High-Resolution NMR Techniques in Organic Chemistry

High-Resolution NMR Techniques in Organic Chemistry

Third Edition

Timothy D.W. Claridge
Chemistry Research Laboratory
Department of Chemistry
University of Oxford
Oxford, United Kingdom

AMSTERDAM • BOSTON • HEIDELBERG • LONDON • NEW YORK • OXFORD • PARIS
SAN DIEGO • SAN FRANCISCO • SINGAPORE • SYDNEY • TOKYO

Elsevier
Radarweg 29, PO Box 211, 1000 AE Amsterdam, Netherlands
The Boulevard, Langford Lane, Kidlington, Oxford OX5 1GB, UK
50 Hampshire Street, 5th Floor, Cambridge, MA 02139, USA

First edition 1999
Reprinted 2004, 2005, 2006
Second edition 2009

Notices
Knowledge and best practice in this field are constantly changing. As new research and experience broaden our understanding, changes in research methods, professional practices, or medical treatment may become necessary.

Practitioners and researchers must always rely on their own experience and knowledge in evaluating and using any information, methods, compounds, or experiments described herein. In using such information or methods they should be mindful of their own safety and the safety of others, including parties for whom they have a professional responsibility.

To the fullest extent of the law, neither the Publisher nor the authors, contributors, or editors, assume any liability for any injury and/or damage to persons or property as a matter of products liability, negligence or otherwise, or from any use or operation of any methods, products, instructions, or ideas contained in the material herein.

British Library Cataloguing-in-Publication Data
A catalogue record for this book is available from the British Library

Library of Congress Cataloging-in-Publication Data
A catalog record for this book is available from the Library of Congress

ISBN: 978-0-08-099986-9

For information on all Elsevier publications
visit our website at https://www.elsevier.com/

Typeset by Thomson Digital

Contents

Preface ix

1 Introduction

1.1 The Development of High-Resolution NMR 1
1.2 Modern High-Resolution NMR and This Book 3
1.2.1 What This Book Contains 4
1.2.2 Pulse Sequence Nomenclature 5
1.3 Applying Modern NMR Techniques 7
References 10

2 Introducing High-Resolution NMR

2.1 Nuclear Spin and Resonance 11
2.2 The Vector Model of NMR 14
2.2.1 The Rotating Frame of Reference 14
2.2.2 Pulses 15
2.2.3 Chemical Shifts and Couplings 17
2.2.4 Spin-Echoes 18
2.3 Time and Frequency Domains 20
2.4 Spin Relaxation 22
2.4.1 Longitudinal Relaxation: Establishing Equilibrium 22
2.4.2 Measuring T_1 with the Inversion Recovery Sequence 24
2.4.3 Transverse Relaxation: Loss of Magnetisation in the *x–y* Plane 26
2.4.4 Measuring T_2 with a Spin-Echo Sequence 27
2.5 Mechanisms for Relaxation 31
2.5.1 The Path to Relaxation 32
2.5.2 Dipole–Dipole Relaxation 33
2.5.3 Chemical Shift Anisotropy Relaxation 34
2.5.4 Spin Rotation Relaxation 35
2.5.5 Quadrupolar Relaxation 35
2.6 Dynamic Effects in NMR 38
2.6.1 The Influence of Dynamic Exchange 39
2.6.2 Lineshape Analysis and Thermodynamic Parameters 53
2.6.3 Magnetisation Transfer under Slow-Exchange Conditions 55
References 58

3 Practical Aspects of High-Resolution NMR

3.1 An Overview of the NMR Spectrometer 61
3.2 Data Acquisition and Processing 64
3.2.1 Pulse Excitation 64
3.2.2 Signal Detection 66
3.2.3 Sampling the FID 67
3.2.4 Quadrature Detection 73
3.2.5 Phase Cycling 78
3.2.6 Dynamic Range and Signal Averaging 80
3.2.7 Window Functions 83
3.2.8 Phase Correction 88
3.3 Preparing the Sample 89
3.3.1 Selecting the Solvent 89
3.3.2 Reference Compounds 91
3.3.3 Tubes and Sample Volumes 92
3.3.4 Filtering and Degassing 94
3.4 Preparing the Spectrometer 95
3.4.1 The Probe 95
3.4.2 Probe Design and Sensitivity 97
3.4.3 Tuning the Probe 103
3.4.4 The Field Frequency Lock 105
3.4.5 Optimising Field Homogeneity: Shimming 107
3.4.6 Reference Deconvolution 112
3.5 Spectrometer Calibrations 113
3.5.1 Radiofrequency Pulses 113
3.5.2 Pulsed Field Gradients 122
3.5.3 Sample Temperature 124
3.6 Spectrometer Performance Tests 126
3.6.1 Lineshape and Resolution 127
3.6.2 Sensitivity 128
3.6.3 Solvent Presaturation 130
References 130

4 One-Dimensional Techniques

4.1 Single-Pulse Experiment 133
4.1.1 Optimising Sensitivity 133
4.1.2 Quantitative NMR Measurements and Integration 137
4.1.3 Quantification with an Electronic Calibrant: ERETIC 140
4.1.4 Quantification with an External Calibrant: PULCON 142
4.2 Spin-Decoupling Methods 143
4.2.1 Basis of Spin Decoupling 143
4.2.2 Homonuclear Decoupling 143
4.2.3 Heteronuclear Decoupling 145
4.3 Spectrum Editing with Spin-Echoes 148
4.3.1 J-Modulated Spin-Echo 149
4.3.2 APT 152
4.4 Sensitivity Enhancement and Spectrum Editing 153
4.4.1 Polarisation Transfer 154
4.4.2 INEPT 156
4.4.3 DEPT 162
4.4.4 DEPTQ 165
4.5 Observing Quadrupolar Nuclei 167
References 168

5 Introducing Two-Dimensional and Pulsed Field Gradient NMR

5.1 Two-Dimensional Experiments 172
5.1.1 Generating the Second Dimension 172
5.1.2 Correlating Coupled Spins 176
5.2 Practical Aspects of 2D NMR 177
5.2.1 Two-Dimensional Lineshapes and Quadrature Detection 177
5.2.2 Axial Peaks 181
5.2.3 Instrumental Artefacts 182
5.2.4 Two-Dimensional Data Acquisition 183
5.2.5 Two-Dimensional Data Processing 186
5.3 Coherence and Coherence Transfer 188
5.3.1 Coherence Transfer Pathways 190
5.4 Gradient-Selected Spectroscopy 191
5.4.1 Signal Selection with Pulsed Field Gradients 192
5.4.2 Phase-Sensitive Experiments: *Echo–Antiecho* Selection 195
5.4.3 Pulsed Field Gradients in High-Resolution NMR 196
5.4.4 Practical Implementation of Pulsed Field Gradients 198
5.4.5 Fast Data Acquisition: Single-Scan Two-Dimensional NMR 199
References 201

6 Correlations Through the Chemical Bond I: Homonuclear Shift Correlation

6.1 Correlation Spectroscopy: COSY 203
6.1.1 Interpreting COSY 204
6.1.2 Peak Fine Structure 207
6.1.3 Which COSY Approach? 210
6.1.4 COSY-β 211
6.1.5 Double-Quantum Filtered COSY (DQF-COSY) 212
6.1.6 Long-Range COSY: Detecting Small Couplings 219
6.1.7 Relayed-COSY 220
6.2 Total Correlation Spectroscopy: TOCSY 220
6.2.1 The TOCSY Sequence 221
6.2.2 Applying TOCSY 223
6.2.3 Implementing TOCSY 225
6.2.4 One-Dimensional TOCSY 227
6.3 Correlating Dilute Spins: INADEQUATE 230
6.3.1 Two-Dimensional INADEQUATE 230
6.3.2 One-Dimensional INADEQUATE 233
6.3.3 Implementing INADEQUATE 233
6.4 Correlating Dilute Spins Via Protons: ADEQUATE 235
6.4.1 Two-Dimensional ADEQUATE 236
6.4.2 Enhancements to ADEQUATE 237
References 240

7 Correlations Through the Chemical Bond II: Heteronuclear Shift Correlation

7.1 Introduction 243
7.2 Sensitivity 244
7.3 Heteronuclear Single-Bond Correlations 246
7.3.1 Heteronuclear Single-Quantum Correlation 246
7.3.2 Hybrid HSQC Experiments 253
7.3.3 Heteronuclear Multiple-Quantum Correlation 257
7.4 Heteronuclear Multiple-Bond Correlations 261
7.4.1 HMBC Sequence 263
7.4.2 Applying HMBC 264
7.4.3 HMBC Extensions and Variants 266
7.4.4 H2BC: Differentiating $^{2}J_{CH}$ and $^{3}J_{CH}$ HMBC Correlations 274
7.4.5 Measuring Long-Range $^{n}J_{XH}$ Coupling Constants 275
7.4.6 Long-Range HSQMBC: Interrogating Proton-Sparse Molecules 281
7.5 Heteronuclear X-Detected Correlations 282
7.5.1 Single Bond Heteronuclear Correlations 283
7.5.2 Multiple-Bond Correlations and Small Couplings 285

7.6 **Heteronuclear X–Y Correlations** 286
7.6.1 Direct X–Y Correlations 286
7.6.2 Indirect ^{1}H-Detected X–Y Correlations 288
7.7 **Parallel Acquisition NMR with Multiple Receivers** 291
References 292

8 **Separating Shifts and Couplings: J-Resolved and Pure Shift Spectroscopy**

8.1 **Introduction** 295
8.2 **Heteronuclear J-Resolved Spectroscopy** 295
8.2.1 Measuring Long-Range Proton–Carbon Coupling Constants 298
8.2.2 Practical Considerations 300
8.3 **Homonuclear J-Resolved Spectroscopy** 301
8.3.1 Tilting, Projections and Symmetrisation 302
8.3.2 Applications 303
8.4 **'Indirect' Homonuclear J-Resolved Spectroscopy** 304
8.5 **Pure Shift Broadband-Decoupled ^{1}H Spectroscopy** 306
8.5.1 The Basis of Pure Shift Spectroscopy 307
8.5.2 Pseudo-2D Pure Shift 307
8.5.3 Real-Time Pure Shift 309
8.5.4 Pure Shift Refocussing Elements 309
References 313

9 **Correlations Through Space: The Nuclear Overhauser Effect**

9.1 **Introduction** 315
PART I **THEORETICAL ASPECTS** 317
9.2 **Definition of the NOE** 317
9.3 **Steady-State NOEs** 317
9.3.1 NOEs in a Two-Spin System 317
9.3.2 NOEs in a Multi-Spin System 324
9.3.3 Summary 329
9.3.4 Applications 330
9.4 **Transient NOEs** 335
9.4.1 Nuclear Overhauser Effect Kinetics 335
9.4.2 Measuring Internuclear Separations 336
9.5 **Rotating Frame NOEs** 337
PART II **PRACTICAL ASPECTS** 339
9.6 **Measuring Transient NOEs: NOESY** 339
9.6.1 The 2D NOESY Sequence 339
9.6.2 1D NOESY Sequences 346
9.6.3 Applications 349
9.7 **Measuring Rotating Frame NOEs: ROESY** 353
9.7.1 The 2D ROESY Sequence 353
9.7.2 1D ROESY Sequences 355
9.7.3 Applications 356
9.8 **Measuring Steady-State NOEs: NOE Difference** 359
9.8.1 Optimising Difference Experiments 361
9.9 **Measuring Heteronuclear NOEs: HOESY** 363
9.9.1 2D Heteronuclear NOEs 364
9.9.2 1D Heteronuclear Nuclear Overhauser Effects 365
9.9.3 Applications 366
9.10 **Experimental Considerations for NOE Measurements** 367
9.11 **Measuring Chemical Exchange: EXSY** 368
9.12 **Residual Dipolar Couplings** 371
9.12.1 Measuring RDCs 372
9.12.2 Applying RDCs 375
References 377

10 **Diffusion NMR Spectroscopy**

10.1 **Introduction** 381
10.1.1 Diffusion Coefficients and Molecular Size 382
10.2 **Measuring Self-Diffusion by NMR** 382
10.2.1 The Pulsed Field Gradient Spin-Echo 383
10.2.2 The Pulsed Field Gradient Stimulated-Echo 384
10.2.3 Enhancements to the Stimulated-Echo 385
10.2.4 Data Analysis: Regression Fitting 388
10.2.5 Data Analysis: Pseudo-2D Presentation 389
10.3 **Practical Aspects of Diffusion NMR Spectroscopy** 390
10.3.1 The Problem of Convection 390
10.3.2 Calibrating Gradient Amplitudes 397
10.3.3 Optimising Diffusion Parameters 397
10.3.4 Hydrodynamic Radii and Molecular Weights 401
10.4 **Applications of Diffusion NMR Spectroscopy** 403
10.4.1 Signal Suppression 403
10.4.2 Hydrogen Bonding 405
10.4.3 Host–Guest Complexes 405
10.4.4 Ion Pairing 408
10.4.5 Supramolecular Assemblies 409
10.4.6 Aggregation 411
10.4.7 Mixture Separation 412
10.4.8 Macromolecular Characterisation 413

10.5 Hybrid Diffusion Sequences 414
10.5.1 Sensitivity-Enhanced Heteronuclear Methods 414
10.5.2 Spectrum-Edited Methods 415
10.5.3 Diffusion-Encoded Two-Dimensional Methods (or 3D DOSY) 415
References 418

11 Protein–Ligand Screening by NMR

11.1 Introduction 421
11.2 Protein–Ligand Binding Equilibria 422
11.3 Resonance Lineshapes and Relaxation Editing 424
11.3.1 ^{1}H Relaxation-Edited NMR 426
11.3.2 ^{19}F NMR 428
11.3.3 Paramagnetic Relaxation Enhancement 429
11.4 Saturation Transfer Difference 430
11.4.1 The STD Sequence and Practicalities 432
11.4.2 Epitope Mapping by STD and DIRECTION 436
11.4.3 K_D Measurement by STD 437
11.5 Water-LOGSY 438
11.5.1 The Water-LOGSY Sequence 440
11.5.2 Water-LOGSY Practicalities 441
11.6 Exchange-Transferred Nuclear Overhauser Effects 441
11.7 Competition Ligand Screening 443
11.7.1 Competitive Displacement 444
11.7.2 Reporter Ligand Screening 445
11.7.3 ^{19}F *FAXS* 447
11.8 Protein Observe Methods 448
11.8.1 ^{1}H–^{15}N Mapping 448
11.8.2 ^{1}H–^{13}C Mapping 452
11.8.3 ^{19}F Mapping 452
References 454

12 Experimental Methods

12.1 Composite Pulses 457
12.1.1 A Myriad of Pulses 459
12.1.2 Inversion Versus Refocusing 460
12.2 Adiabatic and Broadband Pulses 461
12.2.1 Common Adiabatic Pulses 462
12.2.2 Broadband Inversion Pulses: BIPs 464
12.3 Broadband Decoupling and Spin Locking 465
12.3.1 Broadband Adiabatic Decoupling 467
12.3.2 Spin Locking 468
12.4 Selective Excitation and Soft Pulses 468
12.4.1 Shaped Soft Pulses 469
12.4.2 Excitation Sculpting 473
12.4.3 Chemical Shift Selective Filters 475
12.4.4 DANTE Sequences 477
12.4.5 Practical Considerations 478
12.5 Solvent Suppression 480
12.5.1 Presaturation 480
12.5.2 Zero Excitation 482
12.5.3 Pulsed Field Gradients 483
12.6 Suppression of Zero-Quantum Coherences 486
12.6.1 The Variable-Delay Z-Filter 486
12.6.2 Zero-Quantum Dephasing 487
12.7 Heterogeneous Samples and Magic Angle Spinning 489
12.8 Hyperpolarisation 491
12.8.1 *Para*-Hydrogen–Induced Polarisation 491
12.8.2 Dynamic Nuclear Polarisation 493
References 496

13 Structure Elucidation and Spectrum Assignment

13.1 ^{1}H NMR 500
13.2 ^{1}H–^{13}C Edited HSQC 501
13.3 ^{1}H–^{1}H COSY and Variants 503
13.3.1 Double-Quantum Filtered COSY 505
13.4 ^{1}H–^{1}H TOCSY and Variants 506
13.4.1 HSQC-TOCSY 508
13.5 ^{13}C NMR 508
13.6 HMBC and Variants 510
13.6.1 ^{1}H–^{13}C HMBC 510
13.6.2 ^{31}P and ^{1}H–^{31}P HMBC 512
13.6.3 ^{1}H–^{13}C HMBC Again 513
13.6.4 ^{19}F and ^{19}F–^{13}C HMBC 515
13.7 Nuclear Overhauser Effects 517
13.7.1 2D NOESY 517
13.7.2 1D NOESY 521
13.7.3 1D ^{19}F HOESY 522
13.8 Rationalization of ^{1}H–^{1}H Coupling Constants 523
13.9 Summary 525

Appendix 527
Subject Index 531

Preface

There can be no doubt that nuclear magnetic resonance (NMR) spectroscopy is now a very well developed analytical tool and one that continues to expand in its capabilities and application. This is manifest in the fact that many NMR sub-disciplines have now established themselves as separate research fields in their own right, with practitioners of NMR often finding themselves specialising in one of these areas. Common, loosely defined fields might include 'small' molecules, bio-macromolecules and metabolomics, aside from solid-state NMR, which itself nowadays largely partitions to materials science and biomolecules. Likewise, NMR texts tend to focus on one of these disciplines as it becomes increasingly difficult to do adequate justice to more than one in a single book. In keeping with the previous two editions, this text continues its focus on solution-state NMR techniques for the study of small molecules. In addition, this third edition extends to medicinal chemistry related applications, reflecting the increasing use of small molecule NMR at the life-sciences interface, and as such includes a completely new chapter on ligand–protein interactions. A second new chapter, presenting a model case study, illustrates how the most common NMR methods may be combined in the structure elucidation of a single compound. The earlier introductory material in the book has also been expanded, notably with regard to dynamic NMR effects, and all chapters have been updated to reflect modern instrument developments, current methodology and new experimental techniques, as outlined subsequently.

The Chapter *Introducing High-Resolution* NMR develops the basic concepts and principles of NMR, including pulse excitation and spin relaxation. This chapter has been extended significantly with a new section on dynamic NMR spectroscopy. This describes the influence of dynamic effects on NMR spectra, introduces the concepts of NMR timescales and defines fast and slow exchange processes. It also describes the use of lineshape analysis and magnetisation transfer experiments for the measurement of equilibrium exchange rates. The chapter describing *Practical Aspects of High-Resolution NMR* focuses on the practicalities of executing NMR experiments and has been updated to include recent technological developments, including nitrogen-cooled cryogenic probeheads. The chapter on *One-Dimensional Techniques* introduces the primary 1D NMR techniques and for this edition discussions on quantitative NMR (qNMR) have been extended to reflect the increasing use of NMR as a primary technique for defining compound purity. The following chapter *Introducing Two-Dimensional and Pulsed Field Gradient NMR* has been revised slightly from that of previous editions and now focuses solely on introducing the principles and practicalities of 2D NMR and of pulsed field gradients (PFGs). The chapter also includes non-uniform sampling for more rapid data collection and concludes by introducing the concept of single-scan NMR, the ultimate in fast data acquisition. The chapter *Correlations Through the Chemical Bond I: Homonuclear Shift Correlation* then focuses on techniques for establishing correlations between homonuclear spins. The theme of through-bond correlations continues with *Correlations Through the Chemical Bond II: Heteronuclear Shift Correlation* which describes heteronuclear correlation techniques. This chapter has been restructured in this edition to emphasise the favoured use of heteronuclear single-quantum correlation (HSQC) over heteronuclear multiple-quantum correlation (HMQC) nowadays, with the later retained partly as a lead into to the very important and closely related heteronuclear multiple-bond correlation (HMBC) technique. The sections covering HMBC have also been extended to consider new variants, including those suitable for measuring long-range, proton-carbon coupling constants and for the investigation of proton-sparse molecules. The chapter concludes with an introduction to the concept of parallel acquisition NMR in which multiple NMR receivers are employed. Whilst uncommon at present, these methods are potentially significant for future developments leading to more efficient data acquisition. The chapter *Separating Shifts and Couplings: J-Resolved and Pure Shift Spectroscopy* introduces classical J-resolved spectroscopy and also contains a substantial new section describing pure shift broadband decoupled proton NMR. The chapter describing *Correlations Through Space: The Nuclear Overhauser Effect* provides an introduction to the basic principles of the NOE. Following this, the practical section has been significantly restructured to emphasise the dominance of the transient NOE (most often measured by NOESY) over the older steady-state difference experiments for measuring proton–proton NOEs. The section on heteronuclear Overhauser effects has been extended to include methods for ^{19}F–^{1}H NOEs, included to match the growing interest in fluorine chemistry. The section on 2D exchange spectroscopy (EXSY) has also been revised to integrate with the new dynamics section found in the chapter *Introducing*

High-Resolution NMR. Finally in this chapter, a new section on residual dipolar couplings (RDCs) and their use in defining stereochemistry has been included, with practical discussions and example applications. The chapter *Diffusion NMR Spectroscopy* describes techniques for studying molecular self-diffusion and has been updated to include the latest consideration of convection and its complicating effects. It also includes coverage of some recent and pragmatic approaches for correlating measured diffusion coefficients with molecular size. The chapter describing *Protein-Ligand Screening by NMR* presents a completely new topic with a focus on methods for studying the interactions of small molecule ligands with macromolecular targets, principally proteins. NMR spectroscopy is increasingly applied in drug discovery programmes and in the development of small molecule probes of biochemical pathways, meaning these techniques are finding increasing use in chemical laboratories that interface with biological science. After an introduction to relevant binding equilibria, various techniques are presented which focus on the response of the small molecule on binding to a target. The most popular NMR methods are described, including relaxation editing, STD, water-LOGSY, exchange-transferred NOEs and the use of competition experiments. Alternative methods employing observation of protein responses are also considered. The chapter on *Experimental Methods* describes elements of NMR that are used to build modern pulse sequences and has been updated to reflect recent developments and to include older methodologies that have become more established in their roles. Descriptions of hyperpolarisation have been extended to include *para*-hydrogen based methods (PHIP) in addition to dynamic nuclear polarisation (DNP). The final chapter *Structure Elucidation and Spectrum Assignment* is also a new addition to this third edition and seeks to exemplify the combined application of some of the primary methods described in the book to the structure elucidation of a moderately complex small molecule. This case study illustrates how these techniques may be applied in a stepwise approach to systematically build a molecular structure and to define its stereochemistry from 1D and 2D NMR data, also highlighting how less-commonly employed techniques might be used in the assignment of more complex data sets. The book concludes with an updated glossary of terms and acronyms that find common use in the field.

In producing the new chapters and sections for this edition, I have once again benefitted from the generous input of many people. From the Chemistry Department in Oxford, I thank my colleagues in the NMR facility Drs Nick Rees and Barbara Odell, and from the University of Auckland Dr Ivanhoe Leung, for providing useful comments and suggestions on new drafts. I am similarly grateful to many group leaders and their research students in the Oxford Chemistry Department for making novel compounds available with which to illustrate the application of various experimental techniques, in particular Profs Stuart Conway, Ben Davis, Philip Mountford, Chris Schofield and Martin Smith. For the provision of original data sets used to prepare figures, I thank Prof Christina Redfield (Department of Biochemistry, University of Oxford), Prof Simon Duckett (Department of Chemistry, University of York), Dr Eriks Kupče (Bruker Biospin UK Ltd), Dr Ignacio Perez-Victoria (formally of Department of Chemistry, University of Oxford) and again Dr Nick Rees. For making their software packages freely available, I thank Prof Hans Reich (University of Wisconsin) for the WinDNMR lineshape simulation program and Prof Alex Bain (McMaster University) for the CIFIT magnetisation transfer fitting program. I am also grateful to Katey Birtcher, Jill Cetel and Anitha Sivaraj of Elsevier for their assistance and patience during the development and production of this new edition.

As previously, those to which I am most grateful are my wife Rachael and daughter Emma for once again demonstrating their continued tolerance and support throughout this project. One might imagine each new edition should be easier and perhaps quicker to complete than the last, but somehow that never seems to be the case, for which I can only apologise.

Timothy D.W. Claridge
Oxford, September 2015

Chapter 1

Introduction

Chapter Outline

1.1 The Development of High-Resolution NMR 1
1.2 Modern High-Resolution NMR and This Book 3
1.2.1 What This Book Contains 4
1.2.2 Pulse Sequence Nomenclature 5
1.3 Applying Modern NMR Techniques 7
References 10

From the initial observation of proton magnetic resonance in water [1] and in paraffin [2], the discipline of nuclear magnetic resonance (NMR) has seen unparalleled growth as an analytical method and now, in numerous different guises, finds application in chemistry, biology, medicine, materials science and geology. The founding pioneers of the subject, Felix Bloch and Edward Purcell, were recognised with a Nobel Prize in 1952 'for their development of new methods for nuclear magnetic precision measurements and discoveries in connection therewith'. The maturity of the discipline has since been recognised through the awarding of Nobel Prizes to two of the pioneers of modern NMR methods and their application, Richard Ernst (1991, 'for his contributions to the development of the methodology of high resolution NMR spectroscopy') and Kurt Wüthrich (2002, 'for his development of NMR spectroscopy for determining the three-dimensional structure of biological macromolecules in solution'). Despite its inception in the laboratories of physicists, it is in chemical and biochemical laboratories that NMR spectroscopy has found greatest use. To put into context the range of techniques now available in the modern chemical laboratory, including those described in this book, we begin with a short overview of the evolution of high-resolution (solution-state) NMR spectroscopy and some of the landmark developments that have shaped the subject.

1.1 THE DEVELOPMENT OF HIGH-RESOLUTION NMR

It is almost 70 years since the first observations of NMR were made in both solid and liquid samples, from which the subject has evolved to become the principal structural technique of the research chemist, alongside mass spectrometry. During this time, there have been a number of key advances in high-resolution NMR that have guided the development of the subject [3–5] (Table 1.1) and consequently the work of chemists and their approaches to structure elucidation. The seminal step occurred during the early 1950s when it was realised that the resonant frequency of a nucleus is influenced by its chemical environment, and that one nucleus could further influence the resonance of another through intervening chemical bonds. Although these observations were seen as unwelcome chemical complications by the investigating physicists, a few pioneering chemists immediately realised the significance of these chemical shifts and spin–spin couplings within the context of structural chemistry. The first high-resolution proton NMR spectrum (Fig. 1.1) clearly demonstrated how the features of an NMR spectrum, in this case chemical shifts, could be directly related to chemical structure and it is from this that NMR has evolved to attain the significance it holds today.

The 1950s also saw a variety of instrumental developments that were to provide the chemist with even greater chemical insight. These included the use of sample spinning for averaging to zero field inhomogeneities which provided a substantial increase in resolution, so revealing fine splittings from spin–spin coupling. Later, spin decoupling was able to provide more specific information by helping the chemists understand these interactions. With these improvements, sophisticated relationships could be developed between chemical structure and measurable parameters, leading to such realisations as the dependence of vicinal coupling constants on dihedral angles (the now well-known Karplus relationship). The inclusion of computers during the 1960s was also to play a major role in enhancing the influence of NMR on the chemical community. The practice of collecting the same continuous wave spectrum repeatedly and combining them with a CAT (computer of average transients) led to significant gains in sensitivity and made the observation of smaller sample quantities a practical realisation. When the idea of stimulating all spins simultaneously with a single pulse of

High-Resolution NMR Techniques in Organic Chemistry. http://dx.doi.org/10.1016/B978-0-08-099986-9.00001-4

TABLE 1.1 A Summary of Some Key Developments that have had a Major Influence on the Practice and Application of High-Resolution NMR Spectroscopy in Chemical Research

Decade	Notable Advances
1940s	First observation of NMR in solids and liquids (1945)
1950s	Development of chemical shifts and spin–spin coupling constants as structural indicators
1960s	Use of signal averaging for improving sensitivity Application of the pulse FT approach The NOE employed in structural investigations
1970s	Use of superconducting magnets and their combination with the FT approach Computer-controlled instrumentation
1980s	Development of multipulse and 2D NMR techniques Automated spectroscopy
1990s	Routine application of pulsed field gradients for signal selection Development of coupled analytical methods (eg LC-NMR)
2000s	Use of high-sensitivity helium-cooled cryogenic probes Routine availability of actively shielded magnets for reduced stray fields Development of microscale tube and flow probes
2010+	Adoption of fast data acquisition methods Use of high-sensitivity nitrogen-cooled cryogenic probes Use of multiple receivers...?

radiofrequency, collecting the time domain response and converting this to the required frequency domain spectrum by a process known as Fourier transformation (FT) was introduced, more rapid signal averaging became possible. This approach provided an enormous increase in the signal-to-noise ratio, and was to change completely the development of NMR spectroscopy. The mid-1960s also saw the application of the nuclear Overhauser effect (NOE) to conformational studies. Although described during the 1950s as a means of enhancing the sensitivity of nuclei through the simultaneous irradiation of electrons, the NOE has since found widest application in sensitivity enhancement between nuclei, or in the study of the spatial proximity of nuclei, and remains one of the most important tools of modern NMR. By the end of the 1960s, the first commercial FT spectrometer was available, operating at 90 MHz for protons. The next great advance in field strengths was provided by the introduction of superconducting magnets during the 1970s, which were able to provide significantly higher fields than the electromagnets previously employed. These, combined with the FT approach, made the observation of carbon-13 routine and provided the organic chemist with another probe of molecular structure. This also paved the way for the routine observation of a whole variety of previously inaccessible nuclei of low natural abundance and low magnetic moment. It was also in the early 1970s that the concept of spreading the information contained within the NMR spectrum into two separate frequency dimensions was proposed in a lecture. However, because of instrumental limitations, the quality of the first 2D spectra were considered too poor to be published, and not until the mid-1970s, when instrument stability had improved and developments in computers made the necessary complex calculations feasible, did the development of 2D methods begin in earnest. These methods, together with the various multipulse 1D methods that

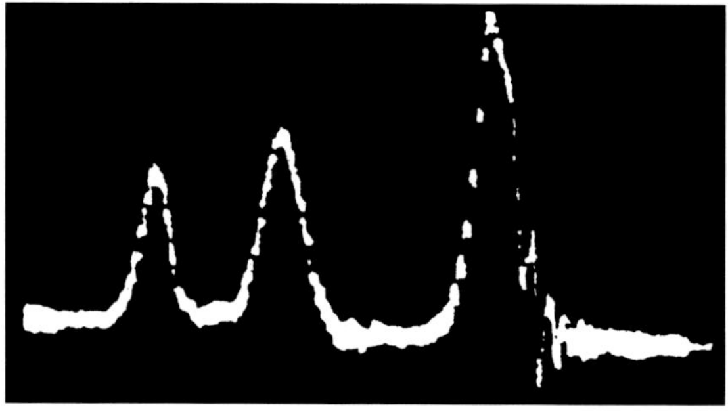

FIGURE 1.1 **The first 'high-resolution' proton NMR spectrum, recorded at 30 MHz, displaying the proton chemical shifts in ethanol.** *(Source: Reprinted with permission from Ref. [6], Copyright 1951, American Institute of Physics.)*

also became possible with the FT approach, were not to have significant impact on the wider chemical community until the 1980s, from which point their development was nothing less than explosive. This period saw an enormous number of new pulse techniques presented which were capable of performing a variety of 'spin gymnastics' and so provided the chemist with ever more structural data, on smaller sample quantities and in less time. No longer was it necessary to rely on empirical correlations of chemical shifts and coupling constants with structural features to identify molecules, instead a collection of spin interactions (through-bond, through-space, chemical exchange) could be mapped and used to determine structures more reliably and more rapidly. The evolution of new pulse methods continued throughout the 1990s, alongside which emerged a fundamentally different way of extracting the desired information from molecular systems. Pulsed field gradient–selected experiments have now become routine structural tools, providing better quality spectra, often in shorter times, than was previously possible. These came into widespread use not so much from a theoretical breakthrough (their use for signal selection was first demonstrated in 1980) but again as a result of progressive technological developments defeating practical difficulties. Similarly, the emergence of coupled analytical methods, such as liquid chromatography and NMR (LC-NMR), came about after the experimental complexities of interfacing these very different techniques was overcome, and these methods have established themselves for the analysis of complex mixtures, albeit as a niche area. Developments in probe technologies over recent decades have allowed wider adoption of probeheads containing coils and preamplifiers that are cryogenically cooled by either cold helium or nitrogen. This reduces system noise significantly and so enhances detection signal-to-noise ratios. Probe coil miniaturisation has also provided a boost in the signal-to-noise ratio for mass limited samples, and the marrying of this with cryogenic technology nowadays offers one of the most effective routes to higher detection sensitivity. Instrument miniaturisation has also been a constant theme, leading to smaller and more compact consoles driven by developments in solid-state electronics. Likewise, actively shielded superconducting magnets with significantly reduced stray fields are now standard, making the siting of instruments considerably easier and far less demanding on space. For example, first-generation unshielded magnets operating at 500 MHz possessed stray fields that would extend to over 3 m horizontally from the magnet centre when measured at the 0.5 mT (5 G) level, the point beyond which disturbances to the magnetic field are not considered problematic. Nowadays, latest-generation shielded magnets have this line sited at somewhat less than 1 m from the centre and only a little beyond the magnet dewar itself. This is achieved through the use of compensating magnet coils that seek to counteract the stray field generated outside the magnet assembly. Other developments have allowed recycling of the liquid cryogens needed to maintain the superconducting state of the magnet through the reliquification of helium and nitrogen gas. Recycling of helium in this manner is already established for imaging magnets but has posed considerable challenges in the context of high-resolution NMR measurements which have only relatively recently been overcome. More extreme miniaturisation of magnets has come about through the use of rare earth metals formed as Halbach array magnets. These ambient temperature devices require no cryogens and provide fields suitable for instruments operating below proton frequencies of 100 MHz. Their compact sizes have allowed the creation of low-field bench-top NMR spectrometers that find use in simple reaction screening or in educational roles, for example. Very recent developments have again taken their inspiration from imaging methodology and have seen the concept of multiple receivers applied to high-resolution NMR. As the name suggests, this allows simultaneous detection of NMR responses from multiple nuclei and has the potential to streamline data acquisition. This is an area still in its infancy and its impact on data collection protocols remains to be seen.

Modern NMR spectroscopy is now a highly developed and technologically advanced subject. With so many advances in NMR methodology in recent years it is understandably an overwhelming task for the research chemist, and even the dedicated spectroscopist, to appreciate what modern NMR has to offer. This text aims to assist in this task by presenting the principal modern NMR techniques and exemplifying their application.

1.2 MODERN HIGH-RESOLUTION NMR AND THIS BOOK

There can be little doubt that NMR spectroscopy now represents the most versatile and informative spectroscopic technique employed in the modern chemical research laboratory, and that an NMR spectrometer represents one of the largest single investments in analytical instrumentation the laboratory is likely to make. For both these reasons it is important that the research chemist is able to make the best use of the available spectrometer(s) and to harness modern developments in NMR spectroscopy in order to promote their chemical or biochemical investigations. Even the most basic modern spectrometer is equipped to perform a myriad of pulse techniques capable of providing the chemist with a variety of data on molecular structure and dynamics. Not always do these methods find their way into the hands of the practising chemist, remaining instead in the realms of the specialist, obscured behind esoteric acronyms or otherwise unfamiliar NMR jargon. Clearly this should not be so and the aim of this book is to gather up the most useful of these modern NMR methods and present them to the wider audience who should, after all, find greatest benefit from their application.

The approach taken throughout is non-mathematical and is based firmly on using pictorial descriptions of NMR phenomena and methods wherever possible. In preparing and updating this work, I have attempted to keep in mind what I perceive to be the requirements of three major classes of potential readers:

1. those who use solution-state NMR as a tool in their own research, but have little or no direct interaction with the spectrometer;
2. those who have undertaken training in directly using a spectrometer to acquire their own data, but otherwise have little to do with the upkeep and maintenance of the instrument; and
3. those who make use spectrometers and are responsible for the day-to-day upkeep of the instrument. This may include NMR laboratory managers, although in some cases users may not consider themselves dedicated NMR spectroscopists.

The first of these could well be research chemists and students in an academic or industrial environment who need to know what modern techniques are available to assist them in their efforts, but otherwise feel they have little concern for the operation of a spectrometer. Their data are likely to be collected under fully-automated conditions, or provided by a central analytical facility. The second may be a chemist in an academic environment who has hands-on access to a spectrometer and has his or her own samples which demand specific studies that are perhaps not available from fully automated instrumentation. The third class of reader may work in smaller chemical companies or academic chemistry departments that have invested in NMR instrumentation but may not employ a dedicated NMR spectroscopist for its upkeep, depending instead on, say, an analytical or synthetic chemist for this. This, it appears (in the United Kingdom at least), is often the case for new start-up chemical companies. NMR laboratory managers may also find the text a useful reference source. With these in mind, the book contains a fair amount of practical guidance on both the execution of NMR experiments and the operation and upkeep of a modern spectrometer. Even if you see yourself in the first of the above categories, some rudimentary understanding of how a spectrometer collects the data of interest and how a sequence produces, say, the 2D correlation spectrum awaiting analysis on your computer, can be enormously helpful in correctly extracting the information it contains or in identifying and eliminating artefacts that may arise from instrumental imperfections or the use of less-than-optimal conditions for your sample. Although not specifically aimed at dedicated spectroscopists, the book may still contain new information or may serve as a reminder of what was once understood but has somehow faded away. The text should be suitable for (UK) graduate-level courses on NMR spectroscopy, and sections of the book may also be appropriate for use in advanced undergraduate courses. The book does not, however, contain descriptions of basic NMR phenomena such as chemical shifts and coupling constants, neither does it contain extensive discussions on how these may be correlated with chemical structures. These topics are already well documented in various introductory texts [7–13] and it is assumed that the reader is already familiar with such matters. Likewise, the book does not seek to provide a comprehensive physical description of the processes underlying the NMR techniques presented; the books by Hore et al. [14] and Keeler [15] present this at an accessible yet rigorous level. Although the chapter on *Protein Ligand Screening by NMR* discusses NMR methods for the study of protein–small-molecule interactions, this text does not discuss techniques for the assignment and structure elucidation of proteins themselves, topics which are also covered in dedicated texts [16–18].

1.2.1 What This Book Contains

The aim of this text is to present the most important NMR methods used for chemical structure elucidation, to explain the information they provide, how they operate and to provide some guidance on their practical implementation. The choice of experiments is naturally a subjective one, partially based on personal experience, but also taking into account those methods most commonly encountered in the chemical literature and those recognised within the NMR community as being most informative and of widest applicability. The operation of many of these is described using pictorial models (equations appear infrequently, and are only included when they serve a specific purpose) so that the chemist can gain some understanding of the methods they are using without recourse to uninviting mathematical descriptions. The sheer number of available NMR methods may make this seem an overwhelming task, but in reality most experiments are composed of a smaller number of comprehensible building blocks pieced together, and once these have been mastered an appreciation of more complex sequences becomes a far less daunting task. For those readers wishing to pursue a particular topic in greater detail, the original references are given but otherwise all descriptions are self-contained.

The following chapter *Introducing High-Resolution NMR* introduces the basic model used throughout the book for the description of NMR methods and describes how this provides a simple picture of the behaviour of chemical shifts and spin–spin couplings during pulse experiments. This model is then used to visualise nuclear spin relaxation, a feature of central importance for the optimum execution of all NMR experiments (indeed, it seems early attempts to observe NMR failed most probably because of a lack of understanding at the time of the relaxation behaviour of the chosen samples). Methods for measuring relaxation rates also provide a simple introduction to multipulse NMR sequences. The chapter concludes with descriptions of how conformational dynamics within molecules can influence the appearance of NMR spectra, and describes how exchange rates may also be measured. The chapter on *Practical Aspects of High-Resolution NMR* describes the practicalities of performing NMR spectroscopy. This is a chapter to dip into as and when necessary and is essentially broken down into self-contained sections relating to the operating principles of the spectrometer and the handling of NMR data, how to correctly prepare the sample and the spectrometer before attempting experiments, how to calibrate the instrument and how to monitor and measure its performance, should you have such responsibilities. It is clearly not possible to describe all aspects of experimental spectroscopy in a single chapter, but this (together with some of the descriptions in the chapter on *Experimental Methods*) should contain sufficient information to enable execution of most modern experiments. These descriptions are kept general and in these I have deliberately attempted to avoid the use of a dialect specific to a particular instrument manufacturer. The chapter covering *One-Dimensional Techniques* contains the most widely used 1D techniques, ranging from optimisation of the single-pulse experiment, through to the multiplicity editing of heteronuclear spectra and the concept of polarisation transfer, another feature central to pulse NMR methods. This includes methods for the editing of carbon spectra according to the number of attached protons. Specific requirements for the observation of certain quadrupolar nuclei that possess extremely broad resonances are also considered. The following chapter *Introducing Two-Dimensional and Pulsed Field Gradient NMR* describes how 2D spectra are generated, the operation of now ubiquitous field gradient pulses and introduces some important concepts (such as coherence transfer between spins) that occur throughout the following chapters. The chapter on *Correlations Through the Chemical Bond I: Homonuclear Shift Correlation* introduces a variety of correlation techniques for identifying scalar (J) couplings between homonuclear spins, which for the most part means protons, as exemplified by the widely employed correlation spectroscopy (COSY) experiment. In addition, less common methods for correlating spins of low-abundance nuclides such as carbon are also discussed. Heteronuclear correlation techniques described in the chapter *Correlations Through the Chemical Bond II: Heteronuclear Shift Correlation* are commonly used to map coupling interactions between, typically, protons and a heteroatom either through a single bond or across multiple bonds. Most attention is given to the modern correlation methods based on proton excitation and detection since these provide for greatest sensitivity. The chapter *Separating Shifts and Couplings: J-Resolved and Pure Shift Spectroscopy* considers methods for separating chemical shifts and coupling constants in spectra, which are again based on 2D methods, and also includes descriptions of the more recently developed pure shift techniques for homonuclear decoupling. Descriptions in *Correlations Through Space: The Nuclear Overhauser Effect* move away from considering through-bond couplings and onto through-space interactions in the form of the NOE. The principles behind the NOE are presented initially for a simple two-spin system, and then for more realistic multispin systems. The practical implementation of both 1D and 2D NOESY experiments is described, as are rotating frame NOE (ROE) techniques which find greatest utility in the study of larger molecules for which the NOE can be poorly suited. The chapter also introduces the use of residual dipolar couplings (RDCs) as an alternative approach to defining molecular stereochemistry. The following chapter on *Diffusion NMR Spectroscopy* considers the measurement of self-diffusion of molecules, a topic now routinely amenable to investigation on spectrometers equipped with standard pulsed field gradient hardware. This describes the operation and application of diffusion NMR methods for the study of molecular interactions, the investigation of mixtures and the classification of molecular size. The theme of molecular interactions is extended in the chapter entitled *Protein-Ligand Screening* by NMR to specifically consider techniques developed for studying the binding of small-molecule ligands with macromolecular targets such as proteins. These methods have become established tools in medicinal chemistry programmes for lead discovery or 'hit' validation and are increasingly widely used in small-molecule NMR laboratories. The chapter describing *Experimental Methods* considers additional experimental elements which do not, on their own, constitute complete NMR experiments but are the tools with which modern methods are constructed. These are typically used within the sequences described in the preceding chapters and include such topics as broadband decoupling, selective excitation of specific regions of a spectrum and solvent suppression. The chapter concludes with a brief overview of specific hyperpolarisation methods for greatly enhancing detection sensitivity. The final chapter entitled *Structure Elucidation and Spectrum Assignment* illustrates application of the most commonly employed techniques for structure elucidation with a worked example defining the structure and stereochemistry of a moderately complex organic molecule. At the end of the book is a glossary of some of the more common acronyms that permeate the language of modern NMR, and, it might be argued, have come to characterise the subject. Although these may provide a convenient shorthand when speaking of pulse experiments, they can confuse the uninitiated and leave them bewildered in the face of such NMR jargon. The glossary provides an immediate breakdown of the acronyms together with a reference to the location in the book of the associated topic.

1.2.2 Pulse Sequence Nomenclature

Virtually all NMR experiments can be described in terms of a *pulse sequence* which, as the name suggests, is a notation which describes the series of radiofrequency or field gradient pulses used to manipulate nuclear spins and so tailor the experiment to provide the desired information. Over the years a largely (although not completely) standard pictorial format

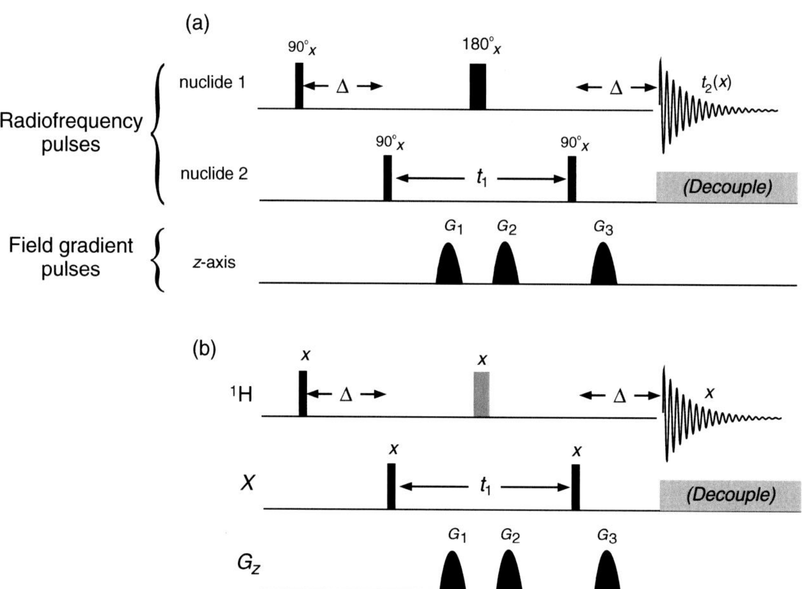

FIGURE 1.2 **Pulse sequence nomenclature.** (a) Complete pulse sequence and (b) the reduced representation used throughout the remainder of the book.

has evolved for representing these sequences, not unlike the way a musical score is used to encode a symphony. Indeed, just as a skilled musician can read the score and 'hear' the symphony in their head, an experienced spectroscopist can often read the pulse sequence and picture the general form of the resulting spectrum. As these crop up repeatedly throughout the text, the format and conventions used in this book deserve explanation. Only definitions of the various pictorial components of a sequence are given here, their physical significance in an NMR experiment will become apparent in later chapters.

An example of a reasonably complex sequence is shown in Fig. 1.2 (a heteronuclear correlation experiment from the chapter *Correlations Through the Chemical Bond II: Heteronuclear Shift Correlation*) and illustrates most points of significance. Fig. 1.2a represents a more detailed account of the sequence, while Fig. 1.2b is the reduced equivalent used throughout the book for reasons of clarity. Radiofrequency (*rf*) pulses applied to each nuclide involved in the experiment are presented on separate rows running left to right, in the order in which they are applied. Most experiments nowadays involve protons, often combined with one (and sometimes two) other nuclides. This is most frequently carbon but need not be, so it is termed the X-nucleus. These *rf* pulses are most frequently applied as so-called 90 or 180 degree pulses (the significance of which is detailed in the following chapter), which are illustrated by a thin black bar and a thick grey bar respectively (Fig. 1.3). Pulses of other angles are marked with an appropriate Greek symbol that is described in the accompanying text. All *rf* pulses also have associated with them a particular phase, typically defined in units of 90 degrees (0, 90, 180, 270 degrees), and indicated above each pulse by x, y, $-x$ or $-y$ respectively. If no phase is defined the default will be x. Pulses that are effective over only a small frequency window and act only on a small number of resonances are differentiated as shaped rather than rectangular bars as this reflects the manner in which these are applied experimentally. These are the so-called *selective* or *shaped* pulses described in the chapter *Experimental Methods*. Pulses that sweep over very wide bandwidths are indicated by horizontal shading to reflect the frequency sweep employed, as also explained

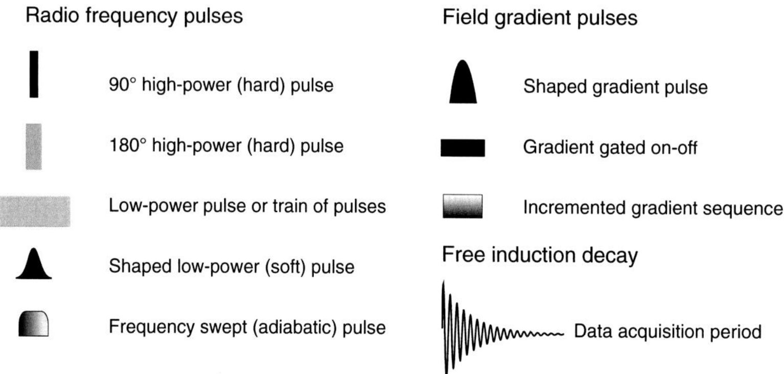

FIGURE 1.3 **A summary of the pulse sequence elements used throughout the book.**

in the chapter *Experimental Methods*. Segments that make use of a long series of very many closely spaced pulses, such as the decoupling shown in Fig. 1.2, are shown as a solid grey box, with the bracket indicating the use of the decoupling sequence is optional. Below the row(s) of radiofrequency pulses are shown field gradient pulses (G_z), whenever these are used, again drawn as shaped pulses where appropriate, and shown greyed when considered optional elements within a sequence.

The operation of very many NMR experiments is crucially dependent on the experiment being tuned to the value of specific coupling constants. This is achieved by defining certain delays within the sequence according to these values, these delays being indicated by the general symbol Δ. Other time periods within a sequence which are not tuned to J values but are chosen according to other criteria, such as spin recovery (relaxation) rates, are given the symbol τ. The symbols t_1 and t_2 are reserved for the time periods which ultimately correspond to the frequency axes f_1 and f_2 of 2D spectra, one of which (t_2) will always correspond to the data acquisition period when the NMR response is actually detected. The acquisition period is illustrated by a simple decaying sine wave in all experiments to represent so-called free induction decay (FID). Again it should be stressed that although these sequences can have rather foreboding appearances, they are generally built up from much smaller and simpler segments that have well-defined and easily understood actions. A little perseverance can clarify what might at first seem a total enigma.

1.3 APPLYING MODERN NMR TECHNIQUES

The tremendous growth in available NMR pulse methods over the last three decades can be bewildering and may leave one wondering just where to start or how best to make use of these new developments. The answer to this is not straightforward since it depends so much on the chemistry undertaken, on the nature of the sample being handled and on the information required of it. It is also dependent on the amount of material and by the available instrumentation and its capabilities. The fact that NMR itself finds application in so many research areas means defined rules for experiment selection are largely intractable. A scheme that is suitable for tackling one type of problem may be wholly inappropriate for another. Nevertheless, it seems inappropriate that a book of this sort should contain no guidance on experiment selection other than the descriptions of the techniques in the following chapters. Here I attempt to broach this topic in rather general terms and present some loose guidelines to help weave a path through the maze of available techniques. Even so, only with a sound understanding of modern techniques can one truly be in a position to select the optimum experimental strategy for *your* molecule or system, and it is this understanding I hope to develop in the remaining chapters. The final chapter serves to illustrate application of the more common techniques to a single compound and also provides something of a guide to their likely use.

Most NMR investigations will begin with analysis of the proton spectrum of the sample of interest, with the usual analysis of chemical shifts, coupling constants and relative signal intensities, either manually or with the assistance of the various sophisticated computational databases now available. Beyond this, one encounters a plethora of available NMR methods to consider employing. The key to selecting appropriate experiments for the problem at hand is an appreciation of the type of information the principal NMR techniques can provide. Although there exist a huge number of pulse sequences, there are a relatively small number of what might be called core experiments, from which most others are derived by minor variation, of which only a rather small fraction ever find widespread use in the research laboratory. To begin, it is perhaps instructive to realise that the NMR methods presented in this book exploit four basic phenomena:

1. Through-bond interactions: scalar (J) spin coupling via bonding electrons.
2. Through-space interactions: the NOE mediated through dipole–dipole coupling and spin relaxation.
3. Chemical exchange: the physical exchange of one spin for another at a specific location.
4. Molecular self-diffusion: the translational movement of a molecule or complex.

When attempting to analyse the structure of a molecule and/or its behaviour in solution by NMR spectroscopy, one must therefore consider how to exploit these phenomena to gain the desired information, and from this select the appropriate technique(s). Thus, when building up the structure of a molecule one typically first searches for evidence of scalar coupling between nuclei as this can be used to indicate the location of chemical bonds. When the location of all bonding relationships within the molecule have been established, the gross structure of the molecule is defined. Spatial proximities between nuclei, and between protons in particular, can be used to define stereochemical relationships within a molecule and thus address questions of configuration and conformation. The unique feature of NMR spectroscopy, and the principal reason for its superiority over any other solution-state technique for structure elucidation, is its ability to define relationships between *specific* nuclei within a molecule or even between molecules. Such exquisite detail is generally obtained by correlating one nucleus with another by exploiting the above phenomena. Despite the enormous power of NMR, there are, in fact, rather few types of correlation available to the chemist to employ for structural and conformational analysis. The principal spin interactions and the main techniques used to map these, which are frequently 2D methods, are summarised

TABLE 1.2 The Principal Correlations or Interactions Established Through NMR Techniques

Correlation	Principal Technique(s)	Comments	Chapter
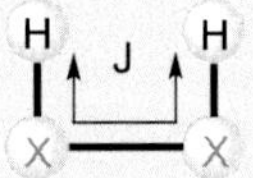	^{1}H–^{1}H COSY	Proton J coupling typically over 2 or 3 bonds.	6
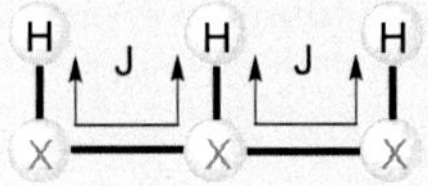	^{1}H–^{1}H TOCSY	Relayed proton J couplings within a coupled spin system. Remote protons may be correlated provided there is a continuous coupling network in between them.	6
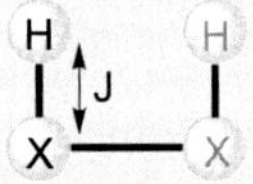	^{1}H–X HSQC ^{1}H–X HMQC	One-bond heteronuclear couplings with proton observation.	7
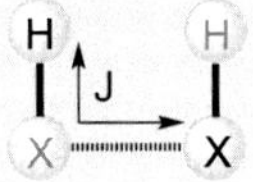	^{1}H–X HMBC	Long-range heteronuclear couplings with proton observation. Typically over 2 or 3 bonds when X = ^{13}C.	7
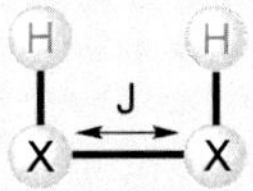	X–X COSY X–X INADEQUATE H–X–X–H ADEQUATE	COSY only used when X-spin natural abundance > 20%. Sensitivity problems when X has low natural abundance; can be improved with proton detection methods.	6
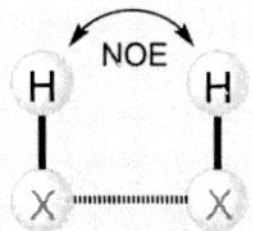	1D/2D NOESY 1D/2D ROESY	Through-space correlations. ROESY applicable to 'mid-sized' molecules with masses of ca. 1–2 kDa.	9
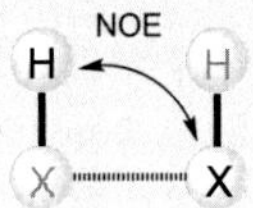	1D/2D HOESY	Sensitivity limited by X-spin observation. Care required to make NOEs specific in presence of proton decoupling.	9
Exchange 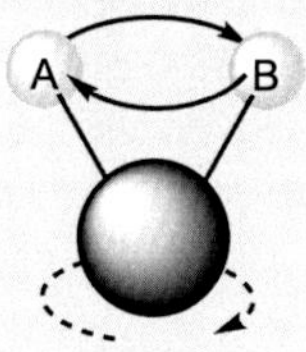	1D saturation or inversion transfer 2D EXSY	Interchange of spins at chemically distinct locations. Exchange must be slow on NMR timescale for separate resonances to be observed. Intermediate to fast exchange requires lineshape analysis.	2 and 9
Diffusion 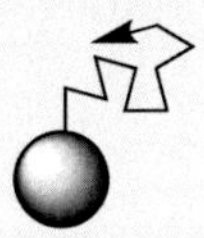	Spin-echo or stimulated-echo methods 2D DOSY	Measurement of molecular self-diffusion using pulsed field gradient technology. Used often in studies of molecular associations.	10
Protein-ligand binding 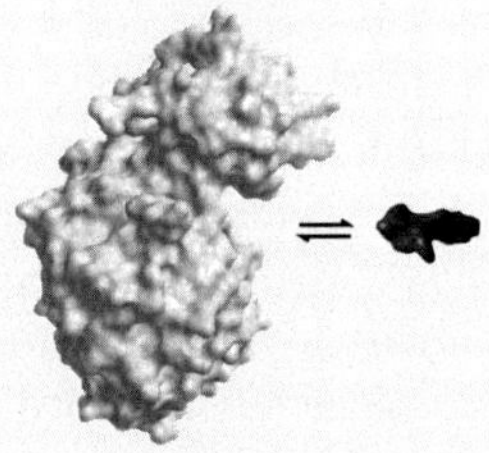	Relaxation editing STD Water-LOGSY Exchange-transferred NOEs Chemical shift perturbation	The qualitative detection of ligand binding to a macromolecular receptor, most often a protein, and the quantitative determination of binding affinities.	11

The correlated spins are labelled in black for each correlation, with X indicating any nuclide other than the proton. The acronyms are explained in the glossary and more fully in subsequent chapters.

in Table 1.2, further elaborated in the chapters that follow, and the use of the primary techniques illustrated in the chapter *Structure Elucidation and Spectrum Assignment*.

The homonuclear correlation spectroscopy (COSY) experiment identifies those nuclei that share a J coupling, which, for protons, operate over two, three and, less frequently, four bonds. This information can therefore be used to indicate the presence of a bonding pathway. The correlation of protons that exist within the same coupled network or chain of spins, but do not themselves share a J coupling, can be made with the total correlation spectroscopy (TOCSY) experiment. This can be used to identify groups of nuclei that sit within the same isolated spin system, such as the amino acid residue of a peptide or the sugar ring of an oligosaccharide. One-bond heteronuclear correlation methods—heteronuclear single quantum correlation (HSQC) or heteronuclear multiple quantum correlation (HMQC)—identify the heteroatoms to which the protons are directly attached and can, for example, provide carbon assignments from previously established proton assignments. Proton chemical shifts can also be dispersed according to the shift of the attached heteroatom, so aiding the assignment of the proton spectrum itself. Long-range heteronuclear correlations over typically two or three-bonds—heteronuclear multiple bond correlation (HMBC)—provide a wealth of information on the skeleton of the molecule and can be used to infer the location of carbon–carbon or carbon–heteroatom bonds. These correlations can be particularly valuable when sufficient proton–proton correlations are absent. Techniques based on the INADEQUATE (incredible natural abundance double quantum transfer) experiment identify connectivity between similar nuclei of low natural abundance. These can therefore correlate directly connected carbon centres, but as they rely on the presence of neighbouring carbon-13 nuclei they suffer from appallingly low sensitivity and thus find limited use. Modern variants that use proton detection, termed ADEQUATE (adequate sensitivity double quantum spectroscopy), have greatly improved performance but are still less used than the above heteronuclear correlation techniques. Measurements based on the NOE are most often applied after the gross structure is defined, and NMR assignments established, to define the 3D stereochemistry of a molecule since these effects map through-space proximity between nuclei. It can also provide insights into the conformational folding of larger structures and on direct intermolecular interactions. The vast majority of these experiments investigate proton–proton NOEs, although heteronuclear NOEs involving a proton and a heteroatom have been applied successfully. Similar techniques to those used in the observation of NOEs can also be employed to correlate nuclei involved in chemical exchange processes that are slow on the NMR timescale and so give rise to distinct resonances for each exchanging species or site. Finally, methods for the quantification of molecular self-diffusion provide information on the nature and extent of molecular associations and provide a complimentary view of solution behaviour. They may also be used to separate the spectra of species with differing mobilities and have potential application in the characterisation of mixtures.

The greatest use of NMR in the chemical research laboratory is in the routine characterisation of synthetic starting materials, intermediates and final products. In these circumstances it is often not so much full structure *elucidation* that is required, rather it is structure *confirmation* or *verification* since the synthetic reagents are known, which naturally limit what the products may be, and because the synthetic target is usually defined. Routine analysis of this sort typically follows a general procedure similar to that summarised in Table 1.3, which is supplemented with data from other analytical techniques, most notably mass spectrometry and infrared spectroscopy. Execution of many of the experiments in Table 1.3 benefits nowadays from the incorporation of pulsed field gradients to speed data collection and to provide spectra of higher quality. Nowadays, the collection of a 2D proton-detected ^{1}H–^{13}C shift correlation experiment such as HSQC requires significantly less time than the 1D carbon spectrum of the same sample, providing both carbon shift data (of protonated centres) and correlation information. This can be a far more powerful tool for routine structure confirmation than the 1D carbon experiment alone. In addition, editing can be introduced to the 2D experiment to differentiate methine from methylene correlations, for example providing yet more data in a single experiment and in less time. Even greater gains can be made in the indirect observation of heteronuclides of still lower intrinsic sensitivity, for example nitrogen-15, and when considering the observation of low-abundance nuclides it is sensible to first consider adopting a proton-detected method for this. The structure confirmation process can also be enhanced through measured use of spectrum prediction tools that are now widely available. Generation of a calculated spectrum from a proposed structure can provide useful guidance when considering the validity of the structure and a number of computational packages are now available that predict at least ^{13}C and ^{1}H spectra, but also the more common nuclides including ^{19}F, ^{31}P and ^{15}N in some cases.

Even when dealing with unknown materials or with molecules of high structural complexity, the general scheme of Table 1.3 still represents an appropriate general protocol to follow, as demonstrated in the chapter *Structure Elucidation and Spectrum Assignment*. In such cases, the basic experiments of this table are still likely to be employed, but may require data to be collected under a variety of experimental conditions (solvent, temperature, pH etc.) and/or may require additional support from other methods or extended versions of these techniques before a complete picture emerges. This book aims to explain the primary NMR techniques and some of their more useful variants, and to describe their practical implementation, so that the research chemist may realise the full potential that modern NMR spectroscopy has to offer.

TABLE 1.3 A Typical Protocol for Routine Structure Confirmation of Synthetic Organic Materials

Procedure	Technique	Information
1D ^{1}H spectrum	1D	Information from chemical shifts, coupling constants, integrals
↓		
2D ^{1}H–^{1}H correlation	COSY	Identify J-coupling relationships between protons
↓		
1D ^{13}C (with spectrum editing)	1D (DEPT or APT)	Carbon count and multiplicity determination (C, CH, CH_2, CH_3). Can often be avoided by using proton-detected heteronuclear 2D experiments.
↓		
1D heteronuclide spectra (eg ^{31}P, ^{19}F, etc.)	1D	Chemical shifts and homonuclear/heteronuclear coupling constants.
↓		
2D ^{1}H–^{13}C one-bond correlation (with spectrum editing)	HSQC (with editing)	Carbon assignments transposed from proton assignments. Proton spectrum dispersed by ^{13}C shifts. Carbon multiplicities from edited HSQC (faster than above 1D approach).
↓		
2D ^{1}H–^{13}C long-range correlation	HMBC	Correlations identified over two and three bonds. Correlations established across heteroatoms (eg N and O). Fragments of structure pieced together.
↓		
Through-space NOE correlation	1D or 2D NOE	Stereochemical analysis: configuration and conformation.

Not all these steps may be necessary, and the direct observation of a heteronuclide, such as carbon or nitrogen, can often be replaced through its indirect observation with more sensitive proton-detected heteronuclear shift correlation techniques.

REFERENCES

[1] Bloch F, Hansen WW, Packard ME. Phys Rev 1946;69:127.
[2] Purcell EM, Torrey HC, Pound RV. Phys Rev 1946;69:37.
[3] Emsley JW, Feeney J. Prog Nucl Magn Reson Spectrosc 2007;50:179–98.
[4] Emsley JW, Feeney J. Prog Nucl Magn Reson Spectrosc 1995;28:1–9.
[5] Shoolery JN. Prog Nucl Magn Reson Spectrosc 1995;28:37–52.
[6] Arnold JT, Dharmatti SS, Packard ME. J Chem Phys 1951;19:507.
[7] Harwood LM, Claridge TDW. Introduction to organic spectroscopy. Oxford: Oxford University Press; 1997.
[8] Anderson RJ, Bendell DJ, Groundwater PW. Organic spectroscopic analysis. Cambridge: Royal Society of Chemistry; 2004.
[9] Silverstein RM, Webster FX, Kiemle DJ, Bryce DL. Spectrometric identification of organic compounds. 8th ed. New York: Wiley; 2014.
[10] Gunther H. NMR spectroscopy: basic principles, concepts and applications in chemistry. 3rd ed. Weinheim: Wiley; 2013.
[11] Akitt JW, Mann BE. NMR and chemistry: an introduction to modern NMR spectroscopy. 4th ed. Cheltenham: Stanley Thornes; 2000.
[12] Balci M. Basic ^{1}H– ^{13}C NMR spectroscopy. Amsterdam: Elsevier; 2005.
[13] Friebolin H. Basic one- and two-dimensional NMR spectroscopy. 5th ed. Chichester: Wiley; 2010.
[14] Hore PJ, Jones JA, Wimperis S. NMR:the toolkit. 2nd ed. Oxford: OUP; 2015.
[15] Keeler J. Understanding NMR spectroscopy. 2nd ed. Chichester: Wiley; 2010.
[16] Zerbe O. BioNMR in drug research, vol. 16. Weinheim: Wiley-VCH; 2003.
[17] Roberts GCK, Lian L-Y. Protein NMR spectroscopy: practical techniques and applications. Weinheim: Wiley; 2011.
[18] Cavanagh J, Fairbrother WJ, Palmer AG, Skelton NJ, Rance M. Protein NMR spectroscopy: principles and practice. 2nd ed. San Diego: Academic Press (Elsevier); 2006.

Chapter 2

Introducing High-Resolution NMR

Chapter Outline

2.1 Nuclear Spin and Resonance 11
2.2 The Vector Model of NMR 14
2.2.1 The Rotating Frame of Reference 14
2.2.2 Pulses 15
2.2.3 Chemical Shifts and Couplings 17
2.2.4 Spin-Echoes 18
2.3 Time and Frequency Domains 20
2.4 Spin Relaxation 22
2.4.1 Longitudinal Relaxation: Establishing Equilibrium 22
2.4.2 Measuring T_1 with the Inversion Recovery Sequence 24
2.4.3 Transverse Relaxation: Loss of Magnetisation in the *x–y* Plane 26
2.4.4 Measuring T_2 with a Spin-Echo Sequence 27
2.5 Mechanisms for Relaxation 31
2.5.1 The Path to Relaxation 32
2.5.2 Dipole–Dipole Relaxation 33
2.5.3 Chemical Shift Anisotropy Relaxation 34
2.5.4 Spin Rotation Relaxation 35
2.5.5 Quadrupolar Relaxation 35
2.6 Dynamic Effects in NMR 38
2.6.1 The Influence of Dynamic Exchange 39
2.6.2 Lineshape Analysis and Thermodynamic Parameters 53
2.6.3 Magnetisation Transfer under Slow-Exchange Conditions 55
References 58

For anyone wishing to gain a greater understanding of modern nuclear magnetic resonance (NMR) techniques together with an appreciation of the information they can provide and hence their potential applications, it is necessary to develop an understanding of some elementary principles. These, if you like, provide the foundations from which the descriptions in subsequent chapters are developed and many of the fundamental topics presented in this introductory chapter will be referred to throughout the remainder of the text. In keeping with the style of the book, all concepts are presented in a pictorial manner and avoid the more mathematical descriptions of NMR. Following a reminder of the phenomenon of nuclear spin, the Bloch vector model of NMR is introduced. This presents a convenient picture of how spins behave in an NMR experiment and provides the basic tools with which most experiments will be described. It is used in the description of spin relaxation processes, after which responsible relaxation mechanisms are described together with methods for measuring relaxation rates. Finally, the chapter considers the appearance of NMR spectra when influenced by the internal dynamics of chemical structures, and describes techniques for measuring the rates of chemical exchange processes.

2.1 NUCLEAR SPIN AND RESONANCE

The nuclei of all atoms may be characterised by a nuclear spin quantum number I, which may have values greater than or equal to zero and which are multiples of ½. Those with $I = 0$ possess no nuclear spin and therefore cannot exhibit NMR, so are termed ‘NMR silent’. Unfortunately, from the organic chemist’s point of view, the nucleus likely to be of most interest, carbon-12, has zero spin, as do all nuclei with atomic mass and atomic number that are both even. However, the vast majority of chemical elements have at least one nuclide that does possess nuclear spin which is, in principle at least, observable by NMR (Table 2.1) and as a consolation the proton is a high-abundance NMR-active isotope. The property of nuclear spin is fundamental to the NMR phenomenon. The spinning nuclei possess angular momentum P and, of course, charge and the motion of this charge gives rise to an associated magnetic moment μ (Fig. 2.1) such that:

$$\mu = \gamma P \tag{2.1}$$

High-Resolution NMR Techniques in Organic Chemistry. http://dx.doi.org/10.1016/B978-0-08-099986-9.00002-6

TABLE 2.1 Properties of Selected Spin-½ Nuclei

Isotope	Natural Abundance (%)	NMR Frequency (MHz)	Relative Sensitivity
^{1}H	99.98	400.0	1.0
^{3}H	0	426.7	1.2[a]
^{13}C	1.11	100.6	1.76×10^{-4}
^{15}N	0.37	40.5	3.85×10^{-6}
^{19}F	100.00	376.3	0.83
^{29}Si	4.70	79.5	3.69×10^{-4}
^{31}P	100.00	161.9	6.63×10^{-2}
^{77}Se	7.58	76.3	5.25×10^{-4}
^{103}Rh	100.00	12.6	3.11×10^{-5}
^{113}Cd	12.16	88.7	1.33×10^{-3}
^{119}Sn	8.58	149.1	4.44×10^{-3}
^{183}W	14.40	16.6	1.03×10^{-5}
^{195}Pt	33.80	86.0	3.36×10^{-3}
^{207}Pb	22.60	83.7	2.07×10^{-3}

Frequencies are given for a 400-MHz spectrometer (9.4-T magnet) and sensitivities are given relative to proton observation and include terms for both intrinsic sensitivity of the nucleus and its natural abundance.
[a]*Assuming 100% ^{3}H labelling. The properties of quadrupolar nuclei are given in Table 2.3.*

FIGURE 2.1 **Nuclear magnetic moments.** A nucleus carries charge and when spinning possesses a magnetic moment μ.

where the term γ is the *magnetogyric ratio* which is constant for any given nuclide and may be viewed as a measure of how 'strongly magnetic' a particular nuclide is. The term gyromagnetic ratio is also in widespread use for γ, although this does not conform to IUPAC recommendations [1–3]. Both angular momentum and the magnetic moment are vector quantities; that is, they have both magnitude and direction. When placed in an external, static magnetic field (denoted B_0, strictly the magnetic flux density) the microscopic magnetic moments align themselves relative to the field in a discrete number of orientations because the energy states involved are quantised. For a spin of magnetic quantum number I there exist $2I + 1$ possible spin states, so for a spin-½ nucleus such as the proton, there are two possible states denoted +½ and −½, while for $I = 1$, for example deuterium, the states are +1, 0 and −1 (Fig. 2.2) and so on. For the spin-½ nucleus, the two states correspond to the popular picture of a nucleus taking up two possible orientations with respect to the static field, either parallel (the α state) or antiparallel (the β state), the former being of lower energy. The effect of the static field on the magnetic moment can be described in terms of classical mechanics, with the field imposing a torque on the moment, which therefore

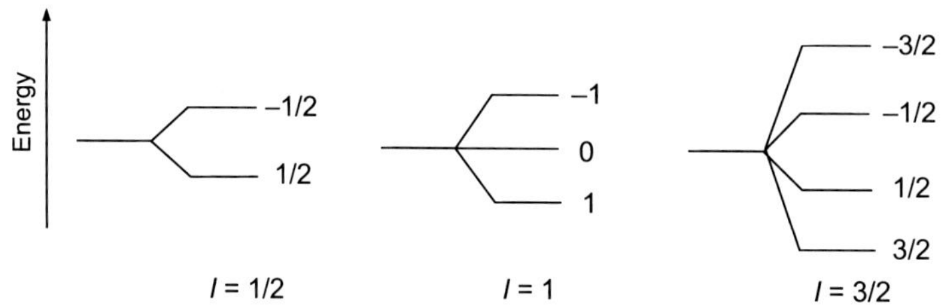

FIGURE 2.2 **Nuclear spin states.** Nuclei with a magnetic quantum I may take up $2I + 1$ possible orientations relative to the applied static magnetic field B_0. For spin-½ nuclei, this gives the familiar picture of the nucleus behaving as a microscopic bar magnet having two possible orientations, α and β.

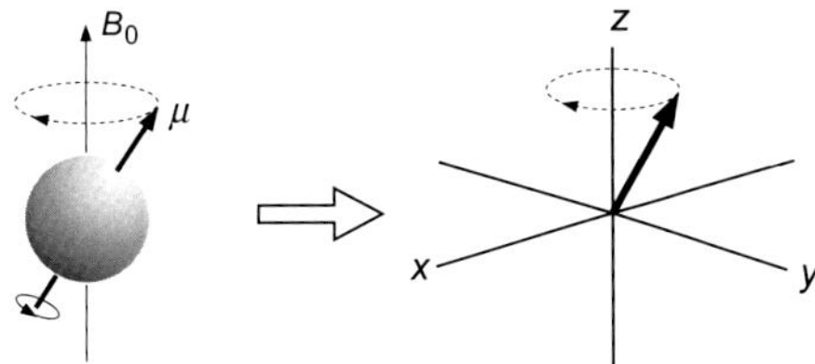

FIGURE 2.3 **Larmor precession.** A static magnetic field applied to the nucleus causes it to process at a rate dependent on the field strength and the magnetogyric ratio of the spin. The field is conventionally applied along the *z*-axis of a Cartesian co-ordinate frame and the motion of the nucleus represented as a vector moving on the surface of a cone.

traces a circular path about the applied field (Fig. 2.3). This motion is referred to as precession, or more specifically *Larmor precession* in this context. It is analogous to the familiar motion of a gyroscope in the Earth's gravitational field, in which the gyroscope spins about its own axis, and this axis in turn precesses about the direction of the field. The rate of precession as defined by the angular velocity (ω rad s^{-1} or ν Hz) is:

$$\omega = -\gamma B_0 \text{ rad s}^{-1} \quad \text{or} \quad \nu = \frac{-\gamma B_0}{2\pi} \equiv -\gamma\!\!\!/ B_0 \text{ Hz} \tag{2.2}$$

(The symbol $\gamma\!\!\!/$ in place of $\gamma/2\pi$ will occur frequently throughout the text when representing frequencies in hertz.)

This is known as the *Larmor frequency* of the nucleus. The direction of motion is determined by the sign of γ and may be clockwise or anticlockwise, but is always the same for any given nuclide. NMR occurs when the nucleus changes its spin state, driven by the absorption of a quantum of energy. This energy is applied as electromagnetic radiation, whose frequency must match that of Larmor precession for the resonance condition to be satisfied, with the energy involved being given by:

$$\Delta E = h\nu = \frac{h\gamma B_0}{2\pi} \tag{2.3}$$

where h is Planck's constant. In other words, the resonant frequency of a spin is simply its Larmor frequency. Modern high-resolution NMR spectrometers currently (2015) employ field strengths up to 23.5 T (tesla) which, for protons, correspond to resonant frequencies up to 1000 MHz that fall within the radiofrequency region of the electromagnetic spectrum. For other nuclei at similar field strengths, resonant frequencies will differ from those of protons (due to the dependence of ν on γ) but it is common practice to refer to a spectrometer's operating frequency in terms of the resonant frequencies of protons. Thus, one may refer to using a '400-MHz spectrometer', although this would equally operate at 100-MHz for carbon-13 since $\gamma_H/\gamma_C \approx 4$. It is also universal practice to define the direction of the static magnetic field as being along the *z*-axis of a set of Cartesian coordinates, so that a *single* precessing spin-½ nucleus will have a component of its magnetic moment along the *z*-axis (the longitudinal component) and an orthogonal component in the *x*–*y* plane (the transverse component) (Fig. 2.3).

Now consider a collection of similar spin-½ nuclei in the applied static field. As stated, the orientation parallel to the applied field α has slightly lower energy than the anti-parallel orientation β, so at equilibrium there will be an excess of nuclei in the α state as defined by the Boltzmann distribution:

$$\frac{N_\alpha}{N_\beta} = e^{\Delta E/k_B T} \tag{2.4}$$

where $N_{\alpha,\beta}$ represents the number of nuclei in the spin orientation, k_B the Boltzmann constant and T the temperature. The differences between spin energy levels are rather small so the corresponding population differences are similarly small and only about one part in 10^4 at the highest available field strengths. This is why NMR is so very insensitive relative to other techniques such as IR and UV, where ground-state and excited-state energy differences are substantially greater. The tiny population *excess* of nuclear spins can be represented as a collection of spins distributed randomly about the precessional cone and parallel to the *z*-axis. These give rise to a resultant *bulk magnetisation vector* **M** along this axis (Fig. 2.4). It is important to realise that this *z*-magnetisation arises because of *population differences* between the possible spin states, a point we return to in Section 2.2. Since there is nothing to define a preferred orientation for the spins in the transverse direction, there exists a random distribution of individual magnetic moments about the cone and hence there is no *net magnetisation* in the transverse (*x*–*y*) plane. Thus, we can reduce our picture of many similar magnetic moments to one of a single bulk magnetisation vector **M** that behaves according to the rules of classical mechanics. This simplified picture

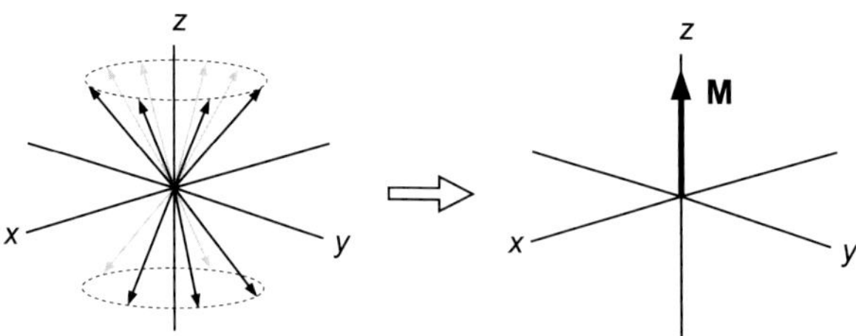

FIGURE 2.4 **Bulk magnetisation.** In the vector model of NMR many like spins are represented by a bulk magnetisation vector. At equilibrium the excess of spins in the α state places this parallel to the $+z$-axis.

is referred to as the Bloch vector model (after the pioneering spectroscopist Felix Bloch), or more generally as the *vector model* of NMR.

2.2 THE VECTOR MODEL OF NMR

Having developed the basic model for a collection of nuclear spins, we can now describe the behaviour of these spins in pulsed NMR experiments. There are essentially two parts to be considered: firstly the application of the radiofrequency (rf) pulse(s) and, secondly the events that occur following this. The essential requirement to induce transitions between energy levels, that is to cause resonance to occur, is the application of a time-dependent magnetic field oscillating at the Larmor frequency of the spin. This field is provided by the magnetic component of the applied rf, which is designated the B_1 field to distinguish it from the static B_0 field. This rf is transmitted via a coil surrounding the sample, the geometry of which is such that the B_1 field exists in the *transverse* plane, perpendicular to the static field. In trying to consider how this oscillating field operates on the bulk magnetisation vector, one is faced with a mind-boggling task involving simultaneous rotating fields and precessing vectors. To help visualise these events it proves convenient to employ a simplified formalism, known as the *rotating frame* of reference, as opposed to the so-called *laboratory frame* of reference described thus far.

2.2.1 The Rotating Frame of Reference

To aid the visualisation of processes occurring during an NMR experiment a number of simple conceptual changes are employed. Firstly, the oscillating B_1 field is considered to be composed of two counter-rotating magnetic vectors in the x–y plane, the resultant of which corresponds exactly to the applied oscillating field (Fig. 2.5). It is now possible to simplify things considerably by eliminating one of these and simultaneously freezing the motion of the other by picturing events in the *rotating frame* of reference (Fig. 2.6). In this, the set of x, y, z co-ordinates are viewed as rotating along with the nuclear precession, in the same sense and at the same rate. Since the frequency of oscillation of the rf field exactly matches that of nuclear precession (which it must for the magnetic resonance condition to be satisfied), the rotation of one of the rf vectors is now static in the rotating frame whereas the other is moving at *twice* the frequency in the *opposite* direction. This latter vector is far from resonance and is simply ignored. Similarly, the precessional motion of the spins has been frozen as these are moving with the same angular velocity as the rf vector and hence the co-ordinate frame. Since this precessional motion was induced by the static magnetic field B_0, this is also no longer present in the rotating frame representation.

The concept of the rotating frame may be better pictured with the following analogy. Suppose you are at a fairground and are standing watching a child going round on the carousel. You see the child move towards you then away from you as the carousel turns, and are thus aware of the circular path the child follows. This corresponds to observing events from the so-called *laboratory frame* of reference (Fig. 2.7a). Now imagine what you see if you step onto the carousel as it turns. You are now travelling with the same angular velocity and in the same sense as the child so the child's motion is no longer

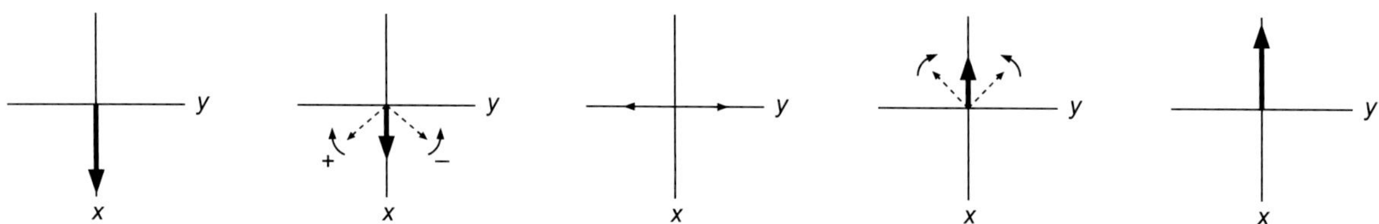

FIGURE 2.5 **The B_1 field.** The rf pulse provides an oscillating magnetic field along one axis (here the x-axis) which is equivalent to two counter-rotating vectors in the transverse plane.

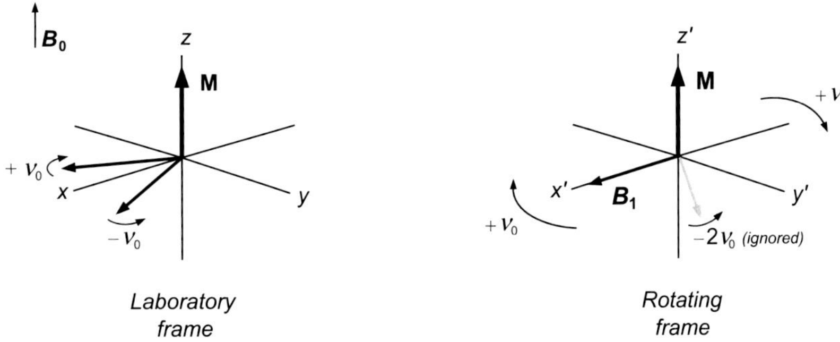

FIGURE 2.6 **Laboratory and rotating frame representations.** In the laboratory frame the coordinate system is viewed as being static, whereas in the rotating frame it rotates at a rate equal to the applied rf frequency ν_0. In this representation the motion of one component of the applied rf is frozen whereas the other is far from the resonance condition and may be ignored. This provides a simplified model for the description of pulsed NMR experiments.

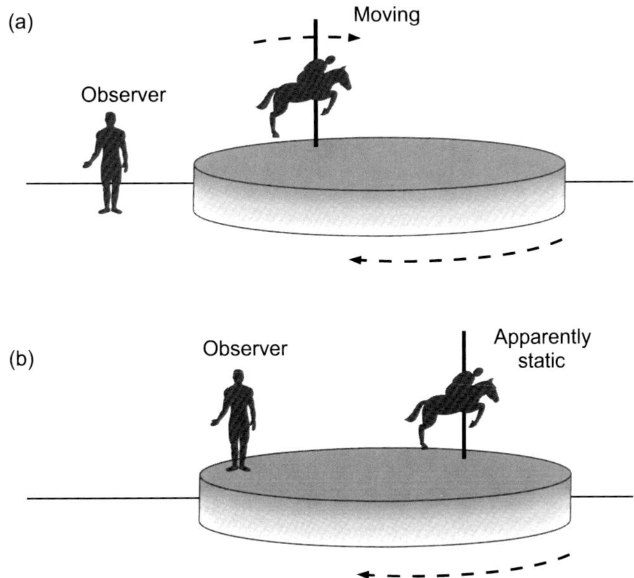

FIGURE 2.7 **Frames of reference.** A fairground carousel can be viewed from (a) the laboratory or (b) the rotating frame of reference.

apparent. The child's precession has been frozen from your point of view and you are now observing events in the *rotating frame* of reference (Fig. 2.7b). Obviously the child is still moving in the 'real' world but your perception of events has been greatly simplified. Likewise, this transposition simplifies our picture of events in an NMR experiment.

Strictly one should use a different co-ordinate labelling scheme for the laboratory and the rotating frames, such as x, y, z and x', y' and z', respectively, as in Fig. 2.6. However, since we shall be dealing almost exclusively with a rotating frame description of events, the simpler x, y, z notations will be used throughout the remainder of the book, and explicit indication provided where the laboratory frame of reference is used.

2.2.2 Pulses

We are now in a position to visualise the effect of applying an rf pulse to the sample. The 'pulse' simply consists of turning on rf irradiation of a defined amplitude for a time period t_p, and then switching it off. As in the case of the static magnetic field, the rf electromagnetic field imposes a torque on the bulk magnetisation vector in a direction perpendicular to the direction of the B_1 field (the 'motor rule') that rotates the vector from the z-axis towards the x–y plane (Fig. 2.8). Thus, applying the rf field along the x-axis will drive the vector towards the y-axis. [Strictly speaking, the sense of rotation is positive about the applied B_1 field so that the vector will be driven towards the $-y$-axis. For clarity of presentation, the vector will always been shown as coming out of the page towards the $+y$-axis for a pulse applied along the $+x$-axis.] The rate at which the vector moves is proportional to the strength of the rf field (γB_1), and so the angle θ through which the vector

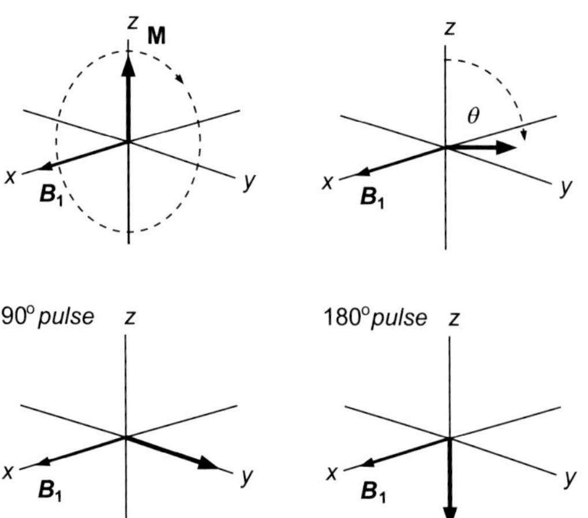

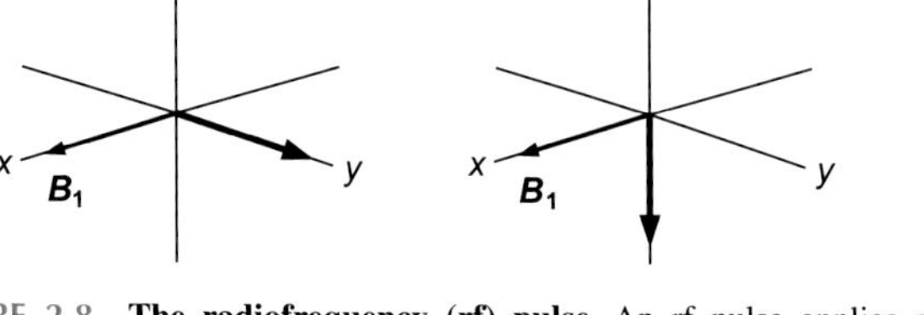

FIGURE 2.8 **The radiofrequency (rf) pulse.** An rf pulse applies a torque to the bulk magnetisation vector and drives it towards the x–y plane from equilibrium. θ is the pulse tip or flip angle which is most frequently 90 or 180 degrees.

FIGURE 2.9 **Phase coherence.** Following a 90 degree pulse, individual spin vectors bunch along the y-axis and are said to possess phase coherence.

turns, colloquially known as the pulse flip or tip angle (but more formally as the nutation angle), will be dependent on the amplitude and duration of the pulse:

$$\theta = 360\,\gamma\,B_1\,t_p \text{ degrees} \tag{2.5}$$

If the rf was turned off just as the vector reached the y-axis, this would represent a 90 degree pulse, if it reached the $-z$-axis, it would be a 180 degree pulse, and so on. Returning to consider the individual magnetic moments that make up the bulk magnetisation vector for a moment, we see that the 90 degrees pulse corresponds to *equalising the populations* of the α and β states, as there is now no net z-magnetisation. However, there *is* net magnetisation in the x–y plane, resulting from 'bunching' of the individual magnetisation vectors caused by application of the rf pulse. The spins are said to possess *phase coherence* at this point, forced upon them by the rf pulse (Fig. 2.9). Note that this equalising of populations is *not* the same as the *saturation* of a resonance, a condition that will be encountered in a variety of circumstances in this book. Saturation corresponds again to equal spin populations but with the phases of the individual spins distributed randomly about the transverse plane such that there exists no net transverse magnetisation and thus no observable signal. In other words, under conditions of saturation the spins lack phase coherence. The 180 degree pulse *inverts the populations* of the spin states, since there must now exist more spins in the β than in the α orientation to place the bulk vector anti-parallel to the static field. Only magnetisation in the x–y plane is ultimately able to induce a signal in the detection coil (see later in the chapter) so that the 90 and 270 degrees pulse will produce maximum signal intensity, but the 180 and 360 degrees pulse will produce none (this provides a useful means of 'calibrating' the pulses, as in the following chapter). The vast majority of the multipulse experiments described in this book, and indeed throughout NMR, use only 90 and 180 degrees pulses.

The example above made use of a 90_x pulse; that is a 90 degree pulse in which the B_1 field was applied along the x-axis. It is however possible to apply the pulse with arbitrary phase, say, along any of the axes x, y, $-x$ or $-y$ as required, which translates to a different starting phase of the excited magnetisation vector. The spectra provided by these pulses show resonances whose *phases* similarly differ by 90 degrees. The detection system of the spectrometer designates one axis to represent the positive *absorption* signal (defined by a receiver reference phase, Section 3.2.2) meaning only magnetisation initially aligned with this axis will produce a pure absorption-mode resonance. Magnetisation that differs from this by +90 degrees is said to represent the pure *dispersion*-mode signal, that which differs by 180 degree is the negative absorption response and so on (Fig. 2.10). Magnetisation vectors initially between these positions result in resonances displaying a mixture of absorption and dispersion behaviour. For clarity and optimum resolution, all resonance peaks are displayed in the favoured absorption mode whenever possible (which is achieved through a process known as phase correction). Note that in all cases the *detected* signals are those *emitted* from the nuclei as described later, and a negative phase signal does not imply a change from emission to absorption of radiation (absorption corresponds to initial excitation of the spins).

The idea of applying a sequence of pulses of different phase angles is of central importance to all NMR experiments. The process of repeating a multipulse experiment with different pulse phases and combining the collected data in an appropriate manner is termed *phase cycling*, and is one of the most widely used procedures for selecting the signals of

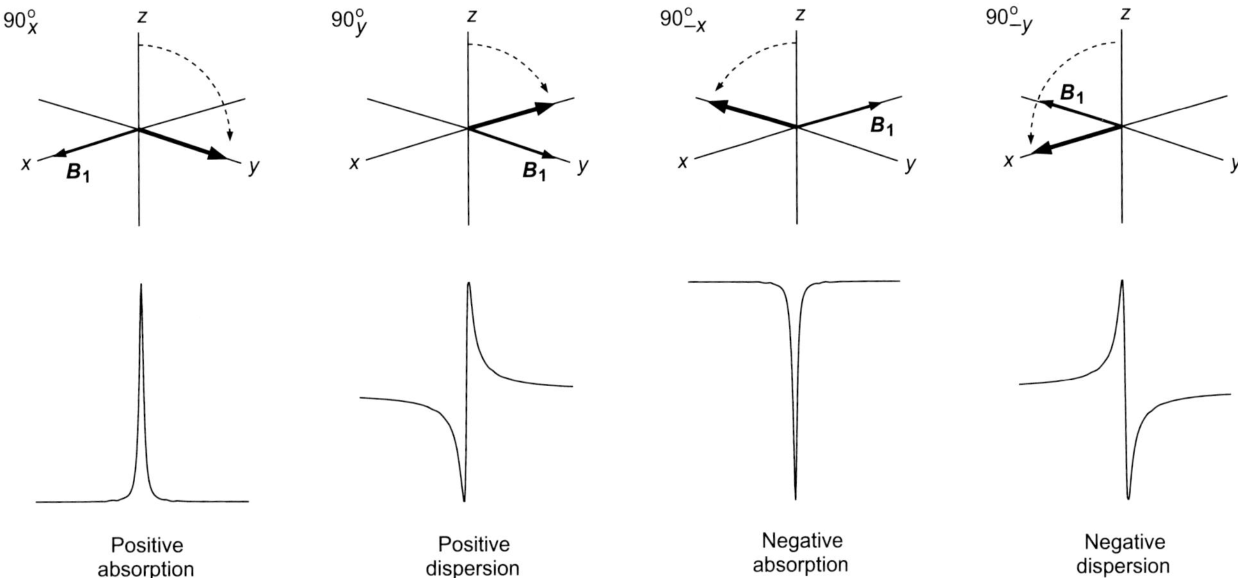

FIGURE 2.10 **Excitation with pulses of varying rf phase.** The differing initial positions of the excited vectors produce NMR resonances with similarly altered phases (here the *y* axis is arbitrarily defined as representing the positive absorption display).

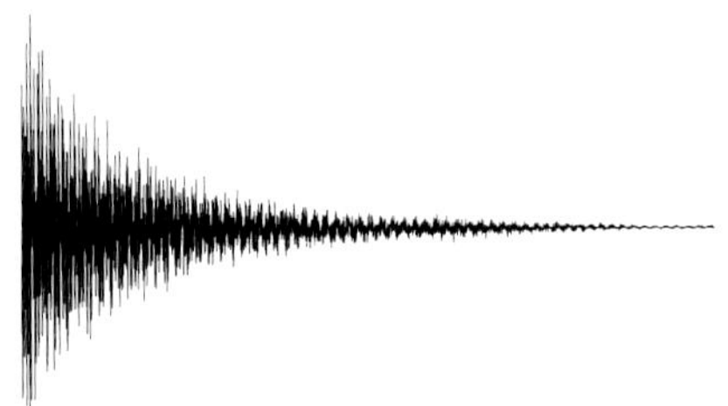

FIGURE 2.11 **The NMR response detected: an FID.** The signal fades as the nuclear spins relax back towards their equilibrium state.

interest in an NMR experiment and rejecting those that are not required. We shall encounter this concept further in the chapter *Practical Aspects of High-Resolution NMR*, and indeed throughout the book.

Now consider what happens immediately after application of, for example a 90°_x pulse. We already know that in the rotating frame the precession of spins is effectively frozen because the B_1 frequency ν_0 and hence the rotating frame frequency exactly match the spin Larmor frequency. Thus, the bulk magnetisation vector simply remains static along the $+y$-axis. However, if we step back from our convenient 'fiction' and return to consider events in the laboratory frame, we see that the vector starts to precess about the z-axis at its Larmor frequency. This rotating magnetisation vector will produce a weak oscillating voltage in the coil surrounding the sample, in much the same way that the rotating magnet in a classic bicycle dynamo induces a voltage in the coils that surround it. These are the electrical signals we wish to detect and it is these that ultimately produce the observed NMR signal. However, magnetisation in the x–y plane corresponds to deviation from equilibrium spin populations and, just like any other chemical system that is perturbed from its equilibrium state, the system will adjust to re-establish this condition, and so the transverse vector will gradually disappear and simultaneously grow along the z-axis. This return to equilibrium is referred to as *relaxation*, and it causes the NMR signal to decay with time, producing the observed free induction decay (FID) (Fig. 2.11). The process of relaxation has wide-ranging implications for the practice of NMR and this important area is also addressed in this introductory chapter.

2.2.3 Chemical Shifts and Couplings

So far we have only considered the rotating frame representation of a collection of like spins, involving a single vector which is stationary in the rotating frame since the reference frequency ν_0 exactly matches the Larmor frequency of the spins (the rf is said to be *on-resonance* for these spins). Now consider a sample containing two groups of chemically distinct but uncoupled spins A and X with chemical shifts of ν_A and ν_X, respectively, which differ by ν Hz. Following

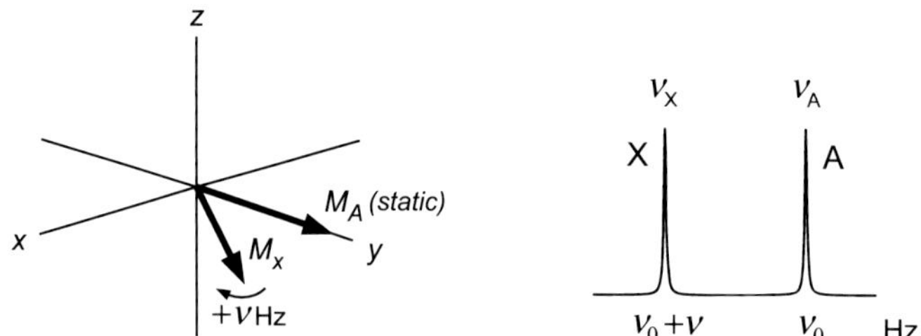

FIGURE 2.12 **Chemical shifts in the rotating frame.** Vectors evolve according to their offsets from the reference (transmitter) frequency ν_0. Here this is on-resonance for spins A ($\nu_0 = \nu_A$) while spins *X* move ahead at a rate of $+\nu$ Hz ($= \nu_X - \nu_0$).

excitation with a single $90°_x$ pulse, both vectors start in the *x–y* plane along the *y*-axis of the rotating frame. Choosing the reference frequency to be on-resonance for the A spins ($\nu_0 = \nu_A$) means these remain along the *y*-axis as before (ignoring the effects of relaxation for the present). If the X spins have a greater chemical shift than A ($\nu_X > \nu_A$) then the X vector will be moving faster than the rotating frame reference frequency by the difference ν Hz so will move *ahead* of A (Fig. 2.12). Conversely, if $\nu_X < \nu_A$ it will be moving more slowly and will lag behind. Three sets of uncoupled spins can be represented by three rotating vectors and so on, such that differences in chemical shifts between spins are simply represented by vectors precessing at different rates in the rotating frame, according to their offsets from the reference frequency ν_0. By using the rotating frame to represent these events we need only consider chemical shift *differences* between the spins of interest, which will be in the kilohertz range, rather than the absolute frequencies, which are of the order of many megahertz. As we shall discover in Section 3.2, this is exactly analogous to the operation of the detection system of an NMR spectrometer, in which a reference frequency is subtracted from the acquired data to produce signals in the kHz region suitable for digitisation. Thus, the 'trick' of using the rotating frame of reference in fact equates directly to a real physical process within the instrument.

When considering the effects of scalar coupling on-resonance it is convenient to remove the effects of chemical shift altogether by choosing the reference frequency of the rotating frame to be the chemical shift of the multiplet of interest. This again helps clarify our perception of events by simplifying rotation of the vectors in the picture. In the case of a doublet, the two lines are represented by two vectors precessing at +J/2 and –J/2 Hz, while for a triplet, the central line remains static and the outer two move at +J and –J Hz (Fig. 2.13). In many NMR experiments it is desirable to control the orientation of multiplet vectors with respect to one another, and, as we shall see, a particularly important relationship is when two vectors are *anti-phase* to one another, that is sitting in opposite directions. This can be achieved simply by choosing an appropriate delay period in which the vectors evolve, which is 1/2J for a doublet but 1/4J for the triplet.

2.2.4 Spin-Echoes

Having seen how to represent chemical shifts and J couplings with the vector model, we are now in a position to see how we can manipulate the effects of these properties in simple multi-pulse experiments. The idea here is to provide a simple introduction to using the vector model to understand what is happening during a pulse sequence. In many experiments, there exist time delays in which magnetisation is simply allowed to precess under the influence of chemical shifts and couplings, usually with the goal of producing a defined state of magnetisation before further pulses are applied or data are acquired. To illustrate these points, we consider one of the fundamental building blocks of numerous NMR experiments: the spin echo.

Consider first two groups of chemically distinct protons A and X that share a mutual coupling J_{AX}, which will be subject to the simple two-pulse sequence in Fig. 2.14. For simplicity we shall consider the effect of chemical shifts and couplings separately, starting with chemical shifts and again assuming the reference frequency to be that of the A spins (Fig. 2.15). The initial $90°_x$ creates transverse A and X magnetisation, after which the X vector precesses during the first time interval Δ. The following $180°_y$ pulse (note this is now along the *y*-axis) rotates the X magnetisation through 180 degrees about the *y*-axis, and so places it back in the *x–y* plane, but now lagging *behind* the A vector. A-spin magnetisation remains along the *y*-axis so is invariant to this pulse. During the second time period Δ, the X magnetisation will precess through the same angle as in the first period and at the end of the sequence finishes where it began and *at the same position as the A vector*. Thus, after the time period 2Δ no phase difference has accrued between the A and X vectors despite their different shifts, and it were as if the A and X spins had the same chemical shift throughout the 2Δ period. We say *the spin-echo has refocused the chemical shifts*, the dephasing and rephasing stages giving rise to the echo terminology.

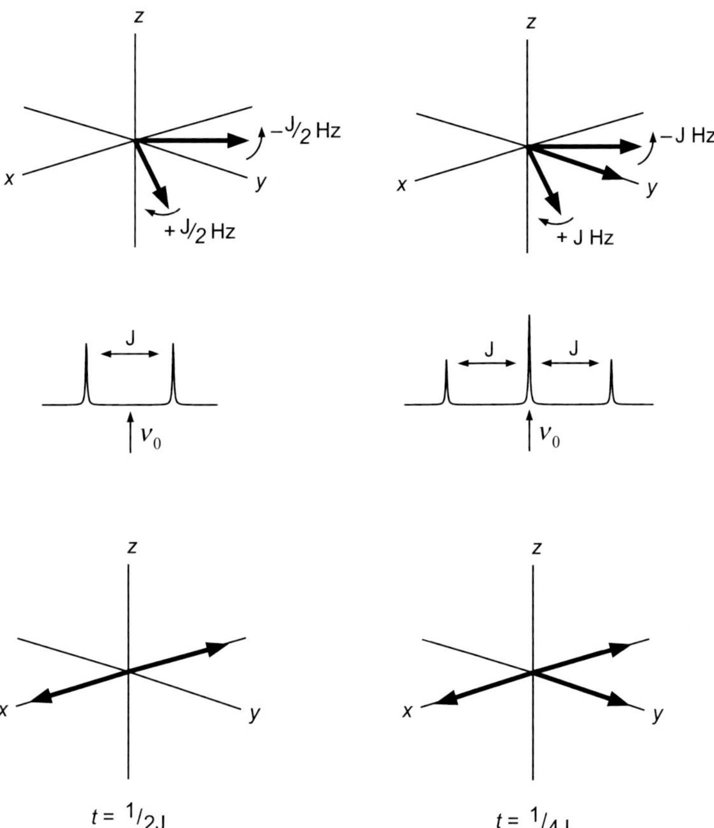

FIGURE 2.13 **Scalar couplings in the rotating frame.** Multiplet components evolve according to their coupling constants. The vectors have an antiphase disposition after an evolution period of 1/2J and 1/4J s for doublets and triplets, respectively.

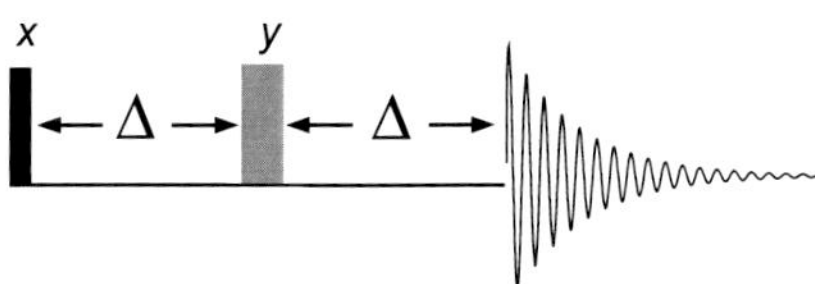

FIGURE 2.14 **The basic spin-echo pulse sequence.**

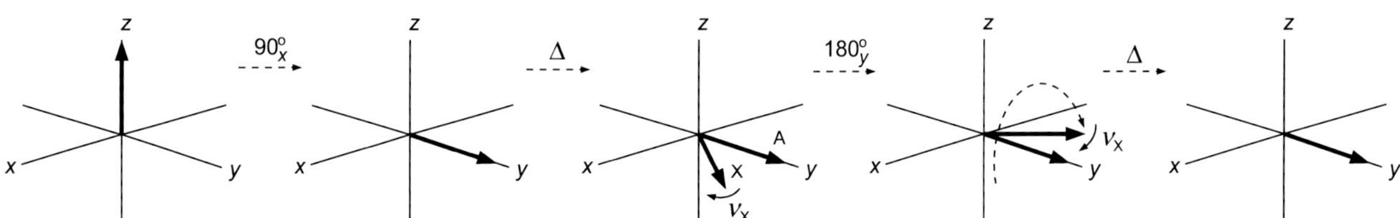

FIGURE 2.15 **Chemical shift evolution is refocused with the spin-echo sequence.**

Consider now the effect on the coupling between the two spins with reference to the multiplet of spin A, safe in the knowledge that we can ignore the effects of chemical shifts. Again, during the first period Δ the doublet components will move in opposite directions, and then have their positions interchanged by application of the $180°_y$ pulse. At this point it would be obvious to assume that the two halves of the doublets would simply refocus as in the case of the chemical shift differences above, but we have to consider the effect of the 180 degrees pulse on the J-coupled partner also; in other words, the effect on the X spins. To appreciate what is happening, we need to remind ourselves of what it is that gives rise to two halves of a doublet. These result from spin *A* being coupled to its partner X, which can have one of two orientations (α or β) with respect to the magnetic field. When spin X has one orientation, spin A will resonate as the high-frequency half of its

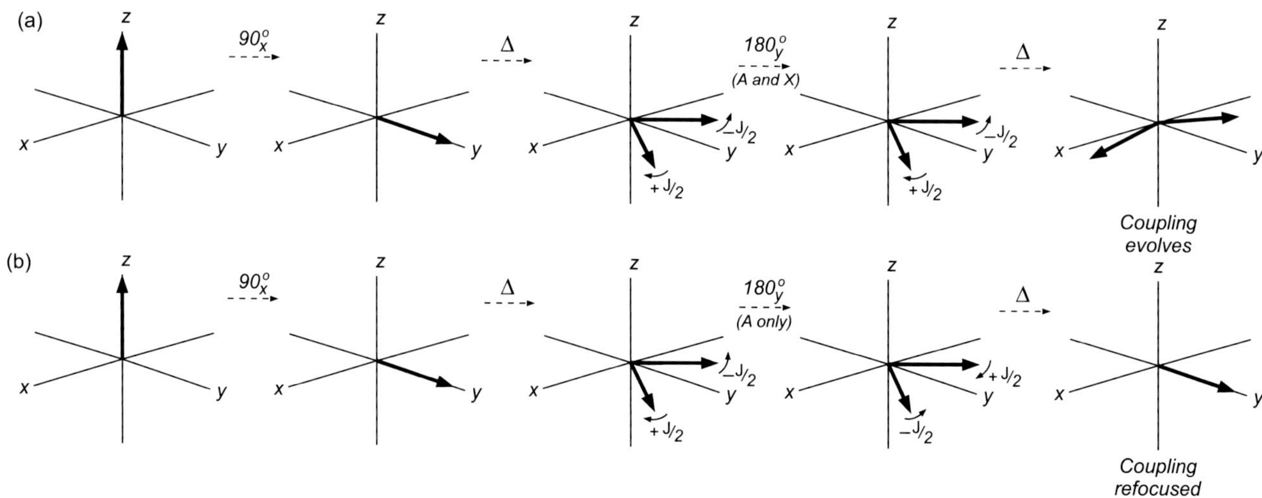

FIGURE 2.16 **The influence of spin-echoes on scalar coupling as illustrated for two coupled spins A and X.** (a) A homonuclear spin echo (in which both spins experience a 180-degree pulse) allows the coupling to evolve throughout the sequence. (b) A heteronuclear spin echo (in which only one spin experiences a 180-degree pulse) causes the coupling to refocus. If both heteronuclear spins experience 180-degree pulses, the heteronuclear coupling evolves as in (a) (see text).

doublet, while with X in the other, A will resonate as the low-frequency half. As there are approximately the same number of X spins in the α and β orientations, the two halves of the A doublet will be of equal intensity (obviously there are not exactly equal numbers of α and β X spins, otherwise there would be no NMR signal to observe, but the difference is so small as to be negligible for these arguments). The effect of applying the 180 degrees pulse on the X spins is to invert the relative orientations, so that any A spin that was coupled to Xα is now coupled to Xβ, and vice versa. This means the faster moving vector now becomes the slower and vice versa, the overall result being represented in Fig. 2.16a. The two halves of the doublet therefore *continue to dephase*, so that by the end of the 2Δ period, the J coupling, in contrast to the chemical shifts, have continued to evolve so that *homonuclear couplings are not refocused by a spin-echo.* The reason for adding the term *homonuclear* to the previous statement is because it does not necessarily apply to the case of heteronuclear spin-echoes, that is when we are dealing with two different nuclides, such as ^{1}H and ^{13}C. This is because in a heteronuclear system one may choose to apply the 180 degree pulse on only one channel, thus only one of the two nuclides will experience this pulse and refocusing of the heteronuclear coupling will occur in this case (Fig. 2.16b). If two simultaneous 180 degree pulses are applied to both nuclei *via* two different frequency sources, continued defocusing of the heteronuclear coupling occurs exactly as for the homonuclear spin-echo above.

The use of the $180°_y$ pulse instead of a $180°_x$ pulse in the above sequences was employed to provide a more convenient picture of events, yet it is important to realise that the refocusing effects on chemical shift and couplings described earlier would also have occurred with a $180°_x$ pulse except that the refocused vectors would now lie along the −*y*-axis instead of the +*y*-axis. One further feature of the spin-echo sequence is that it will also refocus the deleterious effects that arise from inhomogeneities in the static magnetic field, as these may be viewed as just another contribution to chemical shift differences throughout the sample. The importance of the spin echo in modern NMR techniques can hardly be overemphasised. It allows experiments to be performed without having to worry about chemical shift differences within a sample and the complications these may introduce (eg phase differences). This then allows us to manipulate spins according to their couplings with neighbours and it is these interactions that are exploited to the full in many of the modern NMR techniques described later.

2.3 TIME AND FREQUENCY DOMAINS

It was shown in the previous section that the emitted rf signal from excited nuclear spins (the FID) is detected as a time-dependent oscillating voltage which steadily decays as a result of spin relaxation. In this form the data are of little use to us because they are a time domain representation of the nuclear precession frequencies within the sample. What we actually want is a display of the frequency components that make up the FID as it is these we relate to transition energies and ultimately chemical environments. In other words, we need to transfer our *time domain* data into the *frequency domain.*

The time and frequency domains are related by a simple function, one being the inverse of the other (Fig. 2.17). The complicating factor is that a genuine FID is usually composed of potentially hundreds of components of differing frequencies

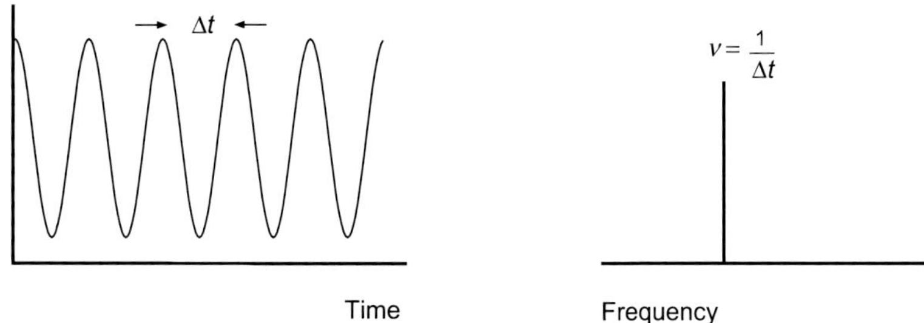

FIGURE 2.17 **Time and frequency domains share a simple inverse relationship.**

and amplitude, in addition to noise and other possible artefacts, and in such cases the extraction of frequencies by direct inspection is impossible. By far the most widely used method to produce the frequency domain spectrum is the mathematical procedure of Fourier transformation, which has the general form:

$$f(\omega) = \int_{-\infty}^{+\infty} f(t)e^{i\omega t}\,\mathrm{dt} \tag{2.6}$$

where $f(\omega)$ and $f(t)$ represent the frequency and time domain data, respectively. In the very early days of pulse–Fourier transform (FT) NMR the transform was often the rate-limiting step in producing a spectrum, although with today's computers and the use of a fast FT procedure (the Cooley–Tukey algorithm) the time requirements are of little consequence. Fig. 2.18 demonstrates this procedure for very simple spectra. Clearly, even for these rather simple spectra of only a few

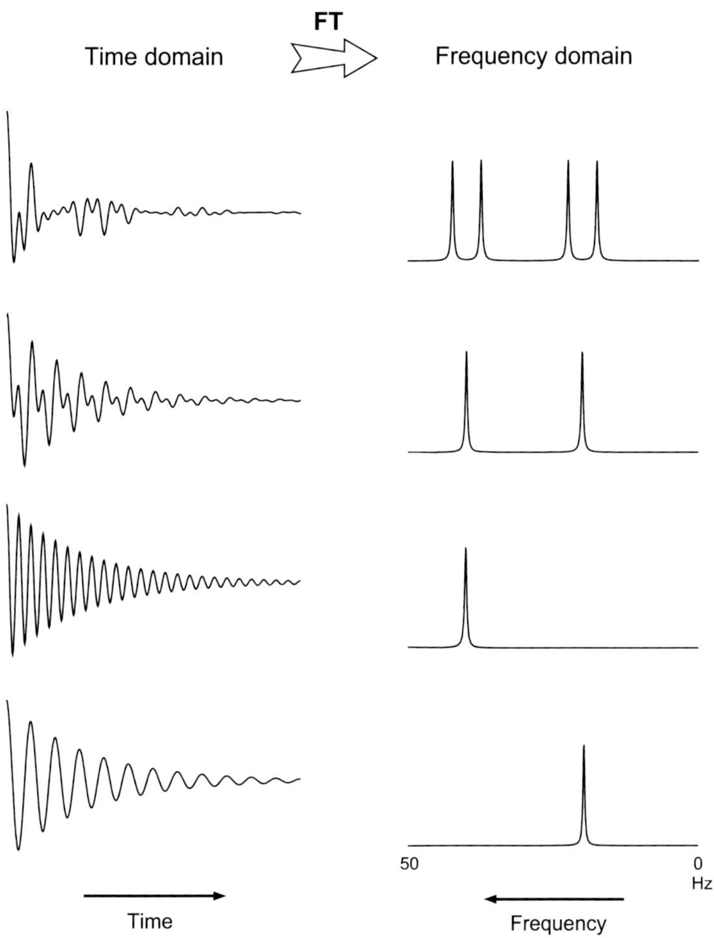

FIGURE 2.18 **Fourier transformation of time domain FIDs produces the corresponding frequency domain spectra.**

lines the corresponding FID rapidly becomes too complex for direct interpretation, whereas this is impossible for a genuine FID of any complexity (see Fig. 2.11).

The details of the Fourier transform itself are usually of little consequence to anyone using NMR, although there is one notable feature to be aware of. The term $e^{i\omega t}$ can equally be written $\cos\omega t + i\sin\omega t$ and in this form it is apparent that the transformation actually results in *two* frequency domain spectra that differ only in their signal *phases*. The two are cosine and sine functions so are 90 degrees out of phase relative to one another and are termed the 'real' and 'imaginary' parts of the spectrum (because the function contains complex numbers). Generally, we are presented with only the 'real' part of the data (although the 'imaginary' part can usually be displayed) and with appropriate phase adjustment we choose this to contain the desired pure absorption-mode data and the imaginary part to contain the dispersion-mode representation. The significance of this phase relationship will be pursued in the chapters *Practical Aspects of High-Resolution NMR* and *Introducing Two-Dimensional and Pulsed Field Gradient NMR*.

2.4 SPIN RELAXATION

The action of an rf pulse on a sample at thermal equilibrium perturbs the nuclear spins from this restful state. Following this perturbation, one would intuitively expect the system to re-establish the equilibrium condition, and lose the excess energy imparted into the system by the applied pulse. A number of questions then arise: Where does this energy go, how does it get there (or in other words what mechanisms are in place to transfer the energy), and how long does all this take? While some appreciation of all these points is desirable, it is the last of the three that has greatest bearing on the day-to-day practice of NMR. The lifetimes of excited nuclear spins are extremely long when compared with, say, the excited electronic states of optical spectroscopy. These may be a few seconds or even minutes for nuclear spins as opposed to less than a picosecond for electrons, a consequence of the low transition energies associated with nuclear resonance. These extended lifetimes are crucial to the success of NMR spectroscopy as an analytical tool in chemistry. Not only do these mean that resonance peaks are rather narrow relative to those of rotational, vibrational or electronic transitions (as a consequence of the Heisenberg Uncertainty principle), but it also provides time to manipulate the spin systems after their initial excitation, performing a variety of spin gymnastics and so modifying the information available in the resulting spectra. This is the world of multi-pulse NMR experiments, into which we shall enter shortly, and knowledge of relaxation rates has considerable bearing on the design of these experiments, on how they should be implemented and on the choice of experimental parameters for optimum results. Even in the simplest possible case of a single-pulse experiment, relaxation rates influence both achievable resolution and sensitivity (as mentioned in the opening chapter, the earliest attempts to observe the NMR phenomenon probably failed because of a lack of understanding at that time of the spin relaxation properties of the samples used).

The relaxation rates of nuclear spins can also be related to aspects of molecular structure and behaviour in favourable circumstances; in particular, internal molecular motions. It is true to say however that the relationship between relaxation rates and structural features are not as well defined as those of the chemical shift and spin–spin coupling constants, and are not used on a routine basis. The problem of reliable interpretation of relaxation data arises largely from the numerous extraneous effects that influence experimental results, meaning empirical correlations for using such data are not generally available and this aspect of NMR will not be further pursued.

2.4.1 Longitudinal Relaxation: Establishing Equilibrium

Immediately after pulse excitation of nuclear spins the bulk magnetisation vector is moved away from the thermal equilibrium $+z$-axis, which corresponds to a change in spin populations. The recovery of magnetisation along the z-axis, termed *longitudinal relaxation*, therefore corresponds to the equilibrium populations being re-established, and hence to an overall loss of energy of the spins (Fig. 2.19). The energy lost by the spins is transferred into the surroundings in the form of heat,

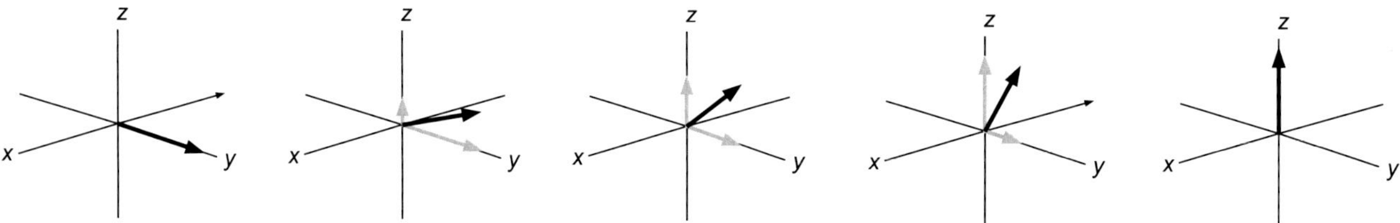

FIGURE 2.19 **Longitudinal relaxation.** The recovery of a magnetisation vector (shown on-resonance in the rotating frame) diminishes the transverse (x–y) and re-establishes the longitudinal (z) components.

although the energies involved are so small that temperature changes in the bulk sample are undetectable. This gives rise to the original term for this process as *spin–lattice relaxation* which originated in the early days of solid-state NMR where the excess energy was described as dissipating into the surrounding rigid lattice.

The Bloch theory of NMR assumes that the recovery of the $+z$ magnetisation M_z follows exponential behaviour described by:

$$\frac{dM_z}{dt} = \frac{(M_0 - M_z)}{T_1} \tag{2.7}$$

where M_0 is magnetisation at thermal equilibrium, and T_1 is the (first-order) time constant for this process. Although exponential recovery was proposed as a hypothesis, it turns out to be an accurate model for the relaxation of spin-½ nuclei in most cases. Starting from the position of no z magnetisation (eg immediately after the sample has been placed in the magnet or after a 90 degree pulse) the longitudinal magnetisation at time t will be:

$$M_z = M_0(1 - e^{-t/T_1}) \tag{2.8}$$

as illustrated in Fig. 2.20. It should be stressed that T_1 is usually referred to as the longitudinal relaxation time throughout the NMR community (and, following convention, throughout the remainder of this book), whereas, in fact, it is a *time constant* rather that a direct measure of the *time* required for recovery. Similarly, when referring to the rate at which magnetisation recovers, $1/T_1$ represents the *rate constant* R_1 (s^{-1}) for this process.

For medium-sized organic molecules (those with a mass of a few hundred), proton T_1s tend to fall in the range 0.5–5 s, whereas carbon T_1s tend to range from a few seconds to many tens of seconds. For spins to relax fully after a 90 degree pulse, it is necessary to wait a period of at least $5T_1$ (at which point magnetisation has recovered by 99.33%) and thus it may be necessary to wait many minutes for full recovery. This is rarely the most time efficient way to collect NMR spectra and Section 4.1 describes the correct approach. The reason such long periods are required lies not in the fact that there is nowhere for the excess energy to go, since the energies involved are so small they can be readily taken up in the thermal energy of the sample, but rather that there is no efficient means for transferring this energy. The time required for *spontaneous* emission in NMR is so long that this has a negligible effect on spin populations, so *stimulated* emission must be operative for relaxation to occur. Recall that the fundamental requirement for inducing nuclear spin transitions, and hence

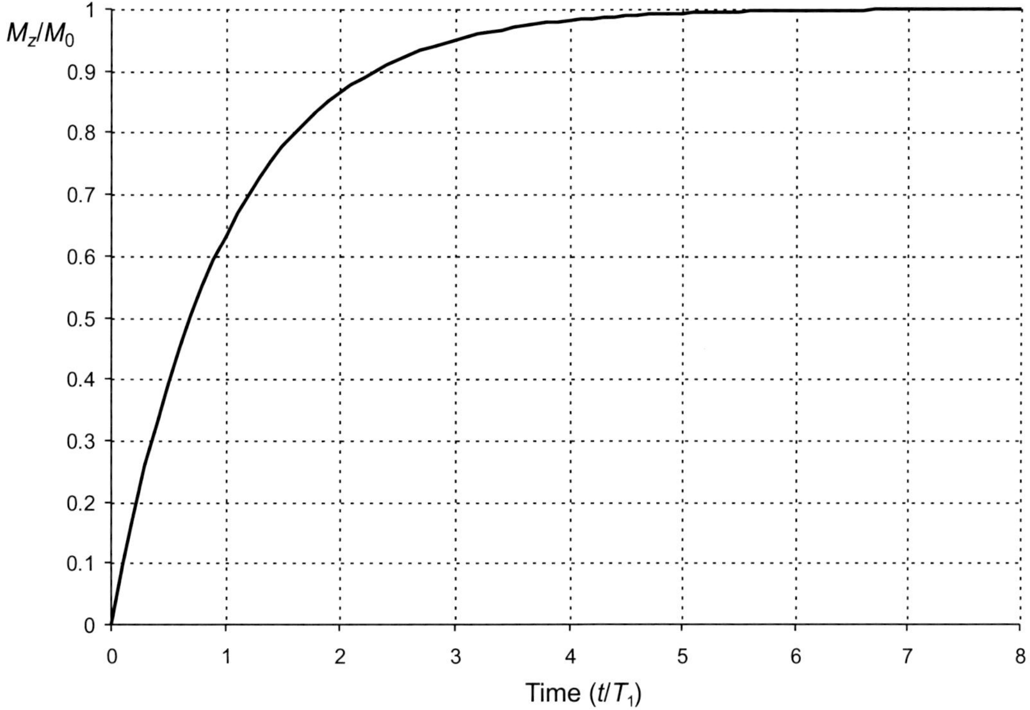

FIGURE 2.20 **Longitudinal magnetisation recovery.** The exponential growth of longitudinal magnetisation is dictated by the time constant T_1 and is essentially complete after a period of $5T_1$.

restoring equilibrium populations in this case, is a magnetic field oscillating at the Larmor frequency of the spins. The long relaxation times suggests such suitable fields are not in great abundance. These fields can arise from a variety of sources with the oscillations required to induce relaxation arising from local molecular motions. Although the details of the various relaxation mechanisms can become rather complex, a qualitative appreciation of these, as in Section 2.5, is important for understanding many features of NMR spectra. At a practical level, some knowledge of T_1s in particular is crucial to optimum execution of almost every NMR experiment, and the simple sequence below offers both a gentle introduction to multipulse NMR techniques as well as presenting a means of measuring this important parameter.

2.4.2 Measuring T_1 with the Inversion Recovery Sequence

There are a number of different experiments devised for determination of the longitudinal relaxation times of nuclear spins and only the most commonly applied method, inversion recovery, will be considered here. The full procedure is described first, followed by the 'quick-and-dirty' approach which is handy for experimental setup.

In essence, all one needs to do to determine T_1s is to perturb a spin system from thermal equilibrium and then devise some means of following its recovery as a function of time. The inversion recovery experiment is a simple two-pulse sequence (Fig. 2.21) that, as the name implies, creates the initial population disturbance by *inverting* the spin populations through application of a 180 degree pulse. The magnetisation vector, initially aligned with the $-z$-axis, will gradually shrink back toward the x–y plane, pass through this and eventually make a full recovery along the $+z$-axis at a rate dictated by the quantity of interest T_1. Since magnetisation along the z-axis is unobservable, recovery is monitored by placing the vector back in the x–y plane with a 90 degrees pulse after a suitable period τ following the initial inversion (Fig. 2.22).

If τ is zero, the magnetisation vector terminates with full intensity along the $-y$-axis producing an inverted spectrum using conventional spectrum phasing; that is, defining the $+y$-axis to represent positive absorption. Repeating the experiment with increasing values of τ allows one to follow the relaxation of the spins in question (Fig. 2.23). Finally, when τ is sufficiently long ($\tau_\infty > 5T_1$) complete relaxation will occur between the two pulses and the maximum positive signal is recorded. The intensity of the detected magnetisation M_t follows:

$$M_t = M_0(1 - 2e^{-\tau/T_1}) \tag{2.9}$$

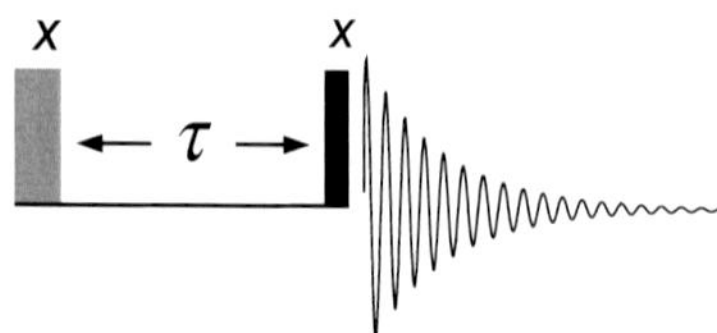

FIGURE 2.21 **The inversion recovery sequence.**

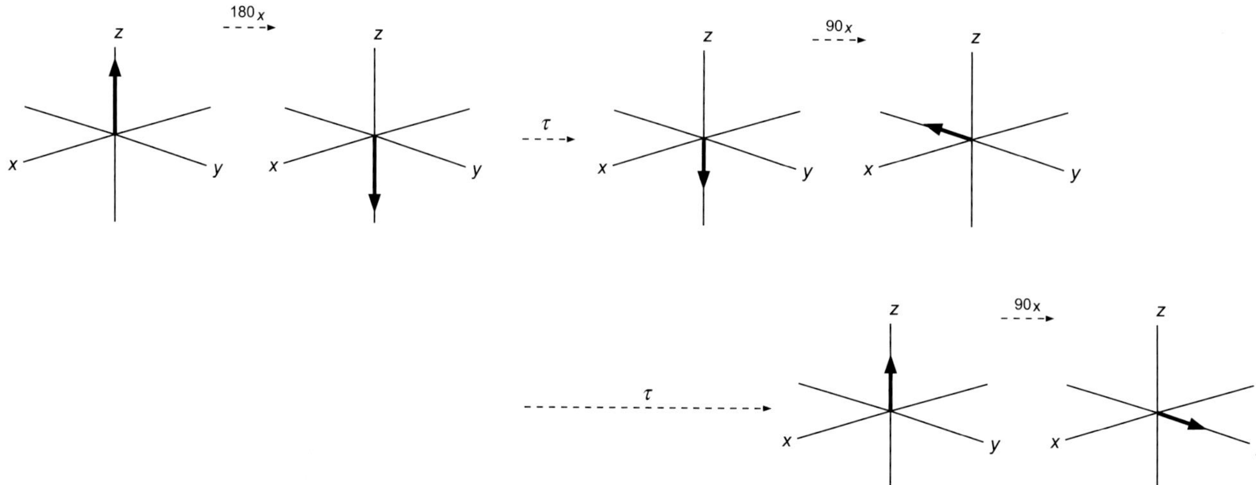

FIGURE 2.22 **The inversion recovery process.** With short recovery periods the vector finishes along the $-y$-axis, so the spectrum is inverted; while with longer periods a conventional spectrum of scaled intensity is obtained.

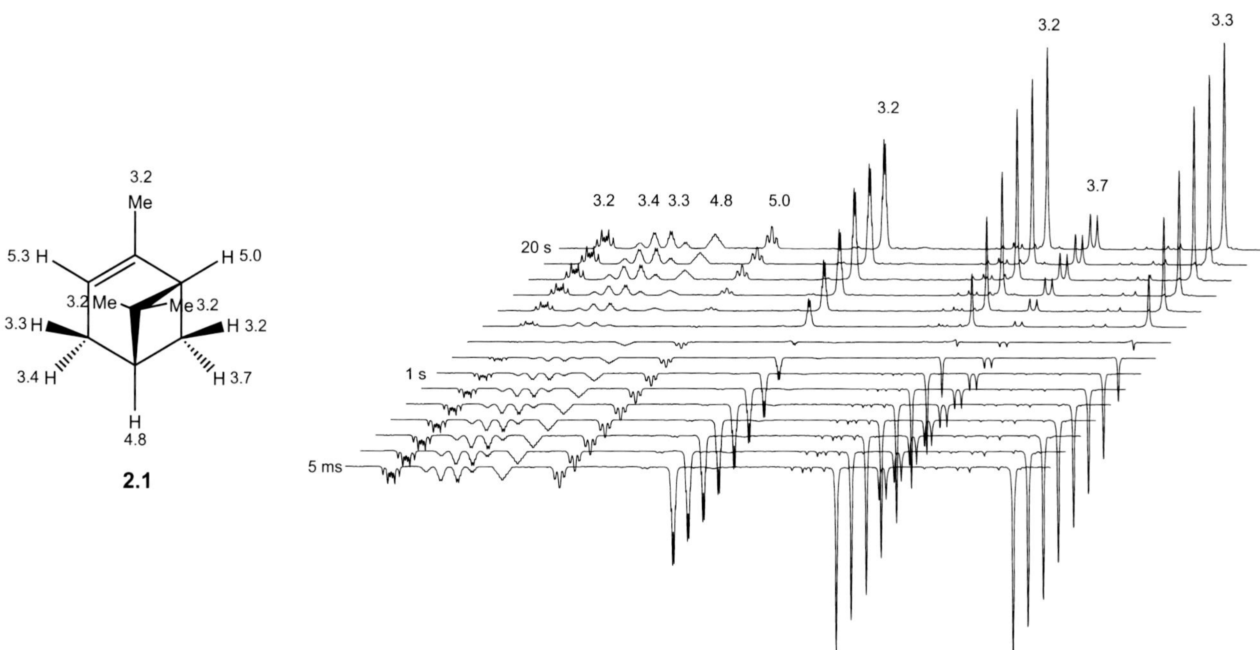

FIGURE 2.23 **The ^{1}H inversion recovery experiment performed on α-pinene 2.1.** Recovery delays in the range 5 ms to 20 s were used and the calculated T_1s shown for each resonance obtained from fitting peak intensities as described in the text. The sample was prepared in non-degassed $CDCl_3$.

where M_0 corresponds to equilibrium magnetisation, such as that recorded at τ_∞. Note here the additional factor of 2 relative to Eq. 2.8 as the recovery starts from inverted magnetisation. The relaxation time is determined by fitting the signal intensities to this equation, algorithms for which are found in many NMR software packages. The alternative traditional method of extracting T_1 from such an equation is to analyse a semi-logarithmic plot of $\ln(M_0 - M_t)$ versus τ whose slope is $1/T_1$.

The most likely causes of error in the use of the inversion recovery method are inaccurate recording of M_0 if full equilibration is not achieved, and inaccuracies in the 180 degree pulse causing imperfect initial inversion. The scaling factor (2 in Eq. 2.9)) can be made variable in fitting routines to allow for incomplete inversion.

2.4.2.1 *Quick* T$_1$ *Estimation*

In many practical cases it is sufficient to have just an estimation of relaxation times in order to calculate the optimum experimental timings for the sample at hand. In these instances the procedure described above is overly elaborate and since our molecules are likely to contain nuclei exhibiting a range of T_1s, accurate numbers will be of little use in experiment setup. This 'quick and dirty' method is sufficient to provide estimates of T_1 and again makes use of the inversion recovery sequence. Ideally, the sample in question will be sufficiently strong to allow rather few scans per τ value, making the whole procedure quick to perform. The basis of the method is the disappearance of signals when the longitudinal magnetisation passes through the x–y plane on its recovery (at time τ_{null}), because at this point the population difference is zero ($M_t = 0$). From the above equation, it can be shown that:

$$\tau_{null} = T_1 \ln 2 \tag{2.10}$$

hence

$$T_1 = \frac{\tau_{null}}{\ln 2} = 1.443\,\tau_{null} \tag{2.11}$$

Thus, the procedure is to run an experiment with $\tau = 0$ and adjust (phase) the spectrum to be negative absorption. After having waited $>5T_1$ repeat the experiment with an incremented τ using the same phase adjustments, until the signal passes through the null condition (Fig. 2.23), thus defining τ_{null}, which may be different for each resonance in the spectrum. Errors may be introduced from inaccurate 180 degree pulses, from off-resonance effects (see Section 3.2) and from waiting for insufficient periods between acquisitions, so the fact that these values are estimates cannot be overemphasised.

One great problem with these methods is the need to know something about the T_1s in the sample even before these measurements. Between each new τ value one must wait for the system to come to equilibrium, and if signal averaging were required one would also have to wait this long between each repetition! Unfortunately, it is the weak samples that require signal averaging that will benefit most from a properly executed experiment. To avoid this it is wise is to develop a feel for the relaxation properties of the types of nuclei and compounds you commonly study so that when you are faced with new material you will have some 'ballpark' figures to provide guidance. Influences on the magnitude of T_1 are considered in Section 2.5.

2.4.3 Transverse Relaxation: Loss of Magnetisation in the *x–y* Plane

Referring back to the situation immediately following a 90 degree pulse in which the transverse magnetisation is on-resonance in the rotating frame, there exists another way in which observable magnetisation can be lost. Recall that the bulk magnetisation vector results from the addition of many microscopic vectors for the individual nuclei that are said to possess phase coherence following the pulse. In a sample of like spins one would anticipate that these would remain static in the rotating frame, perfectly aligned along the *y*-axis (ignoring the effects of longitudinal relaxation). However, this only holds if the magnetic field experienced by each spin in the sample is *exactly* the same. If this is not the case, some spins will experience a slightly greater local field than the mean causing them to have a higher frequency and to creep ahead, whereas others will experience a slightly smaller field and start to lag behind. This results in a fanning-out of the individual magnetisation vectors, which ultimately leads to no *net* magnetisation in the transverse plane (Fig. 2.24). This is another form of relaxation referred to as *transverse relaxation* which is again assumed to occur with an exponential decay now characterised by the time constant T_2.

Magnetic field differences in the sample can be considered to arise from two distinct sources. The first is simply from static magnetic field inhomogeneity throughout the sample volume which is really an instrumental imperfection; it is this one aims to minimise for each sample when optimising or 'shimming' the static magnetic field. The second is from the local magnetic fields arising from intramolecular and intermolecular interactions in the sample, which represent 'genuine' or 'natural' transverse relaxation processes. The relaxation time constant for the two sources combined is designated T_2* such that:

$$\frac{1}{T_2^*} = \frac{1}{T_2} + \frac{1}{T_{2(\Delta B_0)}} \tag{2.12}$$

where T_2 refers to the contribution from genuine relaxation processes, and $T_{2(\Delta B_0)}$ to that from field inhomogeneity. The decay of transverse magnetisation is manifested in the observed FID. Moreover, the widths of resonance peaks are inversely proportional to T_2* since a short T_2* corresponds to faster blurring of the transverse magnetisation which in turn corresponds to a greater frequency difference between the vectors and thus a greater spread (broader line) in the frequency dimension (Fig. 2.25). For (single) exponential relaxation the lineshape is Lorentzian with a half-height line-width $\Delta v_{1/2}$ (Fig. 2.26) of

$$\Delta v_{1/2} = \frac{1}{\pi T_2^*} \tag{2.13}$$

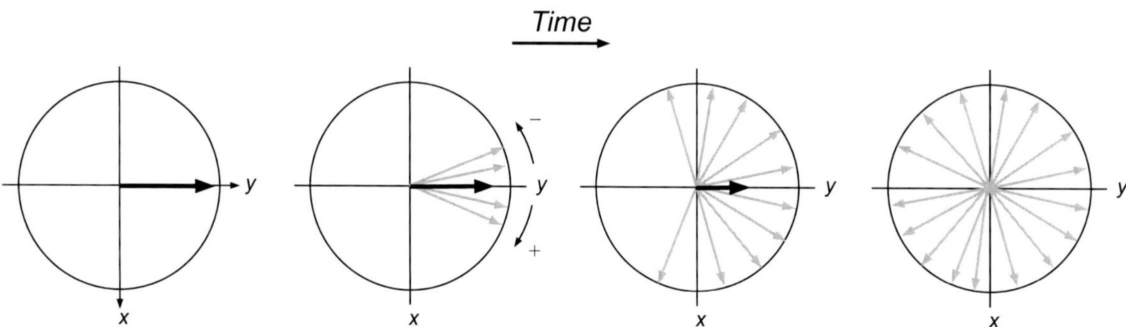

FIGURE 2.24 **Transverse relaxation.** Local field differences within the sample cause spins to precess with slightly differing frequencies, eventually leading to zero net transverse magnetisation.

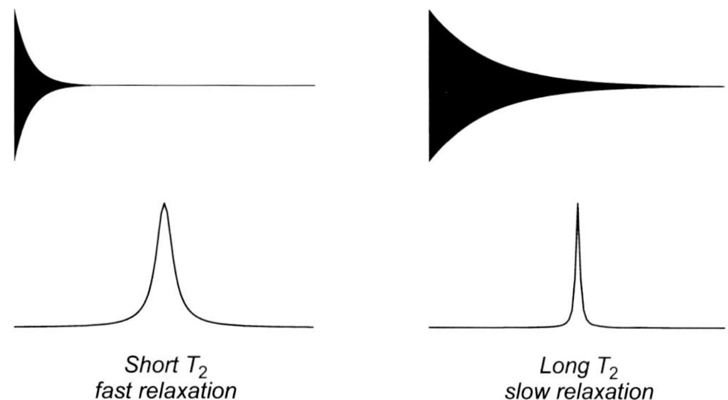

FIGURE 2.25 **Resonance linewidths.** Rapidly relaxing spins produce fast-decaying FIDs and broad resonances, while those which relax slowly produce longer FIDs and narrower resonances.

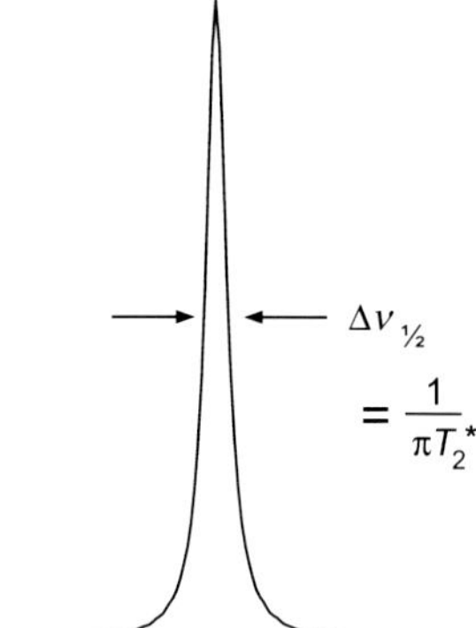

FIGURE 2.26 **Definition of the half-height linewidth of a resonance.**

For most spin-½ nuclei in small, rapidly tumbling molecules in low-viscosity solutions, it is field homogeneity that provides the dominant contribution to observed linewidths, and it is rarely possible to obtain genuine T_2 measurements directly from these. However, nuclei with spin >½ (quadrupolar nuclei) may be relaxed very efficiently by interactions with local electric field gradients and so have broad lines and short T_2s that can be determined directly from linewidths.

Generally speaking, relaxation mechanisms that operate to restore longitudinal magnetisation also act to destroy transverse magnetisation, and since there clearly can be no magnetisation remaining in the *x*–*y* plane when it has all returned to the +*z*-axis, T_2 can never be longer than T_1. However, additional mechanisms may also operate to reduce T_2, so that it may be shorter. Again, for most spin-½ nuclei in small, rapidly tumbling molecules, T_1 and T_2 have the same value, while for large molecules that tumble slowly in solution or for solids, T_2 is often very much shorter than T_1 (see Section 2.5). Whereas longitudinal relaxation causes a loss of energy from the spins, transverse relaxation occurs by mutual swapping of energy *between spins*; for example, one spin being excited to the β state while another simultaneously drops to the α state; the so-called 'flip-flop' process. This gives rise to the original term of *spin–spin relaxation* which is still in widespread use. Longitudinal relaxation is thus an enthalpic process whereas transverse relaxation is entropic. Although the measurement of T_2 has far less significance in routine spectroscopy, methods for this are described below for completeness and an alternative practical use of these is also presented.

2.4.4 Measuring T_2 with a Spin-Echo Sequence

Measurement of the natural transverse relaxation time T_2 could in principle be obtained if the contribution from magnetic field inhomogeneity was removed. This can be achieved, as has been suggested already, by use of a spin-echo sequence. Consider again a sample of like spins and imagine the sample to be composed of microscopically small regions such that within each region the field is perfectly homogeneous. Magnetisation vectors within any given region will precess at the same frequency and these are sometimes referred to as *isochromats* (meaning 'of the same colour' or frequency). In the basic two-pulse echo sequence (Fig. 2.27a) some components move ahead of the mean while others lag behind during

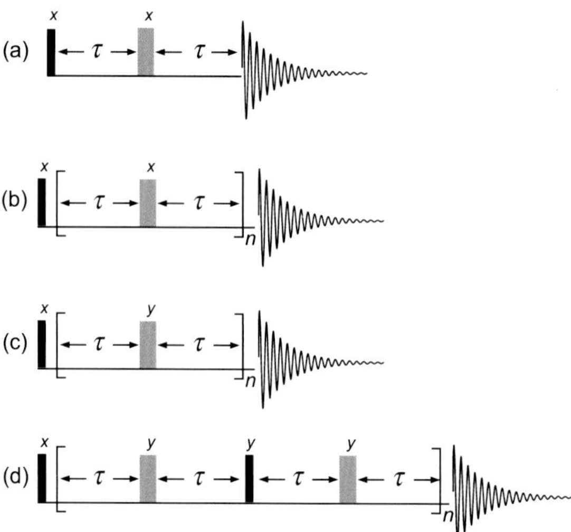

FIGURE 2.27 **Spin-echo sequences for measuring T_2 relaxation times.** (a) A basic spin-echo, (b) the Carr–Purcell sequence, (c) the Carr–Purcell–Meiboom–Gill (CPMG) sequence, and (d) the CPMG–PROJECT sequence.

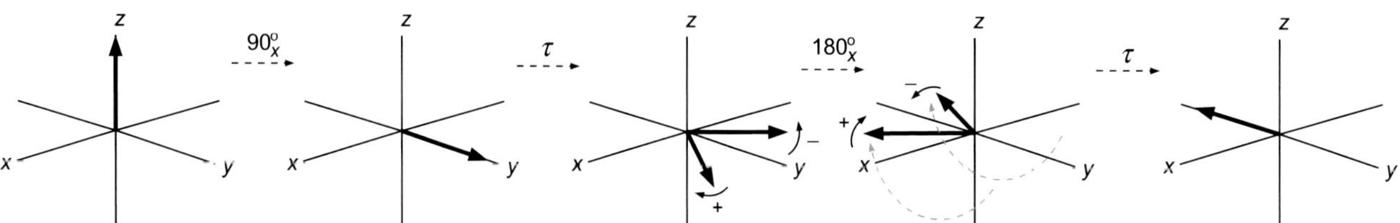

FIGURE 2.28 **The spin echo refocuses magnetisation vectors dephased by field inhomogeneity.**

the time period τ (Fig. 2.28). The 180 degree pulse rotates the vectors toward the $-y$-axis and following a further period τ the faster moving vectors coincide with the slower ones along the $-y$-axis. Thus, the echo has refocused blurring in the x–y plane caused by field inhomogeneities. If one were to start acquiring data immediately after the 90 degree pulse, one would see the FID decay away initially but then reappear after a time 2τ as the echo forms (Fig. 2.29a). However, during the 2τ time period, some loss of phase coherence by natural transverse relaxation also occurs, and this is *not* refocused by the spin-echo since, in effect, there is no phase memory associated with this process to be undone. This means that at the time of the echo the intensity of the observed magnetisation will have decayed according to the *natural* T_2 time constant, independent of field inhomogeneity. This can clearly be seen in a train of spin echoes applied during the acquisition of an FID (Fig. 2.29b).

A logical experiment for determining T_2 would be to repeat the sequence with increasing τ and measure the amplitude of the echo versus time, by analogy with the inversion-recovery method above. However, some care is required when using such an approach as the formation of the echo depends on the isochromats experiencing exactly the same field throughout the duration of the pulse sequence. If any given spin diffuses into a neighbouring region during the sequence it may experience a slightly different field from that where it began, and thus will not be fully refocused. As τ increases, such diffusion losses become more severe and the experimental relaxation data less reliable (although this method does provide the basis for measuring molecular diffusion in solution by NMR; see the chapter *Diffusion NMR Spectroscopy*). Long values of τ will also lead to increased evolution of any homonuclear couplings that may be present, and lead to distortions of multiplet structure.

A better approach to determining T_2, which minimises the effect of diffusion and of J evolution, is to repeat the echo sequence within a single experiment using a short τ to form multiple echoes, the decay of which follows the time constant T_2. This is the Carr–Purcell sequence (Fig. 2.27b) which causes echoes to form alternately along the $-y$ and $+y$-axes following each refocusing pulse. Losses occur from diffusion between the echo peaks, or in other words in the time period 2τ, so if this is kept short relative to the rate of diffusion (typically $\tau < 100$ ms) such losses become negligible. Furthermore, if τ is kept short such that $1/\tau >> \Delta\upsilon$ (the frequency separation between coupled spins), evolution of homonuclear couplings is also suppressed. The intensity of the echo at longer time periods is attenuated by repeating the $-\tau$–180–τ– sequence many

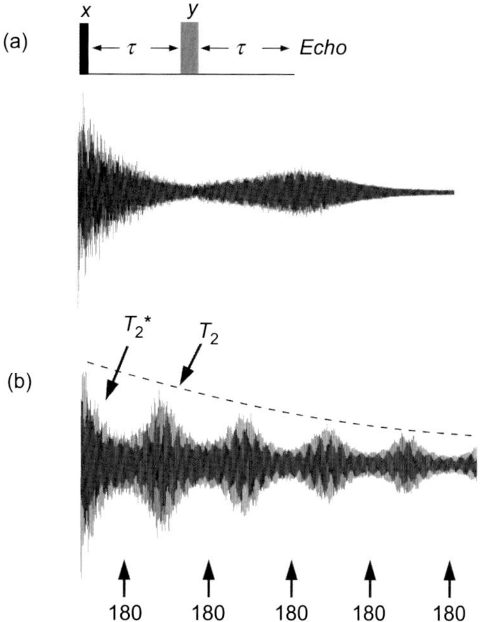

FIGURE 2.29 **Experimental observation of spin echoes.** (a) Signal acquisition was started immediately after a 90 degree excitation pulse and a 180 degree pulse applied to refocus field inhomogeneity losses and produce the observed echo. (b) A train of spin echoes reveals the true T_2 relaxation of magnetisation *(dashed line)*.

times prior to acquisition. The problem with this method is the fact that any errors in the length of the 180 degree pulse will be cumulative leading to imperfect refocusing as the experiment proceeds. A better implementation of this scheme is the Carr–Purcell–Meiboom–Gill (CPMG) sequence (Fig. 2.27c) in which $180°_y$ (as opposed to $180°_x$) pulses cause refocusing to take place in the $+y$ hemisphere for every echo. Here errors in pulse lengths are not cumulative but are small and constant on every odd-numbered echo but will cancel on each even-numbered echo (Fig. 2.30).

T_2 may then be extracted by performing a series of experiments with increasing $2\tau n$ (by increasing n) and acquiring data following the last even echo peak in each case. Application of the CPMG sequence is shown in Fig. 2.31 for a sample with differing resonance linewidths and illustrates the faster disappearance of broader resonances (ie those with shorter T_2s).

In reality, the determination of T_2 by any of these methods is still not straightforward. The most significant problem is likely to be from homonuclear couplings, the evolution of which may not be fully suppressed in the spin echo and hence may impose unwelcome phase modulations on the detected signals (Fig. 2.32c). The requirement that $1/\tau >> \Delta\upsilon$ to suppress such modulation in CPMG can demand very short τ delays and hence rapid pulsing (high-duty cycles) that can lead to undesirable sample heating or can give rise to incomplete modulation suppression. Nevertheless, it is possible to remove the distortions caused by J evolution by use of a more recent CPMG sequence that incorporates so-called 'perfect ' echoes, as in the periodic refocussing of J evolution by coherence transfer (PROJECT) variant (Fig. 2.27d). In this a purging 90 degree pulse is placed at the midpoint of a double echo and serves to effectively reverse the sense of J evolution, such that any evolution occurring in the first echo is then refocused by the end of the second (more formally, the role of the 90 degree pulse is to exchange coherences between coupled spins; see the chapter *Introducing Two-Dimensional and Pulsed Field Gradient NMR*). This approach also allows suppression of J-modulation effects without the constraints of rapid pulsing demanded by the basic CPMG sequence, since now the requirement is for $1/\tau >>$ J rather than $1/\tau >> \Delta\upsilon$, meaning much longer τ values and lower duty cycles can be employed. Fig. 2.32 demonstrates the cleaner multiplet shapes afforded by the CPMG-PROJECT method relative to CPMG alone, even with long echo delays. The benefits of employing perfect echoes are also demonstrated in later sections where spin-echo or CPMG trains find application in other methods.

The complications traditionally associated with use of multiple spin-echo trains has meant that studies involving T_2 measurements are even less widespread than those involving T_1. Fortunately, from the point of view of performing practical day-to-day spectroscopy, exact T_2 values are not important and the value of T_2* (which may be calculated from linewidths as described earlier) has far greater significance. It is this value that determines the rate of decay of transverse magnetisation, so it effectively defines how long a multipulse experiment can be before the system has decayed to such an extent that there is no longer any signal left to detect.

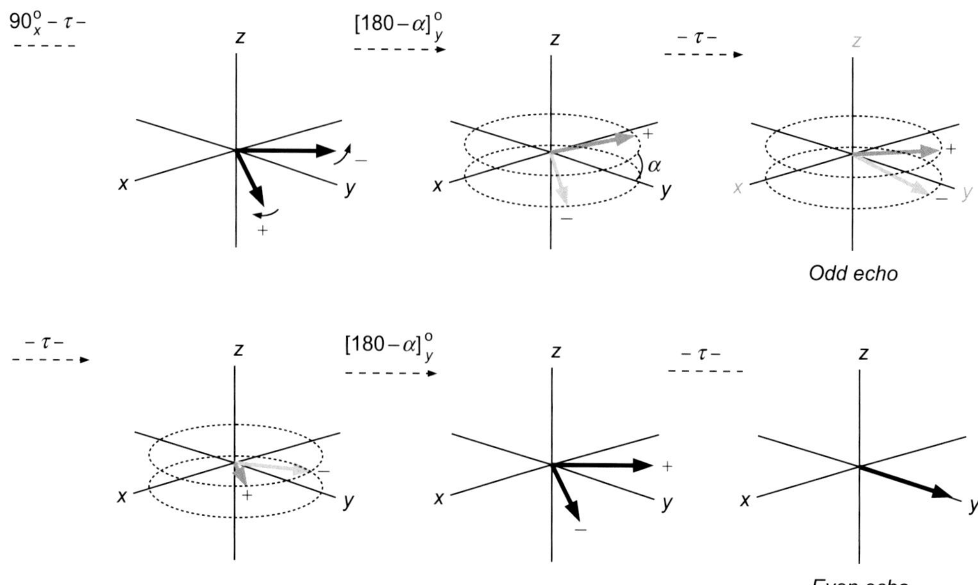

FIGURE 2.30 **Operation of the CPMG sequence in the presence of pulse imperfections.** The 180 degree pulse is assumed to be too short by α degree meaning vectors will fall above (*dark grey*) or below (*light grey*) the *x–y* plane following a single 180 degree pulse and so reduce the intensity of 'odd' echoes. By repeating the sequence, errors are cancelled by the imperfect second 180-degree pulse so 'even' echoes can be used to accurately map T_2 relaxation.

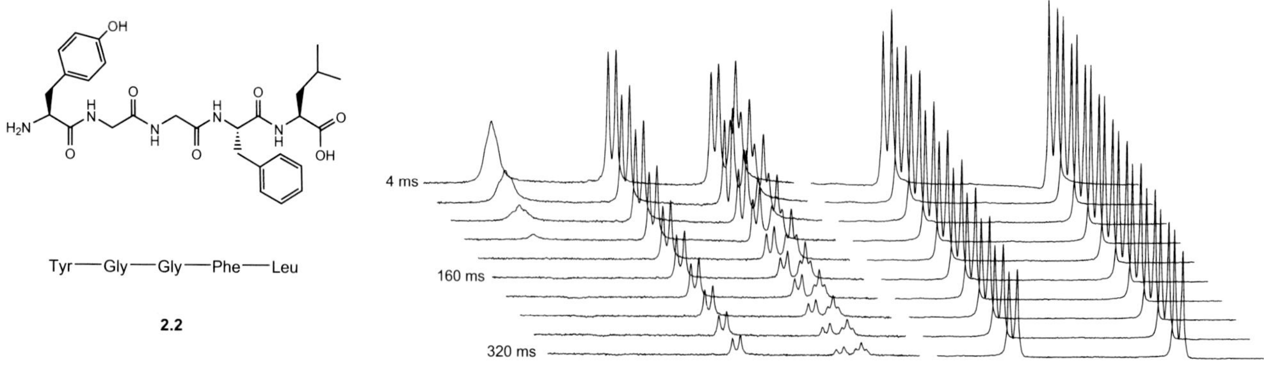

FIGURE 2.31 **The ^{1}H CPMG sequence performed on the pentapeptide leu-enkephalin 2.2 in DMSO.** The faster decay of the amide protons (left) relative to the aromatic protons of the tyrosine residue (right) results from amide protons coupling to quadrupolar ^{14}N (Section 2.5). The very fast decay of the highest frequency amide proton occurs because this is in rapid chemical exchange with dissolved water, broadening the resonance significantly. The numbers show the total T_2 relaxation period $2\tau n$.

2.4.4.1 T_2 Spectrum Editing

One interesting use of these echo techniques lies in the exploitation of gross differences in transverse relaxation times of different species. Larger molecules typically display broader resonances than smaller ones since they possess shorter T_2 spin relaxation times. If the differences in these times are sufficiently large, resonances of the faster relaxing species can be preferentially reduced in intensity with the CPMG echo sequence, while the resonances of smaller, slower relaxing molecules decrease by a lesser amount (Fig. 2.33). This therefore provides a means, albeit a rather crude one, of editing a spectrum according to molecular size, retaining the resonances of the smaller components at the expense of the more massive ones. This approach has been widely used in the study of biofluids to suppress background contributions from very large macromolecules such as lipids and proteins.

Selective reduction of a solvent water resonance can also be achieved in a similar way if the transverse relaxation time of water protons can be reduced (ie the resonance broadened) such that this becomes very much shorter than that of the solutes under investigation. This can be achieved by the addition of suitable paramagnetic relaxation agents (about which the water molecules form a hydration sphere) or by reagents that promote chemical exchange. Ammonium chloride and hydroxylamine have been used to great effect in this way [4,5], as illustrated for the proton spectrum of the reduced arginine

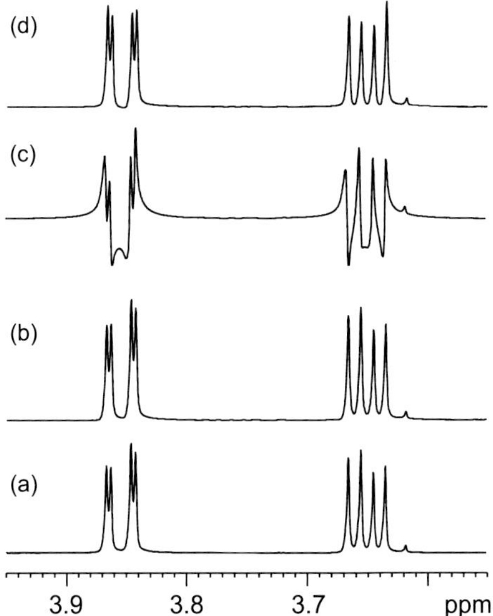

FIGURE 2.32 **Comparison of the ^{1}H CPMG and CPMG–PROJECT echo trains.** (a) ^{1}H spectrum, (b) CPMG with $\tau = 0.5$ ms, (c) CPMG with $\tau = 6$ ms and (d) CPMG–PROJECT with $\tau = 6$ ms. The total echo time was 24 ms in each case.

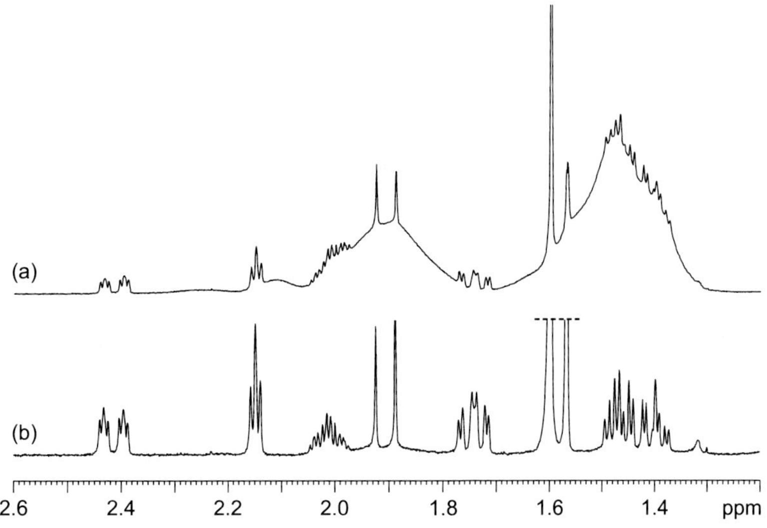

FIGURE 2.33 **Application of the T_2 filter.** The broad resonances of polystyrene ($M_r = 50{,}000$) in (a) have been suppressed in (b) through T_2-based editing with the CPMG sequence, leaving only the resonances of the smaller camphor molecule. The τ delay was 1.5 ms and the echo was repeated 150 times to produce a total relaxation delay period $2\tau n$ of 450 ms.

vasopressin peptide in 90% H_2O [6] (Fig. 2.34). This method of solvent suppression has been termed water attenuation by transverse relaxation (WATR). While capable of providing impressive results it does have limited application; more general solvent suppression procedures are described in the chapter on *Experimental Methods*.

2.5 MECHANISMS FOR RELAXATION

Nuclear spin relaxation is not a spontaneous process, it requires stimulation by a suitable fluctuating field to induce the necessary spin transitions and there are four principle mechanisms that are able to do this: the dipole–dipole, chemical shift anisotropy, spin rotation and quadrupolar mechanisms. Which of these is the dominant process can directly influence the appearance of an NMR spectrum, and it is these factors we consider here. The emphasis is not so much on the explicit details of the underlying mechanisms, which can be found in physical NMR texts [7–10] but on the manner in which the spectra are affected by these mechanisms and how, as a result, different experimental conditions influence the observed spectrum.

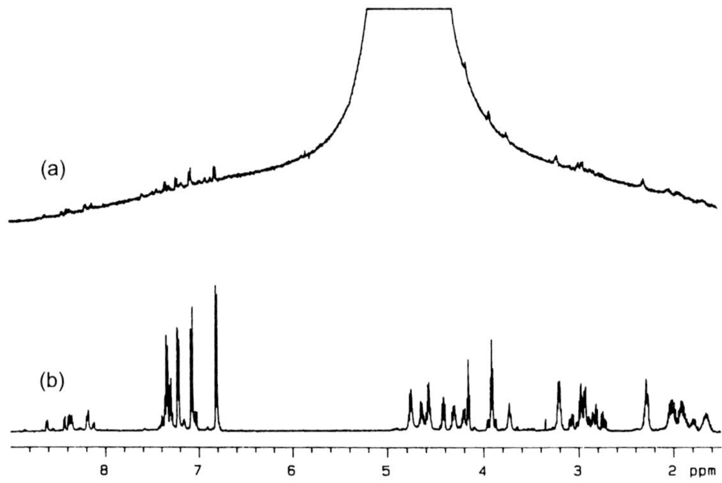

FIGURE 2.34 **Solvent attenuation with the WATR method.** (a) The 1D proton spectrum of 8 mM reduced arginine vasopressin in 90% H_2O/10% D_2O, pH = 2.75 containing 0.2 M NH_2OH. (b) The same sample recorded with the CPMG sequence using a total relaxation delay period of 235 ms. *(Source: Reproduced with permission from Ref. [6] Copyright © 1991, John Wiley & Sons Limited.)*

2.5.1 The Path to Relaxation

The fundamental requirement for longitudinal relaxation of a spin-½ nucleus is a time-dependent magnetic field fluctuating at the Larmor frequency of the nuclear spin. Only through this can a change of spin state be induced or, in other words, can relaxation occur. Local magnetic fields arise from a number of sources, described below, while their time dependence originates in the motions of the molecule (vibration, rotation, diffusion, etc.). In fact, only the chaotic tumbling of a molecule occurs at a rate that is appropriate for nuclear spin relaxation, others being either too fast or too slow. This random motion occurs with a spread of frequencies according to the molecular collisions, associations and so on experienced by the molecule, but is characterised by a rotational *correlation time* τ_c, the average time taken for the molecule to rotate through one radian. Short correlation times therefore correspond to rapid tumbling and vice versa. The frequency distribution of the fluctuating magnetic fields associated with this motion is termed the *spectral density* $J(\omega)$ and may be viewed as being proportional to the probability of finding a component of the motion at a given frequency ω (in radians per second). Only when a suitable component exists at the spin Larmor frequency can longitudinal relaxation occur. The spectral density function has the general form

$$J(\omega) = \frac{2\tau_c}{1 + \omega^2\tau_c^2} \tag{2.14}$$

and is represented schematically in Fig. 2.35a for fast, intermediate and slow molecular tumbling rates (note the conventional use of the logarithmic scale). As each curve represents a probability, the area under each remains constant. For the Larmor frequency ω_0 indicated in Fig. 2.35a, the corresponding graph of T_1 against molecular tumbling rates is also given (Fig. 2.35b). Fast molecular motion has only a relatively small component at the Larmor frequency (point a) so relaxation is slow (T_1 is long). This is the region occupied by small molecules in low-viscosity solvents, known as the *extreme narrowing limit.* As the tumbling rates decrease, the spectral density at ω_0 initially increases (point b) but then falls away once more for slow tumbling (point c) so the T_1 curve has a minimum at intermediate rates. Thus, *for small rapidly tumbling molecules, faster motion corresponds to slower relaxation and hence narrower linewidths*, since longitudinal and transverse relaxation rates are identical ($T_2 = T_1$) under these conditions. A reduction in tumbling rate, such as by an increase in solvent viscosity or reduction in sample temperature, reduces the relaxation times and broadens the resonance peak. The point at which the minimum is encountered and the slow motion regime approached is field dependent because ω_0 itself is field dependent (Fig. 2.35b). Behaviour in the slow motion regime is slightly more complex. The energy-conserving flip-flop processes that lead to transverse relaxation are also stimulated by very low frequency fluctuations and the T_2 curve differs markedly from that for T_1 (Fig. 2.36). Thus, for slowly tumbling molecules such as supramolecular complexes, polymers and biological macromolecules, T_1 relaxation times can again be quite long but linewidths become rather broad as a result of short T_2s.

Molecular motion is therefore fundamental to the process of relaxation, but it remains to be seen how the fields required for this arise and how these mechanisms influence observed spectra.

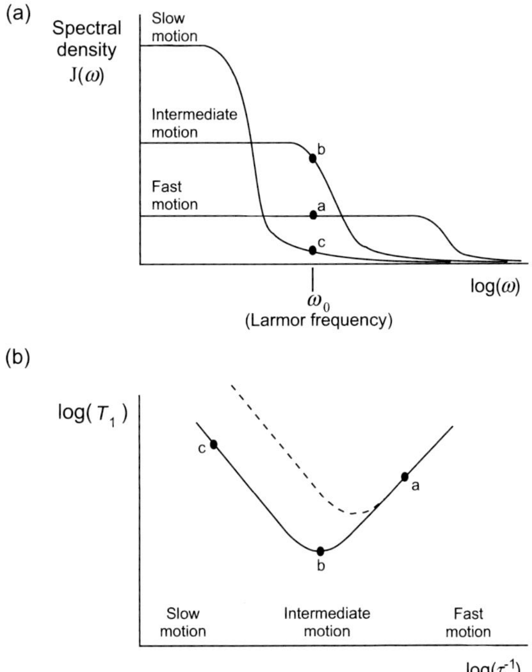

FIGURE 2.35 **Relaxation and spectral density.** (a) Schematic representation of spectral density as a function of frequency shown for molecules undergoing fast, intermediate and slow tumbling. For spins with Larmor frequency ω_0, the corresponding T_1 curve is shown in (b) as a function of molecular tumbling rates (inverse correlation times τ_c). The T_1 curve is field dependent because ω_0 is field dependent and the minimum occurs for faster motion at higher fields [*dashed curve* in (b)].

2.5.2 Dipole–Dipole Relaxation

The most important relaxation mechanism for many spin-½ nuclei arises from the dipolar interaction between spins. This is also the source of the tremendously important nuclear Overhauser effect and further discussions on this mechanism can be found in the chapter *Correlations Through Space: The Nuclear Overhauser Effect*, so are kept deliberately brief here. Dipolar interactions can be visualised using the 'bar magnet' analogy for a spin-½ nucleus in which each is said to possess a magnetic North and South Pole. As two such dipoles approach, their associated magnetic fields interact; they attract or repel depending on their relative orientations. Now suppose these dipoles were two neighbouring nuclei in a molecule that is tumbling in solution. The orientation of each nucleus with respect to the static magnetic field does not vary as the molecule tumbles just as a compass needle maintains its direction as a compass is turned. However, their relative positions in space will alter and the local field experienced at one nucleus as a result of its neighbour will fluctuate as the molecule tumbles (Fig. 2.37). Tumbling at an appropriate rate can therefore induce relaxation.

This mechanism is often the dominant relaxation process for protons which rely on their neighbours as a source of magnetic dipoles. As such, protons which lack near-neighbours relax more slowly (notice how the methine protons in α-pinene (Fig. 2.23) all have longer T_1s than the methylene groups). The most obvious consequence of this is lower than expected

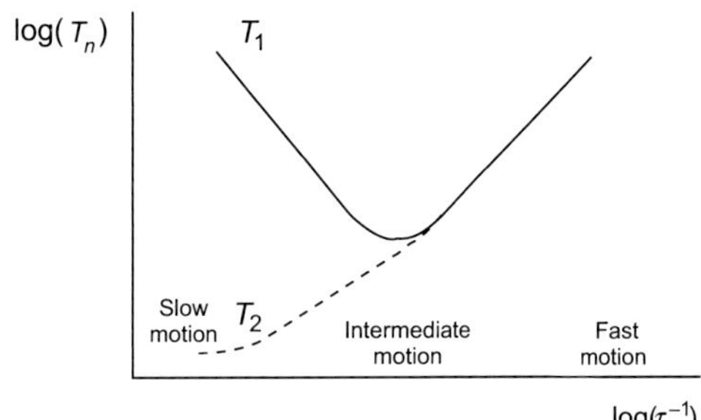

FIGURE 2.36 **Schematic illustration of the dependence of T_1 and T_2 on molecular tumbling rates.** T_1 relaxation is insensitive to very slow motions while T_2 relaxation may still be stimulated by them.

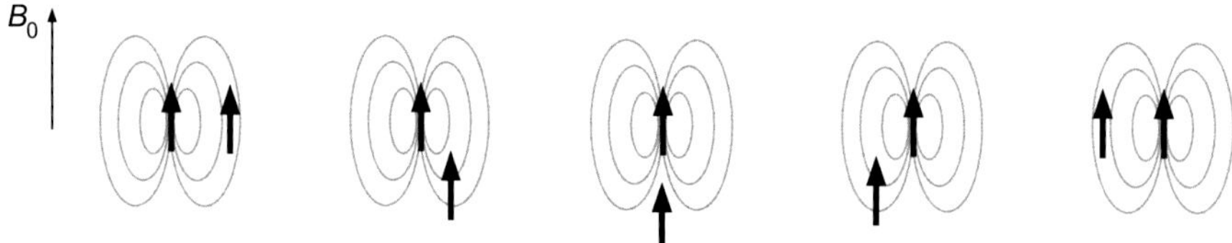

FIGURE 2.37 **Dipole–dipole relaxation.** The magnitude of direct through-space magnetic interaction between two spins is modulated by molecular tumbling and so induces spin transitions and hence relaxation.

integrals in routine proton spectra due to the partial saturation of the slower relaxing spins which are unable to recover sufficiently between each pulse–acquire sequence. If T_1 data are available, then protons with long relaxation times can be predicted to be remote from others in the molecule. Carbon-13 nuclei are also relaxed primarily by dipolar interactions, either with their directly bound protons or, in the absence of these, by more distant ones. In very large molecules and at high field, the chemical shift anisotropy mechanism described below can also play a role, especially for sp and sp^2 centres. Likewise this can be significant for spin-½ nuclei which exhibit large chemical shift ranges. Dipolar relaxation can also arise from the interaction of a nuclear spin with an unpaired electron, the magnetic moment of which is 658 times that of the proton and so provides a very efficient relaxation source. This is referred to as the *paramagnetic relaxation* mechanism. Even the presence of dissolved oxygen, which is itself paramagnetic, can contribute to spin relaxation and the deliberate addition of relaxation agents containing paramagnetic species is sometimes used to reduce relaxation times and so speed data acquisition (see the chapter on *One-Dimensional Techniques*). The most common reagents are chromium(III) acetylacetonate ($Cr(acac)_3$), for organic solvents and manganese(II) chloride or gadolinium(III) chloride for aqueous solutions. Paramagnetic relaxation enhancement is also considered in the chapter *Protein-Ligand Screening by NMR* as a method to aid the detection of protein–ligand binding interactions.

2.5.3 Chemical Shift Anisotropy Relaxation

The electron distribution in chemical bonds is inherently unsymmetrical or *anisotropic* and, as a result, the local field experienced by a nucleus, and hence its chemical shift, will depend on the orientation of the bond relative to the applied static field. In solution, the rapid tumbling of a molecule averages this *chemical shift anisotropy* (CSA) such that one observes only a single frequency for each chemically distinct site, sometimes referred to as the isotropic chemical shift. Nevertheless, this fluctuating field can stimulate relaxation if sufficiently strong. This is generally the case for nuclei which exhibit a large chemical shift range since these possess the greatest shift anisotropy (eg ^{19}F, ^{31}P and, in particular, many metals).

The characteristic feature of CSA relaxation is its dependence on the *square* of the applied field, meaning it has greater significance at higher B_0. For example, the selenium-77 longitudinal relaxation rate in the selone **2.3** was shown to be linearly dependent on B_0^2, indicating CSA to be a significant relaxation mechanism [11] (Table 2.2), while no proton–selenium nuclear Overhauser effect (NOE) could be detected, demonstrating the ^1H–^{77}Se dipole–dipole mechanism to be ineffectual. Nuclei whose relaxation is dominated by the CSA mechanism may show significantly larger linewidths at higher fields if they possess a large shift anisotropy, and any potential benefits of greater dispersion and sensitivity may be lost by line broadening. For this reason, the study of some metal nuclei may be more successful at lower fields. Reducing the correlation time, by warming the sample for example, may attenuate the broadening effect, although this approach clearly has rather limited application. In some cases, enhanced CSA relaxation at higher fields can be advantageous. A moderately enhanced relaxation rate, such as in the ^{77}Se example above, allows for more rapid data collection (Section 4.1), thus providing an improvement in sensitivity (per unit time) above that expected from the increase in magnetic field alone.

TABLE 2.2 ^{77}Se Longitudinal Relaxation Times as a Function of B_0 with the Corresponding Dependence of the Relaxation Rate on the Square of the Applied Field Shown Graphically

B_0 (T)	T_1 (s)	$1/T_1$ (s^{-1})
2.36	2.44	0.41
5.87	0.96	1.04
7.05	0.67	1.49

Adapted with permission from reference [11].

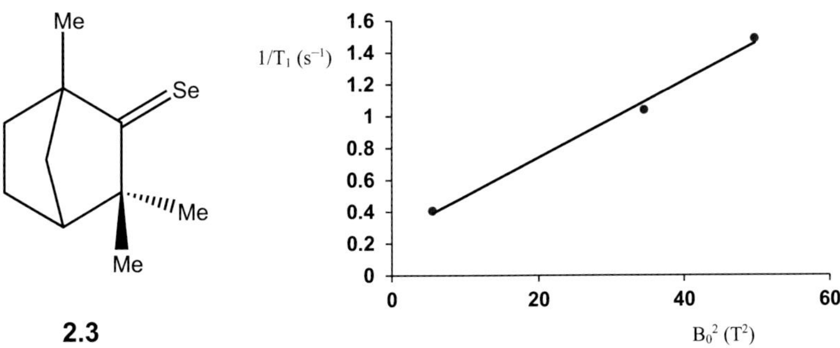

The CSA mechanism can also have a perhaps unexpected influence on the spectra of nuclei that are scalar spin–coupled to the CSA-relaxed spin. If CSA causes very rapid relaxation, the satellites arising from coupling to this spin will broaden and may even disappear altogether. Thus, while the coupling may be apparent at low field, it may vanish at higher values. This effect can be seen in the proton spectra of the platinum complex **2.4** recorded at 80 and 400 MHz [12] (Fig. 2.38). The increased linewidth of the satellites relative to the parent line at higher field scales with the square of the applied field, as expected for the CSA mechanism. To understand why this occurs, consider the origin of the Pt satellites themselves. These doublet components arise from the spin-½ ^{195}Pt nuclei existing in one of two states, α and β, which result in a frequency difference of $J_{Pt\text{–}H}$ Hz for the corresponding proton signals. CSA relaxation induces rapid transitions of the platinum spins between these states causing the doublet components to repeatedly switch positions. As this exchange rate (ie relaxation rate) increases, the satellites first broaden and will eventually merge into the parent line, as for any dynamic process; see Section 2.6. This rapid and repeated change of spin states has a direct analogy with conventional spin decoupling (see the chapter *One-Dimensional Techniques*).

2.5.4 Spin Rotation Relaxation

Molecules or groups which rotate very rapidly have associated with them a *molecular* magnetic moment generated by rotating electronic and nuclear charges. The field due to this fluctuates as the molecule or group changes its rotational state as a result of, for example, molecular collisions and this provides a further mechanism for nuclear relaxation. This is most effective for small, symmetrical molecules or for freely rotating methyl groups and its efficiency *increases* as tumbling rates *increase*. This is in contrast to the previously described mechanisms. Thus, heating a sample enhances spin rotation relaxation, this temperature dependence being characteristic of this mechanism and allowing its presence to be established.

2.5.5 Quadrupolar Relaxation

The quadrupolar relaxation mechanism is only directly relevant for those nuclei that have a nuclear spin quantum number I greater than ½ (quadrupolar nuclei) and is often the dominant relaxation process for these. This can also be a very efficient mechanism and the linewidths of many such nuclei can be hundreds or even thousands of hertz wide. The properties of

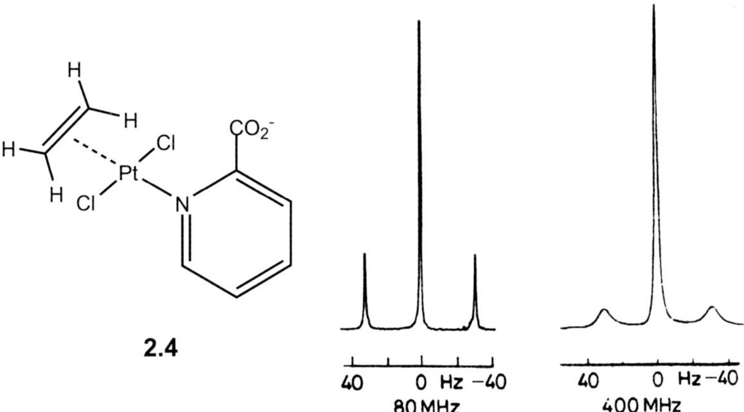

FIGURE 2.38 **Ethene proton resonance of the platinum complex 2.4 in $CDCl_3$.** The spectra at 80 and 400 MHz show broadening of the ^{195}Pt satellites at the higher field. *(Source: Reproduced with permission from Ref. [12].)*

TABLE 2.3 Properties of Selected Quadrupolar Nuclei

Isotope	Spin (*I*)	Natural Abundance (%)	Quadrupole Moment (10^{-28} m^2)	NMR Frequency (MHz)	Relative Sensitivity
^{2}H	1	0.015	2.8×10^{-3}	61.4	1.45×10^{-6}
^{6}Li	1	7.42	-8.0×10^{-4}	58.9	6.31×10^{-4}
^{7}Li	3/2	92.58	-4×10^{-2}	155.5	0.27
^{10}B	3	19.58	8.5×10^{-2}	43.0	3.93×10^{-3}
^{11}B	3/2	80.42	4.1×10^{-2}	128.3	0.13
^{14}N	1	99.63	1.0×10^{-2}	28.9	1.01×10^{-3}
^{17}O	5/2	0.037	-2.6×10^{-2}	54.2	1.08×10^{-5}
^{23}Na	3/2	100	0.10	105.8	9.27×10^{-2}
^{27}Al	5/2	100	0.15	104.2	0.21
^{33}S	3/2	0.76	-5.5×10^{-2}	30.7	1.72×10^{-5}
^{35}Cl	3/2	75.73	-0.1	39.2	3.55×10^{-3}
^{37}Cl	3/2	24.47	-7.9×10^{-2}	32.6	6.63×10^{-4}
^{59}Co	7/2	100	0.38	94.5	0.28

The observation of such nuclei is generally most favourable for those of low quadrupole moment and high natural abundance, which exist in more highly symmetric environments; those listed in italics are considered the least favoured of the available isotopes. The NMR frequencies are quoted for a 400-MHz instrument (9.4-T magnet) and sensitivities are relative to ^{1}H and take account of both the intrinsic sensitivity of the nucleus and its natural abundance.

selected nuclei with $I > ½$ are summarised in Table 2.3. While the direct observation of these nuclei may not be routine for many organic chemists, their observation can at times prove very enlightening for specific problems, and the indirect effects they have on the spectra of spin-½ nuclei should not be overlooked.

Quadrupolar nuclei possess an electric quadrupole moment in addition to a magnetic dipole moment. This results from the charge distribution of the nucleus deviating from the usual spherical symmetry associated with spin-½ nuclei and becoming ellipsoidal in shape. This can be viewed as arising from two back-to-back *electric* dipoles (Fig. 2.39). As such, the quadrupole moment is influenced by electric field *gradients* about the nucleus, but not by symmetric electric fields. The gradient is modulated as the molecule tumbles in solution and again if this occurs at the appropriate frequency it can induce flipping of nuclear spin states and thus stimulate relaxation. This is analogous to the relaxation of nuclear dipoles by time-dependent local magnetic fields, but the quadrupolar relaxation mechanism is the only one that depends on electric rather than magnetic interactions.

The relaxation rates of a quadrupolar nucleus are dictated by two new factors not previously considered. The first is the magnitude of the quadrupole moment itself (Table 2.3). Larger values contribute to more efficient spin relaxation and hence broader linewidths, whereas smaller values typically produce sharper lines. Thus, those nuclei with smaller quadrupole moments are usually more favoured for NMR observation. As before, for the mechanism to be effective, molecular tumbling must occur at an appropriate frequency, so again fast molecular tumbling reduces the effectiveness, leading to longer relaxation times and sharper lines. High temperatures or lower viscosity solvents are thus more likely to produce narrow linewidths. The ultimate in low-viscosity solvents are supercritical fluids which have viscosities more like those of a gas yet solubilising properties more like liquids. These have indeed been used in the study of quadrupolar nuclei [13], but since they are only supercritical at very high pressures they demand the use of single-crystal sapphire NMR tubes so their use cannot be considered routine! The second new factor is the magnitude of the electric field gradient. In highly symmetrical environments, such as tetrahedral or octahedral symmetries, the field gradient is in principle zero and the quadrupolar mechanism is suppressed. In reality, local distortions still arise, if only momentarily, introducing an element of asymmetry and hence enhanced relaxation and line broadening. Nevertheless, a higher degree of electrical symmetry can be correlated with narrower resonances. Thus, for example, the ^{14}N linewidth of $N(Me)_4^+$ is less than 1 Hz whereas that for NMe_3 is nearer to 80 Hz. Linewidth changes in ^{11}B spectra ($I = 3/2$) have been used in the identification of tetrahedral boronic

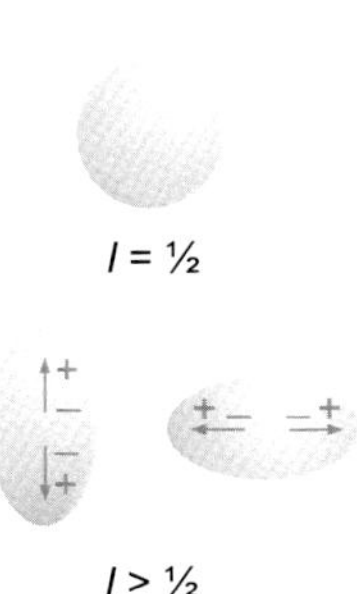

FIGURE 2.39 **Quadrupolar nuclei lack the spherical charge distribution of spin−½ nuclei, having an ellipsoidal shape which may be viewed as arising from pairs of electric dipoles.** Thus quadrupolar nuclei interact with electric field gradients.

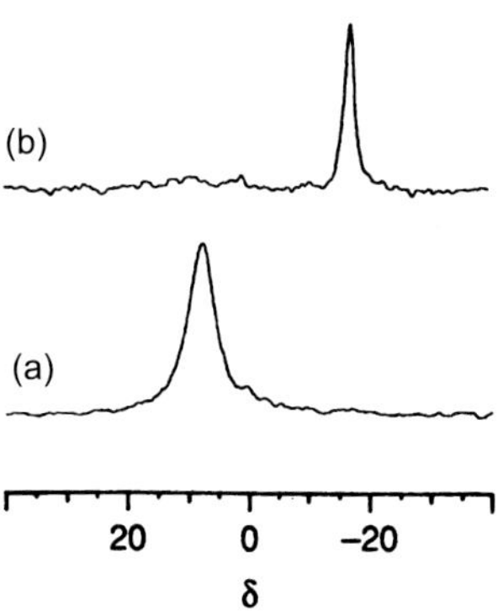

FIGURE 2.40 **Boron-11 NMR.** The ^{11}B NMR spectrum (128 MHz) of dansylamidophenylboronic acid **2.5** (a) as the free trigonal boronic acid ($\Delta\nu_{½}$ 580 Hz) and (b) as the tetrahedral complex with the active site serine of the P99 β-lactamase from *Enterobacter cloacae* ($\Delta\nu_{½}$ 160 Hz). Spectra are referenced to external trimethylborate. (*Source: Adapted with permission from [14].*)

acid complexes at the active site of β-lactamases [14], enzymes responsible for the destruction of β-lactam antibiotics such as penicillins, and part of the defence mechanism of bacteria. Boronic acids, such as 3-dansylamidophenylboronic acid **2.5**, are known to be reversible inhibitors of active site serine β-lactamases and the complexes so formed display significant changes in the ^{11}B chemical shift of the boronic acid together with a reduction in linewidth relative to the free acid (Fig. 2.40). This reduction is attributed to the boron nucleus taking up a more symmetrical tetrahedral environment as it becomes bound by the enzyme's active site serine. This shift and line narrowing can be mimicked by placing the boronic acid in alkaline solution in which the R–B(OH)$_3^-$ ion predominates.

$B(OH)_2$ $N(Me)_2$ $NHSO_2$

2.5

The broad resonance of many quadrupolar nuclei means field inhomogeneity makes a negligible contribution to linewidths so the methods described previously for measuring relaxation times are no longer necessary. For small molecules at least, T_2 and T_1 are identical and can be determined directly from the half-height linewidth. Broad resonances together with the sometimes low intrinsic sensitivity and low natural abundance of quadrupolar nuclei are the principal reasons for their relatively low popularity for NMR studies relative to spin-½ nuclei. The very fast relaxation of certain quadrupolar nuclei can also make their direct observation difficult with conventional high-resolution spectrometers; see Section 4.5.

2.5.5.1 Scalar Coupling to Quadrupolar Nuclei

Probably of more relevance to the practising organic chemist is the influence quadrupolar nuclei have on the spectra of spin-½ nuclei, by virtue of their mutual scalar coupling. Coupling to a quadrupolar nucleus of spin I produces, in theory, $2I + 1$ lines; so, for example, the carbon resonance of $CDCl_3$ appears as a 1:1:1 triplet (^{2}H has $I = 1$) by virtue of the 32-Hz ^{13}C–^{2}H coupling. However, more generally, if the relaxation of the quadrupolar nucleus is rapid relative to the magnitude of the coupling, the splitting can be lost, in much the same way that coupling to a nucleus experiencing rapid CSA relaxation is lost. The carbon resonance of $CDCl_3$ is only a triplet because deuterium has a relatively small quadrupole moment making its coupling

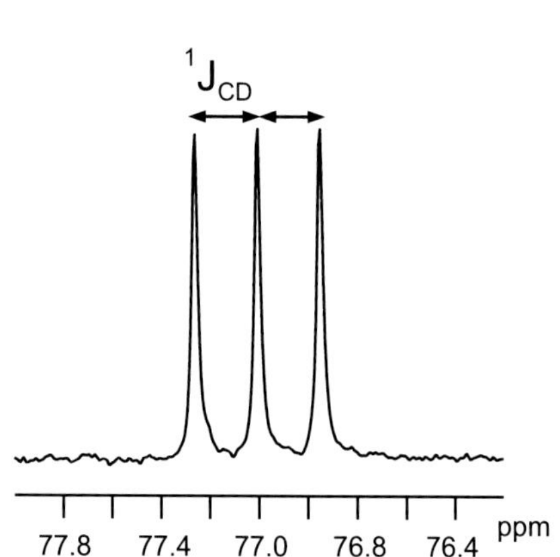

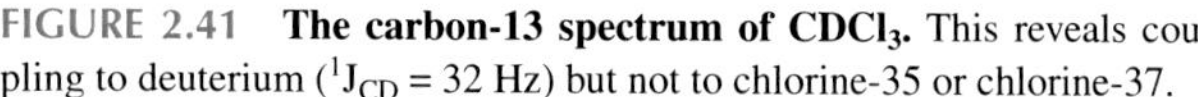

FIGURE 2.41 **The carbon-13 spectrum of $CDCl_3$.** This reveals coupling to deuterium ($^1J_{CD}$ = 32 Hz) but not to chlorine-35 or chlorine-37.

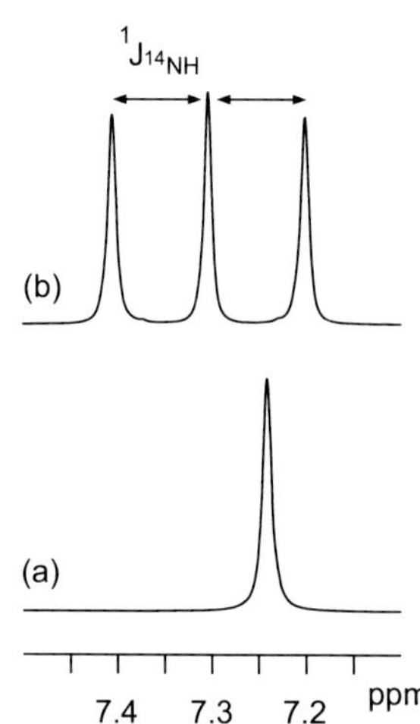

FIGURE 2.42 **The ^1H spectrum of ammonium nitrate in DMSO.** Spectra were recorded (a) before and (b) after acidification. In (b) the NH_4^+ ion dominates and the induced symmetry reduces the ^{14}N quadrupolar relaxation rate and reveals the ^1H–^{14}N one-bond coupling constant (51 Hz).

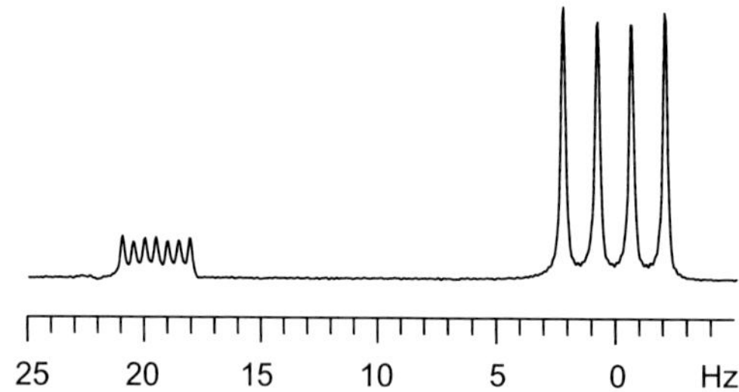

FIGURE 2.43 **The ^{19}F spectrum of sodium borofluoride in D_2O.** The $^{11}BF_4^-$ ion produces the even quartet (^{11}B $I = {}^3/_2$, 80% abundance, J = 1.4 Hz) and the $^{10}BF_4^-$ ion produces the even septet (^{10}B $I = 3$, 20% abundance, J = 0.5 Hz). The frequency difference between the two is due to a $^{10}B/^{11}B$ isotope shift.

apparent whereas all coupling to the chlorine nuclei (^{35}Cl and ^{37}Cl have $I = 3/2$) is quenched by the very rapid relaxation of these spins (Fig. 2.41). Similarly, the proton resonance of $CHCl_3$ is a sharp singlet despite the presence of neighbouring chlorine atoms. A common example of line broadening brought about by a quadrupolar nucleus is seen for the resonances of amino and amido protons due to the adjacent ^{14}N (Fig. 2.31).

The appearance of the spin-½ nucleus spectrum is therefore also influenced by the factors described above which dictate the rate of quadrupolar relaxation. Couplings to quadrupolar nuclei that exist in a highly symmetrical environment are likely to be seen because of the slower relaxation the nuclei experience. For this reason the proton spectrum of $^{14}NH_4^+$ is an unusually sharp 1:1:1 triplet (Fig. 2.42, ^{14}N has $I = 1$) and the fluorine spectrum of $^{11}BF_4^-$ is a sharp 1:1:1:1 quartet (Fig. 2.43, ^{11}B has $I = 3/2$). Increasing sample temperature results in slower relaxation of the quadrupolar nucleus so there is also a greater chance of the coupling being observable. In contrast, reducing the temperature increases relaxation rates and collapses coupling fine structure. This is contrary to the usual behaviour associated with dynamic systems where heating typically leads to simplification of spectra by virtue of resonance coalescence, as described in the following section. The likelihood of coupling fine structure being lost is also increased as the magnitude of the coupling constant decreases. In general, then, the observation of scalar coupling to a quadrupolar nucleus is the exception rather than the rule.

2.6 DYNAMIC EFFECTS IN NMR

Molecules in solution experience many motional processes including the molecular rotations that are responsible for spin relaxation processes and which define the natural linewidths observed in NMR spectra, as discussed earlier. They also experience internal bond vibrations, which occur too rapidly to have directly observable effects on NMR spectra. Here we consider processes occurring at much slower rates that can influence the appearance of spectra by causing lineshape changes

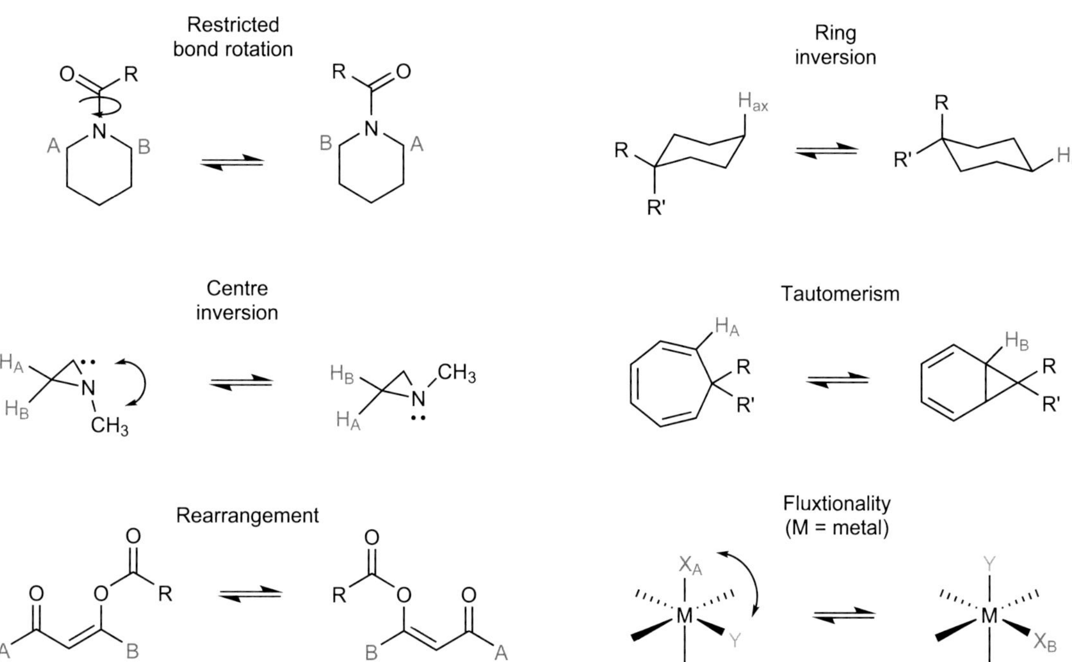

FIGURE 2.44 **Dynamic molecular processes.** Examples of intramolecular dynamic exchange equilibria that may be slow enough to produce directly observable effects in NMR spectra.

or, if slower still, happen at rates that can be measured by magnetisation exchange experiments [15]. The dynamic events considered here lead to a physical interchange between different structural forms arising from a transfer of matter and are referred to as *chemical exchange* processes. Most frequently, these arise from intramolecular conversions such as restricted bond rotations, ring inversions, group interconversions, and valence tautomerism (Fig. 2.44); numerous examples may be found in review papers [16–19] and classic textbooks [20–22]. It is also possible that observable effects may originate from *inter*molecular processes such as proton exchange, as described below, or from the on–off equilibria associated with ligand–protein binding events, a topic discussed in the chapter *Protein-Ligand Screening by NMR*.

The quantitative study of chemical exchange processes to derive exchange rate constants is often termed *dynamic NMR* or simply (and perhaps ambiguously) *DNMR*—not to be confused with ^{2}H (deuterium) NMR. Knowledge of these rates allows relevant thermodynamic parameters to be determined, and descriptions of such analysis are provided below.

Discussions in this section relate to dynamic systems at equilibrium and not to kinetic reaction monitoring in which a reaction pathway is followed towards its endpoint and where initial reactants yield discrete products. Although NMR is well suited to and increasingly used for such monitoring [23,24], it shall not be considered here.

2.6.1 The Influence of Dynamic Exchange

One of the most commonly observed dynamic effects in small-molecule NMR is a broadening or doubling of resonances caused by restricted rotation of a tertiary amide bond (**2.6**) due to its partial double-bond character. This example serves as a useful introduction to the influence on NMR spectra of dynamic chemical exchange within molecules, and to illustrate this we first consider the classic example of *N,N*-dimethylacetamide (DMA, **2.7**). In this, the two *N*-methyl groups may be defined as sitting either *cis* or *trans* to the carbonyl oxygen and if they were to remain in these positions (ie if there were no rotation about the N-CO bond) we would expect the two groups to experience different chemical environments and thus appear with different chemical shifts. Indeed, the ambient temperature spectrum of DMA displays two *N*-methyl resonances because rotation around the amide bond is sufficiently *slow* that the two discrete environments are apparent (Fig. 2.45a). However, at the higher temperature of 420 K only one *N*-methyl resonance is observed which now arises from six equivalent protons. Under these conditions, amide bond rotation causes the *N*-methyl groups to exchange their environments rapidly; moreover, this is sufficiently *fast* that only a single averaged resonance is detected partway between the individual resonance positions (Fig. 2.45c). Between these two extremes the exchange behaviour is said to occur at an *intermediate* rate and the appearance of the spectrum becomes dominated by line broadening that is sensitive to the exchange

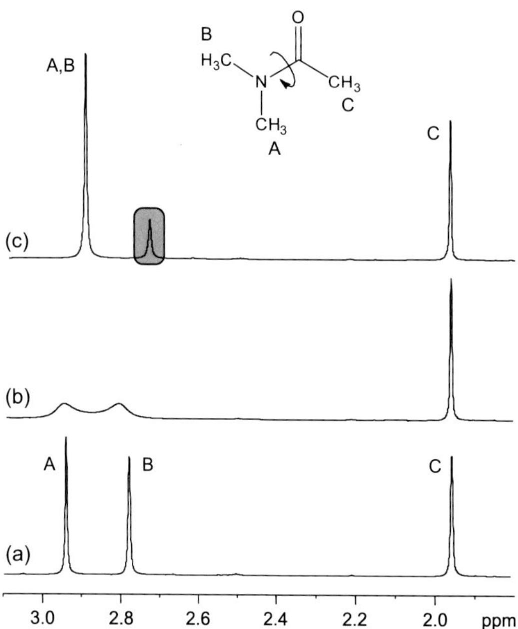

FIGURE 2.45 **The 250-MHz ^{1}H NMR spectra of *N,N*-dimethylacetamide in DMSO.** The spectra show the methyl group behaviour when the amide bond rotation rate is considered to be (a) slow (295 K) (b) intermediate (350 K) and (c) fast (420 K). The box shows the appearance of temperature-dependent water resonance. Methyl resonances B and C are broadened slightly in (a) owing to their mutual, unresolved long-range coupling.

rate (Fig. 2.45b). This regime yields more complex spectra that can inform on exchange rate constants and receives further attention below. Notice that the appearance of the *C*-methyl resonance of DMA does not vary significantly with temperature since its environment remains unchanged following amide bond rotation, although the chemical shifts of all three methyl groups do show a small temperature dependence, a commonly observed effect. To continue, we need to determine what defines a process as being termed slow or fast in the context of NMR spectra, and this is most easily done by again considering exchange between only two discrete sites.

2.6

2.7

2.6.1.1 Two-Site Exchange: Equal Populations

Let us consider the general case of two-site exchange between two equally populated sites A and B for which the equilibrium may be written:

$$A \underset{k}{\overset{k}{\rightleftharpoons}} B \tag{2.15}$$

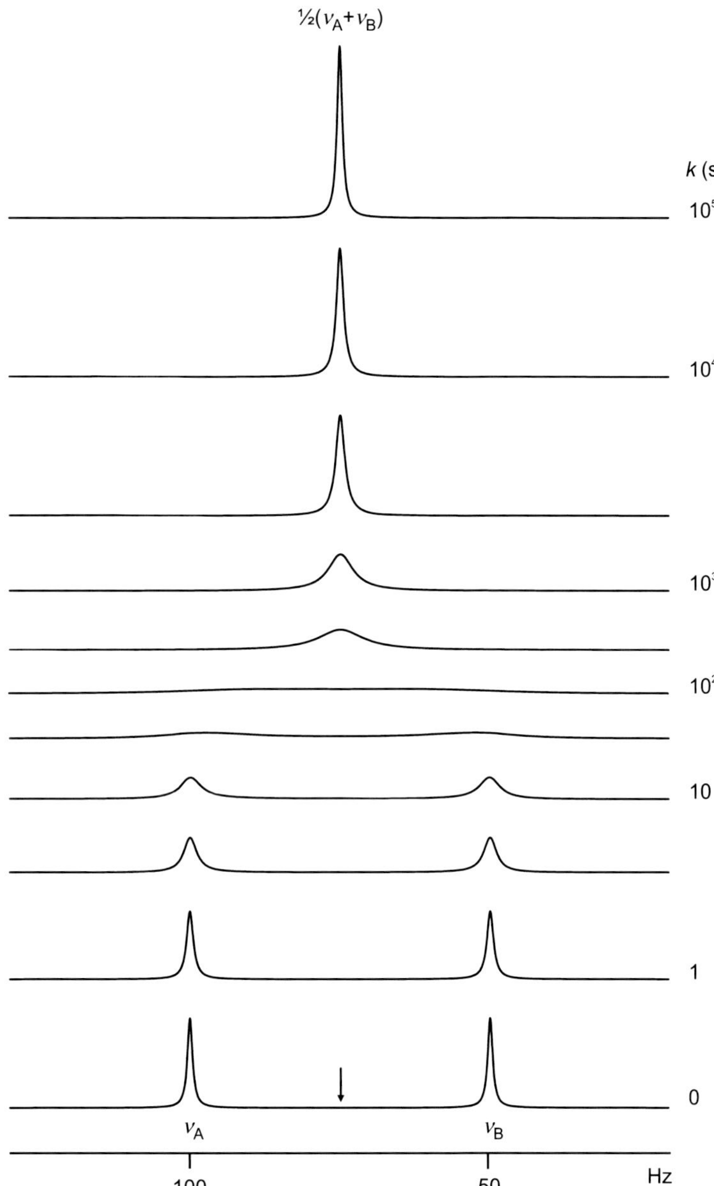

FIGURE 2.46 **Two-site exchange for equal populations.** Calculated spectra for two nuclei undergoing exchange between two equally populated sites *A* and *B* with frequency separation $\Delta\upsilon_{AB}$ of 50 Hz. The *arrow* indicates the position of exchange-averaged resonance. Linewidths in the absence of exchange were 1 Hz.

where the forward and reverse first-order rate constants *k* are necessarily equal. We shall first consider a system which lacks any resolved J coupling; so, we need only consider the resonance frequencies for the two sites υ_A and υ_B (defined in hertz) and the magnitude of their frequency differences $\Delta\upsilon_{AB} = |\upsilon_A - \upsilon_B|$. An example of this might be the two *N*-methyl groups of DMA above representing sites A and B that undergo dynamic interconversion. The exchange process will interchange the locations and hence the environment of sites *A* and *B* such that the resonance frequency of each will jump whenever an exchange event occurs. In the case of DMA, the methyl groups will move between the *cis* and *trans* positions relative to the carbonyl group. The influence of this process on NMR lineshapes as a function of the exchange rate constant *k* is illustrated in Fig. 2.46. Experimentally, changes in exchange rates are typically induced by changes in sample temperature.

When the exchange rate is very slow (ie when the lifetimes of each state τ are very long, $\tau = 1/k$), the resonances for each site are sharp and clearly resolved yielding discrete lines of equal intensity. At the limit of this so-called *slow exchange* regime the dynamic exchange process has a negligible influence on the observed NMR spectrum and what is seen is merely the discrete resonances of each site. Measurement of exchange rates in this regime require the use of magnetisation transfer experiments, as described in Section 2.6.3.

As the rate progressively increases, each resonance initially begins to broaden, reducing in peak height, and the two lines will then appear to move closer; the exchange process is said to lead to *exchange-* or *dynamic-broadening* of resonances. In this *slow–intermediate regime* NMR lineshapes are exquisitely sensitive to the exchange rate, which may be related to the *observed* peak separation $\Delta\upsilon_O$ according to

$$k = \frac{\pi\sqrt{\left[(\Delta\upsilon_{AB})^2 - (\Delta\upsilon_O)^2\right]}}{\sqrt{2}} \tag{2.16}$$

The additional line broadening induced as a result of the dynamic exchange $\Delta\upsilon_{½}^{ex}$ (ie that acting in addition to the natural linewidth in the absence of exchange $\Delta\upsilon_{½}$; see Section 2.4.3) is given by

$$\Delta\upsilon_{½}^{ex} = \frac{k}{\pi} \tag{2.17}$$

Clearly in this regime the resonances becomes broader as the rate becomes faster, although in principle two separate resonances remain. In reality, it may be that the peaks are sufficiently broad that they may not be readily apparent and may even appear to be missing from a spectrum; this can be an even greater problem in situations of exchange between sites of unequal population, as described below. When peak movement and broadening can be measured, exchange rate constants can be determined either from the above expressions or more commonly nowadays through computer lineshape simulation (see Section 2.6.2).

As the rate further increases, the peaks eventually merge into one and are said to have coalesced. The point at which they do so, when the valley between them *just* flattens out, is the so-called *coalescence point*, at which the rate k_c will be:

$$k_c = \frac{\pi\Delta\upsilon_{AB}}{\sqrt{2}} \approx 2.22\Delta\upsilon_{AB} \tag{2.18}$$

Thus, if the frequency separation between peaks is known from the spectrum recorded within the slow exchange limit, the exchange rate at coalescence can be determined quite readily. One caveat here is that chemical shifts may also be temperature dependent meaning the actual shift differences $\Delta\upsilon_{AB}$ may be altered at coalescence temperature; see discussions on the practicalities of these measurements in Section 2.6.2.

At rates above the coalescence point, the system enters the *fast–intermediate regime* and only a single merged peak is seen that reflects the intensity sum of the two merged resonances. This peak appears at the midpoint of the two individual resonances since we now observe the average properties of the species undergoing exchange. In this regime the broadening due to exchange is:

$$\Delta\upsilon_{½}^{ex} = \frac{\pi(\Delta\upsilon_{AB})^2}{2k} \tag{2.19}$$

leading to a reduction in broadening and hence a sharpening of the single resonance as the rate increases. This process is sometimes referred to as *exchange-* or *dynamic-narrowing* of resonances. The rate can again be determined from lineshape fitting procedures provided the frequency separation of the resonances in the absence of exchange is known. In the limit of very fast exchange the contribution to the linewidth becomes negligible relative to its natural linewidth and the process becomes too rapid to influence directly the observed NMR spectrum. Although lineshapes can no longer provide information on exchange rates under these conditions, rapid exchange may still influence spin relaxation rates, allowing exchange rates to be determined through appropriate analysis. However, this finds greatest application in studies of the dynamics and flexibility within macromolecules, and shall not be considered further here.

From the above descriptions, it is apparent that the exchange regimes are determined by the rate of exchange *relative to the frequency separation of the exchanging resonances* $\Delta\upsilon_{AB}$ (Fig. 2.47). In the slow exchange regime ($k << \Delta\upsilon_{AB}$), separate resonances for each interchanging site will be observed. In fact, the term *slow* is somewhat misleading here since the exchange process itself, bond rotation for example, actually occurs rather rapidly. Instead, it is the *intervals* between the exchange events that become lengthy in this regime, and hence the events infrequent. This in turn means the lifetime of each individual state is also long and its resonance position can be accurately determined. For consistency with common

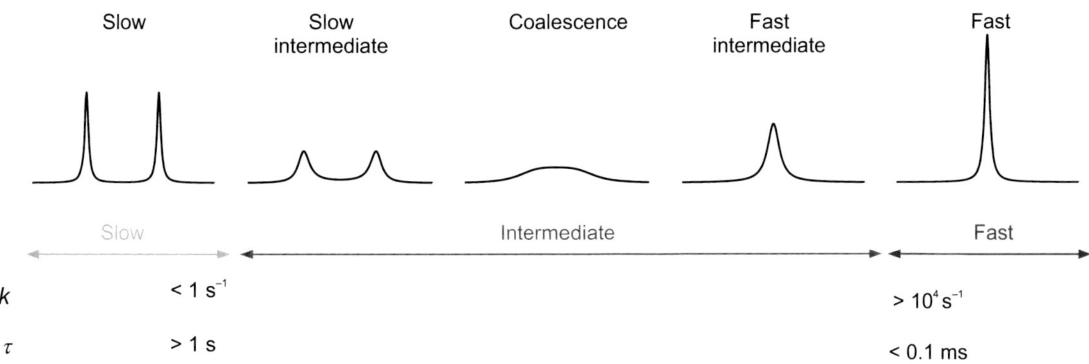

FIGURE 2.47 **Dynamic exchange regimes.** Schematic classification of the dynamic exchange regimes observed in NMR spectra, with approximate rates (k) or lifetimes (τ) associated with these regimes.

parlance, but with this in mind, the term slow will continue to be used throughout these discussions to reflect this condition. In the fast-exchange regime ($k >> \Delta\upsilon_{AB}$), a single resonance is observed representing the averaged properties of each state, which in the case of equally populated sites appears at the midpoint of the two peaks. Here the lifetime of each state is too short for each to be resolved. Between these regimes, within the intermediate-exchange regime ($k \sim \Delta\upsilon_{AB}$), exists the coalescence point when the two signals just merge into a single peak. The importance of discussing exchange regimes relative to an appropriate frequency scale cannot be overemphasised, and descriptions of dynamic effects should be referenced to a corresponding *NMR timescale*, governed by $1/\Delta\upsilon$. In discussions so far we have referred to events as ranging from slow to fast on the *chemical shift timescale* since it is chemical shift differences that defined these exchange regimes. Typically, the phrase 'NMR timescale' will refer to the chemical shift timescale. Since chemical shift separations typically range from a few hertz up to a few kilohertz, dynamic NMR lineshape studies are most often sensitive to relatively slow equilibria occurring on second to millisecond timescales. Processes that are slower than this may be studied through magnetisation transfer techniques, as described in Section 2.6.3. It is equally possible to consider events on a coupling constant timescale where the magnitudes of these provide the frequency reference; see discussions involving scalar coupling below.

Before we proceed further we consider briefly why dynamic exchange should lead to the broadening of resonances at all. To begin, we recall that any spin will have a characteristic precession frequency in the transverse plane during the collection of its FID (Section 2.2.3). We also note that the observed resonance peak for each site arises from the summed intensity of all similar spins in the NMR sample, precessing with a similar frequency. Under conditions of two-site exchange, the change in environment from one site to another leads to the frequency of a spin to jump between two different values whenever an exchange event occurs. If this were to occur so infrequently that a complete FID can be collected before such a jump were to happen, then each spin will exhibit only a single frequency and the exchange process would be too slow to influence the NMR spectrum. Now consider the slow–intermediate exchange situation in which two sites interchange while the FID is being collected. These events are infrequent and random, meaning the jumps will occur at different times for different spins within an ensemble of molecules since they are not correlated events. This will lead to net dephasing of the bulk magnetisation vector arising from the sum of all the spins undergoing frequency jumps, and an associated acceleration of the loss of the net transverse magnetisation beyond that which occurs from natural relaxation processes. As described in Section 2.4.3, faster decay of transverse signal correlates to broader resonance in the transformed NMR spectrum, as is seen in this exchange regime. Still faster exchange rates, that is even shorter lifetimes) give rise to more rapid net magnetisation decay and hence greater resonance broadening. When exchange becomes fast on the shift timescale, the lifetimes of any spin in either of the exchanging sites are so short that little net dephasing of transverse magnetisation can accrue between frequency jumps, meaning the accelerated decay of the NMR response is attenuated and a sharper, averaged resonance is observed, characteristic of the fast-exchange regime.

2.6.1.2 Two-Site Exchange: Unequal Populations

Now consider the result of exchange between two species that exist with an unequal population, a more generalised form of the above equilibrium condition. Continuing with an example of restricted rotation in amide bonds, this is observed for the unsymmetrically substituted tertiary amide **2.8**. Rotation about the N-CO bond will change the environments for all neighbouring groups leading to two different conformational (rotational) forms, commonly known as rotamers, of differing concentrations. Under slow-exchange conditions the ^{1}H spectrum displays the sum of the

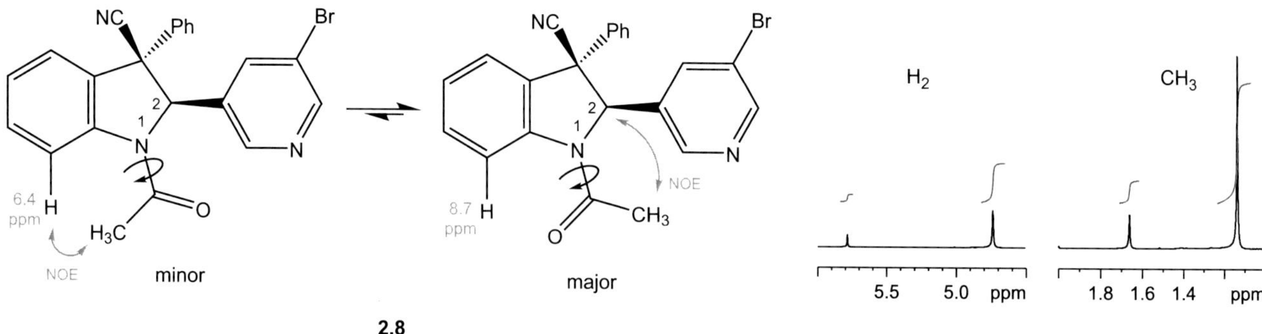

FIGURE 2.48 **The ^1H NMR spectra of the aromatic amide 2.8.** The traces show the acetyl methyl and H_2 resonances under conditions of slow exchange at 230 K. From this, the relative populations of rotamers are 5.3:1.

spectra for each rotamer so that, for example, the acetyl methyl group and the adjacent methane H_2 each give rise to two resonances whose intensities reflect the populations of the interconverting conformers (Fig. 2.48). In this example, it is noteworthy that the carbonyl group exerts a substantial deshielding effect on the adjacent aromatic proton in the major rotamer, giving rise to a shift difference between the rotamers for this proton of a sizeable 2.3 ppm. The fact that these widely separated protons were undergoing exchange was proven by saturation transfer experiments as described in Section 2.6.3, and the assignment of the rotamers was made from NOEs, with all these experiments performed under slow-exchange conditions at 230 K.

In the generic case we may again label the exchanging species as A and B and represent the equilibrium:

$$A \underset{k_B}{\overset{k_A}{\rightleftharpoons}} B \tag{2.20}$$

In this case, the rate constants for the forwards (k_A) and backwards (k_B) processes will differ due to the differing populations such that:

$$P_A \cdot k_A = P_B \cdot k_B \tag{2.21}$$

where P_A and P_B represent the fractional populations of species A and B, respectively, with $P_A + P_B = 1$. The behaviour of a single, uncoupled resonance as a function of the exchange rate is shown in Fig. 2.49 with $k_A > k_B$. The populations of the interchanging species are A:B = 1:2 as reflected in the spectrum at slow exchange. Under slow–intermediate exchange conditions, resonance broadening is apparent and the peaks again tend to move together. However, the extent of broadening is now different for the two resonances, with that of the *less populated* species showing *greater broadening*. This is a consequence of its shorter lifetime associated with its higher rate constant. Following from Eq. 2.17, the terms for exchange broadening may be written:

$$A: \Delta\upsilon_{1/2}^{ex} = \frac{k_A}{\pi} \quad B: \Delta\upsilon_{1/2}^{ex} = \frac{k_B}{\pi} \tag{2.22}$$

quantifying the correlation between the exchange rate constant and the extent of resonance broadening.

As the system moves into the fast–intermediate regime a single resonance is again observed. This now represents a population-weighted average of the exchanging species, meaning the frequency of the averaged peak υ_{AB}^{av} sits towards that of the more populated species such that:

$$\upsilon_{AB}^{av} = P_A\upsilon_A + P_B\upsilon_B \tag{2.23}$$

Any further increase in rate constant leads again to greater exchange narrowing of the single resonance, with residual line broadening arising from exchange of:

$$\Delta\upsilon_{1/2}^{ex} = \frac{4\pi P_A P_B (\Delta\upsilon_{AB})^2}{k_A + k_B} \tag{2.24}$$

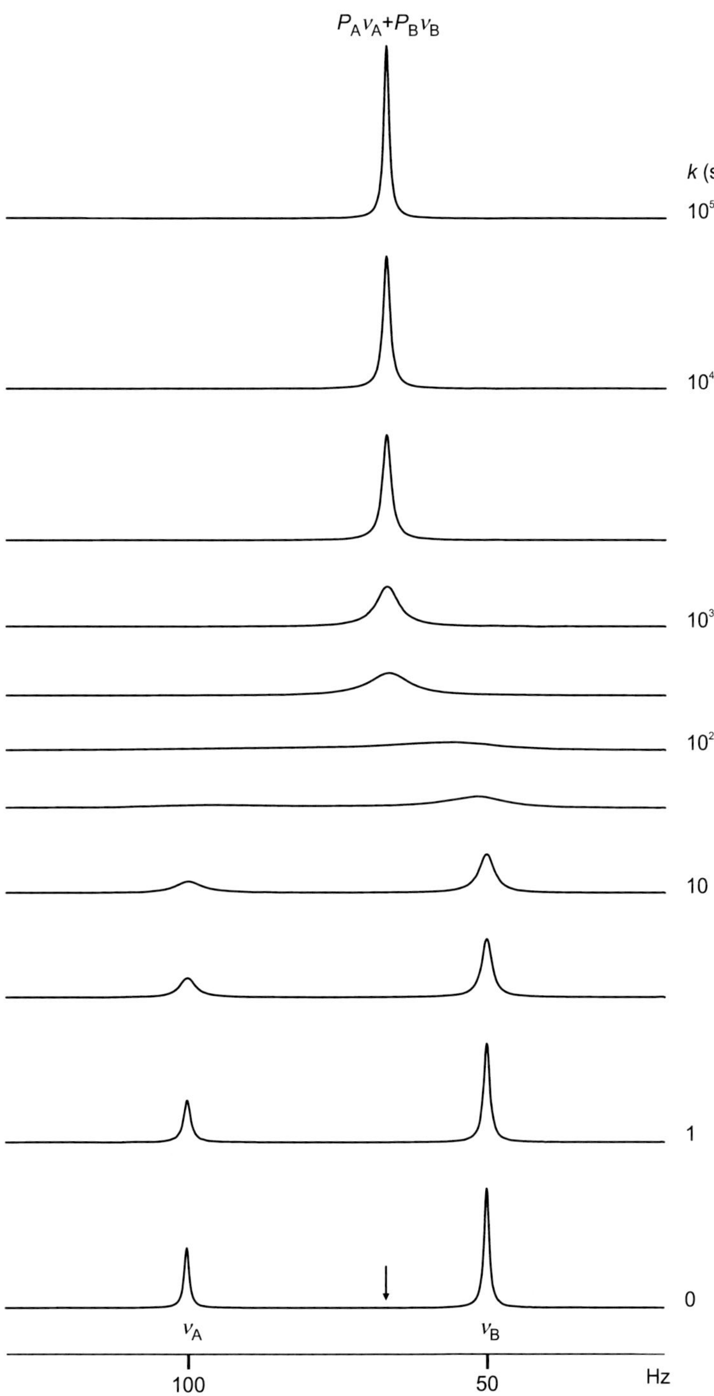

FIGURE 2.49 **Two-site exchange for unequal populations.** Calculated spectra for two nuclei undergoing exchange between two unequally populated sites A and B with frequency separation $\Delta\upsilon_{AB}$ of 50 Hz. The population ratio A:B is 1:2 ($k_A = 2k_B$) and the spectra are labelled according to the mean rate constant (½($k_A + k_B$) s^{-1}). The *arrow* indicates the position of exchange-averaged resonance. Linewidths in the absence of exchange were 1 Hz.

This exchange broadening term can play a very significant role in influencing the linewidths observed for small-molecule ligands experiencing rapid *on–off* exchange equilibria with a protein receptor. Such dynamic exchange broadening may be indicative of ligand binding and so provides a useful indicator of complex formation with a macromolecular target; studies of protein–ligand binding equilibria are considered in the chapter *Protein-Ligand Screening by NMR*.

Under conditions of unequal populations, peak broadening in the intermediate exchange regime can mean the point at which the peaks coalesce is difficult, if not impossible, to define and determination of k_A and k_B is best made through

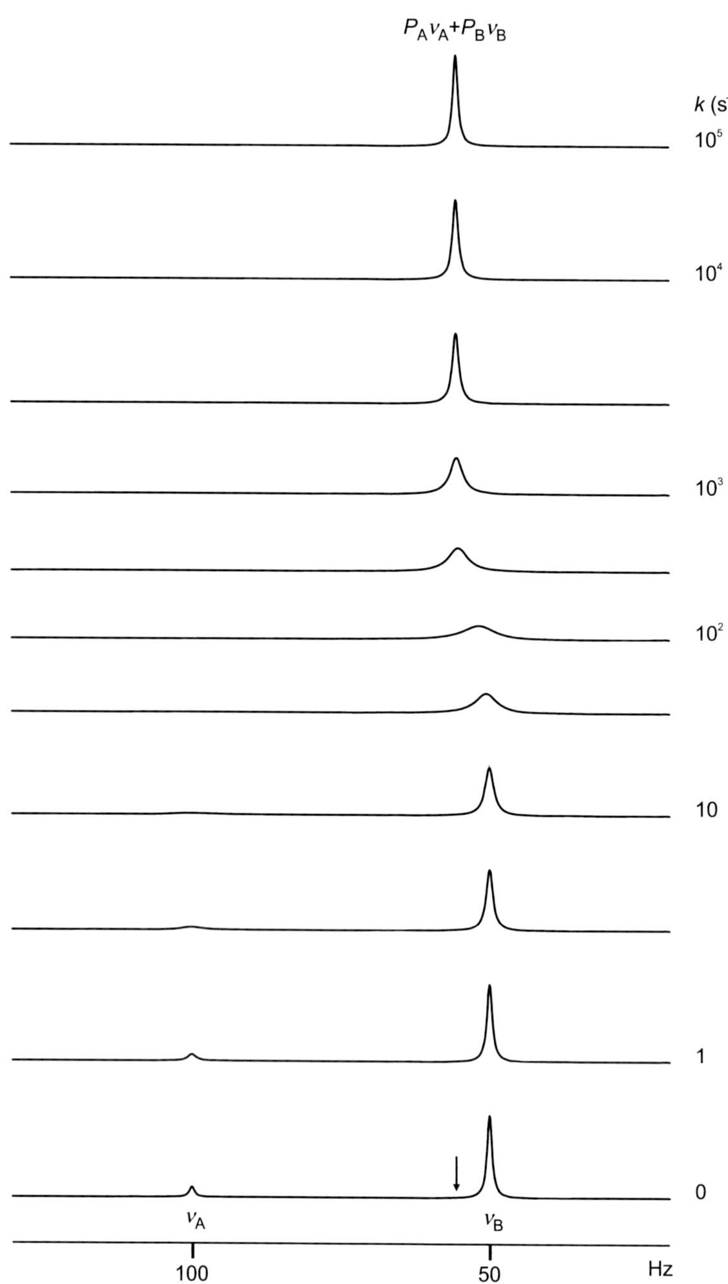

FIGURE 2.50 **Two-site exchange for highly unequal populations.** Calculated spectra for two nuclei undergoing exchange between two sites of greatly differing populations with frequency separation $\Delta\upsilon_{AB}$ of 50 Hz. The population ratio A:B is 1:8 ($k_A = 8k_B$) and the spectra are labelled according to the mean rate constant ($½(k_A + k_B)$ s^{-1}). The arrow indicates the position of exchange-averaged resonance. Linewidths in the absence of exchange were 1 Hz.

lineshape simulation. At a practical level, the greater broadening associated with the more dilute species can mean its resonance becomes very difficult to observe in this regime. The presence of chemical exchange may be revealed only by moderate broadening and shift, followed by progressive sharpening, of the major species resonance as sample temperature, and hence the rate constant, is increased. This is illustrated by the exchange spectra with one species in large excess, shown in Fig. 2.50 for A:B = 1:8. Change in the resonance position of the major species *B* under slow-intermediate exchange and of the averaged peak under fast-intermediate exchange is rather small and only moderate line broadening is observed (compare Fig. 2.46). Moreover, the minor species peak A is barely apparent throughout the intermediate regime and its presence is easily overlooked.

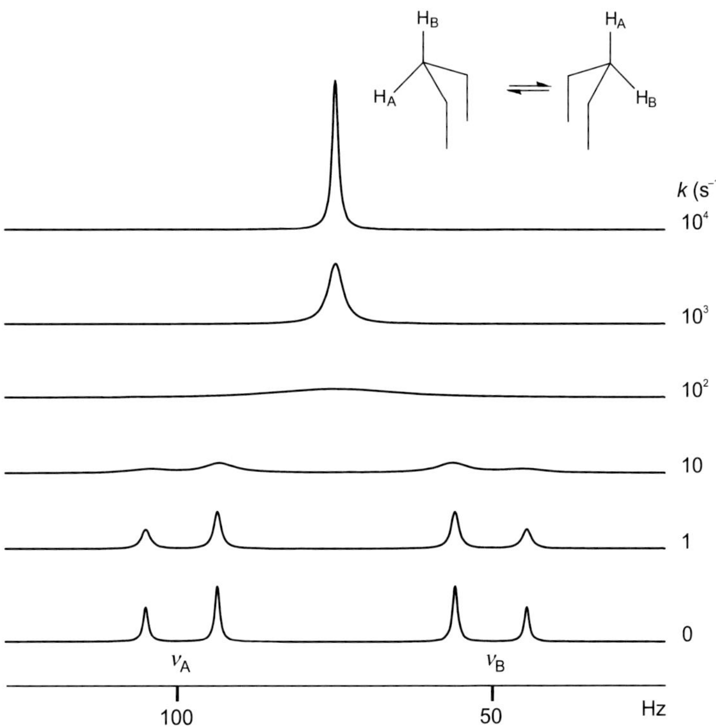

FIGURE 2.51 **Two-site exchange for mutually coupled spins.** Calculated exchange spectra as a function of the rate constant for two mutually J-coupled spins undergoing symmetrical interconversion (J_{AB} = 12 Hz, $\Delta\upsilon_{AB}$ = 50 Hz).

2.6.1.3 Two-Site Exchange between Scalar Coupled Nuclei

We now extend discussions to include scalar (J)-coupled spins experiencing exchange, and how coupling fine structure may be influenced. Let us first consider exchange between two mutually coupled spins A and B that undergo a symmetrical interchange A $\rightleftharpoons$ B and share a coupling J_{AB} but have no other coupled partners (Fig. 2.51). This may be an isolated geminal proton pair undergoing mutual exchange of their environments, for example. Under conditions of slow exchange, geminal coupling is apparent and we see the classical roofed coupling structure. As the exchange rate increases the coupling becomes masked by the resonance broadening until at coalescence this is no longer discernible. At the coalescence point the rate constant is given by:

$$k = \frac{\pi\sqrt{\left[(\Delta\upsilon_{AB})^2 + 6(J_{AB})^2\right]}}{\sqrt{2}} \quad (2.25)$$

having similarity to Eq. 2.18 for the uncoupled two-spin case. As the system tends to the fast-exchange regime the peaks sharpen as the sites interchange more rapidly. However, the resonance lacks coupling structure, appearing as a broad single resonance until in the fast-exchange regime only a single sharp peak exists. The mutual interconversion of A and B leads to a time-averaged system and as the spins have become equivalent the coupling interaction between them no longer gives rise to an observable splitting and the system has been decoupled by the dynamic process. A real example of such behaviour is seen for diphenyldiazetidinone **2.9**, where the nitrogen inversion process leads to the symmetrical interchange of ring CH_2 protons, leading to their eventual coalescence and associated decoupling.

2.9

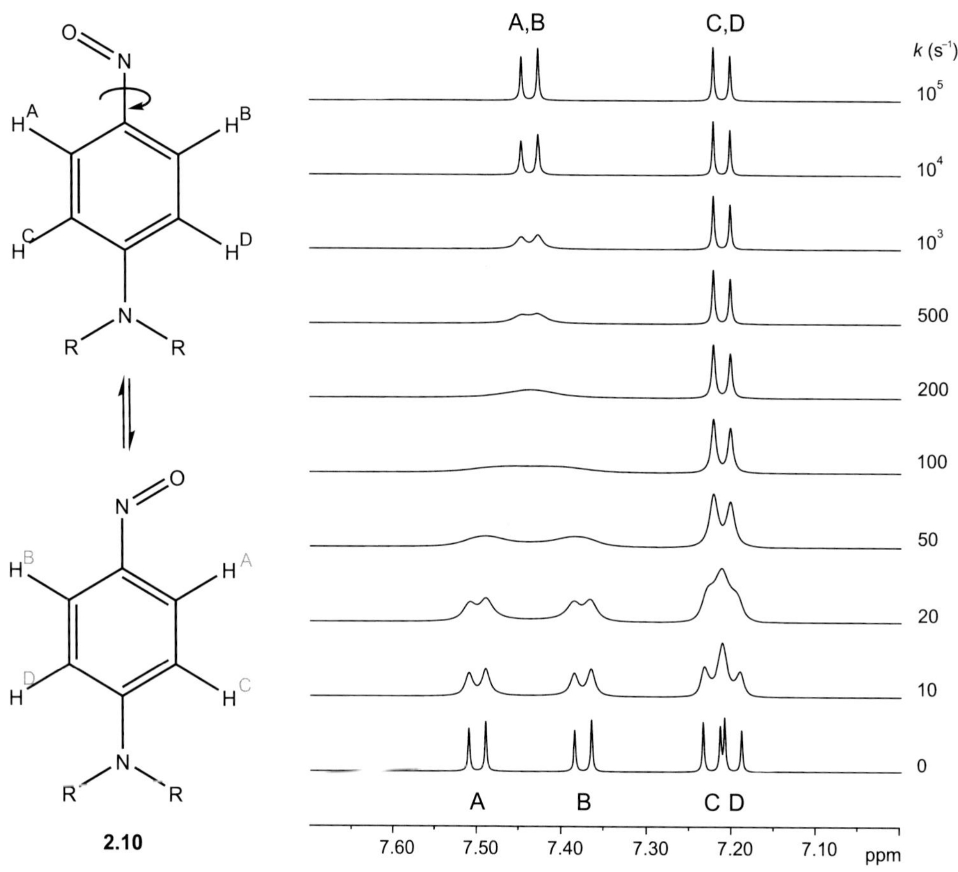

FIGURE 2.52 **Exchange for coupled spins pairs.** Calculated exchange spectra as a function of the rate constant for pairs of *J*-coupled spins undergoing interconversion (^{1}H frequency 400 MHz and $J_{AC} = J_{BD} = 8$ Hz). To aid clarity, each trace has been scaled individually.

Now consider the behaviour of exchanging spins that couple to partners with which they do not undergo mutual exchange. An example of this would be slow exchange of the nitroso group in *N,N*-dialkyl *p*-nitrosoaniline systems, as represented by structure **2.10**. Here the coupled partners are H^A–H^C and H^B–H^D with $J_{AC} = J_{BD}$ and with exchange partners $H^A \rightleftharpoons H^B$ and $H^C \rightleftharpoons H^D$ (for clarity, long-range couplings are not considered here). The slow-exchange condition reveals four doublets for mutually coupled protons, corresponding to the four unique environments these experiences. As the exchange rate increases, these resonances broaden and tend towards coalescence for each exchanging pair (Fig. 2.52). In the intermediate regime, the coupling structure becomes masked, but as the fast–intermediate exchange regime begins the now coalesced peaks of the exchanging pairs once again show evidence for coupling to their vicinal neighbouring partners. In the fast-exchange regime, the environments of the exchanging protons are now fully averaged and chemically equivalent, yielding resolved doublets since the exchange process does not, in this scenario, lead to the dynamic averaging of mutually coupled sites. In this example, it is also important to realise that although the two exchanging conformational forms are equivalent structures (they could be exactly superimposed), the exchange process itself leads to a swap of proton environments and hence the interconversion is still evident in the NMR lineshapes.

2.6.1.4 Scalar Coupling to Exchanging Sites

It is possible that the appearance of the resonance of a spin that is not itself directly influenced by a chemical exchange process may be altered if it shares scalar coupling to one or more spins that are involved in the exchange mechanism. This is most likely to occur when the interchanging sites are chemically identical, so that the environment of the scalar-coupled spin remains unchanged as a result of the exchange event. The influence this has can be appreciated by considering the behaviour of a spin *X* that is coupled to two mutually coupled spins A and B that are themselves undergoing mutual exchange. The symmetrical interchange A $\rightleftharpoons$ B is similar to that described above in Section 2.6.1.3, while the environment of spin X remains invariant to this. However, as may be appreciated from Fig. 2.53, the multiplet structure of X is affected by the exchange, and as A and B average their environments, the coupling of X to these spins is also averaged. In the absence of exchange, spin X

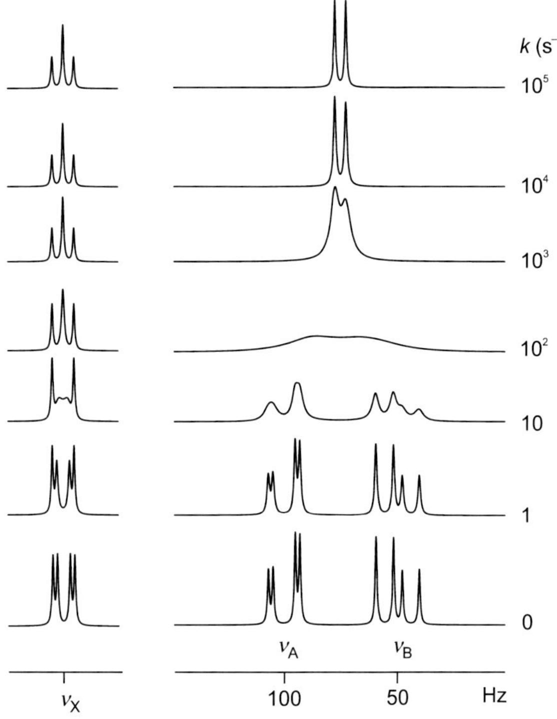

FIGURE 2.53 **Scalar coupling to mutually exchanging sites.** Calculated exchange spectra as a function of the rate constant for three mutually J-coupled spins ABX in which A and B undergo symmetrical interconversion (J_{AB} = 12 Hz, J_{AX} = 2 Hz, J_{BX} = 8 Hz, $\Delta\upsilon_{AB}$ = 50 Hz). The environment of spin X is unchanged by the exchange process, and only its scalar coupling to spins A and B influences its appearance.

shows a typical double-doublet structure, as do spins A and B, each consistent with coupling to two discrete partners. This structure partially collapses in the presence of intermediate exchange until, under conditions of fast exchange, the X-spin multiplet now appears as a triplet and the three spins represent a classical A_2X spin system since the AB environments are fully averaged. The coupling constants are now the average of the initial values and the 2- and 8-Hz couplings to spins A and B, respectively, now collapse and are seen to be 5 Hz for the triplet. Throughout, the chemical shift of spin X is unchanged, but its multiplet structure adopts a rather esoteric appearance at intermediate exchange rates. Notice that the coupling structure of spin X is fully averaged at a rate slower than is required to completely average the chemical environments of spins A and B. This is because the exchange rate need only be fast on the J-*coupling timescale* (not the chemical shift timescale) to fully reduce the multiplet structure to the triplet, and the coupling constants involved are smaller than the A–B chemical shift differences so are more readily averaged. An example in which this behaviour may be observed is represented in the hypothetical fluxional structure **2.11,** where *M* represents a metal centre and *L* a ligand group. Here the interchange of the two fluorine nuclei A and B between axial and equatorial positions will average their environments, while that of the phosphorus nucleus X is unaltered, and only its multiplet structure will exhibit the changes described here.

F^A
L
M
F^B
$(Ar_3)P^X$
L'

2.11

2.6.1.5 Intermolecular Exchange and J Coupling

It is a well-recognised feature of proton NMR spectra recorded in organic solvents, especially $CDCl_3$, that hydroxyl resonances rarely display splitting due to their J-coupling to neighbouring protons, instead most often appearing as broad singlet resonances. This is a further example of dynamic exchange effects influencing the appearance of spectra, and is a direct consequence of intermolecular proton chemical exchange whereby the hydroxyl proton leaves the molecule and is

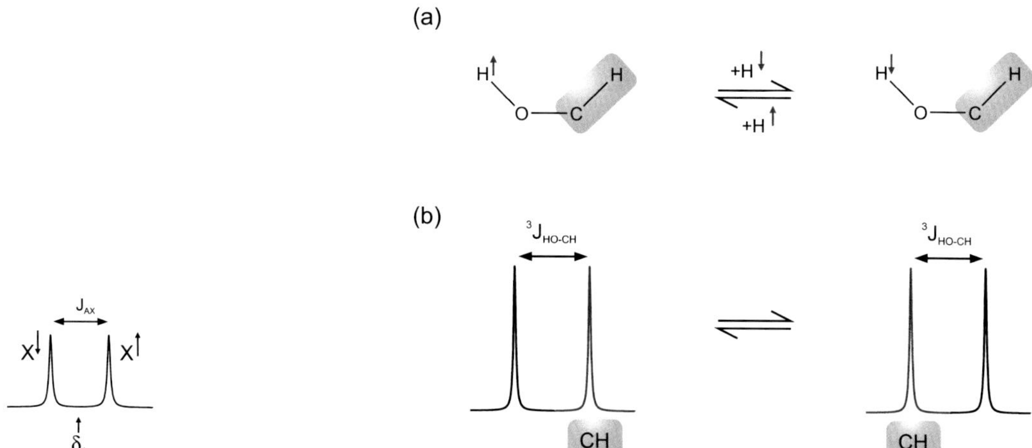

FIGURE 2.54 **The origin of resonance splitting due to scalar coupling with a spin-½ nucleus.** The line frequency of spin A is dictated by the spin state of its J-coupled partner X.

FIGURE 2.55 **The proton exchange process in a coupled CH–OH pair.** (a) Inversion of the OH proton spin state due to proton chemical exchange and (b) its influence on CH proton line frequency.

replaced by another. The incoming proton may have originated from a similar functional group on another solute molecule, may have been from residual solvent water or from acid present in solution. To understand the influence this has on the appearance of coupling fine structure, one must recall that the J-splitting observed for a proton resonance arises from the spin-state orientations adopted by its J-coupled partner. Consider the case for a CH–OH pair sharing a mutual scalar coupling. A doublet structure would be anticipated for the CH group because in half the molecules the neighbouring OH proton exists in the Hα orientation while for the other half the neighbour has the Hβ orientation, thus yielding two lines of equal intensity (Fig. 2.54). During the intermolecular chemical exchange process, it is possible for the incoming proton to have the opposite spin state to that which is leaving (Fig. 2.55a), causing a frequency change of the coupled CH line, meaning it now resonates as the other half of the doublet (Fig. 2.55b). On–off proton exchange therefore leads to an interconversion of the two halves of the doublet (which we may simply consider to be two interchanging sites *A* and *B*) and if this were sufficiently rapid, the lines would average and appear as single resonance (Fig. 2.56). The CH and OH groups no longer display their mutual coupling and they are said to be *exchange decoupled.*

The observed effect is exactly as for the two-site exchange between equally populated states as described above. The significant difference here is that the relevant timescales for the process are now associated with the magnitude of the mutual coupling constant J_{HH}, rather than with chemical shift differences between exchanging sites. Thus, in Fig. 2.56 it is apparent that coalescence of the doublet occurs when $k \approx 2.2$ J s^{-1} or $\sim$18 s^{-1} in this J = 8 Hz example, which follows directly from Eq. 2.18. Above this, the exchange process is said to be fast *on the proton coupling constant timescale.*

For compounds in chloroform, the loss of resonance splitting due to coupling with OH protons (or indeed other acidic protons such as NH or SH) is often associated with traces of hydrochloric acid in commercial deuterated solvent which acts to catalyse proton exchange. Where efforts are made to deacidify solvents with, for example, potassium carbonate, coupling to such protons can be observed. In hygroscopic solvents—most notably, but not exclusively, dimethylsulfoxide (DMSO)—the presence of water can similarly promote exchange, meaning resonances of acidic protons are again often broad and show no coupling structure. This effect is evident in the proton spectrum of the terpene andrographolide **2.12** when freshly prepared in deuterated DMSO and after ageing for some months. The initial spectrum clearly reveals coupling of all hydroxyl resonances whereas after the solvent has absorbed water over time, the proton exchange rate increases, leading to broadening of the hydroxyl protons and a disappearance of coupling structure on neighbouring protons (Fig. 2.57). Thus, the use of dry solvents or suitable drying agents such as molecular sieves, are most likely to reveal coupling to an acidic proton.

2.6.1.6 Some Practicalities Regarding NMR Timescales

From the descriptions so far, it is apparent that the influence on spectra from dynamic exchange correlates with the rate constant for the process, but also that the frequency separation between exchanging sites is critically important. It has already been stressed that dynamic processes may be considered to be *slow* or *fast* only with reference to frequency difference between exchanging sites in the absence of exchange. This has a number of practical consequences on what one might actually observe in spectra according to the conditions used, the instrument employed or the nucleus studied.

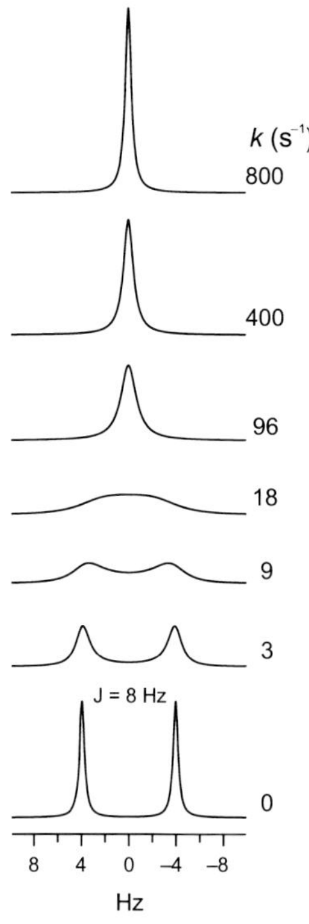

FIGURE 2.56 Calculated spectra showing the loss of coupling fine structure for a CH–OH spin pair as a function of the intermolecular exchange rate constant *k*. The CH–OH coupling constant was 8 Hz.

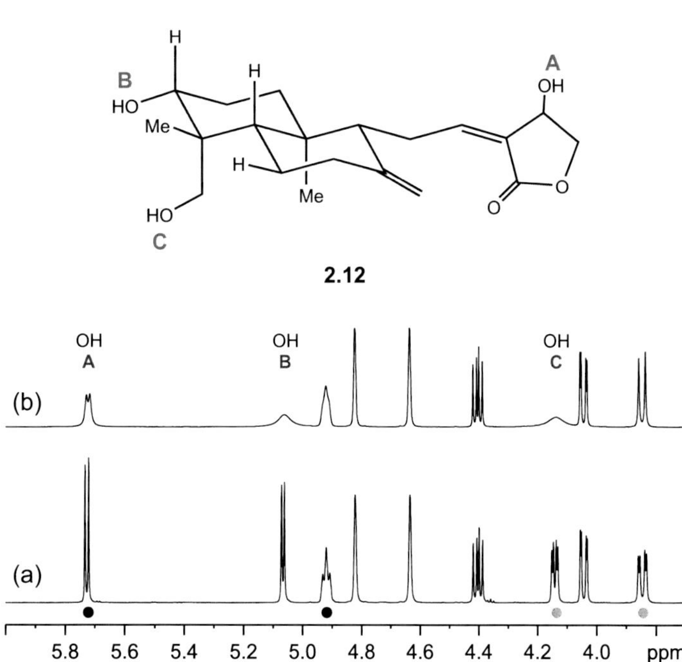

FIGURE 2.57 Exchange broadening and decoupling. The appearance of hydroxyl resonances in andrographolide **2.12** recorded in (a) fresh, dry DMSO and (b) after standing for some months. The *circles* identify mutually coupled protons.

Firstly, when recording a series of spectra at different temperatures to investigate a chemical exchange process, it should be realised that there may be multiple coalescence temperatures associated with a single exchange process. If multiple pairs of nuclei experience interconversion during the exchange it is likely that the frequency separations between each pair would not be identical. As such, the rate constant at which each pair would experience coalescence would differ and hence their coalescence temperatures would be dissimilar, even though they were influenced by the same chemical exchange process. Consider an exchanging system which simultaneously brings about the interconversion of sites A with B and sites C with D which we may represent as $A \rightleftharpoons B$ and $C \rightleftharpoons D$. If the frequency separations between the interconverting pairs were to differ such that $\Delta\upsilon_{AB} > \Delta\upsilon_{CD}$ then for any given rate constant the extent of line broadening due to exchange within the intermediate regime would differ between the pairs, as illustrated in Fig. 2.58. Likewise, the temperature (or rate constant) at which the pairs would reach coalescence would also differ. In practice, this means it is possible for an NMR spectrum to display different exchange regimes for different exchanging nuclei despite only a single exchange process, influenced by one rate constant, being in operation. It is possible that exchanging sites with a small frequency difference may be beyond coalescence and classed as being fast–intermediate whereas those with a greater separation may be yet to coalesce and would be classified as slow–intermediate; see the trace for $k = 40\ s^{-1}$ in Fig. 2.58, for example. It follows that the lowest exchange rate at which the fast-exchange condition exists for *all* exchanging spins will be governed by the pair experiencing the greatest frequency separation in the absence of exchange. This differential behaviour due to differing frequency separations is also observed in Fig. 2.52.

A second consequence of frequency differences being considered in hertz and not parts per million is that the spectra of samples experiencing dynamic exchange broadening will differ according to the magnetic field strength used. This is simply a consequence of shift differences in hertz scaling directly with field; so, for example, a proton frequency separation of 0.2 ppm will correspond to 80 Hz at 400 MHz but to 120 Hz on a 600 MHz spectrometer. It therefore follows that dynamic processes tend to *appear slower* as magnetic fields *increase*. Of course, provided sample conditions including temperature are identical, the rate of exchange itself is unchanged, only the way this is manifested in the spectrum is altered by the change in spectrometer frequency. This effect is illustrated in Fig. 2.59 which shows the dynamic process for the $A \rightleftharpoons B$ and $C \rightleftharpoons D$ pairs described above in ^1H spectra observed over a range of field strengths while all other conditions remain identical. Clearly, at 200 MHz the $C \rightleftharpoons D$ exchange pattern falls within the fast–intermediate regime while at the highest field of

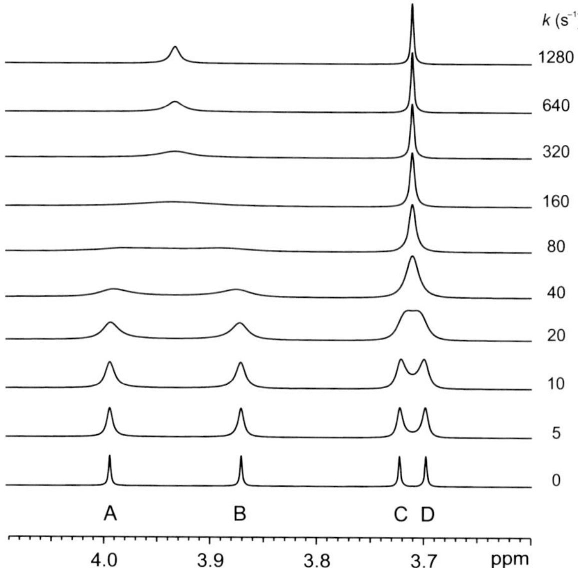

FIGURE 2.58 **The influence of differing frequency separations.** Calculated ^{1}H spectra showing the exchange between two sets of interconverting spin pairs A ⇌ B and C ⇌ D with frequency separations $\Delta\upsilon_{AB}$ = 50 Hz and $\Delta\upsilon_{CD}$ = 10 Hz. To aid clarity, each trace has been scaled individually.

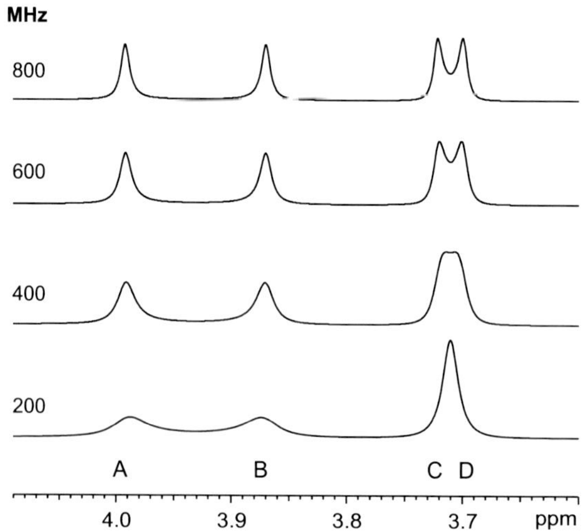

FIGURE 2.59 **The influence of field strength.** Calculated ^{1}H spectra of the exchanging A ⇌ B and C ⇌ D two-pair system of Fig. 2.58 shown as a function of field strength. The rate constant was 10 s^{-1} for all spectra and $\Delta\upsilon_{AB}$ = 50 Hz and $\Delta\upsilon_{CD}$ = 10 Hz.

800 MHz the behaviour of this pair would be classified as slow–intermediate. Due to the larger A–B frequency separation, the exchange behaviour of this pair remains within the slow–intermediate regime at all field strengths.

A final consequence, which follows directly from the above discussion, is that a dynamic process can have quite a different influence on spectra recorded for differing nuclei, sometimes dramatically so. Since chemical shift differences can vary greatly between these, the frequency separation between exchanging spins can likewise show significant variation and are thus influenced to differing degrees by exchange. For example, it is often the case that frequency separations between exchanging carbon centres are significantly larger than those of the corresponding proton sites, due to the greater chemical shift range of carbon-13. This results in carbon-13 resonances exhibiting slower exchange behaviour than proton resonances. In dynamic chemical systems it is not uncommon for the proton spectrum to be well resolved, displaying little or no evidence for exchange since all interconversions exist exclusively in the fast-exchange regime, whereas the carbon-13 spectrum, recorded at the same sample temperature, contains rather broad resonances arising from slow–intermediate exchange behaviour. The study of faster exchange processes may therefore be favoured by ^{13}C NMR rather than ^{1}H NMR, with ^{19}F having the potential to extend the range further due to its very large chemical shift dispersion and high Larmor frequency. Proton-decoupled carbon-13 NMR has also traditionally been popular for the quantitative measurement of exchange rate constants because of the lack of complication associated with J-coupling fine structure that is typical of proton studies. Nowadays, the use of iterative computer lineshape–fitting routines has made analysis of more complex coupled spin systems far more accessible (Section 2.6.2), although the very precise definition of chemical shifts and coupling constants required for such fitting can still present a challenge for complex systems.

2.6.2 Lineshape Analysis and Thermodynamic Parameters

From the discussions above it is apparent that analysis of changes in peak lineshapes in the presence of exchange can allow one to derive rate constants for the exchange processes under study. Nowadays, this is achieved through the computational simulation of spectra and complete lineshape analysis, for which a number of protocols have been developed including DNMR-3 and DNMR-5 [25,26] and MEXICO [27,28]. Gratifyingly, these have also been made accessible in freely (or cheaply) available software [29–33]. Their application involves fitting calculated lineshapes to experimental peaks whereby the rate constant is varied to achieve the optimum match between simulation and experiment. This requires that the spin system under study is accurately defined under all conditions, including chemical shifts and all scalar couplings, where present (see discussions below). Quantitative determination of the rate constant in this manner is often termed DNMR.

From knowledge of the rate constant k, it is then possible to determine the activation energy for the dynamic process, which is most commonly derived from the Eyring relationship:

$$k = \frac{k_B T}{h} \exp\left(-\frac{\Delta G^{\ddagger}}{RT}\right) \tag{2.26}$$

where k_B is the Boltzmann constant, h is Planck's constant and R is the gas constant. From this it is possible to determine the free energy of activation $\Delta G^{\ddagger}$ at absolute temperature T (K) as:

$$\Delta G^{\ddagger} = RT\left\{\ln\left(\frac{k_B}{h}\right) - \ln\left(\frac{k}{T}\right)\right\} \tag{2.27}$$

noting that $\ln(k_B/h) = 23.76$, a common substitution in the above expression found in many texts (note also that in some texts the last term of this expression is given as $+\ln(T/k)$; note the change in sign and swap of numerator and denominator—one should take great care as to which form of the expression is employed). Activation-free energies amenable to determination from NMR lineshape measurements fall typically in the range 20–100 kJ mol^{-1}.

It is further possible to determine the free enthalpy and entropy of activation $\Delta H^{\ddagger}$ and $\Delta S^{\ddagger}$ using the relationship:

$$\Delta G^{\ddagger} = \Delta H^{\ddagger} - T\Delta S^{\ddagger} \tag{2.28}$$

Through the determination of k over a wide range of temperature, one may employ the traditional *Eyring plot* to extract these thermodynamic terms, plotting $\ln(k/\mathrm{T})$ against $1/T$ from which linear correlation gives:

$$\begin{aligned} &\text{Slope}: \frac{-\Delta H^{\ddagger}}{R} \\ &\text{Intercept}: \ln\left(\frac{k_B}{h}\right) + \frac{\Delta S^{\ddagger}}{R} \equiv 23.76 + \frac{\Delta S^{\ddagger}}{R} \end{aligned} \tag{2.29}$$

Alternatively, one may directly plot $\Delta G^{\ddagger}$ obtained from Eq. 2.27 against T to yield a straight line with a slope of $-\Delta S^{\ddagger}$ and an intercept of $\Delta H^{\ddagger}$. This form provides a more useful visual indicator of the magnitudes of these parameters and may be more useful when comparing data from different systems.

2.6.2.1 Practical Considerations in Lineshape Analysis

This approach to determine activation energies appears straightforward, yet demands considerable attention to detail if accurate parameters are to be derived. Firstly, sample temperatures must be accurately known, which requires accurate calibration of probe temperature readings since those from spectrometer temperature units are rarely accurate in the absence of such calibration (see Section 3.5.3 for relevant procedures). Ideally, sample temperatures would be reported directly by

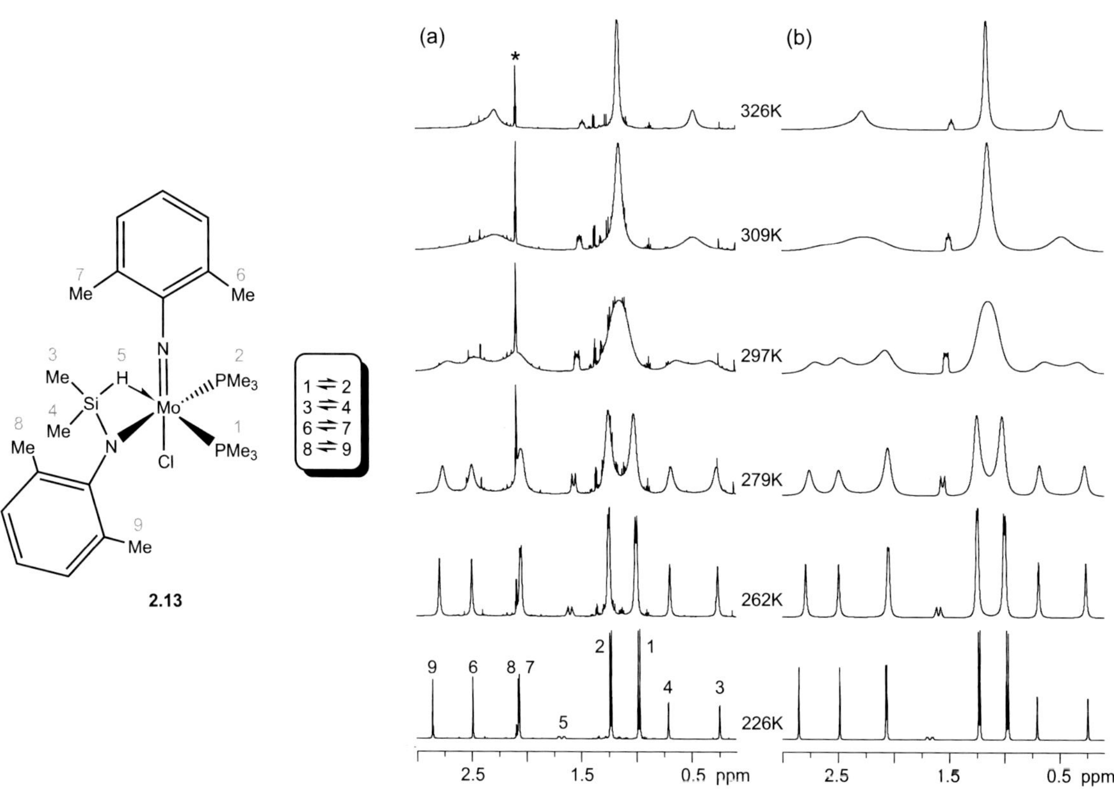

FIGURE 2.60 **Rate constant determination through lineshape analysis.** (a) Experimental ^{1}H NMR spectra showing the methyl region of transition metal complex **2.13** as a function of temperature and (b) simulated ^{1}H spectra (* = solvent toluene). Spectra provided by Dr. N. Rees, University of Oxford.

the response of an internal reference compound. For ^{13}C NMR measurements, tris(trimethylsilyl)methane ($(Me_3Si)_3CH$) has been proposed for this purpose as the CH–CH_3 frequency difference correlates linearly with temperature [34] (this is commercially available, and the synthesis of the ^{13}C labelled analogue has also been described [34]). Secondly, it is often the case that chemical shifts are themselves temperature dependent, meaning it is not sufficient to determine shifts under the condition of slow exchange and assume these will be valid for lineshape simulations of the intermediate-exchange regime. Not accounting for such changes in shift may lead to inaccurate determination of rate constants, even though the simulated spectra may provide a convincing match to the experimental data. The same restriction applies to the estimation of rate constants from peak coalescence, as expressed in Eq. 2.18, where resonance shift differences $\Delta\upsilon_{AB}$ *at the coalescence temperature* must be known. A practical solution to this is to measure chemical shifts accurately over a range of temperatures still within the slow-exchange regime (ie where the separate peaks of the exchanging groups remain clearly resolved and not influenced by line broadening) and from this extrapolate chemical shifts for all temperatures above this. It may also be the case that shift changes are not linear as a function of temperature, and some care is required in performing such extrapolation. Another factor to be determined, as required for simulation, is the natural linewidth of resonances in the absence of exchange. These may be determined by direct inspection of spectra under low-temperature conditions where the influence of exchange is negligible and, for small molecules with narrow resonances, may also be assessed from resonances that are not influenced by the exchange process.

In all cases, it is important to determine rate constants over a wide range of temperatures, especially since the entropy of activation is derived from extrapolation of the Eyring plot to yield the intercept. Indeed, this is generally regarded as the least reliable aspect of Eyring data analysis. Practically, it may not be possible to collect data at sufficiently high temperatures, either due to sample stability or spectrometer limitations, in which case it may be necessary to determine exchange rate constants in the slower exchange regime through use of the magnetisation transfer techniques described in Section 2.6.3.

The lineshape-fitting process is illustrated in Fig. 2.60 for the fluxional dynamics of the transition metal complex **2.13**. Internal rearrangement within the complex interconverts pairs of methyl resonances, as observed in variable temperature ^{1}H spectra, with the mechanism proceeding through opening of the agostic Si—H—Mo bond, followed by rotation about the N—Mo bond, and subsequent ring closure (Fig. 2.61). Although the two forms of the molecule are enantiomeric, the inversion leads to local changes of environment for each methyl group, giving rise to the dynamic behaviour observed.

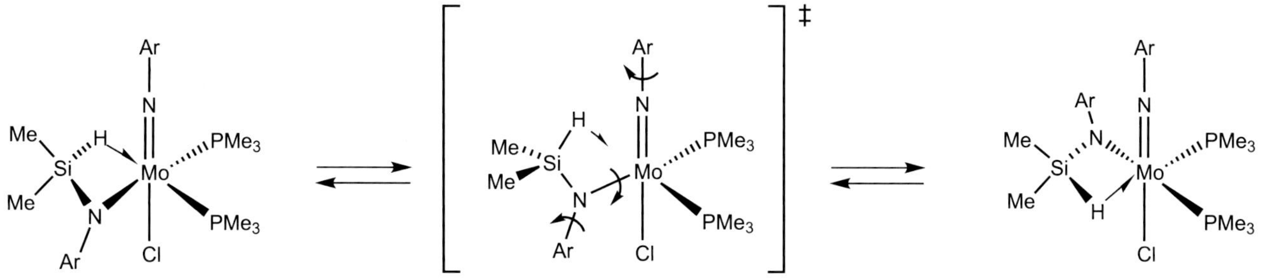

FIGURE 2.61 **The mechanism of inversion for complex 2.13.** This involves weakening of the agostic Si—H—Mo bond followed by concerted bond rotations.

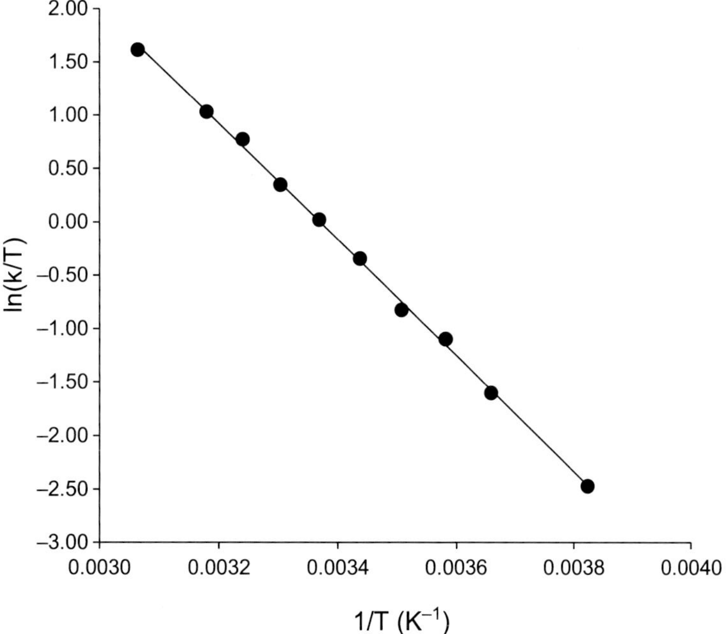

FIGURE 2.62 **Eyring analysis.** The Eyring plot of Me^3–Me^4 exchange rate constants derived from lineshape analysis for the dynamics of 2.13 to yield the activation energies $\Delta H^‡$ and $\Delta S^‡$.

Lineshape analysis demonstrates that all pairwise exchange processes occur at the same rate, indicating correlated motions for all rotational interconversions, including the rotation of aryl groups. This correlated behaviour can be rationalised from steric congestion within the structure, leading to geared dynamics. Analysis of the Eyring plot for Me^3—Me^4 exchange (Fig. 2.62) yields the thermodynamic parameters $\Delta H^‡ = 45.2$ kJ mol^{-1} and $\Delta S^‡ = -46$ J K^{-1} mol^{-1}. A similar analysis for all exchanging pairs indicates similar energy barriers, consistent with the concerted motions [35].

2.6.3 Magnetisation Transfer under Slow-Exchange Conditions

When dynamic exchange processes occur at a rate that is too slow to yield significant influence on lineshapes, it may be possible to interrogate these through the use of magnetisation transfer methods. As the name suggests, these techniques seek to follow the flow of magnetisation from one site to another, carried by the exchanging spin, and can yield rate constants from time-dependent studies. For the exchange to be detectable, the magnetisation transfer process must occur at a rate that is faster than, or at least comparable with, spin relaxation rates so that memory of the initial spin-state perturbation is not lost. This means these methods are suitable where $k \geq (1/T_1)$ s^{-1} which for typical small molecules can extend measurable exchange rates down to $\sim 10^{-2}$ s^{-1}, some two orders of magnitude lower than lineshape analysis. The study of slower relaxing nuclei provides access to slower exchange rates; so, for example, carbon-13 will be favoured over protons for measuring very slow exchange processes due to the relatively slower spin relaxation rates of carbon.

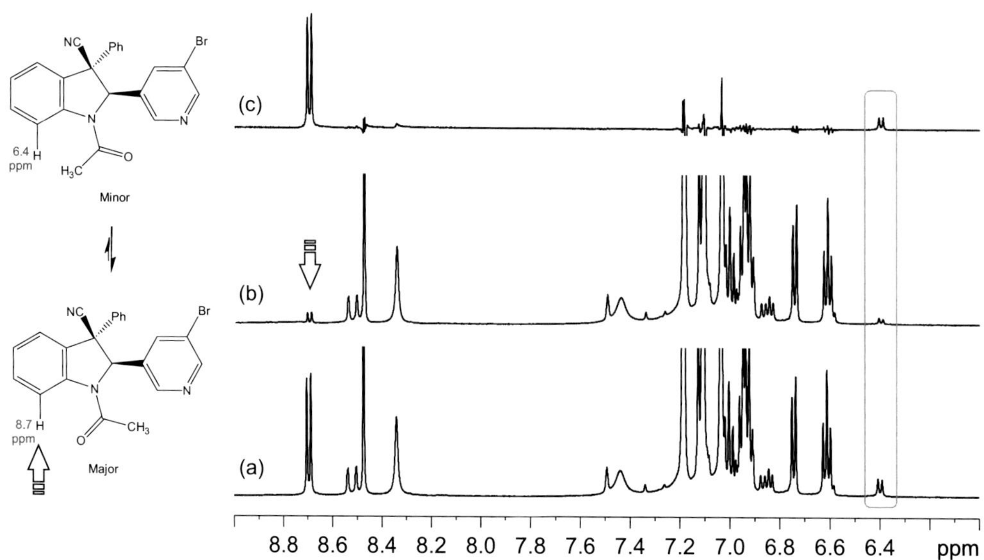

FIGURE 2.63 **Saturation transfer to identify exchanging protons in amide 2.8** (a) the 1D ^{1}H spectrum at 230 K in toluene, (b) the spectrum following saturation of the major resonance at 8.7 ppm (arrowed) and (c) the difference spectrum (a)–(b). The *boxed area* highlights saturation transfer to the minor rotamer.

One of the more simple yet informative applications of magnetisation transfer is for qualitative differentiation of potential conformational or configurational isomers of a molecule [36]. In situations where multiple resonances are observed for a compound, all of which are consistent with a single structure, it may be that these occur because the compound exists as diastereoisomers or because it has two (or more) conformational forms that are undergoing slow chemical exchange. Provided the exchange rate is at least comparable with the spin relaxation rates, then the observation of magnetisation transfer between resonances provides evidence of these being conformers of the compound, rather than stereoisomers, which would show no such interconversion. This is most easily achieved through direct saturation of one resonance with exchange between conformers indicated by resonance attenuation of the exchange partner, a process known as *saturation transfer*. Experimentally, this may be achieved by using a simple presaturation scheme as described in Section 12.5.1 or the 1D NOE spectroscopy (NOESY) experiment described in Section 9.6.2. Similarly, the 2D NOESY experiment also described in the chapter *Correlations Through Space: The Nuclear Overhauser Effect* can also yield cross peaks from chemical exchange and so correlate exchanging partners for the whole molecule. The saturation transfer process is illustrated in Fig. 2.63 for the previously introduced amide **2.8** to identify the exchanging aromatic protons adjacent to the amide group. Direct saturation of the major isomer resonance at 8.7 ppm causes attenuation of the minor resonance at 6.4 ppm (Fig. 2.63b) which is most readily seen in the difference spectrum (Fig. 2.63c). This highlights the sizeable chemical shift difference caused by the carbonyl group deshielding in the major rotamer. To ensure intensity changes arise from exchange and not from accidental spillover of the saturating radiation when interconverting peaks are close, it is also wise to record a control experiment in which the irradiating field is applied equidistant from the exchange partner, avoiding saturation of the target resonance, in which case no saturation transfer should be observed.

Magnetisation transfer techniques have also been employed in the quantitative measurement of slow-exchange rates, most effectively through the use of selective inversion recovery schemes [37,38]. Here, the inversion of a *single* resonance provides an initial perturbation of the spin system, after which its recovery is monitored as a function of time, by analogy with the inversion recovery sequence introduced in Section 2.4.2. Transfer to an exchange partner results in transient reduction of its resonance intensity, after which longitudinal spin relaxation determines that both the inverted spin and its exchange partner return to their equilibrium intensities. This process is illustrated in Fig. 2.64 for amide bond rotation in *N*,*N*-dimethylacetamide. From these data it is possible to extract the rate constant of the exchange process by fitting both the exchange rate and the longitudinal spin relaxation rate, for which programs such as CIFIT [27] have been made available [39]. Since the fitting process requires knowledge of spin relaxation rates, these are typically determined from the conventional non-selective inversion recovery experiment described earlier in this chapter. These rates will be partly averaged by the exchange process, whereas the rates determined from CIFIT analysis will reflect the individual relaxation rates for each resonance. The result of such fitting for the data of Fig. 2.64 is shown in Fig. 2.65, yielding a rate constant for amide rotation of 0.23 s^{-1} at 298 K. A useful description of the procedure for those new to the method has also been published using *N*-methyl formamide as an illustrative example [40].

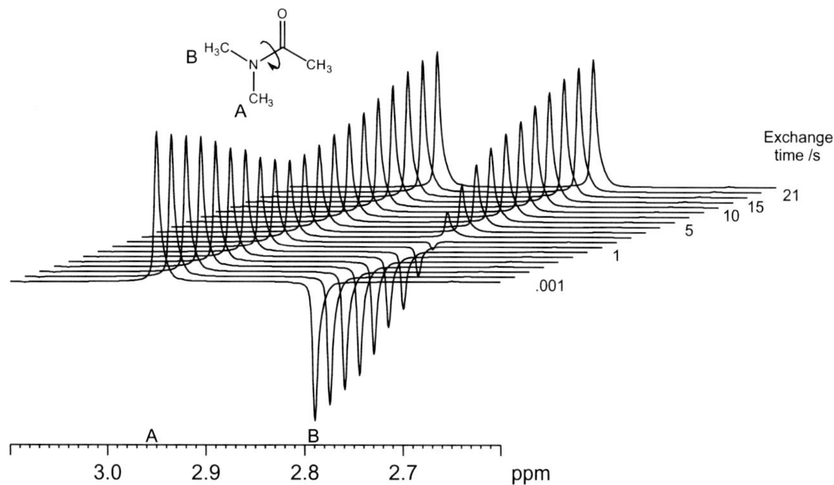

FIGURE 2.64 **The selective inversion recovery experiment.** Spectra were collected for *N,N*-dimethyl acetamide in DMSO at 298 K and were recorded using the sequence of Fig. 2.66b.

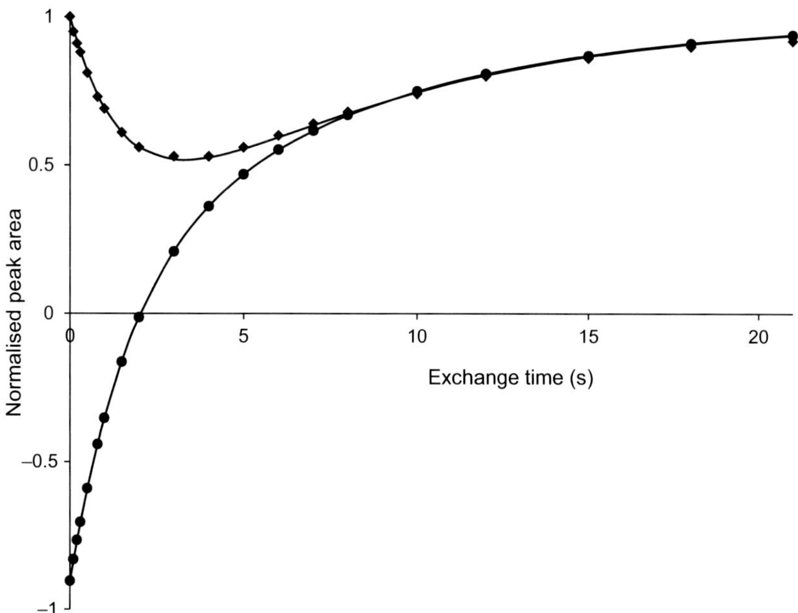

FIGURE 2.65 **Selective inversion recovery analysis.** Fitting the time dependence of magnetisation transfer for the data of Fig. 2.64 to extract the exchange rate constant (0.23 s^{-1}). Data points show normalised experimental peak areas and curves trace calculated points from the CIFIT program.

The ability to measure exchange rate constants for slow processes has practical benefits over the line-shape fitting described earlier. Most significantly, for systems in slow exchange at ambient temperatures, it can avoid the need to apply excessive heat to samples, so reducing the likelihood of sample decomposition. Furthermore, when used in addition to variable temperature lineshape analysis, it provides a means to extend the range of rate constants that can be measured, so improving the reliability of data fitting in the Eyring analysis.

A possible complication to exchange rate measurements relevant to proton studies, in particular, is the potential for the generation of transient NOE effects following initial spin-state perturbation (see Section 9.4). These are likely to arise on a similar timescale to exchange and may contribute to time-dependent resonance intensity changes in addition to those caused by the exchange processes themselves. Although data fitting can include appropriate terms for the NOE, the analysis becomes considerably more complex. Therefore, it may be pragmatic to avoid such complications through the study of alternative nuclei for which time-dependent NOEs will not arise during the magnetisation transfer period, carbon-13 being most commonly employed. In some instances, it may be that proton NOE effects may be considered sufficiently weak to

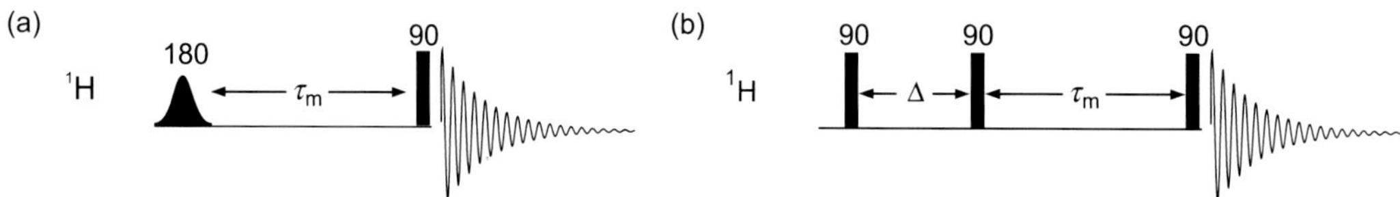

FIGURE 2.66 **Selective inversion recovery methods for studying magnetisation transfer between slowly exchanging spins.** Selective inversion of a target resonance using (a) an on-resonance shaped pulse and (b) a hard-pulse evolution scheme. For the study of heteronuclear systems, broadband ^{1}H decoupling during acquisition times may also be employed.

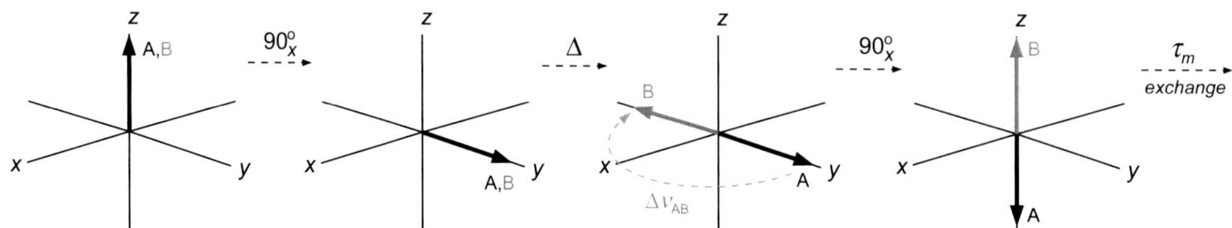

FIGURE 2.67 **Operation of the hard-pulse evolution scheme for selective inversion of spin *A* only.** The transmitter frequency is set on-resonance for spin A causing this to become inverted and the delay $\Delta = 1/2(\Delta\nu_{AB})$ s.

be neglected in the analysis (such as when exchange times are employed that are short relative to NOE buildup timescales) although some caution is required when assessing this.

It is also possible to identify exchange processes for slowing exchanging systems and, in favourable cases, to extract their rate constants, through the use of 2D exchange experiments known commonly as 2D exchange spectroscopy (EXSY); these are considered in the chapter *Correlations Through Space: The Nuclear Overhauser Effect*, after 2D methodology has been introduced. By analogy, the selective 1D inversion recovery methods described here are sometimes also referred to as 1D EXSY experiments.

2.6.3.1 Practical Considerations for Magnetisation Transfer

The magnetisation transfer process requires that a target resonance is subject to selective inversion prior to its exchange behaviour being monitored as a function of the exchange time, often referred to as the *mixing time* τ_m. This inversion may be achieved through the use of a selective 180 degree shaped pulse that acts solely on the resonance of choice, after which the sequence follows that of the standard inversion recovery scheme (Fig. 2.66a). This demands that a suitably selective inversion pulse is defined and calibrated; see Section 12.4 for a description of selective pulses. A simple alternative scheme, well suited to the study of exchange between two uncoupled sites (eg proton singlets or carbon resonances), is to employ the hard-pulse scheme of Fig. 2.66b. Here the transmitter frequency is set on-resonance for the spin to be inverted (A), meaning this remains static in the rotating frame (Fig. 2.67). The delay Δ is chosen so that the magnetisation vector of the exchanging site (B) evolves through exactly one-half of a cycle, which requires that $\Delta = 1/2(\Delta\upsilon_{AB})$ s, where $\Delta\upsilon_{AB}$ is the frequency separation between resonances A and B (in hertz). The second 90 degree pulse then places magnetisation of spin A along $-z$ and so achieves its desired inversion, while that of spin B is returned to $+z$, experiencing no net excitation after the 90 degree–Δ–90 degree element. This method was used to generate the spectra of Fig. 2.64. Note that other resonances in the spectrum may show phase distortions from this scheme due to their differing offsets from spin A, but these should not perturb the measurement of magnetisation transfer between the exchanging sites A $\rightleftharpoons$ B. This method tends to be less well suited to resonances exhibiting scalar coupling since J evolution during the Δ delay can lead to phase distortions for resonances A and B in the resulting spectrum. As for the standard inversion recovery sequence, complete spin relaxation is demanded between each scan, requiring knowledge of T_1 time constants for exchanging spins.

REFERENCES

[1] Harris RK, Kowalewski J, Cabral de Menezes S. Magn Reson Chem 1998;36:145–9.
[2] Harris RK, Becker ED, Cabral de Menezes SM, Goodfellow R, Granger P. Pure Appl Chem 2001;73:1795–818.
[3] Harris RK, Becker ED. J Magn Reson 2002;156:323–6.
[4] Rabenstein DL, Fan S, Nakashima TK. J Magn Reson 1985;64:541–6.
[5] Rabenstein DL, Srivasta GS, Lee RWK. J Magn Reson 1987;71:175–9.

[6] Larive CK, Rabenstein DL. Magn Reson Chem 1991;29:409–17.
[7] Harris RK. Nuclear magnetic resonance spectroscopy. Harlow: Longman; 1986.
[8] Levitt M. Spin dynamics. New York: Wiley; 2001.
[9] Bakhmutov VI. Practical NMR relaxation for chemists. Chichester: Wiley; 2004.
[10] Keeler J. Understanding NMR spectroscopy. 2nd ed. Chichester: Wiley; 2010.
[11] Wong TC, Ang TT, Guziec FS, Moustakis CA. J Magn Reson 1984;57:463–70.
[12] Ismail IM, Kerrison SJS, Sadler PJ. Polyhedron 1982;1:57–9.
[13] Waugh AP, Lawless GA. In: Gielen M, Willem R, Wrackmeyer B, editors. Advanced applications of NMR to organometallic chemistry. Chichester: Wiley; 1996.
[14] Baldwin JE, Claridge TDW, Derome AE, Smith BD, Twyman M, Waley SG. J Chem Soc Chem Commun 1991;573–4.
[15] Bain A. Annu Rep NMR. Spectrosc 2008;63:23–48.
[16] Orrell KG, Šik V, Stephenson D. Prog Nucl Magn Reson Spectrosc 1990;22:141–208.
[17] Orrell KG, Šik V. Annu Rep NMR Spectrosc 1987;19:79–173.
[18] Orrell KG, Šik V. Annu Rep NMR Spectrosc 1993;27:103–71.
[19] Pons M, Millet O. Prog Nucl Magn Reson Spectrosc 2001;38:267–324.
[20] Jackman LM, Cotton FA. Dynamic nuclear magnetic resonance spectroscopy. New York: Academic Press; 1975.
[21] Sandström J. Dynamic NMR spectroscopy. London: Academic Press; 1982.
[22] Oki M. Applications of dynamic NMR spectroscopy to organic chemistry. Weinheim: VCH; 1985.
[23] Clegg IM, Gordon CM, Smith DS, Alzaga R, Codina A. Anal Methods 2012;4:1498–506.
[24] Foley DA, Bez E, Codina A, Colson KL, Fey M, Krull R, Piroli D, Zell MT, Marquez BL. Anal Chem 2014;86:12008–13.
[25] Stephenson DS, Binsch G. J Magn Reson 1978;30:625–6.
[26] Stephenson DS, Binsch G. J Magn Reson 1978;32:145–52.
[27] Bain AD, Cramer JA. J Magn Reson A 1996;118:21–7.
[28] Bain AD, Rex DM, Smith RN. Magn Reson Chem 2001;39:122–6.
[29] Budzelaar, PHM. gNMR. Available from: http://home.cc.umanitoba.ca/~budzelaa/gNMR/gNMR.html
[30] Reich, HJ. WinDNMR. Available from: http://www.chem.wisc.edu/areas/reich/plt/windnmr.htm
[31] Bain, A. MEXICO. Available from: http://www.chemistry.mcmaster.ca/bain/exchange.html
[32] Marat, K. SpinWorks. Available from: http://home.cc.umanitoba.ca/~wolowiec/spinworks/
[33] Nucleomatica. iNMR. Available from: http://www.inmr.net/
[34] Sikorski WH, Sanders AW, Reich HJ. Magn Reson Chem 1998;36:S118–24.
[35] Ignatov SK, Khalimon AY, Rees NH, Razuvaev AG, Mountford P, Nikonov GI. Inorg Chem 2009;48:9605–22.
[36] Hu DX, Grice P, Ley SV. J Org Chem 2012;77:5198–202.
[37] Campbell ID, Dobson CM, Ratcliffe RG, Williams RJP. J Magn Reson 1978;29:397–417.
[38] Led JJ, Gesmar H. J Magn Reson 1982;49:444–63.
[39] Bain, AD. CIFIT. Available from: http://www.chemistry.mcmaster.ca/bain/
[40] Williams TJ, Kershaw AD, Li V, Wu X. J Chem Ed, 2011;88:665–9.

Chapter 3

Practical Aspects of High-Resolution NMR

Chapter Outline

3.1 An Overview of the NMR Spectrometer 61
3.2 Data Acquisition and Processing 64
- 3.2.1 Pulse Excitation 64
- 3.2.2 Signal Detection 66
- 3.2.3 Sampling the FID 67
- 3.2.4 Quadrature Detection 73
- 3.2.5 Phase Cycling 78
- 3.2.6 Dynamic Range and Signal Averaging 80
- 3.2.7 Window Functions 83
- 3.2.8 Phase Correction 88

3.3 Preparing the Sample 89
- 3.3.1 Selecting the Solvent 89
- 3.3.2 Reference Compounds 91
- 3.3.3 Tubes and Sample Volumes 92
- 3.3.4 Filtering and Degassing 94

3.4 Preparing the Spectrometer 95
- 3.4.1 The Probe 95
- 3.4.2 Probe-Design and Sensitivity 97
- 3.4.3 Tuning the Probe 103
- 3.4.4 The Field Frequency Lock 105
- 3.4.5 Optimising Field Homogeneity: Shimming 107
- 3.4.6 Reference Deconvolution 112

3.5 Spectrometer Calibrations 113
- 3.5.1 Radiofrequency Pulses 113
- 3.5.2 Pulsed Field Gradients 122
- 3.5.3 Sample Temperature 124

3.6 Spectrometer Performance Tests 126
- 3.6.1 Lineshape and Resolution 127
- 3.6.2 Sensitivity 128
- 3.6.3 Solvent Presaturation 130

References 130

Nuclear magnetic resonance (NMR) spectrometers are expensive instruments, representing one of the largest financial investments a chemical laboratory is likely to make, and to get the best results from these they must be operated and maintained in the appropriate manner. This chapter explores some of the fundamental experimental aspects of relevance to high-resolution, solution-state NMR spectroscopy, from instrumental procedures through to the preparation of samples for analysis. Later sections also deal with the basics of calibrating a spectrometer and assessing its performance.

3.1 AN OVERVIEW OF THE NMR SPECTROMETER

A schematic illustration of a modern NMR spectrometer is presented in Fig. 3.1. The fundamental requirement for high-resolution NMR spectroscopy is an intense static magnetic field which is provided, nowadays exclusively, by superconducting solenoid magnets manufactured from niobium/tin-based alloy wire. These are able to produce the stable and persistent magnetic fields demanded by NMR spectroscopy. Current technology provides for fields of up to 23.5 T, corresponding to a proton frequency of 1000 MHz (1 GHz). The drive for ever increasing magnetic fields, encouraged by the demands for greater signal dispersion and instrument sensitivity, has continued since the potential of NMR spectroscopy as an analytical tool was first realised over 60 years ago (Fig. 3.2) and still represents an intense research area for magnet manufacturers. Proton observation frequencies beyond 1 GHz demand the use of newer high-temperature superconducting materials in addition to conventional alloy wire to carry the huge currents required. This not only presents substantial technical challenges for the manufacturer but will undoubtedly further increase magnet costs for the end user. However, these ultra high field magnets are currently of most interest to specific research areas such as structural biology, and in the arena of chemistry research the need for such extreme signal dispersion is likely less important, although enhancements in sensitivity remain welcome and valuable developments. In recent years there have been a number of advances in probe technologies that provide the chemist with very significant sensitivity increases without the need for higher field instrumentation and these will also be reviewed later in the chapter.

The magnet solenoid operates in a bath of liquid helium (at or below 4 K), surrounded by a radiation shield and cooled by a bath of liquid nitrogen (at 77 K), itself surrounded by a high vacuum. This whole assembly is an extremely efficient

High-Resolution NMR Techniques in Organic Chemistry. http://dx.doi.org/10.1016/B978-0-08-099986-9.00003-8

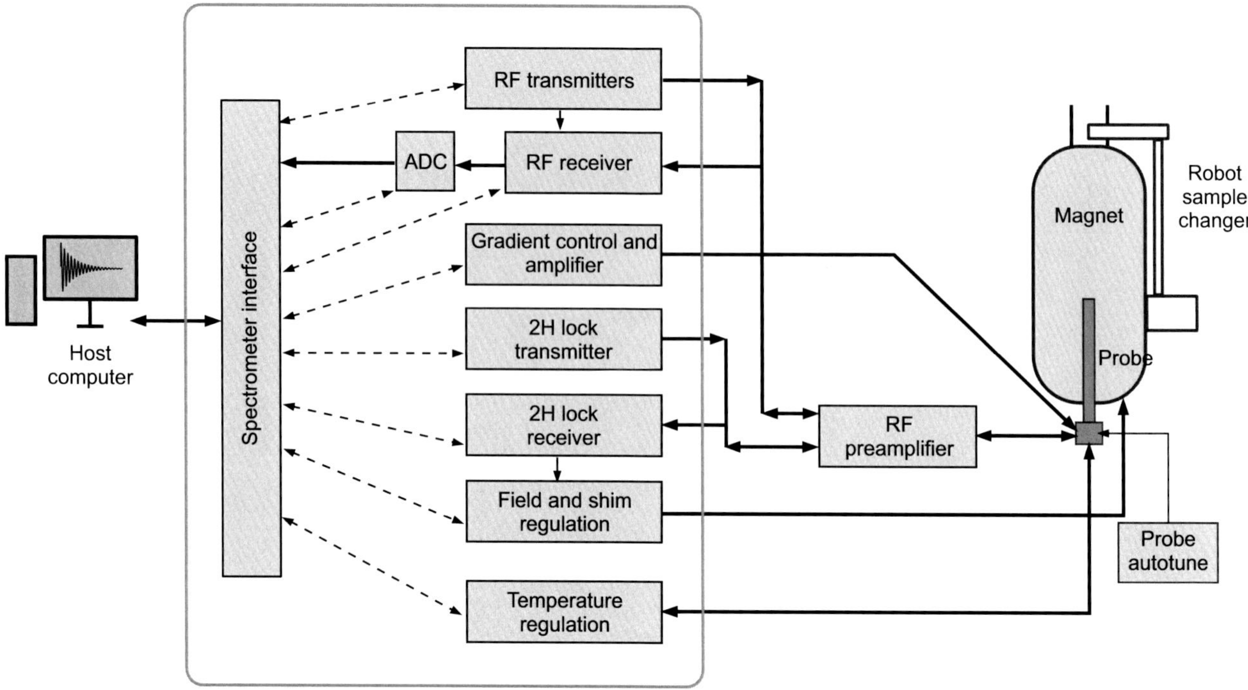

FIGURE 3.1 **Schematic illustration of the modern NMR spectrometer.**

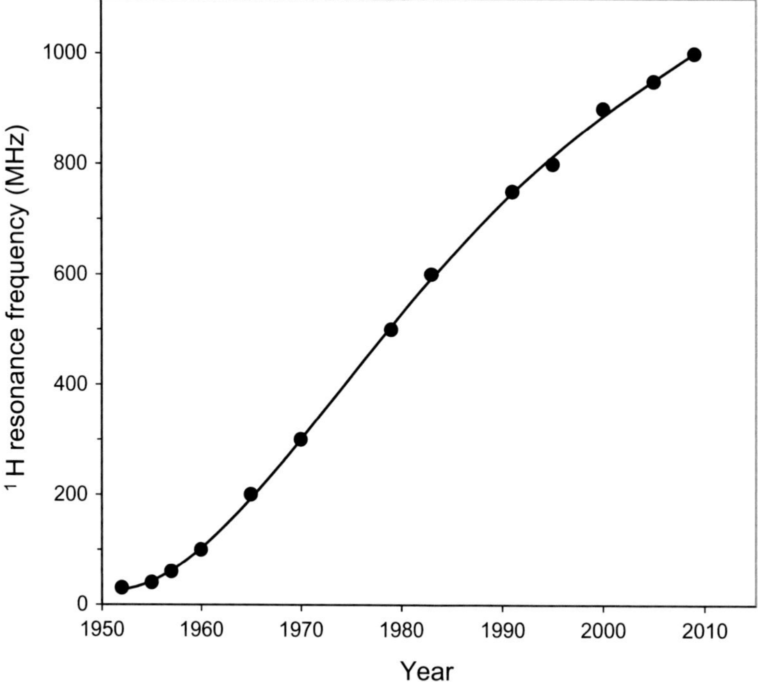

FIGURE 3.2 **The increase in proton resonance frequency since the introduction of NMR spectroscopy as an analytical method.** *(Source: Adapted with permission from Ref. [1] and extended, Copyright 1995, Elsevier).*

system and once energised the magnet operates for many years free from any external power source. It requires only periodic refilling of the liquid cryogens, typically on a weekly or biweekly basis for the nitrogen but only every 2–12 months for helium, depending on magnet age and construction. The central bore of the magnet dewar is itself at ambient temperature, and this houses a collection of electrical coils, known as shim coils, which generate their own smaller magnetic fields and are used to trim the main static field and remove residual inhomogeneities. This process of optimising

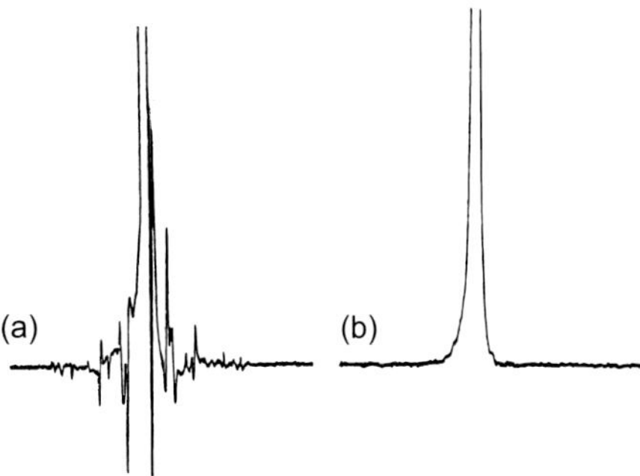

FIGURE 3.3 Floor vibrations can introduce unwelcome artefacts around the base of a resonance (a) which can be largely suppressed by mounting the magnet assembly on an anti-vibration stand (b). *(Source: Courtesy of Bruker BioSpin.)*

the magnetic field homogeneity, known as *shimming*, is required for each sample to be analysed, and is discussed in Section 3.4.5. Within the shim coils, at the exact centre of the magnetic field, sits the head of the probe, the heart of any NMR spectrometer. This houses the radiofrequency coil(s) and associated circuitry that act as antennae, transmitting and receiving electromagnetic radiation. These coils may be further surrounded by pulsed field gradient (PFG) coils that serve to *destroy* field homogeneity in a controlled fashion (this may seem a rather bizarre thing to do, but it turns out to have rather favourable effects in numerous experiments). The sample, usually contained within a cylindrical glass tube, is held in a turbine or 'spinner' and descends into the probe head on a column of air or nitrogen. Alternative configurations may employ fixed flow cells to pass the sample through the detector region of the probe. For routine 1D experiments it may prove beneficial to spin the sample tube held in a turbine about its vertical axis at 10–20 Hz while in the probe to average to zero field inhomogeneities in the transverse (x–y) plane and so improve signal resolution. Sample spinning is often unnecessary on modern instruments and indeed is rarely used for multidimensional NMR experiments since it may induce additional signal modulations and associated undesirable artefacts. This requires that acceptable resolution can be obtained on a static sample and while this is perfectly feasible with modern shim technology, older instruments may still demand spinning for all work.

Probes come in various sizes of diameter and length, depending on the magnet construction, but are more commonly referred to by the diameter of the sample tube they are designed to hold. The most widely used tube diameter is still 5 mm; other available sizes have included 1, 1.7, 2.5, 3, 8 and 10 mm (Section 3.4). Probes may be dedicated to observing one frequency (*selective* probes), may be tuneable over a very wide frequency range (*broadband* probes) or may tune to predefined frequency ranges, for example four-nucleus or quad-nucleus probes. In all cases they will also be capable of observing the deuterium frequency simultaneously to provide a signal for field regulation (the 'lock' signal). A second (outer) coil is often incorporated to allow the simultaneous application of pulses on one or more additional nuclei.

In many locations it is advantageous to mount the whole of the magnet assembly on a vibration damping system as floor vibrations (which may arise from a whole host of sources including natural floor resonances, air conditioners, movement in the laboratory and so on) can have deleterious effects on spectra, notably around the base of resonances (Fig. 3.3). While such artefacts have lesser significance to routine 1D observations, they may severely interfere with the detection of signals present at low levels, for example those in heteronuclear correlation or nuclear Overhauser effect experiments.

Within the spectrometer cabinet sit the radiofrequency transmitters and the detection system for the observation channel, additional transmitter channels, the lock channel and the PFG transmitter. Reference to the 'decoupler channel(s)' is often used when referring to these additional channels, but this should not be taken too literally as they may only be used for the application of only a few pulses rather than a true decoupling sequence. This nomenclature stems from the early developments of NMR spectrometers when the additional channel was only capable of providing 'noise decoupling', usually of protons.

Most spectrometers come in either a two-channel or three-channel configuration, plus the lock channel. The spectrometer is controlled via the host computer (either a Windows or Linux-based PC system) which is linked to the spectrometer via a suitable interface such as Ethernet. Electrical analogue NMR signals are converted to the digital format required by the host computer via the analogue-to-digital converter (ADC), the characteristics of which can have important implications for the acquisition of NMR data (Section 3.4.5). The computer also processes the acquired data, although this may also be performed 'off-line' with one of the many available NMR software packages.

Various optional peripherals may also be added to the instrument, such as variable temperature units which allow sample temperature regulation within the probe, robotic sample changers and so on. The coupling of NMR with other analytical techniques such as high-performance liquid chromatography (HPLC) has become an established method through the development of flowprobes and gained popularity in some analytical areas. The need for this will obviously depend on the type of samples handled and the nature of the experiments employed.

With the hype surrounding the competition between instrument manufacturers to produce ever-increasing magnetic fields, it is all too easy for one to become convinced that an instrument operating at the highest available field strength is essential in the modern laboratory. While the study of biological macromolecules no doubt benefits from the greater sensitivity and dispersion available, problematic small or mid-sized molecules are often better tackled through the use of appropriate modern techniques. Signal dispersion limitations are generally less severe, and may often be overcome by using suitably chosen higher dimensional experiments. Sensitivity limitations, which are usually due to a lack of material rather than solubility or aggregation problems, may be tackled by utilising smaller probe geometries or cryogenically cooled probeheads (Section 3.4.2). So, for example if one has insufficient material to collect a carbon-13 spectrum one could consider employing a proton-detected heteronuclear correlation experiment to determine these shifts indirectly. Beyond such considerations, there are genuine physical reasons largely relating to the nature of nuclear spin relaxation, which mean that certain experiments on small molecules are likely to work *less well* at very high magnetic fields. In particular, this relates to the nuclear Overhauser effect (see the chapter *Correlations Through Space: The Nuclear Overhauser Effect*), one of the principal NMR methods in structure elucidation. For many cases commonly encountered in the chemical laboratory a lower field instrument of modern specification is sufficient to enable the chemist to unleash an array of modern pulse NMR experiments on the samples of interest and subsequently solve the problem in hand. Undoubtedly, a better understanding of these modern NMR methods should aid in the selection of the most appropriate experiments, and subsequent chapters will aim to develop such understanding. In this chapter, we seek to develop an understanding of the NMR spectrometer itself, and the practicalities of applying this to chemical research.

3.2 DATA ACQUISITION AND PROCESSING

This section examines some of the spectrometer procedures that relate to the collection, digitisation and computational manipulation of NMR data, including some of the fundamental parameters that define the way in which data are acquired. Such technicalities may not seem relevant to those who do not consider themselves spectroscopists, but the importance of understanding a few basic relationships between experimental parameters comes from the need to recognise spectrum artefacts or corrupted data that can result from inappropriate parameter settings and to appreciate the limitations inherent in NMR measurements. Only then can one make full and appropriate use of the spectroscopic information at hand.

3.2.1 Pulse Excitation

It is widely appreciated that modern NMR spectrometers use a 'short pulse' of radiofrequency energy to excite nuclear resonances over a range of frequencies. This pulse is supplied as *monochromatic* radiation from the transmitter, yet the nuclear spin transitions giving rise to our spectra vary in energy according to their differing Larmor frequencies, and so it would appear that the pulse will be unable to excite all resonances in the spectrum simultaneously. However, Heisenberg's Uncertainty principle tells us that an excitation pulse of duration Δt has associated with it a frequency uncertainty or spread of around $1/\Delta t$ Hz and so effectively behaves as if it were polychromatic. The duration of the rf pulse Δt is usually referred to as the *pulse width*. A short, high-power pulse provides excitation over a wide frequency window while a longer, low-power pulse (which will provide the same net tip angle) is effective over a much smaller region (Fig. 3.4).

Consider a proton spectrum recorded at 400 MHz with the transmitter frequency placed in the centre of the spectral region of interest. A 10 ppm spectral window corresponds to 4000 Hz; therefore, we need to excite over ±2000 Hz, meaning the pulse duration must be 0.5 ms or less. The observation frequency for carbon-13 on the same instrument would be 100 MHz, so a typical 200 ppm spectral width corresponds to ±10 kHz, requiring a pulse of only 100 μs but with considerably higher power as its energy is now spread over a wider area. If the pulse were to be made very long (say, tens of milliseconds) and weak, it would only excite over a rather small frequency range, giving rise to *selective excitation* of only part of the spectrum. As will become apparent in later chapters, the use of selective excitation methods is now commonplace in numerous NMR experiments. Long, low-power pulses which excite only a selected region of a spectrum are commonly referred to as 'soft pulses' whereas those that are of short duration and high power are termed 'hard pulses'. Unless stated explicitly, all pulses in this book will refer to non-selective hard pulses.

3.2.1.1 Off-Resonance Effects

In practice, modern NMR instruments are designed to deliver high-power 90 degrees pulses closer to 10 μs, rather than the hundreds predicted from the above arguments. This is to suppress the undesirable effects that arise when the pulse rf frequency is *off-resonance*; that is when the transmitter frequency does not exactly match the nuclear Larmor frequency, a situation of considerable practical significance that has been ignored thus far.

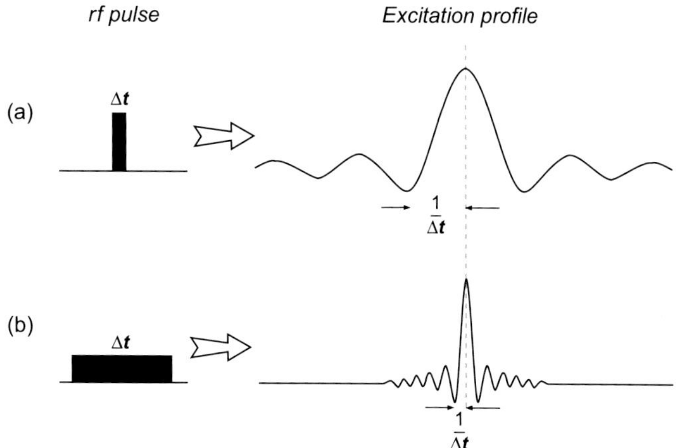

FIGURE 3.4 **Pulse excitation.** A single monochromatic radiofrequency pulse has an effective excitation bandwidth that depends inversely on the duration of the pulse. A short intense pulse is therefore able to excite over a wide frequency window (a), whereas a longer weaker pulse provides a more selective excitation profile (b).

As shown in the previous chapter, spins that are on-resonance have their magnetisation vector driven about the rf B_1 field towards the x–y plane during the pulse. Those spins that are off-resonance will, in addition to this B_1 field, experience a *residual* component ΔB of the *static* B_0 field along the z-axis of the rotating frame for which:

$$\frac{\gamma\Delta B}{2\pi} = \gamma\!\!\!/\Delta B = \Delta\nu \text{ Hz} \tag{3.1}$$

where $\Delta\nu$ represents the offset from the reference frequency. The vector sum of B_1 and ΔB is the *effective field* B_{eff} experienced by an off-resonance spin about which it rotates (or more correctly, nutates) (Fig. 3.5). This is greater than B_1 itself and is tilted away from the x–y plane by an angle θ, where

$$\tan\theta = \frac{\Delta B}{B_1} \tag{3.2}$$

For those spins further from resonance, the angle θ becomes greater and the net rotation towards the x–y plane diminishes until, in the limit, θ becomes 90 degrees. In this case the bulk magnetisation vector simply remains along the $+z$-axis and thus experiences no excitation at all. In other words, the nuclei resonate outside the *excitation bandwidth* of the pulse. Since an off-resonance vector is driven away from the y-axis during the pulse it also acquires a (frequency-dependent) phase difference relative to the on-resonance vector (Fig. 3.6). This is usually small and an approximately linear function of the frequency, so can be corrected by phase adjustment of the final spectrum (Section 3.2.8).

How deleterious off-resonance effects are in a pulse experiment depends to some extent on the pulse tip angle employed. A 90 degree excitation pulse ideally transfers magnetisation from the $+z$ axis into the x–y plane, but when off-resonance the

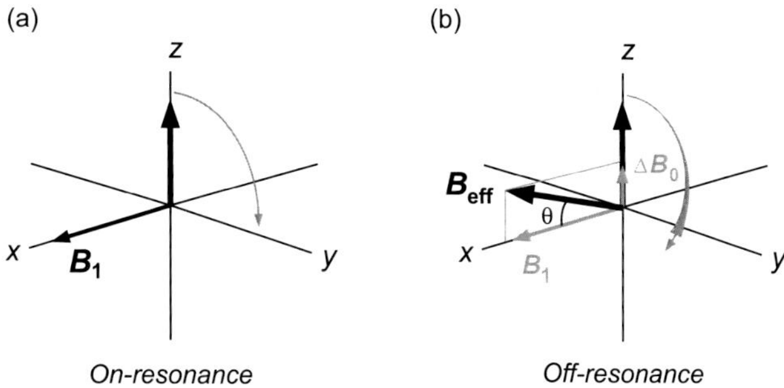

FIGURE 3.5 **On- and off-resonance excitation.** Excitation of magnetisation for which the rf is on-resonance (a) results in the rotation of the bulk vector about the applied rf field B_1. Those spins which experience off-resonance excitation (b) are instead driven about an effective rf field B_{eff}, which is tipped out of the x–y plane by an angle θ, which increases as the offset increases.

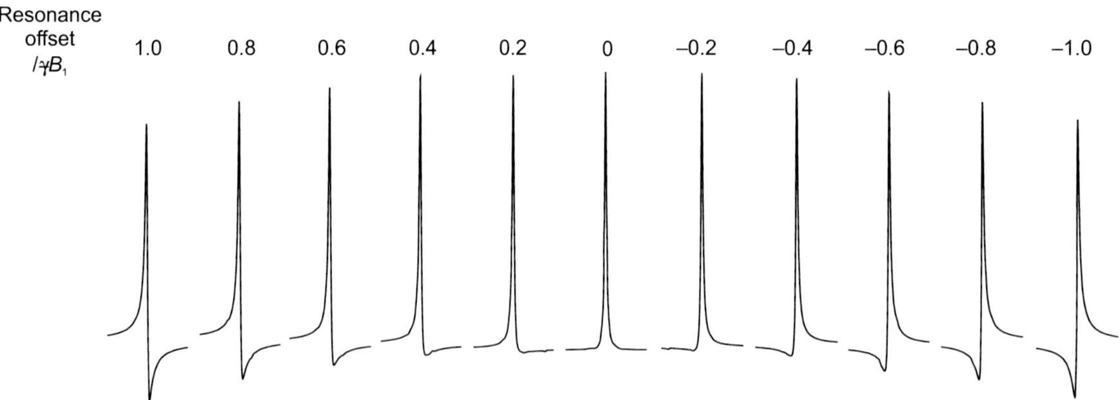

FIGURE 3.6 **Resonance offset effects.** Experimental excitation profiles for a 90 degree pulse as a function of resonance offset relative to the applied rf field strength γB_1. Greater offsets introduce larger phase errors and reduce the amplitude of the resultant transverse magnetisation.

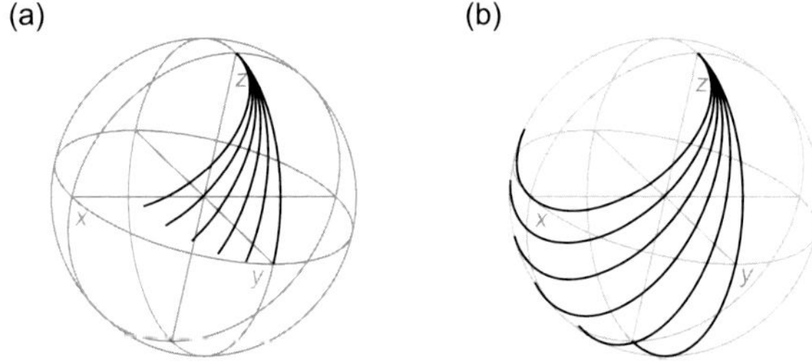

FIGURE 3.7 **Pulse excitation trajectories.** These are shown as a function of resonance offset for (a) a 90 degree pulse and (b) a 180 degree pulse. The offset moves from zero (on-resonance) to $+\gamma B_1$ Hz in steps of $0.2\gamma B_1$ (as in Fig. 3.6). The 90 degree pulse has a degree of 'offset compensation' as judged by its ability to generate transverse magnetisation over a wide-frequency bandwidth. In contrast, the 180 degree pulse performs rather poorly away from resonance, leaving the vector far from the target South Pole and with a considerable transverse component.

tilt of the effective field will act to place the vector above this plane. However, the greater B_{eff} will mean the magnetisation vector follows a longer trajectory and this increased net flip angle offers some compensation, and hence a 90 degree pulse is fairly tolerant to off-resonance effects as judged by the elimination of z-magnetisation (Fig. 3.7a). In contrast, a 180 degree inversion pulse ideally generates pure (-) z-magnetisation leaving none in the transverse plane; but now the increased effective flip angle is detrimental, tending to move the vector further from the South Pole. Thus, 180 degree pulses do not perform well when applied off-resonance (Fig. 3.7b), and can be a source of poor experimental performance. In practice, it is relatively easy to provide sufficiently short pulses (ca. 10 μs) to ensure that excitation and inversion are reasonably uniform over the relatively small frequency ranges encountered in ^{1}H NMR spectroscopy. However, for nuclei that display much greater frequency dispersion, such as ^{13}C or ^{19}F, this is often not the case, and resonance distortion and/or attenuation can occur, and spurious signals may arise in multipulse experiments as a result. Experimental approaches to overcoming these limitations include the use of clusters of pulses, known as *composite pulses,* or of frequency-swept *adiabatic pulses*, both of which aim to compensate for these (and other) defects; see Section 12.2 for further discussion.

3.2.2 Signal Detection

Before proceeding to consider how one collects weak NMR signals, we briefly consider the detection process that occurs within the NMR receiver. The energy emitted by excited spins produces tiny analogue electrical signals that must be converted into a series of binary numbers to be handled by the computer. This *digitisation* process (Section 3.2.3) must occur for all the frequencies in the spectrum. As chemists, we are really only interested in knowing the chemical shift *differences* between nuclei rather than their *absolute* Larmor frequencies since it is from these differences that one infers differences in chemical environments. As we already know, these frequencies are typically tens or hundreds of megahertz while chemical shift ranges only cover some kilohertz. For example a proton spectrum recorded at 400 MHz covers a frequency range of only approximately 4 kHz, and even for carbon (100 MHz on the same instrument) the frequency range is only around 20 kHz. Therefore, rather than digitising signals over many megahertz and retaining only the few kilohertz of interest, a

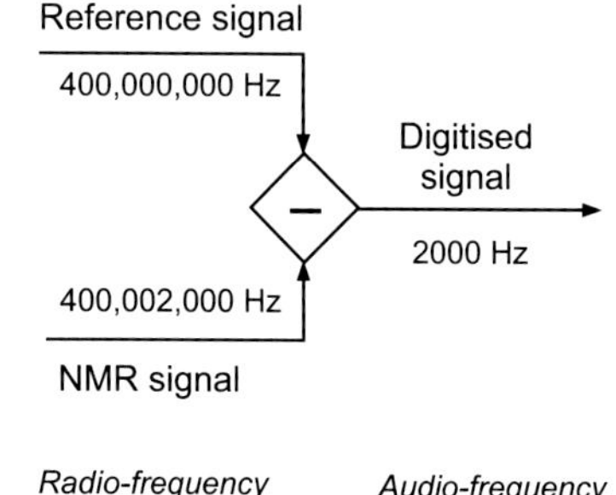

FIGURE 3.8 **The NMR detection process.** A fixed reference frequency is subtracted from the detected NMR signal so that only the frequency *differences* between resonances are digitised and recorded.

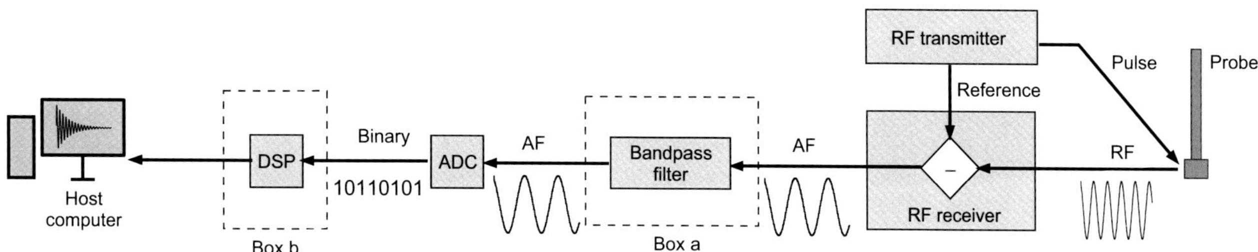

FIGURE 3.9 **Schematic illustration of NMR data collection.** Pulsed rf excitation stimulates the NMR response in the probehead which is then amplified and detected in the receiver. The receiver reference frequency (that of the pulse rf) is subtracted to leave only audio frequencies (AF) which are digitised by the analogue-to-digital converter (ADC) and subsequently stored within the computer. The functions of the boxed sections are described later in the text. DSP, digital signal processor.

more sensible approach is to subtract a reference frequency from the detected signal *before* digitisation, leaving only the frequency window of interest (Fig. 3.8). The resultant signals are now in the *audio* range, which for humans at least, corresponds to frequencies less than 20 kHz (so it is also possible to hear NMR resonances).

The reference frequency is usually chosen to be that of the original pulse used to excite the spin system, and is supplied as a continuous reference signal from the transmitter (Fig. 3.9). This detection process is exactly analogous to the use of the rotating frame representation introduced in the previous chapter where the rotating frame reference frequency is also that of the rf pulse. If you felt a little uneasy about the seemingly unjustified use of the rotating frame formalism previously then perhaps the realisation that there is a genuine experimental parallel within all spectrometers will help ease your concerns. The digitised free induction decay (FID) you see therefore contains only the audio frequencies that remain after subtracting the reference and it is these that produce the resonances observed in the final spectrum following Fourier transformation.

3.2.3 Sampling the FID

3.2.3.1 The Nyquist Condition

To determine the frequency of an NMR signal correctly, it must be digitised at the appropriate rate. The *Sampling* or *Nyquist Theorem* tells us that to characterise a regular, oscillating signal correctly it must be defined by at least two data points per wavelength. In other words, to characterise a signal of frequency F Hz, we must sample at a rate of at least $2F$; this is also known as the *Nyquist condition*. In NMR parlance the highest recognised frequency is termed the *spectral width* (SW) (Fig. 3.10). The time interval between sampled data points is referred to as the *dwell time* (DW), as given by:

$$\mathrm{DW} = \frac{1}{2\mathrm{SW}} \tag{3.3}$$

Signals with frequencies less than or equal to SW will be characterised correctly as they will be sampled at two or more points per wavelength, whereas those with higher frequencies will be incorrectly determined and in fact will appear in

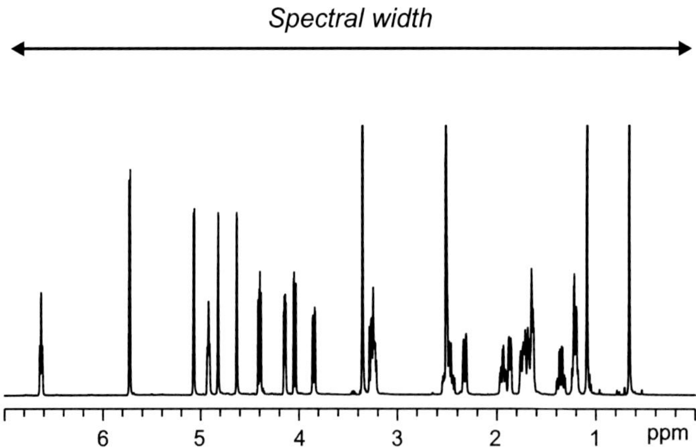

FIGURE 3.10 **The spectral width.** This defines the size of the observed frequency window. Only within this window are the line frequencies correctly characterised.

the spectrum at frequencies which are lower than their true values. To understand why this occurs, consider the sampling process for three frequencies, two within and the third outside the spectral width (Fig. 3.11). The sampled data points of the highest frequency clearly match those of the lower of the three frequencies and this signal will therefore appear with an incorrect frequency within the spectral window. Resonances that appear at incorrect frequencies, because in reality they exist outside the spectral width, are said to be *aliased* or *folded* back into the spectrum. Their location when corrupted in this manner is dependent upon which of the two commonly employed *quadrature detection* schemes is in use, as explained in Section 3.2.4. On modern NMR spectrometers that employ digital signal detection (see Section 3.2.6) digital filters are able to completely eliminate signals that fall outside the defined spectral width, and the folded or aliased signals are therefore not observed. This filtering applies only for the directly sampled (observed) dimension of an NMR experiment, so aliasing of peaks may still be observed in the indirectly detected dimensions of a 2D experiment (see Section 5.2.1).

3.2.3.2 Filtering Noise

One particularly insidious effect of aliasing or folding is that not only will NMR resonances fold into the spectral window but noise will as well. This can seriously compromise sensitivity, since noise can extend over an essentially infinite frequency

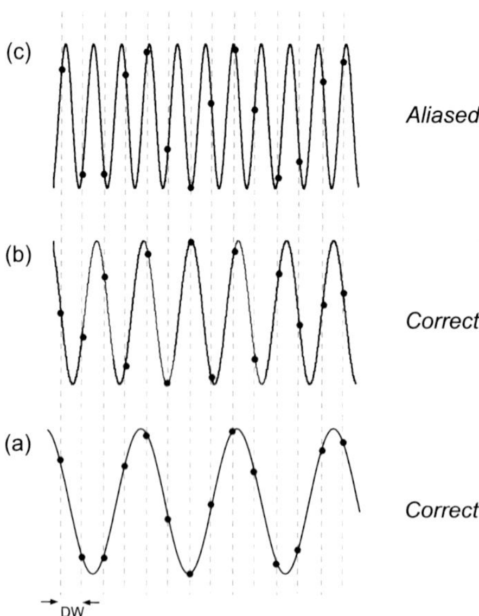

FIGURE 3.11 **The Nyquist condition.** To correctly characterise the frequency of an NMR signal it must be sampled at least twice per wavelength; it is then said to fall within the spectral width. The sampling of signals (a) and (b) meet this criteria. Signals with frequencies too high to meet this condition are aliased back within the spectral width and so appear with the wrong frequency. The sampling pattern of signal (c) matches that of (a) and hence it is incorrectly recorded as having the lower frequency.

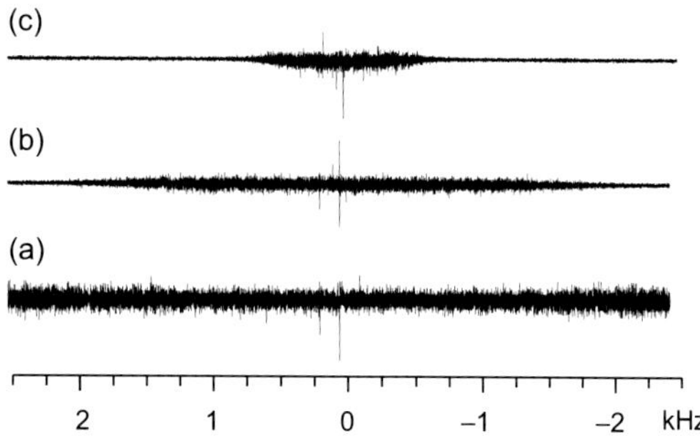

FIGURE 3.12 **Eliminating noise with analogue filters.** Spectra were recorded over a spectral width of 5 kHz with no sample in the probe. The analogue filter window was set to (a) 6 kHz, (b) 3 kHz and (c) 1 kHz and the associated noise attenuation is apparent, although the cutoff point is ill defined.

range (so-called white noise), all of which, potentially, can fold into the spectrum and swamp the peaks. It is therefore essential to prevent this by filtering out all signals above a certain frequency threshold with an audio *bandpass filter* after detection but prior to digitisation (Fig. 3.9, Box a). The effect of the noise filter is demonstrated in Fig. 3.12 and the gain in signal-to-noise in the spectrum employing this is clearly evident in Fig. 3.13. The cutoff bandwidth of the filter is variable, to permit the use of different spectral widths, and is usually automatically set by the spectrometer software to be a little greater than the spectral window. This is necessary because analogue filters are not perfect and do not provide an ideal sharp frequency cutoff, but instead tend to fall away steadily as a function of frequency, as is apparent in Fig. 3.12. A practical consequence is that these filters can attenuate the intensity of signals falling at the extreme ends of the spectrum, so may interfere with accurate signal intensity measurement and it is therefore wise to ensure that resonances do not fall at the edges. Analogue filters can also introduce a variety of distortions to the spectrum and the use of digital filtration methods having superior characteristics is now a standard feature on modern NMR spectrometers. Digital signal processing is considered in Section 3.2.6.

3.2.3.3 Acquisition Times and Digital Resolution

The total sampling period of the FID, known as the *acquisition time* (AQ), is ultimately dictated by the frequency resolution required in the final spectrum. From the Uncertainty Principle, the resolution of two lines separated by $\Delta\nu$ Hz requires data collection for at least $1/\Delta\nu$ s. If one samples the FID for a time that is too short then frequency differences cannot be resolved and fine structure is lost (the minimum linewidth that can be resolved is given approximately by 0.6/AQ). Likewise, if the signal decays rapidly then one is unable to sample it for a long period of time and again can resolve no fine structure. In this case, the rapid decay implies a large linewidth arising from natural (transverse) relaxation processes, or from poor field homogeneity, and one will be unable to recover resolution by extending the acquisition time. Thus, when selecting the appropriate acquisition time, one needs to consider the likely frequency differences that need to be resolved and the relaxation times of the observed spins.

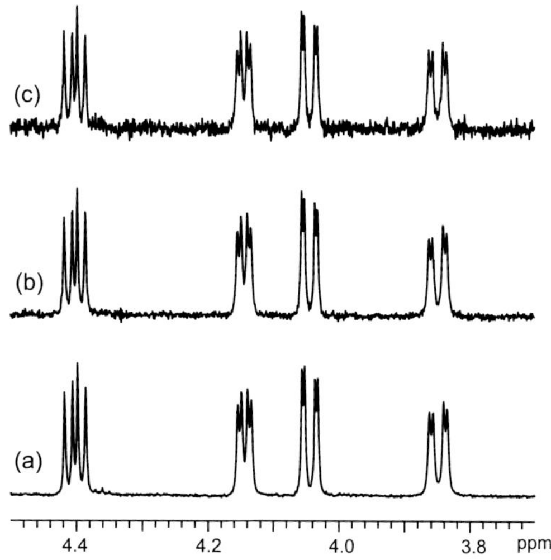

FIGURE 3.13 **The effect of analogue filters on signal-to-noise.** Spectrum (a) was recorded with the correct filter width (1.25 times the spectral width), and (b) and (c) with it increased 10- and 100-fold, respectively.

The acquisition time is defined by the digitisation rate, which is dictated by the spectral width and defines the sampling dwell time, and on how many data points are sampled in total. If the FID contains time domain (TD) data points then:

$$AQ = DW.TD = \frac{TD}{2.SW} \tag{3.4}$$

When dealing with the final spectrum we are concerned with frequency and not time, and what one really needs is some measure of how well the resonances in the spectrum are digitised. The figure of interest is the frequency between adjacent data points in the spectrum, the *digital resolution* (DR) (quoted in hertz per data point or simply Hz/point). It should be stressed that this is not what spectroscopists speak of when they refer to the 'resolution' achieved on a spectrometer as this relates to the homogeneity of the magnetic field. Digital resolution relates only to the frequency window to data point ratio, which is small for a well-digitised spectrum but large when poorly digitised.

Following the Fourier transform, two data sets are generated representing the 'real' and 'imaginary' spectra (Section 2.3), so the real part with which one usually deals contains half the data points of the original FID (in the absence of further manipulation), and its data size (SI) is therefore TD/2. Digital resolution is then:

$$DR = \frac{\text{Total frequency window}}{\text{Total number of data points}} = \frac{SW}{SI} = \frac{2.SW}{TD} = \frac{1}{AQ} \tag{3.5}$$

Thus, DR is simply the reciprocal of AQ, so to collect a well-digitised spectrum one must sample the data for a long period of time; clearly, this is the same argument as presented earlier.

The effect of inappropriate digital resolution is demonstrated in Fig. 3.14. Clearly, with a high value (that is, a small AQ), the fine structure cannot be resolved. Only with extended acquisition times can the genuine spectrum be recognised. For proton spectroscopy, one needs to resolve frequency differences of somewhat less than 1 Hz to be able to recognise small couplings, so AQs of around 2–4 s are routinely used, corresponding to DRs of around 0.5–0.25 Hz/point. This then limits the accuracy with which frequency measurements can be made, including shifts and couplings (even though peak listings tend to quote resonance frequencies to many decimal places).

The situation is somewhat different for the study of nuclei other than protons, however, since they often exhibit rather few couplings (especially in the presence of proton decoupling), and because one is usually more concerned with optimising sensitivity, so does not wish to sample the FID for extended periods and thus sample more noise. Since one does not need to define lineshapes with high accuracy, lower DR suffices to resolve chemical shift differences and 1–2 Hz/point (0.5–1 s AQ) is adequate in ^{13}C NMR. Exceptions occur in the case of nuclei with high natural abundance which may

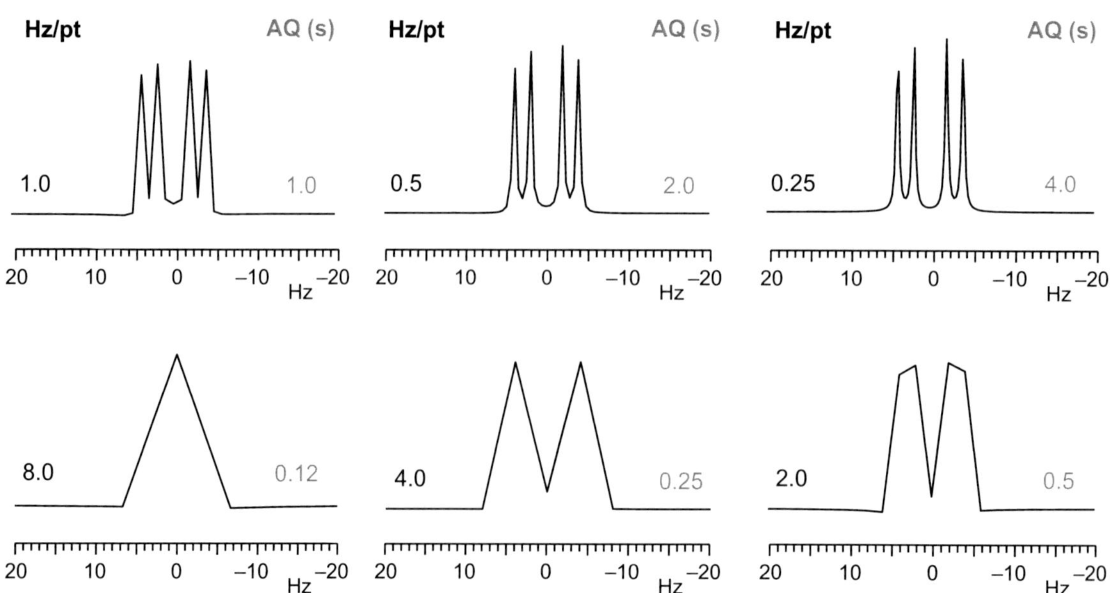

FIGURE 3.14 **The influence of data acquisition times on the ability to resolve fine structure.** Longer acquisition times correspond to higher digitisation levels (smaller Hz/pt) which here enable characterisation of the coupling structure within the double-doublet (J = 6 and 2 Hz).

show homonuclear couplings, such as ^{31}P, or when the spectrum is being used to estimate the relative ratios of compounds in solution. This is true, for example in the use of the chiral derivatising agent known as Mosher's acid (α-methoxy-α-trifluoromethylphenylacetic acid or MTPA) for the determination of enantiomeric excess from ^{19}F NMR spectra. For this the resonances must be well digitised to represent the true intensity of each species. In such cases the parameters that may be used for the 'routine' observation of ^{19}F are unlikely to be suitable for quantitative measurements (see discussions in Section 4.1.2). Similarly, the use of limited DR in routine carbon-13 spectra is one reason signal intensities do not provide a reliable indication of relative concentrations. In contrast, the spectra of rapidly relaxing quadrupolar nuclei contain broad resonances that require only low DR and hence short AQs.

3.2.3.4 Zero-Filling and Truncation Artefacts

While the NMR response decays throughout the FID, the noise component remains essentially constant and will eventually dominate the tail of the FID. At this point there is little advantage in continuing acquisition since this only adds noise to the final spectrum. *Provided the FID has fallen to zero* when acquisition stops, one can artificially improve the DR by appending zeros to the end of the FID. This process is known as *zero-filling* and it interpolates these added data points in the frequency domain and so enhances the definition of resonance lineshapes (Fig. 3.15).

It has been shown [2] that by doubling the number of data points in the TD by appending zeros (a single 'zero-fill') it is possible to improve the frequency resolution in the spectrum. The reason for this gain stems from the fact that information in the FID is split into two parts (real and imaginary) after the transformation, so one effectively loses information when considering only the real (absorption) part. Doubling the SI regains this lost information and is therefore a useful tool in the routine analysis of proton spectra (one could of course simply double the acquired data points but this leads to reduced sensitivity and requires more spectrometer time). Further zero-filling simply increases the *digital* resolution of the spectrum by interpolating data points, so no new information can be gained and the improvement is purely cosmetic. However, because this leads to a better definition of each line it can still be extremely useful when analysing multiplet fine structure in detail or when measuring accurate resonance intensities; in the determination of enantiomeric excess via chiral solvating reagents, for example. In 2D experiments acquisition times are necessarily kept rather short and zero-filling is routinely applied to improve the appearance of the final spectrum. A more sophisticated approach to extending the time domain signal, known as *linear prediction*, is described in the following section.

If the FID has not decayed to zero at the end of the acquisition time, the data set is said to be *truncated* and this leads to distortions in the spectrum after zero-filling and Fourier transformation (FT). The distortions arise from the FT of the sudden step within the FID, the result of which is described by the function (sin x)/x, also known as sinc x. Fig. 3.16 shows

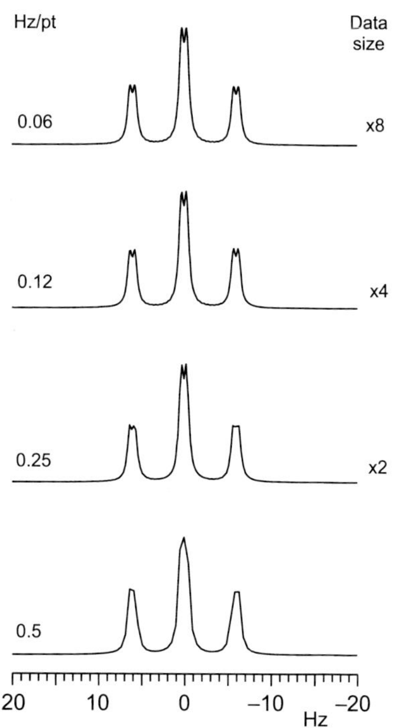

FIGURE 3.15 **Zero-filling.** This can be used to enhance fine structure and improve lineshape definition.

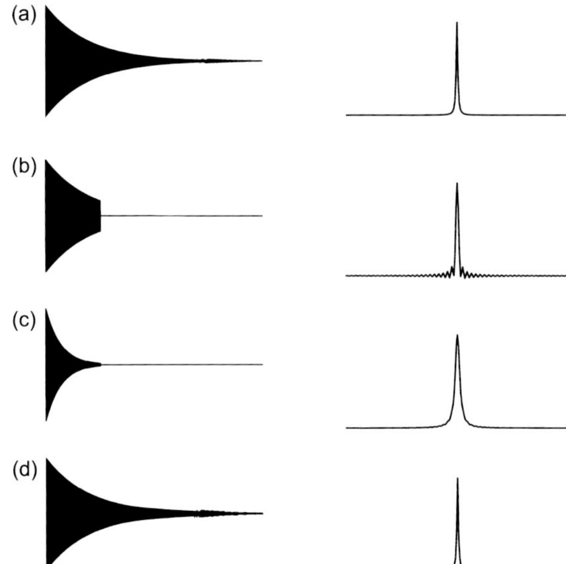

FIGURE 3.16 **Processing a truncated FID.** (a) A complete FID and the corresponding resonance, (b) a truncated FID which has been extended by zero-filling produces sinc wiggles in the spectrum, (c) apodisation of the FID in (b) together with zero-filling and (d) linear prediction of the FID in (b).

that this produces undesirable ringing patterns that are symmetrical about the base of the resonance, often referred to as 'sinc wiggles'. To avoid this problem it is essential to either ensure the AQ is sufficiently long, to force the FID to decay smoothly to zero with a suitable shaping function (Section 3.2.7) or to artificially extend the FID by linear prediction.

In proton spectroscopy, acquisition times are sufficiently long that truncation artefacts are rarely seen in routine spectra, although they may be apparent around the resonances of small molecules with long relaxation times, for example solvent lines. As stated earlier, acquisition times for other nuclei are typically kept short and FIDs are usually truncated. This, however, is rarely a problem as it is routine practice to apply a window function to enhance sensitivity of such spectra, which itself also forces the FID to zero, so eliminating the truncation effects. Try processing a standard ^{13}C spectrum directly with an FT but without the use of a window function; for some resonances you are likely to observe negative responses in the spectra that appear to be phasing errors. In fact, these distortions arise purely from the low DR used (ie the short AQs) and the corresponding truncation of the FID [3] (Fig. 3.17a). Use of a simple decaying function (Section 3.2.7) removes these

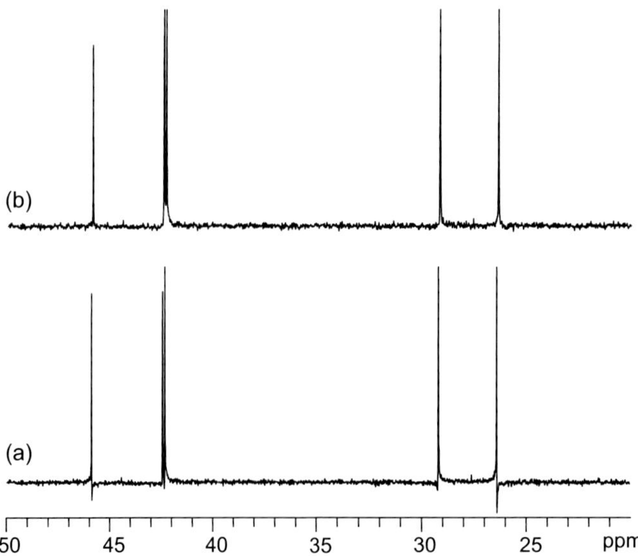

FIGURE 3.17 **Distortions from truncated FIDs.** Carbon spectra often display distortions when transformed directly (a) which appear to be phase errors but which actually arise from a short acquisition time. Applying a line-broadening apodisation function prior to the transform removes these distortions (b) 1 Hz line broadening.

effects (Fig. 3.17b) and when used in this way is often referred to as *apodisation*, literally meaning 'removing the feet'. Similar considerations apply to truncated 2D data sets where shaping functions play an essential role.

3.2.3.5 Linear Prediction

The method of linear prediction (LP) can play many roles in the processing of NMR data [4,5], from the rectification of corrupted or distorted data, through to the complete generation of frequency domain data from an FID, an alternative to the FT. Here we consider its most popular usage, known as forward linear prediction, which extends a truncated FID. Rather than simply appending zeros, this method, as the name suggests, predicts the values of the missing data points by using the information content of the previous points and so genuinely extends the FID (Fig. 3.16d).

In a time sequence of data points the value of a single point d_n can be estimated from linear combination of the immediately preceding values:

$$d_n = a_1 d_{n-1} + a_2 d_{n-2} + a_3 d_{n-3} + a_4 d_{n-4} \dots \tag{3.6}$$

where a_1, a_2, ... and so on represent the LP coefficients. The number of coefficients (referred to as the *order* of the prediction) corresponds to the number of data points used to predict the next value in the series. Provided the coefficients can be determined from the known data, it is then possible to extrapolate beyond the acquired data points. Repetition of this process incorporating the newly predicted points ultimately leads to the extended FID.

This process is clearly superior to zero-filling and produces a much better approximation to the true data than does simply appending zeros. It improves resolution, avoids the need for strong apodisation functions and greatly attenuates truncation errors. Naturally, the method has its limitations, the most severe being the requirement for high signal-to-noise in the FID for accurate estimation of the LP coefficients. Successful execution also requires that the number of points used for the prediction is very much greater than the number of lines that comprise the FID. This may be a problem for 1D data sets with many component signals and linear prediction of such data is less widely used. The method is far more valuable for the extension of truncated data sets in the indirectly detected dimensions of a 2D or 3D experiment (you may wish to return to these discussions when you are familiar with the 2D approach, see the chapter *Introducing Two-Dimensional and Pulsed Field Gradient NMR*). Here individual interferograms contain rather few lines (look at a column from a 2D data set) and are thus better suited to prediction, providing sensitivity is adequate. The use of linear prediction in the routine collection of 2D spectra of organic molecules has been subject to detailed investigation [6]. Typically, twofold to fourfold prediction is used; so, for example $128t_1$ data points may be readily extended to 256 or 512. Apodisation of the data is still likely to be required, although not as severely as for untreated data, and zero-filling may also be applied to further improve digitisation. Fig. 3.18 clearly demonstrates the improved resolution attainable through the use of linear prediction in the indirect dimension of a 2D heteronuclear correlation experiment. The same principles can be used in *backward* linear prediction. Data points at the start of an FID are sometimes corrupted by ringing in receiver circuitry when detection starts resulting in baseline distortion of the spectrum. Replacing these points with uncorrupted predicted points eliminates the distortion; an example of this is found in Section 4.5.

3.2.4 Quadrature Detection

It was described earlier how during the classical NMR signal detection process a reference frequency equal to that of the excitation pulse is subtracted from the NMR signal to produce an audiofrequency signal that is digitised and later subject to FT. The problem with analysing the data produced by this *single-channel detection* is that the FT is intrinsically unable

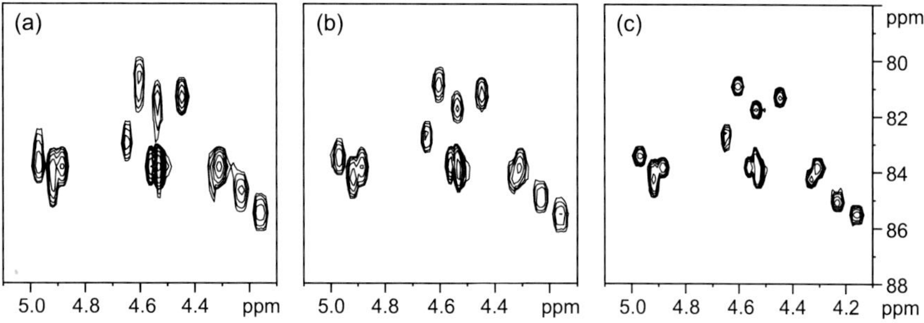

FIGURE 3.18 **Linear prediction of 2D data.** Improved peak resolution in a 2D heteronuclear proton–carbon correlation experiment through linear prediction. The same raw data were used in each spectrum, with the F_1 (carbon) dimension processed with (a) no data extension, (b) one zero-fill and (c) linear prediction in place of zero-filling.

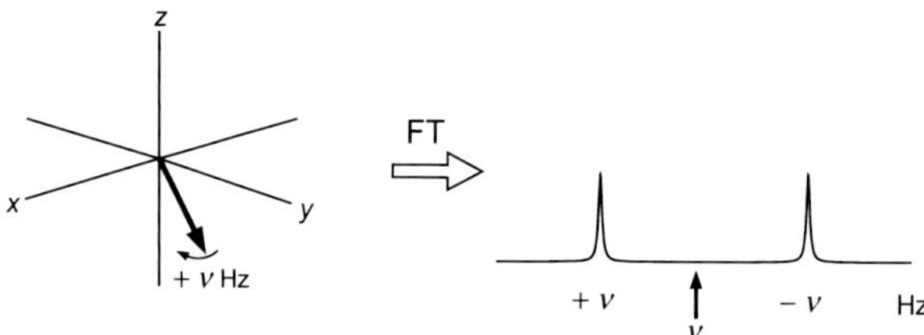

FIGURE 3.19 **Single-channel detection.** This scheme is unable to differentiate positive and negative frequencies in the rotating frame, which results in a mirror image spectrum being superimposed on the true one if the transmitter is placed in the centre of the spectrum.

to distinguish frequencies above the reference from those below it that is it cannot differentiate positive from negative frequencies). This results in a magnetisation vector moving at $+\nu$ Hz in the rotating frame producing two resonances in the spectrum at $+\nu$ and $-\nu$ Hz after FT (Fig. 3.19). The inevitable confusing overlap in a spectrum acquired with the reference positioned in the centre of the spectral width can be avoided by placing the reference at one edge of the spectrum to ensure all rotating frame frequencies have the same sign. While this will solve potential resonance overlap problems, it introduces a number of other undesirable factors. Firstly, although there will be no mirror image of NMR resonances remaining within the spectral window, noise will still be mirrored about the reference frequency and added to that already present. This leads to a decrease in signal-to-noise by a factor of $\sqrt{2}$, or about 1.4 (not by a factor of 2 because noise is random and does not add coherently). Furthermore, this produces the greatest possible frequency separation between the pulse and the highest frequency resonance, thus enhancing undesirable off-resonance effects. The favoured position for the reference is therefore in the centre of the spectrum and two-channel (*quadrature*) detection is then required to distinguish the sign.

To help visualise why single-channel detection cannot discriminate positive and negative frequencies, and how the quadrature method can, consider again a single magnetisation vector in the rotating frame. The use of the single-channel detector equates to being able to observe the precessing magnetisation along *only one axis*; say, the *y*-axis. Fig. 3.20 shows that the resultant signal along this axis for a vector moving at $+\nu$ Hz is identical to that moving at $-\nu$ Hz, both giving rise to a cosinusoidal signal, so the two are indistinguishable (Fig. 3.20a). If, however, one were able to make use of two (phase-sensitive) detection channels whose reference signals differ in phase by 90 degrees (hence the term *quadrature*), one would then be able to observe magnetisation simultaneously along both the *x* and the *y*-axes such that one channel monitors the cosine signal, the other the sine (Fig. 3.20a and b). With the additional information provided by the sinusoidal response of the second channel, the sense of rotation can be determined and vectors moving at $\pm\nu$ Hz can be distinguished (Fig. 3.21). Technically, the FT is then *complex*, with the *x* and *y* components being handled separately as the real and imaginary inputs to the transform, following which the positive and negative frequencies are correctly determined. In the case of the single channel, the data are used as input to a *real* FT.

3.2.4.1 Simultaneous and Sequential Sampling

Quadrature detection is universally employed in all modern spectrometers. However, there exist a number of experimental schemes for implementing this and which of these you are likely to use will be dictated by the spectrometer hardware and perhaps by the age of the instrument (on some modern instruments the operator can choose between these methods). The

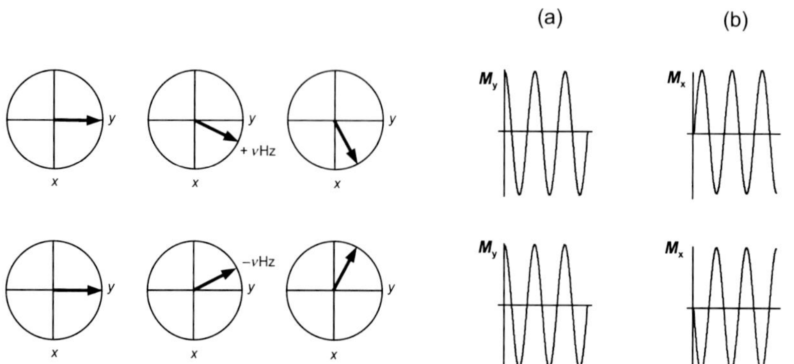

FIGURE 3.20 **Quadrature detection.** A two-channel quadrature detection system monitors magnetisation on two orthogonal axes, providing both (a) cosine and (b) sine–modulated data which ultimately allow the sense of precession to be determined (see text).

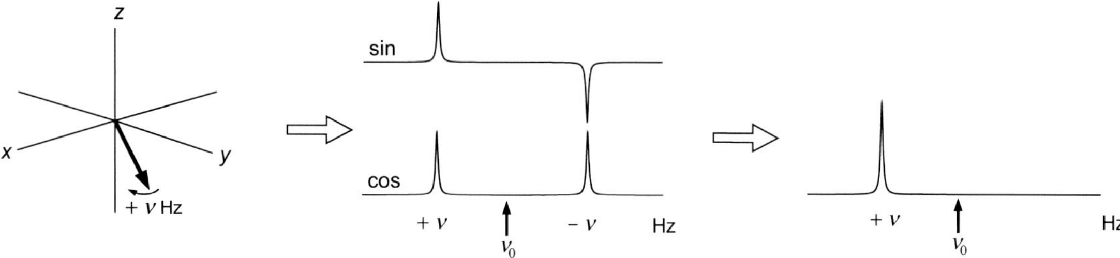

FIGURE 3.21 **Sign discrimination by quadrature detection.** A two-channel detection scheme is able to differentiate positive and negative frequencies in the rotating frame, allowing the transmitter to be placed in the centre of the spectrum without the appearance of mirror image signals.

differing approaches may be divided into the classical *analogue* scheme described here for which there are two widely used methods, and the now standard *digital quadrature detection* scheme which is considered later.

In both analogue implementations the incoming rf signal from the probe and preamplifier is split into two signals and each fed to separate phase-sensitive detectors whose reference frequencies are identical but differ in phase by 90 degrees (Fig. 3.22). The resulting audio signals are then sampled, digitised and stored for subsequent analysis, and it is in the execution of these that the two methods differ. The first relies on *simultaneous* sampling of the two channels (ie the channels are sampled at precisely the same point in time) and the data points for each are stored in separate memory regions (Fig. 3.23a). The two sets of data so generated represent the cosine and sine components required for sign discrimination and are used as the real and imaginary input to a complex FT routine. As this enables positioning of the transmitter frequency in the centre of the spectrum, a frequency range of only ±SW/2 need be digitised. According to the Nyquist criterion the sampling rate

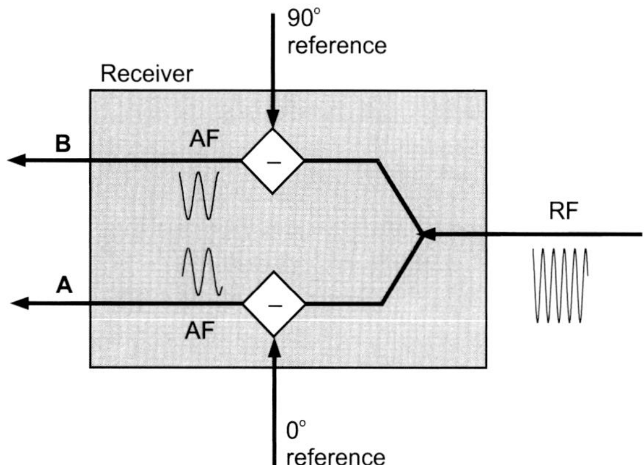

FIGURE 3.22 **Schematic illustration of the experimental implementation of analogue quadrature detection in the NMR receiver.** The incoming rf signal is split in two and the reference signal, differing in phase by 90 degrees in the two channels, subtracted. Channels *A* and *B* therefore provide the required sine and cosine components.

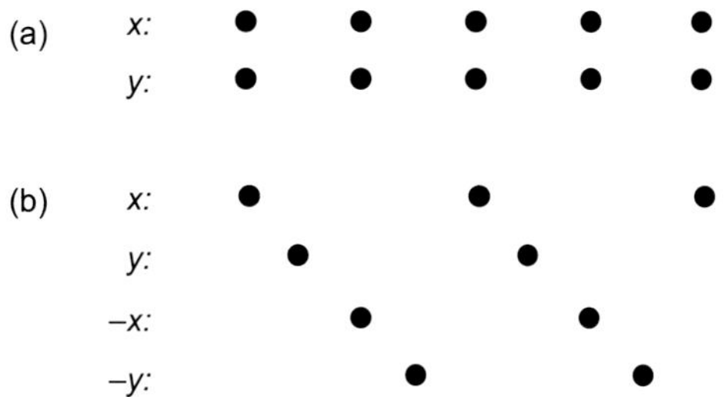

FIGURE 3.23 **Data-sampling schemes for the two common quadrature detection methods.** (a) Simultaneous sampling: the two quadrature channels (representing *x* and *y* magnetisation) are sampled at the same point in time. (b) Sequential sampling: the two channels are sampled alternately at twice the rate of method (a), and the phase inverted for alternate pairs of data points (see text).

now becomes equal to SW (ie 2×SW/2 or half of that required for single-channel detection), so the equations given in the previous section are modified slightly for *simultaneous quadrature detection* to give:

$$DW = \frac{1}{SW} \text{ and } AQ = DW.\frac{TD}{2} = \frac{TD}{2.SW} \tag{3.7}$$

and

$$DR = \frac{SW}{SI} = \frac{2.SW}{TD} \text{ and so again } DR = \frac{1}{AQ} \tag{3.8}$$

Notice that in the calculation of the acquisition time we need only consider *half* the total time domain data points since now two points are sampled at the same time.

The second method for quadrature detection actually strives to eliminate the problems of positive and negative frequencies by effectively mimicking a single-channel detection scheme with the transmitter at the edge of the spectrum, so is sometimes referred to as 'pseudo-quadrature detection'. Although it allows the positioning of the transmitter frequency at the centre of the spectrum, it employs an ingenious sampling scheme to make it appear as if the reference frequency sat at one edge of the spectrum, so that only frequencies of one sign are ever characterised. In this method data points are sampled *sequentially* at a rate suitable for single-channel detection (ie 2.SW or twice that in the simultaneous method), but for each data point sampled the reference phase is incremented by 90 degrees (Fig. 3.23b). The effect on a magnetisation vector is that it appears to have advanced in the rotating frame by 90 degrees more between each sampling period than it actually has, so the signal appears to be precessing faster than it really is. As digitisation occurs at a rate of 2.SW and each 90 degrees phase shift corresponds to an advance of 1/4 of a cycle, the frequency appears to have increased by 2.SW/4 or SW/2. Since the transmitter is positioned at the centre of the spectrum, the genuine frequency range runs from +SW/2 to –SW/2, meaning the artificial increase of +SW/2 moves the frequency window to +SW to 0. Thus, there are no longer negative frequencies to distinguish. Experimentally, the required 90 degrees phase shifts are achieved by alternating between the two phase-sensitive detectors to give the 0 and 90 degrees shifts, and simply inverting the signals from these channels to provide the 180 and 270 degrees shifts, respectively (Fig. 3.23b). Sampled data are handled by a single memory region as *real* numbers which are used as the input to a *real* FT. In this *sequential quadrature detection* scheme, the sampling considerations in Section 3.2.3 apply as they would for single-channel detection. In Section 5.2 these ideas are extended to quadrature detection in 2D spectroscopy.

For either scheme, the total number of data points digitised, acquisition times and spectral widths are identical, so the resulting spectra are largely equivalent. The most obvious difference is in the appearance of aliased signals (that is, those that violate the Nyquist condition), as described below. Experimentally, flatter baselines are observed for the simultaneous method as a result of the *symmetrical* sampling of the initial data points in the FID and this is the recommended protocol for an analogue detection scheme.

3.2.4.2 Aliased Signals

In Section 3.2.3 it was shown that a resonance falling outside the spectral window (because it violates the Nyquist condition) will still be detected but will appear at an incorrect frequency and is said to be *aliased* or *folded* back into the spectrum (if digital signal filters are not employed to eliminate this). This can be confusing if one is unable to tell whether the resonance exhibits the correct chemical shift or not. The precise location of the aliased signal in the spectrum depends on the quadrature detection scheme in use and on how far outside the window it truly resonates. With the simultaneous (complex FT) scheme, signals appear to be 'wrapped around' the spectral window and appear at the *opposite* end of the spectrum (Fig. 3.24b), whereas with the sequential (real FT) scheme signals are 'folded back' at the *same* end of the spectrum (Fig. 3.24c). If you are interested to know why this difference occurs see reference [7].

Fortunately, in proton spectroscopy it is generally possible to detect the presence of an aliased peak because of its esoteric phase which remains distorted when all others are correct. In heteronuclear spectra that display only a single resonance such phase characteristics cannot provide this information as there is no other signal with which to make the comparison. In such cases it is necessary to widen the spectral window and record any movement of the peak. A signal that is not aliased will be unchanged and will appear with the same chemical shift. In contrast, an aliased resonance will move towards the edge of the spectrum and will appear at a different shift. The situation is even worse if digital filters

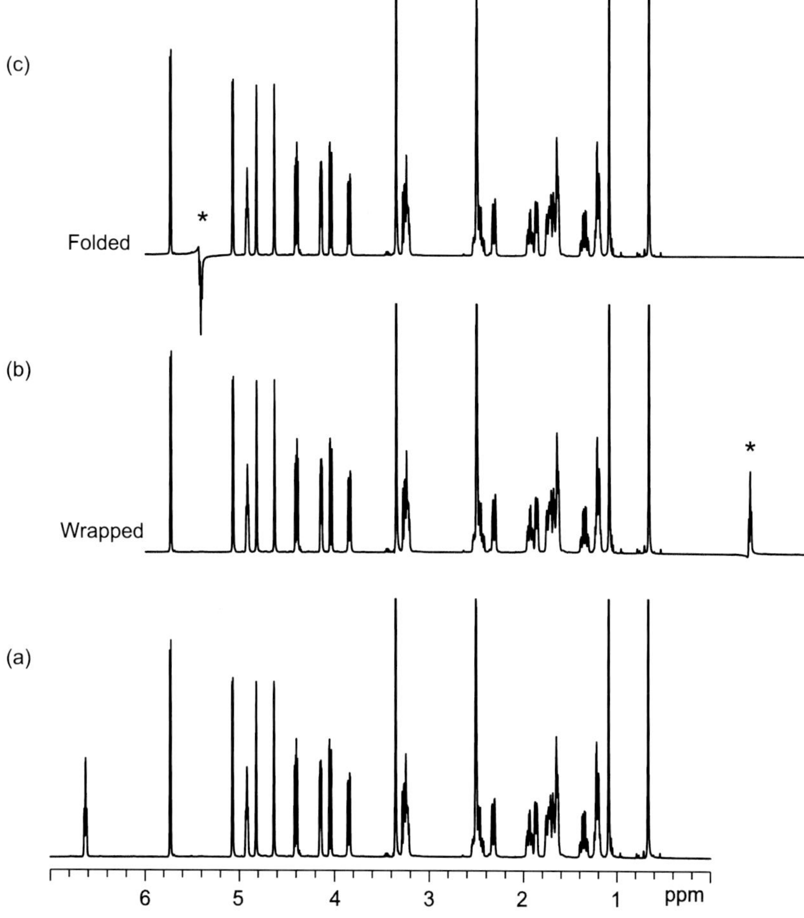

FIGURE 3.24 **Aliasing of resonances.** Spectrum (a) displays all resonances at their correct shifts, while (b) and (c) result from the spectral window being positioned incorrectly. Spectrum (b) shows how the resonance wraps back into the spectrum at the far end when simultaneous sampling is employed, whereas in (c) it folds back in at the near end with sequential sampling. Typically, the phase of the aliased resonance(s) is also corrupted.

are employed as any resonance that falls outside the defined spectral window is simply not observed. If you happen to be searching for a resonance of unknown chemical shift it would be wise to turn off digital filtration to increase the chance of detecting the peak.

3.2.4.3 Quadrature Images

In addition to the corruption of spectra due to signal aliasing, there may also exist unwelcome artefacts that arise from imperfections in analogue quadrature detection schemes. In these one requires two signals to be digitised for subsequent storage in two separate memory blocks that differ in phase by 90 degrees but which are otherwise identical. However, experimentally it is rather difficult to ensure the phase difference is *exactly* 90 degrees and that the signal amplitudes are identical in each channel. Such channel imbalance in phase and amplitude leads to spurious images of resonances mirrored about the transmitter frequency known as *quadrature images* or colloquially *quad images* (Fig. 3.25). Their origin can be realised if one imagines the signal amplitude in one channel to fall to zero. The scheme is then one of single-channel detection, so again positive and negative frequencies cannot be distinguished and thus mirror images appear about the reference (see Fig. 3.19 again). In a correctly balanced receiver these images are usually less than 1% on a single scan, but even these can be significant if you wish to detect weak resonances in the presence of very strong ones, so some means of compensating the imbalance is required. The traditional approach to this is through *phase cycling* of rf pulses and the receiver to cancel these artefacts, and the elimination of these images in this way serves as an introduction to the concept of phase cycling and is described in the section that follows. On modern instruments, these images are avoided through the use of digital quadrature detection in place of the analogue schemes described earlier.

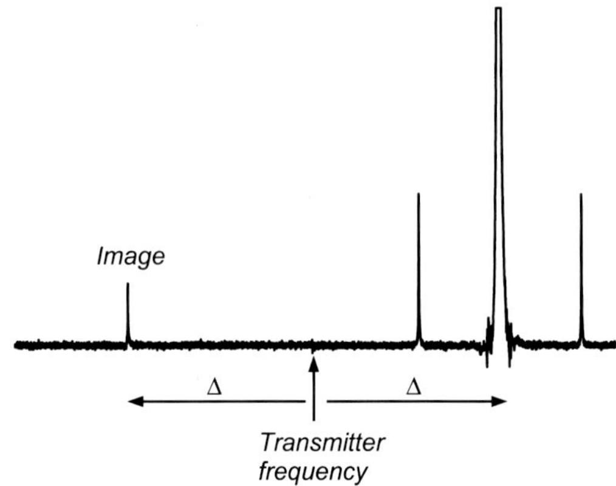

FIGURE 3.25 **Quadrature images.** These are unwanted mirror image artefacts that arise from spectrometer imperfections. Here an image of $CHCl_3$ can be seen at about one half the height of the carbon satellites in a single-scan spectrum. Phase cycling is typically employed to suppress these, although digital quadrature detection can eliminate them completely.

3.2.4.4 Digital Quadrature Detection

In the traditional analogue detection scheme illustrated in Fig. 3.22, the two quadrature signals required for frequency discrimination were generated by mixing with analogue reference signals of (ideally) 90 degrees phase difference prior to independently digitising the data from the two channels. In the digital quadrature detection (DQD) scheme only a single channel is digitised, representing, say, the 0 degrees component. These digital data may then be copied and manipulated numerically to generate a second data set with a mathematically defined 90 degrees phase difference from the originally detected data. This synthesized second-channel data have the same amplitude and a precisely defined phase difference, so the imbalance in the quadrature channels that gives rise to quad images no longer occurs in the DQD scheme and these images are thus eliminated. Since direct digitisation of the rf is not possible (current ADCs do not run fast enough for this) downmixing of the NMR signal to a suitably low *intermediate frequency* (IF) must occur prior to digitisation which nowadays takes place with input frequencies of anything from kilohertz up to tens of megahertz, this made possible by developments coming from telecommunications (Fig. 3.26). With digital schemes it also becomes possible to offset the reference frequency used at the detection stage from that used for the transmitter pulse and this then eliminates any 'zero-frequency' glitch or spike that may be seen at the centre of spectra when analogue receivers are used that produce a residual direct current (DC) offset of the detected FID.

3.2.5 Phase Cycling

As mentioned earlier, the removal of quadrature images can be achieved through the process of *phase cycling*, an experimental method that lies at the very heart of almost every NMR experiment. This process involves repeating a pulse sequence with identical timings and pulse tip angles but with judicious changes to the *phases* of the rf pulse(s) and to the routing of data to the computer memory blocks (often loosely referred to as the receiver-phase cycle). The aim in all this is for the desired signals to add coherently with time averaging whereas all other signals, whether from unwanted NMR transitions or from instrumental imperfections, cancel at the end of the cycle and do not appear in the resulting spectrum.

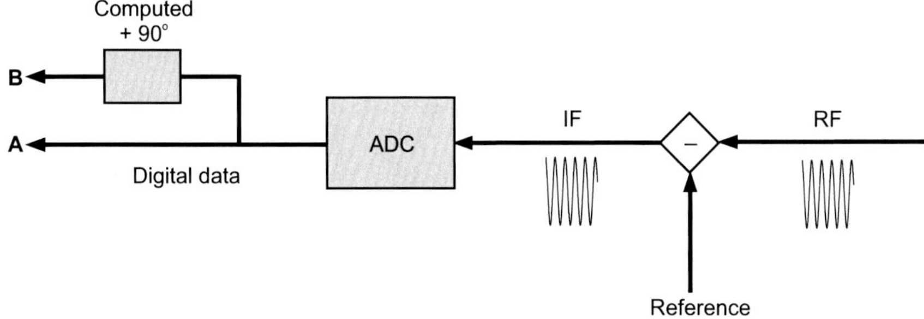

FIGURE 3.26 **Schematic illustration of digital quadrature detection.** A single channel is digitised as a suitable intermediate frequency input (*if*) and the second channel is generated numerically with an identical amplitude and a precise 90 degrees phase shift.

We shall see the importance of phase cycles throughout the remainder of the book, particularly in the sections on multipulse 1D and 2D NMR, but as an introduction we remain with the idea of analogue quadrature detection and the cancellation of quadrature images.

Recall that the quadrature scheme results in two signals differing in phase by 90 degrees that are stored in two separate memory blocks (here designated 1 and 2) and that phase and amplitude imbalance between these gives rise to the images. One solution to the problem is to ensure that data from each channel, *A* and *B*, contribute equally to the two memory blocks (1 and 2). This can be achieved by performing two experiments – the first with a $90°_x$ pulse, the second with $90°_y$ – and adding the data. To keep the cosine and sine components in separate memory regions for use in the FT routine, appropriate data routing must also be used (Fig. 3.27a and b). This is generally handled internally by the spectrometer, being defined as the 'receiver phase' from the operator's point of view, and taking values of 0, 90, 180 or 270 degrees or *x*, *y*, –*x* and –*y*. Note, however, that the phase of the receiver *reference* rf does not alter on sequential scans, only the data routing is changed. For any pulse NMR sequence it is necessary to define the rf phases *and* the receiver phase to ensure retention of the desired signals and cancellation of artefacts.

By use of the two-step phase cycle it is therefore possible to compensate for the effects of imbalance in the two receiver channels. It is also possible to remove extraneous signals that may occur, such as from DC offsets in the receiver, by simultaneously inverting the phase of the rf pulse and the receiver, thus, as shown in Fig. 3.27, *a* steps to *c* and likewise *b* becomes *d*. The NMR signals will follow the phase of the pulse so will add in the memory whereas offsets or spurious signals will

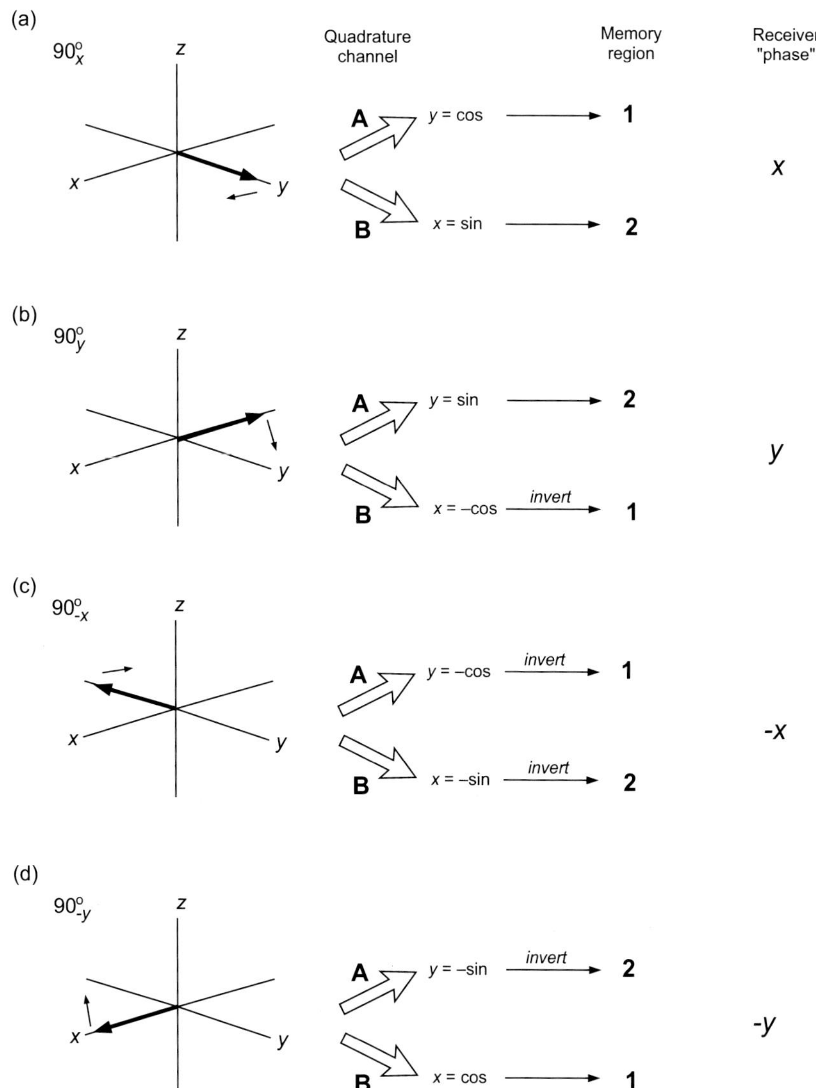

FIGURE 3.27 **Phase cycling.** The CYCLOPS scheme cancels unwanted artefacts while retaining the desired NMR signals. This four-step phase cycle is explained in the text.

TABLE 3.1 The Four-Step CYCLOPS Phase Cycle Illustrated in Fig. 3.27. This Shorthand Notation is Conventionally Used to Describe all Pulse Sequence Phase Cycles

Scan Number	Pulse Phase	Receiver Phase
1	x	x
2	y	y
3	$-x$	$-x$
4	$-y$	$-y$

be independent of this so will cancel. This gives us a second possible two-step phase cycle which, when combined with the first, produces an overall four-step cycle known as the cyclically ordered phase cycle (CYCLOPS) [8] (Table 3.1). This is the standard phase cycle used for one-pulse acquisitions on all spectrometers, and is often nested within the phase cycles of 2D experiments, again with the aim of removing receiver artefacts.

3.2.6 Dynamic Range and Signal Averaging

In sampling the FID, the ADC (or digitiser) limits the frequency range one is able to characterise (ie the SW) according to how fast it can digitise the incoming signal. In addition to limiting the *frequencies*, ADC performance also limits the *amplitudes* of signals that can be measured. The digitisation process converts the electrical NMR signal into a binary number proportional to the magnitude of the signal. This digital value is defined as a series of computer bits, the number of which describes the *ADC resolution*. Typical digitiser resolutions on modern spectrometers operate with 14 or 16 bits. The 16-bit digitiser is able to represent values in the range ±32,767 (ie $2^{15} - 1$) with one bit reserved to represent the sign of the signal. The ratio between the largest and smallest detectable value (the most and least significant bits), 32,767:1, is the *dynamic range* of the digitiser. If we assume the receiver amplification (or *gain*) is set such that the largest signal in the FID on each scan fills the digitiser, then the smallest signal that can be recorded has the value 1. Any signal whose amplitude is less than this will not trigger the ADC; the available dynamic range is insufficient. However, noise will also contribute to the detected signal and this may be sufficiently intense to trigger the least significant bit of the digitiser. In this case the small NMR signal *will* be recorded as it rides on the noise and signal averaging therefore leads to summation of this weak signal, meaning even those whose amplitude is below that of the noise may still be detected. However, the digitiser may still limit the detection of smaller signals in the presence of very large ones when the signal-to-noise ratio is high. Fig. 3.28 illustrates how a reduction in the available dynamic range limits the observation of smaller signals when thermal noise in the spectrum is low. This situation is most commonly encountered in proton studies, particularly of protonated aqueous solutions where the water resonance may be many thousand times that of the solute. Such intensity differences impede solute signal detection, so many procedures have been developed to selectively reduce the intensity of H_2O resonance and ease dynamic range requirements; some of these *solvent suppression* schemes are described in Section 12.5.

If any signal is so large that it cannot fit into the greatest possible value the ADC can record, its intensity will not be measured correctly and this results in severe distortion of the spectrum (Fig. 3.29a). The effect can be recognised in the FID as 'clipping' of the most intense part of the decay (Fig. 3.29b) which results from setting the receiver gain or amplification

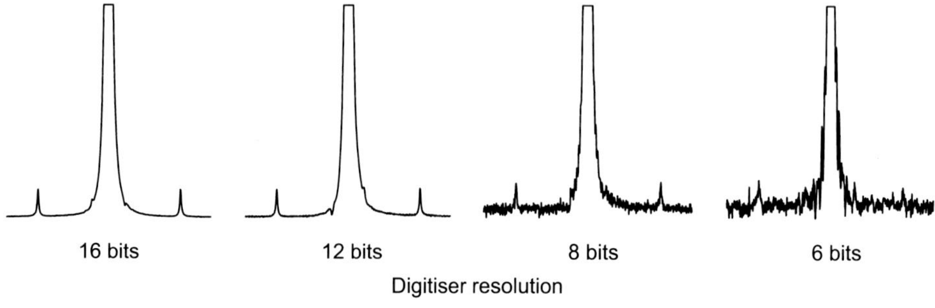

FIGURE 3.28 **Dynamic range and the detection of small signals in the presence of large ones.** As the digitiser resolution and hence its dynamic range are reduced, the carbon-13 satellites of the parent proton resonance become masked by noise until they are barely discernible with only 6-bit resolution (all other acquisition parameters were identical for each spectrum). The increased noise is *digitisation* or *quantisation noise* (see text).

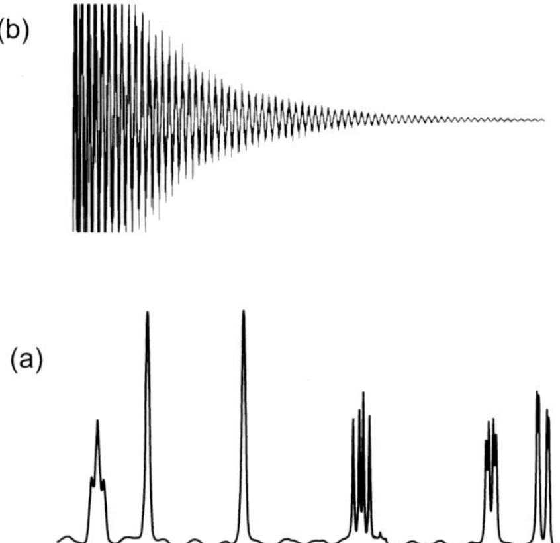

FIGURE 3.29 **Receiver or digitiser overload.** This distorts the spectrum baseline (a). This can also be recognised as 'clipping' of the FID (b).

too high; this must be set appropriately for each sample studied, either by manual adjustment or more conveniently via the spectrometer's automated routines.

3.2.6.1 Signal Averaging

The repeated acquisition and summation of the FID leads to an overall increase in the signal-to-noise ratio as the NMR signals add coherently over the total number of scans (NS), whereas the noise, being random, adds according to √NS. Thus, the signal-to-noise ratio increases according to √NS. In other words, to double the signal-to-noise ratio it is necessary to acquire *four times* as many scans (Fig. 3.30). It is widely believed that continued averaging leads to continuous improvement in the signal-to-noise ratio, although this is only true up to a point. Each time a scan is repeated, a binary number for each sampled data point is added to the appropriate computer memory location. As more scans are collected, the total in each location will increase as the signal 'adds up'. This process can only be repeated if the cumulative total fits into the computer word size; if it becomes too large it cannot be recorded, potentially leading to corruption of the data (although spectrometers generally handle this problem internally by simply dividing the data and reducing the ADC resolution whenever memory overflow is imminent, so preserving the data and allowing the acquisition to continue). The point at which the computer word length

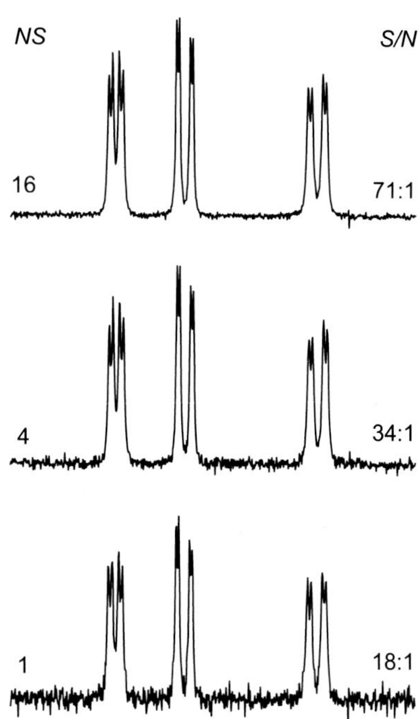

FIGURE 3.30 **Signal averaging produces a net increase in the spectrum signal-to-noise ratio (S/N).** This improves as the *square root* of the number of acquired scans (NS) because noise, being random, adds up *more slowly* than the NMR signal.

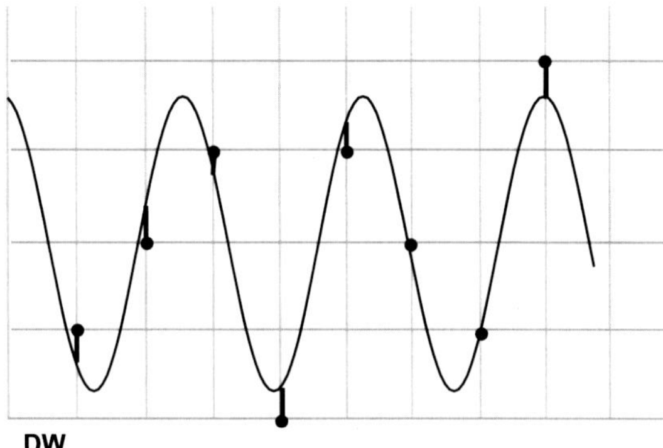

FIGURE 3.31 **Digitisation errors.** Discrete digital sampling of an analogue waveform introduces errors in amplitude measurements, indicated by the vertical bars. These errors ultimately contribute to additional noise in the spectrum.

can no longer accommodate the signal will be dependent upon the number of bits used by the ADC (its resolution), and the number of bits in the computer word. For example if both the ADC and the word length comprise 16 bits then any signal that fills the ADC will also fill the memory, so no further scans may be collected. Thus the word length must be greater than the ADC resolution to allow signal averaging, and the larger the word length or the smaller the ADC resolution the more scans can be collected. Typical word lengths on modern instruments are at least 32 bits, so with a 16-bit ADC it is possible to collect around 66,000 scans ($2^{32} - 16$) if the largest signal fills the digitiser (in fact, more scans than this could be collected since the NMR signal also contains noise which does not add coherently, meaning the memory will not fill as rapidly as this simple estimation suggests). Since signal averaging is most necessary when noise levels are relatively high and because longer word lengths are used on modern instruments, the problem of overflow is now rarely encountered.

3.2.6.2 Oversampling and Digital Filtering

As per standard usage on the latest generation spectrometers, it has long been realised that the effective dynamic range can, in certain circumstances, be improved by *oversampling* the FID [9]. In particular, this applies when the *digitisation noise* (or *quantisation noise*) is significant and limits the attainable signal-to-noise level of the smallest signals in the spectrum. This noise arises from the 'rounding error' inherent in the digitisation process, in which the analogue NMR signal is characterised in discrete 1-bit steps which may not accurately represent the true signal intensity (Fig. 3.31). Although this is not true noise, but arises from systematic errors in the sampling process, it does introduce noise into the NMR spectrum. Fig. 3.28 has already shown how larger errors in the digitisation process, caused by fewer ADC bits, increase the noise level. In realistic situations, this noise becomes dominant when the receiver amplification or gain is set to a relatively low level such that the thermal (analogue) noise arising from the probe and amplifiers is negligible (Fig. 3.32). This is most likely to arise in proton spectroscopy when attempting to observe small signals in the presence of far larger ones and is especially significant with the introduction of low-noise cryogenic probes. High-gain settings are therefore favoured to overcome digitisation noise.

Oversampling, as the name suggests, involves digitising the data at a much faster rate than is required by the Nyquist condition or, equivalently, acquiring the data with a much greater SW than would normally be needed. Digitisation noise may then be viewed as being distributed over a far greater frequency range, such that in the region of interest this noise has reduced intensity so leading to sensitivity improvement (Fig. 3.33). If the rate of sampling has been increased by an oversampling factor of N_{os}, then the *digitisation* noise is reduced by $\sqrt{N_{os}}$. Thus, sampling at four times the Nyquist frequency

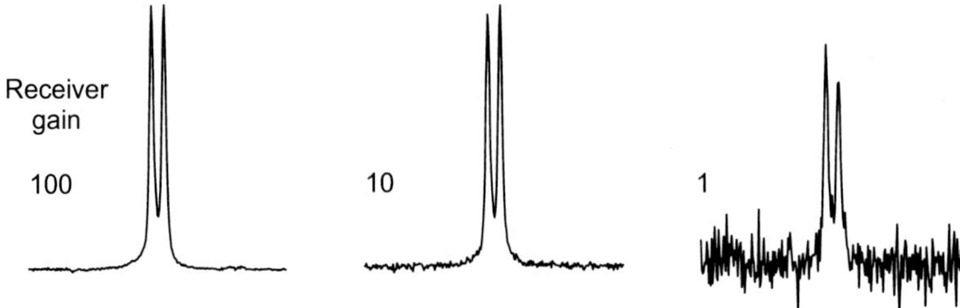

FIGURE 3.32 **Digitisation noise.** At high receiver gain settings the noise in this proton spectrum is vanishingly small, meaning system thermal noise is low. As the gain is reduced, the amplitude of the NMR signal (and thermal noise) is reduced and digitisation noise becomes significant relative to the NMR signal.

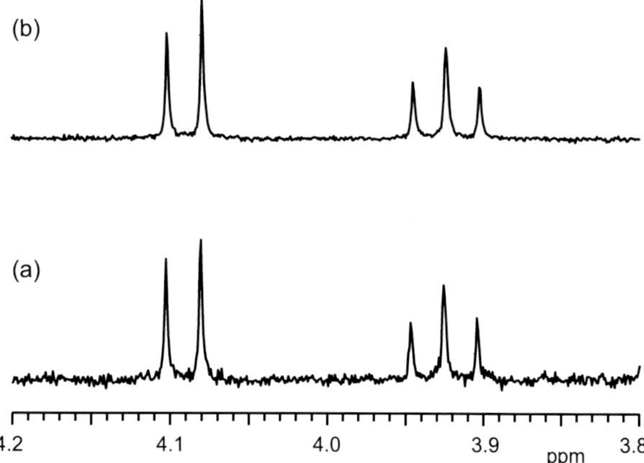

FIGURE 3.33 **Oversampling data.** Enhanced sensitivity can be realised in favourable cases by oversampling the data and hence reducing digitisation noise. Spectrum (a) shows part of a conventional proton spectrum sampled according to the Nyquist criterion. Oversampling the data by a factor of 24 as in (b) provides a sensitivity gain (all other conditions as for (a)).

would theoretically produce a twofold reduction in noise, which in turn equates to an effective gain in ADC resolution of one bit. Likewise, oversampling by a factor of 16 corresponds to a fourfold noise reduction (a resolution gain of 2 bits) and a factor of 64 gives an eightfold reduction or an extra 3 bits. The degree of oversampling that can be achieved is limited by the digitisation speed of the ADC and typical values of N_{os} for proton observation are 16–32, meaning SWs become a few hundred kilohertz and the ADC resolution may be increased by 2–3 bits in favourable circumstances (when thermal noise can be considered insignificant relative to digitisation noise).

To maintain the desired digital resolution in the spectral region of interest when oversampling, data sets would have to be enlarged according to the oversampling factor also, which would demand greater storage capacity and slower data processing. To overcome these limitations, modern spectrometers generally combine oversampling with digital signal–processing methods. Since one is really only interested in a relatively small part of our oversampled data set ($1/N_{os}$ of it) the FID is reduced after digitisation to the conventional number of data points by the process of *decimation* (literally 'removing one tenth of') prior to storage. This, in effect, takes a running average of the oversampled data points, leaving one point for every N_{os} sampled, at intervals defined by the Nyquist condition for the desired spectral window (the peculiar distortions that may be seen at the beginning of a digitally processed FID arise from this decimation process). The resulting FID then has the same number of data points as it would have if it had been sampled normally. These digital signal–processing steps are generally performed with a dedicated processor after the ADC (Fig. 3.9, Box b) and are typically left invisible to the user (short of setting a few software flags perhaps) although they can also be achieved by separate post-processing of the original data or even included in the FT routine itself [10]. The use of fast dedicated processors means the calculations can readily be achieved as the data are acquired, thus not limiting data collection.

One further advantage of digital processing of the FID is the ability to mathematically define frequency filters that have a far steeper and more complete cutoff than can be achieved by analogue filtration alone, meaning signal aliasing can be eliminated. Spectral widths may then be set to encompass only a subsection of the whole spectrum, allowing *selective detection* of NMR resonances [11]. The ability to reduce spectral windows in this way without complications from signal aliasing has considerable benefit when acquiring 2D NMR data, in particular. The principle behind the filtration is as follows (Fig. 3.34). The digital TD filter function is given by the inverse FT of the desired frequency domain window. This function may then be convoluted with the digitised FID such that, following FT, only those signals that fell within the originally defined window remain in the spectrum. One would usually like this window to be rectangular in shape although in practice such a sharp cutoff profile cannot be achieved without introducing distortions. Various alternative functions have been used which approach this ideal, generally with the property that the more coefficients used in the function, the steeper their frequency cutoff. Since these operate only *after* the ADC, the analogue filter (Fig. 3.9, Box a) is still required to reject broadband noise from outside the oversampled SW, but its frequency cutoff is now far removed from the normal spectral window and its performance less critical.

3.2.7 Window Functions

There exist various ways in which the acquired data can be manipulated prior to FT in an attempt to enhance its appearance or information content. Most commonly, one would either like to improve the signal-to-noise ratio of the spectrum to help reveal resonances above the noise level, or to improve the resolution of the spectrum to reveal hidden fine structure. Many

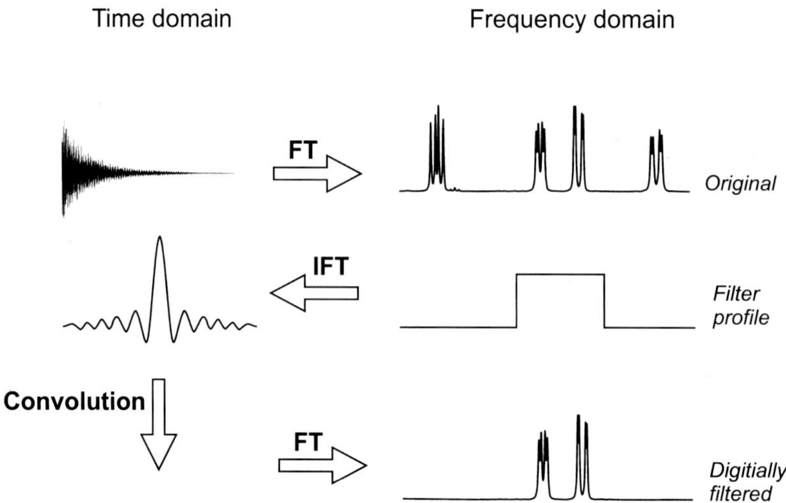

FIGURE 3.34 **Use of a digital filter to observe a selected region of a spectrum.** The desired frequency window profile is subject to an inverse FT and the resulting time domain function convoluted with the raw FID. Transformation of the modified data produces a spectrum containing only a subset of all resonances as defined by the digital filter.

mathematical weighting functions, known as *window functions*, have been proposed to achieve the desired result, but a relatively small number have come into widespread use, some of which are illustrated in Fig. 3.35. The general philosophy behind all these functions is the same. In any FID, NMR signal intensity declines throughout the acquisition time whereas the noise amplitude remains constant, meaning the relative noise level is greater in the latter part. Decreasing the tail of the FID will therefore help reduce the noise amplitude in the transformed spectrum, so enhancing *sensitivity*. Conversely, increasing the middle and latter part of the FID equates to retarding the decay of the signal, thus narrowing the resonance and so enhancing *resolution*.

One word of warning here before proceeding. The application of window functions has also been referred to in the past as 'digital processing' or 'digital filtering' of the data. However, this should not be confused with the digital signal–processing terminology introduced in previous sections and in widespread use nowadays. Adoption of the terms 'window' or 'weighting' function is therefore recommended in the context of sensitivity or resolution enhancement.

3.2.7.1 Sensitivity Enhancement

As suggested earlier, the noise amplitude in a spectrum can be attenuated by de-emphasising the latter part of the FID, and the most common procedure for achieving this is to multiply the raw data by a decaying exponential function (Fig. 3.35a). This process is therefore also referred to as *exponential multiplication*. Because this forces the tail of the FID towards zero, it is also suitable for the apodisation of truncated data sets. However, it also increases the apparent decay rate of the NMR signal, causing lines to broaden, meaning one compromises resolution for the gain in sensitivity (Fig. 3.36). Using too strong a function (that is, one that causes the signal to decay too rapidly) can actually lead to a decrease in signal-to-noise ratio in the resulting spectrum because the broadening of the lines causes a reduction of their peak heights. Spectrometer software usually allows one to define the amount of line broadening directly in hertz (here parameter lb) so exponential multiplication is rather straightforward to use and allows the chemist to experiment with different degrees of broadening to attain a suitable result. The optimum balance between reducing noise and excessive line broadening is reached when the decay of the window function matches the natural decay of the NMR signal, in which case it is known as the *matched filter* [12] which results in a doubling of the resonance linewidth. For the exponential function to truly match the FID, the decay of the NMR signal must also be an exponential (meaning NMR resonances have a Lorentzian lineshape) which on a correctly adjusted spectrometer is usually assumed to be the case (at least for spin-½ nuclei). Despite providing the maximum gain in sensitivity, the matched condition is often not well suited to routine use in proton spectroscopy as the resulting line broadening and loss of resolution may preclude the separation of closely spaced lines, so less line broadening than this may be more appropriate. Furthermore, different resonances in the spectrum often display different unweighted linewidths, so the matched condition cannot be met for all resonances simultaneously. For routine proton work it turns out to be convenient to broaden the line by an amount equal to the digital resolution in the spectrum (typically 0.3-0.5 Hz), as this leads to some sensitivity enhancement but to minimal increase in linewidth. For heteronuclear spectra, resolution is usually a lesser concern and line broadening

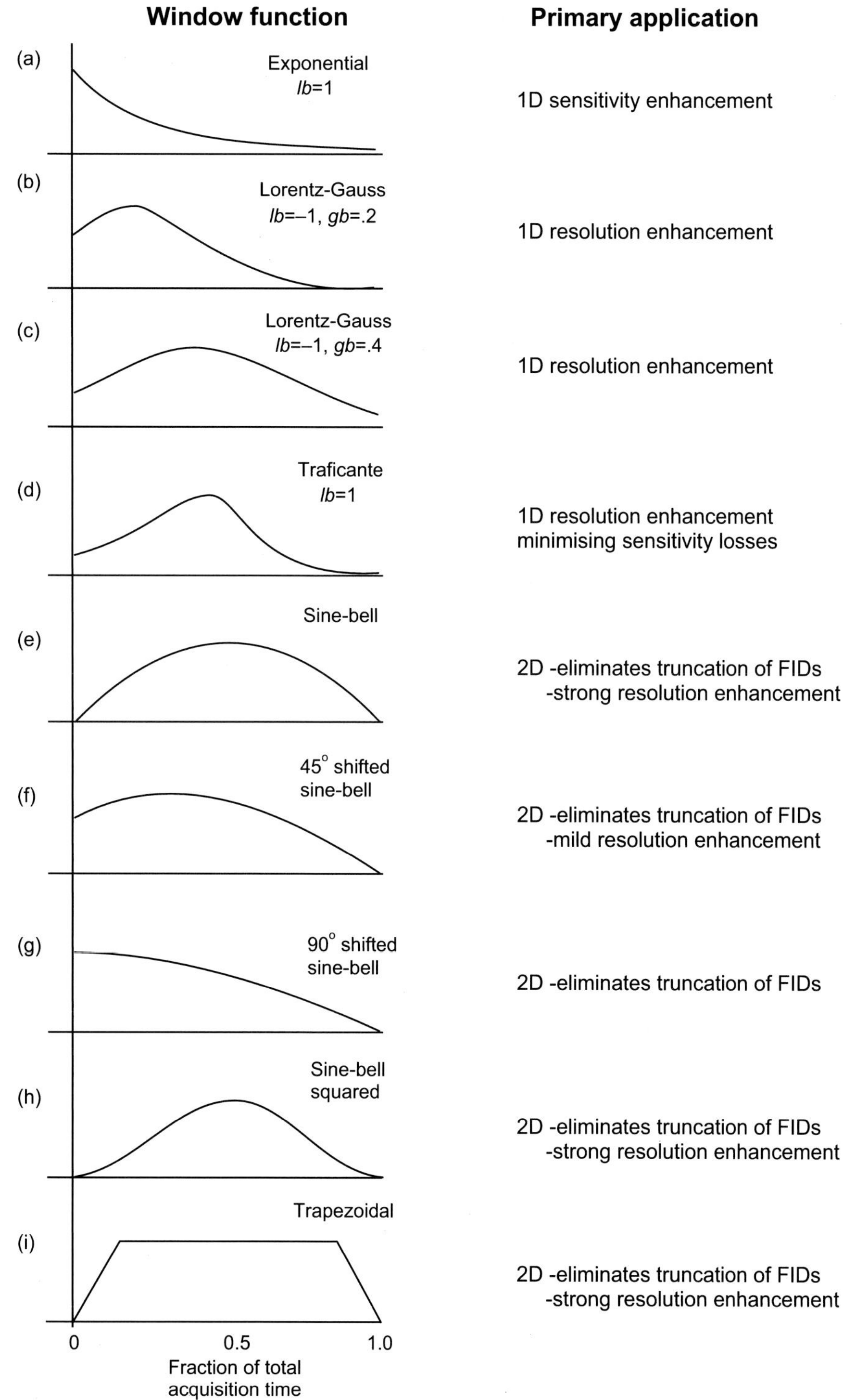

FIGURE 3.35 **Some commonly employed window functions.** These are used to modify the acquired FID to enhance sensitivity and/or resolution (*lb* = line-broadening parameter, *gb* = Gaussian-broadening parameter; ie the fraction of the acquisition time when the function has its maximum value; see text).

of a few hertz is commonly employed (comparable with or slightly greater than digital resolution). This can be considered essential to attenuate the distortions about the base of resonances arising, in part, from truncation of the data set (Section 3.2.3 and Fig. 3.17). For spectra that display resonances that are tens or even hundreds of hertz wide (most notably those of quadrupolar nuclei) the amount of line broadening must be increased accordingly to achieve any appreciable sensitivity gain.

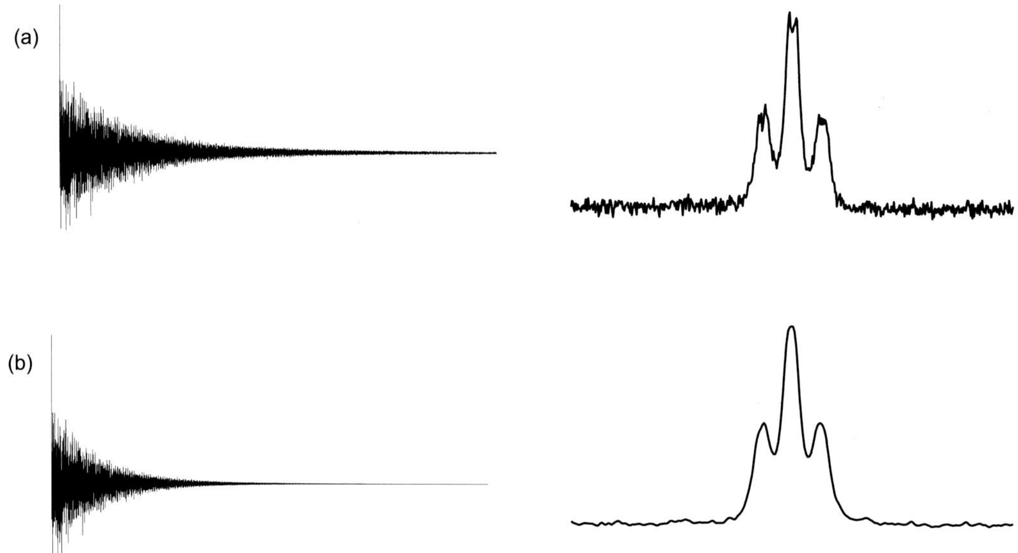

FIGURE 3.36 **Exponential multiplication of the FID.** This can be used to reduce noise in the spectrum. (a) Raw FID and spectrum following Fourier transformation. (b) Results after exponential processing with *lb* = 1 Hz.

3.2.7.2 Resolution Enhancement

One might suppose that to improve the resolution in spectra we should apply a function that enhances the latter part of the FID, so increasing the decay time. The problem with this approach is that by doing this one also increases the noise amplitude in the tail of the FID and emphasises any truncation of the signal that may be present, so increasing the potential for undesirable 'truncation wiggles'. A better approach is to apply a function which initially counteracts the early decay but then forces the tail of the FID to zero to provide the necessary apodisation. The most popular function for achieving this has been the *Lorentz–Gauss transformation* [13] (Fig. 3.35b and c), sometimes loosely referred to as *Gaussian multiplication* (although this strictly refers to yet another mathematical weighting function) and also known as double-exponential multiplication. This transforms the usual Lorentzian lineshape into a Gaussian lineshape which has a somewhat narrower profile, especially around the base (Fig. 3.37), and it is this feature that allows resolution of closely spaced lines.

The shape of the function is altered by two variable parameters which define the degree of line narrowing in the resulting spectra and the point during the acquisition time at which the function reaches its maximum value. These are usually presented to the operator as negative line broadening in hertz (here parameter lb) and as a fraction of the total acquisition time (here parameter gb), respectively. The choice of suitable values for these usually comes down to a case of trial and error, and different optimum values may be found for different groups of peaks within the same spectrum. Modern NMR-processing packages usually allow interactive variation of the parameters during which one can observe both the window function itself and the resulting spectrum, so an optimum can rapidly be determined. More negative line broadening will produce narrower lines, while positioning the maximum further along the decay will enhance this effect. This will also lead to greater distortion of the resonances about the base and to degradation of sensitivity as the early signals are

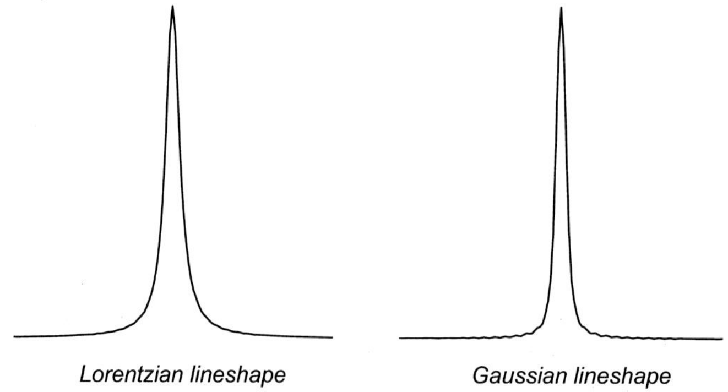

FIGURE 3.37 **Comparison of the Lorentzian and Gaussian lineshapes.**

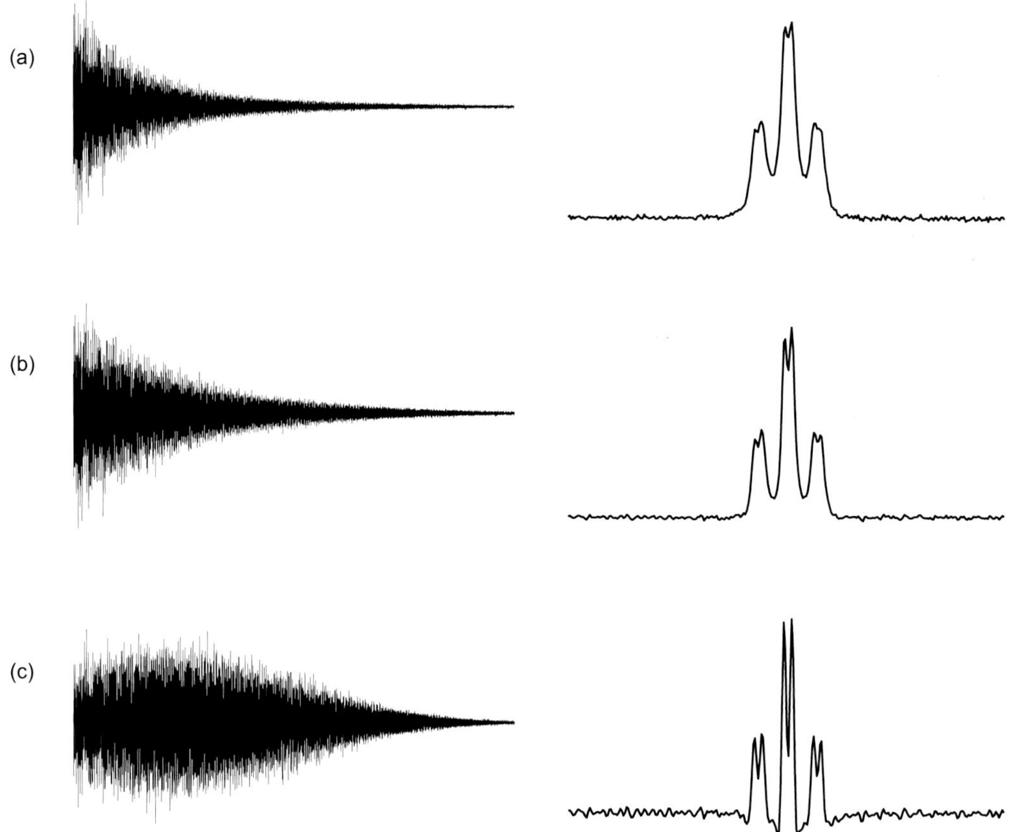

FIGURE 3.38 **The Lorentz–Gauss transformation ('Gaussian multiplication') can be used to improve resolution.** (a) Raw FID and spectrum following Fourier transformation and results after the Lorentz–Gauss transformation with (b) *lb* = −1 Hz, *gb* = 0.2 and (c) *lb* = −3 Hz and *gb* = 0.2.

sacrificed relative to the noise (Fig. 3.38). In any casc, some reduction in sensitivity will invariably result and it is necessary to have reasonable signal-to-noise for acceptable results. It has been suggested [14] that the optimum Gaussian resolution enhancement for routine use on spectra with narrow lines aims for a reduction in linewidth by a factor of 0.66 for which the function maximum should occur at a time of $1/(\Delta v_{1/2})$ s (that is, the inverse of the resonance linewidth in the absence of window functions). The appropriate figures for setting in the processing software can then be arrived at by simple arithmetic. For example a starting linewidth of 1 Hz will require line narrowing of 0.66 Hz and the maximum to occur after 1 s. With a typical proton acquisition time of around 3 s, the function should therefore reach a maximum at 1/3 of the total acquisition. Gaussian multiplication can be used to good effect when combined with zero-filling, thus ensuring the digital resolution is sufficiently fine to define any newly resolved fine structure.

In more recent years new window functions have been introduced [15,16] that are similar to the Lorentz-Gauss window but which aim to improve resolution without a discernible reduction in sensitivity. These so called *TRAF* functions (Fig. 3.35d) generally aim to enhance the middle part of the FID but also use matched filtration of the later part to attenuate noise. Another function, more commonly used in the processing of two-dimensional data sets, is the *sine-bell* window, which has no adjustable parameters. This comprises one half of a sine wave, starting at zero at the beginning of the FID, reaching a maximum half way through the acquisition and falling back to zero by the end of the decay (Fig. 3.35e). The function always has zero intensity at the end, so eliminates the truncation often encountered in 2D data sets. This tends to be a rather severe resolution enhancement function which can introduce undesirable lineshape distortions and also produces severe degradation in sensitivity because the early part of the NMR signal is heavily attenuated, and hence this is rarely used in one-dimensional spectra. A variation on this is the *phase-shifted sine-bell* for which the point at which the maximum occurs is a variable (Fig. 3.35 f and g) and can be moved toward the start of the FID. Again, this is commonly applied in 2D processing and has the advantage that the user has the opportunity to balance the gain in resolution against the lineshape distortion and sensitivity degradation. A final variant is the (optionally shifted) *squared sine-bell* (Fig. 3.35h), which has similar properties to the sine-bell but tails more gently at the edges, which can invoke subtle differences in 2D spectra. Likewise, the *trapezoidal* function (Fig. 3.35i) is sometimes used in the processing of 2D data.

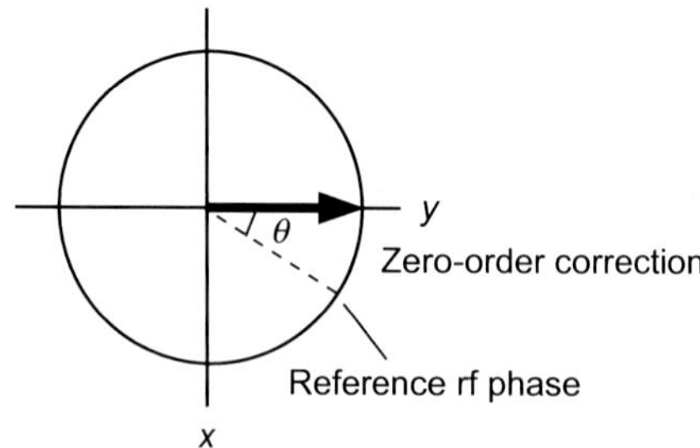

FIGURE 3.39 **Zero-order (frequency-independent) phase errors.** These arise when the phase of the detected NMR signals does not match the phase of the receiver reference RF. All resonances in the spectrum are affected to the same extent.

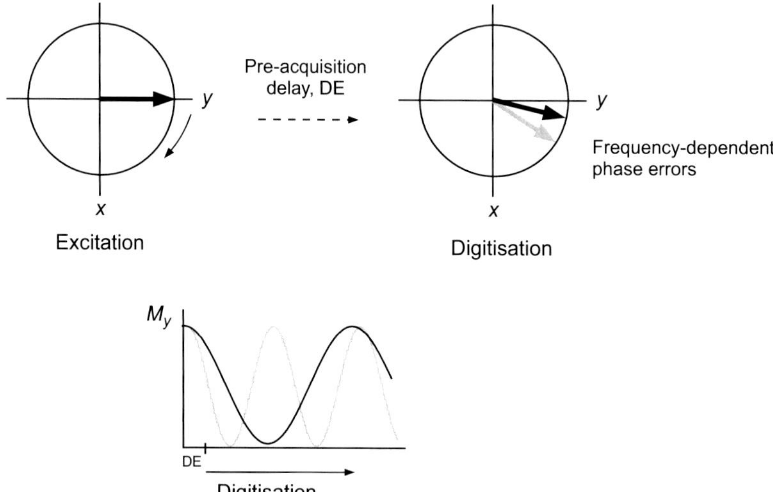

FIGURE 3.40 **First-order (frequency-dependent) phase errors.** These arise from a dephasing of magnetisation vectors during the preacquisition delay which follows the excitation pulse. When data collection begins, vectors with different frequencies have developed a significant phase difference which varies across the spectrum.

3.2.8 Phase Correction

It has already been mentioned in Section 2.2 that the phase of a spectrum needs correcting following FT because the receiver reference phase does not exactly match the initial phase of the magnetisation vectors. This error is constant for all vectors and since it is independent of resonance frequencies it is referred to as 'zero-order' phase correction (Fig. 3.39). Practical limitations also impose the need for a frequency-dependent or 'first-order' phase correction. Consider events immediately after the pulse is applied to the sample. A short period of time, the pre-acquisition delay (DE) is required for the spectrometer electronics to recover from the effect of the pulse before an undistorted FID can be collected. This delay is typically tens of microseconds, during which magnetisation vectors will evolve a little according to their chemical shifts so that at the point digitisation begins they no longer have the same phase (Fig. 3.40). Clearly, those resonances with the greatest shifts require the largest corrections. If the DE is small relative to the frequency offsets, the phase errors have an approximately linear offset dependency and can be removed. If the delay becomes large, the correction cannot be made without introducing a rolling spectrum baseline. The appearance of this zero and first-order phase errors is illustrated in the spectra of Fig. 3.41.

Typically both forms of error occur in a spectrum directly after the FT. The procedure for phase correction is essentially the same on all spectrometers. Zero-order correction is used to adjust the phase of one signal in the spectrum to pure absorption mode, as judged 'by eye' and first-order correction is then used to adjust the phase of a signal far away from the first in a similar manner. Ideally, the two chosen resonances should be as far apart in the spectrum as possible to maximise the frequency-dependent effect. Experimentally, this process of phase correction involves mixing of the real and imaginary parts of the spectra produced by the FT process such that the final displayed 'real' spectrum is in pure absorption mode whereas the usually unseen 'imaginary' spectrum is pure dispersion.

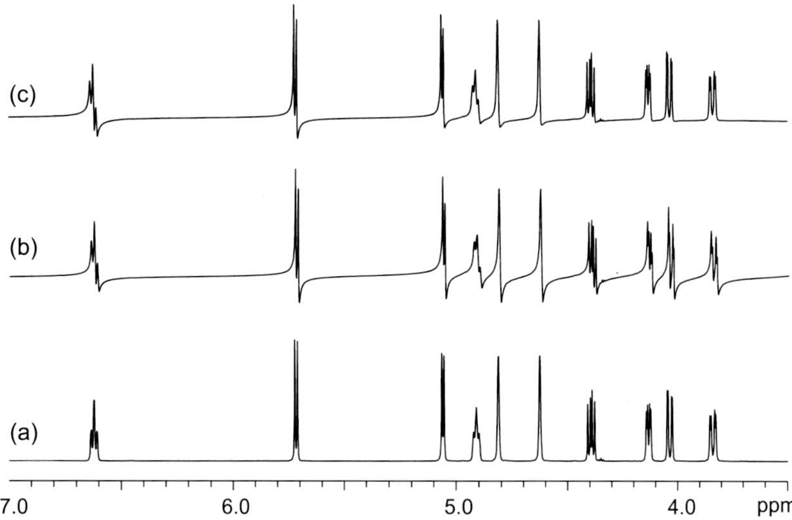

FIGURE 3.41 **Zero and first-order phase errors in a ^{1}H spectrum.** (a) Correctly phased spectrum, (b) the spectrum in the presence of frequency-independent (zero-order) phase errors and (c) the spectrum in the presence of frequency-dependent (first-order) phase errors.

3.3 PREPARING THE SAMPLE

This section describes some of the most important aspects of sample preparation for NMR analysis, a topic that is all too frequently given insufficient consideration in the research laboratory. Even for routine applications of NMR it is wise for the chemist to adopt a systematic strategy for sample preparation that will give consistently good results, so saving instrument time and eliminating frustration.

3.3.1 Selecting the Solvent

Since all modern NMR spectrometers rely on a deuterium lock system to maintain field stability, a deuterated solvent is invariably required for NMR, and these are now available from many commercial suppliers. In a busy organic chemistry laboratory the volume of NMR solvents used can be surprisingly large, and the cost of the solvent may play some part in its selection. Generally, chloroform and water are the most widely used solvents and are amongst the cheapest, although dimethylsulphoxide is also widely employed, notably in the industrial sector, because of its useful solubilising properties. Acetone, methanol and dichloromethane are also in widespread routine use.

The selection of a suitable solvent for the analyte is based on a number of criteria. The most obvious requirement is that the analyte be soluble in it *at the concentration required for the study*. This will be dependent on a diverse range of factors including the fundamental sensitivity of the nucleus, the overall sensitivity of the instrument, the nature of the experiment and so on. If experiments are likely to be performed at very low temperatures then it is also important to know that the solute remains in solution when the temperature is reduced so that a precipitate does not form in the NMR tube and degrade resolution. For experiments at temperatures other than ambient probe temperatures (which may be a few degrees greater than the ambient room temperature), the melting and boiling points of the solvents must also be considered. Table 3.2 summarises some important properties of the most commonly encountered solvents. For work at very high temperatures, dimethylsulphoxide or toluene are generally the solvents of choice, while for very low temperatures dichloromethane, methanol, toluene or the more expensive tetrahydrofuran are most appropriate. Even for experiments performed at ambient temperatures, solvent melting points may be limiting; on a cold day dimethylsulphoxide solutions may freeze in the laboratory. The viscosity of the solvent will also influence the resolution that can be obtained, with the best performance provided by the least viscous solvents, such as acetone (which, for this reason, is used as the solvent for spectrometer resolution tests).

For proton and carbon spectroscopy, the chemical shifts of the solvent resonances need to be anticipated as these may occur in a particularly unfortunate place and interfere with resonances of interest. In proton spectroscopy, the observed solvent resonance arises from residual *protonated* species (NMR solvents are typically supplied with deuteration levels in excess of 99.5%). For routine studies, where a few milligrams of material may be available, lower specification solvents should suffice for which the residual protonated resonance is often comparable in magnitude with that of the solute. Solvents with higher levels of deuteration are beneficial for proton spectroscopy when sample quantities are of the order of tens of micrograms or less

TABLE 3.2 Properties of Common Deuterated Solvents[a]

Solvent	δ_H (ppm)	$\delta_{(HDO)}$ (ppm)	δ_C (ppm)	Melting Point (°C)	Boiling Point (°C)
Acetic acid–d_4	11.65, 2.04	11.5	179.0, 20.0	16	116
Acetone–d_6	2.05	2.0	206.7, 29.9	–94	57
Acetonitrile–d_3	1.94	2.1	118.7, 1.4	–45	82
Benzene–d_6	7.16	0.4	128.4	5	80
Chloroform-d_1	7.27	1.5	77.2	–64	62
Deuterium oxide–d_2	4.80	4.8	—	4	101
Dichloromethane–d_2	5.32	1.5	54.0	–95	40
N,N-dimethyl formamide–d_7	8.03, 2.92, 2.75	3.5	163.2, 34.9, 29.8	–61	153
Dimethylsulphoxide–d_6	2.50	3.3	39.5	18	189
Methanol–d_4	4.87, 3.31	4.9	49.2	–98	65
Pyridine–d_5	8.74, 7.58, 7.22	5.0	150.4, 135.9, 123.9	–42	114
Tetrahydrofuran–d_8	3.58, 1.73	2.4	67.6, 25.4	–109	66
Toluene–d_8	7.09, 7.00, 6.98, 2.09	0.4	137.9, 129.2, 128.3, 125.5, 20.4	–95	111
Trifluoroacetic acid–d_1	11.30	11.5	164.2, 116.6	–15	75
Trifluoroethanol–d_3	5.02, 3.88	5.0	126.3, 61.5	–44	75

[a]*Proton shifts δH and carbon shifts δC are quoted relative to TMS (proton shifts are those of the residual partially* protonated *solvent). The proton shifts of residual HDO/H_2O vary depending on solution conditions.*

and are likely to be used in conjunction with microsample techniques (Section 3.3.3). The solvent proton resonance, with the exception of chloroform and water, will comprise a multiplet from coupling to spin-1 deuterium. For example dichloromethane displays the triplet of $CDHCl_2$ while acetone, dimethylsulphoxide or methanol show a quintuplet from CD_2H (Fig. 3.42). In carbon spectroscopy, the dominant resonance is often that of the deuterated solvent which again will be a multiplet owing to coupling with deuterium. For other common nuclei, interference from the solvent is rarely a consideration, the one notable exception to this being deuterium itself. Unless very large sample quantities are being used (many tens of milligrams) and one can be sure the solvent resonance will not overlap those of the solute, it is usually necessary to record ^{2}H spectra in *protonated* solvent since the huge signals of the deuterated solvent are likely to swamp those of interest. This then precludes the use of the lock system for maintaining field stability. The lack of a suitable lock reference requires a small quantity of the deuterated solvent to be added to the solution to provide an internal chemical shift reference (see later in the chapter).

Aside from the solvent resonance itself, the other significant interference seen in proton spectra is water, which is present to varying degrees in all solvents and is also often associated with solutes. The water resonance is often rather broad and its chemical shift can vary according to solution conditions. All solvents are hygroscopic to some degree (including deuterated water) and should be exposed to the atmosphere as little as possible to prevent them from becoming wet. Very hygroscopic solvents such as dimethylsulphoxide, methanol, and water are best kept under a dry inert gas atmosphere. Solvents can also be obtained in smaller glass ampoules which are particularly suitable for those which are used only infrequently. Molecular sieves may also be used to keep solvents dry, although some care is required in filtering the

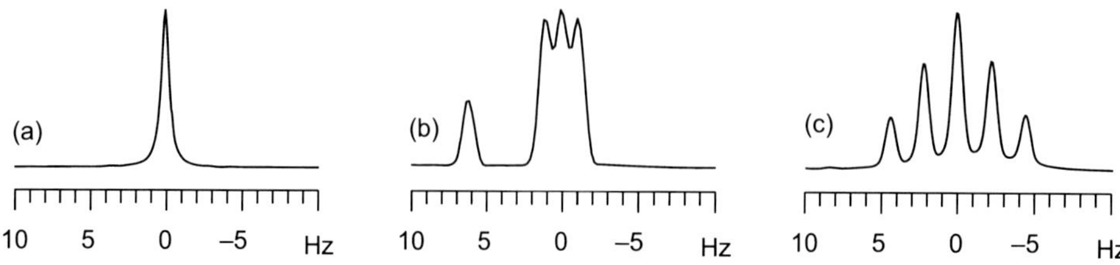

FIGURE 3.42 **Residual protonated resonances of deuterated solvents.** These are shown for (a) $CHCl_3$ in $CDCl_3$, (b) $CHDCl_2$ in CD_2Cl_2 and (c) CHD_2COCD_3 in $(CD_3)_2CO$. The multiplicity seen in (b) and (c) arises from two-bond (geminal) couplings to spin-1 deuterium producing a 1:1:1 triplet and a 1:2:3:2:1 quintet, respectively. The left-hand singlet in (b) is residual CH_2Cl_2 in the solvent, the shift difference arises from the H–D isotope shift of 6 Hz.

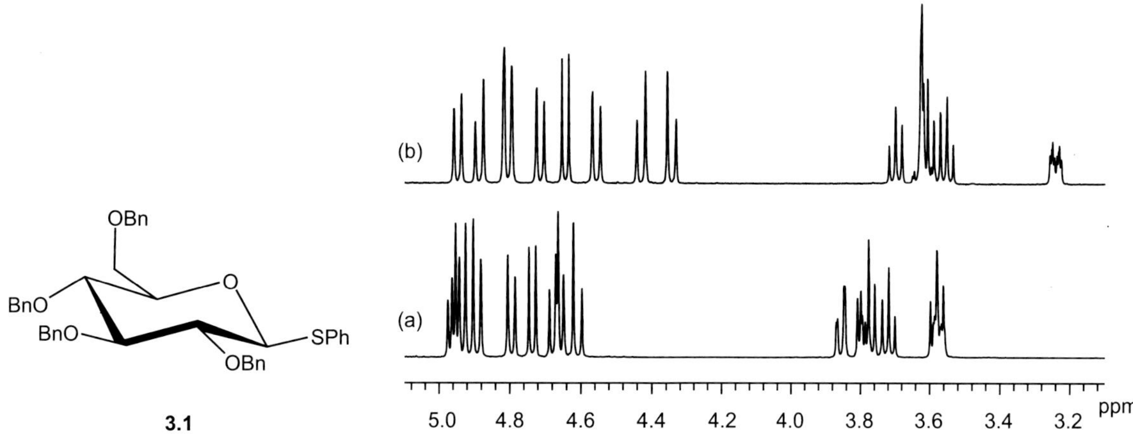

FIGURE 3.43 **Changes in solvent can be used to improve resonance dispersion.** The proton spectrum of the sugar **3.1** is shown in (a) $CDCl_3$ and (b) C_6D_6. Note the appearance of the resonance at 3.2 ppm in (b) that was hidden at 3.6 ppm by another resonance in (a).

prepared solution before analysis to remove fine particulates that may be present. Alternatively, the NMR solution may be passed through activated alumina to remove water, as part of the filtration process.

The solvent may also result in the loss of the resonances of exchangeable protons since these will become replaced by deuterons. In particular, this is likely to occur with deuterated water and methanol. To avoid deuterium replacement altogether one must consider using H_2O containing 5–10% D_2O (to provide a lock signal) or d_3–methanol (CD_3OH). In these cases, a suitable solvent suppression scheme (see Section 12.5) is required to attenuate the large solvent resonance (which may still lead to the loss of the exchangeable protons of the solute by the process known as saturation transfer). Studies in 90% H_2O dominate the NMR of biological macromolecules because solvent exchangeable protons, such as the backbone amide NH protons in peptides and proteins or the imido protons of DNA and RNA base pairs, often play a key role in structure determination of these molecules [17,18].

The nature of the solvent can also have a significant influence on the appearance of the spectrum, and substantial changes can sometimes be observed on changing solvents. Whether these changes are beneficial is usually difficult to predict and a degree of trial and error is required. A useful switch of solvent is from a non-aromatic solvent into an aromatic one, for example chloroform to benzene (Fig. 3.43), making use of the magnetic anisotropy exhibited by the latter. In cases of particular difficulty, where the change from one solvent to another simply produces a different but equally unsuitable spectrum, then titration of the second solvent into the first may provide a suitable compromise between the two extremes. Such changes may prove useful in 1D experiments that use selective excitation of a specific resonance, by revealing a target proton when previously it was hidden, as in the example in Fig. 3.43. Selective removal of a resonance by the addition of another solvent can also prove useful in spectrum interpretation. Adding a drop of D_2O to an organic solution in the NMR tube, mixing thoroughly and leaving the mix to settle removes (or attenuates) acidic exchangeable protons. Acidic protons that are protected from the solvent, such as those in hydrogen bonds, may not fully exchange immediately, but may require many hours to disappear. This can provide a useful probe of H-bonding interactions.

3.3.2 Reference Compounds

When preparing a sample it may be desirable to add a suitable compound to act as an *internal* chemical shift reference in the spectrum, and the selection of this must be suitable for the analyte and solvent. In proton and carbon NMR, the reference used in organic solvents is tetramethylsilane (TMS, 0.0 ppm) which has a number of favourable properties; it has a sharp 12-proton singlet resonance that falls conveniently to one end of the spectrum, it is volatile so can be readily removed and it is chemically inert. In a few cases this material may be unsuitable such as in the study of silanes or cyclopropanes. For routine work it is often not necessary to add any internal reference as the residual lines of the solvent itself can serve this purpose (Table 3.2). For aqueous solutions, the water-soluble equivalent of TMS is partially deuterated sodium 3-(trimethylsilyl)propionate-d_4 (TSP-d_4) which is also referenced to 0.0 ppm. A volatile alternative is 1,4-dioxane (^{1}H 3.75 ppm, ^{13}C 67.5 ppm) which can be removed by lyophilisation. The standard reference materials for some other common nuclides are summarised in Table 3.3. Often these are not added into the solution being studied but are held in an outer, concentric jacket or within a separate axial capillary inside the solution, in which case the reference material may also be in a different solvent to that of the sample.

TABLE 3.3 Spectrum Reference Materials (0 ppm) and Reference Frequencies (Ξ Values) for Selected Nuclides

Nuclide	Primary Reference	Ξ Value (MHz)	Alternative Reference
^{1}H	Me_4Si	100.000 000	TSP–d_4 (aq.)
^{2}H	Me_4Si	15.351	Trace-deuterated solvent
^{6}Li	LiCl (aq)	14.717	
^{7}Li	LiCl (aq)	38.866	
^{10}B	$BF_3O(Et)_2$	10.746	H_2BO_3 (aq.)
^{11}B	$BF_3O(Et)_2$	32.089	H_2BO_3 (aq.)
^{13}C	Me_4Si	25.145 004	1,4-dioxan @ 67.5 ppm, TSP–d_4 (aq.)
^{14}N	CH_3NO_2	7.224	
^{15}N	CH_3NO_2	10.136 767	NH_4NO_3 (aq.), NH_3 (liq.)[a]
^{17}O	H_2O	13.557	
^{19}F	$CFCl_3$	94.094 003	
^{29}Si	Me_4Si	19.867 184	
^{31}P	H_3PO_4 (85%)	40.480 747	
^{119}Sn	Me_4Sn	37.290 665	
^{195}Pt	$K_2[Pt(CN)_6]$	21.414 376	
^{207}Pb	Me_4Pb	20.920 597	

[a] ^{15}N shifts of biomolecular materials are more often quoted relative to external liquid ammonia at 0.0 ppm.

An alternative to adding additional reference materials is to use a so-called external reference. Here the spectrum of a separate reference substance is acquired before and/or after the sample of interest and the spectrum reference value carried over. Identical field settings should be used for both, which, on some older instruments, requires the same lock solvent, or an additional correction to the spectrum reference frequency must be used to compensate any differences. This restriction does not arise on instruments that use shifting of the lock transmitter frequency to establish the lock condition.

The referencing of more 'exotic' nuclei is generally less clearcut than for those in common use and in many cases it is impractical to add reference materials to precious samples, and it is sometimes even difficult to identify what substance is the 'accepted' reference standard. In such cases the IUPAC Ξ-scale can be used, which does not require use of a specific reference material. Instead, this scheme defines the *reference frequency* for the reference material of each nuclide at a field strength at which the proton signal of TMS resonates at exactly 100 MHz. The reference frequencies are scaled appropriately for the magnetic field in use and this then defines the absolute frequency at 0.0 ppm for the nuclide in question. The 'master' reference sample for these measurements is 1% TMS in $CDCl_3$ which is used to determine the absolute frequency of TMS for the instrument in use. The Ξ values for selected nuclei are also summarised in Table 3.3, while more extensive tables are available [19,20].

3.3.3 Tubes and Sample Volumes

Newcomers to the world of practical NMR often find the cost of NMR tubes surprisingly high. The prices reflect the need to produce tubes that conform to strict requirements of straightness (camber) and concentricity, as deviation from these can produce undesirable artefacts in spectra, usually in the form of 'spinning sidebands'. Generally, the higher quality (and more expensive) tubes are required on higher field instruments and this is particularly so for proton work. Some experimenting may be required to decide on a suitable balance between cost and performance for your particular instruments.

The diameter of the tube is generally determined by the dimensions of the probe to be used, or more precisely the probe's rf coils. Tube diameters employed in chemical laboratories (Fig. 3.44) include 1, 1.7, 2.5 or 3 mm (for use with so-called microprobes), 5 mm (the most widely used), or 10 mm (typically used for the observation of low-sensitivity nuclei where the solubility of the material to be studied is limiting or for dedicated polymer studies). Eight millimetre tubes once found use in biological macromolecular work, where solubility and aggregation considerations preclude the use of concentrated samples, but are less common nowadays, having been superseded by cryogenic probes. The benefit of smaller diameter

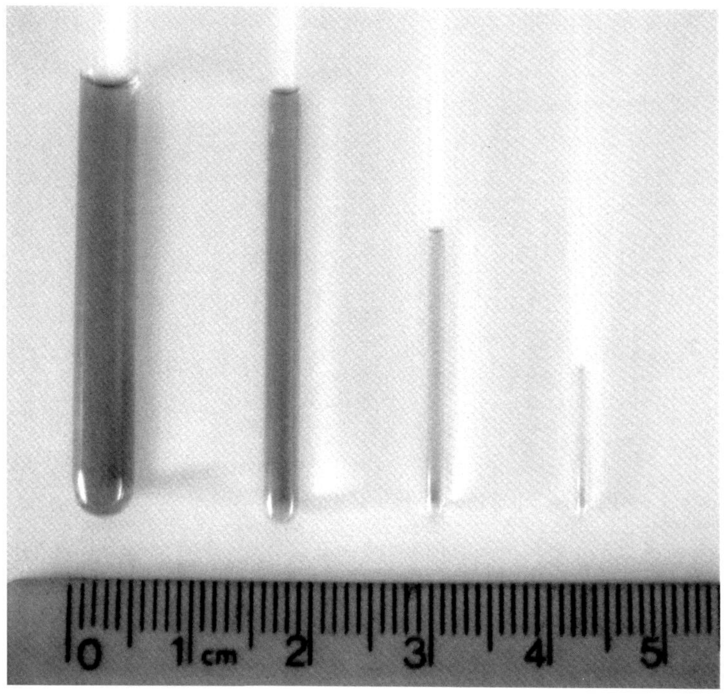

FIGURE 3.44 **NMR tubes of various diameters and their optimum solution volumes.** From left to right: 5 mm (500 μL), 3 mm (150 μL), 1.7 mm (30 μL) and 1 mm (5 μL).

tubes and probes lies in their greater 'mass sensitivity' as will be explained in Section 3.4.2 which describes available probe technologies. As sensitivity is usually of prime importance when performing any NMR experiment, it is prudent to use the correct sample volume for the probe/tube configuration and not to dilute the sample any more than is necessary. An important factor here is the length of the detection coil within the probe. The sample volume should be sufficient to leave a little solution above and below the coil since magnetic susceptibility differences at the solution/air and solution/glass interfaces lead to local distortions of the magnetic field. For a standard 5 mm tube the coil length is typically 16 mm and the required volume will be around 500–600 μL whereas for narrower coils and tubes this may reduce to 150 μL for the 3 mm tube, 30 μL for the 1.7 mm tube or as little as 5–8 μL for the 1 mm microtube (Table 3.4). For 10 mm tubes it may be necessary to use a 'vortex-suppressor' to prevent whirlpool formation in the tube if the sample is spun. These are usually polytetrafluoroethylene (PTFE) plugs designed to fit tightly within the NMR tube which are pushed down to sit on the surface of the sample and thus hold the solution in place. Some care is required when using these plugs in variable temperature work due to the thermal expansion or contraction of the plug, and at low temperatures the plug may even fall into the solution so its use may be inappropriate. Filling the smaller 1 mm and 1.7 mm tubes requires the use of fine needles with a manual syringe or robotic system, or alternatively can be readily filled by placing the solution in the top tube funnel and centrifuging this down.

For the handling of smaller sample quantities when microprobes are not available, the microtubes of small diameter that are designed for use with microprobes may be employed to concentrate the available material within the detection coil.

TABLE 3.4 Sample Tube Diameters and Their Associated Working Sample Volumes and Heights

Tube Dimensions (Outer Diameter)	Solute Volume (μL)	Solute Column Height (mm)
10	2500	40
8	1500	40
5	500	40
3	150	40
1.7	30	20
1	5	10

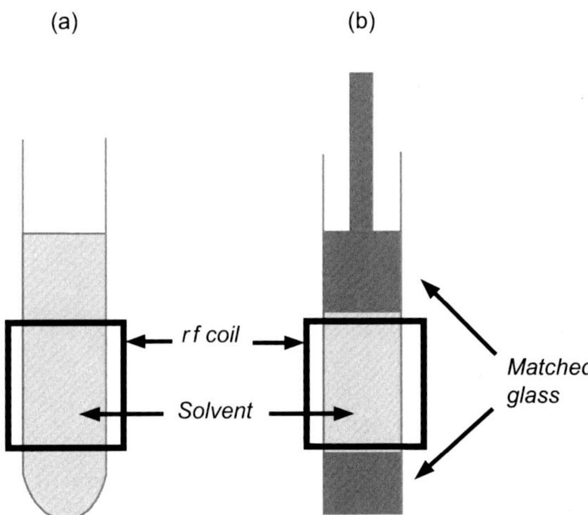

FIGURE 3.45 **Susceptibility-matched NMR tubes.** Schematic comparison of (a) a conventional NMR tube and (b) a Shigemi-matched tube and plug in which the magnetic susceptibility of the glass is matched to that of the solvent. This allows a smaller sample volume and so concentrates more of the analyte within the detection coil.

Thus, using a 3 mm tube in a 5 mm probe has been shown to provide a modest signal-to-noise improvement for conventional probes whereas this approach can provide greater gains when using cryogenically cooled probeheads (see Section 3.4.2). Alternatively, microcells can be obtained that fit within standard 5 mm NMR tubes but which require only tens of microlitres of solution, again allowing greater sample concentration. When using such cells it is advisable to fill the volume around the outside of the cell with solvent, as this minimises susceptibility differences and improves resolution. Additionally, special tubes and plugs ('Shigemi tubes') are available that allow all the sample volume to be held within the sensitive region of the rf coil (Fig. 3.45). These use glasses that have a magnetic susceptibility matched to that of the solvent and so allow shorter sample heights without the introduction of lineshape defects [21]. Typically, tubes are available individually matched to water and to a small selection of organic solvents and are available in 5 and 3 mm sizes. When using these it is imperative that no air bubbles become trapped between the solvent and the upper plug to avoid lineshape perturbations and some practice in their use is highly recommended before embarking on the preparation of your most precious sample. More recently, specialised tubes have been introduced specifically for use with cryogenically cooled probes that have rectangular cross-sections, for example 3 × 6 mm; Section 3.4.2). These have been shown to provide enhanced sensitivity when observing samples of high ionic strength (buffered or salty samples) by reducing pickup of electrical noise arising from the sample solution and so are likely to be of greatest applicability for the study of biomacromolecules or biofluids.

NMR tubes should be kept clean, dry, free from dust and must also be free from scratches on the glass as this distorts the cylinder in which the sample is confined. New tubes may not always be clean when delivered, although this is often assumed to be the case. Organic lubricants used in their manufacture may remain on the glass and become all too apparent when the tube is first used. For routine washing of tubes, rinsing with a suitable organic solvent, such as acetone, or with distilled water a few times is usually sufficient. More stubborn soiling is best tackled by chemical rather than mechanical means and either soaking in detergent or strong mineral acids is recommended. The use of chromic acid must be avoided as this can leave traces of paramagnetic chromium(VI) which will degrade resolution. Tubes should not be dried by subjecting them to high temperatures for extended periods of time as this tends to distort them. A better approach is to keep the tubes under vacuum, at slightly elevated temperatures if possible, or to blow filtered nitrogen into the tubes. If oven drying is used, the tubes must be laid on a flat tray and heated for only 30 minutes, not placed in a beaker or tube rack as this allows them to bend. If the removal of all traces of protiated water from the tube is important, then it is necessary to rinse the tube with D_2O prior to drying to ensure the exchange of water adsorbed onto the surface of the glass. Unused tubes should be stored with their caps on to prevent dust from entering. Finally, it is also important to avoid contamination of the *outside* of the NMR tube (which includes fingerprint oils) as this leads to the transfer of contaminants into the probehead. The accumulation of these contributes to degradation of the instrument's performance over time.

3.3.4 Filtering and Degassing

For any NMR experiment it is necessary to achieve a uniform magnetic field throughout the whole of the sample to obtain a high-resolution spectrum and it is easy to imagine a range of circumstances that detract from this condition. One of the most detrimental is the presence of particulates in the sample since these can distort the local magnetic field, so it is of utmost importance to remove these from an NMR sample prior to analysis. Even very slight changes in the field within the sample can produce a noticeable reduction in resolution. This should not come as a surprise if one recalls that one often wishes to resolve lines whose resonant frequencies are many hundreds of megahertz but which are separated by less than 1 Hz. This corresponds to a resolution of 1 part in 10^9, roughly equivalent to measuring the distance from the Earth to the Sun to an accuracy of 1 km. Commercial sintered filters can be obtained but require constant washing between samples. A convenient and cheap alternative is to pass the solution through a small cotton wool plug in a Pasteur pipette directly into the NMR tube (it is a good practice to rinse the cotton wool with a little of the NMR solvent before this). If cotton wool is not suitable for the sample, a glass wool plug may be used although this does not provide such a fine mesh. If metal ions are likely to be present in solution (paramagnetic ions being the most unwelcome), either from preparation of the sample or from decomposition of the sample itself, then passing the solution through a small column of a metal-chelating resin should remove these.

In situations where very high resolution is demanded, if relaxation studies are to be performed or when handling air-sensitive samples, it may also be necessary to remove all traces of oxygen from the solution. For air-stable samples, the need to remove oxygen may arise because O_2 is paramagnetic and its presence provides an efficient relaxation pathway which leads to line broadening. Essentially, there are two approaches to degassing a sample. The first involves bubbling an inert gas through the solution to displace oxygen. This is usually oxygen-free nitrogen or argon which need be passed through the solution in the NMR tube for about one minute for organic solvents and double this for aqueous solutions. Great care is required when attempting this and a fine capillary must be used to introduce the gas *slowly*. It is all too easy to blow the sample clean out of the tube in an instant by introducing gas too quickly, so it is probably wise to get some feel for this before attempting the procedure on your most precious sample. Note that with volatile solvents a significant loss in volume is likely to occur and TMS, if used, may well also be lost.

The second and more thorough approach is the 'freeze-pump-thaw' technique which is most appropriate for air-sensitive samples. The NMR tube containing the solution is frozen with liquid nitrogen or dry ice and placed under vacuum on a suitable vacuum line. Commercial tube manufacturers produce a variety of specialised tubes and adapters for this purpose but a simple alternative is to connect a standard tube to a vacuum line via a needle through a rubber septum cap on the tube. The tube is then isolated from the vacuum by means of a stopcock and allowed to thaw, during which the dissolved gases leave the solution. The procedure is then repeated typically at least twice more, this usually being sufficient to fully degas the sample. When using an ordinary vacuum pump it may be necessary to place a liquid nitrogen trap between the pump and the sample to avoid the possibility of vacuum oils condensing in the sample tube. Furthermore, when freezing aqueous samples it is easy to crack the tube if this is carried out too fast; holding the sample tube just above the freezing medium while tilting it is usually sufficient to avoid such disasters. An alternative approach for aqueous samples containing involatile solutes is to carefully place the sample directly under vacuum without freezing, for example in a vacuum jar or schlenk and allowing dissolved gas to bubble out of solution.

Following the degassing procedure, the tube should be sealed. If the sample is likely to be subject to short-term analysis (say, over a few hours), then a standard tight-fitting NMR cap wrapped with a *small* amount of paraffin wax film or, better still, use of a rubber septum, is usually sufficient since diffusion into solution of gases in the tube will be rather slow. For longer duration studies, specialised adapters or screw top tubes can be purchased or standard tubes can be flame-sealed, for which NMR tubes with restrictions towards the top are available to make the whole process very much easier. When flame-sealing it is advisable to cool or freeze the solution and then seal while pulling a gentle vacuum since failure to evacuate the tube may cause the pressure within in it to become dangerously high as the sample warms. The seal should be symmetric, otherwise the sample will spin poorly, so practice on discarded tubes is likely to pay dividends.

3.4 PREPARING THE SPECTROMETER

Whenever a new sample is placed within the NMR spectrometer, the instrument must be optimised for this. The precise nature of the adjustments required and the amount of time spent making these will depend on the sample, the spectrometer and the nature of the experiment but in all cases the aim will be to achieve optimum resolution and sensitivity and to ensure system reproducibility. The details of the approaches required to make these adjustments depend on the design of spectrometer in question, so no attempt to describe such detail is made here. They all share the same general procedures, however, which are summarised in Scheme 3.1 and described in the sections that follow.

3.4.1 The Probe

If you are a chemist making use of the NMR facilities available to you, then choosing the appropriate probe for the study in question may not be relevant, as you will likely use that which is available. If, however, you are involved with instrument

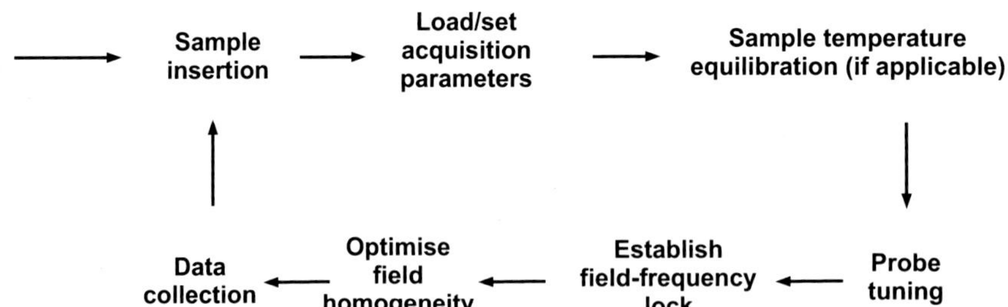

SCHEME 3.1 **The typical procedure followed in preparing a spectrometer for data collection.** All these steps can be automated, although probe tuning demands additional equipment or suitably equipped auto-tuneable probeheads.

purchasing or upgrading or you are fortunate enough to have available a variety of probes for a given instrument, then it is important to be able to make the appropriate selection. Since it is the probe that must receive the very weak NMR signals, it is perhaps the most critical part of the NMR spectrometer and its particular design and construction will influence not only the types of experiments it is able to perform but also its overall performance. There has been considerable progress over the years in the design of probes, resulting in ever increasing performance and a greater array of available probe technologies and these will be reviewed in more detail in Section 3.4.2.

Important considerations when selecting a probe are the frequency range(s) it may be tuned to, since this will dictate the nuclei one is able to observe, and how the rf coil configuration has been optimised. The simplest design is a probe containing a single coil which is designed for the observation of only one nucleus. In fact, it would be 'doubly-tuned' to enable simultaneous observation of deuterium for field frequency regulation via the lock system, although the presence of a deuterium channel is usually implicit when discussing probe configurations. However, many modern NMR experiments require pulses to be applied to two (or more) different nuclides, of which one is most often proton, for which two rf coils are necessary. In this case, two possible configurations are in widespread use depending on whether one wishes to observe the proton or another nuclide, referred to as the X-nucleus. The traditional two-coil design is optimised for the observation of the X-nucleus, with the X coil as the inner of the two allowing it to sit closer to the sample, so offering the best possible sensitivity for X observation; it is said to have the greatest 'filling factor'. This configuration can be described by the shorthand notation X(^{1}H). Nowadays multipulse experiments tend to utilise the higher sensitivity offered by proton observation wherever possible, and benefit from probes in which the proton coil sits closest to the sample with the X coil now the outermost: ^{1}H(X). It is this design of probe that is widely referred to as having the *inverse* configuration because of this switch in geometry. In either case, X coil circuitry can be designed to operate at only a single frequency or can be tuneable over a wide frequency range, such probes being known as *broadband observe* or *broadband inverse* probes. More recent designs of broadband observe probes have incorporated developments that deliver significantly improved proton sensitivity and lineshapes, making them also suitable for work with highly protonated aqueous samples (≥90% H_2O). They also include ^{19}F capability (sharing either the ^{1}H or the broadband channel, according to vendor design) and are now widely regarded as the workhorse probe for chemical research. These probes have now superseded the *quad-nucleus* probe, a once popular configuration in organic chemistry that allowed observation of the four most commonly encountered nuclei ^{1}H, ^{13}C, ^{19}F and ^{31}P. In studies of biological macromolecules, and less frequently in organic chemistry, *triple-resonance* probes are employed, allowing proton observation and pulsing or decoupling of two other nuclei; ^{1}H(X, Y), the most common configurations being ^{1}H(^{13}C, ^{15}N) or ^{1}H(^{13}C, ^{31}P). Alternatively, triple-resonance inverse probes which include one broadband (BB) channel may offer greater flexibility in a chemical laboratory where many nuclei may be of interest with ^{1}H(^{13}C, BB) or ^{1}H(^{31}P, BB) being most common.

A further feature offered is the addition of magnetic field gradient coils to the probehead. These surround the usual rf coils and are designed to destroy static magnetic field homogeneity throughout the sample for short periods of time in a very reproducible manner; PFG–selected techniques are introduced in Section 5.3. While offered as optional features, the inclusion of a single gradient coil in a probe can now be considered as standard.

3.4.1.1 Flow Probes

A fundamentally different approach to sample handling is used in the case of flow probes in which a flow cell sits within the rf coils and solutions are pumped into the cell for analysis, dispensing with the need for NMR tubes altogether. Traditional designs held a fixed cell assembly, meaning these probes could only be used in flow mode, whereas newer versions of both conventional and cryogenic probes (Section 3.4.2) utilise an interchangeable flow cell that allows this to be placed within probes designed to accommodate conventional tubes, so providing greater flexibility. Flow probes are typically used either as the NMR detector at the output of hyphenated systems including HPLC for mixture analysis – as liquid chromatography (LC)-NMR or LC-NMR-MS (mass spectrometry) configurations – or in systems where direct flow injection is employed for automated sample handling often of very large sample numbers. NMR analyses may then be performed in an *on-flow mode* wherein the sample moves continually through the cell, so limiting observation to 1D ^{1}H (or ^{19}F) spectra only, due to the short sample residence time. This mode is typically used for frontline screening of mixtures and has limited applicability to full structural characterisation. Alternative static modes include the *stopped-flow mode* in which the analyte slug is held in the cell for a discrete period by stopping sample elution, so enabling the collection of longer running and 2D NMR spectra for structure identification, or *loop collection mode* where the analyte slug is diverted to a storage loop during a chromatographic run and later parked within the flow cell for NMR analyses. Some of the main issues that arise with flow-based NMR when compared with tube-based NMR include the potential for cross-contamination between samples, requiring appropriate cleaning cycles between these, sample dilution during the transfer process and the fact that samples are not isolated in convenient storage vessels (tubes) should further analysis be required. Hyphenated methods for online separation also tend to be time consuming and technically demanding and procedures employing off-line chromatography followed by tube-based NMR analysis still appear to be widely employed. Developments such as solid-phase extraction (SPE) for sample concentration combined with high-sensitivity microscale or cryogenic probes also provide competitive routes to the analysis of complex mixtures [22]. Indeed, the 30 μL volume of a 1.7 mm NMR tube matches the elution volume of SPE cartridges and the 1.7 mm cryogenic probe design (see below) was no doubt developed with this in mind.

LC-NMR coupling technology has been reviewed [23] and the interested reader is also referred to the detailed text edited by Albert [24]. The miniaturisation of flow probes to the capillary scale has been developed to enhance signal detection sensitivity, as discussed in Section 3.4.2, and even the hyphenation of gas chromatography with microflow cells (GC-NMR) has been demonstrated [25].

3.4.2 Probe Design and Sensitivity

There now exists an ever-increasing array of probe designs and dimensions aimed at delivering improved and optimised detection sensitivity for the analytes of interest and this section sets out some of these key developments and the rationale underlying them. Over the years a number of factors have been employed that have progressively improved probe performance and sensitivity in addition to novel probe designs. One such development has been the material used in the construction of the receiver coil itself. Since this coil sits in very close proximity to the sample, this material may distort the magnetic field within this, compromising homogeneity. Modern composite metals are designed so that they do not lead to distortions of the field (they are said to have *zero magnetic susceptibility*) so allowing better lineshapes to be obtained which ultimately leads to improved signal-to-noise figures. Increasingly, coil materials are also matched to the properties of the cooling gas that flows over the NMR tube, typically either air or pure nitrogen, meaning it may be advantageous to define this when ordering probeheads. A second factor lies in the coil dimensions, with modern coils on standard probes tending to be longer than used previously so that a greater sample volume sits within them. This demands a greater volume in which the magnetic field is uniform, so these changes in probe construction have largely followed improvements in room temperature shim systems (see later in the chapter). Developments in electronics have also contributed to enhanced instrument sensitivity.

3.4.2.1 Detection Sensitivity

In order to appreciate developments in probe and instrument design, we shall first consider some key factors that contribute to the detection sensitivity of an NMR measurement and hence how this may be improved. As an illustration of the progressive enhancements in instrument sensitivity, Fig. 3.46 shows the specified signal-to-noise ratios of commercial ^{1}H observe probes at the time of new magnet launch (as judged by the 0.1% ethyl benzene 'sensitivity test', Section 3.6.2). Although the data reflect in part increasing performance with higher magnetic fields, the improvement in overall instrument and probe performance that paralleled these developments is illustrated by the comparison of 500 MHz data at the time of launch (1979) with that in 2014, which shows an approximately fivefold improvement in S/N (dashed line in Fig. 3.46). Current signal-to-noise specifications for selected field strengths (represented by ^{1}H frequency) are also shown for current (2014) conventional probes and helium and nitrogen cryogenically cooled probes (described later), including those for the current highest commercial field strength of 1 GHz. These data demonstrate that 40 years of combined instrument development has led to over a 10-fold gain in field strength and greater than 600-fold enhancement of ^{1}H signal-to-noise. Nevertheless, sensitivity limitations can still present a barrier when compared against other spectrometric methods and experimental technologies for enhancing this further continue to represent an active research area (see Section 12.8, for example).

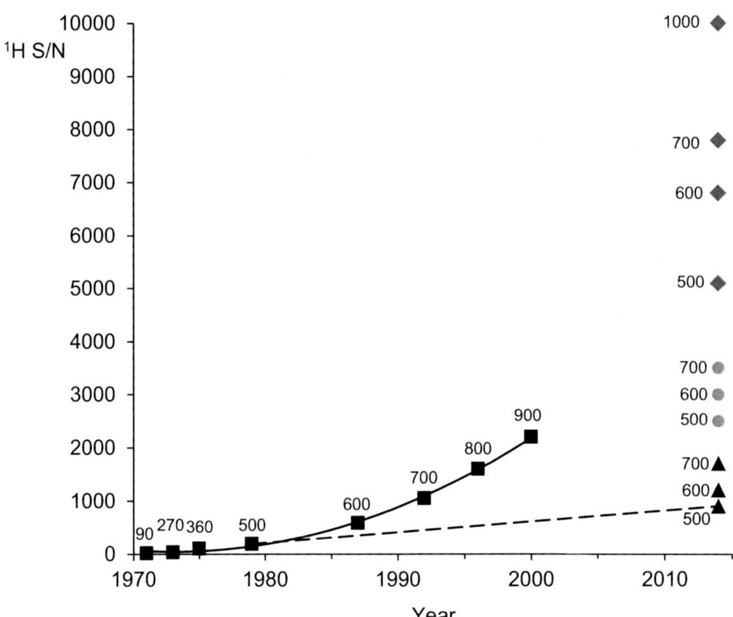

FIGURE 3.46 **Progress in instrument sensitivity.** The specified ^{1}H signal-to-noise ratios for ^{1}H observe probes at the time of magnet launch (indicated by ^{1}H frequency) as a function of year (squares). Also shown are data for current cryogenically cooled probes (helium-cooled: red diamonds, nitrogen-cooled: grey circles), and conventional probes (black triangles) at selected fields. *(Source: Data courtesy of Bruker Biospin in 2014.)*

The signal-to-noise of an NMR measurement depends upon many factors, but improving this ultimately comes down to either boosting the signal intensity or reducing the background noise. A general expression for this parameter for a conventional NMR spectrometer operating at ambient temperature may be formulated thus:

$$\frac{\mathrm{S}}{\mathrm{N}} \propto N A T_{\mathrm{S}}^{-1} B_0^{3/2} \gamma^{5/2} T_2^* (\mathrm{NS})^{1/2} \tag{3.9}$$

where N is the number of molecules in the observed sample volume, A is a term that represents the abundance of the nuclide, T_S is the temperature of the sample and surrounding rf coil, B_0 is the static magnetic field, γ represents the magnetogyric ratio of the nuclide, T_2* is the effective transverse relaxation time and NS is the total number of accumulated scans. Many of these factors are dictated by the properties of the nuclide involved, including the natural abundance, the magnetogyric ratio and relaxation behavior, so are not dictated by instrument design and are considered further in Section 4.4 where pulse techniques that exploit these parameters are introduced. Suffice it to say that the high magnetogyric ratio, near 100% natural abundance and favourable relaxation properties of the proton explains its popularity in high-resolution NMR spectroscopy.

The number of observable molecules N is, of course, dictated by the amount of material available but also by how much of this is held within the observe region of the detection coil, the so-called *active volume* (or *observe volume*) of the coil (Fig. 3.47). Any solution that sits outside this region will not induce a response in the coil and so does not contribute to

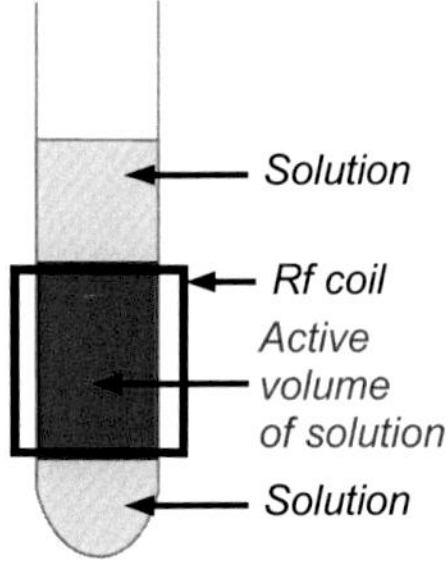

FIGURE 3.47 **The active (observe) volume of an NMR coil.** This is the region of the sample that sits within the coil and so contributes to the detected response.

the NMR signal but may be required to avoid lineshape distortions arising from magnetic susceptibility discontinuities too close to the coil. The susceptibility-matched ('Shigemi') tubes introduced in Section 3.3.3 enhance sensitivity by concentrating more of the sample within the active volume and represent a cost-effective approach to improving signal-to-noise.

From an instrumental point of view, sensitivity scales as $B_0^{3/2}$ (in fact, a more thorough theoretical analysis shows signal-to-noise should scale as $B_0^{7/4}$ [26]) and an increase in field strength has traditionally been one of the drivers of enhanced sensitivity, although is becoming more difficult to justify on these grounds alone given the enormous costs associated with the very highest field magnets. Increasingly, advances in probe design are now employed to provide substantial sensitivity gains at more affordable prices, as illustrated below. In addition, improved electronics have reduced system noise, especially that associated with the NMR signal preamplifier stage. Digital data handling in the form of oversampling reduces so-called quantisation noise in spectra (Section 3.2.6) adding further enhancement. These advances combined with the aforementioned refinements in probe technology have themselves led to progressive increases in signal-to-noise figures for conventional 5 mm probes, as illustrated by the data in Fig. 3.46.

3.4.2.2 Mass and Concentration Sensitivity

When discussing sensitivity in the context of real laboratory samples as opposed to instrument testing, it becomes useful to consider two definitions of this. The classic measurement of instrument sensitivity uses a fixed solution concentration (0.1% ethyl benzene in $CDCl_3$, equivalent to 14 mM) under standard conditions and so measures the *concentration sensitivity* of the system:

$$S_C = \frac{\text{signal-to-noise}}{\text{solution concentration}}$$

Concentration sensitivity may be improved by using greater solution volumes within the active volume of the coil. Thus, a wider or longer detection coil will hold more sample and show an enhanced concentration sensitivity and it follows that the use of wider sample tubes and probes is advantageous when sample solubility is a limiting factor. The use of longer coils in 5 mm probes has contributed to enhanced signal-to-noise performance specifications for certain probe designs. However, such gains may not be realised for *real* mass-limited laboratory samples. Such samples are typically not presented as fixed solution concentrations but are more commonly available as materials of fixed mass and can thus be prepared in any chosen sample volume, to the limit of sample solubility. In such cases it is more useful to refer to the *mass sensitivity* of the probe or system, a measure of signal-to-noise as a function of sample mass or the number of moles of material:

$$S_M = \frac{\text{signal-to-noise}}{\text{moles of material}}$$

A probe of higher mass sensitivity will provide a greater signal-to-noise ratio in a spectrum for a fixed amount of material and this better reflects the intrinsic detection sensitivity of the probe design. When making comparisons between the performance of different probes it is common practice to relate the results to those of a 'conventional' 5 mm probe operating at the same field strength, and this approach will be employed in the discussions that follow. The relative mass sensitivities of some commercially available probes are summarised in Table 3.5. Note that these are conservative figures

TABLE 3.5 Comparative ^1H Mass Sensitivities for Various Probe Configurations

Probe Diameter (Inverse Configuration)	Sample Volume (μL)	Relative Mass Sensitivity
5 mm	500	1
3 mm	150	1.5
1.7 mm	30	2
1 mm	5	4
5 mm cryogenic (He)	500	5
5 mm cryogenic (N_2)	500	2.5
1.7 mm cryogenic (He)	30	14
Capillary microflow	5	10

Cryogenic probes are defined as being cooled by gaseous helium or by liquid nitrogen
Data for entries 1–7 courtesy of Bruker Biospin.

that the manufacturer expects always to meet and it is not uncommon to exceed these, but they nevertheless provide us with an approximate guide to the sensitivity gains afforded by the differing probe configurations.

3.4.2.3 Microprobes

The mass sensitivity of an rf coil scales inversely with the diameter of the coil, d ($S/N \propto 1/d$), to first approximation, meaning that greater intrinsic sensitivity can be achieved by narrowing the coil. This is the basis for the development of so-called *microprobes*. This terminology is used somewhat loosely and there exists no formal definition of a microscale probe, but it is generally taken to mean a probe geometry that is small when compared with a 'standard' probe. Nowadays, the standard remains a probe designed to accept a 5 mm diameter maximum NMR tube, colloquially referred to as a 5 mm *probe*, so commercial microprobes would now include 3.0, 1.7 and 1.0 mm probes.

The mass sensitivity benefits of a microprobe relative to a larger diameter probe can be exploited for a fixed mass of sample through the enhanced signal-to-noise achievable or through reduced data collection times for data sets of equivalent signal-to-noise ratio, or perhaps as a balance between these. This of course demands that the solute can be dissolved in the reduced sample volumes employed (see later in the chapter). As a consequence, such probes enable data to be collected on reduced sample masses and so extend the operating range of NMR experiments for mass-limited samples. For example a microprobe that has a fivefold gain in mass sensitivity relative to a larger probe will provide fivefold improvement in the signal-to-noise ratio, which may be traded for a 25-fold reduction in data collection time for a fixed sample mass. Alternatively, this would allow sample masses to be reduced by a factor of 5 for data of comparable quality with that from the larger probe collected in a similar timeframe. An additional benefit from the reduced solvent volumes of the microprobes is the reduction in background solvent resonances and potentially also those of solvent impurities which can become apparent when dealing with very small sample masses. The smaller sample volumes also mean solvent suppression becomes much easier. The microprobes also provide reduced pulse widths leading to more uniform excitation over wider bandwidths. Indeed, it is generally the case that the length of a 90 degree pulse width for a given pulse power provides a direct indication of the detection sensitivity of the coil with smaller pulse widths correlating with higher sensitivity, and this itself can be a useful way of gauging a probe's performance. They may also demand lower pulse powers to prevent damage to more delicate rf coils, so even greater caution is required with such probes.

The availability of commercial tube-based microprobes came about in 1992 with the introduction of the 3 mm inverse probe, although much earlier work had demonstrated the potential for micro-scale probes [27]. The 3 mm probe with a 150 μL sample volume demonstrated an approximately twofold gain in signal-to-noise in ^{1}H-detected heteronuclear correlation spectra relative to a 5 mm probe with the same sample mass in 600 μL [28]. Subsequent to this in 1999 [29], a 1.7 mm probe [30] utilising a 30 μL sample demonstrated a further ca. twofold sensitivity gain in 2D correlation experiments relative to the 3 mm probe and more recently in 2002 [31] a 1 mm probe using only a 5 μL sample volume demonstrated a fivefold mass sensitivity gain in ^{1}H observation relative to a 5 mm probe. Owing to limitations in sample handling and the delicate nature of 1 mm tubes it seems likely that this will represent the smallest size for tube-based NMR probes, although smaller dimensions are feasible for microflow probes (as described later).

Fig. 3.48 compares the performance capabilities for ^{1}H observation of 5 mm and 1 mm triple-resonance inverse probes. Data were collected under identical conditions (single scan, 90 degree pulse excitation) on the same 500 MHz spectrometer for a 50 μg sample of sucrose in D_2O. The relative signal-to-noise of the anomeric resonance (not shown) is 7.8:1 and 39.5:1 for the 5 mm and 1 mm probes, respectively, indicating a fivefold sensitivity gain for the smaller diameter probe. The reduced solvent volume also attenuates the solvent background and this can also be a significant factor in the selection of smaller diameter tubes for mass-limited samples. Despite the miniscule samples and hence the small volume over which good field homogeneity is required, experience suggests the 1 mm probe resolution tends to be a little poorer than that of the larger geometry

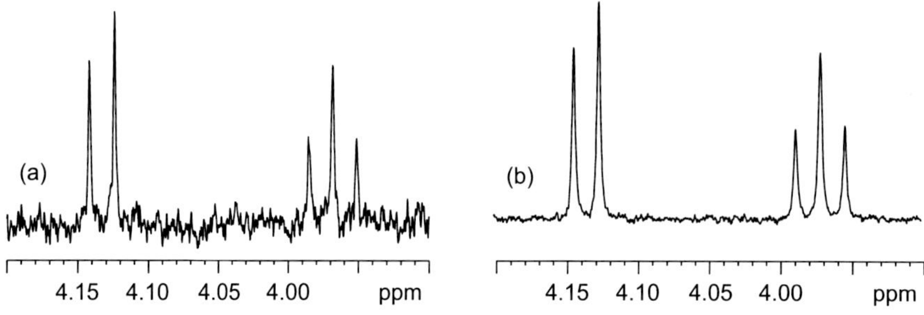

FIGURE 3.48 **Comparing probe mass sensitivity.** Partial ^{1}H spectra of 50 μg sucrose in D_2O recorded with probes of different dimensions: (a) 500 μL in 5 mm probe and (b) 5 μL in 1 mm probe. Spectrum (b) displays a fivefold gain in signal-to-noise relative to spectrum (a).

probes, and vendor probe specifications also support this observation. This most probably is due to the proportionately larger volume of glass present from the NMR tube relative to the solution within the coil and with the associated susceptibility discontinuities; susceptibility-matched 1 mm NMR tubes may yield still greater performance but are not currently available.

3.4.2.4 Micro-Flow Probes

An alternative approach to reducing sample volume requirements in probes is to avoid the use of sample tubes altogether and to use flow technology to place the sample within a fixed coil geometry. In the context of probe miniaturisation this is achieved using capillary flow cells ('capillary NMR') in which the rf coil is wound directly around the capillary tube itself. Such probes have been developed which have active volumes as little as 5 nL [32] although commercially available probes currently utilise active volumes of 2.5 μL that require working sample volumes of 5 μL and have a total *probe* volume of a mere 15 μL [33]. These probes employ rf coils that are surrounded by an inert perfluorinated fluid whose magnetic susceptibility matches that of the coil assembly, as needed to produce acceptable lineshapes. As there is no need to align the coil with the field axis for sample insertion, as is the case for tube-based designs, this allows the use of solenoid coils mounted perpendicular to the field which give an intrinsic sensitivity advantage of ca. twofold over the traditional, vertically aligned saddle coil designs [34]. In accord with this, the data in Table 3.5 indicate the capillary probe to demonstrate a ca. 10-fold mass sensitivity gain over a conventional 5 mm probe (or ca. fivefold if Shigemi tubes are employed [35]). Furthermore, it is possible to tune the single coil to multiple frequencies simultaneously making it possible to use the coil for both ^{1}H and ^{13}C observation without any loss of filling factor.

The flow cell design naturally requires a different approach to sample handling [36] with sample insertion performed via manual syringe injection, robotic injection or via direct hyphenation of the probe with capillary liquid chromatography (CapLC). The narrow capillaries (<100 μm diameter) may be prone to blockage, and effective sample filtration and the routine use of inline filters (2 μm) are essential. Arguably, the flow approach lacks the convenience of NMR tubes which are effective storage vessels should one wish to undertake further analysis of a sample without further preparative steps. The practicalities of employing the flowprobe in conjunction with CapLC has also been considered and contrasted with offline sample separation coupled with tube-based analysis with a cryogenic probe [22].

3.4.2.5 Cryogenic Probes [37,38]

A fundamentally different approach to improving detection sensitivity is to reduce the background noise in spectra by cooling both the probe rf detection coils and the preamplifier. This is achieved either through the use of cold helium gas (helium-cooled probes) or through cooling with liquid nitrogen (nitrogen-cooled probes). Helium-cooled variants utilise a closed-cycle system which cools the rf coils to typically 25 K and the preamplifier (which is housed within the body of the probe rather than being separate from this as with conventional probes) to around 70 K. This concept was proposed [26] and demonstrated experimentally [39,40] some years before the first commercial systems became available (in 1999), a testament to the demanding technical challenges in the development of these systems, not least of which is the need to maintain the detection coil at around 25 K while the adjacent sample, a matter of some millimetres away, remains at ambient temperature. The result of cooling the rf coils leads to a ca. twofold gain in signal-to-noise while cooling of the preamplifier leads to a similar gain, resulting typically in a fourfold signal-to-noise gain relative to a conventional probe of similar dimensions. The dewar assembly required for the cryoprobes actually results in a reduced filling factor that places limitations on the sensitivity improvement achievable through probe cooling; for example, a 5 mm cryogenic probe typically has the coil dimensions of a traditional 8 mm probe. Cooling of the rf coils is achieved through conduction from a cryocooled block on which the coils are mounted rather than by direct cooling of these by flowing He gas so as to avoid disturbance of the coils. Nitrogen-cooled probes are a more recent design and rely on the controlled transfer of liquid nitrogen direct to the probe, allowing the rf coils and the in-built preamplifier to operate at ~85 K. These temperatures mean the overall sensitivity gains do not match those of helium-cooled variants and are more typically two to threefold higher than room temperature probes. This is still a significant and experimentally useful gain, potentially reducing experiment times by ca. 10-fold. They benefit from being cheaper and simpler to install relative to their helium-cooled counterparts.

The gains arising from cryogenic cooling of probes may be described more formally by considering the noise contributions arising within the hardware. This may be summarised as:

$$\mathrm{S/N} \propto \frac{1}{\sqrt{T_c R_c + T_a (R_c + R_s) + T_s R_s}} \tag{3.10}$$

where T_c and R_c represent the temperature and resistance of the coil, T_s is the sample temperature, R_s is the resistance generated in the coil by the sample itself (the 'sample resistance'), and T_a is the effective noise temperature of the preamplifier.

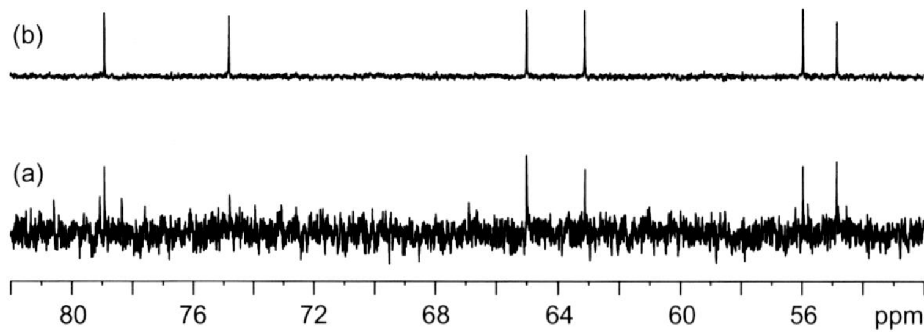

FIGURE 3.49 **Comparison of the carbon-13 sensitivity performance of (a) a conventional room temperature probe and (b) a helium-cooled cryogenic probe operating at 500 MHz (^{1}H).** Data were collected on the same sample under identical experimental conditions using a broadband observe probe tuned to carbon-13 and a cryogenic carbon-13 observe probe, respectively.

By cooling both the coils and the preamplifier the first two terms of the denominator in Eq. 3.10 are reduced and so contribute to the higher sensitivity of cryogenically cooled probes. The third term, representing noise generated by the sample, becomes important for ionic solutions [41], as discussed later, and is of greater significance for cryogenic probes than conventional probes because of the smaller contributions of the first two terms. The comparison of cryogenic and conventional probes in Table 3.5 demonstrates sensitivity gains that are (at least) fourfold, equating to a minimum 16-fold time saving for a given sample mass or the ability to collect data on fourfold less material in a given data acquisition period. This advantage is demonstrated experimentally in Fig. 3.49 for a ^{13}C-optimised helium-cooled cryogenic probe for which there is observed a sixfold gain in signal-to-noise ratio relative to a room temperature probe operating under otherwise identical experimental conditions. While the presence of some resonances is dubious in trace (a) of Fig. 3.49, all are clearly revealed in trace (b). As the performance gain is not dependent upon increased sample concentration, as is the case with microprobes, there is a simultaneous gain in both concentration and mass sensitivity with this technology. Further enhancements of mass sensitivity may thus be achieved by combining miniaturisation with cryogenic technology in the form of microcryogenic probes, as is illustrated by the figures for the 1.7 mm cryogenic probe.

It is also possible to manipulate the influence of sample noise to enhance sensitivity. Consider again the sample resistance R_s, which may be represented in the form:

$$R_s \propto \omega_0^2 l r^4 \sigma \tag{3.11}$$

where ω_0 is the observation frequency, l and r are the sample length and radius, respectively, and σ represents a term for the dielectric properties of the solution. This suggests that the noise contribution from the solution may be reduced by employing tubes of smaller diameter, the most common approach being a 3 mm tube used in a 5 mm probe. While this approach is valid in principle for both conventional and cryogenic probes, the advantages are more pronounced for the cooled probes because the noise contribution from the sample relative to that of the instrument is more significant, as described earlier. Despite the associated reduction in filling factor, the measure of how much of the rf coil volume is filled with sample, a signal-to-noise gain of ca. 1.5-fold has been demonstrated using this approach for a 3 mm versus a 5 mm tube containing identical sample masses when acquired in a 5 mm inverse helium-cooled cryogenic probe, [38] equating to a potential time-saving factor of ~2.2. Personal experience suggests that comparable results to the 3 mm tubes are obtained for identical sample masses in both 2 mm and 1 mm diameter tubes, suggesting further possible gains in noise reduction are offset by the losses in filling factor. Greater gains may be anticipated, however, by the use of 3 mm matched 'Shigemi' tubes. For situations where sample concentration cannot be increased, this method is clearly inappropriate. It has been demonstrated that the contribution from sample noise may also be diminished through the use of NMR tubes with rectangular instead of circular cross-sections with the longer transverse axis of the tube aligned along the rf magnetic field [42]. Such tubes are now commercially available with typical dimensions of 6 × 3 mm and demonstrate a ~20% gain in signal-to-noise versus a 5 mm tube. Such gains are most pronounced for samples of high ionic strength which compromise the performance of cryogenic probes because of the associated increase in the R_s term in Eq. 3.10. The most common examples of this are the buffered aqueous solutions widely employed in biomolecular studies. While higher salt concentrations generally lead to decreased performance, the nature of the buffer itself can also have a significant effect and the behaviour of many common buffers has been assessed in detail in this context [41]. Such adverse effects are generally less pronounced for organic solvents, although losses may still arise for the more polar solvents, especially methanol and dimethylsulphoxide. Finally, note also that Eq. 3.11 predicts that the use of cryogenic probe technology becomes less efficient at higher field strengths and for higher γ nuclides due the to ω_0 dependence of R_s. Thus, despite their high profile for biomolecular studies, the midrange

FIGURE 3.50 **A helium-cooled cryogenic probe installation.** The cryogenic bay (a) contains the cold expansion head that generates cold helium gas which passes along the insulated transfer line (b) to the probe. The vacuum line (c) maintains the cryogenic probe vacuum, and the base of the probe (d) houses the cold preamplifier stages. At the rear can be seen the high-pressure lines (e) that carry helium gas to/from the remote compressor.

frequencies (400–700 MHz) routinely employed for chemistry combined with the prevalence of organic solvents and the importance of ^{13}C NMR might suggest this to be the home territory of cryogenic probe technology.

Despite the obvious benefits of these high-sensitivity probes, there are also, inevitably, some drawbacks in their use. These include high initial purchase costs for the complete system, which can be over five times those of a conventional probe, the running and maintenance costs and the need to accommodate the additional instrumentation to operate the cryogenics. A complete system for the helium-cooled version comprises the cold probe, the cryogenic platform or bay that directly supplies the probe with the cooled helium gas and acts as the master controller for the complete assembly, a remote helium gas compressor unit and some means of dissipating the considerable quantities of heat generated by the closed-cycle cooling system (Fig. 3.50). The cryogenic bay generates the cold helium gas through a compression/expansion cycle, producing the characteristic 'chirping' associated with these cryogenic probes. The heat generated by this refrigeration process is removed either through the use of a chilled water supply or by external air-cooled assemblies as used in air-conditioning systems. Selecting the most appropriate method can be a critical aspect of site-planning these installations. The water-cooled systems may provide the convenience of local access but the instrument is then beholden to the demands on the building-wide installation, while the external compressor systems require suitable routing either out of the building or into internal space with sufficient cooling capacity. Nitrogen-cooled probes are rather simpler in design, requiring a liquid nitrogen vessel and a controller system that regulates the liquid nitrogen supply to the probe, so are easier to install and maintain.

In either case, the probe contains indirectly cooled rf coils and also houses cooled preamplifiers that are fed with the same cold cryogen flow. The internal dewared probe assembly demands continuous pumping of the probe vacuum from the platform and/or the use of internal ion pumps to maintain cryogenic efficiency. The use of these cold probes also imposes different operational protocols, chiefly because of the times required to cool and warm the probes which are typically some hours. This, combined with the more complex nature of the probe and its installation, has led to most cryogenic probe assemblies being considered as permanent fixtures within an instrument, and probe removal being reserved for maintenance procedures only.

In summary, it is apparent that there now exist a range of approaches to improving the sensitivity of NMR measurements using modern probe and tube technologies as an alternative to seeking higher field strengths. These often represent more economical and more practical routes to extending the range of samples that can be studied through probe miniaturisation and cryogenic cooling, through the appropriate use of microtubes and susceptibility-matched tubes, or indeed by a combination of these methods. Clearly, the adoption of appropriate probe technologies offers the chemist exciting new opportunities for the investigation of mass-limited samples.

3.4.3 Tuning the Probe

The NMR probe is a rather specialised (and expensive) piece of instrumentation whose primary purpose is to hold the transmit and receive coils as close as possible to the sample to enable the detection of weak NMR signals. For the coils to be able to transmit rf pulses to the sample and to pick up the NMR signals efficiently, the electrical properties of the coil circuit should be optimised for each sample. The adjustments are made via variable capacitors which sit in the probehead a short distance from the coil(s) and comprise the tuning circuitry. There are two aspects to this optimisation procedure known as *tuning* and *matching*, although the whole process is more usually referred to as 'tuning the probe'. The first of these, as the name implies, tunes the coil to the radiofrequency of the relevant nucleus and is analogous to the tuning of a radio receiver to the desired radio station. A poorly tuned probe will lead to severe degradation in sensitivity, just as a radio broadcast becomes swamped with hissing noise. The second aspect aims to equalise (or match) the impedance (total effective resistance to alternating current) of the coil/sample combination with that of the transmitter and receiver so that the maximum possible rf energy can pass from the transmitter into the sample and subsequently from the sample into the receiver. As electrical properties differ between samples, the optimum tune and match conditions will also vary and so require checking for each new sample. These differences can be largely attributed to differences between solvents with the most significant changes occurring between low-polarity organic solvents and ionic aqueous solutions.

Probe tuning is necessary for a number of reasons. Other than the fundamental requirement for maximising sensitivity, it ensures pulse widths can be kept short which in turn reduces off-resonance effects and minimises the power required for broadband decoupling. A properly tuned probe is also required if previously calibrated pulse widths are to be reproducible, an essential feature for the successful execution of multipulse experiments.

3.4.3.1 Tuning and Matching

The process of probe tuning involves applying rf to the probe, monitoring the response from it by some suitable means and making adjustments to the capacitors in the probehead (via long rods that pass through its base) to achieve the desired response. In the case of broadband probes, which may be tuneable over very wide–frequency ranges, the capacitors may even need to be physically exchanged, either by removing them from the probe altogether or by means of a switching mechanism held within the probe. Various procedures exist for monitoring the response of the probe but the most useful and the one supplied widely on modern instruments uses a frequency sweep back and forth over a narrow region (of typically a few megahertz) about the target frequency during which the probe response is compared with that of an internal 50 Ω reference load (all NMR spectrometers are built to this standard impedance). The display then provides a simultaneous measure of the tune and match errors (Fig. 3.51a), allowing one to make interactive adjustments to the tuning capacitors to achieve the desired result (Fig. 3.51b); the response reaches a minimum when the probe is matched to a 50 Ω load. With this form of display it is possible to see the direction in which adjustments should be made to arrive at the correct tune and match condition. It is important to note that the tune and match controls are not mutually exclusive and adjustments made to one are likely to alter the other, so a cyclic process of tune, match, tune, etc. will be required to reach the optimum.

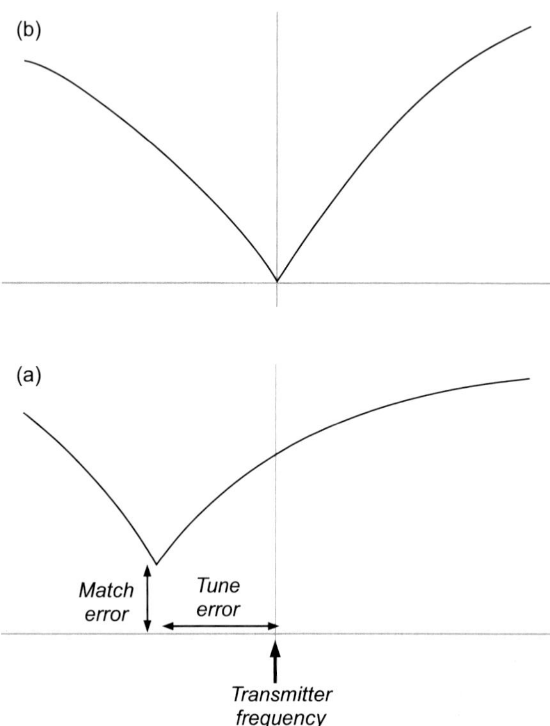

FIGURE 3.51 **Tuning and matching an NMR probehead.** The dark line represents the probe response seen for (a) a mistuned and (b) a correctly tuned probehead.

This process is clearly amenable to automation under computer control, and auto-tune configurations are now widely available through two distinct approaches. The first uses remote motorised units that connect to the probe-tuning rods via flexible cables which drive this process and have the advantage of being compatible with standard probes. The second, more sophisticated, approach employs motors built into the base of the probe itself that perform the optimisation but have the limitation of being incompatible with older probe designs. The fact that the drive units operate within an intense magnetic field also poses rather severe technical challenges that have been overcome by the use of rare earth metals in the motorised units. Whichever approach is adopted the benefits of auto-tune probes means that optimum tuning performance can be obtained under fully automated operation, a wide range of nuclei likewise become accessible without operator intervention (such as with broadband auto-tuneable probes), and for manually operated instruments this provides the ease and convenience of not having to crawl under magnets to tune, a sometimes awkward and potentially hazardous process.

The method for probe tuning on older spectrometers that are unable to produce the frequency sweep display is to place a *directional coupler* between the transmitter/receiver and the probe and to apply rf as a series of very rapid pulses. The directional coupler provides some form of display, usually a simple meter, which represents the total power being reflected back from the probe. The aim is to *minimise* this response by the tuning and matching process so that maximum power is able to enter the sample. Unfortunately with this process, unlike the method described earlier, there is no display showing errors in tune and match separately, and there is no indication of the direction in which changes need be made; one simply has an indication of the overall response of the system. This method is clearly the inferior of the two, but may be the only option available.

When more than one nuclear frequency is of interest, the correct approach is to make adjustments at the *lowest* frequency first and work up to the highest. Tuning will also be influenced by sample and probe temperature and must be checked whenever changes are made. If large temperature changes are required it is wise to quickly recheck the tuning every 10 or 20 degrees so that one never becomes too far from the optimum; this is especially important if using the directional coupler method. Where the spectrometer is used in an open-access environment, where interaction with the spectrometer is kept to a minimum or where the instrument runs automatically, probe tuning for each sample may not be viable (unless an auto-tuneable probe is installed), in which case it is appropriate to tune the probe for the most frequently used solvent, and accept some degradation in performance for the others. Different solvents may require different pulse width calibrations under these conditions.

Finally in this section we consider one particular situation in which it is beneficial to deliberately detune a probe. When performing studies in *protonated* water the linewidth of the solvent resonance is broadened significantly in a well-tuned probe, because of the phenomenon of *radiation damping* [43]. This is where the FID of the solvent decays at an accelerated rate because the relatively high current generated in the coil by the intense NMR signal itself produces a secondary rf field which drives the water magnetisation back to the $+z$-axis at a faster rate than would be expected from natural relaxation processes alone. The rapid decay of the FID in turn results in broadened water resonance. This only occurs for very intense resonances and has greatest effect when the sample couples efficiently with the coil (that is, when the probe is well tuned). Detuning the probe a little provides a sharper (and weaker) water resonance whose lineshape gives a better indication of the field homogeneity. Retuning the probe is essential for subsequent NMR observations employing solvent suppression schemes.

3.4.4 The Field Frequency Lock

Despite the impressive field stability provided by superconducting magnets they still have a tendency to drift significantly over a period of hours, causing NMR resonances to drift in frequency, leading to a loss of resolution. To overcome this problem some measure of this drift is required so that corrections may be applied. On all modern spectrometers the measurement is provided by monitoring the frequency of the deuterium resonance of the solvent. The deuterium signal is collected by a dedicated ^2H observe spectrometer within the instrument that operates in parallel with the principal channels, referred to as the *lock channel* or simply the *lock*.

3.4.4.1 The Lock System

The lock channel regulates the field by monitoring *dispersion*-mode deuterium resonance rather than the absorption mode signal that is usually considered in NMR, and aims to maintain the centre of this resonance at a constant frequency (Fig. 3.52). A drift in the magnetic field alters the resonance frequency and therefore produces an error signal that has both magnitude and sign (unlike absorption-mode resonance which has only magnitude). This then controls a feedback system which adjusts the field setting. The dispersion signal also has the advantage of having a rather steep profile, providing the greatest sensitivity to change. Monitoring the deuterium resonance also provides a measure of the magnetic field

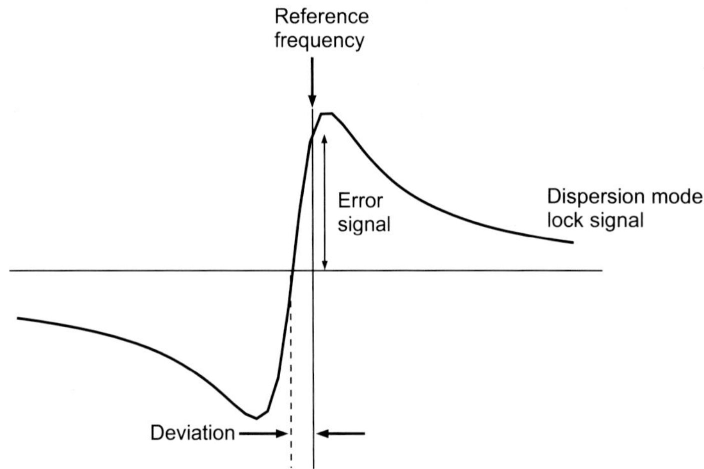

FIGURE 3.52 **The spectrometer lock system.** This monitors the dispersion-mode signal of the solvent deuterium resonance. A shift of the resonance frequency due to drift in the static field generates an error signal that indicates the magnitude and direction of the drift, enabling a feedback system to compensate for this.

homogeneity within the sample since only a homogeneous field produces a sharp, intense resonance. In this manner the lock signal may be used as a guide when optimising (shimming) the magnetic field (Section 3.4.5) where the absorption-mode signal is presented to the operator as the lock display. Since the best field frequency regulation is provided by the strongest and sharpest lock resonance, the more highly deuterated non-viscous solvents provide the best regulation. Water is especially poor since it tends to have a rather broad resonance because of exchange processes which are very dependent on solvent temperature and pH.

3.4.4.2 Optimising the Lock

The first procedure for locking is to establish the resonance condition for the deuterium signal, which involves altering either the field or the frequency of the lock transmitter. Of these two options the latter is preferred since it avoids the need for changing transmitter frequencies and is now standard on modern instruments. Beyond this, there are three fundamental probe-dependent parameters that need to be considered for optimal lock performance. The first of these is the lock transmitter *power* used to excite the deuterium resonance. This needs to be set to the highest usable level to maximise the ^{2}H signal-to-noise ratio but must not be set so high that it leads to lock *saturation*. This is the condition in which more energy is applied to the deuterium spins than can be dissipated through spin relaxation processes, and is evidenced by the lock level drifting up and down erratically. The saturation level can be determined by making small changes to the lock power and observing the effect on the lock signal. If the power is increased the lock signal should also increase, but if saturated it will then drift back down. Conversely, if the power is decreased, the lock level will drop but will then tend to creep up a little. The correct approach is to determine the point of saturation and then to be sure you are operating below this before proceeding. The second feature is the lock *gain*. This is the amplification that is applied to the detected lock signal, and is generally less critical than the power although it should not be so high as to introduce excessive lock noise. The final feature is the lock *phase*. We have already seen that field frequency regulation relies on monitoring the pure dispersion-mode lineshape and this is present only when the resonance has been phased correctly. This is the case when the *observed* lock signal produces the maximum intensity as the lock phase is adjusted since only then does this have the required pure absorption lineshape.

3.4.4.3 Acquiring Data Unlocked

In cases where deuterium is not available for locking it may be necessary to acquire data 'unlocked'. This may be necessary if one wishes to observe deuterium itself or if the optimum solvent is not available in deuterated form, or is simply too expensive. It can also be advantageous when reaction monitoring to directly collect the NMR spectrum of aliquots taken from the reaction vessel without addition of deuterated solvents, as in the no-deuterium proton NMR (no-D NMR) approach [44]. The receiver dynamic range of modern instruments is sufficient to allow for the observation of solute proton resonances even in the presence of solvent resonances that may be two to three orders of magnitude more intense, making the no-D NMR method straightforward to implement. Similarly, running unlocked is perfectly feasible over limited time periods, the duration of which will depend on the drift rate of the magnet. For proton experiments of a few minutes this is negligible. However, for data collection that must run for very many hours for reasons of sensitivity, then a useful approach is to acquire a series of FIDs, each being collected over a shorter time period – say, one hour. Drift within each data set will then be essentially negligible whereas any drift occurring over the whole duration of the experiment can be corrected by adding *frequency domain spectra* for which any frequency drift has been compensated manually by internal spectrum referencing of each. The addition then provides the required enhancement of signal-to-noise without the deleterious effects of field drift. In the absence of lock resonance, it is also necessary to shim samples (see later in the chapter) by an alternative means which is most conveniently done by direct optimisation of the FID or spectrum lineshape observed during acquisition, as described in the following section.

3.4.5 Optimising Field Homogeneity: Shimming

NMR experiments require a uniform magnetic field over the whole of the NMR sample volume that sits within the detection coil. Deviation from this ideal introduces various lineshape distortions, compromising both sensitivity and resolution. Thus, each time a sample is introduced into the magnet it is necessary to 'fine-tune' the magnetic field, and a few minutes spent achieving good resolution and lineshape is time well spent. For anyone actually using an NMR spectrometer, competence in the basic level of field optimisation is essential, but even if you only need to *interpret* NMR spectra, perhaps because someone else has acquired the data or if the whole process is performed through automation, then some understanding of the most common defects arising from remaining field inhomogeneities can be invaluable.

3.4.5.1 The Shim System

Maintaining a stable magnetic field that is uniform to one part in 10^9 over the active volume within modern NMR probes (typically 0.1–1.5 cm^3) is extremely demanding. This amazing feat is achieved through three levels of field optimisation. The first lies in the careful construction of the superconducting solenoid magnet itself, although the field homogeneity produced by these is rather crude when judged by NMR criteria. This basic field is then modified at two levels by sets of 'shim' coils. These coils carry electrical currents that generate small magnetic fields of their own which are employed to cancel remaining field gradients within the sample (in fact, "shims" is the term for small wedges of metal used in engineering to make parts fit together, and were originally used in the construction of iron magnets to modify the position of poles to adjust the field; still in the present day where superconducting magnets dominate, this name permeates NMR, as does the term 'shimming', referring to the process of field homogeneity optimisation.) Note that some (older) texts may refer to shimming as 'tuning', which is now reserved exclusively for processes involving radio frequencies; for example one may *shim* a magnet, but will *tune* a probe. Superconducting shim coils sit within the magnet cryostat and remove gross impurities in the magnet's field. The currents are set when the magnet is first installed and do not usually require altering beyond this. Room temperature shims are set in a former which houses the NMR probe itself, the whole assembly being placed within the bore of the magnet such that the probe coil sits at the exact centre of the static field. These shims (of which there are typically around 20–40 on a modern instrument) remove any remaining field gradients by adjusting the currents through them, although in practice only a small fraction of the total number need be altered on a regular basis (see the following section).

The static field in vertical bore superconducting magnets also sits vertically and this defines, by convention, the z-axis. Shims that affect the field along this axis are referred to as axial or Z shims, whereas those that act in the horizontal plane are known as radial or X/Y shims (Table 3.6). When acquiring high-resolution spectra it is traditional practice to spin the sample (at about 10–20 Hz) about the vertical axis. This has the effect of averaging field inhomogeneities in the X–Y plane, so improving resolution. This averaging means that adjustments to shims containing an X or Y term must be made when the sample is static, hence these shims are also commonly referred to as 'non-spinning shims'. Modern shim sets are capable of delivering non-spinning lineshapes that almost match those when spinning, and it is becoming increasing common *not* to spin samples. For multidimensional studies this is certainly the case, since sample spinning can introduce modulation effects to the acquired data, leading to unwanted artefacts particularly in the form of so-called t_1 noise (see Section 5.2.3).

3.4.5.2 Shimming

In order to achieve optimum field homogeneity, high-quality samples are essential. The depth of a sample also has a considerable bearing on the amount of Z shimming required, which can be kept to a minimum by using solutions of similar depth each time. Most spectrometers possess software that is capable of carrying out the shimming process automatically, and clearly this is essential if an automatic sample changer is used. However, such systems are not infallible and can produce spectacularly bad results in some instances. Here, reproducible sample depths are vitally important for auto-shimming procedures to be successful and to reach an optimum rapidly. It is also crucial for the whole of the sample to be at thermal

TABLE 3.6 Shim Gradients Found on High-Resolution Spectrometers

Shim Gradient	Gradient Order	Principal Interacting Shim Gradients	Shim Gradient	Gradient Order
Z^0 (the main field)	0	—	XYZ^2	4
Z^1 (Z)	1	—	$(X^2–Y^2)Z^2$	4
Z^2	2	Z	X^3Z	4
Z^3	3	Z, [Z^2]	Y^3Z	4
Z^4	4	Z^2, Z^0,[Z, Z^3]	XZ^4	5
Z^5	5	Z^3, Z, [Z^2, Z^4]	YZ^4	5
Z^6	6	Z^4, Z^2, Z^0, [Z, Z^3, Z^5]	XYZ^3	5
X	1	Y, [Z]	$(X^2–Y^2)Z^3$	5
Y	1	X, [Z]	XZ^5	6
XZ	2	X, [Z]	YZ^5	6
YZ	2	Y, [Z]	XYZ^4	6
XY	2	X, Y	$(X^2–Y^2)Z^4$	6
$X^2–Y^2$	2	XY, [X, Y]	XYZ^5	7
XZ^2	3	XZ, [X, Z]	$(X^2–Y^2)Z^5$	7
YZ^2	3	YZ, [Y, Z]		
XYZ	3	XY, [X, Y, Z]		
$(X^2–Y^2)Z$	3	$X^2–Y^2$, [X, Y, Z]		
X^3	3	X		
Y^3	3	Y		
XZ^3	4			
YZ^3	4			

Lower-field instruments (< 500 MHz) may utilise only 20 or so gradients (such as those in the left panel), while higher-field spectrometers may employ in excess of 30. Shims up to third order are those most likely to need periodic optimisation as part of long-term spectrometer maintenance for which the most significant interacting shims are listed. Those shown in square brackets interact less strongly with the listed gradient while those that interact with Z^0 (the main field) may cause momentary disruption of the lock signal when adjusted.

equilibrium so that convection currents do not exist. For aqueous solutions away from ambient this may demand 10–20 min equilibration in the probe.

To provide an indication of progress when shimming, one requires a suitable indicator of field homogeneity. Essentially, there are three schemes that are in widespread use, all of which have their various advantages and disadvantages: (1) the lock level, (2) the shape of the FID and (3) the shape of the NMR resonance. The ultimate measure of homogeneity is the NMR resonance itself, since defects apparent in the spectrum can often be related directly to deficiencies in specific shim currents (as described later). Most often field homogeneity is monitored by the height of the deuterium lock resonance which one aims to maximise. While conceptually this is a simple task, in reality it is complicated by the fact that most shims interact with others. In other words, having made changes to one it will then be necessary to re-optimise those with which it interacts. Fortunately, shims do not influence all others, but can be subdivided into smaller groups which are dealt with sequentially during the shimming process. A detailed account of the shimming procedure has been described [45] and the fundamental physics behind field gradient shims has also been presented [46], but here we shall be concerned more with addressing the lineshape defects that are commonly encountered in the daily operation of an NMR spectrometer.

When shimming, it is not always sufficient to take the simplest possible approach and maximise the lock level by adjusting each shim in turn, as this is may lead to a 'false maximum', in which the lock level appears optimum yet lineshape distortions remain. Instead, shims must be adjusted interactively. As an example of the procedure that should be adopted, the process for adjusting Z and Z^2 shims (as is most often required) should be:

1. Adjust the Z shim to maximise the lock level, and note the new level.
2. Alter Z^2 so that there is a noticeable change in the lock level, which may be up or down, and remember the direction in which Z^2 has been altered.

3. Readjust Z for maximum lock level.
4. Check whether the lock level is greater than the starting level. If it is, repeat the whole procedure, adjusting Z^2 in the same sense, until no further gain can be made. If the resulting level is lower, the procedure should be repeated but Z^2 altered in the opposite sense.

If the magnetic field happens to be close to the optimum for the sample when it is initially placed in the magnet, then simply maximising the lock level with each shim directly will achieve the optimum since you will be close to this already. Here again a reproducible sample depth makes life very much easier.

Shimming is performed by concentrating on one interacting group at a time, always starting and finishing with the lowest order shim of the group. The principal interactions for selected shims are summarised in Table 3.6. Whenever it is necessary to make changes to a high-order shim, it will be necessary to readjust all the low-order shims within the same interacting group, using a similar cyclic approach to that described for the adjustment of Z and Z^2 earlier. Generally, the order of optimisation to be followed will be:

1. Optimise Z and Z^2 interactively, as described earlier. If this is the first pass through Z and Z^2, then adjust the lock phase for maximum lock level.
2. Optimise Z^3. Make a known change, then repeat step 1. If the result is better than previously, repeat this procedure, if not, alter Z^3 in the opposite sense and repeat step 1.
3. Optimise Z^4 interactively with Z^3, Z^2 and Z.
4. Stop the sample spinning (if applicable) and adjust Z to give the maximum response (this is likely to have changed a little as the position of the sample relative to the field will change). Adjust X and Y in turn to give the maximum response.
5. Optimise X and XZ interactively. Adjust Z to give the maximum response.
6. Optimise Y and YZ interactively. Adjust Z to give the maximum response.
7. Optimise XY interactively with X and Y.
8. Optimise X^2–Y^2 interactively with XY, X and Y.
9. Repeat step 1.

The higher the shim order, the greater the changes required and when far from the optimum shim settings, large changes to the shim currents may have only a small effect on the lock level and the shim response will feel rather 'sluggish'. When close to the optimum the response becomes very sensitive and small changes can have a dramatic effect. The above procedure should be sufficient for most circumstances and any field strength, unless the basic shims set have become grossly misset. If lineshape distortions remain then it may be possible to identify the offending shim(s) from the nature of the distortion (see the following section), allowing appropriate corrections to be applied.

3.4.5.3 Common Lineshape Defects

The NMR resonance lineshape gives the ultimate test of field homogeneity, and it is a useful skill to be able to recognise common distortions that are caused by errors in shim settings (Fig. 3.53). Thus, Z shims all influence the width of the NMR resonance, but in subtly different ways; impurities in even-order shims (Z^2, Z^4 and Z^6) will produce unsymmetrical distortions to the lines whereas those in odd-order shims (Z, Z^3, Z^5) will result in symmetrical broadening of the resonance. In any case, the general rule is that the higher the order of the shim, the lower down the resonance the distortions will be seen. Errors in Z^3 usually give rise to a broadening of the base of a resonance and, since a broad resonance corresponds to a rapid decay of the FID, such errors are sometimes seen as a sharp decay in the early part of the FID. Another commonly observed distortion is that of a shoulder on one side of a peak, arising from poorly optimised Z and Z^2 shims (this is often associated with reaching a 'false maximum' simply by maximising the lock level with each shim and is usually overcome by making a significant adjustment to Z^2, and following the procedure described earlier).

Errors in low-order X/Y shims give rise to the infamous 'spinning sidebands' (for a spinning sample). These are images of the main resonance displaced from it by multiples of the spinning frequency. Shims containing a single X or Y term produce 'first-order sidebands' at the spinning frequency from the main line whereas XY and X^2–Y^2 give second-order sidebands at double the spinning frequency. However, unless something has gone seriously amiss, you should not encounter more than first-order sidebands in everyday NMR at most, and these should certainly be no greater than 1% of the main resonance. If there is any doubt as to the presence of sidebands, a simple test is to alter the spinning speed by, say, 5 Hz and re-acquire the data; only the sidebands will have moved. If the sample is not spinning, errors in the low-order X/Y shims contribute to a general broadening of the resonance.

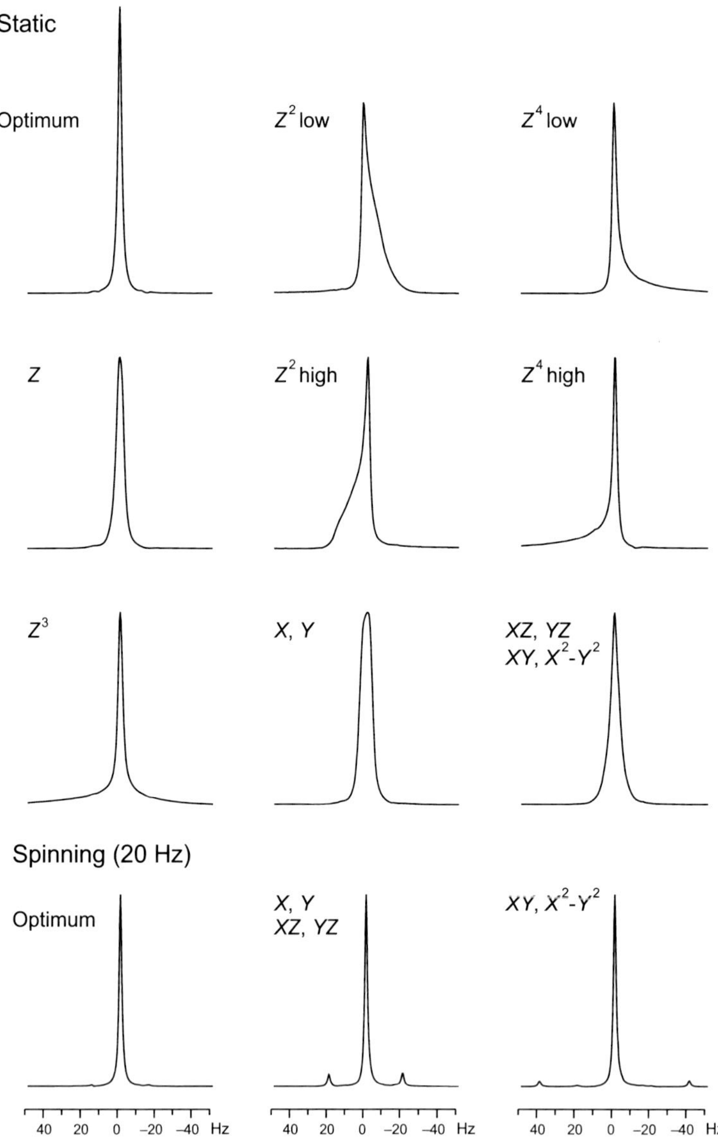

FIGURE 3.53 **Lineshape defects that arise from inappropriate settings of various shims.** These effects have been exaggerated for the purpose of illustration.

3.4.5.4 Shimming Using the FID or Spectrum

Although the lock level is used as the primary indicator of field homogeneity, it is not always the most accurate one and may not always be available (see Section 3.4.4.3). The lock level is dependent upon only one parameter: the height of the deuterium resonance. This, while being rather sensitive to the width of the main part of the resonance, is less sensitive to changes in the broad base of the peak. The presence of such low-level humps can be readily observed in the spectrum (particularly in the case of protons) but, for this to be of use when shimming, the spectrometer must be able to supply a real-time display of a single-scan spectrum so that changes to the shim currents can be assessed rapidly. With modern host computers, the FT and phase correction of a spectrum can be performed very rapidly, allowing one to correct for lineshape distortions in 'real time' as one shims during the repeated collection of single-scan FIDs. Alternatively, the shape and the duration of the FID may be used as a more immediate indicator of homogeneity. This approach works best when a singlet resonance dominates the spectrum (such as for aqueous solutions) for which the shape of the FID should be smooth exponential decay. Since with this method of shimming it is likely that changes to the shim currents will be made *during* acquisition of the spectrum (which will certainly lead to a peculiar lineshape) it is essential that one assesses a later spectrum for which there have been no adjustments during acquisition to decide whether improvements have been made.

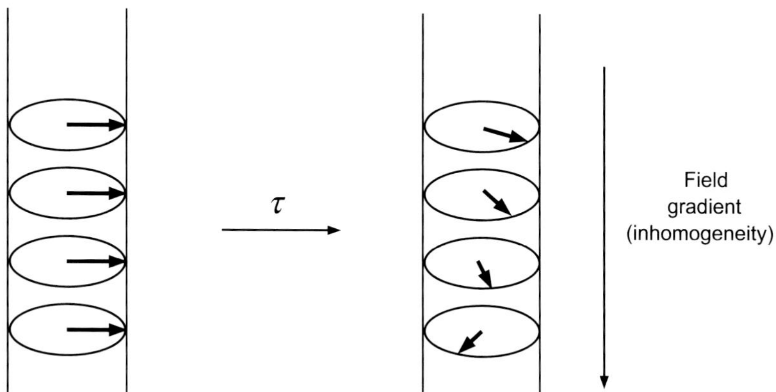

FIGURE 3.54 **Mapping field inhomogeneity.** Errors in the local static field along the length of a sample are encoded as signal phases when magnetisation is allowed to precess in an inhomogeneous static field for time τ. The resulting spatially dependent phase differences are used as the basis of gradient shimming.

3.4.5.5 Gradient Shimming

The most recent approach to field optimisation comes from the world of magnetic resonance imaging and makes use of field gradients to map B_0 inhomogeneity within a sample. This can then be cancelled by calculated changes to the shim settings [47,48]. The results that can be attained by this approach are little short of astonishing when seen for the first time, especially for anyone who has had to endure the tedium of extensive manual shimming of a magnet, and this method now enjoys widespread popularity.

The discussions in this section assume some understanding of the action of PFGs and the reader not familiar with these may wish to return to this section after they are introduced in Section 5.3. In any case, an appreciation of the capabilities of this method should be readily achieved from what follows. Here we shall consider the basis of the method with reference to the optimisation of Z shims which requires z-axis PFGs that are commonly found in modern probeheads (although the use of conventional shim assemblies to generate the appropriate gradients with so-called homospoil pulses has been demonstrated [49], which has the advantage of not requiring specialised gradient hardware). The underlying principle is that all spins throughout a sample contributing to a singlet resonance will possess the same precession (Larmor) frequency *only* if the static field is homogeneous throughout (of course, this is what we aim for when shimming). Any deviation from this condition will cause spins in physically different locations within the sample to precess at differing rates according to their *local* static field. If the excited spins are allowed to precess in the transverse plane for a fixed time period prior to detection, these differing rates simply correspond to different phases of their observed signals (Fig. 3.54). By detecting these signals in the presence of an applied field gradient, the *spatial distribution* of the spins becomes encoded as the *frequency distribution* in the spectrum allowing inhomogeneity (encoded as phase differences) to be mapped along the length of the sample in the case of z-axis gradients (or across the sample for x and y-axis gradients).

A suitable scheme for recording this is the gradient echo of Fig. 3.55 in which spins are first dephased by a PFG and later rephased (after a period of precession τ_1) to allow detection. The resulting spectrum is the 1D spatial profile (or image) of the sample (Fig. 3.56). Recording a second echo with delay τ_2 and taking the difference yields the *phase map* in which only free precession during the period (τ_2–τ_1) is encoded. The phase distribution in this profile therefore directly maps the inhomogeneity along the sample. The necessary corrections to shim currents to remove these inhomogeneities are calculated from a series of reference phase maps recorded with known offsets in each of the Z shims. Once these reference maps have been recorded for a given probe, they can be used for the gradient shimming of all subsequent samples, with the whole process operating automatically. A more recent approach to optimisation (available in Bruker's TOPSHIM routine

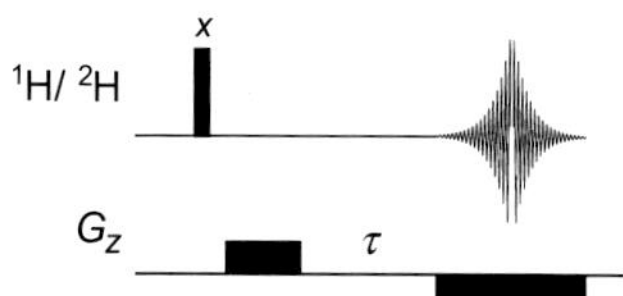

FIGURE 3.55 **A gradient echo sequence suitable for z-axis field gradient shimming.** PFGs provide the spatial encoding while the delay τ encodes static field inhomogeneity as the signal phase.

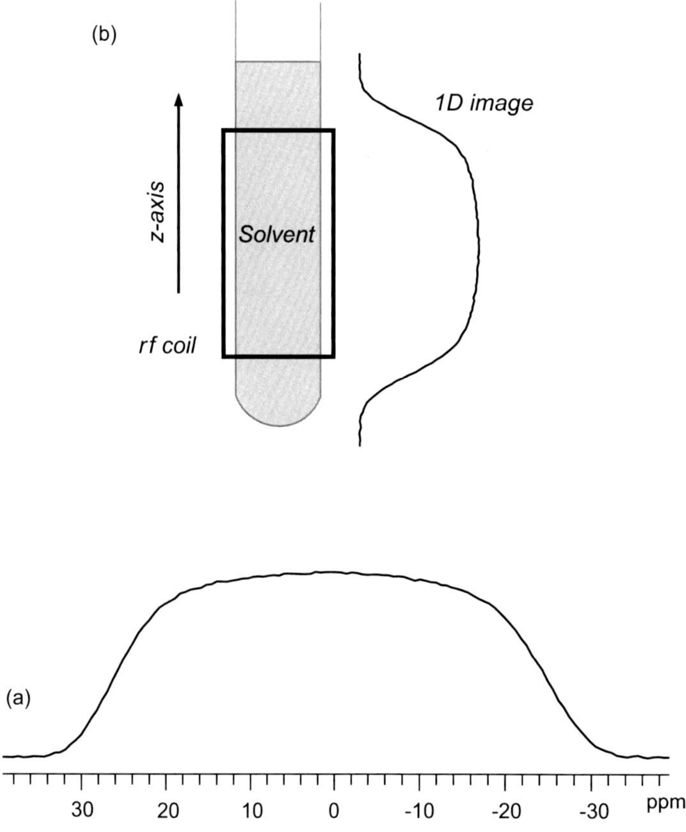

FIGURE 3.56 **1D imaging of a sample.** (a) The 1D z-axis deuterium image of DMSO (magnitude mode). (b) The frequency axis of this image encodes the spatial dimension along the length of the sample.

[48]) seeks to provide the best possible resonance lineshape rather than aiming for minimal B_0 field errors by simulating peak shapes from measured field maps and knowledge of the B_1 (radiofrequency) profile of the rf coil. This is better able to factor in field discontinuities at the ends of sample columns, for example and is claimed to provide a more robust and reliable approach with higher quality results; experience suggests this is indeed case.

The primary experimental requirement for gradient shimming is a sample containing a dominant strong, singlet resonance. A good candidate for proton observation is 90% H_2O, and although this is ideal for biomolecular studies, it is clearly of little use for the majority of solvents used in organic spectroscopy. An alternative in this case is to observe the deuterium resonance of the solvent [50] (which in most cases is also a singlet) using the lock channel of the probe. The potential problem then is one of sensitivity and the need for appropriate hardware to allow deuterium observation on the lock coil without manually recabling the instrument each time a sample is shimmed. The necessary lock channel–switching devices are commercially available or, more recently, dedicated deuterium transmitters are built into lock systems to allow direct ^{2}H observation (and decoupling).

The remarkable power of gradient shimming is illustrated in Fig. 3.57. The lower spectrum was recorded with the z–z^5 shims all set to zero while the upper trace was the result of only three iterations of deuterium gradient shimming using the dimethylsulphoxide solvent resonance. The whole process took less than 2 min without operator intervention. Although a rather extreme example, the capabilities of this approach are clearly evident and explain the now routine use of this methodology. This plays an especially valuable role in automated spectroscopy where irreproducible sample depths can lead to rather poor results with conventional simplex optimisation shim routines. The individual mapping of field errors within each and every sample overcomes these problems in a time-efficient manner.

3.4.6 Reference Deconvolution

Despite efforts to achieve optimum shimming, either manually or through automated procedures, there are invariably instances when the field homogeneity in a sample does not meet expectations, leading to lineshape distortions in the resulting spectrum. Reference deconvolution is a post-processing strategy that aims to correct such distortions and is now

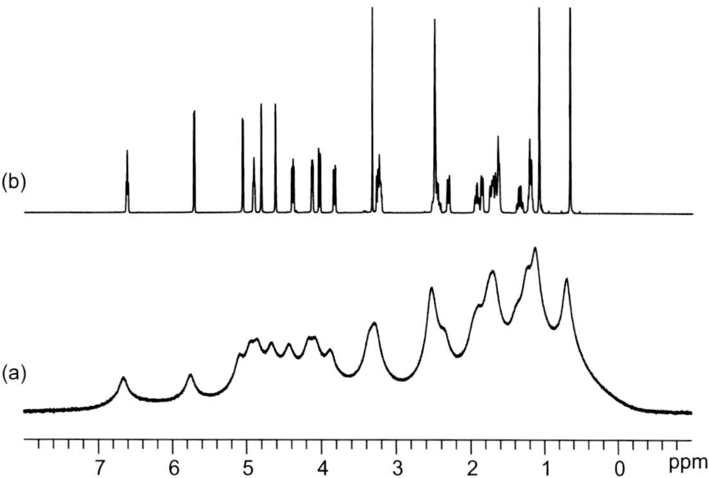

FIGURE 3.57 **Automatic deuterium gradient shimming.** Spectrum (a) was acquired with all z-axis shims set to zero. After less than 2 min gradient shimming of the z–z^5 shims (three iterations) spectrum (b) was obtained. The solvent was DMSO, and one scan was acquired for each deuterium gradient echo collected via the probe lock coil.

a standard feature of some NMR-processing software [51–54]. The process relies on the fact that shim errors will influence all resonances in the distorted spectrum in the same way. If the type and extent of lineshape error can be determined and a correction factor computed for a single resonance this adjustment may be applied to all others to yield a spectrum free from unwanted distortions. The principal limitation of the technique is in the selection of a suitable reference resonance with many common multiplet structures proving unsuitable since they give rise to null points in their free induction decays; a singlet is thus most appropriate, although multiplets have been employed [55]. The reference deconvolution procedure is summarised pictorially in Fig. 3.58. Having identified a suitable reference peak in the spectrum that is free from others, this is extracted (by zeroing all other data points) and its FID generated by inverse Fourier transformation (IFT) of the reduced spectrum. An idealised lineshape is then chosen and the corresponding FID similarly computed. This shape will most likely have a Lorentzian profile, corresponding to exponential decay of the FID, but others may also be selected including a Gaussian profile should resolution enhancement of the spectrum be desired. Comparison of the experimentally derived FID and the idealised FID is then made, a correction function computed and this applied to the original complete FID. This may then be transformed to yield the corrected spectrum.

The reference deconvolution procedure is not limited to addressing lineshape distortions in 1D spectra, having a number of potential applications in high-resolution NMR spectroscopy [53,54] and has been shown to be effective in improving NOE difference spectra [52] for t_1 noise reduction in 2D spectra [56,57] and in the production of high-resolution diffusion-ordered spectroscopy (DOSY) [58].

3.5 SPECTROMETER CALIBRATIONS

This section is primarily intended for those who need to set up experiments or those who have new hardware to install for which new calibrations are required. As with any analytical instrumentation, correct calibrations are required for optimal and reproducible instrument performance. All the experiments encountered in this book are critically dependent on the application of rf and gradient pulses of precise amplitude, shape and duration and the calibrations described below are therefore fundamental to the correct execution of these sequences. Periodic checking of these calibrations, along with the performance tests described in the following section, also provides an indication of the overall health of the spectrometer.

3.5.1 Radiofrequency Pulses

Modern multipulse NMR experiments are critically dependent on the application of rf pulses of known duration (the *pulse width*) that correspond to precise magnetisation tip angles, most frequently 90 and 180 degrees. Pulse width calibrations are normally defined for the 90 degree pulse (PW_{90}), from which all other tip angles may be derived. Those with PW_{90} in the microsecond range generally excite over a rather wide bandwidth and are termed *non-selective* or *hard* pulses while those in the millisecond range are of lower power and are effective over a much smaller frequency window and are thus termed *selective* or *soft* pulses. A third class of pulse is the *frequency-swept* or *adiabatic* pulse which can be effective over far

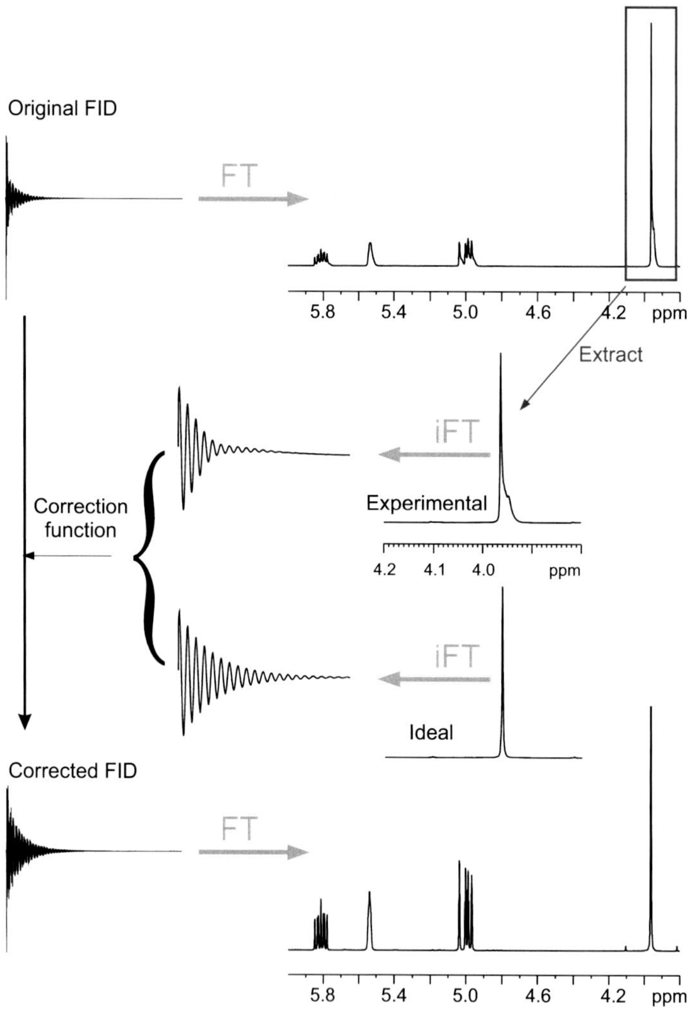

FIGURE 3.58 **Schematic representation of the reference deconvolution procedure.**

wider frequency bandwidths than simple hard pulses and have become especially important with higher field spectrometers. The implementation of selective and adiabatic pulses and their calibration is considered separately in Sections 12.4 and 12.2 respectively, while here we shall concentrate on the more widely used hard pulses, and on weaker rf fields used for decoupling purposes.

3.5.1.1 Rf Field Strengths

Calibration of a pulse width is equivalent to determining the radiofrequency (B_1) field strength of the pulse, and, in fact, it is often more useful to think in terms of field strengths than pulse widths when setting up experiments. For example excitation bandwidths and off-resonance effects are best considered with reference to field strengths, as are decoupling bandwidths. The relationship between pulse width and the rf field strength γB_1 is straightforward:

$$\gamma B_1 = \frac{1}{PW_{360}} \equiv \frac{1}{4PW_{90}} \text{ Hz} \tag{3.12}$$

Thus, a 90 degree pulse width of 10 μs corresponds to a field strength of 25 kHz, a typical value for pulse excitation on a modern spectrometer.

Although spectrometer transmitters are frequently used at full power when applying single pulses, there are many instances when lower transmitter powers are required, for example the application of decoupling sequences or of selective pulses). These lower powers are derived by attenuating the transmitter output, and the units used for defining the level of attenuation are the decibel (dB). This is, in fact, a measure of the ratio between two power levels, P_1 and P_2, as defined by:

$$\mathrm{dB} = 10\ \log_{10}\left(\frac{P_1}{P_2}\right) \tag{3.13}$$

When one speaks of attenuating the transmitter output by so many decibels one must think in terms of a change in power *relative to the original output*. In fact, it is more convenient to think in terms of changes in output voltage rather than power, since the rf field strength and hence pulse widths (our values of interest) are proportional to voltage. Since power is proportional to the *square* of the voltage ($P = V^2/R$, where R is resistance), we may rewrite Eq. 3.13 as:

$$\mathrm{dB} = 20\ \log_{10}\left(\frac{V_1}{V_2}\right) \equiv 20\ \log_{10}\left(\frac{PW_1}{PW_2}\right) \tag{3.14}$$

where PW_1 and PW_2 are the pulse widths for the same net tip angle at the two attenuations. Thus, if one wished to double a pulse width, an additional 20 log 2 dB (6 dB) attenuation is required. An alternative expression has the form:

$$+n\,\mathrm{dB} \Rightarrow PW \times 10^{n/20} \tag{3.15}$$

or in other words, the addition of n dB attenuation increases the pulse width by a factor of $10^{n/20}$. Some example attenuation values are presented in Table 3.7.

An illuminating experiment to perform, if your spectrometer is able to alter the output attenuation internally, is to determine the pulse width over a wide range of attenuations. A plot of pulse widths vs. $\log_{10}$(attenuation) should yield a single straight line over the full range. Discontinuities in the plot may arise when different power amplifiers come into use or when large attenuators are switched in place of many smaller ones. More recent spectrometers make use of linear amplifiers together with calibrated power correction lookup tables, providing pulse output powers that are linear over the whole attenuation range. This means pulse calibrations at different power settings become superfluous since all pulse widths can be calculated with Eq. 3.14 from a single, accurate high-power calibration, so it is worthwhile checking if your spectrometer is equipped for this. This also makes the application of more elaborate pulse profiles whose performance can be critically dependent on accurate pulse calibrations (such as the shaped pulses described in Section 12.4) far more straightforward since their calibrations may also be derived in this way.

3.5.1.2 Observe Pulses: High Sensitivity

We begin with the most basic NMR pulse calibration, that for the observed nucleus. When the spectrum of the analyte can be obtained in a single scan, or only a few scans, it is straightforward to calibrate the pulse width simply by following the behaviour of magnetisation as the excitation pulse is increased, with nulls in the signal intensity occurring with a 180 or 360 degree pulse. To perform any pulse calibration it is essential that the probe is first tuned for the sample, so that the results are reproducible. The transmitter frequency should be placed on-resonance for the signal to be monitored, to eliminate potential inaccuracies arising from off-resonance effects, and a spectrum recorded with a pulse width less than 90 degrees (say, 2–3 μs). The resonance is used to define the phase correction for all subsequent experiments, and is phased to produce the conventional positive absorption signal. The experiment is repeated with identical phase correction but with a progressively larger pulse width (Fig. 3.59). Passing through the 180 degrees null then provides the PW_{180} calibration, while going beyond this yields an inverted resonance, so it becomes clear when you have gone too far. Detecting the null precisely can

TABLE 3.7 Pulse Width Versus Attenuation

Attenuation (dB)	1	3	6	10	12	18	20	24
Pulse width factor	1.1	1.4	2.0	3.2	4.0	7.9	10	15.8

The additional attenuation of transmitter output levels increases the pulse width by the factors shown. Thus, adding 6 dB will double the pulse width whereas removing 6 dB will halve it.

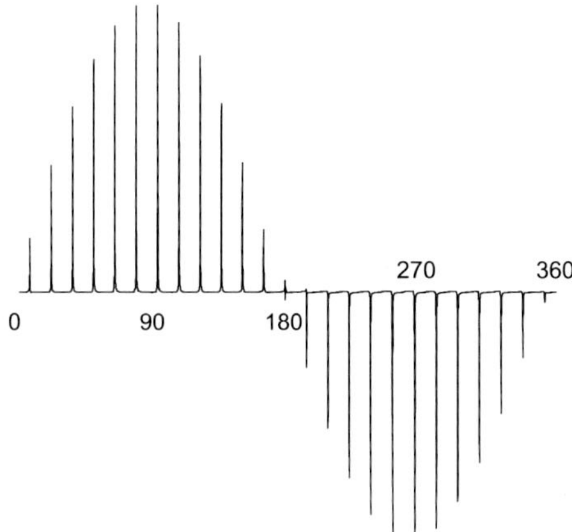

FIGURE 3.59 **Pulse width calibration for the observe channel.** A sequence of experiments is recorded with a progressively incremented excitation pulse. The maximum signal is produced by a 90 degrees pulse and the first null with a 180 degrees pulse. Either the 180 or 360 degrees condition can be used for the calibration (but be sure to know which of these you are observing!).

sometimes be tricky, as there is often some slight residual signal remaining, the exact appearance depending on the probe used, so in practice one aims for the minimum residual signal.

Between acquisitions there must be a delay sufficient for complete relaxation of the spins, and if signal averaging is used for each experiment, this must also be applied between each scan to obtain reliable calibrations. In proton spectroscopy, such delays are unlikely to be too much of a burden, but can be tediously long for slower relaxing spins. If you have some feel for what the pulse width is likely to be, perhaps from similar samples or a 'reference' calibration, then a better approach is to search for the 360 degrees null. Since magnetisation remains close to the +*z*-axis, its recovery demands less time and the whole process can be performed more quickly. The process can be further simplified by monitoring the FID alone, which also has minimum amplitude with the 360 degree pulse. In any case, it is wise to check the null obtained is in fact the one you believe it to be, as all multiples of 180 degrees give rise to nulls.

3.5.1.3 Observe Pulses: Single-Scan Nutation Spectroscopy

A more recent approach to pulse calibration allows this to be performed in a single scan and provides a rapid and convenient method for ^{1}H pulse calibration, in particular, that is also well suited to use under automation [59]. The method has most significance for inverse and cryogenic probes where pulse calibrations may vary significantly between samples. The measurement derives from a single-scan nutation experiment in which rf pulse is applied during each of the dwell periods of an FID acquisition between data point sampling in a manner similar to that used in the homonuclear decoupling experiments described in Section 4.2.2. No initial excitation pulse is required. Provided the rf pulses are applied on-resonance for a single peak the associated magnetisation vector is driven by each of the pulses away from its equilibrium position along the longitudinal axis as the acquisition proceeds; the magnetisation is said to *nutate* about the applied rf field. The transverse magnetisation generated by this process describes an oscillatory behaviour along one axis during data acquisition (Fig. 3.60a). One may imagine this oscillatory behaviour as being equivalent to the *net* magnetisation generated by the evolution of a doublet that is on-resonance in the rotating frame (Fig. 3.60b; see also Fig. 2.13). Indeed, FT of the nutation

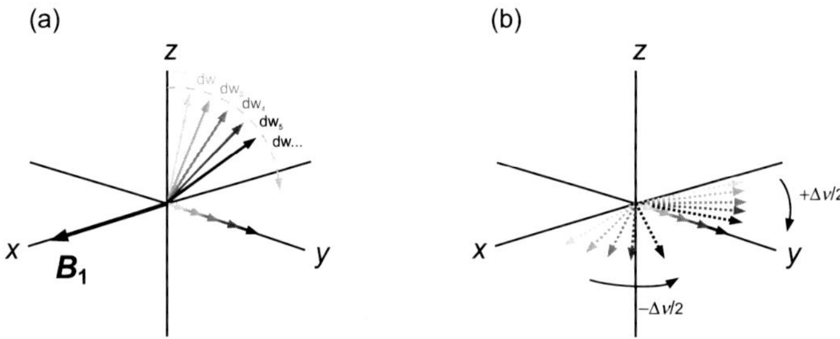

FIGURE 3.60 **The behaviour of an on-resonance magnetisation vector during the nutation experiment.** (a) As the rf field is applied during each of the FID dwell periods (dw_1, dw_2, dw_3, ...) the vector is driven stepwise about the B_1 field (shown in shades of grey). The sampled transverse component (likewise greyed) oscillates along the *y*-axis as data acquisition proceeds and the total nutation angle progressively increments. (b) Equivalent behaviour arises from the evolution of an anti-phase doublet of splitting $\Delta\nu$.

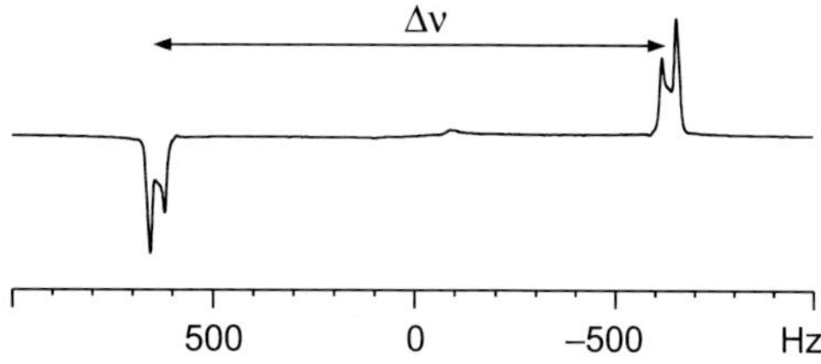

FIGURE 3.61 **The single-scan nutation spectrum of an aqueous sample (12 dB attenuation, 10% duty cycle).** The doublet splitting of 1268 Hz calibrates the high-power 90 degree hard pulse at 9.86 μs.

data leads to the generation of an anti-phase doublet in the resulting nutation spectrum with a splitting of $\Delta\nu$ Hz (Fig. 3.61). The oscillation frequency reflects the mean nutation frequency of the vector during data collection and equates to the frequency of evolution of each half of the hypothetical doublet $\Delta\nu/2$. This provides the mean γB_1 value of the applied rf and hence pulse width calibration under the conditions used.

To avoid possible probe damage and sample heating it is safe practice during the nutation experiment to reduce the rf amplitude from that used for hard pulses by typically 12 dB, corresponding to a fourfold increase in pulse width. It also becomes necessary to account for the fraction of time during which the rf pulse is applied within each dwell period of the FID, the so-called duty cycle *d*. Values may range from 10–80% (d = 0.1–0.8) and the measured rf amplitude must be corrected for this. With the low-power calibration complete the required hard-pulse calibration may be calculated directly when linearised rf amplifiers are in use, as found on modern instruments, otherwise additional calibrations may be required to correct for the power differences. Assuming amplifier linearisation and the use of 12 dB attenuation in the nutation experiment the high-power pulse width may be computed as:

$$t_{360} = \frac{d}{4(\Delta\nu/2)} \text{ or } t_{90} = \frac{d}{8\Delta\nu} \qquad (3.16)$$

The method is most appropriate when the spectrum is dominated by a single resonance, so has obvious application to samples in heavily protonated water but can be equally effective for organic solutions when a single dominant peak can be identified, such as a resolved methyl resonance. This must be placed on-resonance and the smallest frequency separation between the anti-phase peaks in the nutation spectrum is then used in the appropriate calculation. The whole procedure is most conveniently executed through appropriate software routines which may be supplied as standard by your instrument vendor.

3.5.1.4 Observe Pulses: Low Sensitivity

When the sample is too weak to allow its observation within a few scans, one possible approach is to use the method of signal averaging and to search for the 360 degree null as described earlier [60], although this could become a rather laborious affair. An alternative approach [61] requires that only two spectra are collected with the pulse width of the second being exactly double that of the first ($PW_2 = 2PW_1$). Once again, a sufficient delay between scans is required to avoid saturation effects and the spectrometer must be operating in 'absolute intensity' scaling mode so that the relative signal intensities from both experiments can be compared. Heteronuclear spectra should also be acquired without enhancement by the NOE, and the use of the inverse-gated decoupling scheme (Section 4.2.3) allows such data to be collected with broadband decoupling during acquisition to aid sensitivity. Following some simple arithmetic [61], it can be shown that:

$$\theta_1 = \cos^{-1}\left(\frac{I_2}{2I_1}\right) \qquad (3.17)$$

where I_1 and I_2 correspond to the measured signal intensities in the first and second experiment, respectively. The duration of the pulse in the first experiment corresponds to the pulse angle θ_1, from which the 90 degree pulse width can be derived. For optimum results, θ_1 should be greater than 30 degrees and θ_2, the tip angle in the second experiment, should be less than 180 degrees, so again some rough feel for the likely answer is required for this approach to be efficient.

A more convenient approach in many instances is to perform the calibrations not on the sample of interest itself but on a suitable sample made up for the purposes of calibration that is strong enough to permit observation of the 180/360 degrees null directly and quickly. The calibration can then be recorded and used in future experiments, as this should be quite reproducible if similar solvents are used and the probe correctly tuned. The largest discrepancies occur between aqueous and organic solvents, especially when the aqueous solutions are highly ionic, and separate calibrations may be required.

3.5.1.5 Indirect Pulses

In contrast to the above, the calibration of pulses for nuclei other than that being observed must be performed 'indirectly'. This is often referred to as calibration of the 'decoupler' pulses although this is something of a misnomer nowadays since the pulses may have nothing to do with decoupling itself but may be an integral part of the pulse sequence. Most commonly, the indirect channel is used in two modes: the first is to apply short, hard pulses typically at maximum power; the second is to apply a pulsed decoupling sequence for longer periods of time, at significantly less power. It is therefore commonplace to record indirect calibrations at these two power levels, for which the following method is equally applicable, and having performed the calibration at high power, low-power attenuation can be estimated from Eq. 3.14 prior to its accurate determination. Alternatively, the use of accurately linearised amplifiers can avoid the need for low-power calibrations altogether. More recent spectrometers make use of linear amplifiers together with calibrated power correction lookup tables, meaning the pulse output power is linear over the whole attenuation range. This means low-power calibrations become superfluous since all pulse widths can be calculated from single high-power calibration with Eq. 3.14, so it is worthwhile checking if your spectrometer is equipped for this. This also makes the application of more elaborate pulse profiles, such as the shaped and adiabatic pulses described in Sections 12.4 and 12.2, far more straightforward since their calibrations may also be computed for both observe and indirect pulses.

3.5.1.6 Indirect Pulses on High-Abundance Nuclides

Calibrations in which the indirect nuclide exists at high abundance are generally more straightforward because of the clear appearance of heteronuclear coupling. Thus, to relate the behaviour of the indirect spin A to that being observed X, one exploits the mutual heteronuclear scalar coupling which must exist between them J_{AX} for calibration to be possible. The simplest calibration sequence [62,63] suitable for spin-½ nuclei is shown in Fig. 3.62. The A and X pulses should be on-resonance, the X-pulse calibration must already be known, and the value of J_{AX} measured directly from the (coupled) X spectrum. The delay Δ is set to $1/2J_{AX}$ s for an AX group so that the X-spin doublet vectors are anti-phase after this evolution period. The subsequent A pulse renders the X magnetisation *unobservable* when $\theta = 90$ degrees as this generates pure heteronuclear multiple-quantum coherence which is unable to generate a detectable signal in the probe (see Section 5.3 for an explanation of this effect).

The approach is to begin with θ very small and to phase the spectrum so that the doublet lines are in anti-phase. As θ increases the doublet intensity will decrease and become zero when θ is 90 degrees, while beyond this the doublet reappears but with inverted phase (Fig. 3.63). If it is necessary to perform the calibration with an A_2X group, the delay Δ should be 1/4J and it is the outer lines of the triplet that behave as described above while the centre line remains unaffected. When calibrating lower powers for the purpose of broadband decoupling, it is usually more convenient to set the *duration* of the θ pulse according to the decoupler bandwidth required (Section 12.3) and to vary the output attenuation to achieve the null condition. As is often the way in experimental procedures, simplicity is best and indirect pulse calibrations can most conveniently be performed on strong calibration samples using the sequence of Fig. 3.62. For ^1H–^{13}C systems, $CHCl_3$ or (^{13}C-labelled) methanoic acid are convenient materials for organic and aqueous calibrations, respectively.

A readily available alternative that is particularly suitable for (although not restricted to) proton calibration in ^1H–X systems is the distortionless enhancement by polarisation transfer (DEPT) sequence (Section 4.4.3). This improves observation sensitivity by making use of polarisation transfer from ^1H to X and is dependent upon faster relaxing protons for repetition rates in signal averaging. The 90 degree proton pulse is achieved when XH_2 or XH_3 groups pass through a null whereas XH groups have maximum intensity at this point (Fig. 3.64). One caveat here is that for the purposes of pulse

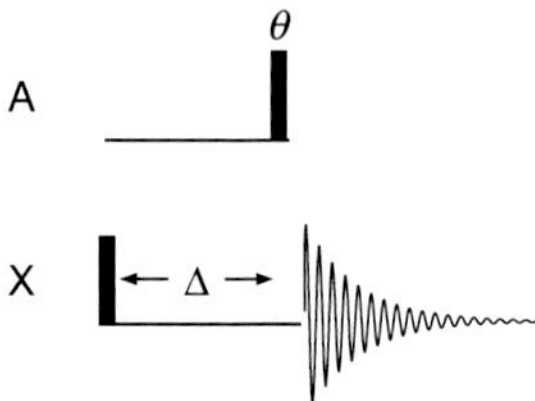

FIGURE 3.62 **Sequence for the calibration of 'indirect' transmitter pulses (those on nuclei other than that being observed).** The period Δ is set to $1/2J_{AX}$.

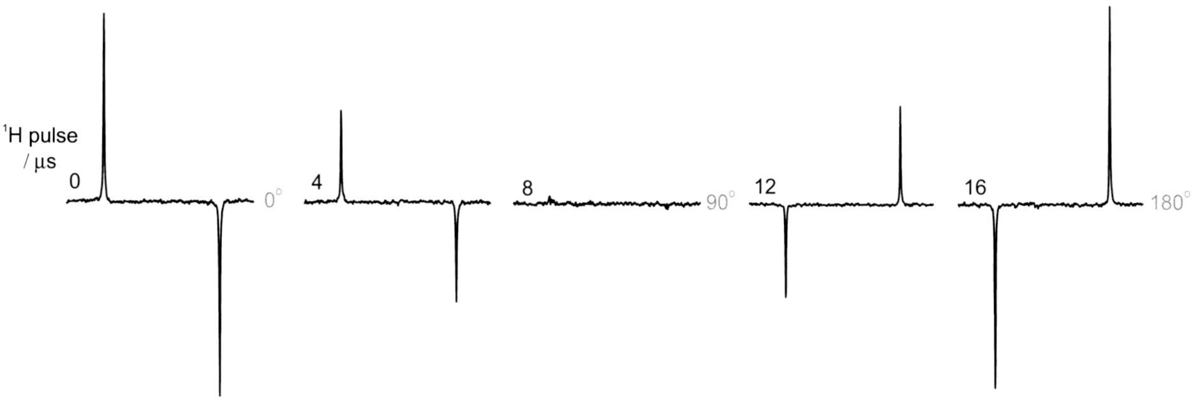

FIGURE 3.63 **Indirect calibration of the proton pulse width with carbon observation using the sequence of Fig. 3.62.** As the proton pulse width increases the carbon signal diminishes until it disappears at the ^{1}H (90 degrees) condition. Going beyond this causes the signal to reappear but with inverted phase (the same phase correction is used for all spectra). The sample is ^{13}C-labelled methanoic acid in D_2O.

calibration, *all* the proton pulses in the sequence need to be altered on each experiment, not just the final θ proton pulse (see the Section 4.4.3 for an explanation of this sequence).

3.5.1.7 Indirect Pulses on Low-Abundance Nuclides

When directly observing protons and indirectly calibrating nuclei of low abundance such as ^{13}C and ^{15}N (the so-called inverse configuration), the dominant resonance will be of no use and may interfere since it is the much less intense coupled satellites that must be employed for the calibration. Thus, the major ^{12}C line of $CHCl_3$ will be apparent as a large dispersive signal between the ^{13}C satellites of interest when using the sequence of Fig. 3.62, but this should not pose too great a problem so long as the sensitivity is sufficient for the satellites to be readily observed (Fig. 3.65). In cases when the major resonance

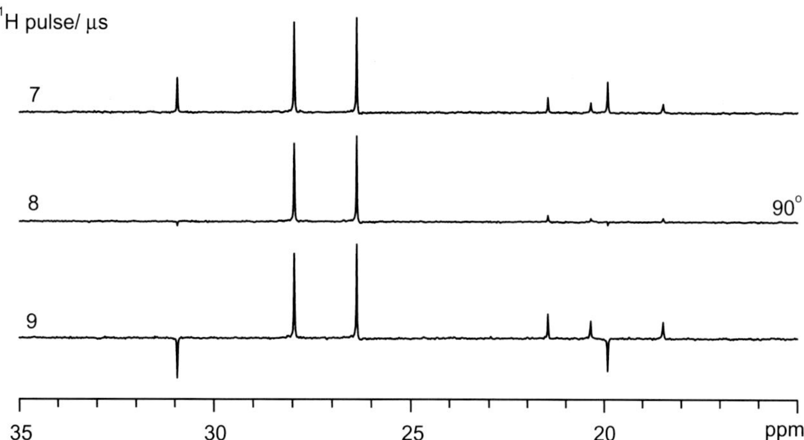

FIGURE 3.64 **Calibration of indirect proton pulses using the DEPT sequence.** When the proton editing pulse in the sequence is exactly 90 degrees, only methine resonances are apparent (see the chapter *One-Dimensional Techniques* also). The sample is menthol in $CDCl_3$.

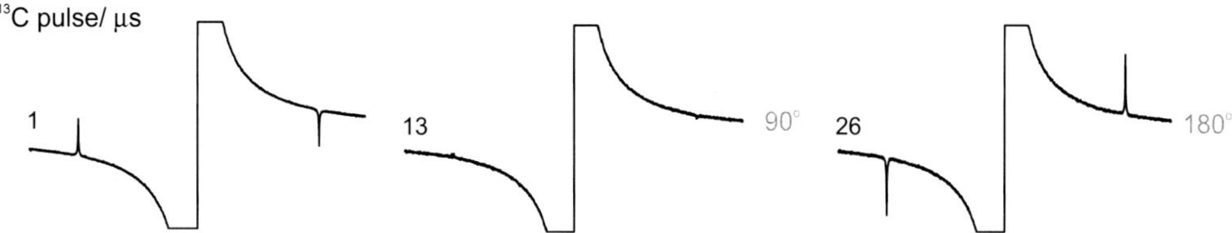

FIGURE 3.65 **Calibration of indirect pulses (here ^{13}C) with proton observation using the sequence of Fig. 3.62.** The parent ^{1}H(^{12}C) resonance appears as a large dispersive signal, but the behaviour of the satellites is clear, again disappearing with a 90 degree pulse on the indirect channel. The sample is $CHCl_3$.

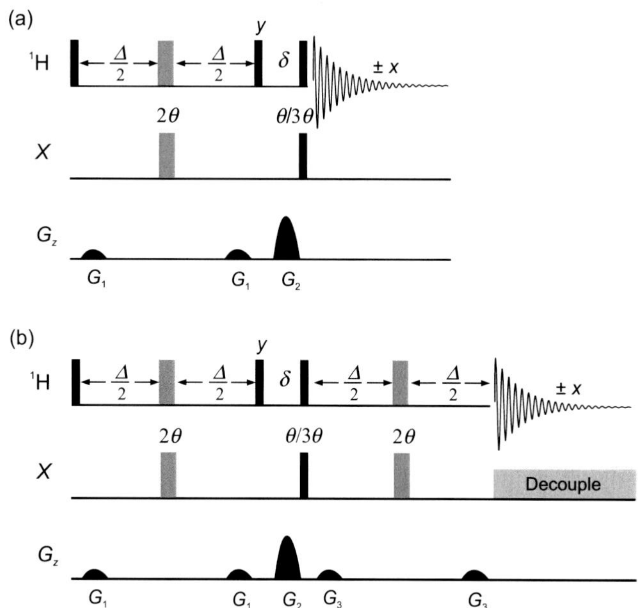

FIGURE 3.66 **The 2CALIS sequences for the calibration of indirectly observed, low-abundance nuclides.** (a) 2CALIS-2 for anti-phase satellite display and (b) the 2CALIS-1 refocused version with *X*-spin decoupling. All pulse phases are *x* unless indicated and the optimum phase cycle is *y*, *y*, −*y*, −*y* on the second proton 90 degree pulse with corresponding alternation of the receiver (*x*, −*x*, *x*, −*x*) as the final *X*-pulse *duration* steps θ, 3θ, 3θ, θ.

hampers observation of the satellites a PFG–based variant may be employed that removes this. The calibration for indirect spins (2CALIS) sequences [64] are shown in Fig. 3.66 and may be most readily understood by considering $\theta = 90$ degrees, where the duration of θ is again the pulse calibration being sought and $\Delta = 1/2J_{XH}$. The initial ^{1}H–X spin-echo generates anti-phase ^{1}H doublet vectors that are then rendered longitudinal by the subsequent ^{1}H $90°_y$ pulse (this is the same initial behaviour is in the INEPT sequence described in Section 4.4.2). The strong purge gradient G_2 then destroys all magnetisation arising from non-X-coupled protons (that is, the ^{1}H–^{12}C resonance in the case of carbon calibration). Following this, the simultaneous 90 degree pulses on both ^{1}H and X transform the longitudinal, anti-phase ^{1}H magnetisation into unobservable multiple quantum coherence and so eliminate the satellites when $\theta = 90$ degrees, as for the simpler sequence of Fig. 3.62. Thus, as θ increases the satellite intensities decay towards this zero-crossing condition (Fig. 3.67). Improved suppression of the unwanted parent resonance may be achieved with additional X-spin filtration by alternating the final X pulse between θ and 3θ (ie 90 and 270 degrees for the ideal calibration condition) combined with alternation of the receiver. The non-X-coupled proton magnetisation remains invariant to these pulses so is cancelled by this receiver cycle. Optionally, an additional refocusing period may be employed, as in 2CALIS-1(Fig. 3.66b), to yield in-phase ^{1}H doublets which may, again optionally, be X-spin decoupled. This approach may prove advantageous when signal overlap makes the presence of anti-phase signals susceptible to accidental cancellation with neighbouring resonances. Note that for these sequences to generate *any* observable signals, 2θ must be close to 180 degrees so that the heteronuclear spin-echo generates anti-phase magnetisation; if no X-spin 180 degree pulse is applied (ie if $\theta = 0$ degrees), the heteronuclear coupling would be fully refocused. Thus, a reasonable estimate of θ is required before the calibration sequence begins as would typically be known from standard calibration samples. 2CALIS would then be employed for accurate calibration on the sample under study.

For the indirect calibration of natural abundance ^{15}N with ^{1}H observation, the dominant ^{14}N-bound proton resonance may be rather broad owing to the quadrupolar ^{14}N nucleus and may mask the small ^{15}N satellites. In such cases, the 2CALIS sequences may prove especially useful. Alternatively, the ^{14}N line can be preferentially suppressed without the need for PFGs, owing to its faster transverse relaxation rate, by application of a spin-echo prior to the sequence of Fig. 3.62 (see [65] for further details). A more convenient alternative is to use ^{15}N-labelled materials for calibration purposes. Calibration over long-range couplings is also possible for example when no one-bond coupling exist, for which the original CALIS-2 sequence [66] is best suited.

3.5.1.8 Homonuclear Decoupling Field Strength

Homonuclear decoupling requires the decoupling (B_2) field to be on during acquisition of the FID and, as described in Section 4.2, this demands the use of rapid gating of the decoupler during this period, so the methods presented above are

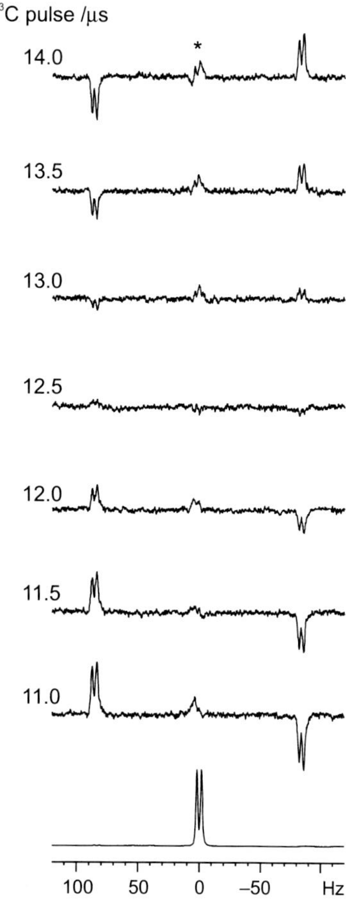

FIGURE 3.67 **^{13}C Indirect calibration with the 2CALIS-2 sequence performed on 50 mM sucrose.** The natural abundance ^{13}C-satellites of the anomeric proton are shown and disappear when $\theta = 90$ degrees, here 12.5 μs. The asterisk indicates the residual ^{12}C parent signal (shown in the lower trace).

therefore no longer applicable. The field strengths used in homonuclear decoupling are considerably less than those required for pulsing and it is sometimes useful to have some measure of these when setting up decoupling experiments, to achieve the desired selectivity. A direct measure of the mean field strength produced during homonuclear decoupling can be arrived at through measurement of the Bloch–Siegert shift [67], which is the change in the resonance frequency of a line on the application of nearby irradiation. Provided the offset of the decoupler from the unperturbed resonance is greater than the decoupler field strength ($\nu_o - \nu_{dec} >> \gamma B_2$), then:

$$\gamma B_2 = ((\nu_{BS} - \nu_o).2(\nu_o - \nu_{dec}))^{1/2}\ \text{Hz} \tag{3.18}$$

where ν_{BS} is the signal frequency in the presence of the irradiating field, ν_o is that in the absence of the irradiating field and ν_{dec} is the frequency of the applied RF. An example of such a calibration is shown in Fig. 3.68.

3.5.1.9 Heteronuclear Decoupling Field Strength

Complementary to the pulse methods described above for the measurement of indirect pulse widths is measurement of the heteronuclear decoupling field strength. This method is applicable to the measurement of medium to low–power outputs, such as those used in broadband decoupling, soft pulses and selective decoupling, and makes use of continuous, rather than pulsed, irradiation. Once again the method exploits the heteronuclear coupling between spins and observes changes to this on the application of off-resonance heteronuclear decoupling [68]. Consider again the AX pair in which continuous A-spin decoupling is applied so far from the A resonance that it has no effect; the X spectrum displays a doublet. As the decoupler frequency moves towards the A resonance the observed coupling begins to collapse until eventually the X spectrum

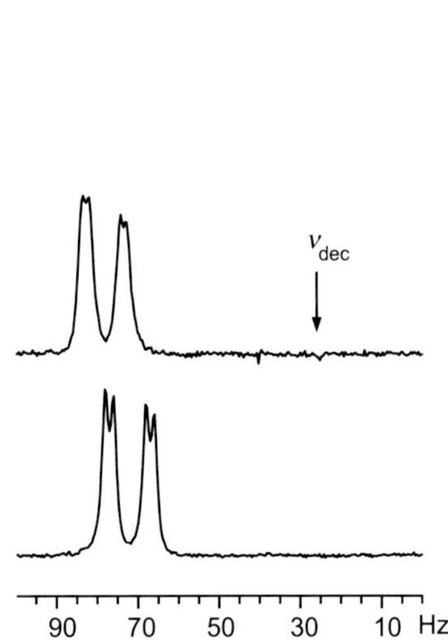

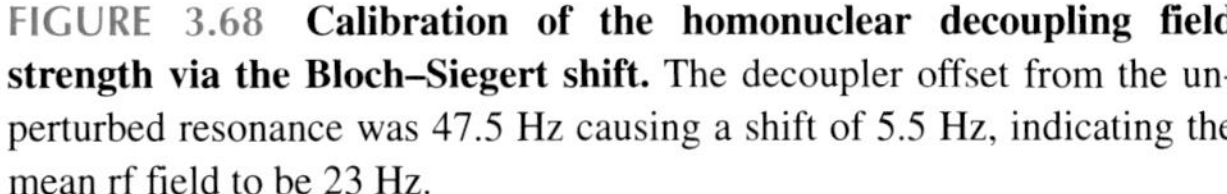

FIGURE 3.68 **Calibration of the homonuclear decoupling field strength via the Bloch–Siegert shift.** The decoupler offset from the unperturbed resonance was 47.5 Hz causing a shift of 5.5 Hz, indicating the mean rf field to be 23 Hz.

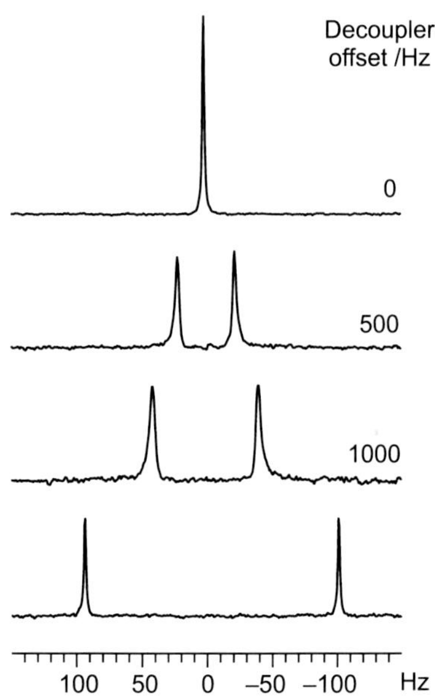

FIGURE 3.69 **Calibration of the heteronuclear decoupler field strength via off-resonance proton decoupling.** From the reduced splitting of the ^{13}C–^{1}H doublet, the proton rf field strength is calculated to be 2.2 kHz, a typical value for broadband proton decoupling.

displays only a singlet when decoupling is on-resonance. Measurement of the reduced splitting in the intermediate stages provides a measure of the field strength according to [68]:

$$\gamma B_2 = \left[\left(\frac{J.\nu_{off}}{J_r}\right)^2 + \frac{(J_r^2 - J^2)}{4} - \nu_{off}^2\right]^{1/2} \text{Hz} \tag{3.19}$$

where J is the AX coupling constant, J_r is reduced splitting in the presence of the decoupling field and ν_{off} is the decoupler offset from the A resonance. When the decoupler field strength is much larger than the decoupler resonance offset, this simplifies to:

$$\gamma B_2 \approx \frac{J.\nu_{off}}{J_r} \text{Hz} \tag{3.20}$$

Thus, by measuring resonance splitting in the absence and presence of irradiation, the field strength is readily calculated. For ^{1}H–^{13}C spectroscopy, one-bond coupling is suitable for medium to strong powers, while for very weak decoupler fields (<50 Hz) it may be necessary to utilise long-range ^{1}H–^{13}C couplings to obtain significant changes in splitting. The calibration of proton field strength with this method is illustrated in Fig. 3.69.

3.5.2 Pulsed Field Gradients

Numerous modern NMR experiments utilise field gradient pulses for signal selection or rejection (see Section 5.4), requiring these to be applied for well-defined durations at known gradient strengths. Experiments presented in the literature will (or should) state the gradient strengths required to achieve the desired results, so it is necessary to have some knowledge of the strengths provided by the instrument if wishing to implement these techniques. Practical procedures for determining gradient strengths and gradient recovery periods are described here, as it is these two parameters that must be defined by the operator when preparing a gradient-selected experiment. Discussions are restricted to static B_0 gradients rather than the far less widely used rf B_1 gradients [69,70]. Despite the claimed advantages of B_1 gradients [71], these have yet to gain widespread acceptance.

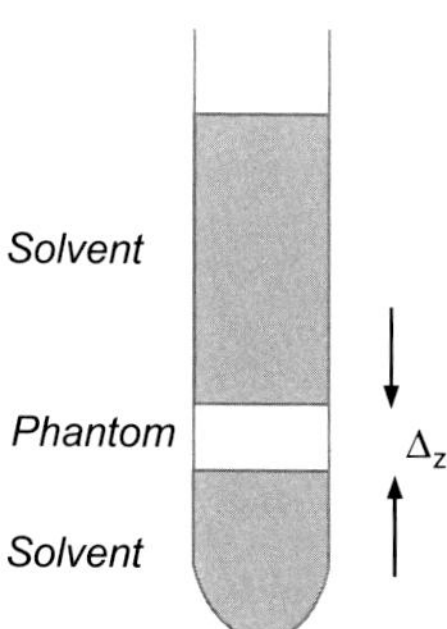

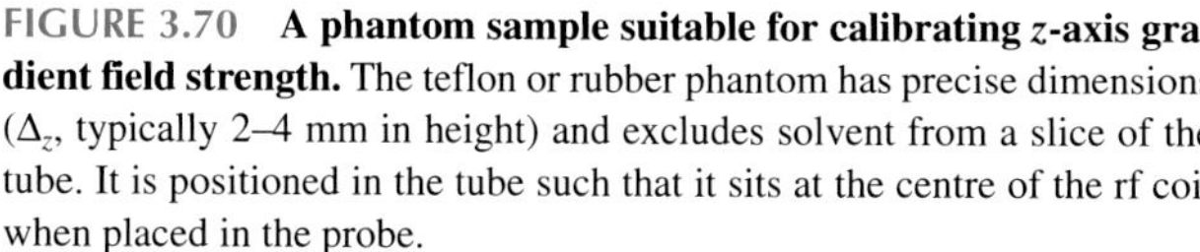

FIGURE 3.70 **A phantom sample suitable for calibrating z-axis gradient field strength.** The teflon or rubber phantom has precise dimensions (Δ_z, typically 2–4 mm in height) and excludes solvent from a slice of the tube. It is positioned in the tube such that it sits at the centre of the rf coil when placed in the probe.

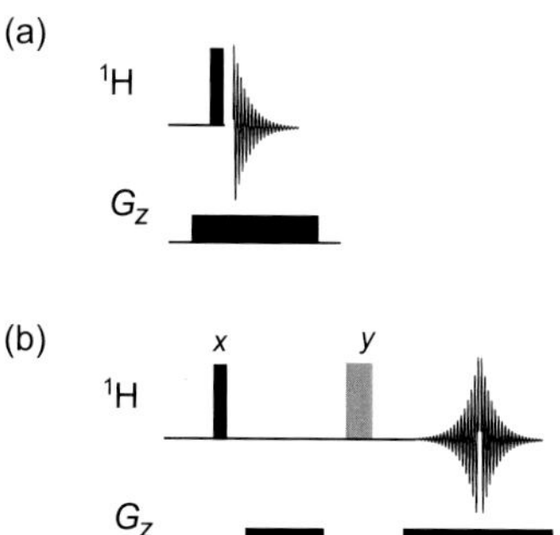

FIGURE 3.71 **Sequences for collecting gradient profiles for calibration of the z-axis gradient field strength.** Sequence (a) collects the FID directly in the presence of the applied gradient while sequence (b) makes use of a gradient echo (see text). In (b) the second gradient has the same strength but twice the duration of the first.

3.5.2.1 Gradient Strengths

For the application of PFGs in high-resolution NMR, knowledge of the exact gradient strengths produced is not, in fact, crucial to the success of most experiments. This is in contrast to the application of rf pulses, where precise pulse widths *are* required. Provided the gradients are *sufficiently strong* for the experiment, the critical factors are pulse reproducibility and the ability to produce gradient pulses of precise amplitude *ratios*. The notable exception to this is in the measurement of self-diffusion coefficients by NMR spectroscopy (see the chapter *Diffusion NMR Spectroscopy*) for which very accurate gradient strength calibrations are required. The procedures described below can be used to calibrate gradient strengths and to check the linearity of gradient amplifier output.

Essentially, there are two approaches to gradient calibration, one based on the measurement of molecular diffusion, the other using an image of a suitable calibration sample. The first requires precise experimental conditions, accurate temperature calibration and a reference solvent or solute with a known diffusion coefficient. Residual H_2O in D_2O is suitable for this, as described in Section 10.3.2 where this is considered in detail. With the relatively low gradient field strengths used in high-resolution NMR, the alternative imaging approach may be rather more straightforward and less dependent on solution conditions. The calibration of z gradients is based on the measurement of a 1D image profile of a phantom of known spatial dimensions placed in a standard NMR tube containing H_2O (Fig. 3.70). The phantom is typically a disk 2–4 mm in length of rubber or teflon (eg a cut down vortex suppresser), which is positioned in the tube such that it sits at the centre of the rf coil. The sample is not spun and need not be locked, in which case the lock field sweep must be stopped. The proton spectrum is first acquired in the absence of the field gradient and the transmitter placed on-resonance. The spectrum is then acquired with a very large spectral width (say, 50 kHz) in the presence of the field gradient. This can, in principle, be achieved by a simple pulse acquire scheme with the gradient turned on throughout (Fig. 3.71a), although this approach has its limitations because the application of rf pulses in the presence of the gradient requires excitation over a very wide bandwidth. Thus, it is preferable to record the spectrum with a gradient spin-echo sequence (Fig. 3.71b), in which the second of the two gradients has the same strength but twice the duration of the first, and is applied throughout the acquisition period during which the spin-echo refocuses. In either case, the acquisition time must be kept short (10 ms or less) and relatively low gradient strengths applied. The resulting data are processed with a *non-shifted* sine bell window function (this matches the shape of the echo FID) and magnitude calculation. The spectrum then displays a dip in the profile corresponding to the region of the sample that contains the phantom, and hence no solution (Fig. 3.72), from which the gradient strength may be calculated as:

$$G_z = \frac{2\pi\Delta\nu}{\gamma\,\Delta_z} \equiv \frac{\Delta\nu}{4358\times\Delta_z}\,\mathrm{G\,cm^{-1}}\ \text{or}\ \frac{\Delta\nu}{4.358\times10^5\times\Delta_z}\,\mathrm{T\,m^{-1}} \tag{3.21}$$

where $\Delta\nu$ is the width of the dip in hertz, and Δ_z is the height of the phantom in centimetres. The SI units for a field gradient are tesla per metre (T m^{-1}), although it is common for values quoted in the literature to be given in gauss per centimetre (G cm^{-1}; 1 G cm^{-1} ≡ 0.01 T m^{-1}). For calibrations of x and y gradients a similar method may be used, in which case the phantom is not required and the internal diameter of the NMR tube defines the sample width, with the areas outside the tube providing the empty region, and the width of the resulting 'peak' used in the above expression.

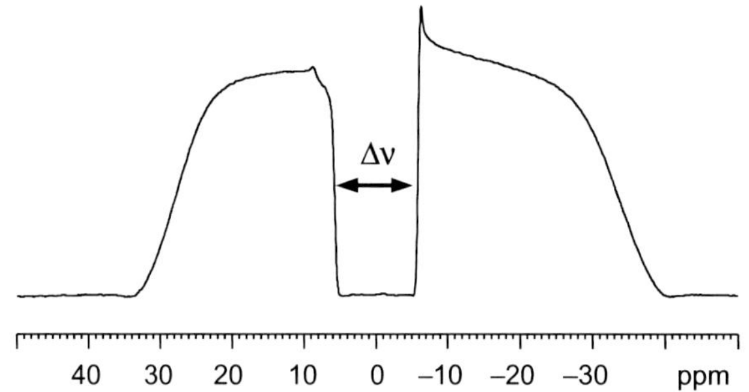

FIGURE 3.72 **A typical gradient profile collected with the sequence of Fig. 3.71b.** Here the dip is 5896 Hz wide, corresponding to a gradient strength of 0.051 T m^{-1} (10% of maximum).

Repeating the above procedure with increasing gradient field strengths indicates the linearity of the amplifier, with larger gradients producing a proportionally wider dip in the profile. Fig. 3.73 shows the result obtained for a 5 mm inverse *z*-gradient probe. Assuming linearity over the whole amplifier range, the maximum gradient strength for this system is 0.51 T m^{-1} (51 G cm^{-1}).

3.5.2.2 Gradient Recovery Times

As discussed in Section 5.4, significant distortions in the magnetic field surrounding a sample will result following application of a gradient pulse, and to acquire a high-resolution spectrum the eddy currents responsible for these field distortions must be allowed to decay before data are collected. A simple scheme for the measurement of the gradient recovery period is shown in Fig. 3.74, in which the initial gradient pulse is followed by a variable recovery period, after which the data are acquired. Distortions of the spectrum are readily seen by observing a sharp proton singlet, such as that of chloroform. Initially, the recovery period τ is set to a large value (1 s) to establish the reference spectrum, and the value progressively decreased until distortions appear in the spectrum (Fig. 3.75). This provides some indication of the recovery period required following a gradient pulse, during which data should not be collected or further gradient or rf pulses applied. The success of high-resolution gradient-selected NMR is critically dependent on being able to use very short recovery periods, which are typically around 100 μs on modern instruments that have actively shielded probeheads. While NMR experiments generally utilise *shaped* gradient pulses (as these help reduce eddy current generation), a more demanding test of hardware performance can be achieved by using *rectangular* gradients, as in Fig. 3.75. Performing this test with a half sine–shaped gradient pulse showed complete recovery of the resonance within 10 μs.

3.5.3 Sample Temperature

Many spectrometers are equipped with facilities to monitor and regulate the temperature within a probehead. Usually, the sensor takes the form of a thermocouple whose tip is placed close to the sample in the gas flow used to provide temperature

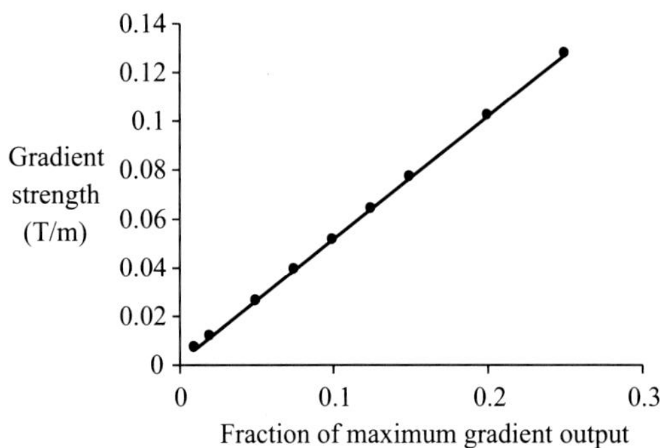

FIGURE 3.73 **Calibration of gradient amplifier linearity using the sequence of Fig. 3.71b.** Assuming system linearity over the full output range and extrapolating the calibration, the maximum gradient strength provided by this system is 0.51 T m^{-1} (51 G cm^{-1}).

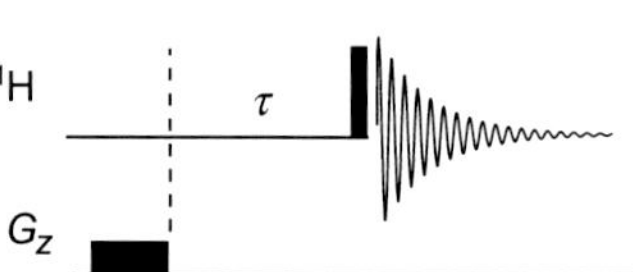

FIGURE 3.74 **A simple scheme for investigating gradient recovery times.**

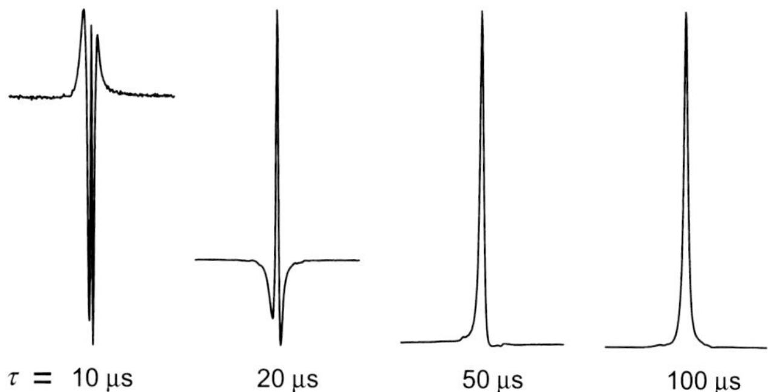

FIGURE 3.75 **Gradient recovery after application of a 1 ms square gradient pulse of 0.25 T m^{-1} (50 % of maximum output).** The sample is chloroform and data were acquired with an actively shielded 5 mm inverse *z*-axis gradient probehead.

regulation. However, the readings provided by these systems may not reflect the true temperature of the sample due to the finite heat capacity of the flowing gas [72] and must be subject to appropriate calibration for accurate temperature recording. One approach to such calibration is to measure a specific NMR parameter that has a known temperature dependence to provide a more direct reading of *sample* temperature. While numerous possibilities have been proposed as reference materials [73], two have become accepted as the standard ^{1}H temperature calibration samples for solution spectroscopy with conventional probeheads. These are neat methanol for the range 175–310 K and 1,2-ethanediol (ethylene glycol) for 300–400 K. Such strong solutions are inappropriate for cryogenic probes, however, and pure perdeuterated methanol offers a better alternative. For use as an *internal* temperature calibrant in ^{13}C measurements, tris(trimethylsilyl)methane is a robust material that may be used in a range of organic solvents, with the shift difference between the ^{13}C resonances of $Si(CH_3)_3$ and CH being indicative of sample temperature; see [74] for details of synthesis and calibration.

For low-temperature ^{1}H calibration, neat MeOH is used, with a trace of HCl to sharpen the resonances, for which the following equation holds [75,76]:

$$T(\mathrm{K}) = 403 - 29.53\Delta\delta - 23.87(\Delta\delta)^2 \tag{3.22}$$

where $\Delta\delta$ is the OH–CH_3 chemical shift difference (ppm). This expression may be approximated to three linear equations of the form:

$$175 - 220\ \mathrm{K}:\quad T(\mathrm{K}) = 537.4 - 143.1\Delta\delta \tag{3.23}$$

$$220 - 270\,\mathrm{K}:\quad T(\mathrm{K}) = 498.4 - 125.3\Delta\delta \tag{3.24}$$

$$270 - 310\,\mathrm{K}:\quad T(\mathrm{K}) = 468.1 - 108.9\Delta\delta \tag{3.25}$$

which are presented as calibration charts in Fig. 3.76. For calibrations with neat ethane-1,2-diol, the appropriate equation is [76]:

$$T(\mathrm{K}) = 466.0 - 101.6\Delta\delta \tag{3.26}$$

where $\Delta\delta$ now represents the OH–CH_2 chemical shift difference (ppm) (Fig. 3.77). For both samples, the peak shift difference decreases as the sample temperature increases due to reduced intermolecular hydrogen bonding, which in turn causes the hydroxyl resonance to move to a lower frequency. Cryogenic probes are limited in their operating temperature range to within typically 280–330 K, but neat solutions produce too strong a signal for these probes, leading to radiation damping which both broadens and shifts the resonances. A more appropriate calibration sample is neat *perdeuterated* methanol for which the following holds [77] (Fig. 3.78):

$$T(\mathrm{K}) = 419.1381 - 52.5130\Delta\delta - 16.7467(\Delta\delta)^2 \tag{3.27}$$

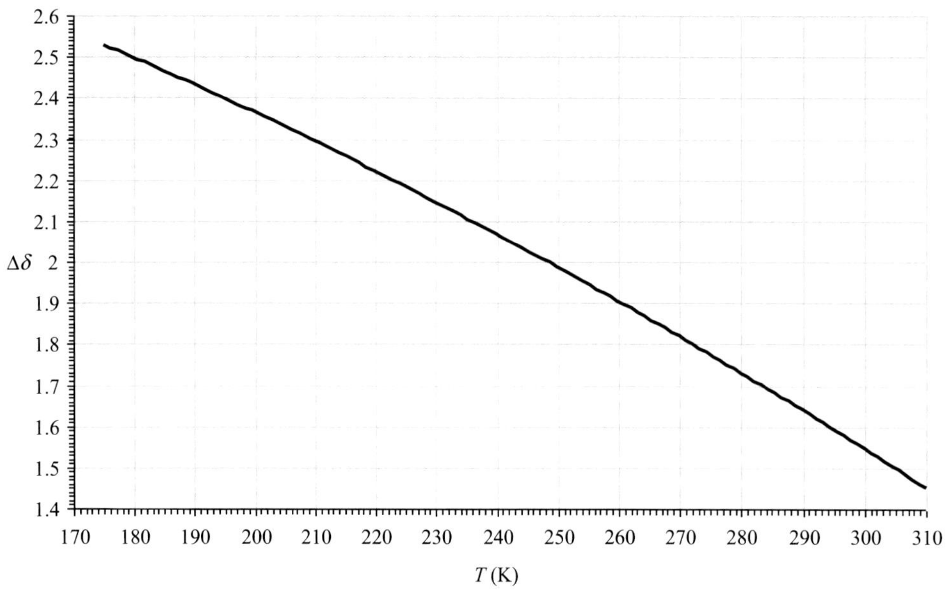

FIGURE 3.76 **Temperature calibration chart for neat methanol.** The shift difference Δδ (ppm) is measured between the CH_3 and OH resonances.

Before making temperature measurements, the sample should be allowed to equilibrate for at least 5–10 min at each new temperature, and it is also good practice to take a number of measurements to ensure they do not vary before proceeding with the calculation.

3.6 SPECTROMETER PERFORMANCE TESTS

There exist a large number of procedures designed to test various aspects of instrument performance, and it would certainly be inappropriate, not to mention rather tedious, to document them in a book of this sort. Instead, I want to introduce briefly only those that are likely to be of use in a routine spectrometer 'health check', and as such this section will be of most interest to anyone with responsibility for upkeep of an instrument. These tests are generally accepted in the NMR community and by instrument manufacturers since they provide a quantitative measure of instrument performance and, in theory at least, should also provide a comparison between the capabilities of different spectrometers. Naturally, this leads

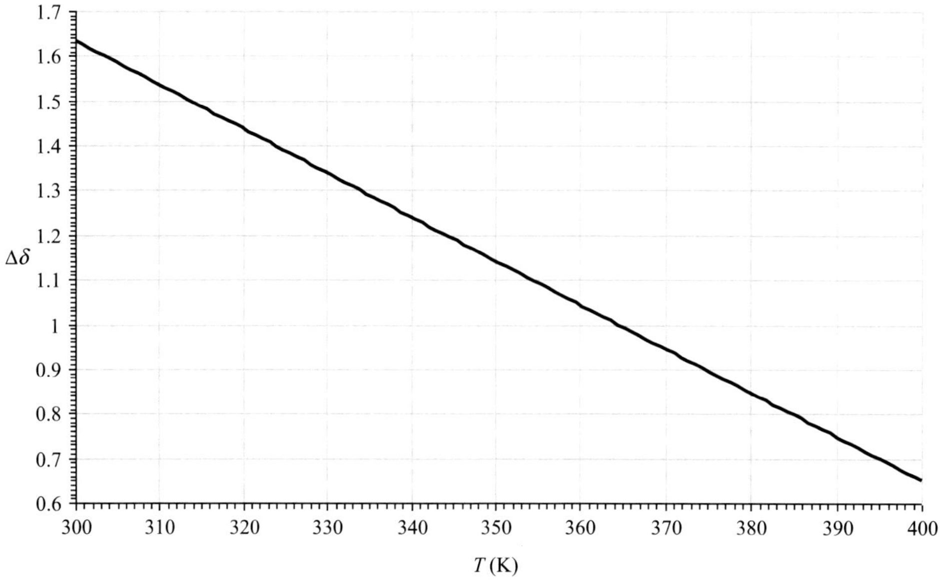

FIGURE 3.77 **Temperature calibration chart for neat ethane-1,2-diol (ethylene glycol).** The shift difference Δδ (ppm) is measured between the CH_2 and OH resonances.

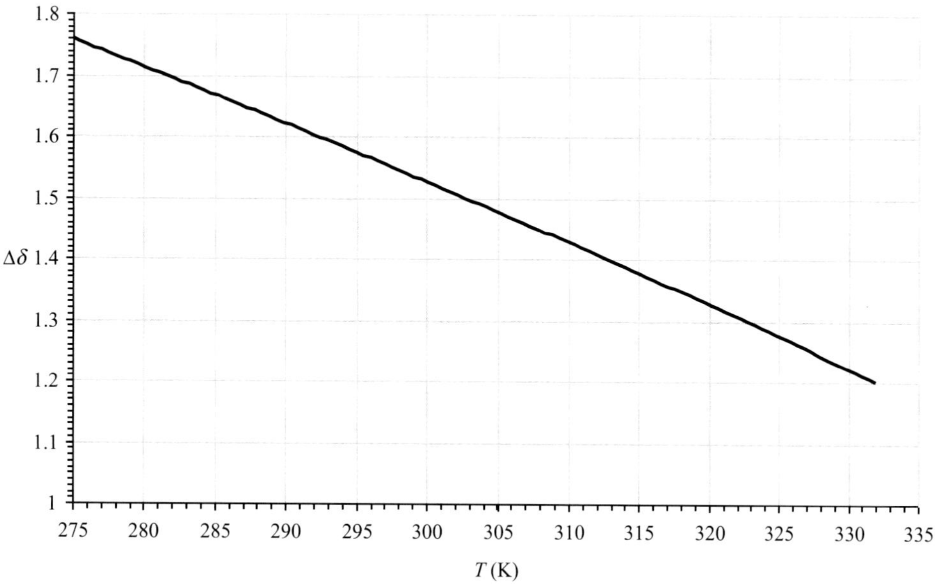

FIGURE 3.78 **Temperature calibration chart suitable for cryogenic probes.** The sample is neat d_4-methanol and the shift difference $\Delta\delta$ (ppm) is measured between the residual CHD_2 and OH resonances.

to aggressive claims by instrument vendors regarding their results and some caution is required when comparing advertised figures. This demands attention to the details of procedures used, for example whether a lineshape test was performed with or without sample rotation or how wide a region of the spectrum was used for estimating the noise level in a sensitivity test. When utilising these tests for in-house checks, consistency in your methodology is the important factor if the results are to be compared with previous measurements. Ideally, these tests should be performed at regular intervals (say, every 6 months or once a year) and appropriate records kept. Additionally, more frequent updating of the 'master' shim settings may also be required for optimal performance, although this tends to be very dependent on instrument stability (as well as that of the surrounding environment). One should also be aware that, perhaps not surprisingly, the installation test results achievable on a modern state-of-the-art instrument will certainly be substantially better than the initial test data of an instrument installed some years ago (eg see the discussions on probes in Section 3.4.2 and see Fig. 3.46). The idea of what is an acceptable test result for an older instrument should therefore be scaled accordingly, and reference to previous records becomes essential if one wishes to gauge instrument performance.

3.6.1 Lineshape and Resolution

The most important test, in may be argued, is that for the NMR resonance lineshape, since only when this is optimised can other quantities also be optimum. The test measures the properties of a suitable singlet resonance which, in a perfectly homogeneous field and in the absence of other distortions, produces a precise Lorentzian lineshape, or, in other words, its FID possesses an exact exponential decay. Naturally the only way to achieve this is through careful optimisation of the magnetic field *via* shimming, which can take many hours to achieve good test results, depending on the initial state of the system. Nowadays, both the lineshape and resolution measurements are often taken from a single set of test data. The *lineshape* is defined by the width of the resonance at 0.55 and 0.11% of the peak height (these numbers having evolved from measurement of proton linewidths at the height of ^{13}C satellites and at one fifth thereof). The *resolution* is defined by the half-height linewidth $\Delta\nu_{1/2}$ of the resonance. In fact, by definition, this cannot be a true measure of resolution (how can one measure resolving power with only a single line?), rather the ability of the instrument to separate neighbouring lines is implied by the narrowness of the singlet. For a genuine Lorentzian line, the widths at 0.55 and 0.11% should be 13.5 and 30 times that at half-height, respectively, although a check for this is often never performed as attempts are often made (erroneously) to simply minimise all measurements. The tests may be performed on both spinning and static samples (be sure you know which if comparing results), the first being of relevance principally for highest resolution proton measurements whereas the second indicates suitability for non-spinning 1D and 2D experiments.

The proton lineshape test uses chloroform in deuteroacetone typically at concentrations of 3% at or below 400 MHz, and 1% at or above 500 MHz. Older instruments and/or probes of lower sensitivity or observations via outer 'decoupler

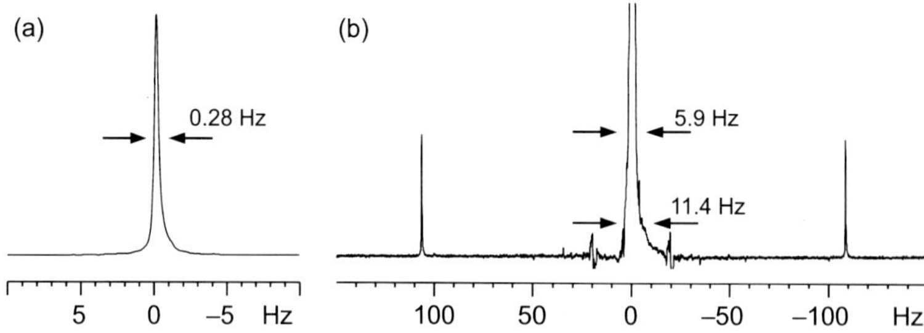

FIGURE 3.79 **Spectrometer proton performance tests for (a) resolution and (b) lineshape.** Spectra are shown for a static 3% $CHCl_3$ sample recorded on a 400 MHz spectrometer equipped with a dual $^{1}H(^{13}C)$ inverse probehead. The ^{13}C satellites are clearly seen in (b) at ±109 Hz, whereas the artefacts at ±20 Hz arise from floor vibrations.

coils' may require 10% at 200 MHz and 3% at 500 MHz to prevent noise interfering with measurements close to the baseline. A single scan is collected and the data recorded under conditions of high digital resolution (acquisition time of 16 s ensuring the FID has decayed to zero) and processed without window functions. Do not be tempted to make measurements at the height of the satellites themselves unless these are confirmed by measurement to be 0.55%. Since these arise from protons bound to ^{13}C, which relax faster than those of the parent line, they may be relatively enhanced if full equilibrium is not established after previous pulses. The test result for a 400 MHz instrument is shown in Fig. 3.79. The traditional test for proton resolution which dates back to the continuous wave (CW) era (o-dichlorobenzene in deuteroacetone) is now no longer employed and has become part of NMR history.

Lineshape and resolution tests on other nuclei follow a similar procedure to that above. Not all nuclei available with a given probe need be tested, and typically only tests for 'inner' and 'outer' coil observations on multinuclear probes are required. This means the second test will often involve carbon-13 for which two samples are in widespread use: the American Society for Testing and Materials (ASTM) test sample (40% p-dioxane in deuterobenzene; also used for the sensitivity test) or 80% benzene in deuteroacetone. In either case on-resonance CW decoupling of protons should be used as this provides improved results for a single resonance relative to broadband decoupling. Rather long (30–40 s) acquisition times will be required for a well-shimmed system.

3.6.2 Sensitivity

A great disadvantage of NMR spectroscopy relative to many other analytical techniques is the intrinsically low sensitivity from which it suffers. This, of course, is greatly outweighed by its numerous benefits, yet is still one of the more likely causes of experiment 'failure' and so deserves serious attention; instrumental approaches to enhancing NMR sensitivity are described in Section 3.4.2. The term 'sensitivity' strictly defines a minimum amount of material that is detectable under defined conditions and so represents the limit of detection for a measurement, but is used rather loosely throughout NMR and often interchangeably with 'signal-to-noise ratio'. The instrument sensitivity test is indeed a signal-to-noise measurement in which the peak height of the analyte is compared with the noise level in the spectrum. For maximum peak height, optimum lineshape is essential, and generally the lineshape and resolution test should be performed first and the shim settings re-optimised on the sensitivity sample (for similar reasons, sensitivity tests are traditionally performed with sample spinning). Likewise, the probe must be tuned for the test sample and the 90 degree pulse determined accurately.

The definition of noise intensity is of prime importance for this measurement since different approaches may lead to differing test results. This has particular significance nowadays as the older method of 'manual' noise estimation has been superseded by automated computational routines provided by instrument vendors. The signal-to-noise measurement compares peak intensity P to twice the root-mean-square (rms) noise level, N_{rms}, in the spectrum. The rms noise is nowadays computed over a defined bandwidth for a region of the spectrum devoid of resonances with a 200 Hz window now most commonly employed. The traditional use of a 2 ppm bandwidth has fallen out of favour with instrument vendors and, perhaps not surprisingly, the narrower bandwidth usually provides a greater signal-to-noise measurement (be sure you know which method is used if comparing results). In cases when computational determination of N_{rms} cannot be made, this may be estimated from the peak-to-peak noise of the baseline N_{pp}, whereby N_{rms} is one fifth N_{pp} as measured directly on screen or from paper plots. The signal-to-noise ratio is then given by:

$$\frac{\mathrm{S}}{\mathrm{N}} = \frac{2.5\,P}{N_{pp}} \tag{3.28}$$

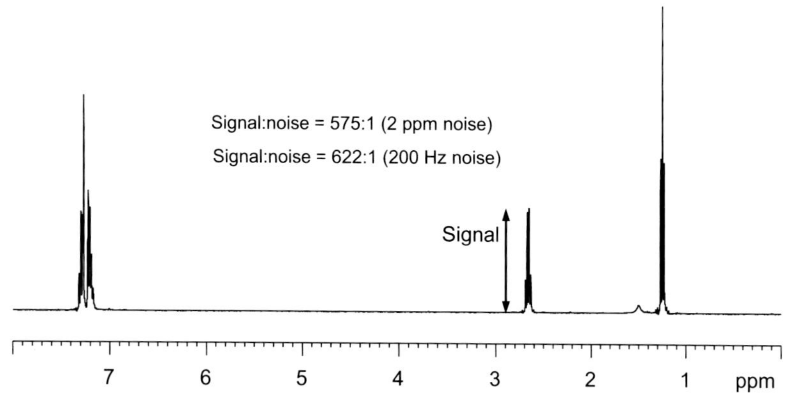

FIGURE 3.80 **Spectrometer proton performance tests for sensitivity.** The spectrum is for a 0.1% ethyl benzene sample recorded on a 400 MHz spectrometer equipped with a dual ^{1}H(^{13}C) inverse probehead.

The peak-to-peak measurement must include all noise bands within the defined frequency window, but is somewhat susceptible to what one might call 'operator optimism' in the exclusion of 'spikes' or 'glitches'. Computational methods for determining noise have the advantage of removing such operator bias, and it is consistency of approach that is important for performance monitoring.

The proton sensitivity test uses 0.1% ethylbenzene in deuterochloroform. As for all sensitivity tests, a single-scan spectrum is recorded following a 90 degree pulse on a fully equilibrated sample (relaxation delay of 60 s in this case). The spectrum is processed with matched filtration corresponding to a line broadening of 1 Hz for the methylene group on which the signal intensity is measured (this resonance is broadened slightly by unresolved long-range couplings to the aromatic protons). Results for a 400 MHz instrument are presented in Fig. 3.80 and show a signal-to-noise ratio of 575:1 or 622:1 with a 2 ppm and a 200 Hz noise bandwidth, respectively.

The carbon sensitivity test comes in two guises: without and with proton decoupling. These give rise to quite different results, so it is again important to be aware of which approach has been used. While the first provides an indication of absolute instrument sensitivity, the second represents a more realistic test of overall performance since it also takes into account the efficiency of proton decoupling and is thus more akin to how one performs genuine experiments. Both approaches acquire carbon spectra under atypical conditions of high digital resolution to correctly define peak shapes, demanding acquisition times of around 4 s. The proton-coupled version makes use of the ASTM sample again (40% p-dioxane in deuterobenzene), but this time measures the peak height of the deuterobenzene triplet. Line broadening of 3.5 Hz is applied, and the noise is measured over the 80–120 ppm window. The proton-decoupled version uses 10% ethyl benzene in deuterochloroform and employs *broadband* composite pulse decoupling usually via the WALTZ-16 sequence (see the chapter *Experimental Methods*). Line broadening of 0.3 Hz is used and the noise recorded over the same region, with the peak height determined for the tallest aromatic resonance. Tuning of the proton coil and appropriate calibration of proton-decoupling pulses are required in this case for optimum results. Test samples for other common nuclei are summarised in Table 3.8. Should you have frequent interests in other nuclei, a suitable standard should be decided upon and used for future measurements.

TABLE 3.8 Standard Sensitivity Test Samples of Some Common Nuclei

Nucleus	Sensitivity Test Sample	Notes
^{1}H	0.1% Ethylbenzene in $CDCl_3$	
^{13}C	10% Ethylbenzene in $CDCl_3$	Use broadband proton decoupling
^{13}C	40% dioxane in C_6D_6 (ASTM sample)	No decoupling, C_6D_6 used for measurement
^{31}P	0.0485 *M* triphenylphosphate in d_6–acetone	No decoupling
^{19}F	0.05% trifluorotoluene in $CDCl_3$	No decoupling
^{15}N	90% formamide in d_6–DMSO	Use inverse gated decoupling to suppress negative NOE
^{29}Si	85% hexamethyldisiloxane in d_6–benzene	No decoupling

ASTM, American Society for Testing and Materials; DMSO, dimethylsulphoxide; NOE, nuclear Overhauser effect.

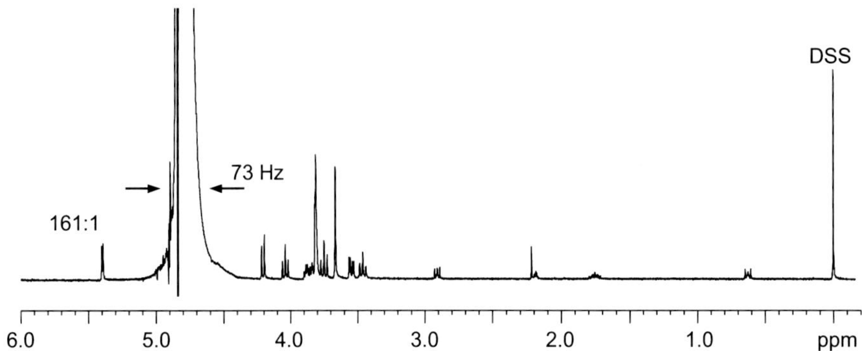

FIGURE 3.81 **The proton solvent presaturation test.** The data were acquired on a 400 MHz spectrometer equipped with a dual $^1H(^{13}C)$ inverse probe, and the sample was 2 mM sucrose and 0.5 mM DSS in 90% H_2O/10% D_2O.

3.6.3 Solvent Presaturation

Interest in biologically and medicinally important materials frequently demands NMR analysis be undertaken on samples in 90% H_2O if solvent-exchangeable protons are also to be observed. The various tests presented so far have all made use of organic solvents, yet the dielectric properties of these and of water are substantially different and a probe that performs well for, say, chloroform may not be optimum for an ionic aqueous solution, particularly with regard to sensitivity. Accurate tuning of the probe and pulse width calibration play an important role here, but protonated aqueous solutions also demand effective suppression of the solvent resonance if the analyte is to be observed (see Section 12.5 for discussions on suppression methods). The solvent suppression test makes use of solvent presaturation and can be used to measure a number of performance characteristics including suppression capability, resolution and sensitivity. Good suppression performance places high demands on both static (B_0) and rf (B_1) field homogeneity. A narrow lineshape down to the baseline, and hence good B_0 homogeneity, is again a prerequisite for good results and ensures all solvent nuclei resonate within a small frequency window. Good B_1 homogeneity means solvent nuclei in all regions of the sample experience similar rf power and are thus suppressed to the same degree. Much of the residual solvent signal that is observed following presaturation arises from peripheral regions of the sample that experience a reduced B_1 field.

The test sample is 2 mM sucrose with 0.5 mM sodium 2,2-dimethyl-2-silapentane-5-sulphonate (DSS) in 90% H_2O/10% D_2O, plus a trace of sodium azide to suppress bacterial growth. In the absence of suppression only the solvent resonance is observed. The test involves on-resonance presaturation of the solvent for a 2 s period, followed by acquisition with a 90 degree pulse. Presaturation power is selected to attenuate the resonance significantly ($\gamma B_1 \approx 25$ Hz), and this setting is used in future comparative tests. Naturally, higher powers will produce greater suppression, but these should not be so high as to reduce neighbouring sucrose resonances. Typically, eight transients are collected following two 'dummy' scans, which achieve a steady state. Suppression performance is judged by measuring the linewidth of the residual solvent signal at 50 and 10% of the height of the DSS resonance. With probes of recent design that have appropriate screening of the rf coil leads, the 50% linewidth should be somewhat less than 100 Hz and reasonably symmetrical. Much older probes may show poorer performance and may be plagued by wide unsymmetrical humps (despite careful attention to shim optimisation) which result from signal pickup in unscreened coil leads [78]. Resolution is judged by the splitting of the anomeric proton doublet at around 5.4 ppm, which should be resolved at least down to 40% of anomeric peak height. Sensitivity may also be measured from this resonance with the noise determined for the 5.5–7.0 ppm region using either of the methods described earlier (baseline correction may be required). Test results for a 400 MHz $^1H(^{13}C)$ probe using moderate presaturation power are shown in Fig. 3.81; further reductions in water resonance could be achieved through the use of greater rf power.

REFERENCES

[1] Shoolery JN. Prog Nucl Magn Reson Spectrosc 1995;28:37–52.
[2] Bartholdi E, Ernst RR. J Magn Reson 1973;11:9–19.
[3] Comisarow MB. J Magn Reson 1984;58:209–18.
[4] Led JJ, Gesmar H. Chem Rev 1991;91:1413–26.
[5] Hoch JC, Stern AS. NMR data processing. New York: Wiley-Liss; 1996.
[6] Reynolds WF, Yu M, Enriquez RG, Leon I. Magn Reson Chem 1997;35:505–19.
[7] Turner CJ, Hill HDW. J Magn Reson 1986;66:410–21.

[8] Hoult DI, Richards RE. Proc Roy Soc (Lond) 1975;A344:311–40.
[9] Delsuc MA, Lallemand JY. J Magn Reson 1986;69:504–7.
[10] Kupče E, Boyd J, Campbell I. J Magn Reson A 1994;109:260–2.
[11] Rosen ME. J Magn Reson A 1994;107:119–25.
[12] Ernst RR. Adv Magn Reson 1966;2:1–135.
[13] Ferrige AG, Lindon JC. J Magn Reson 1978;31:337–40.
[14] Pearson GA. J Magn Reson 1987;74:541–5.
[15] Traficante DD, Nemeth GA. J Magn Reson 1987;71:237–45.
[16] Pacheco CR, Traficante DD. J Magn Reson A 1996;120:116–20.
[17] Evans JNS. Biomolecular NMR spectroscopy. Oxford: Oxford University Press; 1995.
[18] Cavanagh J, Fairbrother WJ, Palmer AG, Skelton NJ, Rance M. Protein NMR spectroscopy: principles and practice. 2nd ed. San Diego: Academic Press; 2006.
[19] Harris RK, Becker ED, Cabral de Menezes SM, Goodfellow R, Granger P. Pure Appl Chem 2001;73:1795–818.
[20] Harris RK, Becker ED. J Magn Reson 2002;156:323–6.
[21] Dykstra RW. J Magn Reson A 1995;112:255–7.
[22] Lewis RJ, Bernstein MA, Duncan SJ, Sleigh CJ. Magn Reson Chem 2005;43:783–9.
[23] Exarchou V, Krucke M, van Beek TA, Vervoort J, Gerothanassis IP, Albert K. Magn Reson Chem 2005;43:681–7.
[24] Albert K. On-line LC-NMR and related techniques. Chichester: Wiley; 2002.
[25] Grynbaum MD, Kreidler D, Rehbein J, Purea A, Schuler P, Schaal W, Czesla H, Webb A, Schurig V, Albert K. Anal Chem 2007;79:2708–13.
[26] Hoult DI, Richards RE. J Magn Reson 1976;24:71–85.
[27] Shoolery JN. In: Levy GC, editor. Topics in carbon-13 NMR spectroscopy, vol. 3. New York: Wiley Interscience; 1979. p. 28–38.
[28] Crouch RC, Martin GE. Magn Reson Chem 1992;30:S66–70.
[29] Martin GE, Hadden CE. Magn Reson Chem 1999;37:721–9.
[30] Martin GE, Crouch RC, Zens AP. Magn Reson Chem 1998;36:551–7.
[31] Schlotterbeck G, Ross A, Hochstrasser R, Senn H, Kuhn T, Marek D, Schett O. Anal Chem 2002;74:4464–71.
[32] Olson DL, Peck TL, Webb AG, Magin RL, Sweedler JV. Science 1995;270:1967–70.
[33] Olson DL, Norcross JA, O'Neil-Johnson M, Molitor PF, Detlefsen DJ, Wilson AG, Peck TL. Anal Chem 2004;76:2966–74.
[34] Webb AG. Prog Nucl Magn Reson Spectrosc 1997;31:1–42.
[35] Gronquist M, Meinwald J, Eisner T, Schroeder FC. J Am Chem Soc 2005;127:10810–1.
[36] Schroeder FC, Gronquist M. Angew Chem Int Ed 2006;45:7122–31.
[37] Kovacs H, Moskau D, Spraul M. Prog Nucl Magn Reson Spectrosc 2005;46:131–55.
[38] Martin GE. Ann Rep NMR Spectrosc 2005;56:1–96.
[39] Styles P, Soffe NF, Scott CA. J Magn Reson 1989;84:376–8.
[40] Styles P, Soffe NF, Scott CA, Cragg DA, Row F, White DJ, White PCJ. J Magn Reson 1984;60:397–404.
[41] Kelly AE, Ou HD, Withers R, Dotsch V. J Am Chem Soc 2002;124:12013–9.
[42] de Swiet TM. J Magn Reson 2005;174:331–4.
[43] Bloembergen N, Pound RV. Phys Rev 1954;95:8–12.
[44] Hoye TR, Eklov BM, Ryba TD, Voloshin M, Yao LJ. Org Lett 2004;6:953–6.
[45] Conover WW. In: Levy GC, editor. Topics in carbon-13 NMR spectroscopy, vol. 4. New York: Wiley; 1983. p. 37–51.
[46] Chmurny GN, Hoult DI. Concept Magn Reson 1990;2:131–59.
[47] van Zijl PCM, Sukumar S, Johnson MO, Webb P, Hurd RE. J Magn Reson A 1994;111:203–7.
[48] Weiger M, Speck T, Fey M. J Magn Reson 2006;182:38–48.
[49] Barjat H, Chilvers PB, Fetler BK, Horne TJ, Morris GA. J Magn Reson 1997;125:197–201.
[50] Sukumar S, Johnson MO, Hurd RE, van Zijl PCM. J Magn Reson 1997;125:159–62.
[51] Morris GA. J Magn Reson 1988;80:547–52.
[52] Morris GA, Cowburn D. Magn Reson Chem 1989;27:1085–9.
[53] Morris GA, Barjat H, Horne TJ. Prog Nucl Magn Reson Spectrosc 1997;31:197–257.
[54] Metz KR, Lam MM, Webb AG. Concept Magn Reson 2000;12:21–42.
[55] Barjat H, Morris GA, Swanson AG, Smart S, Williams SCR. J Magn Reson A 1995;116:206–14.
[56] Horne TJ, Morris GA. J Magn Reson A 1996;123:246–52.
[57] Gibbs A, Morris GA, Swanson AG, Cowburn D. J Magn Reson A 1993;101:351–6.
[58] Barjat H, Morris GA, Smart S, Swanson AG, Williams SCR. J Magn Reson B 1995;108:170–2.
[59] Wu PSC, Otting G. J Magn Reson 2005;176:115–9.
[60] Wesener JR, Günther H. J Magn Reson 1985;62:158–62.
[61] Haupt E. J Magn Reson 1982;49:358–64.
[62] Thomas DM, Bendall MR, Pegg DT, Doddrell DM, Field J. J Magn Reson 1981;42:298–306.
[63] Bax A. J Magn Reson 1983;52:76–80.
[64] Benie AJ, Sorensen OW. J Magn Reson 2006;182:348–52.
[65] Kupče E, Wrackmeyer B. J Magn Reson 1991;94:170–3.

[66] Benie AJ, Sorensen OW. J Magn Reson 2006;180:317–20.
[67] Bloch F, Siegert A. Phys Rev 1940;57:522–7.
[68] Simova SD. J Magn Reson 1985;63:583–6.
[69] Price WS. Ann Rep NMR Spectrosc 1996;32:51–142.
[70] Cory DG, Laukien FH, Maas WE. J Magn Reson A 1993;105:223–9.
[71] Maas WE, Laukien F, Cory DG. J Magn Reson A 1993;103:115–7.
[72] Loening NM, Keeler J. J Magn Reson 2002;159:55–61.
[73] Martin ML, Delpeuch J-J, Martin GJ. Practical NMR spectroscopy. London: Heydon; 1980.
[74] Sikorski WH, Sanders AW, Reich HJ. Magn Reson Chem 1998;36:S118–24.
[75] van Geet AL. Anal Chem 1970;42:679–80.
[76] Raidford DS, Fisk CL, Becker ED. Anal Chem 1979;51:2050–1.
[77] Findeisen M, Brand T, Berger S. Magn Reson Chem 2007;45:175–8.
[78] Dykstra RW. J Magn Reson 1987;72:162–7.

Chapter 4

One-Dimensional Techniques

Chapter Outline

4.1 Single-Pulse Experiment 133
4.1.1 Optimising Sensitivity 133
4.1.2 Quantitative NMR Measurements and Integration 137
4.1.3 Quantification with an Electronic Calibrant: ERETIC 140
4.1.4 Quantification with an External Calibrant: PULCON 142
4.2 Spin-Decoupling Methods 143
4.2.1 Basis of Spin Decoupling 143
4.2.2 Homonuclear Decoupling 143
4.2.3 Heteronuclear Decoupling 145
4.3 Spectrum Editing with Spin-Echoes 148
4.3.1 J-Modulated Spin-Echo 149
4.3.2 APT 152
4.4 Sensitivity Enhancement and Spectrum Editing 153
4.4.1 Polarisation Transfer 154
4.4.2 INEPT 156
4.4.3 DEPT 162
4.4.4 DEPTQ 165
4.5 Observing Quadrupolar Nuclei 167
References 168

The approach to any structural or mechanistic problem will invariably start with the acquisition of one-dimensional (1D) spectra. Since these provide the foundations for further work, it is important that they are executed correctly and full use is made of the data they provide before more extensive and potentially time-consuming experiments are undertaken. This chapter describes the most widely used 1D techniques in the chemistry laboratory, beginning with the simple single-pulse experiment and progressing to consider the various multipulse methods that enhance the information content of our spectra. The key characteristics of these are summarised briefly in Table 4.1. This chapter does not cover the wide selection of techniques that are strictly 1D analogues of two-dimensional (2D) experiments, as these are more appropriately described in association with the parent experiment and are found throughout the following chapters.

4.1 SINGLE-PULSE EXPERIMENT

The previous chapter described procedures for the optimum collection of the FID, and how these were dictated by the relaxation behaviour of the nuclei and by the digital resolution required in the resulting spectrum, which in turn defines the data acquisition time. When setting up an experiment, one also need consider the optimum pulse excitation angle to use and how rapidly pulses can be applied to the sample for signal averaging (Fig. 4.1). These parameters also depend on spin relaxation times. There are two extreme cases to be considered for the single-pulse experiment: in particular, whether one is striving for optimum sensitivity for a given period of data accumulation or whether one requires accurate quantitative data from the sample. The experimental conditions for meeting these criteria can be quite different, so each shall be considered separately.

4.1.1 Optimising Sensitivity

If one were to apply a 90 degree pulse to a spin system at thermal equilibrium, it is clear that the maximum possible signal intensity would result since all magnetisation is placed in the transverse plane. This may therefore appear to be the optimum pulse tip angle for maximising sensitivity. However, one is usually interested in performing signal averaging, so before applying subsequent pulses it becomes necessary to wait many times T_1 for the system to relax and recover the full signal once again; a period of $5T_1$ leads to 99.3% recovery of longitudinal magnetisation (Fig. 2.20), complete for all practical purposes. Such slow repetition is in fact not the most efficient way of signal averaging, and it turns out to be better to do away with the recovery delay altogether and to adjust pulse conditions to maximise the *steady-state* z-magnetisation produced.

High-Resolution NMR Techniques in Organic Chemistry. http://dx.doi.org/10.1016/B978-0-08-099986-9.00004-X

TABLE 4.1 Principal Applications of the Techniques

Technique	Principal Applications
J-modulated spin-echo or APT	Editing of heteronuclide spectra (notably carbon-13) according to resonance multiplicity.
INEPT	Enhancement of a low-γ nuclide through polarisation transfer from a high-γ nuclide (eg ^{1}H, ^{31}P, ^{19}F). Editing of spectra according to resonance multiplicity.
DEPT	Enhancement of a low-γ nuclide through polarisation transfer, with editing according to resonance multiplicity. Experimentally more robust for routine carbon-13 spectra than INEPT. No responses observed for quaternary carbons.
DEPTQ	As for DEPT, but with retention of quaternary carbons.
RIDE or ACOUSTIC	Observation of low-frequency quadrupolar nuclei with very broad resonances. Sequence suppresses 'acoustic ringing' responses from probehead.
ERETIC	Quantification of solute concentration relative to an external rf reference.
PULCON	Quantification of solute concentration relative to an external calibrant.

Under conditions where there is complete decay of transverse magnetisation between scans (ie the FID decays to zero), the optimum tip angle for a pulse repetition time t_r, known as the *Ernst angle* α_e, is given by [1,2]:

$$\cos\alpha_e = e^{-t_r/T_1} \tag{4.1}$$

and is illustrated in Fig. 4.2. As the repetition time decreases relative to the spin relaxation rate, that is, when faster pulsing is used, smaller tip angles produce the optimum signal-to-noise ratio. Proton spectra are typically acquired with sufficient digital resolution to reveal multiplet fine structure, and acquisition times tend to be of the order of $3T_2$*, which is sufficient to enable the almost complete decay of transverse magnetisation between scans ($3T_2$* corresponds to 95% decay). Furthermore, since for small- to medium-sized molecules $T_2 = T_1$ and for proton observation in a well-shimmed magnet $T_2^* \approx T_2$, we have $t_r \approx 3T_1$, and thus, from Fig. 4.2, the pulse angle for *maximising sensitivity* will typically be >80 degrees. However, there will exist a range of T_1 values for the protons in a molecule, and it will not be possible to use optimum conditions for all. In such cases a large pulse angle is likely to lead to significant intensity differences for protons with widely differing relaxation times, with the slower relaxing spins displaying reduced signal intensity [3]. This is exemplified in the signal-to-noise plots of Fig. 4.3. When proton integrals are required to indicate the number of nuclei giving rise to each resonance, as for routine proton acquisitions, it is wise to optimise the pulse tip angle for the longer T_1 values and thus use shorter pulses (see the following section for details of *accurate* quantitative measurements). Assuming the longest T_1 to be around 4 s and an acquisition time of 3 s, the optimum tip angle will be around 60 degrees, although the plots of Fig. 4.3 clearly possess rather broad maxima, meaning the precise setting of the pulse tip angle is not critical. Pulsing too rapidly, that is, using very short repetition times relative to T_1, leads to a substantial decrease in signal-to-noise until in the extreme case magnetisation has no time to recover between pulses and so no signal can be observed. This condition is known as *saturation* and causes resonances to be lost completely which can be nuisance or a bonus depending on your point of view. If you are interested in observing the signal then clearly this must be avoided, whereas if the signal is unwanted, for example, a large solvent resonance, then this is to be encouraged. The deliberate use of resonance (pre)saturation for solvent suppression is described in Section 12.5.

Conditions for the acquisition of heteronuclear spectra are usually rather different from those for proton observations. Often one does not require the spectra to be well digitised when there is little or no fine structure to be observed, therefore acquisition times are kept comparable with T_2*, which is often less than T_2. Under such conditions, transverse magnetisation

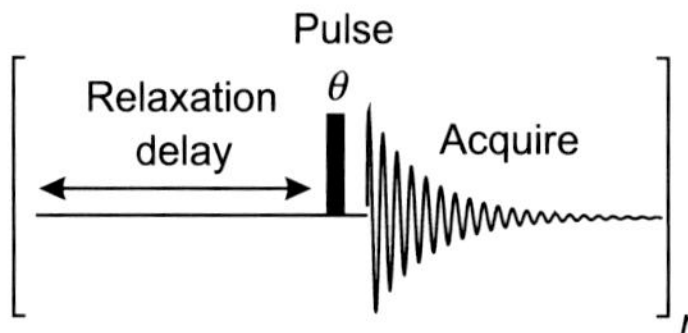

FIGURE 4.1 **The essential elements of the single-pulse NMR experiment.** The relaxation (recovery) delay, pulse excitation and data acquisition may be repeated *n* times for signal averaging.

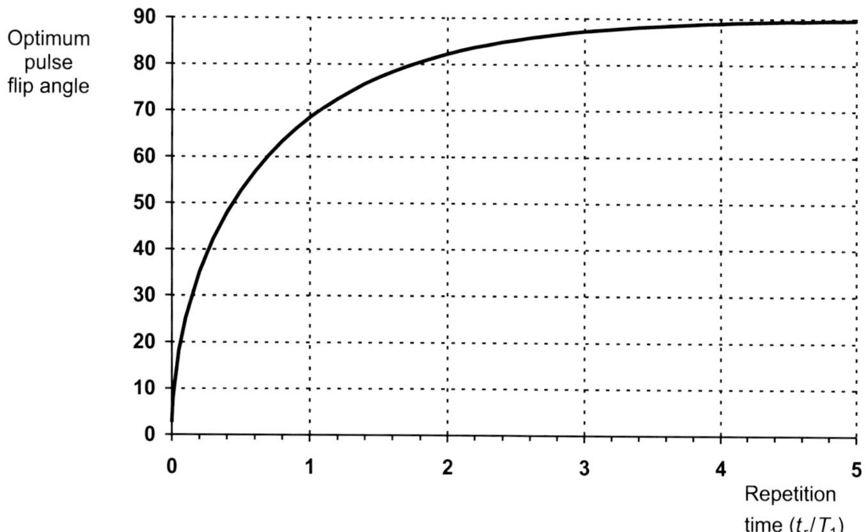

FIGURE 4.2 **The Ernst angle for optimum sensitivity when signal averaging.** The pulse tip angle is dictated by the pulse repetition time, t_r, relative to the longitudinal relaxation time, T_1.

does not decay completely between scans and such rapid pulsing gives rise to steady-state spin-echoes [2,4,5] where transverse magnetisation remaining from one pulse is refocused by subsequent pulses. These echoes can give rise to phase and intensity distortions in resulting spectra which are a function of resonance offset. The use of rapid pulsing, where repetition times are considerably shorter than longitudinal relaxation times (as in the case of routine carbon-13 observations) further contributes to distortions of the relative intensities because of partial saturation effects. This is the principal reason for quaternary carbon centres (which relax rather slowly owing to the lack of attached protons) appearing with often characteristically weak intensities. In extreme cases these can become fully saturated and effectively lost altogether if pulsing is too rapid and/or if the pulse tip angle is too large. Thus, routine carbon spectra are typically acquired with a somewhat reduced pulse angle (<45 degrees) and with a relaxation delay of a second or so. Similar considerations apply to other heteronuclei. For routine observations, there is always some compromise needed between optimum sensitivity and the undesirable

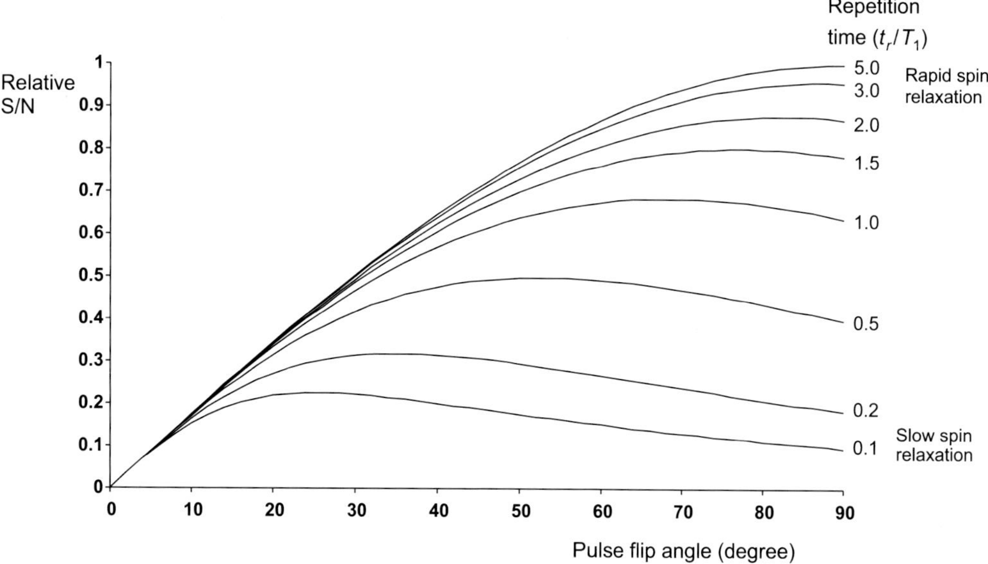

FIGURE 4.3 **The dependence of signal-to-noise ratio on pulse tip angle for different pulse repetition rates.** The maximum of each curve corresponds to the Ernst angle. When a molecule contains nuclei with very different relaxation times (T_1 s), pulsing with a large flip angle produces a large difference in their signal intensities and makes even semi-quantitative measurements unreliable. Optimising the tip angle for the slowest relaxing spins reduces these intensity differences. The plots have been generated from the equations given in [3].

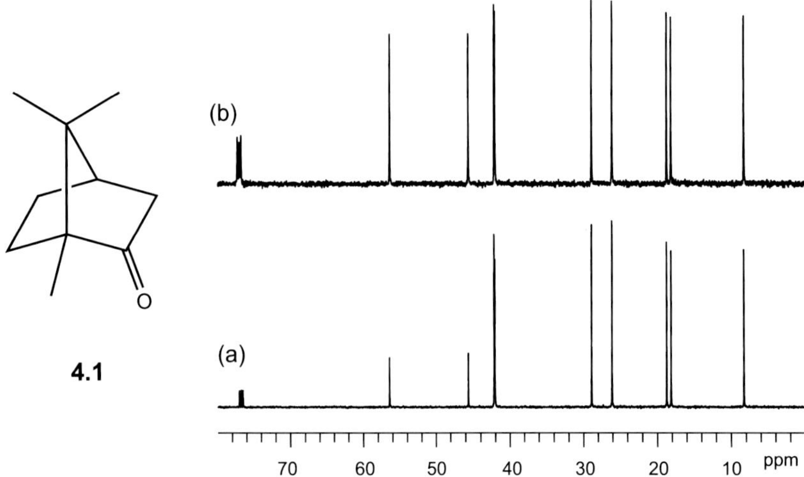

FIGURE 4.4 **Carbon-13 spectra of camphor 4.1.** Spectra were acquired (a) without and (b) with the addition of $Cr(acac)_3$ under otherwise identical standard observation conditions (1 s relaxation delay, 30 degree excitation pulse, 0.5 s acquisition time). In (a) the two quaternary resonances display reduced signal intensities as a result of partial saturation. In (b) this difference is largely eliminated by addition of the relaxation agent, whilst the reduced signal intensity of the protonated resonances arises from suppression of ^{1}H–^{13}C NOE enhancements and the line broadening caused by the relaxation agent.

intensity anomalies that arise for nuclei with widely differing relaxation times. If problematic, one approach towards reducing these anomalies is the addition of paramagnetic relaxation agents such as chromium(III) acetylacetonate, $Cr(acac)_3$, with the aim of reducing and, in part, equalising longitudinal relaxation times to allow faster signal averaging (Fig. 4.4). With $Cr(acac)_3$, around 10–100 mM of the relaxation agent suffices and the solution should take on a slight pink-purple hue.

Finally, we consider the situation where one is compelled by the pulse sequence to use 90 degree pulses in which case the aforementioned arguments no longer apply. This is in fact the case for most multipulse and multidimensional sequences. Under such conditions a compromise is required between acquiring data rapidly and the unwelcome effects of saturation, and it has been shown [6] that for steady-state magnetisation the sensitivity is maximised by setting the repetition rate equal to $1.3T_1$. This leads to a signal-to-noise ratio that is approximately 1.4 times greater than that obtained using a recycle time of $5T_1$ *for a given period of data collection* (Fig. 4.5). Thus, for most sequences a repetition time of $1.3T_1$ is the target.

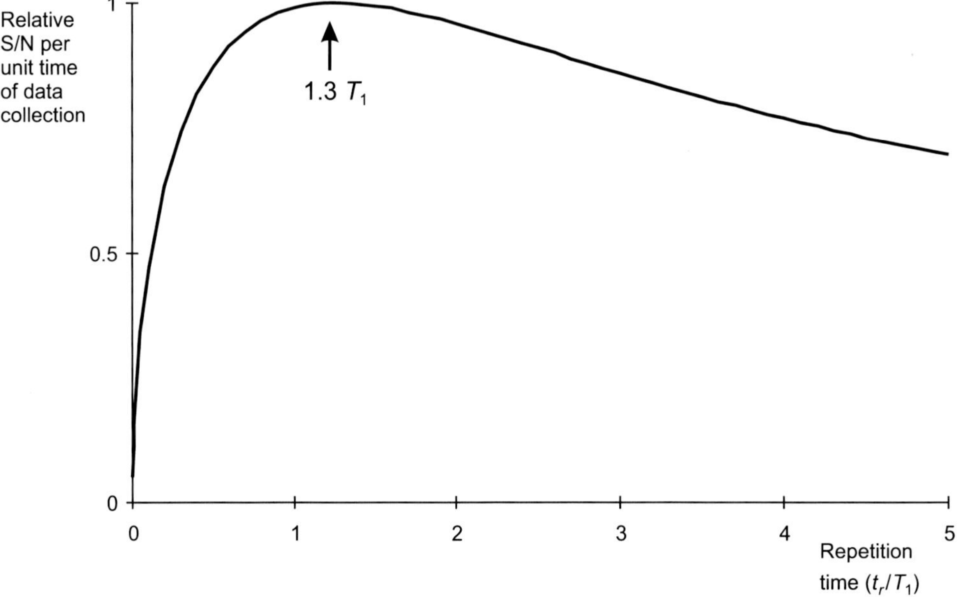

FIGURE 4.5 **Optimum pulse repetition time when using 90 degree pulses.** The curve has been calculated from the equation in Ref. [3].

4.1.2 Quantitative NMR Measurements and Integration

The extensive use of integration in proton NMR arises from the well-used doctrine that the area of an NMR resonance is proportional to the relative number of nuclei giving rise to it. In fact, this is only strictly true under well-defined experimental conditions, and in routine proton spectra integrals may only be accurate to within 10–20% or so, for the reasons given earlier. Whilst this level of accuracy is usually sufficient for estimation of the relative proton count within a molecule, it is clearly inadequate for quantitative measurements where accuracy to within a few percent is required, and indeed this accuracy *is* possible provided appropriate parameters are employed. The fact that integrals are often reported to many decimal places by NMR software can draw one into believing such figures are significant, so inflating one's expectations of what can be derived from routine spectra. Conversely, it is widely taught in introductory NMR lectures that it is not possible to integrate carbon-13 spectra. Whilst this is indeed inappropriate for routine acquisitions, not just for carbon-13 but also for many heteronuclei, it is nevertheless quite possible to obtain meaningful integrals when the appropriate protocols are employed, as described later.

In all these discussions it is important to be conscious of the fact that NMR measurements are always *relative* measures of signal intensities. There exist no NMR parameters equivalent to the extinction coefficients of ultraviolet spectroscopy, so *absolute* sample quantities cannot usually be determined directly. However, in the following sections two protocols will be introduced that do allow sample concentrations to be estimated without the need for an internal calibrant. These make use of either a synthetic electronic calibration signal ERETIC (electronic reference to in vivo concentrations) or a separate *external* calibration sample PULCON (pulse length–based concentration measurements) with which to determine sample concentrations.

4.1.2.1 Quantitative NMR—qNMR

Before considering the general practicalities pertinent to making quantitative measurements, we shall consider the use of NMR as a quantitative analytical method [7]. In this the typical aim is to define, with a high degree of accuracy, the relative levels of analytes present in a sample (such as when determining the percentage of an impurity present with a purity assay) or to determine the absolute concentrations of components in a sample. This approach is often referred to as quantitative NMR or simply *qNMR*, with the most common implementation employing proton spectroscopy termed *qHNMR*. This area has been reviewed in some comprehensive papers to which the interested reader is referred for extensive examples [8–10]. NMR is in fact very well suited as an analytical method, providing high precision and accuracy when properly validated, with the ability to simultaneously quantify multiple analytes within a sample without the need for identical standard compounds for calibration protocols, as are required for chromatographic assays.

The favoured use of proton NMR spectroscopy in qNMR arises from the greater sensitivity of proton detection, allowing an accuracy of 1% in measurements if appropriate experimental protocols are followed. In this, the optimum approach is to keep things simple and employ the single-pulse experiment of Fig. 4.1 (using parameters guided by the considerations discussed further), with an internal calibration standard added when absolute concentrations are to be determined. Herein the term *calibration standard* or *calibrant* refers to a material used to quantify an analyte, whilst the term *reference standard* describes a chemical shift reference material such as tetramethylsilane (TMS) [9]. In some instances it may also be beneficial to employ ^{13}C decoupling during proton acquisition to collapse the ^{13}C satellites within the parent ^{12}C proton resonance such that they do not overlap with neighbouring peaks. This is best achieved using the inverse-gated decoupling scheme of Section 4.2.3, to help attenuate sample heating that may be caused by carbon decoupling.

When assessing concentrations, the appropriate choice of calibrant is critical since its resonance(s) must not overlap those of the analyte(s) or solvent in the sample, meaning compounds with only a single resonance are favoured. The material should optimally be solid for accurate weighing (certainly not of high volatility), and highly soluble in the solvent used. It should also be available with high purity, stable, inert to the solvent and analytes studied, non-hygroscopic and, ideally, cheap. Many such compounds have been employed over the years and a number assessed specifically for use as calibration standards [11]. There is no absolute standard suitable for all studies, but some of the more useful compounds include maleic and fumaric acids, sodium acetate, dimethylsulfone, trimethylsilyl propionic acid (TSP; also a shift reference standard), 1,3,5-trimethoxybenzene, 1,4-dinitrobenzene and 3,4,5-trichloropyridine. Relevant properties of these standards are summarised in Table 4.2. For assessment of the purity of standards themselves, highly pure acetanilide has been proposed as a suitable primary standard [12]. Analyte concentrations may then be determined by comparison of integrated peak intensities against those of the calibrant present at known concentration, taking account of the relative numbers of protons responsible for the integrated peaks. Alternative protocols derived from this procedure are described in Sections 4.1.3 and 4.1.4 and can avoid the need for an internal calibrant. Methods employing quantitative 2D NMR techniques that may be applicable to studies of more complex mixtures have also been reviewed [13], but are not considered further here.

TABLE 4.2 Properties of Selected Calibration Standards

Calibration Standard	Structure	Molecular Weight (g/mol)	^{1}H δ (ppm)	Solvent Solubility			
				D_2O	DMSO–d_6	CD_3OD	$CDCl_3$
Maleic acid	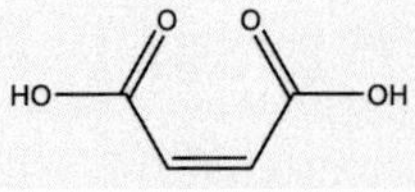	116.07	6.0–6.3	✓	✓	✓	✗
Fumaric acid	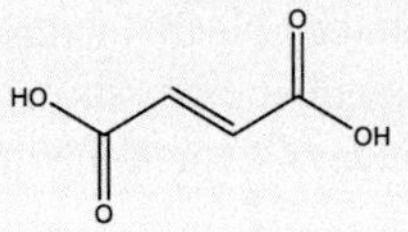	116.07	6.6–6.8	✓	✓	✓	✗
Sodium acetate	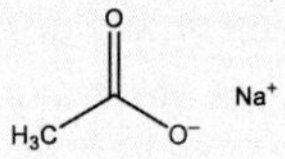	82.03	1.6–1.9	✓	✓	✓	✗
Dimethyl sulfone	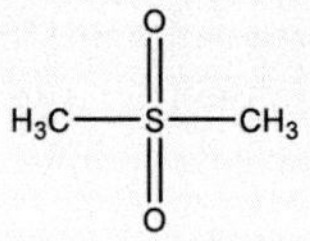	94.13	3.0	✓	✓	✓	✓
TSP–d_4	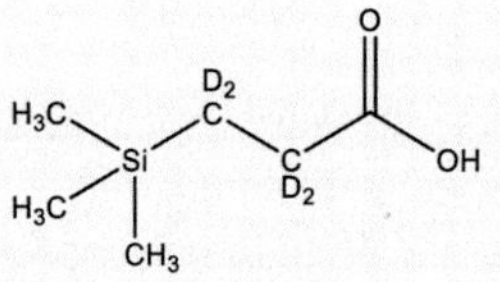	150.29	0.0	✓	✓	✓	✗
1,3,5-Trimethoxybenzene	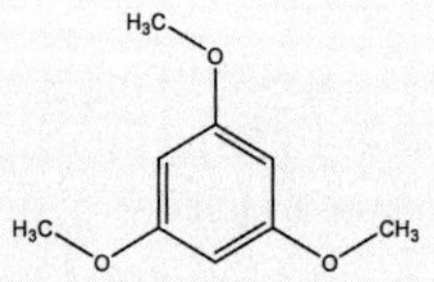	168.19	6.1, 3.7–3.8	✗	✓	✓	✓
1,4-Dinitrobenzene	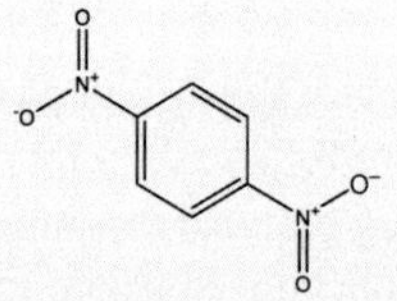	168.11	8.3–8.5	✗	✓	✓	✓
3,4,5-Trichloropyridine	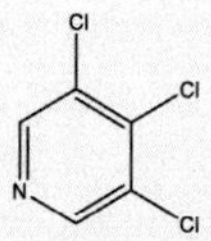	182.43	8.5–8.8	✗	✓	✓	✓
Acetanilide (primary standard)	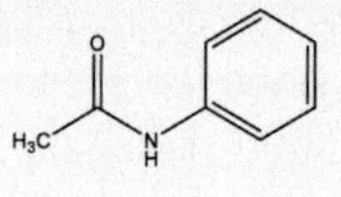	135.17	2.0–2.2, 7.0–7.2, 7.2–7.3, 7.3–7.6, 7–10[a]	✓	✓	✓	✓

^{1}H shift ranges provide a guide but are solvent dependent within these ranges.
[a]The amide NH shift of acetanilide varies greatly with solvent.

4.1.2.2 Data Collection

There are three features of specific importance to quantitative measurements, aside from the obvious need for adequate signal-to-noise in the spectrum. These are the avoidance of differential saturation effects, the need to characterise the NMR resonance lineshape properly and the need to avoid differential nuclear Overhauser effect (NOE) enhancements when decoupling is employed, a situation most significant for carbon-13 observation.

As mentioned in the previous section, acquiring data whilst pulsing rapidly relative to spin relaxation times leads to perturbation of the relative signal intensities in the spectrum, so to avoid this it is *essential* to wait for the spins to fully relax between pulses, demanding recycle times of at least $5T_1$ *of the slowest relaxing nuclei*. This allows the use of 90 degree observation pulses, thus providing the maximum possible signal per transient. Clearly, one requires some knowledge of the T_1 values for the sample of interest which may be determined by the inversion recovery methods described in Section 2.4 or estimates made from prior knowledge of similar compounds. Whilst recycle times of the order of $5T_1$ are usually bearable for proton work, they can be tediously long in the study of heteronuclei which may demand many minutes between scans. Here relaxation reagents can be employed to reduce these periods to something more tolerable.

The second fundamental requirement is for the data to be sufficiently well digitised for the lineshape to be defined properly. To minimise intensity errors it is necessary to have *at least four* acquired data points covering the resonance linewidth, although many more than this are preferable, so it is beneficial to use the minimum spectral width compatible with the sample and to adjust the acquisition times accordingly. For proton spectroscopy, acquisition times of 3–4 s are usually adequate. However, the spectral width should not be too narrow to ensure the receiver filters do not interfere with resonance intensities at the edges of the spectrum.

A further source of intensity distortions in heteronuclear spectra recorded with broadband proton decoupling arises from the NOE produced by proton saturation (see the chapter *Correlations Through Space: The Nuclear Overhauser Effect*). Clearly, differential enhancements will prevent the collection of meaningful intensity data, so it is necessary to take measures to suppress the NOE, yet it is still desirable to collect proton-decoupled spectra for optimum signal-to-noise and minimal resonance overlap. The solution to this apparent dichotomy is to employ the inverse-gated decoupling scheme described in the following section. The lack of the NOE and the need to pulse at a slow rate means quantitative measurements can take substantially longer than would routine observation of the same sample. The addition of a relaxation agent will again speed things along by reducing recycle delays and will also aid suppression of the NOE, as it will eliminate the dipolar relaxation responsible for this enhancement (Fig. 4.4).

4.1.2.3 Data Processing

Having taking the necessary precautions to ensure the acquired data genuinely reflects the relative ratios within the sample, appropriate processing can further enhance results. The spectra of heteronuclei, in particular, benefit from the application of an exponential function that broadens the lines. This helps to ensure the data are sufficiently well digitised in addition to improving the signal-to-noise ratio. For proton spectroscopy, achieving adequate sensitivity is not such a demanding problem, although the use of a matched exponential window will again help to ensure sufficient digitisation. The use of zero-filling will further assist with definition of the lineshape and is highly recommended, although this must not be used as a substitute for correct digitisation of the acquired FID. Careful phasing of the spectrum is also essential. Deviations from pure absorption-mode lineshapes will reduce integrated intensities with contributions from the negative-going components; in the extreme case of a purely dispersive lineshape, the integrated intensity is zero! Another potential source of error arises from distortions of the spectrum baseline, which have their origins in spectrometer receiver stages. These errors mean the regions of the spectrum that should have zero intensity, that is, those that are free from signals, have a non-zero value, and make a positive or negative contribution to measurements. NMR software packages incorporate suitable baseline correction routines for this.

The final consideration when integrating is where the integral should start and finish. For a Lorentzian line, the tails extend a considerable distance from the centre and the integral should, ideally, cover 20 times the linewidth each side of the peak if it is to include 99% of it. For proton observation this is likely to be 10–20 Hz each side. In practice, it may not be possible to extend the integral over such distances before various other signals are met. These may arise from experimental imperfections (such as spinning sidebands), satellites from coupling to other nuclei or from other resonances in the sample. Satellites can be particularly troublesome in some cases as they may constitute a large fraction of a total signal, owing to the high natural abundance of the second nuclide, and one must decide on whether to include them or exclude them *for all measurements*, or alternatively to collapse them entirely through use of the heteronuclear decoupling scheme described above. However, in some instances, satellites can be used to one's advantage in quantitative measurements. One example is in the estimation of enantiomeric or diastereomeric excesses by proton NMR where the minor isomer is present at only a few percent of the major. In such cases, the *ee* or *de* measurement demands the comparison of a very large integral versus a very small one, a situation prone to error. Comparison of similar size integrals can be made if one considers only the carbon-13 satellites of the major species (each present at 0.55%), and scales the calculation accordingly.

An alternative to resonance integration now commonplace in NMR processing software is to use direct lineshape fitting, or deconvolution, to determine peak areas. In this process resonances are decomposed into clusters of singlet peaks of either Lorentzian or Gaussian shape which have known linewidth and area. The sums of these lines should reproduce the original

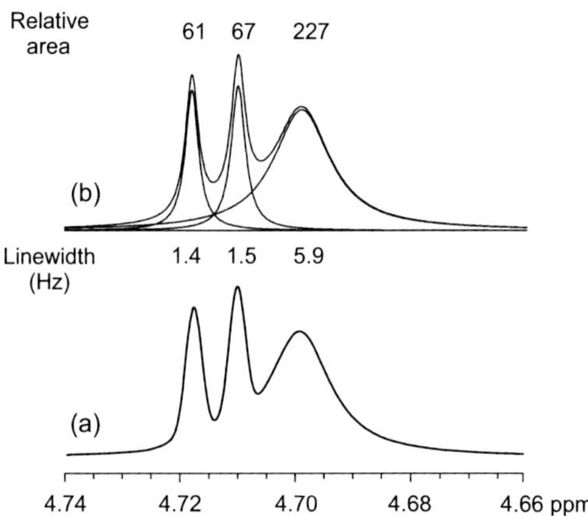

FIGURE 4.6 **Spectrum deconvolution.** The experimental spectrum (a) can be decomposed into singlet Lorentzian lineshapes, (b) the sum of which reproduce the experimental data.

resonance profile, as illustrated in Fig. 4.6. The experimental spectrum comprising an overlapping doublet and broad singlet have been deconvoluted to yield three Lorentizian singlets that represent the relative areas of each component. The use of deconvolution may reduce operator bias in integral definition and may prove to be more reliable in the presence of baseline noise or other low-level artefacts, or when resonances experience overlap with a neighbouring peak and are not sufficiently resolved for integration, as in this example.

4.1.3 Quantification with an Electronic Calibrant: ERETIC

When seeking to determine the concentration of a solute by NMR spectroscopy the established procedure is to measure it against that of a known internal calibrant, as described earlier. However, the use of an internal chemical calibrant may impose constraints that may not be readily met, making this approach undesirable. The calibration material needs to be soluble in the analyte solution, it must be inert to the solute, its resonance(s) must not overlap that of the solute and its relaxation properties should ideally be similar to that of the solute. In addition to these practical considerations, one simply may not wish to add a 'contaminant' to a precious sample. An alternative approach to the use of an internal *chemical* calibrant is to introduce a *synthetic* radio-frequency (rf) calibration signal, as in the Electronic Reference To access In-vivo Concentrations (ERETIC) method [14,15].

The basic approach is to 'leak' a small amount of an rf signal into the NMR detector during the acquisition of the solute FID such that an additional resonance line appears in the final spectrum that arises from the synthesised rf signal. Since this is generated from the NMR spectrometer itself, it is under the control of the operator, meaning the reference line can be placed in any area of the spectrum free from analyte peaks, and its amplitude defined. The application of the method then involves two stages. The first is to *calibrate* the intensity of the synthetic signal itself against a chemical calibrant solution of known concentration. This will define the effective concentration of the ERETIC signal for the second stage of the process, the *quantification* of subsequent samples of unknown concentration. As an illustration, a spectrum with the addition of the ERETIC reference signal at −1.0 ppm is shown in Fig. 4.7.

4.1.3.1 Experimental Implementation

The precise details for implementing the ERETIC method will be dependent upon the make and vintage of your instrument, and here it is possible only to describe the general approach to this; your instrument manufacturer should be able to provide further guidance.

The original approach to introducing the synthetic rf signal is to feed this in via a second rf coil of the probe in use; for the typical case of observing ^{1}H, this would then employ the carbon or broadband (X) channel of the probe. This does not need to be tuned to ^{1}H since we only require a small degree of signal leakage between the coils (Fig. 4.8a). The rf signal does not need to be amplified since the detected signal intensity needs only to be comparable with the NMR response, so is typically taken directly from the second rf signal generation board to the probe, the first being used for the normal pulse-acquire process. To generate a Lorentzian lineshape for the reference signal in the final NMR spectrum, the synthetic signal must exhibit an exponential intensity decay during the data acquisition period (as would the true solute FID) which may be

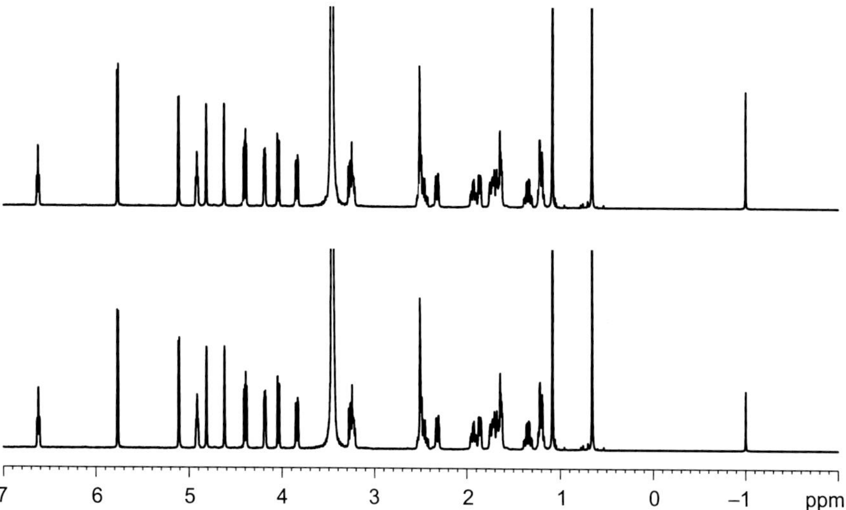

FIGURE 4.7 **Use of the synthetic ERETIC calibration signal for concentration measurement in ^{1}H NMR.** The ERETIC signal appears at −1.0 ppm. The ERETIC intensity in the upper spectrum is precisely twice that in the lower due to a 6 dB increase in ERETIC signal intensity.

possible with the pulse-shaping tools of the spectrometer; that is, the ERETIC pulse can be applied as a single exponentially decaying decoupler pulse throughout the FID acquisition (Fig. 4.8b). The rate of the signal decay will dictate the observed linewidth of the ERETIC calibration signal, so is again under the experimentalist's control. The final parameter that will need optimisation (through trial and error) is the phase of the synthetic rf signal fed to the probe which should be adjusted, preferably under software control, so that the ERETIC line has an identical absorption phase to the solute signals after Fourier transformation and phase correction of the NMR data (Fig. 4.7).

For the method to be reproducible across different samples a number of factors need to remain constant throughout the measurements. The final intensity of the ERETIC signal is dictated by the coupling between the two rf coils, which is influenced by probe tuning and by the dielectric properties of the solvent. Thus, the optimum approach is to use the same

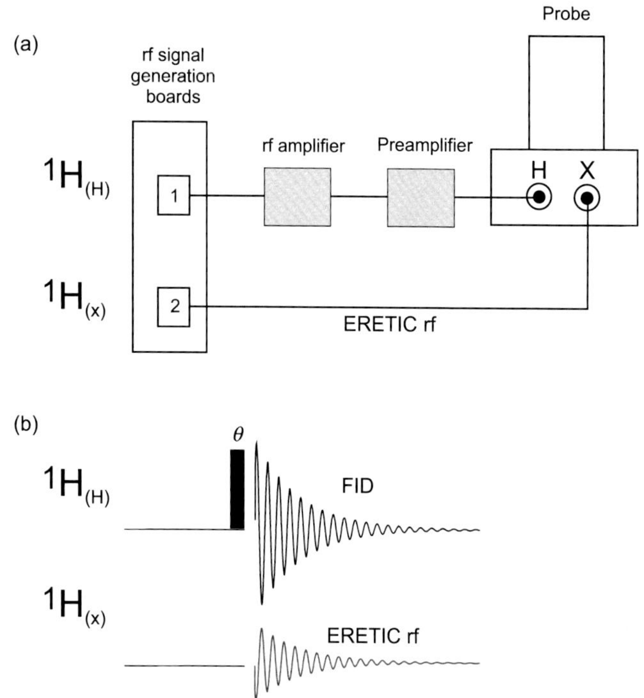

FIGURE 4.8 **Schematic implementation of the ERETIC method.** (a) The rf signal routing and (b) the ERETIC pulse sequence.

solvent throughout, including that of the known compound used to calibrate the ERETIC signal itself, and to keep the probe tuning constant. If a broadband channel has been used to supply the ERETIC signal, this must be tuned to the same nucleus for all measurements, and generally higher frequencies have been observed to give better results, meaning ^{31}P would be favoured if no other constraints apply. The ERETIC method has been demonstrated to have accuracy and precision that is at least as good as that provided by use of an internal calibrant [15,16] and has more recently also been applied to 2D spectroscopy [17].

4.1.4 Quantification with an External Calibrant: PULCON

An alternative procedure for quantification that avoids the need for an internal calibrant, and is simpler to implement than ERETIC, is to employ an external calibration sample. The principle is to compare the detected NMR signal intensity per proton of the analyte against that of the calibrant sample of known concentration and from this determine the analyte concentration directly [18]. In this approach, most parameters are under operator control and can be made identical for both samples or can be accounted for when comparing the analyte and calibrant responses, as described further below. The most critical feature that must be accounted for, however, is the detection sensitivity of the NMR coil for the different samples, which will vary between solvents and solution conditions, and in particular will be sensitive to the ionic strength of aqueous samples. The key to the *Pu*lse-*L*ength based *Con*centration determination (PULCON) method [19] lies in the fact that this variation may be accounted for through the *principle of reciprocity*, which states that the detected NMR signal strength for a coil used for both rf pulsing and detection (the usual scenario in high-resolution NMR) is inversely proportional to the 90 degrees pulse width. This means that, provided the probe is correctly tuned and matched for each sample, the calibrated pulse widths may be used to correct for probe sensitivity variation between samples. The unknown concentration of the analyte C_U may then be related to that of the calibrant C_C from:

$$C_U = C_C f\left(\frac{S_U T_U \theta_U N_C}{S_C T_C \theta_C N_U}\right) \tag{4.2}$$

where S represents the observed signal intensity (ie the absolute integrated intensity of a proton), T the sample temperature (in K), θ is the calibrated 90 degree pulse duration and N is the number of transients collected for each sample. The scaling factor f may be included to accommodate other differences in experimental parameters used for the analyte and calibration samples, as discussed further below. For the common case of routine concentration measurements of mM samples, the sample temperature remains constant and a single transient suffices, reducing the expression to:

$$C_U = C_C f\left(\frac{S_U \theta_U}{S_C \theta_C}\right) \tag{4.3}$$

leading to ready determination of analyte concentration.

4.1.4.1 Experimental Implementation

The PULCON method is applicable only when separate spectra for the analyte and calibrant samples being compared have been acquired with *identical* instrument hardware, including probes and any associated rf filters. The most common protocol employs the single-pulse 1D acquisition scheme of Fig. 4.1, using a 90 degree excitation pulse, although solvent presaturation may also be required for aqueous solutions. The 90 degree pulse duration should be accurately calibrated for each sample after careful probe tuning and matching in each case (see Section 3.5.1; 360 degree pulse calibration is equally valid for this purpose and is often quicker to determine). Other acquisition parameters should ideally remain identical for both samples, although this may not always be possible. Changes in receiver gain settings lead to variations in detected signal intensities, so should be made the same wherever possible. When different values are required, such as to avoid receiver overload, these differences should be included in the scaling factor f of Eqs 4.2 and 4.3. This correction is valid on modern digital spectrometers where gain settings produce a linear response in signal intensity, but this may not be the case with older spectrometers for which identical gain settings are recommended. One should also be aware of internal software scaling factors that come in to play when processing data, since any difference in these may also need to be accounted for in the scaling factor f (vendor algorithms that determine sample concentrations using PULCON will usually take account of these often enigmatic parameters). Finally, when comparing integrals between samples, one must account for the number of protons giving rise to the resonances being compared, and also factor this into the intensity comparisons being made.

Choice of the calibration samples for PULCON is simplified by the fact that resonance overlap with analyte signals is no longer an issue, meaning samples may be selected with attractive sample-handling properties and ready availability, and, once the calibration spectrum is collected, it may be retained for future concentration measurements. Typically, a calibration sample of ~10 mM will allow for analyte concentration determinations of between 0.1 and 100 mM

4.2 SPIN-DECOUPLING METHODS

The use of spin–spin decoupling is no doubt familiar to most regular users of solution NMR spectroscopy, and it has constituted an important tool for the chemist since the early applications of NMR to problems of chemical structure [20]. It is now so widely used in the observation of heteronuclear spectra (when did you last see a fully proton-coupled carbon spectrum?) that one can almost become oblivious to the fact that it is an integral part of numerous pulse experiments. Nowadays, scalar spin–spin decoupling is most often applied with one of two goals in mind: either the *selective* decoupling of a single resonance in an attempt to identify its coupling partner(s), or non-selective (*broadband*) decoupling of one nuclide to enhance and simplify the spectrum of another. Decoupling is usually classified as being *homonuclear*, in which the decoupled and observed nuclides are the same, or *heteronuclear* in which they differ. That which experiences the decoupling rf is conventionally distinguished by placing it in curly brackets; so, for example, $^{13}C\{^{1}H\}$ would indicate carbon observation in the presence of proton decoupling, whilst $^{1}H\{^{13}C, ^{31}P\}$ would represent proton observation in the presence of simultaneous carbon and phosphorus decoupling (or more generally carbon and phosphorus pulsing). One may also find spin decoupling being referred to as one example of a 'double-resonance' experiment, a term originating from CW experiments in which the observe and decouple rf were applied simultaneously. Likewise, the H, C, P example may be termed a 'triple-resonance' experiment as it requires three rf channels.

4.2.1 Basis of Spin Decoupling

A simplified [21] yet convenient description of how spin decoupling operates considers two spin-½ nuclei A and X, that share a mutual scalar coupling of J Hz. The resonant frequency of the X-spins will depend on whether their coupled partners are oriented parallel (α orientation) or anti-parallel (β orientation) to the applied static field. For a spin ensemble we can assume there exist an equal number of A nuclei in the α and β states, owing to the very small energy difference between the two orientations, and thus the X spectrum displays the familiar doublet pattern. Application of an rf field, designated the B_2 field (recall the transmitter field is termed B_1), at the frequency of the A spins causes these to undergo continuous, rapid transitions between the α and β orientations by continually inverting these spins. If this reorientation is fast relative to the coupling constant, the X-spin doublet coalesces into a singlet since the lifetimes of the α and β orientations are no longer sufficient for the coupling to be distinguished. Thus, if the A spins are irradiated during data acquisition with a sufficiently strong field such that $\gamma B_2 > J$ Hz, the X resonance displays no coupling to A and the spins are said to be decoupled.

Whilst the removal of scalar spin–spin coupling is the usual goal of such experiments, a number of additional effects can arise from application of the additional rf field, which may be beneficial or detrimental depending on the circumstances. Incomplete decoupling can introduce residual line broadening or, even worse, leave rather esoteric partially decoupled multiplets, whilst the non-uniform irradiation of a resonance can introduce population transfer effects which cause intensity distortions *within* multiplets (Section 4.4). Population disturbances caused by the rf may also produce intensity changes that arise from the NOE (see the chapter *Correlations Through Space: The Nuclear Overhauser Effect*), which operates quite independently from J coupling. Finally, changes in the resonant positions of signals close in frequency to the applied rf may also be observed, so-called 'Bloch–Siegert shifts'. In many cases it is possible to have some control over these factors, according to the experimental protocol used, as described in later sections.

4.2.2 Homonuclear Decoupling

Homonuclear decoupling traditionally involves selective application of a coherent decoupling field to a target resonance with the aim of identifying scalar spin–coupled partners, and is most often applied in proton spectroscopy (Fig. 4.9). Although the use of this method for identifying coupled ^{1}H–^{1}H spins has been largely superseded by 2D correlation methods (see the Section 6.1), selective decoupling can still be a very useful and convenient tool in the NMR armoury. It is very simple to set-up, providing rapid answers to relatively simple questions, and can be particularly useful in identifying spins that share very small couplings which do not always reveal themselves in 2D correlation experiments. In very crowded

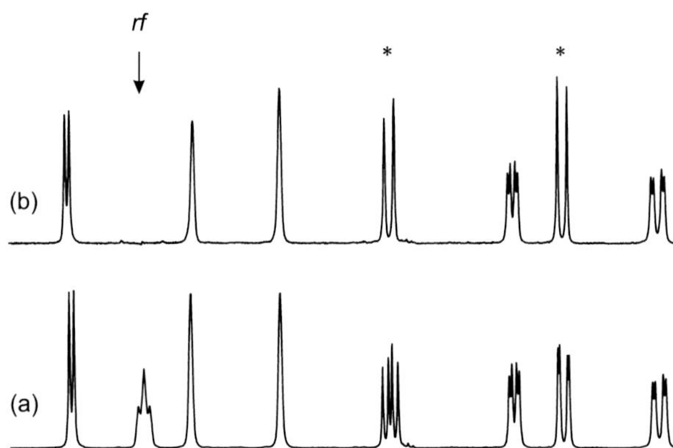

FIGURE 4.9 **Homonuclear decoupling allows rapid identification of coupled partners by removing couplings to the irradiated spin.** (a) Control spectrum and (b) decoupled spectrum.

spectra where only a specific interaction is to be investigated, the affected resonances may not be obvious and the use of *difference spectroscopy* can aid interpretation [22,23]. Here, a control spectrum recorded in the absence of decoupling is subtracted from that collected in the presence of on-resonance decoupling to reveal any changes that arise. (Note that for any form of difference spectroscopy, it is best practice to collect the control FID with the decoupler frequency applied far away from all resonances rather than being turned off completely). In such cases a 2D correlation experiment may however be more appropriate. Recent methods that aim to achieve *broadband* homonuclear decoupling of proton spectra (known as 'pure shift' experiments) are described in Section 8.5.

One of the limitations to the use of selective decoupling lies in the need to irradiate only a single resonance to identify the coupling partners of the desired target. Any saturation spill-over onto neighbouring resonances introduces a degree of ambiguity into the interpretation. Where the target multiplet is free from other resonances, this poses little problem and the decoupler power can be set sufficiently high to ensure complete decoupling ($\gamma B_2 > J$ Hz). When other resonances are close in frequency to the target, it may be necessary to reduce the decoupler power and thus reduce the frequency spread to avoid disturbing neighbouring resonances. The penalty for reducing the decoupler power may be incomplete decoupling of target spins, although changes within the fine structure of other resonances should be sufficient to identify coupled partners.

4.2.2.1 Bloch–Siegert Shifts

Application of an rf field during acquisition of the FID may also move signals that resonate close to the decoupler frequency. This effect is known as the *Bloch–Siegert shift* [24,25] and, more formally, it occurs when γB_2 becomes comparable with the shift difference in hertz between the decoupling frequency and the resonance. It arises because the decoupling field acts on neighbouring spins such that they experience a modified effective field that is inversely dependent upon their resonance offsets from the decoupling frequency, but proportional to $(\gamma B_2)^2$, which causes resonances to move *away* from this decoupling frequency (Fig. 4.10). The effect is principally limited to homonuclear decoupling experiments, where resonances may be very close to the decoupling frequency, but is of no concern in most pulse NMR experiments since the rf pulse is turned off prior to data collection. A notable exception is when solvent presaturation is applied during the evolution time (t_1) of proton homonuclear 2D experiments. This can lead to shifts of f_1 frequencies of resonances close to the solvent, but, since this is not present in f_2, it introduces asymmetry in the shifts of crosspeaks associated with these resonances. In homonuclear decoupling experiments it is rarely a major problem since the requirement for selectivity limits the B_2 field and thus keeps the shifts small. However, such small shifts may still introduce subtraction artefacts into decoupling difference spectra since the reference will not contain Bloch–Siegert shifts, so one should be cautious not to interpret these as evidence of coupling. Caution is also required should one need to measure accurate chemical shifts from decoupled spectra. In Section 3.5.1 the Bloch–Siegert shift is used quantitatively as a means of calibrating decoupler powers.

4.2.2.2 Experimental Implementation

The application of homonuclear irradiation whilst acquiring the FID poses some challenging instrumental problems. Whilst needing to detect the responses of excited spins, one must not have the receiver open when the decoupler is on since this will simply swamp the NMR signal. The solution lies in discrete sampling of the FID, and the application of homonuclear irradiation *only* when data points are *not* being sampled, that is, during the FID dwell time [26]. Spectrometers have

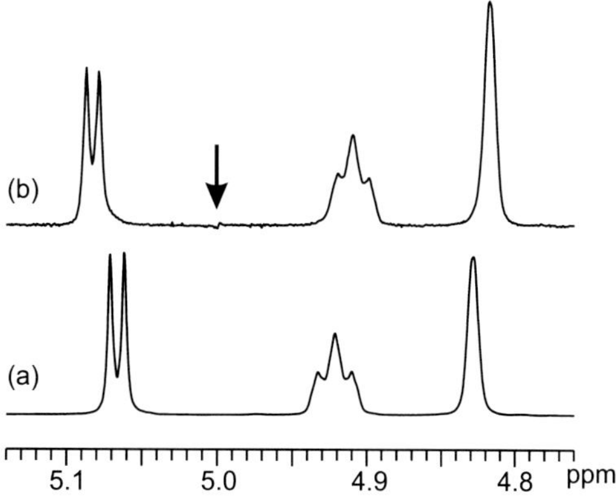

FIGURE 4.10 **The Bloch–Siegert shift.** This causes resonances near an applied rf to move away from its point of application. Plot (a) is the conventional spectrum and (b) is the spectrum acquired with a decoupling field applied at the position of the arrow.

purpose-built homonuclear decoupling modes to handle the necessary gating internally. The time in which the decoupler is gated on is thus only a small fraction of the total acquisition time, this so-called *duty cycle* being typically 20% or less. The low duty cycle means the effective mean decoupler power is somewhat less than the instantaneous B_2 field, hence rf powers are usually greater than those required for presaturation of a resonance where such receiver conflict does not arise. Should decoupler gating-off not be perfect during data collection, a number of spectrum artefacts may be introduced [27,28]. Most notable are a reduction in signal-to-noise, owing to 'leakage' of the decoupler rf into the receiver, and a significant 'spike' occurring at the decoupler frequency, although this is really only a question of aesthetics. The leakage problem varies greatly, it seems, from one instrument to another and if it is a serious problem it may be cured by instrumental modification [28]. The second may be eliminated simply by setting the *transmitter* frequency to match that of the decoupler [29], so that the usual phase-cycling routines employed in 1D acquisitions remove the unsightly 'zero-frequency' spike (Fig. 4.11).

4.2.3 Heteronuclear Decoupling

4.2.3.1 X{^{1}H} Decoupling

The application of broadband proton decoupling during the acquisition of carbon-13 spectra is now universally applied. The removal of all ^{1}H–^{13}C couplings concentrates all the carbon resonance intensity into a single line providing a significant increase in signal intensity and simplification of the spectrum. As an added bonus, the continuous saturation of proton spins provides further enhancement of the signal (by as much as 200% in the case of carbon-13) due to the NOE (see

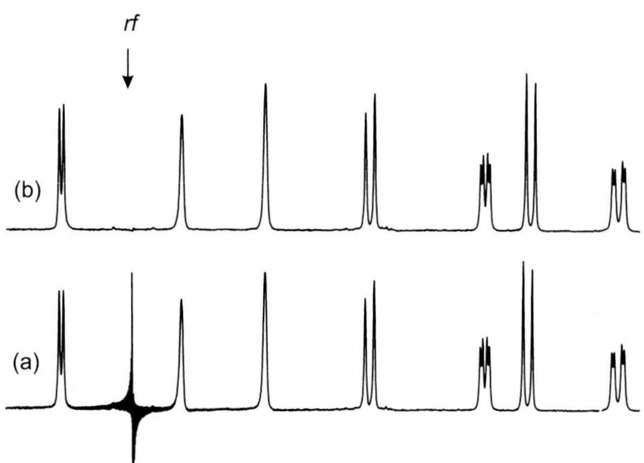

FIGURE 4.11 **Optimising homonuclear decoupling.** The aesthetically unappealing decoupler frequency 'spike' sometimes observed in homonuclear-decoupling experiments, as in (a), can be readily removed by setting the transmitter and decoupler frequencies to be the same, as in (b).

Section 9.3). For these reasons, the use of broadband proton decoupling is standard practice for routine observation of most commonly encountered nuclides.

The principles behind heteronuclear decoupling are no different from the homonuclear case, the major difference being the very much larger frequency separation between the decoupling and observed frequencies, meaning Bloch–Siegert shifts and receiver interference are no longer a problem. The use of broadband heteronuclear decoupling is itself associated with a number of other practical problems. First, there is the need to decouple uniformly over the whole of the proton spectrum, and this requires the applied rf to be effective over a far greater frequency window. Decoupling bandwidths are typically of the order of many kilohertz and are greater at higher field strengths. A 10 ppm proton spectrum covers 4 kHz at 400 MHz but 6 kHz at 600 MHz, for example. Generally, a wider decoupling bandwidth can be achieved by increasing decoupler power, although the continuous application of high-power rf is more likely to destroy the probe and sample than achieve the desired results. High powers also give rise to the problem of rf sample heating, which is most pronounced in ionic samples and can be disastrous for heat-sensitive compounds. The effect is insidious as the heating occurs within the sample itself, so the usual method of sample temperature monitoring (a thermocouple placed within the airflow surrounding the sample) is largely insensitive to its presence, the usual indicator being a drift of the lock level when the pulse sequence is started. To overcome the problem of decoupling over wide bandwidths without the need for excessive powers, specially designed modulated decoupling schemes are employed (which include the so-called *composite pulse decoupling sequences* and more recently *adiabatic decoupling pulses*, described in Section 12.3). The careful design of probes can also assist by reducing sample heating which is caused by the *electric* component of the rf. As with pulse excitation, the required frequency spread can be kept to a minimum by placing the decoupler frequency in the *centre* of the region to be decoupled. If your heteronuclear spectra display unexpectedly broad or even split resonances, this may be due to the decoupler frequency being incorrect, and/or the decoupler channel being poorly tuned or wrongly calibrated (see Section 3.4). If these perturbations occur only at the edges of the spectrum, the decoupling scheme effective bandwidth may be inadequate.

Sample heating can also be reduced by gating off the rf when not essential to the experiment (Fig. 4.12). Gating the decoupler off during recovery delay (*inverse-gated decoupling*, Fig. 4.12b) also removes the NOE and provides a decoupled spectrum *without* NOE enhancements. This is because any NOE that builds up during the acquisition time affects only longitudinal magnetisation, so does not influence the detected (transverse) signals, and is then allowed to decay during recovery delay, so again has no influence. This method has particular value in the observation of nuclei with negative magnetogyric ratios, since the NOE causes a *decrease* in signal intensity for such nuclei (see Section 9.3). It is also employed for accurate quantitative measurements, as discussed in the previous section. Conversely, running the decoupler during relaxation delay but gating it off during the acquisition period (*gated decoupling*, Fig. 4.12a) results in a spectrum that retains spin coupling and is enhanced by the NOE. A further method for reducing sample heating provides a decoupled spectrum with enhancement by the NOE and is known as *power-gated decoupling* (Fig. 4.12c). Here, high decoupler power is applied during acquisition to achieve complete decoupling, but reduced power is applied between transients to maintain a degree of saturation and hence develop the NOE. This approach is particularly applicable to studies at high field where larger bandwidths demand relatively high decoupler powers. Comparison of these various decoupling schemes is illustrated in Fig. 4.13 for the terpene α-pinene **4.2**.

4.2

By analogy with homonuclear decoupling, *selective* heteronuclear decoupling is also possible. The decoupling of a single proton resonance could in principle identify coupled partners through the observation of collapsed multiplet structure in the heterospin spectrum. This however requires the decoupling rf to be effective over the *satellites* of the parent proton resonance, which may be far apart if $^1J_{XH}$ is large. One-bond proton–carbon couplings exceed 100 Hz and may be hard to remove whilst retaining sufficient selectivity, so here a 2D heteronuclear correlation spectrum is likely to be a more efficient approach. The selective removal of long-range (two- or three-bond) proton–carbon couplings is more readily achieved

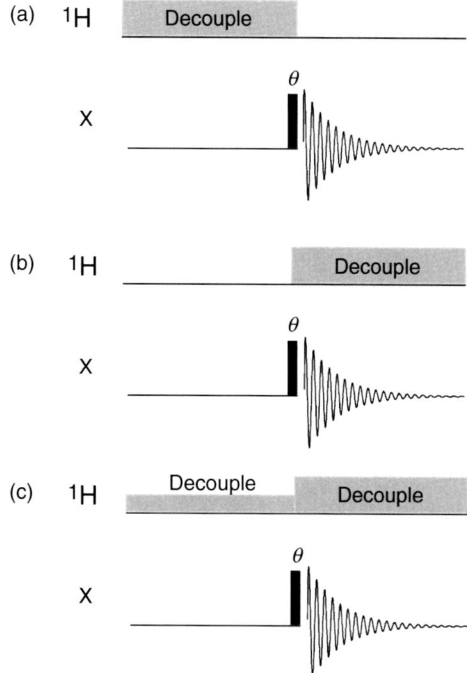

FIGURE 4.12 **Heteronuclear ^{1}H decoupling schemes.** (a) Gated decoupling (coupled spectrum with NOE), (b) inverse-gated decoupling (decoupled spectrum without NOE) and (c) power-gated decoupling where the rf is applied at two different powers (decoupled spectrum with NOE).

since these are typically less than 10 Hz. An example of this is illustrated by identification of configurational isomers of **4.3** through the measurement of three-bond proton–carbon couplings (Fig. 4.14) which share a Karplus-type dependence on dihedral bond angles. The magnitude of proton couplings to carbon atoms of the nitrile and the ester across the alkene would therefore indicate their relative stereochemistry, but whilst the nitrile coupling presented a clear doublet in the proton-coupled ^{13}C spectrum, the carbonyl coupling was masked by the additional three-bond couplings to the ester methyl protons. Selective decoupling of these methyl protons eliminated this interference resulting in a carbonyl doublet, sufficient to identify the nitrile as being *trans* to the alkene proton.

4.2.3.2 ^{1}H{X} Decoupling

Traditionally, FT spectrometers were built with the ability to decouple only protons whilst observing the heteronucleus, as described earlier. Modern instruments now have the capability of providing X-nucleus pulsing and decoupling whilst observing protons, ^{1}H{X} (the 'inverse' configuration). The bandwidths required for broadband decoupling of, for example, carbon-13, far exceed those needed for proton decoupling, and the composite pulse or adiabatic decoupling methods described in Section 12.3 become essential for success. Thus, carbon-13 decoupling over 150 ppm (a typical range for proton-bearing carbons) at 400 MHz–^{1}H requires a bandwidth of 15 kHz and the rf heating problems mentioned earlier are generally more problematic with heteronuclear decoupling, especially at high fields. Broadband X-nucleus decoupling is most frequently applied as part of a multidimensional pulse sequence to simplify crosspeak structures, although it can be a useful tool in interpreting 1D proton spectra when a heteronucleus has high natural abundance and thus makes a significant contribution to coupling fine structure, for example, ^{19}F or ^{31}P. Fig. 4.15 demonstrates simplification of the ^{1}H spectrum of the palladium phosphine **4.4** on application of broadband ^{31}P decoupling, and this procedure often aids the interpretation of multiplet structures or the extraction of homonuclear proton couplings. Likewise, if X-nucleus chemical shifts are known, *selective* X-decoupling can be used to identify coupled partners in the proton spectrum, and in simple cases this may serve as a ready alternative to a 2D heteronuclear correlation experiment. In fact, this approach was used to identify the resonant frequencies of heteronuclei prior to the advent of the pulse–FT methods [20] which made their direct observation possible. Thus, the proton spectrum was recorded whilst applying a second rf field to the heteronucleus. Successively stepping the decoupler frequency and repeating the measurement would ultimately indicate the X-resonance position when the X–^{1}H coupling was seen to disappear from the proton spectrum.

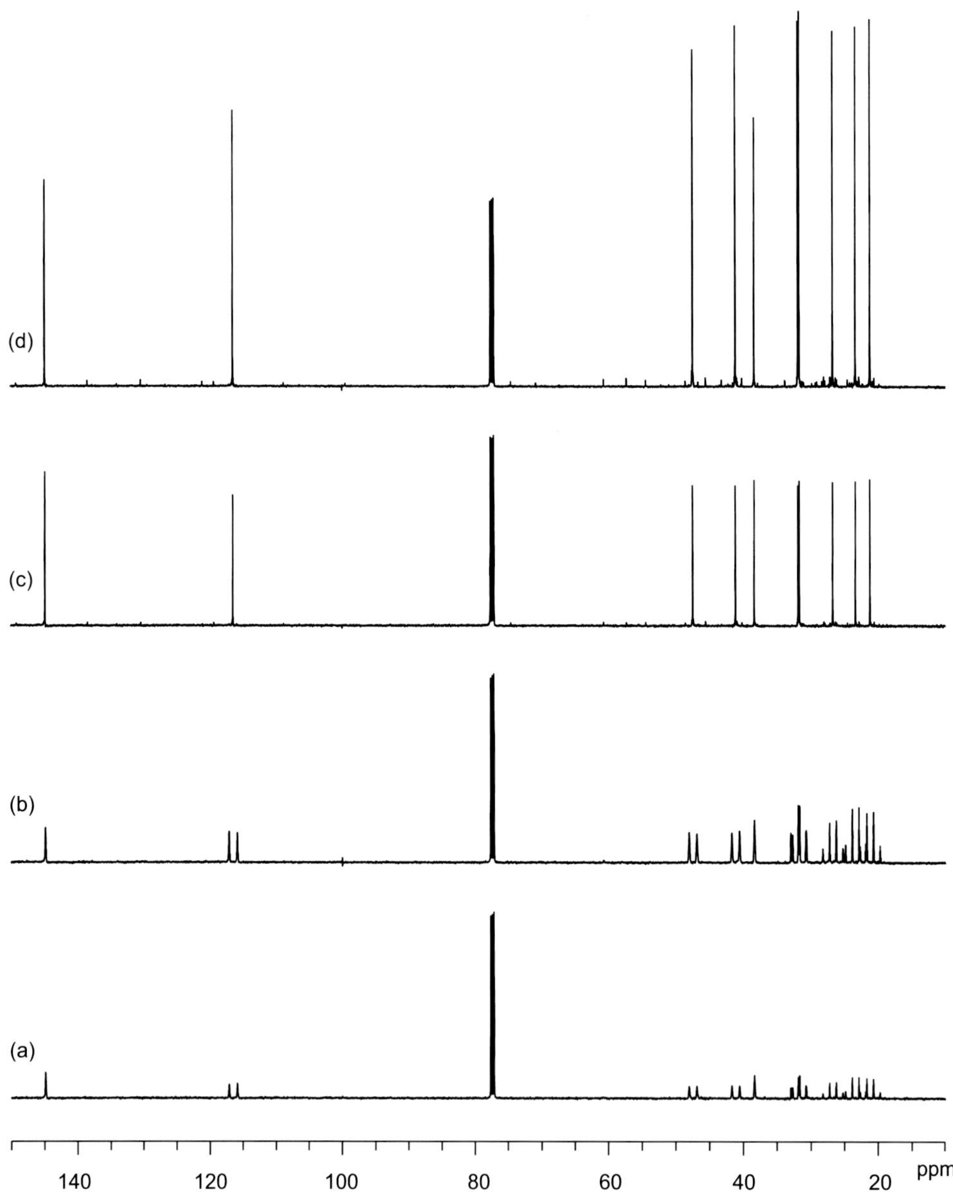

FIGURE 4.13 **The carbon-13 spectrum of α-pinene 4.2 in $CDCl_3$.** Spectra were acquired (a) without proton irradiation at any stage (coupled spectrum without NOE), (b) with gated decoupling (coupled spectrum with NOE), (c) with inverse-gated decoupling (decoupled spectrum without NOE) and (d) power-gated decoupling (decoupled spectrum with NOE). All other experimental conditions were identical and the same absolute scaling was used for each plot.

4.3 SPECTRUM EDITING WITH SPIN-ECHOES

The principal reason behind application of broadband proton decoupling of heteronuclei is removal of the coupling structure to concentrate signal intensity, thereby improving the signal-to-noise ratio and reducing resonance overlap (Fig. 4.13). Additional gains such as signal enhancements from the NOE and clarification of any remaining homonuclear couplings may also arise. Set against these obvious benefits is the loss of multiplicity information present in the proton-coupled spectrum (Fig. 4.13a), meaning there is no way, *a priori*, of distinguishing, for example, a methine from a methylene carbon resonance. It is therefore desirable to be able to record fully proton-decoupled spectra yet still retain valuable multiplicity data. Some of the earliest multipulse sequences were designed to achieve this aim. These were based on simple spin-echoes and provided spectra in which the multiplicities were encoded as signal intensities and signs. Despite competition from 1D and 2D methods based on polarisation transfer described shortly, these techniques still find use in chemistry laboratories. As they are also rather easy to understand, they provide a suitable introduction to the idea of spectrum multiplicity editing.

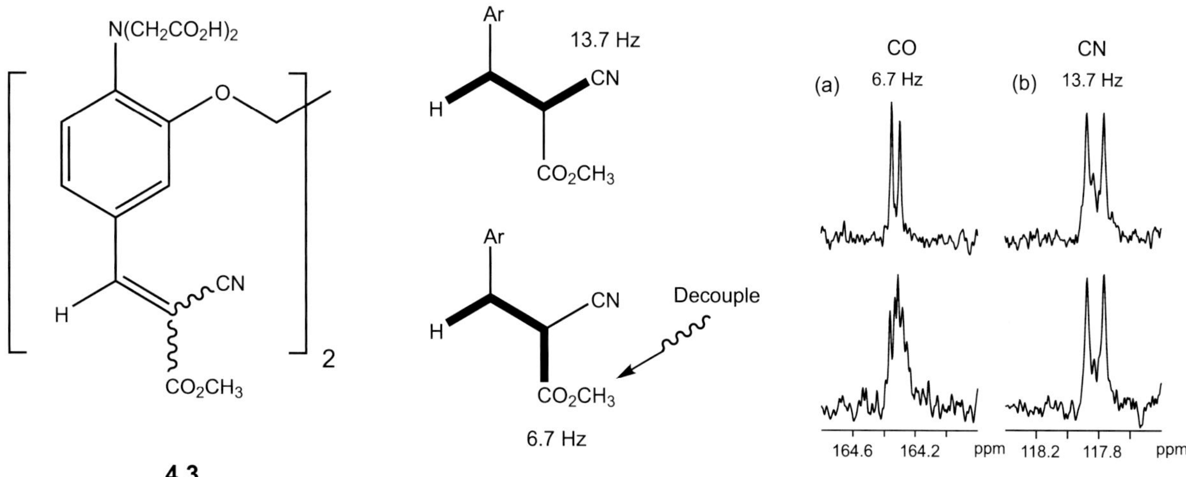

FIGURE 4.14 **Application of selective proton decoupling in the measurement of heteronuclear long-range proton–carbon coupling constants.** Partial carbon-13 spectra are shown for (a) the CO groups and (b) the CN group. Lower traces are from the fully proton-coupled carbon-13 spectrum and the upper traces from that in which the methyl ester protons of **4.3** were selectively decoupled to reveal the three-bond coupling of the carbonyl carbon across the alkene.

The sections that follow utilise the rotating-frame vector model to explain pictorially the operation of the experiments, and familiarity with the introduction in the Section 2.2 is assumed.

4.3.1 J-Modulated Spin-Echo

One of the simplest approaches to editing is the J-modulated spin-echo sequence [30] (Fig. 4.16a, also referred to as the spin-echo Fourier transform (SEFT, [31]) which can be readily appreciated with reference to the vector model. The key to understanding this sequence is the realisation that the evolution of carbon magnetisation vectors under the influence of the $^1J_{CH}$ coupling only occurs when the proton decoupler is gated off, whilst at all other times only carbon chemical shifts are effective. Further simplification comes from ignoring carbon chemical shifts altogether since shift evolution during the first Δ period is precisely refocused during the second by the 180 degree (C) refocusing pulse

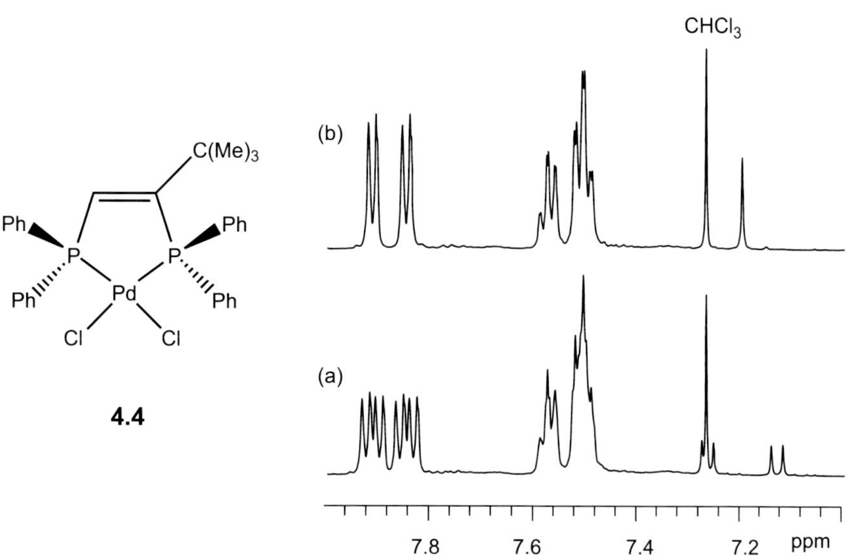

FIGURE 4.15 **Phosphorus decoupling of a proton spectrum.** Simplification of the conventional proton spectrum (a) of the palladium phosphine **4.4** in $CDCl_3$ is possible by application of broadband phosphorus decoupling (b). All long-range ^1H–^{31}P couplings are removed, as is most apparent for the alkene proton (7.2 ppm) and the *ortho* protons of the phenyl rings (above 7.8 ppm).

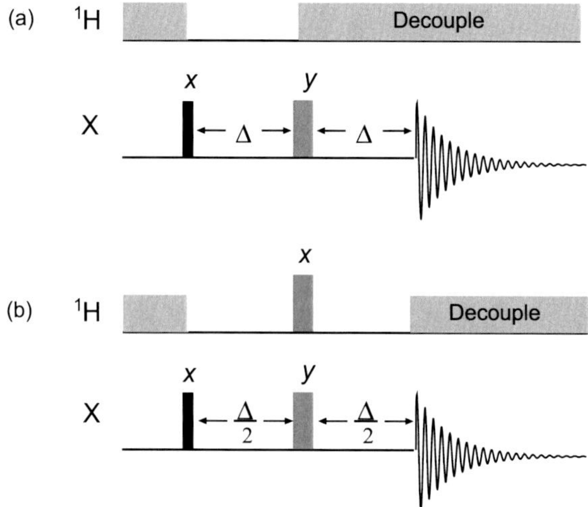

FIGURE 4.16 **J-modulated spin-echo sequences.** (a) The decoupler-gated variant and (b) the pulsed variant.

(see Section 2.2). Thus, to understand this sequence one only need consider the influence of heteronuclear coupling during the first Δ period.

Consider events with Δ = 1/J s, for which the relevant vector evolution for different multiplicities is illustrated in Fig. 4.17. Since chemical shifts play no part, quaternary carbons remain stationary along +*y* during Δ (long-range C—H coupling that may exist will be far smaller than the one-bond coupling and may be considered negligible). Doublet vectors for a methine pair evolve at ±J/2 Hz, so will each rotate through one-half cycle in 1/J s and hence meet once more along −*y*. As these now have a 180 degrees phase difference with respect to the quaternary signals, they will ultimately appear inverted in the final spectrum. Applying these arguments to the other multiplicities shows that methylene vectors evolving

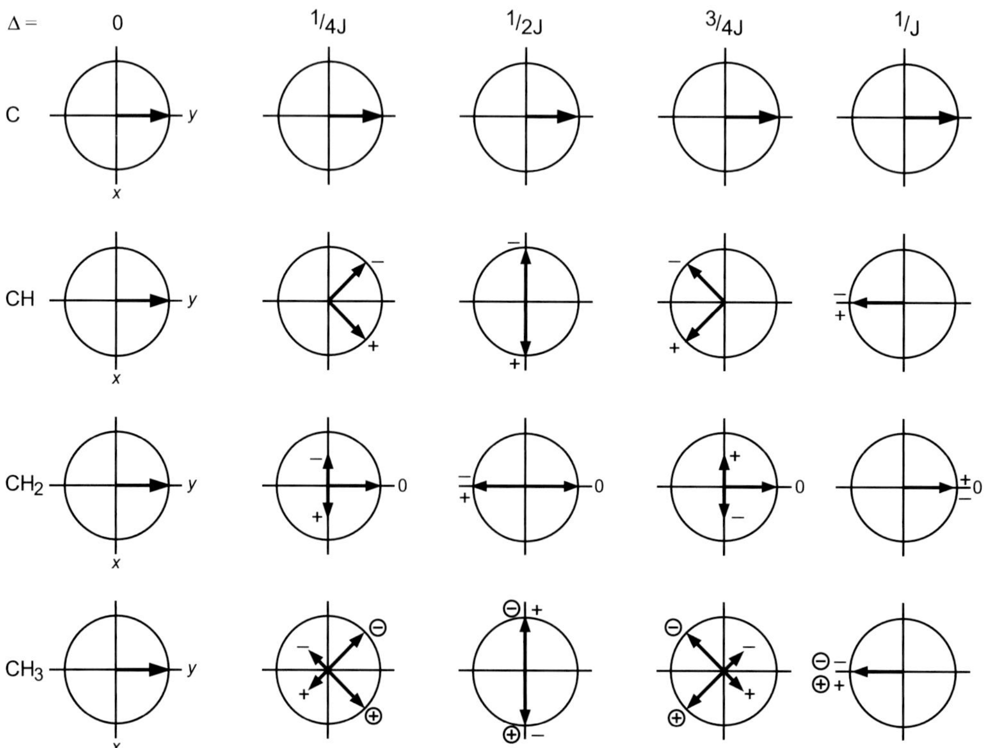

FIGURE 4.17 **Evolution of carbon magnetisation vectors under the influence of proton–carbon couplings.** Vectors are shown for carbon singlets (C), doublets (CH), triplets (CH_2) and quartets (CH_3).

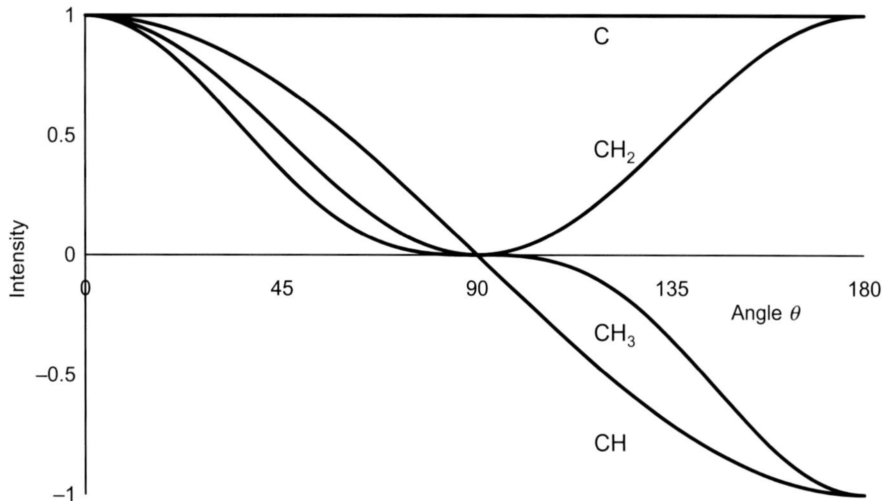

FIGURE 4.18 **The variation of carbon signal intensities in the J-modulated spin-echo as a function of the evolution time Δ ($\theta = 180J\Delta$ degrees).**

at ±J Hz will align with $+y$ whilst methyl vectors evolving at ±J/2 and ±3J/2 Hz will terminate along $-y$. More generally, by defining an angle θ such that $\theta = 180J\Delta$ degrees, the signal intensities of carbon multiplicities I vary according to:

$$\begin{aligned} &C: && I = 1 \\ &CH: && I \propto \cos\theta \\ &CH_2: && I \propto \cos^2\theta \\ &CH_3: && I \propto \cos^3\theta \end{aligned}$$

as illustrated in Fig. 4.18. The spectrum for Δ = 1/J s (θ = 180 degrees) will therefore display quaternaries and methylenes positive with methines and methyls negative if phased as for the one-pulse carbon spectrum (Fig. 4.19b), although edited spectra are often presented with methine resonances positive and methylene negative, so check local convention. The carbon multiplicities therefore become encoded as signal intensities and at least some of the multiplicity information lost in the single-pulse carbon experiment has been recovered, whilst still benefitting from decoupled resonances. The application of proton decoupling for all but a short time period Δ also ensures spectra are acquired with enhancement from the NOE.

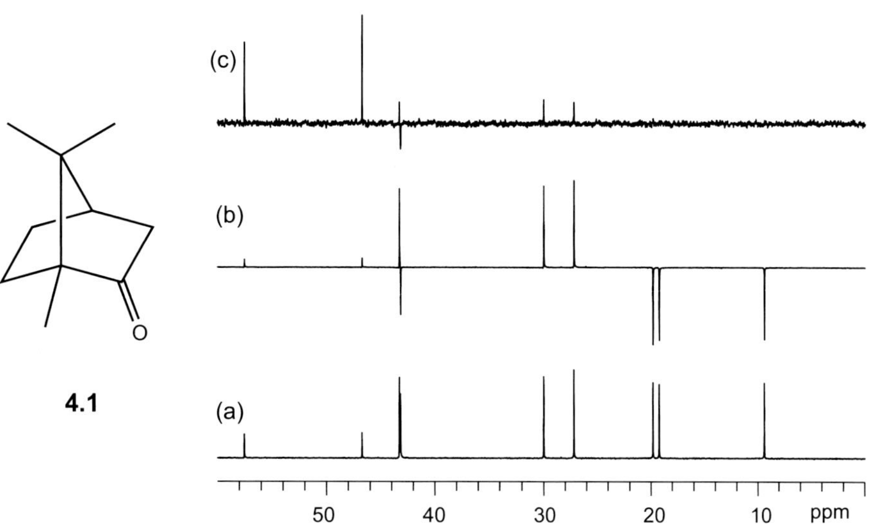

FIGURE 4.19 **Carbon spectrum of camphor 4.1 edited with the J-modulated spin-echo sequence.** (a) Conventional carbon spectrum (carbonyl not shown), and edited spectra with (b) Δ= 1/J (θ = 180 degrees) and (c) Δ = 1/2J (θ = 90 degrees) with J assumed to be 130 Hz. Some breakthrough of protonated carbons is observed in (c) due to variations in coupling constants within the molecule.

TABLE 4.3 Typical Ranges for One-Bond Proton–Carbon Coupling Constants

Carbon Environment	Typical $^1J_{CH}$ Range (Hz)
Aliphatic, CH_n-	125–135
Aliphatic, CH_nX (X = N, O, S)	135–155
Alkene	155–170
Alkyne	240–250
Aromatic	155–165

Setting $\Delta = 1/2J$ s ($\theta = 90$ degrees) corresponds to a null for all protonated carbons (Fig. 4.18), so producing a quaternary-only spectrum (Fig. 4.19c). The accuracy of editing is clearly critically dependent on the correct setting of Δ, which is in turn dependent upon J, and as there are likely to be a wide variety of J values within a sample, one is forced to make some compromise setting for Δ. One-bond proton–carbon couplings range between 125 and 250 Hz, although they are more commonly between 130 and 170 Hz (Table 4.3), so a typical value for Δ would be ≈7 ms ($^1J_{CH} \approx 140$ Hz) when aromatics and carbon centres bearing electronegative heteroatoms are anticipated. If couplings fall far from the chosen value the corresponding resonance can display unexpected and potentially confusing behaviour. Alkyne carbons are particularly prone to this, owing to exceptionally large $^1J_{CH}$ values.

The spin-echo experiment is particularly simple to set-up as it does not require proton pulses or their calibration, a desirable property when the experiment was first introduced but of little consequence nowadays. The same results can, in fact, be obtained by the use of proton 180 degree pulses rather than gating of the decoupler [32] (Fig. 4.16b). In this case the Δ period is broken in two periods of 1/2J separated by simultaneous application of proton and carbon 180 degrees pulses. These serve to refocus carbon chemical shifts but at the same time allow couplings to continue to evolve during the second Δ/2 period (Section 2.2). Hence, the total evolution period in which coupling is active is Δ as in the decoupler-gating experiment, and identical modulation patterns are produced. It is this shorter, pulsed form of the heteronuclear spin-echo that is widely used within numerous pulse sequences to refocus shift evolution yct lcavc couplings to evolve.

4.3.2 APT

The principle disadvantage of the J-modulated spin-echo described above is the use of a 90 degree carbon excitation pulse which, as discussed in Section 4.1, is not optimum for signal averaging and may lead to signal saturation, notably of quaternary centres. The preferred approach using an excitation pulse width somewhat less than 90 degrees requires a slight modification of the J-modulated experiment, giving rise to the attached proton test (APT) sequence [33] (Fig. 4.20). Use of a small tip angle excitation pulse leaves a component of magnetisation along the $+z$-axis, which is inverted by the 180 degree (C) pulse that follows. In systems with slowly relaxing spins, this inverted component may cancel magnetisation arising from relaxation of the transverse components, leaving little or no net signal to observe on subsequent cycles. It is therefore necessary to add a further 180 degree carbon pulse which returns the problematic $-z$ component to $+z$ prior to acquisition, thereby eliminating possible cancellation. Transverse components also experience a 180 degrees rotation prior to detection, but are otherwise unaffected beyond this phase inversion. The APT experiment is commonly used with excitation angles of 45 degrees or less and is thus better suited to signal averaging than the basic echo sequence, but otherwise gives similar editing results.

The poor editing accuracy of spin-echoes in the presence of a wide range of J values and the inability to fully characterise all carbon multiplicities are the major limitations of these techniques. More complex variations on the pulsed J-modulated spin-echo are to be found that do allow complete decomposition of the carbon spectra into C, CH, CH_2 and

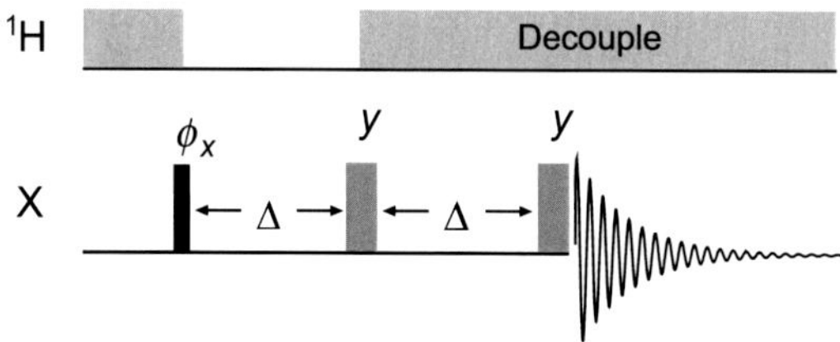

FIGURE 4.20 **The APT sequence.** An excitation pulse (ϕ) less than 90 degrees may be used.

CH_3 sub-spectra [34] and which also show greater tolerances to variations in $^1J_{CH}$ [35]. Likewise, J-compensated APT sequences have been developed for greater tolerance to a spread of J values [36], and for the direct generation of complete sub-spectra [37]. Invariably, the simplest sequences still find widest use.

4.4 SENSITIVITY ENHANCEMENT AND SPECTRUM EDITING

One of the principal concerns for the chemist using NMR spectroscopy is the relatively poor sensitivity of the technique when compared with other analytical methods, which stems from the small energy differences and hence small population differences between spin states. Developments that provide increased signal intensity relative to background noise are a constant goal of research within the NMR community, be it in the design of the NMR instrumentation itself or in novel pulse techniques and data-processing methods. The continued development of higher field magnets clearly contributes to gains in sensitivity, and is paralleled by other instrumental advances such as the probe developments described in Section 3.4.2. Although that section considered the issue of sensitivity with reference to probe design, we now return to this topic in relation to pulse techniques so will again review some of the principal factors that govern this.

Aside from instrumental developments, optimum sensitivity is provided by observing the nuclear species giving rise to the strongest signal, as judged by the intrinsic sensitivity of the nucleus and its natural abundance. The intrinsic sensitivity depends upon the magnetogyric ratio γ of the spin in three ways, with a greater γ contributing to:

- a high resonant frequency, which in turn implies a large transition energy difference and hence a greater Boltzmann population difference;
- a high magnetic moment and therefore a stronger signal; and
- a high *rate* of precession which induces a greater signal in the detection coil (just as cycling faster would cause a bicycle dynamo lamp to glow brighter, should you happen to possess one).

Thus, in general, the strength of an NMR signal is proportional to γ^3 for a single nuclide, but as noise itself increases with the square root of the observation frequency [38] the signal-to-noise ratio scales as $\gamma^{5/2}$. Note also that as two of the aforementioned terms depend upon resonant frequency, the signal-to-noise ratio for a single nuclide would be expected to scale with the static field according to $B_0{}^{3/2}$ (all other things being equal). When more than one nuclide is involved in a sequence, a general expression for the signal-to-noise ratio of a 1D experiment may be derived:

$$\frac{S}{N} \propto N\,A\,T^{-1}\,B_0^{3/2}\,\gamma_{\text{exc}}\gamma_{\text{obs}}^{3/2}\,T_2^*\,(NS)^{1/2} \tag{4.4}$$

where N is the number of molecules in the observed sample volume, A is a term that represents the abundance of NMR active spins involved in the experiment, T is temperature, B_0 is the static magnetic field, γ_{exc} and γ_{obs} represent the magnetogyric ratios of the initially excited and the observed spins, respectively, T_2* is the effective transverse relaxation time and NS is the total number of accumulated scans. The high magnetogyric ratio of the proton, in addition to its 100% abundance (and ubiquity), explains why proton observation is favoured in high-resolution NMR. Indeed, many of the newly developed multipulse heteronuclear experiments utilise direct proton observation to achieve greater sensitivity, whilst heteronuclear spin is observed indirectly, as described in the Section 7.2. This section deals with those 1D pulse methods in widespread use that assist in the *direct* observation of relatively low-γ nuclei, here termed X-nuclei, for example, ^{13}C, ^{15}N, ^{29}Si and so on. These methods are characterised by initial excitation of a high-γ spin, typically ^{1}H, ^{19}F or ^{31}P, followed by *polarisation transfer* onto the low-γ nucleus to which it is scalar spin–spin coupled. This process introduces γ_{high} as the γ_{exc} term in Eq. 4.4 in place of γ_{low}, and so provides sensitivity gains of a factor of $\gamma_{high}/\gamma_{low}$.

It has been alluded to in the previous section, and will be considered in full in the Section 9.3, that significant enhancements of low-γ nuclei may be obtained on saturation of, for example, protons, owing to the NOE. This effect, however, can be ill suited to nuclei that possess a negative γ, since the NOE then causes a *reduction* in signal intensity and may, in the worst case, lead to a complete loss of signal. Polarisation transfer methods do not suffer such disadvantages and, in many cases, are able to provide considerably greater enhancements than the NOE alone. In addition, these experiments provide another means for the multiplicity editing of spectra and enable the differentiation of groups that possess differing numbers of attached nuclei. The ability to edit carbon spectra and hence to distinguish quaternary, methine, methylene and methyl groups offers an alternative to the spin-echo methods of the previous section. These polarisation transfer techniques, the Distortionless Enhancement by Polarisation Transfer (DEPT) experiment in particular, have become routine experiments in the organic chemist's repertoire. Methods that relate to the multiplicity editing of proton spectra are principally dealt with in Section 7.3. More recently developed and experimentally more complex technologies are now also emerging that

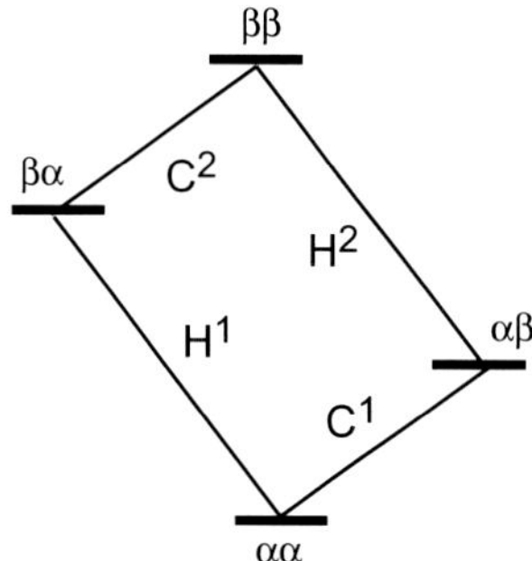

FIGURE 4.21 **Schematic energy-level diagram for a two-spin ^{1}H–^{13}C system.** The two transitions for each nucleus correspond to the two lines in each doublet.

potentially provide very significant sensitivity gains far beyond those attainable with the methods described below. These operate through the generation of very large population differences, examples being *parahydrogen-induced polarisation* (PHIP) and *dynamic nuclear polarisation* (DNP). As these are not widely used technologies, they are introduced briefly in Section 12.8.

4.4.1 Polarisation Transfer

Polarisation transfer methods enhance signal intensity by transferring the greater population difference (or *polarisation*) of a high-γ spin onto its spin-coupled low-γ partner(s). Through this they replace one of the three γ_{low} signal intensity dependencies described earlier with γ_{high}, leading to a signal enhancement by a factor of $\gamma_{high}/\gamma_{low}$. The principles behind all polarisation transfer methods can be understood by considering a spin-½ scalar-coupled pair which for illustrative purposes is here taken to be a ^{1}H–^{13}C pair *both with 100% abundance*, whose energy level diagram is shown in Fig. 4.21. At thermal equilibrium the populations for each of the four spin transitions are governed by the Boltzmann law as represented in Fig. 4.22a. From this, the population *differences* across the transitions are 2ΔH and 2ΔC for the

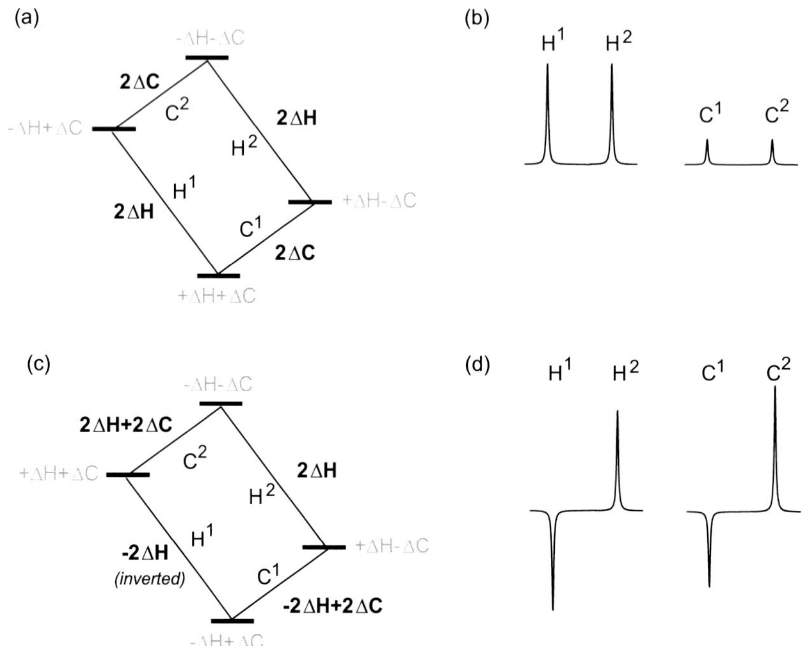

FIGURE 4.22 **Polarisation transfer in a two-spin system.** (a) Populations (in grey) and population differences (in bold) for each transition at equilibrium and (b) the corresponding spectra obtained following pulse excitation illustrating the four-fold population difference (see text). (c) The situation after selective inversion of one-half of the proton doublet and (d) the corresponding spectra showing the enhanced intensity of carbon resonances.

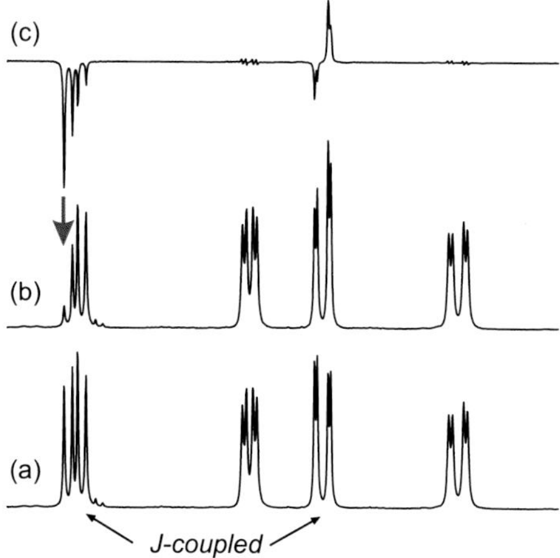

FIGURE 4.23 **Selective population transfer.** (a) The conventional ^{1}H spectrum, (b) the spectrum in which the arrowed multiplet is unevenly saturated, and (c) the difference spectrum (b-a) highlighting the associated distortion.

proton and carbon spins, respectively, and since these scale linearly with γ, the ratio $2\Delta H/2\Delta C$ equates to γ_H/γ_C or ≈ 4. The resonance intensities for proton are thus four times greater than those of carbon when judged by these populations (Fig. 4.22b). Now consider the result of selectively inverting *one* line of the H-spin doublet, say H^1 (eg with a weak selective pulse), which equates to inverting the population difference across the corresponding transition (Fig. 4.22c). As H^2 remains untouched, the population difference across the H^2 transition is no different to that before the inversion. The salient point is that both the C-spin population differences have now been altered by this process. Thus, the C^1 population difference becomes $-2\Delta H + 2\Delta C$ whilst that of C^2 is $2\Delta H + 2\Delta C$ and the population differences previously associated with the protons have been transferred onto the carbons. Since $2\Delta H$ is four times greater than $2\Delta C$, the C-transitions will display relative intensities of $-3{:}5$ so that sampling the carbon magnetisation at this point would produce a spectrum with one half of the doublet inverted and with signal intensities greater than in the single-pulse carbon spectrum (Fig. 4.22d). For a ^{1}H–^{15}N pair, signal intensities following polarisation transfer from protons are $+11$ and -9 ($\gamma_H/\gamma_N \approx 10$) and signal enhancement is even more impressive. Since one half of the H-spin populations have been inverted but the line intensities are otherwise unaffected, there has been no *net* transfer of magnetisation from proton to carbon. The *integrated intensity* of the whole carbon doublet is the same as that in the absence of polarisation transfer, so we say there has been a *differential* transfer of polarisation.

The experiment described is termed *selective population transfer* (SPT) or, more precisely in this case with proton spin inversion, *selective population inversion* (SPI). It is important to note, however, that the complete inversion of spin populations is not a requirement for the SPT effect to manifest itself. Any *unequal perturbation* of the lines within a multiplet will suffice, so, for example, saturation of one proton line would also have altered the intensities of the carbon resonance. In heteronuclear polarisation (population) transfer experiments, it is the heterospin-coupled satellites of the parent proton resonance that must be subject to the perturbation to induce SPT. The effect is not restricted to heteronuclear systems and can appear in proton spectra when homonuclear-coupled multiplets are subject to unsymmetrical saturation. Fig. 4.23 illustrates the effect of selectively but unevenly saturating a double doublet and shows the resulting intensity distortions in the multiplet structure of its coupled partner, which are most apparent in a difference spectrum. Despite these distortions, the integrated intensity of the proton multiplet is unaffected by the presence of the SPT because of equal positive and negative contributions (as in Fig. 4.22d). Distortions of this sort have particular relevance to the NOE difference experiment described in Section 9.8.

The greatest limitation of the SPI experiment is its lack of generality. Although it achieves the desired polarisation transfer, it is only able to produce this for one resonance in a spectrum at a time. To accomplish this for all, one would have to repeatedly step through the spectrum, inverting satellites one by one and performing a separate experiment at each step. Clearly, a more efficient approach would be to invert one half of each proton doublet simultaneously for all resonances in a single experiment and this is precisely what the Insensitive Nuclei Enhanced by Polarisation Transfer (INEPT) sequence achieves.

4.4.2 INEPT

The Insensitive Nuclei Enhanced by Polarisation Transfer (INEPT) experiment [39] was one of the forerunners of many of the pulse NMR experiments developed over subsequent years and still constitutes a feature of some of the most widely used multidimensional experiments in modern pulse NMR. Its purpose is to enable *non-selective* polarisation transfer between spins, and its operation may be readily understood with reference to the vector model. Most often it is the proton that is used as the source nucleus, and these discussions will relate to XH spin systems throughout, although it should be remembered that any high-γ spin-½ nucleus constitutes a suitable source.

The INEPT sequence (Fig. 4.24a) provides a method for inverting one-half of each XH doublet in a manner that is independent of its chemical shift, requiring the use of non-selective pulses only. The sequence begins with excitation of all protons which then evolve under the effects of chemical shift and heteronuclear coupling to the X-spin. After a period $\Delta/2$, the proton vectors experience a 180 degree pulse which serves to refocus chemical shift evolution (and field inhomogeneity) during the second $\Delta/2$ period. Simultaneous application of a 180 degree (X) pulse ensures heteronuclear coupling continues to evolve by inverting the proton vectors' sense of precession. This is once again a spin-echo sequence in which only coupling evolution requires consideration. For an XH pair, a total Δ period of 1/2J ($\Delta/2 = 1/4J$) leaves the two proton vectors opposed or *anti-phase* along $\pm x$, so that the subsequent $90°_y$(H) pulse aligns these along the $\pm z$-axis (Fig. 4.25). This therefore corresponds to the desired inversion of one-half of the proton doublet, as for SPI, but for all spin pairs simultaneously.

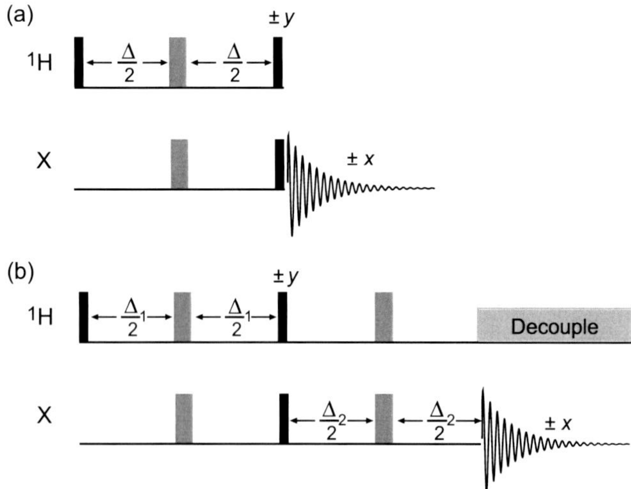

FIGURE 4.24 **The INEPT sequences.** (a) The basic INEPT and (b) the refocused INEPT sequence.

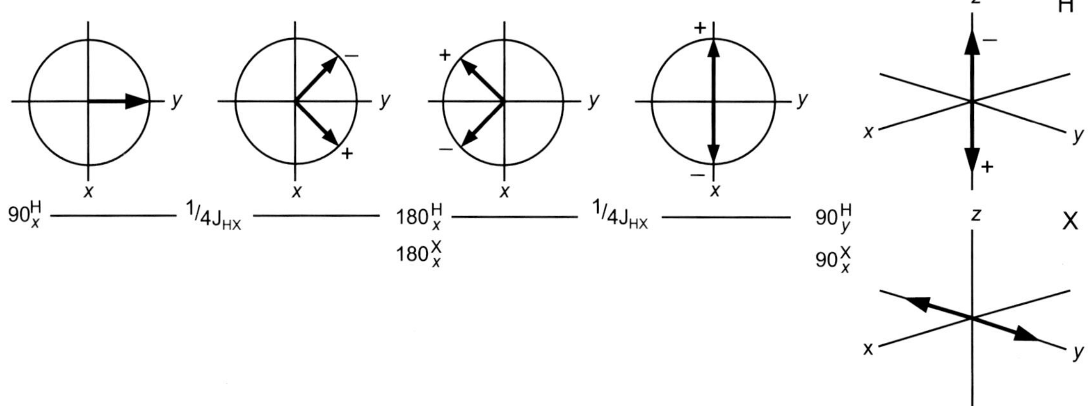

FIGURE 4.25 **Evolution of proton vectors during the INEPT sequence.** Following initial evolution under J_{XH} the 180 degree (H) pulse flips the vectors about the x-axis and the 180 degree (X) pulse inverts their sense of precession. After a total evolution period of $1/2J_{XH}$ the vectors are anti-phase and are subsequently aligned along $\pm z$ by the $90°_y$(H) pulse. This produces the desired inversion of one half of each H–X doublet for all resonances.

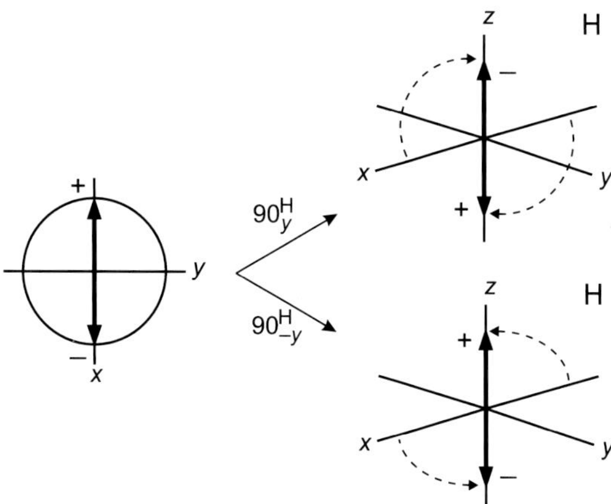

FIGURE 4.26 **INEPT phase cycling.** Inverting the phase of the last proton pulse in INEPT inverts the other half of the proton doublet (compare Fig. 4.25).

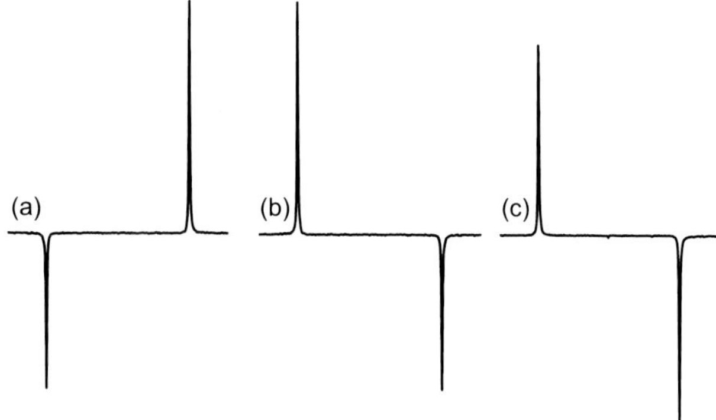

FIGURE 4.27 **Experimental carbon-13 INEPT spectra of methanoic acid.** The spectra were recorded with the phase of the last proton pulse set to (a) *y* and (b) −*y*. Subtracting the two data sets (c) cancels natural magnetisation that has not been generated by polarisation transfer and equalises the intensity of the two lines.

The 90 degree (X) pulse samples the newly created population differences to produce the enhanced spectrum. In practice, these last two pulses are applied simultaneously, although in this pictorial representation it proves convenient to consider the proton pulse to occur first.

The asymmetrical peak intensities produced by the SPI effect, −3 and +5 for a CH pair, arise from the contribution of natural X-spin magnetisation, which may be removed by the use of a simple phase cycle. Thus, repeating the experiment with the phase of the last proton pulse at −*y* causes the anti-phase lines to adopt the inverted disposition relative to the +*y* experiment, that is, what was −*z* now becomes +*z* and vice versa (Fig. 4.26). The resulting X-spin doublets are likewise inverted, although natural X magnetisation, being oblivious of the proton pulse, is unchanged. Subtraction of the two experiments by inverting the receiver phase causes the polarisation transfer contribution to add but cancels natural magnetisation (Fig. 4.27). This two-step phase cycle is the basic cycle required for INEPT. As only the polarisation transfer component is retained by this process, a feature of pure polarisation transfer experiments is the lack of responses from nuclei without significant proton coupling.

4.4.2.1 Refocused INEPT

One problem with the basic INEPT sequence described is that it precludes the application of proton spin decoupling during acquisition of the X-spin FID. Since this removes the J-splitting it will cause the anti-phase lines to become coincident and cancel to leave no observable signal. The addition of a further period Δ_2 after polarisation transfer allows the X-spin magnetisation to refocus under the influence of the XH coupling, giving the *refocused-INEPT* sequence [40] (Fig. 4.24b), the results of which are illustrated in Fig. 4.28. Once again, the refocusing period is applied in the form of a spin-echo to remove chemical shift dependencies and, for the XH system to fully refocus, the appropriate period is again 1/2J. Since it is now the X-spin magnetisation that refocuses, it is necessary to consider coupling to all attached protons. Choosing Δ_2 = 1/2J is not appropriate for those nuclei coupled to more than one proton and does not lead to the correct refocusing of vectors,

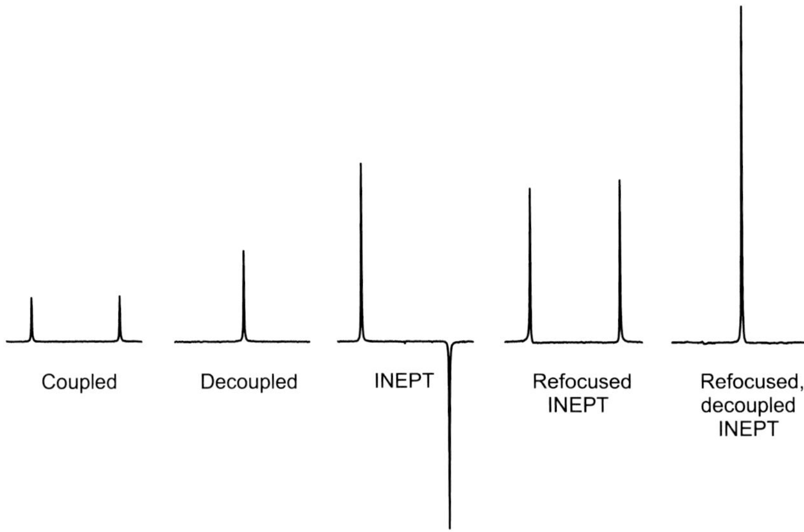

FIGURE 4.28 **Experimental carbon-13 spectra of methanoic acid.** The spectra were acquired without and with INEPT illustrating the intensity gains arising from polarisation transfer.

rather they remain anti-phase and produce no signal on decoupling (this provides a clue as to how one might use INEPT for spectrum editing). It has been shown [41] that for optimum sensitivity a refocusing period for an XH_n group requires:

$$\Delta_2 = \frac{1}{\pi J}\sin^{-1}\left(\frac{1}{\sqrt{n}}\right) \tag{4.5}$$

where the angular term must be calculated in radians. For carbon spectroscopy this corresponds to 1/2J, 1/4J and ≈1/5J for CH, CH_2 and CH_3 groups, respectively. If all multiplicities are to be observed simultaneously, a compromise setting of ≈1/3.3J is appropriate, which for an assumed 140-Hz coupling gives $\Delta_2 = 2.2$ ms. A spectrum recorded under these conditions but without proton decoupling will display significant phase errors since the refocusing period will not be optimal for any multiplicity; such errors are removed with proton decoupling, however.

In cases where the XH-coupling fine structure is of interest and proton decoupling is not applied, the relative line intensities within multiplets will be distorted relative to the coupled spectrum without polarisation transfer, and this is one potential disadvantage of the aforementioned INEPT sequences. Even with suppression of the contribution from natural X-spin magnetisation, intensity anomalies remain for XH_n groups with $n > 1$ (Fig. 4.29). A potentially confusing feature is the disappearance of the central line of XH_2 multiplets. One sequence proposed to generate the usual intensities within multiplets is INEPT$^+$, comprising the refocused-INEPT with an additional proton 'purge' pulse to remove the magnetisation terms responsible for the intensity anomalies [42]. In practice, the DEPT-based sequences described further are preferred as these show greater tolerance to experimental mis-settings.

Finally, note that these INEPT discussions assume the evolution of magnetisation is dominated by the coupling between the X and H spins, with all other couplings being negligibly small. In situations where homonuclear proton coupling becomes significant ($J_{HH} \approx\geq 1/3\ J_{XH}$), it is necessary to modify the Δ_1 period for optimum sensitivity, and analytical expressions for this have been derived [43]. Such considerations are most likely to be significant with polarisation transfer from protons via long-range couplings, that is, when the protons are not directly bound to the heteronucleus. Examples may include coupling to tertiary nitrogens, quaternary carbons or to phosphorus. In some instances, the situation may be considerably simplified if homonuclear couplings are removed by application of *selective* proton decoupling during proton Δ_1 evolution periods [44]. Fig. 4.30 illustrates the acquisition of ^{15}N spectra of adenosine-5′-sulphate **4.5** by polarisation transfer through one-bond and

	(a)	(b)	(c)
CH	1 1	−3 +5	−1 1
CH_2	1 2 1	−7 2 +9	−1 0 1
CH_3	1 3 3 1	−11 −9 15 13	−1 −1 1 1

FIGURE 4.29 **Relative multiplet line intensities in coupled INEPT spectra.** (a) Conventional multiplet intensities and those from INEPT (b) without and (c) with suppression of natural magnetisation.

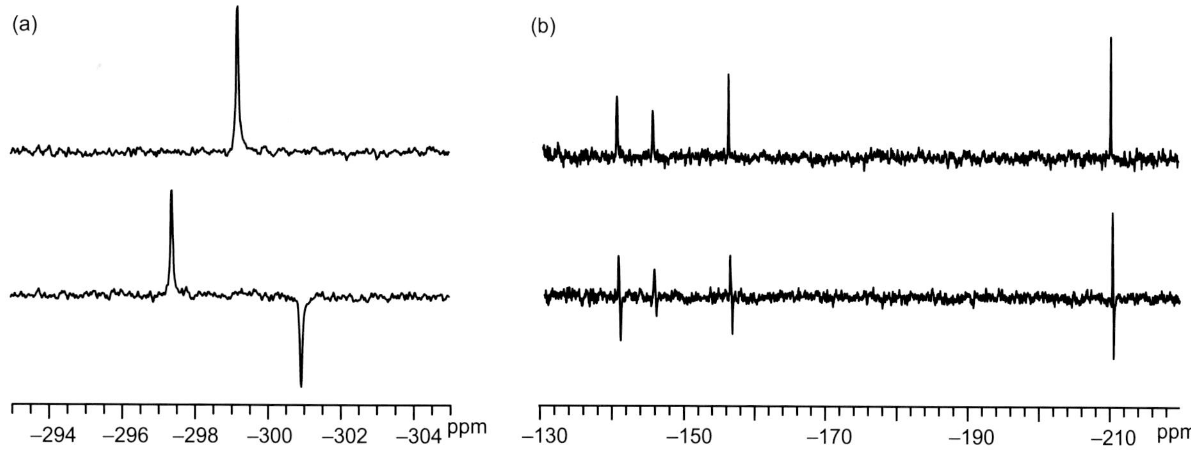

FIGURE 4.30 **Natural abundance ^{15}N INEPT spectra of adenosine-5'-monosulphate 4.5.** Lower traces show results with the INEPT sequence and upper traces with the refocused-INEPT sequence. Delays were calculated assuming (a) J = 90 Hz and (b) J = 10 Hz. For (a) the lower trace displays the 1:0:−1 pattern of the NH_2 group whilst in (b) all resonances display anti-phase two-bond ^1H–^{15}N couplings. Spectra are referenced to nitromethane.

long-range (two-bond) couplings. All nitrogen centres in the purine unit are observed in two experiments in which delays were optimised for J = 90 Hz and 10 Hz, which were assumed values for one- and two-bond couplings, respectively.

NH2
N 1 5 N 7 3 9 N N
SO3−
O
OH OH

4.5

4.4.2.2 Sensitivity Gains

To appreciate the sensitivity gains from application of the INEPT sequence, one should compare the results with those obtained from the usual direct observation of the low-γ species. This invariably means the spectrum obtained in the presence of proton broadband decoupling for which signal enhancement will occur by virtue of the ^1H–X NOE (see Section 9.3). Thus, to make a true comparison, we need to consider the signal arising from polarisation transfer versus that from observation with the NOE, which *for an XH pair* is given by:

$$I_{\text{INEPT}} = I_0 \left| \frac{\gamma_\text{H}}{\gamma_\text{X}} \right| \qquad I_{\text{NOE}} = I_0 \left(1 + \frac{\gamma_\text{H}}{2\gamma_\text{X}} \right) \tag{4.6}$$

where I_{INEPT} is the signal intensity following polarisation transfer, I_{NOE} is that in the presence of the NOE and I_0 represents the signal intensity in the absence of any enhancement. Note, first, that the NOE makes a contribution that *adds* to natural magnetisation and, secondly, because of this the resulting signal intensity is also dependent upon the *sign* of the magnetogyric ratios, whereas polarisation transfer depends only on their magnitudes. The NOE therefore causes a *decrease* in signal intensity for those nuclei with a *negative* magnetogyric ratio, and may cause the observed signal to be inverted if the NOE

TABLE 4.4 Signal Intensities for the X-Spin in ^{1}H–X Pairs Arising From Polarisation Transfer (I_{INEPT}) and From Direct Observation With Maximum NOE (I_{NOE})

X	^{13}C	^{15}N	^{29}SI	^{31}P	^{57}Fe	^{103}Rh	^{109}Ag	^{119}Sn	^{183}W	^{195}Pt	^{207}Pb
I_{INEPT}	3.98	9.87	5.03	2.47	30.95	31.77	21.50	2.81	24.04	4.65	4.78
I_{NOE}	2.99	−3.94	−1.52	2.24	16.48	−14.89	−9.75	−0.41	13.02	3.33	3.39

Intensities are given relative to those obtained by direct observation in the absence of the NOE (I_0).

is greater than natural magnetisation or to become close to zero if comparable with it. Table 4.4 compares the theoretical *maximum* signal intensities that can be expected for polarisation transfer and the NOE from protons to heteronuclear spins in XH pairs. The degree of signal enhancement does not scale linearly with the number of attached source nuclei n used for polarisation transfer. In other words, the enhancement expected for an XH_2 group is not twice that for an XH group. Although greater enhancements do arise with more attached protons when the optimum refocusing delay is used, the gains over that for an XH group are only modest [42].

In practice, owing to experimental imperfections or to other relaxation processes reducing the magnitude of the NOE (chemical shift anisotropy for metals, in particular), the figures in Table 4.4 may not be met, although they provide some guidance as to which method would be the most appropriate. The results for higher-γ heteronuclei such as ^{31}P and ^{13}C are clearly comparable, whereas polarisation transfer provides far greater gains for the lower-γ species. A further important benefit of the polarisation transfer approach not reflected in the figures of Table 4.4 is that the repetition rate of the experiment depends on the longitudinal relaxation times of the *protons*, since the populations of interest originate only from these spins. In contrast, the direct observation experiment, with or without the NOE, depends on the relaxation times of the X-spins, which are typically very much longer. The opportunity for faster signal averaging provides another significant gain in sensitivity per unit time when employing polarisation transfer, and in practice this feature can be as important or sometimes more important than the direct sensitivity gains from the transfer itself.

In short, polarisation transfer methods provide greatest benefits for those nuclei that have low magnetogyric ratios and are slow to relax. It is also the preferred approach to direct observation for those nuclei with negative γs where the NOE may lead to overall signal reduction. Nitrogen-15 [45] (Fig. 4.31) and silicon-29 routinely benefit from polarisation transfer methods, as does the observation of metals; in particular, transition metals which often possess very low γs and are extremely insensitive, despite their sometimes high natural abundance, some examples being ^{57}Fe, ^{103}Rh, ^{109}Ag and ^{183}W.

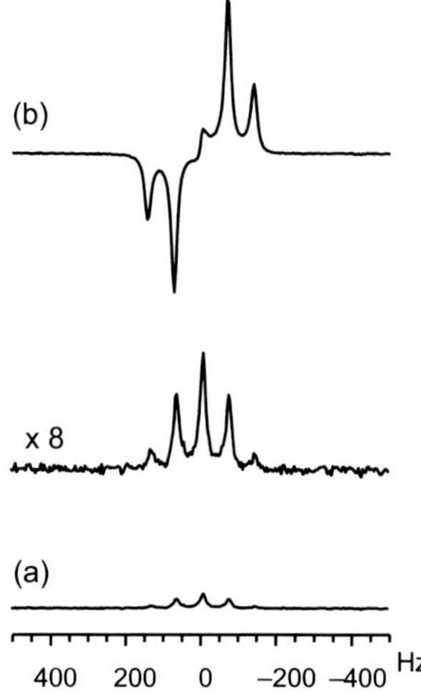

FIGURE 4.31 **Signal enhancement of the ^{15}N spectrum of ammonium nitrate with INEPT.** Direct observation using (a) the Ernst angle optimised for the nitrogen T_1 and (b) INEPT optimised for the proton T_1. Both spectra were collected in the same total accumulation time.

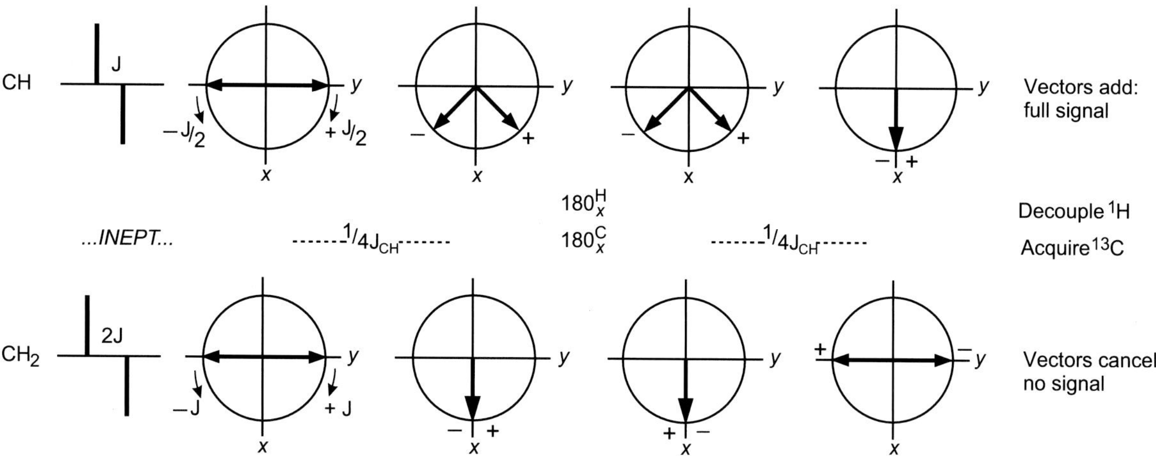

FIGURE 4.32 **Editing with INEPT.** This can be achieved with a judicious choice of refocusing delay Δ_2. A total delay of $1/2J_{CH}$ retains CH resonances but eliminates CH_2 (and CH_3) resonances.

Despite these impressive sensitivity gains when directly observing the X-spin, the more modern approach is to observe the X-spin *indirectly* through the coupled proton when possible, which can be achieved through a number of heteronuclear correlation experiments. As these methods additionally employ proton observation, they benefit from further theoretical gain of $(\gamma_{high}/\gamma_{low})^{3/2}$ over X-observe schemes (see Eq. 4.4). These topics are pursued in Section 7.3, and should be considered as potentially faster routes to X-nucleus data.

4.4.2.3 Editing with INEPT

Returning to selection of the Δ_2 period in the refocused INEPT sequence, it is apparent that whilst a period of 1/2J produces complete refocusing of doublets, the triplets (and quartets) remain anti-phase and will be absent from the spectrum recorded with proton decoupling (Fig. 4.32). Choosing Δ_2 = 1/2J therefore yields a sub-spectrum containing only methine resonances. This idea of editing heteronuclear spectra according to multiplicities is closely related to the editing with spin-echoes described in Section 4.3. Extending this idea to the selection of other carbon multiplicities, it is again convenient to define an angle $\theta = 180J\Delta_2$ degrees from which the signal intensities in the decoupled experiment are:

$$\begin{aligned} &CH: && I \propto \sin\theta \\ &CH_2: && I \propto 2\sin\theta\cos\theta \\ &CH_3: && I \propto 3\sin\theta\cos^2\theta \end{aligned}$$

as presented graphically in Fig. 4.33. To differentiate all protonated carbons, it is sufficient to record three spectra with Δ_2 adjusted suitably to give θ = 45, 90 and 135 degrees. The 90 degrees experiment corresponds to Δ_2 = 1/2J mentioned earlier and hence displays methine groups only, θ = 45 degrees produces all responses whilst θ = 135 degrees has all responses again, but with methylene groups inverted. This process combines signal enhancement by polarisation transfer with multiplicity determination through spectrum editing. The editing attainable here is superior to that provided by the basic spin-echo methods of Section 4.3 since comparison of the three INEPT spectra allows one to determine the multiplicity of all resonances in the spectrum. However, in routine carbon-13 spectroscopy, this information tends not to be derived from INEPT but from the related DEPT sequence, which has better tolerance to experimental imperfections. Alas, DEPT cannot be described fully by recourse to the vector model, and an understanding of INEPT goes a long way to help one appreciate the features of DEPT.

In summary, the INEPT sequence provides signal enhancement for all proton-coupled resonances in a single experiment by virtue of polarisation transfer. This also enables more rapid data collection since the repetition rate is now dictated by faster relaxing protons and not the heteronucleus. By judicious choice of refocusing delays, refocused INEPT can also provide multiplicity editing of, for example, carbon-13 spectra. However, those nuclei that do not share a proton coupling cannot experience polarisation transfer and are therefore missing from INEPT spectra. The INEPT sequence is also used extensively as a building block in heteronuclear 2D sequences, as will become apparent in later chapters.

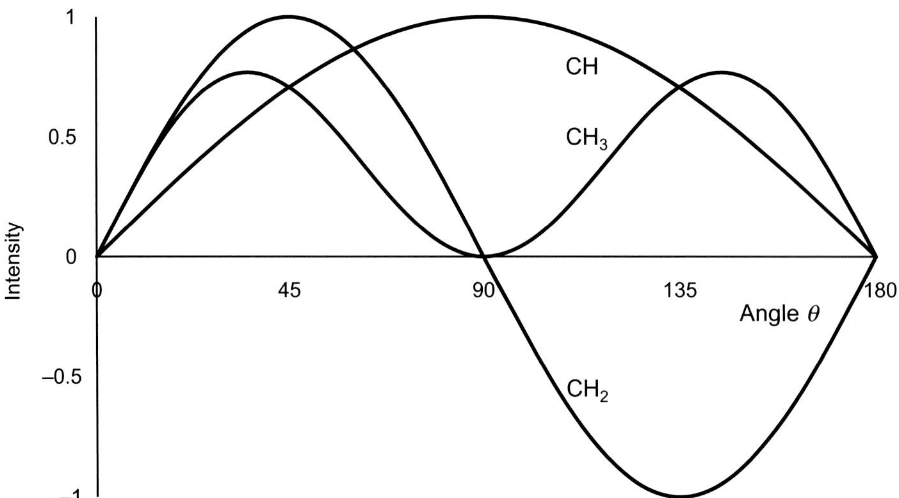

FIGURE 4.33 **The variation of carbon signal intensities in the refocused INEPT experiment as a function of the evolution time Δ ($\theta = 180J\Delta_2$ degrees).** Identical results are obtained for the DEPT experiment for which the angle θ represents the tip angle of the last proton pulse.

4.4.3 DEPT

The Distortionless Enhancement by Polarisation Transfer (DEPT) experiment [46] is the most widely used polarisation transfer editing experiment in carbon-13 spectroscopy, although its application is certainly not limited to the proton–carbon combination. It enables complete determination of all carbon multiplicities, as does the refocused INEPT discussed earlier, but has a number of distinct advantages. One of these is that it directly produces multiplet patterns in proton-coupled carbon spectra that match those obtained from direct observation, meaning methylene carbons display the familiar 1:2:1 and methyl carbons the 1:3:3:1 intensity patterns; this is the origin of the term 'distortionless'. However, for most applications proton decoupling is applied during acquisition and multiplet structure is of no consequence, so the benefits of DEPT must lie elsewhere, as we now discover.

4.4.3.1 DEPT Sequence

The DEPT pulse sequence is illustrated in Fig. 4.34. To follow events during this, consider once more a ^{1}H–^{13}C pair and note the action of the two 180 degree pulses is again to refocus chemical shifts where necessary. The sequence begins in a similar manner to INEPT with a 90 degree (H) pulse after which proton magnetisation evolves under the influence of proton–carbon coupling such that after a period 1/2J the two vectors of the proton satellites are anti-phase. Application of a 90 degree (C) pulse at this point produces a new state of affairs that has not been previously encountered, in which both transverse proton and carbon magnetisation evolve coherently. This new state is termed *heteronuclear multiple quantum coherence* which, in general, cannot be visualised with the vector model, and without recourse to mathematical formalisms it is not possible to rigorously describe its behaviour. However, for these purposes we can imagine this multiple quantum coherence to be a pooling of both proton and carbon magnetisation, to be separated again at some later point. The transverse magnetisation we observe directly in an NMR experiment is known as *single quantum coherence*.

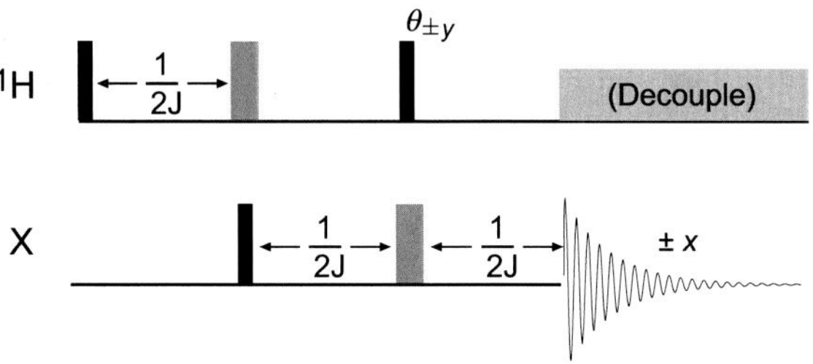

FIGURE 4.34 **The DEPT sequence.**

Multiple quantum coherence, however, cannot be directly observed because it induces no signal in the detection coil. For multiple quantum coherence to be of use to us, it must be transferred back into signal quantum coherence by the action of rf pulses. The concept of coherence is developed further in Section 5.3.

Heteronuclear multiple quantum coherence now evolves for a further time period under the influence of the proton chemical shift but also, simultaneously, under the influence of the carbon chemical shifts, until application of the proton θ pulse. However, it will *not* be influenced by the proton–carbon coupling (this is another interesting property of multiple-quantum coherence, which need not concern us further at this time; see Section 5.3). To remove the effects of proton chemical shifts, a second evolution period is also set to 1/2J and a proton 180 degree pulse is applied between the two delays, coincident with the 90 degree (C) pulse. The action of the θ pulse is to transfer heteronuclear multiple quantum coherence into anti-phase transverse carbon magnetisation, that is, to regenerate observable magnetisation. The details of the outcome of this transfer process depend upon the multiplicity of the carbon resonance. In other words, methine, methylene and methyl groups respond differently to this pulse and this provides the basis of editing with DEPT. In the final 1/2J delay, carbon magnetisation refocuses under the influence of proton coupling but also evolves according to carbon chemical shift, as it did in the second 1/2J period. Thus, simultaneous application of a 180 degree (C) pulse with the proton θ pulse leads to an overall refocusing of carbon shifts during the final 1/2J period. Carbon magnetisation is therefore detected without chemical shift–dependent phase errors, with or without proton decoupling. Phase alternation of the θ pulse, combined with data addition/subtraction by the receiver, leads to the cancellation of natural carbon magnetisation, as for INEPT. The net result is again polarisation transfer from protons to carbon, combined with the potential for spectrum editing.

4.4.3.2 Editing with DEPT

As was hinted at earlier, the key to spectrum editing with DEPT is the realisation that the phase and intensity of carbon resonances depends upon the proton tip angle θ. In fact, the editing results for angle θ are analogous to those produced by refocused INEPT with $\Delta_2 = \theta/180J$, so the previous discussions relating to θ apply equally to DEPT, as does the graph of Fig. 4.33. However, with INEPT the editing delay Δ_2 must be chosen according to the spin-coupling constant whereas with DEPT the editing is achieved through the variable pulse angle θ which has no J dependence. This means the editing efficiency of DEPT tends to be superior to INEPT when a range of J values are encountered, and this is the principal advantage of DEPT in routine analysis. Evolution delays in DEPT do of course depend on J, although it turns out that the experiment is quite tolerant of errors in these settings.

The starting point for determining resonance multiplicities with DEPT is the collection of three spectra with $\theta = 45$, 90 and 135 degrees, noting that the 90 degrees experiment requires twice as many scans to attain the same signal-to-noise ratio as the others. The signs of the responses are then as summarised in Table 4.5. Knowing these patterns, it is a trivial matter to determine multiplicities by direct comparison, whilst quaternaries can be distinguished by their appearance only in the direct carbon spectrum (or the DEPTQ variant described later). Example spectra are shown for the bicyclic terpene andrographolide **4.6** (Fig. 4.35). An extended approach is to combine these spectra appropriately to produce 'sub-spectra' that display separately CH, CH_2 and CH_3 resonances. Personally, I think it is always better to examine the original DEPT spectra directly since this is such a simple matter. The combining of spectra introduces greater potential for 'artefact' generation, and it also becomes more difficult to identify what is going on if editing fails owing to the presence of an unusually high coupling constant or to instrument miscalibration, for example. In practice, the recording of a single DEPT spectrum with $\theta = 135$ degrees is often sufficient to provide the desired information when methyls are absent or are easily recognised on account of their chemical shifts. Alternatively, the DEPTQ experiment has the further advantage of retaining quaternary centres, as described in the following section.

TABLE 4.5 Signs of Multiplet Resonances in DEPT and in DEPTQ Spectra

	DEPT-45	DEPT-90	DEPT-135	DEPTQ
X	0	0	0	–
XH	+	+	+	+
XH_2	+	0	–	–
XH_3	+	0	+	+

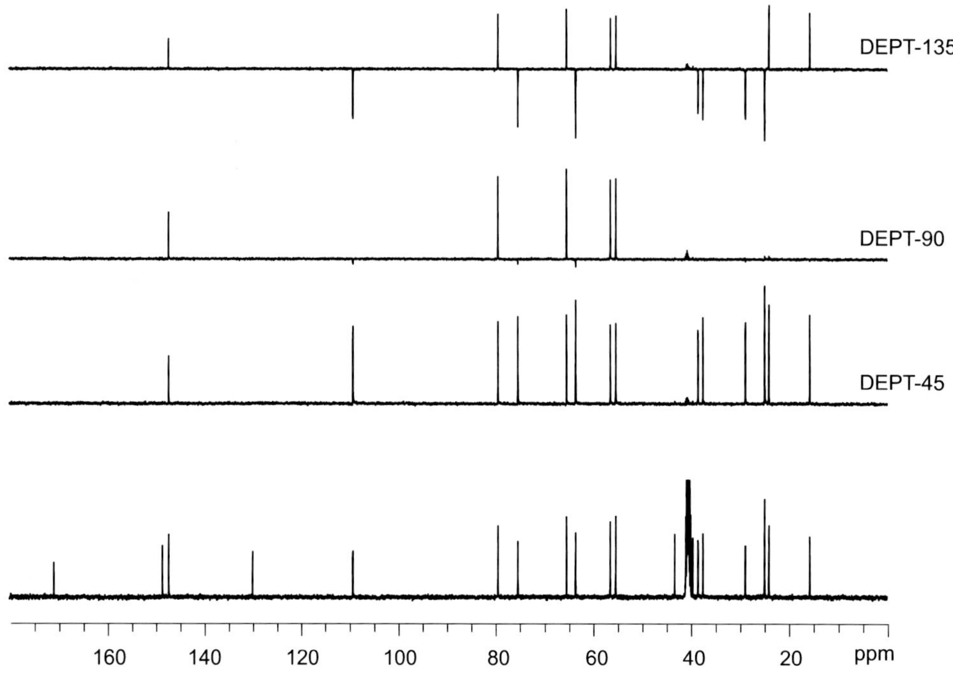

FIGURE 4.35 **The conventional carbon and DEPT-edited spectra of the terpene andrographolide 4.6.**

4.6

Errors in DEPT editing may arise from a number of sources. The most likely is incorrect setting of the proton pulses, especially the θ pulse used for editing, which may often be traced to poor tuning of the proton channel. Even small errors in θ can lead to the appearance of small unexpected peaks in the DEPT-90; if θ is too small it approaches DEPT-45 whilst too big and it approaches DEPT-135. Usually, because of their low intensity, these spurious signals are easily recognised and should cause no problems. Even with correct pulse calibrations, errors can arise when the setting for the delay period is very far from that demanded by $^{1}J_{CH}$ [47] and, in particular, CH_3 resonances may still appear weakly in the CH sub-spectrum. A typical compromise value for $^{1}J_{CH}$ for determining Δ is $\approx$140 Hz, for which $\Delta = 3.6$ ms. Alkyne carbons, in particular, exhibit very high values of $^{1}J_{CH}$ (usually >200 Hz), and may appear as weak signals displaying bizarre sign behaviour. Even more unexpected can be the appearance of the *quaternary* carbon of a terminal alkyne group which experiences polarisation transfer from the alkyne proton two bonds away because of the exceptionally large $^2J_{CH}$ value (~50 Hz).

4.4.3.3 Optimising Sensitivity

It was shown earlier that for the refocused INEPT experiment optimum sensitivity was obtained for a given multiplicity when the refocusing period was set according to the number of coupled protons and the associated coupling constant. When DEPT is used primarily as a means of enhancing the sensitivity of the heteronucleus rather than for spectrum editing, similar considerations apply. The difference here is that it is the θ proton pulse that must be optimised according to the number of spins in the XH_n group according to:

$$\theta_{\text{opt}} = \sin^{-1}\left(\frac{1}{\sqrt{n}}\right) \quad (4.7)$$

Note the similarity to Eq. 4.5 except that the dependence on J is again not relevant for DEPT. For an XH group, θ_{opt} is 90 degrees, whilst this angle decreases with higher multiplicities.

4.4.4 DEPTQ

The polarisation transfer sequences presented earlier provide signal enhancement of insensitive nuclei, but suffer from a lack of responses from those that do not possess directly bound protons. Conversely, the spin echo–based editing sequences do display such responses, but gain enhancement only from the NOE and require signal averaging that is dictated by the typically slower relaxing insensitive spins. As an alternative, the DEPTQ experiment [48] (DEPT retaining Quaternary centres) operates in a similar manner to DEPT, and so benefits from the advantages of polarisation transfer, but also seeks to retain responses from quaternary centres and thus has obvious appeal for routine ^{13}C analysis of small molecules.

In the original DEPT method, it was necessary to remove responses arising from direct carbon excitation by appropriate phase cycling because they would appear 90 degrees out of phase with those arising from polarisation transfer (in other words, they would appear confusingly as dispersive contributions) meaning also that quaternary signals were lost. The DEPTQ sequence (Fig. 4.36a) retains the same polarisation transfer elements as DEPT, and the edited responses for all protonated centres remain the same. In addition, it incorporates a ^{13}C [90 - 1/2J - 180] segment at the beginning of the sequence

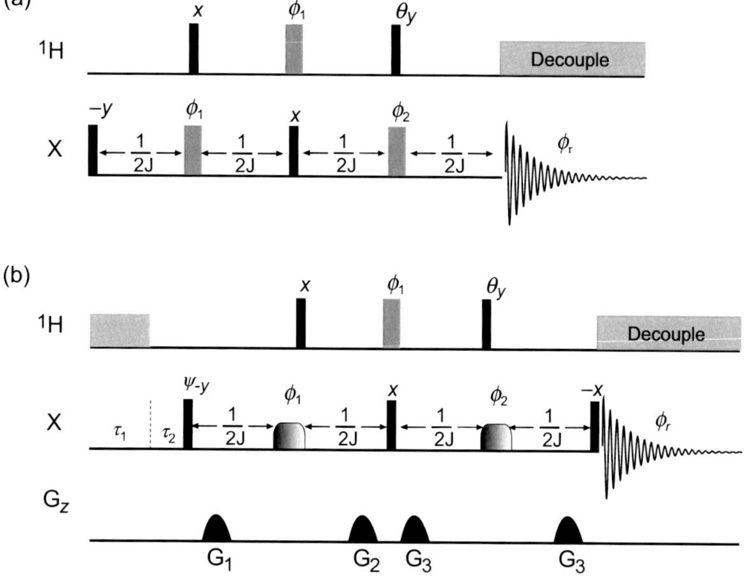

FIGURE 4.36 **DEPTQ sequences.** (a) The original sequence and (b) the enhanced sequence in which the recovery delay is split in two (τ_1 and τ_2, with and without ^1H decoupling, respectively) and the initial X pulse has reduced tip angle (ψ). The part-shaded pulses represent adiabatic 180 refocusing pulses. Phase cycling follows $\phi_1 = x, -x, y, -y$, $\phi_2 = x, -x, x, -x$ and $\phi_r = y, y, -y, -y$.

designed to excite quaternary centres and subsequently manipulate the behaviour of these such that they appear with the same phase response as CH_2 resonances and hence appear of opposite sign to those of CH and CH_3 groups (Table 4.5). For quaternary centres the complete sequence of ^{13}C pulses merely acts as two refocusing spin-echoes and the final spectrum retains all ^{13}C resonances with all the advantages of DEPT.

In practice, this basic implementation of DEPTQ still suffers from limitations in the efficiency of responses arising from quaternary centres meaning these are often weaker than observed in a simple pulse acquire carbon experiment, especially at the far edges of the spectrum. This failing can be traced to three principal causes: (1) the use of a 90 degree ^{13}C excitation pulse, larger than would usually be suggested as the optimum Ernst angle; (2) the lack of potential NOE enhancements onto the quaternary centres since ^{1}H decoupling is gated off during the recovery period to allow for proton relaxation prior to polarisation transfer; and (3) losses due to the off-resonance effects of the ^{13}C pulses, especially the 180 degree refocusing pulses. These issues have been addressed in a modified form of DEPTQ [49] illustrated in Fig. 4.36b. In this the first ^{13}C pulse tip angle is now reduced, requiring incorporation of an additional ^{13}C 90 degree purging pulse immediately prior to data acquisition to remove phase errors in the quaternary responses. The proton recovery delay has also been extended and split in two parts initially maintaining ^{1}H decoupling to generate ^{1}H–^{13}C NOEs, but then gating this off to allow for the necessary ^{1}H magnetisation recovery. Finally, the two ^{13}C 180 degree refocusing pulses are both applied as frequency-swept pulses (see Section 12.2) which are effective over far greater bandwidths and thus seek to overcome the off-resonance losses associated with hard pulses. As a final modification, optional gradient pulses may be incorporated to provide for cleaner signal selection ($G_1 = G_2 = G_3$) or can be used to select only protonated centres ($G_1 = 0$, $G_2 = G_3$) or only quaternaries ($G_1 = G_2 \neq G_3$); see Section 5.4 for a description of pulsed field gradient methodology.

A comparison of DEPT-135 with both the original and enhanced DEPTQ is made in Fig. 4.37 demonstrating the retention of quaternary centres and the improved performance achievable with the enhanced version. This method would now appear to be the optimal approach to spectrum editing with retention of all carbon centres and is attractive as a routine carbon-13 screening tool for the organic laboratory. The use of frequency-swept (adiabatic) 180 degree ^{13}C pulses is equally valid for the standard DEPT sequence, or indeed any using ^{13}C 180 degree pulses, and should also provide enhanced performance when large carbon bandwidths become problematic; see Section 12.2 for more information on these.

Before leaving these sections on the editing of X-nucleus spectra, some comments on the utility of such experiments in the modern NMR laboratory are required. As has already been mentioned, techniques based on proton observation are now used routinely for organic structure elucidation. Developments in this area have evolved techniques in which both 1D and more usefully 2D spectra are edited in such a way as to indicate X-nucleus multiplicities (see Section 7.3.2). The higher sensitivity possible with these methods compared with direct X-nucleus observation, together with the additional correlation data provided by 2D techniques, have led to a reduced dependence on the traditional, more time-consuming X-nucleus observation and editing experiments, most notably where sample masses become limiting. However, for routine analysis of novel synthetic materials, and their associated characterisation to enable publication,

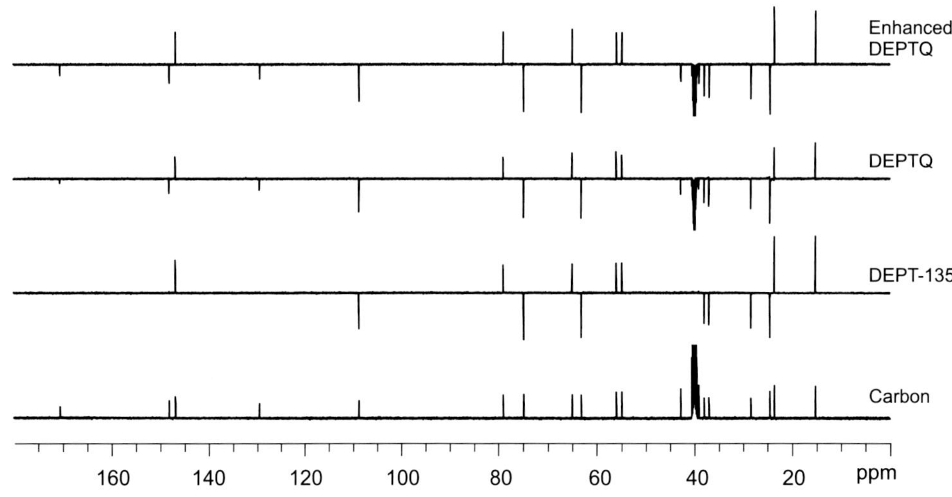

FIGURE 4.37 **The carbon, DEPT-135, DEPTQ and enhanced DEPTQ spectra of the terpene andrographolide 4.6.**

1D carbon observe techniques remain popular with synthetic chemists and are likely to remain so with the availability of cryogenic probes.

4.5 OBSERVING QUADRUPOLAR NUCLEI

A characteristic feature of many quadrupolar nuclei is the broad lines they produce, due to rapid quadrupolar relaxation (Section 2.5.5). The rapid recovery of spins following excitation means they can often be acquired under conditions of very fast pulsing with full excitation by a 90 degree pulse, which is clearly beneficial for signal-averaging purposes. However, the corresponding rapid decay of an FID can make the direct observation of nuclei with linewidths of hundreds or thousands of hertz experimentally challenging. Following an rf pulse, a delay is required for the probe circuits and receiver electronics to recover before the NMR signal can be digitised. One particular concern is so-called *acoustic ringing* in the probehead [50], which is most severe for low-abundance, low-frequency nuclei at lower fields and which may take many tens of microseconds to decay. In the observation of spin-½ nuclei a pre-acquisition delay of this order can be used without problem but in cases of fast-relaxing nuclei, such as ^{17}O or ^{33}S, this can lead to a loss of much of the FID before detection begins (resonance linewidths of 1 and 5 kHz correspond to relaxation times of only c. 320 and 60 μs, respectively). Not only does this compromise sensitivity it also introduces substantial phase errors to the spectrum. The use of shorter delays is therefore essential but leads to a significant contribution to the spectrum in the form of broad baseline distortions from spectrometer transient responses. Numerous sequences to suppress this ring-down contribution have been proposed [50], two of which are illustrated in Fig. 4.38b and c. In both cases, suppression of the acoustic response from the 90 degree excitation pulses is achieved in two scans by inverting the NMR signal phase on the second transient with an additional 180 degree pulse, together with simultaneous inversion of the receiver phase [51] (Fig. 4.38a). Whilst the NMR signals are added by this process, the acoustic response remains unchanged in the two experiments and therefore cancels with receiver inversion. The remaining unwanted feature is now the acoustic response from the additional 180 degree pulse, which is suppressed in a slightly different manner in the two sequences. With Alternate Compound One-eighties Used to Suppress Transients In the Coil (ACOUSTIC, Fig. 4.38b) [52] the whole experiment is repeated with inversion of the 180 degree pulse. This, in effect, inverts the associated acoustic response relative to the first experiment so adding the data from the two experiments cancels this component. The alternative *Ri*ng-down *De*lay sequence (RIDE, Fig. 4.38c) [53] instead repeats the whole process, but with inversion of the NMR response by inverting the phase of both 90 degree pulses, and subtracts this second experiment from the first. Both sequences suffer from the potential problem of off-resonance effects because of the use of 180 degree pulses, but otherwise provide significant reductions in baseline distortions. Fig. 4.39 illustrates the use of RIDE in collection of the ^{17}O spectrum of ethyl acetate, where suppression of baseline distortion is clear.

An alternative approach now available with modern processing software is to collect the distorted FID with the simple one-pulse-acquire sequence and to replace the early distorted data points with uncorrupted points generated through

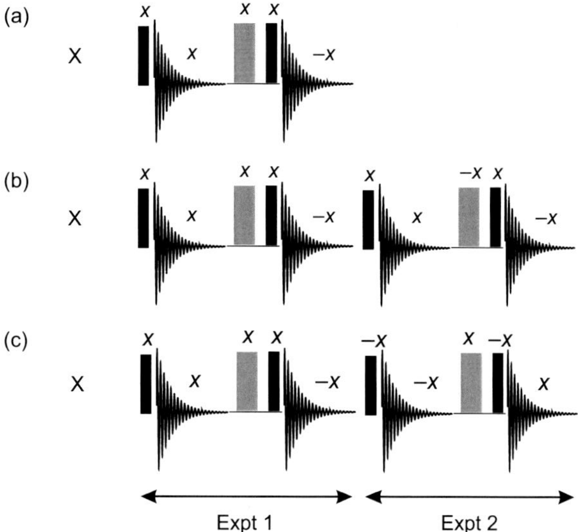

FIGURE 4.38 **Sequences for the observation of quadrupolar nuclei with very broad lines for which acoustic ringing is a problem.** Sequence (a) eliminates ringing associated with the 90 degree pulse whilst ACOUSTIC (b) and RIDE (c) further eliminate that associated with the 180 degree pulses.

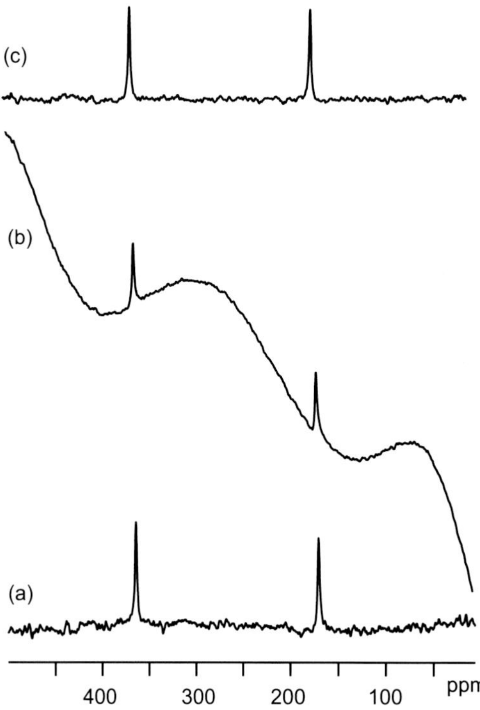

FIGURE 4.39 **^{17}O spectra of ethyl acetate.** Spectra were recorded (a) with and (b) without the RIDE sequence. The severe baseline distortion in (b) arises from acoustic ringing in the probehead. Spectrum (c) was from the same FID of (b) but this had the first 10 data points replaced with backward linear–predicted points, computed from 256 uncorrupted points. The spectra are referenced to D_2O and processed with 100-Hz line broadening.

backward linear prediction. Fig. 4.39c was produced from the same raw data as Fig. 4.39b, but with the first 10 data points of the FID replaced with predicted points. Baseline distortion is completely removed and there is no signal-to-noise loss through experimental imperfections, as occurs for the RIDE data set.

REFERENCES

[1] Ernst RR, Anderson WA. Rev Sci Instr 1966;37:93–102.
[2] Jones DE, Sternlicht H. J Magn Reson 1972;6:167–82.
[3] Becker ED, Ferretti JA, Gambhir PN. Anal Chem 1979;51:1413–20.
[4] Freeman R, Hill HDW. J Magn Reson 1971;4:366–83.
[5] Waldstein P, Wallace WE. Rev Sci Instr 1971;42:437–40.
[6] Waugh JS. J Mol Spectrosc 1970;35:298–305.
[7] Evilia RF. Anal Lett 2001;34:2227–36.
[8] Pauli GF, Jaki BU, Lankin DC. J Nat Prod 2005;68:133–49.
[9] Pauli GF, Gödecke T, Jaki BU, Lankin DC. J Nat Prod 2012;75:834–51.
[10] Simmler C, Napolitano JG, McAlpine JB, Chen S-N, Pauli GF. Curr Opin Biotechnol 2014;25:51–9.
[11] Rundlöf T, Mathiasson M, Bekiroglu S, Hakkarainen B, Bowden T, Arvidsson T. J Pharm Biomed Anal 2010;52:645–51.
[12] Rundlöf T, McEwen I, Johansson M, Arvidsson T. J Pharm Biomed Anal 2014;93:111–7.
[13] Giraudeau P. Magn Reson Chem 2014;52:259–72.
[14] Barantin L, Pap AL, Akoka S. Magn Reson Med 1997;38:179–82.
[15] Akoka S, Barantin L, Trierweiler M. Anal Chem 1999;71:2554–7.
[16] Molinier V, Fenet B, Fitremann J, Bouchu A, Queneau Y. Carb Res 2006;341:1890–5.
[17] Michel N, Akoka S. J Magn Reson 2004;168:118–23.
[18] Burton IW, Quilliam MA, Walter JA. Anal Chem 2005;77:3123–31.
[19] Wider G, Dreier L. J Am Chem Soc 2006;128:2571–6.
[20] Hoffman RA, Forsén S. Prog Nucl Magn Reson Spectrosc 1966;1:15–204.
[21] Anderson WA, Freeman R. J Chem Phys 1962;37:85–103.
[22] Massiot G, Kan SK, Gonard P, Duret C. J Am Chem Soc 1975;97:3277–8.
[23] Sanders JKM, Mersh JD. Prog Nucl Magn Reson Spectrosc 1982;15:353–400.

[24] Bloch F, Siegert A. Phys Rev 1940;57:522–7.
[25] Ramsey NF. Phys Rev 1955;100:1191–4.
[26] Jesson JP, Meakin P, Kneissel G. J Am Chem Soc 1973;95:618–20.
[27] Dykstra RW. J Magn Reson 1990;88:388–92.
[28] Dykstra RW. J Magn Reson 1992;100:571–4.
[29] Dykstra RW. J Magn Reson A 1993;102:114–5.
[30] Brown DW, Nakashima TT, Rabenstein DL. J Magn Reson 1981;45:302–14.
[31] Cocq CL, Lallemand JY. J Chem Soc Chem Commun 1981;150–2.
[32] Pei F-K, Freeman R. J Magn Reson 1982;48:318–22.
[33] Patt SL, Shoolery JN. J Magn Reson 1982;46:535–9.
[34] Bildsøe H, Dønstrup S, Jakobsen HJ, Sørensen OW. J Magn Reson 1983;53:154–62.
[35] Sørensen OW, Dønstrup S, Bildsøe H, Jakobsen HJ. J Magn Reson 1983;55:347–54.
[36] Torres AM, Nakashima TT, McClung RED. J Magn Reson A 1993;101:285–94.
[37] Ollerenshaw J, Nakashima TT, McClung RED. Magn Reson Chem 1998;36:445–8.
[38] Freeman R. A handbook of nuclear magnetic resonance. 2nd ed. Harlow: Longman; 1987.
[39] Morris GA, Freeman R. J Am Chem Soc 1979;101:760–2.
[40] Burum DP, Ernst RR. J Magn Reson 1980;39:163–8.
[41] Pegg DT, Doddrell DM, Brooks WM, Bendall MR. J Magn Reson 1981;44:32–40.
[42] Sørensen OW, Ernst RR. J Magn Reson 1983;51:477–89.
[43] Schenker KV, von Philipsborn W. J Magn Reson 1985;61:294–305.
[44] Mohebbi A, Gonen O. J Magn Reson A 1996;123:237–41.
[45] von Philipsborn W, Müller R. Angew Chem Int Ed 1986;25:383–413.
[46] Doddrell DM, Pegg DT, Bendall MR. J Magn Reson 1982;48:323–7.
[47] Bendall MR, Pegg DT. J Magn Reson 1983;53:272–96.
[48] Burger R, Bigler P. J Magn Reson 1998;135:529–34.
[49] Bigler P, Kümmerle R, Bermel W. Magn Reson Chem 2007;45:469–72.
[50] Gerothanassis IP. Prog Nucl Magn Reson Spectrosc 1987;19:267–329.
[51] Canet D, Brondeau J, Marchal JP, Robin-Lherbier B. Org Magn Reson 1982;20:51–3.
[52] Patt SL. J Magn Reson 1982;49:161–3.
[53] Belton PS, Cox IJ, Harris RK. J Chem Soc Faraday Trans 2 1985;81:63–75.

Chapter 5

Introducing Two-Dimensional and Pulsed Field Gradient NMR

Chapter Outline

5.1 Two-Dimensional Experiments 172
 5.1.1 Generating the Second Dimension 172
 5.1.2 Correlating Coupled Spins 176
5.2 Practical Aspects of 2D NMR 177
 5.2.1 Two-Dimensional Lineshapes and Quadrature Detection 177
 5.2.2 Axial Peaks 181
 5.2.3 Instrumental Artefacts 182
 5.2.4 Two-Dimensional Data Acquisition 183
 5.2.5 Two-Dimensional Data Processing 186
5.3 Coherence and Coherence Transfer 188
 5.3.1 Coherence Transfer Pathways 190
5.4 Gradient-Selected Spectroscopy 191
 5.4.1 Signal Selection with Pulsed Field Gradients 192
 5.4.2 Phase-Sensitive Experiments: *Echo–Antiecho* Selection 195
 5.4.3 Pulsed Field Gradients in High-Resolution NMR 196
 5.4.4 Practical Implementation of Pulsed Field Gradients 198
 5.4.5 Fast Data Acquisition: Single-Scan Two-Dimensional NMR 199
References 201

Having discussed a number of methods in the previous chapter that are considered routine 1D techniques, this chapter provides an introduction to the world of 2D NMR and to the use of pulsed field gradients (PFGs). It aims to explain how 2D NMR methods operate, how and why PFGs are employed in NMR experiments, and seeks to introduce the most important experimental protocols relevant to these techniques. Subsequent chapters describe a range of 2D methods, many of which incorporate PFGs, developed to elucidate the structures and dynamics of molecules in solution by correlating interactions between nuclear spins. If one is most interested in learning how each of these methods may be applied to the investigation of molecular structures, you may choose to skip this chapter on the underlying principles, considering first those that follow. It should be possible to appreciate the information content of the methods presented, although an understanding of their operation follows from the material in this chapter. It is no exaggeration to say that the complete structure characterisation of almost all novel small molecules will involve the application of 2D NMR techniques nowadays, most of which will employ PFGs, and hence these techniques are now ubiquitous in chemistry research laboratories. Further to the following descriptive chapters, the example provided in the chapter *Structure Elucidation and Spectrum Assignment* illustrates how the more commonly employed 2D techniques may be utilised in structure elucidation.

The principles underlying the generation of a 2D spectrum were first presented in a lecture in 1971 [1], although it was a number of years later that the approach found wider application [2]. Throughout the 1980s the world of NMR, and consequently the chemist's approach to structure determination, was revolutionised by the development of numerous 2D techniques, and nowadays many higher dimensionality methods (3D and even 4D) also exist. Methods utilising three or more dimensions find greatest application in the hands of NMR spectroscopists studying biological macromolecular structures and these methods therefore fall beyond the realms of this book, but are considered in texts dedicated to biomolecular NMR [3,4]. Through the 1990s, the incorporation of PFGs in high-resolution NMR methods again revolutionised the discipline, providing cleaner spectra often in less time. The concept of applying magnetic field gradients across samples was taken from the area of magnetic resonance imaging and now lies at the heart of most modern NMR techniques.

It is likely that you have already made some use of 2D methods in your research, quite probably in the form of the correlation spectroscopy (COSY) experiment. Not only is this one of the most widely used 2D experiments, its simple sequence provides a convenient introduction to multidimensional experiments and we shall use this as an illustrative technique with which to introduce the key features of 2D NMR spectroscopy. The approach in this chapter is again a pictorial one. While

High-Resolution NMR Techniques in Organic Chemistry. http://dx.doi.org/10.1016/B978-0-08-099986-9.00005-1

this avoids the need for becoming embroiled in formalisms that rigorously describe the behaviour of magnetisation during multipulse experiments, it has its limitations in that many of the techniques cannot be described completely by such a simplified approach. We have already come up against these limitations when attempting to understand the *distortionless enhancement by polarization transfer* (DEPT) experiment in the previous chapter, where it was not possible to fully describe the behaviour of heteronuclear multiple quantum coherence in terms of the classical vector model. An explanation of this technique was thus derived largely from the more readily understood *insensitive nuclei enhanced by polarisation transfer* (INEPT) sequence. Despite these limitations it is still possible to develop some physical insight into how the experiments operate without resorting to esoteric mathematical descriptions. Section 5.3 introduces graphical formalisms, known as *coherence transfer pathways*, that provide a simple representation of the 'flow' of magnetisation during pulse experiments, which proves particularly enlightening for 2D sequences and for those experiments that utilise PFGs for signal selection. Before any of this however, we shall attempt to develop some understanding of how 2D spectra are generated.

5.1 TWO-DIMENSIONAL EXPERIMENTS

The first point to clarify when discussing 2D techniques is the fact that the two dimensions refer to two *frequency* dimensions, while 1D methods have only one, and in either case there will also be a dimension representing signal intensity. The two frequency dimensions will typically represent any combination of chemical shifts and/or scalar couplings. However, methods have also been developed for which the idea of a 2D representation has been adapted to include one frequency and one 'other' dimension. For example it is possible to disperse the spectra of solutes according to their diffusion properties, in which case the second dimension represents their self-diffusion coefficients (see the chapter *Diffusion NMR Spectroscopy*).

Conventional 2D experiments find wide utility in chemical research because they map interactions within, or sometimes between, our molecules of interest. The interactions that can be interrogated may be broken down into three distinct categories which relate to quite different physical phenomena:

- Through-bond coupling: methods utilise scalar coupling (J) between nuclei with experiments aimed principally at the identification and subsequent piecing together of structural fragments within a molecule (see the chapters *Correlations Through the Chemical Bond I: Homonuclear Shift Correlation* and *Correlations Through the Chemical Bond II: Heteronuclear Shift Correlation*).
- Through-space coupling: methods utilise dipolar coupling (D) between nuclei, providing the basis for the nuclear Overhauser effect which is most often employed to deduce molecular stereochemistry and define conformation (see the chapter *Correlations Through Space: The nuclear Overhauser Effect*).
- Chemical exchange: methods follow dynamic exchange pathways often arising from conformational interconversions or molecular rearrangements, which may be associated with intramolecular or inter-molecular exchange processes (see the chapters *Introducing High-Resolution NMR* and *Correlations Through Space: The Nuclear Overhauser Effect*).

In addition to classification according to nuclear interactions, it also proves useful to divide experiments according to the types of nuclei involved:

- Homonuclear correlations: correlating information between similar nuclides; exemplified by the ^{1}H–^{1}H correlation experiments described in chapters: Correlations Through the Chemical Bond I: Homonuclear Shift Correlation and Correlations Through Space: The Nuclear Overhauser Effect.
- Heteronuclear correlations: correlating information between different nuclides; exemplified by the ^{1}H–^{13}C correlations of chapter: Correlations Through the Chemical Bond II: Heteronuclear Shift Correlation.

It is also possible for heteronuclear correlations to involve more than two nuclides, and although this approach is less widespread in small molecule studies, appropriate methods are also introduced in the chapter *Correlations Through the Chemical Bond II: Heteronuclear Shift Correlation.* In contrast, heteronuclear correlations involving ^{1}H, ^{13}C and ^{15}N are routine in studies of biological macromolecules, for which isotope labelling with ^{13}C and ^{15}N is essentially mandatory; see below and the chapters *Correlations Through the Chemical Bond II: Heteronuclear Shift Correlation* and *Protein-ligand screening by NMR* for related discussions.

5.1.1 Generating the Second Dimension

No matter what the nature of the interaction to be mapped, all 2D sequences have the same basic format and can be subdivided into four well-defined units termed the *preparation*, *evolution*, *mixing,* and *detection* periods (Fig. 5.1). The preparation and mixing periods typically comprise a pulse or a cluster of pulses and/or fixed time periods, the exact details of which

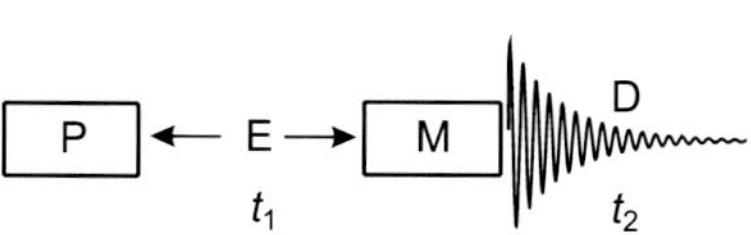

FIGURE 5.1 **The general scheme for any 2D experiment.** P, preparation; E, evolution; M, mixing; D, detection.

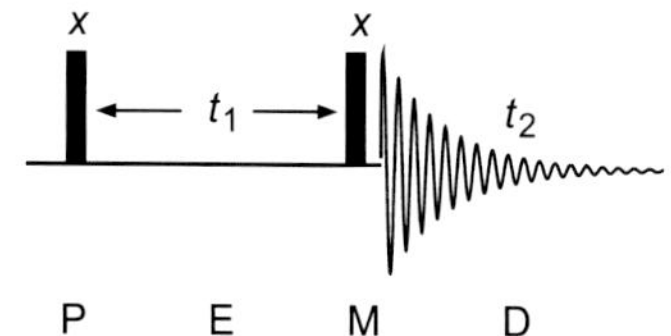

FIGURE 5.2 **An illustrative 2D sequence in which P and M are 90 degree pulses.** This is also the most basic COSY sequence.

vary depending on the nature of the experiment. The detection period is entirely analogous to the detection period of any 1D experiment during which the spectrometer collects the free induction decay (FID) of the observed spins. The evolution period provides the key to the generation of the second dimension. Before proceeding, briefly recall what happens during the acquisition of a 1D FID and the processes that generate a single frequency domain spectrum. The detection process involves sampling (digitising) the oscillating FID at regular time intervals dictated by the Nyquist condition for an appropriate acquisition period. The collected data are then subject to Fourier transformation to produce the required frequency domain spectrum. In other words, the general requirement for creating a frequency dimension is the regular sampling of magnetisation as it varies in some way as a function of time. Extending this idea one can conclude that to generate a spectrum with *two* frequency domains, f_1 and f_2, it is necessary to sample data as a function of *two* separate time variables, t_1 and t_2. Clearly one time domain and hence one frequency domain originates from the usual detection period (t_2 in Fig. 5.1), but how does the other arise?

To illustrate the required procedure consider a simple pulse sequence in which both the preparation and mixing units of Fig. 5.1 are each single $90°_x$ pulses (Fig. 5.2) acting on a sample that contains only a single, uncoupled proton resonance, say chloroform, with a chemical shift offset of ν Hz. Viewing events with the vector model (Fig. 5.3), the initial $90°_x$ pulse places the equilibrium magnetisation in the *x–y* plane along the +*y*-axis, after which it will precess (or *evolve*) according to this chemical shift offset. After a time period, t_1, the vector has moved through an angle of $360\nu t_1$ degrees, and is then subject to the second pulse. To appreciate the influence of this pulse, it is convenient to consider the vector as comprising two orthogonal components, one along the original *y*-axis (M_0.cos $360\nu t_1$) and the other along the *x*-axis (M_0.sin $360\nu t_1$). The second $90°_x$ pulse places the *y*-component along the *z*-axis while the *x*-component is unaffected and remains precessing in the transverse plane and produces the detected FID. Fourier transformation of this FID will therefore produce a spectrum containing a single resonance whose intensity depends upon the factor sin $360\nu t_1$.

Performing this experiment with t_1 set to zero means the two $90°_x$ pulses will add to produce a net $180°_x$ pulse which simply corresponds to inversion of the equilibrium vector. As there exists no transverse magnetisation there is no signal to detect and the spectrum contains only noise. Now imagine repeating the experiment a number of times with the t_1 interval increased by a uniform amount each time, and the resulting FIDs stored separately. As t_1 increases from zero, the resulting signal intensity also increases as *x* magnetisation has time to develop during the evolution period, reaching a maximum when it has evolved through an angle of 90 degrees. Further increases in t_1 cause the *x* component to diminish, pass through a null, then become negative and so on according to the sine modulation. Subjecting each of the acquired t_2 FIDs to Fourier transformation produces a series of spectra containing a single resonance whose intensity (or amplitude) varies as a function of time according to sin $360\nu t_1$ (Fig. 5.4). With longer values of t_1, relaxation effects diminish the intensity of the transverse magnetisation according to T_2*, so the signal intensities show a steady (exponential) decay in addition to the *amplitude modulation*. The intensity of the resonance as a function of time therefore represents another FID for the t_1 time domain (referred to as an interferogram) that has been generated artificially or *indirectly* (Fig. 5.5). The frequency of the amplitude modulation corresponds to the chemical shift offset of the resonance in the rotating frame and we say the

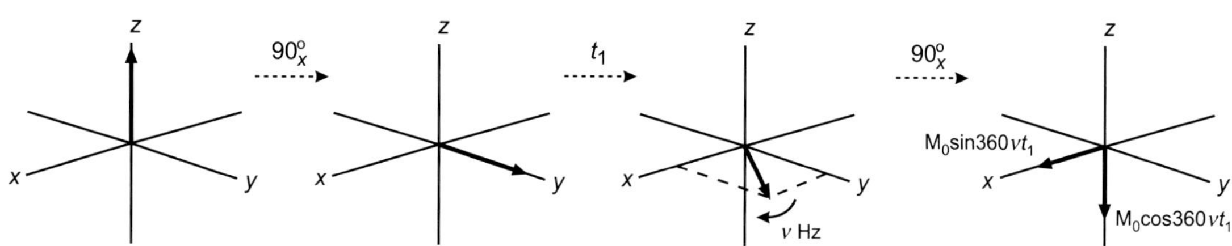

FIGURE 5.3 **The action of the COSY sequence of Fig. 5.2 on a single, uncoupled proton resonance.**

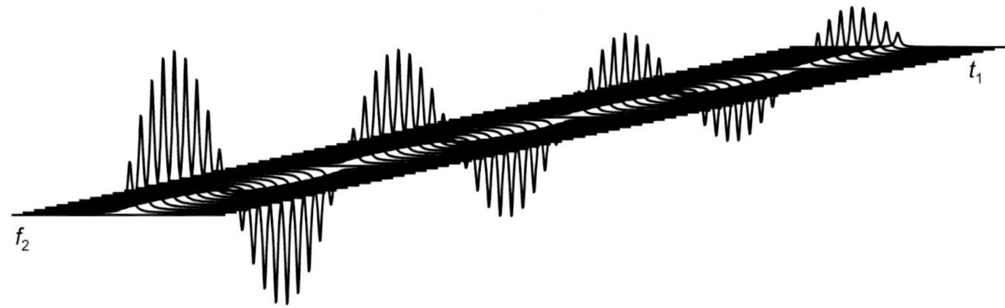

FIGURE 5.4 **Amplitude modulation of a singlet resonance as a function of the evolution period t_1.** At longer values of t_1 the signal intensity is diminished by spin relaxation.

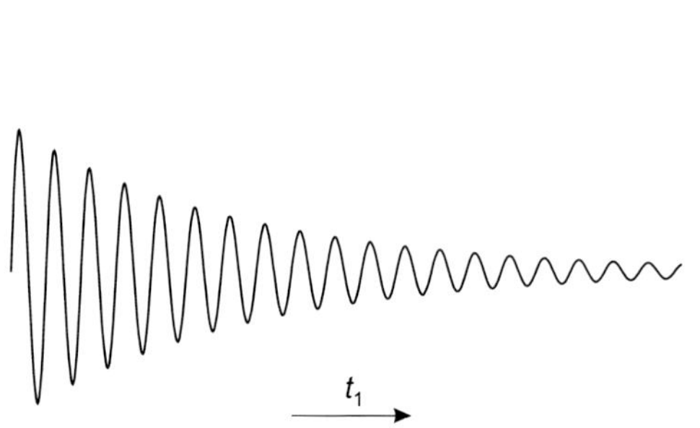

FIGURE 5.5 **The variation in peak intensity of the amplitude modulated resonance of Fig. 5.4 produces a free induction decay (interferogram) for the t_1 domain.**

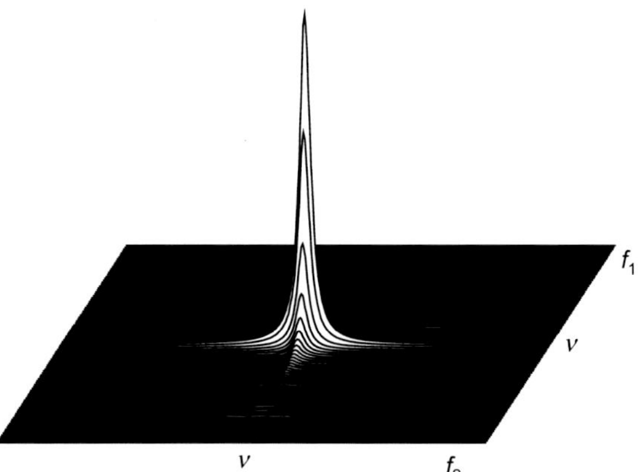

FIGURE 5.6 **The 2D spectrum resulting from the sequence of Fig. 5.2 for a sample containing a single uncoupled spin.** One peak results with a shift of ν Hz in both dimensions.

magnetisation has been *frequency labelled* as a function of t_1. Subjecting this time domain data to Fourier transformation, one would again expect this to produce a single resonance with a chemical shift of ν Hz in the f_1 frequency domain, just as for the f_2 domain. Generalising this process requires monitoring the intensity of every data point in the f_2 domain as a function of t_1, to produce a complete 2D data set. Following Fourier transformation with respect to t_1 another complete frequency domain is generated which, in combination with the conventional domain, produces a 2D spectrum displaying a single resonance at ν Hz in both dimensions (Fig. 5.6). Note that the labelling of the frequency axes as f_1 and f_2 follows from the ordering of the corresponding time domains in the pulse sequence, that is t_1 followed by t_2. The new artificial (t_1) time domain has been sampled discretely by recording a FID for each t_1 time point and storing each separately, in analogy with the sampling of the t_2 data points described earlier. The repeated acquisition of FIDs with systematically incremented t_1 time periods (Fig. 5.7) is fundamental to the generation of all 2D data sets. The philosophy behind the multidimensional approach is that one signal modulated as a function of one time variable (say t_1) is detected at some later point as a function of another (t_2 in the case of a 2D experiment). These general descriptions apply equally to spectra that contain many resonances. Repeating the above experiment for a sample containing just two uncoupled spins, with offsets of ν_A and ν_X Hz, results in a spectrum with two 2D singlets whose frequencies in both dimensions correspond simply to the respective chemical shift offsets of the resonances (Fig. 5.8). This simple idea can be extended to produce a 3D spectrum simply by having three independently incremented time periods, one detected 'directly' and two 'indirectly' (Fig. 5.9); while the technical details may be a little complex, the principle is quite straightforward. As an illustration of this concept, the 3D correlation spectrum of a small protein is shown in Fig. 5.10 in which the chemical shifts of ^{1}H, ^{13}C and ^{15}N nuclei of the protein backbone HN and CO groups are correlated in the 3D cube.

Having gone to the trouble of generating a 2D data set, the spectra of Fig. 5.6 and Fig. 5.8 tell us nothing that we could not derive from the corresponding 1D spectra, simply the chemical shifts of the participating nuclei. This provides no new information because the 2D peaks map (or correlate) exactly the same information in both dimensions ($\nu_1 = \nu_2$), in this case the chemical shifts. The 2D spectra become useful when the peaks they contain correlate *different* information on the two axes, that is when $\nu_1 \neq \nu_2$. This requires that magnetisation evolving at frequency ν_1 during the t_1 time period then evolves at a different frequency ν_2 in the detection (t_2) period. Clearly for this to happen there must be some mechanism by which

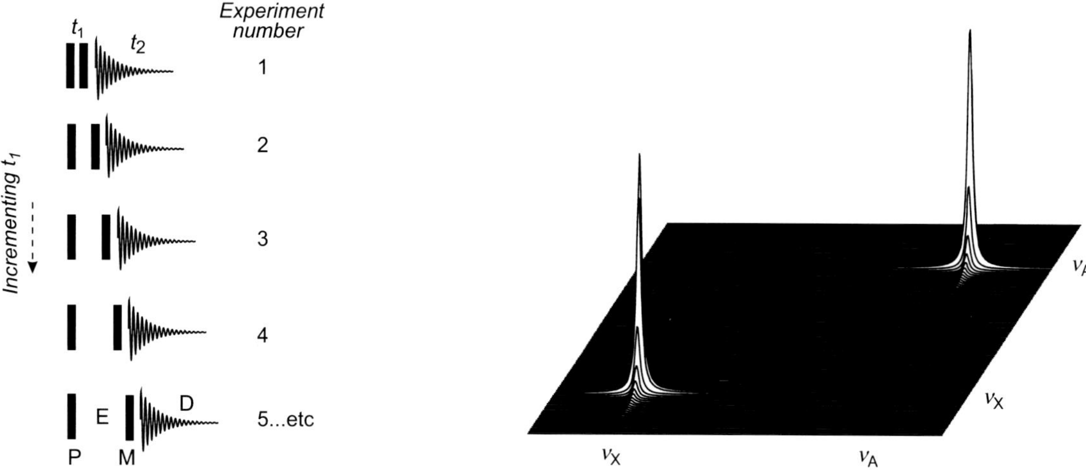

FIGURE 5.7 A generalised scheme for the collection of a 2D data set. The experiment is repeated many times with the t_1 period incremented at each stage and the resulting FIDs stored separately. Following a double Fourier transformation with respect to first t_2 and then t_1, the 2D spectrum results.

FIGURE 5.8 The 2D spectrum resulting from the sequence of Fig. 5.2 for a sample containing two uncoupled spins, A and X, of offsets ν_A and ν_X. Each produces a 2D peak at its corresponding chemical shift offset in both dimensions.

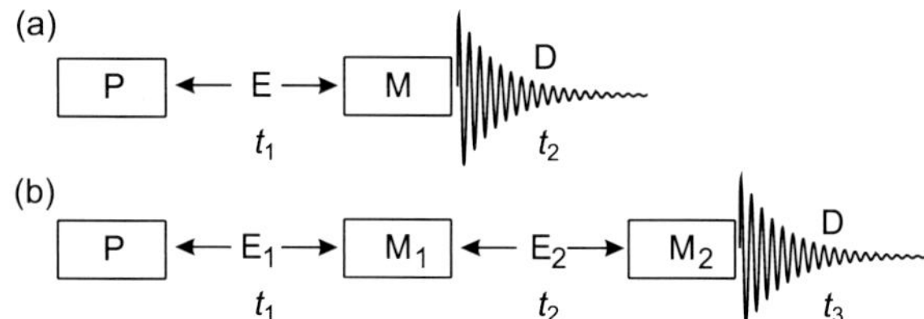

FIGURE 5.9 Schematic representations of (a) 2D and (b) 3D experiments. A 3D experiment is collected by independently varying the t_1 and t_2 intervals to generate the indirect f_1 and f_2 domains respectively, by analogy with a 2D experiment.

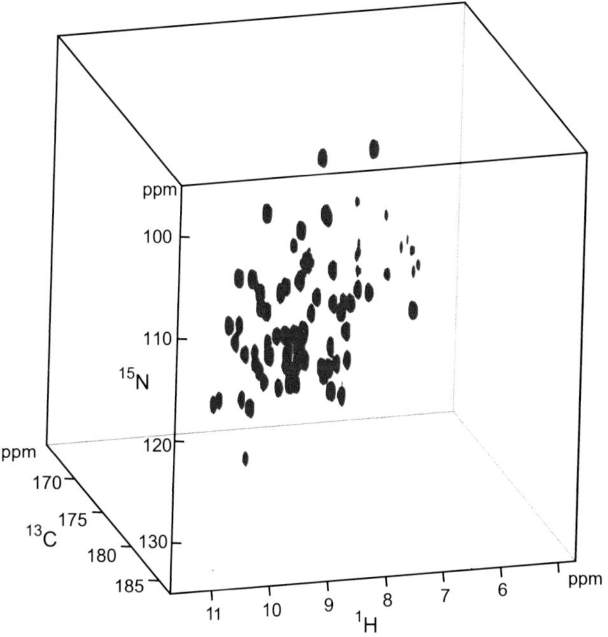

FIGURE 5.10 The 3D heteronuclear correlation spectrum of a small protein linking the HN and CO groups of the backbone (isotopically enriched in ^{13}C and ^{15}N). The three dimensions represent the chemical shifts of ^{13}C, ^{15}N and ^{1}H.

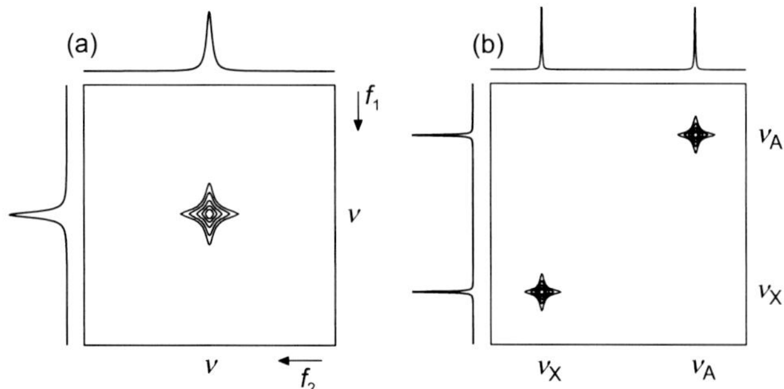

FIGURE 5.11 **The equivalent contour plot representations of (a) Fig. 5.6 and (b) Fig. 5.8.** Alongside the f_1 and f_2 axes are the conventional 1D spectra for reference purposes.

the magnetisation precession frequency changes during the sequence, and more specifically within the mixing period. The details of this period dictate the information content of the resulting spectrum according to the interaction that is exploited, for example coupling, NOE, or chemical exchange. Just how the informative correlation peaks are produced will become apparent as various sequences are examined in the sections that follow.

Before moving on, a comment on the presentation of 2D spectra is required. The spectra of Fig. 5.6 and Fig. 5.8 have been presented in the 'stacked-plot' mode to emphasise the similarity with 1D spectra, and the presence of two frequency axes and one intensity axis. Although these may look aesthetically impressive, this form of presentation is of little practical use, becoming easily cluttered and uninterpretable. The usual way to present 2D spectra is via 'contour plots', in which peak intensities are represented by contours, as a mountain range would be represented on a map. Fig. 5.11 shows the equivalent contour presentation of Fig. 5.6 and Fig. 5.8 and unless stated otherwise, all 2D spectra from now on will make use of this contour mode.

5.1.2 Correlating Coupled Spins

Moving on from the previous discussions concerning uncoupled spins, we now consider the more general case of spins that are scalar coupled. We first note that the frequency labelling component of the experiment described above is in operation for all excited spins so it is unnecessary to consider this further. Of interest now is the mixing part of the sequence, in this case the action of the pulse following t_1. In short, this pulse transfers 'magnetisation' originally associated with one spin onto neighbouring spins to which it is coupled, this process being known formally as *coherence transfer* (Section 5.3), which is only possible because of the J coupling. Coherence transfer has, in fact, already been described in Section 4.4.1 where it was presented in a slightly different guise as polarisation transfer from proton to carbon, which is nothing more than *heteronuclear coherence transfer*. In this, the original spin populations ('magnetisation') associated with the proton were ultimately transferred to carbon. In the heteronuclear case events could be readily pictured by reference to population diagrams (selective population inversion, SPI) and to the vector model (the INEPT sequence) because one could consider the action of pulses on the source proton and the target carbon quite independently. For the homonuclear case things are rather more complicated because all spins experience the same pulses simultaneously and the simple picture of events tends to break down. Nevertheless, similar transfer processes are operative and with COSY we simply have the analogous *homonuclear coherence transfer* taking place between spins. The similarity can be further exemplified if one compares the COSY sequence of Fig. 5.2 with the basic INEPT sequence of Fig. 4.23a. Ignoring the refocusing 180 degree pulses in INEPT, the two sequences are largely identical, both with initial excitation being followed by a period of evolution, after which *both* coupled spins experience a 90 degree pulse which elicits the transfer. The concept of coherence and coherence transfer is central to most NMR experiments and is further considered in Section 5.3.

The key point in all this is that magnetisation transfer occurs between coupled spins. To appreciate the outcome of this in the final COSY spectrum, consider the case of two J-coupled spins, A and X, with a coupling constant of J_{AX} and chemical shift offsets ν_A and ν_X. The magnetisation associated with spin A will, after the initial 90 degree pulse, precess during t_1 according to its chemical shift offset, ν_A. The second 90 degree pulse then transfers some part of this magnetisation to the coupled X spin, while some remains associated with the original spin A (the reason for this segregation is described in Section 5.3). That which remains with A will then precess in the detection period at a frequency ν_A, just as it did during t_1, so in the final spectrum it will produce a peak at ν_A in both dimensions, denoted (ν_A, ν_A). This peak is therefore equivalent to that observed for the uncoupled *AX* system of Fig. 5.8 and because it represents the same frequency in both dimensions it sits on the diagonal of the 2D spectrum and is therefore referred to as a *diagonal peak*. In contrast, the transferred magnetisation will precess in t_2 at the frequency of the new 'host' spin *X*, and will thus produce a peak corresponding to two different chemical shifts in the two dimensions (ν_A, ν_X). This peak sits away from the diagonal and is therefore referred to as an off-diagonal or, more commonly, a *crosspeak* (Fig. 5.12). This is the peak of interest as it provides direct evidence of coupling between spins A and X. The whole

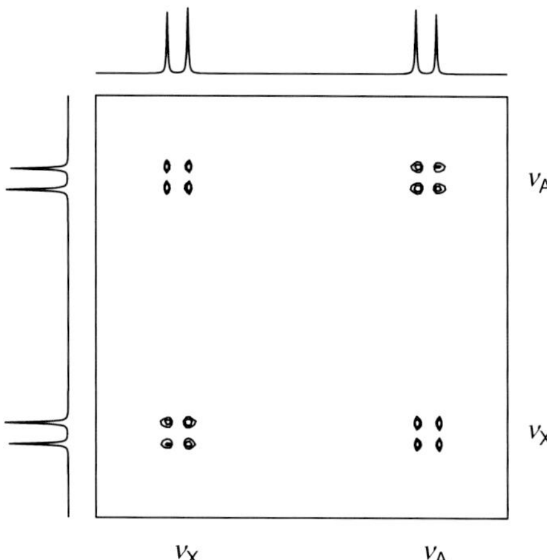

FIGURE 5.12 **The COSY spectrum of a coupled, two-spin AX system.** Diagonal peaks are equivalent to those observed in the 1D spectrum while crosspeaks provide evidence of a coupling between the correlated spins.

process also operates in the reverse direction, that is the same arguments apply for magnetisation originally associated with the X spin, giving rise to a diagonal peak at (ν_X, ν_X) and a crosspeak at (ν_X, ν_A). Thus, the COSY spectrum is symmetrical about the diagonal, with crosspeaks on either side of it mapping the same interaction (Fig. 5.12).

For any 2D experiment it is the crosspeaks that provide the correlation information of interest, whether this is between homonuclear or heteronuclear spin pairs. The chapters *Correlations Through the Chemical Bond I: Homonuclear Shift Correlation* and *Correlations Through the Chemical Bond II: Heteronuclear Shift Correlation* serve to illustrate the principle 2D techniques for correlating spins according to the scalar (J) coupling interactions they share, while the chapter *Correlations Through Space: The Nuclear Overhauser Effect* describes the use of 2D NMR for correlating spins through-space via their dipolar couplings and the resulting nuclear Overhauser effects that are observed.

5.2 PRACTICAL ASPECTS OF 2D NMR

This section introduces the most important experimental aspects relating to 2D data sets, and again uses the COSY experiment to illustrate these. Many of the discussions are simply extensions of what has already been discussed in the chapter *Practical Aspects of High-Resolution NMR* for the 1D experiment and familiarity with this is assumed. No new concepts are introduced here, although a modified approach to experimental set-up and to data processing is necessary if 2D experiments are to be successful.

5.2.1 Two-Dimensional Lineshapes and Quadrature Detection

All the 2D spectra presented so far have made use of quadrature detection in both dimensions, enabling the transmitter frequencies for each to be placed at the centre of the spectrum. Quadrature detection for the f_2 data is achieved by either the simultaneous or sequential sampling schemes described in Section 3.2.4 and is therefore entirely analogous to that used for 1D acquisitions. Quadrature detection in f_1 is also necessary for the same reasons. As with the 1D case, this demands some means of distinguishing frequencies that are higher than that of the reference from those that are lower when evolving during t_1. In other words, it is again necessary to distinguish positive and negative frequencies in the rotating frame so that mirror image signals do not appear either side of $f_1 = 0$. There are two general approaches to this in widespread use, one providing so-called *phase-sensitive* data displays, while the other provides data that are conventionally displayed in the *absolute value* mode in which all phase information has been discarded. The first of these displays lineshapes in which absorption- and dispersion-modes are separated, meaning the preferred absorption-mode signal is available for a high-resolution display. The second approach is inferior in that it produces lineshapes in which the absorptive and dispersive parts are inextricably mixed making it poorly suited to high-resolution work. However, since absolute value spectra are rather easy to process and manipulate, they still find use in some routine, fully automated processes, so are also considered here.

5.2.1.1 Phase-sensitive Presentations

As for 1D data, f_1 quadrature detection requires two data sets to be collected which differ in phase by 90 degrees, thus providing the necessary sine and cosine amplitude-modulated data. Since the f_1 dimension is generated artificially, there is strictly no reference rf to define signal phases so it is the phase of the pulses that bracket t_1 that dictate the phase of the

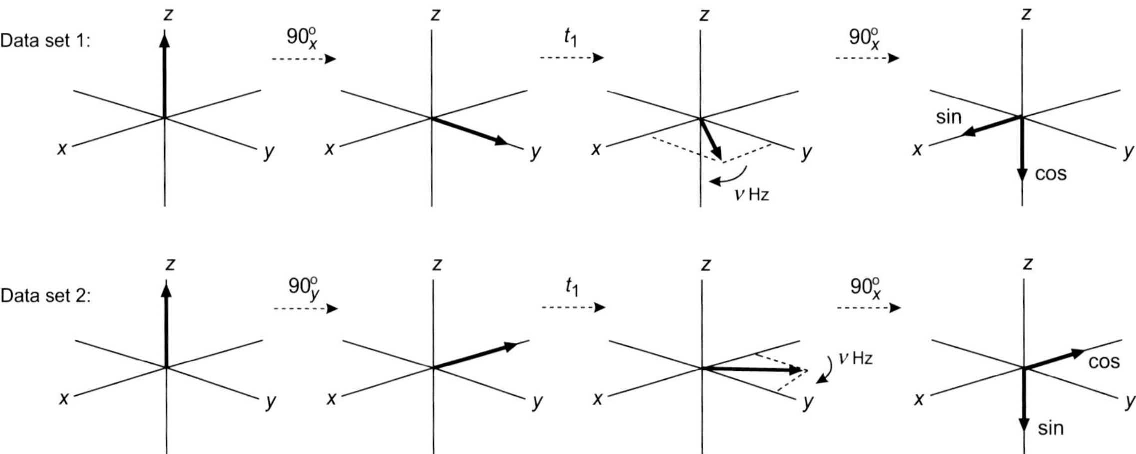

FIGURE 5.13 **The States method of f_1 quad-detection.** This requires two data sets to be acquired per increment to generate separate sine and cosine modulated data sets.

detected signal. Thus, *for each t_1 increment* two data sets are collected, one with a 90_x preparation pulse (t_1 sine modulation), the other with 90_y (t_1 cosine modulation), and both stored separately (Fig. 5.13). These two sets are then equivalent to the two channel data collected with *simultaneous* acquisition which produces the desired frequency discrimination when subject to a complex Fourier transform (also referred to as a hypercomplex transform in relation to 2D data). The rate of sampling in t_1 or, in other words, the size of the t_1 time increment, is dictated by the f_1 spectral width and is subject to the same rules as for the simultaneous sampling of 1D data. This method is derived from the original work of States, Haberkorn and Ruben [5,6] and is therefore often referred to in the literature as the *States* method of f_1 quad-detection.

An alternative approach is analogous to the method of *sequential* sampling introduced in Section 3.2.4. As in the 1D approach the aim is to avoid both positive and negative frequencies arising by shifting the *apparent* frequency range from $\pm\frac{1}{2}SW_1$ to 0 to $+SW_1$ Hz by making the evolution frequencies during t_1 appear faster than they actually are. As for the States method discussed earlier, there is no reference rf phase to shift for this artificial domain, so the equivalent effect is achieved by incrementing the phase of the preparation pulse by 90 degrees for each t_1 increment, and sampling the data twice as fast as for the States method (by halving the t_1 increment). Only one data set is acquired for each t_1 period, but twice as many t_1 increments are collected, so the total t_1 acquisition time, and hence the digital resolution, is equal for both methods. This approach to f_1 quad-detection [7] is now referred to as time proportional phase incrementation (TPPI).

The States and TPPI methods produce equivalent data sets [8] although they differ subtly in the appearance of aliased signals and the artefacts known as axial peaks, which are described more fully below. It turns out that an effective way to deal with axial peaks is to beneficially combine these two approaches to yield the so-called *States–TPPI* method of quadrature detection [9] as described in Section 5.2.2. Selecting between these related approaches is often left to the experimentalist but in general the preferred method is States–TPPI. The most significant point from all this is the 2D lineshapes these methods produce. All methods involve the detection of a signal that is *amplitude modulated* as a function of the t_1 evolution period (Fig. 5.4), and this is the general requirement for producing spectra that have absorption and dispersion parts separated (*pure-phase* spectra) so allowing a *phase-sensitive* presentation. Separate real and imaginary parts of the data exist for both the f_1 and f_2 dimensions, again analogous to the real and imaginary parts of a 1D spectrum. This gives rise to four data quadrants (Fig. 5.14) with only the (real, real) data set being presented to the user as the final 2D spectrum, the others being retained for phase correction. It is usual for the displayed spectrum to contain absorption-mode lineshapes in both dimensions (Fig. 5.6) wherever possible since the double-absorption lineshape affords the highest possible resolution (Fig. 5.14, *RR quadrant*). The phase information contained within crosspeaks can also provide additional information in some circumstances and this is especially true for the phase-sensitive COSY experiment.

Finally we note an alternative approach to quadrature detection known as *echo–antiecho* selection [10] which is applicable only to PFG selected 2D methods and which now finds widespread use. As this involves a quite different procedure it will not be considered further here but will be introduced in Section 5.4.2 after field gradients have been described.

5.2.1.2 Aliasing in Two Dimensions

Resonances that fall outside the chosen spectral width will be characterised with incorrect frequencies and so will appear aliased in the 2D spectrum. For symmetrical data sets such as COSY, the two spectral widths are chosen to be the same, so

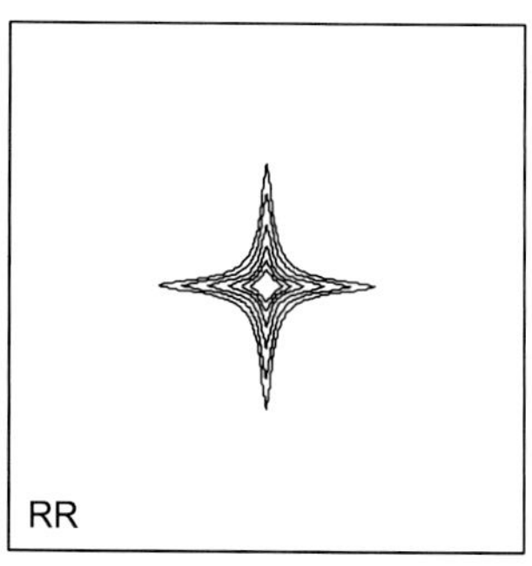

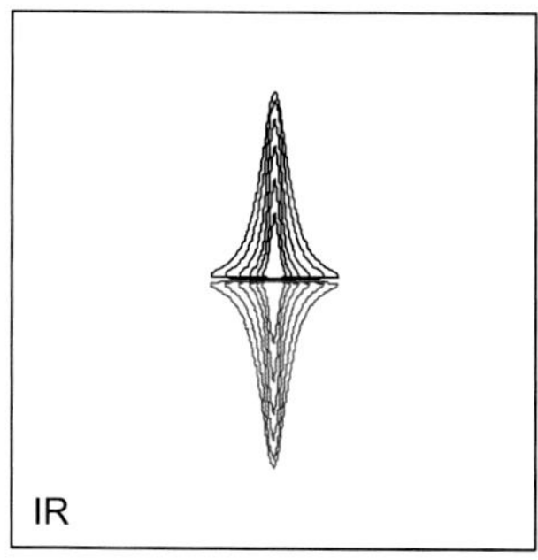

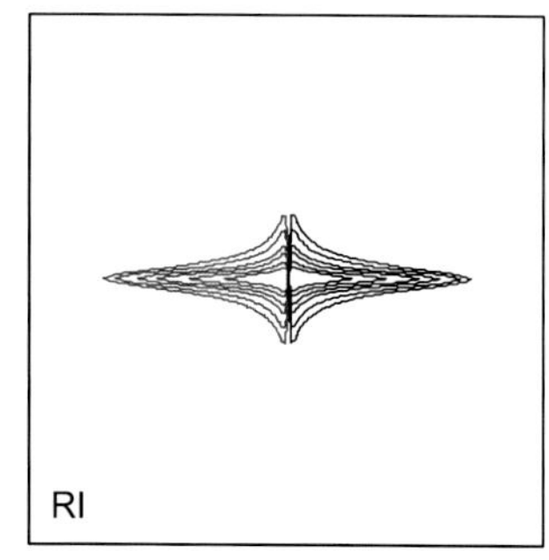

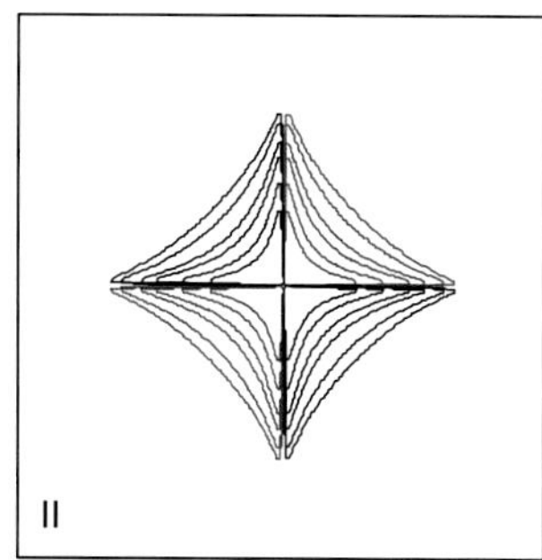

FIGURE 5.14 **The four quadrants of a phase-sensitive data set.** Only the RR quadrant is presented as the 2D spectrum and this is phased to contain absorption-mode lineshapes in both dimensions to provide the highest resolution (R = real, I = imaginary). Positive contours are in black and negative in red.

aliasing will occur in both dimensions, and the position of the aliased signal can usually be predicted from the quadrature detection scheme used for each dimension. Thus, in analogy with 1D spectra, simultaneous or States sampling causes signals to be *wrapped* around, appearing at the far end of the spectrum, while those from sequential or TPPI sampling appear as *folded* signals at the near end. Some confusion can be introduced if different sampling schemes are used for the two dimensions, for example simultaneous sampling in f_2 but TPPI in f_1 [11]. Deliberately aliasing signals can be a useful trick in acquiring 2D data since reduced spectral widths imply reduced acquisition times and smaller data sets, or alternatively, greater resolution. For example it is often feasible to eliminate lone phenyl groups from COSY spectra and concentrate on the aliphatic region only as this is where the interesting correlations most often lie. With modern digital filters aliased signals in f_2 can be eliminated, although those aliased in f_1 remain since there is no equivalent filtration in the indirect dimension.

5.2.1.3 Absolute Value Presentations

Prior to the advent of the above methods that allowed the presentation of phase-sensitive displays, 2D data sets were collected that were *phase-modulated* as a function of t_1 rather than amplitude-modulated. Phase-modulation arises when the sine and cosine modulated data sets collected for each t_1 increment are combined (added or subtracted) by the steps of the phase cycle, meaning each FID per t_1 increment contains a mixture of both parts. Here it is the sense of phase precession that allows the differentiation of positive and negative frequencies. This method is inferior to the phase-sensitive approach because of the unavoidable mixing of absorptive and dispersive lineshapes, so is generally only considered for routine, low-resolution work.

The selection of only one of the two possible mirror image data sets in f_1 is now achieved by suitable combination of the sine and cosine data sets. *Addition* of the two data sets within the phase cycle (Table 5.1) selects those signals that have the *same* sense of precession in t_1 as they have during t_2 (say, both positive) and these are referred to as P-type signals. *Subtraction* within the phase cycle selects those signals that have the *opposite* sense of precession during t_1 and t_2 (say, negative in t_1, positive in t_2) and are referred to as N-type signals. To clarify this point, remember the two senses of precession we speak of here simply represent the two possible signals that would be detected either side of $f_1 = 0$ if one were *not* using quadrature detection. By employing this it becomes possible to select only one of these to appear in the final spectrum while cancelling the mirror image. The information content of the P-type or N-type spectra are equivalent, only their appearance differs by virtue of reflection about $f_1 = 0$. One significant difference arises from the fact that with N-type selection signals are chosen that have opposite senses of precession in the two time periods. This may be thought of as being analogous to a spin-echo where vectors move in opposite senses either side of the refocusing pulse. A similar effect arises during t_2 with N-type selection whereby echoes also occur, known as *coherence transfer echoes* (Fig. 5.15). For this reason, N-type

TABLE 5.1 COSY Phase Cycles to Select the N-Type (Echo) or P-Type (Anti-echo) Signals in f_1

	N-Type			P-Type		
	Pulse 1	Pulse 2	Acquire	Pulse 1	Pulse 2	Acquire
Cycle 1	x	x	x	x	x	x
Cycle 2	y	x	$-x$	y	x	x

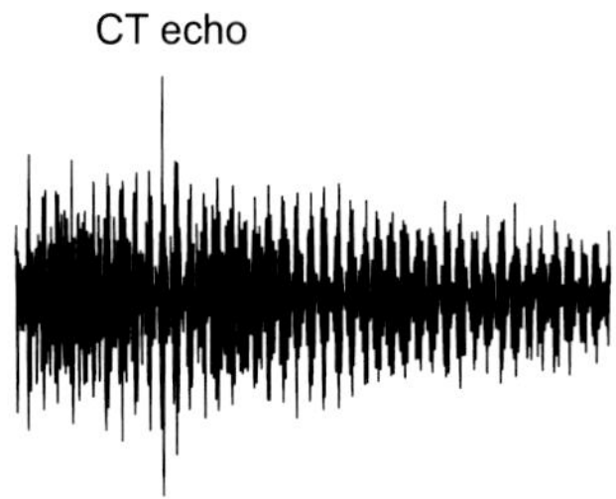

FIGURE 5.15 **The coherence transfer echo is apparent in an FID taken from an N-type COSY data set.**

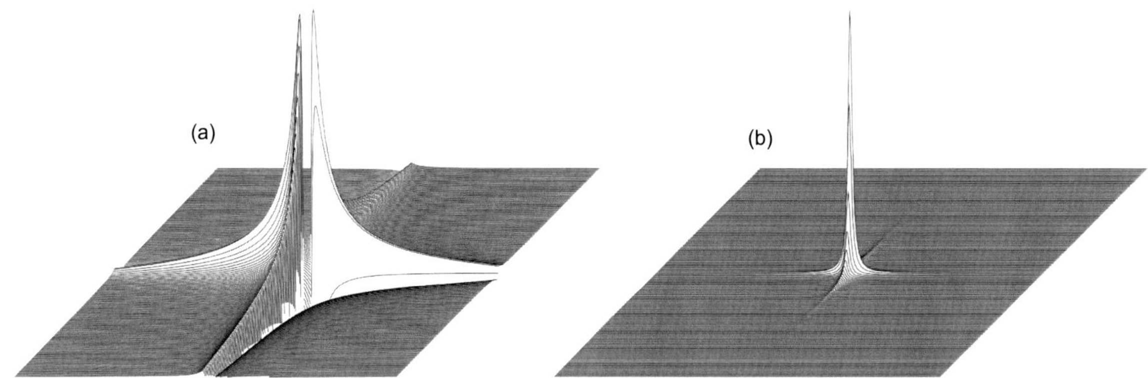

FIGURE 5.16 **2D lineshapes.** A stacked plot illustration of (a) the phase-twisted line shape and (b) the double-absorption lineshape. Clearly the resolution in (b) is far superior and for this reason phase-sensitive methods are preferred.

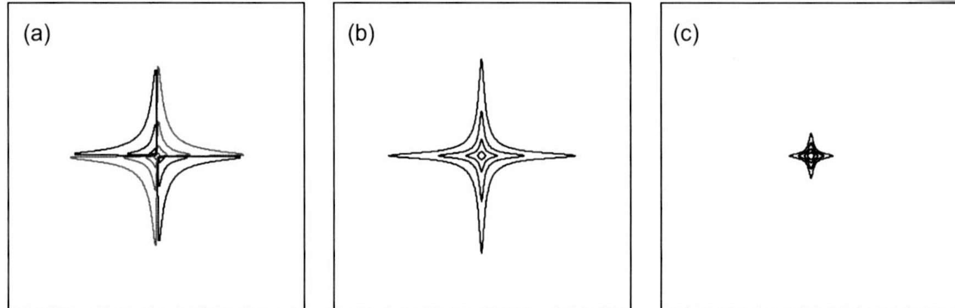

FIGURE 5.17 **Magnitude calculation.** Contour plots of (a) the phase-twist lineshape, (b) the same following magnitude calculation, and (c) the same following resolution enhancement with an unshifted sine bell window and magnitude calculation.

selection is also referred to as *echo selection* and P-type signals as *anti-echo selection*, since the refocusing effect does not arise for signals that precess in the same sense. Refocusing of field inhomogeneity with echo selection together with the fact that P-type signals are subject to more severe distortions, means that N-type selection is preferred when this method of quad-detection is used. Conventionally, these spectra are then presented with the diagonal running from bottom-left to top-right, as in earlier figures.

The greatest drawback with data collected with phase-modulation is the inextricable mixing of absorption and dispersion-mode lineshapes. The resonances are said to possess a *phase-twisted* lineshape (Fig. 5.16a), which has two principle disadvantages. Firstly, the undesirable and complex mix of both positive and negative intensities, and secondly, the presence of dispersive contributions and the associated broad tails that are unsuitable for high-resolution spectroscopy. To remove confusion from the mixed positive and negative intensities, spectra are routinely presented in absolute value mode, usually after a *magnitude calculation* (Fig. 5.17):

$$\mathbf{M} = (\text{real}^2 + \text{imaginary}^2)^{1/2}$$

where **M** represents the resulting spectrum. To improve resolution, severe window functions are also employed to eliminate the dispersive tail from the lineshape. Although a number of shaping functions are suitable [12], the most frequently used are the *sine bell*, or the *squared sine bell* (Section 3.2.7), which are simple to use as they have only one variable, the position of the maximum. When used unshifted, for example when the half sine-wave has a maximum at the centre of the acquisition time, the resulting peaks possess no dispersive component, thus reproducing the desired absorption lineshape (Fig. 5.17c). The sine bell shape is also beneficial in enhancing the crosspeak intensities relative to the diagonal. Since the crosspeaks arise from sine modulations, as described earlier, they initially have zero intensity which builds within the acquisition time. Diagonal peaks are cosine modulated so begin with maximum intensity and are therefore attenuated by the window function. However, the attenuation presents a problem for signals with differing relaxation times (or, in other words, linewidths) as these will be attenuated to different extents and, in particular, broader lines will be notably reduced in intensity by the early part of the sine bell. The moral when interpreting absolute value data processed in this way is to be wary of differing crosspeak intensities and not simply to associate these with smaller coupling constants. The final problem with the use of these extreme resolution enhancing window functions is that much of the early signal is attenuated and the later noise enhanced, leading to a potential loss of sensitivity. It is possible to shift the function maximum to earlier in the FID to help improve this, with the inevitable reintroduction of some dispersive component, although a better approach is to acquire phase-sensitive data which does not demand the application of such severe window functions. Despite this, when sensitivity is not a limiting factor, the magnitude experiment does have the advantage of not requiring phase corrections to be made, so is well suited to routine acquisitions and to automated methods, especially when implemented as the COSY-β variant described in Section 6.1.4.

5.2.2 Axial Peaks

Axial peaks are experimental artefacts that arise from magnetisation that was longitudinal (aligned along the z-axis) during the t_1 evolution period which is subsequently returned to the transverse plane by the mixing pulse and detected in t_2. Since this magnetisation does not evolve during t_1 it is not frequency labelled and therefore has zero f_1 frequency. The peaks can be associated with all resonances in the spectrum and therefore appear as a band of signals across the spectrum at $f_1 = 0$ Hz. For the States or absolute value methods, this occurs at the centre of the f_1 dimension (Fig. 5.18) and, unless eliminated, may interfere with genuine crosspeaks. If TPPI is used for quadrature detection in f_1, the peaks are shifted toward one edge of the spectrum which, as described earlier, is where $f_1 = 0$ appears to be, and although aesthetically undesirable, are less of an interference. This advantage has led to the combined approach to quadrature detection known as the *States–TPPI* method [9] in which the original States procedure is modified such that axial peaks arise at the edge and not the centre of the f_1 dimension. This is achieved simply by incrementing the phase of the pulse prior to the t_1 period by 180 degrees *on each t_1 increment*, combined with inversion of the receiver phase. For the desired magnetisation components this simultaneous inversion has no net influence and signal modulation remains as for the States procedure. However, magnetisation associated with axial peaks does not experience the initial pulse and so accumulates the receiver phase increment. This is then equivalent to the TPPI procedure *for the axial peaks only* and hence they appear advantageously at the edge of the spectrum.

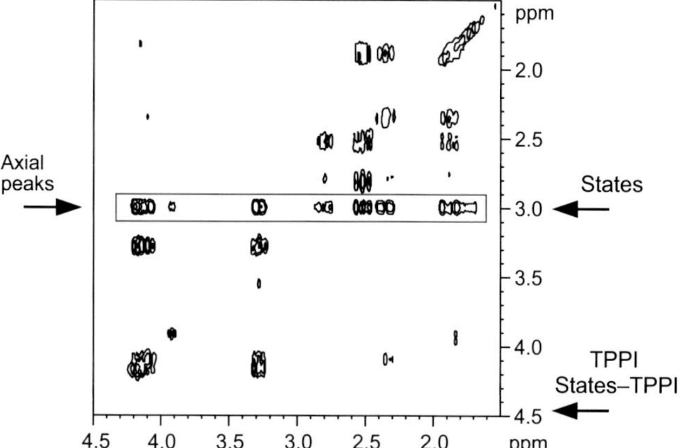

FIGURE 5.18 **Axial peaks in a COSY spectrum.** These form a band of signals along f_2 (boxed) at either the midpoint of f_1 (States or absolute value data sets) or at the high-frequency edge of the spectrum (TPPI or States–TPPI data sets).

TABLE 5.2 The Basic Two-Step Phase Cycle for the Elimination of Axial Peaks

	Pulse 1	Pulse 2	Acquire
Scan 1	x	x	x
Scan 2	$-x$	x	$-x$

The longitudinal magnetisation responsible for axial peaks arises from both experimental imperfection and from relaxation during t_1. Thus if the preparation sequence does not place all magnetisation in the transverse plane, which for COSY means the initial pulse is not exactly 90 degrees, some residual z component remains. Elimination of these unwanted peaks is achieved simply by repeating the experiment with the receiver phase inverted. The phase of the axial peaks will be unchanged, so will cancel when the two data sets are subtracted by the phase cycle. This procedure also has the benefit of cancelling any offset of the FIDs that arises from receiver imperfections. To avoid simultaneous cancellation of the desired signals the phase of the preparation pulse is also inverted to ensure that it matches the receiver phase, leading to the two-step phase cycle of Table 5.2.

5.2.3 Instrumental Artefacts

While there exist a whole host of artefacts that can arise in 2D spectra according to the details of the experiment, the two most likely to be encountered are briefly considered here. Both arise from instrumental imperfections, and how significant they are to you will be somewhat dependent on your instrument and its performance, but in any case it is useful to be able to recognise these artefacts if and when they appear. Nevertheless, the digital architectures of modern NMR instruments contribute significantly to the systematic reduction of many sources of instability and can reduce, or even eliminate, certain artefacts more common to older instrument designs.

5.2.3.1 f_2 Quadrature Artefacts

Artefacts appearing along f_2 rows are in many instances the exact parallel of those that may be observed in 1D acquisitions, and, not surprisingly, the same solutions apply, including their elimination by digital processing methods. If likely to arise, mirror image quadrature artefacts arising from imbalance of the receiver quadrature channels may again be suppressed by integration of the four-step cyclically ordered phase cycle (CYCLOPS) routine (Section 3.2.5) into the existing phase cycle. This involves incrementing the phase of all pulses in the sequence together with the receiver phase in steps of 90 degrees, and leads to a fourfold increase in the duration of the phase cycle. On modern instruments, the intensity of quadrature artefacts are so small (or even non-existent with digital detection schemes) that the addition of CYCLOPS can often be avoided and time savings made when sensitivity is not limiting.

5.2.3.2 t_1-Noise

Digitisation of data during a 2D experiment is subject to the same thermal noise arising from the probe head and preamplifier as in a 1D experiment, and this contributes to the noise baseplane observed in the 2D spectrum. There also exists a particularly objectionable artefact associated with 2D experiments, referred to as t_1 noise (note the t_1 here refers to the evolution period and should not be confused with the T_1 relaxation time constant). This appears as bands of noise parallel to the f_1-axis where an NMR resonance exists and it is sometimes this that limits the observation of peaks in the spectrum rather than the true thermal noise (indeed, it appears that the very earliest work on 2D NMR was unpublished due to excessive t_1 noise present in the spectra). Generally speaking, this is caused by instrument instabilities which lead to random fluctuations in signal intensities and phase from one FID to the next over the course of the 2D experiment. Since these fluctuations relate to perturbations of the NMR resonance, the bands of t_1 noise characteristically appear only at the f_2 shifts of resonances and have intensities proportional to the corresponding resonance amplitude (Fig. 5.19). Thus stronger bands of noise are associated with intense, sharp peaks, most notably singlets.

The instrumental instabilities that contribute to t_1 noise are numerous [13,14], including irreproducible rf pulse phase and amplitude, instability in the field frequency lock and field homogeneity, and so on. These factors are in the hands of the instrument manufacturers, who have made steady progress in reducing various sources of instrumental instabilities over the years. In preparing a 2D acquisition, there are a number of steps that will also help minimise these artefacts. The same arguments apply to any experiment that uses difference spectroscopy to retain selected signals and cancel others and are

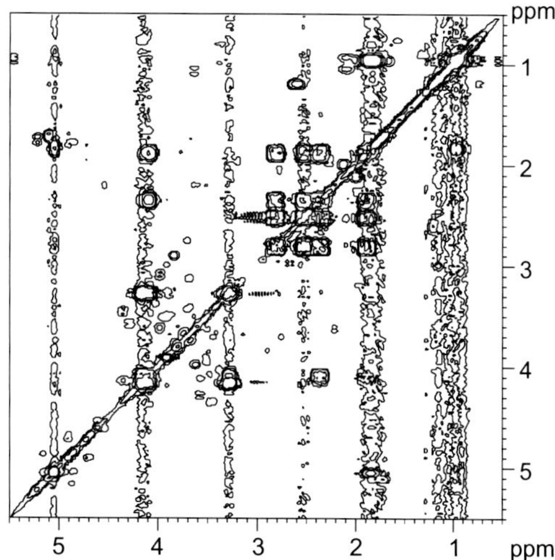

FIGURE 5.19 **Bands of noise parallel to the f_1-axis (t_1 noise) often appear in 2D spectra.**

not exclusive to the world of 2D NMR. Use of a strong, sharp lock signal will provide optimum field frequency regulation, and the use of auto-shim routines on the lower-order shims for long-term acquisitions should also help maintain lineshape reproducibility. Stability of the sample environment is also important. Magnetic field stability is, of course, essential, so movement of any magnetic materials anywhere in the vicinity of the instrument is highly undesirable; this could even include motion (human or otherwise) above or below the magnet or in the laboratory next door. Temperature stability should also be maintained with a suitable temperature control unit, and equilibration of the sample prior to running the experiment may be necessary, especially for aqueous solutions. Finally, it is recommended *not* to spin samples during 2D acquisitions. This typically causes additional modulation of the detected signal as the sample rotates, notably from an imperfect tube or a 'wobbling' sample spinner, further contributing to t_1 noise.

5.2.3.3 Symmetrisation

Unwanted t_1 noise in spectra can also be reduced by a number of post-processing strategies. One of the most common procedures for absolute value COSY experiments is symmetrisation [15]. This involves replacing symmetrically related data points either side of the diagonal with the lesser of the two values so retaining crosspeaks but suppressing unsymmetrical artefacts. This must be applied with caution however, as it may enhance artefacts that have coincident symmetry, such as regions of t_1 noise associated with intense singlets, giving them the appearance of genuine correlation peaks; confusion can usually be avoided by comparison with the non-symmetrised spectrum. This method is not generally suitable for phase-sensitive data as it may introduce distortion of crosspeak structure and can lead to reduced sensitivity, so this symmetrisation finds rather limited use in modern NMR.

Significant reductions in the level of t_1 noise can also be obtained in certain experiments where signal selection is achieved by PFGs, although homonuclear experiments such as COSY show little benefit in this respect, more notable gains being apparent for proton-detected heteronuclear correlation experiments. Alternative post-processing methodologies such as reference deconvolution [16–19] (Section 3.4.6) have also provided impressive reductions in t_1 noise [20,21], and the algorithms required for this are to be found in some commercial processing software.

5.2.4 Two-Dimensional Data Acquisition

The success of any NMR experiment is, of course, crucially dependent on the correct setting of the acquisition parameters. In the case of 2D experiments one has to consider the parameters for each dimension separately, and we shall see that the most appropriate parameter settings for f_2 are rarely optimum for f_1. Likewise, one has to give rather more thought to the setting up of a 2D experiment than is usually required for 1D acquisitions to make optimum use of the instrument time available. Once again, the general considerations below will be applicable to all 2D experiments, although we restrict our discussion at this stage to COSY.

At the most basic level one has to address two fundamental questions when setting up an experiment. (1) Will the defined parameters enable (or limit) the detection of the desired information and exclude the unwanted? (2) How long will it take and do I have enough time? Ideally we would wish to acquire data sets rapidly yet still be able to provide the information we require, so we set up our experiment with these goals in mind.

Firstly, spectral widths, which should be the same in both dimensions of the COSY experiment, should be kept to minimum values with transmitter offsets adjusted so as to retain only the regions of the spectrum that will provide useful correlations. It is usually possible to reduced spectral widths to well below the 10 or so ppm proton window observed in 1D experiments. The use of excessively large windows leads to poorer digital resolution in the final spectrum, or requires greater data sizes, neither of which is desirable. The spectral widths in turn define the sampling rates for data in t_2, in exact analogy with 1D acquisitions, and the size of the t_1 increment, again according to the Nyquist criteria. The acquisition time (AQ_t), and hence the digital resolution ($1/AQ_t$), for each dimension is then dictated by the number of data points collected in each. For t_2, this is the number of data points digitised in each FID, while for t_1 this is the number of FIDs collected over the course of the experiment. The appropriate setting of these parameters is a most important aspect to setting up a 2D experiment, and the way in which one thinks about acquisition times and digital resolution in a 2D data set is, necessarily, quite different from that in a 1D experiment. As an illustration, imagine transferring the typical parameters used in a 1D proton acquisition into the two dimensions of COSY. The acquisition time might be 4 s, corresponding to a digital resolution of 0.25 Hz/pt, with no relaxation delay between scans. On, for example, a 400 MHz instrument with a 10 ppm spectral width, this digital resolution would require 32K words to be collected per FID. The 2D equivalent, with States quad-detection in f_1 and with axial peak suppression, requires 4 scans to be collected for each t_1 increment. The mean acquisition time for each would be 6 s (t_2 plus the mean t_1 value), corresponding to 24 s of data collection per FID. If 16K t_1 increments were to be made for the f_1 dimension (two data sets are collected for each t_1 increment remember) this would correspond to a total experiment time of about 4½ days. I trust you will agree that four days for a basic COSY acquisition is quite unacceptable, so acquiring data with such high levels of digitisation in both dimensions is clearly not possible.

The key lies in deciding on what level of digitisation is required for the experiment in hand. The first point to notice is that adding data points to extend the t_2 dimension leads to a relatively small increase in the overall length of the experiment, so we may be quite profligate with these. Adding t_1 data points on the other hand requires that a complete FID of potentially many scans is required *per increment*, which makes a far greater increase to the total data collection time. Thus one generally aims to keep the number of t_1 increments to a minimum that is consistent with resolving the correlations of interest, and increasing t_2 as required when higher resolution is necessary. For this reason, the digital resolution in f_2 is often greater than that in f_1, particularly in the case of phase-sensitive data sets. The use of smaller t_1 acquisition times (AQ_{t1}) is, in general, also preferred for reasons of sensitivity since FIDs recorded for longer values of t_1 will be attenuated by relaxation and so will contribute less to the overall signal intensity. The use of small AQ_{t1} is likely to lead to truncation of the t_1 data, and it is then necessary to apply suitable window functions that force the end of the data to zero to reduce the appearance of truncation artefacts.

For COSY in particular, one of the factors that limits the level of digitisation that can be used is the presence of intrinsically anti-phase crosspeaks, since too low a digitisation will cause these to cancel and the correlation to disappear (see Section 6.1.5 for further discussions). The level of digitisation will also depend on the type of experiment and the data one expects to extract from it. For absolute value COSY one is usually interested in establishing where correlations exist, with little interest in the fine structure within these crosspeaks. In this case it is possible to use a low level of digitisation consistent with identifying correlations. As a rule of thumb, a digital resolution of J to 2J Hz/pt (AQ of 1/J to 1/2J s) should enable the detection of most correlations arising from couplings of J Hz or greater. Thus for a lower limit of, say, 3 Hz a digital resolution of 3–6 Hz/pt (AQ of ca. 300–150 ms) will suffice. The acquisition time for t_1 is typically half that for t_2 in this experiment, with one level of zero-filling applied in t_1 so that the final digital resolution is the same in both dimensions of the spectrum (as required for symmetrisation).

For phase-sensitive data acquisitions one may be interested in using the information contained within the crosspeak multiplet structures, and a higher degree of digitisation is required to adequately reflect this, a more appropriate target being around J/2 Hz/pt or better (AQ of 2/J s or greater). Again, digitisation in t_2 is usually 2 or even 4 or 8 times greater than that in t_1. In either dimension, but most often in t_1, this may be improved by zero filling, although one must always remember that it is the length of the time domain acquisition that places a fundamental limit on peak resolution and the effective linewidths after digitisation, regardless of zero-filling. The rule as ever is that high resolution requires long data-sampling periods. An alternative approach for extending the time-domain data is to use forward linear prediction when processing the data (Section 3.2.3). It is now also possible to reduce the sampling requirements in t_1 through the use of non-uniform sampling, as described in the following section, potentially leading to significant time savings. Following recent developments, the acquisition of 2D spectra in a single scan is also possible and the simultaneous acquisition of multiple nuclei (requiring dual instrument receiver channels) also has the potential to accelerate data acquisition; these topics are introduced briefly in Sections 5.4.5 and 7.7, respectively.

TABLE 5.3 Illustrative Data Tables for COSY Experiments

Experiment	Spectral Width (ppm)	$N(t_2)$	$N(t_1)$	Hz/pt (t_2)	Hz/pt (t_1)	Experiment Time
(a) Phase sensitive	10 × 10	32K	32K	0.25	0.25	4.5 day
(b) Phase sensitive	6 × 6	2K	1K	2.3	2.3	55 min
(c) Absolute value	6 × 6	1K	256	4.6	4.6	22 min

Scenario (a) transplants acquisition parameters from a typical 1D proton spectrum into the second dimension leading to unacceptable time requirements, whereas (b) and (c) use parameters more appropriate to 2D acquisitions. All calculations use phase cycles for f_1 quad-detection and axial peak suppression only and, for (b) and (c), a recovery delay of 1 s between scans. A single zero-filling in f_1 was also employed for (b) and (c).

Having decided on suitable digitisation levels and data sizes, one is left to choose the number of scans or transients to be collected per FID and the repetition rates and hence relaxation delays to employ. The minimum number of transients is dictated by the minimum number of steps in the phase cycle used to select the desired signals. Further scans may include additional steps in the cycle to suppress artefacts arising from imperfections. Beyond this, further transients should only be required for signal averaging when sensitivity becomes a limiting factor. Since most experiments are acquired under 'steady-state' conditions, it is also necessary to include 'dummy' scans prior to data acquisition to allow the steady-state to establish. On modern instruments dummy scans are required only at the very beginning of each experiment so make a negligible increase to the total time required. On older instruments it is necessary to add dummy scans for each t_1 increment, and these may then make a significant contribution to the total duration of the experiment. The repetition rate will depend upon the T_1 s of the excited spins (protons in the case of COSY and many of the heteronuclear correlation experiments) and since the sequence uses 90 degree pulses, the optimum sensitivity is achieved by repeating every 1.3 T_1 s.

Returning to the example 400 MHz acquisition discussed earlier, we can apply more appropriate criteria to the selection of parameters. Table 5.3 compares the result from above with more realistic data, and it is clear that under these conditions COSY becomes a viable experiment, requiring only minutes to collect, rather than days. The introduction of PFGs to high-resolution spectroscopy (Section 5.4) allows experiments to be acquired with only one transient per FID where sensitivity is not limiting so further reducing the total time required for data collection. Although illustrated for COSY spectra, the general line of reasoning presented here is applicable to the set-up of any 2D experiment. These issues are briefly considered with reference to different classes of techniques in the following chapters describing other 2D methods.

5.2.4.1 Non-Uniform Sampling

The classical sampling of 2D (and more generally multi-dimensional) NMR experiments requires the uniform sampling of data in the indirect dimension(s) that allows for the processing of the data by the discrete Fourier transform. This means a sequential, stepwise increment of the t_1 period of a 2D data set is made to the limit $t_{1(max)}$ which dictates the resolution in this dimension. The number of such t_1 increments employed ultimately defines the total duration of the experiment. The method of non-uniform sampling (NUS) seeks to reduce the number of data points collected in the indirect dimension(s) and so reduces the total experiment time. The development of NUS in NMR spectroscopy has been largely driven by the widespread use of 3D experiments in biomolecular NMR, where single experiments may last days with conventional data sampling and where time savings through NUS can be very significant. Nevertheless, NUS may also be applied to 2D experiments and can yield time savings that become significant when many samples are being analysed, and it is in this context that we consider this approach.

A typical approach to collecting NUS experiments is to randomly acquire only a subset of the usual t_1 data traces. Most often only 50–25% of data points may be sampled (Fig. 5.20), leading to a time saving of two to fourfold. Such sampling means, however, that it is then not possible to process the t_1 time domain data with the conventional Fourier transform, and

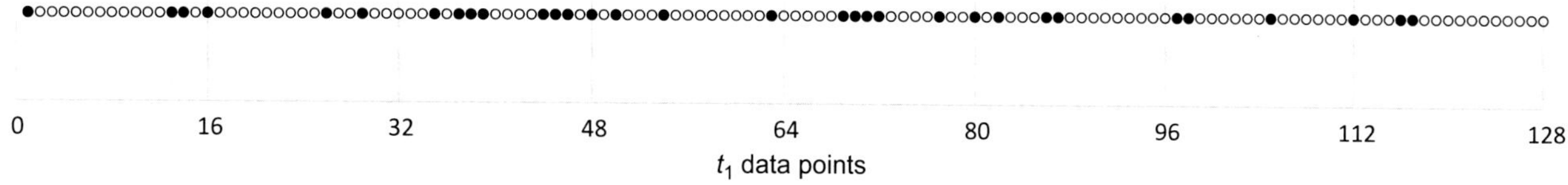

FIGURE 5.20 **Non-uniform data sampling.** Filled circles show the randomly distributed t_1 data points of a 2D data set sampled at 25%, open circles represent those additional points that would also be collected for conventional t_1 data sampling (128 complex points in total).

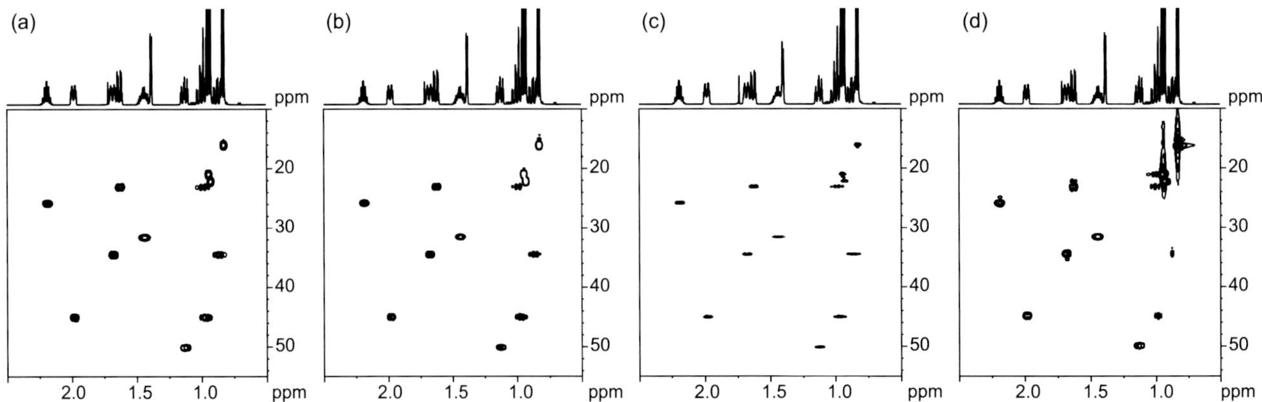

FIGURE 5.21 **NUS sampling of 2D ^{1}H–^{13}C heteronuclear correlation spectra (HSQC).** (a) HSQC with conventional sampling of 128 complex t_1 points (10 min), (b) NUS sampling at 25% (32 complex points as in Fig. 5.20, 2.5 min), (c) NUS sampling at 25% (128 complex points, 10 min), and (d) NUS at 10% (13 complex points, 1.25 min).

reconstruction of the missing data points is required. Many approaches to the processing of non-uniformly sampled data have been proposed [22], although in reality you are likely to employ that provided by your software vendor (for which specific licences may be required). The critical element required for these to succeed is adequate signal-to-noise in the collected data points, and NUS spectra are most likely to contain artefacts when this is too low, or when too few t_1 data points have been sampled.

The benefits of using NUS for 2D experiments may be viewed in two ways. Firstly, it may allow the collection of a spectrum in a shorter time when compared to a conventionally sampled data set with the same f_1 resolution. This is illustrated in the comparison of the 2D heteronuclear correlation spectra (see Section 7.3) in Fig. 5.21. The conventionally sampled (128 complex points) and the 25% sampled (32 complex points) NUS spectra appear indistinguishable, although that later required only one-quarter of the time for data acquisition; this sample provided high ^{1}H signal to noise, allowing such sparse sampling. Alternatively the gain from NUS could be to increase resolution in the indirect dimension when compared to a conventionally sampled data set of the same overall duration. The heteronuclear single quantum correlation (HSQC) spectrum of Fig. 5.21c was collected in the same total time as that for Fig. 5.21a but has fourfold greater points in the reconstructed t_1 domain (512 vs. 128 complex points) due to the use of 25% NUS. This may appear to be gaining resolution for no penalty, but as ever, the compromise here is sensitivity, since the non-uniformly sampled points are now spread over a larger total t_1 duration, meaning the later points will suffer reduced intensity from greater relaxation losses. Some sampling schemes, especially those used for protein samples, are therefore optimised to concentrate the NUS points towards the earlier time points to mitigate such sensitivity losses, although this is likely to be less of a problem for small molecules where transverse relaxation rates are low compared to those of macromolecules. Too great a reduction in the number of sampled points will ultimately lead to corruption of the final spectrum, as illustrated for the HSQC spectrum of Fig. 5.21d, for which only 10% sampling was employed. Due to the requirement of sufficiently high signal-to-noise ratios, NUS methods may be less satisfactory for data sets with intrinsically low sensitivity, such as 2D nuclear Overhauser effect spectroscopy (NOESY) (see Section 9.6). When sample quantities are not restrictive, anecdotally they appear more robust for heteronuclear correlation experiments such as HSQC and possibly heteronuclear multiple bond correlation (HMBC).

5.2.5 Two-Dimensional Data Processing

The general procedure for processing any 2D data set is the same and essentially follows that outlined in the flow chart of Fig. 5.22. The details of the parameters selected for each stage will depend upon the technique used, the acquisition conditions, and the nature of the sample being studied, so only the general principles that are to be considered during data processing are considered here. All of these are extensions of the ideas already introduced for the handling of 1D data and this section assumes familiarity with these (see Sections 3.2.3 and 3.2.8). Processing techniques have even greater importance for 2D work where the operator is able to tailor the appearance of the final spectrum to a large degree according to the chosen parameters. The choice of window functions has a major impact on the final spectrum and this selection differs

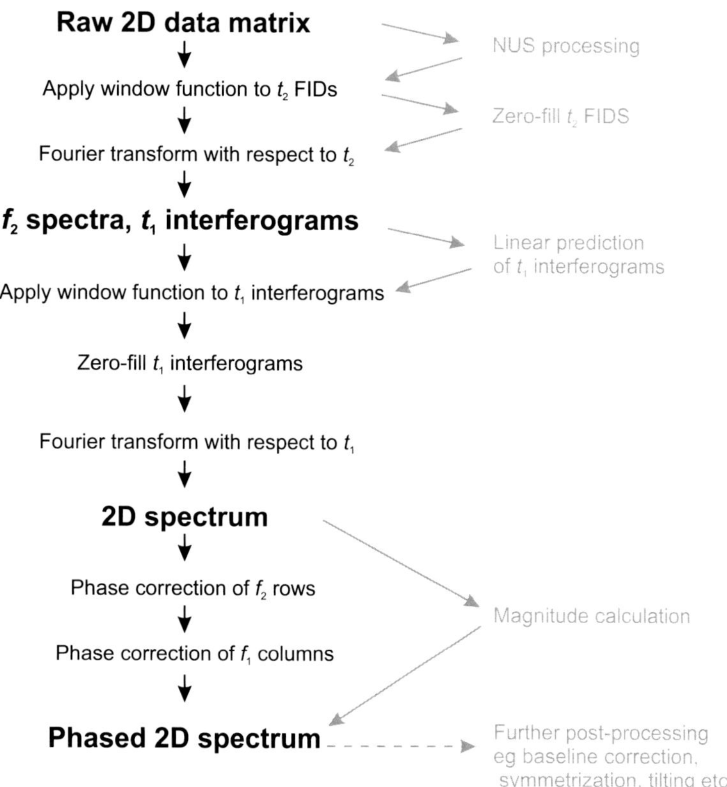

FIGURE 5.22 **The typical scheme followed in the processing of a 2D data set.** The items shown to the right in grey are additional or alternative procedures that may be executed.

markedly for phase-sensitive and absolute value data sets, so initially we restrict discussion to the phase-sensitive case. These functions are applied typically with three goals in mind:

- To ensure truncation of the time-domain data and thus attenuate truncation artefacts.
- To enhance resolution in the spectrum.
- To provide optimum signal-to-noise.

Owing to the short acquisition times used, data truncation typically occurs in both dimensions. Particularly when zero-filling is applied, as is nearly always the case in t_1, the window function must force the FID to zero if truncation wiggles are to be avoided. The longer t_2 acquisition times often means zero-filling is not required in this dimension and similar considerations apply to those for handing a 1D FID. Generally some enhancement of resolution without degradation of signal-to-noise is of primary concern for which any of the resolution enhancement functions of Section 3.2.7 can be used, the most popular being the Lorentz–Gauss transformation, or the shifted sine bell, or squared sine bell. The latter two guarantee forcing the decay to zero, so are also suitable for the shaping of t_1 data prior to zero-filling, which is typically applied at least once. In either dimension, excessively strong enhancement leads to the appearance of negative-going excursions about resonances and trial and error is generally the best approach to achieving optimum results. Often the first FID of a 2D data matrix can be used as a convenient test 1D spectrum to visualise the effects different window functions will have on the final f_2 data. Modern software now allows interactive changes of shaping parameters along with a display of the final spectrum making such experimenting rather straightforward. An alternative to the zero-filling of t_1 data for high signal-to-noise data sets is the extension of t_1 interferograms by linear prediction to typically two or four times their original size. Not only does this extend the decay, it reduces the need for apodisation of highly truncated data, both of which contribute to improved resolution in f_1. Further zero-filling beyond this may also be applied to improve the digitisation of the final spectrum if desired.

Absolute value data sets (typically presented following a magnitude calculation) demand stronger resolution enhancement functions in both dimensions to suppress the undesirable dispersion-mode contributions to the phase-twisted lineshapes. This is most often achieved with the sine bell or squared sine bell functions. These are well suited to the anti-phase peaks that give rise to COSY correlations since these start with zero intensity (being initially anti-phase they have no net magnetisation) and grow during the course of the FID. However, most 2D experiments produce in-phase resonances which provide full intensity at the start of the FID and if used non-shifted these functions lead to attenuation of much of the signal and can severely degrade sensitivity (try processing a 1D proton spectrum with a non-shifted sine bell).

5.2.5.1 Phase Correction

Phase correction of a 2D spectrum is undertaken independently for the two frequency domains. Where the first FID of a 2D matrix contains signals, this can be processed and phased as for a 1D spectrum and the resulting zero- and first-order phase constants applied to the f_2 data. The frequency independent phase correction for f_1 will depend on the details of the phase cycling used in the pulse sequence, but should be close to 0 degrees or 90 degrees. In theory, a frequency dependent (1st order) phase correction in f_1 should not be necessary since the time delays prior to data sampling that produce these errors do not arise for the artificially generated time domain. In practice some small correction is often required to both domains for optimum results and this is undertaken interactively as for a 1D spectrum. Thus, one aims to observe signals at either end of the spectrum on which to make the corrections. This involves independently selecting f_2 rows (or f_1 columns) from the 2D spectrum. To provide the necessary frequency distribution of resonances, 2 or more from each dimension are usually required, and the resulting phase constants applied to all other rows (or columns).

5.2.5.2 Presentation

It has already been stated that the most informative and convenient way to visualise 2D data is as a contour plot, in which the horizontal and vertical axes represent the frequency dimensions and the peak intensity is indicated with appropriate contours. The lower contour setting defines the cut-off below which no signals are presented. Typically this will be set some way above the baseplane noise level but if set too high may eliminate potentially useful information such as weak correlations. In some instances a cut-off level suitable for revealing weak signals will be too low for more intense ones, revealing artefacts such as t_1 noise, which does not make for an aesthetically appealing spectrum. Interrogating expansions of selected regions will likely prove beneficial, especially in crowded spectra or where the details of crosspeak structures are to be analysed, and is best performed interactively on screen whenever possible, with the use of cursors and crosshairs aiding accurate analysis.

5.3 COHERENCE AND COHERENCE TRANSFER

The notion of coherence has already been touched upon in the introduction to COSY in this chapter and has been mentioned briefly in Section 4.4. This lies at the foundation of every NMR experiment, and to appreciate many of the topics that follow in this book it is useful to develop some understanding of this and the notion of coherence transfer. More rigorous discussions of this topic may be found in the excellent texts of Hore et al. [23] and Keeler [24].

Coherence is a generalisation of the notion of transverse magnetisation yet it is important to appreciate some specific features of coherence, which may be illustrated with reference to the energy level diagram of a coupled two-spin AX system (Fig. 5.23a). Following the application of a 90 degree pulse on this system at equilibrium, transverse magnetisation exists which, according to the vector model, may be represented as four vectors in the rotating frame, one for each of the transitions in Fig. 5.23a. These give rise to the four lines of the two doublets in the resulting spectrum. These transitions are termed single-quantum transitions because they are associated with a change in magnetic quantum number M of one ($\Delta M = 1$). The magnetisation associated with each is therefore referred to as a *single-quantum coherence* which is said to possess a *coherence order p* of 1. Single-quantum coherence is the conventional transverse magnetisation detected in any NMR experiment as this is able to induce a voltage in the NMR detector. In this respect it is useful to consider the difference between a transition excited by a 90 degree pulse, as discussed earlier, and one that is saturated by weak rf irradiation. Both situations correspond to equal α and β populations and hence to zero population differences across the transitions yet only the first produces a detectable signal. The saturated transition has no *net* magnetisation to detect because the spins are distributed about the *x*–*y* plane with random phase. The spins excited by the pulse, however, are bunched together along one axis in the transverse plane and so produce a net component that is detectable; they have *phase coherence* imposed on them by the pulse (Fig. 5.24). In general we may distinguish a single-quantum coherence as a phase coherence between two states that have $\Delta M = \pm 1$.

Now imagine the application of a selective 90 degree pulse to the two-spin system at equilibrium but to just one of the transitions, say A_1. This then creates a single-quantum coherence for this transition alone (Fig. 5.23b). Next imagine applying a selective 90 degree pulse to the transition X_2 and hence equalising the populations across this transition. This

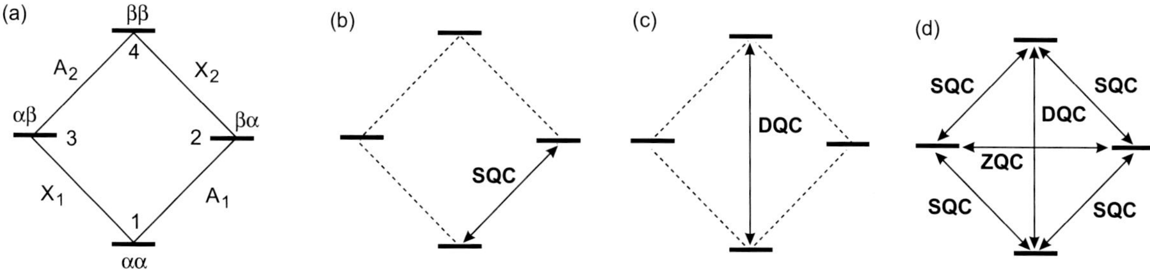

FIGURE 5.23 **Schematic energy level diagrams for a coupled two-spin AX system (see text).** SQC, single-quantum coherence; DQC, double-quantum coherence; ZQC, zero-quantum coherence.

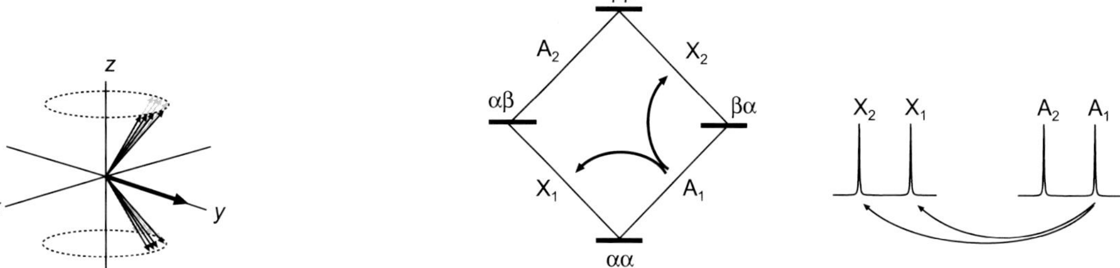

FIGURE 5.24 **Phase coherence.** Net transverse magnetisation is produced from the bunching of individual magnetic moments which gives rise to an observable NMR signal. These spins are said to possess *phase coherence*, and because only single-quantum spin transitions (α ↔ β) are involved in generating this state, it is termed *single-quantum coherence*.

FIGURE 5.25 **The coherence transfer process responsible for generating crosspeaks in correlation spectroscopy, illustrated for the A_1 transition.** Coherence of spin-A during t_1 of a 2D sequence becomes coherence of spin-X during t_2, thus correlating the two spins in the resulting spectrum.

process transfers some of the spins associated with the βα state to the ββ state and some part of the phase coherence originally associated with A_1 is now associated with the 1–4 transition (αα − ββ) (Fig. 5.23c). This is a new form of coherence not previously considered in any detail. It corresponds to a change in magnetic quantum number of 2 (ΔM = 2) so is therefore termed two-quantum or, more usually, *double-quantum coherence* ($p = 2$) which cannot readily be represented with the vector model notation used thus far. In the COSY experiment a similar state of affairs arises through the use of non-selective 90 degree pulses applied simultaneously to all transitions. Thus, during t_1 the single-quantum coherences (transverse magnetisation) generated by the first 90 degree pulse evolve under the influence of spin–spin coupling to give the anti-phase magnetisation components required for *coherence transfer*. On application of the second 90 degree pulse, the coherence originally associated with the A_1 single-quantum transition is distributed amongst the other transitions within the spin system (X_1, X_2, A_2, 1–4 and 2–3) while some part remains with A_1 (Fig. 5.23d). The first two again correspond to single-quantum coherence but now of the X spin, so this coherence transfer from A to X (Fig. 5.25) gives rise to the COSY crosspeak. Coherence remaining with A_1 or that transferred to A_2 ultimately produces the diagonal peak multiplet that we have already seen. The 1–4 coherence is the double-quantum coherence described above while the 2–3 coherence corresponds to a change in magnetic quantum number of zero, so is termed *zero-quantum coherence* ($p = 0$). Longitudinal magnetisation, although not a state of coherence, also has $p = 0$ since it does not correspond to transverse magnetisation. All coherences for which $p \neq \pm 1$ are referred to as *multiple-quantum coherences*. Quantum mechanical selection rules dictate that only those coherences corresponding to ΔM = ±1 are able to induce a signal in the detection coil or, in other words, *only single-quantum coherences may be observed directly* since all others have no *net* magnetisation associated with them. Unfortunately, simply stating this fact leaves a slightly uncomfortable air of mystery surrounding the 'invisible' multiple-quantum coherences. A simple physical picture of these states views them as combinations of pairs of evolving vectors that are *always* anti-phase and hence never produce observable *net* magnetisation (Fig. 5.26). The evolution frequency of these states is dictated by the chemical shifts of the participating nuclei and the coherence order. Whereas single-quantum coherences evolve in the rotating frame according to the chemical shift offset of the spin from the transmitter, double-quantum coherence evolves according to the *sums* of the chemical shift offsets of the two spins ($\nu_A + \nu_X$) and zero-quantum coherences according to their offset *differences* ($\nu_A - \nu_X$). In a system containing more than two coupled spins, higher orders of multiple-quantum coherence may also be excited, so, for example at least three coupled spins are required for the generation of triple-quantum coherence, and so on. Although again these higher orders cannot be observed *directly* their presence

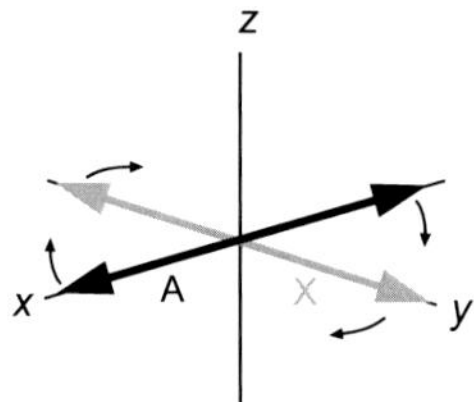

FIGURE 5.26 **A simplified picture of multiple-quantum coherence.** In an AX system this may be viewed as being composed of groups of evolving anti-phase vectors which have zero net magnetisation, and hence can never be directly observed.

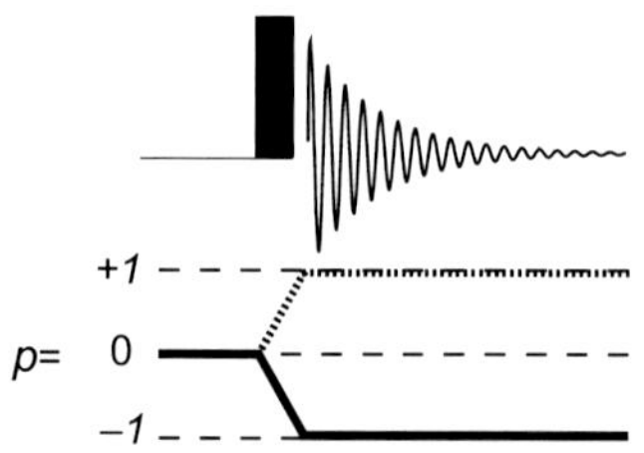

FIGURE 5.27 **The coherence level diagram for a single-pulse 1D experiment.** The solid line represents the coherence transfer pathway of the detected magnetisation. The dotted line represents the mirror image pathway that is rejected by the quadrature detection scheme.

may be detected *indirectly* if they are reconverted to single-quantum coherences prior to detection. For example adding a third pulse to COSY would regenerate single-quantum coherence from the otherwise invisible multiple-quantum coherences that were generated; an example of this is found in the double-quantum filtered COSY described in the following chapter. Just as spin–spin relaxation produces a loss of observable signals by destroying the phase coherence of transverse (single-quantum) magnetisation, multiple-quantum coherences are also dephased by such relaxation processes and similarly have a finite lifetime in which they can be manipulated.

It should also be noted that a coherence order has sign associated with it, so single-quantum coherence may have an order of $p = +1$ or $p = -1$, whereas double-quantum coherences will have $p = \pm 2$ and so on. The positive and negative signs represent hypothetical vectors that precess at the same rate but in opposite senses in the rotating frame. For single-quantum coherence, these simply correspond to the two sets of mirror-image signals that would be observed either side of the reference frequency in the absence of quadrature detection, so have a realistic physical manifestation. By using quadrature detection when collecting data we choose to keep one of these sets of signals and eliminate the other, to avoid possible confusion.

5.3.1 Coherence Transfer Pathways

The existence of the different levels of coherence present during any pulse sequence may be illustrated by means of coherence-level diagrams. These were originally introduced to NMR for the purpose of designing pulse sequence phase cycles [25–27] but can also provide an extremely effective yet simple graphical means of following the flow of magnetisation through a pulse sequence. This formalism proves to be especially powerful when considering the pathway taken by magnetisation from the start of a 2D pulse sequence to its arrival at the detector. It also provides a means of following multiple-quantum coherences which cannot be represented by the vector model, so overcoming some of the limitations of this without resorting to mathematical formalisms. The coherence level diagram for a single-pulse 1D acquisition is presented in Fig. 5.27. A single pulse generates only coherences with $p = \pm 1$, and these are represented in this figure as solid and dashed lines. As stated above, one of these pathways is selected by the hardware quadrature detection while the other is rejected, and purely by convention the $p = -1$ pathway is retained (note that in the original paper [25] the opposite convention was chosen). Coherence level diagrams are used to indicate the pathway followed by the desired magnetisation that is selected by the phase cycle or PFGs in use. All other possible pathways (of which there could be very many) are not shown and are assumed to be eliminated, thus the dashed pathway of Fig. 5.27 would not usually be presented.

To introduce the idea of coherence-transfer pathways for 2D experiments, the COSY sequence is again considered and the P- and N-type signals introduced in Section 5.2.1 are described. The desired pathways for both these experiments are shown in Fig. 5.28 with the only difference between these two being the relative sense of precession of (single-quantum) coherence in the t_1 and t_2 time periods, this being the same for the P-type and opposite for the N-type signals. This difference is then clearly illustrated by the coherence-transfer pathways followed, both of which begin with equilibrium longitudinal magnetisation which has coherence order zero. Since coherence order $p = -1$ is selected in t_2, the corresponding coherence in t_1 is therefore $p = -1$ for the P-type signal but $p = +1$ for the N-type. These pathways are selected by the phase cycle described previously and illustrated in Table 5.1. During the t_1 period, and in general any period of free precession under the influence of chemical shifts and/or couplings, the coherence orders remain unchanged; only rf pulses are able to alter coherence levels (although relaxation will ultimately regenerate longitudinal magnetisation and hence coherence-level zero).

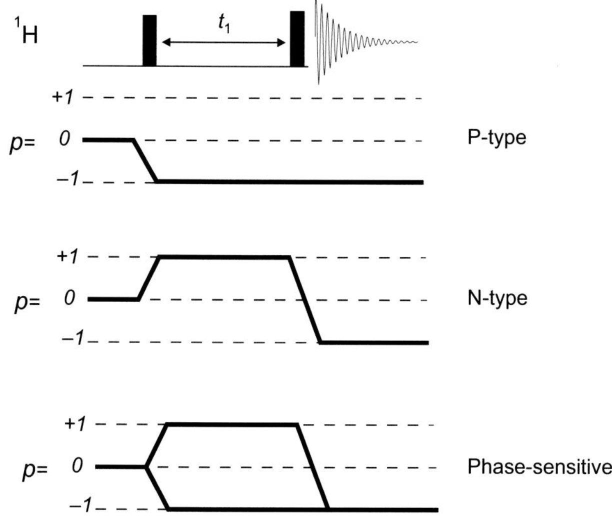

FIGURE 5.28 **Coherence level diagrams.** Coherence transfer pathways are shown for P-type, N-type and phase-sensitive COSY.

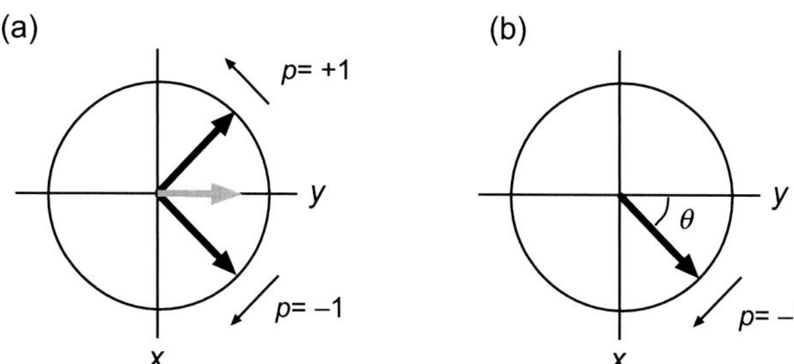

FIGURE 5.29 (a) Amplitude modulation requires both counterrotating vectors ($p = \pm 1$ coherences) be retained during t_1. The resultant signal (greyed) is modulated in amplitude along the y-axis as the vectors precess. (b) Phase modulation results if only one vector ($p = 1$ or $p = -1$) is retained; the phase angle θ varies as the vector precesses.

For the phase-sensitive COSY experiment both $p = \pm 1$ pathways must be preserved during t_1. However, for a pure-phase 2D spectrum the signal detected in t_2 *must* be amplitude-modulated as a function of t_1. Such a signal can only be obtained if both of the counter-rotating coherences are retained since the retention of only one of these unavoidably leads to a phase-modulated signal (Fig. 5.29) and hence the phase twist lineshape, as for the P- or N-type COSY. However, the presence of both pathways implies the detection of the NMR resonance plus its mirror (quadrature) image in f_1, so the use of a suitable quadrature detection scheme (States or TPPI) to eliminate one set of these responses at the data processing stage is essential. The need to retain both pathways during t_1 for the acquisition of pure phase spectra is fundamental to all 2D experiments. This has prime importance when utilising PFGs for coherence pathway selection.

5.4 GRADIENT-SELECTED SPECTROSCOPY

It may be argued that the most significant fundamental development in experimental methodology in high-resolution NMR during the 1990s was the introduction of PFGs, which extend the scope of numerous NMR techniques [28–30]. The concept of applying magnetic field *gradients* across a specimen or sample has long been used in (nuclear) magnetic resonance imaging [31] and although the potential for their use in high-resolution NMR was realised some time ago [32,33], technical limitations precluded their widespread use for some years [34]. Nowadays, PFGs find application in numerous one and multidimensional NMR techniques as a means of selecting those signals deemed interesting and suppressing those which are not, and have proved hugely beneficial to many of the routine methods used for structural organic chemistry. This section introduces the general principles behind the use of PFGs, and although presented alongside an introduction to 2D NMR techniques, the discussions are quite general and applicable to all gradient-selected experiments. The majority of methods that use PFGs are often simple variations of the original non-gradient equivalent (hence the use of terms such as *gradient-enhanced* or *gradient-accelerated* experiments in the literature) and thus require no greater understanding to enable one to apply the experiment and interpret the data. In fact the advantages endowed by PFGs over traditional methods largely relate to the practicalities of data collection, often allowing the collection of superior quality data in shorter times.

The term 'traditional' in this context, refers to those experiments that make use of phase cycling to select the desired signals and suppress all others. The notion of phase cycling has already been encountered in previous sections; the point to recall at this stage is that this procedure involves the repetition of a pulse sequence with the phases of the rf pulse(s) adjusted on each transient and the data from each combined such that the desired signals add constructively while those that are unwanted cancel. Phase-cycling procedures are thus all based on difference spectroscopy and as a result are subject to problems of imperfect subtraction of the undesired signals which leads to artefacts in the final spectrum. The dependence on a difference procedure also means that, on each transient, data are collected which are ultimately discarded; this clearly does not make optimum use of the receiver's dynamic range. These methods also require that a full phase cycle be completed to achieve the required selection regardless of signal strength, thus placing a fundamental limit on how quickly an experiment can be collected. Experiments which make use of PFGs for signal selection do not suffer from these limitations because *only* the *desired signal* is collected *on each transient*, so leading to cleaner spectra that are free from difference artefacts and, when sensitivity is not limiting, to faster data collection because one is not limited by the number of steps in a phase cycle.

5.4.1 Signal Selection with Pulsed Field Gradients

5.4.1.1 Defocusing and Refocusing with Pulsed Field Gradients

Before subjecting any sample to NMR analysis, the magnetic field must be shimmed to remove inhomogeneities. This ensures that all spins in the same chemical environment experience the same applied field and hence possess the same Larmor frequency regardless of their location within the sample. If, however, one were to deliberately impose a linear magnetic field *gradient* (known as a B_0 gradient) across the sample, say along the *z*-axis, chemically equivalent spins would now experience a different applied field according to their position in the sample, in this case, along its length. This may be understood by considering the sample to be composed of microscopically thin disks in the *x–y* plane with each experiencing a different local magnetic field according to their physical location (Fig. 5.30). If the field gradient were imposed for a discrete period of time as a *gradient pulse* (Fig. 5.31a), then during this the spins, having been previously excited with an rf pulse, precess with differing frequencies and will thus rotate through differing angles (Fig. 5.30b). Provided the gradient pulse is of sufficient strength and duration, the *net* magnetisation remaining, which is the sum of the transverse magnetisation from all disks in the sample, will be zero and there would be no detectable NMR signal. The gradient pulse is said to have *defocused* (or *dephased*) the magnetisation. The effect of applying such a sequence with increasing gradient strengths is shown in Fig. 5.32.

Although there is now no *detectable* signal, magnetisation has not been lost from the transverse plane and may be recovered. Applying a second gradient pulse immediately after the first, of equal intensity and duration but in the opposite sense (now along the -*z*-axis, Fig. 5.31b), causes the dephasing of the first gradient to be exactly 'unwound' by the action of the second (Fig. 5.30c) and the magnetisation vectors from each disk would *refocus* (or *rephase*) to produce a *gradient*

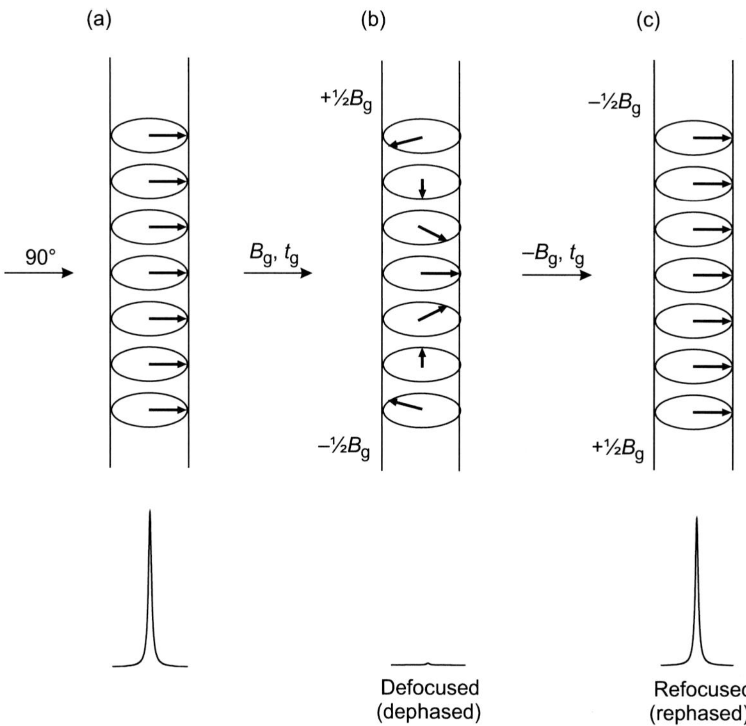

FIGURE 5.30 **The action of field gradient pulses.** (a) In a homogeneous magnetic field all chemically similar spins possess the same Larmor frequency within every microscopically thin disk throughout the sample causing their individual magnetic moments to add and to produce a sharp resonance. (b) After the action of a linear field gradient of strength B_g for a duration τ_g, the spins have evolved through different angles according to their local magnetic fields, adding to give zero net magnetisation. (c) Application of a second, identical gradient but in the opposite direction refocuses the individual vectors once more and recovers the resonance. The magnetisation is said to have been first defocused and then refocused by the action of the gradients.

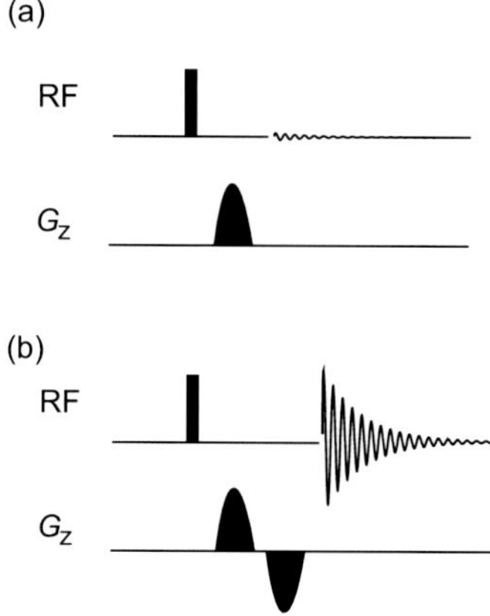

FIGURE 5.31 **Pulsed field gradients.** The pulse sequence representation of (a) a single *z*-axis field gradient pulse and (b) two *z*-axis gradient pulses applied in opposite directions. Gradient pulses typically have a shaped rather than square profile, such as the sine profile illustrated here (see Section 5.4.4). RF represents the radiofrequency pulse channel.

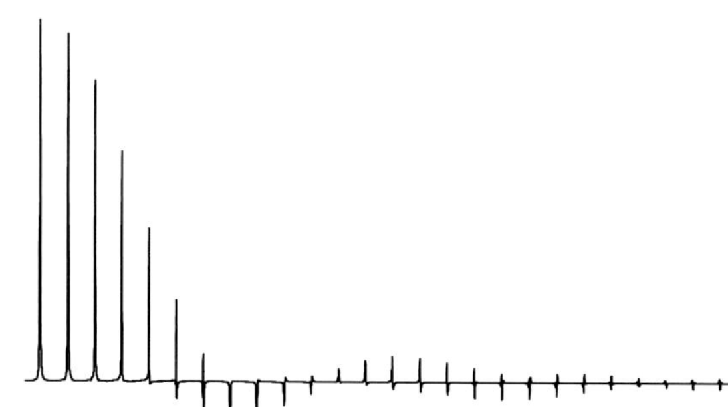

FIGURE 5.32 **The destruction of an NMR resonance with a PFG.** The sequence of Fig. 5.31a was employed with a progressively stronger gradient for each experiment.

echo and an observable net signal (Fig. 5.33). The concept of defocusing all unwanted signals and *selectively* refocusing only those that one desires lies at the heart of every gradient selected experiment but to understand how this refocusing can be made selective, we must consider the action of the gradient pulse on transverse magnetisation in a slightly more formal context.

5.4.1.2 Selective Refocusing

In the absence of a field gradient pulse, the Larmor frequency ω of a spin depends upon the applied static field B_0 such that $\omega = \gamma B_0$. When the gradient pulse is applied there is an additional spatially dependent field $B_{g(z)}$ associated with the gradient giving rise to a spatially dependent Larmor frequency $\omega_{(z)}$:

$$\omega_{(z)} = \gamma(B_0 + B_{g(z)})\,\text{rad}\,s^{-1} \tag{5.1}$$

If the gradient is applied for a duration τ_g seconds, the magnetisation vector rotates through a spatially dependent phase angle $\Phi_{(z)}$ of

$$\begin{aligned}\Phi_{(z)} &= \gamma(B_0 + B_{g(z)})\tau_g \\ &= \gamma B_0 \tau_g + \gamma B_{g(z)}\tau_g\ \text{rad}\end{aligned} \tag{5.2}$$

The first term in the final expression simply represents the Larmor precession in the absence of the field gradient which is constant across the whole sample and as such shall not be considered further at this point, except to note that this precession has bearings on the practical implementation of field gradients, as described later. The second term represents the spatially dependent phase caused by the gradient pulse itself, and it is this quantity that is of interest. Generalising the above expression which is only applicable to transverse single-quantum magnetisation, it is necessary to include a term that represents the coherence order p of the magnetisation, so that:

$$\Phi_{(z)} = p\gamma B_{g(z)}\,\tau_g \tag{5.3}$$

This modification reflects the fact that a p-quantum coherence dephases at a rate proportional to p. Thus, double-quantum coherences dephase twice as fast as single-quantum coherences yet zero-quantum coherences are insensitive to field gradients. When the coherence involves different nuclear species, allowance must be made for the magnetogyric ratio and coherence order for each, such that:

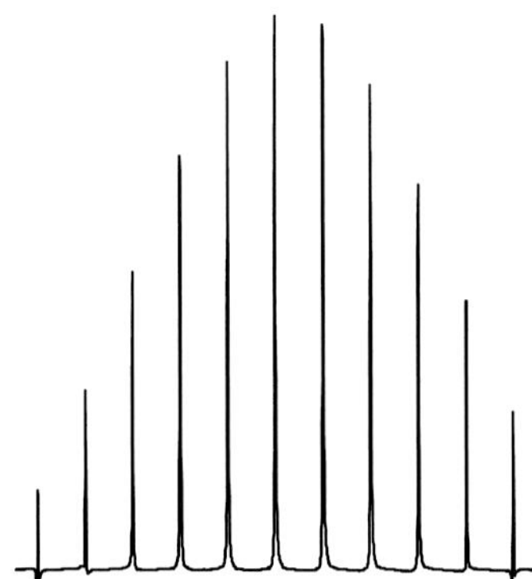

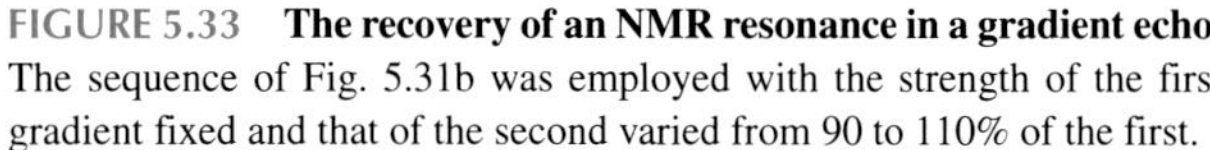

FIGURE 5.33 **The recovery of an NMR resonance in a gradient echo.** The sequence of Fig. 5.31b was employed with the strength of the first gradient fixed and that of the second varied from 90 to 110% of the first.

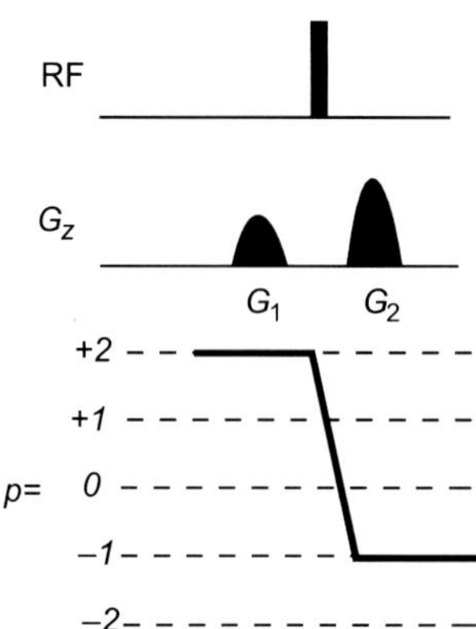

FIGURE 5.34 **An illustration of signal selection by PFG.** Using two gradients with ratio G_1:G_2 of 1:2 selects only the coherence transfer pathway shown, leaving all others defocused and unobservable.

$$\Phi_{(z)} = \mathrm{B}_{g(z)}\,\tau_g \sum p_1\gamma_1 \qquad (5.4)$$

It is therefore apparent that the degree of defocusing caused by a gradient is dependent upon the coherence order and the magnetogyric ratio of the spins involved, and these two features provide the key to signal selection with field gradients. The purpose of gradients in the majority of experiments is the selective refocusing of magnetisation that has followed the desired coherence transfer pathway during the experiment while leaving all other pathways defocused and hence unobservable. While PFGs themselves are unable to alter coherence orders, rf pulses can, and it is a combination of rf pulses to generate the appropriate coherence orders together with the PFGs to select these that ultimately provides the desired result.

This general process is illustrated for a homonuclear spin system in the scheme of Fig. 5.34 in which only double-quantum coherence existing prior to the pulse is to be retained. Thus, prior to the rf pulse $p = 2$ coherence exists which experiences a gradient of strength B_{g1} and duration τ_1 and thus obtains a spatially dependent phase:

$$\Phi_1 = 2\gamma \mathrm{B}_{g1}\tau_1 \qquad (5.5)$$

Following the rf pulse, the coherence is transformed into observable single-quantum magnetisation of order $p = -1$ which experiences a second pulsed gradient of strength B_{g2} and duration τ_2 which encodes a phase:

$$\Phi_2 = -\gamma \mathrm{B}_{g2}\tau_2 \qquad (5.6)$$

For coherence following a defined transfer pathway to be refocused and hence observable, its *total spatially dependent phase from all gradient pulses must be zero*:

$$\sum \Phi_i = \sum \gamma \mathrm{B}_{gi}\tau_i = 0 \qquad (5.7)$$

or, in other words, the refocusing induced by the last gradient must exactly undo the defocusing caused by all earlier gradients in the coherence transfer pathway. Thus, in this illustration Φ_2 must equal $-\Phi_1$ for this condition to be satisfied. This may be achieved by either altering the duration of the second gradient relative to the first, while keeping the amplitude the same, or by altering the amplitude and retaining the duration. Nowadays it is uniform practice to use gradient pulses of the same duration throughout a pulse sequence and to alter their amplitudes. Hence, setting $B_{g2} = 2B_{g1}$ in this example (remembering the coherence order has sign as well as magnitude) selects the desired pathway (Fig. 5.34). In contrast, single-quantum magnetisation originating from, for example any triple-quantum coherence ($p = 3$) that may have originally existed, acquires a net phase of:

$$\Phi = (3\gamma \mathrm{B}_{g1}\tau_1) + (-1\gamma 2\mathrm{B}_{g1}\tau_1) = \gamma \mathrm{B}_{g1}\tau_1 \qquad (5.8)$$

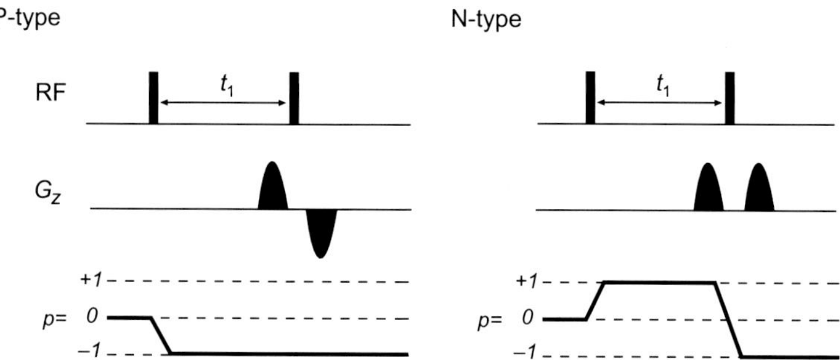

FIGURE 5.35 **The necessary gradient combinations to select P- and N-type COSY data.** Also shown are the corresponding selected coherence transfer pathways.

and hence remains dephased and does not contribute to the detected signal. The gradient pair has therefore selected only the desired pathway. For *homonuclear* systems with gradients of equal duration, it is sufficient to consider only the *coherence orders* and *ratios of gradient amplitudes* involved in the selection of the coherence transfer pathway, as illustrated in the COSY examples below. For *heteronuclear* systems, *magnetogyric ratios* of the participating spins must also be included, as will be illustrated in Section 7.3. Since in all following discussions gradients within a sequence are assumed to have the same shape and duration, a gradient pulse will simply be denoted as G_i in shorthand form from now on, which is taken to represent a pulse of strength B_{gi} and duration τ_i.

5.4.1.3 Gradient-Selected Correlation Spectroscopy

To illustrate the incorporation of PFGs into conventional experiments, the basic COSY sequence is again used as an example. It has already been shown that for the absolute value COSY experiment those signals to be retained in the final COSY spectrum follow the coherence transfer pathway illustrated in Fig. 5.35, with N-type selection being the favoured option. For signal selection it is sufficient to phase-encode coherences before the second 90 degree pulse and decode these after the pulse, immediately prior to detection. For N-type signals, the desired coherence orders are +1 and −1 at these points and thus the phase-encoding induced by each gradient may be represented using shorthand notation as:

$$\begin{aligned} \Phi_1 &= +1G_1 \\ \Phi_2 &= -1G_2 \end{aligned} \tag{5.9}$$

Since in COSY one is dealing with a *homonuclear* spin system the magnetogyric ratios need not be included as these are constant throughout, so the only variables are the gradient strengths. Hopefully it is quite apparent that for these two expressions to cancel one another, one simply selects $G_2 = G_1$, that is both gradients should have the same strength applied in the same direction. Conversely, for P-type selection (coherence orders −1 either side of the pulse) one must select $G_2 = -G_1$, that is similar gradients applied in opposite directions). In either case, the desired pathway is selected *in a single scan* without the need for phase cycling because gradients that refocus N-type signals cannot simultaneously refocus P-type; hence quadrature detection in f_1 is inherent with this scheme. Axial peaks are also suppressed by this gradient selection because, by definition, these arise from magnetisation that was along the z-axis during t_1 and was therefore unaffected by the first gradient. When this magnetisation is made transverse by the second 90 degree pulse, the final gradient dephases it and it is simply not observed. Thus, the original four steps of the phase-cycled absolute value COSY (two for f_1 quad detection and two for axial peak suppression) can be replaced by a single step with PFGs, allowing the 2D data set to be collected with only one transient per t_1 increment and leading to a fourfold reduction in time. This assumes that sensitivity is sufficient to allow the collection of only one scan per increment, which in many routine applications is the case. As an educational exercise, try running a gradient COSY sequence with first N-type and then P-type selection as described above and compare the results. One should simply be the mirror image of the other about the midpoint of f_1.

5.4.2 Phase-Sensitive Experiments: *Echo–Antiecho* Selection

One problem with the gradient approach described above for COSY is that it necessarily precludes the acquisition of high-resolution phase-sensitive data sets by selecting only one pathway in t_1. Since this leads to phase-modulated data it provides

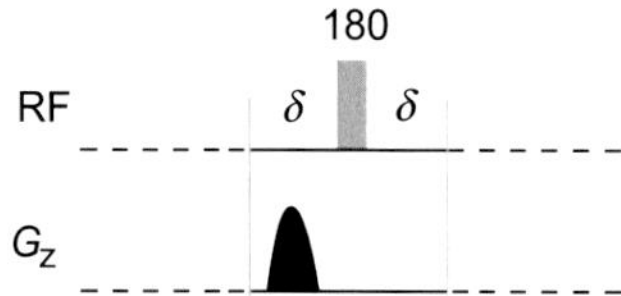

FIGURE 5.36 **Gradient pulses are often applied within a spin-echo.** This serves to refocus chemical shift evolution occurring during the pulse and thus avoids phase errors in spectra.

only phase-twisted lineshapes. Recall that phase-sensitive acquisitions require both $p = \pm 1$ pathways to be retained during t_1, but this is impossible when a gradient is placed within the evolution period since only one of these pathways may be refocused prior to detection. The most general solution to collecting phase-sensitive data is to avoid placing gradients within t_1 altogether, as this then allows conventional States or TPPI f_1 quadrature detection to be employed. The details of how this is done depend on the sequence in use, and examples of this approach will be illustrated for various techniques in the following chapters.

However, there also exists an ingenious approach to f_1 quadrature detection [10] which provides absorption-mode phase-sensitive data and allows gradients to be placed conveniently within the evolution time. The procedure involves collecting both P- and N-type data sets alternately and storing them *separately* for each t_1 value. The two pathways are selected typically by inverting the sign of the final refocusing gradient, as for the P versus N selection above. These data sets contain mirror image signals in the f_1 dimension when processed conventionally, so mirroring one data set about $f_1 = 0$ and adding it to the second will retain only one set of signals and hence provide the required f_1 frequency discrimination. This procedure also exactly cancels the dispersive contributions otherwise apparent within the phase-twist lineshapes of each data set and so produces the favoured absorption-mode lineshapes. The data handling required for this procedure is now found within conventional processing software, and is often referred to as the *echo–antiecho* approach, which is now very widely used.

Nevertheless, this is not all that is required to produce pure-phase spectra. Following Eq. (5.2), the evolution of chemical shifts under the influence of the static B_0 field during the application of a gradient pulse was ignored for convenience as this played no part in determining signal selection. Such shift evolution during the gradient pulse (which is typically of the order of milliseconds) leads to appreciable frequency dependent phase errors in spectra if not eliminated. These errors are of no consequence for absolute value presentations and hence were not considered for the COSY examples above, but prove disastrous for phase-sensitive data sets. To avoid these failings, gradient pulses are either applied within existing J-evolution periods or must be placed within a spin-echo to refocus the chemical shift evolution (Fig. 5.36). Examples of both these approaches are to be found in the techniques described in the following chapters.

5.4.3 Pulsed Field Gradients in High-Resolution NMR

This section summarises some of the key features of PFGs when applied to high-resolution NMR spectroscopy, highlighting some of the benefits they provide and also warning of potential limitations. Many of these features will be made apparent when describing the gradient selected techniques in the remainder of the book, so this merely provides an overview of these topics. Broadly speaking, the role of PFGs in high-resolution NMR experiments may be grouped into four classes:

- Selection of a coherence transfer pathway: The examples discussed earlier have already provided an illustration of how gradients may be used in this manner, the aim being to select magnetisation that has followed one desired pathway and suppress that which has followed any other. This application represents the most widespread use of gradients in NMR spectroscopy, the resulting experiments providing alternatives to the conventional phase-cycled equivalent.
- Suppression of a solvent resonance: The ability of PFGs to completely destroy an NMR resonance makes them ideally suited for methods of solvent suppression, where the aim is to selectively remove the solvent resonance but retain all others. The most widespread use of these methods is in the studies of molecules in *protonated* water where gradient-based suppression methods prove most effective (see Section 12.5).
- Purging of unwanted magnetisation: Purging in this context means the elimination of responses that arise from experimental imperfections such as inaccurate pulse widths. For example the use of an imperfect 180 degree inversion pulse on $+z$ magnetisation will leave an unwanted residual component in the transverse plane. This may be destroyed by the application of a single purging z gradient (sometimes referred to as a 'homospoil' pulse), ensuring that it will not contribute to the final spectrum. The use of purging z gradients requires all wanted magnetisation to sit along the z-axis and so be invariant to the gradient pulse. One of the earliest applications of a gradient pulse within a high-resolution experiment was as a purge pulse in a 2D exchange sequence [35].

- Mapping translational motion: The measurement of self-diffusion of molecules (see the chapter *Diffusion NMR Spectroscopy*) requires that their spatial location can be defined. This is made possible through the use of PFGs since resonance frequency correlates with location of a molecule in the solution in the presence of a field gradient. Similar principles provide the basis of (nuclear) magnetic resonance imaging (MRI).

5.4.3.1 Advantages of Field Gradients

The principle benefits arising from the use of field gradients for coherence pathway selection as opposed to conventional phase cycling may be summarised thus:

- Quality: On each transient only the desired signal is detected, avoiding the need for signal addition/subtraction in the steps of a phase cycle. Difference artefacts are thus not encountered, and the t_1-noise often associated with these in 2D spectra is reduced, providing cleaner spectra.
- Speed: As there is no requirement to complete a phase cycle, experiment times are dictated by sensitivity and resolution considerations only. When sensitivity is not limiting, 2D experiments can be acquired with a single transient per t_1 increment, leading to significant time savings.
- Dynamic range: As only the desired signal is ever detected, all others being destroyed in the probehead, optimum use can be made of the receiver's dynamic range and greater receiver gains can be employed. This compensates to some degree for the sensitivity losses sometimes associated with gradient selection (see later).
- Signal suppression: Very high signal suppression ratios can be achieved. The elimination of protons bound to carbon-12 in proton-detected ^{1}H–^{13}C correlations becomes trivial (suppression ratio of ≈1:100) and even the selection of natural abundance ^{1}H–^{13}C–^{13}C fragments becomes feasible [36] requiring suppression ratios of 1:10,000 (Section 6.4). Likewise the suppression of large solvent resonances becomes very much easier with gradient methods [37,38] (Section 12.5).
- Ease of use: In contrast to the use of rf pulses, the accurate calibration of gradient strengths is not required and the absolute strengths used are not critical for the success of the experiment, provided they are sufficient to dephase unwanted magnetisation (the exception is in the case of diffusion measurements, where field gradient strengths must be known accurately). More important is the fact that gradient *ratios* must be precise and the gradients reproducible. These demands can be quite readily met with appropriate instrument design, making gradient-selected techniques easy to implement and experimentally robust.

5.4.3.2 Limitations of Field Gradients

Despite their undoubted advantages, PFGs have a number of fundamental limitations that should also be appreciated. These have important consequences for the design of gradient experiments and although one may not be concerned with experiment design, these points help explain some of the features that are common to many of the gradient-selected sequences.

- Sensitivity: When used for coherence selection, gradients are able to refocus only one of two $\pm p$ coherence orders. In some applications this means that only one-half of the available signal is detected in contrast to phase-cycled experiments in which both pathways may be retained. Experiments which use PFGs for coherence selection may therefore exhibit lower sensitivity than the phase-cycled equivalent by a factor of typically 2 or $\sqrt{2}$ depending on the precise experimental details [30,39]. Purging gradients, however, do not cause such a sensitivity loss.
- Quadrature detection: As described earlier, some care is required when employing gradients in phase-sensitive experiments. This either requires that gradients are not placed within time domain evolution periods or that the echo–antiecho approach be used.
- Phase errors: Chemical shifts also evolve during the application of a gradient pulse due to the static B_0 field. Therefore, gradient pulses are generally applied within a spin-echo to refocus this evolution and thus avoid phase distortions when phase-sensitive displays are required.
- Diffusion losses: The detected signal can be attenuated by losses due to diffusion. The refocusing condition requires that the refocusing gradient negates the spatially dependent phase caused by all previous gradients. If a molecule were to move along the direction of the gradient (along the length of the NMR tube for a z-gradient) between the dephasing and rephasing pulses the gradient strengths would not be perfectly matched and incomplete refocusing will result, leading to a loss of signal intensity (it is exactly this phenomenon that is used to study diffusion in solution by NMR; see the chapter *Diffusion NMR Spectroscopy*). This will be more of a problem for small, fast-moving molecules in low-viscosity solvents which can be best avoided by keeping the defocusing and refocusing gradient pulses close together in the sequence. The degree of signal attenuation also depends on the *square* of the gradient strengths so the use of weaker gradients, which must still be sufficient to dephase magnetisation, will also help minimise such losses.

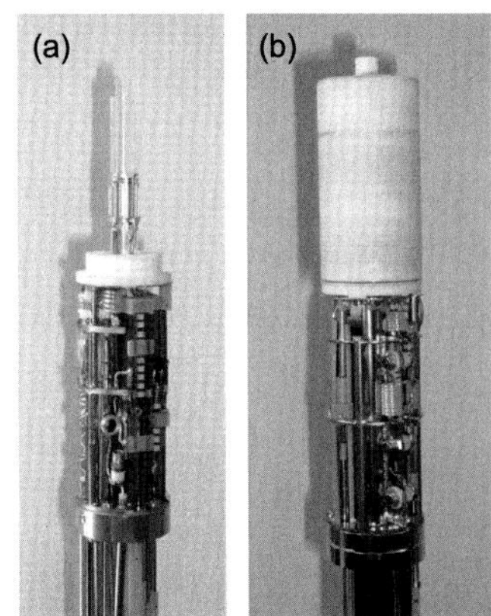

FIGURE 5.37 **A high-resolution probe head (cover removed).** This shows (a) the rf coils and (b) the gradient coil block surrounding the rf coils in which the tracks holding the coils can be seen.

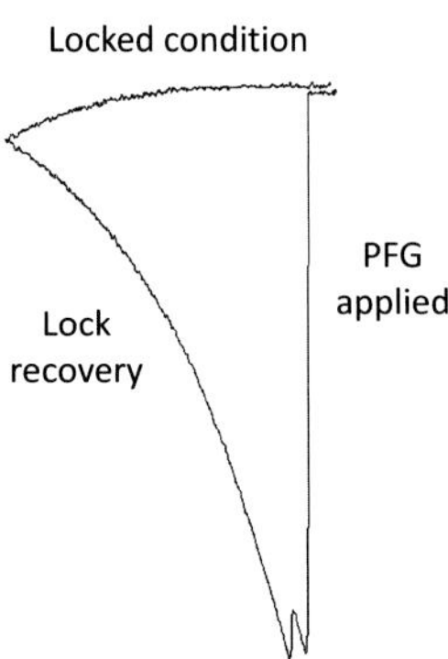

FIGURE 5.38 **The instantaneous destruction of the ^{2}H lock signal caused by a PFG, followed by its recovery.** *(Source: From a Bruker spectrometer.)*

5.4.4 Practical Implementation of Pulsed Field Gradients

The simplest way to generate a field gradient through a sample is to offset one of the shim currents from its optimum value and in principle it is possible to generate a PFG by momentarily driving the Z-shim at maximum current (on older spectrometers this may be referred to as a 'homospoil' pulse). However, such a set-up is only suitable for providing a basic purge gradient since it produces only relatively weak fields, offers no control over amplitude, and would be plagued by the dreaded eddy currents described later. For performing the full range of gradient-selected experiments a probe equipped with a dedicated gradient coil surrounding the usual rf coils is required (Fig. 5.37) together with an appropriate gradient amplifier; these are found routinely on modern NMR spectrometers. The field gradient across the sample is then generated by applying a current to the gradient coils, and inverting the sign of a gradient corresponds to reversing the applied current. Typical maximum gradient strengths are around 0.5 T m^{-1} (50 G cm^{-1}) for routine work, although in practice the actual gradient strengths used will often be somewhat less than this. A method for calibrating gradient strengths is described in Section 3.5.2.

As simple as this scheme may sound, it is plagued by a number of technical problems that would make the collection of high-resolution data impossible if left unchecked. The worst of these are the eddy currents that are generated on application of the gradient pulse. These are currents in the conducting components that surround the sample, such as the probe case, magnet bore, and even the shim coils themselves, which in turn generate further spurious magnetic fields within the sample. Since these can last for hundreds of milliseconds, they prevent the acquisition of a high-resolution spectrum. The most effective way to suppress the generation of eddy currents is the use of so-called actively shielded gradient coils, which are now used in all commercial gradient probes. These consist of a second gradient coil surrounding the first and driven by the same current. The outer coil is designed in such a way that the field it generates *outside of the active sample region* cancels that produced by the inner coil. The net result is a field gradient within the sample, but not outside of it and hence no eddy currents are stimulated in surrounding structures. Since eddy currents are caused by the rapid change in the local magnetic field on application of the gradient, they can also be attenuated by using a shaped gradient profile with a gentle rise and fall, rather than a simple rectangular pulse that has very steep leading and trailing edges. Commonly used gradient profiles are half sine-waves or smoothed rectangular shapes. With these steps it is possible to acquire an unperturbed spectrum within tens of microseconds of the gradient pulse; a method for gauging gradient recovery periods is also given in Section 3.5.2.

One final point regarding field gradients is worthy of note. Since a gradient pulse induces a spatially dependent phase for all magnetisation in the sample, this also leads to the destruction of the deuterium lock signal itself, a rather alarming sight when first encountered (Fig. 5.38). How then does the spectrometer retain the field frequency stability over extended experiments? The key point stems from the fact that the regulation system integrates the lock error signal over a period of typically many seconds so is little effected by the brief disturbances caused by the gradient pulse. The signal presented to the operator is not exactly that used for the field frequency regulation and is integrated over much shorter time periods so presents a more dramatic picture of events to the operator (which itself can be a useful indicator of whether the gradient system is operating).

5.4.5 Fast Data Acquisition: Single-Scan Two-Dimensional NMR

In recent years a number of approaches have been developed that aim to reduce the time required for recording multidimensional data sets, most notably when detection sensitivity is not limiting. Many of these methods are tailored toward high-dimensionality experiments (≥3D) or to spin systems associated with the common amino acids so are of most relevance to biomolecular NMR spectroscopy [40]. However, one ingenious method stands out as having wider applicability for very fast data acquisition and will be considered briefly here. It makes use of spatial encoding of spins with PFGs, borrowing from methods developed for magnetic resonance imaging, and produces a complete 2D spectrum *in a single scan*, meaning typically less than one second [41,42]. Not surprisingly, this approach has been termed *ultrafast 2D NMR* and has obvious appeal for the monitoring of fast processes and for the characterisation of transient species. Early examples have demonstrated its use in the reaction monitoring of relatively simple chemical transformations [43–45], enabling short-lived intermediates to be identified. Despite this, the methods are yet to find widespread use, although they do have potential for wider acceptance. The reasons for their limited uptake lie primarily in the highly complex setup protocols required when compared with the established 2D NMR procedures introduced above, and the stringent demands they place on spectrometer hardware and field gradient systems in particular. Should you wish to employ these methods, the practical guidance provided in reference [46] is highly recommended. Here, the basic operating concepts are introduced and the role of PFGs explained [47,48].

As described above, the collection of conventional 2D data sets relies upon the *sequential* acquisition of experiments collected as a series of 1D experiments of increasing t_1 evolution times according to the preparation–evolution–mixing–detection paradigm. In contrast, the single-scan approach collects these *simultaneously* and it is this parallelisation that affords the significant time savings. The approach taken is to excite different ‘slices’ through a sample at slightly different times such that the start point of the t_1 evolution period is shifted in time for each one. At the start of the 2D mixing period the total evolution time experienced for spins in different slices will therefore vary such that responses from each slice correspond to experiments recorded with differing t_1 periods. The ‘slicing’ of the sample is achieved by the imposition of a field gradient along the z-axis so that resonant frequencies vary along the sample due to differences in the local magnetic field. A sequence of selective pulses (Section 12.4) is then applied with each progressively stepped in frequency so as to cause excitation of only those spins within a discrete slice of the sample, this being referred to as *spatially selective excitation*. Since these pulses are applied sequentially, the start of t_1 evolution is correspondingly stepped in time between each slice, resulting in parallel evolution periods (Fig. 5.39). The number of t_1 increments is therefore equal to the number of selective pulses applied. To refocus chemical shift evolution differences imposed by the field gradient, a second must be applied in the opposite sense so that only chemical shifts imposed by the external B_0 field give rise to net evolution during t_1. This gives rise to the general preparation–evolution (encode) segment of Fig. 5.40, following which any standard mixing element may be applied according to the correlations being sought.

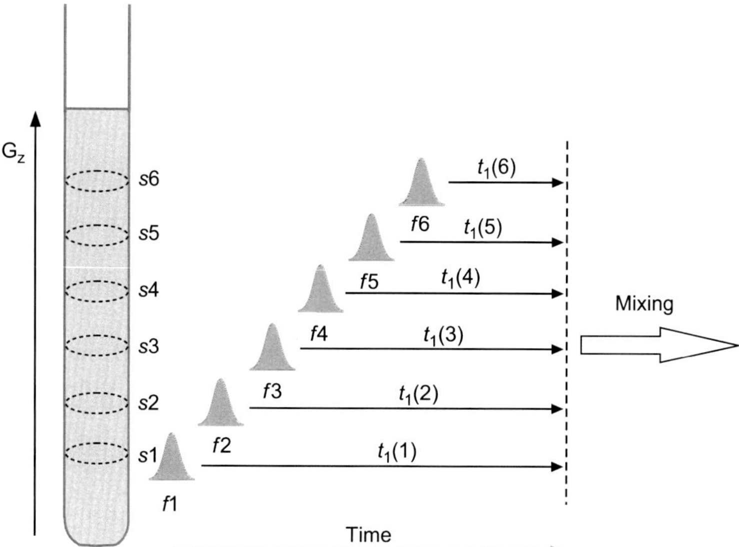

FIGURE 5.39 **Spatially selective excitation for single-scan 2D spectra.** This representation shows a hypothetical experiment of 6 t_1 increments, each initiated by selective excitation applied at frequencies f_1–f_6 corresponding to ‘slices’ s_1–s_6.

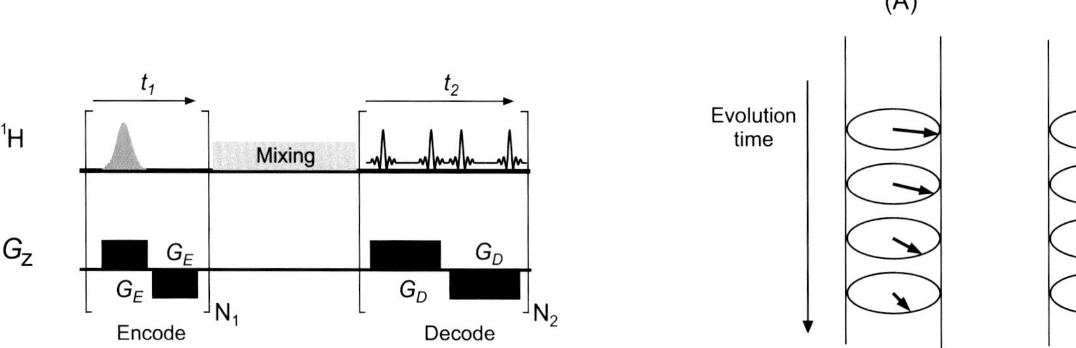

FIGURE 5.40 **A generalised scheme for the collection of single-scan 2D spectra.**

FIGURE 5.41 **The spatially encoded evolution of spins A and B result in their vectors giving helices of differing pitch.**

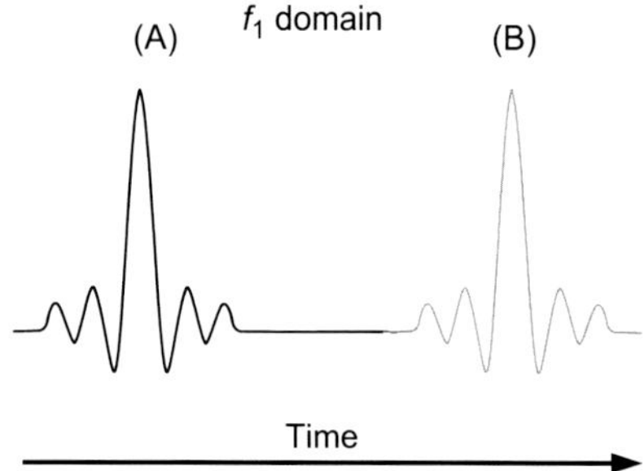

FIGURE 5.42 **Echo formation due to unwinding (refocussing) of the helices.** The position of each echo reflects the f_1 chemical shifts for spins A and B.

Data acquisition requires that the spatial encoding used to define the discrete t_1 frequency labelled segments is now decoded and this may be achieved through the application of a series of read gradients during the acquisition period (Fig. 5.40). The action of the decoding module may be understood by considering the results from a sample giving rise to only two resonances of chemical shift offsets ν_A and ν_B. The net chemical shift evolution experienced during t_1 for spins A with offset ν_A imposes a helical distribution of magnetization vectors along the sample length due to the differing effective t_1 periods experienced by each slice (Fig. 5.41). For spins B of greater chemical shift offset, the pitch of the resulting helix is correspondingly larger. The application of the read gradient causes these spatially encoded vectors to refocus and produce an echo within the acquisition period at a time dictated by how long it takes each helix to unwind and thus by the magnitude of ν_A and ν_B. Spins with the smallest offset ν_A will refocus first while those with greater offset ν_B will take longer to refocus and hence will generate an echo at a later time. In each case the initial chemical shift evolution occurring during t_1 has been forcibly refocused by the applied read gradient and the position of each echo reflects the corresponding chemical shifts encoded in f_1 (Fig. 5.42). A second read gradient is also applied in the opposite sense to refocus the echoes once more and the corresponding data sets generated for the positive and negative gradients are stored separately. These may be considered as the echo and antiecho counterparts giving rise to data sets reflected in f_1 and can be added in a similar manner to the conventional protocol to improve signal-to-noise. The read gradient module is repeated N_2 times to define the total t_2 acquisition period, during which the evolution of the echo responses reflects the chemical shifts of the detected spins (Fig. 5.43). Fourier transformation of these data yields the f_2 dimension. Perhaps surprisingly, it is the case that the complete 2D spectrum arises from this single transformation since the locations of the detected echo peaks define the f_1 chemical shifts directly. Most significantly, the whole 2D data set is acquired in only a single transient and data collection is complete within a fraction of a second. This is exemplified by the single-scan 2D total correlation spectroscopy (TOCSY) spectrum of *n*-butyl chloride recorded in a mere 200 ms (Fig. 5.44).

More recent implementations of single-scan techniques have replaced the sequential application of many frequency selective pulses in the initial encoding phase with more efficient single frequency swept pulses (Section 12.2). This enables t_1 encoding with only a single gradient pair and so reduces the demands on instrument hardware quite significantly. Nevertheless, beyond the obvious requirement for good signal-to-noise in a single transient, the principle limitations with

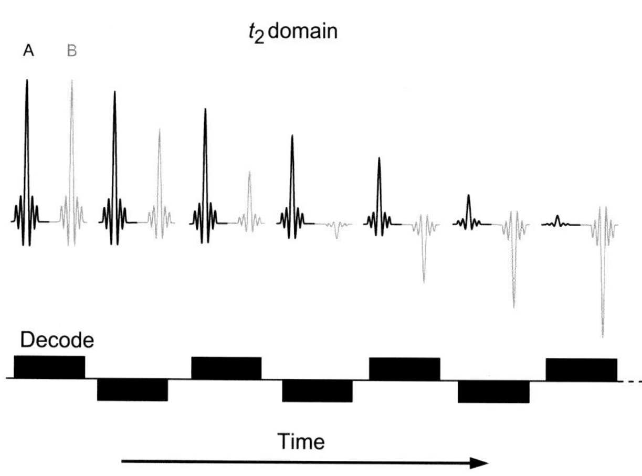

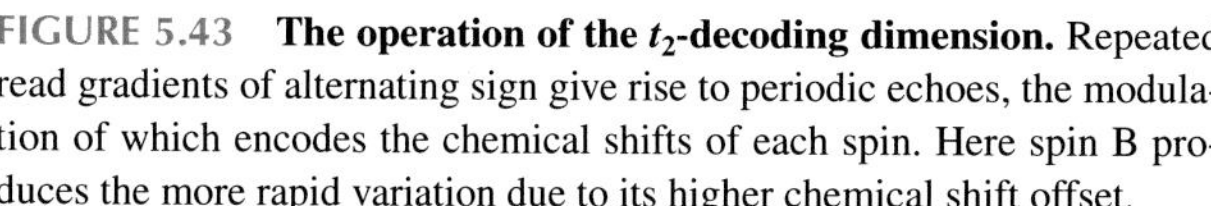

FIGURE 5.43 **The operation of the t_2-decoding dimension.** Repeated read gradients of alternating sign give rise to periodic echoes, the modulation of which encodes the chemical shifts of each spin. Here spin B produces the more rapid variation due to its higher chemical shift offset.

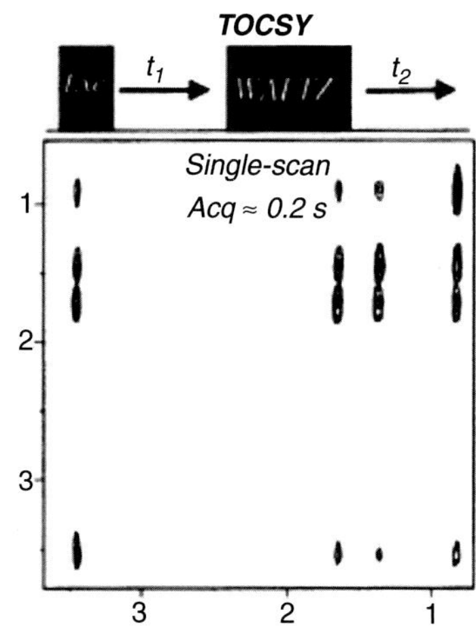

FIGURE 5.44 **A single-scan 2D TOCSY spectrum of *n*-butyl chloride.** This was recorded with a 74-ms mixing period and a total data collection time of ~200 ms. *(Source: Reproduced with permission from Ref. [42], Copyright 2003, American Chemical Society.)*

this approach still lie in the need for high-power gradients with fast switching times for signal decoding. These must be applied repetitively without intensity loss or imbalance between gradients of opposite sign. Current gradient systems may struggle to satisfy all these requirements, and further technical developments to overcome these and simplify experimental setup are likely required before these methods gain wider application. Nevertheless, the potential time savings have obvious appeal.

This then finishes laying the foundations for understanding 2D (or more generally *n*D) NMR and the operation of PFGs. Throughout this, the basic two-pulse COSY sequence has been used as an illustrative 2D sequence. Not only is COSY a simple sequence, but the spectra are rather easy to interpret and require relatively little explanation. Subsequent chapters describe the most important 2D homonuclear and heteronuclear correlation techniques, and examples of the application of the most commonly employed variants may be found in the chapter *Structure Elucidation and Spectrum Assignment.*

REFERENCES

[1] Jeener, J, 1971. Ampère International Summer School, Basko Polje, (former) Yugoslavia.
[2] Aue WP, Bartholdi E, Ernst RR. J Chem Phys 1976;64:2229–46.
[3] Evans JNS. Biomolecular NMR spectroscopy. Oxford: Oxford University Press; 1995.
[4] Cavanagh J, Fairbrother WJ, Palmer AG, Skelton NJ, Rance M. Protein NMR spectroscopy: principles and practice. 2nd ed. San Diego: Academic Press (Elsevier); 2006.
[5] States DJ, Haberkorn RA, Ruben DJ. J Magn Reson 1982;48:286–92.
[6] Ohuchi M, Hosono M, Furihata K, Seto H. J Magn Reson 1987;72:279–97.
[7] Marion D, Wüthrich K. Biochem Biophys Res Commun 1983;113:967–74.
[8] Keeler J, Neuhaus D. J Magn Reson 1985;63:454–72.
[9] Marion D, Ikura M, Tschudin R, Bax A. J Magn Reson 1989;85:393–9.
[10] Davis AL, Keeler J, Laue ED, Moskau D. J Magn Reson 1992;98:207–16.
[11] Turner CJ, Hill HDW. J Magn Reson 1986;66:410–21.
[12] Bax A, Freeman R, Morris GA. J Magn Reson 1981;43:333–8.
[13] Mehlkopf AF, Korbee D, Tiggleman TA, Freeman R. J Magn Reson 1984;58:315–23.
[14] Morris GA. J Magn Reson 1992;100:316–28.
[15] Baumann R, Wider G, Ernst RR, Wüthrich K. J Magn Reson 1981;44:402–6.
[16] Morris GA. J Magn Reson 1988;80:547–52.
[17] Morris GA, Cowburn D. Magn Reson Chem 1989;27:1085–9.
[18] Morris GA, Barjat H, Horne TJ. Prog Nucl Magn Reson Spectrosc 1997;31:197–257.
[19] Metz KR, Lam MM, Webb AG. Concept Magn Reson 2000;12:21–42.
[20] Horne TJ, Morris GA. J Magn Reson A 1996;123:246–52.

[21] Gibbs A, Morris GA, Swanson AG, Cowburn D. J Magn Reson (A) 1993;101:351–6.
[22] Mobli M, Hoch JC. Prog Nucl Magn Reson Spectrosc 2014;83:21–41.
[23] Hore PJ, Jones JA, Wimperis S. NMR: the toolkit. 2nd ed. Oxford: Oxford University Press; 2015.
[24] Keeler J. Understanding NMR spectroscopy. 2nd ed. Chichester: Wiley; 2010.
[25] Bain AD. J Magn Reson 1984;56:418–27.
[26] Bodenhausen G, Kogler H, Ernst RR. J Magn Reson 1984;58:370–88.
[27] Kessler H, Gehrke M, Griesinger C. Angew Chem Int Ed 1988;27:490–536.
[28] Keeler J, Clowes RT, Davis AL, Laue ED. Method Enzymol 1994;239:145–207.
[29] Berger S. Prog Nucl Magn Reson Spectrosc 1997;30:137–56.
[30] Parella T. Magn Reson Chem 1998;36:467–95.
[31] Lauterbur PC. Nature 1973;242:190–1.
[32] Bax A, de Long PG, Mehlkopf AF, Smidt J. Chem Phys Lett 1980;69:567–70.
[33] Barker P, Freeman R. J Magn Reson 1985;64:334–8.
[34] Hurd RE. J Magn Reson 1990;87:422–8.
[35] Jeener J, Meier BH, Bachmann P, Ernst RR. J Chem Phys 1979;71:4546–53.
[36] Weigelt J, Otting G. J Magn Reson A 1995;113:128–30.
[37] Sklenár V, Piotto M, Leppik R, Saudek V. J Magn Reson A 1993;102:241–5.
[38] Hwang TL, Shaka AJ. J Magn Reson A 1995;112:275–9.
[39] Kontaxis G, Stonehouse J, Laue ED, Keeler J. J Magn Reson A 1994;111:70–6.
[40] Freeman R, Kupče E. . J Biomol NMR 2003;27:101–13.
[41] Frydman L, Scherf T, Lupulescu A. Proc Natl Acad Sci USA 2002;99:15858–62.
[42] Frydman L, Lupulescu A, Scherf T. J Am Chem Soc 2003;125:9204–17.
[43] Herrera A, Fernández-Valle E, Martínez-Álvarez R, Molero D, Pardo ZD, Sáez E, Gal M. Angew Chem Int Ed 2009;48:6274–7.
[44] Queiroz LHK, Giraudeau P, dos Santos FAB, de Oliveira KT, Ferreira AG. Magn Reson Chem 2012;50:496–501.
[45] Pardo ZD, Olsen GL, Fernández-Valle ME, Frydman L, Martínez-Álvarez R, Herrera A. J Am Chem Soc 2012;134:2706–15.
[46] Pathan M, Charrier B, Tea I, Akoka S, Giraudeau P. Magn Reson Chem 2013;51:168–75.
[47] Freeman R. Concept Magn Reson A 2011;38A:1–6.
[48] Mishkovsky M, Frydman L. Ann Rev Phys Chem 2009;60:429–48.

Chapter 6

Correlations Through the Chemical Bond I: Homonuclear Shift Correlation

Chapter Outline

6.1 **Correlation Spectroscopy: COSY** **203**
6.1.1 Interpreting COSY 204
6.1.2 Peak Fine Structure 207
6.1.3 Which COSY Approach? 210
6.1.4 COSY-β 211
6.1.5 Double-Quantum Filtered COSY (DQF-COSY) 212
6.1.6 Long-Range COSY: Detecting Small Couplings 219
6.1.7 Relayed-COSY 220
6.2 **Total Correlation Spectroscopy: TOCSY** **220**
6.2.1 The TOCSY Sequence 221
6.2.2 Applying TOCSY 223
6.2.3 Implementing TOCSY 225
6.2.4 One-Dimensional TOCSY 227
6.3 **Correlating Dilute Spins: INADEQUATE** **230**
6.3.1 Two-Dimensional INADEQUATE 230
6.3.2 One-Dimensional INADEQUATE 233
6.3.3 Implementing INADEQUATE 233
6.4 **Correlating Dilute Spins via Protons: ADEQUATE** **235**
6.4.1 Two-Dimensional ADEQUATE 236
6.4.2 Enhancements to ADEQUATE 237
References **240**

This chapter focuses on methods for establishing through-bond correlations between similar nuclides, the so-called 'homonuclear shift correlation experiments'. By far the most common implementation of these methods exploit spin-coupling interactions between protons, and these represent some of the most widely used correlation techniques for the identification of chemical structures. In the previous chapter the basic correlation spectroscopy (COSY) experiment was employed to illustrate how 2D NMR spectroscopy operates, and in this chapter we shall consider the application of COSY and of some variants that display specific beneficial characteristics. Subsequently, the related technique of total correlation spectroscopy (TOCSY) is introduced as it provides a more informative alternative to COSY for establishing correlations within more complex molecules. This provides efficient transfer of information along a network of coupled spins, a feature that can be extremely powerful in the analysis of more complex spectra. We then consider INADEQUATE (Incredible Natural Abundance Double-Quantum Transfer Experiment) which establishes homonuclear correlations between nuclides of low natural abundance. This may be used to directly correlate neighbouring carbon centres, and in this form is perhaps the ultimate experiment for defining the molecular skeleton of an organic molecule although, alas, it is also about the least sensitive. More pragmatic methods known as ADEQUATE (adequate sensitivity double-quantum spectroscopy) that are based on proton excitation and detection offer significant sensitivity gains and are considered in the final part of the chapter. An overview of the main features of these techniques may be found in Table 6.1.

6.1 CORRELATION SPECTROSCOPY: COSY

In the structure characterisation of organic molecules, the two most widely used 2D methods are COSY for defining ^{1}H–^{1}H correlations and the heteronuclear single-quantum correlation (HSQC) for defining ^{1}H–^{13}C correlations (described in Section 7.3). Both these techniques exist in many variant forms designed to manipulate their information content to better assist with structure definition, and below the most useful forms of COSY are described and their application illustrated. We shall begin by considering the simplest COSY sequence (Fig. 6.1), as described in the previous chapter and used to illustrate the principles underlying the generation of 2D NMR data sets. Here we review the basic principles regarding the appearance of COSY spectra and the interpretation of 2D COSY data, before considering their application and that of other variants.

High-Resolution NMR Techniques in Organic Chemistry. http://dx.doi.org/10.1016/B978-0-08-099986-9.00006-3

TABLE 6.1 The Principal Applications of the Main Techniques Described in This Chapter

Technique	Principal Applications
COSY-90	Correlating coupled homonuclear spins. Typically used for correlating protons coupled over 2- or 3-bonds, but may be used for any high-abundance nuclide. The basic COSY experiment.
COSY-β	Correlating coupled homonuclear spins. Typically used for correlating protons coupled over two or three bonds, but may be used for any high-abundance nuclide. Reduced 2D peak structure over basic COSY. Vicinal and geminal coupling relationships can be differentiated in some cases.
DQF-COSY	Correlating coupled homonuclear spins. Typically used for correlating protons coupled over two or three bonds. Higher-resolution display than basic COSY. Additional information on magnitudes of coupling constants may be extracted from 2D peak fine structure. Singlets suppressed.
Long-range COSY	Correlating coupled homonuclear spins through small couplings. Often used to identify proton correlations over many bonds (>3).
TOCSY	Correlating coupled homonuclear spins and those that reside within the same spin system, but which may not share mutual couplings. Employs the propagation of magnetisation along a continuous chain of spins. Powerful technique for analysing complex proton spectra.
INADEQUATE	Correlating coupled homonuclear spins of low natural abundance (<20%). Typically used for correlating adjacent carbon centres at natural abundance, but has extremely low sensitivity.
ADEQUATE	Correlating coupled homonuclear spins of low natural abundance, primarily ^{13}C, but employing ^{1}H excitation and detection for sensitivity improvement.

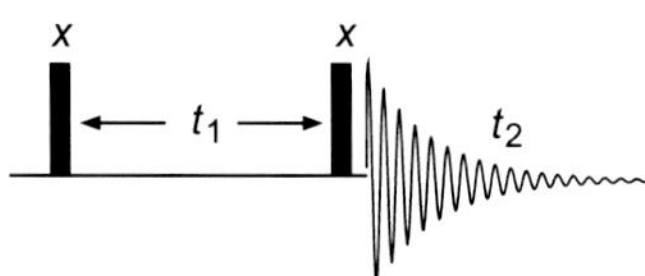

FIGURE 6.1 **The basic COSY sequence.**

6.1.1 Interpreting COSY

Consider first the COSY spectrum of β-methyl glucose **6.1** (Fig. 6.2). The spectrum displays two identical proton chemical shift axes with matching shifts represented by a hypothetical diagonal, shown as a dashed line in this spectrum. The peaks that sit along this line are termed the *diagonal peaks*. These provide reference chemical shifts matching those seen in the conventional 1D proton spectrum, here shown alongside each axis, and present no new information. The peaks of interest are the off-diagonal or *crosspeaks* which correlate differing shifts in the two dimensions and provide direct evidence that the correlated protons share a mutual scalar (J) coupling. The correlated protons are labelled for each crosspeak and have a symmetry-related partner about the diagonal. The crosspeaks observed for **6.1** correlate neighbouring protons via three-bond (vicinal) couplings and two-bond (geminal) couplings (here between the diastereotopic H6 protons). Assignment of the proton resonances of β-methyl glucose can be made starting from H1, which can be identified due to its higher chemical shift, as expected for this anomeric proton. Stepping from diagonal to crosspeak identifies the shift of the coupled H2 proton, and repetition of this from the diagonal peak of H2 and so on provides a stepwise assignment for all protons around the ring. It is noteworthy that the H3–H4 and H4–H5 correlations overlap due to similar H3 and H5 chemical shifts, but may still be identified through careful consideration of crosspeak shifts. Note also that although the methyl resonance appears on the diagonal it has no associated crosspeaks since it shares no coupling to neighbouring protons, and consequently also appears as a singlet in the proton spectrum. It is also worth noting that although correlations over two- or three-bonds are most frequently observed in COSY, since these give rise to scalar couplings of significant size, 'long-range' crosspeaks may also become apparent, occurring over four or even five bonds when favourable bonding pathways enhance the magnitudes of these longer range coupling constants. A variant of COSY described below (Section 6.1.6) seeks to specifically enhance correlations arising from long-range couplings should these be deemed valuable. In the example of **6.1**, the similar multiplet structures and coupling constants for H2, H3 and H4, due to their similar axial positions on the ring, makes their direct assignment from the 1D spectrum alone difficult to assign unambiguously. However, their identification from COSY crosspeaks becomes trivial and unambiguous and can be made without regard to multiplet information, exemplifying the utility of COSY, and of 2D correlation methods more generally. These methods accelerate and make more reliable the task of spectrum assignment and hence play a critical role in modern structure characterisation.

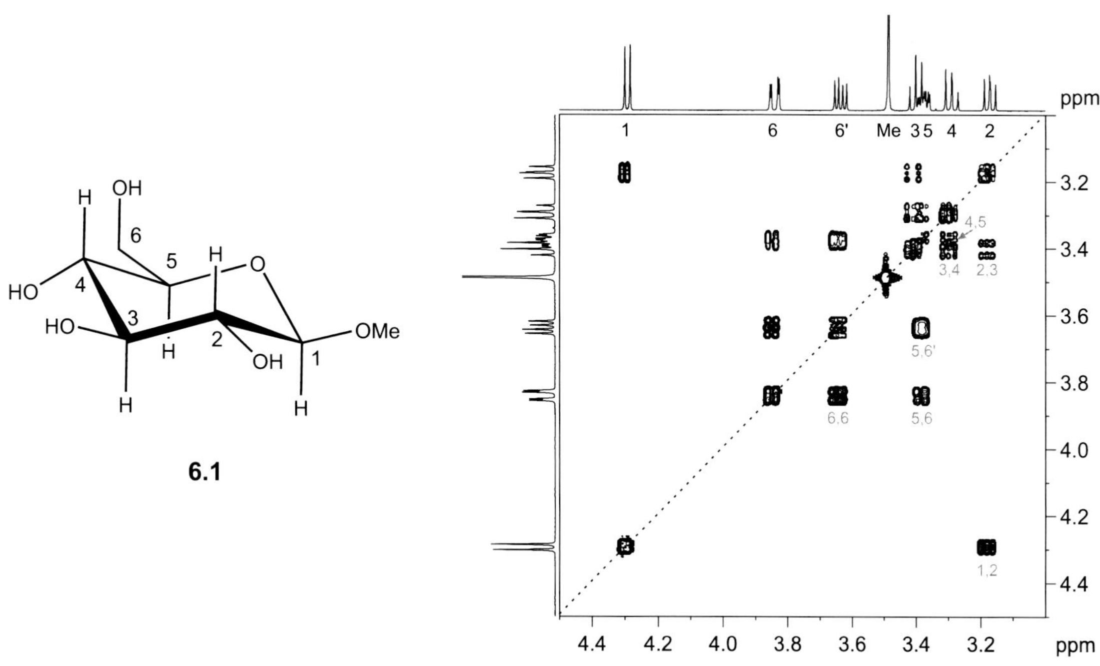

FIGURE 6.2 **The ^1H–^1H COSY spectrum of β-methyl glucose 6.1.** The dashed line represents the hypothetical spectrum diagonal and each crosspeak is labelled with the mutually coupled protons it correlates.

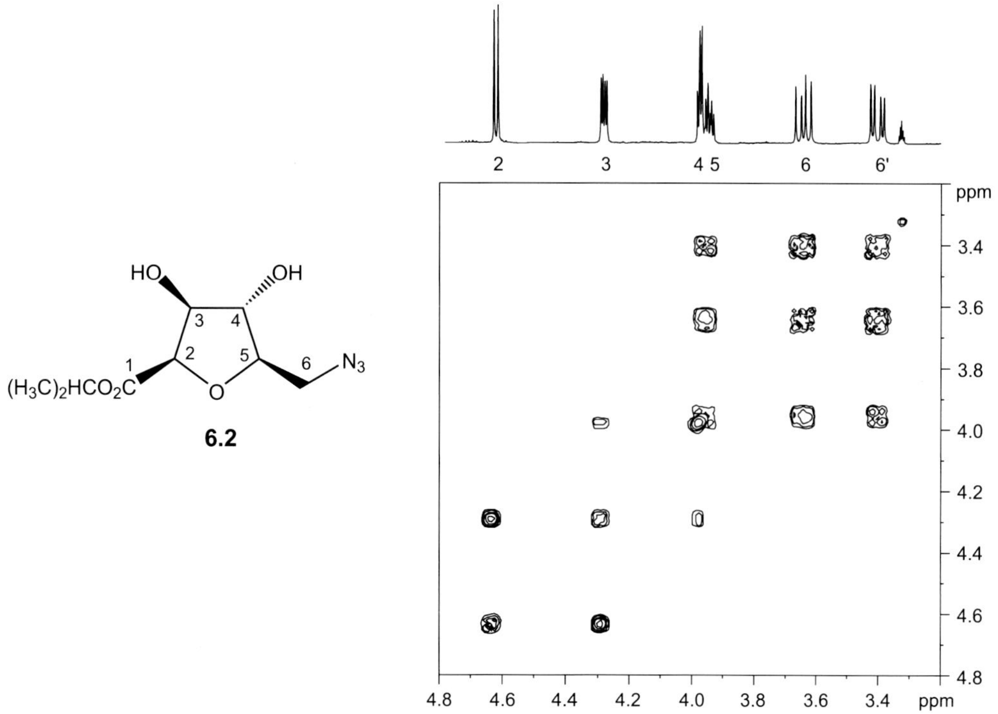

FIGURE 6.3 **The 400 MHz COSY spectrum of the azo-sugar 6.2 in MeOD.** Shown above is the conventional 1D spectrum.

The COSY spectrum of the azo-sugar **6.2**, a monomer used in the synthesis of oligomeric carbopeptoids, further demonstrates both the power of the COSY technique as well as some of its limitations (Fig. 6.3). In this relatively simple example, proton assignments could be accomplished through considered analysis of the multiplet fine structure of each resonance, although more generally this process fails when spectra display significant resonance overlap. In such cases the greater dispersion available in two dimensions can often overcome such limitations. Naturally, one may return to multiplet

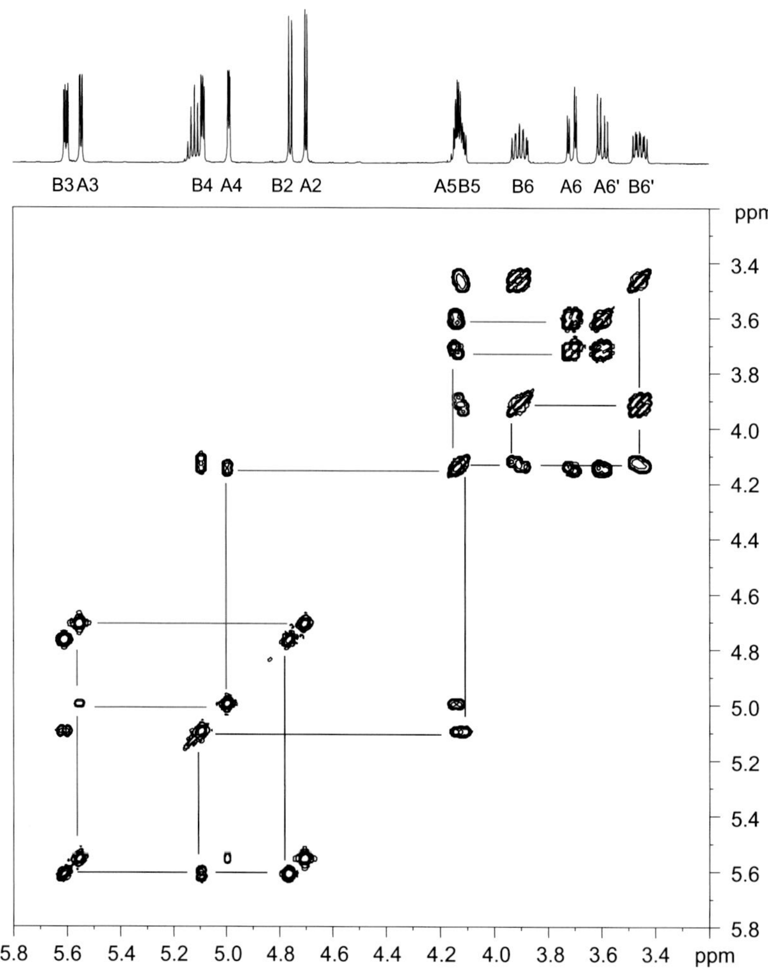

FIGURE 6.4 **The 500 MHz COSY spectrum of the carbopeptoid dimer 6.3 in $CDCl_3$ shown with the 1D spectrum above.** Traced in red above the diagonal are the assignments for sugar ring A and in black below the diagonal are those for sugar ring B (this plot excludes correlations to the amide proton of residue B). Both can be assigned starting from the H2 proton of each sugar. The rings are labelled from the N-terminus of the carbopeptoid and the sugar ring numbering follows IUPAC recommendations for carbohydrate nomenclature.

analysis in detail when assignments have been made and the gross structure identified to glean additional structural data (eg from coupling constants), but COSY can very rapidly provide evidence in support of a structure without the need for this. The first step in the analysis of **6.2** is again the matching of diagonal peaks to the equivalent resonances in the 1D spectrum, as this provides the orientation for all subsequent assignments. One then needs to identify a suitable start point from which to trace correlations, preferably a resonance that can be assigned based on its chemical shift, multiplet pattern or integral. Assignments for **6.2** begin with the characteristic doublet of H2. Correlation to its vicinal partner in the sugar is made via either of the symmetrically related crosspeaks which thus assigns H3, and so on, as seen for **6.1** earlier. This basic procedure provides the basis on which all COSY spectra are analysed. The power and simplicity of the experiment is further illustrated by the carbopeptoid dimer **6.3** (Fig. 6.4). In this case, two discrete spin systems exist within the molecule on adjacent sugar rings, and the differentiation and assignment of these follow readily from the characteristic doublets of H2 in each. The two-ring systems are traced in Fig. 6.4, above and below the diagonal in red (ring A) and black (ring B) for clarity. Further correlations from the H6 protons of ring B to the adjacent amide NH are not shown in this figure, but their presence readily differentiates rings A and B. Indeed, correlations from the lone amide resonance would also provide a suitable starting point to map ring B.

The COSY experiment, despite its simplicity, is however subject to a number of limitations that one should also be aware of. The correlation between H4 and H5 in Fig. 6.3 is almost lost due to the near degeneracy of these resonances, and when coupled spins overlap in this manner the stepwise assignment process can break down since it may not be clear from which resonance subsequent correlations should be traced. Some caution is required in such instances to avoid incorrect assignments, and more sophisticated methods may ultimately be required to provide unambiguous results; the TOCSY experiment introduced in Section 6.2 is often helpful in such cases. With complex structures, one may be left with groups of coupled spins identified from COSY spectra which cannot be linked through homonuclear correlations. These pieces of the molecular 'jigsaw' require additional experimental data if they are to be joined together, such as from the heteronuclear correlation experiments or nuclear Overhauser effect (NOE) measurements described in subsequent chapters. As noted earlier, most proton–proton couplings operate over two or three bonds and the proton COSY spectrum typically identifies vicinal and geminal relationships. However, a general limitation of COSY is the inability a priori to differentiate between these two types of correlation; for example in Fig. 6.3 the vicinal H5–H6 crosspeaks appear similar to the geminal H6–H6′ crosspeak. When analysing COSY these two possible relationships should be borne in mind, particularly when diastereotopic geminal protons may be anticipated, notably in cyclic systems or those with stereogenic centres. A variant on the basic COSY sequence (COSY-β, Section 6.1.4) can, in favourable circumstances, differentiate vicinal from geminal crosspeaks, and the one-bond heteronuclear correlation experiments of Section 7.3 can often aid in the identification of geminal proton pairs. Structural features, such as unsaturation or the well-known *w* geometry, can also enhance long-range coupling constants, and COSY can be surprisingly effective at revealing these correlations. As such, it is not safe to automatically assume *all* crosspeaks correspond to either vicinal or geminal coupling relationships. Geometrical factors can also cause vicinal couplings to be close to zero, in which case no COSY crosspeak is produced despite the presence of neighbouring protons. Thus, the *absence* of a crosspeak between protons does *not* always exclude them from being adjacent in a structure.

The COSY experiment is not limited to proton spectroscopy, but is suitable for establishing homonuclear correlations for any high-abundance nuclide, the most common examples being ^{19}F, ^{31}P or ^{11}B. Fig. 6.5 illustrates the ^{31}P–^{31}P COSY spectrum of the P_7Me_5 cluster **6.4** for which the assignment procedure is exactly as described for the proton spectra earlier. In this case, single-bond P–P couplings are identified directly, along with a lone two-bond coupling between P^2 and P^7 which arises from the sizeable $^2J_{PP}$ coupling.

6.1.2 Peak Fine Structure

It may be apparent on closer inspection that the spectrum of Fig. 6.2 reveals fine structure within each diagonal and crosspeak that is equivalent to the structure seen within 1D multiplets. This is even more apparent in the spectrum of the two-spin system shown in Fig. 5.12 of the previous chapter. The level to which such fine structure is resolved depends on the experimental settings used in the acquisition of the data, and the fine structure in Figs 6.3 and 6.4 is barely resolved, as is often the case. Interpretation of multiplet fine structure can, under the appropriate experimental conditions, provide valuable information, a topic addressed in Section 6.1.5 where we encounter a version of the COSY experiment better suited to this purpose. To realise the origin of this fine structure in the 2D spectrum, it is necessary to modify slightly the description of magnetisation transfer given in Section 5.1.2. It was stated that this transfer occurs between coupled spins, but more precisely one should say it occurs between the *transitions* associated with these spins. Reference to the energy-level diagram for a coupled two-spin system should clarify this (Fig. 6.6). Magnetisation associated with, say, transition A_1, which gives rise to one half of the A-spin doublet, is redistributed to the other three transitions A_2, X_1 and X_2 by the second 90 degrees pulse of the COSY sequence. That associated with A_2 during the detection period gives rise to the fine structure *within* the diagonal peak, while that now present as X_1 and X_2 provides the structure of the crosspeak. The result is that, in addition to modulation due to chemical shifts during t_1 and t_2, there is additional modulation arising from J couplings, and this produces the coupling fine structure in both dimensions.

Recall also that following the second pulse, 'some magnetisation remains associated with the original spin'. Thinking back to the discussions of polarisation transfer in the insensitive nuclei enhanced by polarisation transfer (INEPT) experiment, it was shown that the basic requirement for the transfer of polarisation was an anti-phase disposition of the doublet vectors of the source spin, which for INEPT was generated by a spin-echo sequence. Magnetisation components that were

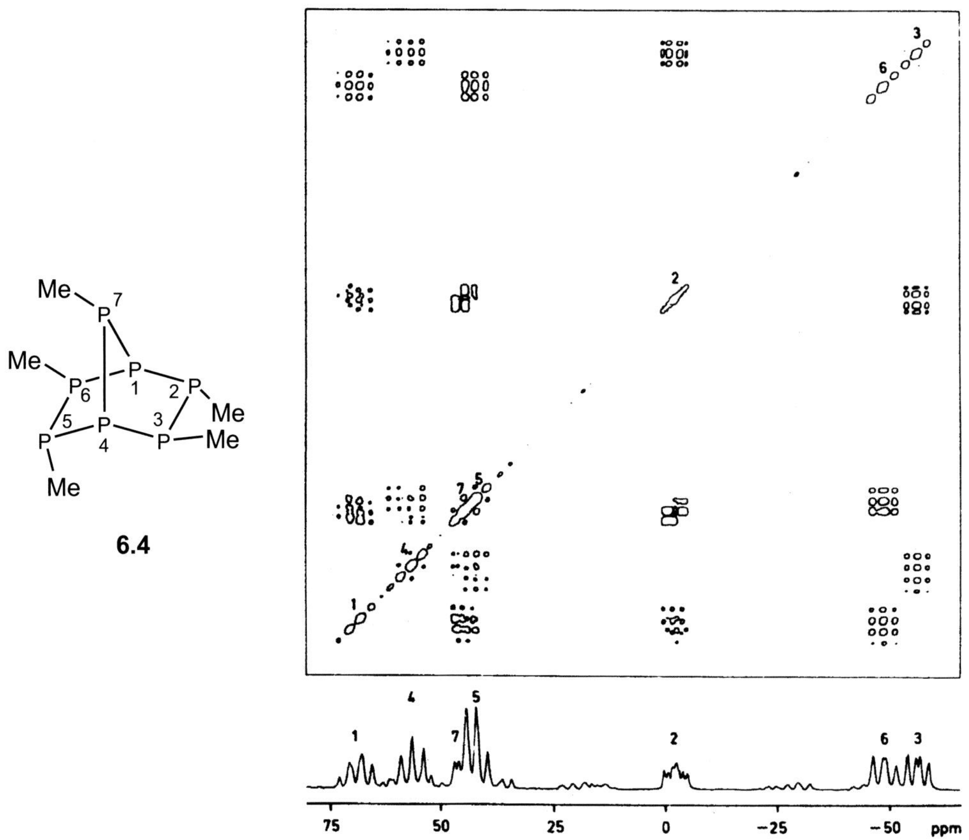

FIGURE 6.5 **The ^{31}P–^{31}P COSY spectrum of the P_7Me_5 cluster 6.4 with the ^{31}P 1D spectrum below.** *(Source: Reproduced with permission from Ref. [1], Copyright Wiley-VCH Verlag GmbH & Co. KGaA.)*

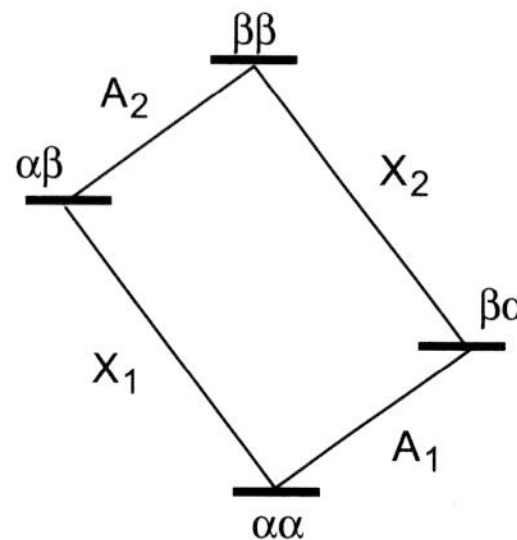

FIGURE 6.6 **A schematic energy-level diagram for the coupled two-spin AX system.**

in-phase just before the second 90 degree pulse would not contribute to the transfer, hence the Δ period was optimised to maximise the anti-phase component. The same condition applies for magnetisation transfer between two protons in the COSY experiment. This requires that the proton–proton coupling be allowed to evolve to give a degree of anti-phase magnetisation that may be transferred by the second pulse, while the in-phase component remains associated with the original spin (Fig. 6.7). The coupling evolution period for COSY is the t_1 period, so that the amount of transferred magnetisation detected in t_2 is also modulated as a function of t_1 (sin $180Jt_1$); this is the modulation mentioned at the end of the last paragraph that ultimately characterises the crosspeak coupling fine structure in f_1. Likewise, the amplitude of the in-phase, non-transferred component is also modulated in t_1 by the coupling (cos $180Jt_1$), and this produces the coupling fine structure of the diagonal peak in f_1. Multiplet structure is further considered in Section 6.1.5 with the aim of extracting coupling constants from 2D crosspeaks.

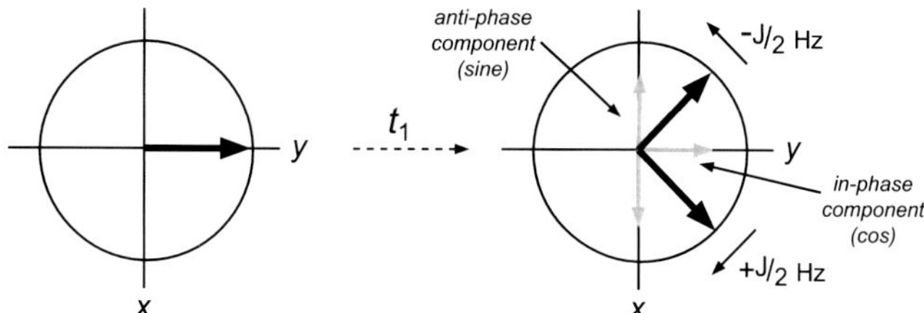

FIGURE 6.7 **Coupling evolution during t_1 produces in-phase and anti-phase magnetisation components.** Only the anti-phase component contributes to magnetisation transfer and hence to crosspeaks in the 2D spectrum.

6.1.2.1 Signal Phases in Phase-Sensitive COSY

When discussing 2D spectra, it is important to consider the relative phases and lineshapes for all resonances present. For the COSY spectrum, as explained earlier, diagonal peaks arise from the in-phase component of magnetisation produced by evolution under spin–spin coupling. This is not transferred to the coupled spin by the mixing pulse, and diagonal peaks therefore have in-phase multiplet fine structure in both dimensions. The crosspeaks, in contrast, arise from the transferred anti-phase component, so have anti-phase multiplet structure in both dimensions. The initial phase of these two sets of signals also differs by 90 degrees since the diagonals arise from magnetisation that was cosine modulated while the crosspeaks arise from that which was sine modulated. Thus, in the final spectrum, the phases of the diagonal and crospeaks will also differ by 90 degrees, and since the crosspeaks are of most interest these are phased so as to have (anti-phase) absorption-mode lineshapes in both dimensions, meaning the diagonal peaks possess (in-phase) double-dispersion lineshapes (Fig. 6.8). The unfortunate presence of dispersion-mode lineshapes is an unavoidable consequence of the COSY sequence with the wide tails of the diagonals potentially masking peaks that sit close by. One variant of the experiment, the double-quantum filtered COSY (DQF-COSY), removes dispersive contributions, so is often favoured for high-resolution work. The widely used magnitude-mode presentation of COSY data masks the signal phase and also removes dispersive tails through the application of harsh window functions.

The anti-phase character of COSY crosspeaks can also be troublesome when couplings are small relative to resonance linewidths, since cancellation of the lines occurs and the crosspeak disappears. This is an important factor in determining whether correlations due to small couplings can be detected with this experiment, a topic addressed in Section 6.1.5.

There exist many experiments that are variations on the simple COSY sequence described here. Below we take a look at a selection of these and examine the benefits these modified sequences provide. Many of the COSY variants proposed in

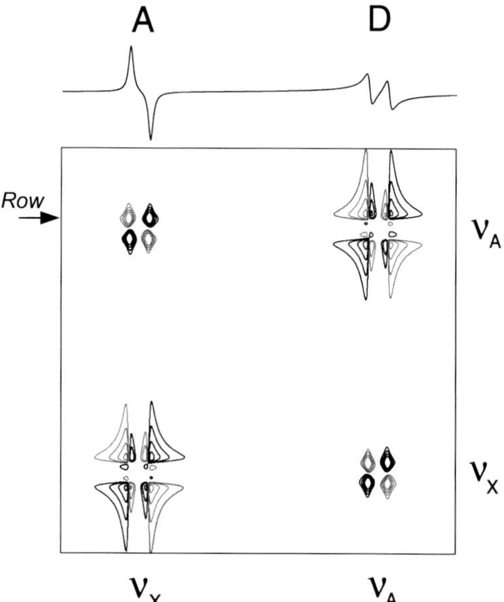

FIGURE 6.8 **The phase-sensitive COSY for a coupled two-spin AX system.** Diagonal peaks have broad, in-phase double-*dispersion* lineshapes (D) whereas crosspeaks have narrow, anti-phase double-*absorption* lineshapes (A), as further illustrated in the row extracted from the spectrum (red and black lines reflect positive and negative intensities respectively).

the literature offer little gain over simpler sequences while being rather more complex to execute or process, or are beneficial for a rather limited class of compounds and lack wide applicability. While the experiments presented below represent only a fraction of the proposed sequences, they have established themselves over the years as the most widely used and informative methods.

6.1.3 Which COSY Approach?

Before moving on, we briefly examine the key characteristics of the different experiments and consider why these might be of interest to the research chemist. Table 6.2 summarises the most significant attributes, some of which have already been introduced while others are expanded in the sections that follow. While the TOCSY experiment is not strictly a member of the COSY family, its information content is so closely related to that of COSY it has also been included in the table.

Pre-empting what is to follow, some general recommendations can be made at this stage. For establishing correlations in relatively well–dispersed spectra, magnitude-mode COSY offers fast turnover and simplicity of data handling, so may be favoured when fully automated generation of spectra is required. The magnitude-mode COSY-β variant, where the mixing pulse β is typically 45 or 60 degrees, provides the most compact peak structure for both diagonal and crosspeaks and is best suited for routine use. The COSY experiments with the greatest information content are the phase-sensitive versions. For solving spectra of any complexity the most useful, by virtue of its high-resolution and informative crosspeak structure, is the phase-sensitive DQF-COSY. As described later the theoretical loss in sensitivity associated with this variant is not as detrimental as it may at first sound, and this method is well suited to the elucidation of complex structures. To genuinely benefit from the information contained within crosspeak structures, it requires high digital resolution which invariably corresponds to longer experiments, and phasing of the resulting 2D spectrum can be something of a black art to the uninitiated. For these reasons it is not so well suited to the rapid sample throughput demanded in busy synthetic chemistry laboratories. In the analysis of complex spectra where considerable peak overlap occurs, the TOCSY experiment can be extremely informative. It provides additional correlation information by relaying magnetisation along networks of coupled spins. This has been particularly favoured in the analysis of peptides and oligosaccharides, since the molecules are comprised of discrete monomer units which themselves represent isolated spin networks. This method has all but replaced the older

TABLE 6.2 A Summary of the Characteristic Features of the Principal COSY Experiments and of the Related TOCSY Experiment

Sequence	Advantages	Potential Drawbacks
Absolute value COSY-90 (magnitude mode)	Simple and robust. Magnitude processing well suited to automated operation.	Phase-twisted lineshapes produce poor resolution and require strong-resolution enhancement functions. Crosspeak fine structure not usually apparent.
COSY-β	Simple and robust. Magnitude processing well suited to automated operation. Simplification of crosspeak structures reduces peak overlap. Vicinal and geminal couplings can be distinguished in some cases from tilt of peaks.	Usually requires magnitude-mode presentation as phase-sensitive variant has mixed-phase lineshapes.
Phase-sensitive DQF-COSY	High-resolution display due to absorptive lineshapes. Crosspeak fine structure apparent; J measurement possible. Diagonals also have absorptive lineshapes. Singlets suppressed.	Theoretical sensitivity loss by a factor of 2 relative to the COSY-90 variant. Requires high digital resolution to reveal multiplet structures.
Long-range COSY	Enhances detection of small and long-range couplings (<2 Hz) such as between protons in allylic systems or those in *w* relationships.	Requires magnitude-mode presentation. Crosspeaks due to larger couplings can be significantly attenuated.
Relayed-COSY	Provides two (or more) step transfers and can reduce ambiguities arising from crosspeak overlap.	Typically has low sensitivity and responses show mixture of lineshapes, so magnitude-mode presentations may be required. TOCSY preferred.
TOCSY	Provides multistep (relayed) transfers to overcome ambiguities arising from crosspeak overlap. High sensitivity. In-phase lineshapes can provide correlations even in the presence of broad resonances.	Number of transfer steps associated with each crosspeak not known *a priori*. In-phase lineshapes tend to mask crosspeak fine structure and may preclude J measurement.

and inferior relayed-COSY experiment that elicits stepwise transfers between coupled spins. The long-range (or delayed) COSY experiment tends to be reserved for addressing specific questions regarding the presence of small couplings and is typically less used.

6.1.4 COSY-β

A common modification of the basic COSY experiment described earlier is one in which the 90 degree mixing pulse is replaced with one of shorter tip angle β, usually of 45 or 60 degrees (Fig. 6.9). These experiments are typically acquired and presented as magnitude-mode experiments, since, strictly speaking, the use of a pulse angle less than 90 degrees does not produce purely amplitude-modulated data.

The use of a reduced mixing pulse largely restricts coherence transfer between transitions that are *directly* connected or, in other words, those that share an energy level, for example A_1 and X_2 in Fig. 6.10. Coherence transfer between *remotely* connected transitions, for example A_1 and A_2, is attenuated. This results in a reduction in intensity of certain lines *within* the multiplet structures of both diagonal and crosspeaks, which is particularly noticeable when spins experience more than one coupling. Diagonal peaks have a somewhat smaller 'footprint', increasing the chances of resolving crosspeaks that sit close to this, while crosspeaks themselves also have a simplified structure which can take on a distinctive 'tilted' appearance (Fig. 6.11). In favourable cases, this tilting effect can be used to differentiate between peaks arising from active couplings of opposite sign, such as geminal or vicinal couplings. Consider the situation for a three-spin AMX system in which all spins couple to each other. The AM crosspeak, for example will have a positive slope and its tilt will be approximately parallel to

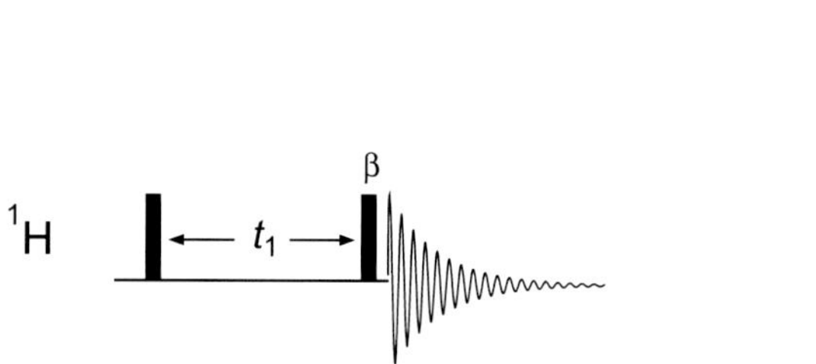

FIGURE 6.9 **The COSY-β experiment.** The β mixing pulse is usually set to 45 or 60 degrees.

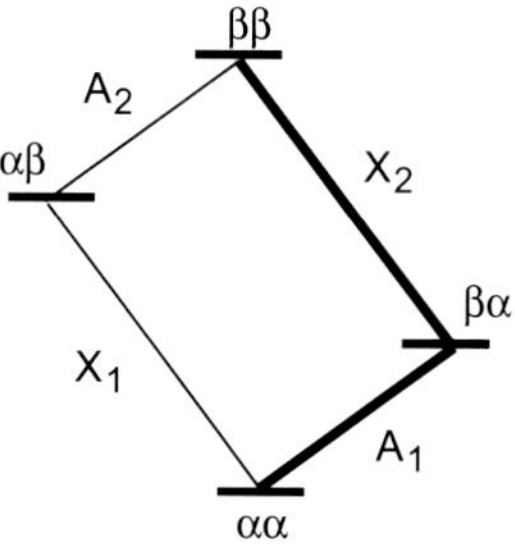

FIGURE 6.10 **The schematic energy-level diagram for a two-spin AX system highlighting the directly connected A_1–X_2 transitions.**

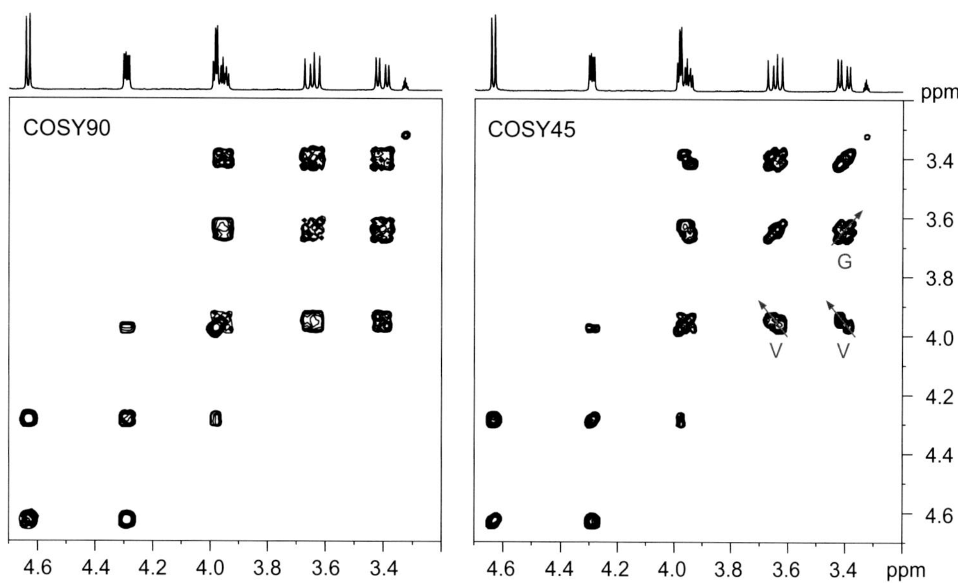

FIGURE 6.11 **The absolute value COSY-45 spectrum of 6.2 compared with the equivalent COSY-90 spectrum.** The tilting apparent on some COSY-45 crosspeaks can indicate whether the active coupling arises from protons sharing geminal (G) or vicinal (V) relationships which produce peaks with positive and negative slopes, respectively (see text).

the diagonal if the two *passive* couplings, J_{AX} and J_{MX} have the *same* sign. In contrast, the peak will display a negative slope and lie anti-parallel to the diagonal if the signs of the passive couplings differ. In proton spectroscopy, vicinal couplings are typically positive and geminal couplings negative, meaning a crosspeak between a proton and its *geminal* partner will possess a *positive slope* whereas that with the *vicinal* partner will display a *negative slope*. This effect is apparent in the COSY-45 spectrum of **6.2** in Fig. 6.11. Couplings to additional spins further split the crosspeak structure and in such cases the tilting may not be distinguished. Nevertheless, when this effect is observed it provides valuable additional evidence by identifying geminal pairs.

The choice between β = 45 and 60 degrees is something of personal taste. Setting β = 45 degrees causes the greatest intensity ratio between direct and remote transitions (a factor of about 6 for 45 degrees vs 3 for 60 degrees), and thus gives the greatest reduction in peak complexity, while setting β = 60 degrees provides slightly higher sensitivity. Since the primary goal of this variant is the reduction of the diagonal and crosspeak footprint, the COSY-45 experiment is probably the one to choose. Due to the more compact peak structures and simplicity of processing, the absolute value COSY-45 experiment still finds widespread use in routine analyses and automated procedures in the chemistry laboratory.

6.1.5 Double-Quantum Filtered COSY (DQF-COSY)

Previous sections have already made the case for acquiring COSY data such that it may be presented in phase-sensitive mode. The pure absorption lineshapes associated with this provide the highest possible resolution and allow one to extract information from the fine structure within crosspeak multiplets. However, it was also pointed out that the basic COSY-90 sequence suffers from one serious drawback in that diagonal peaks possess dispersion-mode lineshapes when crosspeaks are phased into pure absorption mode. The broad tails associated with these can mask crosspeaks that fall close to the diagonal, so there is potential for useful information to be lost. The presence of dispersive contributions to the diagonal may be (largely) overcome by the use of DQF-COSY [2], and for this reason DQF-COSY is the experiment of choice for recording phase-sensitive COSY data.

6.1.5.1 The DQF Sequence

The DQF-COSY sequence (Fig. 6.12) differs from the basic COSY experiment by the addition of a third pulse and the use of a modified phase cycle or gradient sequence to provide the desired selection. Thus, following t_1 frequency labelling, the second 90 degree pulse generates multiple-quantum coherence which is not observed in the COSY-90 sequence since it remains invisible to the detector. This may, however, be reconverted into single-quantum coherence by application of the third pulse, and hence subsequently detected. The required phase cycle or gradient combination selects only signals that existed as double-quantum coherence between the last two pulses, while all other routes are cancelled, hence the term 'double-quantum filtered COSY'.

The rules for filtering multiple-quantum coherences of order p are simple; all pulses prior to the p-quantum coherences must be cycled in steps of 180/p degrees with associated alternation of the receiver phase on each step, [3] so a suitable phase cycle (although not the only one) for a double-quantum filter would be that presented in Table 6.3. More effective filtration can be obtained through gradient selection [4]. Coherence order +2 within the filter may be selected by using gradient amplitude ratios of +1:+2 either side of the last 90 degree pulse, as in the example of Section 5.4.1, while order −2 is retained with ratios of +1:−2. Thus, it is possible to select either the +2 or −2 pathway, *but not both at once.* This is in contrast to the phase-cycled experiment in which both pathways are retained, so the gradient experiment detects only *one*

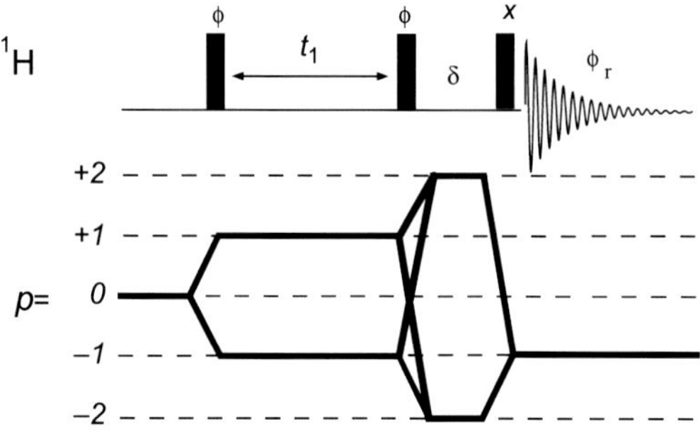

FIGURE 6.12 **The DQF-COSY experiment and coherence transfer pathway.** The pulses are phase-cycled as described in the text to select the pathway shown with quad-detection observing the p = −1 magnetisation. The period δ allows for rf phase changes and is typically of only a few microseconds.

TABLE 6.3 A Suitable Phase Cycle for Double-Quantum Filtration

φ	ϕ_r
x	x
y	$-x$
$-x$	x
$-y$	$-x$

Phase φ is used for all pulses prior to double-quantum coherence and ϕ_r for the receiver. The phase of the final pulse remains unchanged.

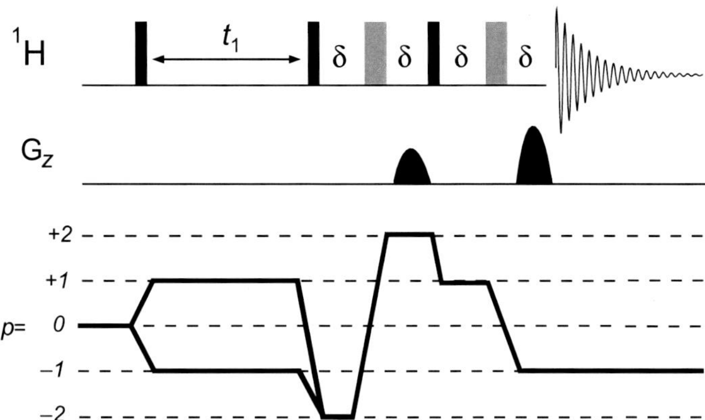

FIGURE 6.13 **The gradient-selected DQF-COSY experiment and coherence transfer pathway.** No phase-cycling is needed as the required pathway is selected with gradient ratios of 1:2. Both gradient pulses are applied within spin-echoes for phase-sensitive presentations. Note only one pathway is retained from the double-quantum filter.

half of the available signal and thus has twofold poorer sensitivity relative to its phase-cycled cousin. This fundamental limitation of being able to refocus only one of the two possible $\pm p$ coherence order pathways on each transient is one of the disadvantages of using pulsed field gradients (PFGs) for coherence selection, but must be balanced against the cleaner suppression of unwanted signals. Since these gradients are not required within t_1 in this sequence, conventional quadrature detection—States or time proportional phase incrementation (TPPI)—can be employed. However, simply inserting the appropriate gradients before and after the final pulse would introduce large phase errors to the spectrum because of chemical shift evolution during the gradient pulses. This must therefore be refocused if phase-sensitive data are required, so both gradients are applied within spin-echoes, producing the sequence of Fig. 6.13 for the gradient-selected, phase-sensitive DQF-COSY experiment.

The result from the filtration step, and the principal reason for its use, is that the diagonal peaks now possess anti-phase absorption-mode lineshapes, as do the crosspeaks which are unaffected by the filtration. Strictly speaking, for spin systems of more than two spins the diagonal peaks still possess some dispersive contributions, but these are now anti-phase, so cancel and tend to be weak and rarely problematic. The severe tailing previously associated with diagonal peaks therefore is removed, providing a dramatic improvement in the quality of spectra (Fig. 6.14).

An additional benefit from filtration is that singlet resonances do not appear in the resulting spectrum because they are unable to create double-quantum coherence, so cannot pass through the filter. Since sharp singlets produce the most intense t_1 noise bands, their suppression alone can be beneficial. This filtration could also be used to suppress large solvent resonances that would otherwise dominate the spectrum. While some success may be achieved through phase cycling, with suppression ratios of the order of a few hundred, far greater suppression can be provided by gradient selection, where suppression ratios can reach 10,000:1. Editing through filtration is exemplified in the 1D double-quantum filtered spectrum of the peptide Leu-enkephalin **6.5** in CD_3OD (Fig. 6.15), produced with the 1D sequence of Fig. 6.16. The singlet solvent resonances are removed whereas those from all coupled spins are retained, and appear with the characteristic anti-phase structure.

The potential disadvantage of using double-quantum filtration is the theoretical reduction in signal-to-noise by a factor of 2 due to losses associated with the generation of double-quantum coherence. However, the benefits arising from the removal of the dispersive contributions to the diagonal usually compensate for the reduction in sensitivity.

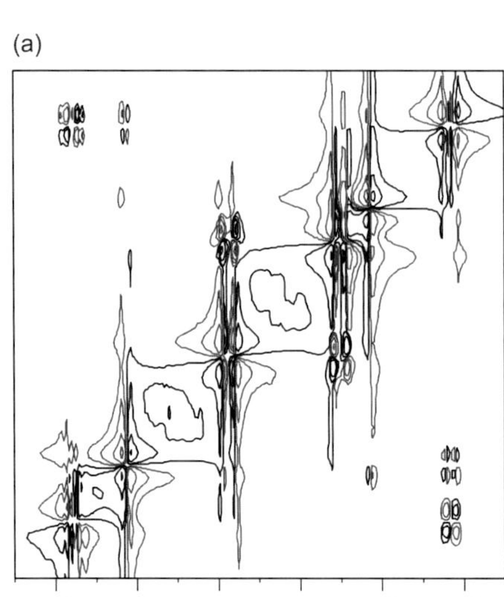

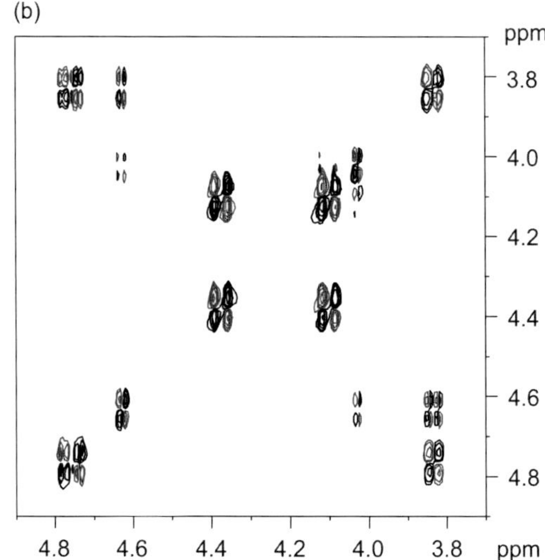

FIGURE 6.14 **Phase-sensitive COSY spectra.** The double-quantum filtered COSY spectrum (b) provides greater clarity close to the diagonal peaks than the basic phase-sensitive COSY (a) as it does not suffer from broad, dispersive diagonal peaks.

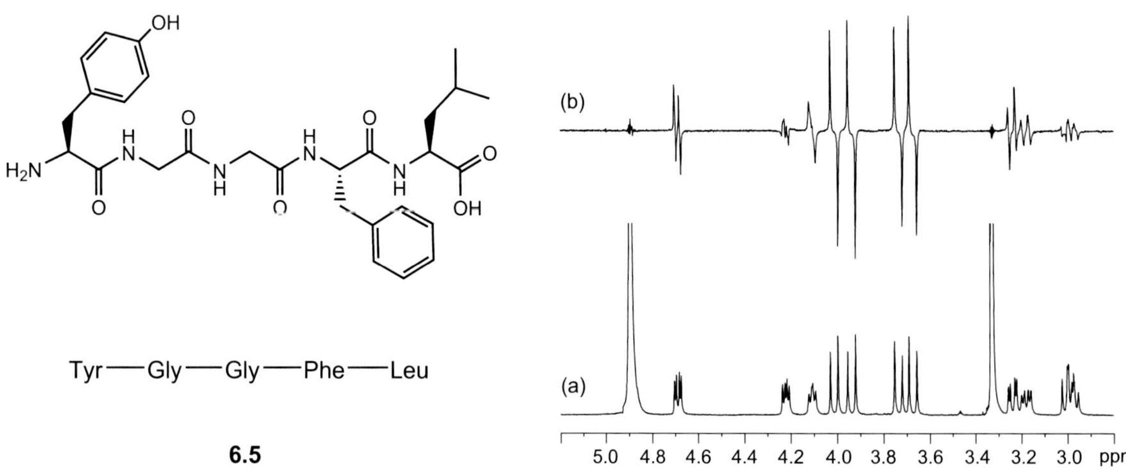

FIGURE 6.15 **1D double-quantum filtration of the spectrum of the peptide Leu-enkephalin 6.5 in CD_3OD.** The singlet resonances of the solvent, truncated in the conventional 1D spectrum (a), have been filtered out in (b). The remaining peaks in (b) display the characteristic anti-phase multiplet structure (which may be masked by magnitude calculation if desired).

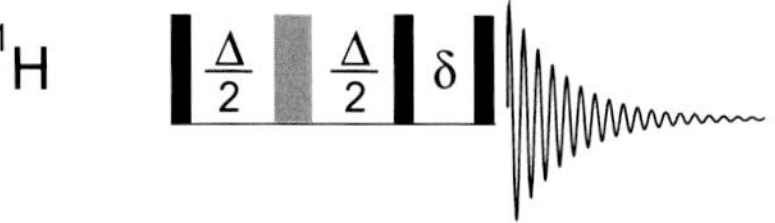

FIGURE 6.16 **The 1D double-quantum filter.** The sequence is derived from the 2D experiment by replacing the variable t_1 period with a fixed spin-echo optimised to produce anti-phase vectors ($\Delta = 1/2J$) as required for the generation of double-quantum coherence. Signal selection is then as for the 2D experiment, and gradient selection may be implemented as in Fig. 6.13.

6.1.5.2 Interpreting Multiplet Structure

A major part of 1D spectrum analysis consists of measuring coupling constants within multiplets to gain as much information as possible on relationships between spins within the molecule. While this is at least plausible for well-resolved multiplets the task becomes rapidly more difficult and the data obtained less reliable for resonances that overlap. In such cases it would be desirable to use the greater dispersion available in the 2D spectrum to reveal coupling patterns from otherwise intractable regions, and it has already been shown that the COSY crosspeak structure reflects that of corresponding 1D multiplets. This section examines how one interprets the data within crosspeaks with the ultimate aim of extracting coupling constants.

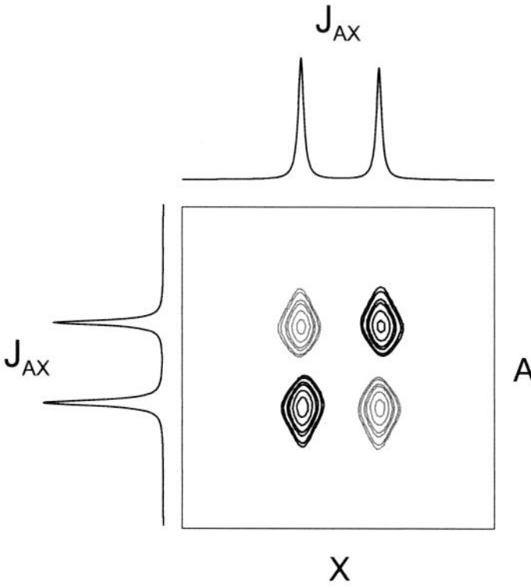

FIGURE 6.17 **The anti-phase square coupling pattern reflects the active coupling J_{AX} between two correlated spins A and X.** This pattern provides the basis for all COSY crosspeak structures.

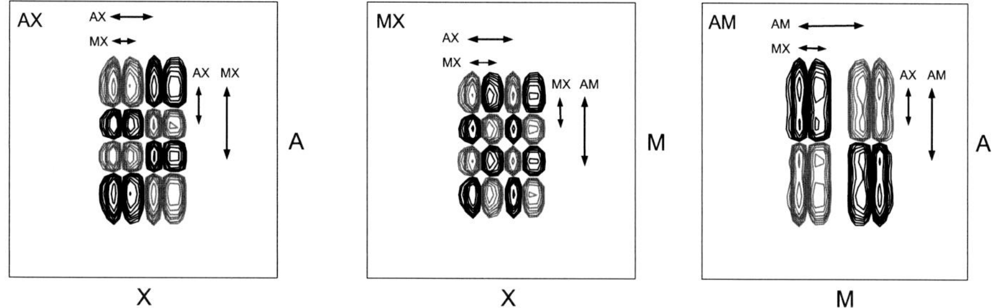

FIGURE 6.18 **The crosspeak structures from the phase-sensitive COSY spectrum of a three-spin AMX system.** The arrows indicate splittings due to the labelled couplings. The spectrum was simulated with J_{AM} = 18, J_{AX} = 12 and J_{MX} = 6 Hz and the final digital resolution was 2.3 Hz/pt in each dimension.

To do this, it is first necessary to differentiate between active and passive couplings within a crosspeak. The *active* coupling is that which *gives rise to the crosspeak* that correlates the coupled spins. As described previously, the coupling responsible for the crosspeak appears as an anti-phase splitting in both dimensions, and so produces the basic anti-phase square array from which all COSY crosspeaks are ultimately derived (Fig. 6.17). *Passive* couplings are those to *all other spins*, and appear in the spectrum as in-phase splittings. These replicate the anti-phase array of the active coupling without introducing further changes of sign. These features and the basic principles behind analysing multiplet structures may be illustrated by reference to the crosspeaks of a three-spin AMX system in which $J_{AM} > J_{AX} > J_{MX}$ (Fig. 6.18). Consider first the MX crosspeak, in which the anti-phase square pattern of the active J_{MX} coupling is clearly apparent in the upper left corner. This whole pattern is then reproduced with the same phase by the larger passive couplings J_{AX} in f_2 and J_{AM} in f_1 to produce the final 4 × 4 structure. This stepwise splitting is the precise equivalent to that used in the analysis of 1D multiplets via coupling 'trees'. The f_2 structure of the AX multiplet arises from the large anti-phase J_{AX} coupling being further split by the smaller passive J_{MX} coupling to yield the + + − − structure. In f_1 the active coupling is now the smaller of the two, so is simply repeated to yield the + − + − pattern. Finally, the AM multiplet is dominated by the very large active J_{AM} coupling in both dimensions which initially yields a large square array, with each part then split by the smaller passive couplings to spin X. The rules for interpreting these structures therefore exactly parallel those for interpreting 1D multiplets, aside from the distinction between active and passive couplings. When coupling occurs between one spin and *n* equivalent spins, the resulting peak structure may be derived by imagining only one of the *n* spins to be active and all others passive. One can then construct the anti-phase analogue of the familiar Pascal triangle for coupling with equivalent spins (Fig. 6.19). Thus, for an AX_3 system the crosspeak displays a doublet for X and a +1 +1 −1 −1 quartet for A (Fig. 6.20).

In realistic systems multiplet structures may derive from very many couplings, and resolution of all of these may not be possible. This leads to overlap within the multiplet with associated line cancellation and/or superposition. In general, crosspeak structures tend towards the simplest pattern as neighbouring lines with like-sign merge. Often in such cases smaller couplings may not be resolved, as is nearly the case for the passive MX and AX couplings within the AM crosspeak of Fig. 6.18. When analysing fine structure in detail it is often advantages to examine both sets of equivalent crosspeaks on either side of the diagonal since these may have differing appearances according to the different digitisation levels used, meaning a coupling may be unresolved in one dimension but quite apparent along the other.

n	(a)	(b)
0	1	1
1	1 1	1 −1
2	1 2 1	1 0 −1
3	1 3 3 1	1 1 −1 −1
4	1 4 6 4 1	1 2 0 −2 −1
5	1 5 10 10 5 1	1 3 2 −2 −3 −1
6	1 6 15 20 15 6 1	1 4 5 0 −5 −4 −1

FIGURE 6.19 **Pascal's triangles.** (a) The conventional triangle for 1D multiplets and (b) the anti-phase equivalent for predicting COSY multiplet structures from coupling to *n*-equivalent spins.

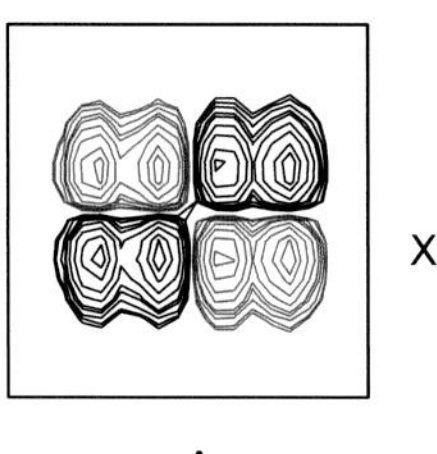

FIGURE 6.20 **The COSY crosspeak for an AX_3 group illustrating the structure predicted from the Pascal triangle.**

6.1.5.3 Measuring Coupling Constants

Provided multiplet structures are sufficiently resolved and free from overlap with other peaks, it is in principle possible to measure coupling constants directly from these [5]. Some caution is required, however, since one needs to consider whether cancellation within the multiplet could lead to erroneous measurement. For example if the central lines of a double-doublet were to interfere destructively and cancel, the measured splitting would then be greater than the true coupling constant. Digital resolution must also be adequate to properly characterise the coupling constant, with the highest resolution being found in the f_2 dimension. To improve the accuracy of the measurement, it is advantageous to extract 1D traces from the 2D spectrum and to subject these to *inverse* Fourier transformation, zero-filling (or better still, linear prediction) and then Fourier transformation to reproduce the 1D row with increased digital resolution. This process is illustrated in Fig. 6.21 for

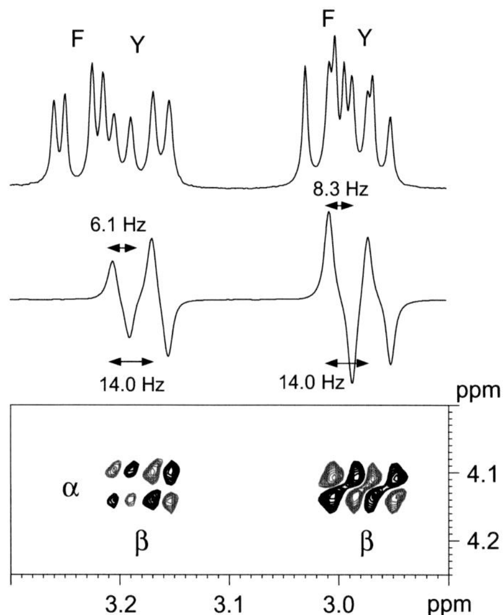

FIGURE 6.21 **Sections from the DQF-COSY spectrum of the pentapeptide Leu-enkephalin 6.5.** The 2D crosspeaks and the f_2 1D trace taken through these correspond to correlations within the tyrosine (Y) residue. The upper trace is taken from the conventional 1D spectrum in which the β-proton resonances partially overlap with those of phenylalanine (F). The original 2D data had an f_2 resolution of 1.8 Hz/pt, but the 1D trace was treated as described in the text to yield a final resolution of 0.4 Hz/pt.

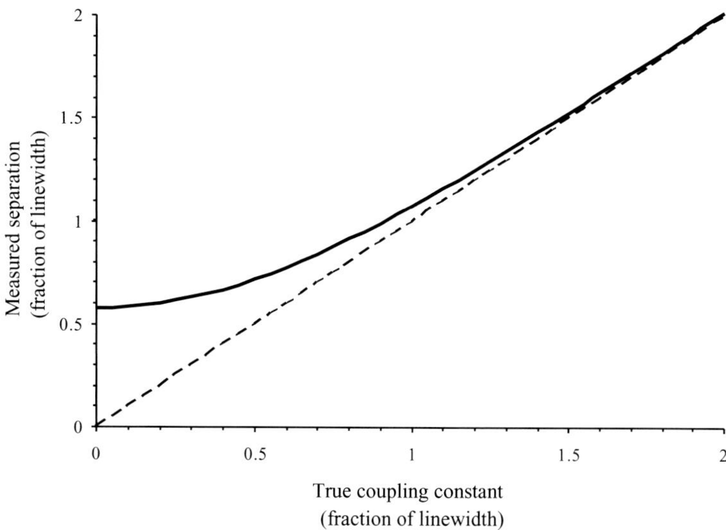

FIGURE 6.22 **Peak separations of anti-phase lines as a function of coupling constant and linewidth.** As coupling constants become relatively smaller, the measured splitting *(solid line)* deviates from the true value *(dashed line)*, producing measurements that are too large.

the αβ coupling constants in the tyrosine residue of the peptide Leu-enkephalin **6.5**. Above the 2D crosspeaks is the 1D trace through these, which has been treated as described here and zero-filled to produce the same digital resolution as the conventional 1D spectrum shown on top (0.4 Hz/pt). From the anti-phase and in-phase splittings the two active αβ couplings and the passive ββ coupling are readily measured as 6.1, 8.3 and 14.0 Hz.

The measurement of coupling constants in this manner is, however, limited by finite linewidths, which in 2D spectra of small to mid-sized molecules are usually dictated by the digital resolution of the spectrum rather than natural linewidths. Under such conditions, lines that are in phase tend to merge together to produce a single maximum which may mask the coupling whereas anti-phase lines remain separated but may produce apparent splittings that are greater than the true couplings. As coupling constants become smaller the splitting of anti-phase lines in fact tends to a *minimum* of ≈0.6 times the linewidth [6] (Fig. 6.22), after which further reductions in the true separation act only to reduce peak intensities until complete cancellation occurs and the crosspeak is lost. One may conclude therefore that it is not possible to measure reliably coupling constants that are significantly smaller than the observed linewidths after digitisation. A rule of thumb is that couplings should be greater than 1.5 times the *digitised* linewidth for measurements to be made, and spectra recorded for the measurement of coupling constants are typically acquired with higher digital resolution in f_2 where one can afford to be profligate with data points. These considerations also indicate that crosspeaks will disappear from COSY spectra if linewidths are large or if digital resolution is low relative to the active coupling constant. Thus, crosspeaks due to small couplings or to broad lines are more likely to be missing. This can be a serious problem for very large molecules that have naturally broad resonances, and in such cases the TOCSY experiment provides superior results since crosspeaks have only in-phase structures. More sophisticated methods also exist which enable the measurement of coupling constants in complex multiplets. For example the DISCO method [7–9] of post-processing relies on the addition and subtraction of rows from COSY spectra to produce simplified traces containing fewer signals, thus making measurements more reliable. Details of the various methods available shall not be addressed here, but have been reviewed [5].

Even when the data at hand preclude the measurement of coupling constants, analysis of the relative magnitudes of active and passive couplings within multiplets can prove enlightening when interpreting spectra. For example it is possible in favourable cases to determine the relative configuration of protons within substituted cyclohexanes from consideration of the crosspeak structures. Since *axial–axial* couplings in chair conformers are typically far greater than *axial–equatorial* or *equatorial–equatorial* couplings (*ax–ax* ≈ 10–12 Hz, *ax–eq*/*eq–eq* ≈ 2–5 Hz), they appear as large anti-phase splittings within crosspeak multiplets, indicative of the diaxial relationship between the correlated spins. This is illustrated in Fig. 6.23 which shows a region of the DQF-COSY spectrum of andrographolide **6.6**, a hepatoprotective agent found in traditional Indian herbal remedies. Three of the crosspeaks of proton H_C display large active couplings consistent with either geminal (H_d) or diaxial (H_a, H_e) relationships, while the much smaller active coupling to H_b limits these to being *ax-eq* or *eq-eq*. Such detail can be extracted from the 2D map even when multiplets are buried or are too complex for direct analysis in the 1D spectrum. For example correlations of H_d identify its geminal partner and indicate three vicinal *ax-eq* or *eq-eq*

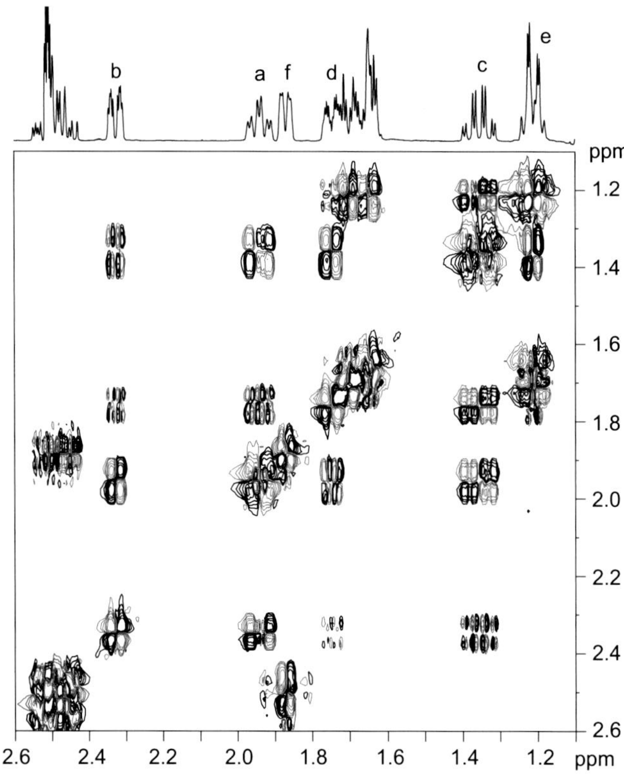

FIGURE 6.23 **A region of the DQF-COSY spectrum of andrographolide 6.6.** The data were collected under conditions of high f_2 resolution (1.7 Hz/pt) to reveal the coupling fine structure within the crosspeaks.

relationships with H_e, H_a and H_b. This type of information, which can only be determined reliably from phase-sensitive presentations, can often be used to good effect in stereochemical assignments, particularly when used in conjunction with the NOE.

HO Me H_e H_f H_a OH H_d HO Me H_b O O H_c

6.6

6.1.5.4 Higher Order Multiple-Quantum Filters

Using the same sequences as for DQF-COSY, but with modified phase cycles or gradient combinations, it is possible to filter for higher orders of multiple-quantum coherence between the last two pulses, so, for example a gradient ratio of 1:3 in the sequence of Fig. 6.13 selects triple-quantum coherence only. The triple-quantum filtered (TQF) COSY experiment eliminates responses from all singlets and two-spin systems leading to potential simplification. More generally, a p-quantum filter can be used to remove peaks arising from spin systems with fewer than p-coupled spins [3,10]. Such experiments have potential utility in the study of molecules containing well defined spin systems that are isolated from each other in the molecule, notably amino acids in peptides and proteins [11]. However, the great disadvantage of higher-order filtering is the reduction in sensitivity by a factor of 2^{p-1} for a p-quantum filter. While loss of a factor of 2 for the DQF-COSY relative to its unfiltered cousin might be tolerable, a factor of 4 for the TQF-COSY may make it untenable. Largely for this reason the TQF-COSY is rarely used and higher order filtering is essentially unheard of. The rules for interpreting multiplet fine structure in COSY recorded with high-order filtration also require some modification [11,12].

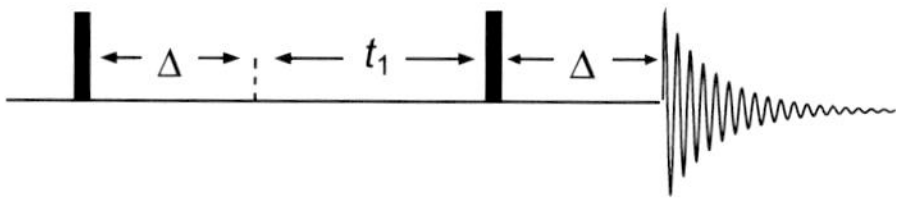

FIGURE 6.24 **The long-range (or delayed) COSY sequence.** Additional fixed delays are inserted into the basic COSY sequence to enhance the appearance of correlations from small couplings (see text).

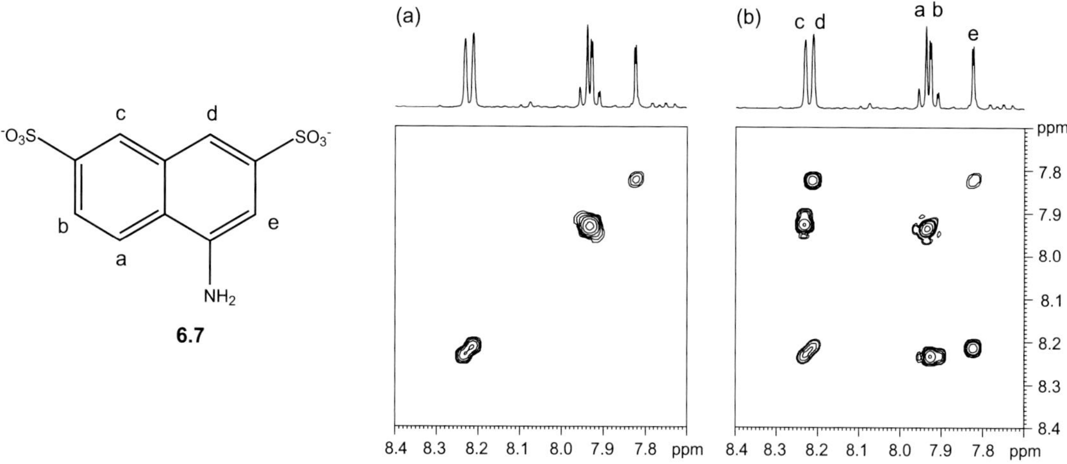

FIGURE 6.25 **Long-range COSY.** (a) The conventional COSY-90 spectrum and (b) the long-range COSY spectrum of **6.7.** Additional Δ delays of 200 ms were used in (b) while all other parameters were as for (a). The small (1 Hz) long-range couplings are not apparent in the conventional COSY experiment, but the correlations in (b) unambiguously provide proton assignments.

6.1.6 Long-Range COSY: Detecting Small Couplings

The COSY sequences encountered thus far are surprisingly effective at revealing small couplings between protons. However, such crosspeaks are typically rather weak, and their identification as genuine correlations may be difficult. The long-range COSY experiment (Fig. 6.24) enhances the intensity of correlations due to couplings that are smaller than natural linewidths such as those often encountered for coupling pathways over many bonds. Correlations over 4 or 5 bonds in rigid or unsaturated systems, such as those from *w*-, allylic or homoallylic couplings, can be observed readily with this experiment even though the coupling may not be resolved in the 1D spectrum (Fig. 6.25).

Previous discussions have stressed the need for an anti-phase disposition between multiplet vectors for coherence transfer to occur. In COSY, this arises from the evolution of spin–spin couplings during t_1 and develops at a slower rate the smaller the coupling. For the case of very small coupling constants, there is insufficient time for this to occur under the typical conditions employed for COSY acquisitions, and hence the crosspeaks from these couplings are weak or undetectable. It is possible to enhance the intensity of these peaks by adjusting the t_1 and t_2 domains such that they are optimised for detection of the coherence transfer signal from these smaller couplings. While this could in principle be achieved by extending both time periods and hence increasing digital resolution in both dimensions, this calls for greatly protracted experiments. The approach used in the long-range COSY sequence is to insert a *fixed* delay after each pulse to increase the *apparent* evolution times yet retain the same level of digitisation as in a standard COSY experiment. The experiment is therefore also known as *delayed-COSY*. The use of the additional delays precludes the use of phase-sensitive presentations because of the large phase distortions that arise, so absolute value processing is necessary.

For optimum detection of small couplings, the maximum of the coherence transfer signals should occur at the *midpoint* of each time domain. It may be shown that this maximum arises at a time t, where:

$$t = \frac{\tan^{-1}(\pi J T_2)}{\pi J} \tag{6.1}$$

When J is small such that $JT_2 << 1$ the approximation $t = T_2$ holds, and to ensure that the midpoint of each time domain coincides with this maximum, the additional delay Δ is therefore:

$$\Delta = T_2^{-1/2}\mathrm{AQ} \tag{6.2}$$

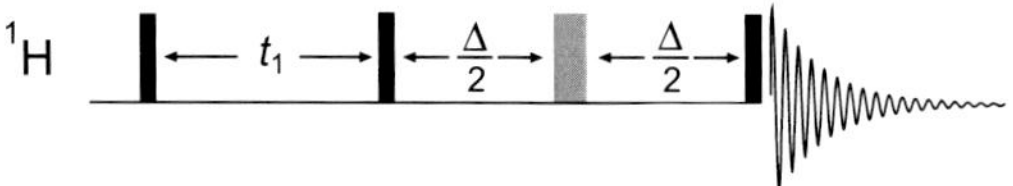

FIGURE 6.26 **The relayed-COSY sequence.** In this an additional transfer step has been appended to the basic COSY experiment, where $\Delta = 1/2J_{HH}$.

where AQ is total t_1 or t_2 acquisition time. For typical COSY acquisition parameters and typical proton T_2 values, Δ falls in the range 50–500 ms. Since the actual values of T_2 are not usually known in advance, and there will in any case exist a spread within the molecule, a compromise setting of around 200 ms should suffice in most cases. The crosspeaks arising from couplings of greater magnitude can be severely attenuated in this experiment, so it is invariably also necessary to record the standard COSY to establish the usual coupling relationships.

6.1.7 Relayed-COSY

The relayed-COSY experiment shall be considered only very briefly because it has essentially been superseded by the far superior TOCSY experiment described below, although one may still encounter references to this experiment in older literature. The relayed-COSY experiment (Fig. 6.26) attempts to overcome problems caused by coincidental overlap of crosspeaks in COSY that can lead to breakdown in the stepwise tracing of coupling networks within a molecule. It incorporates an additional coherence transfer step in which magnetisation transferred from a spin A to its partner M, as in the standard COSY, is subsequently relayed onto the next coupled spin X in the sequence. This produces a crosspeak in the spectrum between spins A and X *even though there exists no direct coupling between them*, by virtue of them sharing the common coupling partner M. Ambiguities from coincidental overlap of crosspeaks associated with spin M are therefore removed by providing direct evidence for A and X existing within the same spin system.

Relayed coherence transfer [13] is achieved by inserting a fixed time period Δ after the second pulse, again in the form of a spin-echo to remove chemical shift effects. During this, M–X coupling evolves to produce anti-phase multiplet vectors (the M–X coupling was passive in the initial A–M transfer, so these vectors are in-phase after the second 90 degree pulse). The third 90 degree pulse then elicits the M–X coherence transfer prior to detection. The fixed delay Δ should be set optimally to 1/2J to maximise this transfer, so a compromise setting of J = 7 Hz gives Δ = 70 ms. The direct responses observed in the usual COSY experiment may also be present, but with varying intensities.

6.2 TOTAL CORRELATION SPECTROSCOPY: TOCSY

The principal feature of all the COSY experiments described earlier is the direct correlation of homonuclear spins that share a scalar coupling. In proton spectroscopy, this typically provides the chemist with evidence for geminal and vicinal relationships between protons within a molecule. TOCSY [14] yields homonuclear proton correlation spectra based on scalar couplings, but is also able to establish correlations between protons that sit within the *same spin system*, regardless of whether they are themselves coupled to one another. In other words, provided there is a continuous chain of spin–spin coupled protons, A–B–C–D–, etc., the TOCSY sequence transfers magnetisation of spin A onto spins B, C, D, etc, by relaying coherence from one proton to the next along the chain. In principle, this can correlate all protons within a spin system, and this feature earns the title *total* correlation spectroscopy. The ability to relay magnetisation in this manner provides an enormously powerful means of mapping correlations by making even greater use of the additional dispersion found in two dimensions. This is particularly advantageous in cases of severe resonance overlap, for which COSY spectra can often leave ambiguities. The TOCSY spectrum for **6.2** is shown in Fig. 6.27, and has been recorded so as to provide multistep transfers around the carbohydrate ring. This should be compared with the COSY spectrum of the same material in Fig. 6.3. The additional relayed crosspeaks seen in TOCSY provide correlations between *all* protons within the ring system, and this ability to fully map spin systems is particularly advantageous when discrete units exist within a molecule, as illustrated shortly. A second significant feature of TOCSY that contrasts with COSY is that it utilises the net transfer of *in-phase* magnetisation, so does not suffer from cancellation of anti-phase peaks under conditions of low digital resolution or large linewidths. In these instances, this feature makes TOCSY the more sensitive of these two methods.

An essentially identical experiment has also been referred to as homonuclear Hartmann–Hahn (HOHAHA) spectroscopy [15,16] (the two differ only in some technical details in the originally published sequences). This name arises from its similarity with methods used in solid-state NMR spectroscopy for the transfer of polarisation from proton to carbon nuclei (so-called cross-polarisation), which are based on the Hartmann–Hahn match described later. For the same reason, the transfer of magnetisation during the TOCSY sequence is sometimes referred to as *homonuclear cross-polarisation*. Throughout this text the original TOCSY terminology is used, although TOCSY and HOHAHA have been used synonymously in the chemical literature.

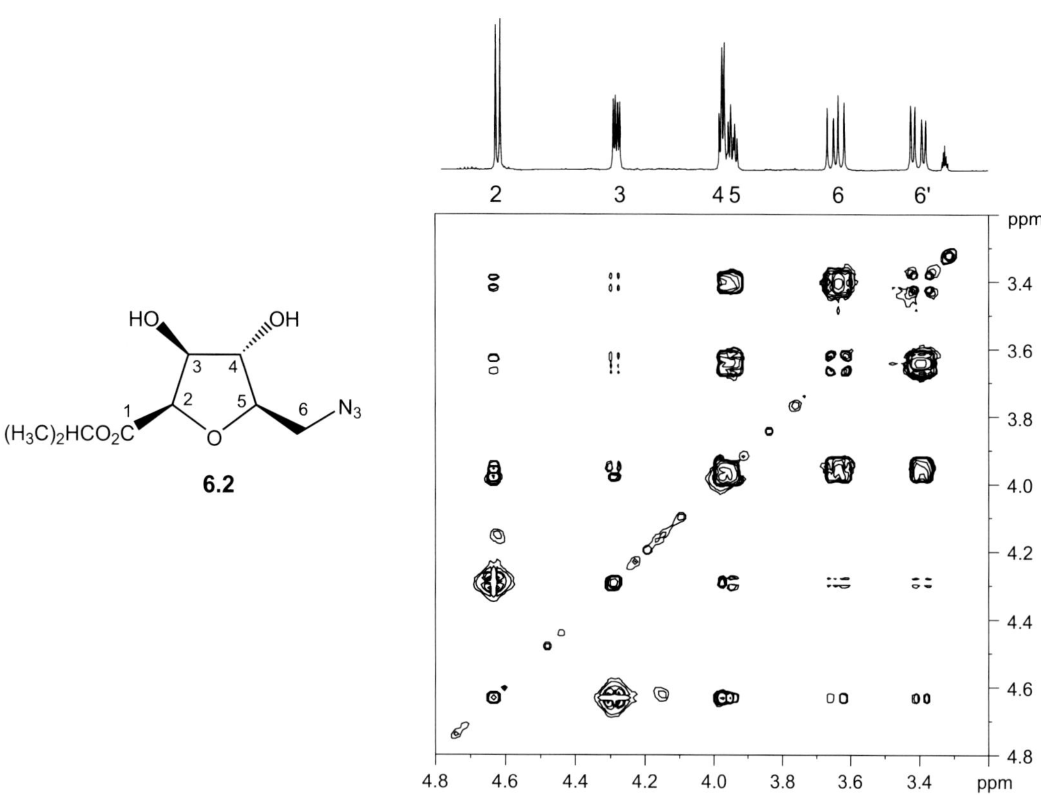

FIGURE 6.27 **The TOCSY spectrum of 6.2.** Relayed crosspeaks are apparent between spins that lack direct scalar couplings, but which exist within the same spin system and arise from the propagation of magnetisation along the chain of coupled spins. Compare this with the COSY spectrum in Fig. 6.3.

6.2.1 The TOCSY Sequence

The TOCSY sequence, represented in its most general form in Fig. 6.28, is rather similar to the COSY sequence already described, the only difference being the use of a mixing sequence in place of a single mixing pulse. This is known as a *spin lock* or *isotropic mixing* sequence (this terminology is explained below) and its purpose is to execute the relayed magnetisation transfer mentioned earlier. Details such as quadrature detection and axial peak suppression parallel those for COSY, so to understand the operation of the TOCSY experiment it only becomes necessary to appreciate the influence of the spin lock.

6.2.1.1 The Spin Lock and Coherence Transfer

To understand the action of the spin lock period in the TOCSY experiment, consider the sequence of events during the first transient of the 2D experiment (ie that with $t_1 = 0$). The experiment begins by exciting the spins with a $90°_x$ pulse such that they all lie along the +*y*-axis in the rotating frame. At this point the spin lock rf field is applied, this time along the +*y*-axis, parallel to the nuclear vectors. In its simplest form the spin-lock is a continuous low-power pulse of constant phase applied for a period of typically tens of milliseconds. It is convenient to imagine this to be composed of a continuous sequence of closely spaced $180°_y$ pulses bracketed by infinitely small periods δ (Fig. 6.29). Each δ–180–δ period constitutes nothing more than the homonuclear spin-echo described in Section 2.2.4 and, to picture the evolution of chemical shifts and

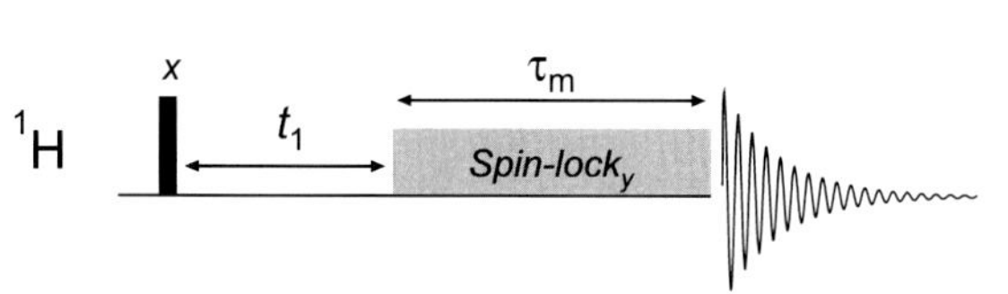

FIGURE 6.28 **The TOCSY sequence.** The spin-lock mixing time, τ_m, replaces the single mixing pulse of the basic COSY experiment.

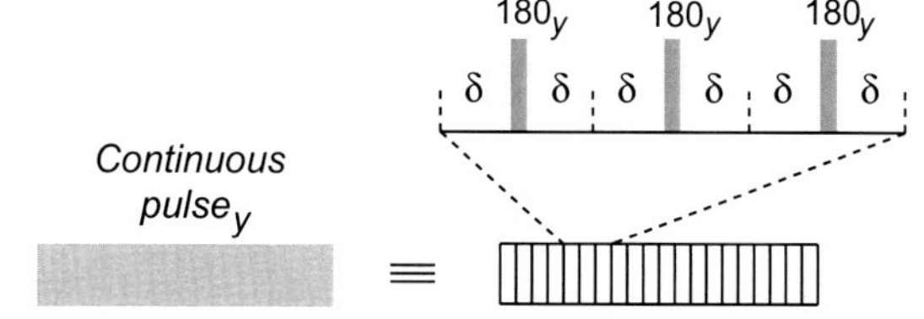

FIGURE 6.29 **The spin lock in its simplest form is a single, long, low-power pulse.** This can be viewed as a continuous sequence of closely spaced 180 degree pulses bracketed by infinitely small periods δ.

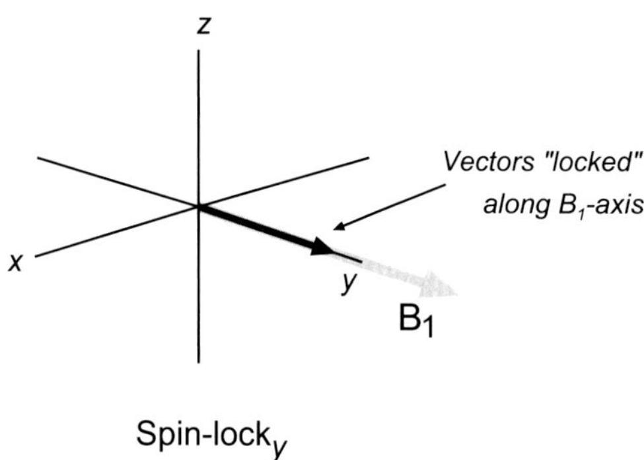

FIGURE 6.30 **The role of the spin-lock.** Vectors are held along the B_1 axis by the long rf pulse and are said to be *spin-locked* in the rotating frame.

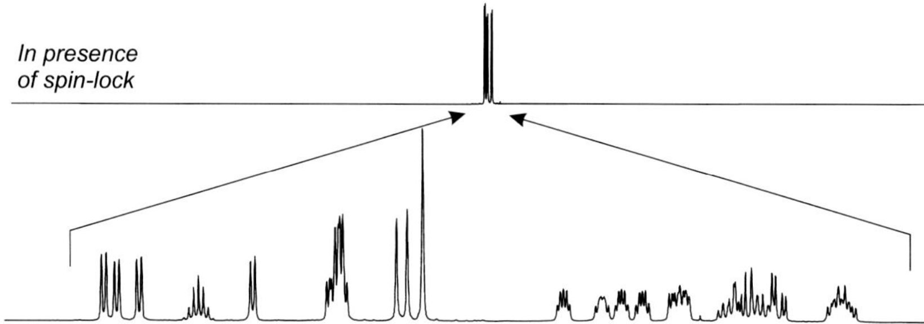

FIGURE 6.31 **A schematic illustration of events during spin lock mixing.** All chemical shift differences between spins are eliminated, yet all spin–spin couplings between them remain. This forces the *strong-coupling* condition on all spins (see text).

homonuclear spin–spin couplings during the spin lock, it is sufficient to consider events within a discrete δ–180–δ period. It has already been shown in Section 2.2.4 that a 180 degree pulse refocuses the evolution of chemical shifts, so after each δ–180–δ period no shift evolution has accrued and all spin vectors remain along the +*y*-axis. Extending this argument for the whole of the mixing period, it is apparent that no net chemical shift evolution occurs during τ_m, and hence the nuclear vectors are said to have been *spin-locked* in the rotating frame along, in this case, the *y*-axis (Fig. 6.30). During spin lock mixing all protons have experienced the same effective field and hence the same chemical shift offset (ie zero) in the rotating frame. Section 2.2.4 has also shown that, in contrast, *homonuclear* spin–spin couplings *continue* to evolve following a 180 degree pulse, so that throughout the spin lock J couplings behave as they would for a period of free precession. In summary, during spin lock mixing all chemical shift differences in the rotating frame are removed yet spin–spin couplings remain active (Fig. 6.31).

Nuclei which share a coupling and have very similar (or coincident) chemical shifts relative to their coupling constant are said to be *strongly coupled*, and under such conditions they lose their unique identity and ultimately become indistinguishable. Thus, for a system containing a coupled AB pair, it is not possible to consider the interactions of spin A independently from those of its strongly coupled neighbour B and vice versa. Such a strong coupling condition is *forced* upon all spin systems by application of the spin lock field since the protons remain coupled yet all possess the same effective chemical shift. The protons therefore lose their 'unique' identity during the mixing period, and this provides the mechanism by which coherence may be shared over all spins within the same spin system. Under these conditions, there exists an oscillatory exchange of coherence between protons during the spin lock which, for an AX system, results in complete transfer from A to X after a period of $1/2J_{AX}$ s [14], and a return to A after $1/J_{AX}$ s (Fig. 6.32). Although events become more complex for larger systems, the general idea of an oscillatory exchange of coherence between spins still holds, leading to a propagation of magnetisation along the chain of coupled spins [17] (Fig. 6.33). For short mixing times in proton experiments of around 20 ms, only single-step transfers have significant intensity and the correlations seen are equivalent to those seen in COSY. Longer mixing periods enable magnetisation to propagate further along the coupled chain and relay peaks arising from multistep transfers appear, as seen in Fig. 6.27. Small couplings lead to poor transfer efficiencies and can lead to a breakdown in the transfer and to loss of relay peaks. Since magnetisation may travel in either direction along a spin chain, 2D TOCSY spectra are again symmetrical about the diagonal.

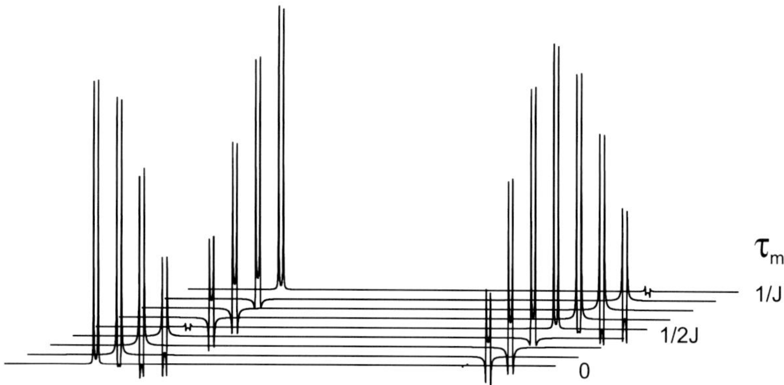

FIGURE 6.32 The oscillatory transfer of magnetisation between a coupled proton pair under the influence of a spin lock. The transfer was simulated using an MLEV-17 spin lock (see text) of increasing duration in each experiment. In each case this was applied immediately after selective excitation of the high-frequency doublet with a pure-phase EBURP-2 selective 90 degree pulse (see the chapter *Experimental Methods*).

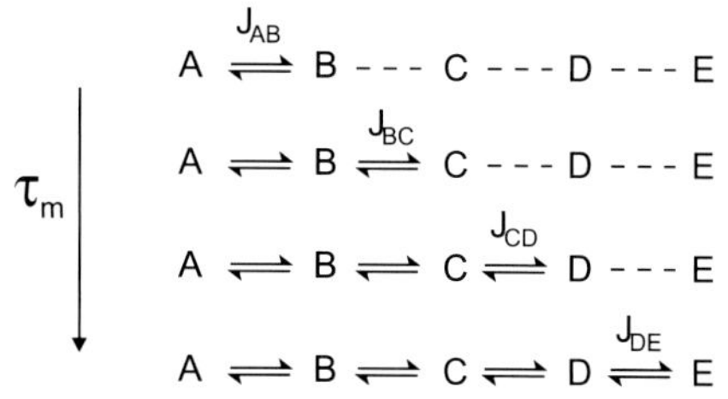

τ_m

$A \overset{J_{AB}}{\rightleftharpoons} B - - - C - - - D - - - E$

$A \rightleftharpoons B \overset{J_{BC}}{\rightleftharpoons} C - - - D - - - E$

$A \rightleftharpoons B \rightleftharpoons C \overset{J_{CD}}{\rightleftharpoons} D - - - E$

$A \rightleftharpoons B \rightleftharpoons C \rightleftharpoons D \overset{J_{DE}}{\rightleftharpoons} E$

FIGURE 6.33 Schematic illustration of the propagation of magnetisation along a chain of coupled spins (A–E) as a function of the spin lock mixing time (τ_m). The arrows indicate the cyclic exchange of coherence between nuclei.

The requirement for nuclei to experience identical local fields during the mixing time for transfer to occur between them is also referred to as the *Hartmann–Hahn match*. More formally and more generally, this may be expressed as:

$$\gamma_A B_{1A} = \gamma_X B_{1X} \tag{6.3}$$

where B_{1A} and B_{1X} are the B_1 fields experienced by spins A and X and γ_A and γ_X are their magnetogyric ratios, respectively. Inclusion of the γs in this means the nuclides involved in the exchange need not be the same, so, for example transfer can be initiated between proton and carbon if simultaneous rf fields that satisfy Eq. (6.3) can be applied to each nuclide. This, by analogy, leads to hetero-TOCSY or, more amusingly, heteronuclear Hartmann–Hahn (HEHAHA) spectroscopy as a possible means of establishing heteronuclear correlations [18,19], although this is rarely used in solution-state NMR.

6.2.2 Applying TOCSY

The ability of TOCSY to propagate magnetisation along a spin system has meant it has become a popular tool for studying molecules that are comprised of discrete and often well-defined units. Obvious examples of this are proteins, peptides and oligosaccharides, for which it is often possible to trace the whole amino acid or sugar ring system from a single resolved proton. This is illustrated in the TOCSY spectrum of gramicidin-S, **6.8**, a cyclic decapeptide (Fig. 6.34). The aliphatic region only is shown, in which all sidechain protons of each residue can be traced from the associated α-proton (with the exception of the phenylalanine aromatic ring), as illustrated for the proline and ornithine residues. The relaying of magnetisation along a coupled chain of spins can often overcome problems due to close or coincident crosspeaks in COSY by effectively moving related correlations into regions of the spectrum that would otherwise be devoid of crosspeaks, therefore utilising redundant 'space' in the 2D spectrum. This approach was used in the sidechain assignment of **6.9**, an intermediate in the biomimetic synthesis of members of the manzamine and keramaphidine alkaloids [20,21], found naturally in marine sponges. Due to extensive crowding in some areas of the aliphatic region, sidechain assignments were more readily derived from relayed correlations involving resolved alkene protons such as H23 (Fig. 6.35). Similarly, the numerous correlations from H4 provided confirmation of assignments derived from a DQF-COSY spectrum.

The somewhat indiscriminate propagation of magnetisation found with TOCSY can also be disadvantageous, however, since it is generally not possible to state explicitly the number of relay steps involved in the production of any given

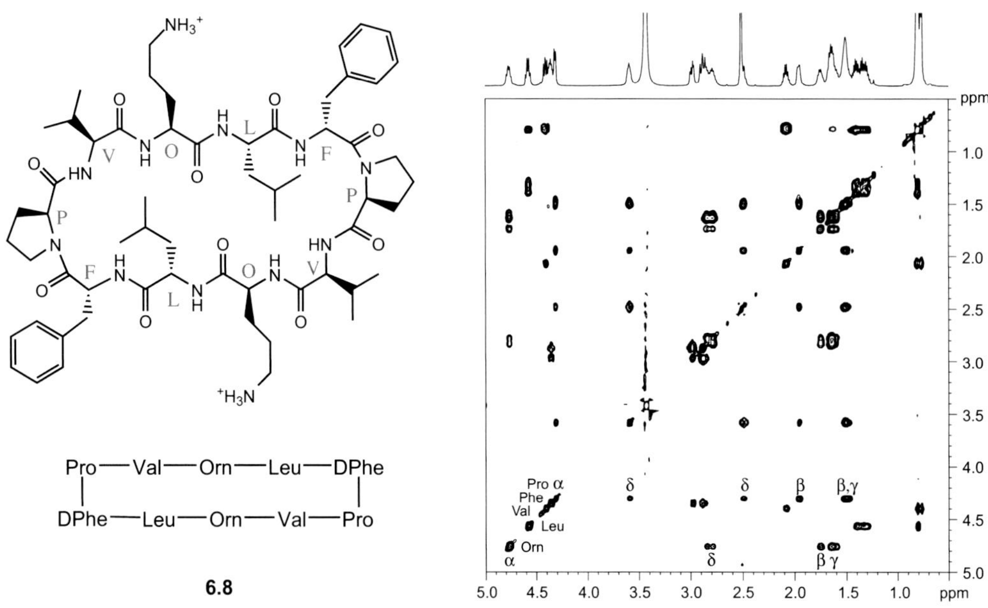

FIGURE 6.34 **The aliphatic region of the TOCSY spectrum of the cyclic decapeptide gramicidin-S 6.8.** An 80 ms spin lock was used, providing complete transfer along the aliphatic sidechains of each residue. Correlations from the α-protons of proline and ornithine are labelled, with those out to the δ-protons corresponding to three-step transfers.

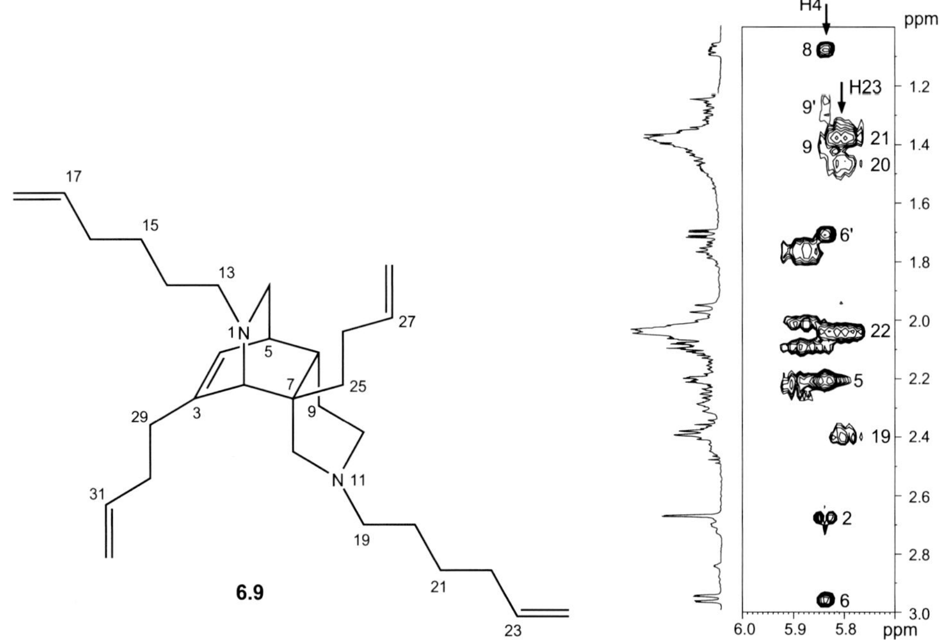

FIGURE 6.35 **A section of the TOCSY spectrum (τ_m = 80 ms) of the biomimetic intermediate 6.9 alongside the conventional 1D spectrum.** Correlations from protons H4 and H23 are labelled. Despite extensive overlap in some regions of the 1D spectrum, the numerous relayed correlations in TOCSY occur in an otherwise clear region of the 2D spectrum and provide ready identification of proton shifts.

correlation. A time dependence of crosspeak buildup from a number of TOCSY spectra recorded with increasing τ_m (in, eg 20 ms steps; Fig. 6.36) can provide an approximate indication of whether the peak arises from one, two or more steps, although it is often not possible to be more precise than this. When using TOCSY on molecules that are not composed of well-characterised monomer units, it is often beneficial to use this in addition to COSY spectra from which more immediate coupling neighbours can be identified. The use of very long spin lock periods for promoting multistep transfer also leads to sensitivity losses due to relaxation (in the rotating frame) of spin-locked magnetisation.

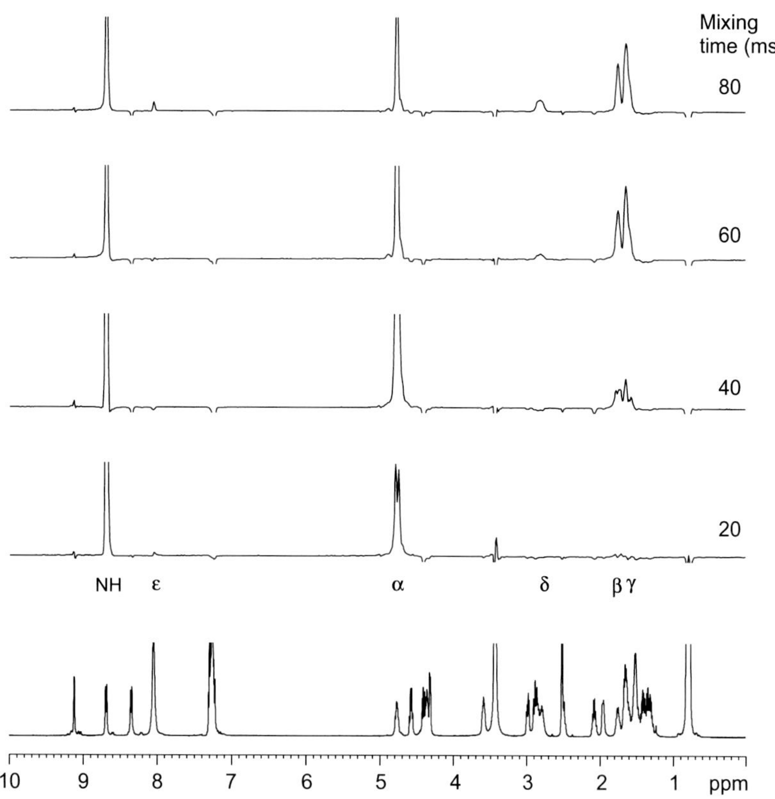

FIGURE 6.36 **The time-dependent propagation of magnetisation along a proton spin system.** This is illustrated for the ornithine residue of gramicidin-S 6.8 where are extracted from 2D TOCSY spectra at the amide NH proton shift and show progressive transfer along the sidechain as the mixing time increases.

It has also been mentioned that TOCSY results in the net transfer of *in-phase* magnetisation, meaning the cancellation effects from *anti-phase* multiplet fine structure associated with COSY are not a feature of TOCSY. Such cancellation can be problematic for molecules that possess large natural linewidths for example (bio)-polymers, but may also prevent the observation of COSY peaks in the spectra of small molecules that have complex multiplet structures which may cancel under conditions of poor digital resolution. In these cases the TOCSY experiment may be viewed as the more sensitive option because of the greater crosspeak intensities. The lack of anti-phase structure also means spectra may be recorded very rapidly under conditions of rather low digital resolution without a loss of peak intensity. Finally, since both diagonal and crosspeaks possess the same phase, both produce double-absorption lineshapes meaning high-resolution phase-sensitive spectra can be obtained.

One side-effect of the spin lock is that it also enables *incoherent* magnetisation transfer in the rotating frame. In other words, through-space effects known as rotating frame NOEs (ROEs) may, in principle, also be observed between spins. These tend to be far weaker than the TOCSY peaks, particularly for small molecules and short mixing times, so are rarely problematic. They may be more of a problem for very large molecules in which they are stronger and faster to build up. For such molecules so-called 'clean-TOCSY' sequences have been developed [22–24] which eliminate ROE peaks by cancelling them with equal but opposite NOE peaks; such methods are only applicable to very large molecules where this difference in sign exists. A far greater problem is the appearance of unwelcome TOCSY peaks in ROE spectroscopy (ROESY), a topic also addressed in Section 9.7.

6.2.3 Implementing TOCSY

Although in principle the simple scheme presented in Fig. 6.28 should produce TOCSY spectra, its suitability for practical use is limited by the effective bandwidth of the continuous-wave spin lock. Protons whose chemical shifts are off-resonance from the applied frequency will experience a reduced rf field causing the Hartmann–Hahn match to break down and transfer to fail. This is analogous to the poor performance of an off-resonance 180 degree pulse (Section 3.2.1). The solution to these problems is to replace the continuous-wave spin lock with an extended sequence of 'composite' 180 degree pulses which extend the effective bandwidth without excessive power requirements. Composite pulses themselves are described in Section 12.1

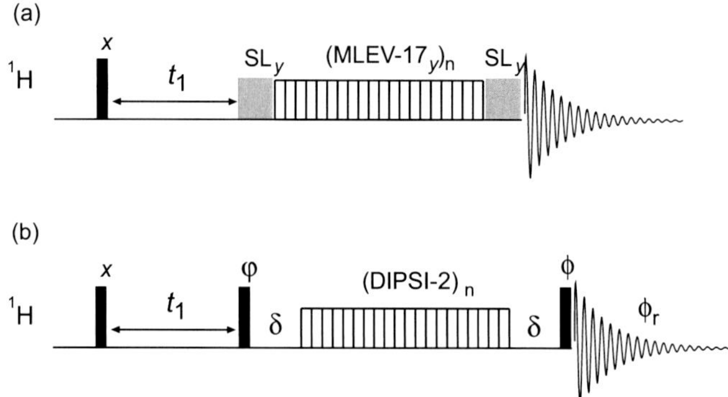

FIGURE 6.37 **Practical schemes for implementing TOCSY.** These are based on (a) the MLEV-17 mixing scheme and (b) the DIPSI-2 isotropic mixing scheme. The MLEV sequence is bracketed by short, continuous wave, spin lock trim pulses (SL) to provide pure-phase data. In scheme (b) this can be achieved by phase-cycling the 90 degree *z*-filter pulses that surround the mixing scheme. This demands the independent inversion of each bracketing 90 degree pulse with coincident receiver inversion, thus $\varphi = x, -x, x, -x$; $\phi = x, x, -x, -x$ and $\phi_r = x, -x, -x, x$. The δ periods allow for the necessary power switching.

alongside the common mixing schemes employed in TOCSY, so shall not be discussed here. Suffice it to say at this point that these composite pulses act as more efficient broadband 180 degree pulses within the general scheme of Fig. 6.29.

There are essentially two approaches based on composite pulse methods in widespread use for the practical implementation of the TOCSY experiment (Fig. 6.37). The first of these [16] (Fig. 6.37a) is based on the so-called 'MLEV-17' spin lock, in which an even number of cycles through the MLEV-17 sequence are used to produce the desired total mixing period. To ensure the collection of absorption-mode data, only magnetisation along a single axis should be retained, so it is necessary to eliminate magnetisation not parallel to this before or after the transfer sequence. In this implementation, this is achieved by the use of 'trim pulses' applied for 2–3 ms along the chosen axis. These are periods in which a single, low-level pulse is applied, during which magnetisation not parallel to this axis will dephase due to inhomogeneity of the rf field (just as transverse magnetisation rapidly dephases in an inhomogeneous static field). This process eliminates dispersive contributions to the spectrum.

The alternative approach is to regenerate longitudinal magnetisation following the evolution period and use a suitable *isotropic mixing* scheme to transfer this *z* magnetisation between spins [25,26] ('isotropic' here means that magnetisation transfer is equally effective along the *x*-, *y*- or *z*-axes, so the overall sequence has a similar effect to that described earlier). The *z* magnetisation is then reconverted to pure-phase transverse magnetisation following the mixing, enabling the collection of absorption-mode data. The idea of generating and subsequently selecting *z* magnetisation with the view to collecting pure-phase spectra has been termed '*z* filtration' [27]. This also has the advantage for older instruments that the proton 'decoupler' may be used for generating the low-power mixing scheme [25] if the main proton transmitter is incapable of suitable power-level control; this restriction does not arise on modern instruments. A number of isotropic mixing schemes may be used in this approach, including MLEV-16, WALTZ-16 [28,29] and DIPSI-2 [30,31], the last of these being particularly well suited for transfer in ^{1}H–^{1}H TOCSY.

In either of these two general schemes, a greater effective bandwidth is achieved with composite pulse mixing and significantly lower powers are required for this than for the 90 degree preparation pulse. For MLEV, the rf field strength need be about twice the desired bandwidth, while for the more efficient DIPSI-2 it may be about equal to it. Thus, a 10 ppm window at 400 MHz requires $\gamma B_1 \approx 8$ kHz (90 degree pulse ≈ 30 μs) or 4 kHz (90 degree pulse ≈ 60 μs), respectively, either of which can be achieved comfortably.

6.2.3.1 Gradient-Selected TOCSY

For samples of sufficient strength to require only a single scan per increment, gradient versions of TOCSY may be attractive alternatives for the rapid recording of spectra. By analogy with the previous COSY discussions, the absolute value TOCSY sequence simply requires equal gradients to be placed either side of the spin lock sequence (Fig. 6.38a) to selectively refocus the N-type pathway to provide f_1 frequency discrimination. Trim pulses are not required since pure-phase spectra are not produced by this method. As for absolute value COSY, this experiment may be well suited to automated acquisition schemes because of the simplicity of processing, although in general the phase-sensitive version is preferred for reasons of resolution. This may be performed by placing the gradients within spin-echoes to refocus shift evolution. Quad-detection is then afforded by collecting P- and N-type data separately for each t_1 increment via gradient inversion, and processing these according to the echo–antiecho procedure of Section 5.4.2 (Fig. 6.38b).

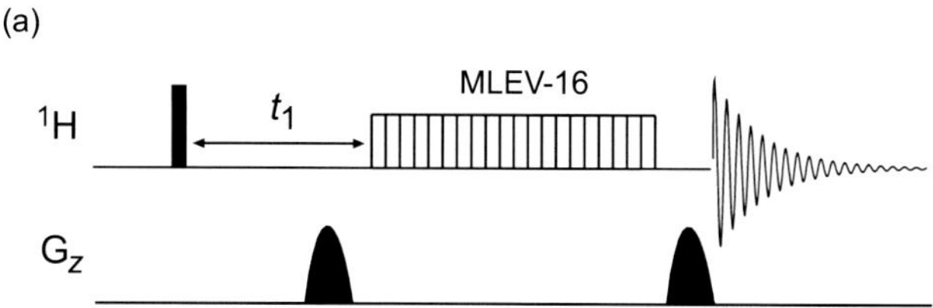

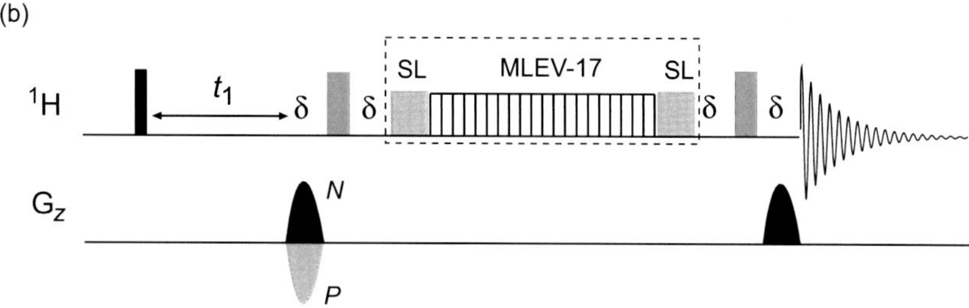

FIGURE 6.38 **Gradient-selected TOCSY.** Sequence (a) is suitable for absolute value presentations with a 1:1 gradient combination selecting the N-type spectrum. Sequence (b) provides phase-sensitive data sets via the echo–antiecho method for which separate P- and N-type data are collected through inversion of the first gradient. The boxed section may contain any suitable mixing element including the 90 degree–DIPSI2–90 scheme described above.

6.2.4 One-Dimensional TOCSY

As an alternative to the collection of a full 2D data set, the 1D analogue can prove advantageous in some circumstances. In general, most 2D sequences can be adapted to produce the 1D equivalent [32,33], meaning the experiments can be quicker to acquire and can be recorded under conditions of higher digital resolution. These may prove particularly attractive when only specific information is required, as is often the case for small- to medium-sized molecules, or when small sample quantities demand many scans for adequate signal-to-noise and so preclude the collection of 2D data sets. All 1D analogues of multidimensional experiments begin by selectively exciting one resonance in the spectrum and using this as the source for all subsequent magnetisation transfer. The resulting spectra may therefore be viewed as being equivalent to a high-resolution row through the related 2D spectrum at the shift of the excited spin. Selective excitation methods are described in Section 12.4, so are not addressed here. Numerous methods for generating 1D TOCSY spectra without [34–37] or with [38–41] gradient assistance have been presented, a generalised scheme using as an example the MLEV-17 sequence being shown in Fig. 6.39. Following the excitation of the chosen spins, magnetisation transfer is initiated immediately with the mixing scheme since there is now no need for the evolution period. The only resonances appearing in the resulting spectra are those that have 'received' magnetisation from the source spin.

The experiment produces 1D subspectra for discrete spin systems within the molecule, potentially revealing multiplet structures that were otherwise overlapped or buried. Furthermore, the higher digital resolution afforded by the 1D analogue may resolve ambiguities arising from crosspeak overlap in 2D spectra and may also provide a more rapid method for following magnetisation transfer along a chain of spins as a function of mixing time. Fig. 6.40 shows 1D TOCSY spectra for the carbopeptide **6.10** in which transfer is from the selected amide protons, with the individual ring systems resolved in each case. This approach is often applicable to the study of oligosaccharides where the anomeric protons can provide a convenient starting point from which to establish intraresidue assignments.

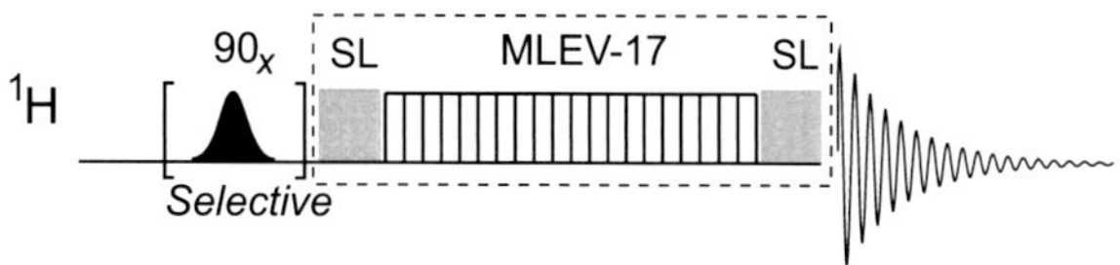

FIGURE 6.39 **The general sequence for 1D selective TOCSY.** Any suitable selective 90 degree pulse scheme (see Section 12.4) can be used to selectively excite the target resonance from which transfer is initiated. The boxed section can contain any suitable mixing element.

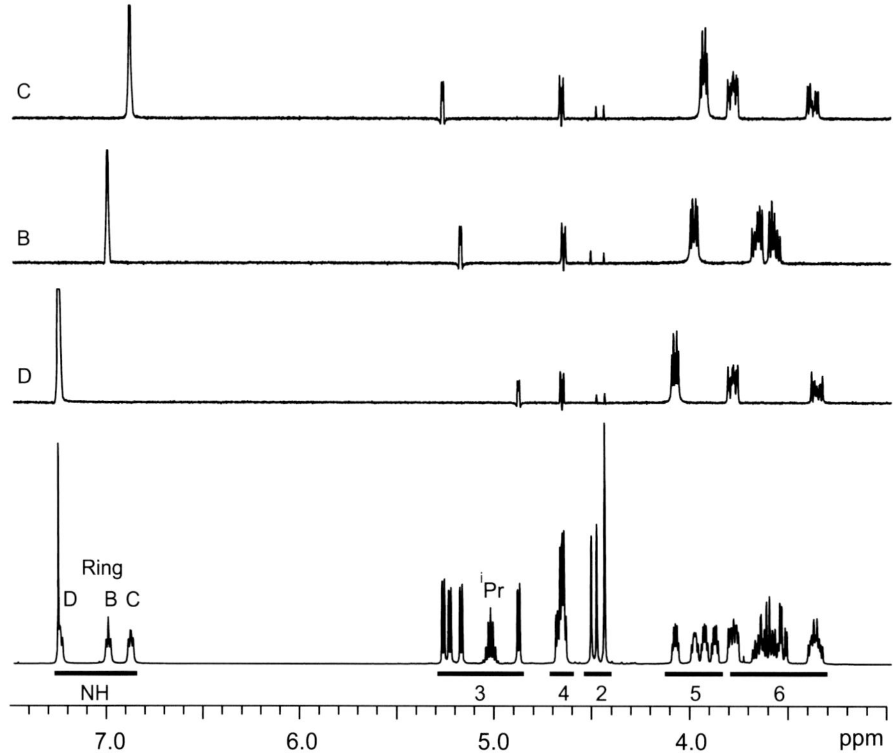

FIGURE 6.40 **1D TOCSY spectra of the tetrameric carbopeptoid 6.10 in $CDCl_3$.** Each amide proton was selectively excited and used as the starting point for coherence transfer. Selective excitation was achieved with the excitation sculpting method of Section 12.4 and mixing used a 97-ms spin lock.

6.2.4.1 Eliminating Zero-Quantum Artefacts

The ability to reveal otherwise hidden multiplet structures with the 1D TOCSY experiment can prove to be extremely beneficial to spectrum analysis, but a faithful representation of multiplet structures can be hampered by undesirable artefacts arising from zero-quantum coherences (a form of transverse magnetisation; Section 5.3) generated prior to and/or during the isotropic mixing scheme employed. While unobservable, these coherences can be transferred into observable signals by subsequent rf pulses, and produce responses in spectra that have an anti-phase coupling structure (akin to those in DQF-COSY described in Section 6.1.5) and are orthogonal to the desired TOCSY peaks, meaning these also appear as dispersive contributions when the desired peaks are phased to have the usual absorption lineshape. The zero-quantum contributions lead to lineshape distortions of TOCSY peaks in both 2D and 1D spectra and, while undesirable in both, these can present a greater nuisance in 1D spectra where the high-resolution analysis of multiplet structure is most useful (Fig. 6.41b). Thus, some effort has gone into designing methods to eliminate these contributions, and two methods are presented here in the context of the TOCSY experiment. The first does not require pulsed field gradients while the second,

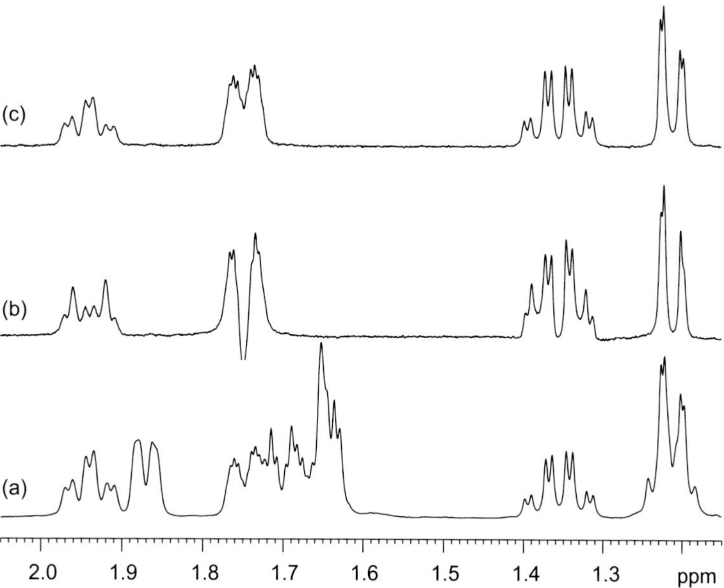

FIGURE 6.41 **Zero-quantum interference in selective 1D TOCSY spectra.** (a) Partial ^{1}H spectrum and the corresponding regions of 1D TOCSY recorded with 80 ms DIPSI-2 (b) without zero-quantum suppression and (c) with the zero-quantum dephasing scheme shown in Fig. 6.43 employing adiabatic smoothed CHIRP pulses with 40 kHz sweep widths and durations of 15 and 10 ms.

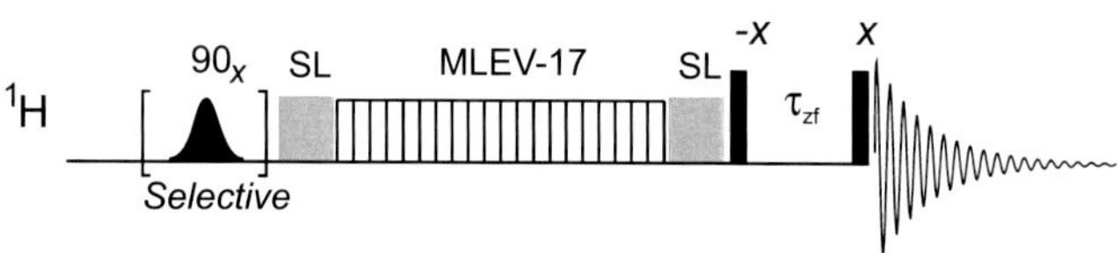

FIGURE 6.42 **The *z*-filter scheme for suppressing zero-quantum contributions in 1D TOCSY.** The delay τ_{zf} is randomly varied between experiments and the resulting spectra co-added. When using DIPSI-2 the two δ periods either side of the isotropic mixing scheme can act directly as independent τ_{zf} periods and may be randomised in a similar manner.

the most recent and successful approach, does. The elimination of these responses is a complex task given that zero-quantum coherences are insensitive to both phase-cycling routines and PFGs. The available approaches are, however, quite general and can be applied to any sequence in which these coherences are troublesome, so the techniques themselves are described more fully in Section 12.6.

The simplest approach to zero-quantum suppression employs a *z*-filter scheme in which the desired TOCSY magnetisation is placed along the *z*-axis for a defined period τ_{zf} before being returned to the transverse plane for detection (Fig. 6.42). Repetition of the experiment with random variation of the τ_{zf} period leads to differential evolution of the zero-quantum coherences such that the summation of many experiments leads to a destructive addition of the zero-quantum responses, as described in Section 12.6. The desired magnetisation, being longitudinal during the *z* filter, is not influenced by this variation and so adds constructively. Typically, the τ_{zf} delay varies over a 2–20 ms range and requires the addition of approximately 10 experiments for acceptable suppression. The need for this repetition is clearly a major limitation of this technique, meaning it may be applicable to selective 1D TOCSY experiments, but essentially impractical for 2D experiments. Even then, it may demand extended acquisition periods beyond what is required to satisfy signal-to-noise requirements.

A more sophisticated approach achieves suppression *in a single scan* and thus has wider applicability, being equally useful for 1D and 2D experiments [42]. This incorporates a method for zero-quantum dephasing based on the simultaneous application of a PFG and a swept frequency 180 degree inversion pulse during a *z*-filter delay, as described in Section 12.6. Implementation with the DIPSI-2 isotropic mixing scheme is illustrated in Fig. 6.43, again for the 1D version. The dephasing element is applied either side of the isotropic mixing scheme with different total durations such that the dephasing generated by the first is not accidentally refocused by the second. The effectiveness of the scheme is illustrated in Fig. 6.41 where appreciable lineshape distortions are seen without the use of the dephasing element whereas clean multiplet structures matching those observed for the 1D spectrum are revealed with this included.

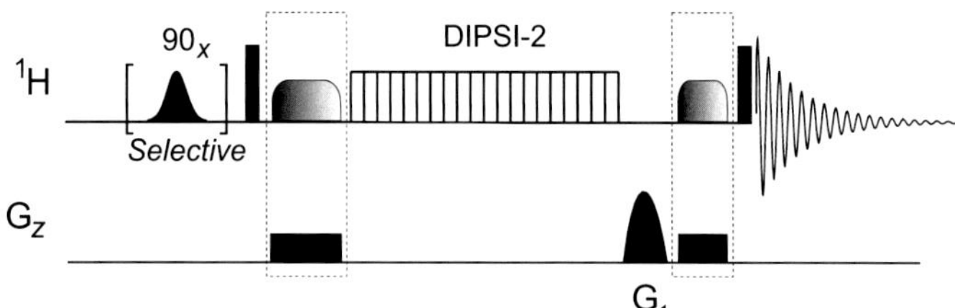

FIGURE 6.43 **The zero-quantum dephasing scheme applied to TOCSY.** The boxed regions contain the dephasing elements in which the gradients are applied during the swept inversion pulse; the two elements are of different durations to avoid accidental refocusing. G_1 represents a purge gradient.

6.3 CORRELATING DILUTE SPINS: INADEQUATE

One of the major goals in determining the structure of an organic molecule is the unambiguous identification of the carbon skeleton of the compound. Carbon–carbon connectivities are, however, rarely determined directly, rather we imply their presence from the correlations observed between protons, such as in COSY-type experiments, or between proton and carbon nuclei, such as from the long-range heteronuclear correlation experiments described in Section 7.4. The need for one to take this indirect approach lies in the relatively poor sensitivity and low natural abundance (1.1%) of the carbon-13 isotope. The abundance of molecules containing two (adjacent) ^{13}C spins is only 0.01%, so in attempting to directly identify ^{13}C–^{13}C linkages, we are forced to observe only 1 in every ~10,000 molecules in our sample (in the absence of isotopic enrichment). In other words, one must detect the ^{13}C satellites in the ^{13}C spectrum itself, and clearly larger sample quantities and extended acquisition periods are required for this. Despite these limitations, and because the ability to directly trace the carbon skeleton of a molecule is the ultimate approach to organic structure elucidation, considerable effort has gone into developing methods to identify C–C connectivities directly, all generally based on the incredible natural abundance double quantum transfer experiment (INADEQUATE) sequence [43,44]. These techniques are suitable for correlating nuclides that exist with relatively low natural abundance levels of around 1–20%, including ^{29}Si, ^{119}Sn and ^{183}W (4.7, 8.6 and 14.4%, respectively) [44]. More recent developments have incorporated proton detection to enhance sensitivity, principally within the realm of organic structure elucidation, and these methods will also be described later.

Despite the power of these techniques for structure elucidation, they are certainly not the first to turn to when considering a structural problem because of the large sample quantities typically required, but in favourable cases may prove effective, and with molecules that possess very few or even no protons, one may have few other options. In addition, the more widespread availability of cryogenic probes is likely to extend the range of applications for these techniques for both the ^{13}C and ^{1}H–detected approaches.

6.3.1 Two-Dimensional INADEQUATE

The principal complication when attempting to correlate low-abundance spins (commonly referred as *dilute* or *rare* spins) is the interference caused by the dominant parent resonances which lack homonuclear couplings and cannot therefore provide connectivity data. The more popular correlation techniques such as COSY, while suitable in principle, are not well suited in these cases and the INADEQUATE sequence is more appropriate. The basis of INADEQUATE is double-quantum filtration to suppress the uninformative parent resonances [45], this filtration being analogous to that which we have already encountered in the DQF-COSY experiment (Section 6.1.5). Natural abundance carbon-13 spins are very well suited to such filtration because they are restricted to AX or AB two-spin systems only (because a molecule containing more than two ^{13}C centres will exist at such low abundance as to be non-existent) and because $^{1}J_{CC}$ couplings cover a limited range, meaning the sequence can be reasonably well optimised for the generation of double-quantum coherence. Furthermore, one-bond C–C coupling constants are typically an order of magnitude greater than those operating over two or three bonds, meaning the experiment can also be optimised to detect only the short-range couplings indicative of direct carbon–carbon connectivity.

The two-dimensional correlation spectrum [46] is generated by allowing the double-quantum coherences associated with coupled spins to evolve during a variable t_1 period, following which they must be reconverted to single-quantum carbon magnetisation for observation. The complete INADEQUATE sequence is given in Fig. 6.44.

The generation of double-quantum coherence requires an anti-phase disposition of coupling vectors, which here develop during a period $\Delta = 1/2J_{CC}$. This is provided in the form of a homonuclear spin-echo to make the excitation independent of chemical shifts. The t_1 period represents a genuine double-quantum evolution period in which these coherences evolve at the *sums* of the rotating frame frequencies of two-coupled spins, that is at the sums of their *offsets* from the transmitter frequency. Thus, for two-coupled spins A and X that have offsets of υ_A and υ_X, the f_1 crosspeak frequency will be $(\upsilon_A + \upsilon_X)$.

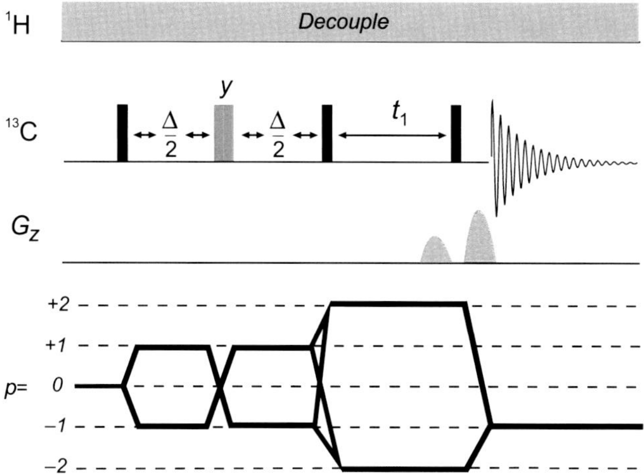

FIGURE 6.44 **The INADEQUATE sequence and the corresponding coherence transfer pathway.** The experiment selects double-quantum coherence during the evolution period with a suitable phase cycle or optionally via gradient selection (shown greyed). In doing so it rejects all contributions from all uncoupled spins.

Following the reconversion of double-quantum into single-quantum coherence, crosspeaks in f_2 will be observed with the conventional chemical shift frequencies of A and X. Throughout the sequence, broadband proton decoupling can be applied to prevent evolution of heteronuclear couplings and to generate the ^{1}H–^{13}C nuclear Overhauser enhancement to aid sensitivity. PFGs may also be incorporated for more complete suppression of the parent ^{12}C resonances for which a 1:2 ratio provides double-quantum filtration.

The resulting spectrum, here illustrated for *n*-butanol **6.11** (Fig. 6.45), differs considerably in appearance from the other correlation spectra encountered in this chapter, although again the f_2 dimension represents the chemical shifts of each participating spin. Since two-coupled spins share the same double-quantum frequency in f_1, correlations are made by following horizontal traces parallel to f_2. Carbon connectivity is therefore established by a sequence of vertical and horizontal steps. If the double-quantum excitation sequence is optimised for one-bond C–C couplings, then each step identifies adjacent carbon centres in the molecule. The spectra display no diagonal peaks, although the midpoints of correlated spin pairs appear along a 'pseudo' diagonal (the double-quantum diagonal) of slope 2, that is where $f_1 = 2f_2$. This characteristic symmetry can be useful in differentiating genuine responses from artefacts. Each crosspeak is composed of a doublet with a splitting of J_{CC} arising from the AX (or AB) fine structure.

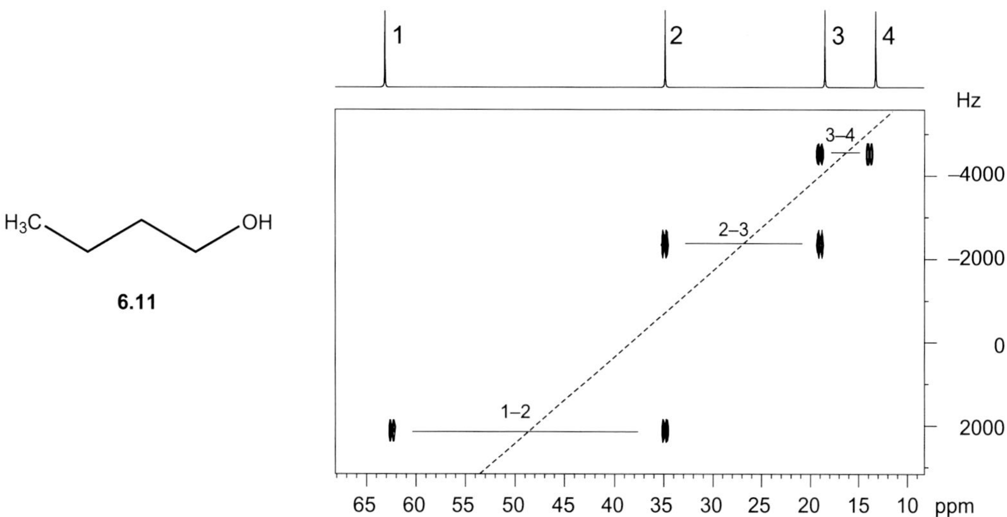

FIGURE 6.45 **The 2D INADEQUATE (magnitude mode) and conventional carbon spectrum of *n*-butanol 6.11.** The dotted line indicates the double-quantum 'pseudo' diagonal.

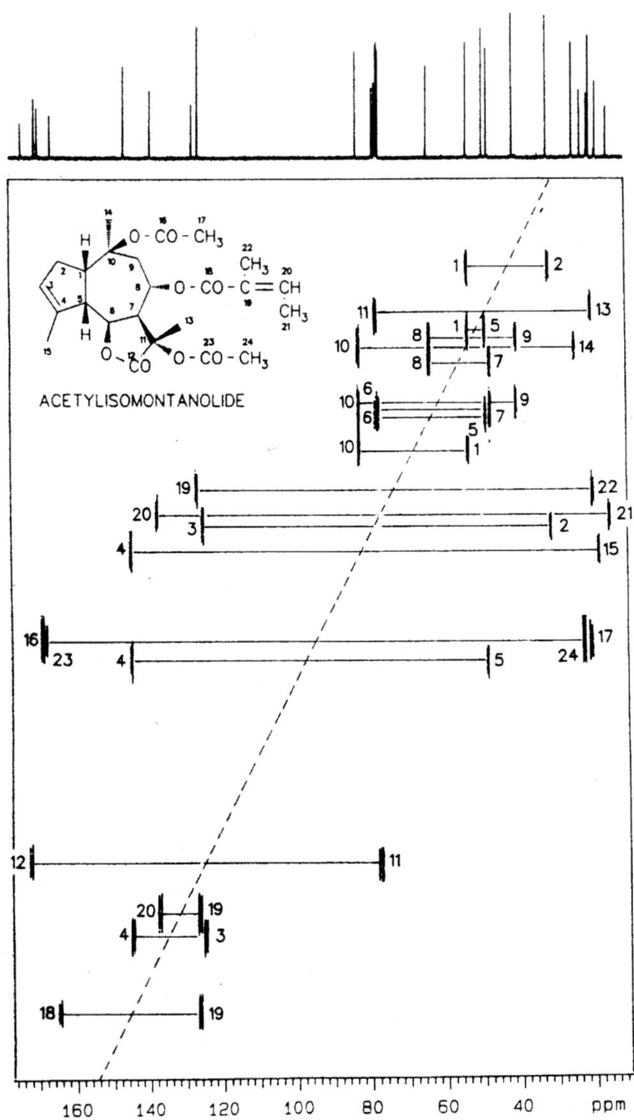

FIGURE 6.46 **The 2D ^{13}C INADEQUATE and conventional carbon spectrum of the sesquiterpene lactone acetylisomontanolide in $CDCl_3$.** *(Source: Reproduced with permission from Ref. [47].)*

An example of the power of this method in structure elucidation is shown for the sesquiterpene lactone acetylisomontanolide (Fig. 6.46). The complete carbon skeleton of the molecule can be traced directly in the spectrum, with breakdown occurring only when heteroatom linkages arise. Unfortunately, it is rare indeed to have sufficient sample quantities at hand to consider employing this technique, and still it remains of limited application in its original form. Nevertheless, in very proton-sparse systems it may present the only option, as in the identification of the nitration product of **6.12a**. The product was to undergo reduction of the NO_2 to NH_2 followed by further derivatisation of the NH_2 and SO_2 groups, so served as a template in the parallel synthesis of potential antibiotics. As such, knowledge of the core template structure was critical prior to further elaboration, but carbon shifts alone were not considered definitive enough to differentiate 4- from 5-nitro products and methods based on long-range proton–carbon correlations (heteronuclear multiple-bond correlation, HMBC; see Section 7.4) would also not differentiate these. The ability to trace carbon connectivities directly in 2D INADEQUATE (Fig. 6.47) provided evidence for the CH at the 4 position and hence unambiguous proof of the nitration site at the 5 position as in **6.12b**.

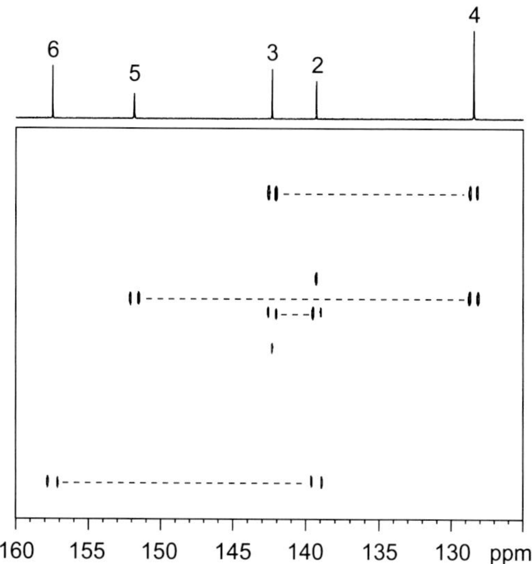

FIGURE 6.47 **The 2D ^{13}C INADEQUATE spectrum of the thiophene 6.12b in $CDCl_3$.** Data were collected on ~150 mg of sample using a ^{13}C optimised cryogenic probe with the experiment optimised for $^1J_{CC} = 70$ Hz.

ClO_2S, MeOOC, S, 2, 3, 4, 5, 6 — **6.12a** → ClO_2S, MeOOC, S, NO_2 — **6.12b** *or* ClO_2S, NO_2, MeOOC, S — **6.12c**

6.3.2 One-Dimensional INADEQUATE

The one-dimensional INADEQUATE experiment was in fact the original version and was proposed as a means of measuring carbon–carbon coupling constants [45]. The use of double-quantum filtration removes the parent resonances which dominate the spectrum and mask the desired splittings, especially when smaller two- and three-bond couplings are to be measured. This can also suppress resonances from low-level impurities that may otherwise interfere. Replacing the t_1 evolution period by a short, fixed delay (sufficient only to allow spectrometer phase changes or the application of gradient pulses) results in a 1D double-quantum filtered experiment in which carbon–carbon couplings appear as anti-phase doublets (Fig. 6.48). From these patterns the values of J_{CC} can be determined directly if sufficiently resolved and have value in studies of stereochemistry and conformation [48–50]. An alternative application that has been used extensively is in the tracking of doubly ^{13}C-labelled precursors in the study of metabolic or biosynthetic pathways [51]. Double-quantum filtration is used to suppress background signals of those species that do not contain the original ^{13}C–^{13}C fragment (to the limit of natural abundance), while retaining only those that have survived the pathway intact, and is nowadays achieved most effectively with PFGs.

6.3.3 Implementing INADEQUATE

The basic components of the INADEQUATE phase cycle comprise double-quantum filtration and f_1 quad-detection. The filtration may be achieved as for the DQF-COSY experiment described previously; that is, all pulses involved in double-quantum excitation (those prior to t_1 in this case) are stepped x, y, $-x$, $-y$ with receiver inversion on each step (an equivalent scheme found in spectrometer pulse sequences is to step the *final* 90 degrees pulse x, y, $-x$, $-y$ as the receiver steps in the opposite sense x, $-y$, $-x$, y; other possibilities also exist). This simple scheme may not be sufficient to fully suppress singlet

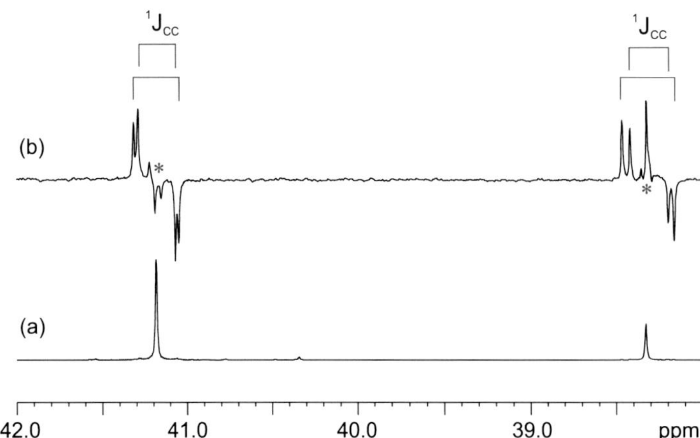

FIGURE 6.48 **The 1D INADEQUATE experiment.** Spectrum (b) reveals the carbon-13 satellites in the conventional carbon-13 spectrum (a) and provides a means of measuring carbon–carbon coupling constants (here marked as anti-phase doublets). *Some residual parent resonances from lone ^{13}C centres remain in (b).

contributions, which appear along $f_1 = 0$ as axial peaks and are distinct from genuine C–C correlations. Extension with the four-step phase cycle for spin-echoes (EXORCYCLE) sequence (Section 8.2.2) on the 180 degree pulse together with the cyclically ordered phase sequence (CYCLOPS) (Section 3.2.5) may improve this. Cleaner suppression could also be achieved by the use of PFGs in a 1:2 ratio, which for sensitivity reasons benefits from gradient probes optimised for ^{13}C observation and even more so from a cryogenically cooled probehead.

Quad-detection requires the observed signals to exhibit 90 degrees phase increments on each t_1 step. Since the desired signals pass through double-quantum coherence, they are *twice* as sensitive to rf phase shifts, meaning the excitation pulses prior to t_1 must themselves be incremented in 45 degrees steps only. Such phase increments are readily achieved on modern spectrometers, so the usual States or TPPI approaches to phase-sensitive data collection can be applied. For a pure-phase display, the implementation of PFGs as in Fig. 6.44 requires that each of these are applied within ^{13}C spin-echoes to provide chemical shift refocusing. During the development of the INADEQUATE sequences the available hardware was incapable of producing small phase shifts and a number of alternative schemes were developed that avoided 45 degrees rf phase shifts altogether. One popular approach used to generate absolute value spectra was to use pulse width dependence to enhance the intensity of N-type signals relative to the P-type [52]. Increasing the final pulse angle to 135 degrees produces a 6:1 intensity ratio of N:P, so the P-type signal essentially disappears and frequency discrimination results, albeit with mixed-mode lineshapes which demand magnitude calculation.

The success of INADEQUATE may be compromised by deleterious off-resonance effects which are particularly troublesome for nuclei such as carbon-13 which have a relatively large chemical shift range. Variations employing composite pulses to counteract offset effects have therefore been developed to minimise losses [53,54]. An alternative approach is to replace the ^{13}C 180 degree pulses with composite adiabatic refocusing pulses (see Section 12.2).

6.3.3.1 Experimental Setup

There is no doubt that the deciding factor when considering the use of ^{13}C INADEQUATE is sensitivity, and far greater sample quantities are required for this experiment than for other 2D sequences. As a rule of thumb, one should be able to obtain a conventional 1D ^{13}C spectrum *in a single scan* following a 90 degree excitation pulse, and this should display a signal-to-noise ratio of *at least* 20:1. For realistic sample quantities encountered in the chemical laboratory, this is rarely achievable and, even in favourable cases, overnight acquisitions at least are typically required. The post-processing algorithm CCBond [55,56] (available as a feature in some software) has proved extremely successful at automatically analysing INADEQUATE spectra even when the signal-to-noise was too poor for the human eye to recognise connectivity patterns. Using such an approach, spectra with an order of magnitude poorer signal-to-noise than is conventionally required may be subject to reliable analysis [57,58], clearly enhancing the utility of this method (for other experimental approaches to improving sensitivity, see the following section).

Maximum sensitivity requires optimum generation of double-quantum coherence. For weakly coupled systems Δ should be equal to $1/2J_{CC}$, while the presence of strong coupling ($\Delta\delta/J_{CC} < 3$) requires optimum periods of $3/2J_{CC}$ [59]. As ever, the choice of Δ represents some compromise over anticipated J_{CC} values. One-bond couplings range typically from 35 to 70 Hz [43,49,60,61] (Table 6.4), but are more often less than 50 Hz in the absence of C–C unsaturation, so Δ of

TABLE 6.4 Typical Ranges for One-Bond C–C Coupling Constants

C–C Group	$^1J_{CC}$ (Hz)
C–C	35–40
C–C–OH (R)	40–60
C=C (alkene)	70–80
C=C (aromatic)	55–70
C≡C	170–220

The principal factor in determining the magnitudes of these is carbon hybridisation.

TABLE 6.5 Typical Values for Long-Range C–C Coupling Constants

J Coupling	C–C Group	$^nJ_{CC}$ (Hz)
$^2J_{CC}/^3J_{CC}$	**C**...**C** (aliphatic)	≤5
$^3J_{CC}$	**C**–C=C–**C** (*trans*)	~7
$^3J_{CC}$	**C**–C=C–**C** (*cis*)	~3
$^2J_{CC}$	**C**–(CO)–**C**	~15
$^2J_{CC}$	**C**=C–**C**	3–8
$^4J_{CC}$	Any	<1

around 10–14 ms is a reasonable choice. Couplings over multiple bonds are considerably smaller, often ≤5 Hz for $^{2/3}J_{CC}$ (Table 6.5), and correspondingly longer delays are required if these correlations are being sought. Since the optimum repetition rate is 1.3 times the ^{13}C T_1s, the use of relaxation agents to speeddata acquisition should also be considered, particularly when quaternary centres may be present. For organic solutions, chromium acetylacetonate is widely used for this and even dissolved oxygen has been shown to be advantageous [62]. Gadolinium triethylene-tetraamine-hexaacetate, Gd(TTHA) [63], has been recommended for aqueous solutions and experience suggests concentrations up to 10 mM produce acceptable results without significantly affecting the carbon resonances, although this can substantially broaden the *proton* spectrum of the sample.

Due to the large spectral widths involved, spectra are collected with low resolution in both dimensions; around 10–20 Hz/pt in f_2 and 50–100 Hz/pt in f_1. Processing then requires window functions that minimise distortions arising from the truncated FIDs, such as that shifted sine bell or squared sine bell. To minimise the f_1 spectral width, the transmitter frequency should be placed in the centre of the 1D ^{13}C spectrum so that double-quantum frequencies fall either side of this, that is both positive and negative double-quantum frequencies appear in f_1. It is also possible to further reduce the f_1 spectral width by deliberately folding peaks in the f_1 dimension by making this equal to that in f_2. This causes no ambiguity in interpretation because both participating crosspeaks fold in together and connectivities can still be identified in horizontal traces.

6.4 CORRELATING DILUTE SPINS VIA PROTONS: ADEQUATE

A number of variations on the basic INADEQUATE sequence have been presented over the years, particularly with reference to ^{13}C experiments, all primarily aimed at improving the appallingly low sensitivity of the original technique. Proton–carbon polarisation transfer prior to the INADEQUATE sequence via INEPT [64] or DEPT [65] may be used as a means of enhancing ^{13}C population differences and of allowing more rapid repetition rates since these would be dictated by faster proton spin relaxation. However, these preclude the use of broadband proton decoupling and hence gain no enhancement from the NOE, so only marginal improvements may obtained. Conversely, magnetisation may be transferred onto protons directly attached to ^{13}C sites by a reverse-INEPT step *after* the INADEQUATE sequence. This variant, referred to as INSIPID [66], benefits from the higher sensitivity of proton detection. Alas, it also suffers severely from the need to suppress the signals of protons attached to lone ^{13}C centres as well as those attached to ^{12}C, which possess intensities that are, respectively, 100 and ~10,000 times greater than those of the desired ^{1}H–^{13}C–^{13}C moieties. More recent experiments, referred to originally as proton-detected INEPT-INADEQUATE [67] or more boldly in one development as ADEQUATE [68], tackle the sensitivity problem by utilising both polarisation transfer and proton observation, and overcome the proton suppression problems by exploiting PFGs for signal selection. In principle, this leads to sensitivity gains by a factor of 32 ($(\gamma_H/\gamma_C)^{5/2}$), although a factor of around 8–16 seems more realistic when allowing for experimental losses [67]. The resulting 2D spectrum presents a stepwise correlation of protons that sit on neighbouring ^{13}C centres, and so maps direct carbon connectivities within the molecular framework. This powerful technique should prove valuable in the structure elucidation of unknown materials, although the obvious limitation with this approach is the need for proton-bearing carbons in the molecule, which may limit its utility for proton-sparse systems. Furthermore, these techniques are likely to be employed only after more sensitive methods, such as those based on proton–proton and proton–carbon correlations, have failed to solve the problem at hand.

6.4.1 Two-Dimensional ADEQUATE

The operation of proton-enhanced sequences is best understood by reference to the original INEPT-INADEQUATE sequence shown in simplified form in Fig. 6.49 in which individual component parts have been identified and may be followed in a stepwise manner. Thus, initial transfer from proton to carbon is via the INEPT sequence introduced in Section 4.4.2 followed by the generation of ^{13}C double-quantum coherence as for the INADEQUATE mentioned earlier. This again evolves to encode t_1 after which gradient selection begins. The ^{13}C double-quantum then returns to single-quantum and the ^{13}C–^{13}C coupling is refocused by the spin-echo. The final step is INEPT in reverse to transfer from carbon back onto proton, followed by gradient refocusing of the desired responses and then observation in the presence of ^{13}C decoupling. The sequence can thus be viewed as a variation on the HSQC technique of Section 7.3 in which double-quantum ^{13}C coherence evolves during t_1 instead of the single-quantum ^{13}C coherence utilised in HSQC. The resulting spectrum has carbon double-quantum frequencies along f_1 with proton shifts along f_2 in which correlated protons are indicative of H–C–C–H fragments. As an illustration, the spectrum in Fig. 6.50 was recorded on a 50-mg sample of unlabelled menthol **6.13** in 16 h at 400 MHz, although 'adequate' results would have to be achieved in less than half this time.

Since the original sequence, a more refined family of experiments (known collectively as the ADEQUATE techniques) have been presented [68], offering the ability to correlate over single and multiple bonds by employing both one-bond and long-range J_{CH} and J_{CC} couplings, as summarised in Table 6.6. The types of molecular fragments interrogated by these sequences are summarised in Fig. 6.51 correlating, in principle, nuclei up to six bonds apart and potentially providing a wealth of structural data. The sequence for the 1,1- and 1,*n*- ADEQUATE experiments is illustrated in Fig. 6.52 and features a number of enhancements including the use of a 60 degree ^{13}C pulse for optimum transfer of double-quantum to single-quantum ^{13}C magnetisation and the inclusion of the preservation of equivalent pathways (PEP) cluster for optimum signal transfer from carbon back to proton (see Section 7.3.1). The 1,*n*- version correlates remotely coupled carbons by optimising the Δ_2 coupling evolution delays for the considerably smaller long-range J_{CC} coupling constants (Table 6.5) rather than

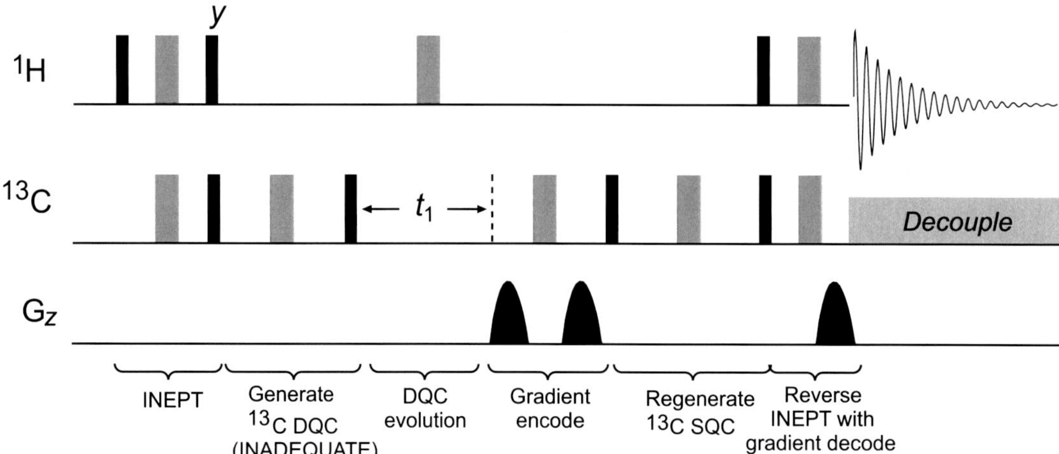

FIGURE 6.49 **The gradient-selected INEPT-INADEQUATE sequence for identifying carbon–carbon connectivities through proton observation.** The sequence selects for ^{1}H–^{13}C–^{13}C fragments. Despite the complexity of the sequence, its operation may be understood by recognising discrete, simpler segments within it that have defined roles to play, which in this case are INEPT and INADEQUATE steps (time delays in the sequence have been removed for clarity).

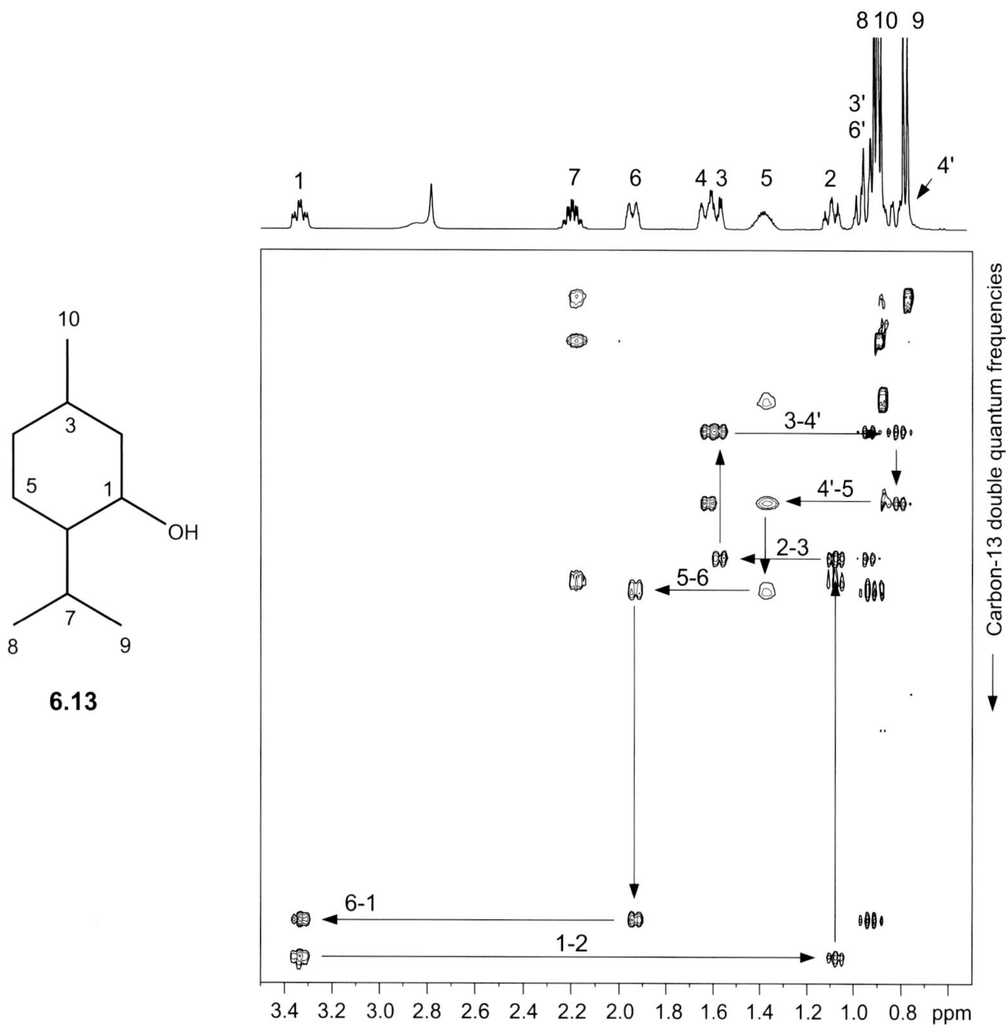

FIGURE 6.50 **The proton-detected 1,1- ADEQUATE spectrum of menthol 6.13 (50 mg in $CDCl_3$) recorded at 400 MHz.** Stepwise tracing of the carbon skeleton of the cyclohexane ring is illustrated in which carbon connectivities are identified in horizontal traces (alternative 'routes' involving the diastereotopic 3, 4 and 6 partners also exist but, for clarity, are not shown).

${}^1J_{CC}$ values. The *n*,1- and *m*,*n*- ADEQUATE techniques employ long-range J_{CH} couplings for initial transfer from proton to carbon and may thus be viewed as the HMBC analogues of the above; see [68] for sequence details. The use of long-range couplings in this manner will further compromise sensitivity, so their application is likely to be further restricted despite their potentially rich information content. An alternative approach to detecting very long–range correlations that utilises only long-range J_{CH} and thus does not depend on carbon–carbon couplings is the long-range HSQMBC method described in Section 7.4.6.

6.4.2 Enhancements to ADEQUATE

In recent years there have been further developments that have sought to simplify the appearance, improve the performance or enhance the content of the ADEQUATE techniques. As described in the original publication [68], one modification of the 1,1- and 1,*n*-sequence includes an additional element which is executed when ${}^{13}C$ single-quantum coherence has been regenerated to refocus the chemical shift of the carbon to which the observed proton is attached. This results in the f_1 dimension displaying the conventional ${}^{13}C$ chemical shift axis (rather than double-quantum frequencies) such that the spectrum appearance matches that of ${}^1H$-detected heteronuclear correlation experiments such as HSQC or HMBC. The observed correlations then relate the observed proton to the *remote* carbon centres which, in the case of the 1,1-experiment, equates to those carbon atoms that are two bonds away. This correlation is equivalent to that which would be observed for ${}^2J_{CH}$ correlations in the HMBC long-range correlation experiments (Section 7.4), even though the couplings exploited differ in these two techniques

TABLE 6.6 The Scalar Couplings Exploited in the ADEQUATE Family of Sequences

ADEQUATE Version	Couplings Exploited	Bonds Correlated
1,1-	$^{1}J_{CH}$, $^{1}J_{CC}$	2
1,*n*-	$^{1}J_{CH}$, $^{n}J_{CC}$	4
n,1-	$^{n}J_{CH}$, $^{1}J_{CC}$	4
m,*n*-	$^{m}J_{CH}$, $^{n}J_{CC}$	6

The bond correlations indicate the *maximum* number of bonds over which correlations may be identified.

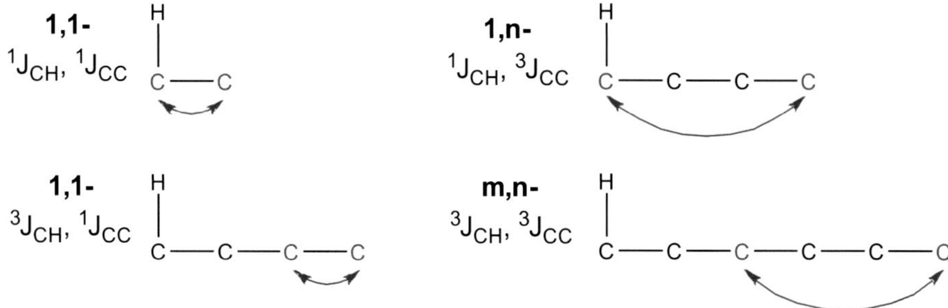

FIGURE 6.51 **Schematic illustration of the correlations explored with the ADEQUATE sequences.** The maximum number of bonds are indicated for typical J_{CH} and J_{CC} values but shorter range (two-bond) correlations and those crossing heteroatoms will also be detected. The linked carbon centres coloured red are those that generate the double-quantum coherence and are thus ultimately correlated.

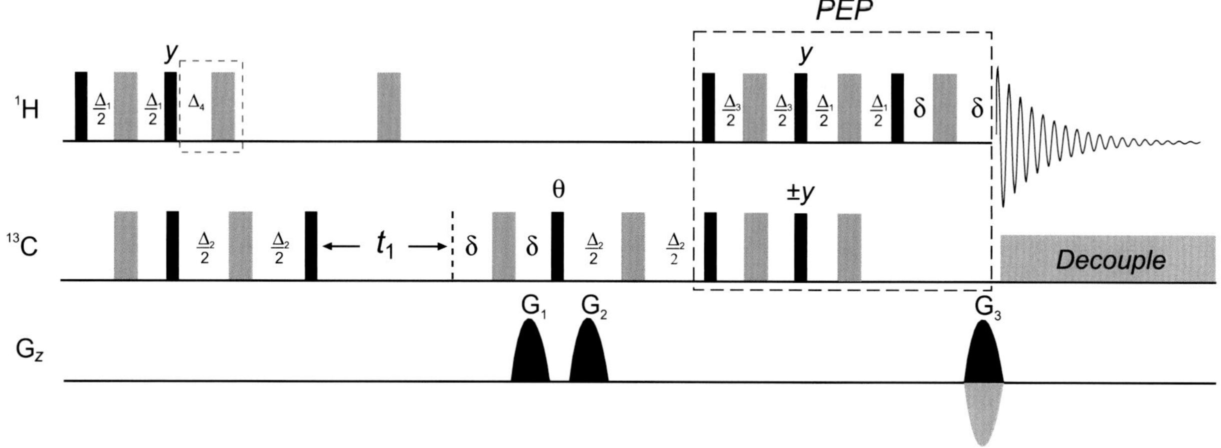

FIGURE 6.52 **The 1,1- ADEQUATE sequence.** Delays are $\Delta_1 = 1/(2^1J_{CH})$, $\Delta_2 = 1/(2^1J_{CC})$, $\Delta_3 = 1/(6^1J_{CH})$, δ = gradient encoding and ^{13}C pulse angle $\vartheta = 60$ degrees. In the 1,*n*- experiment the Δ_2 J_{CC} evolution period is optimised for long-range carbon–carbon couplings. Gradient pulses together with alternation of the final carbon 90 degree pulse select echo–antiecho signals for f_1 quad-detection. For the multiplicity-edited version the additional red-boxed element is employed with $\Delta_4 = 1/(2^1J_{CH})$. For details of the PEP element see Section 7.3.1.

(ie $^{1}J_{CH} + {}^{1}J_{CC}$ vs. $^{2}J_{CH}$). Indeed, the refocused 1,1- ADEQUATE has been suggested as one means of differentiating $^{2}J_{CH}$ from $^{3/4}J_{CH}$ correlations observed in HMBC experiments [69]; other potentially more sensitive methods also exist including heteronuclear two-bond correlation (H2BC) (Section 7.4.4). As an example, the 1,1-experiment was used to support HMBC data in the identification of C–C connectivity and assignment of $^{2}J_{CH}$ correlations in the proof of structure of the synthetic indol-3-yl-2-oxo acetamide anti-tumour agent known as D-24851 **6.14** [70]. The appearance of the H8–C9 and the lack of H8–C3 correlation provided proof of substitution position. Similarly, the 1,1- and 1,*n*-experiments were used in support of ^{1}H–^{13}C and ^{1}H–^{15}N HMBC data in the identification of oxepinamide A **6.15** isolated from marine fungi [71]. The 1,*n*- experiment has also been applied to the assignment of proton-sparse, highly substituted aromatic compounds for which HMBC affords ambiguous results [72]. Thus, the 'pseudo-$^{4}J_{CH}$' correlations observed in this experiment were able to distinguish between the two possible synthetic regioisomers **6.16a** and **b** and identify the product as the former by virtue of the four-bond correlations from the two O-methyl groups to the aromatic methine carbon.

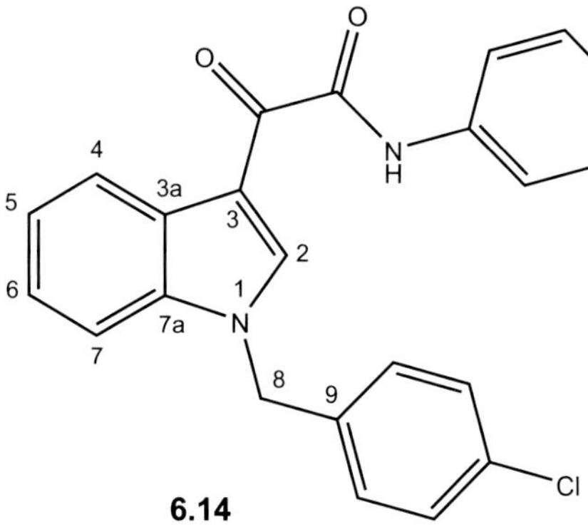

6.14

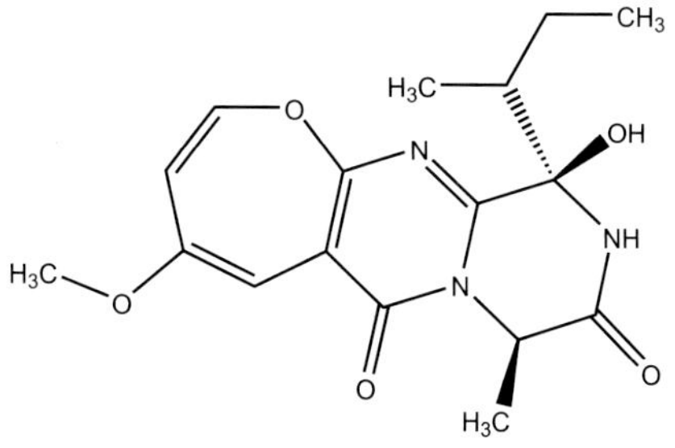

6.15

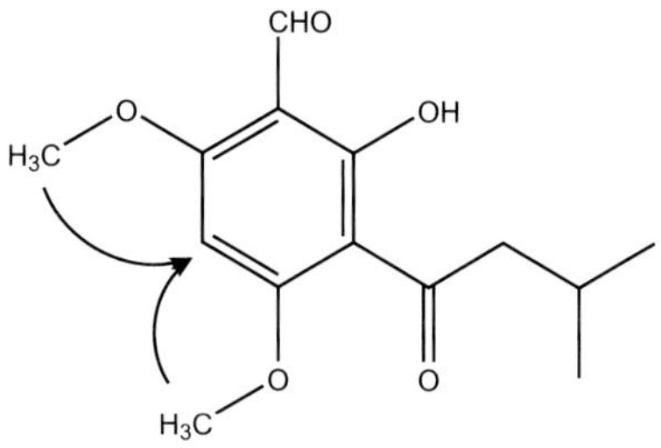

6.16a

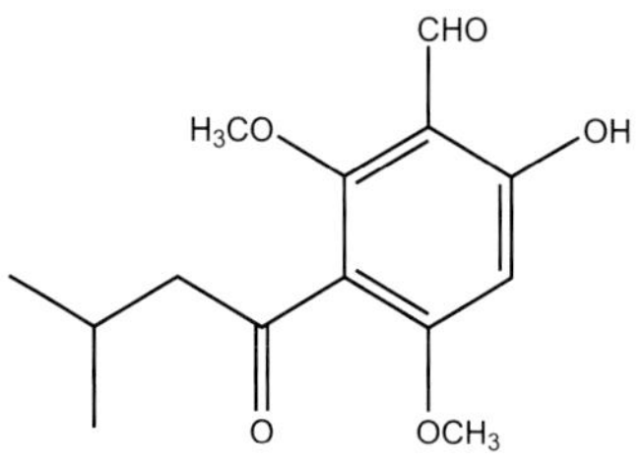

6.16b

The performance of ADEQUATE experiments can be significantly compromised by deleterious off-resonance effects associated with the many ^{13}C 180 degree pulses employed, and the use of frequency-swept adiabatic pulses (Section 12.2) has been shown to greatly improve spectral quality [73]. The first and last ^{13}C 180 degree pulses of Fig. 6.52 may both be replaced with smoothed CHIRP pulses, where carbon inversion is required, and the remaining four may be substituted for composite-smoothed CHIRP pulses to effect the desired carbon chemical shift refocusing. For situations in which the molecules of interest are likely to contain a wide range of $^1J_{CC}$ values, notably when both aliphatic and olefinic or aromatic moieties exist, an ACCORDION-optimised version of ADEQUATE has been presented that seeks to detect correlations over a wider range of J_{CC} coupling constants [74], in an analogous manner to the approach employed in HMBC experiments for the detection of $^nJ_{CH}$ couplings (Section 7.4.3). In a separate development, a simple modification of the experiment, illustrated by the boxed area in Fig. 6.52, provides for multiplicity editing of the spectrum whereby the observed correlations arising from protons belonging to CH_2 groups are inverted relative to those from CH or CH_3 (as in the multiplicity-edited HSQC variants of Section 7.3.2), so further enhancing information content. Finally, we note that further variants of the ADEQUATE experiment have been proposed for the measurement of one-bond and long-range carbon–carbon coupling constants [75–77] which, although rarely used to date, have potential application in stereochemical and conformational analysis. [48–50,78]

Finally, we finish with a small digression. The original sequence for identifying ^1H–^{13}C–^{13}C connectivities, shown in Fig. 6.49, serves to illustrate a feature common to many modern pulse sequences. Although at first sight it may appear to be frighteningly complex and is undoubtedly one of the most elaborate sequences in this book, it can be broken down into more readily digestible fragments, as illustrated. While the details of this sequence may appear somewhat esoteric, an appreciation of its operation is rather more accessible when viewed in this stepwise manner as it becomes apparent that this is built up from simpler sequences or elements already encountered (the aforementioned INEPT and INADEQUATE). As already mentioned, the sequence can then simply be viewed as a variation on the HSQC technique in which carbon double-quantum rather than single-quantum coherence is employed. When faced with trying to follow a new or unfamiliar pulse sequence, a good approach is to try to recognise smaller fragments within it that are derived from shorter, more familiar or recognisable sequences and to piece these elements together to follow the flow of magnetisation through the experiment. We shall encounter further examples of this in the following chapter.

REFERENCES

[1] Hahn J. In: Verkade JG, Quin LD, editors. Phosphorus-31 NMR spectroscopy in stereochemical analysis. Florida: VCH; 1987.
[2] Rance M, Sørensen OW, Bodenhausen G, Wagner G, Ernst RR, Wüthrich K. Biochem Biophys Res Commun 1983;117:479–85.
[3] Piantini U, Sørensen OW, Ernst RR. J Am Chem Soc 1982;104:6800–1.
[4] Davis AL, Laue ED, Keeler J, Moskau D, Lohman J. J Magn Reson 1991;94:637–44.
[5] Eberstadt M, Gemmecker G, Mierke DF, Kessler H. Angew Chem Int Ed 1995;34:1671–95.
[6] Neuhaus D, Wagner G, Vasák M, Kági JHR, Wüthrich K. Eur J Biochem 1985;151:257–73.
[7] Kessler H, Oschkinat H. Angew Chem Int Ed 1985;24:690–2.
[8] Kessler H, Müller A, Oschkinat H. Magn Reson Chem 1985;23:844–52.
[9] Oschkinat H, Freeman R. J Magn Reson 1984;60:164–9.
[10] Shaka AJ, Freeman R. J Magn Reson 1983;51:169–73.
[11] Müller N, Ernst RR, Wüthrich K. J Am Chem Soc 1986;108:6482–92.
[12] Boyd J, Redfield C. J Magn Reson 1986;68:67–84.
[13] Eich G, Bodenhausen G, Ernst RR. J Am Chem Soc 1982;104:3731–2.
[14] Braunschweiler L, Ernst RR. J Magn Reson 1983;53:521–8.
[15] Davis DG, Bax A. J Am Chem Soc 1985;107:2820–1.
[16] Bax A, Davis DG. J Magn Reson 1985;65:355–60.
[17] Cavanagh J, Chazin WJ, Rance M. J Magn Reson 1990;87:110–31.
[18] Morris GA, Gibbs A. J Magn Reson 1991;91:444–9.
[19] Gibbs A, Morris GA. Magn Reson Chem 1992;30:662–5.
[20] Baldwin JE, Bischoff L, Claridge TDW, Heupel FA, Spring DR, Whitehead RC. Tetrahedron 1997;53:2271–90.
[21] Baldwin JE, Claridge TDW, Culshaw AJ, Heupel FA, Lee V, Spring DR, Whitehead RC, Boughtflower RJ, Mutton IM, Upton RJ. Angew Chem Int Ed 1998;37:2661–3.
[22] Griesinger C, Otting G, Wüthrich K, Ernst RR. J Am Chem Soc 1988;110:7870–2.
[23] Briand J, Ernst RR. Chem Phys Lett 1991;185:276–85.
[24] Cavanagh J, Rance M. J Magn Reson 1992;96:670–8.
[25] Rance M. J Magn Reson 1987;74:557–64.
[26] Bazzo R, Campbell ID. J Magn Reson 1988;76:358–61.
[27] Sørensen OW, Rance M, Ernst RR. J Magn Reson 1984;56:527–34.
[28] Shaka AJ, Keeler J, Freeman R. J Magn Reson 1983;53:313–40.
[29] Shaka AJ, Keeler J, Frenkiel T, Freeman R. J Magn Reson 1983;52:335–8.
[30] Rucker SP, Shaka AJ. Mol Phys 1989;68:509–17.
[31] Shaka AJ, Lee CJ, Pines A. J Magn Reson 1988;77:274–93.
[32] Kessler H, Mronga S, Gemmecker G. Magn Reson Chem 1991;29:527–57.
[33] Parella T. Magn Reson Chem 1996;34:329–47.
[34] Davis DG, Bax A. J Am Chem Soc 1985;107:7197–8.
[35] Kessler H, Anders U, Gemmecker G, Steuernagel S. J Magn Reson 1989;85:1–14.
[36] Kessler H, Oschkinat H, Griesinger C, Bermel W. J Magn Reson 1986;70:106–33.
[37] Subramanian S, Bax A. J Magn Reson 1987;71:325–30.
[38] Adell P, Parella T, Sánchez-Ferrando F, Virgili A. J Magn Reson B 1995;108:77–80.
[39] Fäcke T, Berger S. J Magn Reson A 1995;113:257–9.
[40] Dalvit C, Bovermann G. Magn Reson Chem 1995;33:156–9.
[41] Xu GZ, Evans JS. J Magn Reson B 1996;111:183–5.
[42] Thrippleton MJ, Keeler J. Angew Chem Int Ed 2003;42:3938–41.
[43] Buddrus J, Bauer H. Angew Chem Int Ed 1987;26:625–42.
[44] Buddrus J, Lambert J. Magn Reson Chem 2002;40:3–23.
[45] Bax A, Freeman R, Kempsell SP. J Am Chem Soc 1980;102:4849–51.
[46] Bax A, Freeman R, Frenkiel TH. J Am Chem Soc 1981;103:2102–4.
[47] Budesinsky M, Saman D. Ann Rep NMR Spectrosc 1995;30:231–475.
[48] Marshall JL. Carbon–carbon and carbon–proton NMR couplings: applications to organic stereochemistry and conformational analysis. Orlando, FL: VCH; 1983.
[49] Krivdin LB, Kalabin GA. Prog Nucl Magn Reson Spectrosc 1989;21:293–448.
[50] Krivdin LB, Zinchenko SV. Curr Org Chem 1998;2:173–93.
[51] Vederas JC. Nat Prod Rep 1987;4:277–337.
[52] Mareci TH, Freeman R. J Magn Reson 1982;48:158–63.
[53] Lambert J, Kuhn HJ, Buddrus J. Angew Chem Int Ed 1989;28:738–40.
[54] Torres AM, Nakashima TT, McClung RED, Muhandiram DR. J Magn Reson 1992;99:99–117.
[55] Dunkel R, Mayne CL, Curtis J, Pugmire RJ, Grant DM. J Magn Reson 1990;90:290–302.

[56] Dunkel R, Mayne CL, Pugmire RJ, Grant DM. Anal Chem 1992;64:3133–49.
[57] Dunkel R, Mayne CL, Foster MP, Ireland CM, Li D, Owen NL, Pugmire RJ, Grant DM. Anal Chem 1992;64:3150–60.
[58] Harper JK, Dunkel R, Wood SG, Owen NL, Li D, Cates RG, Grant DM. J Chem Soc Perkin Trans 1996;2. 191-100.
[59] Bax A, Freeman R. J Magn Reson 1980;41:507–11.
[60] Breitmaier E, Voelter W. Carbon-13 NMR spectroscopy. 3rd ed. Weinheim: VCH; 1987.
[61] Kamienska-Trela K. Ann Rep NMR Spectrosc 1995;30:131–230.
[62] Mattiello DL, Freeman R. J Magn Reson 1998;135:514–21.
[63] Lettvin J, Sherry AD. J Magn Reson 1977;28:459–61.
[64] Sørensen OW, Freeman R, Frenkiel TA, Mareci TH, Schuck R. J Magn Reson 1982;46:180–4.
[65] Sparks SW, Ellis PD. J Magn Reson 1985;62:1–11.
[66] Keller PJ, Vogele KE. J Magn Reson 1986;68:389–92.
[67] Weigelt J, Otting G. J Magn Reson A 1995;113:128–30.
[68] Reif B, Kock M, Kerssebaum R, Kang H, Fenical W, Griesinger C. J Magn Reson A 1996;118:282–5.
[69] Kock M, Reif B, Fenical W, Griesinger C. Tetrahedron Lett 1996;37:363–6.
[70] Knaack M, Emig P, Bats JW, Kiesel M, Müller A, Günther E. Eur J Org Chem 2001;3843–7.
[71] Belofsky GN, Anguera M, Jensen PR, Fenical W, Köck M. Chem Eur J 2000;6:1355–60.
[72] Kock M, Reif B, Gerlach M, Reggelin M. Molecules 1996;1:41–5.
[73] Kock M, Kerssebaum R, Bermel W. Magn Reson Chem 2003;41:65–9.
[74] Williamson RT, Marquez BL, Gerwick WH, Koehn FE. Magn Reson Chem 2001;39:544–8.
[75] Reif B, Kock M, Kerssebaum R, Schleucher J, Griesinger C. J Magn Reson B 1996;112:295–301.
[76] Kövér KE, Forgó P. J Magn Reson 2004;166:47–52.
[77] Pham TN, Kövér KE, Jin L, Uhrín D. J Magn Reson 2005;176:199–206.
[78] Bose B, Zhao S, Stenutz R, Cloran F, Bondo PB, Bondo G, Hertz B, Carmichael I, Serianni AS. J Am Chem Soc 1998;120:11158–73.

Chapter 7

Correlations Through the Chemical Bond II: Heteronuclear Shift Correlation

Chapter Outline

7.1 Introduction 243
7.2 Sensitivity 244
7.3 Heteronuclear Single-Bond Correlations 246
7.3.1 Heteronuclear Single-Quantum Correlation 246
7.3.2 Hybrid HSQC Experiments 253
7.3.3 Heteronuclear Multiple-Quantum Correlation 257
7.4 Heteronuclear Multiple-Bond Correlations 261
7.4.1 HMBC Sequence 263
7.4.2 Applying HMBC 264
7.4.3 HMBC Extensions and Variants 266
7.4.4 H2BC: Differentiating $^2J_{CH}$ and $^3J_{CH}$ HMBC Correlations 274
7.4.5 Measuring Long-Range $^nJ_{XH}$ Coupling Constants 275
7.4.6 Long-Range HSQMBC: Interrogating Proton-Sparse Molecules 281
7.5 Heteronuclear X-Detected Correlations 282
7.5.1 Single-Bond Heteronuclear Correlations 283
7.5.2 Multiple-Bond Correlations and Small Couplings 285
7.6 Heteronuclear X–Y Correlations 286
7.6.1 Direct X–Y Correlations 286
7.6.2 Indirect ^{1}H-Detected X–Y Correlations 288
7.7 Parallel Acquisition NMR with Multiple Receivers 291
References 292

7.1 INTRODUCTION

This second chapter on establishing correlations through the chemical bond concentrates on techniques which correlate different nuclides, so-called *heteronuclear* shift correlations. In chemistry this primarily means establishing connectivities between proton and carbon nuclei and as such the techniques encountered in the following sections are concerned mainly with these. That is not to say the techniques are not suitable for correlating other nuclides, and combinations such as ^{1}H with ^{15}N or ^{31}P, and ^{19}F or even ^{31}P with ^{13}C are widely employed. Indeed many of the modern techniques used routinely in the chemical laboratory were originally implemented as methods for ^{1}H–^{15}N correlations in proteins and peptides. The principal techniques described in the sections that follow are summarised in Table 7.1.

A variety of methods have already been described in the chapter *One-Dimensional Techniques* that allow the editing of the one-dimensional (1D) spectrum of the heteronuclear spin, for example, those based on spin-echoes or polarisation transfer, so providing valuable information on the numbers of attached protons. They do not, however, provide any *direct* evidence for which protons are attached to which heteronucleus (X-spin) in the molecule and for this heteronuclear two-dimensional (2D) correlations are widely used to transfer previously established proton assignments onto the directly bonded heteronucleus or, on occasions, vice versa. This may then provide further evidence to support or reject the proposed structure as being correct, or may provide assignments that can be used as the basis of further investigations. In addition, the experiment may be used as a means of spreading the resonances of a complex proton spectrum according to the chemical shift of the directly attached nucleus, utilising the typically greater dispersion of the X-spin chemical shifts to assist with proton interpretation. The high sensitivity of modern correlation techniques often provides a fast method for determining indirectly chemical shifts of the X-nucleus and avoids the need for its direct observation altogether, offering considerable time savings. Thus, the ^{1}H–^{1}H COSY and the ^{1}H–^{13}C correlation experiments (preferably the heteronuclear single-quantum correlation, HSQC) represent the primary 2D techniques in structural chemistry.

For more complex problems there exist methods that combine the features of two, otherwise separate, techniques, which shall be referred to here as 'hybrid' experiments. Whilst a wide range of combinations have been devised, essentially limited by the imagination of the spectroscopist, two principal features are of considerable utility. The first is editing of the correlation spectrum such that it contains both shift and multiplicity data, by analogy with the 1D editing methods presented in Sections 4.3 and 4.4. The second feature is the relaying of correlations to enhance the

High-Resolution NMR Techniques in Organic Chemistry. http://dx.doi.org/10.1016/B978-0-08-099986-9.00007-5

TABLE 7.1 Principal Applications of the Main Techniques

Technique	Principal Applications
HSQC	Correlating coupled heteronuclear spins across a single bond and hence identifying directly connected nuclei. Employs detection of high-sensitivity nuclides, for example, ^{1}H, ^{19}F, ^{31}P (an 'inverse technique'). Favoured over HMQC as it can provide improved resolution.
HMQC	Correlating coupled heteronuclear spins across a single bond and hence identifying directly connected nuclei, most often ^{1}H–^{13}C. Employs detection of high-sensitivity nuclides, for example, ^{1}H, ^{19}F, ^{31}P (an 'inverse technique').
HMBC	Correlating coupled spins across multiple bonds. Employs detection of high-sensitivity nuclides, for example, ^{1}H, ^{19}F, ^{31}P (an 'inverse technique'). This is HMQC tuned for the detection of small couplings. Most valuable in correlating ^{1}H–^{13}C over two- or three-bonds. Powerful tool for linking together structural fragments.
H2BC	Used to identify HMBC peaks that equate to two-bond ^{1}H–^{13}C correlations and so distinguish them from three-bond correlations. Limited to correlations to protonated carbon centres only.
HSQMBC	Used for measurement of the magnitudes of long-range heteronuclear coupling constants, most often between proton and carbon centres.
LR-HSQMBC	Used for the detection of very long–range heteronuclear correlations. For ^{1}H–^{13}C systems this may provide correlations over four, five and even six bonds and will likely prove most useful for proton-sparse structures.
HETCOR	Correlating coupled heteronuclear spins across a single bond. Employs detection of the lower-γ nuclide, typically ^{13}C, so has significantly lower sensitivity than inverse techniques. Benefits from high resolution in the ^{13}C dimension, so may find use when this is critical, otherwise superseded by the aforementioned methods.
H–X–Y	Triple-resonance methods for correlating protons with two heteroatoms X and Y. One heteroatom may be used to either relay correlations or to edit the correlation spectrum.

information content of the spectrum by providing additional neighbouring group information. Common implementations of these schemes are found in the form of the edited HSQC and the HSQC-TOCSY experiment, both described in this chapter.

Beyond these methods that utilise one-bond heteronuclear couplings, correlations with a heteronucleus over more than one bond, so-called *long-range* or *multiple-bond* correlations, can provide a wealth of connectivity information on how molecular fragments are linked together. These methods have become increasingly important in recent years and the techniques for establishing such correlations have been embellished in various ways to enhance information content, providing the chemist with a range of complimentary techniques with which to define structure. The final class of techniques that will be considered seek correlations between, or involving, two heteronuclei, these generally being referred to as 'X–Y correlations', which may be employed when ^{1}H–X correlations provide insufficient information or when these simply do not exist. Those methods that also utilise coupling to protons (^{1}H–X–Y correlations) may be derived conceptually from those developed for establishing ^{1}H–^{13}C–^{15}N correlations in biological macromolecules and mostly find application in the realm of organometallic chemistry where suitable NMR-active X/Y spin pairs exist.

7.2 SENSITIVITY

The original methods for determining heteronuclear shift correlations were based on the observation of the low-γ X-nucleus, with the proton being indirectly detected and consequently appearing along the f_1 dimension of the 2D experiment. This approach was adopted because, as we shall see, the original 2D sequences were derived from early polarisation transfer experiments, such as INEPT (Section 4.4), which were themselves designed to enhance the sensitivity of low-γ observations, and because early pulsed NMR instruments were designed with this mode of operation in mind. During the last two decades, the approach to data collection has fundamentally changed to one in which the high-γ nucleus, most frequently the proton, is observed, with the heteronucleus now detected indirectly. This switch has given rise to a body of experiments frequently referred to as 'inverse' shift correlations, the motivation for change being improved sensitivity.

The dependence of the strength of an NMR signal on magnetogyric ratio has been discussed previously in Section 4.4 where the concept of polarisation transfer was introduced as a means of sensitivity enhancement. From the qualitative arguments presented in that section a more formal expression for the signal-to-noise ratio of a 1D experiment involving spin-½ nuclei was given as:

$$\frac{S}{N} \propto NAT^{-1} B_0^{3/2} \gamma_{exc} \gamma_{obs}^{3/2} T_2^{*} (NS)^{1/2} \qquad (7.1)$$

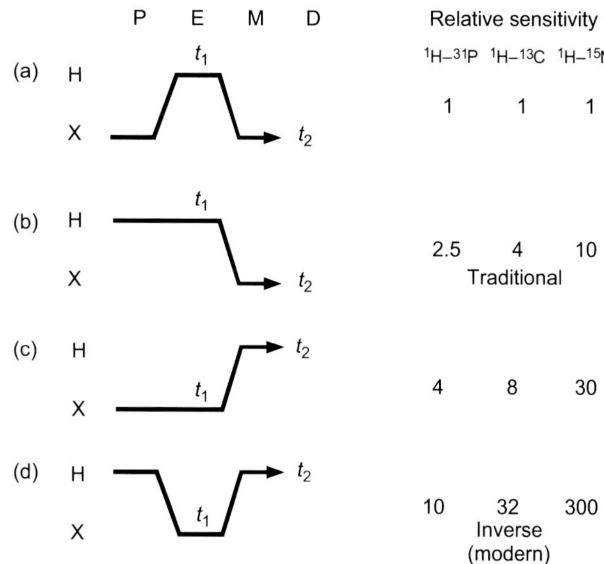

FIGURE 7.1 **The four general schemes to produce 2D heteronuclear shift correlation spectra.** (a) X–H–X, (b) H–H–X, (c) X–X–H and (d) H–X–H. The relative sensitivities of these approaches are compared for proton correlation experiments with phosphorus-31, carbon-13 and nitrogen-15.

where N is the number of molecules in the observed sample volume, A is a term that represents the abundance of the NMR active spins involved in the experiment, T is temperature, B_0 is the static magnetic field, γ_{exc} and γ_{obs} represent the magnetogyric ratios of the initially excited and the observed spins, respectively, $T_2{}^*$ is the effective transverse relaxation time, and NS is the total number of accumulated scans. When choosing how to perform a heteronuclear shift correlation experiment, four general schemes may be devised, as represented in Fig. 7.1, according to which spin is used as the initial magnetisation 'source' and which is detected (represented in the γ factors in Eq. 7.1). Alongside each schematic is summarised the relative theoretical sensitivities expected for an H–X pair based on the participating γ values. The scheme in Fig. 7.1b represents the approach adopted in the traditional shift correlation experiments, in which initial proton magnetisation is frequency labelled in t_1 and then transferred onto the X-spin for detection. The scheme Fig 7.1d is the modern 'inverse' approach in which the proton nucleus is used both as the source and observed spin. Clearly the predicted sensitivity is significantly greater than that of the traditional method, and the dependence on the magnetogyric ratio of the low-γ spin has been completely removed, although the natural abundance of this spin is still an important factor for all approaches (term A in the equation). The lower the γ of the X-spin, the greater the gains arising from proton detection; compare the figures for ^{31}P and ^{15}N in Fig. 7.1, for example. Additional theoretical gains by a factor of 2 or 3 can be anticipated for XH_2 or XH_3 groups, respectively, with proton-detected experiments.

The impressive gains illustrated in Fig. 7.1 for inverse versus traditional methods may not be met in practice when the details of a particular sequence are considered. Significant factors may include the different relaxation behaviour of participating spins or the presence of multiplet splittings that spread a resonance and so reduce the signal-to-noise ratio. However, even a realistic gain of a factor of four in ^{1}H–^{13}C correlations corresponds to a time saving of a factor of 16, meaning experiments that once required an overnight acquisition with ^{13}C detection can be collected in about an hour, whilst those traditionally requiring a couple of hours can be completed within a matter of minutes. Furthermore, studies on very dilute samples that would have been considered intractable become viable targets with proton detection. Such considerations have led to the universal adoption of proton-detected inverse correlation methods whenever possible in both chemical and biological spectroscopy, and it is these techniques that are focused upon later, along with only a relatively brief consideration of the more traditional X-detected methods that may still have utility in specific circumstances.

The adoption of the inverse approach also has implications for the design of the NMR instrument. Conventional probes were constructed so as to optimise the sensitivity for observation of the low-γ X-nucleus, which entails placing the X-nucleus coil closest to the sample and positioning the proton coil outside this. Inverse probes have this configuration switched such that the proton coil sits closest to the sample for optimum sensitivity, thus providing a greater *filling factor*. However, even with conventional probes, the proton detected experiments can still be performed, albeit with less than optimum sensitivity, and may still provide a faster approach than the former X-observe experiments.

7.3 HETERONUCLEAR SINGLE-BOND CORRELATIONS

There are two techniques in widespread use that provide single-bond heteronuclear shift correlations, known colloquially as heteronuclear single-quantum correlation (HSQC) and heteronuclear multiple-quantum correlation (HMQC). The correlation data provided by these two methods are essentially equivalent, the methods differing only in finer details which, for routine spectroscopy, are often of little consequence. HSQC has now become the dominant technique as it has more favourable characteristics for very high-resolution work and is more flexible with regard to modification and extension of the sequence. HMQC will be described as it lays the foundations for the widely used and closely related long-range correlation experiment known as heteronuclear multiple bond correlation (HMBC) that is also described in this chapter. Historically, the HMQC experiment was favoured by the chemical community and HSQC by biological spectroscopists. This was, at least in part, due to the manner in which the early experiments were presented, viz HMQC for ^{1}H–^{13}C correlations in small molecules and HSQC for ^{1}H–^{15}N correlations in proteins. Both techniques employ the optimum approach to establishing heteronuclear connectivity utilising proton detection and follow the general scheme of Fig. 7.1d. However, this approach demands the suppression of the parent resonance arising from protons bound to nuclides with $I \neq ½$, ^{1}H–^{12}C and ^{1}H–^{14}N most commonly. This is the dominant line observed in 1D proton spectra but is merely a source of interference in heteronuclear correlations since it is only the low-intensity *satellites* that can give rise to the desired correlations, that is, the ^{1}H–^{13}C or ^{1}H–^{15}N protons in these examples. The necessary suppression is nowadays most effectively executed by the application of pulsed field gradients (PFGs) which have had an enormous impact on heteronuclear correlation spectroscopy in particular.

7.3.1 Heteronuclear Single-Quantum Correlation

The 2D HSQC spectrum provides a simple map of connectivities in which a crosspeak correlates two attached nuclei, as seen in the ^{1}H–^{13}C correlation spectrum of menthol **7.1** (Fig. 7.2). This illustrates three of the most significant features of the experiment when applied to routine structural problems. The first is the ability to transfer known proton assignments, determined with the methods described in the previous chapter. The second is the dispersion of the proton resonances according to the heteronuclear shift, which itself can aid the initial interpretation of the proton spectrum. For example, the

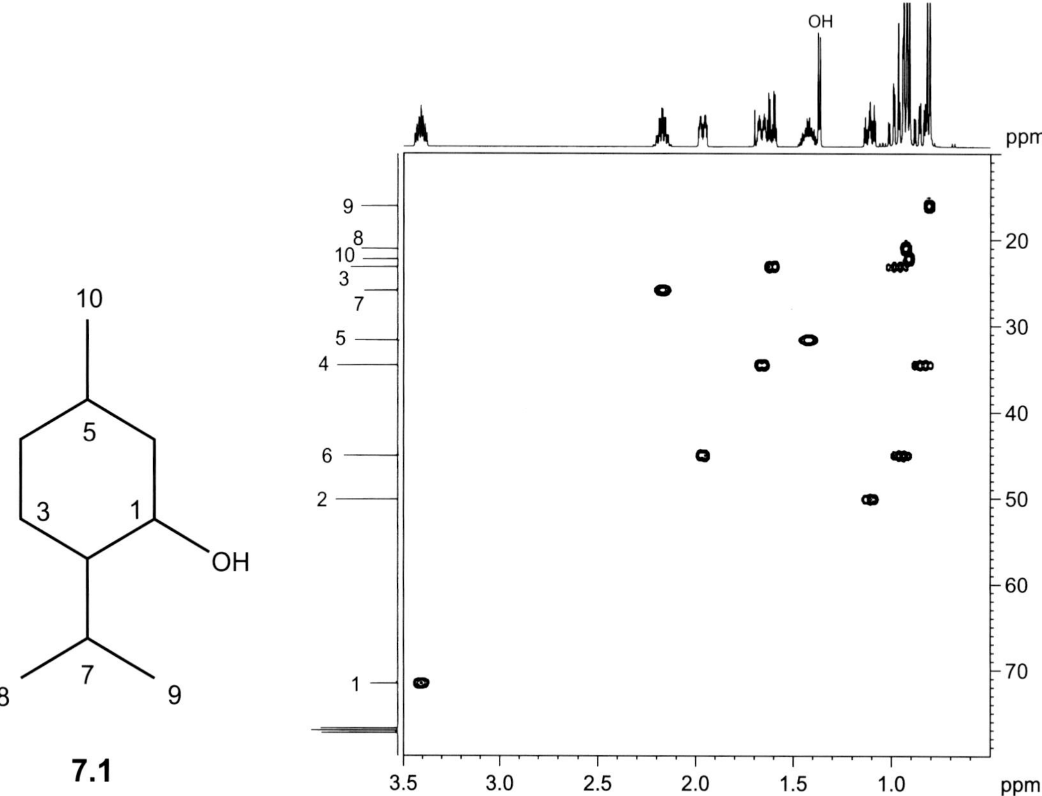

FIGURE 7.2 **The 500 MHz HSQC single-bond correlation spectrum of menthol 7.1.** The conventional 1D proton and carbon spectra are also shown for reference.

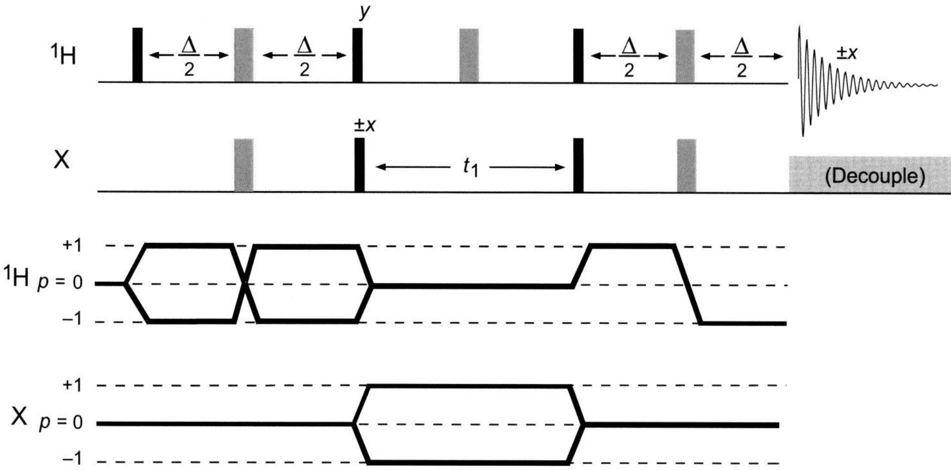

FIGURE 7.3 **The HSQC sequence and associated coherence transfer pathway.** The experiment uses the INEPT sequence to generate transverse X magnetisation which evolves and is then transferred back to the proton by an INEPT step in reverse ($\Delta = 1/2J_{XH}$).

region between 0.7 and 1.2 ppm of Fig. 7.2 contains a number of overlapped proton resonances which defy direct analysis even at 500 MHz. A clearer picture of events in this region emerges when these are dispersed along the ^{13}C dimension, revealing the presence of seven distinct groups. This property becomes increasingly valuable as the complexity and overlap within proton spectra increases. The final feature which can prove surprisingly useful in structural assignment is the ability to identify *diastereotopic* geminal pairs. These are not always readily identifiable in COSY spectra owing to the lack of differentiation of geminal and vicinal couplings, and can lead to ambiguity in proton assignment. Only for these geminal pairs will two correlations to a single carbon resonance be observed, as seen for the correlations of C6, C4 and C3 in Fig. 7.2. In contrast to previous *homonuclear* 2D spectra encountered in this book, *heteronuclear* shift correlation spectra lack a diagonal and are not symmetrical about $f_1 = f_2$, a simple consequence of there being different nuclides represented in the two dimensions.

7.3.1.1 HSQC Sequence

The HSQC experiment [1] follows the scheme of Fig. 7.1d in which transverse (single-quantum) magnetisation of the heteronuclear spin evolves during the t_1 period ($\Sigma p = \pm 1$) (Fig. 7.3). The transverse heteronuclear magnetisation is generated by polarisation transfer from the attached protons via the INEPT sequence, exactly as described for the 1D experiment in Section 4.4.2 (compare the first part of Fig. 7.3 with that of Fig. 4.24a). Thus, during the INEPT period Δ, anti-phase proton magnetisation develops with respect to $^1J_{XH}$ and is maximised by setting Δ to $1/2J_{XH}$. Accepting a compromise value of around 145 Hz for $^1J_{CH}$ (Table 7.2), Δ is therefore typically set to ~3.4 ms for carbon-13 correlations. For nitrogen-15 correlations, $^1J_{NH}$ is typically 90 Hz and Δ therefore 5.6 ms. X-nucleus magnetisation evolves during t_1 with the proton 180 degree pulse at its midpoint refocusing ^{1}H–X coupling evolution, thus decoupling the ^{1}H–X interaction, so only heteronuclear chemical shifts remain in f_1. Following t_1, heteronuclear magnetisation is transferred back onto the proton by an INEPT step in reverse to produce, once again, in-phase proton magnetisation for detection. Since the proton coupling is refocused, it is possible to employ X-spin decoupling during the acquisition to collapse the doublet satellite structure, thus

TABLE 7.2 Typical Ranges for One-Bond Carbon–Proton Coupling Constants

Proton Environment	Typical $^1J_{CH}$ range (Hz)
Aliphatic, CH_n–	125–135
Aliphatic, CH_nX (X = N, O, S)	135–155
Alkene	155–170
Alkyne	240–250
Aromatic	155–165

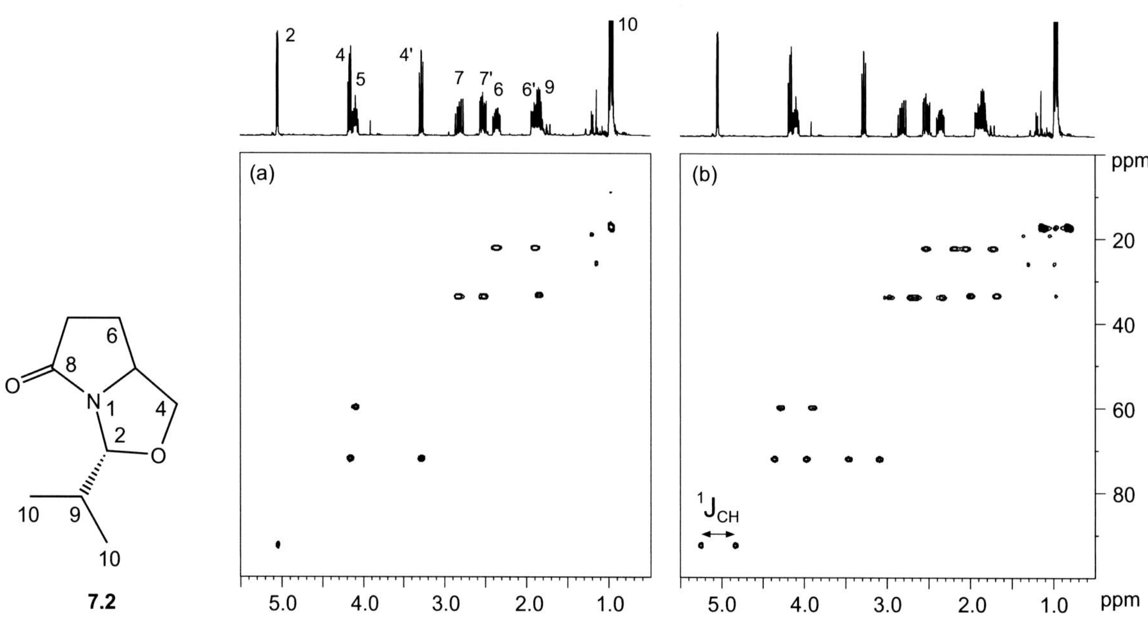

FIGURE 7.4 **The 400 MHz heteronuclear correlation spectra of 7.2.** These spectra were recorded (a) with and (b) without carbon decoupling during data collection. In the absence of decoupling, each crosspeak appears with doublet structure along f_2 arising from $^1J_{CH}$. These doublets are merely the usual ^{13}C satellites observed in the 1D proton spectrum. Data were collected as 256 t_1 increments of 2 transients each. These were processed with $\pi/2$-shifted sine bells in both dimensions and presented in phase-sensitive mode. Zero-filling once in t_1 resulted in digital resolutions of 4 and 40 Hz/pt in f_2 and f_1, respectively.

doubling the signal-to-noise ratio and producing a single crosspeak per correlated spin pair. The doublet structure for each correlation seen in the absence of X-spin decoupling reflects the $^1J_{XH}$ coupling constant and provides a direct measure of this parameter (Fig. 7.4). The overall transfer process may be summarised using a shorthand notation indicating these key transfer and encoding stages thus:

$$^1\text{H} \xrightarrow[\text{INEPT}]{^1J_{XH}} \text{X} \xrightarrow[t_1]{\delta_X} \text{X} \xrightarrow[\text{R-INEPT}]{^1J_{XH}} {}^1\text{H} \Rightarrow \text{Detect}_{(\text{X-dec.})}$$

A key feature of HSQC is the presentation of only heteronuclear chemical shifts along f_1 with all proton coupling interactions removed. This allows high resolution to be obtained in this frequency dimension, this being a notable advantage over the HMQC experiment, as explained later. Its potential drawback is the larger number of pulses it employs, especially 180 degree pulses on the heteronucleus, potentially leading to intensity losses from rf inhomogeneity, pulse miscalibration or off-resonance excitation. This has greater significance for ^{13}C than ^{15}N owing to the greater frequency spread. Such losses may be minimised by careful probe tuning and by the use of frequency swept adiabatic pulses in place of hard 180 degree pulses. These are effective over wider bandwidths than single high-power pulses and now find widespread use in heteronuclear experiments (see descriptions in Section 12.2). The other critical requirement for HSQC is the suppression of the dominant parent resonance which is most effectively achieved via PFGs, as described later.

One subtle point to be aware of is that although this sequence (and HMQC later) nominally detects single-bond correlations, in exceptional circumstances these may be missing and longer range correlations may appear. This occurs when the actual coupling constant involved is far from the value assumed when calculating Δ, a situation most likely to occur for alkynes where the one-bond and two-bond (**H–C≡C**) couplings are unusually large (Table 7.2). Thus, one-bond correlations could well be weak or absent whilst a large two-bond coupling (>50 Hz) can be sufficient to produce a crosspeak, meaning care should be taken not to confuse this with a single-bond correlations if dealing with these systems.

7.3.1.2 Interference From Parent ^1H–^{12}C/^1H–^{14}N Resonances

The HSQC sequence aims to detect only those protons that are bound to a spin-½ heteronucleus, or in other words only the satellites of the conventional proton spectrum. In the case of ^{13}C, this means that only one in every 100 proton spins contributes to the 2D spectrum (all others being attached to NMR-inactive ^{12}C) whilst for ^{15}N with a natural abundance of a mere 0.37%, only one in ~300 contributes. When the HSQC FID is recorded, all protons will induce a signal in the receiver

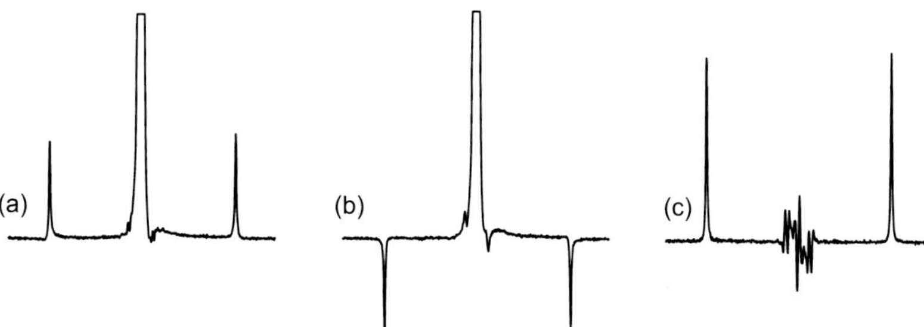

FIGURE 7.5 **Selection of satellite resonances through phase-cycling.** The phase of the carbon-13 satellites can be inverted by inverting the phase of the first 90 degree carbon pulse (a vs. b). Subtraction of these two data sets, by inverting the receiver phase also, cancels the parent ^{1}H–^{12}C resonance but reinforces the satellites (c).

on each scan and the unwanted resonances, which clearly represent the vast majority, must be removed if the correlation peaks are to be revealed. In principle, this can be achieved through phase cycling by inverting the first ^{13}C 90 degree pulse on alternate scans, whereby the phase of the ^{13}C satellites are themselves inverted whilst the ^{12}C-bound protons remain unaffected (Fig. 7.5). Simultaneous inversion of the receiver will thus lead to cancellation of the unwanted resonances with corresponding addition of the desired satellites.

The problem with this scheme is that clean suppression of the unwanted resonances is unlikely to be achieved by phase cycling alone, with residual signals contributing to undesirable bands of t_1 noise in the resulting spectra which may mask the genuine correlations. Nowadays, signal selection in HSQC is achieved almost exclusively through the use of PFGs which attenuate the parent proton resonance in the probe *before* data collection begins, and thus represent the ultimate approach to signal suppression (Fig. 7.6c). Attenuation ratios in excess of 1000:1 are readily achieved in a single scan, meaning suppression of the parent signals is complete and conveniently implemented [2]. Phase cycling is also not essential because signal selection is achieved solely by gradient refocusing, and in situations where sample quantities are not limiting these experiments may be performed within a matter of minutes.

The gradient-selected sequence (Fig. 7.7) employs a suitable combination of gradients that refocus only those responses that have followed the desired transfer pathway. To understand the signal selection process, consider the action of each gradient in turn, paying due attention to coherence orders p (as represented in the coherence transfer pathway) and magnetogyric ratios of the participating spins. The coherence transfer pathway represents only the pathway we wish to preserve, others are not shown since they will not be selected. Assuming the gradients have the same profile and are of the same duration but differ only in their strengths G_n, it is straightforward to summarise the phase induced by the gradients using the shorthand notation of Section 5.4. Thus, for ^{1}H–^{13}C HSQC, the first gradient of Fig. 7.7 acts when carbon has coherence order $p = +1$ (single-quantum coherence), so the effect of the gradient is written $G_1(\gamma_C)$. The second gradient acts on single-quantum proton magnetisation ($p = -1$), so the term becomes $G_2(-\gamma_H)$. To preserve this pathway, the overall phase induced by the gradients must be zero:

$$G_1(\gamma_C) + G_2(-\gamma_H) = 0 \tag{7.2}$$

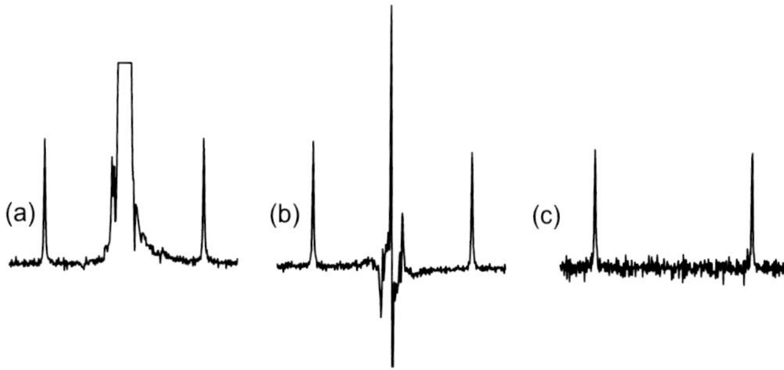

FIGURE 7.6 **A comparison of signal suppression methods used in proton-detected heteronuclear correlation experiments (see descriptions in text).** Spectrum (a) is taken from a conventional 1D proton spectrum without suppression of the parent resonance and displays the required ^{13}C satellites. Other spectra are recorded with (b) phase-cycling and (c) PFGs to remove the parent line. All spectra were recorded under otherwise identical acquisition conditions and result from two transients. Complete suppression can be achieved with gradient selection, but at some cost in sensitivity in this case (see text).

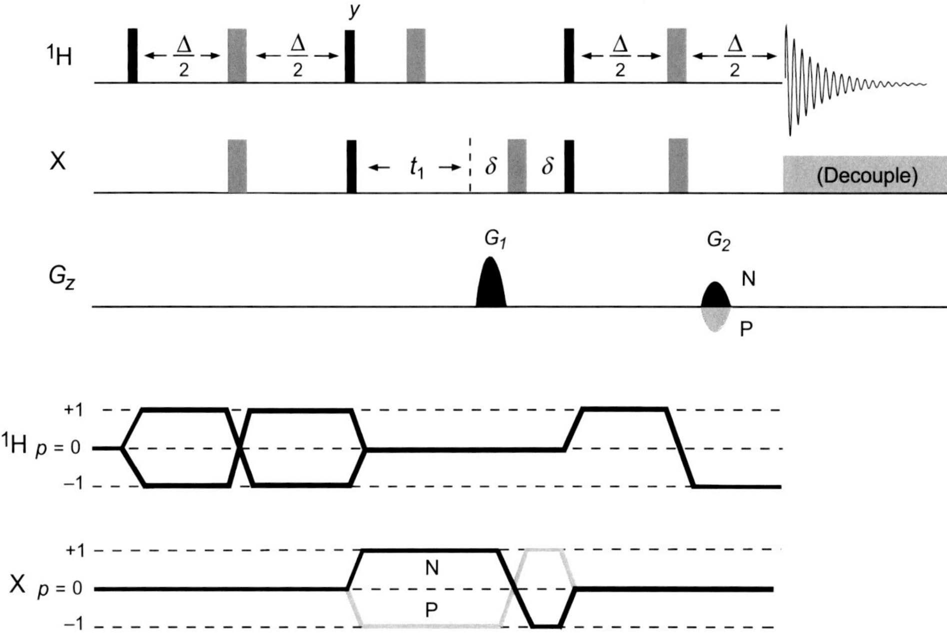

FIGURE 7.7 **A gradient-selected, phase-sensitive HSQC sequence using the echo–antiecho approach.** The N- and P-type pathways are selected by the last gradient.

Note that only the *ratios* of the γ values become important in heteronuclear experiments, and since $\gamma_H/\gamma_C = 4$ the expression is simplified to:

$$G_1(1) + G_2(-4) = 0 \tag{7.3}$$

A suitable gradient ratio that will satisfy this expression and lead to the desired signal selection would therefore be $G_1{:}G_2 = 4{:}1$ for ^{1}H–^{13}C HSQC. By analogy, the appropriate combination for ^{1}H–^{15}N correlations would be 10:1 since $\gamma_H/\gamma_N = 10$. With the appropriate gradient ratio, it remains to ensure the gradients are of sufficient strength to achieve the desired suppression.

To achieve a phase-sensitive presentation, it is necessary to retain both N- and P-type ^{13}C pathways during t_1. This, however, is fundamentally impossible in a single scan if gradients are applied during the t_1 period since no gradient combination is able to refocus *simultaneously* signals from both the +1 and −1 pathways. For example, the net effect of the 4:1 gradient combination on the ^{13}C P-type pathway (coherence order −1 during t_1) will be:

$$G_1(-\gamma_C) + G_2(-\gamma_H) = 4 \times (-1) + 1 \times (-4) = -8 \tag{7.4}$$

and the signal will remain dephased. There are two basic approaches to overcoming this limitation: either to avoid the use of gradients during t_1 altogether [3,4] so allowing use of the conventional States or TPPI methods of quad-detection, or to collect the N- and P-type signals on alternate scans and combine them via the echo–antiecho method of processing [5–7] (Section 5.4.2). Fig. 7.7 shows the more common scheme based on the echo–antiecho method. To obtain pure-phase spectra it is necessary to refocus carbon-13 chemical shift evolution that occurs during the gradient pulses, so the first is placed within a spin-echo immediately following t_1, whilst the second can be applied during the usual INEPT refocusing period. The two different pathways may be collected by inverting the sign of the last gradient, so, for example, the N-type is refocused with gradient ratios 4:1 and the P-type with 4:−1. The experiment may be summarised using the reduced notation as:

$$^1\mathrm{H} \xrightarrow[\text{INEPT}]{^1J_{XH}} \mathrm{X} \xrightarrow[t_1]{\delta_X} \mathrm{X}_{(\text{grad.encode})} \xrightarrow[\text{R-INEPT}]{^1J_{XH}} {}^1\mathrm{H}_{(\text{grad.decode})} \Rightarrow \mathrm{Detect}_{(\mathrm{X-dec.})}$$

The ability to completely suppress the parent ^{1}H–^{12}C or ^{1}H–^{14}N resonances produces spectra that are largely devoid of the bands of t_1 noise that may otherwise plague the experiment. This is illustrated in Fig. 7.8 which shows a section of the gradient-selected ^{1}H–^{15}N HSQC spectrum of the carbopeptoid **7.3** recorded with ^{15}N decoupling, plotted conventionally and with the contour levels reduced to show baseline thermal noise.

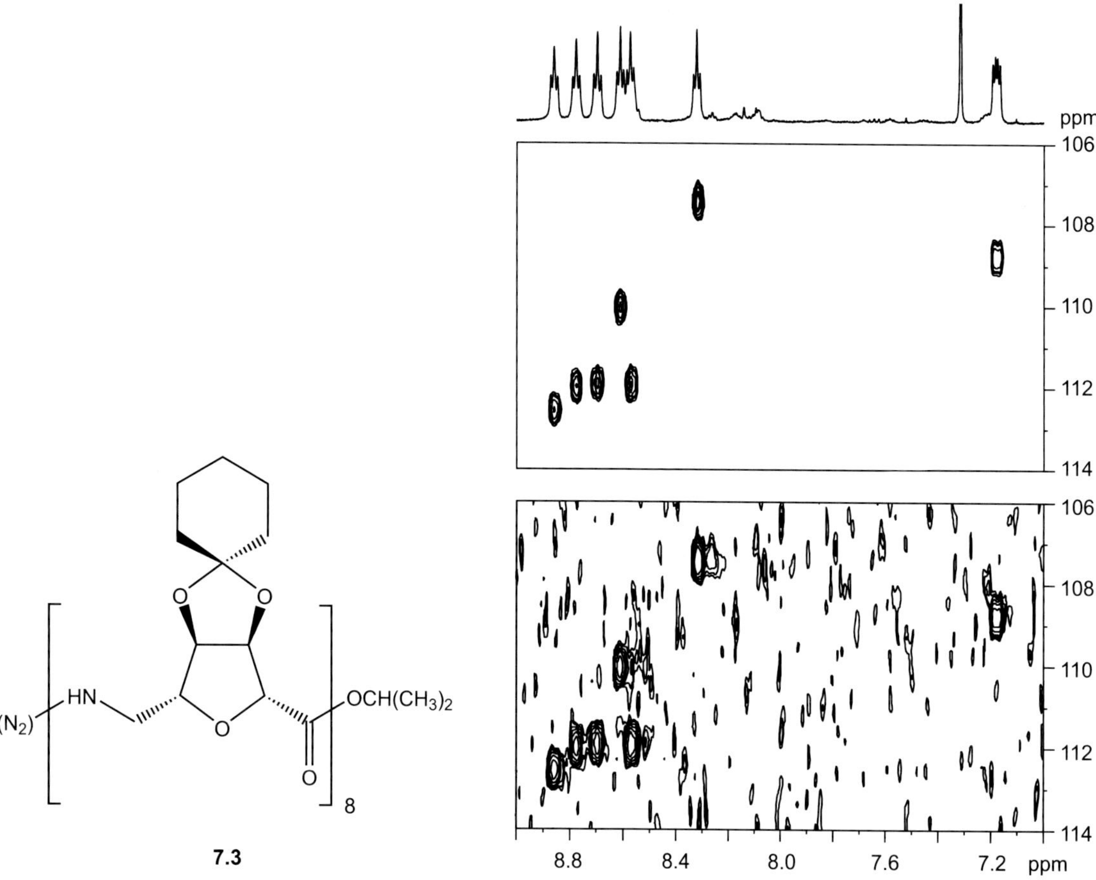

FIGURE 7.8 **A gradient-selected HSQC spectrum of the carbopeptoid 7.3 at natural ^{15}N abundance.** Spectra are plotted at high- and at low-contour levels to show the thermal noise floor. No t_1-noise artefacts remain from the parent ^{1}H–^{14}N resonances (^{15}N chemical shifts are referenced to external liquid ammonia).

In the absence of PFGs, the traditional approach for large molecules [8] has been to apply a strong spin-lock period during the first INEPT sequence of the HSQC experiment. This destroys all magnetisation not aligned with the spin-lock axis and should suppress unwanted parent signals. This tends to enjoy only partial success, and if gradients are not available the BIRD-HMQC experiment described in Section 7.3.3 will likely yield better results.

7.3.1.3 Sensitivity Improvement: PEP

The signal losses associated with gradient selection that arise from selective refocusing of a single N- or P-type pathway can be ameliorated to some extent by the use of the sensitivity improvement scheme now known as the *preservation of equivalent pathways* (PEP) [9–11]. This can yield sensitivity gains up to a factor of 2 relative to the standard sequences by recovering one of the magnetisation components that is usually lost in conventional HSQC methods and is applicable to both non-gradient and gradient-selected experiments. In the HSQC experiment, heteronuclear single-quantum magnetisation that exists at the end of the t_1 period may be decomposed into two orthogonal components that lie along the x- and y-axes. The subsequent 90 degree(H,X) pulses are able to transfer only one of these back into single-quantum proton magnetisation whilst the other gives rise to ^{1}H–X multiple-quantum coherence that remains unobservable, meaning only one half of the initial proton magnetisation contributes to the observed signal. The PEP method extends the HSQC sequence (Fig. 7.9) so as to reconvert the unobservable component into observable single-quantum coherence for simultaneous detection with its orthogonal partner. Thus, at point A in the sequence of Fig. 7.9 (the point at which detection would begin in the standard HSQC experiment) the refocused ^{1}H single-quantum magnetisation produced by the reverse INEPT step is placed along the z-axis by the 90 degree(^{1}H) pulse for storage. Subsequently, the multiple-quantum signals are transferred into single-quantum proton magnetisation by the simultaneous 90 degree(X) pulse, effectively initiating a second reverse INEPT transfer. Within this, the anti-phase magnetisation so generated undergoes refocusing of the heteronuclear $^1J_{XH}$ coupling (as ultimately required for detection in the presence of X-spin decoupling) at which point the component held in storage along

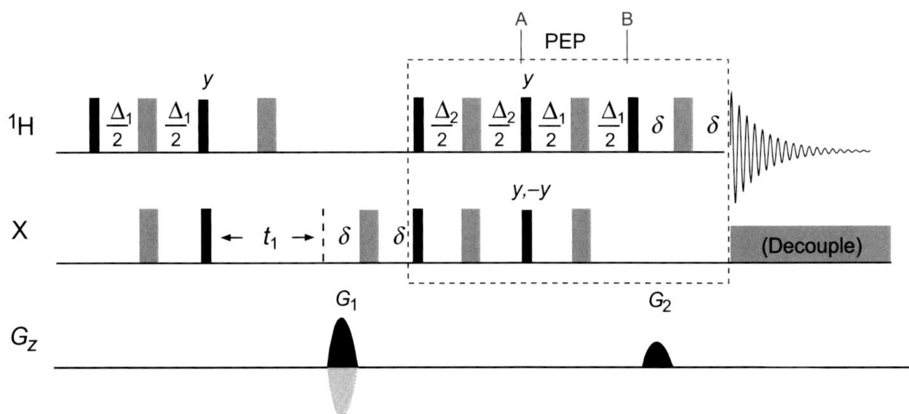

FIGURE 7.9 **The PEP-HSQC sequence for sensitivity improvement; see text for discussions.**

the *z*-axis is again placed into the transverse plane by the final 90 degree(^{1}H) pulse (point B, Fig. 7.9). Subsequent gradient refocusing then allows detection of *both* components of proton magnetisation. Since these two components are again orthogonal they would produce phase-twisted lineshapes in the resulting spectrum, so it becomes necessary to record two data sets per t_1 increment in which the phase of one of the 90 degree(X) pulses after t_1 is inverted and the data sets stored separately (this may also be combined with gradient inversion according to the echo–antiecho protocol for f_1 quadrature detection). This inverts the phase of one of the two orthogonal components such that appropriate addition and subtraction of the data sets then separates these components and allows for quadrature detection with absorption-mode lineshapes.

The maximum theoretical gain in signal intensity provided by the PEP extension when using the echo–antiecho scheme is a factor of 2 although experimental gains are likely to be less than this due to losses through relaxation and pulse imperfections arising from the PEP extension. Furthermore, the whole process operates to give retention of both pathways most efficiently only in the case of XH pairs but to a lesser extent when the heteroatom is bound to more than one proton. In situations when only XH pairs are to be observed, the INEPT delays should be optimised for $\Delta_1 = \Delta_2 = 1/2J_{XH}$ whereas for the detection of XH_2 or XH_3 as well the Δ_2 period is best reduced to $1/4J_{XH}$ [11]. In this case the theoretical enhancement factors become 1.7 (XH), 1.4 (XH_2) and 1.2 (XH_3). Fig. 7.10 compares the signal intensities for a disaccharide anomeric CH resonance observed for HSQC with and without the PEP extension and shows a 1.4-fold sensitivity gain using the PEP methodology when selecting for all multiplicities.

7.3.1.4 Practical Set-Up

The high crosspeak dispersion typically associated with heteronuclear shift correlation experiments alongside the lack of any requirement for well-defined crosspeak fine structure means HSQC experiments can be recorded with rather low digital resolution for routine applications, enhancing their time efficiency. Acquired digital resolutions of 5 Hz/pt in the proton f_2 dimension and only 50 Hz/pt in the heteronucleus f_1 dimension are generally sufficient to resolve correlations. Improved f_1 resolution can be achieved by linear prediction of the FIDs when sensitivity allows, and/or the digital resolution enhanced by zero-filling

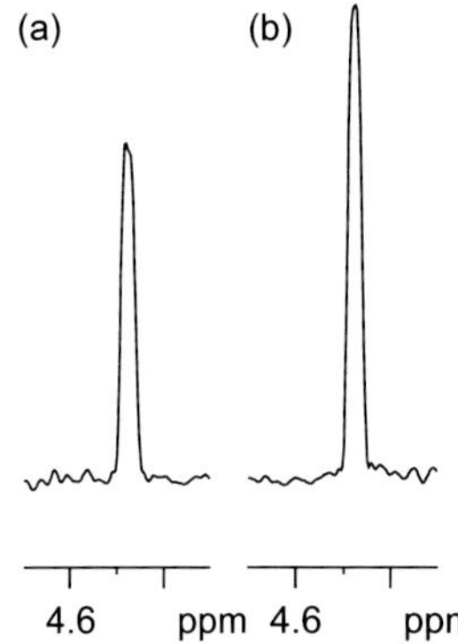

FIGURE 7.10 **Traces through an anomeric CH resonance of a disaccharide taken from (a) HSQC and (b) PEP-HSQC.** Both spectra were collected with PFGs with delays $\Delta_1 = 1/2J_{CH}$ and $\Delta_2 = 1/4J_{CH}$ optimised for $^1J_{CH} = 145$ Hz. Trace (b) demonstrates a 40% signal increase.

(usually at least once). The use of linear prediction can lead to significant time savings when high sample quantities are available, by reducing the number of t_1 increments that must be recorded and computationally regenerating those missing t_1 data points to provide adequate f_1 resolution [12,13]. A more recent alternative to increase f_1 resolution (or alternatively to save time by recording fewer t_1 increments) is to employ non-uniform sampling (NUS), as described in Section 5.2.4. Sampling of typically 50–25% data points is often adequate provided signal-to-noise is sufficient. Beyond this, processing requires only simple apodisation in both dimensions and either shifted squared sine bells or Gaussian windows function well for phase-sensitive data sets.

Proton–carbon HSQC experiments can be surprisingly fast to acquire for routine organic samples, especially when PFGs and probes optimised for proton detection are employed. Repetition rates are dictated by the shorter relaxation times of ^{13}C-bound protons and can therefore be faster than homonuclear COSY spectra, for example, where it is the ^{12}C-bound protons that are monitored. They are usually quicker to obtain and often more informative than the 1D carbon spectrum, often accessible in a matter of minutes, and are now routine experiments in the chemical laboratory. Combining these methods with the spectrum-editing techniques described in the following section further extends their utility as structural tools, so making one less reliant on direct carbon-observe 1D experiments such as APT and DEPT in structural investigations.

7.3.2 Hybrid HSQC Experiments

As was alluded to earlier, HSQC spectra may be further enhanced by combining them with other pulse techniques to produce spectra with altered characteristics or increased information content. These sections describe two of the more useful methods for structure elucidation and also illustrate the manner in which different techniques may be concatenated to produce 'new' experiments. They also illustrate how heteronuclear spin may be used in the editing or simplification of 1D proton spectra.

7.3.2.1 2D Multiplicity Editing

Short of the ability to identify diastereotopic XH_2 groups, the HSQC experiments described so far provide no direct evidence for the multiplicities of XH_n groups giving rise to each correlation, analogous to the way in which broadband-decoupled heteronuclear spectra provide no multiplicity information. Various methods are now available which incorporate editing *within* the 2D heteronuclear correlation experiment itself. These provide more information in routine analysis and require less time than conventional 1D editing methods. Furthermore, such experiments provide access to heteronuclear multiplicity information when small sample quantities preclude the direct observation of X-spin edited spectra, so further extending the range of sample quantities one can consider accessible to structural studies.

Editing may be readily introduced to the HSQC sequence [4,14,15] by the simple addition of a spin-echo after t_1. During this, only the heteronuclear coupling evolves, exactly as in the pulsed 1D J-modulated spin-echo sequence described in Section 4.3.1. Hence, setting $\Delta = 1/2J_{XH}$ produces a 2D correlation spectrum in which XH_2 responses are inverted relative to those of XH and XH_3. In one example of this approach, concatenating this echo with the existing t_1 period produces the modified HSQC scheme of Fig. 7.11 [2,16]. The edited HSQC spectrum of the substituted disaccharide **7.4** is shown in Fig. 7.12 and clearly differentiates CH from CH_2 correlations. A potential problem with these editing approaches is the cancellation of overlapping correlations of opposite phase in crowded regions of 2D spectra, so one need be aware of this caveat for groups having similar chemical shifts in both dimensions.

7.3.2.2 Utilising X-Spin Shift Dispersion

When analysing molecules that display very crowded proton spectra, the 2D homonuclear shift correlation experiments discussed in Sections 6.1 and 6.2 may still provide spectra which are too overlapped to allow complete interpretation. In such

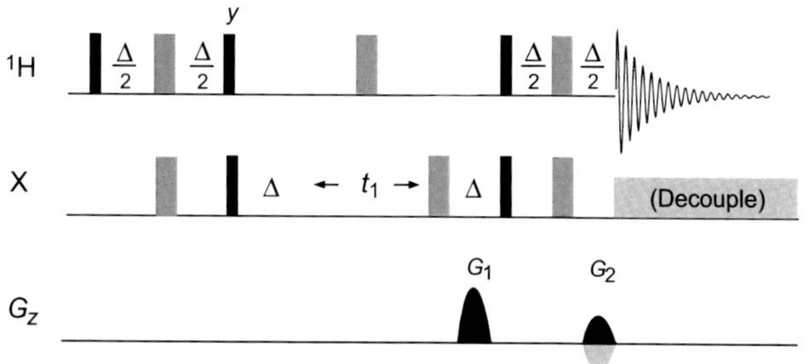

FIGURE 7.11 **The gradient-selected, spin-echo HSQC sequence for multiplicity editing within the 2D correlation experiment.** Setting $\Delta = 1/2J_{XH}$ inverts XH_2 responses relative to those of XH and XH_3.

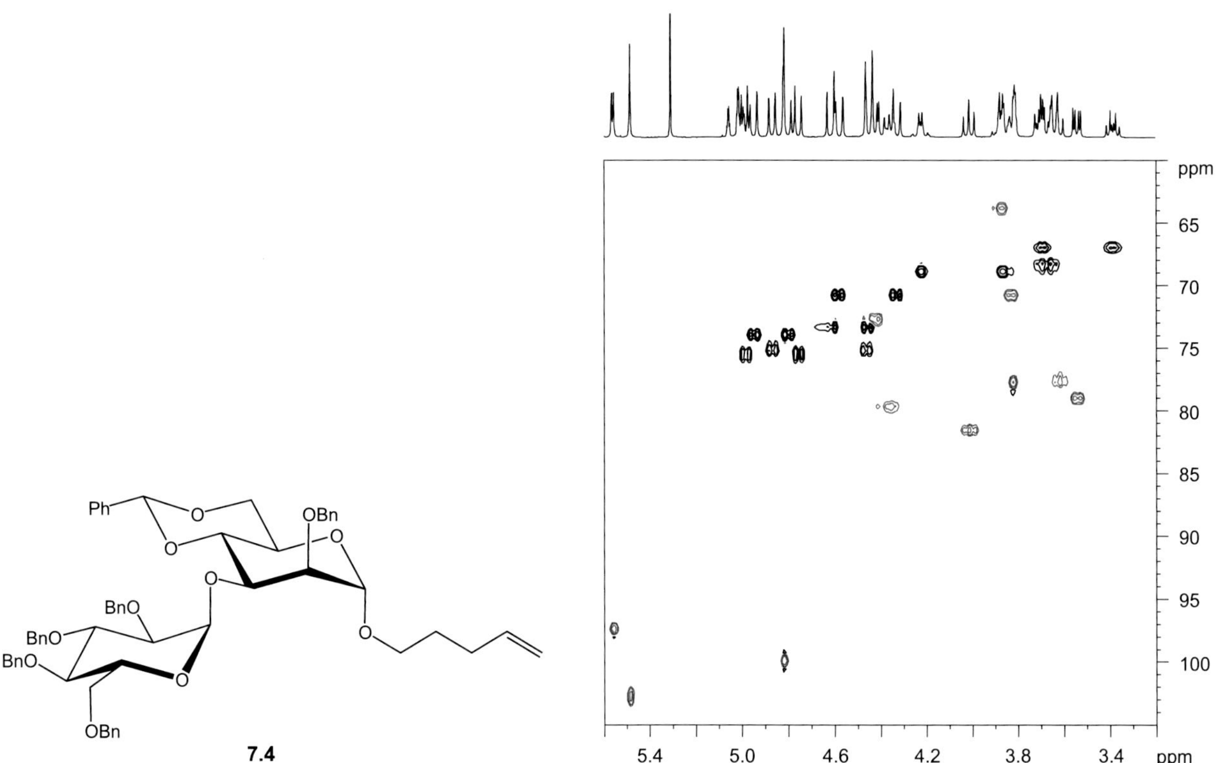

FIGURE 7.12 **The multiplicity-edited HSQC spectrum of the disaccharide 7.4.** In this the positive CH correlations (red) are distinguished from negative CH_2 correlations (black).

instances, the potentially greater dispersion of parent heteroatom chemical shifts can be used as an additional means for separating proton–proton correlations. Furthermore, in cases of exactly overlapping *proton* resonances, one-bond heteronuclear correlation experiments do not provide unambiguous identification of the parent heteroatom, and one approach to overcoming such problems lies in the transfer of heteronuclear correlation information onto neighbouring protons. Thus, adding a TOCSY spin-lock mixing period after the HSQC sequence and immediately prior to data collection [17,18] transfers magnetisation that has returned to the proton from which it originated onto neighbouring J-coupled protons (Fig. 7.13). The process may again be summarised for HSQC-TOCSY in shorthand notation as:

$$^{1}H \xrightarrow[\text{INEPT}]{^{1}J_{XH}} X \xrightarrow[t_1]{\delta_X} X_{(\text{grad.encode})} \xrightarrow[\text{R-INEPT}]{^{1}J_{XH}} {}^{1}H \xrightarrow[\text{SPIN-LOCK}]{^{n}J_{HH}} {}^{1}H_{(\text{grad.decode})} \Rightarrow \text{Detect}_{(X-\text{dec.})}$$

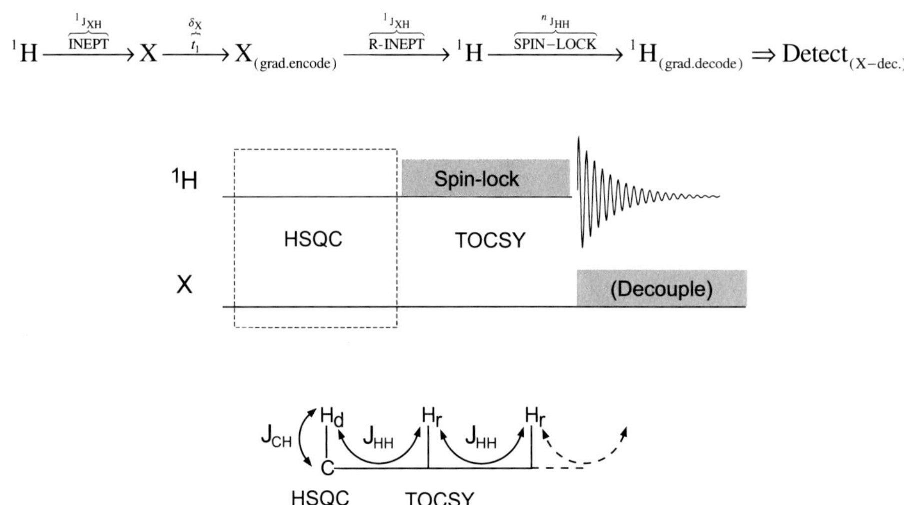

FIGURE 7.13 **The schematic HSQC-TOCSY sequence and the coupling pathway it maps.** Direct correlations are produced for the proton (H_d) bound to the X-heteroatom as in the basic shift correlation sequence, and further relayed correlations are produced for those protons receiving magnetisation through TOCSY transfer (H_r).

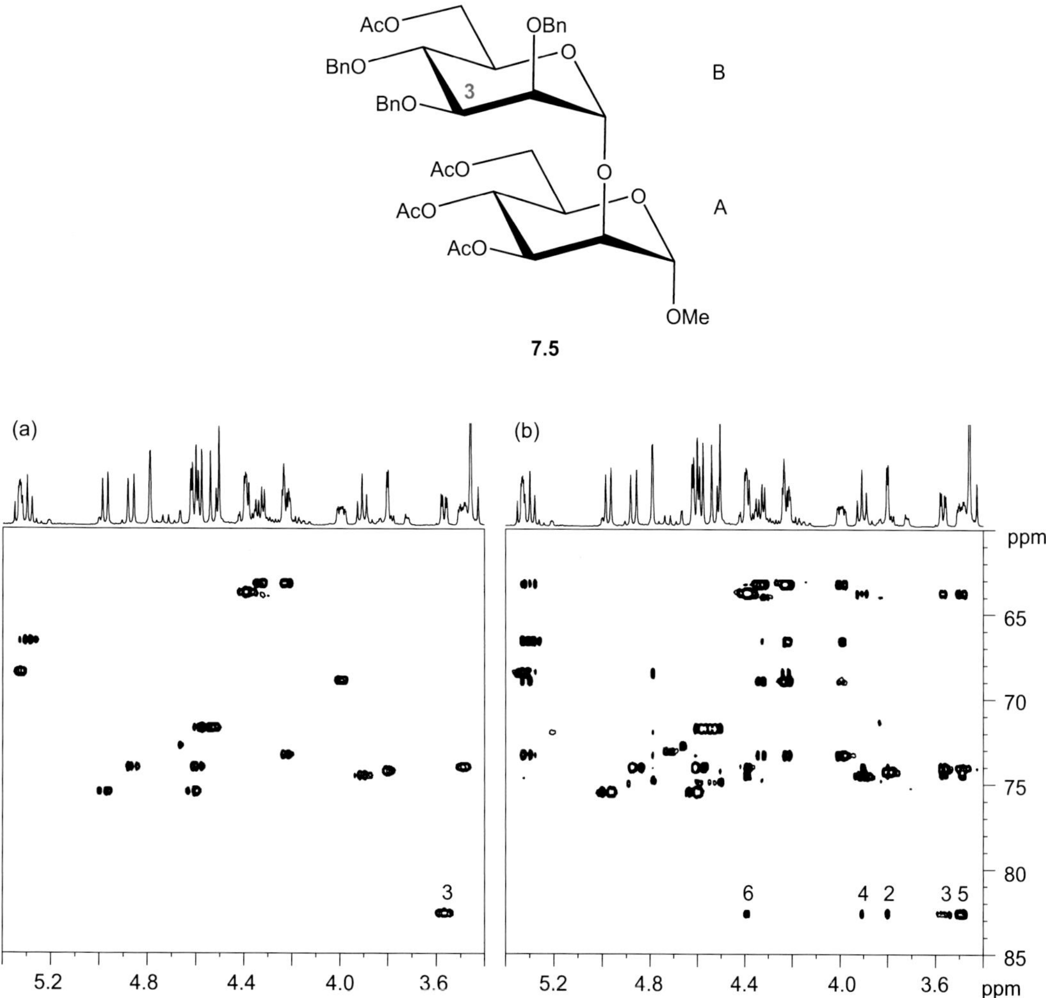

FIGURE 7.14 **Selected regions of the (a) HSQC and (b) HSQC-TOCSY spectra of the disaccharide 7.5.** The proton correlations originating from H3 of ring B are labelled and provide an almost complete map of protons within this ring (only the correlation to H1 is missing).

Extended mixing periods can again be used to *relay* magnetisation along the proton network, potentially providing a complete proton sub-spectrum of the molecular fragment to which the heteroatom belongs. The result may be viewed as an X–spin edited 2D TOCSY experiment, the proton correlations now appearing along rows taken at the hetero-spin f_1 chemical shifts. With sufficient shift dispersion in the X-spin dimension, overlap present in the 2D TOCSY is removed. The HSQC-TOCSY spectrum of the disaccharide **7.5** (Fig. 7.14b) contains numerous additional correlations over the HSQC spectrum (Fig. 7.14a), and can be used to map proton-coupling pathways. This is most clearly seen at the C3 shift of ring B, from which the H3 proton produces TOCSY correlations to all other protons in the ring, with the exception of H1 (which in this case had zero coupling with H2, so causing a breakdown in transfer).

The enormous simplification of crowded spectra in this manner makes this a very powerful technique. However, a point to bear in mind if considering applying these methods is that crosspeaks in the ^{13}C-edited experiment originate only from ^{13}C satellites because of the initial HSQC step, and the experiment is therefore of significantly lower sensitivity than homonuclear 2D TOCSY, in which all protons may participate. The experiment will generally find use after initial stages of investigation of complex spectra using the techniques already presented and where ambiguities remain. The use of long-range heteronuclear correlations described shortly should also be considered in such cases.

Further modifications (the addition of a spin-echo) allow the direct and relayed peaks to be differentiated through inversion of the direct correlations [19] and gradient-selected versions of these sequences without [20,21] and with [4,22] such editing have also been proposed. Although the addition of TOCSY transfer is probably the most useful extension, any homonuclear mixing scheme can, in principle, be added, including a COSY, NOESY or ROESY step.

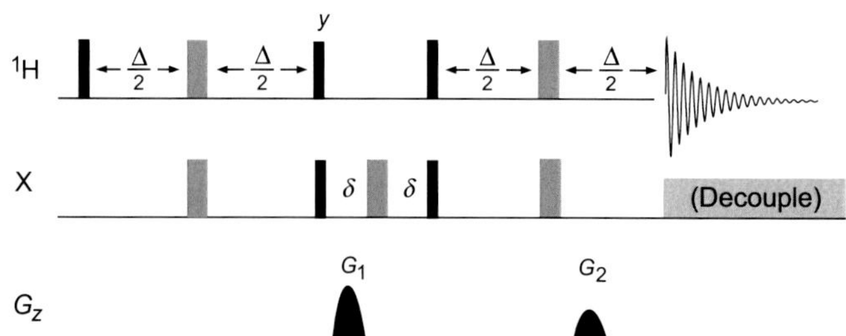

FIGURE 7.15 **The 1D HSQC sequence using gradient selection.** This can be used for editing spectra by selecting only those protons bound to NMR-active heteroatoms.

7.3.2.3 Editing and Filtering 1D Proton Spectra

The idea of using the heteroatom to edit the 2D spectrum as described earlier is equally applicable to the editing of 1D proton spectra. Generally speaking, there are two reasons one may consider doing this. Firstly, one may wish to selectively observe only the protons attached to a heteroatom isotope label. In this way, one uses the label to *filter* the proton spectrum, transferring the selectivity associated with the hetero-label onto the more sensitive proton. Whilst the natural abundance proton satellites of unlabelled positions will also pass the filter, they will be present at significantly lower intensity. Secondly, one may wish to *edit* the proton spectrum according to heteroatom multiplicities to produce subspectra that are the proton analogues of, for example, APT- or DEPT-edited carbon spectra. These analogues provide an alternative and possibly more sensitive route to identifying group multiplicities within a molecule, and allow the analysis of smaller sample quantities by virtue of proton detection.

The HSQC sequence may be transformed into its 1D equivalent to act as a heteronuclear filter simply by removing the incremental t_1 time period (Fig. 7.15). Only magnetisation that has passed via the X-spin will be observed in the final spectrum, and again the suppression of all unwanted signals is greatly improved by the use of PFGs. The selective observation of ^{13}C-labelled glycine in an aqueous mixture is illustrated in Fig. 7.16. One must use caution in this approach if decoupling of the heteroatom is employed during proton detection, due to the application of high rf decoupling powers. To protect the probe, it is therefore necessary to keep acquisition times much shorter than for conventional proton detection and so limit digital resolution, or to avoid the use of decoupling altogether.

A readily implemented approach to generating multiplicity-edited proton spectra employs the 1D variant of the edited HSQC described earlier. An example of this is shown in Fig. 7.17 which distinguishes methine from methylene groups and also eliminates hydroxyl protons. Relative to the standard 1D proton spectrum there is a price to pay in sensitivity because one is again forced to observe the carbon-13 satellites of the conventional proton spectrum, and again caution is required if using heteroatom decoupling. Nevertheless, this editing may prove useful for rapid characterisation of simple molecules and may provide valuable multiplicity data when small sample quantities preclude the use of carbon-detected 1D editing methods.

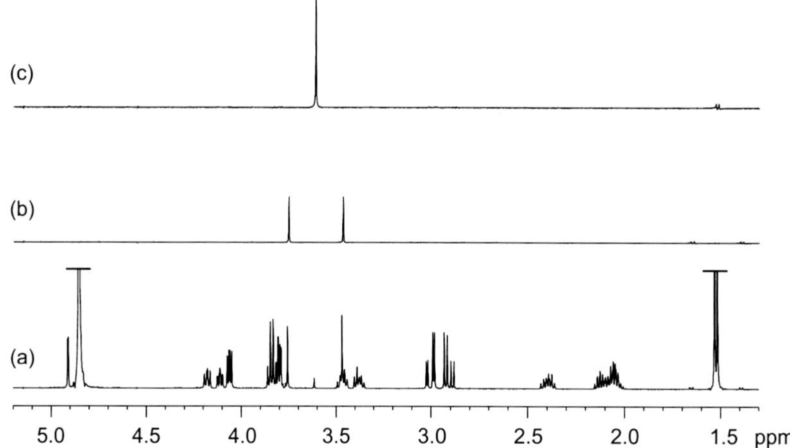

FIGURE 7.16 **The selective observation of protons bound to a carbon-13 label (2–^{13}C–glycine) with a gradient-selected 1D heteronuclear filter.** (a) The 1D proton spectrum and the filtered spectrum recorded (b) without and (c) with carbon-13 decoupling during acquisition.

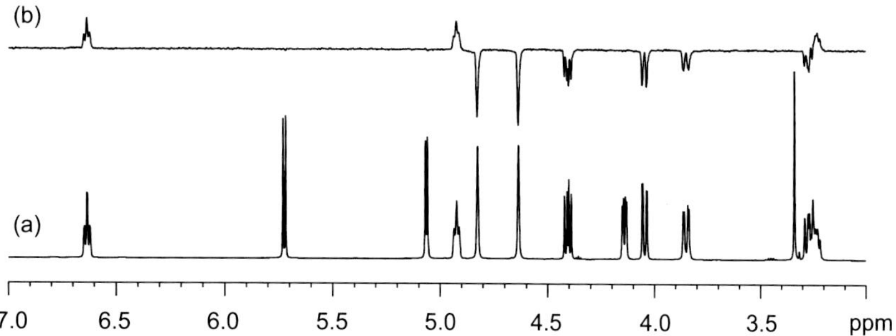

FIGURE 7.17 **Editing of a proton spectrum according to carbon multiplicities.** In (b) methylene proton multiplets are inverted and hydroxyl resonances fully suppressed with PFGs.

7.3.3 Heteronuclear Multiple-Quantum Correlation

The heteronuclear multiple-quantum correlation (HMQC) experiment provides an alternative route to detect single-bond correlations, and once dominated organic NMR spectroscopy. Suggested many years ago [23,24], the experiment gained widespread use when a scheme was presented [25] that was able to overcome the technical difficulties associated with parent proton suppression described earlier, before selection through PFGs became commonplace. Nowadays, HSQC has become the primary heteronuclear correlation technique and HMQC finds less use. However, the basic sequence has close similarities with the multiple-bond heteronuclear correlation technique known as HMBC that finds extensive use in structure identification, so for this reason alone is worthy of description. It is also the case that the bilinear rotation decoupling (BIRD)-HMQC experiment can be a more useful technique than HSQC when PFGs are *not* available, as described later.

7.3.3.1 HMQC Sequence

Despite the slightly foreboding title, the basic HMQC sequence is rather simple, comprising only four rf pulses (Fig. 7.18), the operation of which is considered here for a simple ^{1}H–^{13}C spin pair. The sequence starts with proton excitation followed by evolution of proton magnetisation under the influence of one-bond carbon–proton coupling. During a period Δ, anti-phase proton magnetisation develops with respect to $^{1}J_{CH}$, and to maximise this Δ is again set to $1/2J_{CH}$ (it will be shown shortly that proton chemical shift evolution during Δ is later refocused, so need not be considered here). As for HSQC, this anti-phase magnetisation may be transferred to the coupled partner by the action of a subsequent rf pulse and the role of the first carbon pulse in HMQC is to generate proton–carbon multiple-quantum coherence (hence the title of the experiment). Multiple-quantum coherence was described in Section 5.3 as a pooling of the transverse magnetisation of coupled spins, in this case a proton and its directly bound carbon, that evolves coherently but which cannot be directly observed. If one were to start data collection of the proton signal directly after this carbon pulse there would be nothing to detect, provided the Δ delay was set precisely to $1/2J_{CH}$. Note that this is exactly the procedure described in Section 3.5.1 for the calibration of

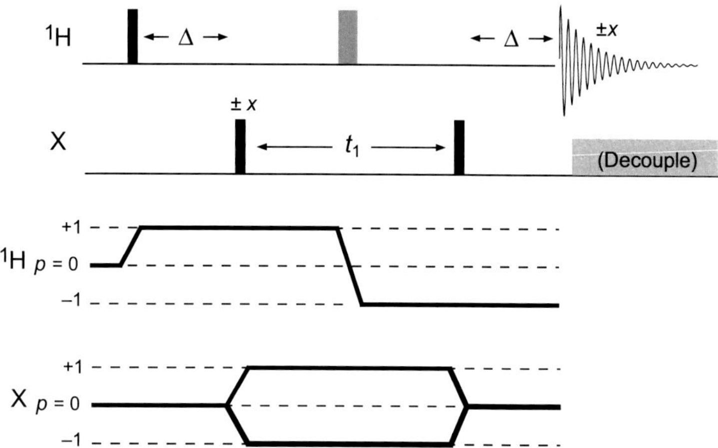

FIGURE 7.18 **The HMQC sequence and associated coherence transfer pathway.** The Δ periods are set to $1/2J_{XH}$ defocusing and subsequent refocusing of the one-bond heteronuclear coupling.

pulses on the indirectly observed spin and, allowing for the proton 180 degree pulse, is also closely related to the start of the DEPT sequence (Section 4.4.3) which similarly relies on the generation of multiple-quantum coherence.

This coherence is, in fact, a combination of both heteronuclear double- and zero-quantum coherence, as represented on the coherence transfer pathway of Fig. 7.18. Thus, for example, $^{1}H_{p} = +1$ and $^{13}C_{p} = +1$ corresponds to *double*-quantum coherence ($\Sigma p = 2$) whilst $^{1}H_{p} = +1$ and $^{13}C_{p} = -1$ is *zero*-quantum coherence ($\Sigma p = 0$). The salient point at this stage is how such coherences evolve during the subsequent t_1 period. Since they contain terms for both transverse proton and carbon magnetisation, they will evolve under the influence of *both proton and carbon chemical shifts* (although a feature of multiple-quantum coherences is that they will *not* evolve under the 'active' coupling, so J_{CH} need not be considered at this point). What is ultimately required, however, is frequency labelling according to *carbon* shifts only, since this is what one wishes to characterise in the indirectly detected f_1 dimension. To remove the effect of proton shifts during t_1, a spin-echo is incorporated by placing a proton 180 degree pulse at the midpoint of t_1, so by the end of the evolution period these shifts have refocused and thus have no influence in f_1. Evolution of the carbon shifts is unaffected by the proton pulse, so these remain to produce the desired frequency labelling. The final carbon pulse then reconverts the multiple-quantum coherence back to observable single-quantum proton magnetisation which is once again anti-phase with respect to $^{1}J_{CH}$. To enable application of carbon decoupling during data acquisition, heteronuclear coupling is again refocused during a period Δ, by analogy with HSQC. Conventional quadrature detection in f_1 is implemented by incrementing the phase of the carbon pulse prior to t_1 according to either the States or TPPI procedures to yield a phase-sensitive display.

Gradient selection in HMQC is also desirable and a scheme utilising the echo–antiecho approach to phase-sensitive data collection is shown in Fig. 7.19. Using the approach described for HSQC, the influence of the gradients may be understood with consideration of the fact that they may act on both proton and carbon magnetisation present as multiple-quantum coherence. Thus, the first gradient of Fig. 7.19 acts when proton and carbon have coherence orders $p = -1$ and $p = +1$, respectively (heteronuclear double-quantum coherence) in the N-type pathway, so the effect of the gradient is written $G_1(-\gamma_H + \gamma_C)$. For the second gradient this becomes $G_2(-\gamma_H - \gamma_C)$ and for the third $G_3(-\gamma_H)$. To preserve this pathway, the overall phase induced by the gradients must again be zero:

$$G_1(-\gamma_H + \gamma_C) + G_2(-\gamma_H - \gamma_C) + G_3(-\gamma_H) = 0 \tag{7.5}$$

which simplifies to:

$$G_1(-3) + G_2(-5) + G_3(-4) = 0 \tag{7.6}$$

There are a number of gradient ratios that will satisfy this expression and lead to the desired signal selection, one example being 2:−2:1 and −2:2:1 for N-type and P-type pathways, respectively. To obtain pure-phase spectra, it is again necessary to refocus carbon-13 chemical shift evolution that occurs during gradient pulses, so an additional spin-echo containing these pulses follows t_1.

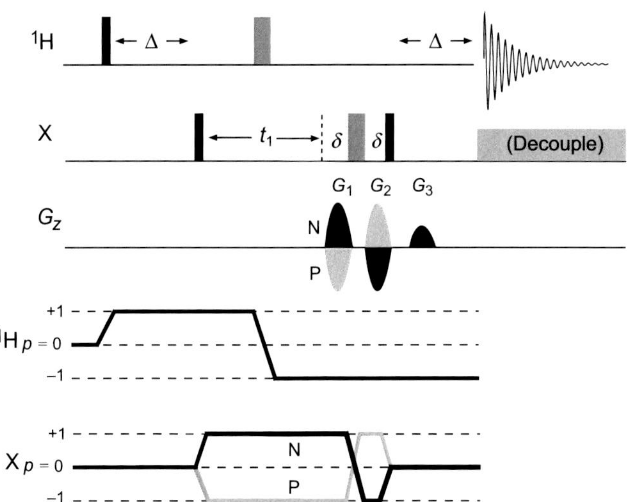

FIGURE 7.19 **The gradient-selected HMQC sequence yielding phase-sensitive data via the echo–antiecho procedure.** The N- and P-type pathways are selected by inversion of the first two gradients which are placed within a spin-echo to refocus shift evolution.

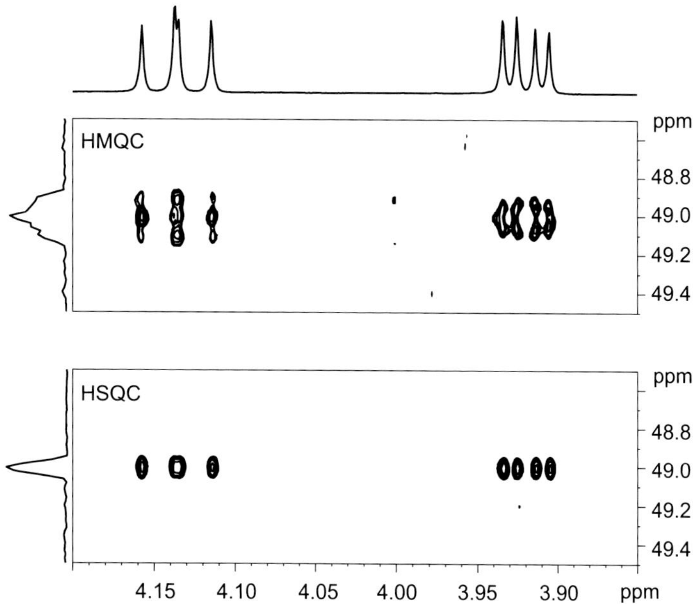

FIGURE 7.20 **A comparison of experimental crosspeaks taken from HMQC and HSQC spectra acquired under identical conditions of high f_1 resolution (2.5 Hz/pt).** The upper 1D trace is taken from the conventional 1D proton spectrum, and the vertical traces are f_1 projections from the 2D spectra. The additional broadening in the HMQC spectrum arises from unresolved homonuclear proton couplings in f_1.

7.3.3.2 Influence of Homonuclear Proton Couplings

It was noted earlier that proton magnetisation will also evolve according to its chemical shift during Δ after initial excitation. However, this is exactly refocused during the second Δ period because of the presence of the proton spin-echo, so does not give rise to phase errors in the proton dimension. Of greater concern is the evolution of proton magnetisation in the two Δ periods and of heteronuclear multiple-quantum coherence during t_1, under the influence of homonuclear proton–proton couplings. Since these homonuclear couplings are *not* refocused by a spin-echo they will evolve in both Δ periods and potentially contribute to unwanted phase errors in the proton dimension. However, Δ is set according to $^1J_{CH}$ which is typically at least an order of magnitude greater than J_{HH}, so that in practice the degree of evolution due to proton–proton coupling is rather small. In other words, Δ is too short for significant evolution to occur and the small phase errors that may arise are rarely troublesome (as described later, this is not the case when seeking correlations through long-range heteronuclear couplings that are comparable in size with homonuclear proton couplings). In contrast, multiple-quantum coherence evolves during t_1 under *passive* J_{HH} couplings without being refocused, and thus the final carbon resonances are spread by proton –proton couplings along f_1. This may seem a little odd, but it is a consequence of the fact that during t_1 both proton and carbon coherences evolve; we simply choose to remove proton chemical shifts with the spin-echo. In fact, because of the rather low digital resolution used in the carbon-13 dimension, these proton couplings are rarely resolved, but do contribute to undesirable *broadening* of the resonance along f_1. These homonuclear couplings *do not* appear in the f_1 dimension of HSQC, and this feature is the principal difference between the two spectra (Fig. 7.20).

7.3.3.3 BIRD-HMQC: Suppressing Parent Resonances Without Gradient Pulses

Not every NMR spectrometer or probehead is equipped with PFG capabilities, yet it may still be possible to execute the basic heteronuclear shift correlation experiments described earlier. In such cases it is desirable to employ an alternative scheme to remove interfering parent resonances and produce spectra devoid of objectionable artefacts. In the BIRD variant (Fig. 7.21), unwanted ^1H–^{12}C resonances are suppressed prior to the HMQC sequence by an ingenious presaturation scheme. This commences with the inversion of *only carbon-12 bound protons* by the so-called 'BIRD pulse' (the action of which is described later), leaving the carbon-13 bound protons unaffected. Following the inversion, a recovery period τ allows the magnetisation vectors to relax back towards the equilibrium +*z*-axis, until they pass through the *x*–*y* plane (Fig. 7.22). At this point, the HMQC sequence itself begins and because there exists no longitudinal ^1H–^{12}C magnetisation, no transverse component is ever generated for these spins and hence the desired suppression of the resonances is achieved. During detection and the next inversion recovery period, ^1H–^{13}C magnetisation simply relaxes back towards its equilibrium value in readiness for the sequence to begin once more since it is not inverted by the BIRD pulse.

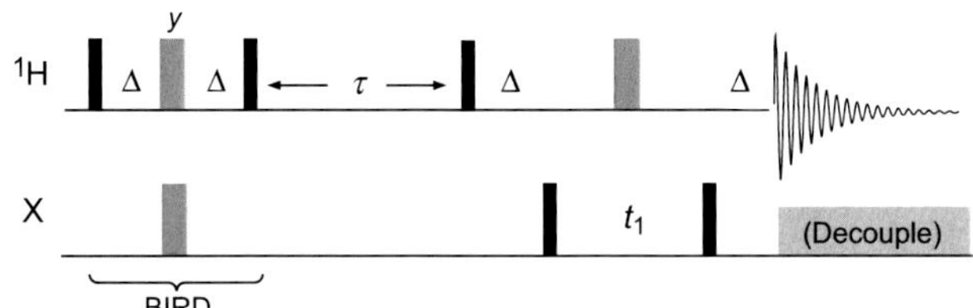

FIGURE 7.21 **The BIRD variant of the HMQC experiment.** The conventional HMQC sequence is employed, but is preceded by the BIRD inversion element and an inversion recovery delay, τ. This procedure ultimately leads to saturation of unwanted parent resonances.

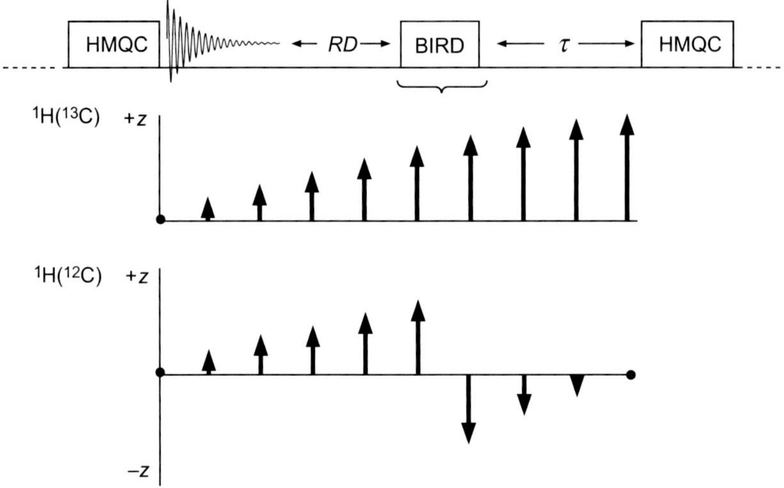

FIGURE 7.22 **Elimination of parent ^{1}H–^{12}C proton resonances through the BIRD inversion recovery sequence.** At the start of data collection no longitudinal proton magnetisation exists but this reappears during the acquisition period and subsequent recovery delay (*RD*) through spin relaxation. The BIRD element selectively inverts only those protons attached to carbon-12, which then continue to relax during inversion recovery delay τ. With an appropriate choice of τ, the ^{1}H–^{12}C magnetisation has no longitudinal component when the HMQC sequence starts, so does not contribute to the detected FID.

The BIRD pulse [26] is in fact a cluster of pulses (Fig. 7.21) used as a tool in NMR to differentiate spins that possess a heteronuclear coupling from those that do not. The effect of the pulse can vary depending on the phases of the pulses within the cluster, so we concentrate here on the desired selective inversion. For illustrative purposes, proton pulse phases of *x*, *y*, *x* will be considered as this provides a clearer picture with the vector model, although equivalent results are achieved with phases *x*, *x*, −*x*, as in the original publication. The scheme (Fig. 7.23) begins with a proton excitation pulse followed by a spin-echo. Since carbon-12 bound protons have no one-bond heteronuclear coupling, only their chemical shifts evolve during Δ. These

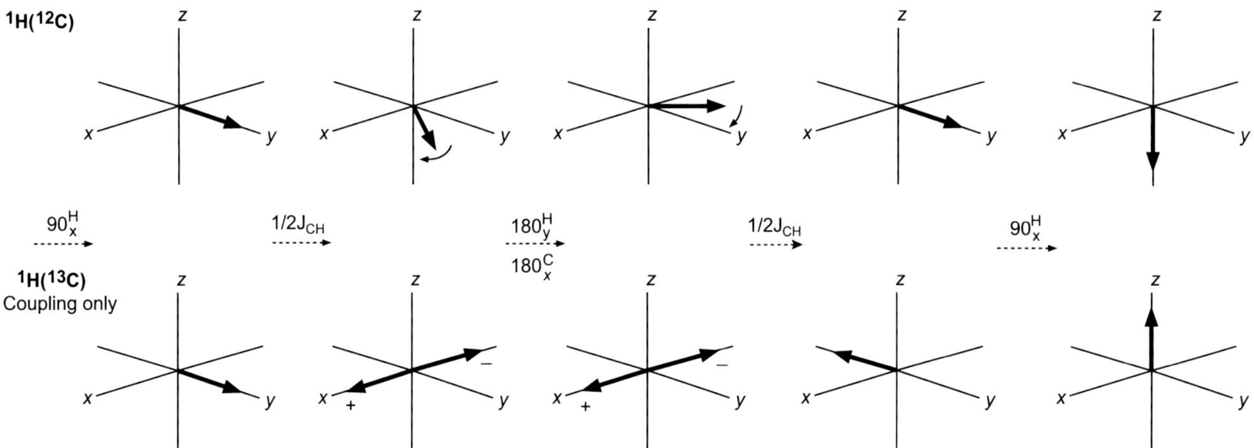

FIGURE 7.23 **Selective inversion of carbon-12 bound protons with the BIRD element.** Proton chemical shift evolution is refocused in the sequence but heteronuclear one-bond couplings evolve throughout.

are subsequently refocused by the 180 degree(^{1}H), so that at the end of the second Δ period, the second 90 degree(^{1}H) places the vector along the $-z$-axis and produces the desired inversion of the ^{1}H–^{12}C resonances. For those spins that possess a heteronuclear coupling, chemical shifts will also refocus as mentioned earlier so we need only consider the effect of the coupling itself. If Δ is set to $1/2J_{CH}$, the two proton vectors will lie along $\pm x$ immediately prior to the 180 degree(^{1}H, ^{13}C) pulses, after which the coupling will *continue* to evolve. Thus, after a total period 2Δ, the doublet vectors lie along $-y$, and are then returned to $+z$ by the action of the final proton pulse, and for these spins it is as if the BIRD had never appeared.

The success of the BIRD suppression scheme depends crucially on the correct setting of recovery delays and repetition rates, according to the proton T_1 values of the molecule. In practice, one is unlikely to know precisely what these are, and in any case there will exist a spread of these within the molecule, so a best guess compromise must be made. It turns out [25] that optimum sensitivity is achieved by setting timings according to the *shortest* T_1 value in the molecule, and once again setting the recycle time of the experiment (the time between the start of data acquisition of one experiment and the beginning of the HMQC sequence of the next; t_2 + RD + τ of Fig. 7.22), to around $1.3T_1$. The inversion recovery period τ must be set to approximately $0.5T_1$ for efficient suppression, and the proton acquisition time t_2 set according to the desired digital resolution. The relaxation delay (RD) is then chosen to make up the desired recycle time (t_2 + RD ≈ $0.8T_1$). For example, assuming the smallest anticipated T_1 in a molecule to be 600 ms and that t_2 is set to 200 ms (corresponding to an f_2 digital resolution of 5 Hz/pt), the aforementioned guidelines suggest one selects τ = 300 ms and RD = 280 ms. Such rapid pulsing may seem excessively fast for the resonances one wishes to observe, especially for those protons with longer relaxation times. However, one should note that the proton T_1s generally measured relate to carbon-12 bound protons, whereas in HMQC one is interested in carbon-13 bound protons. This nucleus acts as an additional dipolar relaxation source for the directly attached proton leading to shortening of the proton T_1 value, consistent with the rapid repetition. Since one is generally forced to make a best guess at the T_1 values, the experimental settings are fine-tuned to give optimum suppression of unwanted resonances. This is most conveniently achieved by running the experiment in an interactive set-up mode that allows real-time adjustment of parameters, and altering τ to produce the *minimum* FID.

The BIRD scheme is remarkably efficient at suppressing troublesome ^{1}H–^{12}C resonances and associated t_1 noise, and has been widely employed in the study of small molecules. However, this method is not suitable for the study of very large molecules because during the τ period the *negative* NOE generated from the inverted protons causes a *reduction* in the signal intensity of the observed protons and therefore compromises sensitivity.

7.4 HETERONUCLEAR MULTIPLE-BOND CORRELATIONS

The ^{1}H–X heteronuclear correlation methods presented so far all depend upon the presence of a proton bound to the heteroatom, relying on the presence of $^1J_{XH}$ couplings, so are therefore unable to provide assignments for non-protonated centres. They also do not identify atom connectivity through longer range coupling pathways. This section considers methods designed to establish correlations between heteroatoms and neighbouring protons over more than one bond, so-called *long-range* or *multiple-bond* correlations, now most commonly achieved with the proton-detected HMBC experiment. In the vast majority of cases, this will involve proton–carbon connectivities through couplings over two or three bonds ($^nJ_{CH}$, $n = 2, 3$), since those over greater distances are often vanishingly small. The ability to identify ^{1}H–^{13}C correlations across carbon–carbon or carbon–heteroatom linkages presents a wealth of information on the molecular skeleton, providing one of the most powerful approaches to defining an organic structure, perhaps second only to the more difficult to realise direct ^{13}C–^{13}C correlations provided by the INADEQUATE-based methods of Sections 6.3 and 6.4. The HMBC spectrum itself closely resembles that of HSQC in appearance, with the long-range correlations of each proton represented in the column taken at its chemical shift. The abundance of information in such spectra is illustrated in the HMBC spectrum of **7.2** (Fig. 7.24) which should be compared with the single-bond correlations of Fig. 7.4. The complete set of correlations observed in this spectrum are summarised in Table 7.3, but the salient features of the experiment may be appreciated by considering the correlations of proton H2 alone. Firstly note the breakthrough of the one-bond correlation appearing as the arrowed doublet at 93 ppm, the presence of which is explained in Section 7.4.1. This is equivalent to the correlation seen in Fig. 7.4b, and here serves as a reference point, although these are most often an unwelcome complicating factor, and attempts are usually made to remove or filter these from HMBC spectra. Nevertheless, the possibility of such peaks appearing should always be borne in mind when interpreting these spectra. Secondly, notice that the correlations to C4, C5 and C8 arise from couplings across heteroatoms (N and O), a feature which sometimes seems to surprise those new to the experiment. Such data can be particularly valuable when proton–proton couplings are absent. Finally, note that H2 shows a correlation to the carbonyl carbon C8, this centre being unobservable in one-bond correlation experiments. The ability to observe non-protonated centres in this way not only allows their chemical shifts to be determined but also provides valuable connectivity data. Nowadays, the combination of COSY, HSQC and then HMBC typically represent the primary techniques to turn to when addressing problems of molecular connectivity in small organic molecules. As for the one-bond correlation methods described earlier, a variety of extensions to the basic HMBC experiment also exist to enhance or modify information content, and the most significant of these are also presented in the sections that follow.

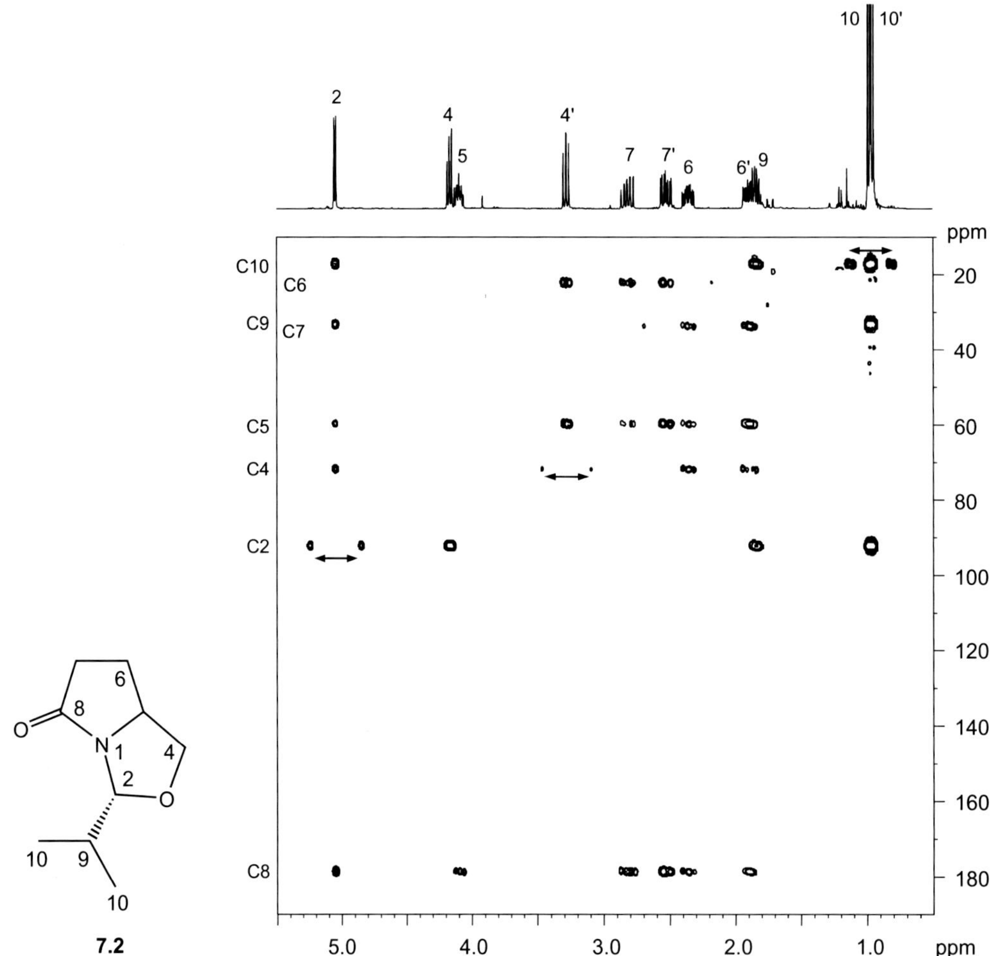

FIGURE 7.24 **The HMBC long-range correlation spectrum of 7.2.** This was recorded with Δ_{LR} = 60 ms and with gradient selection. The sequence used the most basic first-order low-pass J-filter (Section 7.4.3) to attenuate breakthrough from one-bond correlations [which appear with $^1J_{CH}$ doublet structure along f_2 (arrowed)]. 1K data points were collected for 256 t_1 increments of 8 transients each and the data processed with unshifted sine bells in both dimensions, followed by magnitude calculation. After zero-filling once in t_1 the digital resolution was 4 and 80 Hz/pt in f_2 and f_1, respectively.

TABLE 7.3 Summary of the Long-Range Correlations Observed in the HMBC Spectrum (Δ_{LR} = 60 ms) of 7.2

Proton	Correlated Carbon
2	4, 5, 8, 9, 10
4	2, 5(w), 6(w)
4′	2(w), 5, 6
5	4(w), 7(w), 8
6	4, 5, 7, 8
6′	4, 5, 7, 8
7	5, 6, 8
7′	5, 6, 8
9	2, 10
10	2, 9, 10′
10′	2, 9, 10

Weaker correlations, corresponding to smaller coupling constants, are identified with (w). Not all these are observed in Fig. 7.24 at the contour levels shown.

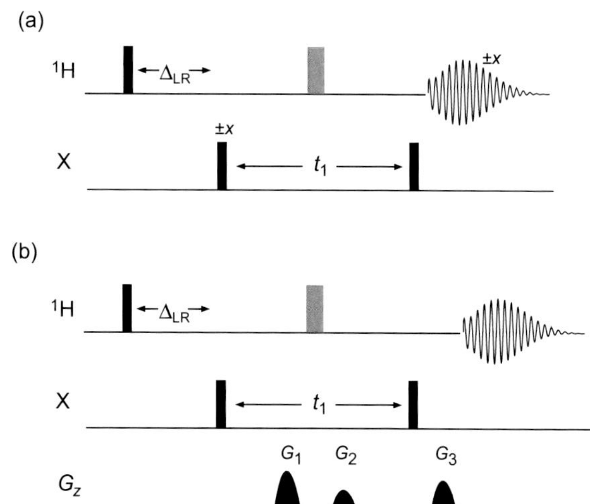

FIGURE 7.25 **The HMBC sequence (a) without and (b) with incorporation of PFGs.** The sequence is closely related to the HMQC experiment and follows a similar coherence transfer pathway but has the coupling evolution period Δ_{LR} optimised for long-range coupling constants.

7.4.1 HMBC Sequence

The HMBC experiment [27,28] establishes multiple-bond correlations by again taking advantage of the greater sensitivity associated with proton detection, and in essence is the HMQC sequence 'tuned' to detect correlations via small couplings (Fig. 7.25). Owing to the close similarity of the two, only differences pertinent to HMBC will be presented here, with application of the experiment being considered in the following section.

The tuning of the experiment is achieved by setting the Δ_{LR} preparation period to a sufficiently long time to allow the small long-range proton–carbon couplings to evolve to produce the anti-phase displacement of vectors required for subsequent generation of heteronuclear multiple-quantum coherence. Since long-range ^{1}H–^{13}C couplings are at least an order of magnitude smaller than one-bond couplings (often <5 Hz), Δ_{LR} should, in principle, be at least 100 ms ($1/2^n J_{CH}$), although shorter delays are often used routinely to avoid relaxation losses. During this long Δ_{LR} period, homonuclear ^{1}H–^{1}H couplings, which are of similar magnitude to long-range heteronuclear couplings, also evolve and introduce phase distortions to the observed crosspeaks (these distortions are small enough to be ignored in HMQC only because of the much smaller Δ periods used). Absolute value presentations are therefore widely used for HMBC spectra to mask these phase errors, and it has therefore been commonplace to acquire HMBC data sets that are phase-modulated (N-type selection) as a function of t_1. For sensitivity reasons, the refocusing Δ period of HMQC is omitted in HMBC so that long-range heteronuclear couplings are anti-phase at the start of t_2, precluding the application of ^{13}C-decoupling. Furthermore, it is generally the case that the use of decoupling in *refocused* HMBC spectra provides no significant benefit and can result in signal reduction from incomplete refocusing of long-range couplings, so its use is not generally recommended.

7.4.1.1 Phase-Sensitive HMBC

Although traditionally acquired as phase-modulated data, the echo–antiecho procedure described in Section 5.4.2 can be applied to HMBC [29] to produce spectra that are phase sensitive in f_1 (Fig. 7.26). This approach can yield higher f_1 resolution by producing absorption-mode lineshapes, so avoiding the need for harsh apodisation associated with magnitude calculation [30]. It has also been shown to provide a $\sqrt{2}$ sensitivity gain relative to the phase-modulated experiment so is gaining in popularity and finds use in other HMBC variants. Due to the appearance of anti-phase heteronuclear coupling and complex proton phase behaviour in f_2, it remains advantageous to apply magnitude calculation to this dimension for routine analysis of HMBC correlations. Alternatively, the retention of these anti-phase peaks may enable measurement of the long-range heteronuclear coupling constants themselves from the f_2 traces; see Section 7.4.5.

7.4.1.2 Suppressing Parent Resonances

Without any doubt, the greatest problem associated with the HMBC sequence lies in suppression of the parent ^{1}H–^{12}C signals which may otherwise mask long-range satellites. Unlike HMQC, the BIRD sequence is not well suited to the removal of these resonances since this is also likely to lead to attenuation of the desired signals. Traditionally, the sequence has relied on phase cycling alone to cancel the intense parent resonance, requiring a very stable spectrometer to be effective. Even so, bands of residual t_1 noise routinely plagued the original HMBC experiment and limited its use as a routine tool

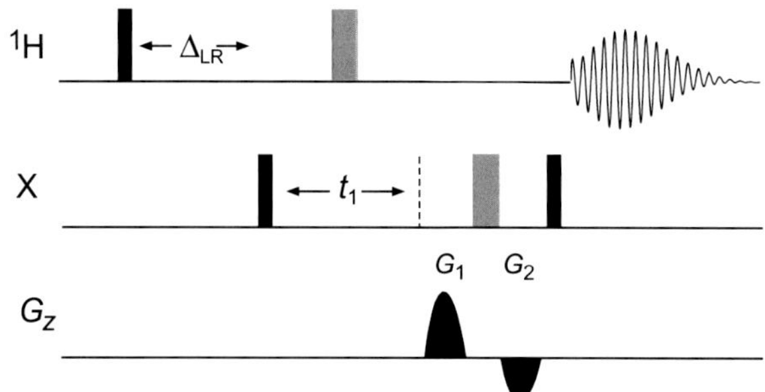

FIGURE 7.26 **The phase-sensitive HMBC experiment with echo–antiecho selection.** The gradient ratios G_1:G_2 are 5:–3 and 3:–5 for odd and even experiments, respectively, when selecting carbon-13.

in organic chemistry. The introduction of PFGs [4,31] has revolutionised the applicability of HMBC, since complete suppression of the parent line is readily achieved, along with a dramatic reduction in the associated t_1 noise. The benefits of the gradient-selected approach are clearly demonstrated in the spectra of Fig. 7.27. Spectrum (a) was recorded with gradient selection and (b) with conventional phase cycling under otherwise identical conditions (other than a 256-fold increase in receiver amplification in (a) over that in (b)). Four transients were collected per increment, corresponding to the minimum phase cycle (two steps for signal selection and two for suppression of axial peaks). Long-range correlations are clearly identified in (a) but can barely be observed above the t_1 noise bands in (b). Improved results for the phase-cycled version can be achieved by collecting far more transients at the expense of instrument time.

Long-range correlations based on the HSQC sequence have traditionally been less widely used. Significant evolution of ^{1}H–^{1}H couplings during the Δ period leads to unwanted COSY-type transfers among protons by the second 90 degree proton pulse of the INEPT sequence, a problem not found with HMBC. More recently, multiple-bond HSQC sequences have become more popular, especially for the measurement of long-range heteronuclear coupling constants and these are considered in Section 7.4.5.

7.4.2 Applying HMBC

The presence of long-range correlations in HMBC spectra is influenced by many factors, both experimental and structural, and an awareness of these points is important for the optimum application and interpretation of the experiment, so are addressed here before proceeding to some illustrative applications. Crosspeak intensities depend upon, among other things, both the magnitude of the long-range coupling and on the value selected for Δ_{LR}, which should optimally be set to $1/2^nJ_{CH}$ (crosspeak intensity $\propto \sin \pi^nJ_{CH}\,\Delta_{LR}$). Long-range proton–carbon couplings over two or three bonds rarely exceed 25 Hz, and in the absence of unsaturation are more often less than 5 Hz [32,33] (Table 7.4), indicating Δ_{LR} should be 100 ms or more. Such long delays can lead to relaxation losses prior to detection, especially for larger molecules, so in practice a compromise is met with Δ_{LR} being set to around 60 ms for routine applications, that is, optimised for ~8 Hz coupling. Since small organic molecules in low-viscosity solvents tend to have slower relaxation rates, longer delays can be used successfully in the search for more connectivities through smaller couplings, with Δ_{LR} taking values of up to 200 ms. The faster relaxation associated with large

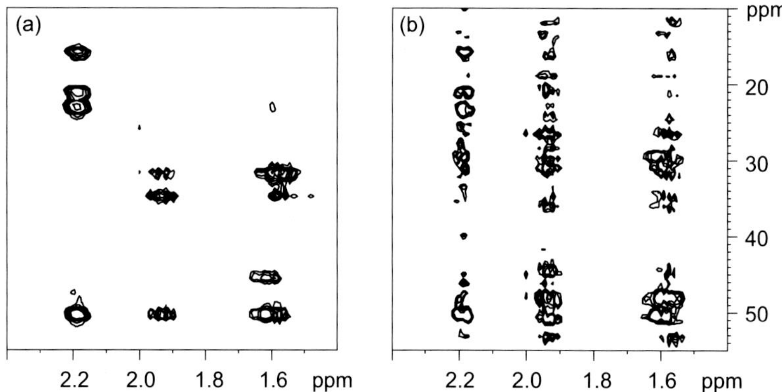

FIGURE 7.27 **Reducing t_1-noise in HMBC spectra.** These spectra were recorded under identical conditions but with signal selection through (a) PFGs and (b) phase-cycling alone.

TABLE 7.4 Typical Values of Long-Range Carbon-Proton Coupling Constants

Coupling Pathway	$^2J_{CH}$	Coupling Pathway	$^3J_{CH}$	Coupling Pathway	$^4J_{CH}$
H–C–**C**	(±) ≤ 5	H–C–C–**C**	≤5	H–C=C–C=**C**	(±) ≤ 1
H–C=**C**	≤10	H–C=C–**C**	≤15[a]	H–C–C–C–**C**[b]	≤1
H–C≡**C**	40–60	H–C≡C–**C**	≤5		
H–C(=O)–**C**	20–25				

Heteroatoms such as O, N, etc., may also be included in the coupling pathways illustrated in place of C.
[a]Trans > cis (*cis usually <10 Hz*).
[b]*w-Configuration favoured.*

molecules benefits from reduced Δ_{LR} periods, perhaps as small as 40 ms. The potentially wide range of $^nJ_{CH}$ values present in a molecule can mean correlations are not observed in HMBC when Δ_{LR} is poorly matched to the optimum delays, so presenting an incomplete set of correlations; more sophisticated HMBC variants aimed at increasing the number of observed correlations are described in Section 7.4.3. When using longer delays it also becomes possible to detect peaks that arise from four-bond correlations which are most likely to occur when the coupling pathway contains unsaturation or when it has the planar zig-zag (*w*-coupling) configuration, as commonly observed in long-range proton–proton couplings also. One should also recognise that three-bond couplings can be, and often are, greater in magnitude than two-bond couplings, displaying a Karplus-type relationship with the dihedral angle [33–35]. Indeed, one further limitation when using HMBC data is the lack of differentiation between two- and three-bond connectivities, a topic also addressed in Section 7.4.3. Furthermore, lack of a correlation cannot on its own be taken as evidence for the nuclei in question being distant in the structure, since a variety of factors can contribute to $^nJ_{CH}$ being close to zero (for more details see discussions in [32]). The most intense correlations are typically observed for methyl groups since magnetisation is detected on three protons simultaneously and because they display simpler coupling structure, if any, whereas the weakest correlations are generally associated with poorly resolved, complex proton multiplets.

Despite the lack of discrimination, protons can potentially correlate to a great many carbon neighbours within two or three bonds, providing a mass of structural data on an unknown molecule. Connectivities may also be traced across heteroatom linkages (as seen in Fig. 7.24) where proton–proton couplings are usually negligibly small, and in this the experiment is extremely effective at piecing together otherwise uncorrelated molecular fragments. Thus, the HMBC experiment was able to define the regiochemistry of a dipeptide antibiotic, Tü 1718B, isolated from *Streptomyces* cultures [36]. The two possible structures **7.6a** and **7.6b** were proposed from biosynthetic arguments, but neither proton correlation nor carbon chemical shift data could unambiguously identify which was correct. Confirmation was gained from analysis of the long-range correlations of protons H4 and H5, and in particular those to the carbonyl groups (Fig. 7.28). H5 was observed to correlate to both whereas H4 correlated only to C6, these data being consistent with **7.6a** only.

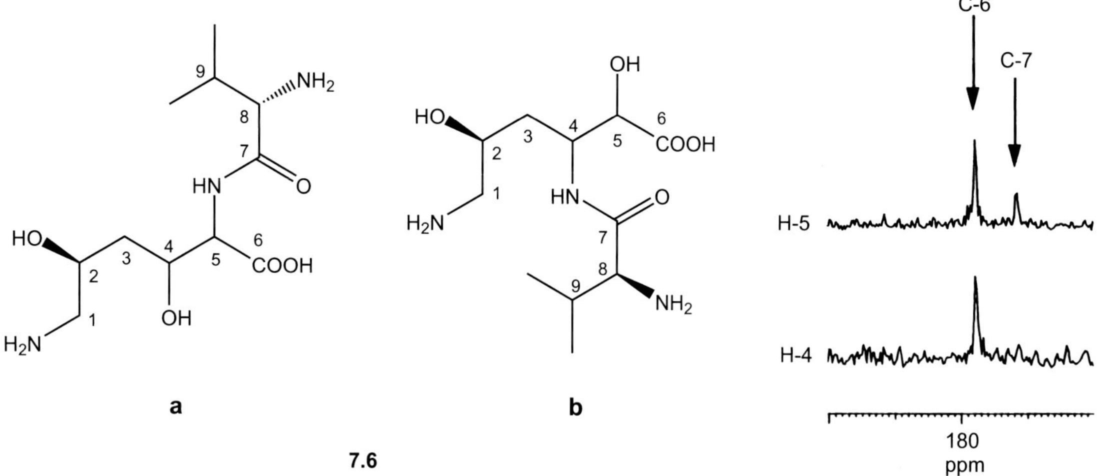

FIGURE 7.28 **Columns taken from the HMBC spectrum of the antibiotic Tü 1718B at proton shifts H4 and H5.** Only the key correlations to the carbonyl resonances are shown. *(Reproduced with permission from reference [36], Copyright 1993, Elsevier.)*

Connectivities across the oxygen atom linking neighbouring sugar residues in oligosaccharides likewise provides a useful means of identifying neighbouring residues in these compounds. Naturally, the sequence is not limited to proton–carbon connectivities, but can be tailored to any spin-½ pair. Thus, long-range proton–silicon and proton–carbon correlations established through HMBC were used to confirm the structure of an unexpected product from the rearrangement of epoxydisilanes [37] such as the silanol **7.7a** rather than **7.7b**. Additional NOE studies identified the stereochemistries across the alkene.

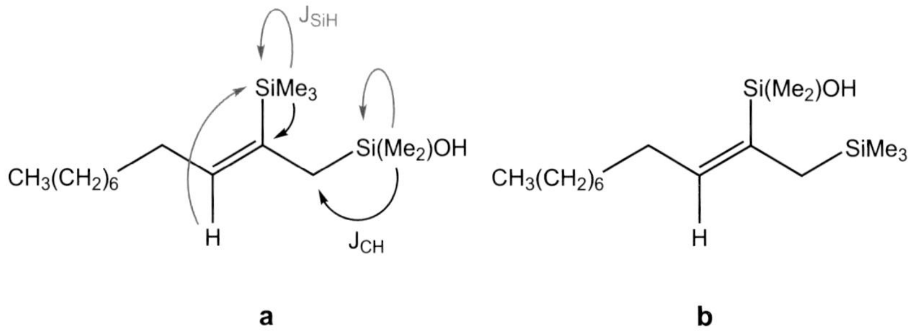

J_{SiH} SiMe$_3$ Si(Me$_2$)OH CH$_3$(CH$_2$)$_6$ H J_{CH} **a**

Si(Me$_2$)OH SiMe$_3$ CH$_3$(CH$_2$)$_6$ H **b**

7.7

7.4.2.1 Practical Set-Up

Owing to the relatively small size of long-range couplings and because of the need for long and often non-optimal Δ_{LR} delay periods, the sensitivity of the HMBC experiment is somewhat less than that of its HSQC or HMQC cousins. In the absence of PFGs, it is necessary to acquire many scans for each increment in an attempt to suppress the intense ^{1}H–^{12}C signal and reveal the correlations of interest, making the experiment time consuming even when large sample quantities are available (spectral quality being highly dependent on spectrometer stability). Gradient-selected versions are considerably quicker to acquire, being dictated by sensitivity and resolution arguments alone since the suppression of the parent signal is no longer an issue and better use is made of the receiver dynamic range. As a rule of thumb when using gradient selection, the HMBC experiment will typically take around four times as long to acquire as the corresponding HSQC experiment to provide acceptable data. From what was once typically an overnight phase-cycled experiment, high-quality data can now be obtained within a few hours or even tens of minutes; the impact of PFGs on the HMBC experiment has been profound.

The setting of Δ_{LR} has been described earlier and the choice of digital resolution for the two dimensions follows similar arguments as for the one-bond correlation experiments; around 50 Hz/pt in f_1 and around 5 Hz/pt in f_2 should provide acceptable results in most cases. In HMBC the acquired t_2 FID begins as anti-phase magnetisation with respect to $^nJ_{CH}$ since the refocusing period is omitted, and the signal builds as refocusing occurs during t_2 itself. Optimal window functions providing close to matched filtering are therefore the unshifted sine bell or squared sine bell. In contrast, t_1 interferograms decay from their maximum values and require only sufficient apodisation to avoid truncation errors that may appear under conditions of high signal-to-noise, so a simple exponential decay or shifted sine bell function suffices. For the phase-modulated experiment it is commonplace to execute a magnitude calculation on the f_1 dimension for which an unshifted sine bell function will improve resolution. Such harsh apodisation is not required for the echo–antiecho phase-sensitive version although f_2 magnitude calculation may be employed to sum anti-phase signal intensity. Improved *digital* resolution of the f_1 dimension can be achieved through zero-filling (at least once is recommended) or the resolution increased by the use of forward linear prediction. Finally, the repetition time of the experiment is dictated by the T_1s of protons *directly* bound to carbon-12 centres. Repetition times are therefore estimated from relaxation times as they would be for *homonuclear* correlation experiments, and are longer than those optimal for HSQC.

7.4.3 HMBC Extensions and Variants

The ability to detect two- and three-bond heteronuclear correlations with HMBC plays a hugely important role in structure elucidation and has led to the widespread and routine use of this experiment. Unsurprisingly then, this has also spawned the development of a range of methods that improve the appearance of the correlation spectrum or which add additional information to it. The basic HMBC experiment described so far suffers from a number of limitations ranging from complications arising from the appearance of one-bond coupling correlations, through limited detection of correlations in the presence of a wide range of $^nJ_{CH}$ values and the inability to distinguish two-bond from three-bond correlations. In this section some of the more significant developments that address these and other issues are presented. We start with methods for the removal of one-bond correlations from HMBC, that is, those correlations observed in HSQC, since it is now standard practice to incorporate this filtration in almost all HMBC experiments. Such filtering is applicable to the other HMBC variants discussed later and its use would be expected, but for reasons of clarity will not be shown in the sequences presented hereafter.

7.4.3.1 Low-Pass J Filtration: Removing One-Bond Correlations

Despite setting the Δ_{LR} period of HMBC according to long-range coupling constants, this may also happen to be a multiple of the appropriate value for selecting one-bond correlations, additionally causing these to appear in the spectrum. Since the FID is acquired without ^{13}C decoupling, these crosspeaks possess a distinctive doublet structure in f_2 (as seen in Figs 7.4b and 7.24) which aids their identification. These may be considered useful additions or unwanted interferences, depending on your point of view; while they simultaneously provide one-bond correlation data, they may obscure or become confused with long-range correlations. Their suppression may be achieved, to a greater or lesser degree, by incorporation of a *low-pass J filter*, so-called because it retains or passes only those peaks arising from couplings that are *smaller* than a chosen cut-off value, here the one-bond coupling constant [27,38]. Typically, these elements are placed at the start of the HMBC sequence prior to the transfer of polarisation onto carbon.

Fig. 7.29 illustrates the simplest first-order low-pass filter. Here, the Δ_1 period is set to $1/2^1J_{CH}$, whilst Δ_{LR} is set according to the long-range coupling, as earlier. The first carbon pulse will generate multiple-quantum coherence for the anti-phase one-bond ^1H–^{13}C pairs only since Δ_1 is too short for significant evolution of the long-range couplings. Alternation of the phase of this first carbon pulse *without* changing that of the receiver causes these coherences to cancel on alternate scans and so removes one-bond correlations from the final spectrum. In practice, the suppression is often not complete, particularly when a wide range

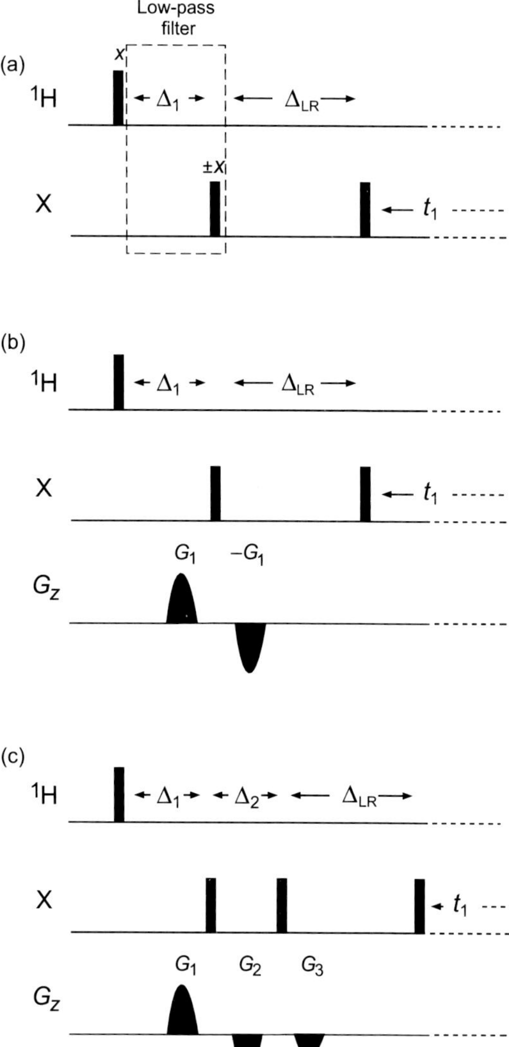

FIGURE 7.29 **The low-pass filter for removing spurious one-bond correlation peaks from HMBC spectra.** (a) The basic first-order filter in which inverting the phase of the first X pulse on alternate scans cancels unwanted one-bond contributions, (b) the gradient purged first-order filter (G_1:G_2 = 1:−1) and (c) the gradient-purged second-order filter (G_1:G_2:G_3 = 3:−2:−1). Δ_{LR} is set to $1/2^nJ_{CH}$ whilst Δ_1 and Δ_2 are set for one-bond couplings as described in the text.

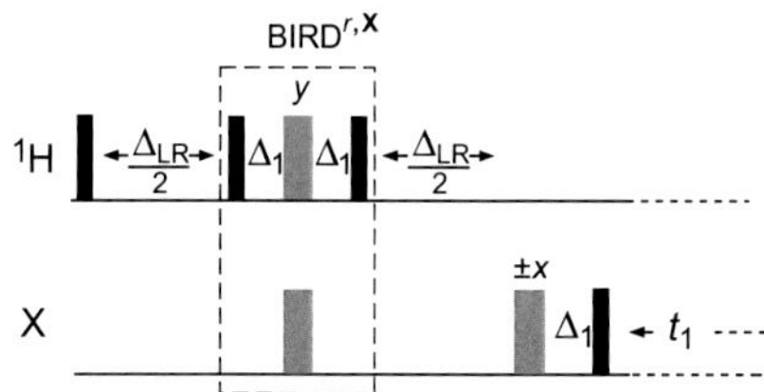

FIGURE 7.30 **An alternative low-pass filtering scheme employing selective refocusing of $^1J_{CH}$ couplings with a BIRD^{r,X} pulse cluster.** The 180 degree X pulses may be replaced advantageously by composite or adiabatic pulses. Δ_1 is set to $1/2^1J_{CH}$ and Δ_{LR} to $1/2^nJ_{CH}$.

of $^1J_{CH}$ values are present, and improved results can be achieved with a second-order low-pass filter incorporating PFGs. In this a second $^1J_{CH}$ evolution delay Δ_2 is included with three purging gradients of ratio 3:−2:−1. For long-range coupled ^{1}H–^{13}C pairs, which generate negligible anti-phase components during the short Δ_1 and Δ_2 periods, the three gradient pulses cancel and have no net effect allowing the $^nJ_{CH}$ couplings to act during the subsequent Δ_{LR} delay. In contrast, the multiple-quantum coherence arising from anti-phase $^1J_{CH}$ couplings remains dephased by these gradient pulses and the corresponding correlations are suppressed. The filter delays are set typically to $\Delta_1 = 1/2^1J_{CH(min)}$ and $\Delta_2 = 1/2^1J_{CH(max)}$ for the smallest and largest anticipated one-bond couplings, typically 125 and 165 Hz. More elaborate values of $\Delta_1 = 1/2[^1J_{CH(min)} + 0.146(^1J_{CH(max)} - {}^1J_{CH(min)})]$ and $\Delta_2 = 1/2[^1J_{CH(max)} - 0.146(^1J_{CH(max)} - {}^1J_{CH(min)})]$ have been suggested as optimum values [39].

An alternative approach to low-pass filtering that is highly effective even without gradient pulses is to selectively refocus one-bond couplings whilst letting the long-range couplings develop to become anti-phase for transfer to ^{13}C [40]. This can again be achieved by exploiting the large difference in coupling constants by use of a suitable BIRD pulse cluster at the midpoint of Δ_{LR} (Fig. 7.30). The so-called 'BIRD^{r,X} pulse' [41] serves to invert only X-spins (here ^{13}C) and those protons remotely (long-range) coupled to these, but not the protons directly attached to the heteronucleus. This achieves the desired refocusing of $^1J_{CH}$ (since only the ^{13}C experiences a 180 degree pulse) but defocusing of $^nJ_{CH}$ (for which both ^{1}H and ^{13}C experience net 180 degree pulses). The range of potential $^1J_{CH}$ values can again reduce the efficiency of this approach and the addition of a subsequent 180 degree ^{13}C pulse followed by a short $1/2^1J_{CH}$ period Δ_1 aims to refocus those one-bond couplings that have 'passed' through the BIRD filter. Being refocused, these are no longer able to generate multiple-quantum coherence and so do not contribute to the final spectrum. Phase alternation of the 180 degree pulse then serves to cancel contributions from those components initially refocused by the BIRD that would subsequently become defocused in the following Δ_1 period, giving rise to effective suppression of one-bond correlations. Extension to a second-order filter by addition of a second 180 degree(X)–Δ_1 period (with the X-pulse phase alternated independently of the first) provides still greater suppression. Application of this filter is also found in the long-range heteronuclear single-quantum multiple-bond correlation (HSQMBC) experiment described in Section 7.4.6 for the detection of very long–range heteronuclear correlations.

7.4.3.2 Constant Time HMBC: Eliminating Proton–Proton Coupling in f1

One of the less desirable features of the HMBC (and HMQC) experiment is the contribution from proton–proton couplings in the X-spin f_1 dimension, a consequence of the evolution of ^{1}H–X multiple-quantum coherence during t_1. This leads to peak broadening in the f_1 dimension and may limit peak resolution so that, as described earlier, typical HMBC experiments are recorded with low f_1 resolution so as not to resolve this additional fine structure. In cases where high f_1 resolution is desired, such as in highly crowded ^{13}C spectra and when broadband sampling of long-range coupling constants is employed (described later), the contribution from proton–proton couplings can be more troublesome. These arise because proton coupling evolution modulates the detected signal *as a function of the t_1 period* meaning these can be eliminated from the f_1 dimension if this t_1-dependent modulation of each FID can be suppressed. This may be achieved through introduction of the *constant-time* principle [42] in which the *total* period in which proton couplings evolve is held constant throughout the experiment *regardless* of the value of t_1. In the HMBC experiment, proton–proton coupling evolves throughout the whole sequence following the initial proton excitation, meaning its duration must remain fixed as t_1 increases. One possible approach to this based on the echo–antiecho scheme is illustrated in Fig. 7.31 in which the total period ($\Delta_{CT}-t_1$) decreases as t_1 is incremented, so defining the constant time period. The delay Δ_{CT} is set equivalent to the maximum value that the t_1 period will attain, $t_{1(max)}$, ($\Delta_{CT} = t_{1(max)}$), so that the variable period ($\Delta_{CT}-t_1$) contracts from Δ_{CT} to zero as t_1 increments from zero to $t_{1(max)}$, with the net result that proton–proton coupling evolution no longer modulates the detected signal.

Owing to the low-resolution conditions appropriate for most routine uses of HMBC, the constant-time approach is arguably of limited use here. However, this can prove more valuable in the band-selective experiment described in the following section, where substantially higher f_1 resolution is employed and where any proton coupling structure may limit crosspeak resolution.

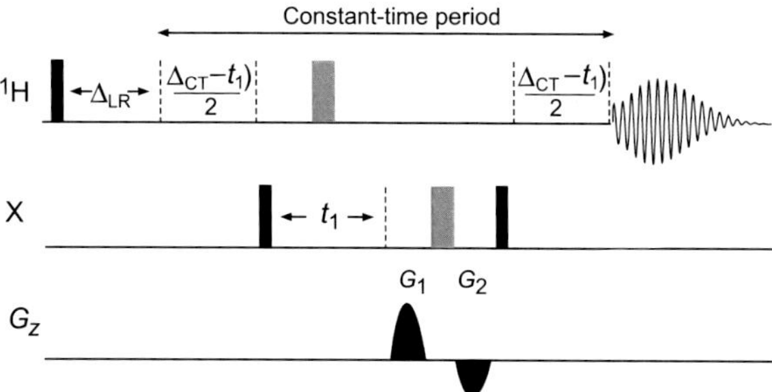

FIGURE 7.31 **The constant time HMBC experiment based on the echo–antiecho f_1 selection scheme of Fig. 7.26 ($\Delta_{CT} = t_{1(max)}$).** The combined period (Δ_{CT}–t_1) is decremented as t_1 increases, so defining a constant-time period for proton coupling evolution that is invariant with changes in t_1.

7.4.3.3 Band-Selective HMBC: Optimising f_1 Resolution

A further potential problem with ^{1}H–^{13}C HMBC, in particular, is the large spectral width that must be indirectly digitised in f_1. This tends to be a greater problem than with one-bond correlation experiments because of the need to include the carbonyl region in many instances, meaning, as already mentioned, the data are usually collected with rather low f_1 resolution. In regions of the ^{13}C spectrum where many resonances fall close together, such low resolution may prove insufficient and result in crosspeak overlap. One solution to this is to select only the region of the f_1 dimension that contains the most crowded resonances, meaning only these appear in the final spectrum and hence only a small f_1 window need be digitised at high resolution. In the most basic approach, this may be achieved by selectively exciting only the ^{13}C resonances of interest with a shaped carbon pulse (Section 12.4) in place of the first non-selective carbon pulse of the conventional sequence [43]; selectively acquiring a smaller region of a 2D data set gives rise to so-called *band-selective* or *semi-selective* variants [44]. However, in such cases of high f_1 resolution the removal of proton–proton coupling fine structure from f_1 has considerable benefits and the band-selective, constant-time HMBC experiment is then recommended [45]. This may be derived from the constant time HMBC of Fig. 7.31 simply by replacing the X-spin 180 degree pulse in the gradient echo with a band–selective shaped 180 degree pulse. Fig. 7.32 compares the carbonyl region of **7.8** from the conventional HMBC with that from the band-selective and the band-selective, constant-time variants. The aim was to obtain sequence-specific assignments for each carbohydrate amino acid residue by identifying long-range ^{1}H–^{13}C correlations to the carbonyl carbons, thus providing a link between adjacent residues. For Fig. 7.32b and c the carbonyl region was selected with a 0.5-ms 180 degrees Gaussian pulse and only a 6 ppm ^{13}C window digitised (centred at 172 ppm), providing a 15-fold increase in f_1 resolution relative to the non-selective experiment. The crosspeak skew observed in Fig. 7.32b reveals the presence of the J_{HH} coupling fine structure which is suppressed in the constant time variant of Fig. 7.32c. This resolves peaks sufficiently to identify the required H3^i, H2^i to CO^i and CO^i to H6^{i+1} long-range correlations and so establish connectivities across the amide linkages.

7.8

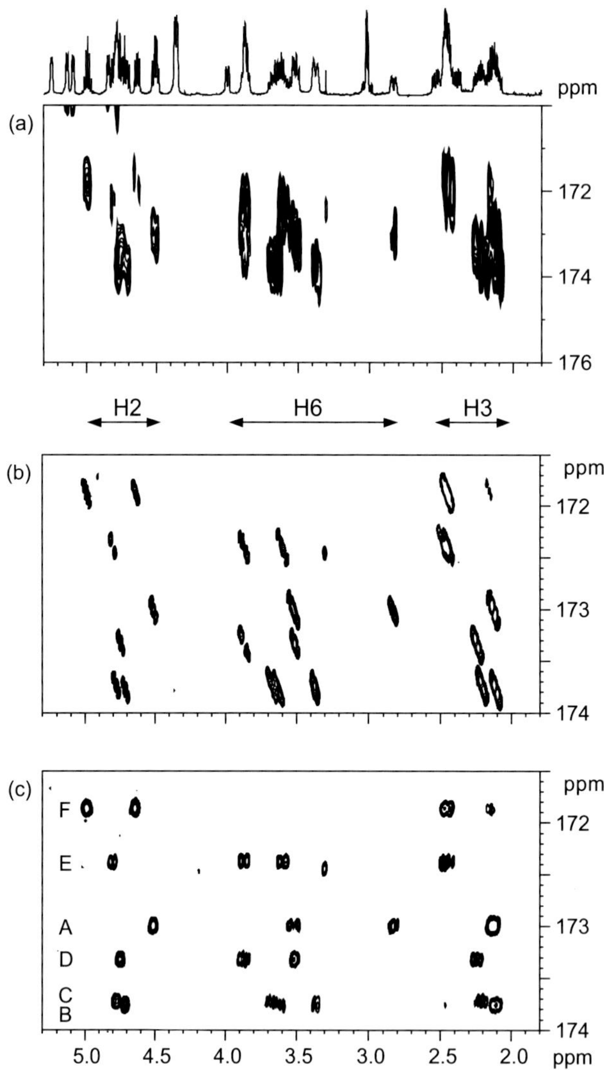

FIGURE 7.32 **Carbonyl regions of the HMBC spectra of 7.8.** These spectra were recorded with (a) the conventional HMBC, (b) the band-selective HMBC and (c) the band-selective, constant-time HMBC sequences. The small carbonyl shift dispersion causes considerable crosspeak overlap with the low f_1 resolution of 45 Hz/pt in (a), whereas the higher resolution in (b) and (c) of 3 Hz/pt removes this limitation. In (c) the proton–proton multiplet structure is also suppressed for optimum peak dispersion. *(Reproduced with permission from reference [45], The Royal Society of Chemistry.)*

7.4.3.4 ACCORDION Optimisation: Broadband $^nJ_{XH}$ Detection

One of the limiting features of the HMBC experiment is its inability to sample simultaneously and effectively a wide range of coupling constants, potentially leading to missing long-range correlations in the spectrum. The full range of $^nJ_{CH}$ values is approximately 1–25 Hz whereas the typical HMBC experiment would be optimised for a coupling of around 8 Hz. When dealing with complex and/or unknown structures it can be advantageous to seek a large number of correlations to aid their identification, and the simplest approach here is to collect multiple HMBC spectra with differing values of Δ_{LR}. As mentioned previously, the HMBC signal intensity scales with the amount of anti-phase ^{1}H–X magnetisation generated during Δ_{LR} and so has a $\sin(\pi^nJ_{XH}\Delta_{LR})$ dependence (ignoring relaxation losses) as illustrated in Fig. 7.33. Thus, collecting two experiments optimised for $^nJ_{CH}$ of 8 and 4 Hz, corresponding to delays of ~62 and 125 ms, respectively, should provide reasonable sampling over a range of 2–14 Hz, which is appropriate for observing many long-range ^{1}H–^{13}C correlations. For larger molecules where relaxation losses may be more substantial, it is advantageous to employ delays somewhat shorter than those predicted here, especially when seeking correlations from smaller couplings. The use of the simplest HMBC sequence here has its attractions in that it uses rather few

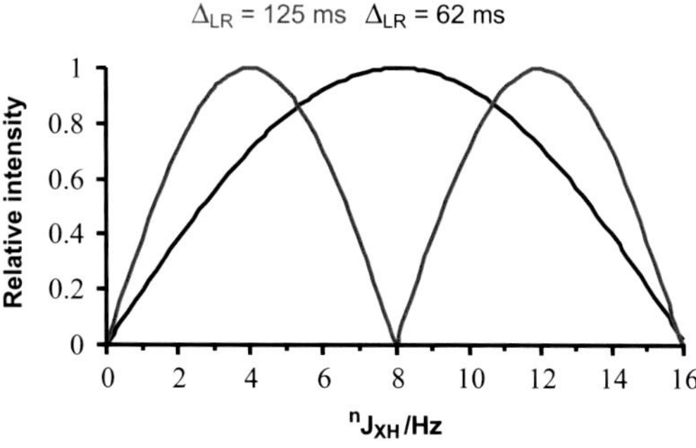

FIGURE 7.33 **The intensity of HMBC correlations as a function of $^nJ_{XH}$ for different values of the Δ_{LR} period.** Two experiments optimised for coupling constants of 8 and 4 Hz (Δ_{LR} delays of 62 and 125 ms, respectively) provide reasonable sampling over the range 2–14 Hz.

pulses and so suffers minimal signal loss from pulse imperfections. Nevertheless, it is generally undesirable to have to collect multiple experiments, especially when samples are limited in quantity or even unstable, and a number of sequences have been developed to provide more efficient 'broadband' detection over a range of coupling constants in a single experiment, which we shall now consider. Whilst it appears to have been little used, we shall begin with the originally proposed sequence since it is from this that the more useful variants are derived, many of which have been reviewed [46].

The original ACCORD-HMBC sequence [47] is illustrated in Fig. 7.34 and differs from the simple HMBC in that the usual 'static' delay for the evolution of long-range couplings Δ_{LR} is replaced by a variable delay Δ_{VD} that is also repeated prior to data acquisition and which serves to sample over a range of coupling constants from $^nJ_{XH(min)}$ to $^nJ_{XH(max)}$. Thus, during the experiment, as t_1 is incremented sequentially, the period Δ_{VD} is decremented in a stepwise manner from $\Delta_{VD(max)}$ to $\Delta_{VD(min)}$ in steps of $(\Delta_{VD(max)} - \Delta_{VD(min)})/n_{t_1}$, where n_{t_1} represents the total number of t_1 increments used to define the f_1 dimension. The delay $\Delta_{VD(max)}$ defines the evolution period for the smallest coupling to be sampled ($1/2^nJ_{XH(min)}$) whilst $\Delta_{VD(min)}$ is that for the largest ($1/2^nJ_{XH(max)}$). The stepwise sampling of all intermediate values introduces this new variable time period (in addition to t_1 and t_2) and is termed the ACCORDION approach [48]; it is this that gives rise to the enhanced detection of long-range correlations. Fig. 7.35 shows a region of the 8–Hz optimised HMBC and the 2–16 Hz sampled ACCORD-HMBC spectra of the diterpene **7.9** and clearly illustrates the larger number of long-range correlations observed in the latter. The sequence is held symmetric by inclusion of a second Δ_{VD} period which allows for optional X-spin decoupling, but is in any case essential to refocus the ^{1}H chemical shift evolution that occurs during the first Δ_{VD} period. Whilst this ensures proton chemical shifts do not modulate the f_1 responses as a function of Δ_{VD}, proton–proton couplings do contribute and are detrimentally scaled in this dimension, giving rise to a characteristic tilting or *skew* of the crosspeak that can compromise peak dispersion, as seen in Fig. 7.35b. The scaling arises from the way in which the Δ_{VD} period alters with each increase in t_1 and is best understood by considering the situation in which Δ_{VD} is *increased* on each t_1 increment. Thus, as t_1 increases, the total evolution time in which J_{HH} is active is increased by t_1 *plus* the associated increase in Δ_{VD} leading to a greater net coupling evolution than would occur if Δ_{VD} did not alter, that is, if it had remained static as in conventional HMBC. The effect of this greater evolution is to make it *appear* as if larger proton–proton coupling constants had

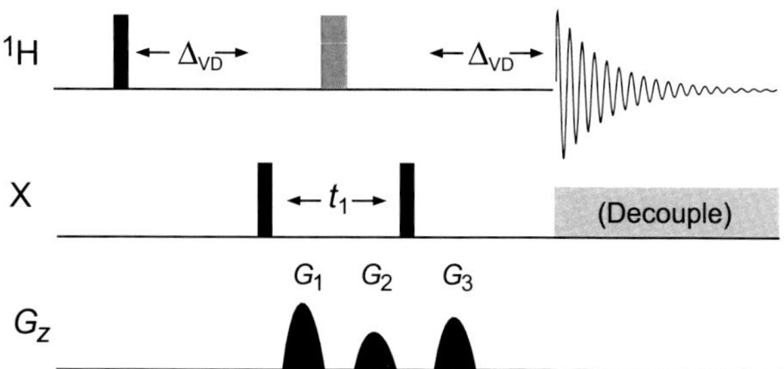

FIGURE 7.34 **The ACCORD-HMBC sequence.** The usual delay for the evolution of long-range couplings Δ_{LR} is replaced by a variable delay Δ_{VD} that samples a range of $^nJ_{XH}$ values as t_1 increases.

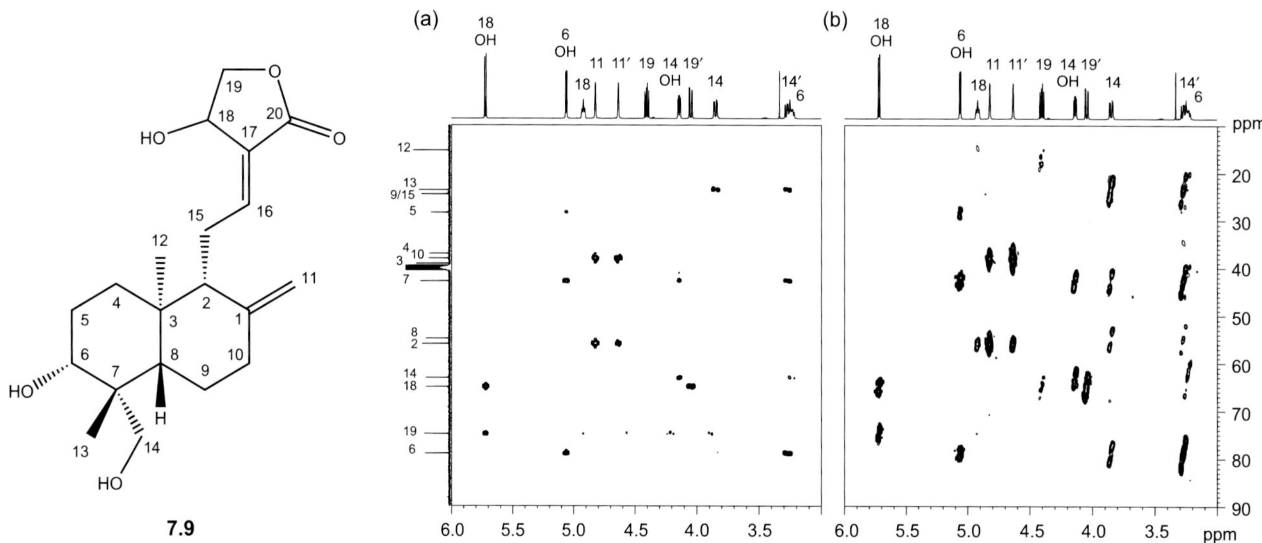

FIGURE 7.35 **The ACCORD-HMBC experiment.** A comparison of regions from (a) an 8-Hz HMBC and (b) a 2–16-Hz ACCORD-HMBC spectrum of **7.9**; see text for an explanation of the peak skew observed in (b).

been operative during t_1 and hence the proton homonuclear couplings appear to be amplified or *scaled* in f_1. The magnitude of the amplification may be defined by a *scaling factor* J_{scale} thus:

$$J_{scale} = \frac{2(\Delta_{VD(max)} - \Delta_{VD(min)})}{n_{t_1} \cdot \delta_{t_1}} \tag{7.7}$$

where n_{t_1} again represents the number of t_1 increments and δ_{t_1} is the size of each increment (t_1 dwell time) [49]. From this it is clear that sampling a wider range of $^nJ_{XH}$ values will lead to increased scaling of J_{HH} in the X-spin dimension. Whilst sampling of the long-range heteronuclear coupling may be improved, this is achieved at the expense of poorer peak separation (Fig. 7.36a). Conversely, the scaling may be diminished by the use of more t_1 increments to characterise f_1 (Fig. 7.36b)

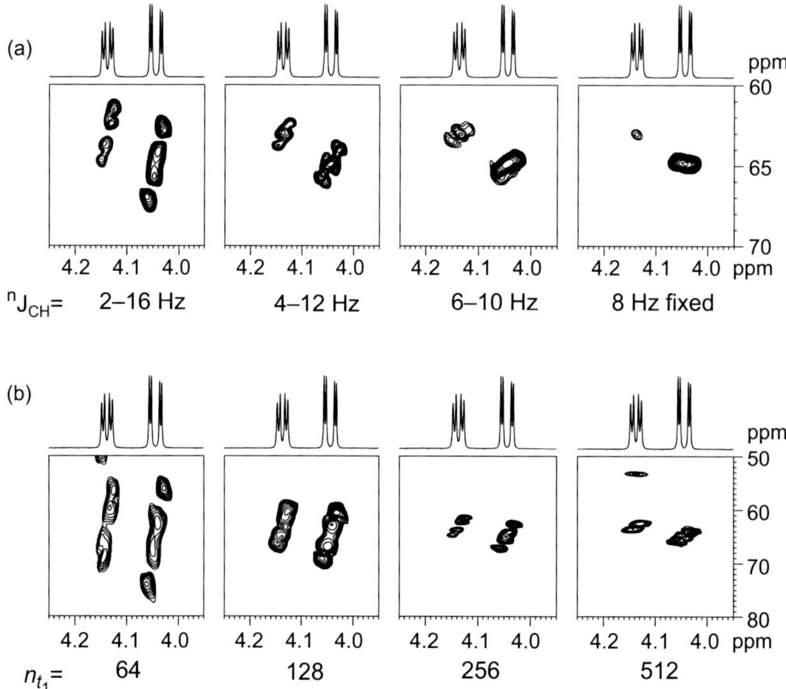

FIGURE 7.36 **The variation in crosspeak skew in ACCORD-HMBC.** This is shown as a function of (a) the $^nJ_{CH}$ range sampled and (b) the number of t_1 increments n_{t_1} used to characterise f_1. For (a) n_{t_1} was 256 and for (b) the sampling range was 2–16 Hz.

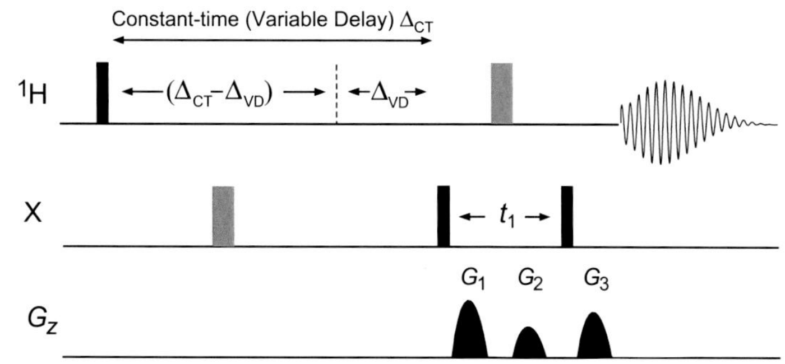

FIGURE 7.37 **The IMPEACH-MBC sequence shown without refocusing of long-range couplings.**

since the change in Δ_{VD} per increment is then reduced, but this will unavoidably increase the total duration of the experiment. Finally, the need to retain two Δ_{VD} periods necessarily leads to a longer sequence and hence reduced sensitivity relative to the basic HMBC scheme. Thus, whilst the ACCORDION sampling has potential benefits for increased detection of long-range correlations, it suffers from a number of undesirable features that limit its utility in this form.

A more practical implementation of this concept is to be found in the improved performance ACCORDION heteronuclear multiple-bond correlation (IMPEACH-MBC) experiment [50] (Fig. 7.37). In this the variable time delay that permits the broadband sampling of heteronuclear coupling constants Δ_{VD} is preceded by an additional variable period ($\Delta_{CT}-\Delta_{VD}$) such that the sum of these delays remains constant, giving rise to the so-called *constant-time* (*variable delay*) element. Thus, as Δ_{VD} is decremented on each t_1 increment in the ACCORDION style, the compensating delay is increased by an equivalent amount following the constant time concept. As described for the constant-time HMBC earlier, this results in no modulation as a function of t_1 of the detected FIDs by the homonuclear proton–proton couplings and so removes the exaggerated skew that is observed with the ACCORD-HMBC experiment. The only period that modulates the observed signals is t_1 itself, in which homonuclear proton couplings remain active, meaning the IMPEACH-MBC experiment displays J_{HH} structure in f_1 in the manner of the standard HMBC, without further scaling. To ensure broadband sampling of the long-range heteronuclear couplings occurs only in the Δ_{VD} period, the compensating element ($\Delta_{CT}-\Delta_{VD}$) is split centrally by a 180 degree X pulse to refocus the evolution of these heteronuclear couplings during this delay. In the original presentation of this experiment, an equivalent constant time period was included after t_1 to allow refocusing and X-spin decoupling. However, this extension typically reduces sensitivity and the non-refocused variant presented here may be preferred. Fig. 7.38 compares regions of the 8-Hz HMBC with two IMPEACH-MBC experiments. With moderate sampling of 4–12 Hz an increased number of correlations is observed, such as those highlighted in the red boxes, although the more aggressive 2–16 Hz sampling also leads to the loss of some signals (dashed red box). These losses occur for the broader resonances and are associated with the longer delays required for the sampling of very small couplings.

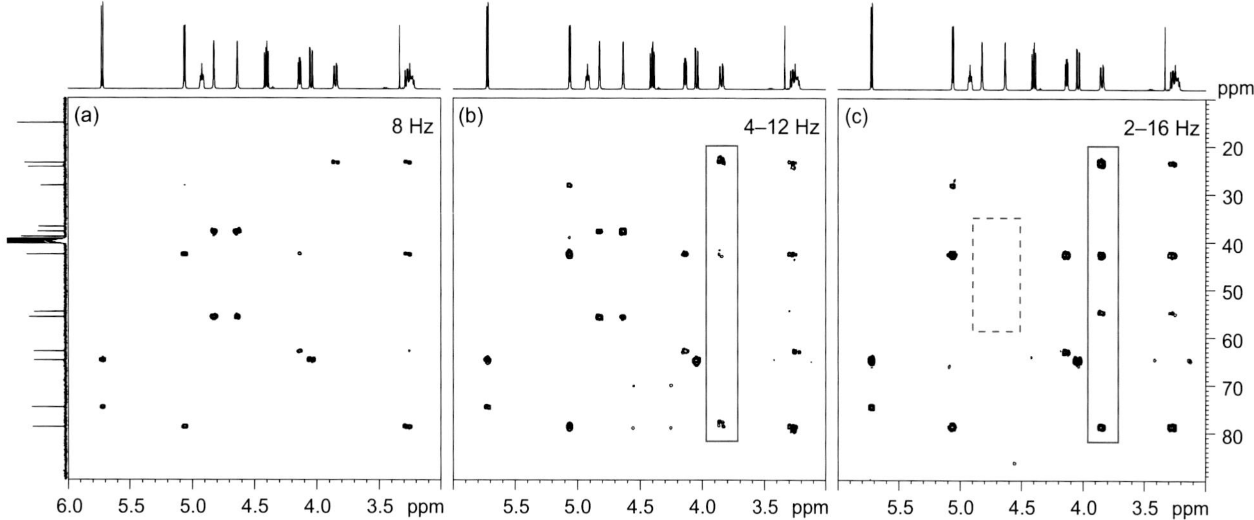

FIGURE 7.38 **The IMPEACH-HMBC experiment of 7.9.** The spectra show a comparison of regions from (a) an 8-Hz HMBC and (b and c) the IMPEACH-MBC recorded with the sequence of Fig. 7.37.

So, how might one decide upon a suitable approach when the use of a single HMBC employing a static Δ_{LR} period is deemed insufficient for the sample under scrutiny? Arguably the most direct and pragmatic approach is simply to acquire multiple spectra with differing values of Δ_{LR} and so make best use of the simplicity and relatively high sensitivity of the basic HMBC sequence. Preferably, this would employ the f_1 phase-sensitive version of Section 7.4.1. When this approach is undesirable and a single–broadband optimised experiment is attractive, then the non-refocused IMPEACH-MBC of Fig. 7.37 presents a relatively simple technique. When using the ACCORDION-based sequences, one needs to be cautious in how aggressive one is in sampling for small long-range coupling constants since this may encourage the appearance of correlations arising from couplings over four or possibly more bonds. In proton-sparse structures this may of course be a desirable feature [51] whilst for routine structure elucidation it may introduce additional complication or ambiguity. In cases when two- and three-bond correlations are most desirable a limited sampling of around 4–12 Hz seems most appropriate; methods aimed at distinguishing two-bond and three-bond correlations in HMBC spectra are the subject of the following section. As indicated earlier, the sequences presented here have not, for reasons of clarity, explicitly included low-pass J filters for the removal of one-bond correlations whereas in most applications these would be incorporated prior to the sampling elements described earlier, usually as gradient-optimised filters as in the original publications or, alternatively, employing the $\text{BIRD}^{r,X}$ filter also described earlier [52].

Whilst most obviously relevant to establishing ^{1}H –^{13}C correlations, broadband J_{XH} optimisation has also been suggested as beneficial in the acquisition of ^{1}H–^{15}N long-range correlations [53,54], a topic of increasing interest [55]. Finally, note that ACCORDION optimisation has also been suggested for the HMQC [56] and HSQC [57] experiments when one-bond couplings are expected to vary widely, which is more often the case for nuclides other than ^{13}C and ^{15}N, this having been demonstrated for ^{1}H–^{113}Cd correlations [56], for example.

7.4.4 H2BC: Differentiating $^{2}J_{CH}$ and $^{3}J_{CH}$ HMBC Correlations

One of the features of HMBC that may lead to ambiguity in structure elucidation is its inability *a priori* to differentiate correlations arising over two bonds from those over three bonds. The coupling constants giving rise to such correlations are similar in magnitude and so cannot be separated on the basis of size alone, unlike one-bond correlations that are at least an order of magnitude larger. Recent methods seek to utilise the presence of homonuclear $^{3}J_{HH}$ vicinal couplings to generate spectra with the appearance of HMBC in which *only* crosspeaks indicative of two-bond pathways appear. These methods exploit quite different transfer pathways by tracing $^{1}J_{CH}$ and $^{3}J_{HH}$ couplings and conceptually follow the scheme of a hybrid HMQC-COSY experiment in seeking *pseudo*-two-bond correlations. Provided $^{4}J_{HH}$ couplings are negligible, crosspeaks corresponding to three-bond correlations in HMBC remain absent. Since this approach does not exploit the often small $^{2}J_{CH}$ couplings, it can offer the further benefit of enhanced intensity for these correlations relative to those observed in HMBC, so acting to complement this, as illustrated in the example presented shortly. Note, however, that correlations to non-protonated carbons cannot be observed due to the lack of the necessary vicinal proton–proton coupling pathway.

The most recent and refined incarnation of this scheme is the heteronuclear two-bond correlation (H2BC) [58,59] experiment illustrated in Fig. 7.39. The core of the sequence comprises a constant time period Δ_{CT} during which proton homonuclear couplings evolve after which the $90°_y$ proton pulse initiates COSY-type transfer to neighbouring protons. Embedded within this period is the HMQC sequence to provide encoding of carbon-13 shifts in the f_1 dimension and so yields a similar appearance to HMBC. Here, the Δ_2 delay is optimised for one-bond heteronuclear couplings in the usual way and the sequence employs gradient encoding for echo–antiecho signal selection. To ensure $^{1}J_{CH}$ coupling evolution

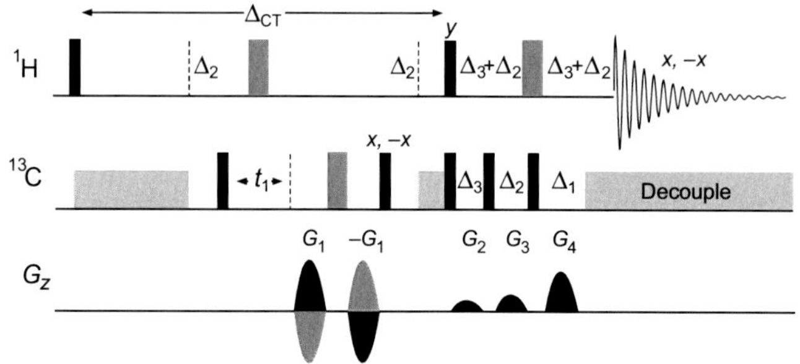

FIGURE 7.39 **The H2BC sequence incorporating a third-order low-pass J filter.** The constant time period for J_{HH} evolution Δ_{CT} is set typically to 16–22 ms, and the heteronuclear delays are $\Delta_1 = 1/2^1J_{CH(min)}$, $\Delta_2 = 1/(^1J_{CH(min)} + {}^1J_{CH(max)})$, $\Delta_3 = 1/2^1J_{CH(max)}$. Gradient selection follows the echo–antiecho protocol, and the gradient ratios for the low-pass filter are G_2:G_3:G_4 = 0.075G_1:0.175G_1:0.75G_1.

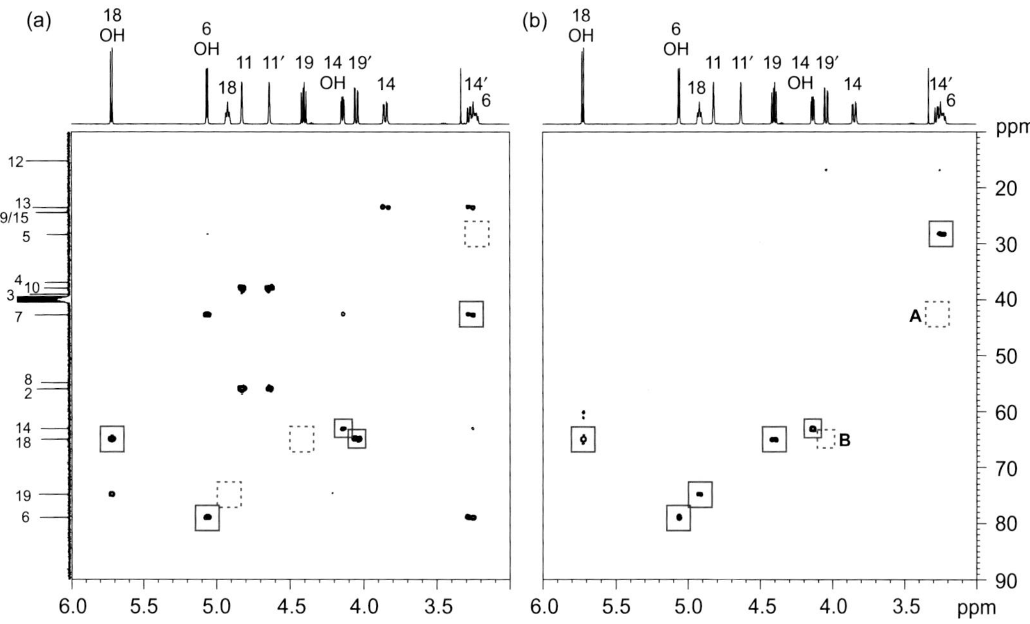

FIGURE 7.40 **The H2BC spectrum of 7.9.** The spectra show a comparison of regions from (a) the 8-Hz HMBC and (b) the H2BC spectra (Δ_{CT} = 22 ms); see text for discussions.

occurs only in the Δ_2 delays within the total Δ_{CT} period, two additional X-spin decoupling periods are included in this; these decrease in duration as t_1 increases to maintain the constant time period. The use of a constant time approach also ensures that J_{HH} couplings do not appear in f_1. The final element of the sequence is a third-order low-pass J filter to efficiently suppress one-bond correlations prior to signal detection in the presence of carbon-13 decoupling. Optionally, the addition of a ^{13}C spin-echo within the HMQC segment yields an edited H2BC experiment [59,60] in which CH_2 responses are inverted relative to those of CH or CH_3 in a manner reminiscent of the edited HSQC. In all cases the constant time period Δ_{CT} is set to allow the formation of anti-phase homonuclear proton magnetisation arising from the evolution of $^3J_{HH}$ couplings and values in the region of 16–22 ms are recommended [58]. Comparison of the HMBC and H2BC spectra of **7.9** is presented in Fig. 7.40 and illustrates the specificity afforded by the H2BC experiment in identifying two-bond correlations. The red boxes highlight all two-bond correlations observed within this region, with those shown dashed indicating missing peaks in either spectrum. Clearly the H2BC experiment identifies a number of additional two-bond correlations that were unobservable in HMBC due to vanishingly small long-range heteronuclear couplings, although it is also missing two correlations that are apparent in HMBC. That marked **A** cannot be observed in H2BC since it would represent a correlation to a quaternary carbon (H14–C7) whereas that marked **B** is not seen because of the small $^3J_{HH}$ coupling involved (H19′–H18, <2 Hz). Note that correlations are also seen from the hydroxyl protons to adjacent carbons two bonds away.

A final route to the identification of two-bond correlations is through the exploitation of one-bond ^{1}H–^{13}C and ^{13}C–^{13}C couplings to explicitly define ^{1}H–^{13}C–^{13}C relationships [61], as in the refocused 1,1-ADEQUATE experiment described in Section 6.4 which has a display format similar to HMBC. Despite the different coupling pathways employed, crosspeaks in the spectrum will equate to those observed in the H2BC experiment with the further beneficial feature that correlations to quaternary centres may also be observed since $^3J_{HH}$ couplings are not involved in the transfer pathway. This will also avoid the potentially misleading three-bond correlation peaks that can arise in H2BC spectra when $^4J_{HH}$ values are non-vanishing, and may also yield additional correlations that are not observed in HMBC in the presence of vanishingly small $^2J_{CH}$ values. The significant drawback to this approach is the undoubtedly poorer sensitivity because of the dependence on $^1J_{CC}$ couplings, but in situations when sample quantities are not limiting it may provide a further useful approach to identifying unambiguously two-bond correlations.

7.4.5 Measuring Long-Range $^nJ_{XH}$ Coupling Constants

In this section we briefly consider experiments for the measurement of the magnitudes of long-range heteronuclear coupling constants, a topic also considered in Section 8.2.1. These can be of considerable value in the stereochemical assignment of organic molecules [62] owing to the Karplus-like dependence of vicinal couplings on dihedral angles [35,63,64]; see discussions in [65] and [46] and the so-called 'J-based configurational analysis' method [66,67] for the definition of stereochemistry, as examples. However, these have traditionally found limited use in the laboratory despite their considerable potential, principally due to the difficulty in measuring these often-small coupling constants. In addition to the low natural abundance of carbon-13, the difficulty originates in the potential for many long-range couplings to exist with a single carbon which, combined with their small magnitude, means they are rarely resolved in 1D proton or proton-coupled carbon spectra. This then demands the use of more complex 2D pulse sequences to enable their extraction. There have been numerous methods developed over many years, indeed far too many to list let alone describe here, but many of these have been reviewed and compared [65,68] and are still under development. One of the significant advantages of the HMBC-based approaches is their ability to measure coupling constants to both protonated *and* non-protonated carbon centres, as most often required in stereochemical investigations. In contrast, many alternative approaches are limited to measuring coupling constants to only protonated carbons, due to the transfer processes involved, restricting their applicability; as such, these methods are not considered here. Instead, we shall review only two methods: the first employs the phase-sensitive HMBC experiment and serves to illustrate the potential but also the limitations of the basic HMBC whilst the second presents a more refined technique (HSQMBC) that provides one of the more user-friendly approaches to these measurements.

7.4.5.1 Using HMBC

The HMBC experiment itself encodes the long-range coupling constants within each crosspeak as an anti-phase splitting superimposed on the usual proton multiplet structure in f_2 since the HMBC FID is recorded without refocusing of heteronuclear couplings. Thus, in principle, these crosspeaks contain the desired ${}^{n}J_{CH}$ information. Unfortunately, however, the multiplets also display complex phase behaviour along f_2 owing to the evolution of proton–proton couplings and proton chemical shifts during the Δ_{LR} ${}^{n}J_{CH}$ evolution period prior to t_1, so direct extraction of the heteronuclear coupling constants is all but impossible (Fig. 7.41). The extravagant phase behaviour within these peaks explains why HMBC spectra are traditionally displayed after a magnitude calculation to mask this. Nevertheless, it is possible to determine the magnitudes of the heteronuclear couplings provided one can adequately account for the phase properties arising within each crosspeak structure. The procedure involves the use of a suitable template multiplet that contains the *identical* phase modulation displayed by the HMBC crosspeak, but lacks the additional splitting arising from the heteronuclear long-range coupling [69]. This template is then used in a fitting routine that seeks to reproduce the experimental form of the HMBC multiplet, as illustrated in Fig. 7.42. The offset between the template and its inverted twin in the optimally fitted trial multiplet directly reflects the magnitude of the heteronuclear coupling ${}^{n}J_{CH}$ responsible for the appearance of the crosspeak; see the original reference for details of the fitting procedure [69].

There are a number of approaches to generate suitable template multiplets, the simplest of which is to take these from a 1D proton spectrum recorded with a delay equal to Δ_{LR} immediately after the excitation pulse (Fig. 7.43a) so as to precisely

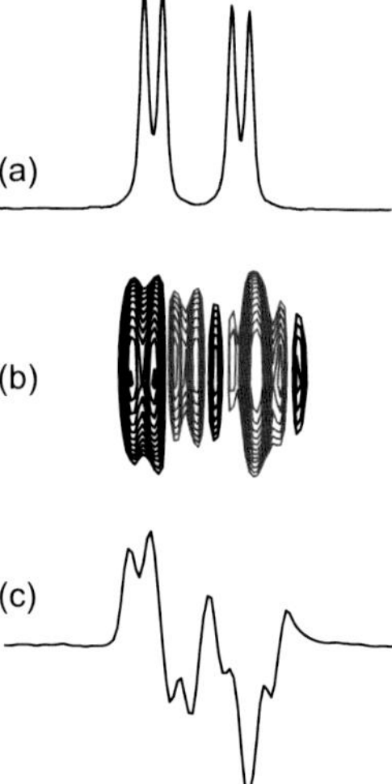

FIGURE 7.41 **Multiplet structure from HMBC.** (a) 1D proton multiplet, (b) HMBC crosspeak and (c) the f_2 (horizontal) trace through the crosspeak displaying the additional anti-phase heteronuclear coupling combined with complex-phase modulation.

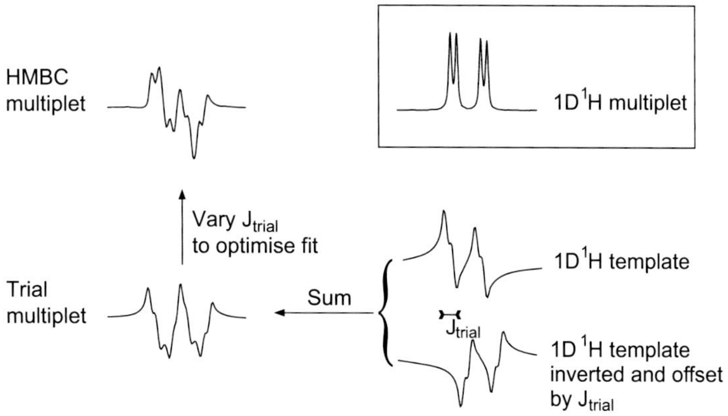

HMBC multiplet
1D ^{1}H multiplet
Vary J_{trial} to optimise fit
Trial multiplet
Sum
1D ^{1}H template
J_{trial}
1D ^{1}H template inverted and offset by J_{trial}

FIGURE 7.42 **Schematic illustration of the template-fitting routine for the extraction of $^nJ_{CH}$ values from HMBC traces.** The trial coupling constant value J_{trial} is varied to produce best fit of the trial multiplet with the experimental HMBC multiplet. The conventional 1D proton multiplet (boxed) is shown for reference purposes only.

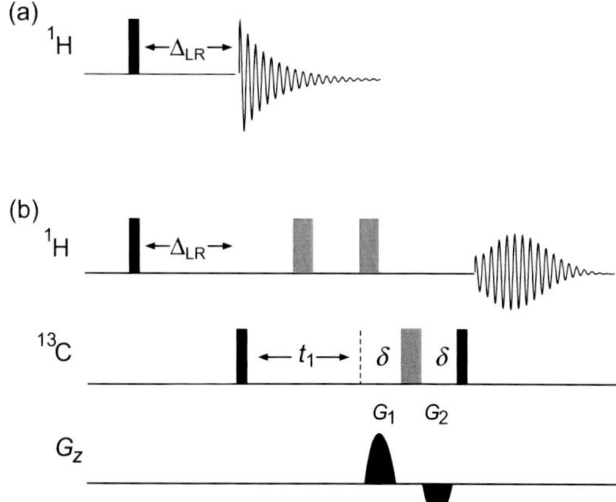

FIGURE 7.43 **The phase-sensitive HMBC sequence for measuring $^nJ_{CH}$ coupling constants.** (a) A possible scheme for producing reference proton template multiplets (see text) and (b) the HMBC experiment.

simulate the phase evolution occurring in the corresponding HMBC delay (experimentally, this can in fact be achieved simply by collecting a standard proton acquisition and removing the first part of the FID equivalent to Δ_{LR} by left-shifting the FID data points). Fig. 7.43b shows an optimised form of the phase-sensitive HMBC containing only one additional proton 180 degree pulse that produces crosspeaks with f_2 phase that precisely match those obtained from the 'delayed' 1D proton spectrum of Fig. 7.43a (strictly, a delay of $\Delta_{LR} + 2\delta$ must then be used in the 1D template when gradient pulses are employed for signal selection in HMBC). Alternatively, the one-bond correlation peaks that often appear in HMBC spectra recorded without low-pass filtration also provide template multiplets since each satellite peak of the $^1J_{CH}$ doublets has the identical phase properties of long-range correlations, but cannot itself contain the long-range heteronuclear coupling (since the associated proton is attached directly to carbon-13 in this case). These may have the advantage of greater separation when multiplets are buried in the 1D proton spectrum and thus not available as templates. Sequences optimised for the retention of these satellites have also been presented [69] and software suitable for performing automatic template-fitting procedures has also been made available [70]. This may also be achieved in most software packages through manual offsetting of the template multiplet and its inverted partner and observing their sum. Typically, the HMBC experiment is recorded with higher digital resolution in the proton dimension than would be employed for routine HMBC acquisitions, so that multiplet fine structure is better defined; a corresponding increase in the t_2 acquisition time by a factor of 4 is usually sufficient. If using multiplets from the 1D spectrum as templates, the 1D digital resolution should match that of the HMBC traces to enable the fitting procedure to operate.

Fig. 7.44 illustrates the application of HMBC to measure three-bond coupling constants to carbonyl C20 in **7.9**, as highlighted in fragment **7.9a**. The complexity of the HMBC multiplet structures precludes the direct measurement of long-range

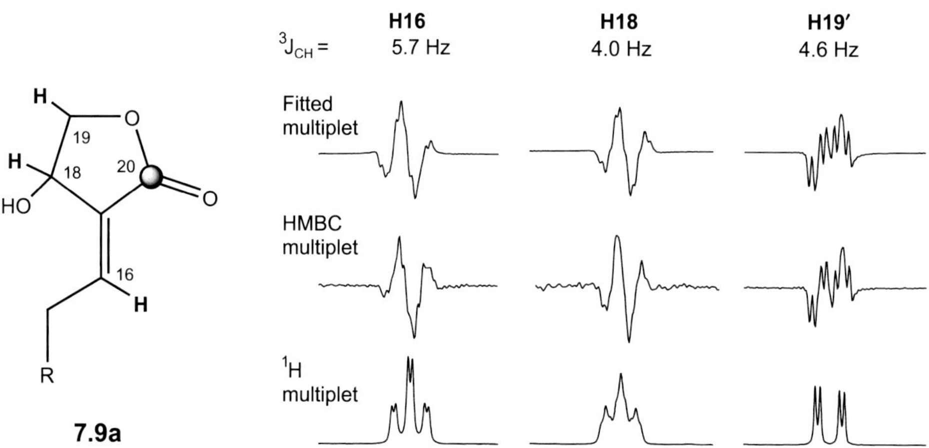

FIGURE 7.44 **Extracting $^nJ_{CH}$ from HMBC spectra.** Example experimental and fitted multiplet structures for the long-range ^1H–^{13}C correlations to carbonyl C20 occurring in the structural fragment **7.9a**. The HMBC spectrum was acquired using the sequence of Fig. 7.43b and the fitting performed as described in [69].

heteronuclear couplings, but these may be extracted via the multiplet-fitting procedure. Nevertheless, it is desirable to avoid such complications in the measurement of these coupling constants, and the improved methods described in the following section nowadays prove more attractive.

7.4.5.2 Using HSQMBC

More recent techniques for the measurement of coupling constants are derived from the long-range optimised HSQC experiment and aim to produce multiplet structures devoid of the complex phase properties associated with the HMBC method [68,71]. In the ideal case, crosspeak multiplets contain only pure, in-phase proton homonuclear *and* heteronuclear couplings such that $^nJ_{CH}$ values can be measured directly from these splittings, without recourse to complex fitting procedures. One of the more attractive methods now available for this is the long-range optimised version of HSQC, known as HSQMBC (heteronuclear single-quantum multiple-bond correlation). The benefit of HSQMBC over HMBC is the fact that only single-quantum magnetisation is used throughout the sequence, simplifying crosspeak structures in f_1 and enabling additional modifications of the sequence. Specifically, this allows for methods that remove the deleterious effects of proton–proton coupling evolution which would otherwise complicate the phase properties of the resulting long-range crosspeaks, as noted earlier.

One approach is to suppress all evolution of J_{HH} couplings during the INEPT transfer steps. An effective method is to use selective 180 degree proton refocusing pulses in the long-range optimised INEPT steps, as in the ^{1}H-selective *clean in-phase* (CLIP)-HSQMBC method (Fig. 7.45) [72]. In this, only a single-proton resonance experiences the selective pulse and thus only this produces long-range correlations. For all other protons, heteronuclear coupling evolution is refocused by the single 180 degrees ^{13}C pulse and, hence, they do not participate in the INEPT transfer. Since protons coupled to the target proton do not experience the selective pulse (they must reside outside its effective bandwidth), all homonuclear

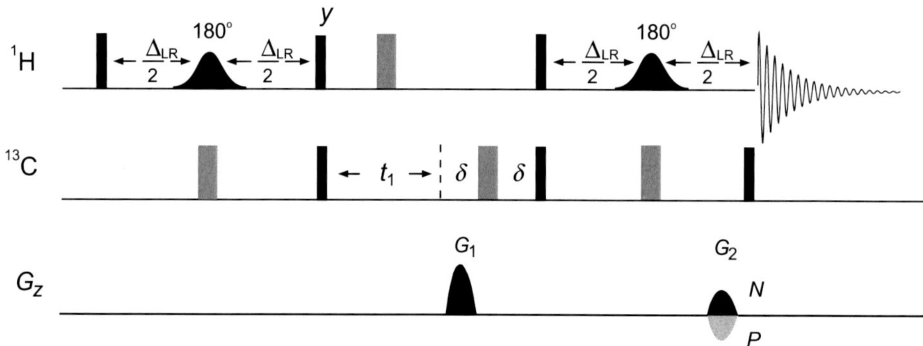

FIGURE 7.45 **The CLIP-selHSQMBC experiment for measurement of long-range heteronuclear coupling constants.** Selective ^{1}H 180 degree pulses are used within the long-range optimised INEPT blocks of HSQC ($\Delta_{LR} = 1/2^nJ_{CH}$).

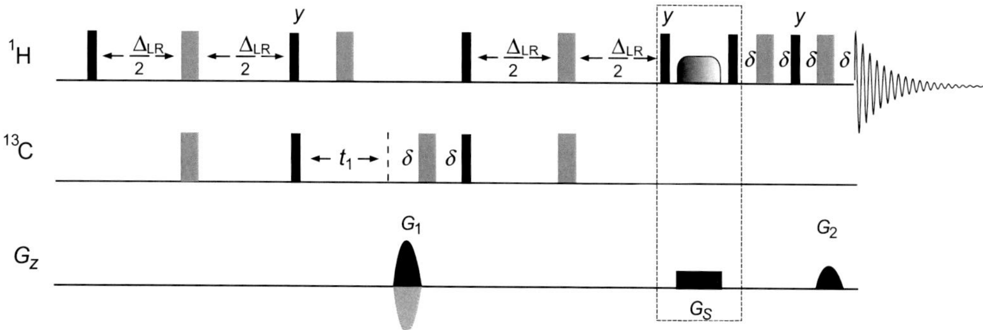

FIGURE 7.46 **The PIP-HSQMBC sequence for measurement of long-range heteronuclear coupling constants.** The purging element (boxed) is applied immediately after the long-range optimised HSQC stage ($\Delta_{LR} = 1/2^nJ_{CH}$) and the final refocusing gradient G_2 is placed within a perfect echo to retain pure, in-phase multiplet structures.

proton coupling evolution in the INEPT periods is refocused, yielding pure proton multiplet structure in the observed crosspeaks. One final refinement is to purge any phase distortions associated with incomplete refocusing of the long-range heteronuclear couplings in the second INEPT step which, as usual, is exact for only a single $^nJ_{CH}$ coupling constant. These residual anti-phase components of the heteronuclear coupling can be removed by applying a final 90 degree carbon pulse immediately before data acquisition starts. This transforms anti-phase magnetisation to invisible heteronuclear multiple-quantum coherences and thus removes them from the crosspeaks, so yielding the CLIP peak structures, exemplified later. In addition to the original proton–proton coupling structure, these crosspeaks contain the desired long-range heteronuclear coupling, allowing for its direct measurement without the need for fitting procedures.

CLIP-HSQMBC does require that a target proton can be selected whilst *all* of its J-coupled proton partners are not, so may fail when coupled protons resonate close to each other, that is, when they experience strong coupling. A further disadvantage is that a complete 2D experiment is collected for each targeted proton, and only correlations for this proton are observed (strictly, for any proton resonating within the selected bandwidth), so the method lacks time efficiency. If desired, the method can be reduced in dimensionality to the 1D variant by replacing the 180 degree proton pulse applied during t_1 with a selective 180 degree *carbon* pulse and by keeping the t_1 period fixed [73]. This then yields 1D proton spectra showing only a single correlation peak between the selected proton *and* carbon spins. To be effective, clean selection of both ^{1}H and ^{13}C multiplets is required.

An alternative approach that allows for the collection of long-range correlations from all protons simultaneously is implemented in the *pure in-phase* (PIP)-HSQMBC [74] technique (Fig. 7.46). This again follows the classical HSQC scheme with INEPT delays Δ optimised for long-range coupling constants ($\Delta_{LR} = 1/2^nJ_{CH}$). The unwanted proton–proton and proton–carbon anti-phase terms that remain after refocusing of the long-range heteronuclear coupling in the second INEPT step are destroyed through application of a combined adiabatic inversion and gradient pulse (G_S: the action of this purging element is described in Section 12.6.2 for the suppression of zero-quantum coherences, so is not considered further here). This leaves only PIP crosspeak structures that are free from distortion, thus cleanly revealing homonuclear and heteronuclear couplings. To ensure the pure-phase character of the multiplet is retained after this purging, the final refocusing gradient G_2 is placed within a perfect echo, which suppresses any further J_{HH} evolution (Section 2.4.4).

Application of the HSQMBC method is illustrated in Fig. 7.47 for the differentiation of *E* and *Z* isomers of **7.10** present as a mixture. Fig. 7.47a shows the HSQMBC spectrum highlighting the alkene proton of the minor isomer at 6.67 ppm. The correlation of interest is that to the ester carbonyl at δ_C 163 ppm (boxed), and the 1D trace through this crosspeak is shown in Fig. 7.47b, above the reference proton multiplet which displays a 7.8-Hz triplet structure. The additional coupling in the HSQMBC crosspeak reveals the presence of the $^3J_{CH}$ 11.0 Hz heteronuclear coupling, consistent with a *trans* relationship. For comparison, the equivalent trace for the major isomer is shown in Fig. 7.47c, revealing the smaller 4.8-Hz coupling arising from the *cis* geometry. Note that the one-bond satellites that may also appear, as in Fig. 7.47a, may provide a useful reference proton structure should the ^{1}H multiplet itself be hidden by other resonances in the 1D spectrum, as noted later. A second example illustrates the definition of a stereocentre in a substituted carbohydrate **7.11**, for which the lack of appropriate protons precluded resolution by other NMR techniques (Fig. 7.48). Here, identification of the axial H3 proton could be made through the large 12.2-Hz triplet structure, reflecting one geminal and one diaxial ^{1}H–^{1}H coupling of similar size. The further large 8.2-Hz heteronuclear coupling to the ester carbonyl (168 ppm) present in the HSQMBC crosspeak, establishes the ester as occupying the axial position. By contrast, no correlation from

Minor

Major

7.10

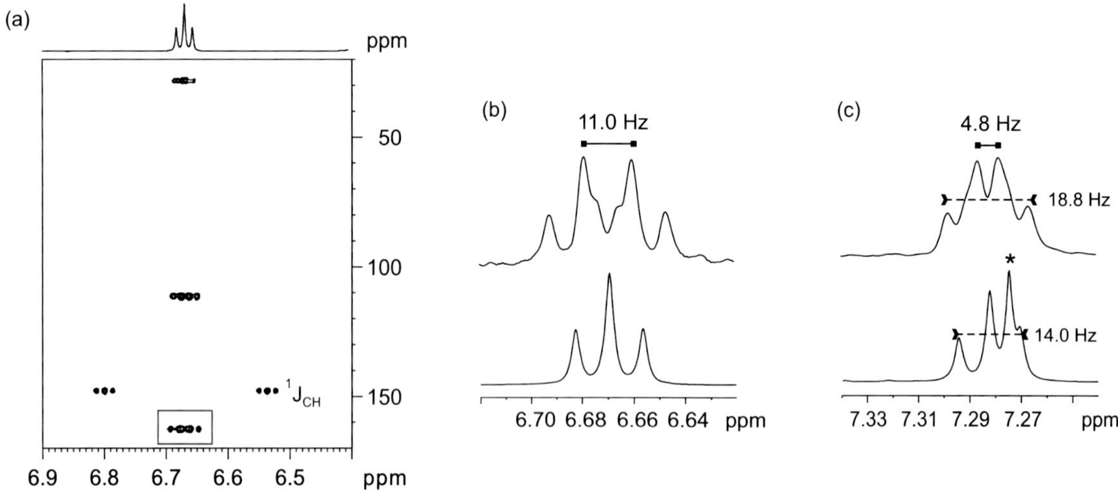

FIGURE 7.47 **Application of the HSQMBC method.** (a) The experiment performed on **7.10** highlighting the alkene proton of the minor isomer at 6.67 ppm, (b) a slice taken through the crosspeak (boxed) in (a), shown above the reference proton 1D multiplet and (c) an equivalent slice for the major isomer resonance at 7.28 ppm, above its reference multiplet (* = residual $CHCl_3$). The difference in multiplet widths (dashed lines) in (c) also yields the heteronuclear coupling constant (see text).

the equatorial H3 proton to the ester carbonyl could be detected, consistent with the smaller dihedral angle and negligible coupling.

As described for HMBC earlier, the HSQMBC experiment is typically collected with a large number of data points in f_2 to enable resolution of the small heteronuclear couplings being sought, leading to acquisition times of ~0.5 s. Once proton traces through the crosspeak of interest have been extracted, it is also possible to back (inverse) Fourier-transform these to produce single FIDs that can be subject to zero-filling and then further transformed to improve the digital resolution in the resultant spectra (in these examples, data sizes were increased four-fold in this manner). In situations where the multiplet structure has significant complexity and direct measurement of the heteronuclear coupling is not possible, one solution is to measure the total width of the multiplet and compare this with the equivalent measurement for a reference peak. This most conveniently would be the ^1H multiplet from the 1D proton spectrum. In situations where the multiplet is hidden in the 1D proton spectrum, a one-bond satellite for the multiplet of interest or the equivalent trace from an HSQMBC spectrum recorded with broadband carbon decoupling may be used. The increased width of the crosspeak trace relative to the reference corresponds to the magnitude of the heteronuclear coupling constant, as exemplified in the multiplet of Fig. 7.47c. In the measurement of a very small heteronuclear coupling (<2–3 Hz) it is likely that this may not be resolved in the

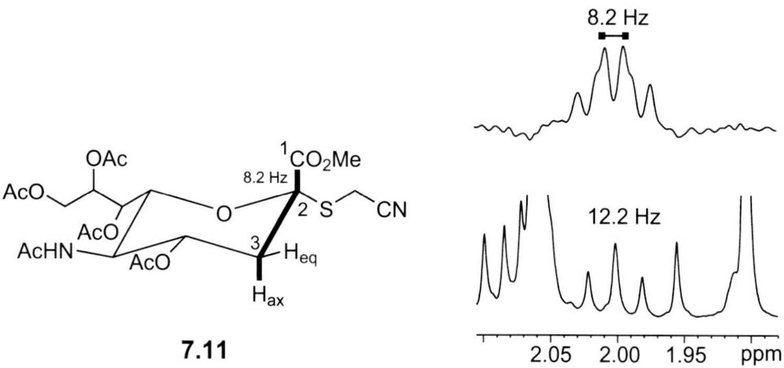

FIGURE 7.48 **Stereochemical definition of 7.11 with HSQMBC.** A trace through the HSQMBC crosspeak, shown above the reference ^{1}H multiplet, reveals the 8.2 Hz $^{3}J_{CH}$ coupling to the ester carbonyl at 168 ppm.

crosspeak fine structure, placing an upper limit on its value. If accurate determination of this is required an extended variant of CLIP-HSQMBC that employs an additional TOCSY transfer to neighbouring protons provides an effective solution and may allow coupling constants as small as 0.5 Hz to be measured [75]. In conclusion, the PIP-HSQMBC currently appears to be one of the more pragmatic and user-friendly techniques for the direct measurement of the magnitudes of long-range heteronuclear coupling constants.

7.4.6 Long-Range HSQMBC: Interrogating Proton-Sparse Molecules

Despite the significant information content provided by the HMBC experiment and its widespread role in defining molecular connectivity, it is hampered by the fact that it typically detects heteronuclear ^{1}H–^{13}C correlations over only two or three bonds, occasionally extending to four when bonding pathways allow for enhanced coupling constants. Although the experiment may be set-up for the detection of very small coupling constants by extending the heteronuclear evolution delay Δ_{LR}, HMBC may not prove effective in this because the complex phase behaviour within its crosspeak structures can lead to cancellation when couplings are small. This restriction can be particularly limiting in studies of proton-sparse molecules, in which heteroatoms of interest may be too remote from protons to be correlated and where the detection of even longer range pathways may be desirable. Examples of this are often to be found in heavily substituted polycyclic aromatic natural products, as exemplified later. To address this need, a variant of the multiple–bond optimised HSQC has been developed specifically to detect very small coupling constants and has been named *long-range* HSQMBC [76]. It has been demonstrated to detect significantly more four-, five-, and even six-bond ^{1}H–^{13}C correlations than HMBC and has proved similarly effective at revealing typically very small four-bond ^{1}H–^{15}N couplings [77].

The sequence (Fig. 7.49) follows the non-selective HSQMBC described earlier with some specific modifications not present in HMBC, but designed to enhance the appearance of crosspeaks originating from very small couplings (ie, $\ll$2 Hz). In particular, these couplings are allowed to be refocused in the final INEPT step, with any remaining anti-phase components purged by the final 90 degree pulse, as in the CLIP element described earlier. These features provide two significant benefits relative to HMBC. Firstly, by detecting only in-phase multiplets, one avoids peak cancellation

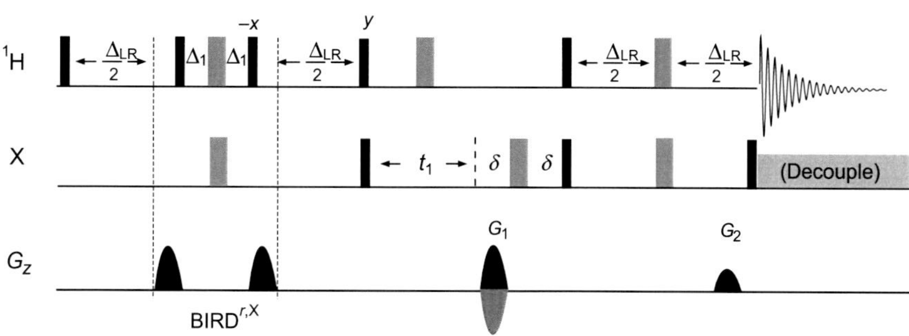

FIGURE 7.49 **The long-range HSQMBC sequence for the detection of very small heteronuclear couplings.** The coupling evolution delays are $\Delta_{LR} = 1/2^{n}J_{CH}$ and $\Delta_{1} = 1/2^{1}J_{CH}$.

that would arise from small anti-phase heteronuclear couplings, especially as these are likely to be comparable with, or less than, the peak linewidths. Secondly, it allows application of heteronuclear decoupling and so helps to increase the intensity of the unavoidably weak very long–range crosspeaks. Attenuation of crosspeaks originating from one-bond couplings is achieved through the use of a gradient-enhanced BIRD inversion element during the first INEPT transfer, as described for low-pass filtration in Section 7.4.3. The INEPT stages themselves are optimised for very small couplings, typically 2 Hz rather than the more common 8 Hz employed for routine HMBC, leading to a very long sequence. Magnetisation losses due to proton transverse relaxation throughout the sequence may therefore be considerable, suggesting the method may be optimal for smaller molecules in low-viscosity solvents which display rather narrow resonance linewidths.

By way of example, correlations observed for quercetin **7.12** in DMSO in a 2-Hz optimised long-range HSQMBC experiment that were not apparent in the standard 8-Hz optimised HMBC are highlighted on the structure. Significantly, a long-range five-bond correlation could be seen to the isolated carbon to which no HMBC correlations were detected; the appearance of this is likely favoured by the 'zig-zag' geometry of the bonding pathway. Similarly, the sharp phenolic OH proton also displayed a four-bond correlation to the adjacent ketone only in the long-range HSQMBC experiment. It is also noteworthy that the typically small $^{2}J_{CH}$ coupling constants observed in aromatic ring systems may also yield enhanced crosspeaks with this technique, and one should be aware that not all new correlations will necessarily correspond to very long–range coupling pathways.

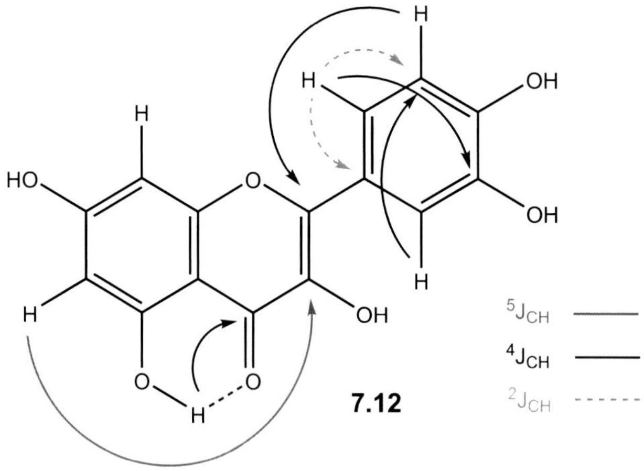

7.12

The very long–range couplings detected with this method are invariably weak, meaning high detection sensitivity is required for success, and access to cryogenic probes will likely prove beneficial when sample quantities are limiting. Nevertheless, this method is more attractive than the ADEQUATE family of experiments (Section 6.4), that are also capable of detecting very long–range connectivities within molecules, because it does not rely on long-range carbon–carbon couplings and so avoids the more severe sensitivity penalty this imposes when working at natural isotopic abundance. As such, the long-range HSQMBC complements the HMBC in defining molecular connectivity, especially for fragments which are substantially devoid of protons.

7.5 HETERONUCLEAR X-DETECTED CORRELATIONS

Traditional methods of heteronuclear shift correlation follow the scheme of Fig. 7.1b, relying on direct observation of the X-spin and with the source nucleus (typically protons) detected indirectly. As discussed at the beginning of this chapter, this general approach is less sensitive than that involving the direct observation of the proton and this has led to the general decline in popularity of X-spin observed techniques. However, one should not shun their existence completely as they have one major advantage over their proton-detected counterparts and that is the ability to record the X-spin with high digital resolution. With proton-detected methods, the often-large X-spin chemical shift range is measured in the indirect dimension and recording this with high resolution entails collecting many t_1 incremented data sets, making the experiments time consuming. With X-spin detection, high resolution may be achieved simply by collecting more data points per FID, making a minimal difference to the total duration of the experiment. When heteronuclear correlations are required for a crowded carbon spectrum, complex aromatic systems, for example, the ^{13}C-detected approach may thus prove superior. The sensitivity of these methods can also be increased by the use of microcell or microtube techniques [78] in place of standard NMR tubes and are aided considerably by cryogenic probes with enhanced carbon sensitivity. The following experiments are described in the context of $^1H–^{13}C$, but again can be tailored to other spin pairs.

7.5.1 Single-Bond Heteronuclear Correlations

Methods for establishing heteronuclear connectives are based on the general idea of polarisation transfer from the proton to carbon and as such can be understood with reference to the previously encountered 1D INEPT experiment of Section 4.4.2. The refocused INEPT sequence allows the transfer of proton populations (polarisation) onto the attached carbon by application of simultaneous proton and carbon pulses after heteronuclear coupling has evolved for a period Δ_1. Following the transfer, the carbon magnetisation so created is anti-phase with respect to $^1J_{CH}$, so is allowed to evolve under the influence of heteronuclear coupling for a period Δ_2 until the carbon vectors have realigned. At this point, proton decoupling is applied and the carbon FID recorded (Fig. 7.50a; here the 180 degree pulses at the midpoint of Δ_1 and Δ_2 which serve to remove chemical shift evolution are omitted for simplicity). To produce the 2D shift correlation experiment [79], the variable t_1 evolution period is added immediately after the initial proton excitation and prior to the polarisation transfer step so that the detected ^{13}C signal becomes modulated by the proton chemical shift as a function of t_1 (Fig. 7.50b). During this, $^1H–^1H$ and $^1H–^{13}C$ coupling will also evolve, leading to the appearance of both homonuclear and heteronuclear splittings in f_1, thereby reducing signal intensities. The removal of $^1H–^{13}C$ coupling can be achieved by refocusing $^1J_{CH}$ with the insertion of a 180 degree carbon pulse at the midpoint of t_1, preferably applied as a composite or adiabatic pulse to reduce resonance offset effects (Fig. 7.50b). Because there is no net CH coupling evolution in t_1, the anti-phase magnetisation required for polarisation transfer only develops during the subsequent Δ_1 period, which is therefore optimised for $\Delta_1 = 1/2^1J_{CH}$ precisely as for INEPT (typically, 3.4 ms for $^1H–^{13}C$ correlations). The resulting spectrum displays crosspeaks correlating carbon chemical shifts in f_2 and proton shifts in f_1 which are further spread by homonuclear proton couplings in f_1. Fig. 7.51 displays a part of the carbon–proton shift correlation spectrum of the palladium complex **7.13**. Despite extensive crowding in the aromatic region, the carbon shifts are sufficiently dispersed to resolve all correlations (note some resonances are broadened by restricted dynamic processes within the molecule and some are split by coupling to phosphorus).

The scheme of Fig. 7.50b has been widely used to produce absolute-value shift correlation spectra, and is often referred to as HETCOR or hetero-COSY. Conversion to the preferred phase-sensitive equivalent (of which various forms have been investigated [80]) requires reintroduction of the simultaneous 180 degree(^1H, ^{13}C) pulses into the midpoints of both Δ_1 and Δ_2 to remove chemical shift evolution during these periods, exactly as in the fully refocused INEPT. In addition, incorporation of the States or TPPI phase cycling of the 90 degree proton pulse of the polarisation transfer step is required. Suppression of axial peaks is through phase alternation of the final proton pulse together with the receiver (which is equivalent to

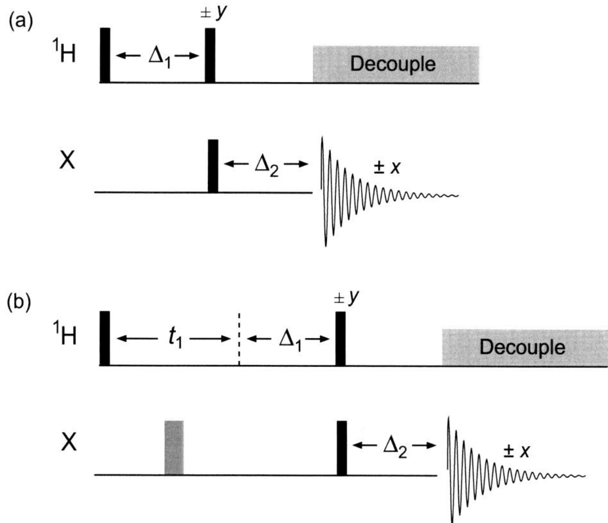

FIGURE 7.50 **Heteronuclear correlations with X-spin detection.** (a) The 1D refocused INEPT experiment, shown in simplified form with the refocusing pulses at the midpoints of Δ_1 and Δ_2 removed for clarity. The 2D shift correlation experiment (HETCOR) in (b) is derived from INEPT by the addition of the t_1 evolution period to encode proton chemical shifts prior to polarisation transfer. The 180 degree X-spin pulse at the midpoint of t_1 refocuses heteronuclear coupling evolution and thus provides X-spin decoupling in f_1.

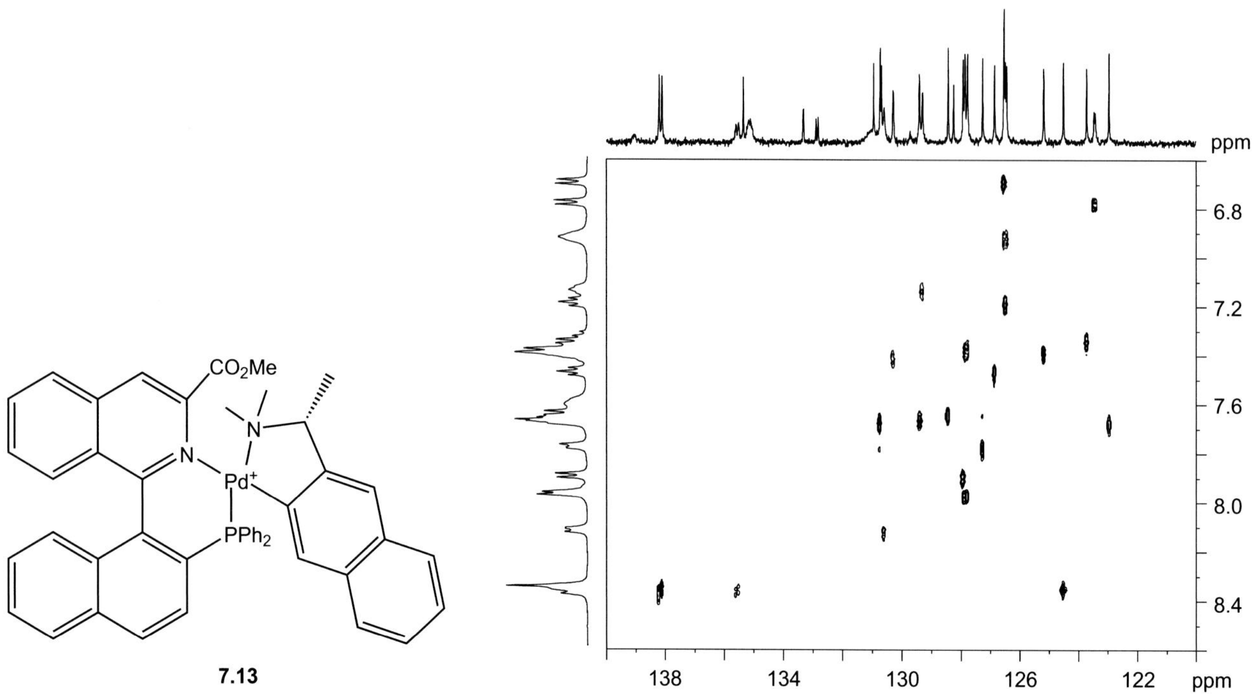

FIGURE 7.51 **The aromatic region of the absolute value ^{13}C–^{1}H HETCOR spectrum (500 MHz) of the palladium complex 7.13.** 2K data points were collected for 256 t_1 increments of 128 transients each for a 40 × 2.6-ppm window. Unshifted sine bells were applied in both dimensions and, after zero-filling once in each, the digital resolution was 2.5 and 5.0 Hz/pt in f_2 and f_1, respectively.

that used in INEPT to suppress contributions from the natural X-spin magnetisation). As for the INEPT sequence, editing of the phase-sensitive spectrum is possible by judicious choice of the Δ_2 refocusing delay, for example, setting $\Delta_2 = 1/1.5J_{CH}$ produces a spectrum in which responses from CH_2 groups are inverted. For routine application, selecting $\Delta_2 = 1/3J_{CH}$ (typically 2.3 ms for ^{1}H–^{13}C) provides positive intensities for all protonated carbons.

Since polarisation transfer is employed, the experiment repetition rate is dictated by recovery of the faster relaxing proton spins, and repetition times should be around 1.3 times the proton T_1s. Resolution in the proton dimension can be quite low for routine applications since one does not usually wish to resolve the proton fine structure, and because the homonuclear coupling is in-phase there is no fear of signal cancellation; f_1 digital resolution may therefore be as low as 10 Hz/pt or so, requiring rather few t_1 increments. The number of scans per increment should be set such that the carbon resonances of interest can just be observed in the spectrum of the first recorded FID (which is equivalent to the 1D INEPT experiment).

7.5.1.1 Homonuclear Decoupling in f_1

By introducing a slight modification to the sequences presented earlier, it is possible to achieve (almost complete) removal of homonuclear proton couplings from f_1 also, so removing the multiplet spread and increasing both sensitivity and f_1 resolution. To do this, a similar approach to that for removal of the *heteronuclear* coupling in f_1 described earlier is employed; that is, for each coupled spin pair one must invert, at the midpoint of t_1, *only one* of the two spins so that the coupling evolution is refocused at the end of t_1. For the heteronuclear case this is trivial, since one can apply a 180 degree carbon pulse confident in the knowledge that its proton partner will be unaffected. However, applying a single proton 180 degree pulse will fail since both protons experience this pulse and the coupling will continue to evolve, as for a homonuclear spin-echo. Worse still, proton chemical shifts will be refocused during t_1, destroying the proton shift dimension altogether (although undesirable in this application such an approach can be useful if one wishes to analyse only coupling patterns in f_1; see Section 8.2). What is required then is a means of inverting only the *remote* J-coupled neighbours of a *directly* ^{13}C-bound proton (which is ultimately involved in the polarisation transfer step). The solution to this seemingly impossible task lies in the application of the BIRD sequence [26] already introduced in Section 7.3.3 as a means of selectively inverting ^{12}C-bound protons whilst leaving ^{13}C-bound protons unaffected [81,82]. The differentiation of protons in the context of f_1 decoupling relies on the very different magnitudes of the one-bond and long-range couplings ($^{1}J_{CH} \gg {}^{2/3}J_{CH}$) such that the long-range couplings can be considered negligible. Hence, placing the BIRD cluster at the midpoint of t_1 refocuses proton–proton coupling whilst the carbon 180 degree pulse of BIRD also serves to refocus proton-carbon coupling, as earlier (hence bilinear rotational *decoupling*). The f_1 resonances then appear as singlets with the exception of those from non-equivalent geminal protons which retain their mutual coupling (as neither proton experiences the inversion since both are bound to ^{13}C) [83].

7.5.2 Multiple-Bond Correlations and Small Couplings

Numerous X-detected sequences have been presented over the years for establishing long-range ^{1}H–^{13}C correlations [84] prior to the widespread adoption of proton-detected counterparts. This section presents the most widely used sequence—correlation through long-range coupling (COLOC)—and briefly mentions more recent sequences that have superior performance in most instances.

A direct way to obtain long-range correlations is to optimise the J-delay periods found in the one-bond HETCOR sequence for much smaller long-range couplings. The potential drawback to this approach is the total length of the resulting sequence because of the small couplings and the associated signal losses due to spin relaxation. COLOC is an early modification that places the t_1 period *within* the Δ_1 delay and thus reduces the overall length of the sequence (Fig. 7.52) [85,86]. The t_1 period is now defined by a pair of 180 degree(H, C) pulses that move through Δ_1 as t_1 is incremented, such experiments being another example of a constant-time experiment. The detected signal is modulated by proton shifts because, although these are refocused during t_1, they continue to evolve in the *remainder* of the Δ_1 period which itself is dependent on t_1, so allowing characterisation of proton chemical shifts. In contrast, homonuclear proton couplings evolve for the whole of the fixed Δ_1 period regardless of the position of the moving 180 degree proton pulse, meaning these couplings cause no modulation as a function of t_1 and the f_1 dimension conveniently displays only proton chemical shifts. Transfers due to one-bond couplings may also appear in the COLOC spectrum if the Δ_1 and Δ_2 periods happen to be multiples of the appropriate $^1J_{CH}$ values, and, if desired, the inclusion of a low-pass J filter, as described for HMBC earlier, may help attenuate these [87].

A greater problem with this sequence is the influence of one-bond couplings during the Δ_2 refocusing period, which can lead to the disappearance of a long-range crosspeak even when the Δ_1 and Δ_2 delays are optimised for the long-range couplings [88]. Following the polarisation transfer step, the vectors associated with the *one-bond* couplings of the carbon nuclei are in-phase (they are *passive* couplings since they have not been responsible for the polarisation transfer process) and during the subsequent Δ_2 period they will pass in and out of phase a number of times under the influence of $^1J_{CH}$. If these happen, coincidentally, to have an anti-phase disposition when the acquisition starts and proton decoupling is applied, the vectors will cancel and the crosspeak will be lost. In any case, the detected long-range correlation intensity will be modulated by the one-bond H–C coupling of the carbon. To remove these effects, one must ensure these vectors are returned to their original in-phase disposition at the end of Δ_2, but at the same time allow the initially anti-phase vectors due to *long-range* couplings to continue to evolve and become in-phase. Once again, this can be achieved by positioning a BIRD sequence at the midpoint of Δ_2, which inverts the carbon and its long-range coupled proton but not the directly coupled proton [89,90]. The use of the BIRD cluster in this manner is recommended for all X-detected long-range correlation methods.

In general, optimisation of the Δ_1 and Δ_2 delays for the COLOC experiment can be problematic. In addition to the one-bond modulation effects already described, transfer efficiency is also influenced by homonuclear proton couplings evolving during Δ_1. The fixed Δ_1 period also limits $t_{1(max)}$ ($t_{1(max)}/2 \leq \Delta_1$) which may dictate fewer t_1 increments be collected than would otherwise be desirable, so reducing the digital resolution of the f_1 dimension. These problems have been addressed in some detail and 1D sequences presented for parameter optimisation [91]. For routine use, it has been suggested [84] that optimising delays for $^nJ_{CH} = 10$ Hz should provide most responses of interest, particularly when using the BIRD sequence within the Δ_2 refocusing delay. This corresponds to delays of 50 ms ($1/2^nJ_{CH}$) and 33 ms ($1/3^nJ_{CH}$) for Δ_1 and Δ_2, respectively.

Various other sequences have been proposed over the years which generally show improved performance over COLOC. If heterospin-detected methods are best suited to the study of your molecules, two sequences to note are XCORFE [92] and FLOCK [93], details of which can be found in the original literature.

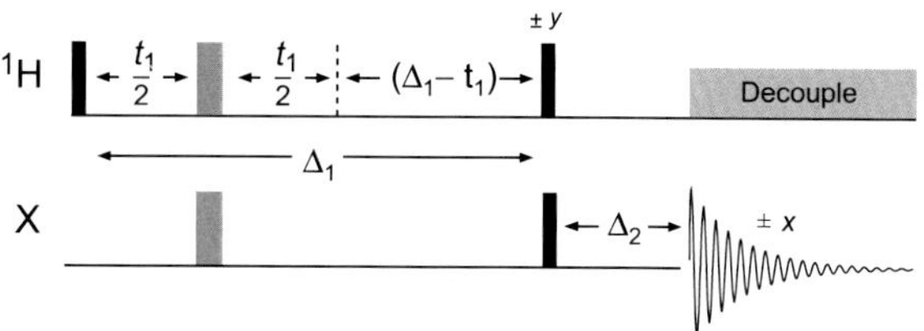

FIGURE 7.52 **The COLOC sequence for establishing long-range correlations through X-spin observation.**

7.6 HETERONUCLEAR X–Y CORRELATIONS

The methods presented so far in this chapter have focused principally on the correlation of two nuclei, ^{1}H and a single heteronucleus X. There are, however, areas of chemistry in which it proves advantageous to correlate, or at least utilise, coupling interactions between two heteronuclei, such experiments being often referred to as *heteronuclear X–Y correlations*. Most examples of this come from studies of organometallic chemistry, in which correlations with a metal centre are sought, or from non-metallic organoelement compounds, in which multiple NMR-active heteroatoms exist to be correlated. Likewise, most examples have both X and Y as spin-½ nuclei.

Methods suitable for such studies may be classified into two groups: those that employ the direct observation of one of the heteroatoms and those that employ proton detection, most often for reasons of sensitivity, both of which will be considered briefly in this section. Both classes require the spectrometer to possess (at least) three rf channels and a triple-resonance probe that can be tuned to all the nuclei involved and hence are not accessible on standard two-channel instruments. However, with increasing access to three-channel spectrometers and the wider range of triple-resonance probes available, such methods can be expected to provide additional routes to structure determination and the study of chemical transformations. Probe configurations such as ^{1}H–^{13}C–broadband, ^{1}H–^{31}P–broadband or custom-designed ^{1}H–X–Y combinations are suitable for such work. Some of the practical considerations of these methods and the range of techniques adapted for such studies have been reviewed extensively and numerous examples given [94–96].

Of increasing interest are also methods in which fluorine is utilised, either in place of proton, such as for ^{19}F–X correlations, or in addition to this, as in ^{1}H–^{19}F–Y correlations that again require triple-resonance probes, here with separate proton and fluorine channels. Experiments in which fluorine is correlated with one other nuclide can be performed on two-channel probes where the ^{1}H channel can be tuned down to ^{19}F. An example of this can be seen in the ^{19}F–^{13}C correlation spectrum of dexamethasone **7.14** (Fig. 7.53). One potential complication with fluorine correlations is the potentially very wide fluorine bandwidths due to the large chemical shift dispersion of this nuclide. This can mean that efficient excitation and, in particular, refocusing over the full bandwidth can be poor for standard correlation sequences employing hard pulses. To address this, HSQC experiments optimised for ^{19}F–^{13}C correlations have been proposed that enable the use of broadband adiabatic pulses on the fluorine channel for refocusing, greatly improving crosspeak detection [97]. Alternative commercial two-channel probe designs place ^{19}F on the broadband channel, so will instead allow ^{1}H–^{19}F methods to be employed, and the heteronuclear sequences presented in earlier sections have been compared specifically for ^{1}H and ^{19}F correlations [98,99].

7.6.1 Direct X–Y Correlations

Direct observation of correlations between two different nuclides can, in principle, be achieved using the 2D methods already described in this chapter [100] and do not necessarily require proton detection, so long as the observed nucleus exists at high abundance. Indeed, such methods may become necessary in proton-sparse systems and may even provide greater

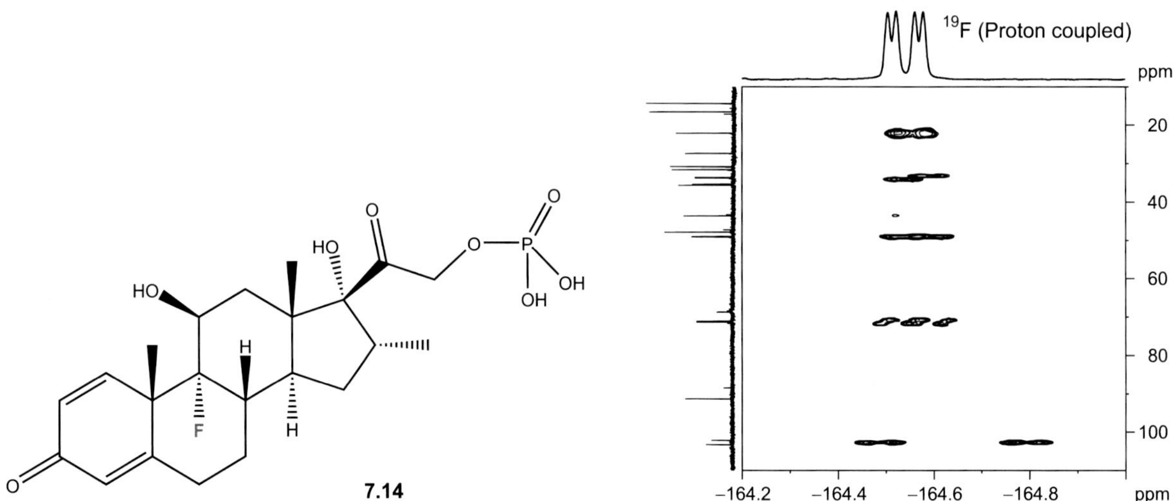

FIGURE 7.53 **The ^{19}F–^{13}C HMBC spectrum of dexamethasone phosphate 7.14.** This was recorded on a two-channel broadband probe with the ^{1}H channel tuned down to ^{19}F. Offsets for the crosspeaks relative to the 1D ^{19}F reference multiplet arise from ^{13}C isotope shifts on the ^{19}F nucleus.

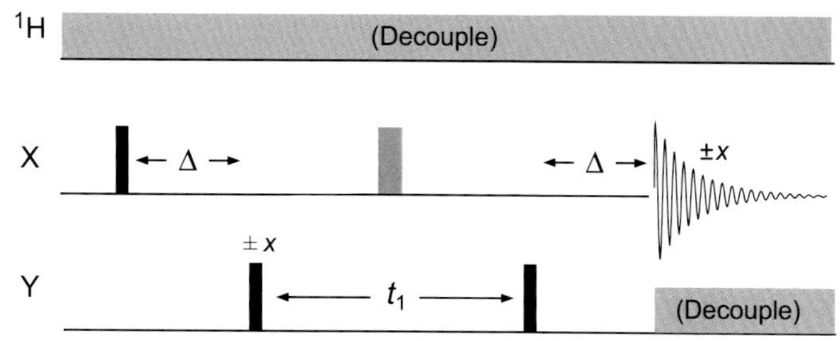

FIGURE 7.54 **The HMQC sequence for X–Y correlations in the presence of broadband proton decoupling.** PFGs may be incorporated as for the standard 2D HMQC experiment.

sensitivity by utilising large J_{XY} coupling constants in situations when only small, long-range proton couplings exist to the heteroatoms. In such cases, the HMQC experiment may be employed with observation of the nuclide of greater sensitivity X (combined with optional Y-spin decoupling), with indirect detection of nucleus Y and, typically, with broadband proton-decoupling operating throughout the sequence (Fig. 7.54). As an illustration, Fig. 7.55 shows a phosphorus-detected ^{31}P–^{13}C correlation of **7.15** acquired in this manner, but without carbon-13 decoupling during the detection period; the correlation thus appears as a $^{1}J_{CP}$ doublet in f_2. The use of ^{31}P as the observed nucleus X is well documented due to its favourable properties, and many examples of this may be found, including ^{31}P correlations with ^{103}Rh, ^{109}Ag, ^{119}Sn and ^{183}W to name a few [101]. Inverse correlation methods can therefore provide access to chemical shift data of nuclei that are otherwise too insensitive to be observed directly. This has particular significance for the indirect observation of metals, many of which have very low intrinsic sensitivity but can nevertheless be detected through their spin-coupled neighbours on adjacent ligands [96]. The indirect observation of ^{57}Fe, a notoriously difficult nucleus to observe directly, is one more example [102–104] (Fig. 7.56).

The HMQC sequence in particular is favoured for such correlations because of its robustness and because the phase-cycled version can be collected without the need for 180 degree pulses on the Y channel. This is beneficial when the Y-nucleus chemical shifts cover a wide range, notably the case for metals again, where off-resonance losses from the 180 degree Y pulse could be highly detrimental. When PFGs are desirable for optimal signal suppression, the use of composite or adiabatic 180 degree pulses within the gradient echoes may prove beneficial for both HSQC and HMQC. When the heteronuclear couplings vary widely in magnitude, the optimal detection of all correlations can be problematic and the ACCORDION-optimised HMBC sequences described previously may then be adapted for X–Y correlations. One such example has been application of the IMPEACH-MBC experiment in the observation of ^{19}F–^{15}N correlations where the $^{2/3}J_{FN}$ couplings ranged between 4 and 50 Hz [105].

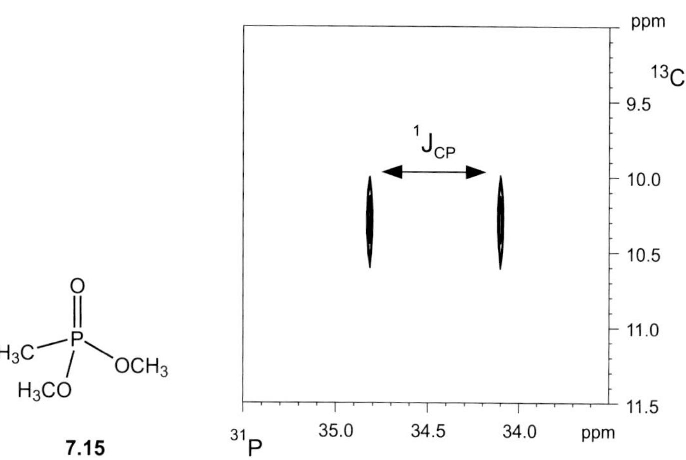

FIGURE 7.55 **Phosphorus-detected ^{31}P–^{13}C HMQC spectrum of 7.15.** Proton decoupling was applied throughout but no carbon-13 decoupling was used during detection of the phosphorus FID. These data were acquired using a home-built third channel and a probe modified for triple-channel operation.

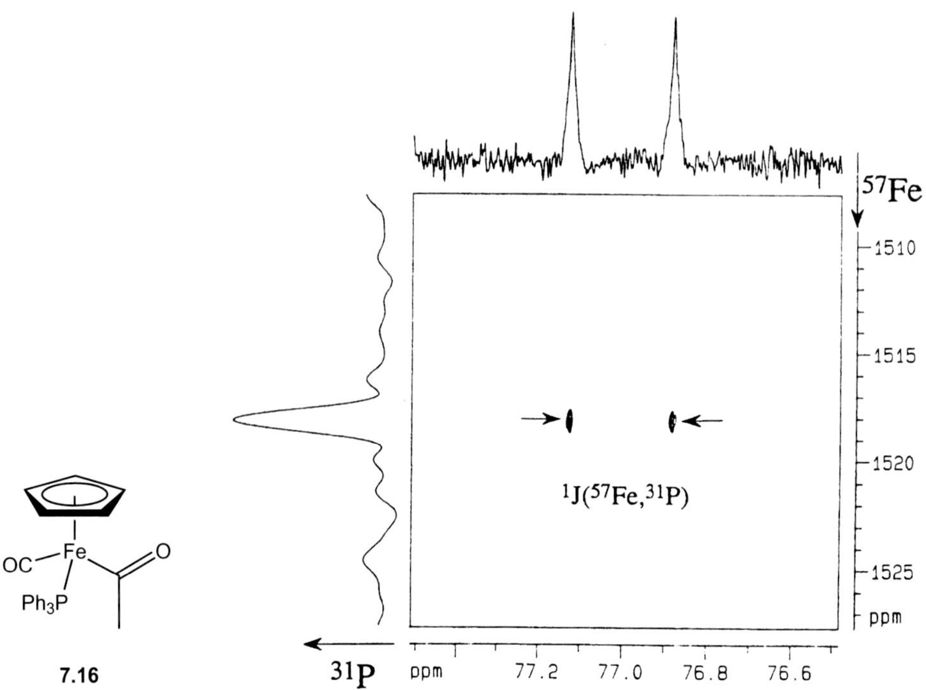

FIGURE 7.56 **Phosphorus-detected, proton-decoupled ^{31}P–^{57}Fe HMQC spectrum of 7.16.** The traces are projections taken from the 2D spectrum. *(Reproduced with permission from reference [104].)*

7.6.2 Indirect ^{1}H-Detected X–Y Correlations

It is perhaps not surprising that more recent developments for identifying X–Y correlations have incorporated proton detection and PFGs for signal selection, with many of the techniques being derived from the triple-resonance ^{1}H–^{13}C–^{15}N techniques originally designed for the study of biological macromolecules [106]. These were conceived as three-dimensional (3D) methods which can be derived conceptually from their 2D counterparts by the incorporation of another variable evolution time period and associated mixing steps (Fig. 7.57). The additional evolution period is incremented in a stepwise manner independently of the others, thus yielding three independent time domains t_1, t_2 and t_3 which ultimately transform to three frequency domains f_1, f_2 and f_3. Such methods result in 3D data cubes in which crosspeaks are dispersed throughout according to the three chemical shifts of the participating spins (Fig. 5.10). These provide clarity by dispersing the numerous correlation peaks that arise in the spectra of isotopically labelled proteins. Whilst 3D sequences have been presented for ^{1}H–^{13}C–^{31}P correlations, these find little use in the study of smaller organic molecules or their complexes for two principal reasons. Firstly, the need to sample two *indirect* time domains for 3D spectra invariably leads to longer experiment times and, secondly, since there are typically rather few heteroatom centres in small molecules there is little need for the additional dispersion provided by 3D experiments. Rather, most ^{1}H–X/Y correlation experiments are recorded as 2D spectra correlating ^{1}H and Y with the X heteroatom acting in a spectrum-editing or magnetisation relay capacity, signified by the notation ^{1}H–(X)–Y. Such methods may be derived from their 3D parent simply by making static one of the incremented time delays, thus using the corresponding nucleus for the relaying of coherence rather than the indirect sampling of its chemical shift.

One such approach that illustrates this principle is shown in Fig. 7.58 which may be viewed as an extended HSQC experiment now employing two transfer steps between nuclei [107,108]. In this case the transfers are made with INEPT stages sequentially from ^{1}H to X and then onto Y for encoding in t_1. Gradient selection is made on the Y-nucleus (which may also

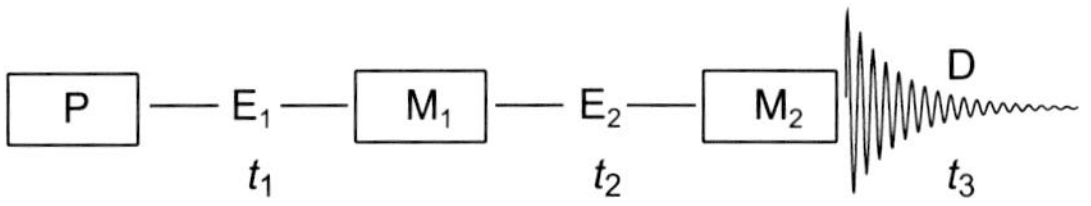

P — E_1 — M_1 — E_2 — M_2 D
t_1 t_2 t_3

FIGURE 7.57 **The general scheme for 3D pulse sequences.** In the 3D techniques there exist two indirectly measured time domains t_1 and t_2 and a third t_3 in which the FID is recorded (see Section 5.1 also).

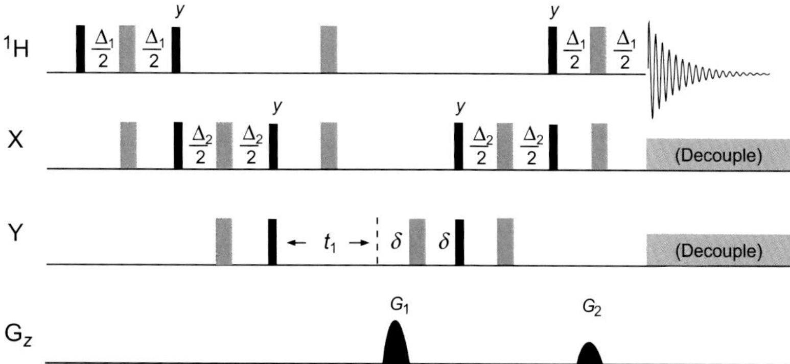

FIGURE 7.58 **A triple-resonance sequence for recording 2D ^{1}H–(X)–Y heteronuclear correlations.** The delays are set to $\Delta_1 = 1/2^nJ_{XH}$ and $\Delta_2 = 1/2^nJ_{XY}$ with gradient strengths set as $G_2 = G_1(\gamma_{(Y)}/\gamma_{(H)})$.

provide for echo–antiecho selection for f_1 quad-detection) and then reverse INEPT stages to transfer magnetisation back onto X and finally onto ^{1}H for gradient decoding and then detection, optionally with X- and/or Y-spin decoupling. Following the notation used for HSQC in Section 7.3.1, the overall transfer process may be summarised thus:

$$^1\text{H}\xrightarrow[\text{INEPT}]{^nJ_{XH}}\text{X}\xrightarrow[\text{INEPT}]{^nJ_{XY}}\text{Y}\xrightarrow[t_1]{\delta_Y}\text{Y}_{(\text{grad.encode})}\xrightarrow[\text{R-INEPT}]{^nJ_{XY}}\text{X}\xrightarrow[\text{R-INEPT}]{^nJ_{XH}}{}^1\text{H}_{(\text{grad.decode})}\Rightarrow\text{Detect}_{(\text{X,Y-dec.})}$$

The 2D spectrum therefore correlates ^{1}H with the Y-spin via a pathway involving both $^nJ_{XH}$ and $^nJ_{XY}$ couplings, where *n* is 1 or greater according to the available coupling pathways. The 'relay' nucleus is frequently ^{13}C or ^{31}P with the Y-nucleus representing a more exotic spin including, for example, ^{15}N, ^{183}W [108], ^{195}Pt [107] and ^{109}Ag [109]. Fig. 7.59 illustrates the ^{1}H–(^{13}C)–^{195}Pt correlation spectrum of a mixture of the *E*- and *Z*- isomers of the thiourea platinum complexes **7.17** recorded using the sequence of Fig. 7.58. Magnetisation transfer steps made use of $^1J_{CH}$ and $^4J_{CPt}$ couplings and the assignment of the *E*- and *Z*- isomers was based on the observation that significant $^4J_{CPt}$ couplings of ~30 Hz existed only when the carbon of the *N*-alkyl branch existed in a favourable '*w*' configuration with the Pt centre, as represented in bold in the

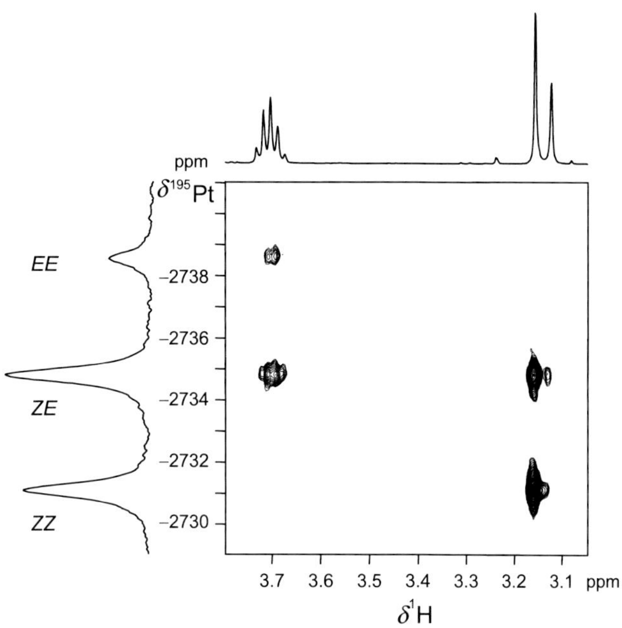

FIGURE 7.59 **The ^{1}H-detected ^{1}H–(^{13}C)–^{195}Pt correlation spectrum of *E*- and *Z*-mixtures of the thiourea platinum complexes of 7.17.** *(Reproduced with permission from reference [107].)*

structures. Any potential $^{5}J_{HPt}$ couplings were not observed and thus could not be employed in HMBC-based experiments. From the 2D spectrum it was possible to directly assign the platinum resonances of the different isomers and hence determine the isomer ratios. These data were collected on a triple-channel instrument equipped with a ^{1}H–^{13}C–broadband probe for which the broadbanded channel had been tuned to ^{195}Pt.

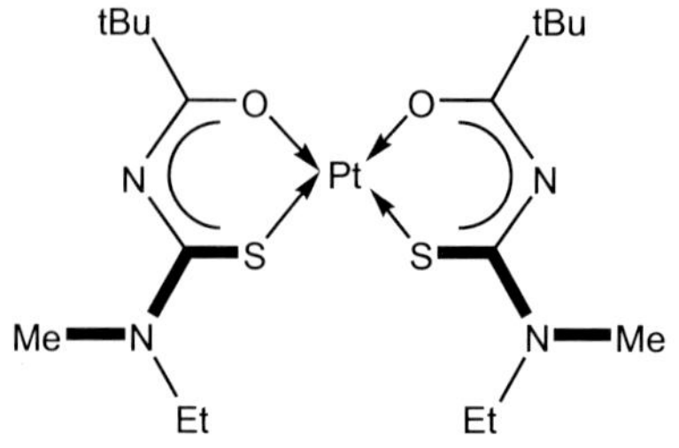

Z,Z-isomer

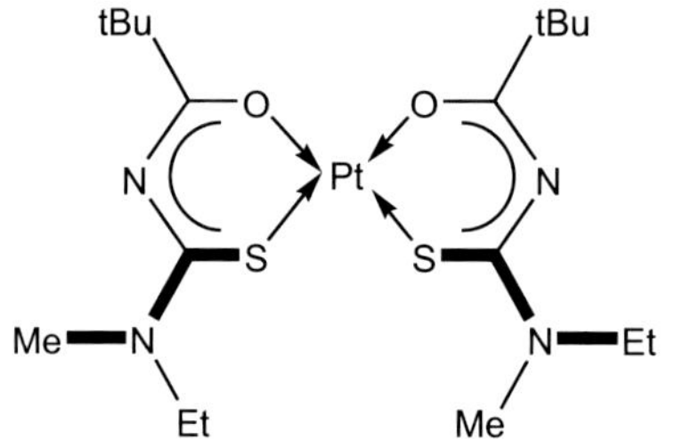

Z,E-isomer

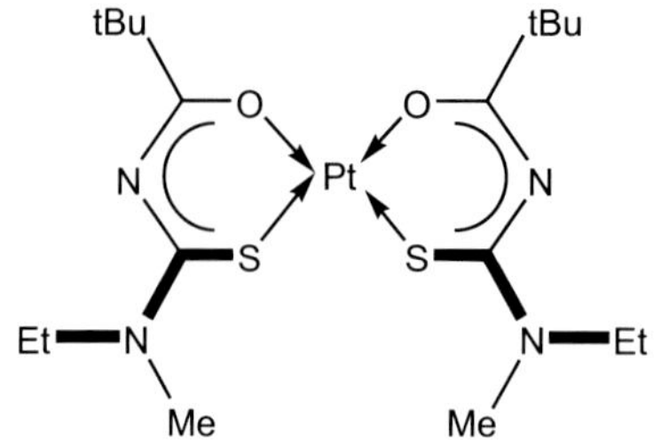

E,E-isomer

7.17

An alternative approach—gradient spectroscopy selective triple resonance (gsSELTRIP) [110]—correlates ^{1}H and X spins with the Y-spin serving to *edit* the resulting 2D spectrum according to the presence of X–Y coupling. Thus, one may produce a ^{1}H–^{13}C correlation spectrum (X = ^{13}C) with the same appearance as a standard ^{1}H–^{13}C HSQC but in which correlations appear only for those carbons that are themselves *also* coupled to ^{31}P (Y = ^{31}P). The transfer process for this may be summarised:

$$^{1}\mathrm{H}\xrightarrow[\text{INEPT}]{\overbrace{\;^{n}J_{XH}\;}}\mathrm{X}\xrightarrow[t_1]{\widehat{\delta_X}}\mathrm{X}\xrightarrow{^{n}J_{XY}}\mathrm{X,Y_{DQF}}\xrightarrow{^{n}J_{XY}}\mathrm{X}\xrightarrow[\text{R-INEPT}]{\overbrace{\;^{n}J_{XH}\;}}\,^{1}\mathrm{H}_{(\text{grad.decode})}\Rightarrow \text{Detect}_{(\text{X,Y-dec.})}$$

where X, Y_{DQF} represents a gradient-based filter element that selects for double-quantum X–Y coherence. This provides one potential approach to simplifying the often-crowded aromatic region of the ^{1}H–^{13}C correlation spectra of aromatic organometallic phosphorus complexes, for example. In this regard, it should also be noted that the ^{1}H–^{31}P HMBC experiment provides a very sensitive method to aid in the assignment of proton spectra by utilising long-range proton–phosphorus couplings and benefiting from the high natural abundance of phosphorus-31.

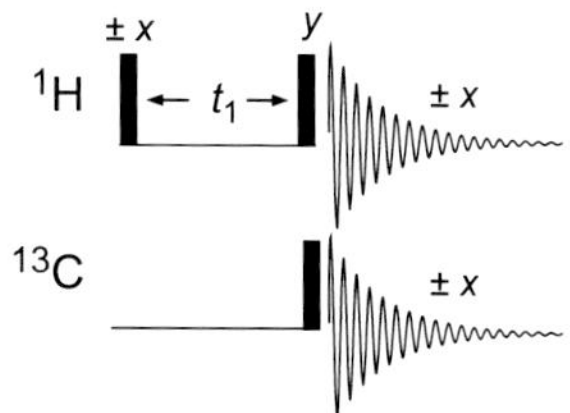

FIGURE 7.60 **A basic scheme for the parallel acquisition of ^{1}H–^{1}H and ^{1}H–^{13}C 2D correlation spectra using dual spectrometer receiver channels.**

7.7 PARALLEL ACQUISITION NMR WITH MULTIPLE RECEIVERS

In this final section we consider briefly a recent addition to high-resolution NMR methodology that has once again found its inspiration in magnetic resonance imaging methods, as with the initial implementation of PFGs. The concept is simple yet revolutionary: the use of multiple NMR receivers in a spectrometer so that it can collect responses from different nuclei *in parallel,* thus providing time-efficient schemes for recording multiple heteronuclear correlation spectra [111]. The method has been termed *parallel acquisition NMR spectroscopy* (PANSY) [112] and offers the potential for more rapid data collection. Although the presence of multiple receivers on NMR spectrometers remains relatively rare, modern consoles are built with the capability to accommodate these, meaning new techniques based on the multiple receiver concept are to be expected and may yet become commonplace.

The philosophy can be readily understood by considering the simplest approach to the simultaneous acquisition of both ^{1}H–^{1}H and ^{1}H–^{13}C 2D correlation spectra (Fig. 7.60). In this, the proton channel operates as for the conventional 2D COSY sequence incorporating the usual ^{1}H detection. In parallel to this, the carbon pulse applied simultaneously with the second proton pulse initiates ^{1}H to ^{13}C polarisation transfer which is then followed by ^{13}C detection. This magnetisation has been frequency-labelled in t_1 according to the ^{1}H chemical shifts and thus provides a ^{13}C-detected 2D heteronuclear shift correlation spectrum. The detected ^{13}C signal is acquired in the absence of ^{1}H decoupling in this case, so although this does not represent an optimum approach it does serve to illustrate the principles behind parallel acquisition spectroscopy.

More sophisticated and practical sequences have been developed, such as for the simultaneous acquisition of ^{1}H–^{13}C and ^{19}F–^{13}C 2D correlation spectra, for which dual ^{1}H and ^{19}F receivers and a triple-resonance ^{1}H–^{19}F–^{13}C probe were employed [113]. The combination of ^{1}H and ^{19}F is especially beneficial for parallel acquisition due to the similar detection sensitivities of these nuclides. The sequence is illustrated in Fig. 7.61 and shares some similarity with the triple-resonance sequence for X–Y heteronuclear correlations described in the preceding section. The scheme utilises independent but parallel INEPT transfers from both ^{1}H and ^{19}F onto ^{13}C and employs simultaneous ^{1}H and ^{19}F detection in the presence of carbon decoupling. To allow for the larger $^1J_{CF}$ over $^1J_{CH}$ couplings, the ^{19}F INEPT segment is shorter than that of the ^{1}H INEPT, so is nested symmetrically within this. Example HSQC spectra collected simultaneously are shown in Fig. 7.62 for a mixture of fluorinated benzoic acids and highlight the potential time savings from parallel acquisitions. A related approach employing dual receivers provides HSQC one-bond and HMBC long-range ^{1}H–^{13}C correlation spectra, a ^{13}C–^{13}C INADEQUATE spectrum and a 1D proton-decoupled ^{13}C spectrum in a single experiment, and has been termed PANACEA

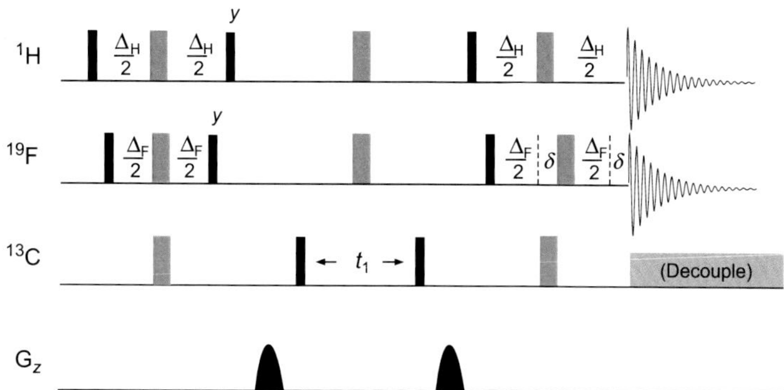

FIGURE 7.61 **The PANSY sequence for recording simultaneous ^{1}H–^{13}C and ^{19}F–^{13}C 2D HSQC spectra.** The delays $\Delta_H = 1/2J_{CH}$ and $\Delta_F = 1/2J_{CF}$, and the additional delay δ ensures that ^{1}H and ^{19}F signals refocus at the same point prior to detection.

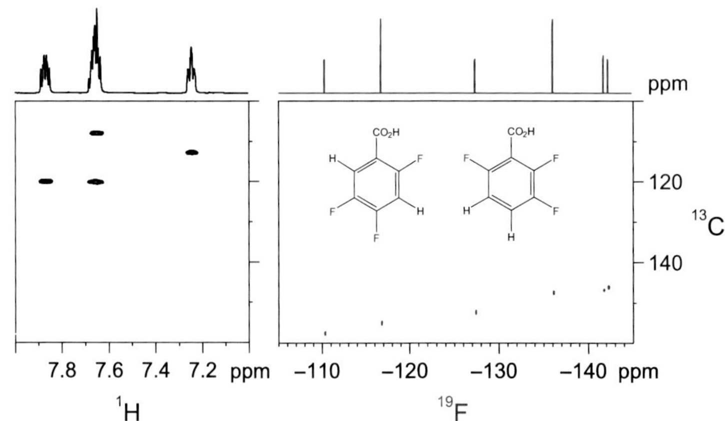

FIGURE 7.62 **Dual-receiver ^{1}H–^{13}C and ^{19}F–^{13}C HSQC spectra.** These are shown for a mixture of fluorobenzoic acids and were recorded in parallel using the sequence of Fig. 7.61, collected for $^{1}J_{HC}$ = 170 Hz and $^{1}J_{FC}$ = 255 Hz. *(Data from Bruker Biospin.)*

(*parallel acquisition NMR, an all-in-one combination of experimental applications*) [114]. The addition of a third receiver allows for the simultaneous acquisition of a ^{15}N spectrum also, and a similar sequence has also been described for ^{1}H–^{29}Si experiments [115]. The detection of high-sensitivity nuclides has also allowed parallel acquisitions to be combined with the single-scan ultrafast approach described in Section 5.4.5 to yield *parallel ultrafast spectroscopy* (PUFSY) [116], possibly the fastest route to multiple 2D experiments so far developed. Whilst rarely employed to date, clearly there are potential savings to be made from instruments equipped with multiple receivers, which will also provide for new areas of exploration for those engaged in pulse sequence development.

REFERENCES

[1] Bodenhausen G, Ruben DJ. Chem Phys Lett 1980;69:185–9.
[2] Parella T. Magn Reson Chem 1998;36:467–95.
[3] Vuister GW, Ruiz-Cabello J, van Zijl PCM. J Magn Reson 1992;100:215–20.
[4] Willker W, Leibfritz D, Kerrsebaum R, Bermel W. Magn Reson Chem 1993;31:287–92.
[5] Tolman JR, Chung J, Prestegard JH. J Magn Reson 1992;98:462–7.
[6] Davis AL, Keeler J, Laue ED, Moskau D. J Magn Reson 1992;98:207–16.
[7] Boyd J, Soffe N, John B, Plant D, Hurd R. J Magn Reson 1992;98:660–4.
[8] Otting G, Wüthrich K. J Magn Reson 1988;76:569–74.
[9] Palmer AG, Cavanagh J, Wright PE, Rance M. J Magn Reson 1991;93:151–70.
[10] Kay LE, Keifer P, Saarinen T. J Am Chem Soc 1992;114:10663–5.
[11] Schleucher J, Schwendinger M, Sattler M, Schmidt P, Schedletzky O, Glaser SJ, Sørensen OW, Griesinger C. J Biomol NMR 1994;4:301–6.
[12] Reynolds WF, Yu M, Enriquez RG, Leon I. Magn Reson Chem 1997;35:505–19.
[13] Reynolds WF, Enriquez RG. Magn Reson Chem 2003;41:927–32.
[14] Tate S-I, Masui Y, Inagaki F. J Magn Reson 1991;94:625–30.
[15] Parella T, Sánchez-Ferrando F, Virgili A. J Magn Reson 1997;126:274–7.
[16] Parella T, Belloc J, Sánchez-Ferrando F, Virgili A. Magn Reson Chem 1998;36:715–9.
[17] Lerner L, Bax A. J Magn Reson 1986;69:375–80.
[18] Davis DG. J Magn Reson 1989;84:417–24.
[19] Domke T. J Magn Reson 1991;95:174–7.
[20] John BK, Plant D, Heald SL, Hurd RE. J Magn Reson 1991;94:664–9.
[21] Mackin G, Shaka AJ. J Magn Reson A 1996;118:247–55.
[22] Crouch RC, Davis AO, Martin GE. Magn Reson Chem 1995;33:889–92.
[23] Müller L. J Am Chem Soc 1979;101:4481–4.
[24] Bax A, Griffey RH, Hawkins BL. J Magn Reson 1983;55:301–15.
[25] Bax A, Subramanian S. J Magn Reson 1986;67:565–9.
[26] Garbow JR, Weitekamp DP, Pines A. Chem Phys Lett 1982;93:504–8.
[27] Bax A, Summers MF. J Am Chem Soc 1986;108:2093–4.
[28] Summers MF, Marzilli LG, Bax A. J Am Chem Soc 1986;108:4285–94.

[29] Cicero DO, Barbato G, Bazzo R. J Magn Reson 2001;148:209–13.
[30] Bax A, Marion D. J Magn Reson 1988;78:186–91.
[31] Rinaldi PL, Keifer PA. J Magn Reson A 1994;108:259–62.
[32] Hansen PE. Prog Nucl Magn Reson Spectrosc 1981;14:179–296.
[33] Marshall JL. Carbon–carbon and carbon–proton NMR couplings: applications to organic stereochemistry and conformational analysis. Orlando, FL: VCH; 1983.
[34] Bystrov VF. Prog Nucl Magn Reson Spectrosc 1976;10:41–81.
[35] Contreras RH, Peralta JE. Prog Nucl Magn Reson Spectrosc 2000;37:321–425.
[36] Baldwin JE, Claridge TDW, Goh KC, Keeping JW, Schofield CJ. Tetrahedron Lett 1993;34:5645–8.
[37] Hodgson DM, Comina PJ. Tetrahedron Lett 1996;37:5613–4.
[38] Kogler H, Sørensen OW, Bodenhausen G, Ernst RR. J Magn Reson 1983;55:157–63.
[39] Meissner A, Sørensen OW. Magn Reson Chem 2000;38:981–4.
[40] Burger R, Schorn C, Bigler P. Magn Reson Chem 2000;38:963–9.
[41] Uhrin D, Liptaj T, Kover KE. J Magn Reson A 1993;101:41–6.
[42] Furihata K, Seto H. Tetrahedron Lett 1998;39:7337–40.
[43] Kessler H, Schmeider P, Köck M, Kurz M. J Magn Reson 1990;88:615–8.
[44] Gaillet C, Lequart C, Debeire P, Nuzillard J-M. J Magn Reson 1999;139:454–9.
[45] Claridge TDW, Pérez-Victoria I. Org Biomol Chem 2003;1:3632–4.
[46] Martin GE. Ann Rep NMR Spectrosc 2002;46:37–100.
[47] Wagner R, Berger S. Magn Reson Chem 1998;36:S44–6.
[48] Bodenhausen G, Ernst RR. J Am Chem Soc 1982;104:1304–9.
[49] Martin GE, Hadden CE, Crouch RC, Krishnamurthy VV. Magn Reson Chem 1999;37:517–28.
[50] Hadden CE, Martin GE, Krishnamurthy VV. J Magn Reson 1999;140:274–80.
[51] Araya-Maturana R, Delgado-Castro T, Cardona W, Weiss-Lopez BE. Curr Org Chem 2001;5:253–63.
[52] Furrer J. Magn Reson Chem 2006;44:845–50.
[53] Martin GE, Hadden CE. Magn Reson Chem 2000;38:251–6.
[54] Kline M, Cheatham S. Magn Reson Chem 2003;41:307–14.
[55] Martin GE, Hadden CE. J Nat Prod 2000;63:543–85.
[56] Zangger K, Armitage IM. Magn Reson Chem 2000;38:452–8.
[57] Hadden CE, Angwin DT. Magn Reson Chem 2001;39:1–8.
[58] Nyberg NT, Duus JO, Sorensen OW. J Am Chem Soc 2005;127:6154–5.
[59] Nyberg NT, Duus JØ, Sørensen OW. Magn Reson Chem 2005;43:971–4.
[60] Petersen BO, Vinogradov E, Kay W, Würtz P, Nyberg NT, Duus JØ, Sørensen OW. Carb Res 2006;341:550–6.
[61] Kock M, Reif B, Fenical W, Griesinger C. Tetrahedron Lett 1996;37:363–6.
[62] Bifulco G, Dambruoso P, Gomez-Paloma L, Riccio R. Chem Rev 2007;107:3744–79.
[63] Palermo G, Riccio R, Bifulco G. J Org Chem 2010;75:1982–91.
[64] Rafet Aydin HG. Magn Reson Chem 1990;28:448–57.
[65] Marquez BL, Gerwick WH, Williamson RT. Magn Reson Chem 2001;39:499–530.
[66] Matsumori N, Kaneno D, Murata M, Nakamura H, Tachibana K. J Org Chem 1999;64:866–76.
[67] Sharman GJ. Magn Reson Chem 2007;45:317–24.
[68] Parella T, Espinosa JF. Prog Nucl Magn Reson Spectrosc 2013;73:17–55.
[69] Edden RAE, Keeler J. J Magn Reson 2004;166:53–68.
[70] Keeler J. 2004. Available from: http://www-keeler.ch.cam.ac.uk
[71] Williamson RT, Marquez BL, Gerwick WH, Kover KE. Magn Reson Chem 2000;38:265–73.
[72] Sauri J, Parella T, Espinosa JF. Org Biomol Chem 2013;11:4473–8.
[73] Espinosa JF, Vidal P, Parella T, Gil S. Magn Reson Chem 2011;49:502–7.
[74] Castañar L, Saurí J, Williamson RT, Virgili A, Parella T. Angew Chem Int Ed 2014;53:8379–82.
[75] Saurí J, Espinosa JF, Parella T. Angew Chem Int Ed 2012;51:3919–22.
[76] Williamson RT, Buevich AV, Martin GE, Parella T. J Org Chem 2014;79:3887–94.
[77] Williamson RT, Buevich AV, Martin GE. Tetrahedron Lett 2014;55:3365–6.
[78] Dykstra RW. J Magn Reson A 1995;112:255–7.
[79] Bax A, Morris GA. J Magn Reson 1981;42:501–5.
[80] Carpenter KA, Reynolds WF. Magn Reson Chem 1992;30:287–94.
[81] Bax A. J Magn Reson 1983;53:517–20.
[82] Wilde JA, Bolton PH. J Magn Reson 1984;59:343–6.
[83] Rutar V. J Magn Reson 1984;58:306–10.
[84] Martin GE, Zektzer AS. Magn Reson Chem 1988;26:631–52.
[85] Kessler H, Griesinger C, Zarbock J, Loosli HR. J Magn Reson 1984;57:331–6.
[86] Kessler H, Griesinger C, Lautz J. Angew Chem Int Ed 1984;23:444–5.

[87] Salazar M, Zektzer AS, Martin GE. Magn Reson Chem 1988;26:28–32.
[88] Quast MJ, Zektzer AS, Martin GE, Castle RN. J Magn Reson 1987;71:554–60.
[89] Bauer C, Freeman R, Wimperis S. J Magn Reson 1984;58:526–32.
[90] Krishnamurthy VV, Casida JE. Magn Reson Chem 1987;25:837–42.
[91] Perpick-Dumont M, Enriquez RG, McLean S, Puzzuoli FV, Reynolds WF. J Magn Reson 1987;75:414–26.
[92] Reynolds WF, Hughes DW, Perpick-Dumont M, Enriquez RG. J Magn Reson 1985;63:413–7.
[93] Reynolds WF, McLean S, Perpick-Dumont M, Enriquez RG. Magn Reson Chem 1989;27:162–9.
[94] Gudat D. Ann Rep NMR Spectrosc 1999;38:139–202.
[95] Gudat D. Ann Rep NMR Spectrosc 2003;51:59–103.
[96] Iggo JA, Liu J, Overend G. Ann Rep NMR Spectrosc 2008;63:191–274.
[97] Adams B. Magn Reson Chem 2008;46:377–80.
[98] Marchione AA, Dooley RJ, Conklin B. Magn Reson Chem 2014;52:183–9.
[99] Howe PWA. Magn Reson Chem 2012;50:705–8.
[100] Berger S, Facke T, Wagner R. Magn Reson Chem 1996;34:4–13.
[101] Lopez-Ortiz F, Carbajo RJ. Curr Org Chem 1998;2.
[102] Benn R, Brenneke H, Frings A, Lehmkuhl H, Mehler G, Rufunska A, Wildt T. J Am Chem Soc 1988;110:5661–8.
[103] Meier EJM, Kozminski W, Linden A, Lustenberger P, von Philipsborn W. Organometallics 1996;15:2469–77.
[104] Nanz D, Bell A, Kozminski W, Meier EJM, Tedesco V, von Philipsborn W. Bruker Report 1996;143:29–31.
[105] Russell DJ, Hadden CE, Martin GE, Krishnamurthy K. Magn Reson Chem 2002;40:207–10.
[106] Cavanagh J, Fairbrother WJ, Palmer AG, Skelton NJ, Rance M. Protein NMR spectroscopy: principles and practice. 2nd ed. San Diego: Academic Press (Elsevier); 2006.
[107] Argyropoulos D, Hoffmann E, Mtongana S, Koch KR. Magn Reson Chem 2003;41:102–6.
[108] Gudat D. Magn Reson Chem 2003;41:253–9.
[109] Weske S, Li Y, Wiegmann S, John M. Magn Reson Chem 2015;53:291–4.
[110] Wagner R, Berger S. J Magn Reson A 1996;120:258–60.
[111] Kupče E. In: Heise H, Matthews S, editors. Modern NMR methodology, 335. Berlin: Springer; 2013. p. 71–96.
[112] Kupče E, Freeman R, John BK. J Am Chem Soc 2006;128:9606–7.
[113] Kupče E, Cheatham S, Freeman R. Magn Reson Chem 2007;45:378–80.
[114] Kupče E, Freeman R. J Am Chem Soc 2008;130:10788–92.
[115] Kupče E, Wrackmeyer B. Appl Organomet Chem 2010;24:837–41.
[116] Donovan KJ, Kupče E, Frydman L. Angew Chem Int Ed 2013;52:4152–5.

Chapter 8

Separating Shifts and Couplings: J-Resolved and Pure Shift Spectroscopy

Chapter Outline

8.1 Introduction 295
8.2 Heteronuclear J-resolved Spectroscopy 295
8.2.1 Measuring Long-Range Proton–Carbon Coupling Constants 298
8.2.2 Practical Considerations 300
8.3 Homonuclear J-resolved Spectroscopy 301
8.3.1 Tilting, Projections and Symmetrisation 302
8.3.2 Applications 303
8.4 'Indirect' Homonuclear J-resolved Spectroscopy 304
8.5 Pure Shift Broadband-Decoupled ^{1}H Spectroscopy 306
8.5.1 The Basis of Pure Shift Spectroscopy 307
8.5.2 Pseudo-2D Pure Shift 307
8.5.3 Real-Time Pure Shift 309
8.5.4 Pure Shift Refocussing Elements 309
References 313

8.1 INTRODUCTION

Unlike the techniques encountered in the previous two chapters that exploit scalar (J) couplings to *correlate* the chemical shifts of interacting spins, those described in this chapter aim to remove the influence of these couplings from spectra. It is widely recognised that the analysis of crowded proton spectra is complicated by the overlap of neighbouring multiplets, making the extraction of coupling constants, accurate measurement of chemical shifts or assessment of resonance intensities difficult or even impossible. J-resolved experiments aim to separate (or resolve) chemical shifts from scalar couplings by presenting these parameters in separate frequency dimensions, so allowing the chemist to examine one parameter without complications arising from the other. Pure shift experiments are a more recent development which seek to provide broadband proton-decoupled proton spectra. As the name suggests, these contain resonances devoid of homonuclear-coupling structure and offer a substantial gain in effective spectral resolution.

Although J-resolved methods seek the separation of chemical shifts from couplings, in practice these can be subject to a number of technical difficulties that may limit their effectiveness. As such, the methods find rather less use in routine structural work than shift correlation experiments. The methods are based on 2D spectroscopy in which shift and coupling parameters are resolved by presenting chemical shifts in f_2 and only spin couplings in f_1, which may be heteronuclear or homonuclear couplings depending on the details of the experiment.

Broadband proton-decoupled ^{1}H spectroscopy has long been recognised as a means to greatly simplify proton spectra since this yields singlet resonances similar to those of conventional carbon spectra. However, the simultaneous removal of *all* proton-coupling structure while observing protons presents considerable experimental challenges, meaning robust and readily implemented methods have largely remained elusive. There now exists a family of pure shift experiments that are able to achieve this aim. As these relatively recent methods are likely to find wider use within 1D and 2D experiments, their operation will be described after J-resolved spectroscopy has been introduced.

The principal techniques of this chapter are summarised in Table 8.1. J-resolved techniques, in particular, were the most widely studied methods in the early development of 2D NMR spectroscopy and may be understood with reference to the vector model, being based on simple spin-echoes. Pure shift spectroscopy follows along similar principles and, as such, I would recommend familiarity with spin-echoes before proceeding (see Section 2.2).

8.2 HETERONUCLEAR J-RESOLVED SPECTROSCOPY

In the heteronuclear version of the J-resolved experiment, the chemical shift of the X spin is presented in f_2 while couplings to a second nucleus, typically protons, are presented in f_1. The f_1 dimension therefore enables an analysis of resonance multiplicity as well as measurement of heteronuclear-coupling constants (J_{XH}), as described later in these sections.

High-Resolution NMR Techniques in Organic Chemistry. http://dx.doi.org/10.1016/B978-0-08-099986-9.00008-7

TABLE 8.1 The Principal Applications of the Main Techniques Described in this Chapter

Technique	Principal Applications
Heteronuclear J resolved	Separation of heteronuclear couplings (usually ^{1}H–X) from chemical shifts. Used to determine the multiplicity of the heteroatom or to provide direct measurement of heteronuclear coupling constants.
Homonuclear J resolved	Separation of homonuclear couplings (usually ^{1}H–^{1}H) from chemical shifts. Used to provide direct measurement of homonuclear coupling constants or to display resonance chemical shifts without homonuclear coupling fine structure ('proton-decoupled' proton spectra).
'Indirect' homonuclear J resolved	Separation of proton homonuclear couplings according to the chemical shift of an attached heteroatom centre. Used to provide direct measurement of homonuclear coupling constants.
Pure shift	Production of proton spectra that are devoid of homonuclear proton couplings ('proton-decoupled' proton spectra). Yields very high resolution in the proton dimension(s) and may be employed in 1D or 2D experiments.

To reduce the f_1 information content of the 2D spectrum to only couplings, it is necessary to make the free induction decays detected insensitive to chemical shift evolution during the t_1 period, which is readily achieved by the use of a spin-echo during t_1 (Fig. 8.1a) [1,2]. Thus, following initial X-nucleus excitation, simultaneous 180 degree proton and X-spin pulses are applied at the midpoint of t_1, such that X-spin chemical shifts will refocus, but heteronuclear coupling will continue to evolve. There is no 'mixing' step in J-resolved experiments because magnetisation or coherence transfer between nuclei is not employed. The sequence is the exact 2D analogue of the J-modulated 1D editing sequence presented in Section 4.2, with the fixed coupling evolution period Δ here being replaced with variable period t_1. Since proton decoupling will invariably be applied during the detection period, heteronuclear X–H couplings do not appear in f_2. The detected signals therefore experience the desired pure amplitude modulation according to evolution of the heteronuclear couplings in t_1 (Fig. 8.2).

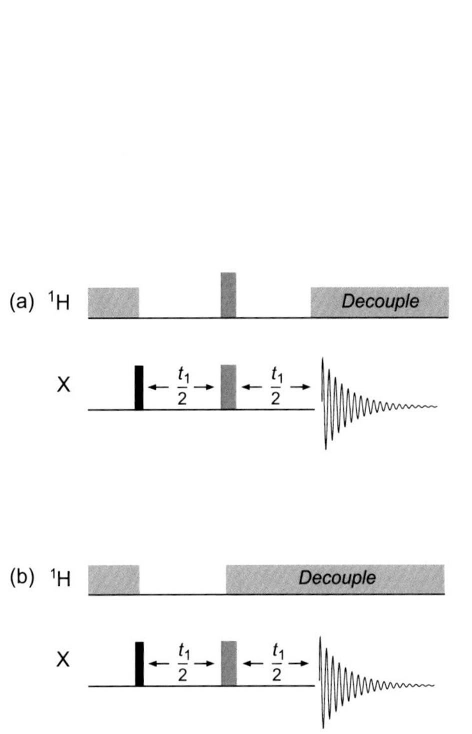

FIGURE 8.1 **Heteronuclear J-resolved sequences.** The (a) spin-flip and (b) gated decoupling schemes. In (b) coupling evolution occurs for half of the t_1 period only, so splittings observed in f_1 appear with half their true $^1J_{XH}$ values.

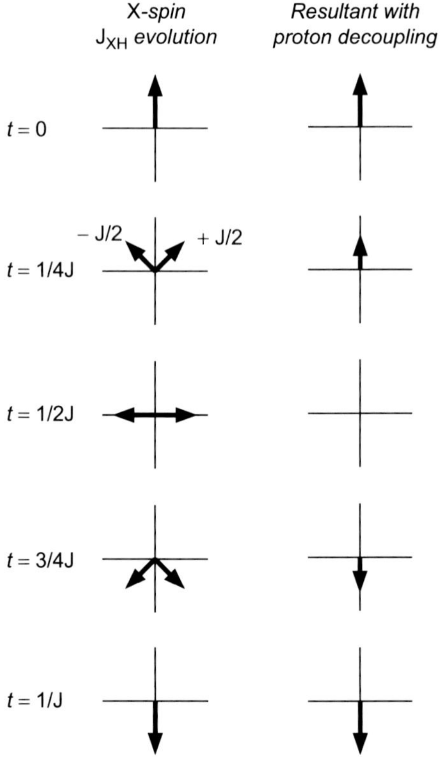

FIGURE 8.2 **Amplitude modulation of the X-spin signal by heteronuclear-coupling evolution during t_1, illustrated for a coupled X–H pair.** The doublet vectors evolve throughout t_1 but are frozen during t_2 by the application of proton decoupling, so the resultant X-spin signal displays pure amplitude modulation according to the heteronuclear-coupling constant.

In addition, the spin-echo refocusses field inhomogeneity effects since these are merely another source of chemical shift differences within the sample, meaning the resonances in f_1 possess *natural* linewidths ie they are governed by the T_2 rather than the T_2* of the heterospin [3].

The scheme of Fig. 8.1a is referred to as the 'spin-flip' or 'proton-flip' method because of use of the 180 degree proton pulse. An alternative is the 'gated decoupler' method [4] (Fig. 8.1b) in which this pulse is omitted and instead proton decoupling is applied for one half of the evolution period (again analogous to the 1D spectrum-editing method). Chemical shifts are again refocussed by the X pulse, but the heteronuclear couplings now evolve for only half the total evolution time, resulting in the f_1 splittings being reduced by a factor of 2 (be sure which version you are using if you intend to measure J_{XH}). Note that if neither the proton-decoupling or proton 180 degree pulse were applied in t_1, the effect of the lone X(180 degree) would be to refocus the heteronuclear coupling in addition to the chemical shifts, so preventing operation of the experiment altogether. The attraction of the gated decoupler method lies in the simplicity of implementation since no ^{1}H(180 degree) calibrations are required (a point of greater significance when these methods were initially developed), because the results are better behaved in the presence of strong coupling (see later in these sections), and because no artefacts are introduced by inaccuracies in the proton pulse.

Fig. 8.3 shows the ^{1}H–^{13}C J-resolved spectrum of menthol **8.1**. Projection onto the f_2 axis produces the usual proton-decoupled 1D carbon spectrum in which resonances appear as singlets, while in f_1 the carbon multiplicities from $^1J_{CH}$ are clearly delineated. Such a spectrum represents a straightforward way of determining multiplicities, but this approach is rarely used nowadays since the 1D spectrum editing methods of Section 4.4 offer a more rapid alternative and the edited heteronuclear shift correlations of Section 7.3.2 are also more informative. The J-resolved spectrum may prove useful when unusually large $^1J_{CH}$ values exist and produce ambiguous results from spectrum-editing methods. Recall that these rely on delays being set according to estimated J values, and may fail when the timings used are far from optimum, whereas J-resolved methods have no such requirement. A more attractive use for the J-resolved experiment nowadays lies in the measurement of heteronuclear coupling constants themselves and, in particular, the measurement of long-range proton–carbon couplings, since these can be of considerable use in conformational or configurational studies. While, in principle, it is possible to record the f_1 dimension with sufficient digital resolution to resolve small long-range couplings in the presence of the far greater one-bond couplings (which are typically at least an order of magnitude larger), a more efficient approach is to eliminate $^1J_{CH}$ from f_1 and retain only $^{2/3}J_{CH}$, so reducing the f_1 spectral width and allowing finer digitisation for more accurate measurements. A number of approaches towards this are introduced in the following section.

Refinements to the basic sequences of Fig. 8.1 include the use of polarisation transfer sequences prior to the spin-echo to prepare transverse X-spin magnetisation. INEPT and DEPT [5] have been proposed for this purpose. In addition to sensitivity gains, polarisation transfer sequences allow repetition rates to be dictated by the T_1s of the faster relaxing protons.

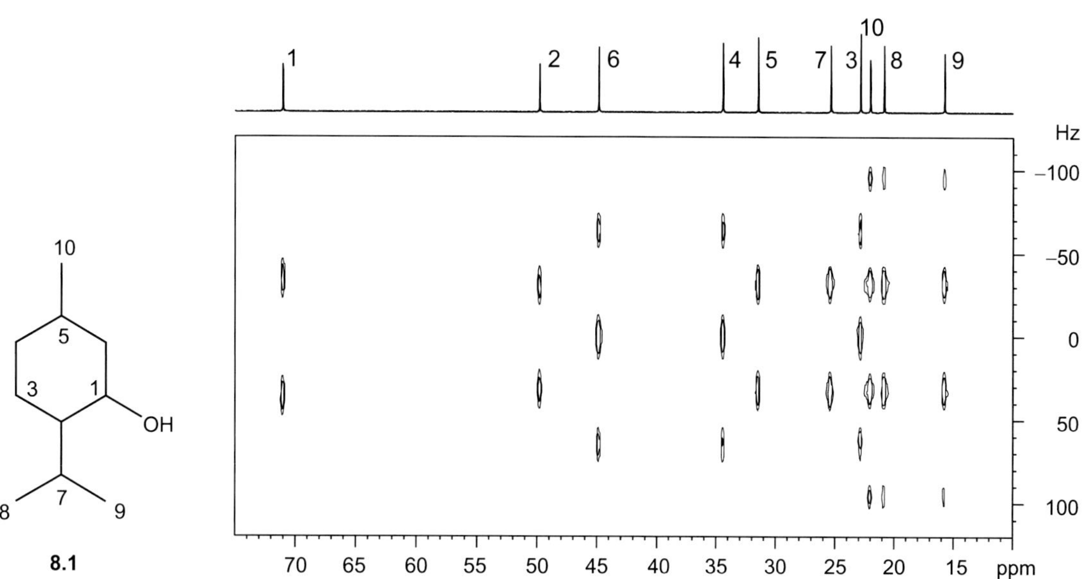

FIGURE 8.3 **The 125 MHz ^{1}H–^{13}C heteronuclear J-resolved spectrum of menthol 8.1, above which is the conventional 1D carbon spectrum.** The gated decoupler method of Fig. 7.1b was used, and the splittings in f_1 are therefore half their true $^1J_{CH}$ values.

They may also be favoured when studying nuclei that experience a loss of intensity from the ^{1}H–X nuclear Overhauser effect (NOE) generated by proton decoupling; that is, those X spins that have negative magnetogyric ratios (eg ^{15}N or ^{29}Si). Avoiding the use of proton decoupling during the relaxation delay minimises NOE buildup and hence signal losses, while the addition of the polarisation transfer step enhances sensitivity.

8.2.1 Measuring Long-Range Proton–Carbon Coupling Constants

The use of long-range ^{1}H–^{13}C coupling constants in the definition of molecular configuration or conformation is an increasingly active area, with numerous methods for measuring these developed in recent years [6,7]. The techniques based on the HMBC and HSQMBC experiments have already been presented in Section 7.4.5. In principle, J-resolved methods are well suited to such measurements, but tend to suffer from the presence of the far greater $^1J_{CH}$ couplings, as mentioned above. The common aim of the sequences described below is to eliminate these less informative one-bond ^{1}H–^{13}C couplings from f_1 and so reduce the corresponding spectral width. Since $^1J_{CH}$ values are greater than c. 125 Hz, while long-range couplings are typically less than 10 Hz, substantial reductions in the f_1 spectral width can be made which allows higher digital resolution and provides accurate characterisation of the remaining $^nJ_{CH}$ values without the need to collect time-demanding data sets. The descriptions are deliberately brief, but serve to illustrate the most useful approaches.

8.2.1.1 Semiselective

By placing a BIRD cluster (Section 7.3.3) with proton pulse phases x, x, $-x$ at the midpoint of t_1 in place of the simultaneous 180 degree pulses of the spin-flip method (Fig. 8.4a), it is possible to invert only those protons that share long-range couplings, while leaving those directly bound to a ^{13}C spin unaffected. This results in selective refocussing of the one-bond couplings by the ^{13}C 180 degree pulse, and thus their removal from f_1, whereas the long-range couplings remain [8,9]. The f_1 traces display *all* the $^nJ_{CH}$ values associated with each carbon, and may thus possess quite complex fine structure since each carbon is likely coupled to many protons. Such complexity may itself preclude measurement of the coupling constants, and no information on which spin pair gives rise to a specific coupling is available.

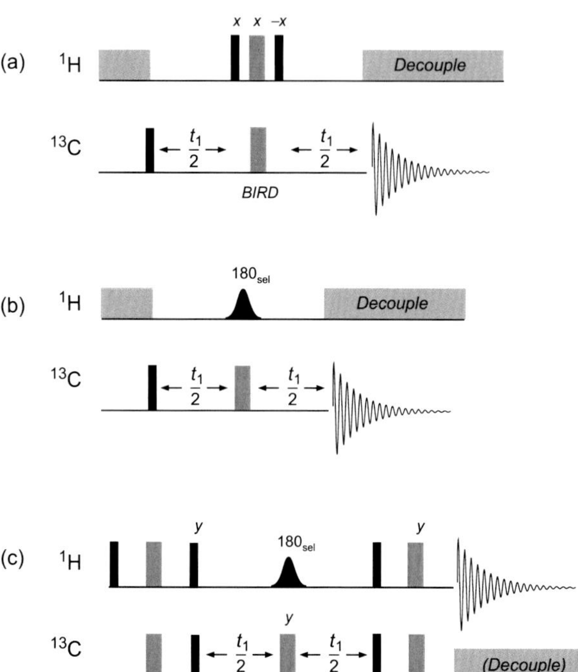

FIGURE 8.4 **Heteronuclear J-resolved sequences for the measurement of long-range proton–carbon coupling constants.** These show the (a) semiselective, (b) selective and (c) selective with proton observation variants (see text).

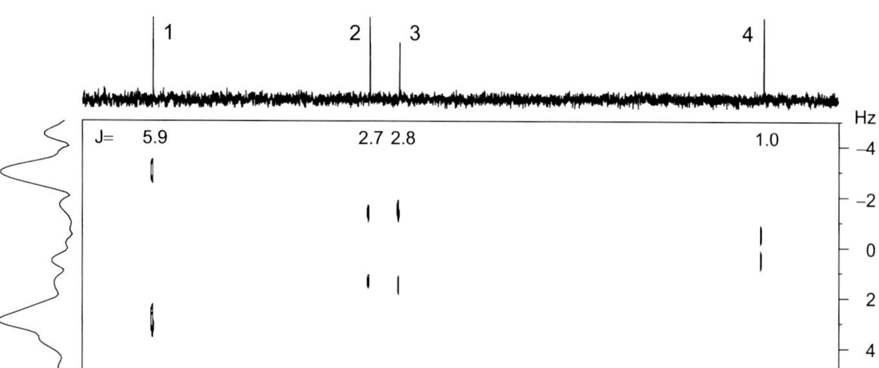

FIGURE 8.5 **A region of the selective heteronuclear J-resolved spectrum of 8.2.** Long-range heteronuclear couplings to the selected proton appear as splittings in f_1, the values of which are shown. The vertical trace is taken through the resonance at 199 ppm. 4K t_2 data points were collected for 32 t_1 increments over spectral widths of 100 ppm and 14 Hz, respectively. A 100-ms Gaussian 180 degree pulse was used to select the proton resonance. The final f_1 resolution after zero-filling was 0.1 Hz/pt.

8.2.1.2 Selective

To simplify the fine structure appearing in f_1, it is possible to selectively invert only *one* proton resonance during t_1, taking care to avoid inverting its one-bond ^{13}C satellites. This requires application of a selective ^1H 180 degree pulse in place of the usual 'hard' pulse in the spin-flip method (Fig. 8.4b). In this case, all couplings will be refocussed and therefore removed from f_1 with the exception of those to selectively inverted proton(s) [10]. The f_1 dimension therefore displays only simple multiplets (doublets, triplets or quartets depending on the proton group selected) and the coupling pairs giving rise to each are readily identified. This was one approach used for measuring long-range couplings to the bridgehead proton in a series of structures related to **8.2** in an attempt to define the unknown bridgehead stereochemistry and thus differentiate *endo* and *exo* products. A lack of NOEs to the bridgehead protons did not provide an unambiguous definition, so long-range couplings in the unknown products were compared with those in structures of known configuration. Fig. 8.5 shows a section of one such spectrum in which these couplings were measured from f_1 doublet splittings. This requires the proton(s) of interest to be sufficiently resolved, and numerous experiments will be required if many $^nJ_{CH}$ values are to be determined. This has, nevertheless, been employed extensively in the measurement of long-range coupling constants in monosaccharide units [11], and other variants have incorporated the INEPT [12] and DEPT [13] sequences for sensitivity enhancement.

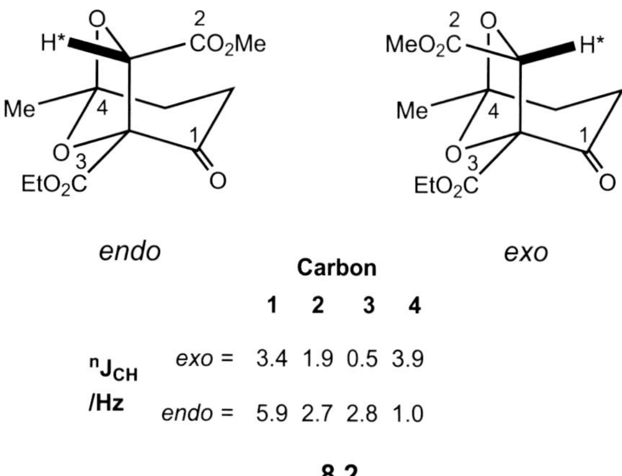

$^nJ_{CH}$ /Hz	Carbon 1	2	3	4
exo =	3.4	1.9	0.5	3.9
endo =	5.9	2.7	2.8	1.0

8.2.1.3 Selective with Proton Detection

Long-range couplings can also be observed through observation of the participating proton, leading to significant sensitivity gains. An elegant way of performing the proton-detected experiment is based on a simple modification of the HSQC sequence presented in Section 7.3.1, and may be performed as a phase-sensitive experiment with or without pulsed field gradients (PFGs) [14] (Fig. 8.4c). The non-selective ^1H 180 degree inversion pulse within t_1 (which serves to refocus

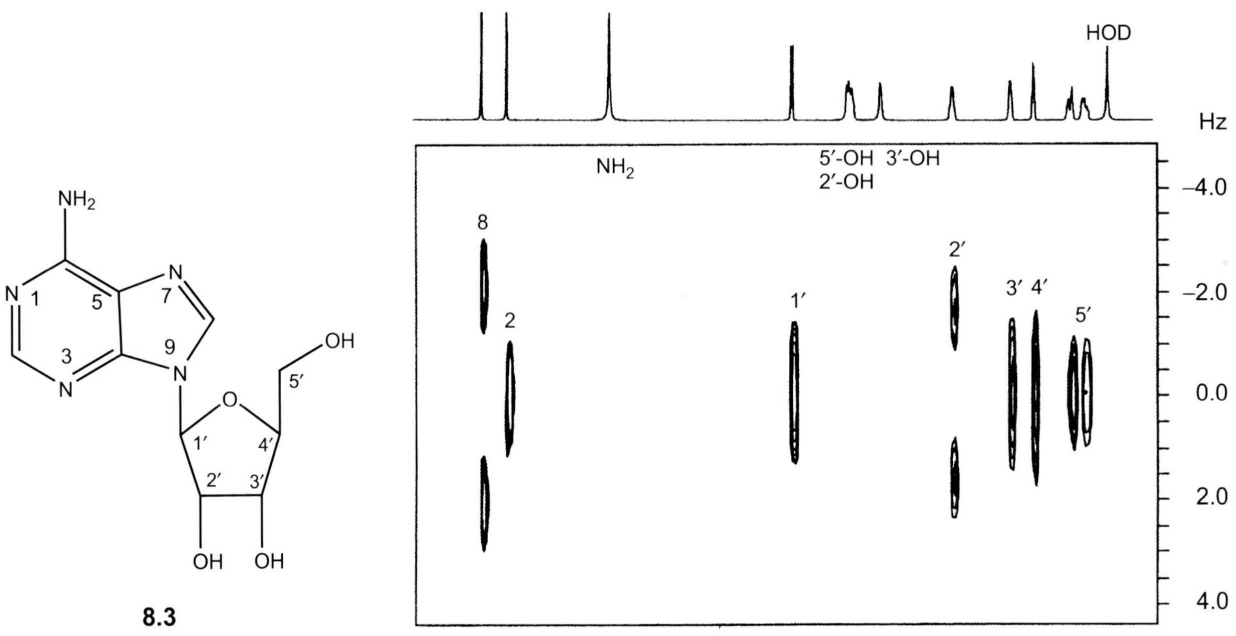

FIGURE 8.6 **Proton-detected selective heteronuclear J-resolved spectrum of adenosine 8.3.** The H1′ proton has been selectively inverted, and the doublet splittings in f_1 record its long-range couplings to C8 and C2′. *(Source: Reproduced with permission from reference [14], Copyright 1995, Elsevier.)*

heteronuclear couplings in HSQC) is here made selective, and a simultaneous non-selective ^{13}C(180 degree) pulse added, producing a spin-echo during t_1 analogous in design and operation to the selective experiment above. The polarisation transfer step following t_1 transfers magnetisation back onto protons for detection through their one-bond proton–carbon coupling. The resulting spectrum thus displays the normal ^{1}H spectrum in f_2 and $^nJ_{CH}$ doublets along f_1. These couplings are from the selectively inverted proton to the carbon attached to the proton at the f_2 chemical shift (Fig. 8.6), so measurements are restricted to long-range couplings to protonated carbons only. Once again, the spin pair giving rise to the coupling can be identified, assuming the target proton is sufficiently well resolved to be inverted selectively.

8.2.2 Practical Considerations

The requirements for sign discrimination in f_1 for the J-resolved experiments of Fig. 8.1 are readily met in *a first-order system* because the multiplets are symmetrical about their midpoints and thus folding about $f_1 = 0$ causes no confusion. Phase-cycling for f_1 quad-detection is, therefore, unnecessary. Since, in the presence of broadband decoupling, the detected signals experience only amplitude modulation, pure-phase spectra may also be obtained. However, when the system is no longer first order (ie in the presence of strong coupling *between protons*—more precisely, between those protons giving rise to ^{13}C satellites and others), the H–X multiplets lose their symmetry and the detected signals experience some phase modulation which in turn introduces dispersion-mode contributions to the spectrum. In this case, the 2D spectrum may be presented in the absolute value mode to mask the phase distortions or, alternatively, following the f_2 transform, columns may be extracted from the 2D data set at the chemical shift of a resonance of interest and processed separately as 1D traces. Despite the overall dispersive contribution to the 2D lineshape, a single column displays absorption-mode characteristics. Further complications arise in the presence of strong coupling for the spin-flip method [1].

The quality of J-resolved spectra can be seriously compromised by deficiencies in the accuracy of the pulses (especially the 180 degree pulses), arising from pulse miscalibration, radiofrequency inhomogeneity and off-resonance effects. The result is the appearance of a variety of additional weak resonances at esoteric positions within the spectrum, referred to as ‘ghosts’ and ‘phantoms’. The ghosts arise from imperfections in the 180 degree refocussing pulse, such that some transverse magnetisation fails to experience the effect of the pulse and thus fails to be refocussed. Phantoms arise from combined deficiencies in the 90 and 180 degree pulses causing residual longitudinal magnetisation existing after the 90 degree pulse to become transverse following the imperfect 180 degree pulse. These spurious responses may be eliminated with the four-step phase cycle for spin-echoes (EXORCYCLE) [15], which is widely employed in sequences that utilise spin-echoes. This involves stepping the phase of the X-spin refocussing pulse through *x*, *y*, –*x*, –*y* while the receiver inverts (ie steps *x*, –*x*, *x*, –*x*). Stepping the refocussing pulse by 90 degrees causes the echo to shift in phase by 180 degrees, as explained in Section 2.2.4, hence receiver inversion follows the echo, whereas unwanted responses ultimately cancel. Deficiencies in the proton 180 degree inversion pulse [16] can be avoided by the use of the gated decoupler method, but where this is not possible, as in the case of the homonuclear J-resolved experiments that follow, use of a composite 180 degree pulse such as $90_x240_y90_x$ is advantageous [17] and better still the more recent frequency-swept broadband inversion pulses (BIPs) described in Section 12.2.2.

8.2.2.1 Experimental Setup

Digitisation of the data, in particular the number of t_1 increments that are needed, will be dictated by the information required of the spectrum. If one simply wishes to determine multiplicities, low digital resolution will suffice, allowing few increments and rapid data collection. For carbon-13 the widest multiplets arise from the quartets of methyl groups, so the f_1 spectral width need be about 3.2 times $^1J_{CH}$ (although it would be possible to reduce the spectral width further and deliberately fold in the outer lines of the quartets). This requirement can beneficially be reduced by a factor of 2 for the gated decoupler method to around 200 Hz, assuming a coupling constant of 125 Hz.

To merely characterise the multiplet structure a digital resolution in f_1 of around 20 Hz/pt should suffice, requiring as little as 20 increments. If one wished to *measure* the value of $^1J_{CH}$ from the multiplet structure, a digital resolution of somewhat less than 5 Hz/pt would be more appropriate. Some 200 increments would be required for 2 Hz/pt leading to a significantly longer experiment, particularly if many scans are required per increment for reasons of sensitivity. If even finer digitisation were required for the measurement of *long-range* couplings, this approach becomes impossible. In contrast, when using one of the selective methods for $^nJ_{CH}$, one may only need digitise a 10-Hz window, requiring only 40 increments for 0.5 Hz/pt. Following the f_2 transform, columns may be extracted from the 2D data set and treated as 1D FIDs, including the use of zero-filling, to further enhance measurements.

8.3 HOMONUCLEAR J-RESOLVED SPECTROSCOPY

The homonuclear version of the J-resolved experiment [18] is most frequently applied in proton spectroscopy, although again it is suitable for any abundant nuclide. In principle, the separation of δ and J should reveal proton multiplets in f_1 free from overlap, and singlets in f_2 at the corresponding chemical shifts, such that the f_2 projection represents the 'broadband proton-decoupled proton spectrum'. As already noted, the possibility of generating such a spectrum has obvious appeal, allowing accurate measurement of chemical shifts in even the most heavily crowded proton spectra. However, a number of technical difficulties must be overcome if one is to achieve this goal which, alas, are not readily avoided, although the more recent pure shift methods introduced in Section 8.5 have provided solutions. A second possible application is in the measurement of homonuclear-coupling constants themselves, which is possible within the caveats detailed shortly.

The homonuclear sequence (Fig. 8.7) closely resembles the heteronuclear methods (although restricted to the spin-flip version only), and utilises the EXORCYCLE scheme. The appearance of the homonuclear spectrum is fundamentally different from its heteronuclear equivalent in that both chemical shifts *and couplings* appear in f_2. Thus, rather than lying parallel to the f_1 axis, the proton multiplets appear along a slope of −1 (in units of hertz), or, in other words, sit at 45 degrees to either axis (assuming identical plot scaling for both dimensions, Fig. 8.8). Columns parallel to f_1 do not, therefore, display the expected proton multiplets, and the f_2 projection displays both chemical shifts and scalar couplings. To overcome these

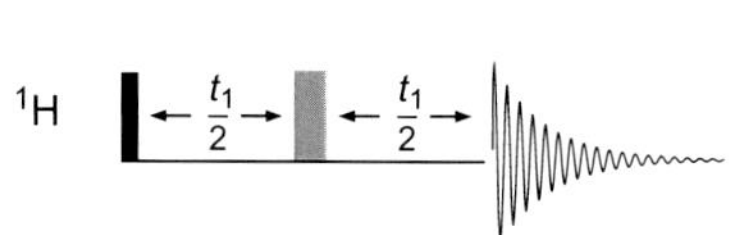

FIGURE 8.7 **The homonuclear J-resolved experiment.** The 180 degree pulse at the midpoint of t_1 refocusses proton shifts but not homonuclear couplings, implying that only these appear in f_1.

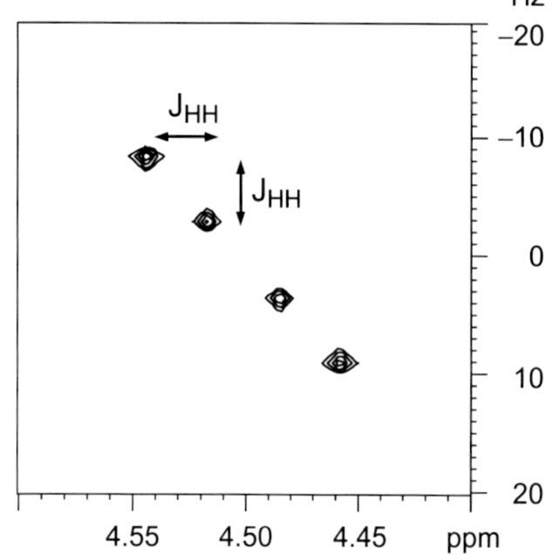

FIGURE 8.8 **A multiplet from the homonuclear J-resolved experiment.** This shows the characteristic tilting brought about by the presence of proton couplings in both f_1 and f_2.

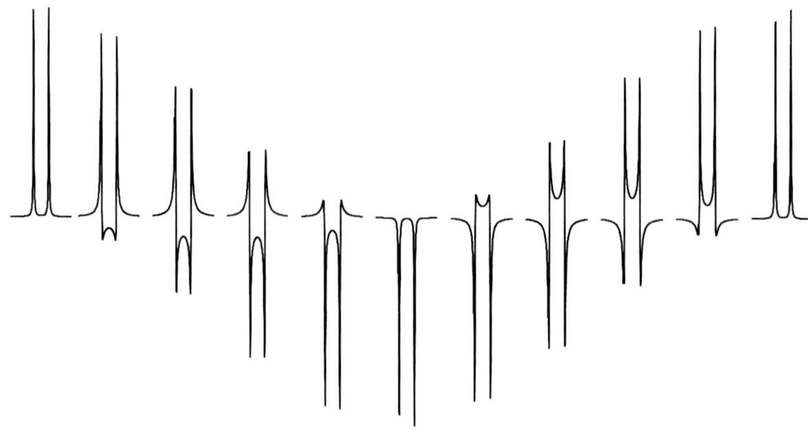

FIGURE 8.9 **Phase modulation of a proton doublet as a function of time.** The signals retain full intensity at all times (ie they do not experience amplitude modulation), but each line experiences a net 360 degree phase change over the sequence. The results are shown for a 10-Hz doublet with evolution times from 0 to 200 ms in 20-ms increments.

shortcomings, postprocessing techniques are routinely applied, as described in the following section. Furthermore, unlike the heteronuclear case, the detected signals unavoidably experience *phase modulation* as a function of t_1 (Fig. 8.9) resulting in phase-twist lineshapes. The spectra are therefore usually presented as absolute value data following strong resolution enhancement and magnitude calculation. Further complications arise in the presence of strong coupling between protons in the form of additional responses (see later in this section), meaning the J-resolved experiment is most suitable for overlapped spectra that are still first order.

8.3.1 Tilting, Projections and Symmetrisation

To reach the ultimate goal of retaining only chemical shifts in f_2, it is possible to eliminate the couplings from this dimension by 'tilting' (or 'shearing') the multiplets through an angle of 45 degrees about their midpoints [19], as illustrated schematically in Fig. 8.10. Software routines for this process are common to NMR processing packages nowadays. The resulting spectrum then has an appearance similar to the heteronuclear analogue, with columns parallel to f_1 reproducing the multiplet structures (providing the magnitude calculation has been performed), and projection onto the f_2 axis producing the broadband-decoupled proton spectrum (Fig. 8.11). Fig. 8.12 compares traces taken from this J-resolved spectrum with the equivalent multiplets from the 1D proton spectrum and illustrates the fine resolution of multiplet structure that can be obtained in the f_1 dimension.

Further improvements in the form of t_1-noise reduction may also be achieved with additional postprocessing. Prior to the tilt procedure, bands of t_1 noise will lie parallel to the f_1 axis, as for all 2D experiments, while following the tilt they will sit at 45 degrees to it. In first- and higher-order systems, the multiplets will themselves be symmetrical about the line $f_1 = 0$ Hz after tilting [20]. If the whole data set were symmetrised about this line (ie the lower intensity point for

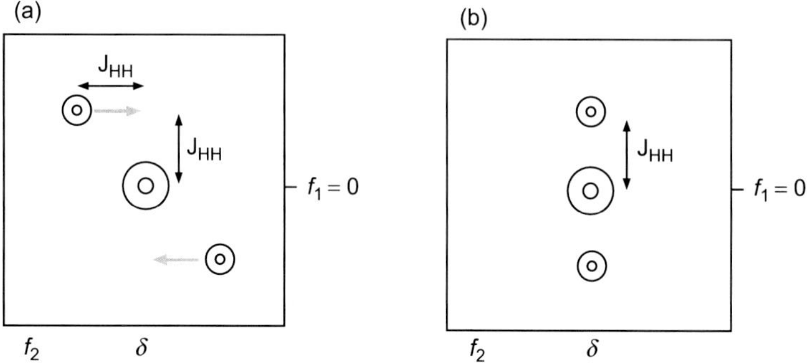

FIGURE 8.10 **Schematic illustration of the tilting procedure for eliminating homonuclear couplings from the f_2 dimension.** (a) The original multiplet structure and (b) that following the tilt procedure.

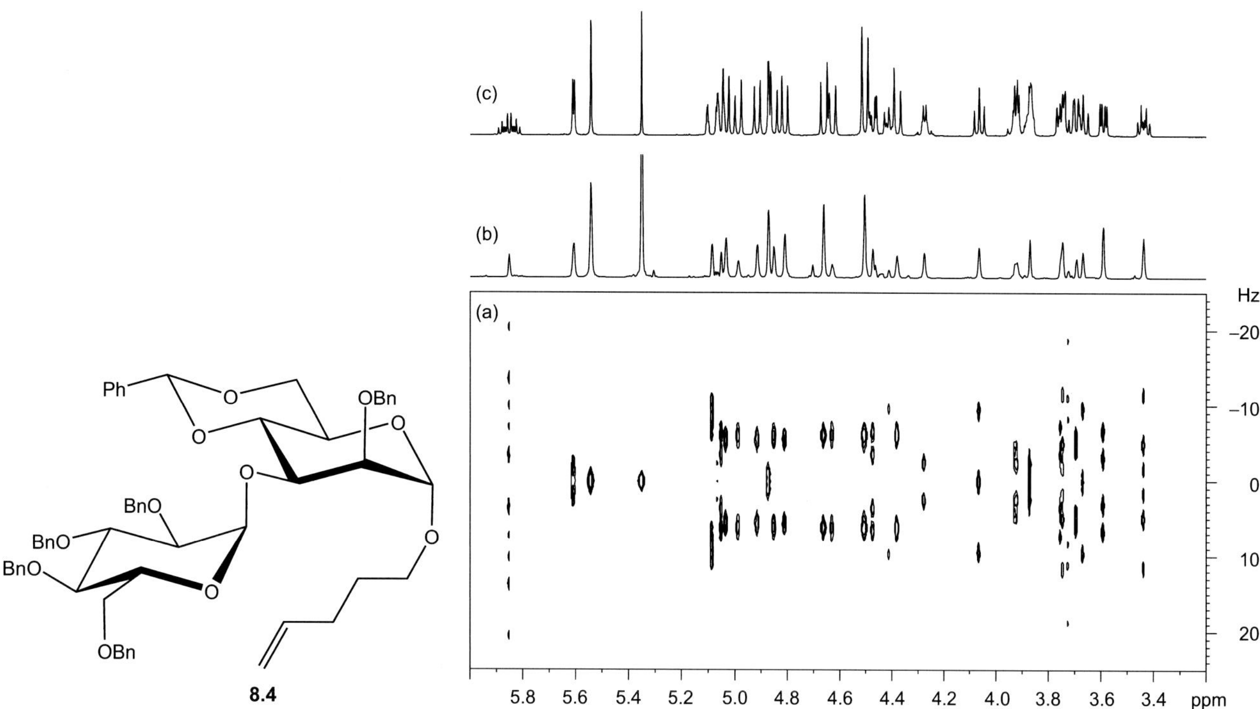

FIGURE 8.11 **The 500 MHz proton homonuclear J-resolved spectrum of 8.4.** (a) The 2D spectrum after tilting and symmetrisation. The f_2 projection (b) approximates to the 'proton-decoupled proton spectrum' and is considerably less complex than the conventional 1D spectrum (c). 4K t_2 data points were acquired for 64 t_1 increments over spectral widths of 5 ppm and 60 Hz, respectively. The final f_1 resolution after zero-filling was 0.5 Hz/pt. Data were processed with unshifted sine bell windows in both dimensions and are presented in magnitude mode.

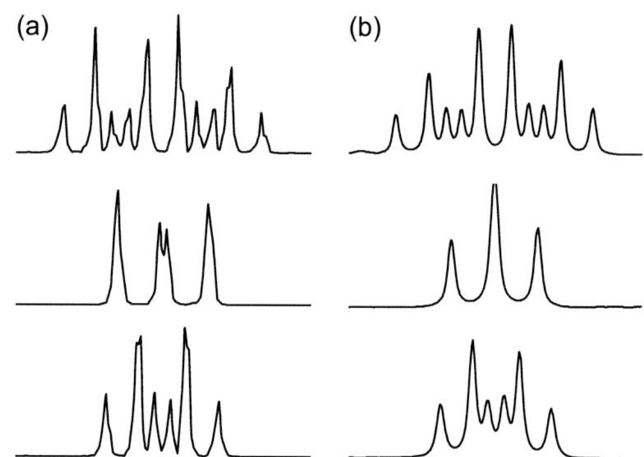

FIGURE 8.12 **J-Resolved spectrum multiplets.** (a) Selected f_1 traces taken from the J-resolved spectrum of Fig. 8.11a and (b) equivalent multiplets from the 1D proton spectrum.

symmetrically related data points replacing the higher), the multiplets would be retained whereas the contributions from the sloping t_1 noise would be diminished [21]. The procedure is similar to that introduced in Section 5.2.3 for t_1-noise reduction in absolute value COSY spectra, although a different symmetrisation procedure is of course required, and again one should be aware of the possibility of introducing artefacts into the spectrum as a result of the symmetrisation routine itself. Again, these routines are common to modern processing packages.

8.3.2 Applications

Perhaps the most obvious application of the homonuclear J-resolved experiment is the separation of overlapping multiplets, so that the fine structure within each may be analysed. There is however a strict requirement for spin systems to be first-order for the separation of shifts and couplings to be successful [20] because strong coupling causes unwanted

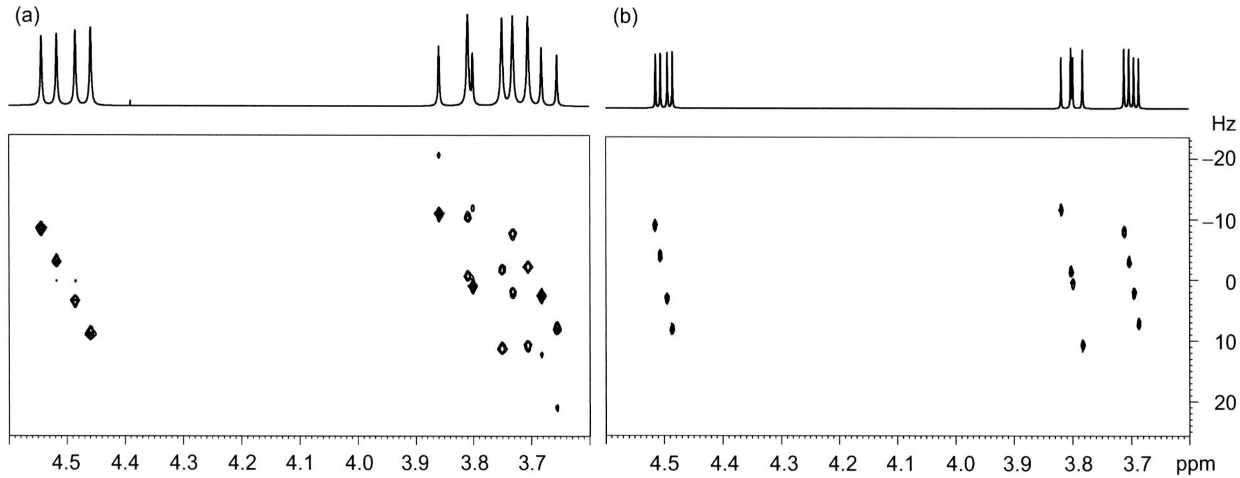

FIGURE 8.13 **Simulated 2D homonuclear J-resolved and 1D spectra for a three-spin proton system.** These are produced for (a) 200 and (b) 600 MHz. The additional artefacts in (a) arise from strong coupling between the two low-frequency protons (no tilting has been applied).

additional responses to appear midway between the shifts of strongly coupled protons. To reduce the degree of strong coupling within spectra, use of the highest available field strength is recommended for J-resolved spectroscopy wherever possible. Fig. 8.13 shows simulated spectra for a three-spin system at field strengths of 200 and 600 MHz. At the lower field (Fig. 8.13a), the coupled protons at 3.7 and 3.8 ppm (J = 12 Hz) experience strong coupling, as evidenced in the 'roofing' of their 1D resonances, which gives rise to additional responses between them in the J-resolved spectrum. No such artefacts are associated with the resonance at 4.5 ppm, which experiences only first-order coupling to its partners. In contrast, the higher field spectrum (Fig. 8.13b) shows no extra responses since all couplings are now (approximately) first order.

The homonuclear J-resolved spectrum can also assist in the measurement of proton *heteronuclear* couplings, notably when the coupling exists to a high-abundance heteronuclide such as ^{31}P or ^{19}F. Since only the proton of a coupled heteronuclear pair will experience the 180 degree pulse, heteronuclear coupling is refocussed in t_1 and therefore absent in f_1. This coupling will nonetheless be operative during detection of the proton FID and will thus appear in f_2. Since this heteronuclear splitting sits *parallel* to the f_2 axis rather than at 45 degrees to it, it is *not* removed from this dimension by the tilting process and may thus be examined without interference from homonuclear proton couplings along f_2. Fig. 8.14 demonstrates this approach for the palladium phosphine complex **8.5**. The f_2 projection of the tilted spectrum contains splittings for the ^{1}H–^{31}P couplings only at each proton shift, as is most clearly seen for phenyl *ortho*-protons at 7.85 and 7.92 ppm. These data can be used to complement the 1D X-spin decoupled proton spectrum in which proton–proton couplings are observed without interference from H–X spin couplings (see Fig. 4.14 of the same complex for comparison, Section 4.2.3). If desired, heteronuclear couplings could also be eliminated from the J-resolved spectrum by application of broadband X-spin decoupling during the acquisition time. For the measurement of heteronuclear couplings, rows taken parallel to f_2 through the multiplet components provide better resolution than the projection itself. The pure shift methods presented later can also be employed to reveal heteronuclear couplings.

8.4 'INDIRECT' HOMONUCLEAR J-RESOLVED SPECTROSCOPY

An alternative approach to resolving proton multiplets in J spectra is to disperse them according to the chemical shift of the carbon nucleus to which the protons are attached, rather than those of the proton themselves [22]. The advantage of this approach lies in the typically greater dispersion of carbon chemical shifts, although one must tolerate the reduced sensitivity of carbon observation.

The sequence that achieves this (Fig. 8.15) is a simple variant on the INEPT-based heteronuclear shift correlation sequence of Fig. 7.50 (HETCOR), so the loss in sensitivity is compensated somewhat by the use of a polarisation transfer step. In fact, the only difference between the two lies in the net evolution of *only shifts* or *only couplings* for the whole of t_1. The addition of a proton 180 degree pulse at the midpoint of t_1 here serves to refocus proton chemical shifts and heteronuclear-coupling constants (so the X-spin 180 degree pulse of HETCOR becomes redundant), but

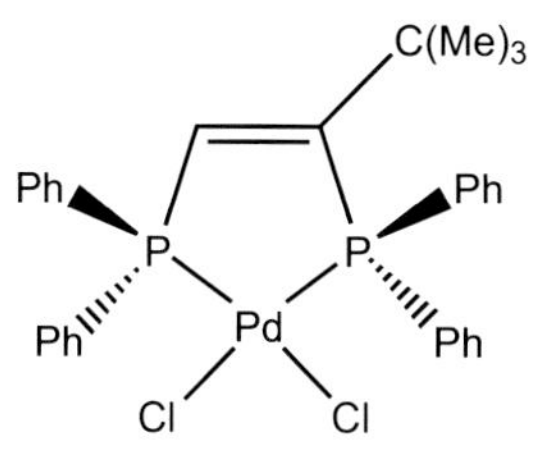

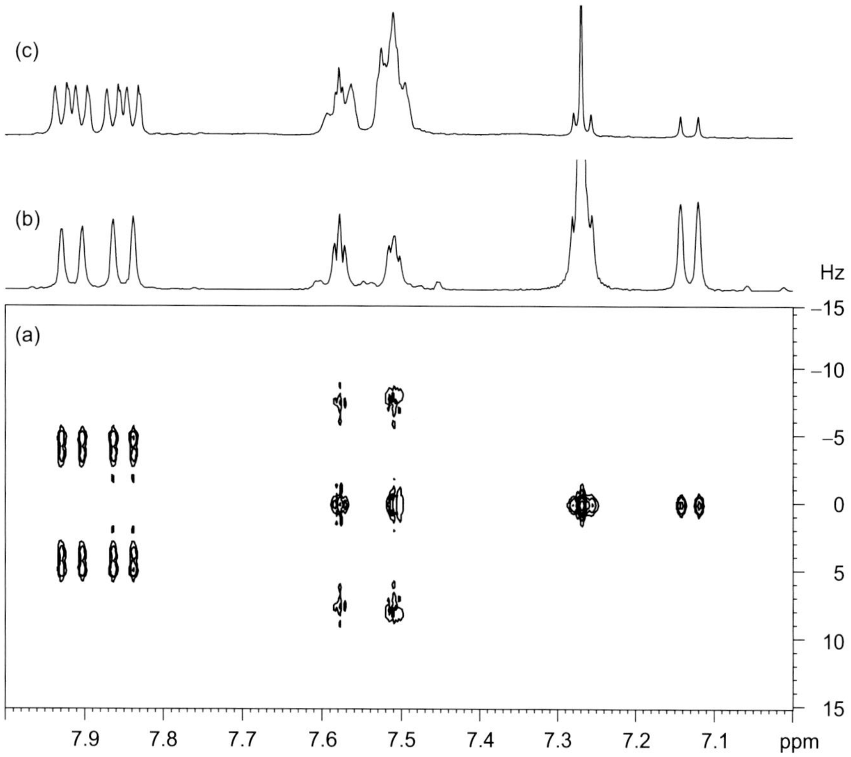

FIGURE 8.14 **Measurement of proton heteronuclear couplings from the tilted homonuclear J-resolved spectrum of 8.5.** The 2D spectrum (a) yields the f_2 projection (b) which displays only proton shifts and $^{1}H-^{31}P$ coupling. Spectrum (c) is the conventional 1D spectrum displaying shifts and both $^{1}H-^{1}H$ and $^{1}H-^{31}P$ couplings.

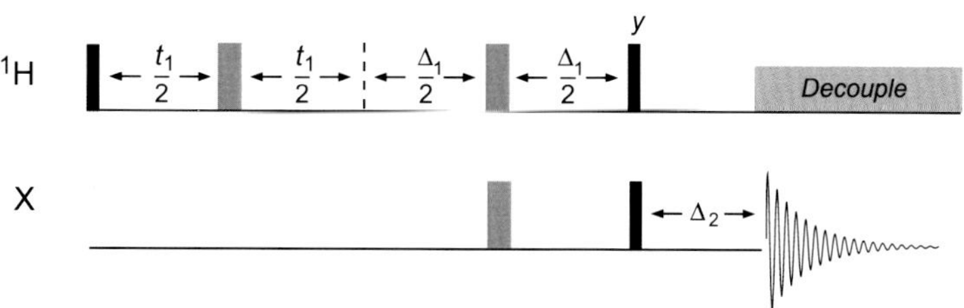

FIGURE 8.15 **The indirect homonuclear J-resolved sequence.** Proton couplings are effective during t_1, after which polarisation transfer leads to X-spin observation.

leaves proton homonuclear couplings free to evolve. The resulting spectrum therefore contains only proton multiplets in f_1 dispersed by the corresponding X-spin shifts in f_2 (Fig. 8.16). Notice the similarity here with the approach used in the sequence of Fig. 8.4c in which the basic HSQC shift correlation sequence was adapted for the measurement of long-range heteronuclear couplings. In both cases the chemical shift evolution period of the shift correlation sequence was converted into one in which only couplings were active (through the use of spin-echoes), so producing the J-resolved experiment. Often, subtle changes involving the addition or removal of one or two strategically placed pulses are all that is required to alter the characteristics of the resulting spectrum, producing a 'new' pulse sequence (which may or may not be useful to the chemist). This, in part, contributes to the plethora of NMR pulse sequences found throughout the chemical literature.

Once again, strong coupling, this time between proton satellites, causes problems. Furthermore, because one must consider the *satellites* of each proton rather than the parent resonance itself (since only the satellites can contribute to polarisation transfer), the multiplet patterns observed for high-order systems may differ from those of the parent resonance in the 1D spectrum. Other complications occur in the case of non-equivalent geminal protons since each may possess different multiplet patterns, yet both will be observed at the same carbon chemical shift, leading to a potentially complex multiplet overlap (this is apparent for the H6 protons in the indirect spectrum of Fig. 8.16) The most likely use of this variant is in

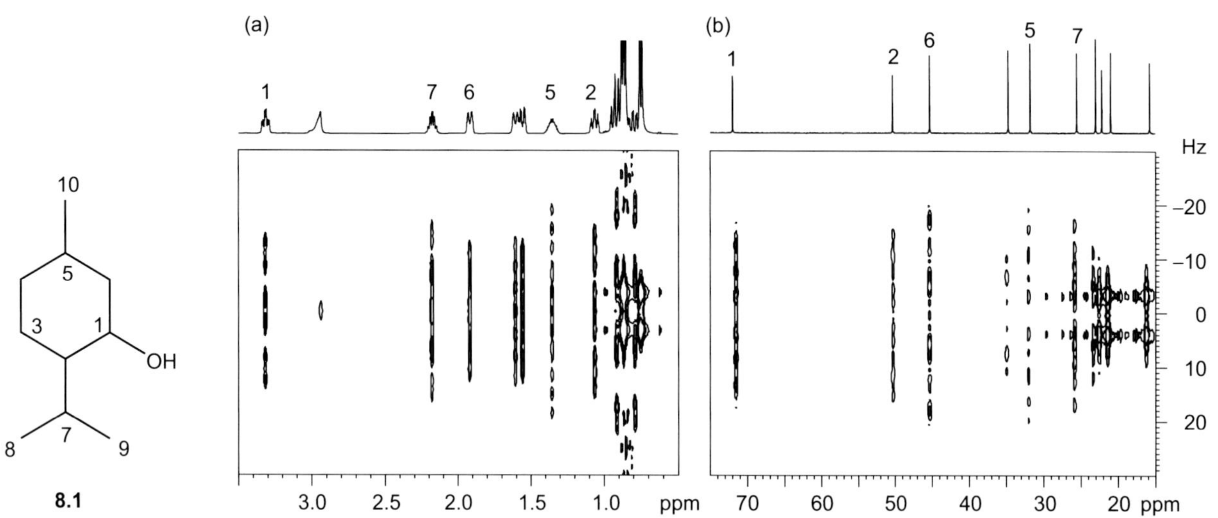

FIGURE 8.16 **J-resolved spectra.** The (a) direct homonuclear J-resolved spectrum and (b) indirect homonuclear J-resolved spectrum of menthol **8.1**. Both experiments present homonuclear couplings in f_1 dispersed by either the proton or the carbon chemical shift, respectively.

separation of the multiplet patterns of protons having coincidentally degenerate chemical shifts which are not therefore separable in the conventional homonuclear version.

8.5 PURE SHIFT BROADBAND-DECOUPLED ^{1}H SPECTROSCOPY

Pure shift spectra, as the name suggests, contain only resonances that portray chemical shifts, but which are devoid of multiplet structure, appearing as singlets only, and is a term that has been applied specifically to broadband homonuclear-decoupled proton spectra, rather than to those that are broadband *heteronuclear* decoupled, such as conventional carbon spectra. The possibility of recording broadband-decoupled proton spectra has been touched on above and the most basic approach illustrated in Fig. 8.11. The simple J-resolved experiment tends to be of limited success in this respect owing to interferences from strong coupling, poor lineshapes and limited resolution obtained and because signal intensities in the projected spectrum bear little relation to signal intensities in the conventional 1D spectrum. In recent years, more sophisticated experimental techniques have been developed that provide practical solutions for achieving fully decoupled absorption-mode lineshapes [23], and hence producing true ultrahigh-resolution decoupled spectra (Fig. 8.17). These pure

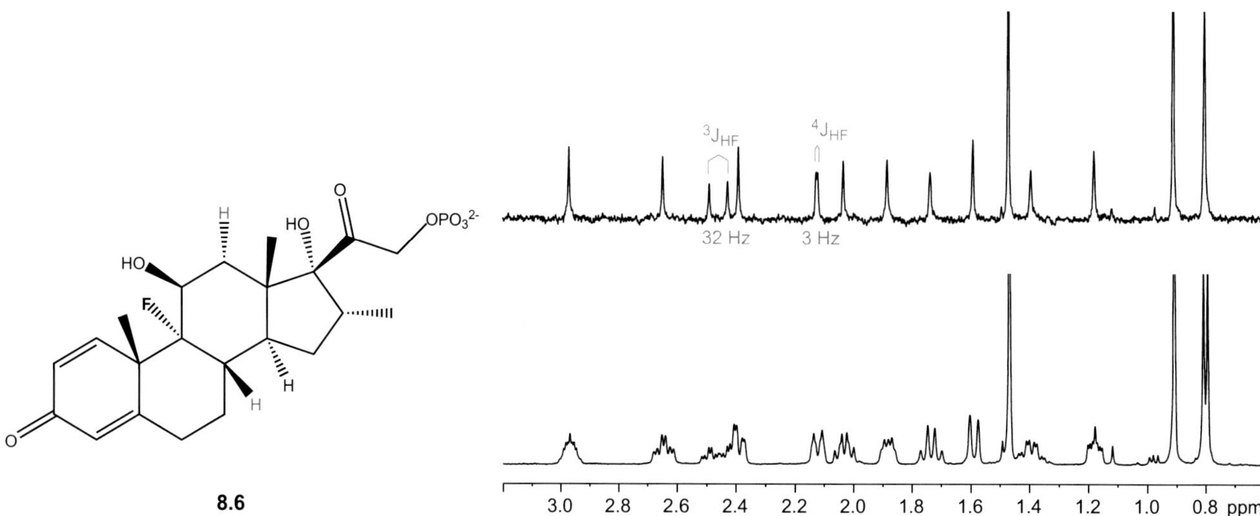

FIGURE 8.17 **The ^{1}H pure shift spectrum of dexamethasone phosphate 8.6.** This is shown above its conventional ^{1}H spectrum, and the remaining heteronuclear ^{1}H– ^{19}F couplings highlighted.

shift methods can yield decoupled 1D experiments [24–29] or can be incorporated into 2D techniques to achieve very high resolution in the proton dimension(s), as demonstrated for HSQC [30,31], total correlation spectroscopy (TOCSY) [28,32] and NOE spectroscopy (NOESY) [33]. The ability to retain only singlet proton resonances may find application where spectral simplification is beneficial, having been exemplified in isomer resolution [34], relaxation measurements[35], the measurement of ^{1}H–^{13}C coupling constants [36,37] and diffusion-ordered spectroscopy [38]. It may also come to prove helpful for peak separation of complex mixtures such as natural product extracts or in metabolomics studies. Alternatively, as described for homonuclear J-resolved spectroscopy above, it can provide a ready means by which to observe and measure *heteronuclear* couplings that will remain in pure shift proton spectra (eg with ^{19}F or ^{31}P [39]). Such ^{19}F couplings may be observed in the ^{1}H pure shift spectrum of dexamethasone phosphate **8.6** (Fig. 8.17). Pure shift methods may also prove to be useful for accurate automated peak-picking protocols for both 1D and 2D spectra, so may come to play a role in automated structure elucidation routines. The primary drawback with most current pure shift methods is a very significant sensitivity penalty, as will be explained later, and this feature will likely play a significant role in defining their practical utility.

8.5.1 The Basis of Pure Shift Spectroscopy

While there exist a number of related methods to achieve the goal of broadband proton decoupling, they are all based on the concept of *refocussing* proton–proton coupling evolution while at the same time allowing chemical shifts to evolve. This leads to homonuclear couplings being absent from the resulting proton spectrum and only chemical shifts evident. The underlying process may be understood with reference to a pair of mutually coupled protons H^A and H^X whose magnetisation is evolving during a time period τ. This period forms a spin-echo in which a 180 degree refocussing pulse is applied at the midpoint ($\tau/2$) selectively to H^A *only*. Considering the behaviour of proton H^X, we would expect its chemical shift to evolve throughout the whole τ period since this proton does not experience the refocussing pulse. In contrast, J_{AX} coupling evolution *will* be refocussed at the end of the τ period since only the H^A proton has been inverted by the selective pulse, thereby inverting the sense of precession of coupling evolution (Fig. 8.18; this is analogous to heteronuclear coupling being refocussed in the heteronuclear spin-echo described in Section 2.2.4). Thus, after the period τ, there is no evidence for coupling evolution for proton X and it is as if the J_{AX} coupling had never existed during this interval. The idea behind all pure shift methods is for this refocussing effect to occur repeatedly for the whole of the proton data acquisition period, so again evidence of homonuclear coupling is, in effect, hidden. As in this example, the requirement for this to succeed is that for every proton in the molecule that is to be decoupled (retained active spin), its J-coupled partners (suppressed passive spins) must be inverted while the observed active proton is left untouched. This must occur for every proton in the molecule at the same time to achieve a fully broadband-decoupled spectrum. The solution to this seemingly impossible task lies in the details of the refocussing element employed at the midpoint of the echo period, and a number of solutions now exist for this, as will be described below. Before considering these, we shall consider the two current approaches for incorporating repeated spin-echoes into the proton detection period such that homonuclear coupling remains unseen. The first of these builds a conventional proton FID from a series of spectra collected as a pseudo-2D data set while the second approach directly incorporates the refocussing echoes into the proton FID in real time.

8.5.2 Pseudo-2D Pure Shift

The pseudo-2D approach to generating a pure shift spectrum employs a scheme in which short FID data 'chunks' are collected during which the evolution of homonuclear proton couplings is effectively negligible [25,40]. A series of these

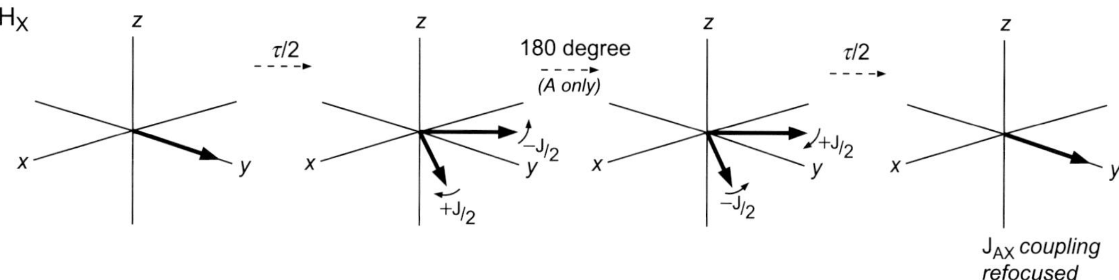

FIGURE 8.18 **J evolution during a spin-echo for a H^A–H^X coupled spin pair.** The behaviour of proton X is represented during the echo in which a selective 180 degree pulse is applied *only* to proton A.

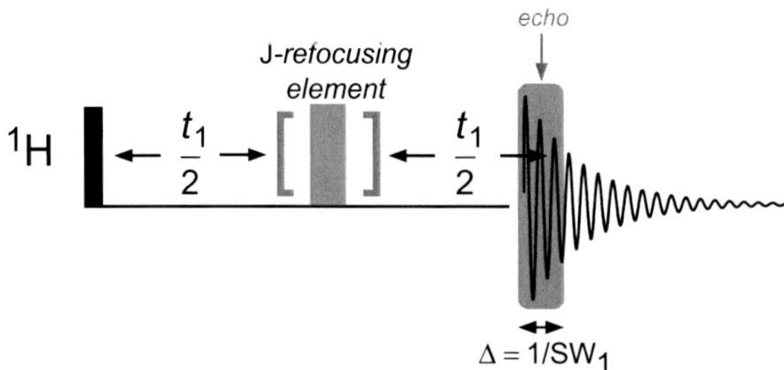

FIGURE 8.19 **The pseudo-2D sequence for generating pure shift ^{1}H spectra.** At the centre of the t_1 period sits the J_{HH} refocussing element with the *shaded box* indicating the retained data chunk centred at t_1.

chunks is then concatenated to build a complete FID, transformation of which yields a spectrum devoid of proton J-coupling structure.

In this method, a series of complete FIDs is collected as increments of a 2D data sequence made up of the selective refocussing element within a spin-echo, as described above (Fig. 8.19). For each increment of the 2D experiment, homonuclear couplings are refocussed at the end of the echo period and each acquired data chunk is centred at this point. Provided the duration of each sampled data chunk is kept short ($\Delta \ll 1/J_{HH}$), then proton J evolution occurring within this period may be considered negligible. Subsequent data chunks corresponding to later parts of the final FID are collected by incrementing the t_1 evolution period while ensuring the centre of each data chunk corresponds with the end of the echo period. Sequential data chunks are then extracted from each of the collected FIDs (Fig. 8.20) and subsequently reassembled (via a suitable processing script) to yield the complete pure shift FID. Throughout the sequence, chemical shifts continue to evolve and are therefore apparent in the final spectrum whereas homonuclear couplings are not. Since proton couplings are small and hence evolve rather slowly, data chunk durations of ~20 ms can be acquired without coupling structure being resolved. The chunk durations are themselves defined as being equal to $1/SW_1$ s, where SW_1 is the spectral width defined for the t_1 dimension. Thus, data chunks of 20 ms collected from

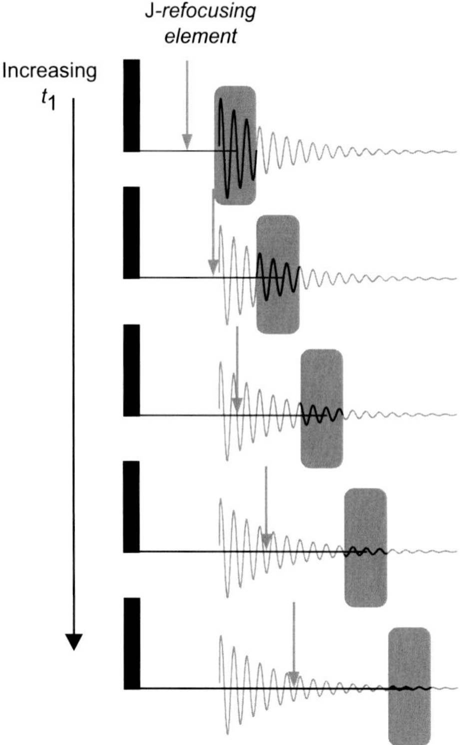

FIGURE 8.20 **Schematic illustration of the pure shift data chunking method.** As t_1 is incremented, sequential data chunks (shaded region) are retained and subsequently concatenated to regenerate a complete FID devoid of J-coupling modulation.

40 t_1 increments will yield a final FID with an acquisition time of 800 ms. While somewhat shorter than conventional proton FIDs (~3 s), such acquisition times are usually adequate in these applications since coupling fine structure does not exist to be resolved. Artefacts arising from residual coupling appear as sidebands at offsets of 1/Δ Hz from the parent peaks, but these may be attenuated by adjusting SW_1 to ensure data chunks remain sufficiently short and hence decoupling effective.

The pseudo-2D approach may be employed using any of the refocussing elements described in Section 8.5.4, and its application will likely be constrained by their performance characteristics. Fig. 8.17 shows a pseudo–2D derived pure shift spectrum recorded using the Zangger–Sterk element introduced below.

8.5.3 Real-Time Pure Shift

An alternative approach to collecting pure shift spectra is to directly include the refocussing element repeatedly within a conventional 1D FID acquisition so as to continually refocus homonuclear proton-coupling evolution [28]. As discussed earlier, proton couplings will not appear in the resulting spectrum provided each FID chunk Δ remains appropriately short (Fig. 8.21). The principal advantage of this 'real-time' or 'instant' approach is speed, since it avoids the need to collect a 2D data set from which to build a single FID. It does however place more significant demands on instrument hardware since rapid receiver gating is required to allow repeated application of refocussing elements and because these must be applied frequently relative to J_{HH} so that coupling evolution remains negligible. The refocussing elements themselves also have finite durations (possibly tens of milliseconds, see below) during which spin relaxation occurs, leading to an effective increase in the signal decay rate and hence introducing residual broadening of resonance linewidths. In principle, the real-time approach can be included in the FID acquisition period of any 1D or 2D experiment without increasing its dimensionality, making it a more general approach to pure shift spectroscopy than that described above. The sensitivity of the method will however remain constrained by the characteristics of the refocussing element employed, as considered below.

8.5.4 Pure Shift Refocussing Elements

The critical aspect in all pure shift experiments is the action of the refocussing element within the spin-echoes. Recall that the net effect of this element must be to invert only passive spins so as to refocus their homonuclear couplings with the observed active spins. A number of approaches to this have been developed and current methods will be introduced here, yet this remains an active field of research.

8.5.4.1 Band Selective

We shall begin with the simplest refocussing element to understand, which is in fact designed to achieve homonuclear proton decoupling over a selected region of the whole spectrum, with other resonances outside this region removed completely. Thus, the method actually yields band-selective decoupling and is referred to either as homodecoupled band selective (HOBS) NMR [26] or band selective homonuclear (BASH) decoupling [27]. The refocussing element employs a pulse pair comprising a hard non-selective 180 degree pulse followed by a shaped band-selective 180 degree pulse, each bracketed by purging gradient pulses (Fig. 8.22). Those protons that resonate within the bandwidth of the selective pulse (active spins) experience no net rotation from combined application of hard and soft 180 degree pulses, while all spins outside the selected region (passive spins) experience only the hard inversion pulse at the midpoint of the echo. This combination therefore leads to the desired refocussing of all homonuclear couplings experienced by active spins, provided their coupled partners do no resonate within the selected region also. This requirement places limitations on how this particular scheme

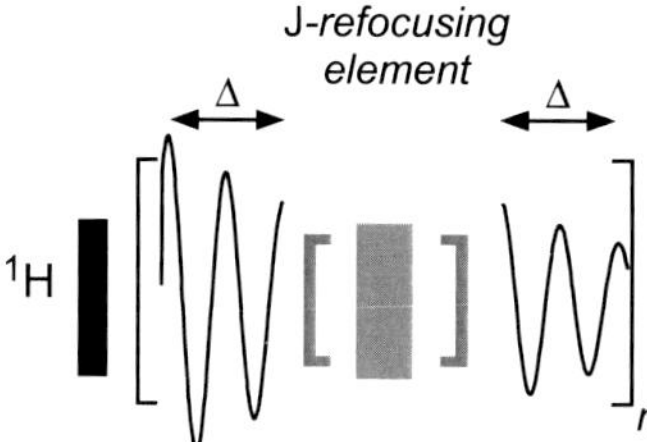

FIGURE 8.21 **Schematic illustration of real-time pure shift data collection.** The refocussing element is applied repeatedly throughout the proton acquisition to suppress J-coupling evolution.

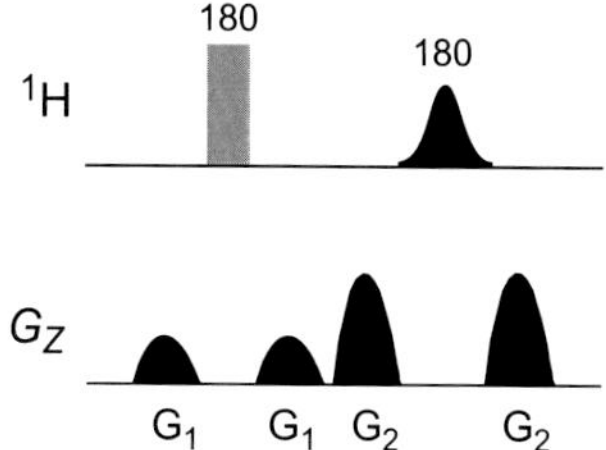

FIGURE 8.22 **The refocussing element employed in band-selective homonuclear decoupling.** The selective 180 degree pulse defines the spectrum region for which decoupled proton resonances are observed.

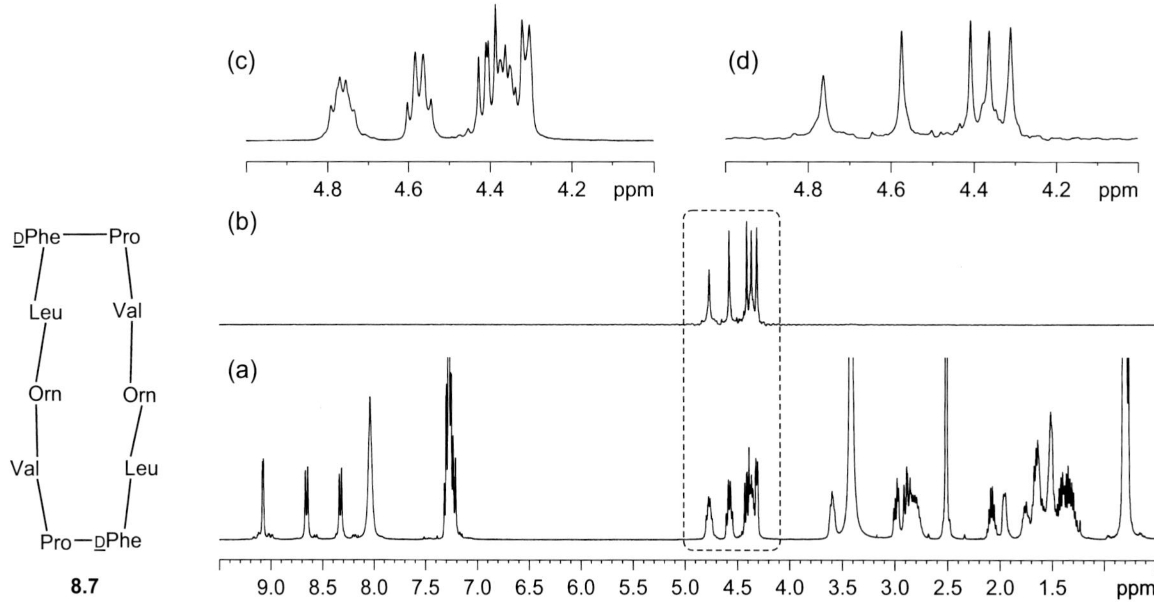

FIGURE 8.23 **Band-selective, real-time homonuclear decoupling of the peptide gramicidin-S 8.7 collected in a single transient.** (a) ^{1}H spectrum, (b) HOBS spectrum selecting Hα region, (c and d) expansion of the Hα region (boxed) from (a) and (b), respectively. An 18-ms selective REBURP pulse was employed and a 700-ms FID collected from 40 data chunks of 17.5 ms each.

may be employed and the types of samples to which it is suited. Structures such as peptides have spectral regions in which groups of protons resonate that are not mutually coupled (amide NH or Hα protons) and for which this approach is well suited. Similarly, oligonucleotides possess isolated clusters of resonances that are also amenable to HOBS decoupling [41]. The selective decoupling of smaller spectral regions can also prove useful for small organic molecules, as demonstrated by the resolution of isomers with overlapping resonances [34]. The effect of such decoupling may be seen in Fig. 8.23 for the cyclic decapeptide gramicidin-S **8.7**. The Hα protons are selected and therefore appear as fully resolved singlets, with all other passive protons destroyed by an initial PFG echo selection element. The selective pulse may be any shaped inversion pulse such as Gaussian, REBURP or RSNOB (see Section 12.4), the duration of which defines the selective bandwidth. While the band-selective method is restricted to decoupling of specific regions, it does have the significant advantage over the alternative methods described below that it does not suffer a severe sensitivity penalty since all selected protons contribute to the observed spectrum.

8.5.4.2 Zangger–Sterk

This implementation seeks to achieve decoupling across the *whole* proton spectrum and is based on one of the earliest approaches to *broadband* homonuclear decoupling [25,28,40]. The refocussing element again uses a non-selective 180 degree pulse to act on all spins followed by a selective 180 degree pulse applied simultaneously with a weak PFG (Fig. 8.24). This shaped pulse gradient combination was first proposed by Zangger and Sterk for broadband decoupling [40], and this refocussing element now takes their name, herein referred to as the Z–S block. This element acts to invert only the active spins which contribute to the final spectrum but not their mutually coupled passive partners, which are removed from the final spectrum by PFGs that bracket the Z–S block.

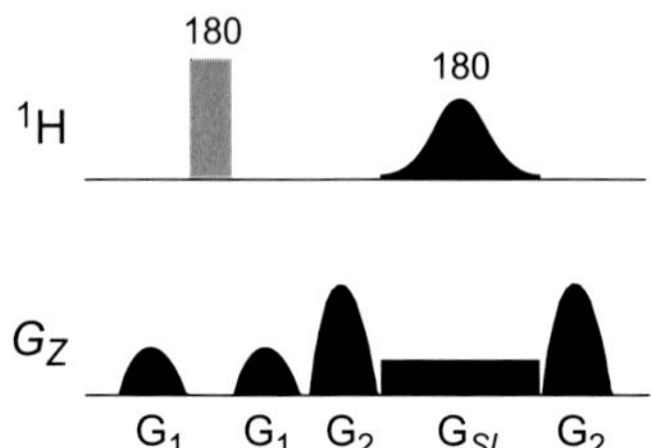

FIGURE 8.24 **The Zangger-Sterk refocussing element during which a selective 180 degree ^{1}H pulse is applied simultaneously with a weak PFG G_{SL}.**

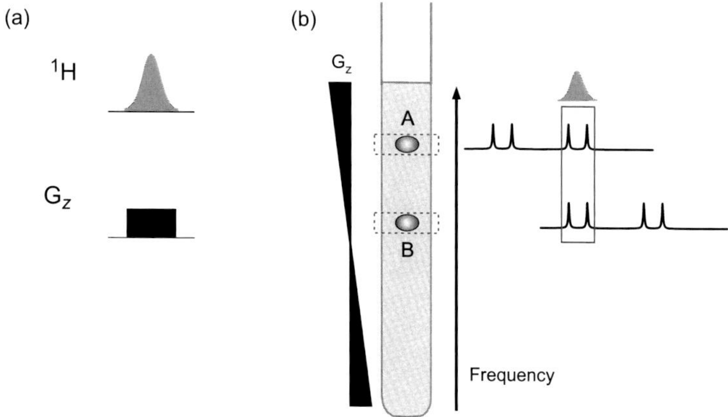

FIGURE 8.25 **Operation of the Z–S pulse element.** (a) The selective inversion pulse/field gradient combination and (b) schematic illustration of selective inversion of only a single spin per molecule by this element; see text for discussion.

Operation of the Z–S block is subtle. The weak *z*-gradient pulse imposes a frequency spread of resonances along the length of the sample due to variations in the local field. This means that when the frequency-selective 180 degree pulse is applied, different regions of the proton spectrum are inverted for molecules that sit in differing locations over the length of the NMR sample. This concept is illustrated schematically in Fig. 8.25 for two molecules A and B of identical structure and demonstrates how the inversion of only a single spin per molecule may be achieved simultaneously for all resonances in the spectrum. Since molecule A experiences a greater local field than B due to the applied field gradient, its spins will resonate at relatively higher frequencies than the equivalent nuclei in molecule B; this is shown as a horizontal offset of the corresponding spectra. Application of the frequency-selective inversion pulse will thus act simultaneously on *different* spin multiplets for the two molecules, as illustrated by the red box which contains the active spin resonances. However, this pulse will have no influence on the coupled passive spin resonances that sit outside the inversion bandwidth (those outside the red box). When all signals from the selectively excited region are retained, each active spin has experienced the required inversion while none of its passive partners have done so. The combined effect of the hard inversion pulse and the Z–S block is therefore to yield no net inversion for active spins but full inversion for all passive spins, thus leading to the required refocussing of proton-coupling evolution.

The Z–S block leads to what is known as *slice selection* since the observed resonance of each active spin derives only from a narrow slice of the whole NMR sample over which the selective pulse was effective (dashed red boxes in Fig. 8.25). This leads to a substantial decrease in the sensitivity of the method since most of the NMR sample no longer contributes to the detected signal, and this is by far the most significant weakness of this method (as apparent in Fig. 8.17). The greater the selectivity of the pulse, the narrower the contributing sample slices and the weaker the detected signal. The frequency bandwidth of the selective 180 degree pulse is itself dictated by the frequency separation of the most closely spaced, mutually coupled spins to be decoupled, so one is likely to require some compromise between complete decoupling of neighbouring multiplets and sensitivity losses. Invariably, strongly coupled resonances are less likely to be fully decoupled because of their similar resonant frequencies and the greater likelihood that both will fall within the bandwidth of the selective pulse. The strength of the gradient pulse is itself chosen such that the frequency spread imposed on the sample is comparable with the width of the proton spectrum. The frequency spread $\Delta\nu$ across a sample of active length l within the probe RF coil is $\Delta\nu = \gamma G l$ for a gradient strength G, from which this may be defined. Thus, at 500 MHz a 10-ppm spectrum covers 5 kHz, so for a typical RF coil length of 1.6 cm a gradient strength of ~0.7 G cm^{-1} is appropriate ($\gamma_H = 4258$ Hz G^{-1}). The sensitivity of the method may then be determined from the bandwidth of the selective pulse since this defines the fraction of the NMR sample that contributes to the observed signal. For a bandwidth of 50 Hz, the retained signal is only ~1% ($50/\Delta\nu$) and, as is so often the case, this sensitivity loss is the price to pay for the accompanying resolution gain in the decoupled spectrum.

8.5.4.3 BIRD

An alternative approach to differentiate passive from active spins within the refocussing element is to use isotope editing, as in the bilinear rotation decoupling (BIRD)-based method. The BIRD pulse cluster was introduced in Section 7.3.3 as a block for manipulating the behaviour of protons bound to ^{13}C separately from those bound to ^{12}C. In this pure shift application, the BIRD cluster serves to invert *only* those ^{13}C-bound protons, leaving those bound to ^{12}C unaffected. When combined with a non-selective 180 degree proton pulse (Fig. 8.26), the desired net inversion of only the passive ^{12}C-bound protons results. The full sequence implementation incorporating the BIRD-refocussing element [24] ensures that only ^{13}C-bound protons are detected with all others suppressed. Sequence timings are such that chemical shifts remain active throughout the echo period and heteronuclear ^{1}H–^{13}C couplings are refocussed at the end of the echo. This then allows the use of ^{13}C decoupling during detection of the proton FID so that the detected ^{13}C satellites are both heteronuclear (^{13}C) and homonuclear (^{1}H) decoupled. Since it is the natural abundance carbon-13 satellites that are retained in this approach, net signal retention is fixed at ~1% so is comparable with the Z–S approach regarding sensitivity losses. However, the method is well suited to the acquisition of homonuclear-decoupled ^{1}H–^{13}C HSQC spectra which inherently select these same satellites, thus imposing no additional sensitivity penalty (indeed, sensitivity gains may be realised from collapse of the proton multiplet structure) [30,31]. The BIRD pure shift experiments are also superior to the above methods when strongly coupled proton multiplets are to be decoupled, because natural isotope abundance will ensure coupled protons will not both be bound to ^{13}C, meaning only one spin of each coupled pair will be inverted. This approach does fail however for mutually coupled geminal protons since both are ^{13}C bound and will experience the BIRD inversion, leaving geminal proton couplings apparent in the pure shift spectrum.

8.5.4.4 PSYCHE

The pure shift yielded by CHIRP excitation (PSYCHE) method [29,42] provides a further approach to manipulating active and passive spin sets separately (Fig. 8.27) and is based on the little-known '*z*-COSY' experiment which has the sequence $90–t_1–\beta–\tau–\beta–$*acquire*. In this the traditional 90 degree mixing pulse of COSY has been replaced by two pulses of small tip angle $\beta \ll 90$ degree, which serve to define the active spin population as a subset of all spins that experience excitation by these pulses. The two frequency-swept pulses each of net tip angle β serve a similar role in PSYCHE. When applied simultaneously with a weak PFG G_S, their effect is to refocus responses from these active spins and suppress other possible coherence transfer pathways involving homonuclear couplings, including COSY-like transfers and zero-quantum coherences (the swept-pulse/field gradient combination parallels that used for zero-quantum suppression described in Section 12.6.2). When combined with the preceding hard 180 degree pulse, the net effect is again to refocus proton couplings but not their chemical shifts. It is noteworthy that despite the apparent similarity with the Z–S method above, the simultaneous swept-pulse/gradient combination is not utilised for slice selection here but only for coherence selection, so does not dictate the sensitivity of the method. Instead, in PSYCHE the population of active spins, and hence signal intensity in the decoupled spectrum, is proportional to $\sin^2\beta$, while the population of passive spins scales as $\cos^2\beta$. This means the operator has some control over signal intensity, albeit constrained by the need to maintain decoupling by retaining a high population of passive spins. It is this that demands that β is small, and in practice a compromise of $\beta \approx 20$ degrees provides for reasonable signal intensity with adequate decoupling. This is achieved by reducing the peak RF power of swept CHIRP pulses (~15 ms each) from that required by the more usual 180 degree inversion pulses (Section 12.2). The potential advantage of the PSYCHE method over those described above (with the exception of the band-selective approach) is a less severe sacrifice in sensitivity, although this remains substantial.

It is clear that pure shift methods have the ability to provide significant resolution gains for ^{1}H spectroscopy, in particular, and may offer effective solutions where overlapping resonances prove problematic, provided associated sensitivity

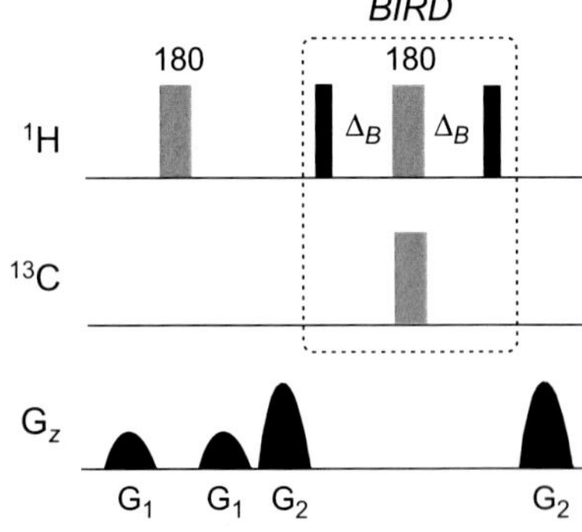

FIGURE 8.26 **The BIRD-based refocussing element using ^{13}C isotope selection ($\Delta_B = 1/2^1J_{CH}$).**

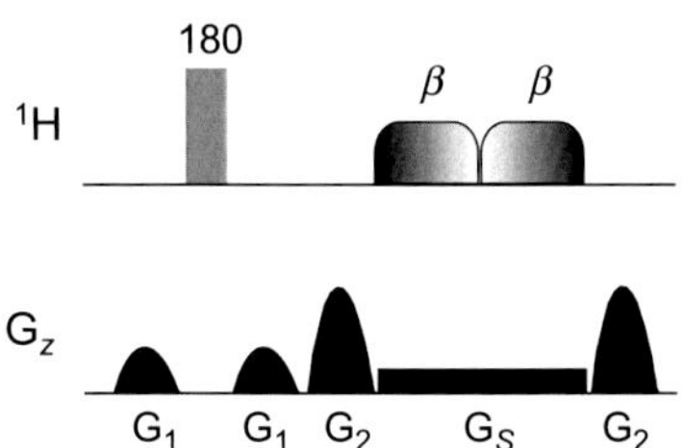

FIGURE 8.27 **The PSYCHE refocussing element.** The two low-power frequency-swept CHIRP pulses each have a net tip angle $\beta \approx 20$ degree but sweep in opposite directions, while the simultaneous weak PFG G_S serves to suppress unwanted transfer pathways.

losses can be tolerated. In this regard, band-selective pure shift methods (BASH or HOBS) have little signal loss and may be most practical where the decoupling of isolated regions containing no mutually coupled spins is desired. For true broadband-homodecoupled proton spectra, PSYCHE currently appears to offer the greatest potential to minimise sensitivity loss, and further developments to this area are to be anticipated.

REFERENCES

[1] Bodenhausen G, Morris GA, Freeman R, Turner DL. J Magn Reson 1977;28:17–28.
[2] Müller L, Kumar A, Ernst RR. J Magn Reson 1977;25:383–90.
[3] Bodenhausen G, Freeman R, Niedermeyer R, Turner DL. J Magn Reson 1976;24:291–4.
[4] Freeman R, Morris GA, Turner DL. J Magn Reson 1977;26:373–8.
[5] Rutar V, Wong TC. J Magn Reson 1983;53:495–9.
[6] Marquez BL, Gerwick WH, Williamson RT. Magn Reson Chem 2001;39:499–530.
[7] Parella T, Espinosa JF. Prog Nucl Magn Reson Spectrosc 2013;73:17–55.
[8] Bax A. J Magn Reson 1983;52:330–4.
[9] Rutar V. J Magn Reson 1984;56:87–100.
[10] Bax A, Freeman R. J Am Chem Soc 1982;104:1099–100.
[11] Morat C, Taravel FR, Vignon MR. Magn Reson Chem 1988;26:264–70.
[12] Jippo T, Kamo O, Nagayama K. J Magn Reson 1986;66:344–8.
[13] Uhrín D, Liptaj T, Hricovíni M, Capek P. J Magn Reson 1989;85:137–40.
[14] Liu ML, Farrant RD, Gillam JM, Nicholson JK, Lindon JC. J Magn Reson (B) 1995;109:275–83.
[15] Bodenhausen G, Freeman R, Turner DL. J Magn Reson 1977;27:511–4.
[16] Bodenhausen G, Turner DL. J Magn Reson 1980;41:200–6.
[17] Freeman R, Keeler J. J Magn Reson 1981;43:484–7.
[18] Aue WP, Karhan J, Ernst RR. J Chem Phys 1976;64:4226–7.
[19] Nagayama K, Bachmann P, Wüthrich K, Ernst RR. J Magn Reson 1978;31:133–48.
[20] Bodenhausen G, Freeman R, Morris GA, Turner DL. J Magn Reson 1978;31:75–95.
[21] Mersh JD, Sanders JKM. J Magn Reson 1982;50:171–4.
[22] Morris GA. J Magn Reson 1981;44:277–84.
[23] Castañar L, Parella T. Magn Reson Chem 2015;53:399–426.
[24] Aguilar JA, Nilsson M, Morris GA. Angew Chem Int Ed 2011;50:9716–7.
[25] Aguilar JA, Faulkner S, Nilsson M, Morris GA. Angew Chem Int Ed 2010;49:3901–3.
[26] Castañar L, Nolis P, Virgili A, Parella T. Chem Eur J 2013;19:17283–6.
[27] Ying J, Roche J, Bax A. J Magn Reson 2014;241:97–102.
[28] Meyer NH, Zangger K. Angew Chem Int Ed 2013;52:7143–6.
[29] Foroozandeh M, Adams RW, Meharry NJ, Jeannerat D, Nilsson M, Morris GA. Angew Chem Int Ed 2014;53:6990–2.
[30] Sakhaii P, Haase B, Bermel W. J Magn Reson 2009;199:192–8.
[31] Paudel L, Adams RW, Király P, Aguilar JA, Foroozandeh M, Cliff MJ, Nilsson M, Sándor P, Waltho JP, Morris GA. Angew Chem Int Ed 2013;52:11616–9.
[32] Morris GA, Aguilar JA, Evans R, Haiber S, Nilsson M. J Am Chem Soc 2010;132:12770–2.
[33] Aguilar JA, Colbourne AA, Cassani J, Nilsson M, Morris GA. Angew Chem Int Ed 2012;51:6460–3.
[34] Castañar L, Pérez-Trujillo M, Nolis P, Monteagudo E, Virgili A, Parella T. Chem Phys Chem 2014;15:854–7.
[35] Castañar L, Nolis P, Virgili A, Parella T. J Magn Reson 2014;.
[36] Timári I, Kaltschnee L, Kolmer A, Adams RW, Nilsson M, Thiele CM, Morris GA, Kövér KE. J Magn Reson 2013;239:130–8.
[37] Castañar L, Saurí J, Nolis P, Virgili A, Parella T. J Magn Reson 2014;238:63–9.
[38] Nilsson M, Morris GA. Chem Commun 2007;933–5.
[39] Aguilar JA, Morris GA, Kenwright AM. RSC Advances 2014;4:8278–82.
[40] Zangger K, Sterk H. J Magn Reson 1997;124:486–9.
[41] McKenna JM, Parkinson JA. Magn Reson Chem 2015;53:249–55.
[42] Foroozandeh M, Adams RW, Nilsson M, Morris GA. J Am Chem Soc 2014;136:11867–9.

Chapter 9

Correlations Through Space: The Nuclear Overhauser Effect

Chapter Outline

9.1 Introduction 315
9.2 Definition of the NOE 317
9.3 Steady-State NOEs 317
9.3.1 NOEs in a Two-Spin System 317
9.3.2 NOEs in a Multi-Spin System 324
9.3.3 Summary 329
9.3.4 Applications 330
9.4 Transient NOEs 335
9.4.1 Nuclear Overhauser Effect Kinetics 335
9.4.2 Measuring Internuclear Separations 336
9.5 Rotating Frame NOEs 337
9.6 Measuring Transient NOEs: NOESY 339
9.6.1 The 2D NOESY Sequence 339
9.6.2 1D NOESY Sequences 346
9.6.3 Applications 349
9.7 Measuring Rotating Frame NOEs: ROESY 353
9.7.1 The 2D ROESY Sequence 353
9.7.2 1D ROESY Sequences 355
9.7.3 Applications 356
9.8 Measuring Steady-State NOEs: NOE Difference 359
9.8.1 Optimising Difference Experiments 361
9.9 Measuring Heteronuclear NOEs: HOESY 363
9.9.1 2D Heteronuclear NOEs 364
9.9.2 1D Heteronuclear Nuclear Overhauser Effects 365
9.9.3 Applications 366
9.10 Experimental Considerations for NOE Measurements 367
9.11 Measuring Chemical Exchange: EXSY 368
9.12 Residual Dipolar Couplings 371
9.12.1 Measuring RDCs 372
9.12.2 Applying RDCs 375
References 377

9.1 INTRODUCTION

The previous three chapters in this book have all been concerned with *scalar couplings* between nuclei, that is, the indirect couplings that are transmitted through intermediate electron spins in chemical bonds. We have seen that by application of the appropriate techniques the chemist is able to exploit this coupling information and piece together molecular fragments and, ultimately, gross molecular structures. In this chapter we shall be concerned with a fundamentally different form of interaction between nuclear spins: the direct, through-space magnetic interactions (*dipolar couplings*) that give rise to the nuclear Overhauser effect (NOE). This brings about changes in resonance intensities and is, as we shall discover, intimately related to nuclear spin relaxation. The NOE is typically employed during the later stages of a structural investigation when the gross structure of the molecule has been (largely) defined through application of the various techniques described in the preceding chapters. The NOE is then able to provide the chemist with information on three-dimensional (3D) molecular geometry. Such information can be obtained because the NOE depends upon, amongst other factors, internuclear separations such that only those spins that are 'close' in space are able to demonstrate this effect.

The NOE also finds widespread use as a means of sensitivity enhancement of low-γ spin-½ nuclei, so widespread in fact that it is often taken for granted and its contribution to experiments often overlooked. As described in Sections 4.4 and 9.3, the use of broadband proton decoupling during the recording of carbon spectra contributes as much as a threefold increase in resonance intensity by virtue of the proton to carbon NOE. However, the principal aim of this chapter is to develop an appreciation of the NOE as a tool in structural analysis where it has a unique role to play. The interpretation of NOE measurements does however require more care than for those methods that exploit scalar couplings, and is generally more susceptible to erroneous conclusions being drawn. As part of this, it is also important to be conscious of the nature of the NOE measurement being taken, and in particular whether it is a *transient* or a *steady-state* protocol that is used. The first of these is exemplified by the widely used NOE spectroscopy (NOESY) technique (executed as either the selective 1D or the 2D variants) while the second is best known as the older NOE difference experiment that finds less use nowadays. The

High-Resolution NMR Techniques in Organic Chemistry. http://dx.doi.org/10.1016/B978-0-08-099986-9.00009-9

nature of the measurement has fundamental implications for how the data should be *interpreted* and indeed *reported*. As such these two fundamentally different approaches to NOE measurements will be treated separately for much of the chapter, although they both share the same underlying theory. Pre-empting what is to follow, it will be shown that steady-state experiments are only appropriate for molecules that tumble 'rapidly' in solution (we shall also see what defines 'rapidly' in this context). Such measurements have traditionally been the home territory of small organic molecules in relatively non-viscous solutions, but have been largely superseded by transient NOE methods. In contrast, very much larger molecules that tumble 'slowly' in solution (or smaller molecules in very viscous solutions) can only be meaningfully studied with the transient NOE techniques, which are also suitable for small-molecule studies and are now widely applied in this area. Between these two extremes of molecular tumbling rates the conventional NOE can become weak and vanishingly small, a condition most likely to occur for those molecules with masses of around 1000–2000 Da. It is here that rotating frame NOE spectroscopy (ROESY) measurements play a vital role, and these shall also be described.

Included in this chapter also are 2D techniques that may be employed to study slow chemical exchange, a topic introduced in Section 2.6. These are considered here because similar sequences, in the form of exchange spectroscopy (EXSY) and NOESY, may be applied to study slow chemical exchange and transient NOEs, respectively, since these are in fact identical in their most basic forms. Finally, the quantitative measurement of dipolar coupling constants is introduced as these provide an alternative means of defining stereochemical relationships between nuclei that complement scalar coupling and the NOE. These methods rely on placing analytes in suitable media that inhibit the isotropic motion of molecules and leads to so-called *residual dipolar couplings* (RDCs) between nuclei becoming apparent as resonance splitting in spectra. These couplings in turn reflect the orientation of intermolecular vectors relative to the applied magnetic field and so contain valuable structural information. While the use of RDCs cannot yet be considered routine, they have been applied to the assignment of configuration when this has proved intractable by conventional methods and currently represent an active research area in small-molecule NMR.

The principal techniques described in this chapter are summarised in Table 9.1. The chapter is presented in two parts, the first covering the essential theory that underlies the NOE, and the second addressing the practicalities of how one measures NOE enhancements, the experimental steps required to optimise such measurements and how to correctly interpret the data. In keeping with the style of this book, mathematical equations are kept to a minimum and are introduced only when they serve to illustrate a point of fundamental importance. Likewise, the equations are generally presented rather than being derived, and the interested reader is encouraged to read dedicated texts on these topics for further elaboration [1].

TABLE 9.1 Principal Applications of the Main Techniques Described in This Chapter

Technique	Principal Applications
NOESY[a]	Establishing NOEs and hence spatial proximity between protons. Suitable for small ($M_r \ll 1000$) and large ($M_r > 2000$) molecules for which NOEs are positive and negative, respectively, but may fail for mid-sized molecules (zero NOE). Observes *transient* NOEs generated from *inversion* of a target resonance. Estimates of internuclear separations can be obtained in favourable cases.
ROESY[a]	Establishing NOEs and hence spatial proximity between protons. Suitable for any molecule but often essential for mid-sized molecules; NOEs are positive for all molecular sizes. Observes *transient* NOEs in the rotating frame, but is prone to interference from other mechanisms, so requires cautious interpretation. Estimates of internuclear separations can be obtained in favourable cases.
NOE difference	Establishing NOEs and hence spatial proximity between protons. Suitable only for small molecules ($M_r \ll 1000$), for which NOEs are positive. Observes *steady-state* or *equilibrium* NOEs generated from *saturation* of a target resonance. Above methods favoured nowadays.
HOESY[a]	Establishing heteronuclear NOEs and hence spatial proximity between different nuclides (eg ^{1}H–^{13}C or ^{1}H–^{19}F). Can provide useful stereochemical information when homonuclear NOEs prove inadequate. Often suffers from low sensitivity, but ^{1}H-detected variants can help.
EXSY[a]	Qualitative mapping of exchange pathways in dynamic systems when exchange rates are slow on the NMR chemical shift timescale, meaning separate resonances are observed for each exchanging species. Quantitative data on exchange kinetics can be obtained in favourable cases.
RDCs	RDCs provide information on relative bond vector orientations within a molecule and may be used to define the relative configurations and conformations of molecules. These methods require that the sample be weakly aligned in an appropriate medium for RDCs to be apparent.

The molecular masses mentioned provide only approximate ranges over which the experiments are applicable (see main text).

[a]*These methods may be executed as 1D or 2D experiments.*

PART I THEORETICAL ASPECTS

9.2 DEFINITION OF THE NOE

The NOE may be defined as the change in intensity of one resonance when the spin transitions of another are somehow perturbed from their equilibrium populations. The perturbation of interest usually corresponds to either *saturating* a resonance, that is equalising the spin population differences across the corresponding transitions, or to *inverting* it (ie inverting the population differences across the transitions). Magnitude is expressed as a relative intensity change between the equilibrium intensity I_0 and that in the presence of the NOE I, such that:

$$\eta_{\mathrm{I}}\{\mathrm{S}\}=\frac{\mathrm{I}-I_0}{I_0}\times 100(\%) \tag{9.1}$$

where $\eta_{\mathrm{I}}\{\mathrm{S}\}$ indicates the NOE observed for spin I when spin S is perturbed, which shall also be referred to as the NOE *from* spin S *to* spin I. Use of the symbols I and S stems from the original publications on the phenomenon (which, in fact, predicted the Overhauser effect from electron spins S to nuclear spins I in a metal) and have become the recognised nomenclature when describing the NOE. However, both S and I have been used to define the *perturbed* spin over the years, and even across modern texts both definitions are encountered, so one should always be clear as to the terminology in use. Herein, S will always refer to the perturbed (or source) spin and I to the enhanced (or interesting) spin. The intensity changes brought about by the NOE can be both positive (an increase) or negative (a decrease), as dictated by the motional properties of the molecules and by the signs of the magnetogyric ratios of the participating spins. Throughout, the term 'enhancement' will be used to refer to intensity changes.

9.3 STEADY-STATE NOEs

In laying down the background to how the NOE arises and what factors dictate its sign and magnitude, we shall focus our discussion on steady-state NOEs in which perturbation is brought about by *saturating* S-spin transitions by selective application of weak radiofrequency (rf) irradiation to the S resonance (Fig. 9.1). It is this form of the NOE that is observed with the once popular NOE difference method, which has had such an enormous impact on structural organic chemistry. Further discussions relating to other forms of NOE measurement then follow logically from this background material, including transient NOEs that are now more commonly measured in 1D or 2D NOESY experiments. We begin by considering the simple case of a homonuclear two-spin system then progress to consider more realistic multi-spin systems.

9.3.1 NOEs in a Two-Spin System

9.3.1.1 Origin of the NOE

Consider a system comprising only two homonuclear spin-½ nuclei I and S that exist in a rigid molecule which tumbles isotropically in solution, that is, it has no preferred axis about which it rotates. The two nuclei do not share a scalar coupling (J_{IS} = 0), but are sufficiently close to share a dipolar coupling (D_{IS}). This is the direct, through-space magnetic interaction between the two spins such that one spin is able to sense the presence of its dipolar-coupled partner. This coupling may be viewed as being analogous to the interaction one witnesses when two bar magnets are brought close together and follows the idea introduced in Section 2.1 that nuclear magnetic dipoles can be viewed as microscopic bar magnets.

The energy-level diagram for an ensemble totaling 4N molecules is shown in Fig. 9.2. Since we are considering a homonuclear system the energies of the I and S transitions will be essentially identical (chemical shift *differences* are negligible relative to Larmor frequencies), and we can therefore assume that the populations of the αβ and βα states are equal at equilibrium. According to the Boltzmann distribution, there will then exist an excess of nuclei in the lower energy αα orientation and a deficit in the higher energy ββ state. We shall ultimately be interested in the population *differences* across

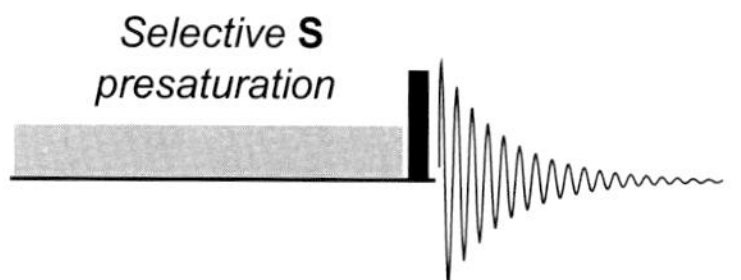

FIGURE 9.1 **The general experimental scheme for observing steady-state NOE enhancements.**

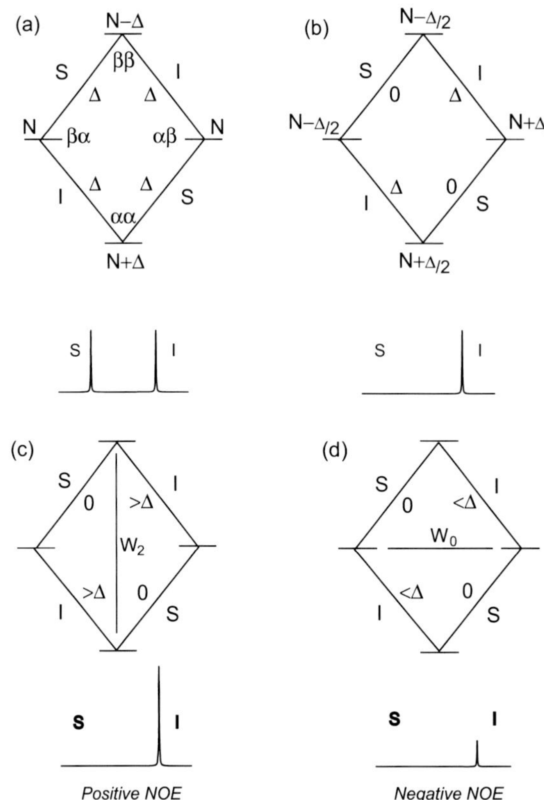

FIGURE 9.2 **Schematic energy-level diagrams and population differences for two spins S and I, which share a dipolar coupling.** (a) At equilibrium, (b) after instantaneous saturation of the S spins, (c) after relaxation via W_2 processes and (d) after relaxation via W_0 processes. Below each are the corresponding schematic spectra.

transitions, as it is these that dictate the intensity of observed resonances, so we shall simply symbolise the population excess as $+\Delta$ and the deficit as $-\Delta$ relative to those of the $\alpha\beta$ and $\beta\alpha$ states. Note that dipolar couplings do not produce observable splittings in solution spectra (see later discussions), so the two transitions associated with each spin are of identical energy. The spectrum in the absence of perturbation therefore contains two singlet resonances of equal intensity (Fig. 9.2a).

Now suppose we instantaneously saturate the S resonance forcing the population differences across the S transitions to zero. The new spin populations are indicated in Fig. 9.2b. Clearly the system has been forced away from equilibrium population differences, so will attempt to regain this by altering its spin populations. The changes of spin states required to achieve this are brought about by longitudinal spin relaxation processes, so we need to consider which relaxation pathways are now available to the spins. Ignoring for the moment the mechanism by which these changes may occur, we see that six possible pathways can be identified for a two-spin system (Fig. 9.3). Four of these correspond to the single-quantum transitions, involving the flip of a single spin, for example $\alpha\alpha$–$\beta\alpha$. The W labels represent the 'transition probabilities' for each or, in other words, the rates at which the corresponding spin-flips occur, and the subscripts represent the magnetic quantum number of the transition. The two other transitions $\alpha\beta$–$\beta\alpha$ and $\alpha\alpha$–$\beta\beta$ involve the simultaneous flipping of both S and I spins. Although these transitions do occur, they cannot be directly *observed* in an NMR experiment, unlike single

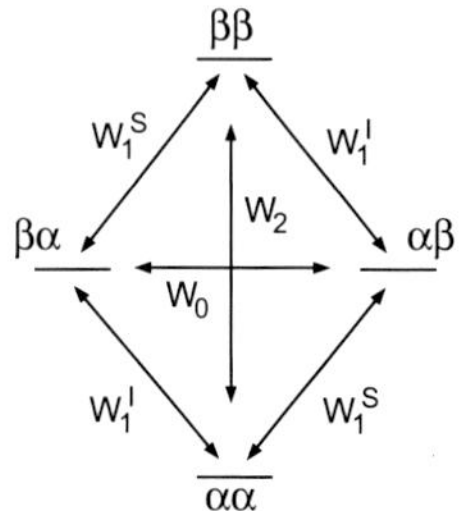

FIGURE 9.3 **The six possible transitions in a two-spin system.**

spin-flips, because the overall change in magnetic quantum ΔM does not equal one. They are said to be 'forbidden' by quantum-mechanical selection rules. The αβ–βα W_0 process is referred to as the zero-quantum transition ($\Delta M = 0$) while the αα–ββ W_2 process is the double-quantum transition ($\Delta M = 2$) (note the same terminology was used when describing the transitions of scalar-coupled spins in Section 5.3). These are both able to act as relaxation pathways and, in fact, it is only these two that are responsible for the NOE itself. Collectively, they are referred to as *cross-relaxation* pathways, a term suggestive of the simultaneous participation of both spins.

Returning to Fig. 9.2 we are now in a position to consider how the various relaxation pathways may be used to re-establish the equilibrium condition noting that, throughout, the $W_1{}^S$ transitions remain saturated by continuous rf energy. The population differences across the I-spin transitions are still Δ, as they were at equilibrium, so the $W_1{}^I$ processes will play no part in re-establishing equilibrium and thus have no role to play in producing the NOE. The W_2 process will act to remove spins from the ββ state and transfer them to the αα state in an attempt to recover the population differences across the S transitions. In doing so, this will *increase* the population *difference* across the two I transitions (Fig. 9.2c). Thus, relaxation via the W_2 process will result in a net increase in the I-spin resonance intensities in the spectrum; this is then a *positive NOE*. Likewise the W_0 process will act to transfer spins from the βα to the αβ state, again in an attempt to recover the population differences across the S transitions. In this case the result will be a *decrease* in the population *difference* across the two I transitions (Fig. 9.2d), so that relaxation via the W_0 process will result in a net reduction in the I-spin resonance intensities in the spectrum; this is then a *negative NOE*.

From these qualitative considerations, we can already say a fair amount about how we might expect the NOE to appear. Clearly, the W_2 and W_0 cross-relaxation processes compete with one another, with the dominant pathway dictating the sign of the observed NOE. In addition, the $W_1{}^I$ pathways will act to re-establish the equilibrium population differences for the I transitions as soon as the NOE begins to develop, so will tend to act against the buildup of the NOE. Thus, if relaxation mechanisms for the $W_1{}^I$ pathways happen to be rather more efficient than those of the W_2 and W_0 pathways, then a measurable NOE may never develop; it is, in effect, bypassed altogether. This can have a significant bearing on the experimental measurement of NOEs, as we shall see in due course. The NOE therefore results from a balance between a number of competing relaxation pathways. Saturating the S transitions for a period of time that is long relative to the relaxation times allows a new *steady state* of populations to arise as a result of this competition, and it is these one eventually measures. A full consideration of the various rate processes involved in the population changes leads to the so-called Solomon equation, which for the steady-state NOE can be used to derive the expression:

$$\eta_I\{S\} = \frac{\gamma_S}{\gamma_I}\left[\frac{W_2 - W_0}{W_0 + 2W_1^I + W_2}\right] \equiv \frac{\gamma_S}{\gamma_I}\left[\frac{\sigma_{IS}}{\rho_{IS}}\right] \tag{9.2}$$

where σ_{IS} represents the *cross-relaxation rate constant* for the two spins, and ρ_{IS} is the total *dipolar longitudinal relaxation rate constant* of spin I. The magnetogyric ratios (γ_S and γ_I) are included to take account of the different equilibrium populations that would exist for spins with differing γs; for a homonuclear spin system, as considered thus far, these values would obviously be equal and may be ignored. This fundamental expression contains within it the qualitative arguments arrived at above; $W_2 - W_0$ dictates the sign of the NOE whereas $W_1{}^I$ processes make no contribution to this but serve to reduce its magnitude. To appreciate the size and sign of the NOE, how this relates to molecular motion and indeed how this can be related in any way to internuclear distances, it is necessary to define what factors influence the participating rate constants, and for this one needs to consider the spin relaxation processes involved.

9.3.1.2 Spin Relaxation and Dipolar Coupling

The NOE arises as a result of the redistribution of spin populations and hence flips between spin states. Such redistributions occur as a result of longitudinal spin relaxation (see Section 2.4) which does not occur spontaneously but requires a suitable stimulus to induce the transitions. This stimulus is a magnetic field fluctuating at the frequency of the corresponding transition (both here and below 'frequency' corresponds to the energy of the transition rather than the rate at which the spin-flips occur). Here exists an analogy with the pulse excitation of a spin system initially at equilibrium. The time-dependent magnetic component of the electromagnetic rf radiation interacts with nuclear magnetic moments and is thus able to tip the bulk magnetisation vector into the transverse plane. Only if the rf has magnetic components oscillating at the Larmor frequencies of the spins does excitation occur; this is why one is able to apply excitation pulses to protons while leaving, say, carbon spins unaffected.

The magnetic field of relevance to the NOE is the local field experienced by a spin as a result of dipolar interactions with neighbouring magnetic nuclei. These interactions may be visualised using the microscopic bar magnet analogy for spin-½ nuclei in which they are considered to possess a magnetic North and a South Pole (Fig. 9.4). Depending on the relative

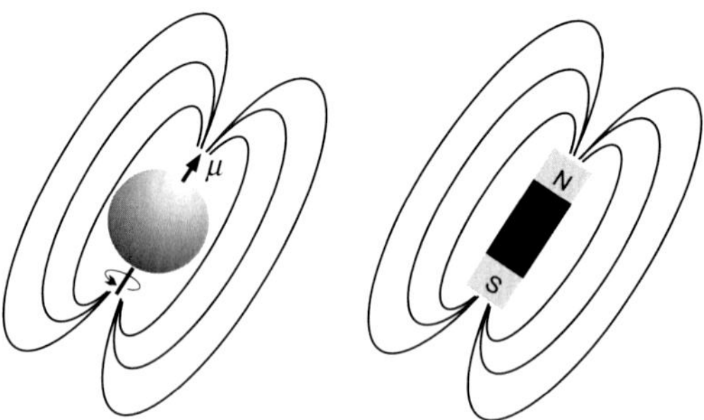

FIGURE 9.4 **The bar magnet analogy for a spin-½ nucleus in which the magnetic dipole is viewed as possessing a magnetic North and South Pole.**

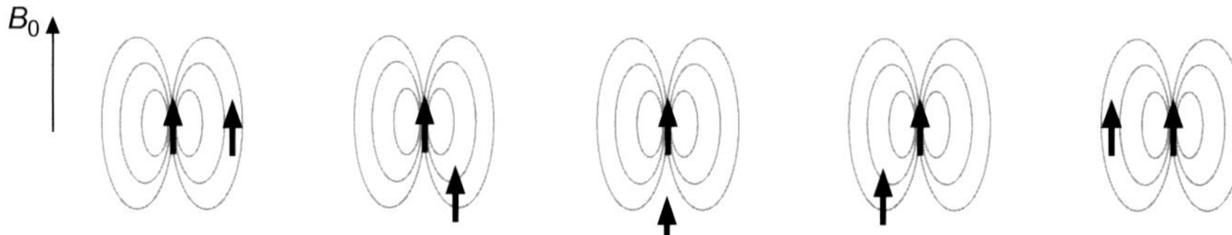

FIGURE 9.5 **The direct, through-space interaction between two near spin-½ nuclei (the dipolar interaction).** This fluctuates as the molecule tumbles in solution and can provide a time-dependent field capable of inducing spin transitions.

orientation of the two nuclei to one another, the field generated by a neighbouring spin will either reinforce or counteract the applied magnetic field, with the time-dependent fluctuation produced by the rotational motion of the molecule in which these nuclei sit (Fig. 9.5). It is precisely this that provides the mechanism for relaxation, known as *longitudinal dipole–dipole relaxation*, and this requirement for an interaction with a neighbouring spin intuitively fits with the arguments presented above for the NOE arising from *mutual* spin-flips (the W_2 and W_0 processes). The magnitude of dipolar coupling between spins is acutely sensitive to the internuclear separation *r*, being proportional to r^{-3}, and it is in this that the NOE itself ultimately has a distance dependence.

One should note here that, despite providing an important relaxation mechanism, *dipolar couplings do not usually produce observable splittings in solution state NMR spectra*. This is because, although the couplings have a finite value at any instant in time, they are averaged precisely to zero on the NMR timescale by the rapid isotropic tumbling of a molecule. In situations where this motion is inhibited, it is possible that some proportion of the dipolar coupling is apparent and these *residual* dipolar couplings can themselves provide useful structural information, as discussed in Section 9.12.

To induce the spin transitions we have been considering, the molecule must tumble at the appropriate frequency to provide a suitable fluctuating field. The rate at which a molecule tumbles or rotates in solution is typically defined by its rotational *correlation time* τ_c. This is usually taken to define the average time required for the molecule to rotate through an angle of one radian about any axis, meaning rapidly tumbling molecules possess small correlation times while slowly tumbling molecules have large correlation times. A *very rough* estimate of this time for a molecule of mass M_r may be obtained from the relationship:

$$\tau_c \approx M_r \times 10^{-12}\ \text{s} \tag{9.3}$$

The power available within a molecular system to induce transitions by virtue of its molecular tumbling is referred to as the *spectral density* J(ω) (Section 2.5) and this provides a measure of how the relaxation rates W_0, W_1 and W_2 vary as a function of tumbling rates. This is illustrated schematically in Fig. 9.6 for three different correlation times. An alternative description of the spectral density is that it represents the probability of finding a fluctuating magnetic component at any given frequency as a result of the motion, and as such the area under each of the curves of Fig. 9.6 must then be equal. Thus, for a molecule with a short τ_c (rapid tumbling) there exists an almost equal but comparatively small chance of finding components at both high and low frequencies, up to about $1/\tau_c$ at which point the probability falls away rapidly. Conversely, there is only a very small probability that molecules which tumble slowly (*on average*) will generate rapidly oscillating fields, so the corresponding spectral density is concentrated into a smaller frequency window. These curves therefore predict how the relaxation rates will vary with correlation time. For a transition of frequency ω_t the spectral density will be rather low if the molecular tumbling rate is far greater than this *(point a)* and thus the relaxation rates will also be low. As the rate of tumbling slows and approaches ω_t, the spectral density and hence relaxation rates increase *(point b)*, only to decrease once more as the tumbling rate falls below ω_t. Thus, the dependence of relaxation rates on τ_c may be represented

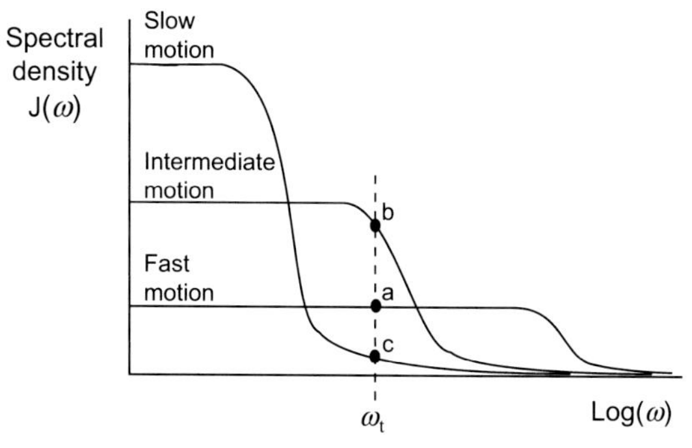

FIGURE 9.6 **Schematic spectral densities.** These are shown for molecules tumbling in three motional regimes as a functßion of frequency ω, where ω_t represents the frequency of the spin transition.

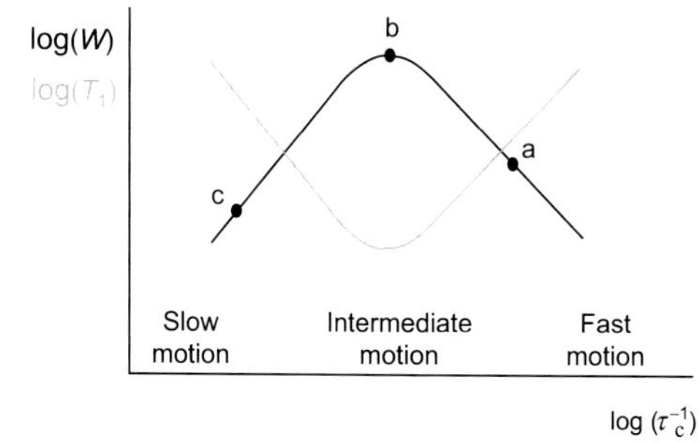

FIGURE 9.7 **Schematic variation in relaxation rates, and hence relaxation times (grey), as a function of molecular tumbling rates.** Points a, b, and c are as in Fig. 9.6.

as in Fig. 9.7, with the fastest relaxation occurring when the correlation rate $1/\tau_c$ matches the frequency of the transition. The point of fastest W_1 relaxation also corresponds to the T_1 minimum and vice versa, so T_1 values themselves are also dependent on the rates of molecular motion.

Using these arguments it is possible to predict when the W_0 or W_2 processes will be dominant. For zero-quantum W_0 processes, the energy differences involved are rather small, being the differences between the I and S frequencies, so $\omega_t = |(\omega_I - \omega_S)|$. These transitions will therefore be strongly favoured for a molecule that tumbles slowly in solution. Double-quantum W_2 transitions correspond to the sum of the I and S frequencies, so $\omega_t = (\omega_I + \omega_S)$ and these will be stimulated by rapidly tumbling molecules. As an example of the transition frequencies involved, consider a homonuclear proton system at an observation frequency of 400 MHz. Single-quantum W_{1I} (and W_{1S}) transition frequencies will correspond to approximately 400 MHz. The W_0 transition frequencies are given by the frequency differences of I and S, that is, their chemical shift differences, which will be in the hertz or kilohertz region. The W_2 transition frequencies are the sums of ω_I and ω_S which will equate to around 800 MHz. Clearly the frequency spread of molecular motions to which the NOE is sensitive is extremely large.

Qualitatively then, one can predict from the previous arguments that *molecules which tumble rapidly in solution* are likely to favour the higher energy W_2 process and hence *exhibit positive NOEs* while *those that tumble slowly* will favour the W_0 process and thus *display negative NOEs*; indeed, this is what is observed in practice.

Quantitative expressions for the relaxation rates in a dipolar-coupled two-spin system have been derived, thus:

$$W_{1I} \propto \gamma_I^2 \gamma_S^2 \left[\frac{3\tau_c}{r^6(1+\varpi_I^2 \tau_c^2)} \right] \tag{9.4}$$

$$W_0 \propto \gamma_I^2 \gamma_S^2 \left[\frac{2\tau_c}{r^6(1+(\varpi_I - \varpi_S)^2 \tau_c^2)} \right] \tag{9.5}$$

$$W_2 \propto \gamma_I^2 \gamma_S^2 \left[\frac{12\tau_c}{r^6(1+(\varpi_I + \varpi_S)^2 \tau_c^2)} \right] \tag{9.6}$$

where the constant of proportionality is the same for each.

Note that, in addition to dependence on correlation times, these expressions contain a term for the internuclear separation r between spins S and I. Here, at last, one starts to see the origins of the famed 'r^{-6}' distance dependence, widely and sometimes dangerously associated with NOE interpretations [2]. This distance term is manifested in the degree of dipolar coupling between the two spins. An important point to note at this stage is that this distance dependence actually lies in the relaxation *rates* and, as we shall see, this can have enormous implications for the way in which NOE data are interpreted. The inverse sixth relationship also means the NOE falls away very rapidly with distance, so in practice significant NOEs will only develop between protons that are within c. 0.5 nm of each other (naturally, this will be influenced by how sensitive and stable the spectrometer is and hence how small an enhancement one is able to 'see'). Note also the dependence of Eqs. 9.4–9.6 upon the square of the magnetogyric ratios of the two spins, so very different rates may occur in heteronuclear systems, depending on the participating spins.

When a molecule tumbles so rapidly in solution such that $\omega\tau_c \ll 1$, all terms in these expressions containing ω become negligible and the rates simplify to:

$$W_{1I} \propto \gamma_I^2 \gamma_S^2 \frac{3\tau_c}{r^6} \tag{9.7}$$

$$W_0 \propto \gamma_I^2 \gamma_S^2 \frac{2\tau_c}{r^6} \tag{9.8}$$

$$W_2 \propto \gamma_I^2 \gamma_S^2 \frac{12\tau_c}{r^6} \tag{9.9}$$

This condition is referred to as the *extreme narrowing limit* since all broadening effects attributable to dipolar interactions are fully averaged to zero under these conditions. This regime typically applies only to small molecules in low-viscosity solvents, and the point at which this condition breaks down depends on the correlation time of the molecule as well as the field strength of the spectrometer (through ω).

9.3.1.3 NOEs and Molecular Motion

Having taken the trouble to see how the relaxation rates in a two-spin system depend upon molecular motion, we are now in a position to predict the behaviour of the NOE itself as a function of this motion and of internuclear separation. Taking the rate constants Eqs. 9.4–9.6 and substituting these into that for the NOE (Eq. 9.2) produces the curve presented in Fig. 9.8 for the theoretical variation of the *homonuclear* NOE as a function of molecular tumbling rates as defined by $\omega_0\tau_c$ (where ω_0 is the spectrometer observation frequency, approximately equal to ω_I and ω_S). Note this is for a two-spin system which relaxes solely by the dipole–dipole mechanism, and as such represents the theoretically maximum possible NOE. The curve has three distinct regions to it, which we shall loosely refer to as the fast-, intermediate- and slow-motion regimes. For those molecules that tumble rapidly in solution (short τ_c, those in the extreme narrowing limit) the NOE has a maximum possible value of +0.5 or 50%. Smaller organic molecules in low-viscosity solvents typically fall within this fast-motion regime which is traditionally the home ground of steady-state NOE measurements. At the other extreme, molecules that tumble very slowly in solution experience negative NOEs, as shown above. The maximum enhancement in this motional regime is obtained when W_2 and W_{1I} are both zero, which from Eq. 9.2 can be seen to be –1 or –100%. For NOE measurements in this region, it becomes essential to use transient experiments since those based on steady-state measurements become uninformative, as explained later. This is the region inhabited by (biological) macromolecules, and it is studies of these systems that have traditionally made widespread use of transient NOE measurements, principally through the 2D NOESY experiment. Between these two extremes is the intermediate region in which the NOE changes sign and even becomes zero when $W_2 = W_0$. Within this region the magnitude and sign of the NOE is highly sensitive to the rate of molecular motions and can be rather weak, possibly too weak to be observed, clearly a major hindrance to structural studies. The

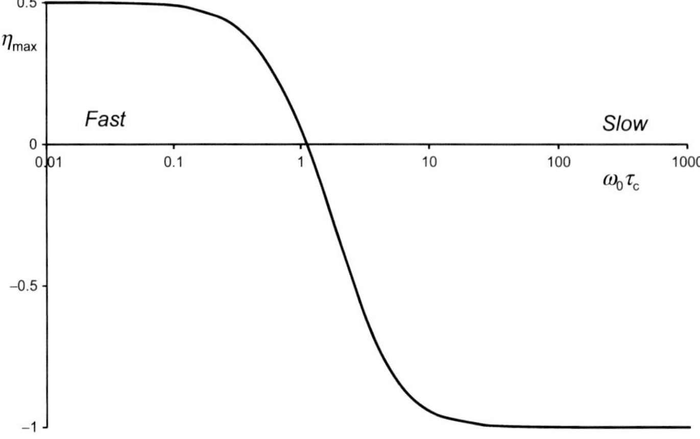

FIGURE 9.8 **Variation in the maximum theoretical homonuclear steady-state NOE in a two-spin system as a function of molecular tumbling rates (defined by the dimensionless product $\omega_0\tau_c$).** The region of fast motion is the extreme narrowing limit and that of slow motion is the spin diffusion limit.

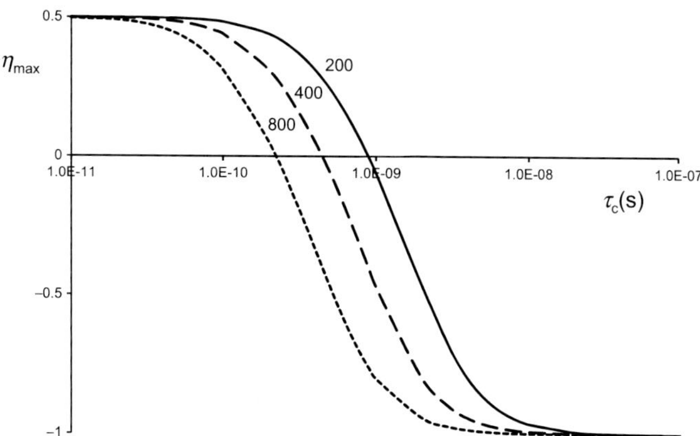

FIGURE 9.9 **Variation in the maximum theoretical homonuclear steady-state NOE in a two-spin system.** This is shown as a function of molecular tumbling rates shown for three spectrometer observation frequencies (MHz).

point at which this region is entered will be dependent on a number of factors: the size and shape of the molecule, solution conditions (viscosity, temperature, possibly pH, etc.) and spectrometer field strength. As a rule of thumb, molecules with a mass of 1000–2000 Da are likely to fall within this intermediate regime. Increasing interest in larger synthetic molecules in many areas, for example supramolecular chemistry, is likely to mean more molecules routinely handled by the research chemist will fall into this potentially troublesome region. The actual zero-crossover point occurs when

$$\omega\tau_c = \sqrt{(5/4)} = 1.12 \tag{9.10}$$

or, in other words, when the molecular tumbling rate ($1/\tau_c$) approximately matches the spectrometer observation frequency. This is therefore field dependent, as illustrated in Fig. 9.9, which now shows the variation of the NOE as a function of τ_c itself for three different field strengths. The use of a higher field strength increases the likelihood of a relatively 'small' molecule falling within the intermediate regime or perhaps a mid-sized molecule passing from this into the slow-motion regime. In some cases, therefore, the use of a higher field instrument, often regarded a panacea for all chemists' woes, may even prove detrimental to the NOE experiment as the 'zero NOE' condition is approached.

The field dependence of the intermediate regime suggests one solution to the problem of 'zero' NOEs: trying a different field strength. This is often an impossible or impractical answer, so an alternative approach is to alter the solution conditions and hence the rate of molecular tumbling, one of the simplest approaches being to vary the sample temperature. A technically different experimental approach to the problem is to measure NOEs in the *rotating frame* instead. This is described further in Sections 9.5 and 9.7, suffice it to say here that these NOEs remain positive for all molecular tumbling rates one is likely to encounter so avoids the 'zero-crossover' problem altogether.

9.3.1.4 NOEs and Internuclear Separation

It is now possible to consider how the separation between two spins influences steady-state NOE enhancement. Assuming the molecule exists in the extreme narrowing regime, then substituting the simplified rate (Eqs. 9.7–9.9) into Eq. 9.2 we obtain:

$$\eta_I\{S\} = \left[\frac{\left(\dfrac{12\tau_c}{r^{-6}}\right) - \left(\dfrac{2\tau_c}{r^{-6}}\right)}{\left(\dfrac{2\tau_c}{r^{-6}}\right) + 2\left(\dfrac{3\tau_c}{r^{-6}}\right) + \left(\dfrac{12\tau_c}{r^{-6}}\right)}\right] = \left[\frac{12-2}{2+6+12}\right] = \frac{1}{2} \tag{9.11}$$

As in Fig. 9.8 for the maximum NOE above, the enhancement is predicted to be 50%. However, it is also predicted to be *independent* of the internuclear distance. Thus, at least for the hypothetical isolated two-spin system considered here, the magnitude of the steady-state enhancement provides no distance information whatsoever. The important point here is that it is too bold a statement to say that differing NOE enhancements within a molecule scale directly with r^{-6}. In realistic chemical systems various 'other' factors must also be taken into account before any distance dependence is reintroduced; this is further pursued in the following sections. Although the *magnitude* of the steady-state enhancement is predicted to be

TABLE 9.2 Theoretical Maximum Steady-State Heteronuclear NOE Enhancements in the Presence of Proton Saturation

X	^{6}Li	^{7}Li	^{13}C	^{15}N	^{19}F	^{29}Si	^{31}P	^{57}Fe	^{103}Rh	^{109}Ag	^{119}Sn	^{183}W	^{195}Pt	^{207}Pb
$\eta_X\{^1H\}$ (%)	339	129	199	−494	53	−252	124	1548	−1589	−1075	−141	1202	233	239

These numbers assume relaxation exclusively via dipole–dipole interactions, although for the metals, in particular, chemical shift anisotropy may also be a significant mechanism. The lithium isotopes are somewhat anomalous in that they are quadrupolar yet can still demonstrate NOEs. ^{6}Li in particular has the smallest quadrupole of all such nuclei, so the dipole mechanism still makes a significant contribution to relaxation.

independent of distance in this system, the *rate* at which this is reached is not because of the dependence of relaxation rates on distance (as expressed in Eqs. 9.4–9.6) meaning NOEs between closer spins develop more rapidly; this is the basis of the transient NOE measurements described in Section 9.4. This also implies that longer range NOEs will only have significant intensities when long NOE build-up periods are employed, a point of considerable practical importance.

9.3.1.5 Heteronuclear NOEs

The equivalent of Eq. 9.11 for a heteronuclear pair experiencing extreme narrowing is the more general expression:

$$\eta_I\{S\}=\frac{\gamma_S}{2\gamma_I} \tag{9.12}$$

For the common situation of carbon-13 observation in the presence of proton saturation (broadband decoupling), $\gamma_H/\gamma_C \approx 4$, and NOE enhancements can be as much as 200%, equating to a threefold intensity increase. Since the relaxation of carbon nuclei is largely dominated by proton dipolar interactions, this maximum is almost met in practice. This is clearly a valuable route to sensitivity enhancement and, at least for the case of ^{13}C, compares favourably for routine acquisitions with the factor of 4 attainable with the ^{1}H to ^{13}C polarisation transfer sequences of Section 4.4. The maximum NOE enhancements for a variety of nuclei in the presence of proton saturation, denoted X{^{1}H}, are summarised in Table 9.2.

In heteronuclear systems the observed NOE also depends on the signs of the magnetogyric ratios of cross-relaxing spins, so NOEs from protons to nuclei with negative γs will display negative NOEs *even if the molecule is within the extreme narrowing regime*. The most common examples of this are for ^{15}N and ^{29}Si for which a *reduction* of signal intensity occurs on proton saturation, so much so that the observed resonance can itself become negative. If less than the full (negative) NOE is generated, the resonance may disappear altogether as the NOE cancels the natural signal, and because of this it is usual to record the spectra of negative γ species in the absence of the NOE, either by use of the inverse-gated decoupling scheme (Section 4.2.3), by polarisation transfer methods (Section 4.4) or, less commonly, by the addition of a paramagnetic relaxation reagent to quench the NOE. In some instances the heteronuclear NOE may be used more specifically for structural assignment also; see Section 9.9.

Beyond the extreme narrowing condition, the absolute magnitudes of heteronuclear X{H} NOEs decrease and, in the case of ^{13}C, ^{15}N and ^{29}Si, closely approach zero, although, with the exception of ^{19}F, do not change sign. For very large molecules there is then little sensitivity gain, or loss, arising from the heteronuclear NOE.

9.3.2 NOEs in a Multi-Spin System

The previous section considered the NOE for the hypothetical case of a two-spin system in which the spins relax exclusively via mutual dipole–dipole relaxation. In progressing to consider more realistic multi-spin systems two key issues will be addressed: how the presence of other spins affects the magnitudes of steady-state NOEs and how these reintroduce distance dependence to the NOE. These considerations lead to the conclusion that steady-state NOE measurements must be used in a comparative way to provide structural data, and that they do not generally provide estimates of internuclear distances *per se*.

9.3.2.1 Additional Relaxation Pathways

The NOE arises as a result of dipolar cross-relaxation between two nuclei, and hence only the dipole–dipole relaxation mechanism is able to generate the NOE. All other competing mechanisms (with the subtle exception of scalar relaxation in a strongly coupled system, which is rarely seen; see Section 9.6.1.1) serve to dilute the overall influence of the W_2 and W_0 pathways by stimulating the W_1 relaxation pathway only, and hence reduce the magnitude of the NOE. These various

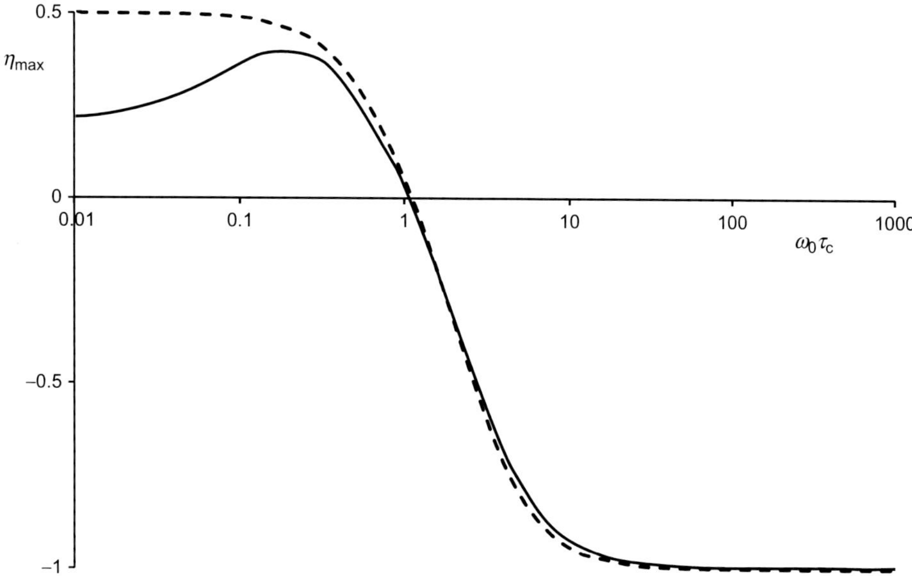

FIGURE 9.10 **Schematic illustration of the maximum homonuclear steady-state NOE in the presence (solid line) and absence (dotted line) of external relaxation sources that compete with cross-relaxation.**

other contributions to spin relaxation may be grouped together and represented by the term ρ_I*, the external relaxation rate for spin I. Adding this to Eq. 9.2 gives:

$$\eta_I\{S\} = \frac{\gamma_S}{\gamma_I}\left[\frac{\sigma_{IS}}{\rho_{IS} + \rho_I*}\right] \tag{9.13}$$

This illustrates the diluting effect of ρ_I* on the magnitude of the NOE, which is sometimes referred to as the 'leakage' term. The effect of this turns out to be of somewhat greater significance for molecules in the extreme narrowing limit than those in the negative NOE regime. For small, rapidly tumbling molecules the absolute magnitude of σ_{IS} is small when compared with that of slowly tumbling molecules, so the relative contribution to Eq. 9.13 of ρ_I* has greater significance. The effect of this is schematically illustrated in Fig. 9.10, which shows that leakage becomes more relevant as molecular tumbling rates increase, meaning very small NOEs may be observed for small molecules. In contrast, leakage effects are less problematic for molecules in the negative NOE regime.

To maximise the size of the NOE for molecules in the extreme narrowing condition, it is necessary to minimise ρ_I*. The most significant contribution to this in routinely prepared solutions is from the paramagnetic oxygen dissolved in solvents. The unpaired spin-½ electron has a magnetic moment that is over 600 times that of the proton, so is able to provide an intense magnetic interaction capable of causing efficient relaxation. Degassing of solutions prior to NOE studies may become necessary when seeking longer range interactions in small molecules, but is otherwise unnecessary for the majority of routine studies (see Section 9.10). Similarly, other paramagnetic impurities, for example some metals, will quench the NOE and must be avoided. Samples to which relaxation agents have been deliberately added to promote relaxation are therefore not suitable candidates for NOE studies. *Intermolecular* dipolar interactions, arising from transient interactions with solute or solvent molecules, are another potential interference. Solvent molecules are deuterated in most cases, so the relaxation arising from these nuclei is rather inefficient when compared with protons (due to the γ^2 dependence of longitudinal dipolar relaxation rates) and interactions with other (protonated) solute molecules are generally only likely to be a problem when very concentrated solutions are used, so these are thus best avoided for NOE studies. Other non-dipole relaxation mechanisms, such as those discussed in Section 2.5, also act to bypass the NOE. Of most significance is the quadrupolar relaxation associated with nuclei with spin >½. This is usually the dominant mechanism for such nuclei, so NOEs on quadrupolar nuclei are very rarely observed (the exception being ^{6}Li, as mentioned in Table 9.2).

9.3.2.2 Internuclear Separations (Again)

Perhaps the most obvious 'other' contribution to the relaxation of spin I in a multispin system arises from neighbouring nuclei other than spin S within the same molecule. Dipolar interactions with these spins act to bring about longitudinal

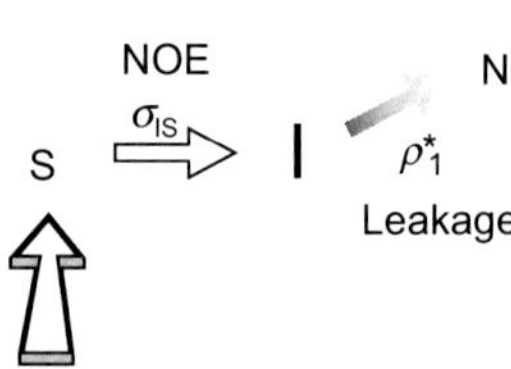

FIGURE 9.11 **Three-spin system in which the I-spin neighbour *N* acts as an external relaxation source for spin-I through their mutual dipolar interaction.**

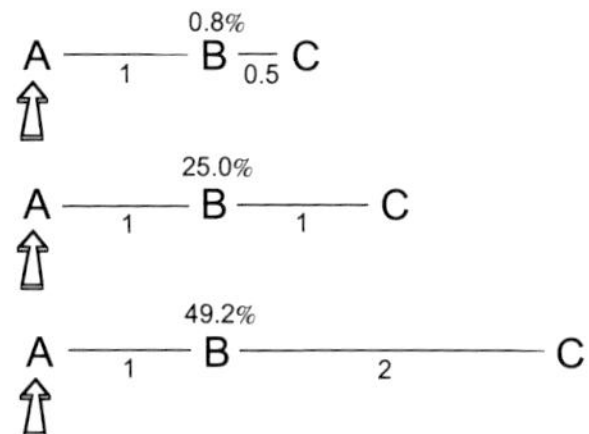

FIGURE 9.12 **Calculated A ⇒ B steady-state NOE enhancements for an isolated three-spin system with the relative internuclear separations as shown.** The *arrow* indicates the saturated spin. Any direct A–C interaction is ignored.

relaxation of spin I *independent* of cross-relaxation between I and S, so diluting the I–S NOE. To illustrate the influence of neighbouring spins, consider a hypothetical homonuclear three-spin system I, S and N (N = neighbour) in which relaxation arises solely from dipolar interactions ignoring, for convenience, all the other possible contributions to ρ_I* described earlier (Fig. 9.11). Assuming, for simplicity, an NOE exists only between I and S, the steady-state NOE may be written:

$$\eta_I\{S\} = \eta_{max}\left[\frac{r_{IS}^{-6}}{r_{IS}^{-6} + r_{IN}^{-6}}\right] \tag{9.14}$$

where η_{max} represents the maximum NOE possible in a homonuclear two-spin system, as previously. Note that if spin N were not present, the NOE would simply be η_{max} and show no distance dependence, exactly as predicted above for a two-spin system. The effect of introducing an additional spin is to reintroduce distance dependence, yet despite this *the magnitude of the steady-state NOE does not scale simply as* r_{IS}^{-6}. In fact, it can be seen that the magnitude of the NOE will be dictated by a balance between the r_{IS} and r_{IN} distances. This is true for all steady-state NOE measurements; the result will always represent a balance between I–S internuclear separation and all other I–N separations (and in a realistic chemical system there may well be a large number of other neighbouring nuclei). To put it another way, the steady-state NOE arises from a competition between I–S cross-relaxation and all other relaxation sources of spin I. Eq. 9.14 also shows that a reduction in the I–S internuclear distance now does indeed contribute to an increased NOE between I and S since the total contribution to the numerator will be relatively more than to the denominator. The arrival of a neighbouring spin has therefore reintroduced the idea that a smaller internuclear distance can be correlated to some degree with larger NOE enhancements.

This statement must still be treated with some caution however, as illustrated in Fig. 9.12, in which $\eta_B\{A\}$ is considered for a three-spin system A, B and C where the B–C distance is varied (and ignoring any direct AC interaction). When C is distant from B it has little influence on its relaxation, allowing AB cross-relaxation to dominate, producing close to the maximum NOE. When both A and C are equidistant from B they play an equal role in relaxing B and the NOE is thus half the maximum possible value. As C becomes very much closer to B than is A, it now dominates B-spin relaxation, and A–B cross-relaxation loses out in the competition resulting in a small NOE. Thus, in general, *despite two spins being 'close' to one another*, in that they share a strong dipolar coupling, *they still may not exhibit a large NOE* if the enhanced spin has other near neighbours.

A further important feature emerges if we consider the results from saturating B and studying the effect on A, $\eta_A\{B\}$ (Fig. 9.13) and compare these with those of Fig. 9.12. With C distant from B, the relaxation of A is essentially completely dominated by B, so it experiences almost the maximum enhancement. As C approaches B the enhancement on A is reduced only a little since its relaxation is still dominated by the much closer spin B. Thus, in general, *steady-state NOEs between spins are not symmetrical* that is $\eta_A\{B\} \neq \eta_B\{A\}$, because the neighbours surrounding A are unlikely to match those surrounding B in both number and proximity.

9.3.2.3 Indirect Effects and Spin Diffusion

The examples discussed above have been restricted to discussing the direct NOE effects between A and B while, for convenience, ignoring effects that may be observed at C itself. If spin C is considered, it is apparent from Fig. 9.14 that it experiences a net decrease in signal intensity when A is saturated. This arises from a relay mechanism in which population changes on B, brought about by the initial A–B NOE, subsequently alters the population of spin C when this also shares a dipolar coupling and hence also cross-relaxes with B. The *negative* NOE seen at C is a result of the *increase* in B-spin population differences generated by the A–B NOE. In the extreme narrowing limit, saturating a resonance (ie *decreasing* the population difference across the corresponding transition) causes a *positive* NOE, so by the same logic an *increase* in population differences for B (the A–B NOE) will in turn generate a *negative* NOE on its dipolar-coupled neighbours. This

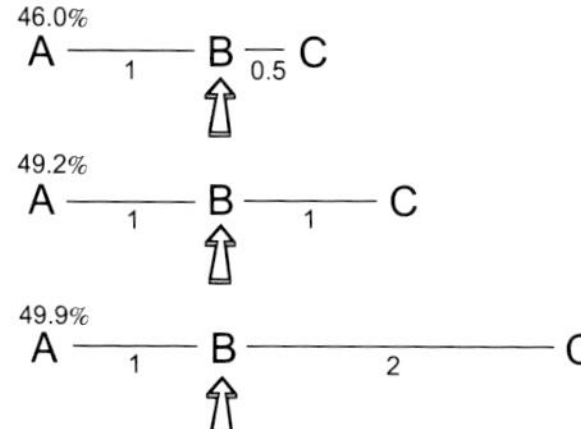

FIGURE 9.13 **Calculated B ⇒ A steady-state NOE enhancements for an isolated three-spin system with the relative internuclear separations as shown.** The *arrow* indicates the saturated spin. Any direct A–C interaction is ignored.

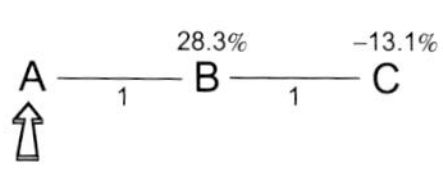

FIGURE 9.14 **Three-spin steady-state effect.** The negative enhancement at C arises from an indirect effect via spin B when spin A is saturated. Altered C-spin populations also contribute to the enhanced NOE at B (see Fig. 9.12).

indirect effect, often referred to as the *three-spin effect* should not be confused with direct negative NOEs observed for slowly tumbling molecules.

The magnitudes of negative three-spin enhancements are usually rather small since they rely on the buildup of a sizeable NOE on a neighbouring spin suitable for relaying. Similarly, they also tend to be slow to develop and show a characteristic lag period before appearing (Fig. 9.15), so tend to be observed only when longer NOE build-up periods are employed. Note that the illustrations used above produce unrealistically high values for all enhancements since they assume pure dipolar relaxation throughout. Experimental NOEs are more often somewhat less than 20% and three-spin effects are rarely more than a few percent at most. They appear most commonly when B and C are a diastereotopic geminal pair, since the B–C distance is then constrained to be rather short so facilitating the relay. They are also favoured when the three spins have an approximately linear relationship and their appearance can be diagnostically useful when observed. The reason for this geometry being particularly favourable arises from the balance between the negative indirect three-spin effect on C and the positive direct effect between A and C (Fig. 9.16). At small A–B–C angles the direct effect dominates that which is relayed via B, while when linear the opposite applies. In between these extremes the two effects cancel, so that *even though* A *and* C *may be close in space*, in that they share a strong dipolar coupling, *an NOE between them may be rather small*. This is another important point to be aware of (particularly when longer build-up times are employed) since in any realistic system there may well be a number of competing indirect pathways present. The curves of Fig. 9.16 demonstrate that the zero-crossover point varies with internuclear separations and that the slopes of the curves are large at this point, so although the NOE is unlikely to cancel to precisely zero, it may well be less than expected from simple geometrical considerations. Although in principle, four-spin relay effects can be predicted to give rise to positive NOEs, they are very rarely observed in practice simply because they are so very weak.

The influences of the various factors described earlier are further illustrated by reference to the four-spin system of Fig. 9.17, which lays testament to the need to consider all neighbouring spin interactions to correctly interpret steady-state NOE data. Clearly the NOE enhancements between B and C differ dramatically despite these effects arising over identical internuclear separations. Furthermore, the enhancements B ⇒ A, B ⇒ C and C ⇒ D are all rather similar despite there being a factor of 4 difference in distance between the largest and smallest separations. The essence of correctly applying steady-state NOE data is to collect a number of measurements and check for self-consistency within the proposed

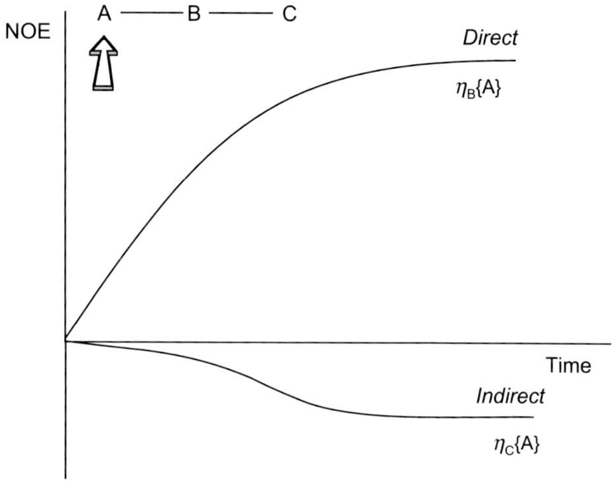

FIGURE 9.15 **Schematic illustration of the buildup of direct and indirect NOEs in a rapidly tumbling three-spin system.**

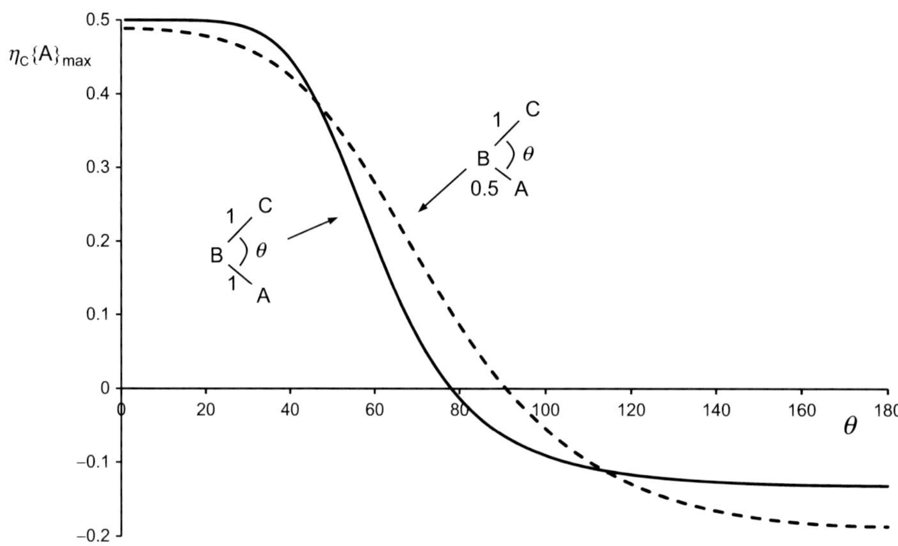

FIGURE 9.16 **Calculated steady-state NOEs for spin C on saturation of spin A, as a function of the A–B–C angle in a rapidly tumbling isolated three-spin system.** The resulting NOE is a balance between direct and indirect A–C effects and the two curves illustrate the dependence on relative internuclear separations.

geometrical arrangement, rather than relying on a single NOE enhancement to provide an answer. If only a single irradiation were performed from C in Fig. 9.17 one might be forced into the erroneous conclusion that D were its closest partner rather than B. A more cautious approach, also measuring the NOEs from both B and D, should lead one to question this conclusion. Again, such spectacular differences are unlikely to occur in reality largely because of the effects of the other numerous ρ_I* relaxation sources we have chosen to ignore, but it should be clear that careful consideration is essential. Generally speaking, only in rather specific cases will a single NOE measurement lead to a definitive structural answer whereas a comparative study of many NOEs within the molecule is more likely to provide a conclusive and correct result.

When a molecule tumbles in solution so slowly that it exhibits negative NOEs, the consequences of indirect effects are more dramatic and far more problematic, so much so that steady-state NOEs become largely useless. The problem is essentially twofold: first, these indirect effects have the same sign as direct effects so cannot be readily distinguished and, second, they grow rapidly and may attain very high intensities. The first of these is a consequence of the fact that a negative NOE (a population decrease) arises from saturation (a forced population decrease), so likewise the indirect effect on the third spin will also be negative, and so on. In the extreme case of extended saturation times, the NOE initially generated between two spins can spread throughout the whole molecule until all nuclei experience the same NOE enhancement (Fig. 9.18). This spreading of information throughout the molecule is often referred to as *spin diffusion* for fairly obvious reasons and may be likened to heat diffusing through a conductive solid. The limit of slow tumbling is also referred to as the *spin diffusion limit*. Because of this, steady-state NOEs in the negative NOE regime fail to provide reliable distance or proximity information. Instead, it becomes necessary to consider the *rate* at which NOEs grow between spins to glean distance information, dictating the use of kinetic measurements in the form of transient NOE experiments (Section 9.4).

9.3.2.4 Saturation Transfer

Complications arising from the transfer of saturation from one resonance to another by means of chemical exchange (Fig. 9.19) also differ in the two extreme motional regimes. Exchange peaks always display the same sign behaviour as the originally saturated resonance, so have opposite sign to positive NOEs but the same as negative NOEs. The identification

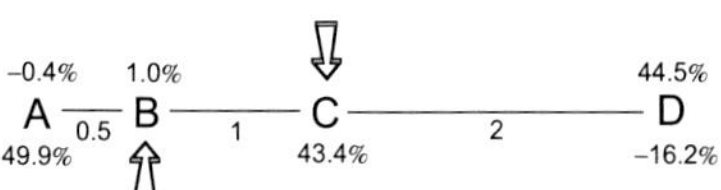

FIGURE 9.17 **Calculated steady-state NOE enhancements in an isolated four-spin system with relative internuclear separations as shown.** These enhancements were derived from the equations presented in Table 3.1 of [1].

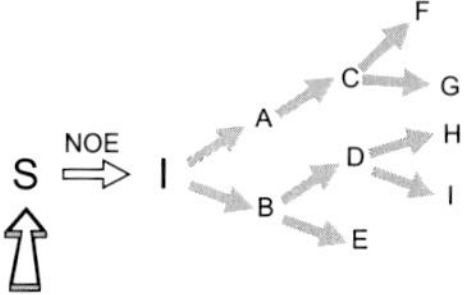

FIGURE 9.18 **Schematic illustration of the spin diffusion process.** In this the original S–I NOE is efficiently relayed onto neighbouring nuclei and propagated throughout the molecule.

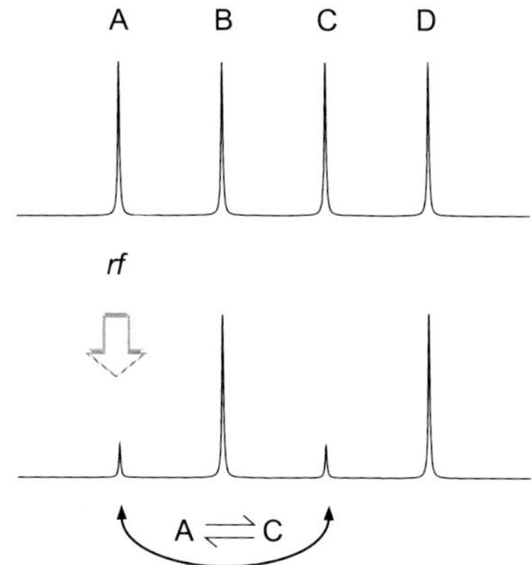

FIGURE 9.19 **Schematic illustration of the saturation transfer process.** Saturation of resonance A will lead to the simultaneous saturation of resonance C if nuclei A and C experience mutual chemical exchange during the saturation period. This behavior is similar in appearance to the negative NOE.

of responses arising from chemical exchange rather than the NOE is a further problem associated with the negative NOE regime; saturation transfer is considered further in the practical sections that follow and has been described in Section 2.6.3.

9.3.3 Summary

The previous sections have covered the most important principles at the heart of the NOE. Despite having avoided much of the underlying mathematics, instead relying on pictorial models where possible, these sections are necessarily a little more technical than the other chapters in the book. The goal has been to present a reasonably thorough introduction to the NOE that is still compatible with the level of this text, rather than simply presenting the reader with statements of fact with little justification or support. This section reviews the key points that have been presented earlier, allowing the reader a more immediate reminder as to the most significant features of the NOE:

- The NOE is the change in intensity of the resonance of a nuclear spin, I ('interesting'), when the population differences across the transitions of a near neighbour, S ('source'), are perturbed from their equilibrium values, usually by *saturation* or by *population inversion*. It arises as the perturbed spin system alters its spin populations in an attempt to regain the equilibrium condition.
- *Steady-state* NOEs are those measured after a period of continuous S-spin saturation during which a new 'steady-state' equilibrium condition has developed for the I-spin populations. The NOE enhancement is usually denoted $\eta_I\{S\}$ and may be quantified as a percentage.
- Steady-state NOEs can provide information on relative internuclear distances only—not on absolute measurements of internuclear separation.
- The NOE only arises between nuclei that share a mutual *dipolar coupling* (a direct, magnetic through-space interaction) and thus relax each other via the dipole–dipole relaxation mechanism. Only this mechanism contributes to the nuclear spin population changes that produce the NOE, which is intimately related to longitudinal spin relaxation. The dependence on internuclear separation that makes the NOE so useful has its origins in the strength of dipolar coupling between two spins, this being inversely proportional to their separation r_{IS} (as r_{IS}^{-3}).
- Longitudinal spin relaxation requires a stimulus in the form of a magnetic field fluctuating at a frequency equal to the frequency (energy) of the transition. In the case of dipolar relaxation, this arises from the time-dependent field a spin experiences from its dipolar-coupled neighbour as the molecule rotates or tumbles in solution. The NOE in turn is dependent upon the rates of molecular tumbling.
- The magnitude and sign of the NOE results from competition between various relaxation pathways. If the W_2 ($\alpha\alpha \leftrightarrow \beta\beta$) pathway dominates, positive NOEs are observed whereas a dominant W_0 ($\alpha\beta \leftrightarrow \beta\alpha$) pathway leads to the observation of negative NOEs. The W_2 and W_0 relaxation pathways (which involve the mutual flipping of the two spins) are collectively referred to as *cross-relaxation pathways*, and it is only these that contribute to the generation of the NOE. All other contributions to relaxation, in the form of W_1 processes ($\alpha\alpha \leftrightarrow \beta\alpha$ and $\alpha\beta \leftrightarrow \beta\beta$, involving the flip of only a single spin), serve to reduce the overall magnitude of the effect, and are thus referred to as *leakage* contributions.
- Rapid molecular tumbling, corresponding to a short correlation time τ_c, favours the higher energy W_2 process, so small molecules in low-viscosity solvents display positive homonuclear NOEs (the *extreme narrowing limit*). In contrast, slow molecular tumbling (long correlation times) favours the lower energy W_0 process, meaning large molecules or smaller molecules in high-viscosity solvents display negative homonuclear NOEs (the *spin diffusion limit*).

- The maximum possible positive NOE is $\gamma_S/2\gamma_I$; in other words, 50% in a *homonuclear* system. Enhancements in *heteronuclear* systems can be far larger, for example 199 % from ^{1}H to ^{13}C, and can serve as a useful source of sensitivity enhancement in the observation of low-γ I spin. They also become negative if one of the γs is negative, for example −494% from ^{1}H to ^{15}N, and may in some cases lead to severe signal reduction or even disappearance.
- The maximum possible negative NOE in a homonuclear system is –100%.
- Between the extreme narrowing and spin diffusion limits lies the difficult region in which NOEs can become zero (when $\omega_0\tau_c \approx 1$) or at least rather weak, often demanding a change in experimental conditions or the use of rotating frame NOE measurements.
- In an isolated homonuclear two-spin system in the extreme narrowing limit relaxing exclusively via the dipole–dipole mechanism, the steady-state NOE is predicted to be +50% and *independent of internuclear separation* r_{IS}. However, the initial *rate* at which the NOE grows is proportional to r_{IS}^{-6}.
- In a more realistic multi-spin system, neighbouring nuclei N that are close to I can also contribute to its W_1 relaxation pathway. The magnitude of the NOE then becomes dependent on the I–S internuclear separation (inversely as r_{IS}^{-6}), but also has a dependence on the distance(s) between I and its near neighbour(s) (inversely as r_{IN}^{-6}), amongst other factors.
- The direct consequence of this is that to correctly interpret steady-state NOE data, it becomes essential to consider not only the I–S internuclear separation, but also the proximity of all other nuclei (relaxation sources) surrounding I.
- This also means that steady-state NOEs are rarely symmetrical. That is, the NOE observed at spin A on saturating spin B, $\eta_A\{B\}$, is unlikely to equal the reverse measurement from saturating A and observing B, $\eta_B\{A\}$. This is a consequence of the fact that the neighbours surrounding A are unlikely to match those surrounding B in number and in distance.
- The relaxation of a spin with a very near neighbour will be dominated by this neighbour and, as a consequence, NOEs on this spin from a more distant source spin will tend to be small. Conversely, nuclei that have no nearby neighbours will experience relaxation only from distant neighbours and, as a result, NOEs from these neighbours will be large despite the relatively large internuclear distance involved. Thus, it is to be expected that NOEs over similar distances on a methylene and on a methine proton will generally be somewhat smaller for the methylene proton since this will always have at least one near neighbour, its geminal partner.
- Longer range NOEs build up only slowly, due to the r^{-6} rate dependence, so require long saturation periods before becoming appreciable.
- NOE enhancements may also be *relayed* on to neighbouring spins. For example, an NOE from A to B may be further passed onto a spin C that also cross-relaxes with B, such events being referred to as *indirect effects*. The properties of indirect enhancements differ markedly in the positive and negative NOE regimes.
- When direct (A–B) NOEs are positive, indirect effects at C are weak and negative but are favoured when the three spins have an approximately linear relationship (so may provide useful geometrical information), develop only slowly with a characteristic lag time and are thus also favoured by long presaturation periods. These are referred to as *three-spin effects*. Although further relays are theoretically possible, they are generally too weak to be observed.
- It is possible that positive direct effects and negative indirect effects can cancel or act to reduce the magnitude of the NOE. Thus, despite two spins being close, NOEs between them may be rather small or even negligible (particularly when longer saturation periods are employed).
- When direct (A–B) NOEs are negative, indirect effects are also negative so cannot be distinguished, they spread rapidly and have high intensities. This process is referred to as *spin diffusion* and is fatal for the steady-state NOE since it causes a loss of specificity and hence provides no information on molecular geometry. In this regime, it usually becomes necessary to use kinetic (transient) methods based on the measurement of NOE growth rates.
- Taking into account all the subtleties associated with the steady-state NOE presented above, it should be clear that it is unwise to place too much significance on the absolute magnitudes of steady-state NOE enhancements. In reality, differences of a few percent mean little when taken on their own, and it is generally necessary to consider a collection of enhancements when undertaking structural or conformational analysis to be certain of an unambiguous conclusion. A qualitative interpretation of many measurements is the most appropriate approach to interpreting most NOE data.

9.3.4 Applications

To illustrate the issues described above and how NOE measurements can be used to provide unique data in structural analysis, some specific examples are now presented. As stated in the Introduction (Section 9.1), the NOE is most often employed during the later stages of structural investigations when the gross structure of the molecule has, at least to a large extent, been defined. For rather small molecules, this may be possible from knowledge of the chemistry used and from 1D spectra, while larger or structurally complex molecules may demand the application of various correlation techniques before NOE experiments are considered. In either case, confidence in the accuracy of one's proton assignments is of paramount importance since errors in these are as likely (if not more likely) to lead to erroneous stereochemical conclusions being drawn than they are to being unmasked by the NOE studies themselves. The examples presented here make use of steady-state NOEs and provide some indication of the magnitudes of NOE enhancements typically encountered in routine laboratory studies, which often fall short of the theoretical numbers discussed in the previous sections. Additional examples making use of other NOE techniques are presented in later sections.

9.3.4.1 E versus Z Geometry

The differentiation of *E* and *Z* alkene isomers is often possible by direct measurement of vicinal proton–proton couplings across the unsaturation, whereby *cis* and *trans* couplings are usually sufficiently different to allow a distinction to be made (typically J_{cis} 7–11 Hz, J_{trans} 12–18 Hz). When only a single alkene proton exists this method can no longer be used and the NOE then offers an alternative approach provided a protonated group exists across the double bond. One such example is in the differentiation of the *E* and *Z* silanols **9.1a** and **9.1b** [3]. The *Z* isomer was readily identified from the NOEs between the alkene proton and the CH_2Si group and further confirmation was provided by the observation of contrasting NOEs for the other (*E*) isomer. Asymmetry in the magnitudes of NOE enhancements is also clearly apparent in this example and should come as no surprise following the preceding discussions.

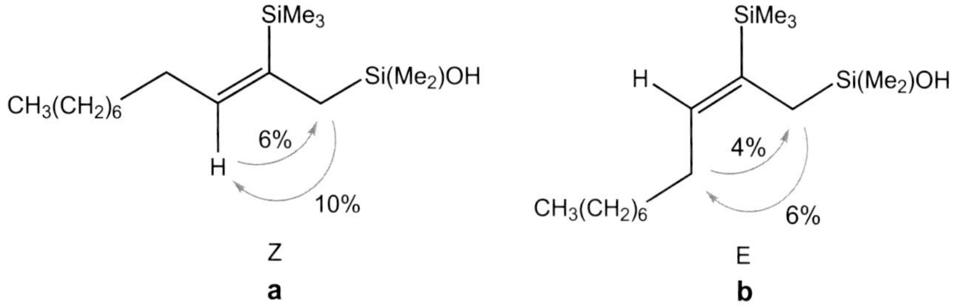

SiMe3
Si(Me2)OH
CH3(CH2)6
H
6%
10%
Z
a

SiMe3
H
Si(Me2)OH
4%
CH3(CH2)6
6%
E
b

9.1

9.3.4.2 Aromatic Substitution Position

Determining the position of substitution within a molecule can also be problematic when no direct proton–proton couplings exist to link the new moiety, which is often the case when the substitution is made on a heteroatom. Particularly when dealing with aromatic systems, the NOE can often provide unambiguous solutions. These systems are particularly favourable because they are restricted to being planar and, hence, it is usually safe to assume that nearest neighbours will be those on adjacent positions in the ring. The differentiation of π- and τ-substituted histidines [4] **9.2a** and **9.2b** provides a simple illustration of this. In the π-substituted systems irradiation of the H5 proton enhances only one of the CH_2 groups bound to the imidazole ring, whereas in the τ-substituted isomers both were enhanced. Prior to the use of the NOE, differentiation was possible only through chemical degradation or through empirical rules based on differences in the small (<1.5 Hz) H2–H5 coupling constant, which could not always be resolved. An alternative approach to consider nowadays in such cases would be to establish connectivity by identifying long-range proton–carbon correlations across the heteroatom via the heteronuclear multiple bond correlation (HMBC) experiment (Section 7.4).

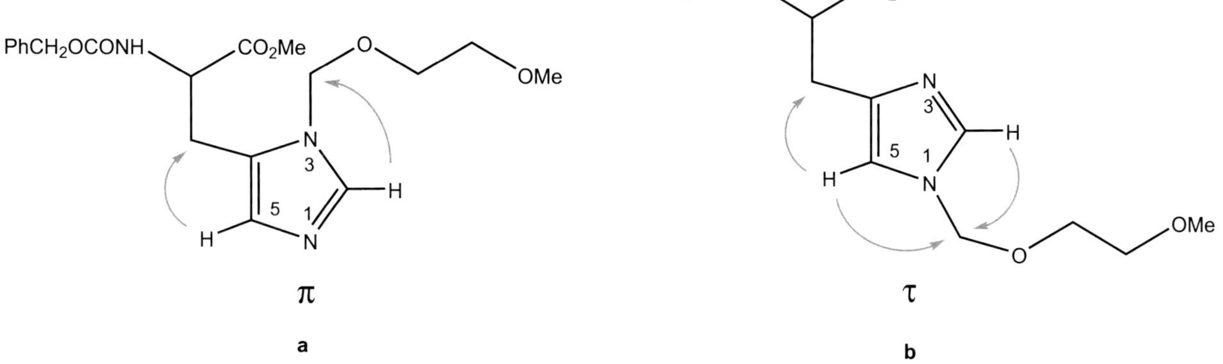

PhCH2OCONH
CO2Me
O
OMe
N
3
H
5
1
H
N
π
a

PhCH2OCONH
CO2Me
N
3
H
5
1
H
N
OMe
O
τ
b

9.2

9.3.4.3 Substituent Configuration

Another widely encountered question, and arguably the most common problem the NOE is used to address in synthetic chemistry, is the relative configuration of substituents on ring systems. This is most often applied to five- and six-membered rings, not only because of their ubiquity but also because larger rings tend to have far greater flexibility, making it more difficult to draw unambiguous conclusions. For six-membered rings in particular, direct analysis of proton-coupling constants within the ring can in itself be informative because of the generally distinct differences in *axial–axial versus axial–equitorial/equitorial–equitorial* coupling constants in chair conformations (eg *ax–ax* ≈ 10–12 Hz, *ax–eq/eq–eq* ≈ 2–5 Hz; see Section 6.1.5 and Fig. 6.23). However, in determination of the orientation of the methyl group in **9.3** these would not have been informative, because of the similarity between *ax–eq* and *eq–eq* coupling constants. A collection of NOE enhancements was able to identify the methyl as occupying the equitorial position, in particular the 1,3-diaxial methine proton NOEs.

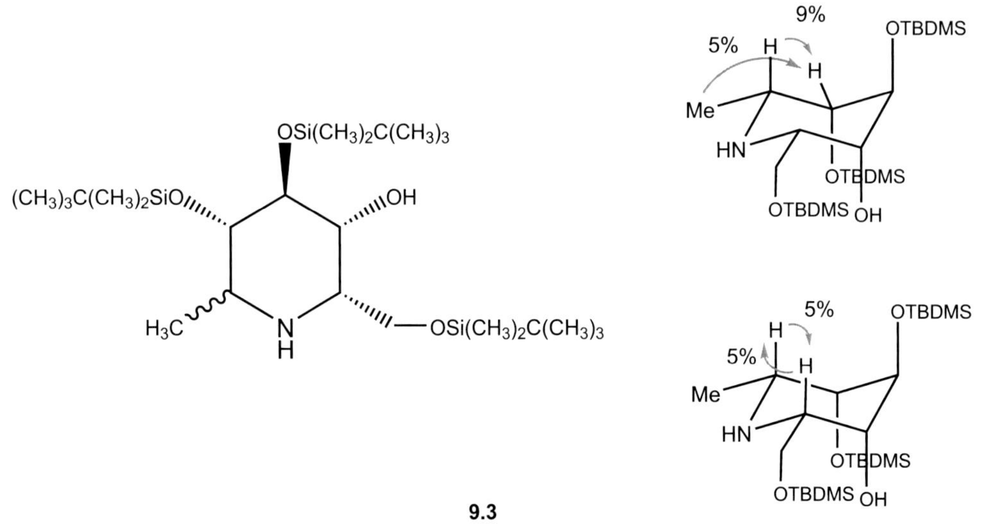

9.3

Vicinal couplings in five-membered rings generally offer greater ambiguity in defining relative configurations and here careful NOE measurements are also required since *cis* and *trans* NOEs between adjacent protons are often of similar magnitude. In these cases it is wise to collect as many NOE enhancements as is possible for the sample and to ensure self-consistency over all of these within the proposed stereochemistry. These points are exemplified by the configurational assignment of the synthesised epoxyprolines [5] **9.4a** and **9.4b**. The negligible J coupling between H2 and H3 in **9.4a** tentatively suggested these protons to be *trans*, and the slightly greater 2.5-Hz coupling in **9.4b** suggested these share a *cis* relationship. Conclusions drawn from interpretation of the NOE data were consistent with this proposal, but were founded on the comparison of NOEs of rather similar magnitude between the H5 and H4 protons in each. The availability of both isomers allowed comparison of their H4–H5 NOE patterns, based on identification of H5β from its NOE with H2, and the differences observed between these provided further support for the assignments. Confirmation of these assignments was provided through additional synthetic structural correlations.

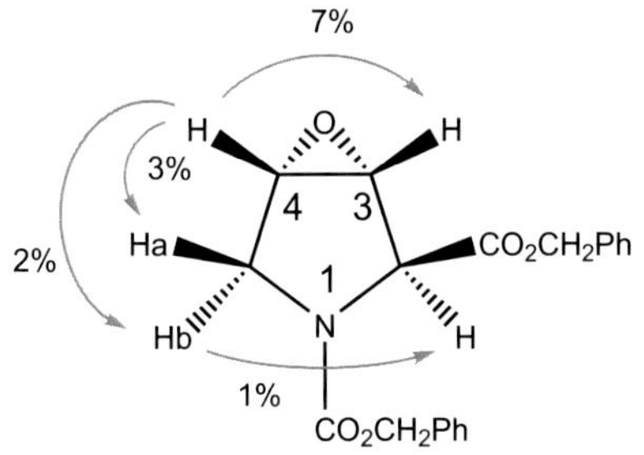

9.4a 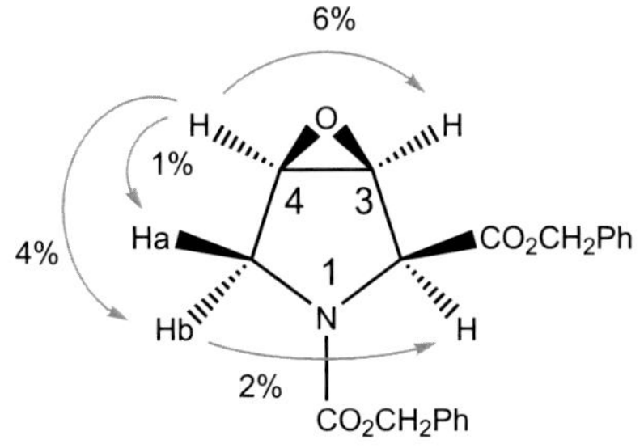 **9.4b**

In the bicyclic lactam **9.5** it was possible to determine the relative configuration at three stereocentres based on the known stereochemistry at only one (C2) through the sequential interpretation of the NOE data. The lack of NOEs between protons on adjacent carbons C5, C6 and C7 suggests the neighbouring protons share *trans* relationships, although the absence of an NOE alone cannot always be considered definitive evidence, as has been stressed in previous discussions. More importantly, this stereochemistry is confirmed by the observation of the additional H6–H4, H6–H2 and H7–H5 NOEs. Again note the asymmetry in the NOEs between H2–H4, H4–H6 and H4*–H5, with the NOE *on to* the methylene proton being always *less* than the NOE from this on to the methine proton, owing to the close proximity of geminal neighbours.

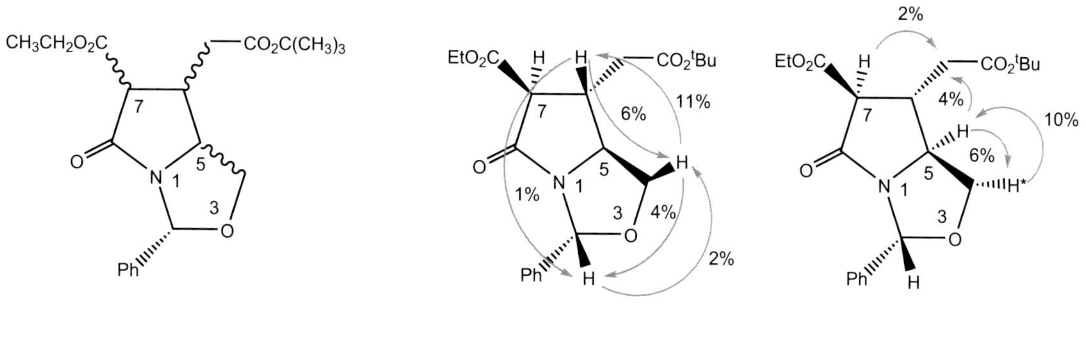

9.5

9.3.4.4 Resonance Assignment

The prerequisite for most NOE studies of configuration or conformation is assignment of the proton spectrum of the molecule, meaning each resonance can be associated with a unique proton within the gross structure. Such assignments are typically derived from 1D spectra and the various correlation methods described in previous chapters, although in some cases the NOE can itself be used for resonance assignment. It is most likely to be of use for assigning resonances of isolated groups that share no proton scalar couplings, although it is by no means limited to this. An example is the assignment of the two methyl resonances in the hepatoprotective agent andrographolide **9.6**, which are distant from one another in the molecule and show quite characteristic NOE patterns. The use of long-range heteronuclear correlation experiments should also be considered when addressing such problems, and indeed may be needed in determination of the gross structure in the first place.

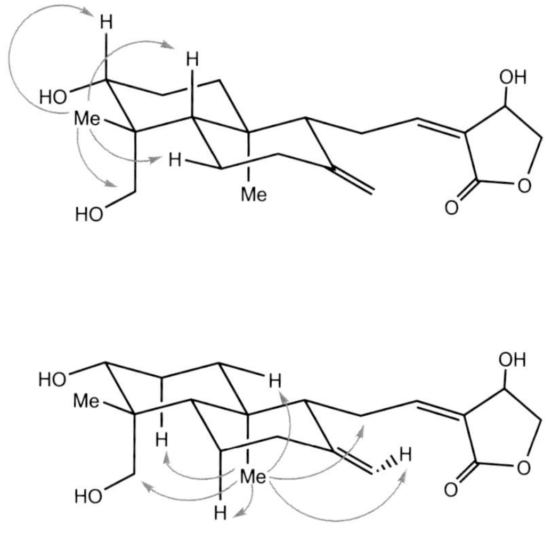

9.6

9.3.4.5 Endo versus Exo Adducts

The need to distinguish between *exo* and *endo* adducts in fused ring systems is another commonly encountered challenge that can often be addressed by the NOE, and determination of the stereochemistry at C5 of the lactam **9.5** above can be viewed as one such example. In such cases the NOE patterns observed between the proton(s) at the junction and those on the adjacent rings can often provide an unambiguous stereochemical assignment. This was the case for the identification of **9.7**, as the *endo* cycloadduct

[6]. The material was synthesised as part of a model study toward the biomimetic synthesis of the manzamine alkaloids, a family of β-carboline alkaloids derived from marine sponges which possess potent antileukaemic and cytotoxic properties. In this, it was important to confirm whether the product stereochemistry was consistent with the proposed biosynthetic hypothesis being investigated, as was shown to be the case. Here the bridgehead proton was in fact too heavily overlapped with an adjacent resonance to be selectively saturated, although NOEs on to this were quite distinct. When bridgehead protons cannot be used at all, or if non-existent, the assignment must then rely on the observation of NOEs between ring protons on either side of the junction.

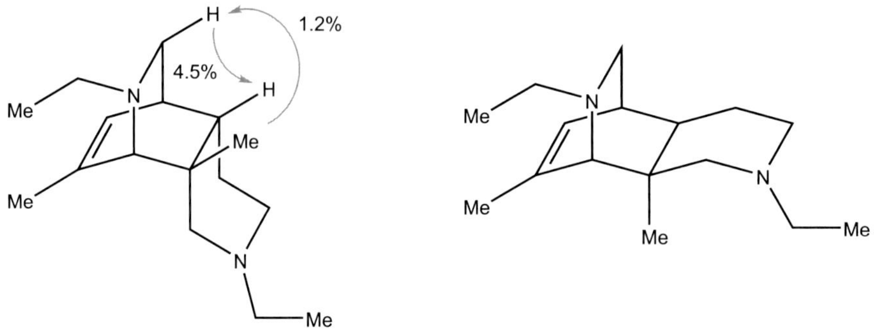

endo exo

9.7

9.3.4.6 Conformational Preference

The definition of a favoured conformation in small, flexible molecules by use of the NOE represents a far greater challenge, notably because of the rapid interchange between many possible conformations and because the NOE itself will represent only a weighted average of internuclear separations present within the conformers. The detailed investigations required to extract meaningful data in these cases is therefore rarely undertaken. However, when restricted conformational processes lead to one conformation being strongly favoured, sufficient NOE data may be available which allow this to be defined. Obvious examples might be the differentiation of chair and boat conformers of cyclohexanes or of slowly interconverting rotamers. Specific conformations may also be favoured in the presence of steric hindrances or strong hydrogen-bonding interactions, for example. An example of a favoured conformation defined by NOE measurements is structure **9.8** (also produced as part of the biomimetic synthesis of the manzamine alkaloids mentioned above [7]) in which the two heterocyclic rings are approximately orthogonal to one another. The limited and specific NOEs observed at the interface of the two rings are not consistent with free rotation about the single bond linking the rings and the structure appears essentially locked, presumably by the presence of the C8 methyl group. This orthogonal relationship is further supported by the almost negligible coupling between the H7 and H3 protons, indicating that the dihedral angle between them is close to 90 degrees. The combined use of NOEs and coupling constants often represents the optimum approach to questions of conformation.

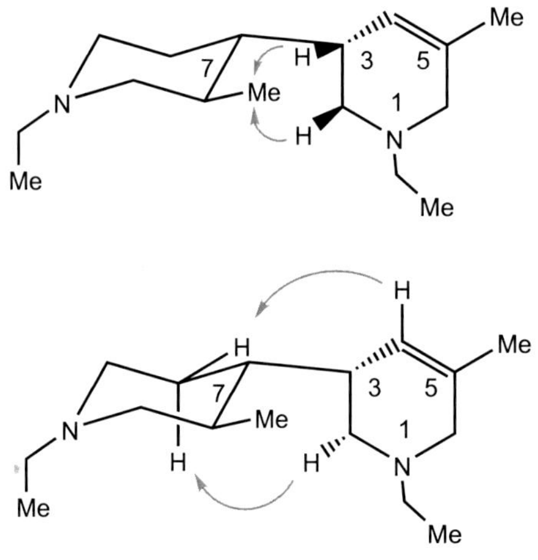

9.8

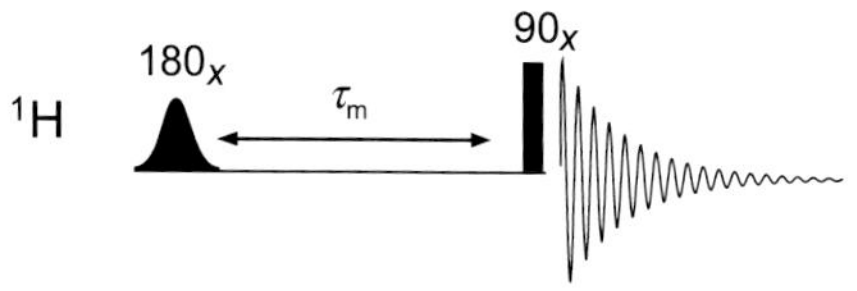

FIGURE 9.20 **General scheme for observing transient NOEs.** Following inversion of a target (source) resonance, the NOE develops during the mixing time τ_m, after which the system is sampled.

9.4 TRANSIENT NOEs

It has been repeatedly stressed in the preceding sections that the steady-state NOEs measured between two nuclei cannot readily be translated into internuclear separations because they result from a balance between the influences of all neighbouring spins. At best, they provide information on *relative* internuclear distances only. It has also been noted however that the *rate* at which the NOE grows towards this steady state can be directly related to these distances under appropriate conditions. It has also been shown that for molecules which exhibit negative enhancements, steady-state measurements may fail to provide any reliable information of spatial proximity and here one is forced to consider the kinetics of the NOE. A logical approach to such measurements would be to follow that taken for the measurement of steady-state effects. Saturation of the target resonance for periods that are far less than those needed to reach the steady state would allow *some* NOE to appear, which is then sampled. Repeating the experiment with progressively incremented saturation periods allows the buildup to be mapped. Owing to the use of shortened saturation periods, the enhancements observed with this method are termed *truncated driven NOEs* (TOEs).

Although once popular, this experimental approach finds little use nowadays and as such shall be considered no further. The more common approach to obtaining kinetic data is to instantaneously perturb a spin system not by saturation but by inverting the target resonance(s) (ie inverting the population differences across the corresponding transitions) and then allowing the NOE to develop in the absence of further external interference. The new populations are then sampled with a 90 degree pulse as usual (Fig. 9.20). In this case the NOE is seen initially to build for some time but ultimately fades away as spin relaxation restores the equilibrium condition; these enhancements are thus termed *transient NOEs*.

The measurement of transient NOEs gained widespread popularity, initially in the biochemical community, in the form of the 2D NOESY experiment, but now also finds widespread use in the analysis of smaller molecules. The 1D transient NOE experiment, generally referred to as 1D NOESY, is also used widely in chemical studies as the gradient-selected sequence is capable of providing quite spectacular, high-quality NOE spectra. Transient experiments, whether 1D or 2D, are more commonly used qualitatively as 'single-shot' techniques, providing an overview of enhancements within a molecule, but may also be employed to map the growth of the NOE in quantitative estimates of internuclear separation. In either case, it is necessary to understand something of the kinetics of the NOE to correctly execute and interpret these experiments.

9.4.1 Nuclear Overhauser Effect Kinetics

Following the (assumed) instantaneous inversion of the S-spin resonance, the *initial* growth rate of the NOE at I depends linearly on the cross-relaxation rate between these two spins *even in multi-spin systems*, such that:

$$\frac{d\mathrm{I}_z}{dt} = 2\sigma_{\mathrm{IS}}\,\mathrm{S}_z^0 \tag{9.15}$$

(the factor of 2 here arises simply from the use of inversion of populations in this case rather than saturation, as considered previously). Using Eqs. 9.4–9.6 and the approximation $\omega_I = \omega_S$ for a homonuclear system, the cross-relaxation rate (W_2–W_0) is given by:

$$\sigma_{\mathrm{IS}} \propto \gamma^4 \left\{ \frac{6}{1+4\omega_0^2\tau_c^2} - 1 \right\} \frac{\tau_c}{r_{\mathrm{IS}}^6} \tag{9.16}$$

Unlike steady-state enhancements, transient enhancements are influenced by only a single internuclear separation as r_{IS}^{-6}, while the so-called *initial rate approximation* is valid. In this the two cross-relaxing spins initially behave as if they were an isolated spin pair and the growth of the NOE has a linear dependence on mixing time. As longer mixing periods are used, the relaxation of spin I begins to compete with cross-relaxation between I and S, so the buildup curve deviates from linearity and the NOE eventually decays to zero (Fig. 9.21). Thus, for the initial rate approximation to be valid, mixing

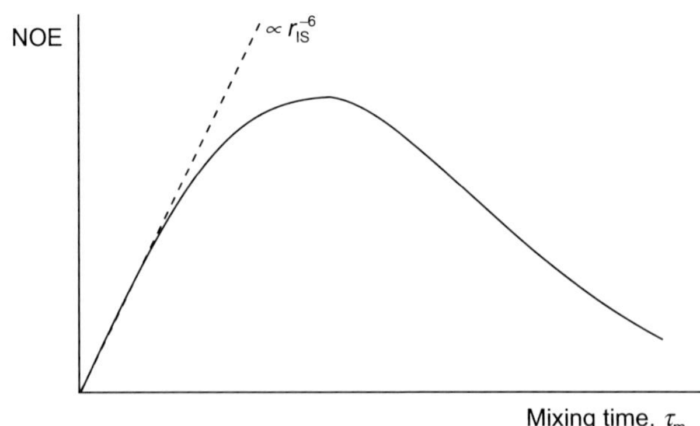

FIGURE 9.21 **Schematic illustration of the development of the NOE between spins I and S as a function of mixing time.**

times significantly shorter than the T_1 relaxation time of spin I must be used. Only under these conditions is meaningful distance measurement possible.

If, on the other hand, the goal is to qualitatively identify through-space correlations, as is more often the case in routine work, mixing periods comparable with T_1 provide maximum enhancements. Since transient NOEs develop in the absence of an external rf field they tend to be weaker than steady-state effects. The maximum theoretical homonuclear enhancement is also reduced to 38% (from 50%) for positive NOEs, so careful choice of timing is crucial to the success of transient experiments. In all other respects, the dependence of transient NOEs on correlation times matches that of steady-state effects, with the maximum homonuclear enhancements for the negative NOE regime again being –100% owing to the domination of efficient cross-relaxation. When dealing with molecules which exist within this regime, it is also necessary to be aware of and if possible avoid potential complications arising from the dreaded spin diffusion (indeed, this is the primary reason transient methods are used) which again demands the use of conservatively short mixing periods.

9.4.2 Measuring Internuclear Separations

Assuming the initial rate approximation to be valid (NOE growth linear), the magnitude of an enhancement between two spins A and B after a period τ will be proportional to the cross-relaxation rate, which in turn depends on $r_{AB}{}^{-6}$:

$$\eta_A\{B\} = k\sigma_{AB}\tau = k' r_{AB}^{-6}\tau \tag{9.17}$$

The constants of proportionality here contain the molecular correlation time τ_c in addition to a number of known physical constants, and {B} is now taken to signify inversion rather than saturation of B. In principle, if τ_c were known, this would directly provide a measure of r_{AB}. While it is possible to determine this (such as from relaxation time measurements) this is rarely done in practice, and it is more common to use a known internal distance as a reference and avoid the need for such laborious measurements. If the NOE between reference nuclei X and Y of internuclear separation r_{XY} is also measured then:

$$\frac{\eta_A\{B\}}{\eta_X\{Y\}} = \frac{r_{AB}^{-6}}{r_{XY}^{-6}} \tag{9.18}$$

Direct comparison of the two NOE intensities thus provides the unknown internuclear distance. This simple relationship has been extensively used to provide measurements of internuclear separations, particularly in biological macromolecules. From a single experiment, distances can be estimated *assuming the initial rate approximation is valid for all interactions*. This relies on all internuclear vectors in question possessing the same correlation time, which may not be the case where internal motion is present. Consideration of these matters lies beyond the scope of this work; further details may be found in [1].

The significance of a single 'internuclear distance' must also be considered carefully. In reality, internuclear separations vary over time with conformational averaging, so the concept of a single distance is simply a convenient model of events. Furthermore, in cases of conformational exchange, the calculated distance tends to be heavily weighted towards shorter separations since the NOE is very much more intense for these because of the r^{-6} factor. Consider the case of rapid averaging on the NMR timescale between two equally populated conformers such that only a single resonance is observed for each chemically distinct site. In one conformer the separation between two spins is 0.25 nm while in the other it is 0.60 nm (Fig. 9.22). The NOE under conditions of fast conformational exchange averages as $\langle r^{-6}\rangle^{-1/6}$ (where $\langle . \rangle$ indicates the

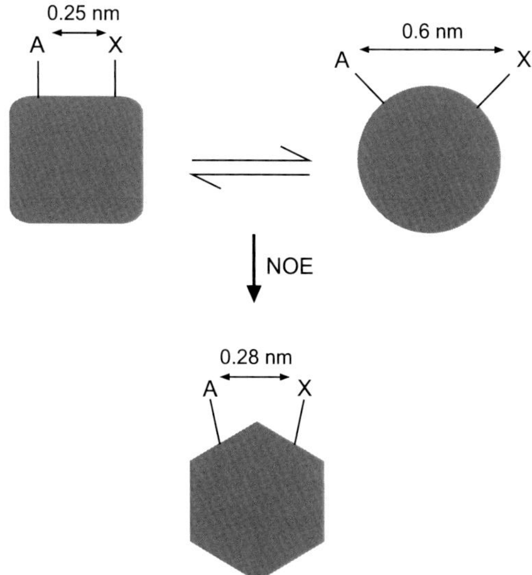

FIGURE 9.22 **Influence of rapid conformational exchange on estimated internuclear separations based on NOE measurements.**

mean value), so that the *apparent* separation, as would be calculated from observed NOE intensities, is 0.28 nm rather than the mean 0.42 nm one may anticipate.

In general then, NOE measurements will tend to underestimate rather than overestimate distances in such cases, which can be problematic for structure calculations. These problems are most severe in the case of small molecules, where extensive conformational averaging is to be expected, and detailed structure calculations based on *quantitative* distance measurements for flexible small molecules remain rather rare. However, this approach has proved enormously successful in the structure calculations of biological macromolecules where, to avoid problems of averaging and errors in intensity measures, semi-quantitative measurements are usually employed (Section 9.6).

9.5 ROTATING FRAME NOEs

The greatest problem associated with the methods described so far is clearly the 'zero-crossing' region around $\omega_0\tau_c \approx 1$ where the conventional (laboratory frame) NOE observed via steady-state or transient techniques becomes vanishingly small. This typically occurs for mid-sized molecules with masses of around 1000–2000 Da, depending on solution conditions and spectrometer frequency. With the increasing interest in larger molecules in many areas of organic chemistry research coupled with the wider availability of higher field instruments, this is likely to be a region visited ever more frequently by our molecules of interest. Other than altering solution conditions in an attempt to escape from this, the measurement of NOEs in the *rotating frame* provides an alternative solution, albeit an experimentally challenging one. These effects are the rotating frame analogues of the transient NOEs described earlier, and many of the discussions in Section 9.4 relating to their application are relevant here also. In this case, however, the cross-relaxation rate between homonuclear spins is given by:

$$\sigma_{\mathrm{IS}} \propto \gamma^4 \left\{ \frac{3}{1+\omega_0^2\tau_c^2} + 2 \right\} \frac{\tau_c}{r_{\mathrm{IS}}^6} \tag{9.19}$$

Unlike the corresponding equation for transient NOEs (Eq. 9.16), this expression remains positive for all values of τ_c, and the undeniable benefit of rotating frame NOEs (ROEs) is quite simply that they remain positive for all realistic molecular tumbling rates. For small molecules, the magnitude of the ROE matches that of the transient NOE, while for larger molecules it reaches a maximum for homonuclear spins of 68%, but under no circumstances does it become zero (Fig. 9.23). Similarly, the NOE and ROE growth rates are identical for small molecules but differ for very large ones. For a small molecule which has $\omega_0\tau_c \ll 1$ both Eqs. 9.16 and 9.19 simplify to:

$$\sigma_{\mathrm{IS}} \propto \frac{5\gamma^4\tau_c}{r_{\mathrm{IS}}^6} \tag{9.20}$$

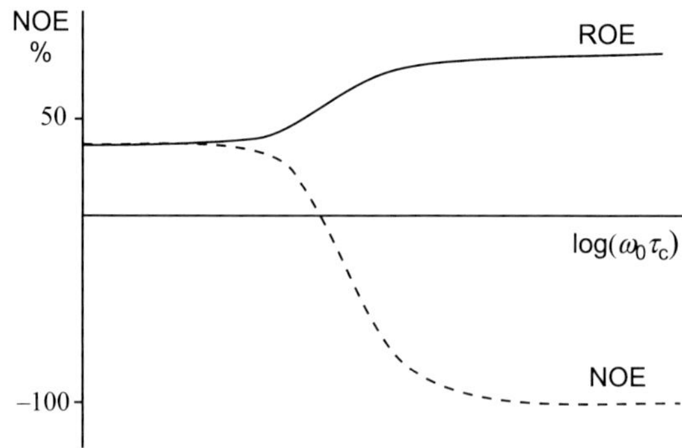

FIGURE 9.23 **Schematic illustration of the dependence of the ROE and transient NOE for an isolated homonuclear two-spin system as a function of molecular tumbling rates.**

In contrast, for very large molecules which have $\omega_0\tau_c \gg 1$, the left-hand expression within the brackets of Eqs. 9.16 and 9.19 becomes negligible, giving:

$$\sigma_{IS}^{NOE} \propto \frac{-\gamma^4\tau_c}{r_{IS}^6} \quad \text{and} \quad \sigma_{IS}^{ROE} \propto \frac{2\gamma^4\tau_c}{r_{IS}^6} \tag{9.21}$$

and hence:

$$\sigma_{IS}^{ROE} = -2\sigma_{IS}^{NOE} \tag{9.22}$$

For very large molecules, the ROE therefore grows twice as fast as the NOE, and has opposite sign [8].

The measurement of ROEs requires a somewhat different experimental approach (Section 9.7). In essence, ROEs develop while magnetisation is held static in the *transverse* plane, rather than along the *longitudinal* axis (hence they are sometimes also referred to as transverse NOEs). To generate the required population disturbance of the source spins, the target resonance is subjected to a *selective* 180 degree pulse prior to the non-selective 90 degree pulse, such that it experiences a net 270 degree flip and is thus inverted relative to all others. Transverse magnetisation is then 'frozen' in the rotating frame by application of a continuous, low-power *spin lock* pulse. This is analogous to the spin-lock described in Section 6.2 for the total correlation spectroscopy (TOCSY) experiment and serves the same purpose, that is to prevent evolution in the rotating frame of chemical shifts. The simplest scheme for a 1D sequence is therefore that of Fig. 9.24. The experiment is more frequently performed as the 2D experiment where it is usually termed rotating frame NOE spectroscopy (ROESY).

The situation during the spin lock may be viewed as the transverse equivalent of events during the transient NOE mixing time (Fig. 9.25). The action of the spin lock is to maintain the opposing disposition of magnetisation vectors which would otherwise be lost through differential chemical shift evolution, and so allows the ROE to develop through cross-relaxation in the transverse plane. Spin relaxation here is characterised by the time constant $T_{1\rho}$ ($\cong T_2$). In utilising the spin lock one has effectively replaced the static B_0 field of the conventional NOE with the far smaller rf B_1 field, and it is this that changes

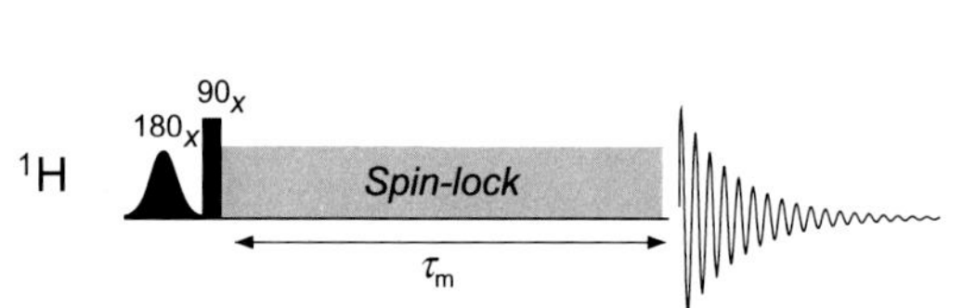

FIGURE 9.24 **General scheme for observing rotating frame NOEs.** The ROE develops during the long spin lock pulse which constitutes the mixing period τ_m.

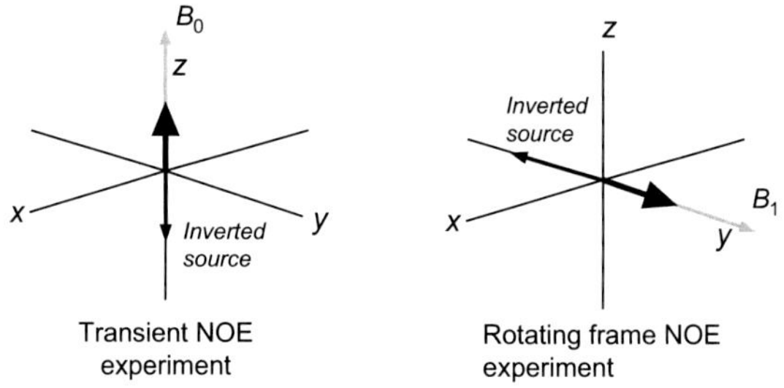

FIGURE 9.25 **The rotating frame NOE experiment can be viewed as the transverse equivalent of the transient NOE experiment.**

the dynamics of the NOE. Whereas γB_0 typically corresponds to frequencies of hundreds of megahertz, γB_1 is typically only a few kilohertz, meaning $\gamma B_1 \ll \gamma B_0$ and hence ω_1(the rotating frame frequencies) $\ll \omega_0$. The consequence of this is that $\omega_1 \tau_c \ll 1$ for all realistic values of τ_c, and *all molecules behave as if they are within the extreme narrowing limit*. Thus, ROEs are positive, any indirect effects have opposite sign to direct effects and tend to be weak and saturation transfer can be distinguished by sign from ROEs, regardless of molecular size and dynamics.

Set against these obvious benefits are a number of experimental problems: principally, TOCSY transfers also occurring during the spin lock and signal attenuation from off-resonance effects. These issues are further addressed in the practical sections that follow, so it is sufficient to note here that even more care is required when acquiring and using ROEs than is needed for NOEs.

PART II PRACTICAL ASPECTS

In this second part of the chapter on the NOE we consider the practicalities of recording and interpreting NOE experiments. The principal techniques used nowadays are based on transient NOE measurements in the form of 1D or 2D NOESY or ROESY experiments, and it is these on which we focus first. The traditional 1D steady-state measurements associated with the NOE difference technique now find less use, generally providing inferior quality data relative to pulsed field gradient (PFG)-selected transient methods. Nevertheless, steady-state methods still have a role when gradient capabilities are not available and may still find use in the measurement of heteronuclear NOEs, so are also considered later in this part.

9.6 MEASURING TRANSIENT NOES: NOESY

Nowadays, two approaches for observing transient NOEs find widespread use: homonuclear 2D NOE spectroscopy (NOESY) and its selective 1D analogue employing PFGs for signal selection. The popularity of the 2D NOESY was driven originally by interest in biological macromolecules, for which NOEs are negative, strong and fast growing, and this area remains the home territory of NOESY. Traditionally, this was less used for smaller molecules principally because the NOE in the extreme narrowing limit is weaker and tends to grow more slowly, and because the steady-state NOE difference experiment provided a viable alternative. However, with improved instrument sensitivity and stability, the NOESY experiment now finds widespread application in routine small-molecule studies. The non-selective nature of the 2D experiment provides the opportunity, at least in principle, to observe all NOEs within a molecule in a single experiment, and the simplicity of setting up NOESY means it is well suited to an automated environment. However, transient enhancements develop solely from the population disturbances brought about by initial resonance inversion so tend to be of low intensity and of fleeting existence, requiring some care in their capture. Furthermore, very small molecules ($M_r < 200$) that tumble very rapidly in solution produce extremely weak NOEs because cross-relaxation tends to be inefficient, and the search for transient enhancements presents still greater challenges in such cases. With larger molecules, more success is to be expected. The popular gradient-selected 1D transient NOE experiments (1D NOESY) described later in these sections have made a significant impact on small-molecule NOE studies and now enjoy widespread use, in particular because of the exceptional quality of spectra they are able to provide.

9.6.1 The 2D NOESY Sequence

The NOESY experiment [9] again follows the principles presented in Section 5.1 for the generation of a 2D data set and has a similar appearance to homonuclear correlation spectroscopy (COSY)-based correlation spectra, although in this case the crosspeaks of the 2D spectrum indicate NOE interactions between the correlated spins. These peaks arise from the *incoherent* transfer of magnetisation between spins during the mixing time via the NOE, so allowing through-space proximities to be mapped directly. The NOESY spectrum of the naturally occurring terpene andrographolide **9.9** in dimethylsulfoxide (DMSO) is shown in Fig. 9.26 and displays a comprehensive map of close contacts within the molecule, some of which are illustrated on the structure. All NOE crosspeaks have opposite phase to the diagonal, indicating these arise from positive NOE enhancements, as anticipated for a molecule of this size ($M_r = 350$ Da) under ambient conditions. The few crosspeaks sharing the same phase as the diagonal in the 3–6 ppm region are attributable to hydroxyl protons and H_2O and arise from chemical exchange of the protons within these groups (see Sections 9.11 and 2.6).

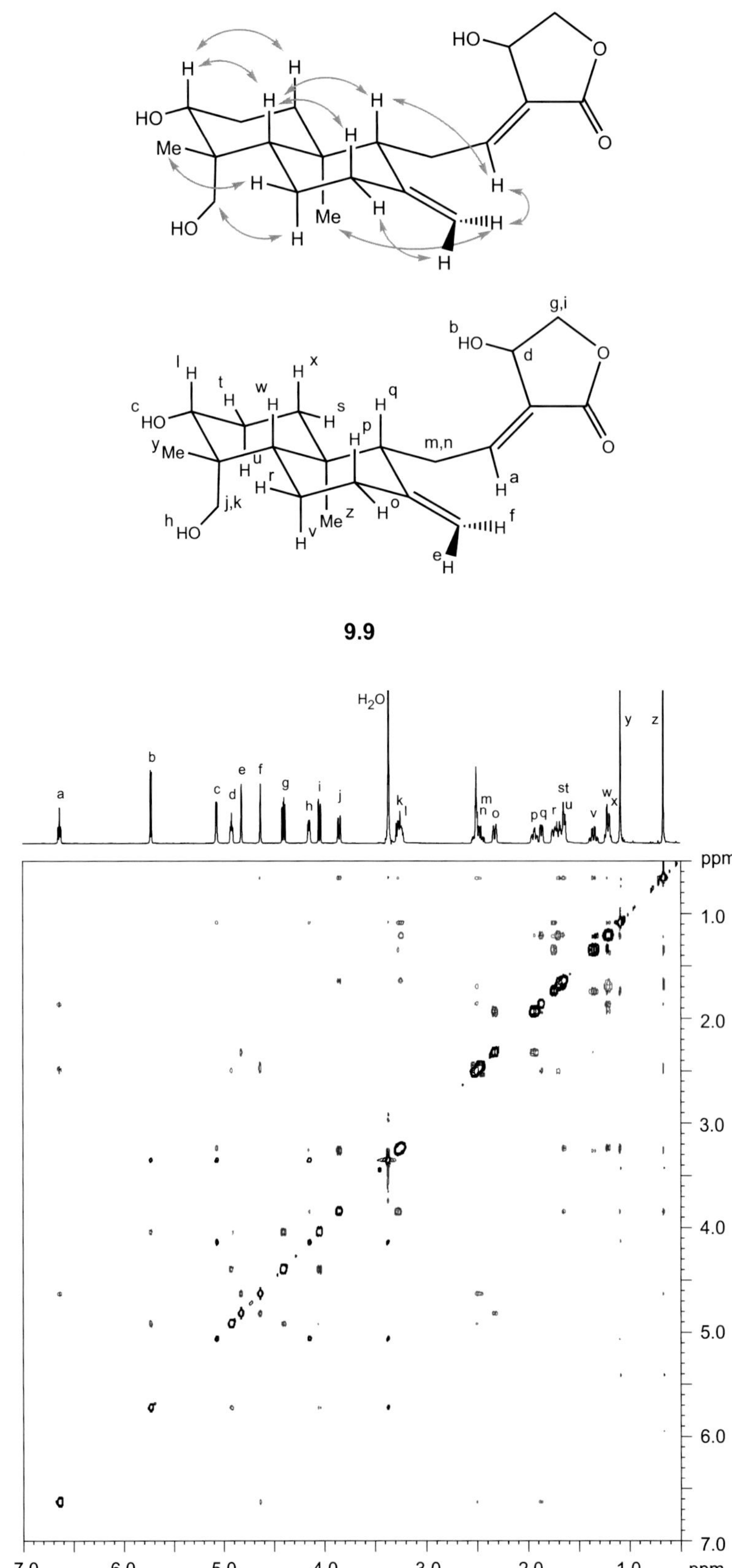

FIGURE 9.26 **The 2D NOESY spectrum of the terpene andrographolide 9.9 in DMSO at 25°C.** The spectrum was recorded with a 600-ms mixing time and a recovery delay of 1.5 s. 2K data points were collected for 512 increments of 16 scans, using time proportional phase incrementation (TPPI) f_1 quadrature detection. Data were processed with a squared cosine bell window in both dimensions with a single zero-fill in f_1.

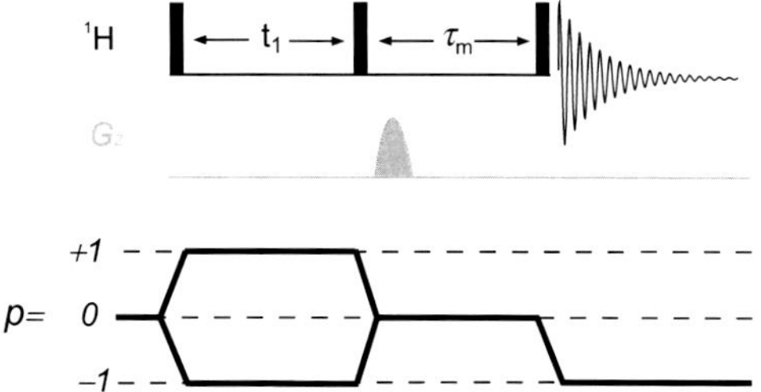

FIGURE 9.27 **The 2D NOESY sequence and the associated coherence transfer pathway.** The optional use of a PFG during the mixing time (grey) is described in the text.

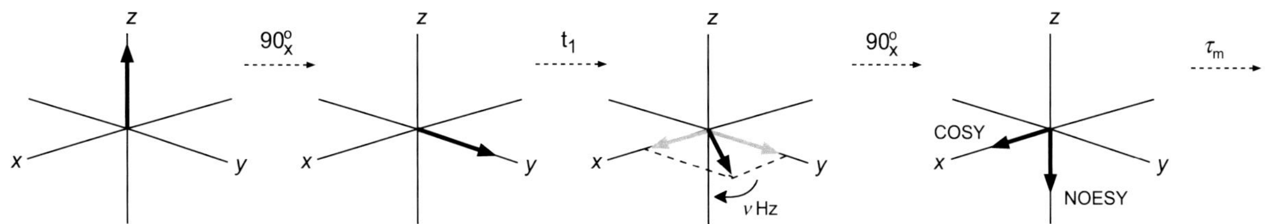

FIGURE 9.28 **Evolution of magnetisation vectors during NOESY.** The inverted magnetisation generates the NOE during the mixing time, while the remaining transverse components are removed by phase cycling or PFGs.

The NOESY sequence (Fig. 9.27) is closely related to COSY, so here we shall consider only the features specific to the NOE experiment. The most significant of these is the mixing period τ_m during which the NOE develops, and to understand the role this plays in a 2D experiment we return to the vector model (Fig. 9.28). Following initial excitation and t_1 evolution, the magnetisation vector exists in the transverse plane. The second 90 degree pulse places one component of this onto the $-z$-axis and has therefore generated the required population inversion that enables transient NOEs to develop during the subsequent mixing period (the components that remain in the transverse plane are those detected in the COSY experiment). After a suitable period τ_m, the new populations are sampled with a 90 degree pulse and the free induction decay (FID) collected; one can readily see a parallel with the 1D 'inversion–mixing–sampling' sequence illustrated in Fig. 9.20. The sequence is repeated to collect sufficient transients for an acceptable signal-to-noise ratio. While, in principle, it should be necessary to leave at least $5T_1$ between transients to allow populations to recover, this would lead to excessively long experiments, particularly in the case of small molecules. In practice, shorter recovery delays of typically 1–3 s for small molecules ($\approx$1–$2T_1$ of the slowest relaxing spins) represents an acceptable compromise [10].

The NOESY sequence, aside from details of the timings used, is identical to the double quantum filtered (DQF)-COSY sequence, so it remains to select the signals arising from NOEs and suppress all others by means of a suitable phase cycle (Table 9.3). Since NOEs are associated only with spin populations, that is z-magnetisation of coherence order zero, they are insensitive to rf phase changes and a four-step cycle is sufficient to cancel signals originating from single- and double-quantum

TABLE 9.3 The Basic NOESY Phase Cycle for the Three Pulses (P_N) and the Receiver (P_R), Including the Suppression of Axial Peaks

Transient	P_1	P_2	P_3	P_R
1	x	x	x	x
2	x	x	y	y
3	x	x	−x	−x
4	x	x	−y	−y
5	−x	x	x	−x
6	−x	x	y	−y
7	−x	x	−x	x
8	−x	x	−y	y

TABLE 9.4 Responses Observed in the NOESY Experiment

Sign of Diagonal	Origin of Crosspeak	Sign of Crosspeak
	Positive NOE	Negative
Positive	Negative NOE	Positive
	Chemical exchange	Positive

By convention the diagonal is phased to be positive and the signs of all other signals are given relative to this.

coherences present during τ_m. The cycle is repeated with additional phase inversion of the preparation pulse and receiver to cancel axial peaks and produces a final eight-step cycle. Since chemical exchange between sites is also associated with longitudinal magnetisation exchange between resonances, this experiment will also detect these processes, in an analogous fashion to saturation transfer described for the 1D experiments above, and for this reason this exact sequence is also termed EXSY when used to study dynamic exchange processes (see Section 9.11). The use of a phase-sensitive presentation is highly recommended for NOESY since this allows discrimination of chemical exchange from positive NOEs on the basis of sign, aids identification of some of the artefacts described later in the sections and allows quantitative measurements [11], in addition to the usual benefits of a higher resolution display. NOESY crosspeaks are then inphase, have pure absorption lineshapes and appear symmetrically about the diagonal [12] (since transient enhancements are symmetrical in nature). Their sign with respect to the diagonal is dictated by the molecule's motional behaviour and hence the sign of the NOE itself (Table 9.4).

Selection of the desired coherences may also be performed via PFGs, so avoiding the need for phase-cycling [13–15]. This may be achieved simply by inserting a single *z*-gradient within the mixing time such that all coherences are dephased and only the desired *z*-magnetisation remains (Fig. 9.27). The time-saving benefits of such an approach are likely to be significant only when large-sample quantities are available and rather few scans per increment provides sufficient sensitivity. When working with aqueous solutions, additional solvent suppression schemes may be added to this basic sequence; these are described in Section 12.5.

9.6.1.1 Chemical Exchange Crosspeaks in NOESY

An additional mechanism which may give rise to correlations in NOESY is the process known as *saturation* or *magnetisation transfer* brought about by chemical exchange processes that are slow–intermediate on the NMR chemical shift timescale (see Section 2.6.3). This may arise from, for example, conformational exchange whereby a resonance belonging to one conformer will correlate with the corresponding resonance in another if exchange of the spins occurs during the mixing period. The principle of saturation transfer is illustrated in Fig. 9.19. This may be advantageous in some circumstances since it may be studied as a means of analysing exchange dynamics, as discussed for the closely related EXSY experiment in Section 9.11. In terms of NOE measurements, it is more often an additional complication to be aware of since any NOEs that develop may also be transferred to other resonances by the exchange process, giving rise to transferred NOE correlations. Exchange processes are often faster than cross-relaxation, and signals arising from exchange are often more intense than NOEs, exhibiting the same sign behaviour as the diagonal peaks. Hence they appear with opposite sign to positive NOE enhancements, allowing them to be distinguished, but with the same sign as negative enhancements. This can be a further hazard if your molecules exhibit behaviour in the negative NOE regime.

A relatively common feature of NOE experiments performed in organic solvents containing traces of water is transfer of saturation between exchangeable protons and the water (Fig. 9.29). Intense exchange crosspeaks are often apparent in such cases between exchangeable protons themselves and with water, which itself can serve as a useful means of identifying acidic proton resonances. Such correlations may be observed in the NOESY spectrum in Fig. 9.26.

Crosspeaks that have an appearance similar to that of exchange peaks but which have no origin in chemical exchange or the NOE can in some circumstances be observed between J-coupled protons when the magnitude of the coupling constant is slowly modulated on the millisecond timescale. The origin of the crosspeaks is a process known as *scalar relaxation of the first kind* and has been reported for nitrogen inversion and is known to occur for proton exchange scenarios [16].

9.6.1.2 Zero-Quantum Interference in NOESY

The principal unwanted signals remaining from the NOESY sequence are those arising from zero-quantum coherences (ZQCs) that existed during the mixing time and which are subsequently transformed into observable signals by the last

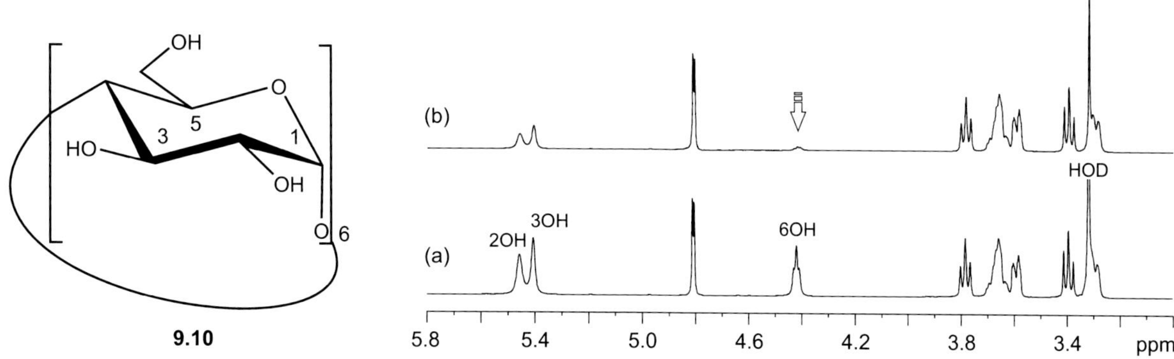

FIGURE 9.29 **Experimental demonstration of saturation transfer.** Direct saturation of the 6-OH resonance of α-cyclodextrin **9.10** in DMSO leads to the simultaneous *indirect* partial saturation of the 2- and 3-OH resonances as well as that of water (truncated in (a)).

90 degree pulse. Since these also possess coherence order zero they are not removed by phase-cycling or gradient selection procedures and thus may contaminate the final spectrum. These coherences arise between J-coupled spins and so give rise to COSY-like peaks which are the zero-quantum analogues of the signals detected in the DQF-COSY sequence and have a similar anti-phase multiplet appearance. Worse still, they appear with dispersive lineshapes if spectra are phased conventionally. If both a ZQC and NOE peak are coincident, which may occur when J-coupled spins also share an NOE, the NOE peak may appear somewhat distorted by this superposition. For large molecules with broad lines, ZQC peaks being anti-phase tend to cancel, but in small or mid-size molecules their active removal is often beneficial.

A number of approaches have been suggested for the suppression of ZQC peaks [9,17,18], the simplest being to introduce a small, random variation in the mixing time between transients or, more commonly, between one t_1 increment and the next. During the mixing time ZQCs oscillate with a frequency equal to the chemical shift difference of the two J-coupled spins ($\nu_A - \nu_X$), but NOEs will grow progressively (Fig. 9.30). Random variation of τ_m will cause the ZQCs to average away while NOE intensities are little affected (in fact, this randomisation of the ZQC signals merely makes them appear as t_1 noise rather than discrete crosspeaks). The degree of variation required must be at least comparable with the inverse of the smallest shift differences $(\nu_A - \nu_X)^{-1}$; so, assuming this to be, say, 50 Hz, a random fluctuation of ±10 ms is required. Even with such an approach, it is not uncommon to observe some residual COSY-like structure within crosspeaks very close to the diagonal. Additional artefact peaks can also arise between J-coupled spins when pulse widths deviate from 90 degree [19], so careful calibration of these is required.

A more direct approach to the elimination of zero-quantum interference is via the use of swept frequency inversion pulses applied in the presence of a field gradient which destroys the zero-quantum contributions *in a single transient.* Application of this has already been described in Section 6.2.4 with a view to obtaining pure lineshapes in TOCSY spectra, and operation of the filter is itself described in Section 12.6.2, so will not be explored here. Implementation of this in the 2D NOESY sequence is illustrated in Fig. 9.31 and requires incorporation of the filter within the usual mixing time, followed

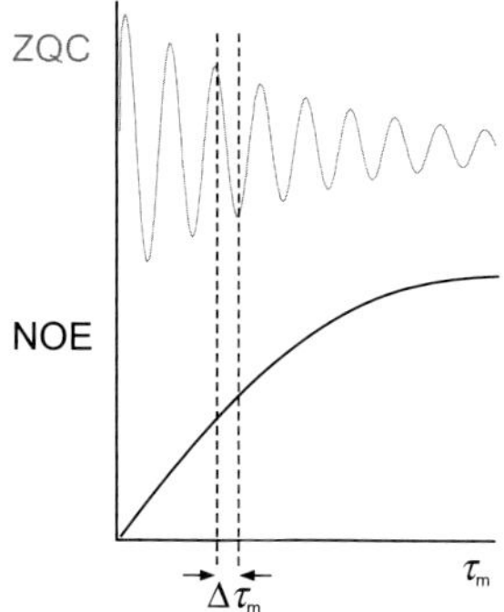

FIGURE 9.30 **Variation of NOE and ZQC intensities during the NOESY mixing time.** ZQC contributions may be suppressed by making small random variations to τ_m, whereas this has negligible effect on NOE intensities.

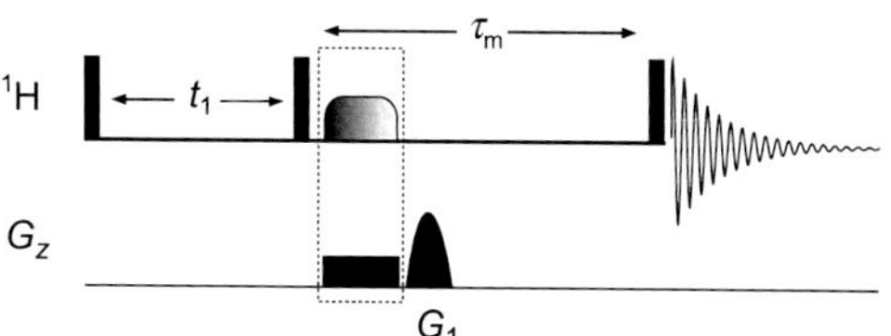

FIGURE 9.31 **NOESY sequence incorporating the frequency sweep/gradient zero-quantum filter.**

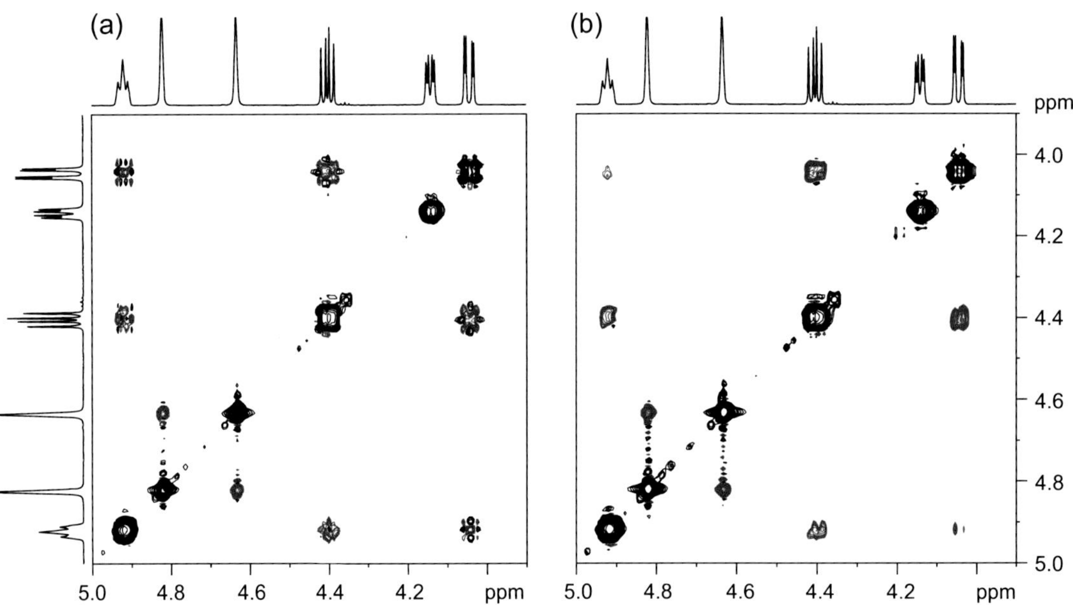

FIGURE 9.32 **Zero-quantum suppression in NOESY.** Regions of 400 ms NOESY spectra recorded (a) without and (b) with the inclusion of the zero-quantum filter shown in Fig. 9.31. ZQC suppression employed a 20 ms adiabatic smoothed CHIRP pulse with a 40 kHz frequency sweep.

by a purging gradient. The effective suppression of zero-quantum interference is illustrated in the high-resolution NOESY spectra of Fig. 9.32 in which spectrum (a) shows clear anti-phase dispersive contributions to crosspeaks between J-coupled spins whereas these cancel to reveal the NOEs more clearly in (b).

9.6.1.3 Optimum Choice of Mixing Time

Appropriate choice of the mixing time τ_m is critical to the success of transient NOE experiments. Incorrect choice of τ_m can cause complete absence of observable signals since too short a value means the enhancements have yet to grow to a detectable level while an excessively long τ_m can mean the enhancements have decayed through relaxation. The motional properties of the molecule and hence the longitudinal relaxation times of the nuclei play the most significant role in dictating NOE growth and in turn dictate the selection of τ_m (Fig. 9.33). This choice is also very dependent on the type of information required of the spectrum. If the desire is to establish a qualitative map of NOEs within a molecule, as is often the case in small-molecule work, then the optimum mixing time will be where the NOEs have their maximum intensities. If the data were to be used for quantitative or semi-quantitative distance measurements for use in structural calculations, then it is necessary to ensure one is working within the linear growth region of the NOE development, and mixing times well short of the NOE maxima will be required.

Selection of τ_m for qualitative studies can be estimated from knowledge of longitudinal relaxation times T_1. Often these are not known prior to running the experiment, so a quick measurement may assist here, or failing that an estimate based on previous knowledge. Precise measurements are little benefit since there will inevitably exist a spread of values within

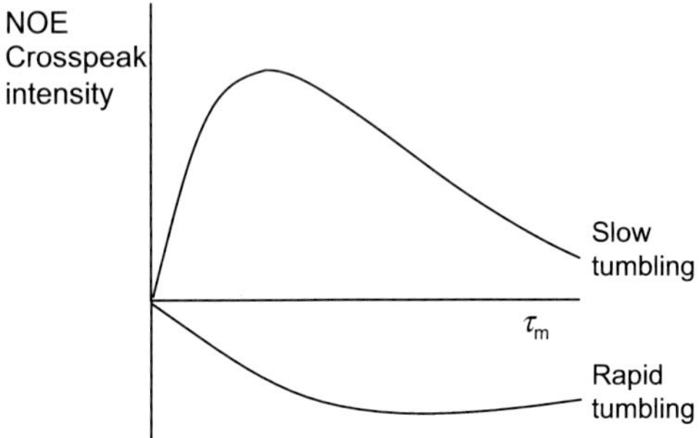

FIGURE 9.33 **Schematic illustration of NOE crosspeak growth for large, slowly tumbling and small, rapidly tumbling molecules.** The diagonal peak is assumed to be positive.

the molecule and a compromise value for τ_m is required in any case. For small molecules a mixing time of around T_1 will produce near maximum intensities, and for many routine organic molecules with masses of <500 Da in low-viscosity solvents, T_1 values typically range from around 0.5 to 2 s. As molecular tumbling rates decrease (eg owing to molecular mass or solution viscosity increases), T_1s likewise decrease and correspondingly shorter mixing times are, in principle, desirable. In such cases one should be aware of approaching the dreaded $\omega_0\tau_c \approx 1$ region where efficacy of the experiment will diminish. For molecules in the negative NOE regime, shorter values of around $0.5T_1$ are more appropriate to avoid complications arising from spin diffusion.

9.6.1.4 Distance Measurement

While a comprehensive discussion of this matter lies beyond the scope of this book, the principles involved are very briefly introduced here, the idea being that the reader should at least be able to follow discussions on structure calculations based on NOE measurements found in the chemical literature. The vast majority of work in this area has been applied to biomolecular structures and the protocols developed with macromolecules in mind. Although these will not translate directly to quantitative measurements in small molecules [20], the general principles remain the same and have been applied to accurate distance measurement in such molecules [21–24].

The most used approach to distance measurements stems from Eqs. 9.17 and 9.18 and relies on a known reference distance r_{XY} from which others may be calculated [25], and the assumption of uniform isotropic molecular tumbling. The basic equation is:

$$\frac{\eta_A\{B\}}{\eta_X\{Y\}} = \frac{r_{AB}^{-6}}{r_{XY}^{-6}} \tag{9.23}$$

The ratio of the NOEs is determined by one of two general approaches. The first (Fig. 9.34a) involves determining the NOE growth rate for both the reference and unknown distances, by recording a series of NOESY spectra over a range of τ_m values and monitoring crosspeak buildup intensities through their volume integrals [26]. The second, simpler approach (Fig. 9.34b) directly compares crosspeak intensities I_{XY} and I_{AB} measured at a single mixing time *that is known to lie within the linear growth regime* for both spin pairs. The success of these approaches relies in part on the insensitivity of the calculated distance and on the accuracy of experimentally measured NOEs due to the r^{-6} dependence. Thus, a factor of 2 error in the growth rate corresponds to only ≈10% error in the final distance estimate.

It should be noted, however, that measurement of a single 'accurate distance' does not allow for internal flexibility, so may have limited significance for a solution-state structure, and a more general and widely used approach has been to make use of semi-quantitative distance measurements in structure calculations. This involves categorising peak intensities as strong, medium and weak relative to the reference peak, which in turn are taken to indicate upper (and perhaps lower) distance bounds on internuclear separations of typically ≤0.25, ≤0.35 and ≤0.5 nm, respectively. The internal reference distance measurement for proton NOEs is typically that between diastereotopic geminal protons (0.175 nm) or between *ortho* aromatic protons (0.246 nm). Although these constraints are lax, combined application of many of them is able to produce well-defined solution conformations, as has been amply demonstrated for numerous protein structures [27,28]. An alternative approach that makes use of a single NOESY spectrum utilises the ratio of crosspeak and diagonal peak intensities in place of growth rates [29]; both methods have been applied to determination of the solution conformation and configuration of a small molecule and shown to produce similar results [30].

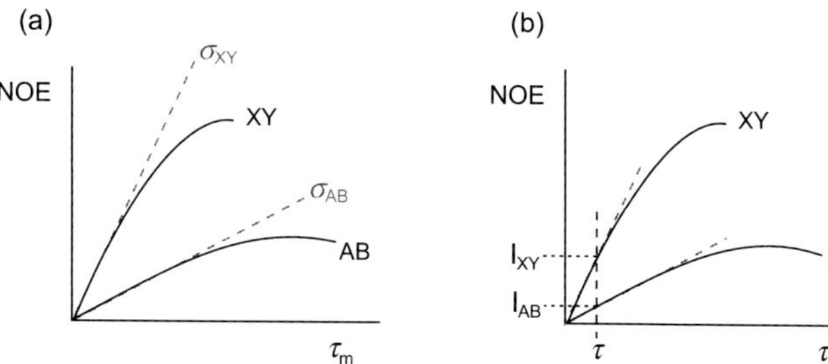

FIGURE 9.34 **Estimating internuclear separations.** Schematic illustration of two approaches in which a reference spin pair XY provides the internal calibration for the unknown pair AB (see text).

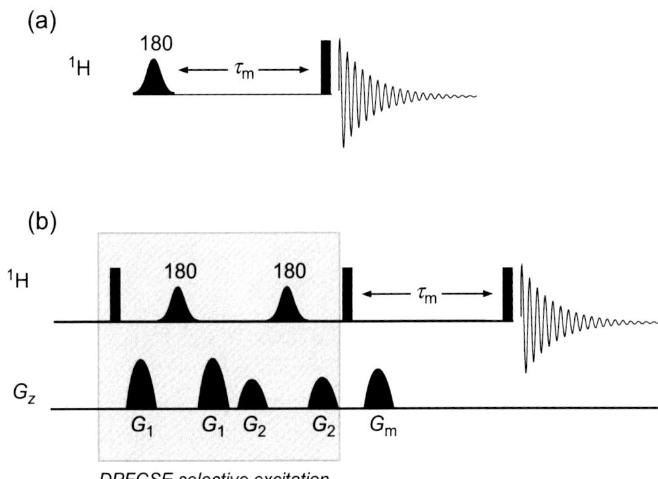

FIGURE 9.35 **1D NOESY sequences.** The basic scheme (a) requires difference spectroscopy to reveal the NOEs whereas the gradient-selected DPFGSE-NOE (b) reveals NOEs without interference from difference artefacts.

9.6.2 1D NOESY Sequences

One-dimensional transient NOE experiments, often referred to as 1D NOESY, now also find widespread use in routine structural work and also find application where quantitative distance measurements are sought with experiments collected over a range of τ_m values. The simplest 1D scheme (Fig. 9.35a) requires the use of difference spectroscopy to reveal NOE enhancements, so suffers from troublesome subtraction artefacts and hence finds little use. However, 1D gradient-selected experiments [31–33] have revolutionised the use of 1D NOESY in small-molecule applications since these experiments do not depend upon difference spectroscopy to reveal NOE enhancements, and cleanly remove those signals not arising from the NOE, so provide very high-quality spectra in shorter times. These are most widely used when a collection of enhancements are used qualitatively to provide structural answers and may also be employed for quantitative measurements through the analysis of NOE buildup rates [21,24,34]. For routine laboratory applications, 1D gradient NOESY experiments have largely superseded the traditional 1D NOE difference experiment for stereochemical investigations. For their application, each target resonance should be well separated from all others, although this criterion cannot always be met. Changes of solvent can be useful here and can sometimes lead to dramatic shifts of resonances which may fortuitously place those of interest in a more exposed position (see the example in Section 3.3.1). If this approach fails, some alternative methods are presented later in this section that enable the targeting of overlapped resonances.

9.6.2.1 1D Gradient NOESY

The most robust experiment to date [32] (Fig. 9.35b) is based upon the 'double PFG spin-echo' (DPFGSE, also referred to as 'excitation sculpting') sequence as a means of selecting a target resonance from which NOEs ultimately develop. This approach to selective excitation is described in Section 12.4.2, so here it is sufficient to note that it serves to selectively refocus only the target spin magnetisation in the transverse plane, leaving that of all other spins dephased and thus unobservable. The subsequent 90 degree pulse places the target magnetisation along the $-z$-axis, after which the NOE enhancements grow, with a purge gradient (G_m) applied to remove any residual transverse components that remain. Finally, magnetisation is sampled in the usual way. The recorded spectrum contains only the inverted target resonance and the NOEs that have developed from it, *without the need for a difference spectrum to be calculated.* High-quality spectra can therefore be collected that are free from difference artefacts; this is the principal benefit of gradient-selected schemes. The experiments also tend to be quick to perform since extensive signal averaging is no longer required for artefact reduction and because optimum use may be made of the receiver dynamic range since only the desired signals are ever acquired.

In practice, relaxation processes operating during τ_m will cause unwanted dephased magnetisation components to recover and produce small observable signals in the final spectrum for all resonances. These can be kept to close to zero intensity by insertion of typically one or two non-selective 180 degree pulses (bracketed by opposing purging gradients) spaced judiciously within the mixing period [32] with any small signals that may remain cancelled by a four-step difference phase cycle (EXORCYCLE). Thus, although difference spectroscopy is being used, in the form of the phase cycle, the signals to be cancelled have close to zero intensity and are readily removed, in contrast to conventional difference spectroscopy where the signals have essentially maximum intensity. For small and mid-sized molecules, experience suggests as a rule of thumb

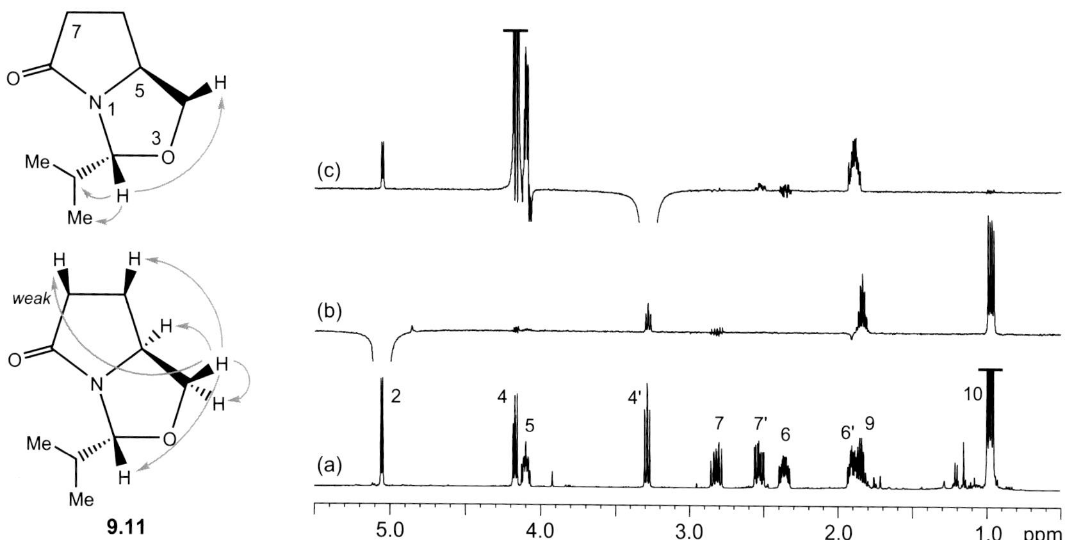

FIGURE 9.36 **Selected 1D gradient NOESY spectra of the bicyclic lactam 9.11 recorded with a mixing time of 800 ms.** Spectrum (a) is the conventional 1D spectrum, and the strong geminal H4′–H4 NOE in (c) has been truncated.

that one inversion pulse during τ_m is sufficient to reduce background signals to undetectable levels with mixing times of up to 500 ms, whereas two inversion pulses are preferred for a τ_m up to 1.5 s. A typical result using the sequence of Fig. 9.35b is shown in Fig. 9.36 for the small bicyclic lactam **9.11**. Each resonance was selected with a 40-ms Gaussian 180 degree pulse with two non-selective 180 degree pulses applied during the 800-ms mixing time. Each spectrum was the result of only 12 min of data collection on a 10-mg sample, yet are perfectly acceptable for qualitative stereochemical assignments. The observed signals may be attributed to NOE enhancements or to scalar coupling artefacts, with an otherwise featureless baseline.

As for the 2D NOESY data described earlier, spectra are vulnerable to interference from undesirable zero-quantum contributions between coupled spins; this is apparent, for example, in trace (c) of Fig. 9.36 between H4′ and H5. This can be especially problematic when shorter mixing times are employed, as in the generation of NOE buildup curves [34], and again the inclusion of the swept frequency/gradient zero-quantum filter should prove beneficial. The complete selective 1D NOESY sequence incorporating this and employing the optimised DPFGSE selection procedure is presented in Fig. 9.37 and an illustration of the improvements provided by the filter may be seen in Fig. 9.38 where the removal of unwelcome anti-phase dispersive contributions between coupled spins is apparent in (b).

9.6.2.2 Chemical Shift Selective Filters and Doubly Selective TOCSY-NOESY

One of the principal limitations to the use of 1D selective NOESY experiments is the need for a well-resolved peak for selective inversion, meaning any that are hidden are not amenable to investigation by these methods. As noted above, one may have the option to change the solvent, yet when such means are ineffective or inappropriate, alternative schemes may be required that allow a specific resonance to be targeted.

One approach is to incorporate a chemical shift selective filter (CSSF) in the 1D NOESY sequence. The CSSF enables selection of a single resonance at a specified chemical shift even when the resonance is overlapped with others; it is described in Section 12.4.3. An alternative approach is to employ a *selective* transfer preparation step from a well-resolved resonance *onto* the overlapped target resonance so as to reveal this. This leads to a concatenated or doubly selective experiment for which a selective TOCSY preparation element is well suited to generating the initial transfer [35], this having been referred to in this context as the selective TOCSY edited preparation (STEP) function [36]. Thus, the TOCSY

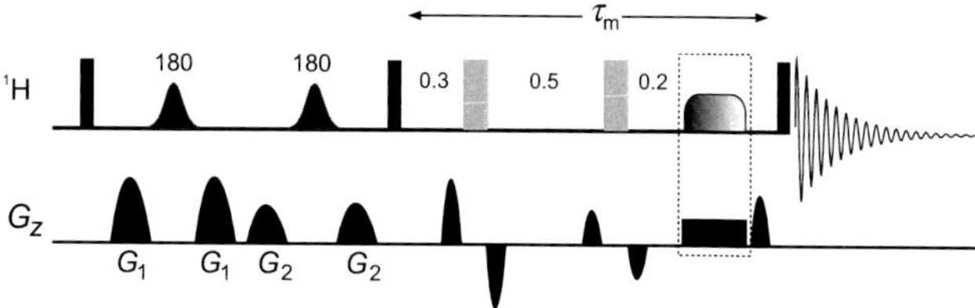

FIGURE 9.37 **The optimised 1D DPFGSE-NOESY sequence incorporating zero-quantum suppression.** The mixing time contains non-selective 180 degree nulling pulses bracketed by opposing purging gradients spaced throughout as shown, followed by the zero-quantum filter.

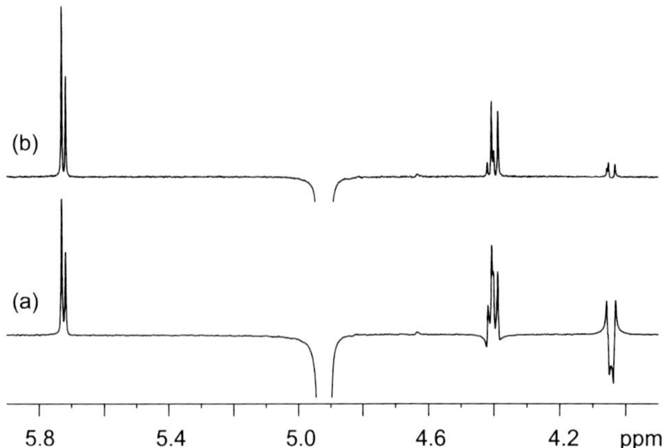

FIGURE 9.38 **1D DPFGSE-NOESY spectra.** These were recorded (a) without and (b) with the zero-quantum suppression scheme of Fig. 9.37. ZQC suppression employed a 20 ms adiabatic smoothed CHIRP pulse with a 40 kHz frequency sweep.

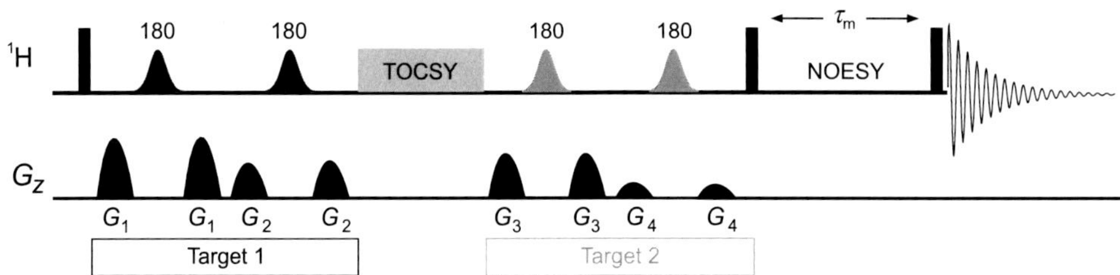

FIGURE 9.39 **A general scheme for implementing the doubly selective 1D TOCSY-NOESY experiment for observing NOEs originating from hidden resonances.** Selectivity of the 180 degree pulses may be optimised independently for the two target resonances 1 and 2.

stage carries selected magnetisation onto the overlapped resonance (with which it shares a spin-coupled network), which is then itself laid bare for a further selective NOESY step, leading to generation of NOEs from only this buried resonance, despite its apparent inaccessibility. A general scheme illustrating this concatenation based upon the excitation sculpting method is shown in Fig. 9.39 in which the selective TOCSY experiment (Section 6.2.4) precedes the selective NOESY element. A more refined and fully optimised sequence incorporates the zero-quantum filter described above into both the TOCSY and NOESY mixing elements and includes 180 degree nulling pulses during the NOE mixing period, such a scheme being referred to as 1D STEP-NOESY [36]. Fig. 9.40 illustrates application of this method in allowing selective observation of NOEs from the H4 proton of ring A across the glycosidic linkage to H1 of ring B in the trisaccharide **9.12**. Although the H4^A proton lies hidden in a cluster of four resonances, its location is revealed clearly in a 1D

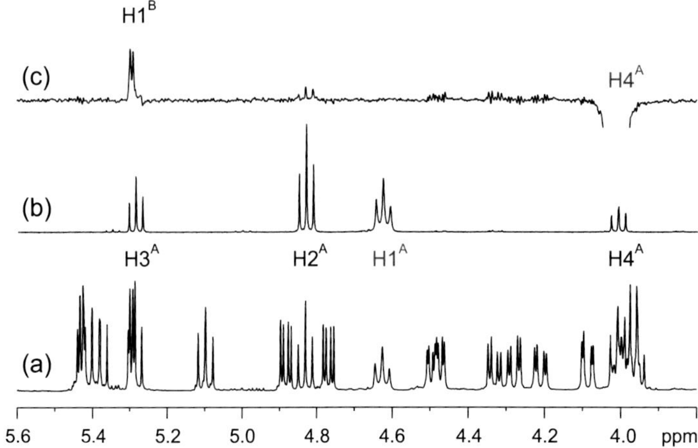

FIGURE 9.40 **Application of a doubly selective 1D TOCSY-NOESY experiment to reveal NOEs from the buried H4^A resonance of 9.12.** Trace (b) shows the 1D TOCSY from the resolved H1^A used in preparation to reveal the shift of H4^A (τ_m = 80 ms) and trace (c) shows the full TOCSY-NOESY combination selecting the target resonances (red) of H1^A and H4^A, respectively (τ_m = 80 ms TOCSY, 400 ms NOESY). Each transfer step included zero-quantum filtration as described above.

TOCSY experiment (Fig. 9.40b). This resonance is then available for clean excitation in the second selective step of the 1D TOCSY-NOESY experiment (Fig. 9.40c).

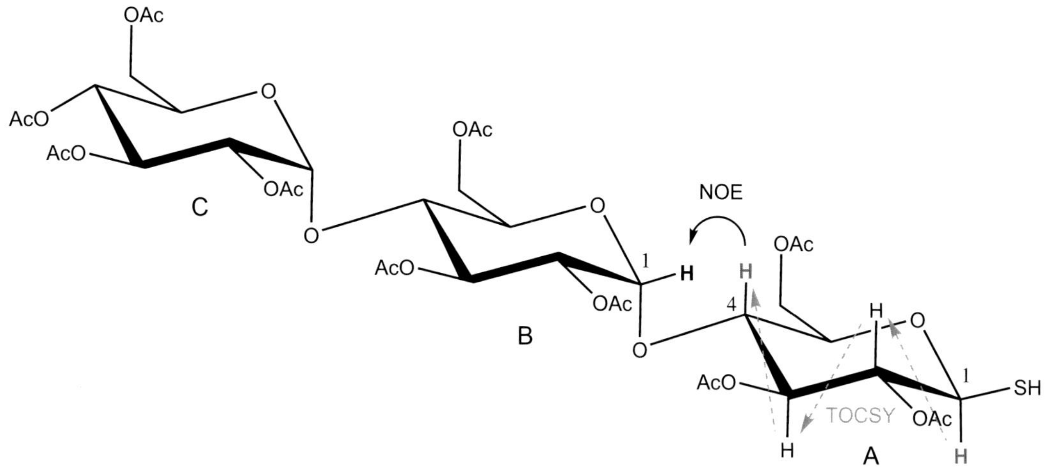

9.12

Both the 1D gradient NOESY and the doubly selective counterpart may be modified through the use of *single* PFG spin-echoes in the selection step(s) to reduce the total sequence duration (Section 12.4.2) which may prove advantageous when dealing with larger molecules that have faster transverse relaxation rates. The selective method may also substitute a ROESY mixing element in place of NOESY mixing when such a scheme is to be preferred; see Section 9.7.

9.6.2.3 *Interpreting One-Dimensional NOESY*

The clear advantages of gradient-selected NOE experiments over the once popular steady-state NOE difference (Section 9.8) means these have become popular tools in small-molecule structural studies. However, despite their similar appearances, there are fundamental differences between the data presented by the two experimental protocols, with *transient* experiments observing *kinetic* NOEs and *steady-state* experiments observing *equilibrium* NOEs. As a consequence, 1D NOESY experiments should be interpreted with the following points in mind:

- NOE enhancement will be acutely sensitive to the choice of mixing time and may vary markedly with changes in this.
- Absolute NOE enhancement will also depend on the degree of target inversion, which may well be less than complete owing to pulse imperfections and other experimental shortcomings.
- Experimentally, the measurement of percentage enhancements is non-trivial. The target resonance cannot be used as an intensity reference (as it conveniently is in steady-state difference spectra) since it is reduced by relaxation during τ_m. A reference spectrum with $\tau_m \approx 0$ s is therefore required from which enhancements can be quantified and the degree of inversion of *each* target resonance must also be determined. Experimentally measured NOE enhancements must then be scaled accordingly for '100% inversion'.
- The *absolute* percentage enhancements from transient experiments will be smaller than steady-state enhancements, even though they may be clearer to see, since they arise only from an initial spin perturbation and are not driven by continual saturation. The perception of what is considered a 'reliable' or 'measurable' enhancement must therefore be adjusted if percentages are reported and compared between differing experimental protocols.

The experimental complexities associated with the correct measurements of percentage enhancements mean that these figures are not routinely determined for the qualitative structural studies commonly undertaken (and anyone purporting to have measured these should give details of the approach taken). A more realistic approach to reporting results would seem to be that adopted in macromolecular NOE studies using semi-quantitative classification of enhancements (small, medium and weak). Assuming mixing times are kept short such that enhancements lie within, or at least close to, the linear growth regime, comparison *within* each trace provides approximate relative distance relationships between the target spin and those exhibiting the enhancements. Piecing together data from a number of experiments to produce a self-consistent argument is again advisable.

9.6.3 Applications

Despite finding its origins largely in the hands (or laboratories) of biological spectroscopists, there is nowadays a widespread use of NOESY methods for structural and conformational analysis of the small to mid-sized molecules encountered in the chemical laboratory. The examples presented here demonstrate a variety of systems to which these experiments have been applied and also serve to illustrate some the benefits of using these in preference to the alternative steady-state experiments.

In the first example, 2D NOESY spectra were used to define the stereochemistry in the synthetic cycloadduct **9.13** [7], a potential biomimetic precursor to the naturally occurring marine-sponge alkaloid keramaphidine *B*, **9.14**. This problem is essentially the same as that addressed for **9.7** using the NOE difference experiment, but in this case the additional unsaturated sidechains caused extensive overlap in the proton spectrum and precluded the use of selective presaturation. Sufficient characteristic NOEs present in a 600-ms NOESY spectrum gave conclusive proof of the *endo* stereochemistry, as shown. Only positive NOEs were observed, consistent with a molecule of mass 436 Da in chloroform. NOESY spectra have also been successfully applied to the structure elucidation of molecules for considerably greater mass and complexity, as illustrated by the cytotoxic macrolide cinachyrolide A, **9.15** [37], also from a marine sponge. The structure of the molecule was determined through extensive 600-MHz 2D NMR experiments, of which NOESY played a crucial role in defining the relative stereochemistry of the six oxane rings, as shown (reproduced with permission from Ref. [37], Copyright 1993, American Chemical Society). Mixing times of up to 700 ms were employed, and the relatively large mass of the molecule, approximately 1200 Da, may have been close to the cutoff for the successful observation of NOEs. Probably for this reason, supporting evidence also came from 200-ms ROESY spectra.

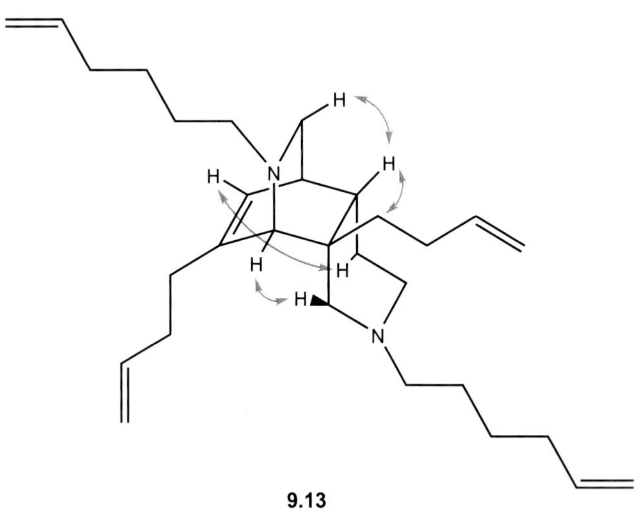

9.13

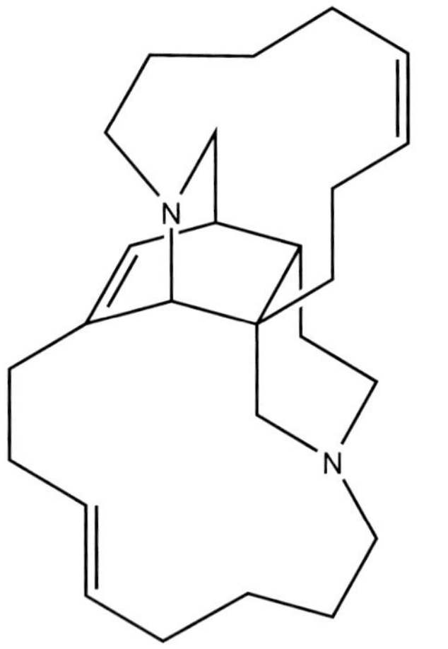

9.14

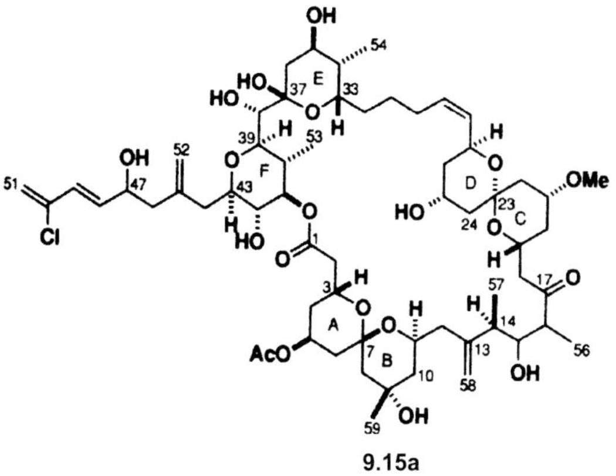

9.15a

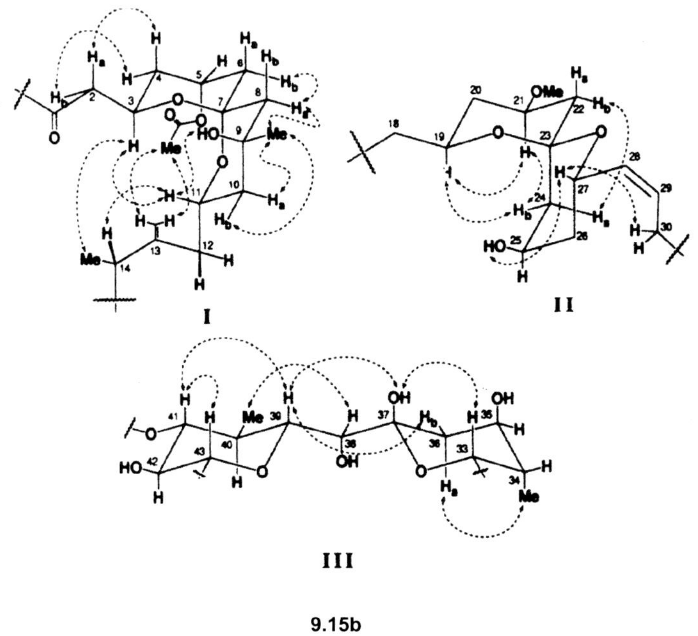

9.15b

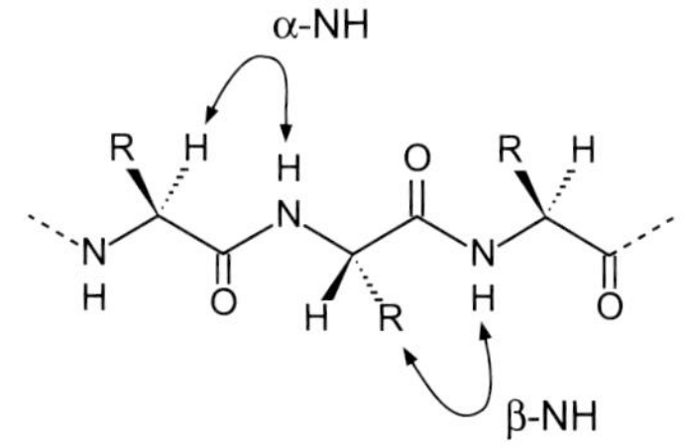

FIGURE 9.41 **Intraresidue NOEs used to identify neighbouring residues in a peptide sequence.**

A common area in which 2D methods have been employed to help provide resonance assignments is in the study of peptides or small proteins, specifically when the same amino acids occur more than once in the peptide sequence. The so-called *sequential assignment* process is used to define the position of a specific amino acid residue within a peptide, and relies upon the observation of NOEs between protons in adjacent residues (Fig. 9.41). Typically, these will be between an alpha proton and the amide NH of the following residue. Identification of neighbouring amino acids in this way can be used to string these units together, which may then be mapped onto the usually known peptide sequence. Once all residues have been sequentially identified, this can provide the basis for conformational studies through longer range NOE contacts [38–40]. As an illustration, stepwise identification of neighbouring resonances is mapped in a NOESY spectrum for the cyclic decapeptide gramicidin-S **9.16** in Fig. 9.42.

While most studies rely on the observation of *intra*molecular NOEs, *inter*molecular NOEs can also be used to define relationships between molecules which share close proximity. This can be of particular relevance in the study of 'host–guest' complexes where one wishes to determine how the guest sits within the host, and here scalar-coupling information can play no direct part. One example of this is the characterisation of rotaxanes, extended conjugated systems sitting within protective encapsulating molecular jackets. The azo-dye rotaxane **9.17** uses α-cyclodextrin (α-CD) as the sheath molecule, which is inherently asymmetric with a smaller cavity entrance at the 5–6 rim of the molecule. The 1D gradient NOESY experiment described above was employed to map long-range host–guest NOEs [41] which differentiated the two ends of the complex and presented a picture of how the α-CD sat. NOEs from both

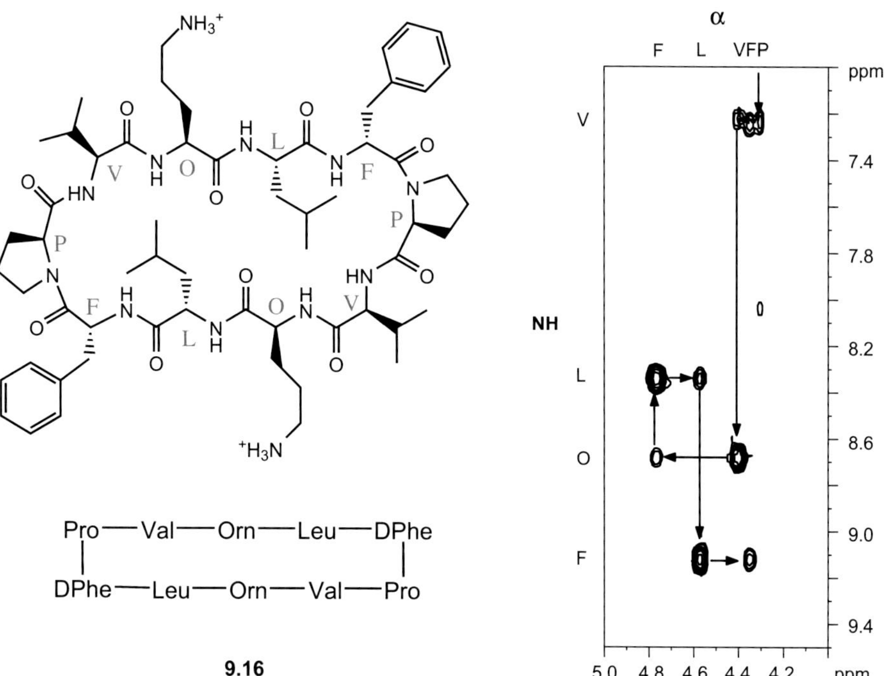

FIGURE 9.42 **Sequential assignment process illustrated for the 300ms NOESY spectrum of gramicidin-S 9.16 in DMSO, starting from the α-proton of proline.** Interresidue αH–NH NOES are often stronger than the intra-residue NOEs in an extended backbone conformation.

HG′ and HI to the 3 and 5 protons on the inner face of the α-CD confirmed that the central portion of the guest was encapsulated within the hydrophobic cavity (structure reproduced with permission from Ref. [41], Copyright Wiley-VCH Verlag GmbH & Co. KGaA.). All NOEs for these molecules in DMSO were negative due to the limited mobility of these bulky complexes.

Both 1D gradient and 2D NOESY experiments have also been used to assess the conformations of anaesthetic steroids [42]. These possess crowded proton spectra requiring the use of 2D experiments, and mean 1D methods can only be applied to a limited set of resonances, which nevertheless prove significant. In **9.18** in particular, long-range NOEs in the 1D experiment were observed from methyl 19 specifically to the H3′ equatorial proton of the attached morpholine ring but not to other protons within this moiety. This indicated free rotation of the ring was restricted (by an intermolecular hydrogen bond) and that it occupied a fixed position relative to the steroid skeleton. Knowledge of such conformations proves useful in elucidating the mechanisms by which these molecules cause anaesthesia.

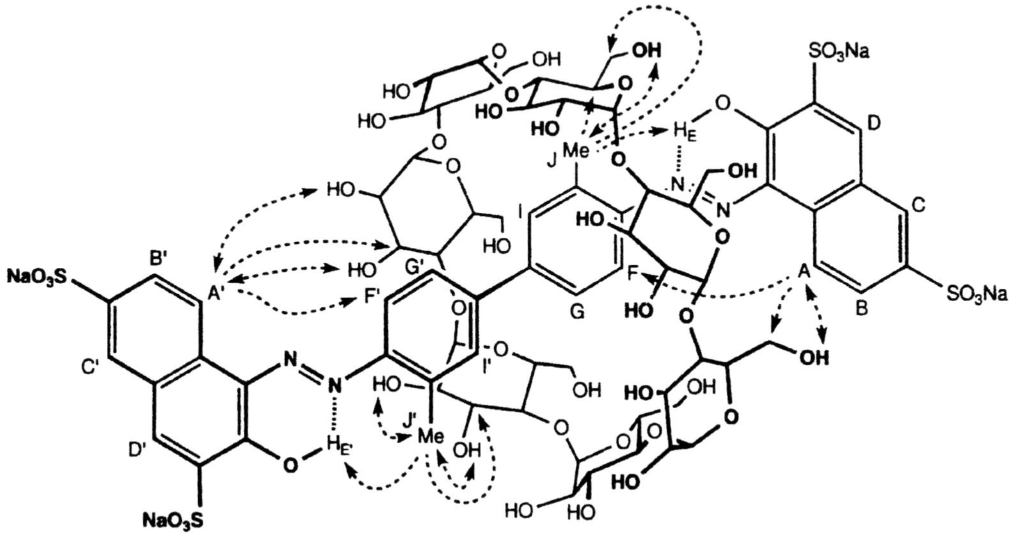

9.17

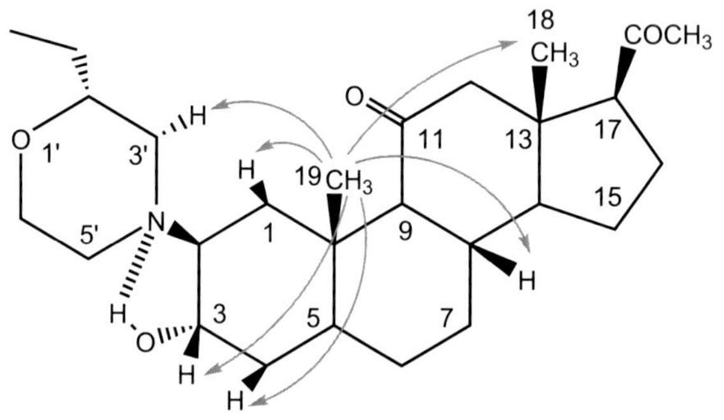

9.18

The final example specifically illustrates some advantages of the 1D gradient NOESY experiment over the once popular NOE difference method. One goal of this work was to compare the solution structure of the diphenylallyl palladium complex **9.19** with the crystal structure and from this derive mechanistic insights [43]. The proton spectrum was assigned through combined analysis of phosphorus-decoupled proton, DQF-COSY and ROESY spectra recorded with a 500-ms mixing time. Specific close proximities were also probed through 1D NOE methods where suitably resolved resonances were available, selected examples of which are shown in Fig. 9.43. Some specific points are worthy of comment, the first of which is the clarity in the NOE spectra which are not confused by difference artefacts,

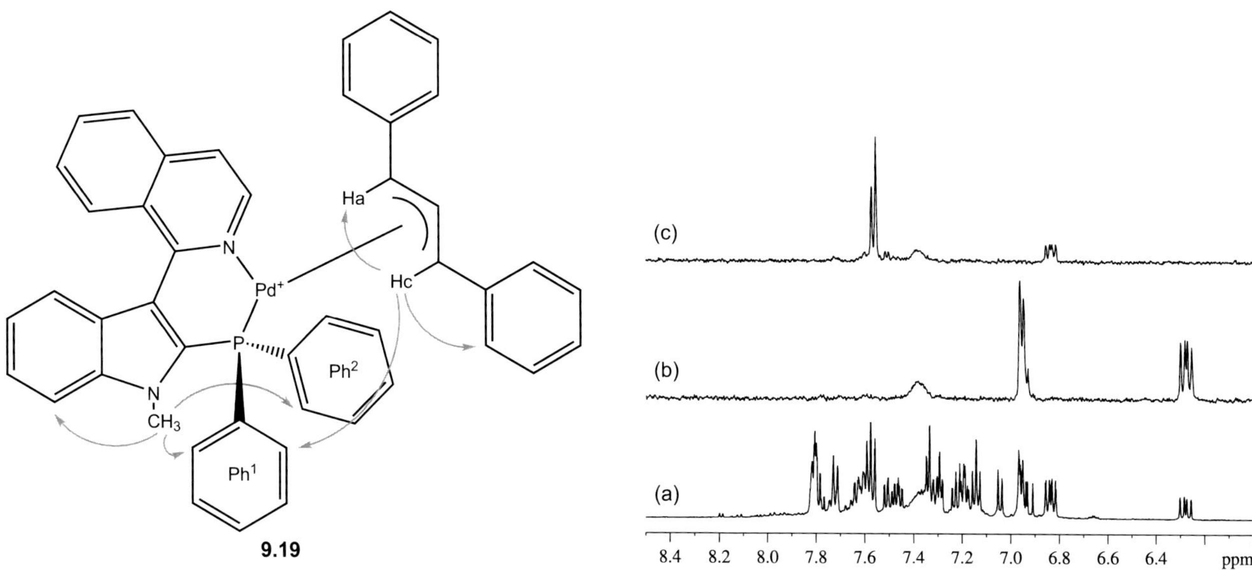

FIGURE 9.43 **Selected 1D gradient NOESY spectra of the palladium complex 9.19 recorded with a mixing time of 800 ms in each case.** (a) Parent 1D spectrum, (b) Hc selected (not shown) and (c) N-Me selected (not shown).

thus giving one confidence that the observed enhancements are genuine. Secondly, these spectra were collected in only 14 min each, while the equivalent difference experiments required data collection over many hours to adequately suppress resonances in the cluttered aromatic region. Finally, note the enhancements of the broad resonance at 7.4 ppm, associated with the *ortho* protons of the dynamically restricted phoshine-Ph^1 group. These were distinctive and highly informative, but could not be reliably distinguished in either the 1D difference or 2D experiments. The ability to extract data of this sort with confidence makes the 1D gradient NOESY experiment an extremely powerful tool in structure elucidation.

9.7 MEASURING ROTATING FRAME NOEs: ROESY

The benefits of recording rotating-frame NOEs have been described in earlier sections and stem from the fact that they are positive for all molecular tumbling rates and hence all molecular sizes, and prove particularly advantageous in the study of mid-sized molecules which exhibit very small conventional NOEs owing to their motional properties ($\omega_0\tau_c \approx 1$). Since the observation of ROEs involves measurement of a transient enhancement, their application and interpretation largely parallels that for NOESY and need not be repeated here. The principal concern with ROESY experiments (originally termed CAMELSPIN [44] owing to the similarity of the motion of a figure skater during this manoeuvre and the behaviour of magnetisation vectors during the experiment) is the fact that they are susceptible to a number of processes other than cross-relaxation and hence may contain a variety of interfering and potentially confusing responses, in addition to the desired ROEs. An appreciation of these effects, and how to deal with them, is therefore mandatory for anyone wishing to employ these methods, and some caution is required in the application of ROE experiments.

9.7.1 The 2D ROESY Sequence

The original and simplest ROESY sequence (Fig. 9.44) has the mixing period defined by the duration of a continuous, low-power rf spin-locking pulse during which the ROE develops. Magnetisation not parallel to the spin-lock axis is dephased by rf field inhomogeneity (Fig. 9.45), so the only phase-cycling required is that for axial peak suppression and f_1 quad-detection. Gradient-selected variants have also been presented [14]. For small and mid-sized molecules the selection of τ_m follows that presented for NOESY with, typically, $\tau_m \leq 600$ ms, whereas for large molecules in the spin-diffusion limit, shorter values are more appropriate since the ROE growth rate is twice that of the NOE (Section 9.5). The 2D spectrum maps through-space interactions through crosspeaks that have *opposite* phase to the diagonal.

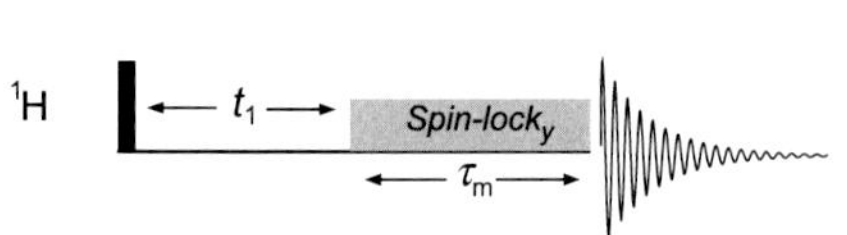

FIGURE 9.44 **2D ROESY sequence.** The mixing time τ_m is defined by the duration of the low-power spin-lock pulse.

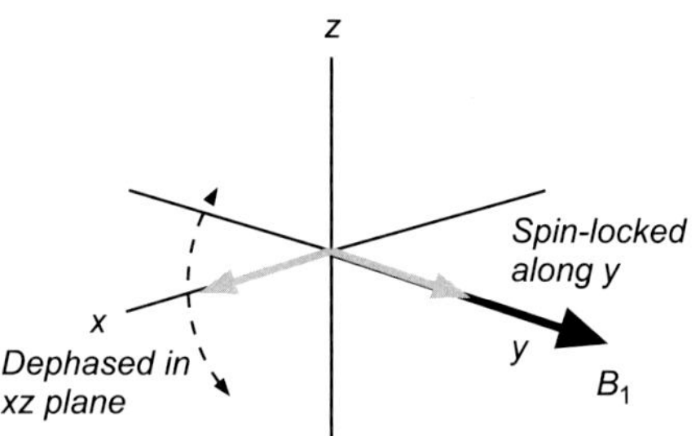

FIGURE 9.45 **ROESY spin-lock.** During this, magnetisation parallel to the B_1 field remains spin-locked whereas orthogonal components are driven about this (here in the *xz* plane) and eventually dephase through B_1 field inhomogeneity.

9.7.1.1 Complications with ROESY

As stated previously, crosspeaks may arise in ROESY spectra as a result of processes that occur during the spin-lock other than cross-relaxation between spins. The principal complications that can arise are: [45]

- COSY-type crosspeaks between J-coupled spins;
- TOCSY transfers within J-coupled spin systems; and
- crosspeak attenuation from rf off-resonance effects.

The first of these arises when the long spin-lock pulse acts in an analogous fashion to the last 90 degree pulse of the COSY experiment, so causing coherence transfer between J-coupled spins. The resulting peaks display the usual anti-phase COSY peak structure and tend to be weak, so are of least concern. A far greater problem arises from TOCSY transfers which arise because the spin-lock period in ROESY is similar to that used in the TOCSY experiment (Section 5.7). This may therefore also induce coherent transfers between J-coupled spins when these experience similar rf fields, that is when the Hartmann–Hahn matching condition is satisfied. Since the ROESY spin-lock is not modulated (ie not a composite pulse sequence), this match is restricted to mutually coupled spins with similar chemical shift offsets or to those with equal but opposite offsets from the transmitter frequency. These conditions are often met in natural products such as carbohydrates, nuclei acids, peptides, alkaloids and steroids for which resonances cluster about the centre of the spectrum. In addition to unwanted direct TOCSY effects, multistep transfers involving both TOCSY and ROE stages can lead to 'false' ROEs, crosspeaks that *a priori* appear to arise from a ROE between two correlated spins but in fact do not [46]. Thus, for example, a TOCSY transfer from spin A to B followed by ROE transfer from B to C in the system of Fig. 9.46 would produce an apparent crosspeak suggesting an ROE between A and C. Likewise the reverse ROE-TOCSY steps would produce a similar signal, but the conclusion would be wrong in either case! Since TOCSY transfer retains signal phase, its involvement in crosspeak generation is not obvious and can lead to drastically incorrect conclusions. Furthermore, if TOCSY and ROE transfers occur simultaneously between two spins, the ROE peak may be reduced in magnitude or even cancelled owing to opposite peak phases. The various transfer pathways that may occur during ROESY are summarised in Table 9.5.

Considerable attention has been given to ways of avoiding TOCSY transfer during ROESY, the simplest of which is through limiting the Hartmann–Hahn match with judicious positioning of the transmitter frequency to ensure coupled spins either side of this are not symmetrically disposed [47]. Alternatively, recording two spectra with differing transmitter offsets leaves genuine ROE peaks little changed, but should significantly alter the intensities of those involving a TOCSY step and so facilitate their identification [46]. Modification of the spin-lock itself also leads to a reduction in TOCSY efficiency (Fig. 9.47). Using low-power rf (Fig. 9.47a) with γB_1 comparable with the maximum resonance frequency offset [45] (typically 2–3 kHz) or a pulsed spin-lock [48] (Fig. 9.47b) offer similar attenuation [49], neither of which is complete in

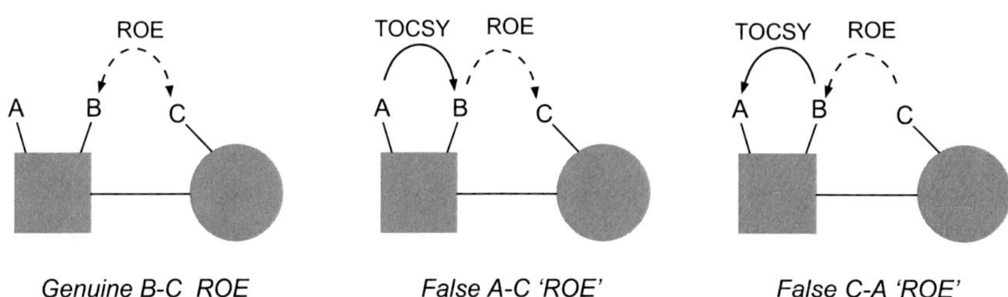

FIGURE 9.46 **Generation of 'false' ROE peaks in ROESY spectra may arise from combined ROE and TOCSY mechanisms.**

TABLE 9.5 Responses Observed in the ROESY Experiment

Sign of Diagonal	Origin of Crosspeak	Sign of Crosspeak
	Direct ROE	Negative
	Indirect ROE (three-spin effect)	Positive (weak)
	TOCSY	Positive
Positive	TOCSY-ROE	Negative (false ROE)
	ROE-TOCSY	Negative (false ROE)
	Chemical exchange	Positive
	COSY type	Anti-phase/mixed phase

By convention the diagonal is phased to be positive and the signs of all other signals are given relative to this.

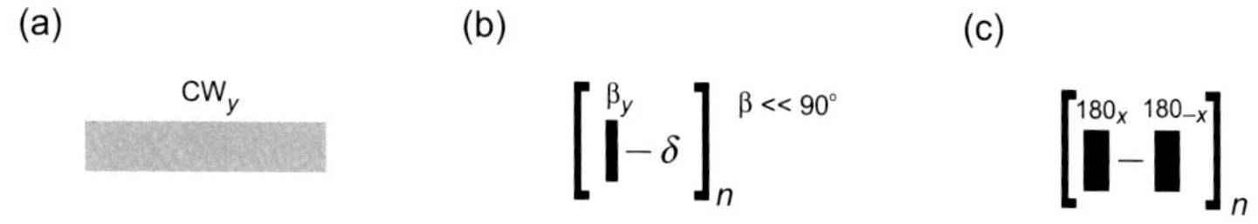

(a) CW_y (b) $[\beta_y - \delta]_n$ $\beta << 90°$ (c) $[180_x\ 180_{-x}]_n$

FIGURE 9.47 **Practical mixing schemes for the ROESY experiment.** (a) A single, low-power pulse, (b) a pulsed spin lock comprising a repeated sequence of a small tip angle pulse followed by a short delay and (c) the Tr-ROESY alternating-phase spin lock.

reality. Considerable care should be taken in the rf power used for the spin-lock as damage to transmitters and/or probes may result if high powers are applied for excessively long periods. More effectively, an alternating phase spin-lock (Fig. 9.47c) has been shown to be effective at reducing TOCSY transfer [50–52] and has been termed Tr-ROESY (transverse ROESY). This approach destroys the Hartmann–Hahn match between coupled spins, so suppressing TOCSY transfer, and measures an average of the ROE and NOE since the magnetisation vectors spend time in both the transverse plane and along the longitudinal axis as they follow a swinging 'tic-toc' trajectory (Fig. 9.48). The drawback is a potential reduction in crosspeak intensity relative to conventional ROESY due to this averaging process. In small molecules ($\omega_0\tau_c \ll 1$) there is no theoretical loss; this becomes a factor of 2 for mid-sized molecules ($\omega_0\tau_c \approx 1$) and a factor of 4 for very large molecules ($\omega_0\tau_c \gg 1$). Hence there is something of a compromise for molecules in the $\omega_0\tau_c \approx 1$ region for which rotating frame measurements prove most beneficial. This method generally requires γB_1 to be twice the maximum resonance frequency offset (γB_1 typically 4–6 kHz), and its success is highly dependent on the use of accurately calibrated 180 degree pulses. The ability of this scheme to suppress interference from the TOCSY mechanism is illustrated in Fig. 9.49 for a tetrameric carbopeptoid **9.20** in which the disappearance of both TOCSY and false ROE peaks alongside the revelation of genuine ROEs in (b) is clear.

The third complicating factor specific to ROESY is the attenuation of crosspeak intensities as a function of resonance offset from the transmitter frequency [53]. Off-resonance spins experience a spin-lock axis that is tipped out of the *x–y* plane (Section 3.2.1) resulting in a reduction in observable transverse signal in addition to a reduction in cross-relaxation rates. This is more of a problem for quantitative measurements, although fortunately mid-sized molecules show the weakest dependence of ROE cross-relaxation rates on offset. The so-called 'compensated ROESY' sequence [53] eliminates these frequency-dependent losses should quantitative data be required.

9.7.2 1D ROESY Sequences

In parallel with the 1D NOESY sequences above, the 2D ROESY experiment also has its 1D equivalent (in fact, this was the original ROE experiment [44]) and gradient-selected analogues [35,54,55], all of which incorporate selective excitation of

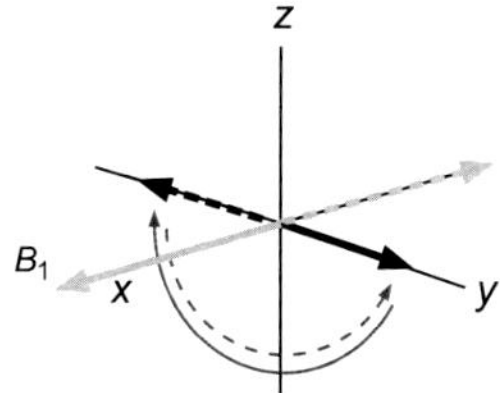

FIGURE 9.48 **Swinging tic-toc motion of spin-locked vectors during the Tr-ROESY mixing sequence.**

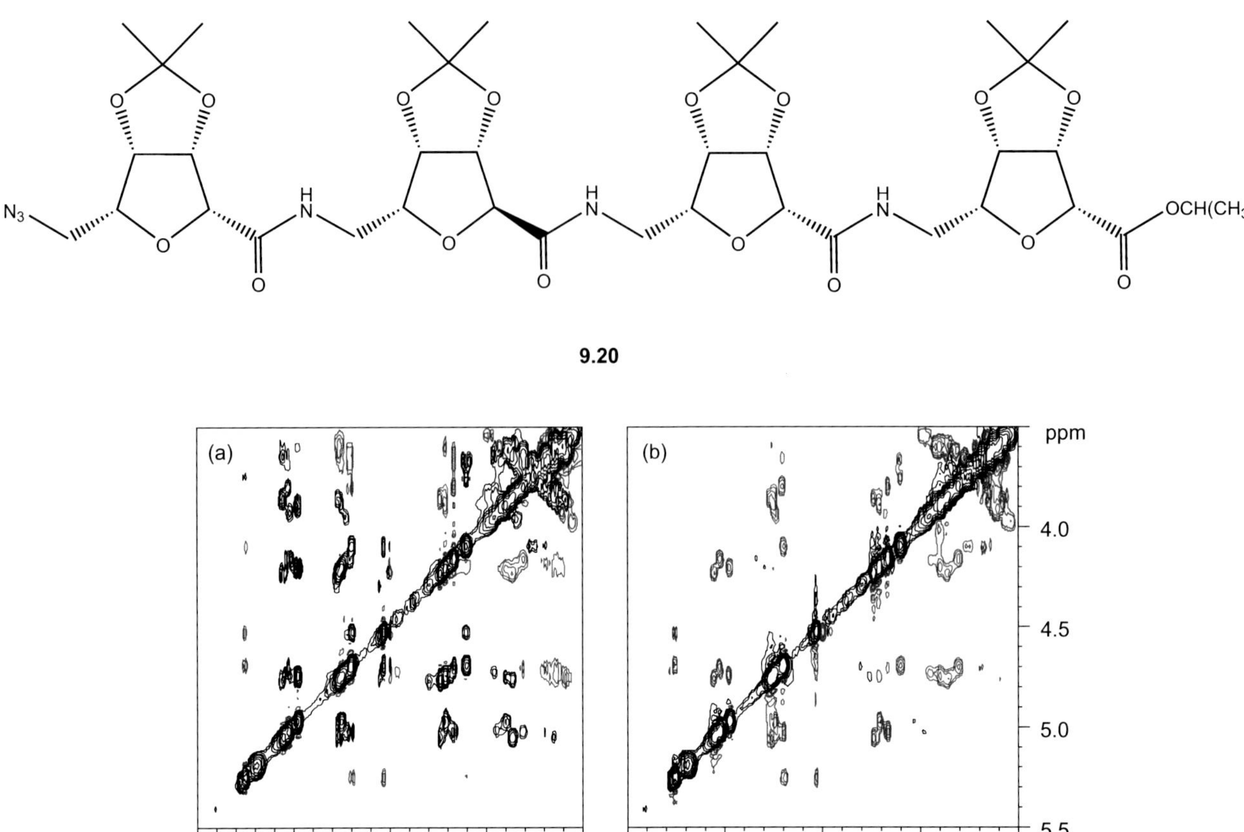

FIGURE 9.49 **ROESY spectra of a tetrameric carbopeptoid 9.20.** These were recorded with (a) a single 2.6 kHz continuous spin-lock pulse and (b) a 3.7 kHz phase-alternating Tr-ROESY spin-lock at 4.5 ppm. Spectrum (a) is dominated by TOCSY peaks which share the same phase as the diagonal, whereas these have been largely suppressed in (b), so revealing the genuine ROE peaks.

the target spin such as via the double PFG spin-echo selection described for 1D NOESY above [56]. These can be derived from 1D NOESY sequences by incorporation of a suitable spin-lock in place of the 90–τ_m–90 segment of NOESY, and thus require no further elaboration.

9.7.3 Applications

In view of the various complicating factors associated with the ROESY experiment, it is perhaps prudent to avoid using the technique initially, and instead select a conventional transient experiment as first choice. Certainly for small molecules within the extreme-narrowing limit there is no theoretical difference between NOEs and ROEs, so there seems little to be gained from undertaking potentially more complex ROE investigations. Nevertheless, situations arise when the use of rotating frame measurements are unavoidable because conventional experiments yield negligible enhancements ($\omega_0\tau_c \approx 1$), and it is not surprising that most applications of ROESY have involved larger molecules, notably macrocyclic natural products, peptides, oligosaccharides and host–guest complexes, as illustrated by these selected examples.

In the structure elucidation of the cytotoxic dimeric steroid crellastatin A [57] **9.21**, the basic skeleton of the molecules was derived principally from numerous long-range proton–carbon correlations observed in HMBC spectra. The relative stereochemistry of the bicyclic system at the junction of the two steroid units was derived from analysis of ROESY spectra recorded at 600 MHz with a 400-ms mixing time, some key enhancements being illustrated. The final configuration at C22′ was determined by comparison of the lowest energy molecular mechanics conformers of the two possible stereoisomers with opposite configurations at this position. Only the C22′–R configuration placed the 21′ and 26′ methyl groups in the proximity required for observation of an ROE between them. The large mass of the molecule, 933 Da, no doubt precluded the use of NOESY measurements, and the literature contains many examples of this sort where rotating frame measurements have played a crucial role in characterising novel natural products. Conformational studies of peptides have also benefited from the ROESY experiment since it is usually larger peptides that possess

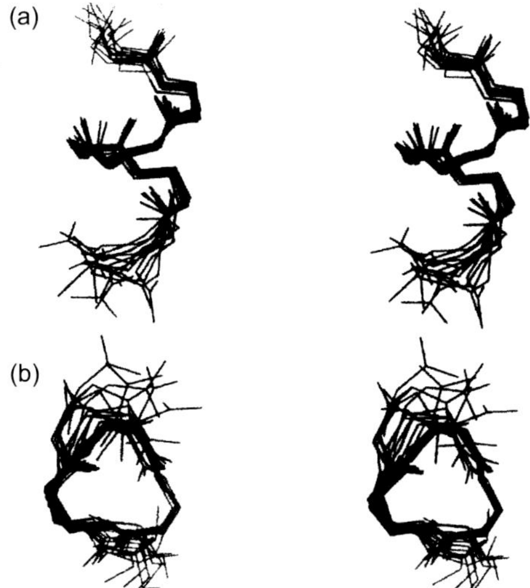

FIGURE 9.50 **Stereo representations of the 14 lowest energy–calculated structures of the hexameric β-peptide 9.22 based on distance restraints derived from ROESY spectra.** Side and top views are shown in (a) and (b), respectively, with the sidechains of the β-amino acids omitted for clarity. *(Source: Reproduced from Ref. [58] with permission from Wiley-VCH.)*

defined conformations and hence potentially interesting pharmocological properties. Studies of oligomers of synthetic β-peptides have now identified various sequences with helical structures in solution that have been extensively characterised by NMR. In the sequence **9.22** distinctive ROEs were observed between the amide proton of residue i and the H^{β} protons of residues $i + 2$ and $i + 3$ along the sequence, and were characteristic of the structure adopted by the peptide [58]. These and additional ROEs were used to define 18 distance constraints for use within molecular dynamics protocols by classifying their intensities as weak, medium and strong (Section 9.6.1), enabling the conformation of the hexapeptide to be calculated. The 14 lowest energy structures derived from this are shown in Fig. 9.50, and clearly illustrate the well-defined helical structure adopted in pyridine. In a second example, ROEs were used in a similar manner to define the structure of a hexameric *cis*-oxetane–derived β-peptide **9.23** as a left-handed helix stabilised by 10-membered intramolecular hydrogen bonds (Fig. 9.51) [59]. Characteristic $i - i + 2$ ROEs were observed along the sequence (only one set is illustrated on the structure) and the hydrogen bonds suggested by these patterns were consistent with the amide NH stretches observed in infrared spectra.

9.21

β-H-Val β-H-Ala β-H-Leu β-H-Val β-H-Ala β-H-Leu

9.22

Host–guest inclusion complexes are also frequently restricted to study by rotating frame measurements because of the requirement for a sufficiently large host to be able to accommodate the guest, and one of the best studied host systems appears to be the cyclodextrins (CD) (see structure **9.10**). One such investigation was into the inclusion of constituents of the natural sex pheromone mix of the olive fruit fly as a means of controlling the release of these volatile substances [60]. Characterisation of the complexes so formed, such as ethyl dodecanoate encapsulated within permethylated α-CD 'cups' **9.24**, was necessarily made from intermolecular ROEs since NOESY enhancements were negligible. Interference from TOCSY transfer was also evident in this work in the form of false ROEs from the guest molecules to

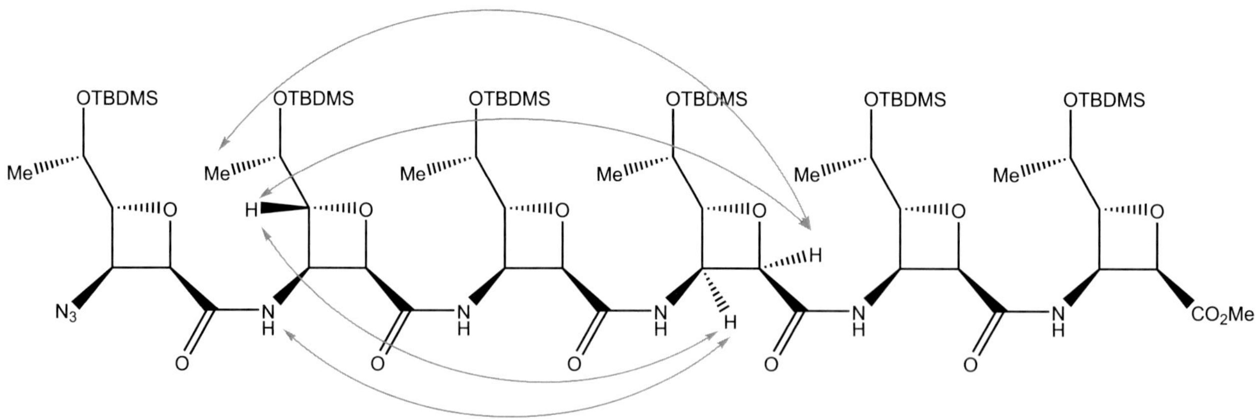

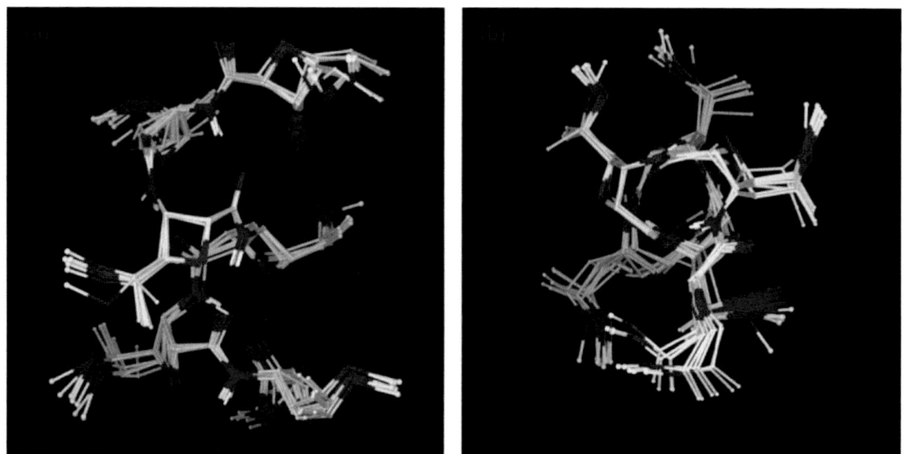

FIGURE 9.51 **Helical structures of 10 overlaid low-energy conformers derived for the hexameric β-peptide 9.23.** These are as side-on (a) and top-down (b) views. The *t*-butyldimethylsilyl (TBDMS) protecting groups are shown truncated at the silicon atom.

the H4 protons of the CD. Since these protons sit on the outer face of the CD cup, the ROEs are unlikely to arise from direct effects but instead are produced by transfer of genuine ROEs between the guest and the inner H3 and H5 protons. Rotating frame measurements have also been used extensively in the characterisation of relatively bulky organometallic complexes, such as **9.19** in Section 9.6.3. A second example is definition of the absolute configuration of the *P*-chiral diphosphine [61] **9.25**. One aspect of this was identification of the complex as either regioisomer **9.25a** or **9.25b** prior to further investigation, and this was shown to be **9.25a** through the observation of ROEs from the phenyl rings, as shown. Analysis of the ROESY data further distinguished between two possible diastereomers and so defined the absolute stereochemisty.

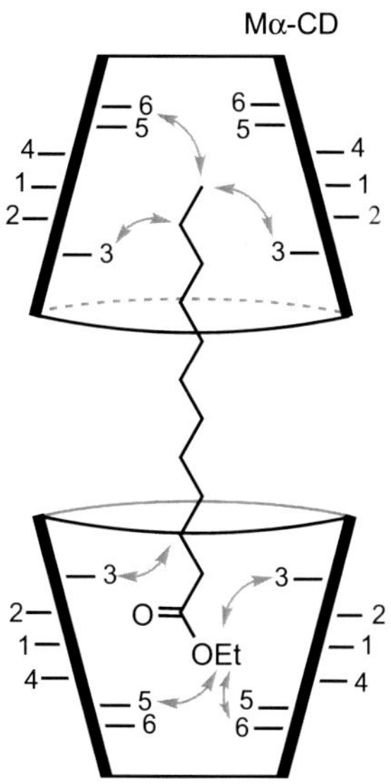

Mα-CD

9.24

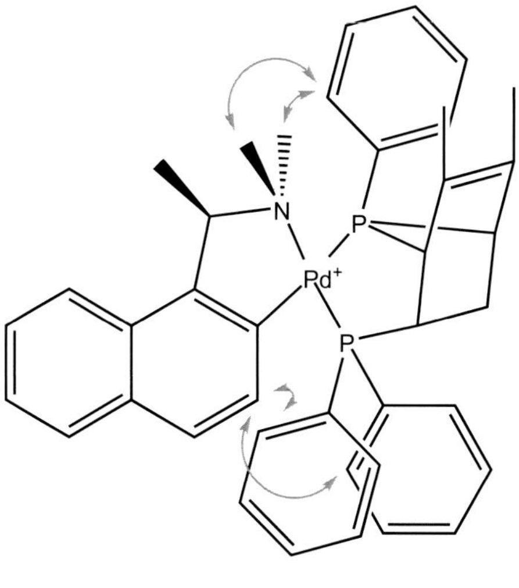

9.25a

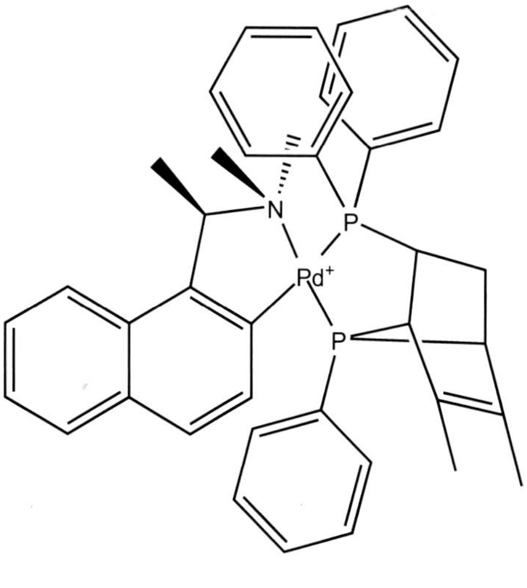

9.25b

9.8 MEASURING STEADY-STATE NOEs: NOE DIFFERENCE

Although once a ubiquitous method for observing ^{1}H–^{1}H NOEs in small molecules, the traditional NOE difference experiment is now less used, having been superseded by the more efficient 1D or 2D NOESY techniques described above. However, the NOE difference experiment may still find use on instruments not equipped with PFGs, and the steady-state

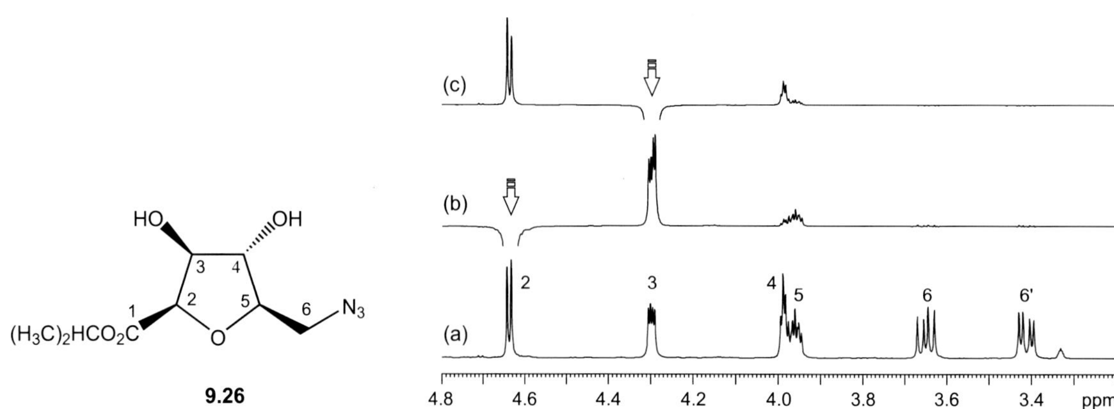

FIGURE 9.52 **The NOE difference spectrum.** Control 1D spectrum (a) and NOE difference spectra (b and c) of **9.26** in MeOD. The difference spectra show only NOE enhancements and the truncated difference signal of the saturated resonance (arrowed). The observed enhancements are consistent with the indicated 2,5-*cis* geometry across the ring oxygen of **9.26**.

methodology has relevance to the observation of heteronuclear NOEs owing to its relatively simple implementation, as described in the following section.

The basic requirement for the observation of steady-state NOEs is a suitable period of presaturation of the target resonance prior to acquisition of the spectrum, as discussed in Section 9.3 and indicated in Fig. 9.1. This dictates that steady-state measurements are derived only from 1D spectra, in which spins experiencing the NOE will produce resonances with altered intensities relative to the conventional 1D spectrum. The question then remains as to how best to display and measure these changes. Although, in principle, this may be achieved by direct integration of resonances relative to a control spectrum, in reality this is a non-trivial exercise when enhancements may only be of a few percent. Instead, the most effective approach is to use 'difference spectroscopy' and to subtract the control spectrum from the NOE spectrum, yielding a difference spectrum in which ideally the only remaining signals are the NOE enhancements and the saturated resonance (Fig. 9.52). This is the so-called 'NOE difference experiment' that played a pivotal role in structural organic chemistry for many years [62]. The purpose of the 'difference' approach is to make the *observation* of enhancements *easier* and *more reliable* but imparts no new information to the difference spectrum that was not within the original NOE spectrum.

The NOE difference experiment is illustrated schematically in Fig. 9.53. The NOE spectrum is generated by applying presaturation to the target resonance for a period τ, after which the presaturating rf is gated off and the 1D spectrum acquired with a 90 degree pulse. Historically, the presaturating rf has been applied via the 'decoupler channel', traditionally the second rf channel of the spectrometer. However, decoupling is a misleading term in the context of the NOE since spectra are acquired fully J-coupled, hence the term 'presaturating rf' is used throughout. The control spectrum is acquired in an *identical* fashion except that presaturation is no longer required. To keep the acquisition conditions for both experiments as similar as possible, which is crucial for a successful difference experiment, the presaturation frequency is moved well away from all resonances (it is placed 'off-resonance') for the control rather than being turned off altogether, typically by placing it at the far edge of the spectrum. Subtracting the resulting spectra—or FIDs followed by Fourier transformation—yields the difference spectrum. The success of this approach is critically dependent on the two spectra being identical in all respects other than those features introduced by on-resonance presaturation; if this is not the case spurious difference responses are introduced (see later in this section). In practice, the perfect subtraction of resonances is experimentally very demanding, so various procedures have been developed to minimise undesirable variations and hence artefacts.

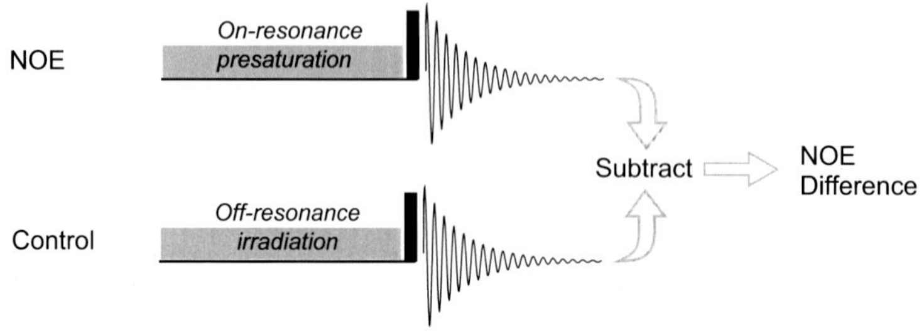

FIGURE 9.53 **Procedure for generating the NOE difference spectrum.**

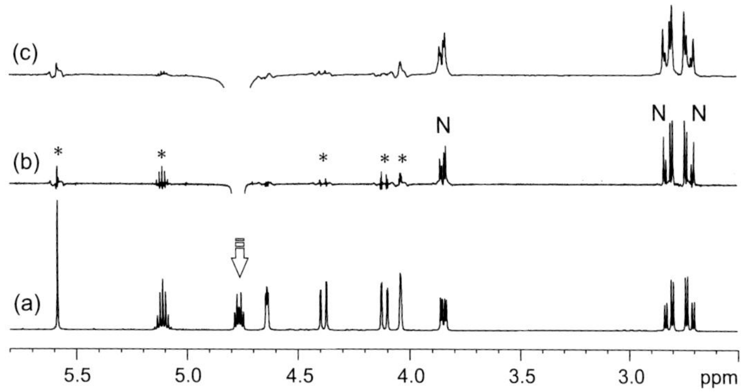

FIGURE 9.54 **NOE difference artefacts.** These are indicated by an asterisk in the difference spectrum (b), shown above the control spectrum (a), and arise when two similar, but not quite identical, signals are subtracted in an attempt to reveal the genuine NOEs (N). The artefacts can be attenuated a little through the use of line-broadening window functions (c, lb = 1 Hz).

9.8.1 Optimising Difference Experiments

9.8.1.1 Minimising Subtraction Artefacts

NOE experiments typically require significant spectrometer time because the enhancements being sought are rather small. The NOE difference experiment represents a stringent test of both short- and long-term spectrometer stability, since changes in rf phase, frequency or magnetic field contribute to *difference artefacts* (Fig. 9.54). These appear as dispersion-like signals which may mask genuine NOEs, in addition to being unsightly. Improvements in spectrometer design have gone a long way to reducing these instabilities although care is still required to achieve optimum results. The far cleaner data that are achieved with the gradient-selected transient experiments described earlier make these preferable for routine analysis and have led to decline in the use of NOE difference.

The first consideration is a stable sample environment, including sample temperature, a point of particular concern for aqueous solutions and for solutes with temperature-dependent shifts. Activity in the vicinity of the magnet has long been considered a contributing factor to poorer difference spectroscopy. How significant this is will depend on the physical location of the instrument, the field strength and on activities in adjacent laboratories or corridors, but in any case it is clearly wise to minimise disturbances of the field. Acquiring data overnight or at weekends may prove beneficial if such interferences prove to be detrimental. Optimal performance can also be achieved by selecting solvents with a strong and sharp lock signal where possible, for example acetone. Solvents with weak deuterium signals such as $CDCl_3$ or broad deuterium resonances such as D_2O, may perform less well in this respect, depending on instrument performance.

Short-term random instrumental instabilities are suppressed by acquiring a large number of scans, just as random noise is similarly reduced by signal averaging, and the improved signal-to-noise in the resulting spectra also aids reliable identification of enhancements. Longer term instabilities, such as from small field or temperature drifts, are addressed by interleaving acquisitions between the control and the NOE experiments, so that over the course of the experiment all spectra experience the same net variation. A typical approach when multiple NOE measurements are required of a sample is to acquire 8 transients for each experiment and cycle round all irradiation frequencies. For further signal averaging, the appropriate data sets are co-added to acquire the desired total number of transients which must be identical for each data set collected. For multiple NOE experiments a single control is usually sufficient for all difference calculations. Also included in each acquisition must be two or four ‘dummy scans’ (in which the recorded data are discarded) to ensure the complete decay of saturation effects from the previous irradiation. Since presaturation is required for each acquisition, the usual relaxation delay between transients is not necessary and is wholly replaced by the presaturation period.

Subtraction artefacts can be further reduced by suitable processing of the data. Difference spectra can be generated by subtracting two FIDs (the NOE and the control) and Fourier transforming the resulting FID, or alternatively by directly subtracting the NOE and the control spectra, in which case they must both be processed with *digitally identical* phase corrections prior to subtraction. The results of these methods are equivalent except when spectra exhibit a large dynamic range (ie where very large signals exist in the presence of very small ones) in which case subtracting FIDs may reduce spectrum noise [63]. In either approach, applying mild line-broadening to the spectra, typically of a few hertz, helps reduce difference artefacts (Fig. 9.54c). The rationale behind this is that intensity of the residual signal obtained by subtracting closely overlapping Lorentzian lineshapes is inversely proportional to the resonance linewidths [64].

9.8.1.2 Optimising Presaturation

Selection of the optimum presaturation time τ is highly dependent on the information required of the molecule and on its size and structure, so general guidelines only can be presented here. Since steady-state measurements are only of use in the positive NOE regime, these considerations are limited to small- to medium-sized molecules and, in fact, most of what we need to know has already been presented in the previous sections since the choice of τ is dictated by the kinetics of the NOE. Thus, if measurements of genuine quantitative steady-state enhancements are required, then τ must be greater than $5T_1$ of the slowest relaxing spins, which could make signal averaging very time consuming. However, most structural work does not require the full steady-state to be attained so long as a measurable NOE has been able to develop, allowing more transients to be collected in the time available, which itself facilitates the observation of NOEs. The use of shorter presaturation times then favours the observation of short-range interactions since these are quickest to develop. In contrast, long-range interactions and indirect effects are slow to build and are enhanced with extended presaturation. For the majority of structural studies, information on immediate neighbours is sufficient, so relatively short presaturation times of around 4 or 5 s (approximately three times the longest T_1) tend to be used in routine NOE investigations of small organic molecules ($M_r < 500$). Under such conditions, one should not be surprised to find that longer range interactions cannot be observed, and if these are expected to be informative, or if indirect effects may provide useful geometrical information, further experiments with extended presaturation ($\gg 5T_1$, possibly tens of seconds) will be required in order to detect these. Clearly, it is advantageous to have some prior knowledge of the approximate T_1 values of protons in the molecule, and these may be measured by the quick inversion recovery method of Section 2.4. This additionally provides supporting data since remote protons that are isolated from dipolar relaxation sources will exhibit unusually long T_1s as well as slow-to-develop NOEs.

9.8.1.3 Selective Saturation and Selective Population Transfer

To ensure the integrity of NOE data it is essential that only a single resonance is subject to presaturation at any one time. Even a small degree of saturation of a neighbouring resonance caused by spillover of the presaturating rf can be detrimental for structural studies. Such spillover can usually be readily observed in the difference spectrum and should call for great caution when interpreting data. Naturally, it is more useful to spot this failing before the experiment is left running for many hours and to take measures to alleviate it. An effective approach to this is to directly overlay the NOE and control spectra obtained after the first experiment cycle (eg after eight transients for each). While NOE enhancements are unlikely to be apparent at this stage (unless very large), saturation effects close to the target resonance are usually clear and may suggest changes to the experimental setup are required.

To maintain selectivity, each target resonance should be well removed from all others and, as for 1D NOESY, changes of solvent may be useful here. Beyond this, a variety of experimental procedures can help. The most direct way to reduce the frequency window over which the rf is effective and hence improve selectivity is to attenuate its power. This also reduces the degree of saturation of the target resonance which in turn reduces the *absolute* magnitude of the NOE enhancement, so a compromise must be sought between the two, usually with the emphasis on selectivity. Additional artefacts may also arise when using low presaturation powers if multiplets are subject to *unequal* perturbation. This arises from so-called *selective population transfer* (SPT) which has already been described in Section 4.4. This is a manifestation of polarisation transfer between J-coupled spins and is related to the process by which crosspeaks are generated in COSY spectra. The responses appear as *anti-phase* multiplets for those spins J-coupled to the saturated spin (Fig. 9.55). The integrated intensity of such anti-phase lines is zero if correctly phased, so should not interfere with NOE quantification, although the potentially intense SPT responses could be distracting and may mask genuine NOE responses, so are well worth suppressing.

One approach to achieving even saturation of a multiplet when using weak rf powers to maintain selectivity is to cycle the presaturation frequency between individual lines *within* each multiplet [65]. The process involves irradiating each line for a short period in turn, then repeating the sequence a number of times to achieve the desired total presaturation period. For each irradiation the aim is to saturate only a single line rather than the whole multiplet, so considerably lower rf powers may be used, while the cycling ensures approximately equal suppression across the whole multiplet. An important consideration here is the saturation period used for each line; too short and the saturation may be ineffectual (and unwelcome frequency modulation artefacts introduced) while too long a period will enable relaxation to become effective leading to uneven saturation. The relaxation behaviour of the spins is again important, and in general periods of 50–300 ms work well for most small molecules; again trial and error is the best approach to optimisation (Fig. 9.56). Thus, a five-line multiplet with 200-ms saturation of each line requires five cycles to achieve a total presaturation period of 5 s. Such sequences can be readily programmed to operate automatically on modern spectrometers.

An additional approach to suppressing SPT distortions is to collect spectra with an *exact* 90 degree observation pulse. This effectively spreads saturation evenly throughout all multiplet components and so removes the source of these

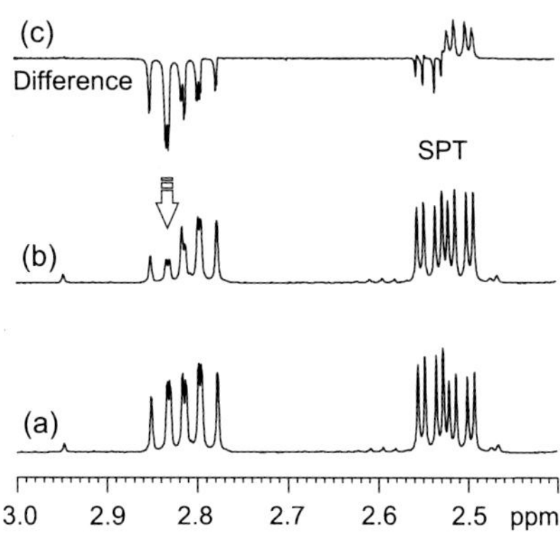

FIGURE 9.55 **SPT artefacts.** Unequal saturation of a resonance (b) can cause SPT intensity distortions to appear at its J-coupled neighbours. These appear as anti-phase patterns in the difference spectrum b–a (c) which have zero net integral but may mask genuine NOEs.

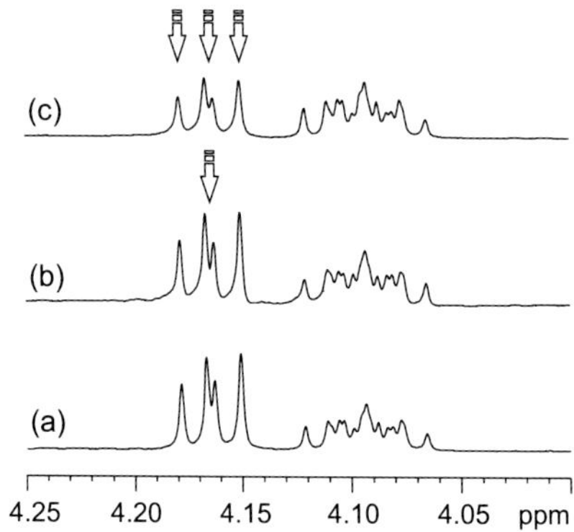

FIGURE 9.56 **Selective multiplet saturation through frequency cycling.** (a) Control spectrum, (b) single-frequency, low-power presaturation applied to the centre of the multiplet and (c) presaturation using the same power but with frequency cycling over the four lines. The low power ensures the neighbouring multiplet is untouched.

distortions. The accuracy of the 90 degree pulse can be improved by the use of a composite pulse (Section 12.1) for which the $270_x360_{-x}90_y$ sequence has been suggested [66]. The 90 degree acquisition pulse also leads to maximum signal, and since long presaturation periods are used between transients, this represents a suitable choice for optimum sensitivity and is standard for the NOE difference method. In situations where a multiplet is overlapped by another resonance, the frequency-cycling method is unsuitable for uniform multiplet suppression, but use of an accurate 90 degree pulse means only a single line from the multiplet need be available for saturation for the experiment to work, although the absolute magnitude of the NOE(s) will be small in such cases because of the small degree of population perturbation.

9.8.1.4 Quantifying Enhancements

The quantification of percentage NOE enhancements can be made most economically by direct analysis of the NOE difference spectrum alone, rather than the perhaps more obvious option of directly comparing integrals between the control and NOE spectra. The saturated peak may be used as an internal reference for these measurements with the peak of a single resonance referenced to –100% or that of a methyl group to –300% and so on (the minus signs arise from the *convention* of plotting enhancements as positive-going responses which defines the saturated peak as having negative intensity when NOEs themselves have positive sign). All NOE enhancements are then integrated relative to this reference and the percentages obtained directly. In doing this, one is at the same time compensating for incomplete saturation that may have been brought about by using lower presaturation powers; by scaling the reference peak to assume complete saturation, we likewise scale all enhancements. This procedure is justifiable since absolute NOE enhancements scale in direct proportion to the degree of saturation. The careful use of baseline correction of the difference spectra prior to integration should be considered since baseline errors can make dramatic differences when measuring small enhancements. Even so, very accurate measurements of enhancements are usually of limited use in the final analysis since such small differences cannot be interpreted meaningfully. Quoting results to 1% or perhaps 0.5% is sufficient for most qualitative interpretations where the emphasis should be on interpreting a collection of enhancements rather than relying on a single percentage measurement. Note that use of the targeted resonance as a reference intensity for quantifying enhancements in this manner is not possible for 1D transient NOE experiments (perhaps the greatest limitation of this approach), as discussed in Section 9.6.2.

9.9 MEASURING HETERONUCLEAR NOEs: HOESY

Beyond its general use as a means of sensitivity enhancement of spin-½ nuclei with low magnetogyric ratios, the specific heteronuclear NOE has occasionally been used as a tool in structural studies [67] and is capable of providing a unique source of structural information in favourable circumstances. Techniques for its observation largely parallel those for homonuclear experiments as both 1D and 2D transient or steady-state experiments, so there is nothing fundamentally new to understand here. The main limitation with these approaches is often the low sensitivity associated with observation of the low-γ spin, meaning heteronuclear NOEs tend to be far less used than their homonuclear proton counterparts. The most widespread applications have involved the ^{1}H–^{13}C NOE, in which the proton is saturated and the carbon observed, the ^{1}H–^{6}Li and ^{1}H–^{7}Li NOE and increasingly the ^{1}H–^{19}F NOE, in the context of structure definition and ion pair association. In the case of carbon-13, the relaxation of a proton-bearing carbon is dominated by its dipolar interactions with this proton, meaning only quaternary carbons tend to show useful specific long-range NOEs. Furthermore, the selective irradiation of a proton resonance is often restricted to the parent ^{12}C line leaving the ^{13}C satellites in the spectrum unaffected, which means the ^{13}C centre associated with the target proton usually does *not* show enhancement under these conditions. The lithium isotopes are somewhat exceptional in that they are both quadrupolar, yet their small quadrupole moments allow dipolar relaxation and hence the NOE to remain operative. Lithium-6 in particular exhibits the lowest quadrupole moment of any quadrupolar nucleus, with caesium-133 being the only other quadrupolar nucleus to give measurable NOEs [68]. The growing interest in fluorine chemistry and the wider availability of instruments equipped for $^{19}F/^{1}H$ NMR means heteronuclear fluorine NOEs are more readily measurable, and optimised sequences for this are presented later.

9.9.1 2D Heteronuclear NOEs

Often when heteronuclear NOEs are employed their observation is through the use of 2D experiments for the detection of transient enhancements, avoiding the need for any form of selective irradiation. The traditional approach utilised observation of the lower-γ heteroatom (X) in detecting NOEs *from* the proton *onto* X, denoted $X\{^{1}H\}$, whereas more recent sequences have adopted the 'inverse' approach and have made use of proton detection in particular, $^{1}H\{X\}$. This latter approach offers two possible benefits. Firstly there is the potential for enhanced sensitivity by observing the proton. However, in this it is the $X \rightarrow {}^{1}H$ NOE that is detected which is intrinsically weaker than the equivalent ${}^{1}H \rightarrow X$ transfer (with the exception of ^{19}F–^{1}H NOEs for which there is little difference for small molecules). The actual sensitivity gain expected from proton detection arises from a balance between these effects (see discussions in Ref. [69]) although some sensitivity advantage is expected in most cases. Secondly, and often more importantly, there is the ability to acquire the typically more crowded proton dimension at higher resolution by observing this directly and frequency labelling the heteroatom in the indirect dimension.

The classic 2D sequence [70,71], referred to as heteronuclear Overhauser effect spectroscopy (HOESY, Fig. 9.57a) parallels that of homonuclear NOESY, with the additional 180 degree pulse during t_1 serving to refocus the $^{1}J_{XH}$ coupling evolution and hence provide X-nucleus decoupling in f_1. The use of an optional purging gradient during the mixing time τ_m serves to eliminate residual transverse components during this period. Equations have been presented to allow estimation of the optimum τ_m for observing structurally informative long-range NOEs to non-protonated nuclei [72]. For the case of the ^{1}H–^{13}C NOE where carbon relaxation times are considerably longer than those of the proton, maximum crosspeak intensity is achieved when $\tau_m \approx 2T_1$ of the proton. Negative three-spin ${}^{1}H \rightarrow {}^{1}H \rightarrow {}^{13}C$ effects may also become apparent [73].

The proton-detected sequences, loosely referred to as inverse-HOESY, depend critically for their success on the incorporation of PFGs. In experiments involving nuclei of low natural abundance, gradients are required to suppress the intense background resonances arising from those protons in molecules that do not contain the heteroatom at the site of interest and so prevent the required NOEs being masked by intense t_1 artefacts [74,75]. A general scheme employing gradient selection is illustrated in Fig. 9.57b [76] and again includes the option of a purging gradient during τ_m. The gradient ratios G_1:G_2 are

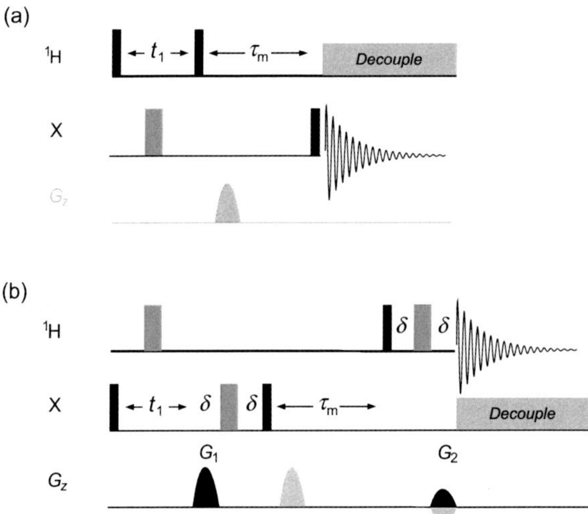

FIGURE 9.57 **2D sequences for measuring transient heteronuclear NOEs.** (a) Classic HOESY and (b) gradient-selected inverse-HOESY employing the echo–antiecho scheme. Both sequences may employ the optional purging gradient during the mixing period (grey) and for (b) the gradient ratio G_1:G_2 is set according to γ_H:γ_X.

set according to the magnetogyric ratios of the spins γ_H:γ_X and G_2 alternated in sign according to the echo–antiecho protocol for f_1 quad-detection. One potential weakness with many gradient-selected sequences proposed to date is the potential for signal attenuation arising from molecular diffusion during the NOE mixing period (see the chapter *Diffusion NMR Spectroscopy*), a particular problem if NOEs are slow to develop, requiring τ_m to be long, and where molecular diffusion is rapid, such as in the case of smaller molecules.

9.9.2 1D Heteronuclear Nuclear Overhauser Effects

Traditionally, 1D heteronuclear NOEs have been recorded as steady-state experiments, and the heteronuclear sequence (Fig. 9.58) is essentially identical to the homonuclear equivalent, except for the addition of (optional) broadband proton decoupling during acquisition. Since this generates non-specific NOE enhancements, a sufficient recovery period must be left between transients to allow this to decay if this is employed. The presaturation period is now dictated by the potentially long relaxation times of the heteroatom, but beyond this consideration similar steps are required for optimisation as for the homonuclear experiment. Problems associated with selective presaturation of the target proton resonance may be handled in a similar fashion, although here SPT interferences are suppressed by the use of broadband decoupling. A 1D gradient-selected transient heteronuclear NOE experiment has also been presented that makes use of selective excitation in the generally better dispersed carbon dimension and which uses proton detection for enhanced sensitivity [69]. This also demonstrates superior signal suppression through gradient selection and provides a useful route to the detection of these transient NOEs. In this approach it is NOEs *from* the carbon-13 centres *onto* the observed protons that is employed, and thus does not exactly parallel the 1D steady-state experiment for which the protons are the source of the NOE.

With growing interest in fluorine chemistry for novel pharmaceuticals and agrochemicals, and the associated drive to develop versatile fluorine-labelling methods, NOE techniques have been also developed to benefit from the high abundance of ^{19}F in the observation of NOEs from fluorine to proton [77]. While the 2D HOESY methods described above may be employed, it is often the case that 1D variants can provide the required information more rapidly and with greater resolution. This is especially true when only a single fluorine label exists within the molecule. In these cases the sequence of Fig. 9.59 provides for the clean observation of NOEs from ^{19}F to ^{1}H, using proton observation, denoted ^{1}H{^{19}F}. The sequence begins with the application of sequential 90 degree pulses on fluorine to invert the target resonance, from which the heteronuclear NOE develops. At the same time, proton magnetisation is placed in the transverse plane by a 90 degree pulse and is subsequently dephased by the bipolar gradient pulse pair that bracket 180 degree pulses on both ^{1}H and ^{19}F. These serve to minimise the recovery of proton magnetisation during the mixing time (similar to those used in the 1D ^{1}H NOESY sequence of Fig. 9.37) while allowing the heteronuclear NOE to develop. Finally, the ^{1}H spectrum is recorded following proton excitation. On the subsequent transient, initial fluorine 90 degree pulses act in opposition (phases *x* and –*x*) so that no inversion of the fluorine resonance occurs, with the resultant data subtracted from that of the first transient through receiver inversion. This two-step cycle provides for cancellation of all background proton magnetisation and reveals the desired heteronuclear NOEs. Optionally, and further to this cancellation cycle, a pair of ^{19}F 90 degree pulses may be applied at the end of the mixing period whose net effects add (net 180 degree pulse) or cancel (net 0 degree pulse) on alternate transients. These serve to cancel any anti-phase ^{1}H–^{19}F structure that can arise from heteronuclear J_{FH} couplings, yielding a clean, inphase NOE peak structure. In situations where multiple fluorine centres exist, the sequence may be made ^{19}F selective through the addition of a selective gradient spin-echo between the first two fluorine pulses, allowing individual fluorine resonances to be used as sources of heteronuclear NOEs [77]. As an illustration, the 1D ^{19}F–^{1}H HOESY spectrum of **9.27** is illustrated in Fig. 9.60.

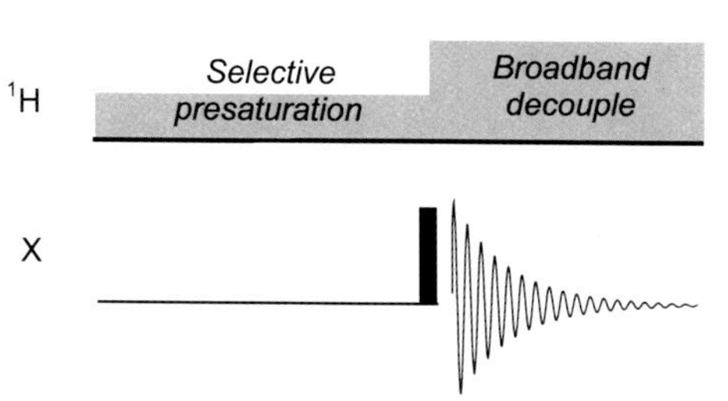

FIGURE 9.58 **The classic sequence for measuring 1D steady-state heteronuclear NOE enhancements.**

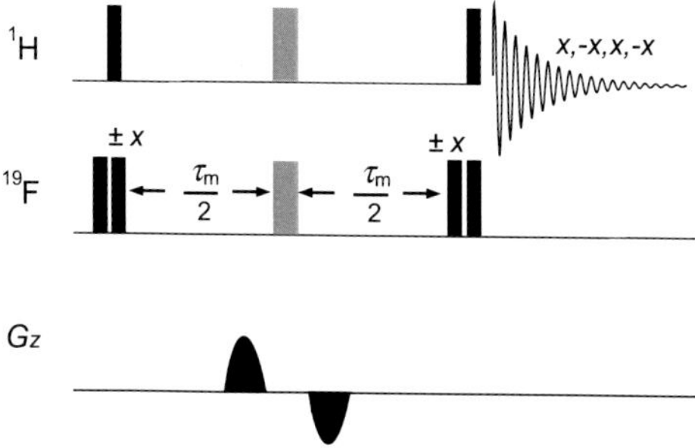

FIGURE 9.59 **1D HOESY sequence for observing ^{19}F–^{1}H NOEs through proton detection.**

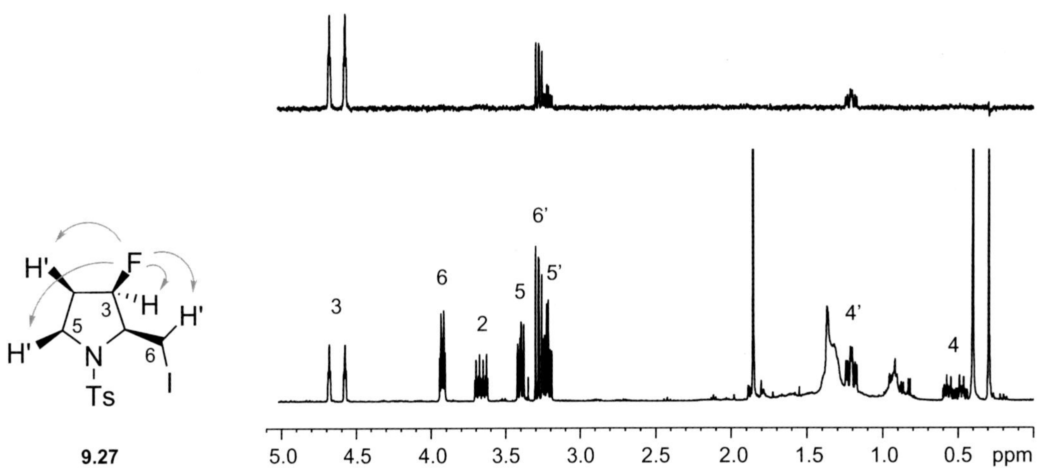

FIGURE 9.60 **Partial 1D ^{19}F–^{1}H HOESY spectrum of 9.27 recorded with a mixing time τ_m of 500 ms, shown above the reference ^{1}H spectrum.**

9.9.3 Applications

Heteronuclear experiments have found use in ^{13}C NMR spectroscopy for the assignment of quaternary carbons and for the determination of stereochemistry in situations where the ^{1}H–^{1}H NOE cannot be applied. Quaternary assignments are nowadays more often made via long-range correlation techniques such as HMBC, although the inability to distinguish *a priori* two- and three-bond correlations may lead to ambiguity. The short-range nature of the NOE means it can prove useful in distinguishing nearby non-protonated centres (ie those two bonds away [78]) from more distant ones [79,80]. The use of heteronucelar NOEs in the determination of stereochemistry is illustrated by the example of compound **9.28** [81]. Homonuclear proton NOEs could not be used to distinguish *E* and *Z* isomers of the oxazolones shown, although this was possible from the heteronuclear NOEs generated at the carbonyl carbon following saturation of the methyl protons.

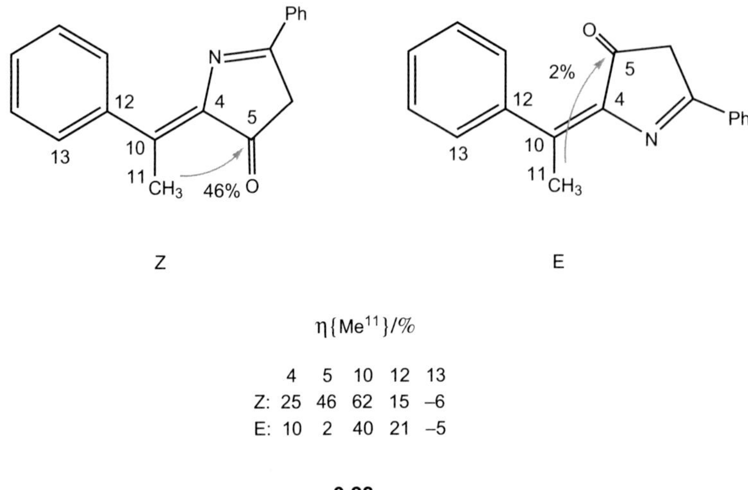

η{Me[11]}/%

	4	5	10	12	13
Z:	25	46	62	15	–6
E:	10	2	40	21	–5

9.28

The observation of $^{19}F\{^{1}H\}$ NOEs have found widespread use in studies of ion pair interactions and have frequently been employed alongside ^{1}H and ^{19}F diffusion measurements (see the chapter *Diffusion NMR Spectroscopy*) [82,83]. Typically, the classic HOESY experiment has been employed with ^{19}F observation when small counterions such as BF_4^-, PF_6^- and $CF_3SO_3^-$ are being considered. For example, in the complex **9.29** $^{19}F\{^{1}H\}$ NOEs located the relative orientation of the ion pair [84] which was subsequently rationalised from computation to arise from an accumulation of positive charge at the CH_2 group. The 1D $^{1}H\{^{19}F\}$ HOESY experiments described above have been used in conformational analysis of variously ring-substituted fluorinated pyrrolidines **9.30** by providing estimates of internuclear fluorine–proton distances from NOE buildup profiles [77]. These were used in combination with vicinal $^{3}J_{FH}$ and $^{3}J_{HH}$ values, which reported on dihedral bond angles, to demonstrate that the weak stereoelectronic fluorine *gauche* effect dominates the ring conformations in all cases, favouring the C^{γ}–*exo* conformation, except when steric interactions were also influential.

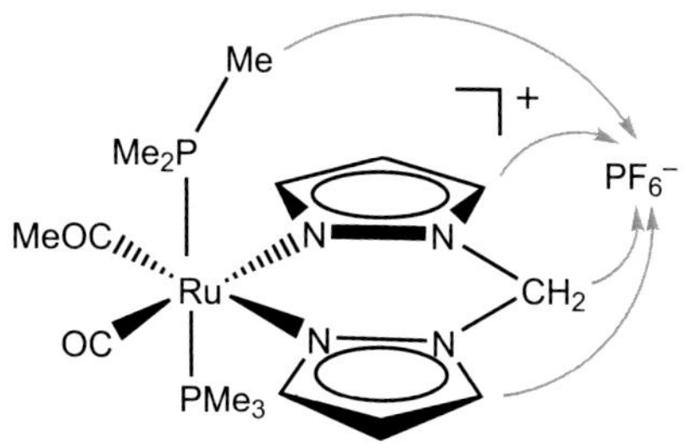

9.29

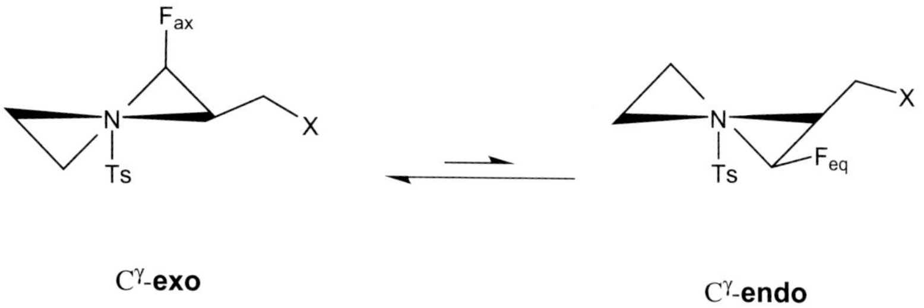

C^{γ}-**exo** C^{γ}-**endo**

9.30

Another area in which heteronuclear NOEs have proved effective is that of organolithium chemistry, where both ^{6}Li and ^{7}Li have been exploited. ^{6}Li has been investigated by the traditional $^{6}Li\{^{1}H\}$ HOESY experiment [85–87], but has been less amenable to the inverse approach [74]. This is most likely due to the longer relaxation times of this nucleus requiring longer mixing times and so leading to unacceptable signal attenuation arising from diffusion processes. In contrast, ^{7}Li appears well suited to inverse detection [74,75,88], and $^{1}H\{^{7}Li\}$ NOEs have been employed to aid in identification of the solution species arising from the dynamic equilibrium observed for the lithium organo-amidocuprate $CuLiMes(N(CH_2Ph)_2)$ **9.31** (where Mes = 2,4,6-trimethylphenyl) [88].

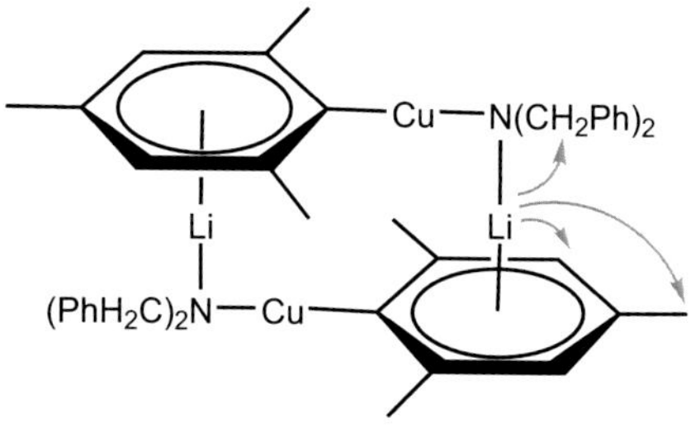

9.31

9.10 EXPERIMENTAL CONSIDERATIONS FOR NOE MEASUREMENTS

The prerequisite for any NMR experiment is, of course, a well-prepared sample (Section 3.3), but in addition to this there are a number of experimental factors that can have particular significance regarding NOE measurements. The choice of solvent and sample temperature has greater influence on the NOE than for most experiments, which can be used to good effect to control molecular tumbling rates (correlation times). This can be particularly useful for mid-sized molecules close to the $\omega_0\tau_c \approx 1$ zero NOE condition since the use of a low-viscosity solvent and/or higher temperatures will increase tumbling rates and move motion towards the extreme-narrowing limit while high-viscosity solvents and/or lower temperatures will encourage a move towards the spin diffusion limit. For NOE experiments it is worth avoiding very high solute concentrations since transient interactions between protonated molecules promote *intermolecular* dipole–dipole relaxation, which competes with the generation of the NOE itself. Another external relaxation source that can quench the NOE is paramagnetic impurities, notably certain metal ions and molecular oxygen. While these are an unwelcome addition to any NMR sample, their presence can be particularly detrimental to NOE studies since paramagnetic relaxation dominates cross-relaxation. Metal ions can be removed by filtration through suitable chelating resins whereas oxygen may be removed with the freeze–pump–thaw method described in Section 3.3. Whether it is worth the effort to remove oxygen traces depends very much on the sample being studied. For very large molecules, efficient cross-relaxation means the NOE builds rapidly to large values and as such oxygen-induced paramagnetic relaxation has relatively little significance and degassing little effect. In contrast, external relaxation for smaller molecules that tumble rapidly causes the NOE to be drastically reduced (see Fig. 9.10), and degassing is likely to be of greater benefit. Likewise, weak, slowly developing NOEs arising from long-range enhancements or indirect effects will also be enhanced if external relaxation is minimised. In general, the smaller the molecule and the further the internuclear separations being investigated, the greater the gain from solvent degassing. In practice, however, *routine* NOE experiments *are* successful in non-degassed solutions for all but the smallest molecules and, with the above caveats in mind, it is not essential to degas for the majority of organic molecules. The use of lock solvents with temperature-dependent shifts, notably D_2O, calls for particular attention regarding temperature stability and will likely benefit from active temperature regulation if experiments are to be acquired over some hours. Drift in the lock resonance frequency arising from sample temperature changes will cause a shift of *all* resonance frequencies in the spectrum as the field follows this drift. If no temperature regulation is available or if it performs inadequately then a small quantity of an additional deuterated solvent (5–10% d_6-acetone is a good choice) may be used for locking and may provide superior results (for the same reason, one should always lock on the CD_3 resonance rather than the OD resonance of d_4-methanol).

Finally, sample spinning can also be detrimental to NOE experiments (or indeed any NMR experiment when small responses are sought), particularly through the process known as 'Q-modulation', the variation in coupling between the rf coil and the sample caused by a slight wobbling of the sample as it rotates in the probe. This is sometimes evident as a modulation or jumping of the probe-tuning profile and by the appearance of responses in 1D spectra that are symmetric about main resonances at frequencies equal to the spinning speed, yet are indifferent to shimming and of random phase, both of which distinguishes them from 'spinning sidebands' caused by field inhomogeneity. Owing to their random properties they tend to be averaged away over many transients but are nonetheless undesirable artefacts. With the improved non-spinning lineshapes attainable with modern shim assemblies, sample spinning can no longer be recommended for either 1D or 2D NOE experiments.

9.11 MEASURING CHEMICAL EXCHANGE: EXSY

The influence of chemical exchange processes on NMR spectra has been described in Section 2.6 in which the concept of *fast*, *intermediate* and *slow* dynamic exchange regimes was introduced. In slow-exchange situations where separate resonances may be observed for interconverting species, the process of exchange may be detected using either 1D or 2D magnetisation transfer experiments. 1D methods have already been described in Section 2.6, so here we consider only the 2D approach, commonly referred to as exchange spectroscopy (EXSY). Although the mechanisms of chemical exchange and the NOE are quite unrelated, they share in common the transfer of longitudinal magnetisation and as such can be detected with the same 1D or 2D NMR experiments. In fact, NOESY and EXSY [89] pulse sequences are, in their most basic forms, identical. Indeed, in the NOESY spectrum of Fig. 9.26, chemical exchange crosspeaks are evident in addition to the NOEs, and arise from the exchange of hydroxyl protons with those of water dissolved in the solvent. Likewise, the 1D transient NOE sequences described above can equally well be applied to measurement of exchange rate constants [90–92] and may also serve to provide qualitative evidence of slow conformational exchange by virtue of magnetisation transfer effects, provided the exchange is rapid relative to the spin relaxation rates (see Fig. 9.19 and associated discussions above). As noted in Section 2.6, this can prove useful in the differentiation of rotamers from diastereoisomers, for example since exchange effects can be observed only between exchanging conformers [93]. The practical difference in all these considerations is that EXSY experiments depend on chemical exchange rates while NOESY experiments are defined by spin cross-relaxation rates. Therefore, the following discussions assume a basic understanding of exchange phenomenon, as presented in Section 2.6. More extensive discussions of 1D and 2D methods for studying dynamic processes may be found in reviews [94–97] and in dedicated texts [98,99].

In parallel with studies of the NOE, exchange experiments can be used both qualitatively to map exchange pathways and quantitatively to determine rate constants. In systems of multi-site exchange, the 2D EXSY experiment proves particularly powerful in the measurement of these for all pathways. However, for either application the exchange processes

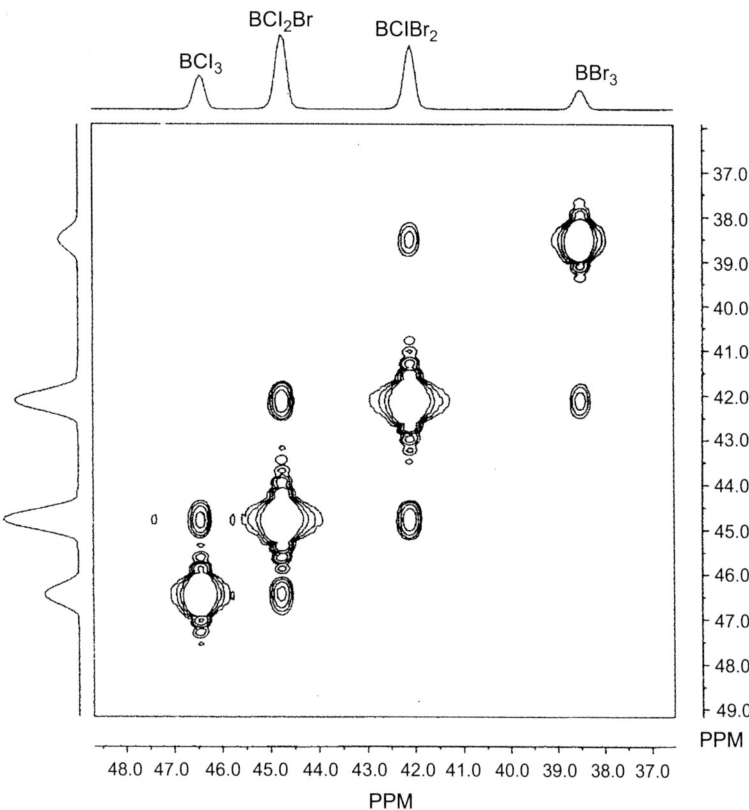

FIGURE 9.61 **The 128-MHz boron-11 2D EXSY spectrum of a 1:1 mix of BCl_3 and BBr_3 at 400 K.** The spectrum was recorded with a mixing time of 50 ms. *(Source: Reproduced with permission from Ref. [100].)*

studied must be slow on the NMR chemical shift timescale since the exchanging resonances must be resolved in order to observe transfer between them. Furthermore, this exchange must not be *too* slow otherwise relaxation processes occurring during τ_m will remove all memory of the exchange process. Thus, magnetisation transfer experiments are suitable for only a limited range of exchange rates and, as a rule of thumb, these must be *at least* comparable with the longitudinal relaxation rates ($k \geq 1/T_1$). In practice, this means these experiments are sensitive to exchange processes with $k \approx 10^2$–10^{-2} s^{-1}. The study of different nuclei provides a window on different rates within this range due to differences in relaxation times, with slower relaxing spins able to probe slower exchange processes. Fast-exchange processes that lead to resonance coalescence can be studied by lineshape analysis (also referred to as 'bandshape' analysis) as described in Section 2.6.

As an illustration of the 2D method, Fig. 9.61 shows the boron-11 EXSY spectrum of a 1:1 mix of BCl_3 and BBr_3. The exchange crosspeaks demonstrate the presence of ligand-scrambling processes and indicates the mechanism involves the exchange of a single halogen atom only [100]. Quantitative analysis of these data also provided (pseudo) first-order rate constants for each exchange process (see later discussions). In a second example, the conformational analysis of substituted ε-caprolactams was studied as part of an investigation into the mode of action of β-lactam antibiotics [101]. Exchange between the ${}^4C_{1,N}$ and ${}^{1,N}C_4$ chair forms of **9.32** was readily detected in a ^{1}H EXSY spectrum recorded at low temperature, where the exchange rate was slow (Fig. 9.62). Inversion is clearly evidenced as crosspeaks between the corresponding H2 and H6 protons of each conformer. As is often the case in proton EXSY studies, NOEs were also detected in the experiment, which in this case were exchange-averaged over the two conformational forms.

The correct choice of mixing time τ_m is again important for the successful application of EXSY and is dependent upon the exchange rate constants. Since these are unlikely to be known with any accuracy (since there would be little point in performing the experiment if they were) and as there may exist a spread of these within the kinetic system being studied, some compromise is again required. In the absence of known rates, an upper limit can be defined (at least for the case of symmetrical two-site exchange) from the knowledge that the rate constant at coalescence k_c is given by $\pi\delta\nu/\sqrt{2}$, where $\delta\nu$ is the shift difference (in hertz) in the slow-exchange limit, and hence $k < k_c$. In qualitative work the aim is to achieve maximum crosspeak intensity, and setting $\tau_m \approx 1/k$ is recommended. Under such compromise conditions, multistep transfers

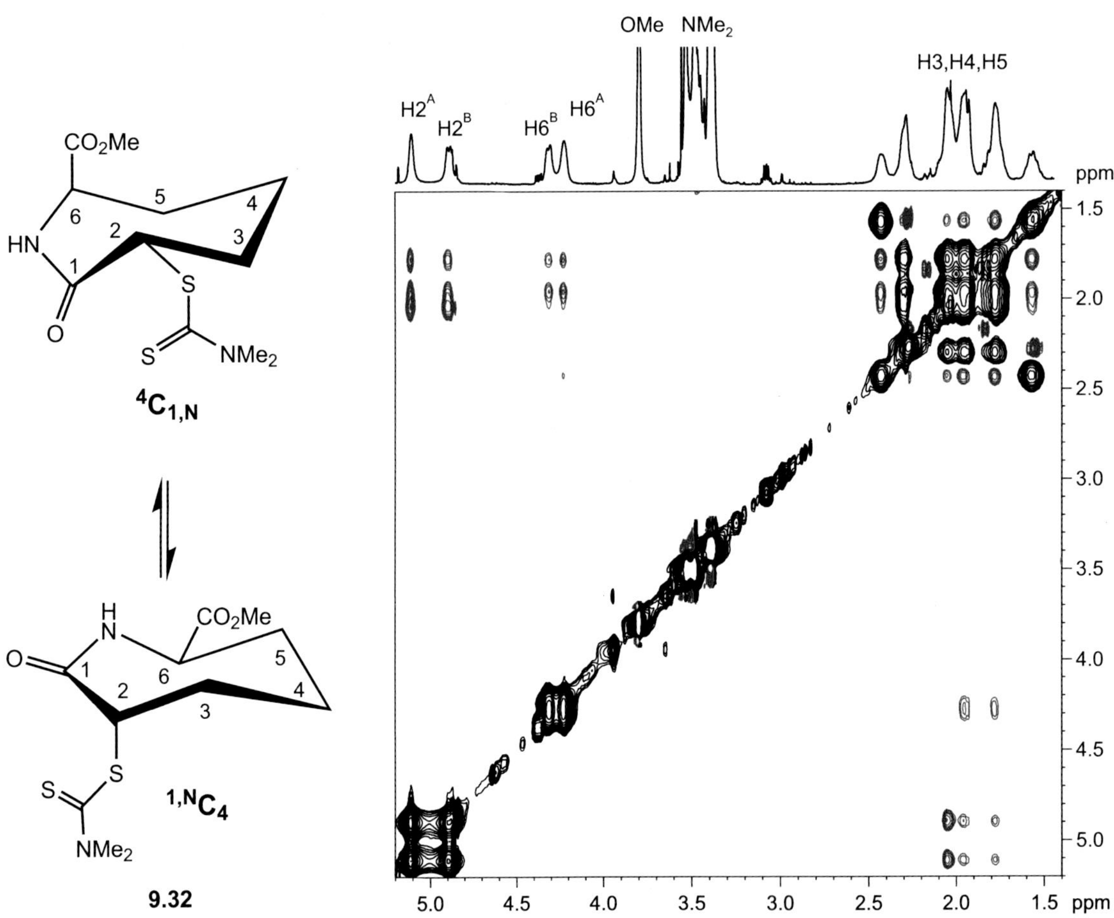

FIGURE 9.62 **The 500-MHz ^{1}H 2D EXSY spectrum of the substituted ε-caprolactam 9.32 at 224 K recorded with a mixing time of 800 ms.** Black crosspeaks arise from the exchange process, red crosspeaks from positive proton NOEs.

may operate when suitable pathways are possible, producing 'indirect' exchange peaks, which may be confused with direct peaks from relatively slow processes. For example this has been observed for the ligand-scrambling reactions of tin halides [102,103], boron halides [100] and of lead(IV) tetracarboxylates [104]. Often one has some element of control over the exchange rates through the use of temperature variation and optimisation of this can also assist in the collection of exchange spectra.

The choice of τ_m for quantitative work will be dictated by the method used to calculate k, of which there are essentially two widely used approaches [100]. The simplest involves collecting EXSY spectra over a range of τ_m values, following the initial linear growth of exchange crosspeaks (the initial rate approximation again, by exact analogy with the approach taken for NOESY analysis). The great disadvantage to this is clearly the need to collect many 2D spectra, which may place unreasonable demands on instrument time. The second approach is to determine rate constants from a single EXSY spectrum and for this the appropriate choice of τ_m is absolutely critical; too small and crosspeaks will have weak intensities which are subject to significant error, too long and the intensities become insensitive to the kinetic parameters. Methods for determining the optimum τ_m in such cases have been described [105,106]. Having obtained suitable data, computational analysis (eg with the program D2DNMR [107]) may be employed to calculate exchange rates for multi-site systems. For the simpler case of equally populated two-site exchange, explicit equations have been presented for the estimation of rate constants from diagonal and crosspeak intensities [108].

Studies involving low natural abundance nuclei such as ^{13}C or those that are present only sparingly in a molecule (eg ^{31}P) have the significant benefit of avoiding potential complications from ^{1}H–^{1}H NOE effects that may be operative. Dynamic proton studies may be subject to such interferences when exchange rates and NOE buildup rates are comparable, and in situations where molecular correlations times are such that negative NOEs are produced the differentiation of exchange and NOE peaks may be problematic owing to them displaying the same sign. In this case the use of the ROESY experiment is advantageous since NOEs will appear with opposite sign to both the diagonal and exchange peaks, although quantification of kinetic parameters can be more complex due to interfering effects from TOCSY transfers and from offset-dependent crosspeak intensities.

1D and 2D magnetisation exchange experiments have found greatest use in the study of inorganic and organometallic systems, for which a high degree of fluxionality often exists, and extensive reviews of these areas have been presented in which a wide variety of example applications may be seen [95,97,109,110]. These methods have also found application to the study of supramolecular complexes [111].

9.12 RESIDUAL DIPOLAR COUPLINGS

Definition of the relative configurations and conformations of small molecules through the combined application of scalar couplings and NOEs is now a very well–established procedure, yet may still lead to ambiguities even for small and apparently simple molecules. These procedures rely on the presence of sufficient relatively short-range spin–spin interactions, either through bonds or through space, and if these are lacking it can prove impossible to determine the relative configurations of relatively remote fragments within a molecule. This section presents an alternative approach to defining relative internuclear orientations that compliments the use of scalar couplings and NOEs [112–118]. The methodology involves the direct measurement of dipolar couplings between nuclei, an approach that is well established in the structure elucidation protocols of biological macromolecules and is gaining popularity in studies of smaller molecules. While these couplings provide the basis for the NOE, they are not usually observed directly in solution spectra owing to the time averaging to zero of the dipolar splittings as molecules tumble rapidly and isotropically in solution (Section 9.3.1). In contrast, such couplings can reach tens or even hundreds of kilohertz in the absence of motional averaging and may dominate the appearance of solid-state spectra. It is, however, possible to reveal dipolar couplings in solution spectra if the molecules are constrained such that they no longer tumble isotropically but instead have some orientational preference imposed upon them. This effect has been known for many decades notably in the study of organic liquid crystals which display very large dipolar couplings and as a result their spectra prove highly complex to analyse. More recent developments have employed weakly aligning media that introduce only small *residual* dipolar couplings (RDCs) that are typically only ~0.05% of their maximum values. In effect, these media very gently 'squeeze' the solute molecules such that their motion becomes weakly anisotropic. In such cases the splittings due to RDCs are typically less than 50 Hz and comparable with scalar couplings so do not complicate spectra excessively. They can be measured directly in spectra since any additional coupling arising from dipole–dipole interactions sums with any scalar J-coupling present. Thus, in the case of a one-bond carbon–proton dipolar coupling ${}^1D_{CH}$ the total observed splitting ${}^1T_{CH}$ will be:

$$ {}^1T_{CH} = {}^1J_{CH} + 2\,{}^1D_{CH} \tag{9.24} $$

The factor 2 here arises because the dipolar splitting adds $+{}^1D_{CH}$ and $-{}^1D_{CH}$ to the multiplet pattern, although some texts simply list the observed splitting in this expression, in which case the factor 2 does not appear. These splittings can be measured using conventional NMR techniques (see later in this section) and comparison with the equivalent experiment performed on a conventional isotropic sample (revealing ${}^1J_{CH}$ coupling constants) yields the magnitude of the residual dipolar couplings.

The value of these couplings lies in their ability to provide information on the relative orientations of internuclear vectors within a molecule. This can, in principle, indicate the relative configuration at different centres within the molecule, even when these centres are remote from each other in the structure. The magnitudes of the dipolar splittings D_{IS} are defined for two isolated spins I and S as:

$$ D_{IS} = \frac{b_{IS}}{2\pi}\left\langle \frac{3\cos^2\theta - 1}{2} \right\rangle \tag{9.25} $$

The angle θ defines the angle between the IS internuclear vector and the static magnetic field (Fig. 9.63a) and is time dependent, the brackets indicating the time average over all molecular orientations as the molecule tumbles. The term b_{IS} is the dipole–dipole coupling constant:

$$ b_{IS} = -\frac{\mu_0 \gamma_I \gamma_S \hbar}{4\pi r_{IS}^3} \tag{9.26} $$

where μ_0 is vacuum permeability, γ_I and γ_S are magnetogyric ratios of spins I and S and r_{IS} is the internuclear distance between them. If this distance is known (the bond length for directly attached nuclei) the size of the dipolar interaction

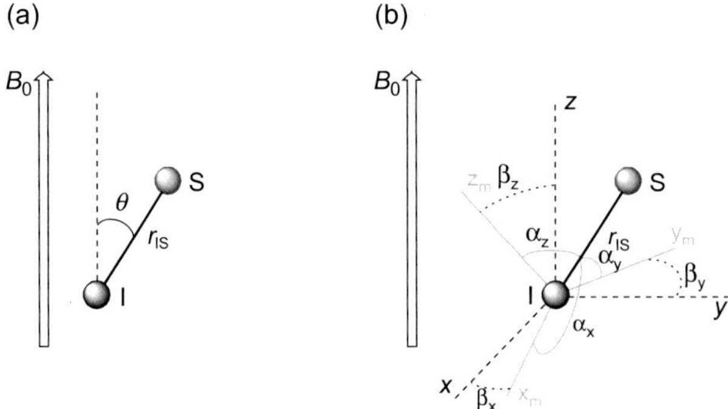

FIGURE 9.63 **Orientational dependence of dipolar couplings.** (a) The internuclear vector between spins I and S defines the angle θ with the static magnetic field B_0. (b) The molecular coordinate system x_m, y_m, z_m rotates with the molecule and defines the time-dependent angles β_x, β_y and β_z with the static magnetic coordinate system.

depends on only one unknown parameter, the angle θ. The r_{IS}^{-3} dependence suggests that dipole–dipole couplings can be measured over greater distances than NOEs (which have a r_{IS}^{-6} dependence) and may provide for the investigation of longer range interactions. More importantly, the RDCs report on angular information relative to an external reference (the applied field) and thus provide information about the relative orientations of *internuclear vectors,* making it possible to correlate information from very distant parts of a structure.

In order to relate the RDCs determined experimentally to intramolecular orientations, it is necessary to define an arbitrary molecular coordinate system that is fixed within the framework of the molecule and rotates with this, against which the angular relationships of internuclear vectors can be measured (angles α_x, α_y and α_z in Fig. 9.63b). This framework is then related to the coordinates of the static magnetic field (angles β_x, β_y and β_z) so as to define an alignment (or order) tensor for the molecule; that is a measure of the extent and direction of ordering of the molecular coordinate system imposed by the alignment media. The alignment tensor may be calculated from a sufficient number of experimentally derived RDCs with at least five independent values required to determine this. Independent values in this context means RDCs from vectors with different (non-parallel) orientations within the molecule. When combined with a proposed structural model, knowledge of the alignment tensor enables the back prediction of RDCs. These may then be compared with experimental values and the model structure refined to optimise the match between predicted and experimental data. Alternatively, the best match to differing configurational isomers may be determined, as in the examples presented later. Software packages are available that allow the calculation of alignment tensors from experimental data and hence the prediction of RDCs for proposed structures. These packages may be freely available (PALES) [119] or produced commercially (MSpin) [120] and a useful practical comparison of these two programs has been made for a model compound [121].

9.12.1 Measuring RDCs

9.12.1.1 Alignment Media

The primary requirement for the measurement of RDCs is that the analyte molecules experience rotational motion that is very slightly anisotropic, and key to this is the availability of suitable molecular alignment media. Ideally, such media are readily (commercially) available, cheap, robust, compatible with many solvents and align solute molecules only weakly. For small molecules, two classes of media have found most use to date which are based on liquid crystals or on polymeric gels [114,115].

Liquid crystalline phases typically comprise helically chiral polymers that align in the NMR tube when sufficiently concentrated in organic solvents, such as chloroform, and orient solute molecules between the polymer chains. The use of chiral liquid crystals such as poly(γ-benzyl-*L*-glutamate) (PBLG) is well proven as a method for distinguishing the spectra of enantiomers in solution by virtue of differential ordering of the R and S isomers of a molecule [122]. However, these tend to impose too great an ordering for the measurement of RDCs, leading to a loss of spectra resolution. Alternative liquid crystalline phases have been demonstrated as being advantageous over PBLG due to their weaker alignment properties and include polyisocyanides [123] and polyacetylenes [124,125], the latter of which are more easily synthesised and appear most favourable for $CDCl_3$ or CD_2Cl_2 solvents. It should also be noted that since these media are inherently chiral they also have the potential to differentiate enantiomers for RDC measurements. A liquid crystal phase compatible with aqueous media

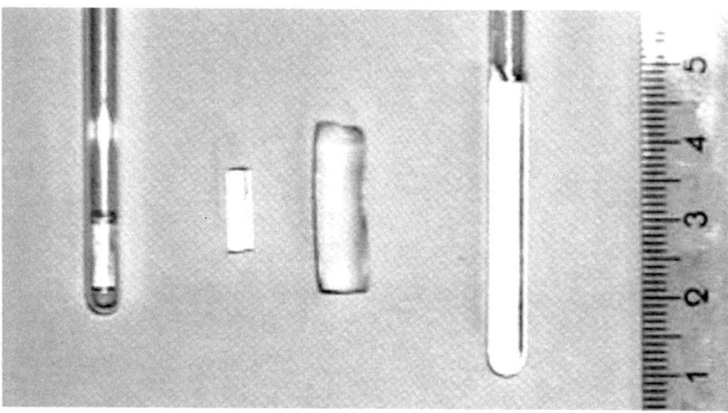

FIGURE 9.64 **The swelling of a cross-linked polystyrene gel confined within a standard NMR tube.** From left to right is shown the unswollen gel in a NMR tube, a polymer stick after polymerisation, the swollen, solvated gel and the solvated gel contained within the tube stretched along the tube axis. *(Source: Reproduced from Ref. [129] with permission from Wiley-VCH.)*

and hence suitable for the study of more polar molecules has also been demonstrated and utilises disodium cromoglycate which self-aggregates as molecular stacks to yield nematic phases comprising columnar regions. While this can strongly orient solute molecules, its alignment properties can be controlled by the addition of brine (NaCl) making it suitable for the measurement of RDCs [126]. The use of ionic liquids has also been shown suitable for water-based studies, employing *N*-dodecyl-*N*-methyl pyrrolidinium bromide in mixtures of D_2O, decanol and DMSO as the alignment media [127].

The alternative use of polymeric gels as alignment media typically involves solvation of a gel within the NMR tube such that as it swells it becomes confined by the tube walls causing it to elongate, introducing anisotropic strain that aligns the analyte molecules contained within the solution (Fig. 9.64). This process has been termed 'strain-induced alignment in a gel' (SAG) [128] and many polymeric materials have been proposed as being suitable for measurements of non-polar organic molecules in particular, although very few of these are commercially available at the time of writing (2015). Thus, cross-linked polymers such as polystyrene [128,129] and poly(dimethylsiloxane) [130] were some of the first gels to prove effective, yet are costly to produce. Polyurethane gels have been shown to be cheap and robust and may be employed without the need for polymer cross-linking [131], with both polyvinyl acetate [132] and polyacrylamide [133] gels having been proposed as alternatives more suitable for polar organic analytes since these gels are compatible with DMSO. As further noted later in this section, one of the most promising gels proposed to date for non-polar molecules appears to be those based on poly(methyl methacrylate) (PMMA) which is well suited to the common chlorinated solvents and can be reused by washing analytes out of the gel [134,135], although in common with most other gels, is available only by synthesis.

One of the great advantages of the gels, however, is the potential for controlling the degree of alignment by altering the strain applied to the gel, for which home-built devices have been described, and commercial systems are now becoming available from NMR peripheral suppliers. One such stretching apparatus employs a perfluorinated tube to allow easily adjustable alignment strength for essentially all polymer gel matrices without any associated proton background from the tube material [136]. An alternative and experimentally rather simple arrangement to control compression of a gel is to use a glass plunger within a standard NMR tube held in place by Teflon tape, such an arrangement being well suited to the use of poly(methyl methacrylate) gels [135]. This apparatus also allows the solute to be effectively washed out of the gel by a series of compression–expansion cycles where it acts as an elastic sponge enabling removal of the solute after multiple rinse cycles so that the gel can be reused. Set against the ability to control alignment is the time required to equilibrate the gels in the solvents and then to diffuse the solutes into the gels. This process typically takes days if not weeks, meaning it is not only slow but may cause issues with sample decomposition. The use of poly(methyl methacrylate) gels with the described compression system allows for the pre-soaking of gels over a few days and then diffusion of the solute into the gel within minutes, enabling a more rapid path to sample measurement. The general lack of readily available alignment media and the length of sample preparation have most likely limited the wider adoption of RDC measurements to date.

For any selected alignment medium, the relative degree of alignment it induces may be assessed by observing the deuterium resonance of the solvent, typically $CDCl_3$ for non-polar molecules. When the solvent (and hence solute) environment becomes anisotropic, the ^2H resonance will split due to the appearance of quadrupolar coupling of the spin-1 deuterium nucleus, a parameter not observed for molecules experiencing isotropic rotation (as for dipolar couplings). The greater the degree of alignment, the larger the quadrupolar splitting $\Delta\nu_Q$, which is typically only tens of hertz with sufficiently weak alignment, so providing a useful internal measure of this.

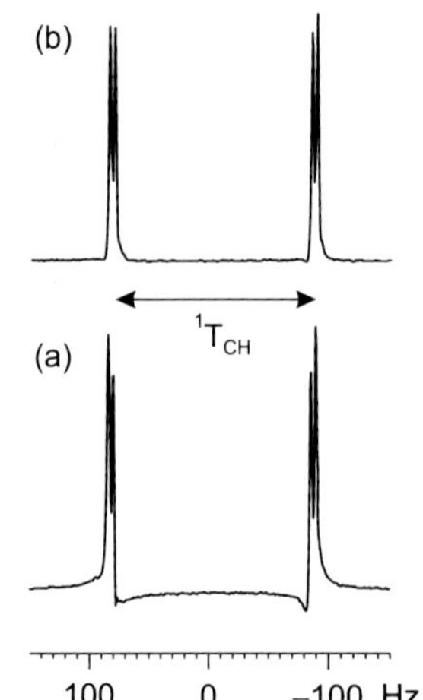

FIGURE 9.65 **Measuring $^1T_{CH}$ from HSQC.** Proton f_2 traces extracted from (a) the standard HSQC and (b) the CLIP-HSQC experiments recorded without carbon decoupling. In each case the HSQC INEPT delays were optimised for $^1T_{CH}$ = 145 Hz whereas the true value was 170 Hz. Only the CLIP-HSQC experiment produces a pure in-phase multiplet structure.

9.12.1.2 Which RDCs to Measure?

While the appearance of RDCs between any neighbouring nuclei may be anticipated for molecules experiencing anisotropic motion, the most readily observed and conveniently employed RDCs of small organic molecules are those between a proton and its directly bound carbon, designated $^1D_{CH}$. Although in a static sample these splittings are ~23 kHz, in a weakly aligned medium these are more typically a few tens of hertz. The corresponding one-bond scalar couplings $^1J_{CH}$ are relatively large in magnitude (>120 Hz), meaning the addition of these residual dipolar splittings can be readily observed according to Eq. 9.24 and will lead to an increase in $^1T_{CH}$ when $^1D_{CH}$ is positive or a decrease when this is negative, so yielding crucial sign information in addition to magnitude. Critically, the H–C distance is also constrained and essentially constant, so does not itself contribute to variations in measured dipolar couplings. Measurement of $^1T_{CH}$ can be made most readily from proton-coupled carbon spectra when sensitivity allows, but given that this is often not the case, it is commonplace to extract these values from proton (f_2) traces taken from carbon-coupled HSQC spectra. In principle, the standard HSQC sequence (Section 7.3.1) without the inclusion of carbon decoupling can be used for this purpose, but in practice phase distortions within the proton traces can lead to erroneous measurements. These distortions arise when the Δ delays in the reverse INEPT step of HSQC do not match the $^1T_{CH}$ values present in the analyte molecule, leading to incomplete refocusing of proton coupling (Fig. 9.65a), a situation more likely to occur when a large spread of $^1D_{CH}$ values exists within a partially aligned molecule. Thus, it is beneficial to use a modified version of the experiment named clean inphase HSQC (CLIP-HSQC) that removes the anti-phase components giving rise to unwanted phase distortions (Fig. 9.66) [137]. In this, an additional 90 degree carbon pulse immediately precedes detection of the proton FID in the absence of carbon decoupling. This purging pulse causes the anti-phase terms to be transformed into invisible multiple-quantum coherence and so eliminates them from the final spectrum, yielding pure phase traces (Fig. 9.65b) that reflect true $^1T_{CH}$ values. Comparison of these values with the corresponding $^1J_{CH}$ magnitudes recorded for the analyte under isotropic

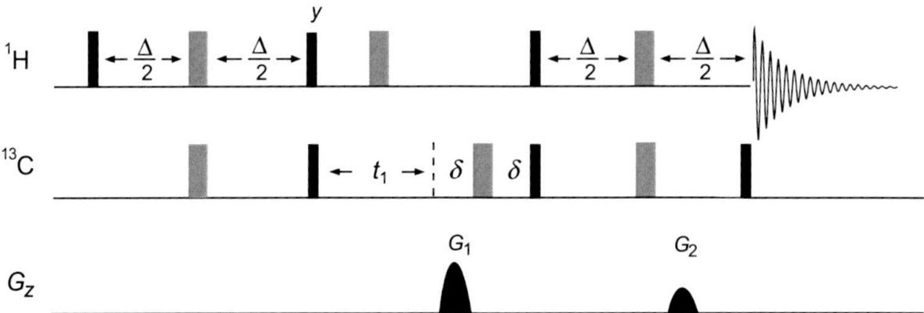

FIGURE 9.66 **The CLIP-HSQC sequence.** An additional 90 degree purging pulse is applied prior to data collection which does not employ carbon decoupling, so that the $^1T_{CH}$ splittings may be observed in the proton dimension.

conditions, that is, in the absence of any aligning media, will yield the required $^1D_{CH}$ RDC values. As an alternative, the D_{HH} splitting between diastereotopic geminal protons may also be employed since their internuclear separation also remains fixed, but these measurements tend to be hampered by strong coupling between the spins so have found significantly less use.

Whichever experimental approach is taken, the resolution of spectra recorded in aligning media will be degraded relative to that of an isotropic solution. This is in part because of the potentially numerous additional dipolar couplings that are introduced between all spatially close nuclei, especially neighbouring protons, leading to net resonance broadening, but also due to the heterogeneous nature of the samples containing the necessary alignment media. The most recent versions of the CLIP-HSQC experiment incorporate pure shift detection described in Section 8.5 to remove all proton–proton couplings and so improve spectral resolution for greater accuracy in RDC measurements [138,139].

9.12.2 Applying RDCs

A simple conceptual illustration of the application of RDCs is in the differentiation of axial and equatorial protons in pyranose sugars **9.33** through the consideration of one-bond $^1D_{CH}$ values [140]. Within the molecular framework all axial protons share the same parallel orientation (bond vectors in black) and thus exhibit RDCs of the same size whereas the equatorial protons point in different directions (vectors in red) and so display different values, enabling their identification. A similar analysis allows definition of the relative configuration of a dihydropyridone **9.34** [141]. In the *trans* isomer the highlighted C–H bond vectors of the stereogenic centres share the same orientation within the molecular frame and hence have similar $^1D_{CH}$ RDC values whereas in the *cis* isomer these differ by virtue of the bond vectors experiencing differing orientations. A more thorough use of RDCs in the assignment of stereochemistry demands the computational analysis indicated above to define molecular alignment tensors and so relate bond geometries, and when this is possible the capabilities of this methodology become highly attractive and uniquely informative, as the examples later in this section illustrate.

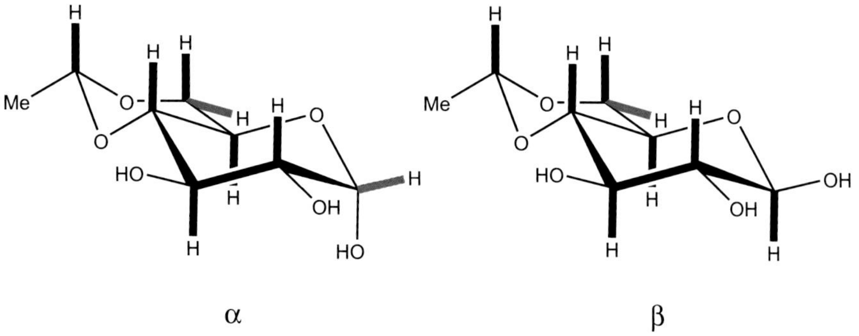

α β

9.33

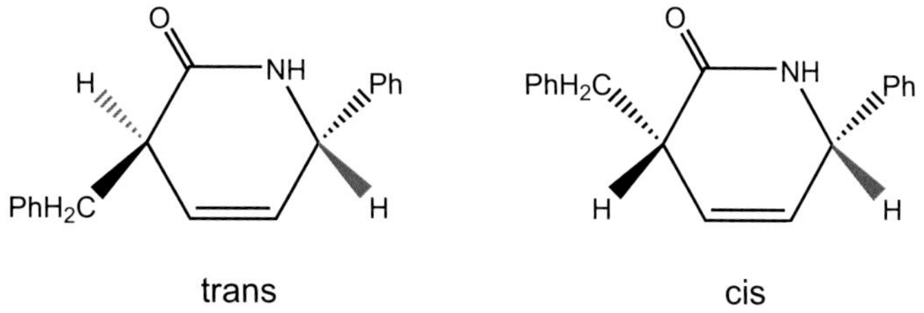

trans cis

9.34

In our first example, we consider the structure of the sesquiterpene lactone ludartin **9.35** whose structure was elucidated through a combination of classical NMR and by chemical transformations which were originally required to unambiguously define the epoxide stereochemistry as **9.35a** rather than **9.35b**. However, it has more recently been shown that this definition could be made on the basis of RDC-based analysis, even though the combined J_{HH} and NOE analyses proved ambiguous due to anticipated similarities for both isomers [134]. For this, the preferred conformations

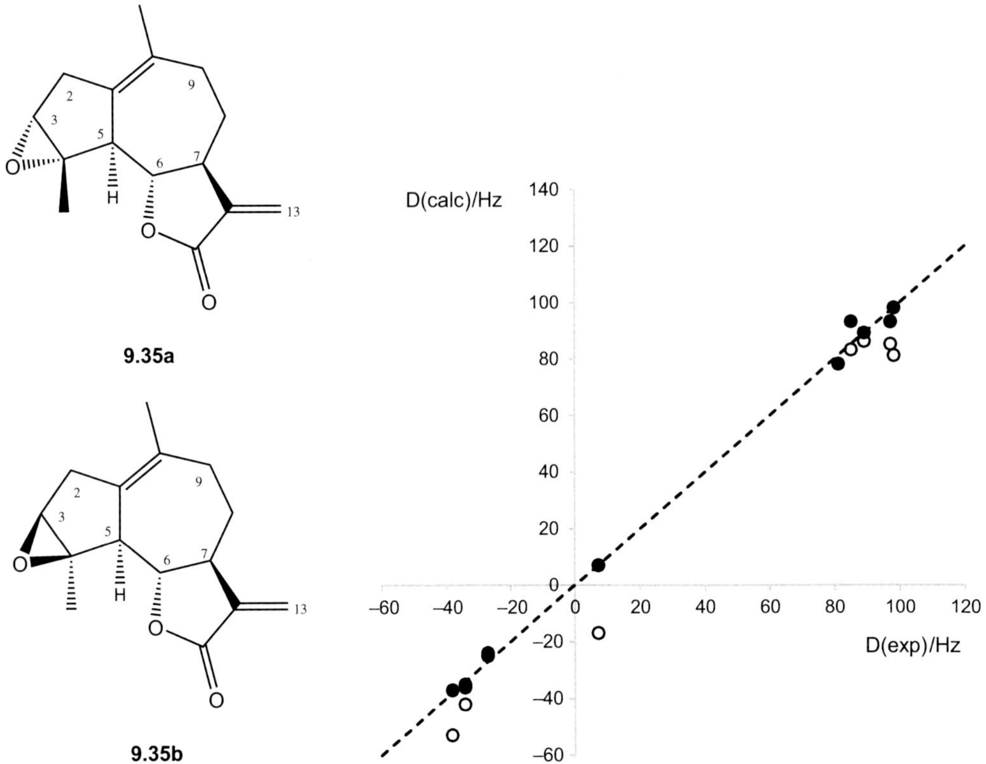

FIGURE 9.67 **RDC fits for ludartin 9.35a (filled circles) and its isomeric epoxide 9.35b (open circles) comparing experimental with back-calculated values.** In the structures shown above, the numbered carbon centres provided the $^1D_{CH}$ values, with two for each of the diastereotopic CH_2 groups (the H8 protons could not be resolved). *(Source: Adapted from Ref. [134] with permission from the American Chemical Society.)*

of the two isomers were established through molecular mechanics calculations for input to the RDC fitting procedure. A sample of ludartin was equilibrated in a polymethyl methacrylate (PMMA) gel solvated with $CDCl_3$, as monitored by the ^2H resonance of the solvent. Ten $^1D_{CH}$ values extracted from HSQC spectra ranging from –40 to +100 Hz were then used to determine the alignment tensors for the two possible structures and the back-calculated RDCs for each were compared against the experimental values (Fig. 9.67). From this, it is apparent that the previously defined structure for ludartin **9.35a** provides a significantly better match than the epoxide isomer **9.35b**, despite the slight structural difference between the two. Note that for the incorrect isomer, poor RDC fits are observed across the whole structure, and not just for the bonds in the vicinity of the epoxide, due to a tilted alignment tensor being calculated for this relative to the correct orientation.

In a second example, RDC analysis was used to define the relative configuration of the fungicidal cyclopentenone 4,6-diacetylhygrophorone A[12] **9.36** derived from mushrooms [142]. Analysis of the few $^3J_{HH}$ and $^4J_{HH}$ coupling constants for this compound suggested the C4 and C5 substituents to be ***trans*** configured by comparison with those of known structures, and the NOE data were consistent with this, including a C4–H to C5–OH enhancement. However, the presence of the quaternary C5 centre and a lack of characteristic NOEs for H6 meant the relative configuration at C6 remained undefined. Thus, the conformational preferences of the four possible diastereoisomers 4*R*,5*R*,6*R*, 4*R*,5*R*,6*S*, 4*R*,5*S*,6*R* and 4*S*,5*R*,6*R* were assessed by molecular mechanics and DFT calculations to provide plausible structures for comparison with experimental data (note that the enantiomeric forms would be indistinguishable as the definition of *absolute* configurations is not possible by RDCs [143]). The sample was weakly aligned in CD_2Cl_2/PBLG ($\Delta\nu_Q$ = 147 Hz) and four single-bond $^1D_{CH}$ and four long-range $^{2/3}D_{CH}$ RDCs were measured. The need to measure long-range RDCs in this case was so that the requisite five independent dipolar couplings could be employed to define the alignment tensor. Due to the r^{-3} distance dependence of RDCs, these values can be an order of magnitude smaller than one-bond RDCs and their measurement requires particular care. The analysis yielded couplings ranging from –15 to +44 Hz which were fitted against computationally optimised isomeric structures. The back-calculated RDCs demonstrated an optimum fit for the 4*R*,5*S*,6*R* diastereoisomer only (Fig. 9.68) and the analysis was therefore able to corroborate the *trans* arrangement proposed by classical J_{HH} and NOE analysis but was also able to define the relative configuration at the exocyclic stereogenic centre. Despite the potential for rotational

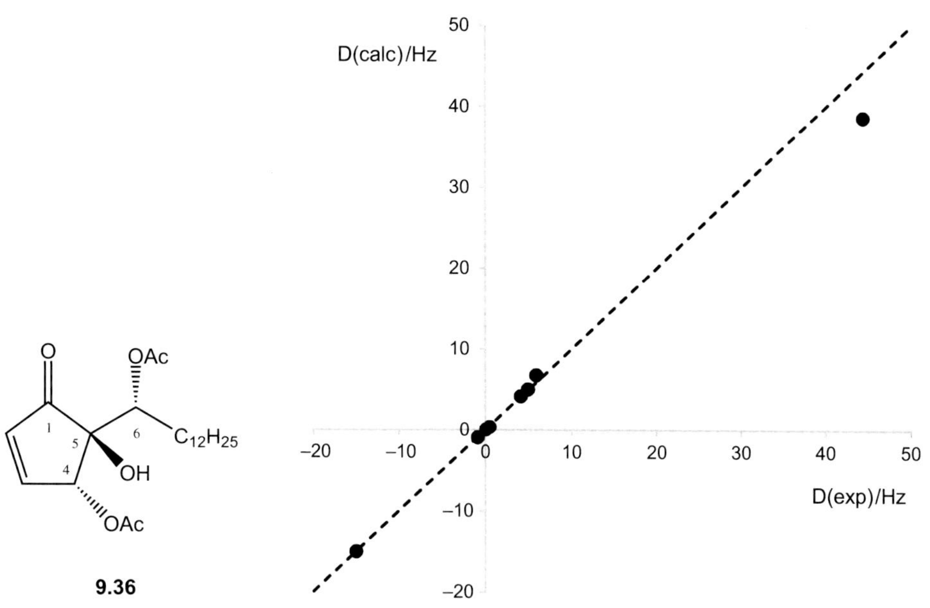

FIGURE 9.68 **RDC fits for the 4*R*,5*S*,6*R* diastereoisomer of 4,6-diacetylhygrophorone A[12] 9.36 showing good agreement between experimental and back-calculated couplings.** *(Source: Adapted from Ref. [142] with permission from the American Chemical Society.)*

freedom about the C5–C6 bond that would average RDCs, the data suggested this not to be the case. The largest RDC of 44 Hz was observed for H6–C6, indicating a largely fixed orientation and enabling RDC fits to a rigid structure. This limited conformational flexibility was believed to arise from intramolecular hydrogen bonding between the OH proton and the C6 acetyl carbonyl group.

As is apparent from these examples, most investigations employing RDCs have to date focused on studies of rigid or semi-rigid organic molecules (see also examples cited in [118] and [121]), and procedures that adequately account for conformational flexibility within molecules are still to be realised. Nevertheless, what these examples also demonstrate is the unique information that may be derived from RDC measurements, suggesting their wider adoption in structure elucidation protocols to be inevitable. This is likely to remain complementary to the interpretation of scalar coupling and NOEs, and will likely find use when these classical methods lead to ambiguities in configuration or conformation.

REFERENCES

[1] Neuhaus D, Williamson MP. The nuclear Overhauser effect in structural and conformational analysis. 2nd ed New York: Wiley-VCH; 2000.
[2] Bell RA, Saunders JK. Can J Chem 1970;48:1114–22.
[3] Hodgson DM, Comina PJ. Tetrahedron Lett 1996;37:5613–4.
[4] Colombo R, Colombo F, Derome AE, Jones JH, Rathbone DL, Thomas DW. J Chem Soc Perkin Trans 1985;1:1811–5.
[5] Robinson JK, Lee V, Claridge TDW, Baldwin JE, Schofield CJ. Tetrahedron 1998;54:981–96.
[6] Baldwin JE, Claridge TDW, Heupel FA, Whitehead RC. Tetrahedron Lett 1994;35:7829–32.
[7] Baldwin JE, Bischoff L, Claridge TDW, Heupel FA, Spring DR, Whitehead RC. Tetrahedron 1997;53:2271–90.
[8] Farmer BT, Macura S, Brown LR. J Magn Reson 1988;80:1–22.
[9] Macura S, Huang Y, Suter D, Ernst RR. J Magn Reson 1981;43:259–81.
[10] Andersen NH, Nguyen KT, Hartzell CJ, Eaton HL. J Magn Reson 1987;74:195–211.
[11] Olejniczak ET, Hoch JC, Dobson CM, Poulsen FM. J Magn Reson 1985;64:199–206.
[12] Williamson MP, Neuhaus D. J Magn Reson 1987;72:369–75.
[13] Dötsch V, Wider G, Wüthrich K. J Magn Reson A 1994;109:263–4.
[14] Parella T, Sánchez-Ferrando F, Virgili A. J Magn Reson 1997;125:145–8.
[15] Wagner R, Berger S. J Magn Reson A 1996;123:119–21.
[16] Kuprov I, Hodgson DM, Kloesges J, Pearson CI, Odell B, Claridge TDW. Angew Chem Int Ed 2015;54:3697–701.
[17] Mitschang L, Keeler J, Davis AL, Oschkinat H. J Biomol NMR 1992;2:545–56.
[18] Otting G. J Magn Reson 1990;86:496–508.

[19] Bodenhausen G, Wagner G, Rance M, Sorensen OW, Wuthrich K, Ernst RR. J Magn Reson 1984;59:542–50.
[20] Andersen NH, Eaton HL, Lai X. Magn Reson Chem 1989;27:515–28.
[21] Butts CP, Jones CR, Towers EC, Flynn JL, Appleby L, Barron NJ. Org Biomol Chem 2011;9:177–84.
[22] Butts CP, Jones CR, Harvey JN. Chem Commun 2011;47:1193–5.
[23] Butts CP, Jones CR, Song Z, Simpson TJ. Chem Commun 2012;48:9023–5.
[24] Jones CR, Butts CP, Harvey JN. Beilstein J Org Chem 2011;7:145–50.
[25] Keepers JW, James TL. J Magn Reson 1984;57:404–26.
[26] Kumar A, Wagner G, Ernst RR, Wüthrich K. J Am Chem Soc 1981;103:3654–8.
[27] Evans JNS. Biomolecular NMR spectroscopy. Oxford: Oxford University Press; 1995.
[28] Cavanagh J, Fairbrother WJ, Palmer AG, Skelton NJ, Rance M. Protein NMR spectroscopy: principles and practice. 2nd ed San Diego: Academic Press (Elsevier); 2006.
[29] Esposito G, Pastore A. J Magn Reson 1988;76:331–6.
[30] Reggelin M, Hoffman H, Köck M, Mierke DF. J Am Chem Soc 1992;114:3272–7.
[31] Stonehouse J, Adell P, Keeler J, Shaka AJ. J Am Chem Soc 1994;116:6037–8.
[32] Stott K, Keeler J, Van QN, Shaka AJ. J Magn Reson 1997;125:302–24.
[33] Stott K, Stonehouse J, Keeler J, Hwang TL, Shaka AJ. J Am Chem Soc 1995;117:4199–200.
[34] Hu H, Krishnamurthy K. J Magn Reson 2006;182:173–7.
[35] Gradwell MJ, Kogelberg H, Frenkiel TA. J Magn Reson 1997;124:267–70.
[36] Hu H, Bradley SA, Krishnamurthy K. J Magn Reson 2004;171:201–6.
[37] Fusetani N, Shinoda K, Matsunaga S. J Am Chem Soc 1993;115:3977–81.
[38] Wüthrich K. NMR of proteins and nucleic acids. New York: Wiley; 1986.
[39] Dyson HJ, Wright PE. Annu Rev Biophys Biophys Chem 1991;20:519–38.
[40] Williamson MP, Waltho JP. Chem Soc Rev 1992;21:227–36.
[41] Anderson S, Claridge TDW, Anderson HL. Angew Chem Int Ed 1997;36:1310–3.
[42] Fielding L, Hamilton N, McGuire R. Magn Reson Chem 1997;35:184–90.
[43] Claridge TDW, Long JM, Brown JM, Hibbs D, Hursthouse MB. Tetrahedron 1997;53:4035–50.
[44] Bothner-By AA, Stephens RL, Lee J, Warren CD, Jeanloz RW. J Am Chem Soc 1984;106:811–3.
[45] Bax A, Davis DG. J Magn Reson 1985;63:207–13.
[46] Neuhaus D, Keeler J. J Magn Reson 1986;68:568–74.
[47] Chan TM, Dalgarno DC, Prestegard JH, Evans CA. J Magn Reson 1997;126:183–6.
[48] Kessler H, Griesinger C, Kerssebaum R, Wagner K, Ernst RR. J Am Chem Soc 1987;109:607–9.
[49] Bax A. J Magn Reson 1988;77:134–47.
[50] Hwang TL, Kadkhodaei M, Mohebbi A, Shaka AJ. Magn Reson Chem 1992;30:S24–34.
[51] Hwang TL, Shaka AJ. J Am Chem Soc 1992;114:3157–9.
[52] Hwang TL, Shaka AJ. J Magn Reson B 1993;102:155–65.
[53] Griesinger C, Ernst RR. J Magn Reson 1987;75:261–71.
[54] Adell P, Parella T, Sánchez-Ferrando F, Virgili A. J Magn Reson B 1995;108:77–80.
[55] Dalvit C, Bovermann G. Magn Reson Chem 1995;33:156–9.
[56] Bauer W, Soi A, Hirsch A. Magn Reson Chem 2000;38:500–3.
[57] D'Auria MV, Giannini C, Zampella A, Minale L, Debitus C, Roussakis C. J Org Chem 1998;63:7382–8.
[58] Seebach D, Overhand M, Kühnle FNM, Martinoni B, Oberer L, Hommel U, Widmer H. Helv Chim Acta 1996;79:913–41.
[59] Claridge TDW, Goodman JM, Moreno A, Angus D, Barker SF, Taillefumier C, Watterson MP, Fleet GWJ. Tetrahedron Lett 2001;42:4251–5.
[60] Botsi A, Yannalopoulou K, Perly B, Hadjoudis E. J Org Chem 1995;60:4017–23.
[61] Aw B-H, Selvaratnam S, Leung P-H, Rees NH, McFarlane W. Tetrahedron Asymm 1996;7:1753–62.
[62] Sanders JKM, Mersh JD. Prog Nucl Magn Reson Spectrosc 1982;15:353–400.
[63] Lindon JC, Ferrige AG. Prog Nucl Magn Reson Spectrosc 1980;14:27–66.
[64] Neuhaus D, Wagner G, Vasák M, Kági JHR, Wüthrich K. Eur J Biochem 1985;151:257–73.
[65] Kinns M, Sanders JKM. J Magn Reson 1984;56:518–20.
[66] Shaka AJ, Bauer C, Freeman R. J Magn Reson 1984;60:479–85.
[67] Kövér KE, Batta G. Prog Nucl Magn Reson Spectrosc 1987;19:223–66.
[68] Bauer W. Magn Reson Chem 1991;29:494–9.
[69] Stott K, Keeler J. Magn Reson Chem 1996;34:554–8.
[70] Rinaldi PL. J Am Chem Soc 1983;105:5167–8.
[71] Yu C, Levy GC. J Am Chem Soc 1983;106:6533–7.
[72] Kövér KE, Batta G. J Magn Reson 1986;69:344–9.
[73] Kövér KE, Batta G. J Magn Reson 1986;69:519–22.
[74] Bauer W. Magn Reson Chem 1996;34:532–7.
[75] Alam TM, Pedrotty DM, Boyle TJ. Magn Reson Chem 2002;40:361–5.

[76] Walker O, Mutzenhardt P, Canet D. Magn Reson Chem 2003;41:776–81.
[77] Combettes LE, Clausen-Thue P, King MA, Odell B, Thompson AL, Gouverneur V, Claridge TDW. Chem Eur J 2012;18:13133–41.
[78] Sánchez-Ferrando F. Magn Reson Chem 1985;23:185–91.
[79] Leeper FJ, Staunton J. J Chem Soc Chem Commun 1982;911–2.
[80] Aldersley MF, Dean FM, Mann BE. J Chem Soc Chem Commun 1983;107–8.
[81] Cativiela C, Sánchez-Ferrando F. Magn Reson Chem 1985;23:1072–5.
[82] Macchioni A. Eur J Inorg Chem 2003;195–205.
[83] Pregosin PS, Kumar PGA, Fernandez I. Chem Rev 2005;105:2977–98.
[84] Stahl NG, Zuccaccia C, Jensen TR, Marks TJ. J Am Chem Soc 2003;125:5256–7.
[85] Bauer W. In: Sapse A-M, Schleyer PVR, editors. Lithium chemistry. Wiley: New York; 1995. p. 125–72.
[86] Gschwind RM, Rajamohanan PR, John M, Boche G. Organometallics 2000;19:2868–73.
[87] Hilmersson G, Malmros B. Chemistry 2001;7:337–41.
[88] Davies RP, Hornauer S, Hitchcock PB. Angew Chem Int Ed 2007;46:5191–4.
[89] Jeener J, Meier BH, Bachmann P, Ernst RR. J Chem Phys 1979;71:4546–53.
[90] Campbell ID, Dobson CM, Ratcliffe RG, Williams RJP. J Magn Reson 1978;29:397–417.
[91] Led JJ, Gesmar H. J Magn Reson 1982;49:444–63.
[92] Bain A. Prog Nucl Magn Reson Spectrosc 2003;43:63–103.
[93] Hu DX, Grice P, Ley SV. J Org Chem 2012;77:5198–202.
[94] Bain A. Ann Rep NMR Spectrosc 2008;63:23–48.
[95] Orrell KG, Šik V. Ann Rep NMR Spectrosc 1993;27:103–71.
[96] Orrell KG, Šik V, Stephenson D. Prog Nucl Magn Reson Spectrosc 1990;22:141–208.
[97] Perrin CL, Dwyer TJ. Chem Rev 1990;90:935–67.
[98] Sandström J. Dynamic NMR spectroscopy. London: Academic Press; 1982.
[99] Oki M. Applications of dynamic NMR spectroscopy to organic chemistry. Weinheim: VCH; 1985.
[100] Derose EF, Castillo J, Saulys D, Morrison J. J Magn Reson 1991;93:347–54.
[101] Gruber T, Thompson AL, Odell B, Bombicz P, Schofield CJ. New J Chem 2014;38:5905–17.
[102] Ramachandran R, Knight CTG, Kirkpatrick RJ, Oldfield E. J Magn Reson 1985;65:136–41.
[103] Pianet I, Fouquet E, Pereyre M, Gielen M, Kayser F, Biesemans M, Willem R. Magn Reson Chem 1994;32:613–7.
[104] Buston JEH, Claridge TDW, Moloney MG. J Chem Soc Perkin Trans 1995;2:639–41.
[105] Perrin CL. J Magn Reson 1989;82:619–21.
[106] Dimitrov VS, Vassilev NG. Magn Reson Chem 1995;33:739–44.
[107] Abel EW, Coston TPJ, Orrell KG, Šik V, Stephenson D. J Magn Reson 1986;70:34–53.
[108] Bodenhausen G, Ernst RR. J Am Chem Soc 1982;104:1304–9.
[109] Orrell KG, Šik V. Ann Rep NMR Spectrosc 1987;19:79–173.
[110] Orrell KG. Ann Rep NMR Spectrosc 1999;37:1–74.
[111] Pons M, Millet O. Prog Nucl Magn Reson Spectrosc 2001;38:267–324.
[112] Gschwind RM. Angew Chem Int Ed 2005;44:4666–8.
[113] Yan J, Zartler ER. Magn Reson Chem 2005;43:53–64.
[114] Thiele CM. Concept Magn Reson A 2007;30:65–80.
[115] Thiele CM. Eur J Org Chem 2008;5663.
[116] Kummerlöwe G, Luy B. Ann Rep NMR Spectrosc 2009;68:193–232.
[117] Kummerlöwe G, Luy B. Trac-Trend Anal Chem 2009;28:483–93.
[118] Gil RR. Angew Chem Int Ed 2011;50:7222–4.
[119] Zweckstetter M. Nat Protoc 2008;3:679–90.
[120] Navarro-Vázquez A. Magn Reson Chem 2012;50:S73–9.
[121] Tzvetkova P, Luy B, Simova S. Magn Reson Chem 2012;50:S92–S101.
[122] Surfati M, Lesot P, Merlet D, Courtieu J. Chem Commun 2000;2069–81.
[123] Dama M, Berger S. Org Lett 2012;14:241–3.
[124] Dama M, Berger S. Tetrahedron Lett 2012;53:6439–42.
[125] Meyer N-C, Krupp A, Schmidts V, Thiele CM, Reggelin M. Angew Chem Int Ed 2012;51:8334–8.
[126] Troche-Pesqueira E, Cid M-M, Navarro-Vázquez A. Org Biomol Chem 2014;12:1957–65.
[127] Dama M, Berger S. Carbohydr Res 2013;377:44–7.
[128] Luy B, Kobzar K, Knor S, Furrer J, Heckmann D, Kessler H. J Am Chem Soc 2005;127:6459–65.
[129] Luy B, Kobzar K, Kessler H. Angew Chem Int Ed 2004;43:1092–4.
[130] Freudenberger JC, Spiteller P, Bauer R, Kessler H, Luy B. J Am Chem Soc 2004;126:14690–1.
[131] Kaden P, Freudenberger JC, Luy B. Magn Reson Chem 2012;50:S22–8.
[132] Freudenberger JC, Knör S, Kobzar K, Heckmann D, Paululat T, Kessler H, Luy B. Angew Chem Int Ed 2005;44:423–6.
[133] Haberz P, Farjon J, Griesinger C. Angew Chem Int Ed 2005;44:427–9.

[134] Gil RR, Gayathri C, Tsarevsky NV, Matyjaszewski K. J Org Chem 2008;73:840–8.
[135] Gayathri C, Tsarevsky NV, Gil RR. Chem Eur J 2010;16:3622–6.
[136] Kummerloewe G, McCord EF, Cheatham SF, Niss S, Schnell RW, Luy B. Chem Eur J 2010;16:7087–9.
[137] Enthart A, Freudenberger JC, Furrer J, Kessler H, Luy B. J Magn Reson 2008;192:314–22.
[138] Reinsperger T, Luy B. J Magn Reson 2014;239:110–20.
[139] Timári I, Kaltschnee L, Kolmer A, Adams RW, Nilsson M, Thiele CM, Morris GA, Kövér KE. J Magn Reson 2014;239:130–8.
[140] Yan J, Kline AD, Mo H, Shapiro MJ, Zartler ER. J Org Chem 2003;68:1786–95.
[141] Aroulanda C, Boucard V, Guibé F, Courtieu J, Merlet D. Chem Eur J 2003;9:4536–9.
[142] Schmidts V, Fredersdorf M, Luebken T, Porzel A, Arnold N, Wessjohann L, Thiele CM. J Nat Prod 2013;76:839–44.
[143] Berger R, Courtieu J, Gil RR, Griesinger C, Köck M, Lesot P, Luy B, Merlet D, Navarro-Vázquez A, Reggelin M, Reinscheid UM, Thiele CM, Zweckstetter M. Angew Chem Int Ed 2012;51:8388–91.

Chapter 10

Diffusion NMR Spectroscopy

Chapter Outline

10.1 Introduction 381
- 10.1.1 Diffusion Coefficients and Molecular Size 382

10.2 Measuring Self-Diffusion by NMR 382
- 10.2.1 The Pulsed Field Gradient Spin-Echo 383
- 10.2.2 The Pulsed Field Gradient Stimulated-Echo 384
- 10.2.3 Enhancements to the Stimulated-Echo 385
- 10.2.4 Data Analysis: Regression Fitting 388
- 10.2.5 Data Analysis: Pseudo-2D Presentation 389

10.3 Practical Aspects of Diffusion NMR Spectroscopy 390
- 10.3.1 The Problem of Convection 390
- 10.3.2 Calibrating Gradient Amplitudes 397
- 10.3.3 Optimising Diffusion Parameters 397
- 10.3.4 Hydrodynamic Radii and Molecular Weights 401

10.4 Applications of Diffusion NMR Spectroscopy 403
- 10.4.1 Signal Suppression 403
- 10.4.2 Hydrogen Bonding 405
- 10.4.3 Host–Guest Complexes 405
- 10.4.4 Ion Pairing 408
- 10.4.5 Supramolecular Assemblies 409
- 10.4.6 Aggregation 411
- 10.4.7 Mixture Separation 412
- 10.4.8 Macromolecular Characterisation 413

10.5 Hybrid Diffusion Sequences 414
- 10.5.1 Sensitivity-Enhanced Heteronuclear Methods 414
- 10.5.2 Spectrum-Edited Methods 415
- 10.5.3 Diffusion-Encoded Two-Dimensional Methods (or 3D DOSY) 415

References 418

10.1 INTRODUCTION

The study of molecular diffusion in solution potentially offers insights into a range of physical molecular properties including molecular size, shape, aggregation, encapsulation, complexation and hydrogen bonding, as illustrated later in this chapter, and NMR-based measurements have been applied to many areas of chemistry for over four decades [1-4]. Diffusion measurements are now routinely accessible with conventional high-resolution NMR spectrometers equipped with actively shielded pulsed field gradient (PFG) probeheads. All modern NMR-based diffusion measurements rely on the application of field gradients to encode the physical location of a molecule or complex in solution and so characterize its diffusion along the direction of the applied field gradient, typically the *z*-axis of conventional gradient probeheads. The ability to use standard commercial instrumentation for such measurements on routine samples has obvious advantages over alternative methods for measuring diffusion, such as radioactive tracer studies. NMR techniques allow relatively fast measurements over a range of temperatures and do not require specialised handling of radioactive isotopes.

In this chapter we shall focus on the use of high-resolution NMR methods for studying diffusion in physically homogeneous (isotropic) solutions. Such self-diffusion arises from the random translational (Brownian) motion of molecules driven by the thermal energy of the system and may be characterised quantitatively by the *self-diffusion coefficient D*. While in an isotropic system the average displacement of all molecules over time in three dimensions is zero (the sample remains in the tube after all), the mean square displacement of a single molecule is non-zero such that the distance travelled by a molecule in a single direction during a period t is given by:

$$z_{\mathrm{rms}} \equiv (2Dt)^{1/2} \qquad (10.1)$$

where z_{rms} represents the root-mean-square distance travelled as a time average for many molecules. The self-diffusion coefficient is therefore a measure of the rate of mean square displacement of a molecule and consequently has units of square metres per second. It is this physical parameter we aim to measure or utilise when performing diffusion NMR experiments which from this point on we shall refer to simply as the *diffusion coefficient.*

Characterisation of the rates of diffusion of molecules or their complexes may be related intuitively to, amongst other things, their size and shape and so provides a different, and often complementary, insight into molecular structures and

High-Resolution NMR Techniques in Organic Chemistry. http://dx.doi.org/10.1016/B978-0-08-099986-9.00010-5

their solution behaviour over that provided by the familiar parameters of chemical shift and through-bond or through-space spin–spin coupling. Different mobility rates may also be used as the basis for separation of the spectra of mixtures of compounds in solution, this procedure often being referred to as *diffusion ordered spectroscopy* (DOSY) [5]. As we shall see, this typically produces 'pseudo'-2D spectra with one axis representing chemical shifts and the second diffusion coefficients, so dispersing the NMR resonances according to the diffusion properties of the parent molecules. The application of high-resolution NMR diffusion measurements to the separation of small-molecule mixtures in this way remains a developing area that has potential use in the analysis of complex matrices such as synthetic and biosynthetic mixtures, biofluids, foods and beverages. However, it is worth noting that the experimental techniques for the direct quantitative measurement of diffusion coefficients and for the production of 2D DOSY spectra are identical, with the different formats for presenting results arising only from different approaches taken in data processing, as will be explained in the sections that follow.

10.1.1 Diffusion Coefficients and Molecular Size

It is often the case that diffusion measurements are performed so as to gain information on (relative) molecular sizes, and the diffusion coefficient D may itself be related to molecular dimensions through the relationship:

$$D = \frac{k_B \mathrm{T}}{f} \qquad (10.2)$$

where k_B is the Boltzmann constant, T is the absolute temperature and f represents the hydrodynamic friction coefficient or frictional factor, a term that reflects, amongst other things, the size and shape of the molecular species. For the idealised case of a diffusing sphere, f may related to the hydrodynamic or Stokes radius of the sphere r_S and the solution viscosity η through the Stokes equation:

$$f = 6\pi\eta r_S \qquad (10.3)$$

Combining these gives the widely quoted *Stokes–Einstein equation* which for a sphere is:

$$D = \frac{k_B T}{6\pi\eta r_S} \qquad (10.4)$$

This demonstrates that the diffusion coefficient is inversely related to the size (radius) of the diffusing species (sphere), as one might expect; that is larger molecules or complexes will tend to exhibit smaller diffusion coefficients, all other conditions being comparable. Whether Eq. 10.4 is valid for your systems will depend upon how well their shape can be approximated to that of a sphere, and suggests caution when translating the diffusion coefficient to a hydrodynamic radius (although this has been done in a number of cases; see Section 10.4). Furthermore, in the study of smaller molecules, notably where the hydrodynamic radius of the solute tends towards that of the solvent, the scaling factor 6 in Eq. 10.4 becomes invalid and must be corrected; see Section 10.3.4. For larger and more irregular molecular shapes, more complex descriptions of the friction coefficient must be developed which may become a non-trivial exercise, meaning it is more commonly the case that the diffusion coefficients themselves are reported. Note also that these depend not only on molecular size and shape, as reflected in the friction coefficient, but also on solution temperature and inversely on its viscosity which has important bearings on how one should *interpret* the magnitudes of diffusion coefficients. This will be discussed further below, along with other very important practical considerations required when undertaking these measurements. Indeed, while it is technically rather straightforward to *execute* diffusion experiments with modern NMR spectrometers, obtaining *meaningful* answers free from interferences can be rather more challenging, as we shall discover. Before discussing the practicalities of such measurements and presenting examples of their applications, we shall consider how one goes about measuring diffusion coefficients.

10.2 MEASURING SELF-DIFFUSION BY NMR

In this section we shall consider the operation of more widely used diffusion sequences and their variants, and consider methods for processing the acquired data. This essentially falls into two possible approaches: either direct regression analysis to provide values of the diffusion coefficients or treatments that present the pseudo-2D DOSY spectra mentioned

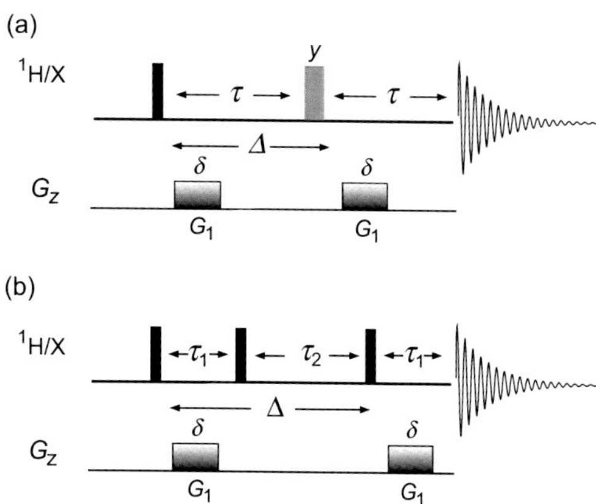

FIGURE 10.1 **Basic sequences for measuring molecular diffusion.** These are based on (a) the PFG spin-echo and (b) the PFG stimulated-echo. The diffusion during the period Δ is characterised by a series of measurements with increasing gradient strengths, as represented by the faded gradient pulses.

above. To understand the process of measuring diffusion coefficients by NMR we shall first consider the simplest gradient spin-echo experiment.

10.2.1 The Pulsed Field Gradient Spin-Echo

The basic scheme for the characterisation of diffusion is the PFG spin-echo [6] (Fig. 10.1a). In the absence of gradient pulses, this will act simply as a standard spin-echo and so refocus chemical shift evolution such that the detected signal is attenuated only by transverse relaxation during the 2τ period. When PFGs are employed, the first gradient pulse will impose a spatially dependent phase on the magnetisation vectors which can be refocused by a second gradient of equal duration and magnitude; although these pulses are applied in the same sense, their effects will cancel due to application of the 180 degree pulse between them. However, *complete* refocusing of the signal will only occur when the local field experienced by a spin is identical during the two gradient pulses. Since a field *gradient* is imposed this refocusing condition is only met if the spin remains in the same physical location when the two PFG pulses are applied. If the molecule were to diffuse away from its initial position during the diffusion delay Δ, then the local field experienced during the second PFG would not exactly match that of the first and only partial refocusing of the signal would occur (Fig. 10.2). The detected signal would therefore be attenuated by an amount dependent upon how far the molecule moved during the period Δ, and hence by its diffusion coefficient.

To characterise diffusion rates, it is possible to progressively increase the diffusion period Δ, the length of the gradient pulses δ, or the strength of the gradient pulses G and to monitor the corresponding change in NMR signal intensity. Thus, increasing Δ provides more time for the molecules to move through the bulk sample, whereas increasing δ or G imposes a greater degree of signal dephasing across the sample; in all cases the amount of signal refocused by the second gradient will have decreased in the presence of self-diffusion and net NMR signal intensity will have diminished. However, changes made to the overall length of the echo sequence by increasing Δ will introduce additional attenuation arising from increased relaxation losses, so it is now universal practice to increase the diffusion–encoding/decoding gradient strengths G while keeping all time periods invariant. While T_2 relaxation losses still occur relative to a single pulse acquisition, they are constant for all echo experiments and thus do not contribute to the progressive signal attenuation that is monitored (Fig. 10.3).

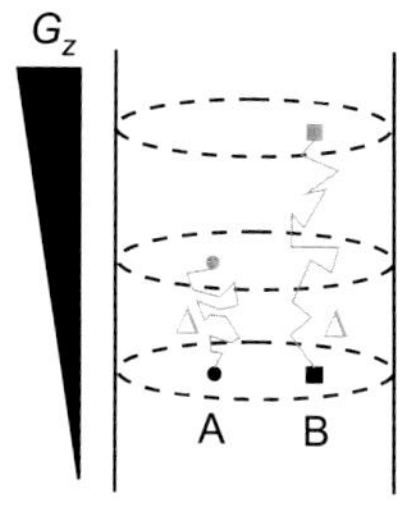

FIGURE 10.2 **Schematic representation of signal attenuation through molecular diffusion.** The local field experienced by molecule A during the first gradient pulse (molecules in black) does not precisely match that experienced during the second gradient pulse (molecules in grey) due to diffusion during the delay Δ. The signal of A does not fully refocus and its response is attenuated. Greater attenuation is observed for the faster moving molecule B due to the greater difference in local fields it experiences during the two gradient pulses.

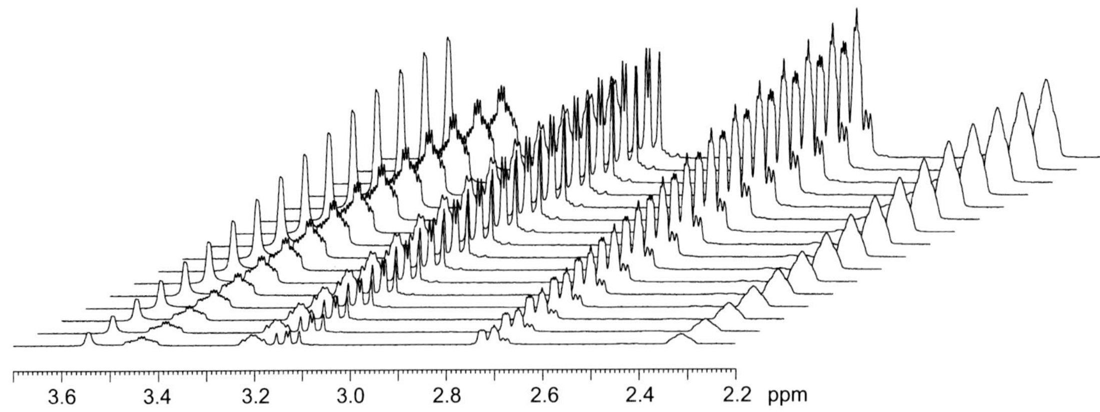

FIGURE 10.3 **Experimental 1D ^{1}H diffusion traces.** These show a progressive intensity decay as a function of increasing gradient strength (from back to front).

The observed signal intensity at the end of the spin-echo I_G for the basic PFG spin-echo experiment is given by the so-called Stejskal–Tanner expression:

$$I_G = I_0 \exp\left(-\frac{2\tau}{T_2}\right)\exp\left[-(\gamma\delta G)^2 D\left(\Delta - \frac{\delta}{3}\right)\right] \tag{10.5}$$

where I_0 is signal intensity in the absence of the gradient spin-echo (ie that after a single 90 degree pulse), G is gradient strength, D is the diffusion coefficient, the delays τ, δ and Δ are as in Fig. 10.1, γ is the magnetogyric ratio of the observed nuclide and T_2 is the transverse relaxation rate constant. For the typical case where the total echo time 2τ is held constant and the gradient strength G varied, a more useful form is:

$$I_G = I_{G=0} \exp\left[-(\gamma\delta G)^2 D\left(\Delta - \frac{\delta}{3}\right)\right] \tag{10.6}$$

where $I_{G=0}$ represents the signal intensity of the spin-echo at zero gradient strength. As the constants γ, δ and Δ are known, the diffusion coefficient may be calculated directly from appropriate linear or non-linear regression and, optionally, the data presented in the DOSY format; see Section 10.2.5.

10.2.2 The Pulsed Field Gradient Stimulated-Echo

The PFG spin-echo sequence described above is limited in practice by the aforementioned relaxation losses and is now less used for diffusion measurements. Because magnetisation is transverse during the diffusion period, these losses are dictated by the transverse (T_2) relaxation rates which themselves increase with molecular size. Since larger molecules tend to require longer diffusion periods to translate significant distances, the use of long Δ delays (which are typically in the region of 10 ms to 1 s) can lead to unacceptable signal-to-noise degradation. The spin-echo sequence may also be subject to J modulation (coupling evolution) when magnetisation is transverse which can lead to undesirable signal distortion, especially for strongly coupled spin systems. In the alternative stimulated-echo (STE) sequence [7] (Fig. 10.1b) magnetisation is transverse only during the relatively short gradient-encoding periods τ_1 (1–10 ms) but longitudinal during the remainder of the diffusion period τ_2 by virtue of the second 90 degree pulse, meaning signal loss is dictated more by the potentially slower longitudinal (T_1) relaxation rates instead. Following the diffusion period, magnetisation is returned to the transverse plane by the third 90 degree pulse for refocusing and detection. The observed signal intensity for the stimulated-echo I_G is now given by:

$$I_G = \frac{I_0}{2} \exp\left(-\frac{2\tau_1}{T_2} - \frac{\tau_2}{T_1}\right)\exp\left[-(\gamma\delta G)^2 D\left(\Delta - \frac{\delta}{3}\right)\right] \tag{10.7}$$

or again when varying gradient strength alone:

$$I_G = I_{G=0} \exp\left[-(\gamma\delta G)^2 D\left(\Delta - \frac{\delta}{3}\right)\right] \tag{10.8}$$

In addition to the dependence on T_1 it may be seen in Eq. 10.7 that signal intensity is now reduced by a factor of 2 relative to that of the spin-echo sequence. This is a consequence of the second 90 degree pulse placing only one of the two orthogonal components of the transverse x–y magnetisation back along the longitudinal z-axis during the diffusion period, the remaining component being removed by phase cycling or additional gradient purging. This signal loss is a necessary evil of this STE sequence, but is often tolerated because of its improved behaviour; the more recently introduced diffusion sequences described below tend to be derived from this basic STE sequence. One further feature of the STE that has consequences for data analysis is that the intensity of the echo is also dependent on the chemical shift offset of each resonance *when the applied gradient strength is zero* (ie with no gradient pulses applied; see Johnson [2] for details). This dependence is removed for non-zero gradients, so in practical implementations it is usual either to ignore spectra with $G = 0$ for data analysis or, more typically, to begin the experiment with a small but non-zero value for G.

10.2.3 Enhancements to the Stimulated-Echo

A number of enhancements to the basic STE sequence have been suggested over the years aimed at minimising lineshape distortions and reducing experiment time by minimising phase-cycling requirements. Resonance perturbations arise from the application of strong field gradient pulses, causing eddy currents in metal work surrounding the sample and disturbance of the field frequency lock. The use of actively shielded field gradient probeheads combined with shaped gradient pulses significantly reduces eddy current generation but diffusion experiments can remain sensitive to small residual effects. Some of the more commonly employed sequences are presented below that address these issues and further variants are discussed in Section 10.3.1 when the thorny issue of convection, the bane of diffusion measurements, is addressed.

10.2.3.1 The BPP-STE Sequence

The key feature of the bipolar pulse pair STE (BPP-STE) sequence (Fig. 10.4a) is the use of bipolar gradient pulse pairs. In this a single gradient pulse in the conventional STE sequence is replaced by two pulses of half the duration and opposing signs that are separated by a non-selective 180 degree pulse [8] (Fig. 10.5). The two equal but opposite gradient pulses impose a net additive dephasing effect for all spins experiencing the central 180 degree pulse and so mirror the effect of a single pulse. However, they generate less eddy current distortions since the two pulses are applied in opposite senses. Furthermore, since deuterium nuclei do not experience the 180 degree rf pulse, the bipolar pair has a net zero effect on the lock system unlike a single gradient pulse that would also dephase the lock resonance. Thus, perturbations of the lock circuitry are also reduced with the bipolar pair (Fig. 10.6), all of which can lead to reduced lineshape disturbances in spectra (Fig. 10.7).

(a)

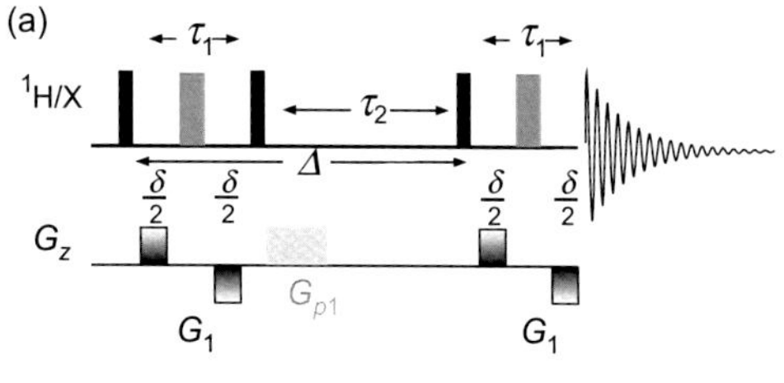

(b)

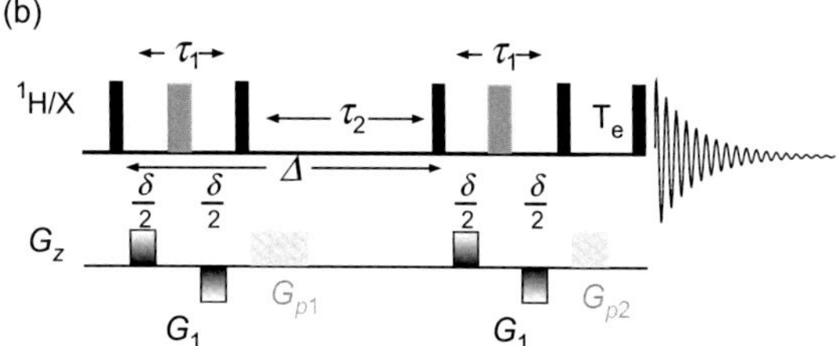

FIGURE 10.4 **Diffusion sequences incorporating bipolar gradient pulses.** The (a) BPP-STE and (b) BPP-LED sequences. The encoding gradients of the STE are applied as symmetrical BPPs of total duration δ and the LED sequence is extended with an eddy current delay period T_e. Optional purging pulses G_{p1} and G_{p2} may also be employed.

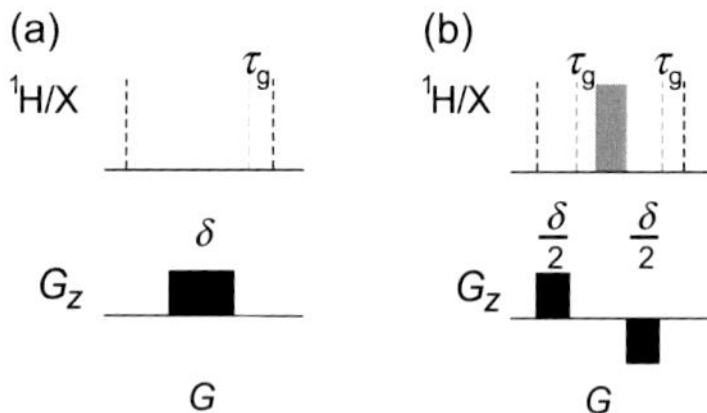

FIGURE 10.5 **Gradient pulse elements.** The (a) monopolar gradient pulse and (b) the equivalent biploar pulse pair. Dashed vertical lines indicate time points and the period τ_g is gradient recovery delay.

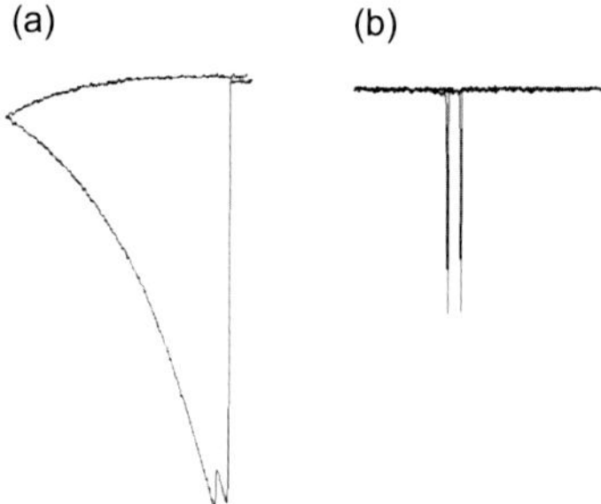

FIGURE 10.6 **Deuterium lock signal behaviour during the STE sequence.** These traces were recorded with (a) monopolar and (b) bipolar gradient pulses. Lock signal refocusing is apparent in (b) leading to a minimal and only momentary perturbation from the two gradient pulse pairs.

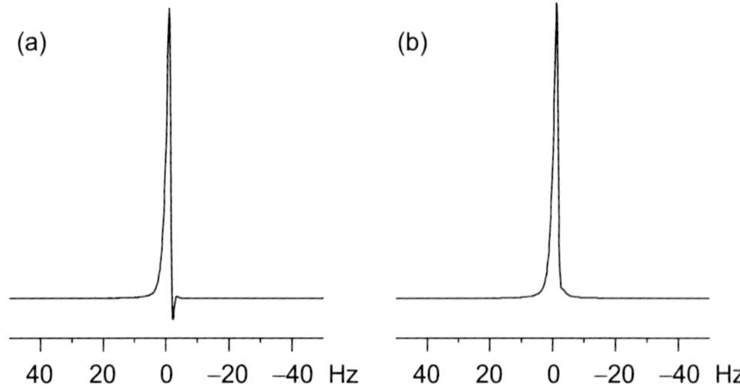

FIGURE 10.7 **^{1}H lineshapes observed for the STE sequence.** These spectra were recorded with (a) monopolar and (b) bipolar gradient pulses.

To allow for gradient pulse recovery delays τ_g in bipolar pairs, Eq. 10.8 for defining the signal amplitude of the STE is modified slightly:

$$I_G = I_{G=0} \exp\left[-(\gamma\delta G)^2 D\left(\Delta - \frac{\delta}{3} - \frac{\tau_g}{2}\right)\right] \tag{10.9}$$

One consequence of employing bipolar pulse pairs is the need for suitable phase cycling of pulses to select only the magnetisation that experiences the 180 degree pulses and thus exhibits the correct diffusion weighting [9]. This requirement for extended phase-cycling may be reduced somewhat by the use of additional gradient purge pulses applied when the desired magnetisation is longitudinal (Fig. 10.4a) which leads to a minimum eight-step cycle for the BPP-STE sequence (Table 10.1). In this, the first four steps select transverse magnetisation produced by the 90 degree pulse of the last spin-echo, while the next four compensate for errors in the final 180 degree pulse.

10.2.3.2 The BPP-LED Sequence

The BPP longitudinal eddy current delay (BPP-LED) sequence [10,11] (Fig. 10.4b) has been one of the most commonly used sequences and employs the BPP-STE with an additional eddy current delay period T_e in which the magnetisation is

TABLE 10.1 A Suitable Minimum Phase Cycle for the Gradient-Purged BPP-STE Experiment

Transient	P_1	P_2	P_3	P_4	P_5	ϕ_R
1	x	x	x	x	x	x
2	x	x	x	y	x	$-y$
3	x	x	x	$-x$	x	x
4	x	x	x	$-y$	x	y
5	x	$-x$	x	x	$-x$	x
6	x	$-x$	x	y	$-x$	$-y$
7	x	$-x$	x	$-x$	$-x$	x
8	x	$-x$	x	$-y$	$-x$	y

Pulses P_1–P_5 represent the five rf pulses of the sequence in the order they are applied and ϕ_R the receiver phase.

TABLE 10.2 The Minimum Phase Cycle for the BPP-LED Experiment

Transient	P_1	P_2	P_3	P_4	P_5	ϕ_R
1	x	x	x	x	x	x
2	x	x	x	$-x$	x	$-x$
3	x	$-x$	x	x	x	$-x$
4	x	$-x$	x	$-x$	x	x
5	x	x	$-x$	$-x$	$-x$	$-x$
6	x	x	$-x$	x	$-x$	x
7	x	$-x$	$-x$	$-x$	$-x$	x
8	x	$-x$	$-x$	x	$-x$	$-x$

Pulses P_1–P_5 represent the five 90 degree pulses of the sequence in the order they are applied and ϕ_R the receiver phase.

stored longitudinally prior to detection while eddy currents are allowed to decay. Following a suitable recovery period the longitudinal magnetisation is placed transverse by the final 90 degree pulse for detection. The duration T_e will depend on the hardware in use, and the LED extension may indeed be unnecessary on modern, well-shielded probeheads operating with relatively low gradient strengths. Otherwise T_e may typically be some milliseconds or tens of milliseconds, but in any case must be short relative to the longitudinal relaxation times of the nuclei being observed so as to retain signal intensity.

The bipolar pulses here lead to a minimum eight-step (preferably 16-step) cycle [11] for the BPP-LED sequence (Table 10.2). As for the BPP-STE experiment, an additional element in common usage is the incorporation of purging gradient pulses to destroy any residual transverse magnetisation during the diffusion period and potentially also during the eddy current recovery period where the desired magnetisation is longitudinal. This will however invalidate the refocusing effects on the lock resonance since these purge pulses are not applied as bipolar pairs, so their use may in fact be detrimental on some instruments. In this case one must also be aware of the possibility of accidental refocusing by these purge pulses of unwanted coherences that may interfere with peak intensities and, if used, some trial and error in the selection of purge gradient strengths and durations may become necessary. An improved implementation of purging gradients is as balanced pulses, as in the one-shot sequence below.

10.2.3.3 The 'One-Shot' Sequence

To avoid the need for extensive phase-cycling yet retain the benefits of BPPs, Morris et al. have suggested the 'one-shot' sequence [12] which requires only a single transient per gradient value, signal-to-noise permitting, so allowing the very rapid collection of diffusion data. A key feature of the sequence is the use of balanced asymmetrical bipolar gradient pairs to minimise the potential for resonance distortion arising from eddy currents and lock disturbance (Fig. 10.8).

The bipolar gradient pairs themselves are unbalanced with an intensity ratio of $1 + \alpha{:}1 - \alpha$ to dephase any magnetisation that is *not* refocused by the 180 degree pulse (which would otherwise require suppression through phase cycling of the

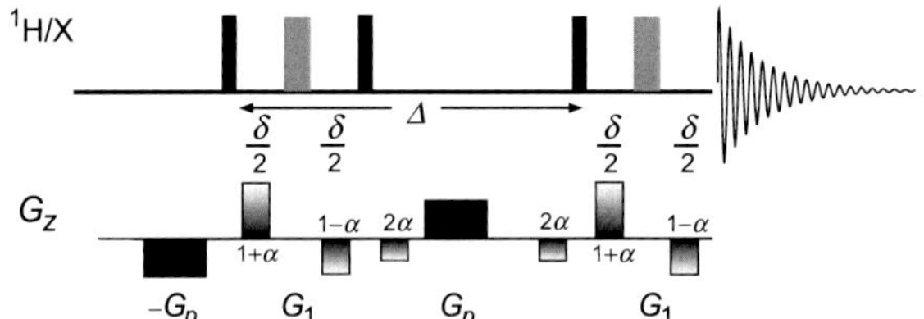

FIGURE 10.8 **The 'One-shot' diffusion sequence.** The bipolar pulses G_1 are applied as unbalanced pairs in a ratio $1 + \alpha:1 - \alpha$ with additional balancing pulses of relative intensity 2α. An intense purge gradient G_p is applied during the diffusion period Δ and is balanced by the gradient pulse $-G_p$ during the relaxation delay.

TABLE 10.3 Eight-Step Phase Cycle for the One-Shot Experiment

Transient	P_1	P_2	P_3	ϕ_R
1	x	x	x	x
2	$-x$	x	x	$-x$
3	x	x	$-x$	$-x$
4	$-x$	x	$-x$	x
5	y	x	x	y
6	$-y$	x	x	$-y$
7	y	x	$-x$	$-y$
8	$-y$	x	$-x$	y

P_1–P_3 represent the three 90 degree pulses of the sequence in the order they are applied and ϕ_R the receiver phase.

180 degree pulses as for a standard spin-echo; Section 8.2.2). To counteract the resultant imbalance in these opposing gradient pulses, and thus counter destruction of the lock, additional balancing pulses each of intensity 2α are applied at the start and end of the diffusion period in which magnetisation is longitudinal. Finally, an additional intense purge pulse G_p is applied also during this diffusion period to select pure longitudinal magnetisation, and this is similarly balanced with an equal but opposite gradient during relaxation delay. Thus, the resultant is retention of the desired magnetisation through the sequence by gradient selection in a single transient but with no net destruction of the lock signal owing to the balancing throughout of positive with negative gradient pulses. If signal-to-noise considerations demand more than a single transient per experiment, phase cycling may also be employed advantageously, for which a basic eight-step cycle is presented in Table 10.3; this may be further extended as in Ref. [12]. The use of the imbalance factor α (typically in the region 0.1–0.2) demands a slight modification of the expression for signal attenuation which is also described in Ref. [12].

10.2.4 Data Analysis: Regression Fitting

Having collected the diffusion data set consisting of a series of 1D traces recorded with increasing gradient amplitudes, it remains to extract from this the diffusion coefficients for the species of interest. These data will take the form of typically 8–32 1D spectra and for all sequences described above the data-fitting procedures rely on measurement of the decay in signal intensity (recorded either as peak height or integrated area) as a function of applied gradient strength. This process relies on the resonances being monitored having arisen from a single species only and thus being free from overlap with those of other molecules such that each decay profile reflects the behaviour of only one molecular component. As such, maximum resonance dispersion is essential for reliable measurements of individual diffusion coefficients. It is also vitally important to ensure baselines are flat and have no offset prior to measuring peak amplitudes, meaning baseline correction routines are likely to prove beneficial.

Diffusion coefficients may be extracted directly from regression fits of the peak intensities versus gradient strengths. This typically takes the form of plotting I_G against G for which a Gaussian decay profile is obtained (Fig. 10.9a) or I_G against G^2 for which an exponential profile is obtained (Fig. 10.9b) and D extracted by an appropriate non-linear fit since all other constants γ, δ and Δ (eg in Eq. 10.8) are known. In this approach, $I_{G=0}$ may also be a variable in the fit since this is not usually determined (Section 10.2.2). An alternative approach favoured by some is to plot $\ln(I_G/I_{G=0})$ against G^2 for which a linear least-squares fit is appropriate (Fig. 10.9c) with the slope proportional to $-D$.

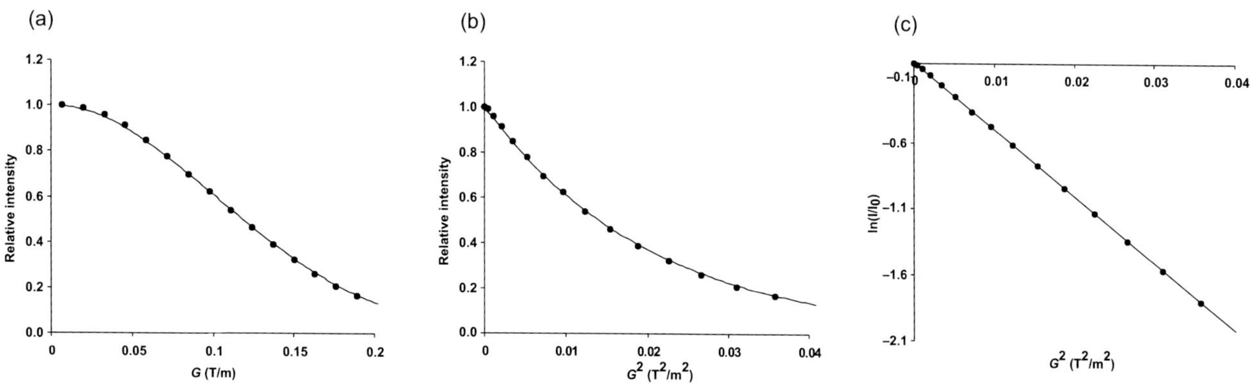

FIGURE 10.9 **Regression analysis to obtain the diffusion coefficient.** This may be achieved by fitting to the (a) Gaussian decay profile (I_G vs. G), (b) exponential decay profile (I_G vs. G^2) or (c) linear decay ($\ln(I_G/I_{G=0})$ vs. G^2); see text.

In situations where two peaks cannot be resolved but arise from separate molecules with differing diffusion behaviour, an apparent diffusion coefficient will result from fitting for a single value of D that has a magnitude intermediate between those of the overlapping components. In such cases it is theoretically possible to perform bi-exponential fits to extract the two diffusion coefficients although resolving superimposed exponential decays is notoriously difficult and achieving reliable results can be problematic; see Ref. [13]. Coincidental overlap of even more components would demand multi-exponential fits making extraction of meaningful diffusion coefficients all but impossible for routine applications; an experimental alternative in such cases is the use of hybrid diffusion sequences to improve resonance dispersion as described in Section 10.5.

10.2.5 Data Analysis: Pseudo-2D Presentation

An aesthetically different approach to presenting the results of diffusion experiments is as a pseudo-2D contour presentation which has been termed 'diffusion-ordered spectroscopy' (DOSY). In this, one dimension represents chemical shifts and the other the diffusion coefficients of each component as exemplified in Fig. 10.10. The most general and widely used data-processing scheme is direct exponential fitting of decay intensities, as described above, from which the 2D display is synthesised such that the centre of a crosspeak in the diffusion dimension represents the magnitude of the coefficient itself and the peak width in the same dimension reflects the magnitude of the fitting error, presented typically as a Gaussian profile. More generally, procedures suitable for extracting diffusion coefficients are equally applicable to the generation of DOSY plots, and there exist a range of more complex mathematical procedures for transforming diffusion decays into

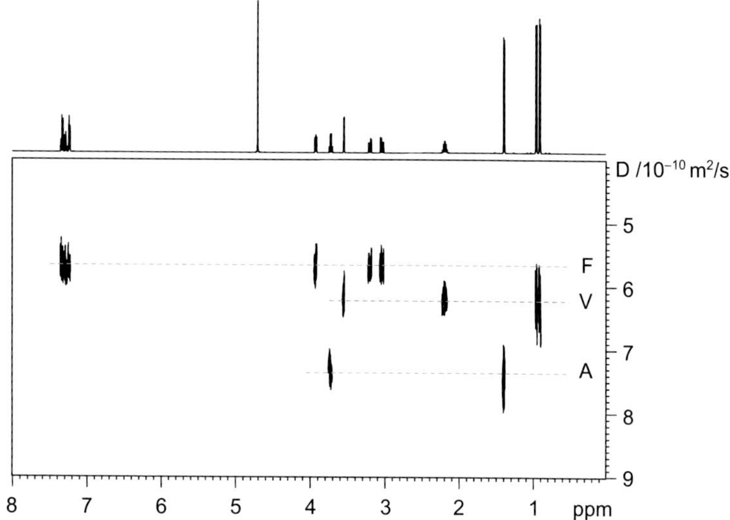

FIGURE 10.10 **2D DOSY presentation.** The spectrum illustrates the separation of resonances from an equimolar mixture of three *L*-amino acids alanine (A), valine (V) and phenylalanine (F) in aqueous solution at 298 K according to their diffusion coefficients, plotted along the vertical axis. Diffusion coefficients obtained from regression analysis of the data indicated values of 5.67 (F), 6.25 (V) and 7.37 (A)$\times10^{-10}$ m^2/s.

2D DOSY spectra when peak overlap is present including CONTIN, SPLMOD, DECRA, CORE and GIFA, and the interested reader is referred to selected review papers [2,14,15]. In all cases, high-quality data are a prerequisite for successful analysis. In practice, one is most likely to make use of the tools available in commercial NMR processing packages or the freely available and platform-independent *DOSY Toolbox* [16]. The DOSY presentation clearly lends itself to experiments on multi-component systems for which it shares some parallels in chromatographic separation of mixtures, although the underlying physical processes giving rise to the separation are, of course, very different.

10.3 PRACTICAL ASPECTS OF DIFFUSION NMR SPECTROSCOPY

In this section we shall explore some of the procedures needed for the recording of reliable diffusion data including instrument calibration, sample preparation and equilibration requirements. We shall also address some of the most important factors that can lead to the collection of inaccurate data, something to which diffusion measurements can be especially prone. Indeed, the setup of a diffusion sequence is no more demanding than most other 2D experiments, but the potential for erroneous results is considerably greater in most cases and, worst still, is not always readily apparent if one is not aware of the potential pitfalls and how to avoid them.

10.3.1 The Problem of Convection

There can be little doubt that the greatest barrier to reliable determination of diffusion coefficients on a modern high-resolution NMR spectrometer stems from thermal convection arising in the NMR sample. This typically takes the form of convective flow along the length of the NMR tube which is driven by temperature gradients in the sample that cause the upward flow of the warmer, less dense fluid, this process being known as Rayleigh–Bénard convection. This becomes significant only when temperature gradients exceed a threshold level, as described below, and becomes progressively more problematic at higher temperatures if not addressed. More recently, it has also been demonstrated that small horizontal temperature gradients also occur in NMR tubes and that these also drive convection [17]. This process, known as Hadley convection, is not dependent on a temperature gradient threshold, so contributes to convection both above and below ambient temperatures and is likely present during all solution-phase NMR experiments to some degree. Sample temperature gradients are generated by the temperature regulation methods used in NMR probeheads. The most widely adopted method on commercial instrumentation involves use of a gaseous flow, typically air or nitrogen, entering through the base of the NMR probe and passing over the NMR sample tube. To regulate the resulting sample temperature, the gas is usually heated prior to the sample with the heater current controlled via a feedback mechanism, as monitored by a thermocouple placed close to the base of the NMR tube. Even though this may maintain a very stable mean sample temperature, it remains the case that temperature gradients exist within the sample that may initiate convective flow. Since this provides an alternative mechanism to diffusion for the translational motion of molecules, its presence will lead to enhanced attenuation of signal decay in diffusion experiments and hence to larger but erroneous (apparent) diffusion coefficients.

Problems arising from convection are expected to be greatest for low-viscosity solvents and at sample temperatures furthest from the ambient probe temperature. As discussed in the sections that follow, convection can be significant at temperatures only very slightly higher or lower than ambient for commonly used low-viscosity solvents [17]. Indeed, for a solvent such as chloroform, a temperature gradient as small as 0.2 K along a typical NMR sample can be sufficient to induce convection [18]. Conversely, there is anecdotal evidence that convection can be less severe although still present when operating at very low temperatures; this is most likely because of increases in solution viscosity and the fact that cold rather than hot gas is flowing over the base of the tube. In any case, it is vitally important that one can recognise the influence of convection in an NMR sample before diffusion measurements are undertaken and, if present, take measures to either avoid this or to compensate for its influence in the techniques used.

10.3.1.1 Diagnosing Convection

In the case of severe thermal gradients (and assuming no other potentially complicating factors are present such as resonance overlap of different components or chemical exchange processes) the influence of convection may be seen as distortions to the usual signal amplitude Gaussian decay profiles. In cases where convection dominates over diffusion it may also be that signal amplitudes become negative before again displaying positive intensities since convection introduces what approximates to a cosine modulation into the diffusion signal attenuation [19] thus:

$$I_G \approx I_{G=0} \underbrace{\cos(\gamma\delta G\Delta v)}_{Convection} \exp\left[\underbrace{-(\gamma\delta G)^2 D\left(\Delta-\frac{\delta}{3}\right)}_{Diffusion}\right] \qquad (10.10)$$

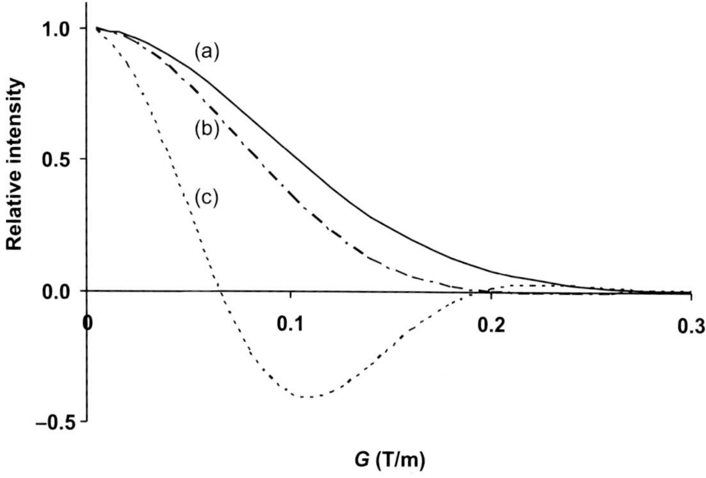

FIGURE 10.11 **Simulated diffusion attenuation profiles illustrating the influence of convection.** Trace (a) illustrates signal decay for a diffusion coefficient of 10×10^{-10} m^2/s in the absence of convection, while trace (b) shows how a moderate flow velocity (0.1 mm/s) accelerates signal attenuation. In trace (c) a relatively high flow velocity (0.3 mm/s) induces pronounced negative-going transitions.

where v represents the convective flow velocity. This effect may be seen in Fig. 10.11 where signal attenuations are presented without and with such contributions from convection. Although in the case of significant and dominant flow (Fig. 10.11c) the distortion is obvious and hence indicative of convection, the situation with a more moderate contribution from flow is far less clear (Fig. 10.11b) owing to similarities between the Gaussian and cosine function profiles. Hence, convection-moderated data can appear to fit the usual exponential function characterising diffusion rather well, but the apparent diffusion coefficients D_{app} so determined will be greater than the true value D and may be approximated as [19]:

$$D_{app} = D + \frac{v^2 \Delta}{2} \tag{10.11}$$

Note that flow velocities as little as 0.05 mm/s can be sufficient to lead to an erroneous apparent diffusion coefficient for a species whose actual value of D is 1×10^{-10} m^2/s.

A more sensitive and reliable test for the presence of convection is to record two or more diffusion experiments with differing values of the diffusion period Δ under otherwise identical conditions. In the absence of convection (or where its influence has been suppressed) the value of D obtained should not differ between data sets. In contrast, the *apparent* diffusion coefficients measured in the presence of convection will vary with Δ, as indicated by Eq. 10.11, and produce progressively larger values of D_{app} with longer diffusion periods. This influence is readily apparent for quinine **10.1** in $CDCl_3$ recorded at the slightly elevated temperature of 313 K, but may also be observed at a much reduced level at 298 K where probe temperature regulation is employed (Fig. 10.12).

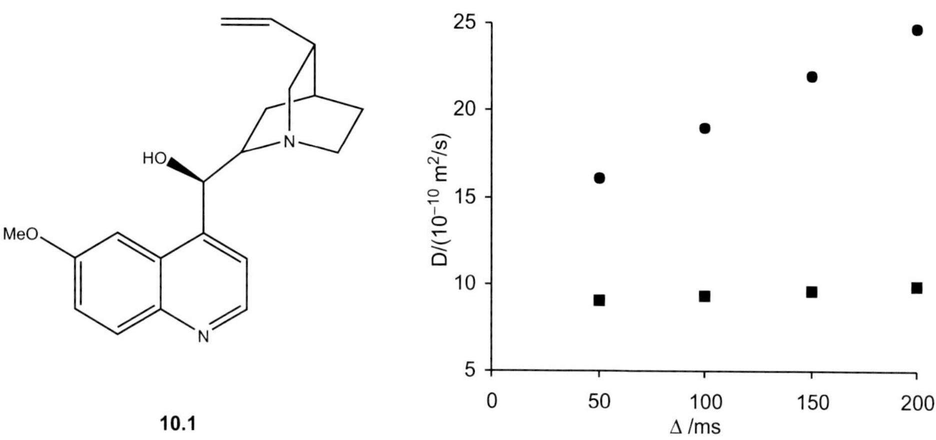

FIGURE 10.12 **Diffusion coefficients for quinine 10.1 in $CDCl_3$ as a function of diffusion time Δ.** Data were recorded at 298 K (squares) and 313 K (circles) demonstrating the influence of convection.

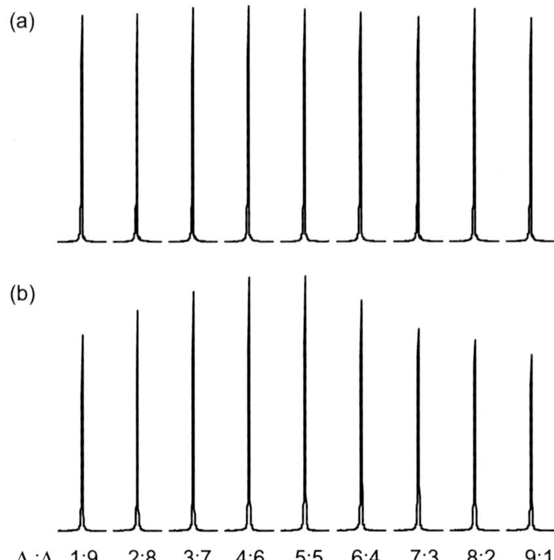

FIGURE 10.13 **Signal profiles recorded with the asymmetrical double STE (aDSTE) sequence as a test for convection as the ratio Δ_1:Δ_2 is varied.** (a) A uniform profile indicates the absence of convection whereas in (b) the asymmetrical profile demonstrates convection to be present.

A further approach to establishing the presence of convection is to record a series of 1D diffusion traces with an *asymmetrical* version of the double STE experiment introduced to compensate for convection (a description of which is given below that you may wish to consider before proceeding). In the compensated sequence the total diffusion period Δ is split into two *equal* periods (here denoted Δ_1 and Δ_2 with $\Delta_1 + \Delta_2 = \Delta$) such that contributions from convection arising during the first echo are exactly cancelled by those in the second. It then follows that an asymmetrical double echo in which $\Delta_1 \neq \Delta_2$ will lead to only partial cancellation of these contributions according to the imbalance in the ratio Δ_1:Δ_2. Thus, a sequence of 1D experiments in which Δ is held constant but Δ_1 and Δ_2 are systematically varied will present spectra with increasing signal attenuation in the presence of convection as the imbalance in the two diffusion periods increases. This is due to the progressively larger contributions to signal attenuation from uncompensated convection. Conversely, uniform intensity will be observed for all spectra in the absence of convection since the total diffusion period Δ remains invariant (Fig. 10.13). This method may also be employed to determine flow velocities within NMR tubes [17,18].

10.3.1.2 Dealing with Convection

As indicated earlier, when convection can be shown to be present within a sample there are essentially two approaches for negating this: one may seek to suppress convection itself or one can employ methods that are designed to compensate for it. The optimum approach will depend on a number of factors relating to the nature of the sample and the behaviour of the instrumentation, so both these approaches will be considered before some general recommendations are presented.

Before proceeding, it is instructive to consider the conditions under which convection arises. The onset of Rayleigh–Bénard convection, as driven by temperature gradients along the length of the NMR sample, is characterised by the Rayleigh number R, which for a long cylindrical sample may be defined as:

$$R = \frac{g\alpha}{\kappa\nu} r^4 T^* \tag{10.12}$$

where g is acceleration due to gravity, α is thermal expansivity, κ is thermal diffusivity, ν is kinematic viscosity, r is the (inner) radius of the NMR tube and T^* is the temperature gradient over the length of the sample (equivalent to $\Delta T/l$ where ΔT is the temperature difference across a sample height l). Thermal expansivity is a measure of the expansion of a sample as a result of heating, thermal diffusivity is a measure of the rate at which temperature disturbance applied to one part of a sample travels to another and kinematic viscosity reflects the viscosity–density ratio. According to theory, convection will arise when the Rayleigh number for the solution exceeds a critical threshold, which itself depends upon the thermal conductivity of the sample container walls (ie the NMR tube walls). Although in practice the onset of convection may deviate from this idealised expression, it does serve to suggest how one may at least deter this onset, as discussed below. As noted earlier, Hadley convection, which is driven by horizontal temperature gradients, has no threshold and will be present to some degree whenever such gradients exist.

10.3.1.3 Temperature Gradients and Decoupling

It is apparent from the above discussions that one option to delay the onset of convection is to reduce the temperature gradients throughout the sample tube. A crude approach to this is to not use active temperature regulation of the sample at all; that is, have no gas flow over the sample and to ensure it is left to fully equilibrate, although this is clearly somewhat limiting, likely impractical and certainly not safe for cryogenically cooled probes. Undoubtedly, it is critical to allow samples to equilibrate fully prior to diffusion measurements and the timescales for this are considerably greater than would typically be considered sufficient for most high-resolution NMR experiments; equilibration periods of 15–30 min are appropriate.

When active temperature regulation is employed, any temperature gradients generated are, at least in part, dictated by probe design, for which most of us have no immediate control. However, it has been demonstrated that temperature gradients along NMR samples may be substantially reduced by using high gas flow rates over the sample [18,20,21]. While higher flow rates may be better, these may induce undesirable sample vibrations or even lift the sample out of the probehead if too large, and a good-quality, heavy, sample turbine may be beneficial in this respect. An additional insidious cause of temperature gradients arises from non-uniform radiofrequency heating of a sample through the use of broadband decoupling, as would be required, for example in the case of diffusion measurements of heteronuclear species benefiting from proton decoupling. Indeed, even conventional proton decoupling using composite decoupling schemes such as WALTZ-16 (see Section 12.3) can induce measurable changes in diffusion coefficients, even at ambient temperatures for low-viscosity solvents. In such cases, reduced decoupler powers may be required, with correspondingly longer decoupler pulse widths. The use of extended relaxation delays during which the decoupler is gated off also prove beneficial.

Additional methods to eliminate temperature gradients include the use of coaxial inserts to hold the solution within a conventional NMR tube so as to jacket the sample, with the intermediate space filled with a perfluorinated hydrocarbon oil [22], an NMR silent solvent [23] or simply just empty space [24], the last of these having been demonstrated specifically for low-temperature (~230 K) studies. Improved heat transfer into the NMR solution is also possible through the use of thick-walled NMR tubes, and the use of these or of small-diameter tubes (see later discussions) is recommended for accurate diffusion coefficient measurements. Although a less practical solution, it appears the ultimate approach is to employ sapphire NMR tubes [17]. The thermal conductivity of sapphire is approximately 25 times that of standard borosilicate glass, leading to significantly reduced temperature gradients, but also to a poorer lineshape due to a reduction in magnetic field homogeneity caused by such tubes.

10.3.1.4 Tube Diameters

One of the most effective and pragmatic approaches to minimising the influence of convection is to reduce the diameter of the NMR tube [21] since the Rayleigh number scales with r^4 (as indicated by Eq. 10.12), and because the Hadley convection rates scale as r^3 for an NMR tube. Thus, reducing the tube diameter from 5 to 3 mm will require temperature gradients to be in excess of an order of magnitude greater for convection to arise in the narrower tube as compared with those in the wider tube. The disadvantage of such an approach is the reduction in signal intensity with the smaller sample volume (for a fixed sample concentration), so may only be appropriate when sample strength is not a limiting factor. A related approach is to employ a reduced sample column height to reduce the total temperature gradient across the sample, and purpose-designed, susceptibility matched tubes have been developed for this [25]. While these may be impractical for routine work, ensuring solution depths are not excessive appears a prudent experimental procedure.

10.3.1.5 Solvent Viscosity

Eq.10.12 further indicates that a solvent with greater viscosity will be less likely to suffer from convection, as one may anticipate. Thus, lower viscosity solvents (Table 10.4) such as acetone and acetonitrile may suffer from significant convection at temperatures within a few degrees of ambient (unregulated) probe temperatures, whereas more viscous solvents such as dimethylsulfoxide (DMSO) and water are more robust in this respect (although the low coefficient of thermal expansion of water also contributes to it being less prone to convection). However, solvent selection is more usually dictated by the solubility and stability of the compounds of interest and may offer limited scope for dealing with convection.

10.3.1.6 Sample Spinning

The rotation of an NMR sample tube has also been demonstrated to reduce convection owing to the induced stabilising forces acting to retard the onset of laminar flow [26,27] and to the averaging of transverse temperature gradients [17]. This effect is illustrated in Fig. 10.14 which shows the measured diffusion coefficients of quinine **10.1** in $CDCl_3$ recorded at 313 K for a static sample and sample spinning at 20 Hz, as a function of Δ. Spinning the sample produces a reliable measure

TABLE 10.4 Viscosities of Some Common Protonated Solvents at 298 K

Solvent	Viscosity (mPa/s)
Acetone	0.31
Acetonitrile	0.35
Benzene	0.60
Chloroform	0.54
Dichloromethane	0.41
Dimethylsulfoxide	1.99
Methanol	0.54
Tetrahydrofuran	0.46
Toluene	0.56
Water	0.89

The SI units of millipascals per second are equivalent to the widely quoted centipoise (cP).

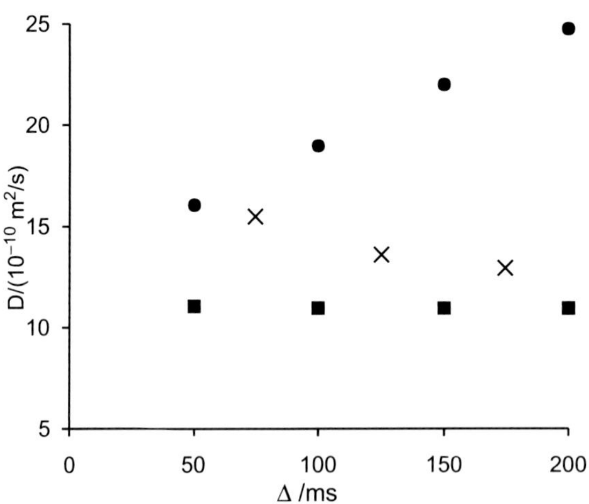

FIGURE 10.14 **Observed diffusion coefficients of quinine 10.1 in $CDCl_3$ at 313 K recorded as a function of Δ for a static sample (circles) and for the sample spinning at 20 Hz (squares).** When the spin rate is not synchronised with the diffusion time the apparent diffusion coefficients may be erroneous and dependent on Δ (crosses).

of the diffusion coefficient that is independent of Δ. However, an absolute requirement for this approach to be effective is that *the spinning rate and diffusion periods must be synchronised* such that the sample tube prescribes an integer number of rotations during the diffusion period. Thus, for a sample spinning at 20 Hz, appropriate diffusion periods would be 50 ms or multiples thereof. This is to prevent any net displacement of the solute molecules along the direction of the applied gradient due to the physical relocation of the particle through rotation. If such a displacement were to occur in the period between the defocusing and refocusing gradient pulses, this would contribute to signal attenuation in addition to that arising from diffusion itself and so result in an erroneously high apparent diffusion coefficient. This process is illustrated in exaggerated form in Fig. 10.15; a lack of perfect concentricity between the sample tube and the gradient coils means the solute particle effectively travels along the gradient direction as the sample rotates, whereas no net displacement arises following a complete revolution. A similar attenuation may arise in the presence of transverse inhomogeneities in the field gradient, which again may be countered by integer tube rotations. Since NMR diffusion measurements are sensitive to translational motions on the micrometer scale, only a miniscule displacement of a particle by rotation is sufficient to interfere with diffusion measurements. The effect of mismatched sample spinning is also illustrated in Fig. 10.14 wherein erroneously high apparent diffusion coefficients are measured when the diffusion period Δ and the spin rate are no longer synchronised; in this case where Δ is not a multiple of 50 ms. Thus, while sample spinning can be an effective means of suppressing convection, it demands cautious implementation, stable sample rotation and benefits from high-quality tubes. Sample spinning

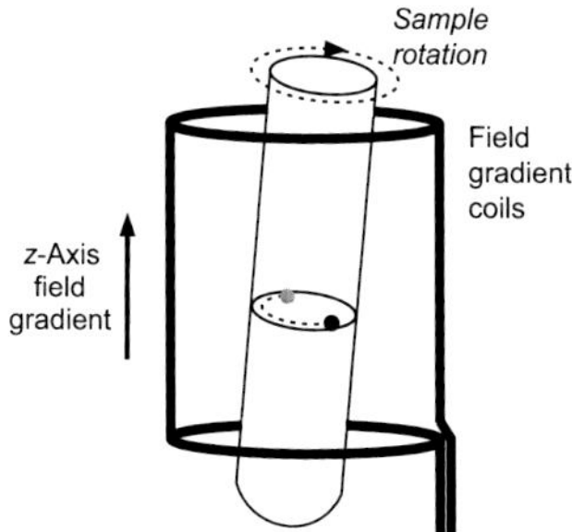

FIGURE 10.15 **Schematic (and exaggerated) illustration of molecular translation brought about by sample spinning in cases where the axes of the sample tube and gradient coils are not perfectly aligned.**

may not therefore provide a complete solution to the problem of convection when very accurate quantitative measurements of diffusion coefficients are required. Furthermore, at temperature extremes in particular, spinning may not lead to complete convection suppression and a check for the presence of this, as described above, is still desirable.

10.3.1.7 Convection-Compensating Sequences

A quite different yet very effective approach to removing the deleterious effects of convection is to employ sequences that directly compensate for this effect by cancelling the influence of flow in the sample: the so-called 'double-echo' experiment. It has already been stated that convection imposes an attenuation factor on the NMR response in addition to that arising from diffusion itself, as was expressed in Eq. 10.10. More precisely, the presence of linear flow during a gradient echo sequence imposes a *velocity*-dependent phase on the NMR signal that arises because the spins move during the gradient pulse and so experience a varying magnetic field. This velocity-dependent term acts in addition to the spatially-dependent phase imposed by the field gradient pulse. In any convecting NMR sample, molecules will be moving in both upward and downward directions and will exhibit a spread of flow velocities such that the net effect of the various imposed phases is the aforementioned attenuation of the detected NMR response.

In a basic gradient echo–refocusing experiment such as that in Fig. 10.16a, the *spatially* dependent phase imposed by the first gradient may be refocused simply by applying a second equal but opposite gradient to produce the echo. However, in the presence of convective flow there remains a velocity-induced phase after the second gradient pulse which depends upon the duration of the gradient pulses, the time between these and the velocity itself. In the typical situation where the gradient durations are small relative to the time between them the remaining phase is primarily due to the linear flow that occurs between the gradient pulses, meaning different local fields are experienced by spins during these pulses. This unwanted residual phase may however be cancelled by repeating the echo sequence but with inversion of the sign of the two gradient pulses (Fig. 10.16b). The spatially dependent phase is again refocused within this second gradient pair but, critically, the residual flow-induced phase is now equal but opposite to that produced by the first gradient pair. Thus, the velocity-dependent phase arising from the first pair of gradients is exactly cancelled by that arising from the second pair and the desired convection compensation is achieved using a double-echo.

In this, it has been assumed that the coherence order p of magnetisation (Section 5.3) has remained constant (at either +1 or −1) throughout the echo sequence. In fact, the velocity-induced phase is also sensitive to the sign of the coherence, suggesting an equivalent approach to convection compensation is to repeat the echo with gradients of the same sign but with an associated change of coherence order (brought about by appropriate rf pulses) between the two echoes (eg $p = +1$ to $p = -1$) as in Fig. 10.1c. This approach has been incorporated in the mostly widely used convection-compensated sequences. In the

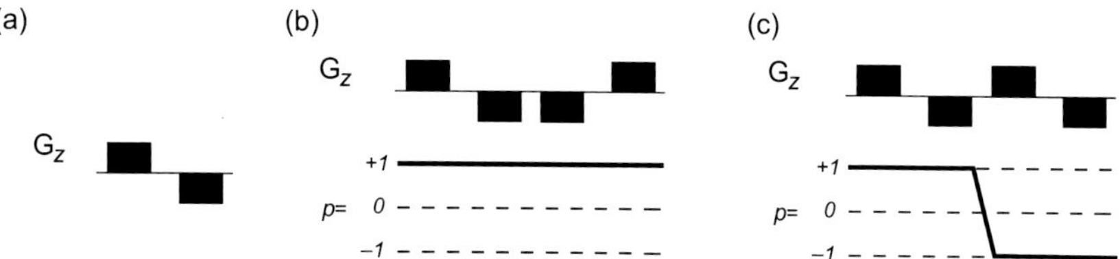

FIGURE 10.16 **Generic gradient schemes for convection compensation.** (a) A basic refocusing gradient echo scheme, (b) a double-echo scheme to refocus the velocity-dependent phase (with retention of coherence order at +1) and (c) a double-echo scheme to refocus the velocity-dependent phase (with inversion of coherence order +1 to −1). The change in coherence order is only possible by the application of suitable rf pulses (not shown).

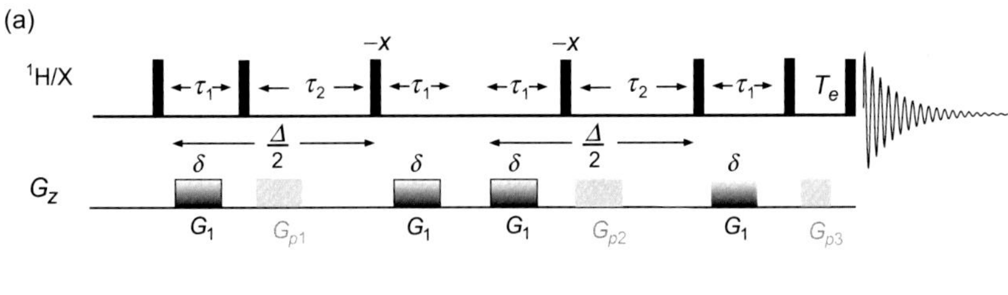

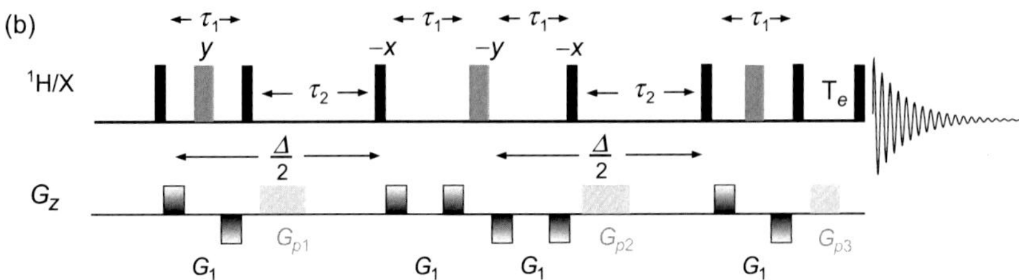

FIGURE 10.17 **Convection-compensated diffusion sequences.** (a) DSTE, (b) BPP-DSTE. Delays are as for the STE scheme but the total diffusion period Δ is split and applied as two echo sequences. Optional gradient pulses (denoted G_p) and the LED scheme are also shown; see. [28] for details of the necessary phase-cycling.

double STE (DSTE) sequence (Fig. 10.17a) the total diffusion period Δ is now divided in two such that each half comprises the basic STE sequence. The gradient pulses are applied with the same sign throughout, demanding the coherence orders selected by phase cycling are [28] $p = -1, +1, +1, -1$ when these are applied to give the desired convection compensation. As for the non-compensated sequences described above, the use of an eddy current delay period may also be advantageous, and additional purging gradients may be employed during the diffusion or LED periods when magnetisation is longitudinal. A further refinement is the incorporation of bipolar pulses (Fig. 10.17b); in this scheme the two adjacent sets of BPPs at the centre of the sequence have been combined such that only a single 180 degree refocusing pulse is necessary. Fitting to the Stejskal–Tanner equation is as for the STE.

Fig. 10.18 compares the diffusion decay profiles of the uncompensated BPP-STE and the compensated BPP-DSTE sequences for a sample of quinine **10.1** in $CDCl_3$ at 313 K. The apparent diffusion coefficient from the STE data of 17.1×10^{-10} m²/s is erroneously high due to the contribution from convection whereas a lower value of 11.1×10^{-10} m²/s is provided by both the DSTE method and for the STE with sample rotation. While providing effective compensation for convection, the DSTE sequences suffer from the disadvantage of reduced sensitivity as each echo sacrifices half of the

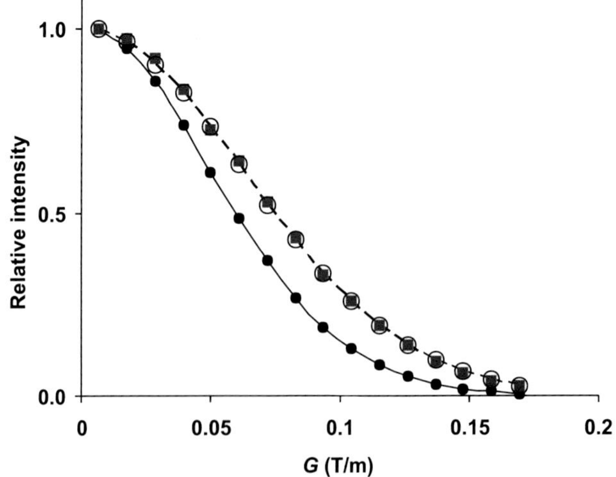

FIGURE 10.18 **Diffusion decay profiles for quinine 10.1 in $CDCl_3$ at 313 K recorded without and with convection compensation.** Experiments used the BPP-STE sequence (closed circles), the double-BPP-STE sequence (red squares) and the BPP-STE sequence with the sample spinning at 20 Hz (open circles). The total diffusion time Δ in each case was 100 ms with gradient pulses δ totalling 4 ms each.

available signal which may make the sequences less attractive in some instances. In contrast, double-echo sequences based on spin-echoes rather than STEs do not suffer such sensitivity losses and have been employed in some combined 3D DOSY sequences [29].

10.3.1.8 Summary

When executing diffusion experiments it is critically important that convection does not contribute to the observed signal decay intensities and its presence should always be considered a possibility. This will be most significant for lower viscosity solvents or when operating at elevated or reduced sample temperatures, although will likely be present only a few degrees above or below ambient. A simple approach to test for this is to perform the experiment with two or more diffusion periods Δ and to compare the measured diffusion coefficients. In the absence of convection these will be the same. When convection is present it may be suppressed either by physically reducing the convective flow or compensating for it. A most effective and readily implemented approach is to employ narrower NMR tubes (typically 3 mm diameter) to greatly reduce the flow, or thick-walled NMR tubes which also reduce temperature gradients. Alternatively, the influence of convection on recorded data may be eliminated by use of suitable convection-compensated pulse sequences, which may also be combined with the practical measures described in this section.

10.3.2 Calibrating Gradient Amplitudes

A fundamental requirement for accurate determination of diffusion coefficients is a measure of the gradient strengths employed, and there are essentially two procedures that find widespread use for calibrating the gradient amplitudes produced by an instrument. This amplitude will represent the combined capabilities of the gradient amplifier and the gradient coils themselves and thus may vary between consoles and between probes. In principle, the actual field gradients delivered by a probe can be calculated from the coil geometry and knowledge of the gradient amplifier output, and vendors often quote the maximum field gradient a probe can produce based on this. Nevertheless, a basic calibration is recommended if only to confirm this value to be correct for your system.

The first method is based on taking a 1D image of a 'phantom' in an NMR tube and from this calculating the applied gradient strength; this procedure is described in Section 3.5.2. The second approach is to determine the strength from a diffusion experiment performed at a defined temperature by using a suitable calibration sample of known diffusion coefficient. In this, single diffusion traces are acquired with a nominal fixed gradient strength applied (often defined as some fraction of the maximum gradient amplifier output) and the resonance intensity of the calibrant compared with that in the absence of gradient pulses. The simple gradient spin-echo scheme of Section 10.2.1 is well suited for this. The applied gradient amplitude G may then be back-calculated from the Stejskal–Tanner equation (Eq. 10.6) since the diffusion coefficient of the calibrant is known. A commonly used sample for this is H_2O in D_2O for which the diffusion coefficient at 298 K is given by [23,30]:

$$D_{\mathrm{HDO(298K)}} = 2.30 - 0.4652x + 0.0672x^2 \times 10^{-9}\,\mathrm{m^2/s} \tag{10.13}$$

where x represents the mole fraction of D_2O in the sample. Thus, the residual HDO in 'pure' D_2O (99.9% *D*) has a diffusion coefficient of 19.02×10^{-10} m^2/s and provides a suitable calibrant for the back calculation of G. While this method may appear more straightforward than the use of a phantom, it does require that the sample temperature has also been accurately calibrated (see Section 3.5.3) and regulated, and that convection does not hamper amplitude intensities. Its notable advantage is that, provided calibration uses the same sequence as subsequent measurements, this procedure allows for any non-ideal behaviour in the field gradients produced. Reference data for a selection of calibration samples for nuclei other than ^1H have also been reported should these ever be required and include ^7Li, ^{13}C, ^{19}F and ^{31}P [23].

10.3.3 Optimising Diffusion Parameters

Diffusion coefficients depend upon many parameters including the size and shape of the solute under investigation, properties of the solvent such as its viscosity, sample temperature and concentration and so on, meaning it is most often the case that optimum parameters for diffusion experiments have to be determined on a sample-to-sample basis. When the physical conditions for the experiment have been decided (solvent, temperature, etc.) the principal parameters that remain to be defined are the diffusion period Δ, the gradient pulse duration δ (and possibly its shape) and the range of gradient strengths to be employed G_{min} to G_{max}. The aim is to obtain a well-characterised decay profile to allow reliable

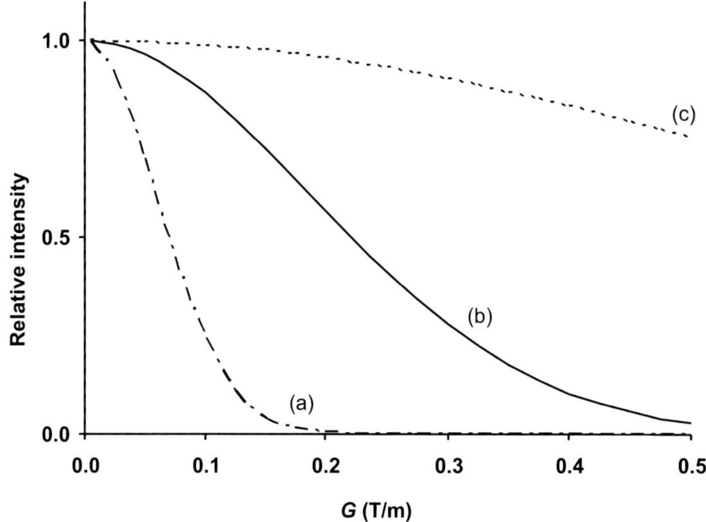

FIGURE 10.19 **Optimizing diffusion decay profiles.** When signal attenuation is too fast (trace (a)) the later data points do not contribute to the profile, when it is too slow (trace (c)) there is insufficient attenuation to provide an accurate determination of D. Trace (b) shows an optimum profile where all data points would contribute to characterisation of the decay.

data fitting (Fig. 10.19). Optimisation may be done following an iterative process along the following lines for a standard high-resolution probehead:

- Select a value for Δ based on prior knowledge or literature precedence; 50–100 ms being a typical value for 'small' molecules.
- Select a value for δ using similar guidance; 2–4 ms is often appropriate.
- Set the gradient output amplitude to a small value, for example ~2% of maximum output.
- Collect a single 1D diffusion trace as a reference point (trace A) using sufficient transients to clearly see all peaks of interest. This would equate to the first data point of the full diffusion experiment.
- In a second data set, repeat the above acquisition with the gradient amplitude set close to the maximum possible value (~90% of maximum output) but with otherwise identical acquisition parameters to provide a test spectrum (trace B).
- Compare trace B with the reference trace A; one should expect to observe a significant reduction in signal intensity.

The aim is for the signal intensity in trace B to be ~10% of reference trace A for the solutes of interest, as this will represent the final data point in the full diffusion experiment (Fig. 10.20). This process is iterated with suitable parameter modification to achieve the desired attenuation. Thus, if the signal has been completely attenuated in trace B, then the upper gradient amplitude and/or the diffusion time may be reduced. Shortening Δ may be advantageous as this reduces relaxation

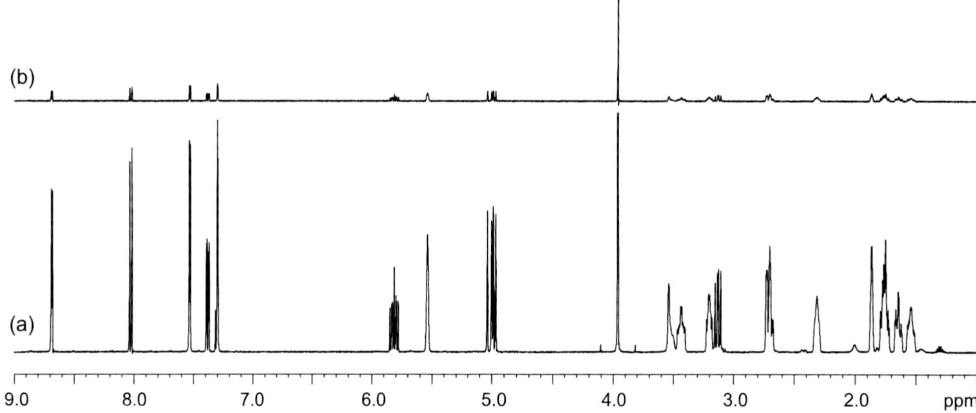

FIGURE 10.20 **Diffusion parameter optimization.** Experimental 1D ^{1}H diffusion traces for quinine **10.1** showing gradient optimisation with (a) an initial minimum gradient value G_{min} of 2% (of maximum output) and (b) a maximum gradient value G_{max} of 70%. The residual signal in (b) is 7% of that in (a).

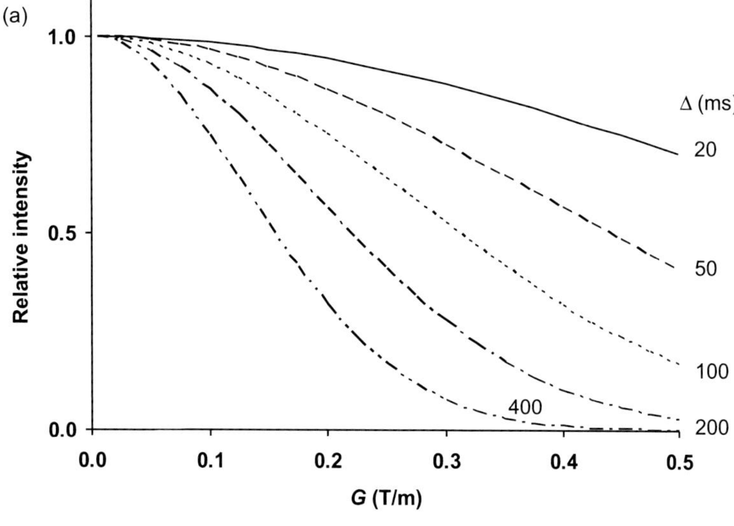

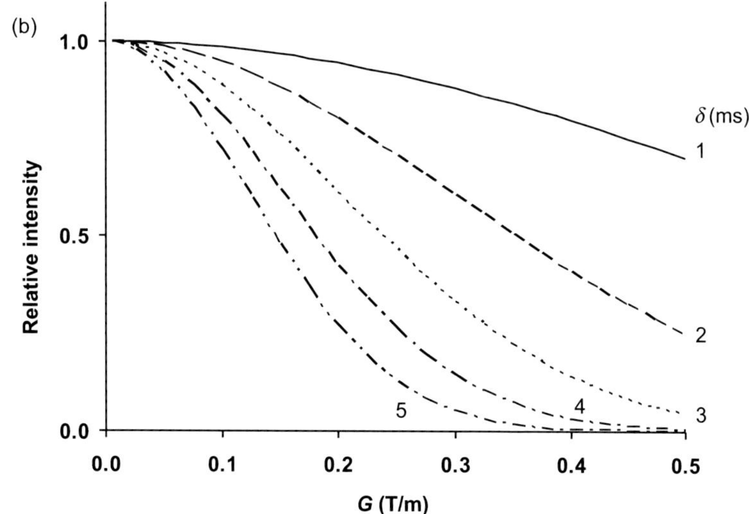

FIGURE 10.21 **Diffusion decay profiles illustrating the influence of the key experimental parameters Δ and *δ*.** The graphs show (a) variation as a function of Δ ($D = 10 \times 10^{-10}$ m^2/s and $\delta = 1$ ms) and (b) variation as a function of δ ($D = 10 \times 10^{-10}$ m^2/s and $\Delta = 20$ ms).

losses, although its selection may be constrained if sample rotation is employed; see Section 10.3.1. If the signal attenuation in trace B is insufficient, then a longer diffusion period may be employed, although this must remain shorter than or comparable with the T_1 relaxation time constant (or T_2 for spin-echo–based sequences), so that signal intensity is not overly sacrificed prior to further diffusion-based attenuation. An alternative approach in such cases is to increase the net gradient power applied by increasing δ, noting here that signal attention scales with δ^2 but is linear in Δ. The profiles of Fig. 10.21 are simulated for $D = 10 \times 10^{-10}$ m^2/s and illustrate the influence of the Δ and δ parameters on signal attenuation and may offer some guidance on parameter selection. Ultimately, one may need to alter Δ, δ and G_{max} to achieve optimum signal decay profiles. The simulation of diffusion decay profiles can be usefully employed when setting up diffusion experiments on new systems by suggesting appropriate initial parameters prior to optimisation. Typical values for Δ are tens to hundreds of milliseconds, and those for δ are in the range 1–10 ms.

During the setup procedure, and depending on the nature of the systems under study, one should remain alert to the possibility of damaging hardware if the gradient amplifier is driven too high for too long, and it is wise to be guided by vendor specifications appropriate to your system for selecting δ and G_{max}. Indeed, for the study of very slowly moving particles, such as large macromolecular species, purpose-built gradient amplifiers and water-cooled probeheads are commercially available that are able to deliver far larger gradient strengths than conventional high-resolution probeheads. You may also

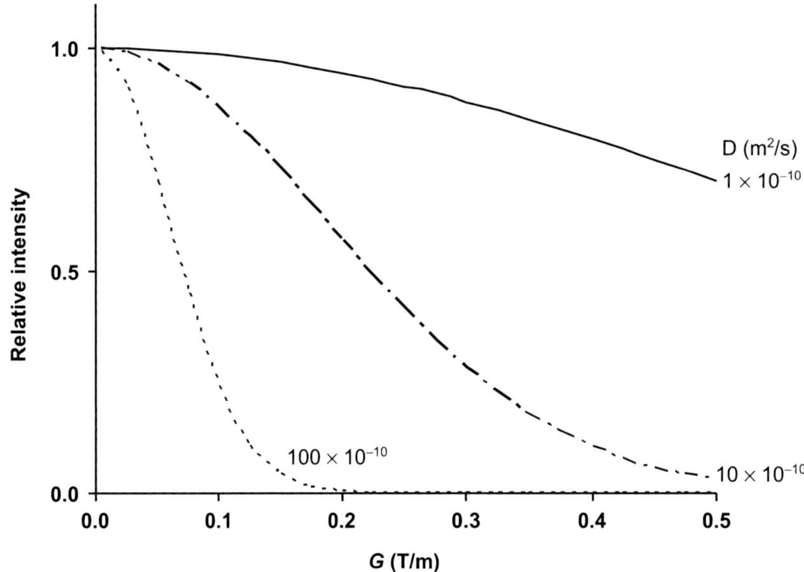

FIGURE 10.22 **Diffusion decay profiles for three diffusion coefficients spanning two orders of magnitude.** Diffusion parameters were $\Delta = 50$ ms, $\delta = 2$ ms, and D of 1, 10 and 100 $\times 10^{-10}$ m^2/s.

choose to use non-rectangular gradient profiles to reduce eddy current generation in which case the resultant gradient amplitudes will be less than those for rectangular pulses of equivalent duration. An appropriate gradient shape factor must then be employed in fitting the Stejskal–Tanner equation to account for this. Equivalently, one can simply apply a correction factor to the gradient amplitudes used in the fit to allow for the difference in integrated intensities of the shape profiles; the shape factor for the commonly employed half-sine bell shape is $2/\pi$ or 0.6366. Of course, the chosen gradient shape must also be employed during optimisation of the parameters for the diffusion experiment. Finally, if using the LED extension, it remains to determine what duration is needed to remove residual lineshape distortions. This is best done through trial and error by examining 1D diffusion traces recorded with a high gradient strength (ie equating to the parameters for the final spectrum of the diffusion experiment) and increasing T_e from, say, 1 ms, to find a suitable value. Periods of 5–10 ms are most commonly reported in the literature, although the use of modern shielded probeheads and shaped gradients may make this stage superfluous.

When selecting parameters, one may also need to be aware of the possibility of solutes exhibiting a range of diffusion coefficients if dealing with sample mixtures, and it may not be possible to characterise reliably the diffusion of all components with a single diffusion parameter set. Thus, for example if one were looking to characterise both macromolecules and small molecules in the same sample, it may be that the attenuation of the small molecules is far too rapid to characterise adequately when diffusion parameters are optimised for the macromolecules. In practice, it is easily possible to measure diffusion coefficients varying over an order of magnitude from a single data set, as illustrated in Fig. 10.22, and possibly over two orders for good-quality data, whereas larger differences are likely to demand multiple experiments with appropriately optimised parameters. The diffusion coefficients of many small molecules and their complexes often fall in the region of 1–20 $\times 10^{-10}$ m^2/s and may therefore be readily characterised by ^{1}H diffusion experiments using standard gradient hardware.

10.3.3.1 Diffusion Measurements with Nuclides Other than ^{1}H

Besides ^{1}H measurements there are potential benefits in the observation of other nuclides for diffusion measurements. There may be fewer problems with resonance overlap of different species or less interference from solvent or impurity resonances, for example, or the compound of interest may simply lack hydrogen as is often the case with counterions. In this case one should be aware that optimum parameters for ^{1}H measurement may not be suitable for nuclides of lower magnetogyric ratio γ. This is because the total *effective* gradient strengths employed also depend upon this parameter and, as can be seen from the Stejskel–Tanner equation (Eq. 10.6), the degree of attenuation of resonance intensity will be reduced as γ becomes smaller. This is illustrated in Fig. 10.23 where the decay profiles for four different nuclides are shown, having been calculated with identical diffusion parameters but with the appropriate γ value employed. Clearly, while the profiles for ^{1}H and ^{19}F are perfectly adequate, those for ^{31}P and ^{13}C are less so. This means that larger diffusion

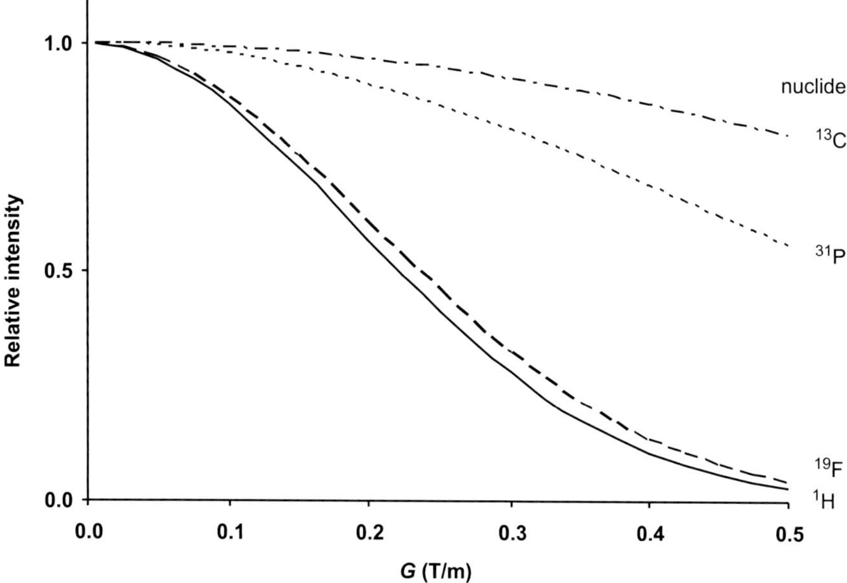

FIGURE 10.23 **Diffusion decay profiles calculated for when differing nuclides are being observed.** The parameters are optimum for ^{1}H observations ($D = 10 \times 10^{-10}$ m^2/s, $\Delta = 50$ ms and $\delta = 2$ ms) but the profiles change according to the magnetogyric ratio (γ) of the observed nuclide, becoming suboptimal for smaller values of γ.

times and/or increased gradient pulse durations or amplitudes will be required relative to those suitable for ^{1}H observations when lower γ nuclides are studied, and may thus demand a separate optimisation procedure for each nuclide to be observed. As noted above, care will also be required to avoid sample heating from proton decoupling when observing heteronuclides.

A problem that is likely to be relevant only to proton observation, although is likely to be rarely encountered, is that of radiation damping caused by very concentrated samples. This has been shown to interfere severely with diffusion measurements, although the simple expedient of using a reduced excitation flip angle can greatly reduce the problem [31].

10.3.4 Hydrodynamic Radii and Molecular Weights

Having made careful measurements of diffusion coefficients it may be inviting to use these to calculate hydrodynamic radii, molecular volumes and even to estimate molecular weights, and there are many reports of such studies in the literature, as highlighted in the following section. Diffusion measurements may be used to address commonly asked questions pertaining to molecular size, including whether species exist as discrete monomers, dimers or higher oligomers, whether ion pairs remain tightly bound in solution, or whether molecules have a tendency to aggregate, and in this section we consider briefly appropriate methods for interpreting diffusion coefficients.

In analyses of this type, there are a number of issues to be addressed, some of which may influence the experimental protocol one employs. The hydrodynamic radius of a molecule or complex may be derived directly from the diffusion coefficient using the Stokes–Einstein equation (Eq. 10.4) which itself includes terms for the sample temperature T and the solution viscosity η. While the temperature may be regulated following appropriate calibration, knowledge of the solution viscosity may be less readily available, and there are two general approaches to obviating the need for direct viscosity measurements. The first is simply to assume the solution viscosity is that of the pure solvent under the conditions employed, which may only be applicable for dilute solutions, perhaps low millimoles. The second is to utilise a calibrant molecule of known hydrodynamic radius placed in the solution and use the measured diffusion coefficient of this to calculate η or, alternatively, to negate the need for this parameter. This requires that the internal reference material is of similar size to the species being investigated such that the same Stokes–Einstein scaling factor applies (see below). In this case the hydrodynamic radius of the solute r_{sol} may be determined from that of the reference r_{ref} by comparison of the respective diffusion coefficients D_{sol} and D_{ref}:

$$\frac{D_{sol}}{D_{ref}} = \frac{r_{ref}}{r_{sol}} \tag{10.14}$$

Both tetramethylsilane (TMS) and terakis(trimethylsilyl)silane (TMSS) have been suggested as internal calibrants no doubt owing to their being chemically inert and in approximating well to the desired spherical shape. The use of an internal reference such as this can also be useful in indicating viscosity changes on a sample-to-sample basis even if r_{sol} itself is not required and may also act as an indicator of convective flow if diffusion coefficients differ greatly.

If solution viscosity is known, then r_{sol} may be calculated directly from the Stokes–Einstein equation, thus:

$$r_{sol} = \frac{k_B T}{6c\pi\eta D_{sol}} \tag{10.15}$$

where the correction factor in the denominator c is now included since the scaling factor 6 is only applicable for molecules or complexes that are significantly larger than the solvent molecules surrounding the particle, falling to around 4 when these become comparable. It has been suggested that a scaling of 6 is applicable when the solute-to-solvent size ratio exceeds 5, but requires correction below this. Thus, the factor 6 may be appropriate for studies of supramolecular systems [4], but is unlikely to be so for studies of isolated small molecules in common organic solvents such as chloroform. Studies on friction theory [32] determined the correction factor c may be expressed in terms of the solute and solvent hydrodynamic radii r_{sol} and r_{solv}, respectively, as:

$$c = \left(\frac{3r_{solv}}{2r_{sol}} + \frac{r_{sol}}{r_{sol} + r_{solv}} \right)^{-1} \tag{10.16}$$

This expression was modified semi-empirically [33] from experimental measurements of variously sized crown ethers, to yield the commonly used term:

$$c = \left(1 + 0.695 \left(\frac{r_{solv}}{r_{sol}} \right)^{2.234} \right)^{-1} \tag{10.17}$$

Nevertheless, for the small molecules commonly encountered in many organic chemistry laboratories, this expression tends to overestimate the values calculated for diffusion coefficients and more appropriate empirically derived correlations have been presented based on a generalised model suitable for small molecules in a range of common solvents [34]. The approach recognises that it is possible to relate the hydrodynamic radius of a particle to its partial specific volume, $\overline{v}$, which in turn allows one to correlate a solute self-diffusion coefficient D_{sol} with its molecular weight M_{sol}, as frequently required in the laboratory. For the case of a spherical particle the following relationship holds [35]:

$$r_{sol} = \sqrt[3]{\frac{3M_{sol}\overline{v}}{4\pi N_A}} \tag{10.18}$$

in which N_A is Avogadro's number. The model then makes use of the notion of a single reduced effective density of a small molecule (meaning both solute and solvent) ρ_{eff}, as a pragmatic route to account for solute flexibility, non-spherical dimensions and solvation effects, yielding an expression relating D_{sol} to M_{sol}:

$$D_{sol} = \frac{k_B T \left(\frac{3\alpha}{2} + \frac{1}{1+\alpha} \right)}{6\pi\eta \sqrt[3]{\frac{3M_{sol}}{4\pi\rho_{eff} N_A}}} \quad \text{for which} \quad \alpha = \sqrt[3]{\frac{M_{solv}}{M_{sol}}} \tag{10.19}$$

and where M_{solv} is the molecular weight of the solvent. Experimental measurements for a range of solutes and solvents yielded an optimised value for ρ_{eff} of 619 kg/m^3 for molecules containing only light atoms (ie sulphur or lighter). Conveniently, this expression has been implemented in an easily used spreadsheet available online [36] that is able to estimate diffusion coefficients from input molecular weights, or vice versa. The correlation between the experimental and calculated diffusion coefficients for 44 small molecules with masses ranging from 19 to 1135 g/mol recorded in five deuterated solvents is shown in Fig. 10.24 and provides sufficient accuracy to address many common questions asked of small-molecule diffusion studies. Deviations arise most notably for very rapidly diffusing solutes in chloroform, due primarily to the

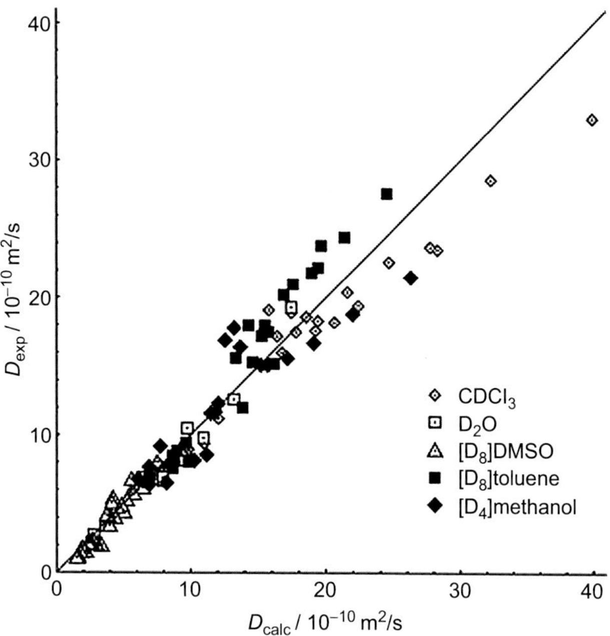

FIGURE 10.24 **A plot of experimentally determined diffusion coefficients against those calculated using Eq. 10.19 for 44 small molecules in five different solvents.** *(Source: Reproduced with permission from Ref. [34] copyright Wiley-VCH.)*

presence of heavy chlorine atoms in the solvent biasing its molecular weight. Parameter optimisation explicitly for chloroform suggests a ρ_{eff} of 647 kg/m^3 and an effective solvent molecular weight of 83 g/mol (rather than the true 119 g/mol) provides for slightly better estimates of D_{sol} in such cases.

More generally, Eq. 10.18 indicates the diffusion coefficient has a reciprocal cube root dependence on molecular weight. This may be used in a relative sense to relate the diffusion coefficients of two molecules D_{sol} and D_{ref} to their molecular weights M_{sol} and M_{ref}:

$$\frac{D_{sol}}{D_{ref}} = \sqrt[3]{\frac{M_{ref}}{M_{sol}}} \tag{10.20}$$

Again, this applies to idealised spherical molecules but may nonetheless provide a useful approximation for comparison between solutes, especially if these belong to a homologous series, whereas deviations from such idealised geometry may demand the use of corrective shape factors; see reference [35] and references therein.

10.4 APPLICATIONS OF DIFFUSION NMR SPECTROSCOPY

In the sections discussed earlier, we have reviewed the most widely used diffusion sequences and have explored the important experimental parameters and procedures for recording meaningful diffusion data. This section will illustrate some of the applications of diffusion NMR spectroscopy and while not exhaustive is meant to give sufficient insight to allow one to realise opportunities where these methods may be applicable to your own work. A number of extensive reviews are available that deal with such applications [3,4,37–39], and the interested reader is encouraged to refer to these. It is noteworthy that many of these reviews deal simultaneously with diffusion and the nuclear Overhauser effect (NOE) and stands testament to the manner in which these methods may be combined in the elucidation of intermolecular interactions in particular, a recurrent theme in the examples discussed in the following sections.

10.4.1 Signal Suppression

A rather simple but sometimes highly effective application of diffusion NMR methods is in the editing of the NMR spectrum to suppress unwanted resonances, whereby use is made of the differing diffusion properties of species in the solution even though a measurement of diffusion coefficients themselves is not necessary. A typical application might be selective

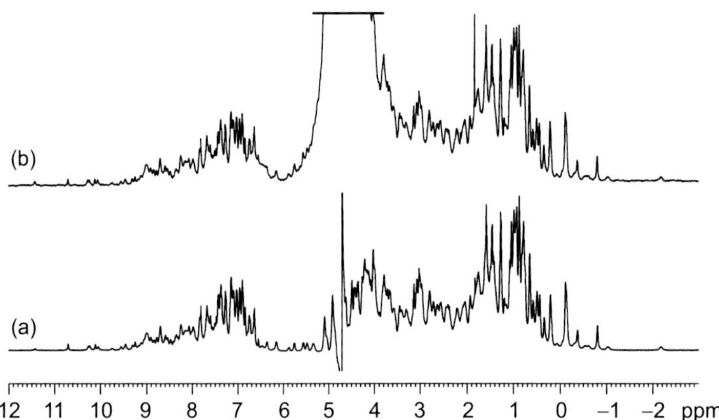

FIGURE 10.25 **Solvent signal suppression by a diffusion filter.** The sample is 1 mM lysozyme in 50:50 H_2O:D_2O and suppression in (a) was achieved using a ^{1}H STE-based sequence employing a diffusion delay Δ of 50 ms, bipolar gradient pairs of 4 ms total duration (δ) and gradient strengths of 0.43 T/m. The standard ^{1}H spectrum is shown in (b).

removal of the resonance(s) of a solvent. In this, a set of 1D diffusion parameters are established in which the faster moving solvent has been completely attenuated whereas the slower diffusing solutes of interest are themselves also attenuated *but to a lesser degree* and thus remain in the resulting 1D spectrum. This approach is illustrated in Fig. 10.25 for an aqueous protein solution (lysozyme, M_r 14.4 kDa, in 50:50 H_2O:D_2O) in which the water is essentially fully suppressed but the protein resonances remain visible, even those that are completely masked by the solvent in the absence of suppression (Fig. 10.25b; see Section 12.5 for discussions of alternative solvent suppression schemes).

As a solvent suppression method, this requires there to be a sufficiently large difference in the diffusion coefficients of solvent and solute and may therefore be limited to investigations of larger systems or complexes. It does, however, have the notable advantage relative to more conventional signal suppression schemes of being able to suppress multiple signals irrespective of chemical shift and signal overlap. A slightly unusual example in this context is the use of DOSY editing for NMR spectroscopy in neat ionic liquids [40], systems of increasing commercial interest as 'green' solvents. In these systems the use of deuterated solvents is often impractical or too expensive, meaning the solvents often present multiple resonances and are thus poorly suited to conventional selective suppression schemes. However, application of the DOSY approach allows extraction of a trace from the pseudo-2D plot through the solute alone (which may itself diffuse *faster* than the ionic solvent) to reveal a 1D spectrum devoid of the ionic solvent peak. This concept is illustrated in Fig. 10.26 for ethyl bromide in an imidazolium ionic solvent system. This has the added advantage over coefficient measurements that precise diffusion coefficients are not required so that accurate gradient calibrations are not needed, making implementation straightforward.

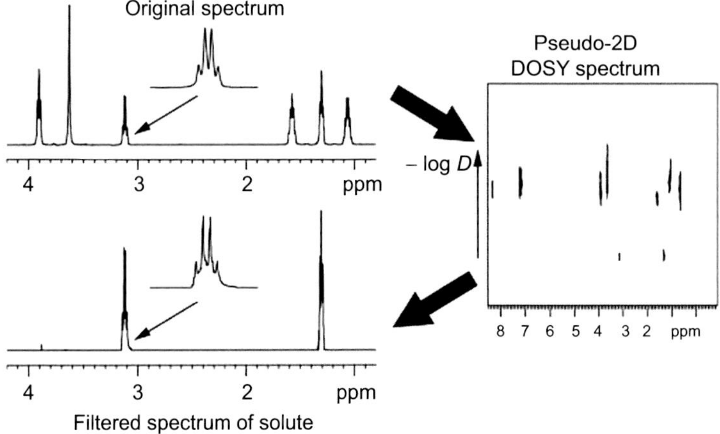

FIGURE 10.26 **The process of signal editing with 2D DOSY.** The ^{1}H spectrum of the solute ethyl bromide can be extracted free from contributions from the neat ionic liquid solvent *(Source: Reproduced with permission from Ref. [40] copyright Wiley-VCH.)*

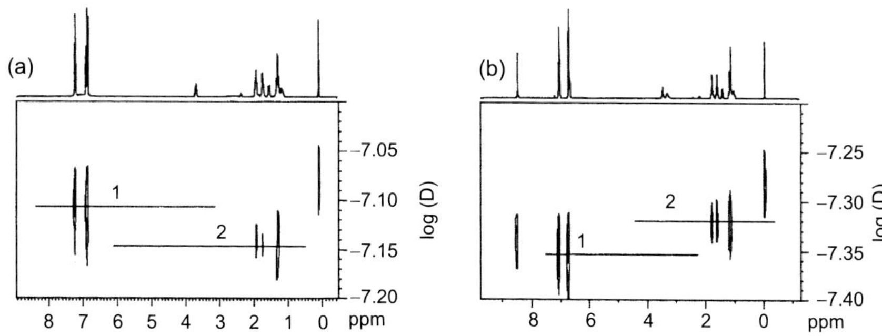

FIGURE 10.27 **Investigating hydrogen bonding with DOSY.** ^{1}H DOSY spectra of phenol (1) and cyclohexanol (2) in (a) $CDCl_3$ and (b) $CDCl_3$ containing 1 molar equivalent DMSO–d_6. In $CDCl_3$, phenol diffuses more rapidly than cyclohexanol whereas stronger hydrogen bonding with DMSO causes this to be retarded more than the alcohol *(Source: Reproduced with permission from Ref. [41] copyright Elsevier.)*

10.4.2 Hydrogen Bonding

The presence of intermolecular hydrogen bonding has also been investigated successfully with diffusion methods and the relative strength of such bonds probed in a qualitative manner. Thus, in a mixture of phenol and cyclohexanol in $CDCl_3$ the cyclohexanol exhibits a lower diffusion coefficient than phenol (Fig. 10.27a) owing to the larger bulk of this system and potentially to differences in intermolecular hydrogen bonding. On addition of the hydrogen bond acceptor DMSO, the phenol is observed to be retarded by a greater extent than the alcohol (Fig. 10.27b), suggesting a greater propensity for the phenol to exist in a H-bonding complex with the DMSO, attributable to its greater acidity relative to the alcohol [41]. In the same report, ^{31}P DOSY was used in a similar manner to study hydrogen bonding of various phosphorus-containing compounds.

In a similar approach, and as part of a larger study [42], hydrogen bonding between *p*-cresol and piperazine has been shown to occur in $CDCl_3$ with the changes in diffusion coefficients relative to the pure, dilute materials in the same solvent (Table 10.5) being consistent with the 2:1 binding stoichiometry observed in the crystal structure (Fig. 10.28). The similarity of diffusion coefficients for both molecules in the complex suggest they exist predominantly in the hydrogen-bonded state. In such studies of intermolecular interactions it has been suggested [43] that TMS offers a suitable internal reference material to allow corrections for viscosity changes to be made as different components are mixed, and so isolate those changes in diffusion coefficients arising from hydrogen-bonding interactions.

10.4.3 Host–Guest Complexes

Host–guest complexes, in which a smaller guest molecule becomes encapsulated within a larger host, present another form of intermolecular association that is well suited to study by diffusion NMR methods. One may observe a measurable change in the diffusion coefficient of the guest molecule as it is taken up by the host providing not only direct evidence for the

TABLE 10.5 Diffusion Coefficients for Pure *p*-Cresol and Piperazine in $CDCl_3$ and in the Hydrogen-Bonded Complex

	D (10^{-10} m^2/s)	
	Pure	Complex
p-Cresol	16.7	9.70
Piperazine	25.3	9.91

Adapted with permission from Ref. [42] copyright Elsevier.

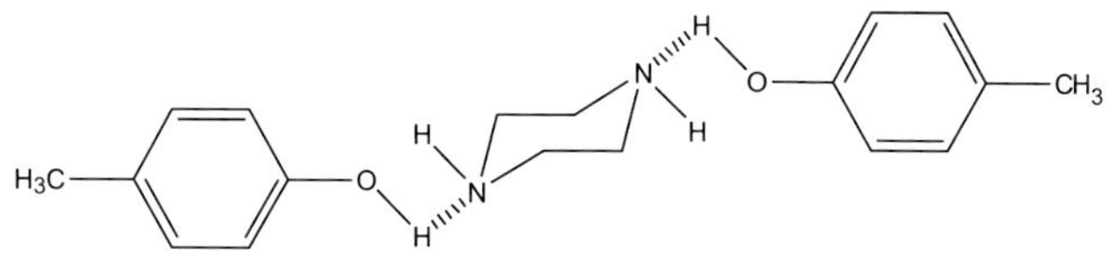

FIGURE 10.28 **Structure of the *p*-cresol:piperazine hydrogen-bonded complex derived from X-ray data.** *(Source: Adapted with permission from Ref. [42] copyright Elsevier.)*

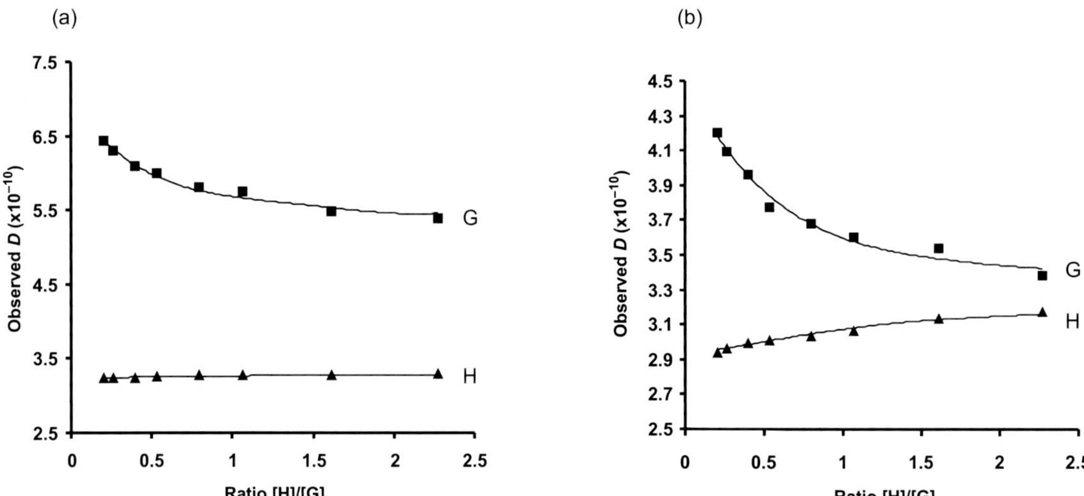

FIGURE 10.29 **Investigating host-guest complexes with DOSY.** Diffusion titration curves for the binding of the host β-cyclodextrin **10.4** with the guest (a) cyclohexylacetic acid **10.2** (K_a 1800 M^{-1}) and (b) cholic acid **10.3** (K_a 5900 M^{-1}). *(Source: Adapted with permission from Ref. [45] copyright American Chemical Society.)*

binding event but also, in favourable situations, allowing the determination of association constants from the diffusion data. In a fast-exchange process between the free and bound states of the guest, its observed diffusion coefficient represents the population-weighted average of the diffusion properties of these two states. Monitoring the change in diffusion coefficient D as a function of the host:guest ratio allows the determination of association constants K_a; the use of D replaces the more usual chemical shift as the NMR observable in such titrations [44].

In the world of host–guest complexes, the cyclodextrins represent one of the most studied host systems by virtue of their open hydrophobic cavities, and not surprisingly have been studied by diffusion NMR. One example of the determination of association constants by diffusion measurements is the binding of cyclohexylacetic acid **10.2** and the bile acid cholic acid **10.3** to β-cyclodextrin **10.4** [45]. Fig. 10.29 shows the variation in diffusion coefficients as the hosts are titrated against differing guest concentrations. For the smaller guest **10.2** (Fig. 10.29a) there is an appreciable change in D as the guest concentration varies, consistent with the binding process, whereas the diffusion behaviour of the cyclodextrin host itself is largely invariant to the binding as it is able to consume the small guest. In contrast, addition of the larger steroid guest **10.3** produces a significant change in both the guest and host diffusion profiles (Fig. 10.29b). The host variation must also be accounted for when determining K_a since the simpler assumption of the host behaviour being unaffected by the presence of the guest is clearly no longer valid. In both cases, the solid curves represent the fitted values for the association constants determined from the simultaneous fitting of both the host and guest diffusion coefficients.

OH O OH OH H H H HO H OH

10.2 **10.3**

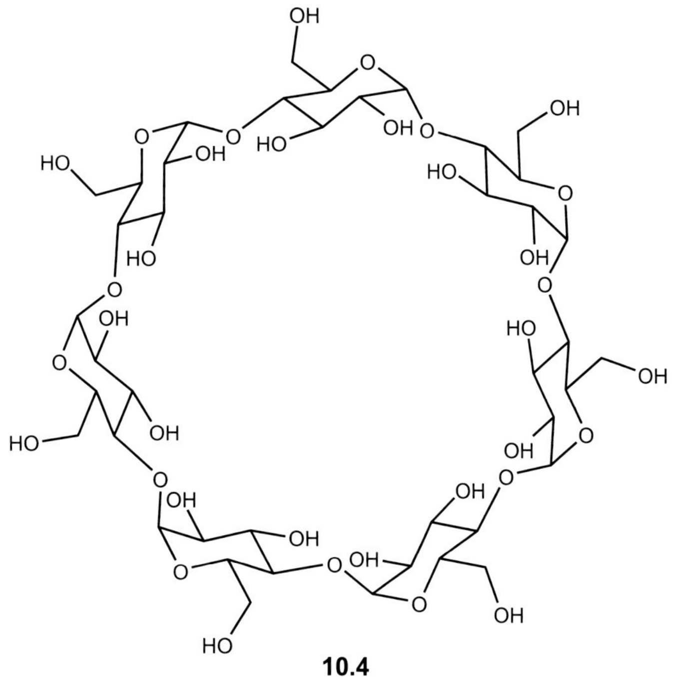

10.4

TABLE 10.6 Diffusion Coefficients for the Free Diaminoalkanes 10.5–10.9 and Their α-CD Complexes in D_2O at 298 K

		Free	Complex			
System		D_{amine} (10^{-10} m^2/s)	D_{amine} (10^{-10} m^2/s)	$D_{\alpha\text{-CD}}$ (10^{-10} m^2/s)	X	K_a (M^{-1})
	α-CD	—	—	3.0		
10.5	$n = 1$	7.6	7.1	2.9	0.11	15
10.6	$n = 2$	6.5	4.9	3.0	0.46	65
10.7	$n = 3$	6.0	4.7	2.9	0.42	680
10.8	$n = 4$	5.5	3.8	2.9	0.65	13,500
10.9	$n = 5$	5.0	3.2	2.6	0.75	7080

Bound molar fractions X are determined from the measured diffusion coefficients as weighted averages of the values for the bound and free states, and association constants K_a are derived from changes in diffusion coefficients as a function of the host:guest ratio.
Adapted with permission from Ref. [47] copyright American Chemical Society.

In a study of camphor binding with cyclodextrins both chemical shift and diffusion titrations were shown to be equally suitable for the determination of binding constants [46]. However, chemical shift changes are not always resolvable or of sufficient magnitude to be reliable for K_a determination. Thus, the diffusion-based approach was taken in the investigation of the binding of α,ω-diaminoalkanes of various lengths **10.5–10.9** in α-CD cavities [47] with a view to determining whether protonation of the nitrogen prevents the diamines from escaping from the cavity and so forming rotaxanes from the initially formed pseudorotaxanes (in which the diamines exchange between the free and complexed states). Owing to the small chemical shift changes on binding, diffusion coefficients were considered a favoured parameter with which to monitor binding events. Table 10.6 shows the variation in diffusion coefficients and association constants as a function of chain length with a general trend of increased binding for longer chains. The **10.5** $n = 1$ diamine demonstrates very weak binding with the α-CD whereas the exceptional binding for the **10.8** $n = 4$ diamine is attributed to additional favourable interactions between the terminal amine and hydroxyls of the cyclodextrin. Further studies on these systems demonstrated that protonation of the amines resulted in weakened binding with the α-CD.

H_2N–[$CH_2CH_2CH_2$]$_n$–NH_2

n = 1 **10.5**
n = 2 **10.6**
n = 3 **10.7**
n = 4 **10.8**
n = 5 **10.9**

10.4.4 Ion Pairing

In the field of organometallic chemistry in particular, diffusion measurements have been used extensively by a number of pioneering groups in various studies of complex coordination and ion pairing [37,38]. The notion is that when held in an ion pair the observed diffusion of the ions is retarded relative to that of the dissociated states and in the usual case of fast exchange between the coordinated and dissociated ions a population-weighted average of individual diffusion coefficients will be observed. In the first application of diffusion NMR to the simultaneous observation of cations and anions as ion pairs [48] it was demonstrated that tetrabutylammonium (TBA^+) and tetrahydroborate (BH_4^-) ions exist as a tight pair in chloroform solution as evidenced by their similar diffusion coefficients and the fact that these were significantly smaller than that of the non-aggregating internal reference tetrabutylsilane. The concentration dependence of the diffusion coefficients indicated the ion pairs formed aggregates, and from comparisons with the diffusion data for $Si(Bu)_4$ it was possible to estimate the aggregate size derived from estimated molecular volumes (Table 10.7).

In a number of examples it has also been shown that ion pairing is more strongly favoured in chloroform solutions for many organometallic salts as illustrated by the diffusion data collected for the two complexes **10.10** and **10.11** (Table 10.8) [49]. Thus, in chloroform the anions and cations of both complexes exhibit very similar diffusion coefficients, consistent with them existing as tight ion pairs. However, the different coefficients recorded in dichloromethane suggest the pairing to be weakened

TABLE 10.7 Diffusion Coefficients for Tetrabutylammonium Tetrahydroborate in $CDCl_3$ as a Function of Concentration with Tetrabutylsilane as Internal Reference

	D (10^{-10} m^2/s)			
Concentration (mM)	TBA^+	BH_4^-	$Si(Bu)_4$	Aggregation Number
200	4.99	5.34	8.64	5.2
40	6.83	7.35	10.3	3.4
8	8.10	8.42	10.5	3.3
2	8.48	8.73	10.6	2.0
0.5	8.34	8.64	10.5	2.0

Adapted with permission from Ref. [48] copyright American Chemical Society.

TABLE 10.8 Diffusion Coefficients for the Ionic Complexes 10.10 and 10.11 as a Function of Solvent

		$CDCl_3$		CD_2Cl_2	
		D (10^{-10} m^2/s)	r_s (Å)	D (10^{-10} m^2/s)	r_s (Å)
10.10	Cation	6.25	6.3	8.74	6.2
	PF_6^{-1}	6.27	6.3	10.17	5.3
10.11	Cation	6.64	6.0	9.14	5.9
	OTf^-	6.45	6.1	11.69	4.7

Coefficients for the metal complexes and their fluorinated counterions were determined from ^1H and ^{19}F diffusion experiments, respectively and the hydrodynamic radii r_s calculated from these values assuming solution viscosities matched those of pure solvents.
Adapted with permission from Ref. [49] copyright Wiley-VCH.

and incomplete with the anions now moving significantly faster on average than the larger cations. Nevertheless, comparisons with the hydrodynamic radii of c. 2.6 Å (PF_6^-) and 3.2 Å (OTf^-) for the two anions determined in methanol (a strongly coordinating solvent likely to disrupt ion pairing) suggest a significant amount of pairing still exists in dichloromethane. These data also serve to illustrate the combined application of ^{1}H and ^{19}F diffusion studies, as required for cations and anions, respectively. In a similar multinuclear approach to chiral recognition it has even been suggested that diffusion measurements are sensitive enough to distinguish diastereomeric complexes through their differing degrees of ion pairing [50].

PF_6^- Ru Bu_3P Bu_3P Cl

10.10

P Pd P Ph Ph $CF_3SO_3^-$

10.11

In a separate study on a series of zirconium-based metallocenes complexed with $[MeB(C_6F_5)_3]^-$ there has been shown a good correlation between their hydrodynamic volumes estimated from diffusion measurements [using Si(p-tolyl)$_4$ as internal reference in benzene] and the van der Waals volumes determined for the 1:1 ion pairs from their crystal structures [51] (Fig. 10.30), suggesting that appropriate application of the diffusion NMR methodology can be an effective means of determining the sizes of molecular complexes in solution.

10.4.5 Supramolecular Assemblies

The investigation of large supramolecular complexes, cages and capsules has been one of the areas to have benefited most from diffusion measurements, with numerous examples to be found [4]. As a simple yet powerful illustration of the capabilities of diffusion methodology, the encapsulation of benzene within the cage of the tetraurea calix[4]arene **10.12** was

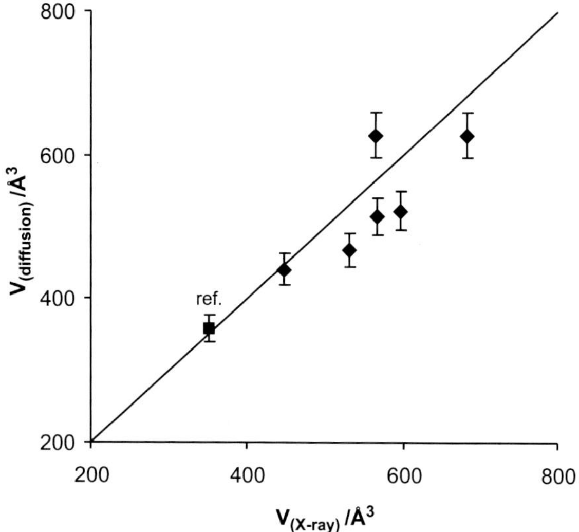

FIGURE 10.30 **Hydrodynamic volumes determined from diffusion NMR.** The plot of hydrodynamic volumes and van der Waals volumes computed from 1:1 ion pairs in the corresponding crystal structures for a series of metallocenes. The point labelled ref. is for the internal diffusion reference Si(p-tolyl)$_4$. *(Source: Adapted with permission from Ref. [51] copyright American Chemical Society.)*

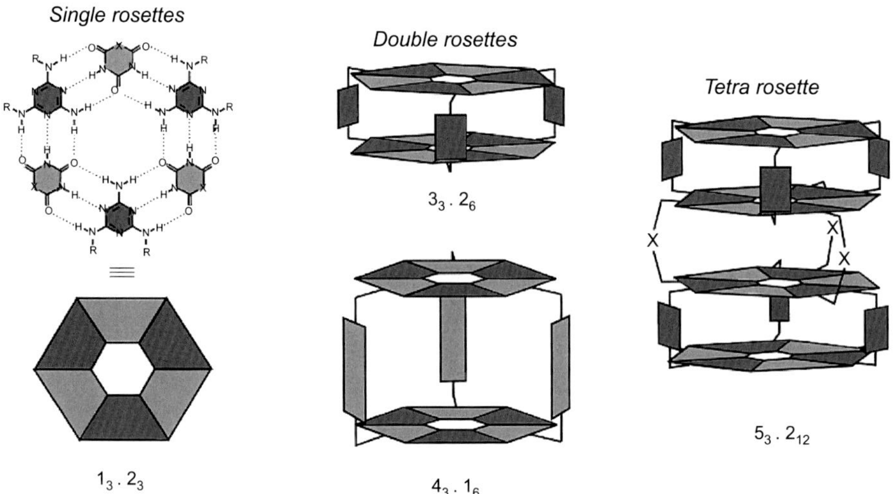

FIGURE 10.31 **Structure and schematic representation of the single, double and tetra rosette supramolecular hydrogen-bonded assemblies characterised by diffusion NMR spectroscopy.** *(Source: Reproduced with permission from Ref. [53] copyright Royal Society of Chemistry.)*

proven when the resonance at 4.4 ppm believed to arise from the encapsulated solvent was shown to possess an identical diffusion coefficient to the dimeric assembly itself (3.4×10^{-10} m^2/s) and that this was considerably less than that of free benzene in the solution (22.1×10^{-10} m^2/s) [52]. The characterisation of hydrogen-bonded supramolecular assemblies known as rosettes has also been undertaken, and single, double and tetra rosette complexes have been classified according to their diffusion behaviour (Fig. 10.31) [53]. Moreover, these experiments were able to demonstrate that the single rosettes are dynamic assemblies in which the components exchange rapidly on the diffusion timescale whereas some of the double rosettes were more robust and undergo slow exchange in the presence of an excess of one of the constituent components.

10.12

In a similar approach, double-stranded helicates of various lengths that were derived from copper-and silver-based metallosupramolecular architectures have also been classified by their diffusion properties and estimates of the molecular sizes made [54]. Owing to the ellipsoidal structures it was necessary to introduce appropriate shape factors to translate the hydrodynamic radii determined directly from the unmodified Stokes–Einstein equation into dimensions that were meaningful for these assemblies. Thus, knowledge of the width of the helicates (determined from the X-ray structure of a single complex in this case) allowed determination of their lengths from the hydrodynamic radii. The results for a series of these helicates is summarised in Table 10.9. It was further shown that 2D DOSY spectra could be employed to differentiate the helicates of different lengths when present simultaneously in a mixture.

TABLE 10.9 Diffusion Coefficients and Calculated Molecular Dimensions of a Series of Cu(I) Helicates

Cu(I) helicate	D (10^{-10} m²/s)	r_s (Å)	Width[a] (Å)	Length (Å)
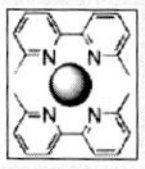	15.5	4.1	11.0	6.5
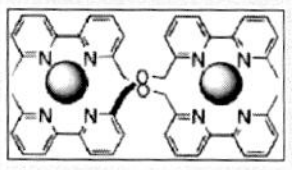	12.6	5.0	11.0	9.5
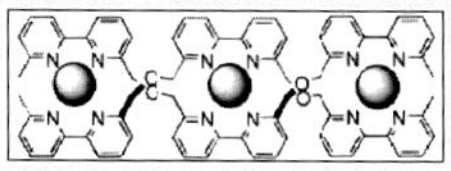	10.1	6.3	11.0	16.0
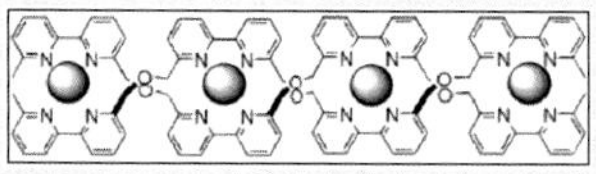	9.5	6.7	11.0	18.0
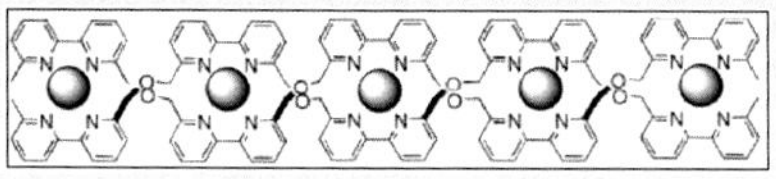	8.3	7.6	11.0	24.5

The *spheres* identify the Cu(I) centres.
[a]*The widths of the assemblies were used as reference distances for calculation of their lengths from the hydrodynamic radii r_s when corrected for the ellipsoidal shapes of the structures.*
Adapted with permission from Ref. [54] copyright Wiley-VCH.

10.4.6 Aggregation

The phenomenon of aggregation presents an obvious area in which one would intuitively anticipate that diffusion measurements could provide insights, and indeed there have been a number of such examples. The π–π stacking of the hydrophobic herbicide Metolachlor **10.13** in D_2O has been detected by diffusion NMR measurements [55] and has been used to explain the higher than expected solubility of this compound in water through the reduced exposure of the aromatic rings to the solvent. In relatively concentrated solution, two species are evident. Those diffusion coefficients differ by two orders of magnitude and suggest an ordered polymeric aggregate ($D = 2.8 \times 10^{-12}$ m²/s) to be present in addition to the monomer ($D = 4.9 \times 10^{-10}$ m²/s) under conditions of slow exchange.

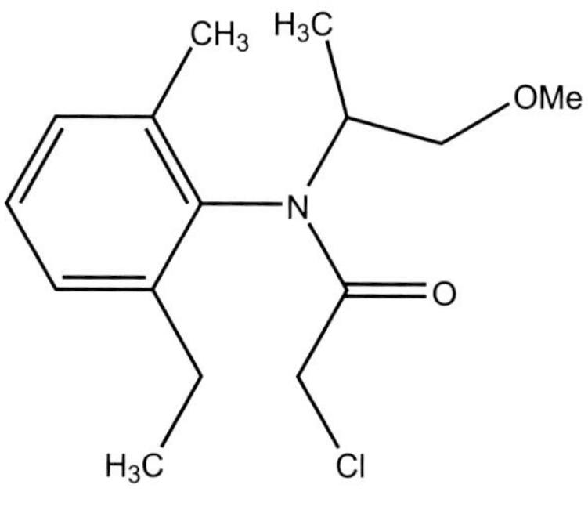

CH3 H3C OMe N O H3C Cl

10.13

Both ^{1}H and ^{7}Li diffusion have been used in the discrimination of previously characterised aggregates of *n*-butyllithium in tetrahydrofuran (THF) solvent, a widely used reagent in synthetic chemistry. Diffusion measurements at 189 K (using the DSTE sequence to overcome convection) were able to distinguish the ^{1}H α-CH_2 resonances of dimeric and tetrameric aggregates at -1.12 and -1.00 ppm, respectively [56]. While ^{7}Li diffusion measurements were possible for the tetramer, the broader resonance of the faster relaxing dimer prevented its observation. The experimental diffusion coefficients were in good agreement with those predicted from both the available X-ray structures and the optimised gas-phase computed structures (Table 10.10). Organocuprates are widely used synthetic reagents for C–C bond-forming reactions, and in an

TABLE 10.10 Experimentally Determined Diffusion Coefficients for the Aggregates of *n*-Butyllithium in THF at 189 K Together With the Values Predicted From Solid-State Structures or Gas–Phase Optimised Structures

	Calculated D (10^{-10} m^2/s)		Experimental D (10^{-10} m^2/s)	
	X-Ray	Gas-Phase Model	^{1}H	^{7}Li
Tetramer	0.89	0.85	0.88 ± 0.05	0.86 ± 0.01
Dimer	0.99	0.95	1.01 ± 0.03	—

The resonances of the dimer could not be observed in the diffusion experiment.
Adapted with permission from Ref. [56] copyright American Chemical Society.

extensive study [57] the identification of aggregates of alkylcuprates in diethyl ether has also been achieved by diffusion NMR spectroscopy and the information used to explain the reactivity of methyl cuprates. Thus, the presence of dimethylcuprate aggregates larger than dimers (so-called higher aggregates) has been correlated with reduced reactivity and the dimer identified as the reactive species.

10.4.7 Mixture Separation

Despite much of the original attention on diffusion NMR methods having focused on the potential for mixture separation in the form of 'NMR chromatography' via the DOSY presentation format, this tends to find limited application in the field of chemistry. This is most likely due to difficulties associated with the efficient separation of species when resonance overlap occurs, a situation most likely to arise in complex mixtures where, unfortunately, the technique has greatest potential. In such cases, the use of higher magnetic fields may be beneficial owing to enhanced signal dispersion. The application of extended diffusion sequences to separate resonances represents an alternative or additional way forward and a number of these methods are presented in Section 10.5. The use of pure shift methods to yield better resolved singlet peaks (Section 8.5) has also been demonstrated.

In one of the first applications of this sort, the concept of mixture separation based on molecular size was demonstrated with a mixture of glucose, adenosine 5′-triphosphate (ATP) and sodium dodecyl sulphate (SDS) in D_2O (Fig. 10.32), with the large differences in size providing clear separation of the constituent parts [58]. A more demanding application was the separation of the metabolite components of a perchloric acid tissue extract containing many small

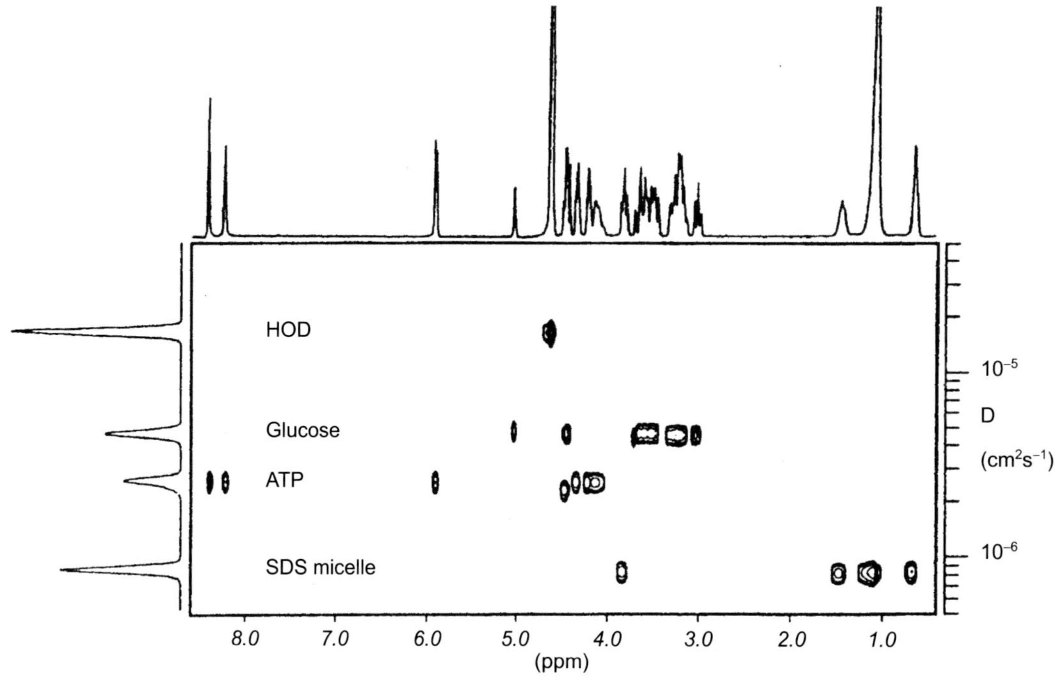

FIGURE 10.32 **The 2D diffusion-ordered (DOSY) spectrum of a mixture of glucose, ATP and SDS micelles in D_2O demonstrating the potential for mixture separation.** *(Source: Reproduced with permission from Ref. [58] copyright American Chemical Society.)*

molecules of more similar molecular masses [59]. DOSY has been applied to the analysis of crude column fractions arising from methanol extracts of *Agelas* sponges which were shown to contain unknown bromopyrrole alkaloids [60]. The use of DOSY in this manner to separate components of the partially purified fractions provided evidence for a novel species being present and indicated the need for additional purification and subsequent characterisation of the novel isomers agesamides A and B **10.14**.

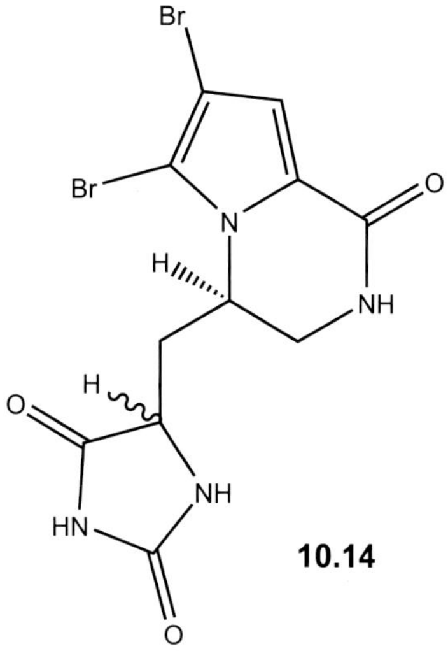

10.4.8 Macromolecular Characterisation

In this final section on illustrative applications we stray into the realm of biomacromolecules with some examples of the characterisation of the molecular masses of such compounds. In the area of oligosaccharide analysis, DOSY has been proposed as a complementary method to size exclusion chromatography for molecular weight (MW) determinations [61]. In a study of a series of *N*-acetyl-chitooligosaccharides $(GlcNAc)_{1-6}$ a linear correlation was obtained for log MW versus log D and a readily interpretable strip form of the DOSY data presented, akin to that of polyacrylamide gels in which the smaller, faster moving components appear towards the bottom of the strips (Fig. 10.33). This provided a calibration against

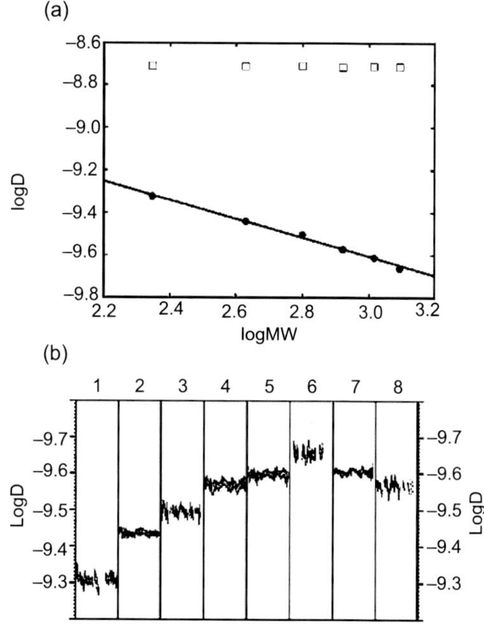

FIGURE 10.33 **Macromolecular characterization with DOSY.** (a) Plot of log *D* versus log MW for a series of *N*-acetyl chitooligosaccharides $(GlcNAc)_{1-6}$ and (b) an alternative presentation of DOSY strips in a format similar to gel electrophoresis. The GlcNAc oligomers appear in *lanes 1 to 6* while *lanes 7 and 8* are the traces for the chitoligosacchraides **10.15** and **10.16**. *(Source: Reproduced with permission form Ref. [61] copyright Oxford University Press.)*

which the molecular masses of two *Escherichia coli*–expressed chitoligosaccharides **10.15** and **10.16** were estimated. In a similar approach, uncharged water-soluble polysaccharides with molecular masses in the range 5000–850,000 have also shown a similar linear log–log plot [62]. Thus, through the use of appropriate standards as diffusion calibrants it becomes possible to use measured diffusion coefficients as direct indicators of molecular mass even for very large oligomeric or polymeric systems.

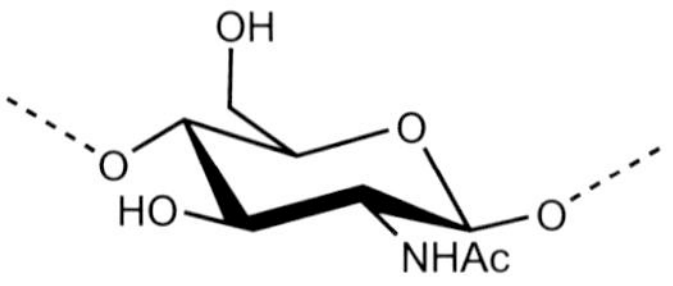

GlcNAc

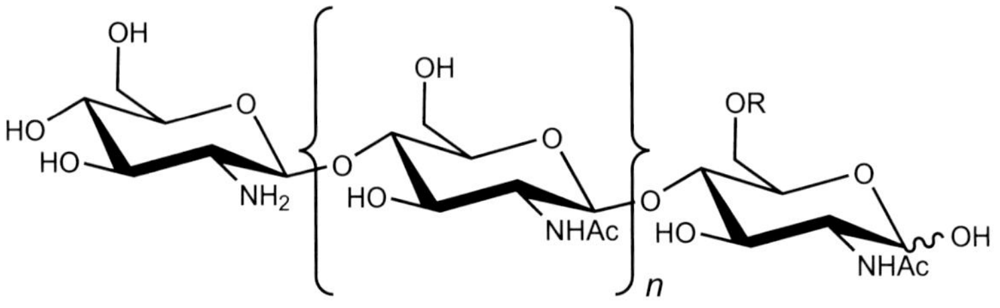

10.15: R = H, n = 3
10.16: R = SO_3Na, n = 2

10.5 HYBRID DIFFUSION SEQUENCES

Having reviewed a range of applications primarily employing the standard techniques described previously, we finish this chapter by considering alternative sequences for determining diffusion coefficients when those above prove limiting. It has been emphasised in previous discussions that an important requirement for the success of diffusion experiments is the ability to resolve the peaks of the species of interest from those of other molecules present in multi-component systems so as to accurately follow their corresponding diffusion-encoded intensity decays. Beyond using the highest available magnetic fields to maximise resonance dispersion there have been a number of diffusion-based sequences proposed which are in essence derived from the usual array of multiple-pulse sequences used in structure elucidation to enhance the information content of spectra. This section provides a brief overview of the principal methods that have been suggested as alternatives to the basic 1D (proton) observation methods described above.

A conceptually attractive approach is to incorporate into diffusion experiments the pure shift methodology described in Section 8.5 to remove all homonuclear proton couplings from diffusion spectra and in so doing greatly increase resonance separation. While this has been demonstrated for DOSY by use of the Zangger–Sterk pure shift methodology [63], the severe sensitivity loss imposed reduces the utility of this otherwise appealing combination, so is not considered further, although a J-resolved DOSY variant is described below.

10.5.1 Sensitivity-Enhanced Heteronuclear Methods

A direct approach to improving resonance dispersion is to observe a heteronucleus rather than the more typical proton, with carbon-13 an attractive alternative for organic systems since resonance overlap is rare. However, it is well known that the direct observation of low-γ nuclides suffers from reduced sensitivity so the use of signal enhancement by polarisation transfer from protons may be considered mandatory. This has been demonstrated with the INEPT-DOSY sequence for both ^{13}C [64] and ^{29}Si [65] and with DEPT-DOSY for ^{13}C [66]. In the first approach, an INEPT transfer was appended onto a bipolar STE sequence such that diffusion encoding is performed on proton magnetisation with signal detection on the heteroatom being combined with proton decoupling [67] (Fig. 10.34a). In this, it becomes possible to include the diffusion-decoding bipolar gradients within the INEPT transfer step. The DEPT variant makes use of a spin-echo rather than the STE and so benefits from twice the signal intensity (Fig. 10.34b). This also employs gradient encoding of the carbon spin rather than

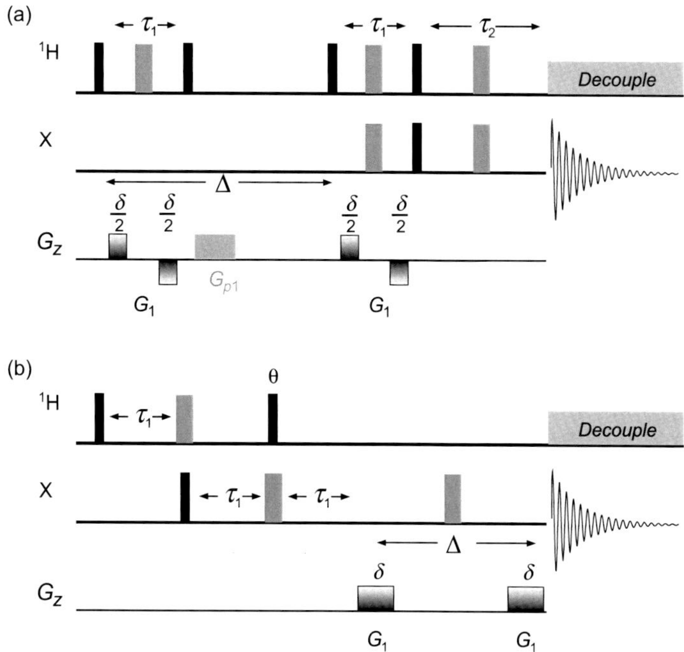

FIGURE 10.34 **Heteronuclear DOSY methods.** (a) INEPT-DOSY sequence and (b) the DEPT–spin-echo–DOSY sequence. The delays τ_1 are set to $1/2J_{XH}$ as required for polarisation transfer and the delay τ_2 set as necessary for the INEPT refocusing stage to enable the use of ^{1}H decoupling. The DEPT ϑ pulse is typically set to 45 degree.

the proton and so demands gradient pulses to have a fourfold greater net intensity than would be required for proton encoding. The experiment would typically be acquired with $\vartheta = 45$ degree to retain all protonated carbon resonances. In both experiments, the use of polarisation transfer allows for more rapid data collection since recycle times are dictated by proton relaxation rates and not those of the heteronucleus.

An even more efficient approach, although applicable only for polarisation transfer across small, long-range heteronuclear couplings, is to use the necessarily long J-evolution delays of the INEPT sequence to encode molecular diffusion within the spin-echo [65]. This approach further benefits from convection compensation since gradient encoding–decoding can both be performed either side of the 180 degree pulse of the initial spin-echo. This scheme has been demonstrated for INEPT transfer using small two-bond ^{1}H–^{29}Si couplings (~6 Hz) in silicon-observed DOSY, and represents one example of the 'IDOSY' approach described in Section 10.5.3.

10.5.2 Spectrum-Edited Methods

For situations when proton sensitivity remains desirable but resonance dispersion is hampered by peak overlap, direct editing of the proton spectrum to observe selectively the signals of interest may offer a solution. An effective approach to this is to combine a selective 1D TOCSY experiment with diffusion encoding such that only magnetisation originating from the selectively excited proton, and subsequently transferred to others within the same spin system, will contribute to the final diffusion traces. The net result is a 2D DOSY spectrum of a subset of the complete proton spectrum. One implementation of this, termed 'selective TOCSY edited preparation' (STEP)-DOSY [68], employs excitation sculpting (Section 12.4.2) and TOCSY mixing with clean zero-quantum suppression (Section 6.2.4) as preparation for the BPP-LED diffusion experiment (Fig. 10.35). Thus, resonances that would otherwise be buried become cleanly resolved for diffusion encoding providing they share a suitable spin-coupling network with a neighbouring spin; note how this approaches parallels that used for the 1D STEP-NOESY experiment described in Section 9.6.2 for the observation of NOEs originating from buried proton resonances.

10.5.3 Diffusion-Encoded Two-Dimensional Methods (or 3D DOSY)

A further extension to improve resonance separation is to employ the additional signal dispersion afforded by 2D NMR techniques and to combine these with diffusion encoding. In essence, there are two general approaches to this: the first is to concatenate a conventional 2D experiment with a diffusion sequence; the second is to somehow embed the diffusion

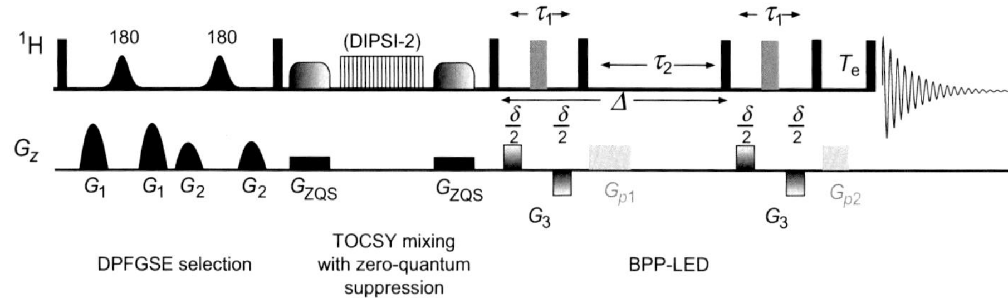

FIGURE 10.35 **The 2D STEP-DOSY sequence for editing of ^{1}H diffusion spectra.** Magnetisation of the DPFGSE-selected resonance(s) is transferred via isotropic mixing and subsequently encoded for diffusion measurements with the BPP-LED scheme. Details of each step labelled in the sequence may be found in the relevant chapter sections.

encoding within the 2D experiment itself. Both cases result in a pseudo-3D data set, that is, with one diffusion and two frequency dimensions. Planes from the 3D cube extracted for a given diffusion coefficient then present the parent 2D spectrum for only the corresponding species. Concatenation itself may take two forms in which the 2D experiment '*X*' precedes the diffusion encoding (*X*-DOSY) or is appended to this (DOSY-*X*) with both approaches having been advocated. Embedding the diffusion period within the 2D method requires that suitably long periods already exist within the sequence in which to encode molecular diffusion with both J-evolution periods and TOCSY mixing times having been utilized for this purpose leading to the so-called 'internal-DOSY' (IDOSY) family of experiments that will be exemplified later.

Data acquisition demands a full 2D data set be collected for every diffusion-encoding gradient amplitude applied and this may then lead to extended experiment times; clearly, high-sensitivity 2D methods and/or relatively concentrated samples are best suited to such an approach. With one exception (mentioned later) data analysis of such experiments involves the fitting of the diffusion-encoded intensity decays of 2D crosspeak volumes to extract diffusion coefficients. Alternatively, a 3D DOSY display may be generated by direct analogy with the fitting of 1D peak areas described in Sections 10.2.4 and 10.2.5. Planes taken through a 3D DOSY spectrum should then contain only the signals arising from species that share the same diffusion coefficients. Examples of these experimental approaches are described briefly in the sections that follow.

10.5.3.1 Diffusion-Encoded Correlation Spectroscopy

The most widely used 2D COSY experiment may be combined with the BPP-LED sequence simply by replacing the final 90 degree read pulse of BPP-LED with the two-pulse COSY sequence to give the BPP-LED-COSY experiment [69]. However, a more direct and efficient approach is simply to incorporate the diffusion period directly into the two-pulse COSY sequence by the appropriate use of gradient pulses, as in the COSY-IDOSY sequence [70] (Fig. 10.36). The gradient pulses now cleverly serve a dual purpose providing both coherence selection, as in gradient-selected COSY, and diffusion encoding–decoding. Here, the extended Δ period employed for diffusion encoding would be expected to produce severe phase distortions in the proton traces, but these errors are masked completely with the usual absolute value (magnitude-mode) processing of COSY. Note, however, that this will allow for extended J evolution, meaning the sequence is in fact more akin to the delayed or long-range COSY experiment (Section 6.1.6) and would be expected to enhance the appearance of long-range correlations arising from small proton–proton couplings. Alternatively, and advantageously, this feature can reduce the total number of t_1 increments required to resolve the COSY crosspeaks.

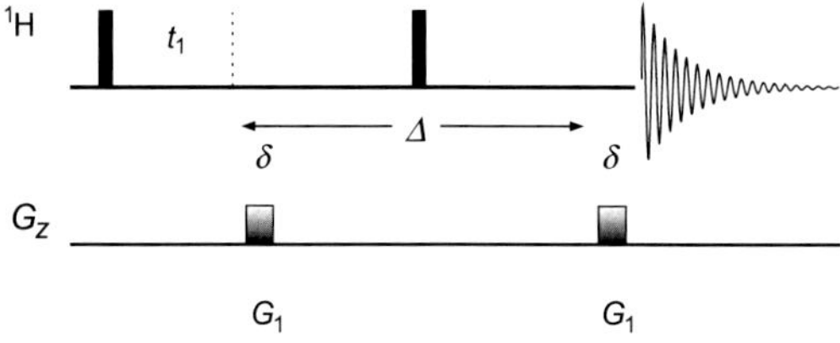

FIGURE 10.36 **COSY-IDOSY sequence.** Gradient pulses select for the conventional N-type COSY pathway in addition to providing diffusion encoding (selection of the P-type pathway would use gradient pulses of opposite sign).

10.5.3.2 Diffusion-Encoded Total Correlation Spectroscopy

By direct analogy with the COSY-based experiment above, diffusion encoding with the TOCSY experiment may be achieved by direct concatenation of the parent sequences to give the DOSY-TOCSY experiment, with most sequences reported employing LED or BPP-LED diffusion encoding [71-73]. However, the mixing time employed in the TOCSY step can equally lend itself to acting as the diffusion-encoding period, so following the IDOSY concept [74]. Thus, encoding and decoding gradients can be placed either side of the spin lock period which is often many tens of milliseconds and thus potentially of sufficient duration to map diffusion. Indeed, one limitation with some early implementations of gradient-selected sequences was signal intensity losses arising from molecular diffusion when gradient pulses were placed either side of lengthy mixing or evolution periods, and many more recent sequences are designed with closely spaced gradient pulses to avoid such losses.

10.5.3.3 Diffusion-Encoded J-Resolved Technique

While the COSY- and TOCSY-derived methods described earlier offer the potential for crosspeak dispersion, they become less effective when analysing mixtures of similar compounds since crosspeak overlap may still be a problem. In such cases an alternative approach may be to use the homonuclear J-resolved technique in an attempt to eliminate multiplet overlap by placing J-coupling fine structure orthogonal to the shift axis (see Section 8.3). In this approach the intensity decay of the 2D peaks may be fitted to extract diffusion coefficients or, alternatively, the projections of the tilted 2D data set may be treated as a homonuclear-decoupled 1D spectrum and the peak intensities fitted as for conventional diffusion experiments [75].

Direct concatenation results in either the J-BPP-LED [76] or, alternatively, the BPP-LED-J [75] sequences. However, as the basic J-resolved experiment is itself derived from a simple spin-echo this can be diffusion-encoded directly yielding the more time-efficient and sensitive J-IDOSY technique [77] (Fig. 10.37). In this, the diffusion period Δ straddles the refocusing 180 degree pulse of the spin-echo and splits equally the t_1 evolution period. Although the minimum t_1 period is now defined by Δ, this does not present a problem in the final spectrum since, as for COSY-IDOSY above, any unwelcome phase modulation this produces is masked by the absolute value presentation usually employed for J-resolved spectroscopy. The gradient-encoding pattern avoids the need for phase cycling, meaning only a single transient per t_1 increment is required, sensitivity permitting.

10.5.3.4 Diffusion-Encoded HSQC

Arguably the most effective means of avoiding resonance overlap is through the application of heteronuclear correlation methods whereby crosspeaks are dispersed by the chemical shifts of (at least) two J-coupled nuclei. Despite the usual penalty of sensitivity loss, heteronuclear correlation–based diffusion methods have been described based on the direct concatenation of HMQC [78] and HSQC [73] or by incorporating these with the IDOSY philosophy [65,79]. A feature common to all of these is the need for extreme caution in the use of broadband decoupling of the heteronucleus since the relatively high powers demanded by conventional composite–pulse decoupling methods can lead to significant sample heating and may cause the onset of the dreaded convection. It is, therefore, essentially mandatory to employ lower power decoupling methods based on adiabatic frequency sweeps so as to minimise this effect (Section 12.3.1).

One example of this approach is the constant time HSQC-IDOSY [79] experiment (Fig. 10.38) that combines the diffusion and constant time periods and incorporates the encoding–decoding bipolar gradient pairs within the INEPT transfer steps. Additional gradients are utilised to provide coherence selection and f_1 quad-detection according to the echo–antiecho

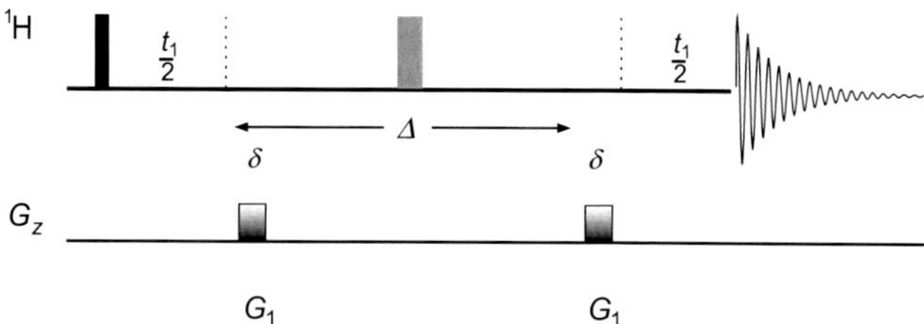

FIGURE 10.37 **The J-IDOSY sequence.** The diffusion period Δ sits at the centre of the incremented t_1 period and the two gradient pulses provide for both diffusion encoding and selection of the spin-echo coherence transfer pathway.

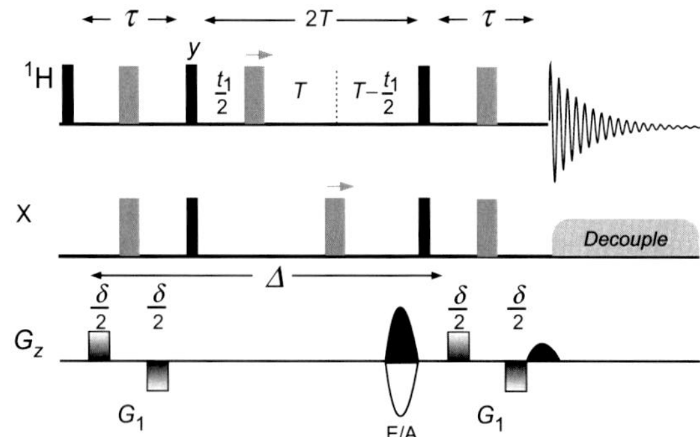

FIGURE 10.38 **Constant time HSQC-IDOSY sequence.** The delays τ are set to $1/2J_{XH}$ as required for INEPT transfer and the constant time period $2T$ remains fixed, its duration being dictated by the desired diffusion time Δ. The effective t_1 evolution time is varied by moving the two 180 degree refocusing pulses within the constant time period (pulses with arrows over) and coherence selection is made with the echo–antiecho (E/A) scheme. The diffusion encoding/decoding gradients are applied as bipolar pairs during the INEPT and reverse-INEPT transfer steps.

approach. The total constant time period $2T$ is defined by the required diffusion period Δ and may lead to lengthy heteronuclear evolution times, and so offers the potential for high f_1 resolution at the expense of extensive t_1 sampling. This may not be practical when the 2D experiment is to be performed for each diffusion gradient amplitude and the use of heteronuclear spectrum aliasing has been suggested as one approach to reduce this requirement [79].

REFERENCES

[1] Stilbs P. Prog Nucl Magn Reson Spectrosc 1987;19:1–45.
[2] Johnson CS. Prog Nucl Magn Reson Spectrosc 1999;34:203–56.
[3] Cohen Y, Avram L, Frish L. Angew Chem Int Ed 2005;44:520–54.
[4] Avram L, Cohen Y. Chem Soc Rev 2015;44:586–602.
[5] Morris KF, Johnson CS. J Am Chem Soc 1992;114:3139–41.
[6] Stejskal EO, Tanner JE. J Chem Phys 1965;42:288–92.
[7] Tanner JE. J Chem Phys 1970;52:2523–6.
[8] Wider G, Dotsch V, Wuthrich K. J Magn Reson A 1994;108:255–8.
[9] Pelta MD, Barjat H, Morris GA, Davis AL, Hammond SJ. Magn Reson Chem 1998;36:706–14.
[10] Gibbs SJ, Johnson CS. J Magn Reson 1991;93:395–402.
[11] Wu D, Chen A, Johnson CS. J Magn Reson A 1995;115:260–4.
[12] Pelta MD, Morris GA, Stchedroff MJ, Hammond SJ. Magn Reson Chem 2002;40:S147–52.
[13] Nilsson M, Connell MA, Davis AL, Morris GA. Anal Chem 2006;78:3040–5.
[14] Antalek B. Concept Magn Reson 2002;14:225–58.
[15] Huo R, Wehrens R, van Duynhoven J, Buydens LMC. Anal Chim Acta 2003;490:231–51.
[16] Nilsson M. J Magn Reson 2009;200:296–302.
[17] Swan I, Reid M, Howe PWA, Connell MA, Nilsson M, Moore MA, Morris GA. J Magn Reson 2015;252:120–9.
[18] Loening NM, Keeler J. J Magn Reson 1999;139:334–41.
[19] Hedin N, Yu TY, Furo I. Langmuir 2000;16:7548–50.
[20] Allerhand A, Addleman RE, Osman D. J Am Chem Soc 1985;107:5809–10.
[21] Goux WJ, Verkruyse LA, Saltert SJ. J Magn Reson 1990;88:609–14.
[22] Hedin N, Furo I. J Magn Reson 1998;131:126–30.
[23] Holz M, Weingartner H. J Magn Reson 1991;92:115–25.
[24] Martinez-Viviente E, Pregosin PS. Helv Chim Acta 2003;86:2364–78.
[25] Hayamizu K, Price WS. J Magn Reson 2004;167:328–33.
[26] Esturau N, Sánchez-Ferrando F, Gavin JA, Roumestand C, Delsuc M-A, Parella T. J Magn Reson 2001;153:48–55.
[27] Lounila J, Oikarinen K, Ingman P, Jokisaari J. J Magn Reson A 1996;118:50–4.
[28] Jerschow A, Muller N. J Magn Reson 1997;125:372–5.
[29] Nilsson M, Morris GA. J Magn Reson 2005;177:203–11.
[30] Longsworth LG. J Phys Chem 1960;64:1914–7.
[31] Connell MA, Davis AL, Kenwright AM, Morris GA. Anal Bioanal Chem 2004;378:1568–73.
[32] Gierer A, Wirtz K. Z Naturforsch A 1953;8:532–8.
[33] Chen H-C, Chen S-H. J Phys Chem 1984;88:5118–21.

[34] Evans R, Deng Z, Rogerson AK, McLachlan AS, Richards JJ, Nilsson M, Morris GA. Angew Chem Int Ed 2013;52:3199–202.
[35] Waldeck AR, Kuchel PW, Lennon AJ, Chapman BE. Prog Nucl Magn Reson Spectrosc 1997;30:39–68.
[36] Evans R. Excel spreadsheet relating diffusion coefficient and molecular weight. Available from: http://nmr.chemistry.manchester.ac.uk/?q=node/290
[37] Pregosin PS, Martínez-Viviente E, Kumar A. Dalton Trans 2004;4007–14.
[38] Pregosin PS, Kumar PGA, Fernandez I. Chem Rev 2005;105:2977–98.
[39] Brand T, Cabrita EJ, Berger S. Prog Nucl Magn Reson Spectrosc 2005;46:159–96.
[40] Giernoth R, Bankmann D. Eur J Org Chem 2005;21:4529–32.
[41] Kapur GS, Cabrita EJ, Berger S. Tetrahedron Lett 2000;41:7181–5.
[42] de Carvalho EM, Velloso MHR, Tinoco LW, Figueroa-Villar JD. J Magn Reson 2003;164:197–204.
[43] Cabrita EJ, Berger S. Magn Reson Chem 2001;39:s142–8.
[44] Fielding L. Tetrahedron 2000;56:6151–70.
[45] Cameron KS, Fielding L. J Org Chem 2001;66:6891–5.
[46] Simova S, Berger S. J Incl Phenom Macro 2005;53:163–70.
[47] Avram L, Cohen Y. J Org Chem 2002;67:2639–44.
[48] Mo H, Pochapsky TC. J Phys Chem B 1997;101:4485–6.
[49] Valentini M, Ruegger H, Pregosin PS. Helv Chim Acta 2001;84:2833–53.
[50] Martinez-Viviente E, Pregosin PS, Vial L, Herse C, Lacour J. Chem Eur J 2004;10:2912–8.
[51] Stahl NG, Zuccaccia C, Jensen TR, Marks TJ. J Am Chem Soc 2003;125:5256–7.
[52] Frish L, Matthews SE, Böhmer V, Cohen Y. J Chem Soc, Perkin Trans 1999;2:669–71.
[53] Timmerman P, Weidmann J-L, Jolliffe KA, Prins LJ, Reinhoudt DN, Shinkai S, Frish L, Cohen Y. J Chem Soc, Perkin Trans 2000;2:2077–89.
[54] Allouche L, Marquis A, Lehn J-M. Chem Eur J 2006;12:7520–5.
[55] Viel S, Mannina L, Segre A. Tetrahedron Lett 2002;43:2515–9.
[56] Keresztes I, Williard PG. J Am Chem Soc 2000;122:10228–9.
[57] Xie X, Auel C, Henze W, Gschwind RM. J Am Chem Soc 2003;125:1595–601.
[58] Morris KF, Johnson CSJ. J Am Chem Soc 1993;115:4291–9.
[59] Barjat H, Morris GA, Smart S, Swanson AG, Williams SCR. J Magn Reson B 1995;108:170–2.
[60] Tsuda M, Yasuda T, Fukushi E, Kawabata J, Sekiguchi M, Fromont J, Kobayashi J. Org Lett 2006;8:4235–8.
[61] Groves P, Rasmussen MO, Molero MD, Samain E, Canada FJ, Driguez H, Jimenez-Barbero J. Glycobiology 2004;14:451–6.
[62] Viel S, Capitani D, Mannina L, Segre A. Biomacromolecules 2003;4:1843–7.
[63] Nilsson M, Morris GA. Chem Commun 2007;933–5.
[64] Schlörer NE, Cabrita EJ, Berger S. Angew Chem Int Ed 2002;41:107–9.
[65] Stchedroff MJ, Kenwright AM, Morris GA, Nilsson M, Harris RK. Phys Chem Chem Phys 2004;6:3221–7.
[66] Botana A, Howe PWA, Caër V, Morris GA, Nilsson M. J Magn Reson 2011;211:25–9.
[67] Wu D, Chen A, Johnson JCS. J Magn Reson A 1996;123:215–8.
[68] Bradley SA, Krishnamurthy K, Hu H. J Magn Reson 2005;172:110–7.
[69] Wu D, Chen A, Johnson JCS. J Magn Reson A 1996;121:88–91.
[70] Nilsson M, Gil AM, Delgadillo I, Morris GA. Chem Commun 2005;1737–9.
[71] Jerschow A, Muller N. J Magn Reson A 1996;123:222–5.
[72] Lin M, Shapiro MJ. J Org Chem 1996;61:7617–9.
[73] Williamson RT, Chapin EL, Carr AW, Gilbert JR, Graupner PR, Lewer P, McKamey P, Carney JR, Gerwick WH. Org Lett 2000;2:289–92.
[74] Birlirakis N, Guittet E. J Am Chem Soc 1996;118:13083–4.
[75] Cobas JC, Martín-Pastor M. J Magn Reson 2004;171:20–4.
[76] Lucas LH, Otto WH, Larive CK. J Magn Reson 2002;156:138–45.
[77] Nilsson M, Gil AM, Delgadillo I, Morris GA. Anal Chem 2004;76:5418–22.
[78] Barjat H, Morris GA, Swanson AG. J Magn Reson 1998;131:131–8.
[79] Vitorge B, Jeanneat D. Anal Chem 2006;78:5601–6.

Chapter 11

Protein–Ligand Screening by NMR

Chapter Outline

11.1 Introduction 421
11.2 Protein–Ligand Binding Equilibria 422
11.3 Resonance Lineshapes and Relaxation Editing 424
11.3.1 ^{1}H Relaxation-Edited NMR 426
11.3.2 ^{19}F NMR 428
11.3.3 Paramagnetic Relaxation Enhancement 429
11.4 Saturation Transfer Difference 430
11.4.1 The STD Sequence and Practicalities 432
11.4.2 Epitope Mapping by STD and DIRECTION 436
11.4.3 K_D Measurement by STD 437
11.5 Water-LOGSY 438
11.5.1 The Water-LOGSY Sequence 440
11.5.2 Water-LOGSY Practicalities 441
11.6 Exchange-Transferred Nuclear Overhauser Effects 441
11.7 Competition Ligand Screening 443
11.7.1 Competitive Displacement 444
11.7.2 Reporter Ligand Screening 445
11.7.3 ^{19}F *FAXS* 447
11.8 Protein Observe Methods 448
11.8.1 ^{1}H–^{15}N Mapping 448
11.8.2 ^{1}H–^{13}C Mapping 452
11.8.3 ^{19}F Mapping 452
References 454

11.1 INTRODUCTION

The focus of this chapter is NMR methods developed to study the binding of small-molecule ligands to biologically relevant macromolecules. In the majority of studies these are proteins, although the methods may also be applicable to nucleic acids and to lipids. Many small molecules are nowadays designed and synthesised as potential leads in drug discovery programs or as protein ligands developed to probe biological pathways. Examples of this include the inhibition of enzyme activity or the blocking of protein–protein interactions to elicit a change in biological function. Accordingly, there now exist a substantial array of NMR methods designed to probe such ligand binding, often referred to loosely as *protein–ligand screening* techniques. These methods can be tailored to address a diverse range of questions including whether a small molecule binds to the macromolecular receptor target of interest, what are the ligand binding interactions at the receptor surface (a process known as epitope mapping) and what is the ligand affinity for the receptor [1–4]. These techniques can also be applied to identify small-molecule binders from a compound mixture or fragment library and are increasingly used as orthogonal techniques in library screening and hit validation, complementing other commonly employed biophysical methods [5,6].

The more commonly employed NMR screening methods can be divided into two classes, so-called *ligand observe* methods that detect the NMR response of the small-molecule ligand in the presence of the macromolecule, or *protein observe* techniques that observe the resonances of the protein receptor itself. Ligand observe methods have wider applicability and are likely to be most relevant to chemists working in the realm of medicinal chemistry, so are the primary focus of this chapter. The main advantage of ligand observation lies in the fact it does not require use of isotopically labelled proteins as is usually demanded in protein-detected methods. These most frequently employ ^{15}N labelling of protein backbone amides and, if site-specific binding information is desired, these also demand protein resonance assignment to be undertaken, a potentially lengthy process requiring significant quantities of soluble protein. Ligand observation also benefits from requiring relatively small quantities of proteins (often micromolar concentrations), by employing 1D experiments that can be rapid to acquire, and it can be applied to the screening of multiple compounds in a mixture. This has the further advantage of not being limited by protein size (and indeed often functions more effectively with larger protein targets) whereas protein observe methods can suffer from poor quality or overly complex spectra when dealing with larger macromolecules, for which the resonance assignment process also becomes more challenging. That said, the ligand observe approach has its own limitations. Most obviously, one gains no direct information on the location of ligand binding on the protein receptor nor the ligand-binding orientation, which can be inferred from protein observe NMR or alternatively through protein crystallography. The detection of high-affinity ligands can also be problematic for reasons explained below

High-Resolution NMR Techniques in Organic Chemistry. http://dx.doi.org/10.1016/B978-0-08-099986-9.00011-7

and interference from non-specific binding (ligand binding at a receptor site separate from the site of interest) can also be potentially misleading. However, both these latter issues can be addressed through the use of competition-based assays, which will also be described in this chapter.

Prior to introducing the most widely employed techniques for protein–ligand screening, it is necessary to introduce some fundamental concepts relating to receptor-binding equilibrium. This is necessary so that the operational basis of the methods introduced below may be understood, and to enable quantitative interpretation of ligand-binding data.

11.2 PROTEIN–LIGAND BINDING EQUILIBRIA

In any ligand observe binding assay it is possible to utilise changes in any NMR parameter to monitor the binding of the small molecule to the receptor, provided the parameter is sufficiently sensitive to the binding event. In most cases any change results from the differences in size between the small-molecule ligand and the macromolecule such that the observable NMR parameters for the ligand change from those associated with rapid rotational and translational motion of the free ligand to the correspondingly slower motions of the macromolecular receptor as the small molecule binds to the target (Fig. 11.1). NMR parameters that are exploited most frequently include chemical shifts, spin relaxation rates and nuclear Overhauser effects (NOEs), and most of the techniques introduced in this chapter will employ these. Although translational diffusion coefficients have also been employed to detect ligand binding [7], these studies generally require long experiments and have been less widely used in this regard. In most cases it is the *free* ligand NMR response that is observed, which requires that the ligand be in rapid on–off exchange with the receptor such that the observed NMR parameter is weighted between that associated with the free and the bound ligand states. This may be expressed:

$$\mathbf{M}_{\mathrm{obs}} = F_{\mathrm{F}}^{\mathrm{L}}\mathbf{M}_{\mathrm{F}}^{\mathrm{L}} + F_{\mathrm{B}}^{\mathrm{L}}\mathbf{M}_{\mathrm{B}}^{\mathrm{L}} \qquad (11.1)$$

where $\mathbf{M}_{\mathrm{obs}}$ represents the observable NMR response, $F_{\mathrm{F}}^{\mathrm{L}}$ and $F_{\mathrm{B}}^{\mathrm{L}}$ are the mole fractions of the free and bound ligand (hence $F_{\mathrm{F}}^{\mathrm{L}} + F_{\mathrm{B}}^{\mathrm{L}} = 1$), and $\mathrm{M}_{\mathrm{F}}^{\mathrm{L}}$ and $\mathrm{M}_{\mathrm{B}}^{\mathrm{L}}$ are the NMR parameters of the free and bound ligand, respectively. Thus, the population (mole fraction) of the bound ligand state $F_{\mathrm{B}}^{\mathrm{L}}$ must be sufficient for it to influence the averaged NMR response $\mathbf{M}_{\mathrm{obs}}$ and hence the protein–ligand binding equilibrium determines one's ability to detect ligand binding via these methods.

For a 1:1 complex PL between the receptor protein P and the free small-molecule ligand L the equilibrium may be represented:

$$\mathrm{P} + \mathrm{L} \underset{k_{\mathrm{off}}}{\overset{k_{\mathrm{on}}}{\rightleftharpoons}} \mathrm{PL} \qquad (11.2)$$

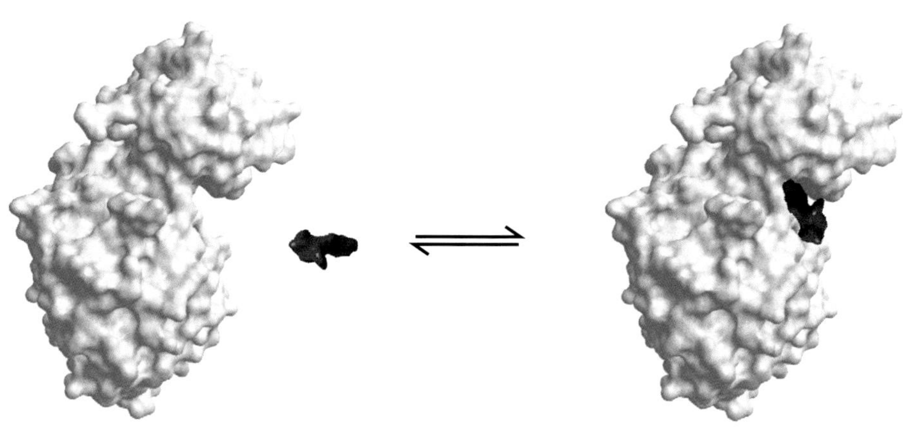

FIGURE 11.1 **Characteristic NMR parameters associated with a ligand (L) when free in solution and when bound to a macromolecular receptor protein (P).** These arise from differences in its rotational and translational motion.

The bimolecular rate constant k_{on} reflects the probability of the free protein and ligand binding to form the PL complex, while the unimolecular rate constant k_{off} is inversely proportional to the lifetime of the bound complex before dissociation. The binding affinity is most often described by the dissociation constant K_D:

$$K_D = \frac{[P][L]}{[PL]} = \frac{k_{off}}{k_{on}} \tag{11.3}$$

Thus, a high-affinity, strong-binding ligand has a small dissociation constant, whereas a relatively weaker binder will display a correspondingly larger K_D. Most NMR screening techniques are sensitive to binding affinities of low millimolar or less. The fraction of protein receptor bound by the ligand is:

$$F_B^P = \frac{[PL]}{[P]+[PL]} \tag{11.4}$$

and substituting for [PL] from Eq. 11.3 yields:

$$F_B^P = \frac{[L]}{[L]+K_D} \tag{11.5}$$

From this it may be seen that K_D equates to the concentration of *free (unbound) ligand* [L] in solution when half the total protein receptor sites are saturated by the ligand, that is when $F_B^P = 0.5$. In situations when the ligand concentration is far less than its K_D, $[L] << K_D$, then the fraction of protein bound to the ligand is directly proportional to the ligand concentration, whereas when the ligand concentration is far greater than K_D, $[L] >> K_D$, the protein becomes fully saturated, meaning all receptor sites are complexed with the ligand. The corresponding term describing the fraction of ligand-bound F_B^L is obtained by substituting [L] with [P] in Eq. 11.5.

It is often most useful to consider the fraction of the bound protein in terms of known *total* concentrations of the protein $[P_T]$ and ligand $[L_T]$ since the free ligand concentration [L] is not usually known nor readily measured. Substituting [P] and [L] in Eq. 11.2 by using the terms $[P_T] = [PL] + [P]$ and $[L_T] = [PL] + [L]$ yields the quadratic solution defining the concentration of the receptor bound by ligand [PL]:

$$[PL] = \frac{([P_T]+[L_T]+[K_D]) - \sqrt{([P_T]+[L_T]+[K_D])^2 - 4[P_T][L_T]}}{2} \tag{11.6}$$

From this the fractions of bound protein or ligand can be determined:

$$F_B^P = \frac{[PL]}{[P_T]} \quad \text{and} \quad F_B^L = \frac{[PL]}{[L_T]} \tag{11.7}$$

Representative curves for the variation in the bound protein fraction as a function of ligand concentration are shown in Fig. 11.2, calculated for a fixed protein concentration of 10 μM for ligand K_D values ranging from 1 μM to 1 mM. Fig. 11.2a presents a typical titration profile, whereas the use of the logarithmic scale in Fig. 11.2b produces the classic 'dose–response' curve. From these curves it is apparent that ligands with lower K_D values will saturate the protein receptor sites at lower concentrations than ligands of higher dissociation constant and thus any exchange-averaged NMR parameter which tends toward its endpoint more rapidly. Conversely, low-affinity ligands of high K_D will have to be present in significant excess over the protein receptor to achieve significant levels of receptor binding and thus cause a measurable change in the observed NMR response of the ligand. Such considerations become important when planning sample preparation for ligand-binding experiments and can allow one to introduce some selectivity in detecting relatively strong from weak binders in a screening assay. The use of low protein concentrations will favour the binding and hence NMR detection of high-affinity ligands, whereas low-affinity ligands may only be detected as binders when higher protein concentrations are employed. These curves also show that for weak affinity binders the *total* ligand concentration $[L_T]$ leading to half saturation of the protein approximates well to K_D since a large ligand excess is required and the majority of the ligand is free in solution, hence $[L_T] \sim K_D$. However, for high-affinity ligands this approximation breaks down since a greater proportion of the total ligand is bound under these conditions and only the relationship $[L] = K_D$ is valid; that is the *free* ligand concentration—not the total ligand concentration—is relevant.

The fraction of ligand bound under similar solution conditions is shown in Fig. 11.3. This indicates that complete occupancy of the ligand in the bound state occurs only with very potent ligands, even when the ligand is sub-stoichiometric

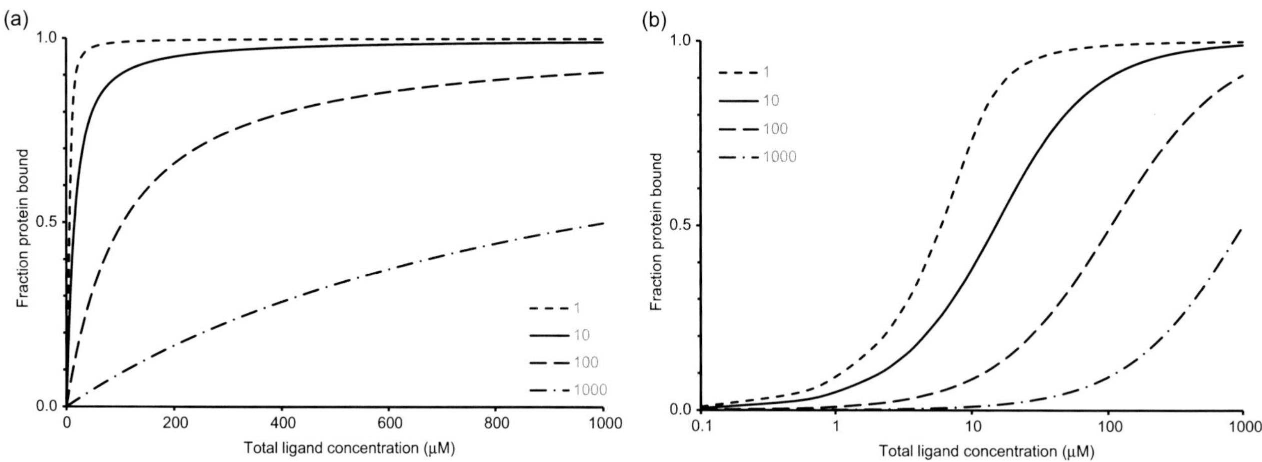

FIGURE 11.2 **Simulations showing the variation of protein receptor bound by ligand as a function of ligand concentration.** Protein concentration $[P_T]$ was 10 μM and graphs are shown for ligands of K_D = 1, 10, 100 and 1000 μM using (a) linear and (b) logarithmic concentration scales.

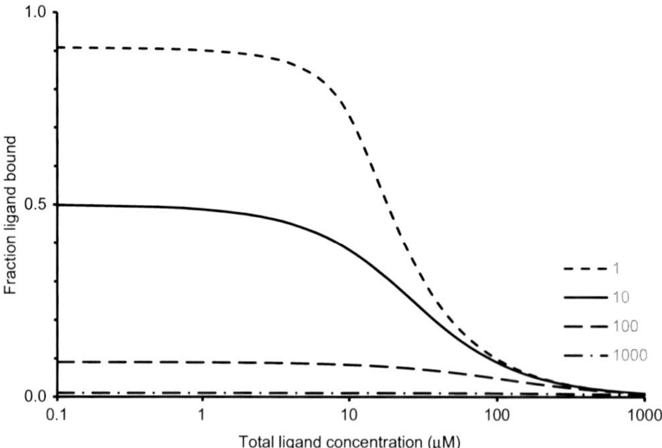

FIGURE 11.3 **Simulations showing the variation of bound ligand as a function of ligand concentration.** Protein concentration $[P_T]$ was 10 μM and graphs are shown for ligands of K_D = 1, 10, 100 and 1000 μM using a logarithmic concentration scale.

with respect to the protein. Thus, when K_D = 1 μM and $[L_T]$ = 1 μM only 90% of the ligand exists in the bound state despite a significant excess of protein (10 μM), this being a consequence of the low total ligand concentration. Only with K_D = 0.1 μM does this increase to 99%, corresponding to essentially complete ligand binding. The significance of these discussions will become more apparent in the sections that follow in which the influence of ligand binding on observable NMR parameters is considered.

11.3 RESONANCE LINESHAPES AND RELAXATION EDITING

The most direct evidence for the binding of a small ligand molecule to a macromolecule can often be seen as changes to the resonance lineshape of the ligand when in the presence of the receptor, observed as resonance broadening and hence peak intensity reduction. The principle here is that when a ligand is in fast or intermediate exchange with the receptor, it will transiently adopt the characteristics of the larger molecule while bound. This includes slow tumbling in solution, and hence it will experience enhanced transverse relaxation rates R_2 (and hence shorter T_2) associated with the longer rotational correlation times of the macromolecule (see Section 2.4). The observed ligand linewidth will be exchange-averaged between that of the bound and uncomplexed states and provided there is a significant contribution from the former then resonance broadening will be observed. The exchange process itself can also contribute to resonance broadening as the ligand environment, and hence its chemical shifts, will likely differ between the free and bound states, giving rise to further dynamic line broadening under conditions of intermediate exchange rates (see Section 2.6). This contribution will increase the greater the chemical

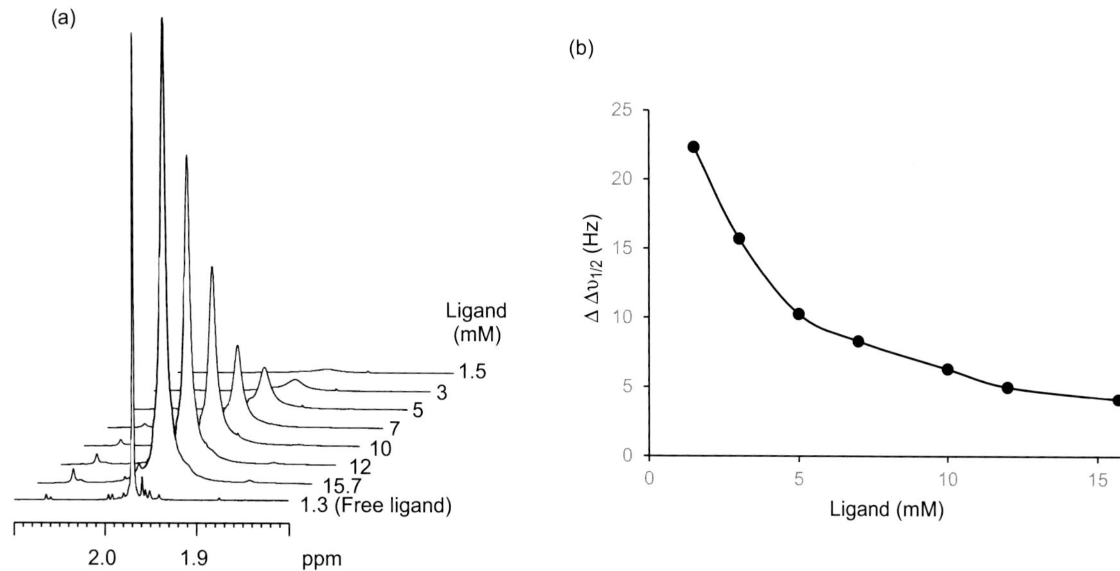

FIGURE 11.4 **Ligand titration.** (a) ^{1}H ligand profile for the NAc resonance of **11.1** titrated against the lectin WGA (fixed at 100 μM) leading to resonance broadening from fast ligand exchange. (b) Change in resonance linewidth $\Delta\Delta\upsilon_{1/2}$ (relative to that of the free ligand) as a function of ligand concentration; data fitting (see text) yields a dissociation constant K_D of 1.6 mM.

shift differences between the two states, a significant feature for ^{19}F observation, in particular; see Section 11.3.2. Thus, under conditions of fast to intermediate on–off exchange the averaged transverse relaxation rates R_2^{av} of the ligand may be expressed:

$$R_2^{av} = F_F^L R_2^F + F_B^L R_2^B + R_2^{ex} \tag{11.8}$$

where the terms in F^L again represent the fractional populations of the free (F) and bound (B) states, the R_2 terms are their corresponding relaxation rates and R_2^{ex} the contribution from exchange-averaging. Typically, the ligand will be present in excess over the receptor protein, yet the significant ligand resonance broadening that can still be observed as relaxation in the bound state is very much faster than that of the uncomplexed ligand ($R_2^B >> R_2^F$), meaning a relatively small population of bound ligand can contribute to measurable line broadening. Fig. 11.4 illustrates this effect in a proton spectrum for binding of the selenium-linked disaccharide **11.1** to the lectin wheat germ agglutinin (WGA) with varying ligand excess. As this excess decreases, the ligand linewidth increases as a greater population is bound to the macromolecule, until the ligand cannot be observed above the resonances of the protein.

HO O Se—Se HN O Me OH NH Me HO HO O OH OH

11.1

Whist such behavior can be expected for ligands that are in fast or intermediate exchange with the protein, ligands of higher affinity that are in slow exchange with the receptor will not show resonance broadening, as the observed free ligand is then not influenced by the properties of the bound state. Furthermore, the resonances of the bound ligand tend to be broad and lost in those of the protein itself, meaning the net result is a decrease in the population of the free ligand (corresponding to the bound population). In situations where the ligand is present in excess, these changes may be too small to observe, meaning high-affinity ligands with low off rates may not be readily detected; this limitation applies to most methods discussed in this chapter that rely on rapid ligand exchange with the receptor. When slow-exchange conditions are suspected, a useful way to establish this is to use the competition-based approach described in Section 11.7.

When fast exchange conditions do exist the NMR observable parameter $\mathbf{M}_{obs}$ (here the resonance linewidth $\Delta\upsilon_{1/2}$ which serves as an indication of the transverse relaxation rate since $\Delta\upsilon_{1/2} \propto 1/T_2$; see Section 2.4.3) is weighted according to the population of the bound state, hence from Eq. 11.5:

$$\Delta\mathbf{M}_{obs} = \Delta\mathbf{M}_{max} F_B^P = \frac{\Delta\mathbf{M}_{max}[L]}{[L] + K_D} \tag{11.9}$$

where $\Delta\mathbf{M}_{max}$ is the maximum change in the observed NMR parameter occurring when the protein is fully saturated by the ligand. Here one may also see that when the free ligand concentration [L] equals K_D the observed parameter change is half the maximum, in accord with previous discussions. Following this expression, a titration of ligand versus protein may be used to derive the dissociation constant for the ligand. This may be performed either through use of the linearized solutions to the binding isotherm that are valid when the ligand is present in high excess, or through the use of Eq. 11.6 to fit for F_B^P. Appropriate curve fitting can be achieved through non-linear regression programs nowadays, although many linear relationships have traditionally been employed [8,9]. In the case of large ligand excess and hence only a small population of the bound ligand L_B, a simplified solution to Eq. 11.3 may be derived assuming $[L_T] >> [L_B]$, from which the linearized form of the binding expression yields the dissociation constant [10]:

$$[L_T] = \frac{[P_T]\Delta\mathbf{M}_{max}}{\Delta\mathbf{M}_{obs}} - K_D \tag{11.10}$$

Plotting ligand concentration $[L_T]$ against the reciprocal of the change in the NMR observable yields an intercept on the ordinate (*y-axis*) of $-K_D$. For **11.1** introduced above, Fig. 11.5 shows this analysis for the data of Fig. 11.4, yielding a dissociation constant of 1.6 mM for this ligand with WGA. Here, the increase in resonance linewidth $\Delta\Delta\upsilon_{1/2}$ is relative to that of the free ligand linewidth ($\Delta\Delta\upsilon_{1/2} = \Delta\upsilon_{1/2(obs)} - \Delta\upsilon_{1/2(free)}$), where the observed linewidth $\Delta\upsilon_{1/2(obs)}$ was that of the *N*-acetyl ^{1}H resonance.

It should also be noted that changes in chemical shift of the ligand as the NMR observable may also be monitored as a function of binding and used to derive dissociation constants. However, for ^{1}H NMR at least, shift changes between bound and free states are often rather small, meaning the averaged shift changes are themselves small and may limit their applicability in K_D measurement. Such limitations are less likely to be encountered for ^{19}F observations, where chemical shift changes can be more significant (Section 11.3.2). Alternatively, direct observation of shift changes of *protein* resonances may be employed to monitor ligand binding, as described in Section 11.8.

11.3.1 ^{1}H Relaxation-Edited NMR

While direct observation of ligand line broadening can be an effective indicator of ligand binding, it is often the case that the degree of broadening can be rather small for ligands of low affinity, with correspondingly low populations of the bound state. The use of low ligand excess to restrict the concentration of free ligand can also lead to ligand resonances becoming obscured by those of the protein, so making an assessment of linewidth changes difficult. In such cases it is useful to apply filtering to the ^{1}H NMR spectrum to attenuate the intensities of broad components. This reduces the potentially troublesome

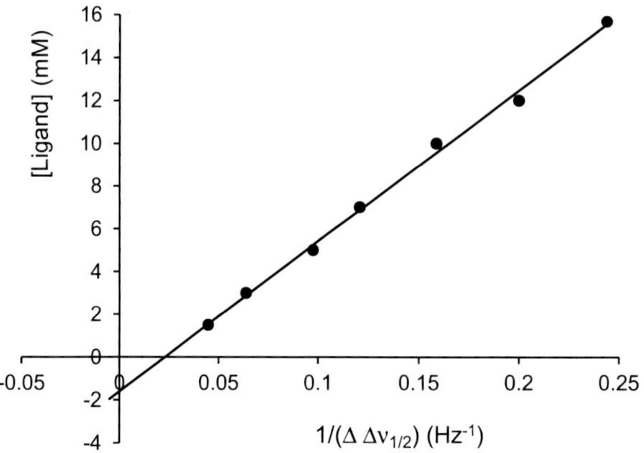

FIGURE 11.5 **Ligand titration data fitting according to Eq. 11.10 to obtain the ligand dissociation constant K_D.** This approach is valid only in cases of large ligand excess over the receptor.

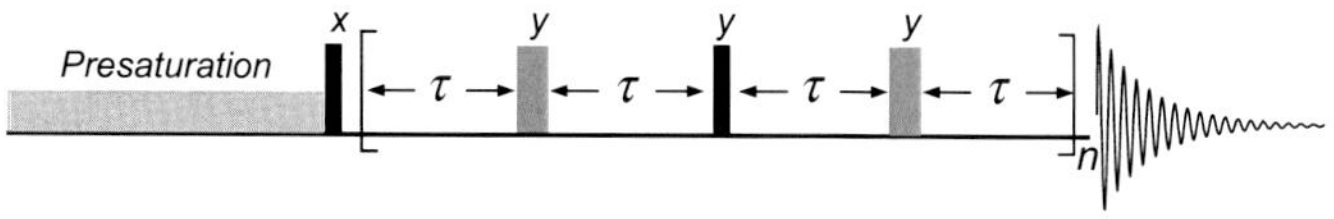

FIGURE 11.6 **The CPMG spin lock filter using the favoured PROJECT sequence with water suppression by presaturation.** The total filter time is achieved by repeating the echo sequence *n* times prior to data acquisition.

protein background and may enhance attenuation of exchange-broadened resonances so that reductions in ligand resonance intensities may be correlated with ligand binding. Such R_2 relaxation-based editing can be achieved through the use of the Carr–Purcell–Meiboom–Gill (CPMG) spin lock filters described in Section 2.4.4; a PROJECT-CPMG sequence employing presaturation of water is shown in Fig. 11.6. Whereas the application of the filter significantly attenuates broad resonances, the sharp resonances of a free ligand experience only a modest attenuation owing to the slower transverse relaxation of the free ligand protons. This process is illustrated for L-tryptophan **11.2** which binds to bovine serum albumin (BSA) in a hydrophobic cavity (known as the Sudlow site II) and for sucrose **11.3**, which does not bind to the protein (Fig. 11.7). The use of long CPMG filter times leads to significant attenuation of the tryptophan ^{1}H resonances, whereas the sucrose is little affected. In the absence of the protein receptor, similarly low levels of attenuation are observed for both compounds. The modest broadening of the tryptophan resonances seen in the ^{1}H spectrum with a short filter would not provide reliable evidence of binding, while substantial attenuation with the longer filter provides greater clarity. Such signal attenuation from binding will usually demand a relatively low ligand excess (<10-fold) as used here, so the influence of the bound ligand population is significant.

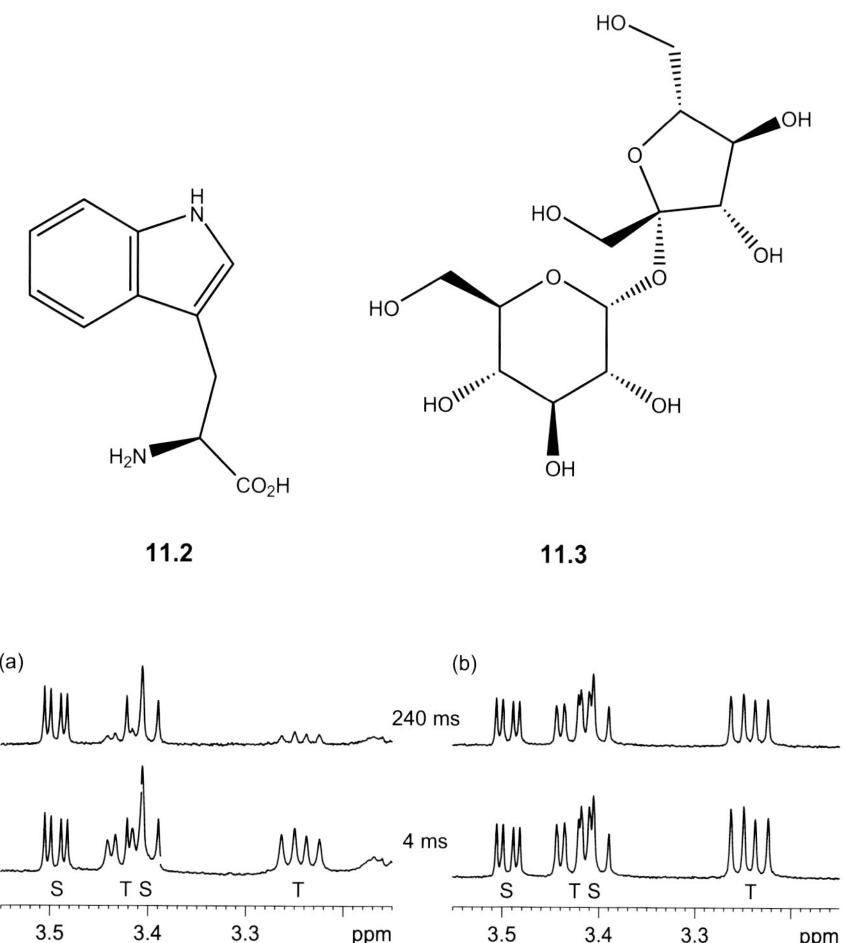

FIGURE 11.7 **Detection of ligand binding enhanced by relaxation editing.** The partial ^{1}H spectra of 300 μM L-tryptophan **11.2** (T) and 200 μM sucrose **11.3** (S) in the presence (a) and absence (b) of protein BSA (100 μM). Spectra were recorded with CPMG filter times as shown under otherwise identical conditions.

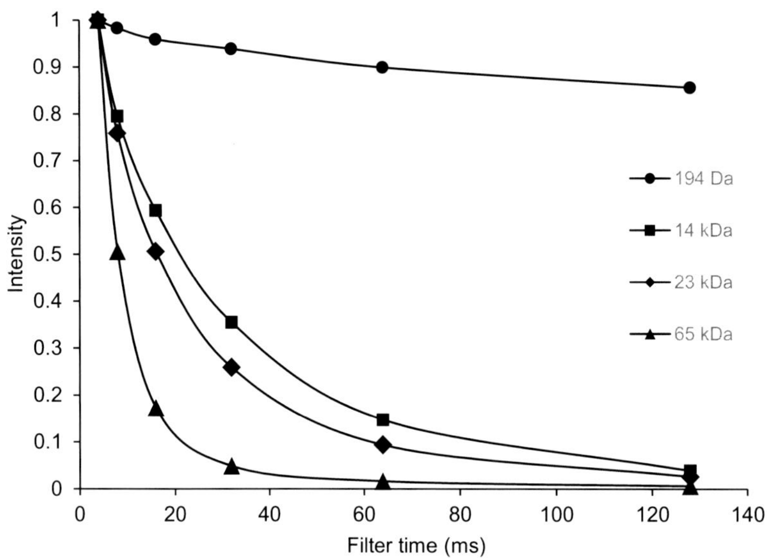

FIGURE 11.8 **^{1}H signal attenuation as a function of CPMG spin lock filter time for a small molecule and for proteins of various sizes.** In each case the CPMG-PROJECT sequence was used with $\tau = 1$ ms, looped to yield the desired total filter time.

The attenuation of ^{1}H resonance intensities of a small molecule and of proteins of differing molecular masses as a function of CPMG filter time is illustrated in Fig. 11.8. It is clear that it is possible to completely attenuate protein resonances, while those of the uncomplexed small molecule retain significant intensity. It is also apparent that resonances of the larger protein are attenuated more rapidly and thus disappear at shorter filter times. In a practical application one would determine through experiment a suitable total filter period sufficient to effectively suppress receptor protein resonances yet retain significant ligand intensity in the absence of the receptor. Filter times typically range from some tens of milliseconds up to hundreds of milliseconds with a filter of 400 ms having been used for ligand screening against a relatively small ~15-kDa protein [11]. Significant attenuation of ligands with higher binding affinities may occur with shorter filter periods, while weak binders may show little signal reduction under such conditions, requiring longer filter periods to experience appreciable attenuation. This may be viewed as problematic as weak binders may be missed if inappropriate conditions are chosen, but can be used to one's advantage to preferentially seek relatively stronger binders, for example when screening mixtures of compounds, provided fast-exchange conditions remain. As described above, peak intensities in CPMG-filtered spectra derived from direct titration can be fitted for determination of ligand dissociation constants. Finally, we note that relaxation-filtered experiments are sometimes also referred to as being $R_{1\rho}$ (or equivalently $T_{1\rho}$) filtered rather than R_2 filtered. The difference between these methods lies in the technical details of the sequences used, with $R_{1\rho}$ experiments typically employing continuous spin locks that hold magnetisation static in the rotating frame as it relaxes (hence the term $R_{1\rho}$ to reflect relaxation in the rotating frame) rather than using CPMG sequences. In terms of their application, there is little difference for editing and henceforth only CPMG filtering will be considered.

11.3.2 ^{19}F NMR

The fluorine nucleus is nowadays widely used in pharmaceuticals and agrochemicals making it an attractive alternative probe nucleus for ligand binding. It has favourable NMR properties well suited to direct ligand observation and, when present in a ligand, is often preferred over the proton for direct observation in screening [12]. Thus, fluorine-19 has high intrinsic detection sensitivity comparable with that of the proton (83%), although available instrument probe geometries may dictate the achievable sensitivities of these two nuclei. Furthermore, fluorine observation benefits from a lack of background fluorine signals, avoiding the need for protein attenuation and solvent suppression that are often required for proton detection. It also allows the experimentalist a greater choice of solvent buffers to be used since again these rarely contain fluorine so will not mask ligand resonances. The high chemical shift dispersion associated with fluorine is also beneficial for screening of a mixture of fluorinated compounds since resonance overlap is likely to be rare [13,14]. The absolute requirement, of course, is for the observed ligand(s) to contain fluorine which is most commonly inserted as a single CF group or as the CF_3 moiety. Nevertheless, even if the ligands to be screened do not contain fluorine, it may be possible to take advantage of ^{19}F-based experiments by using a fluorinated reporter ligand in competition studies, as described in Section 11.7.3.

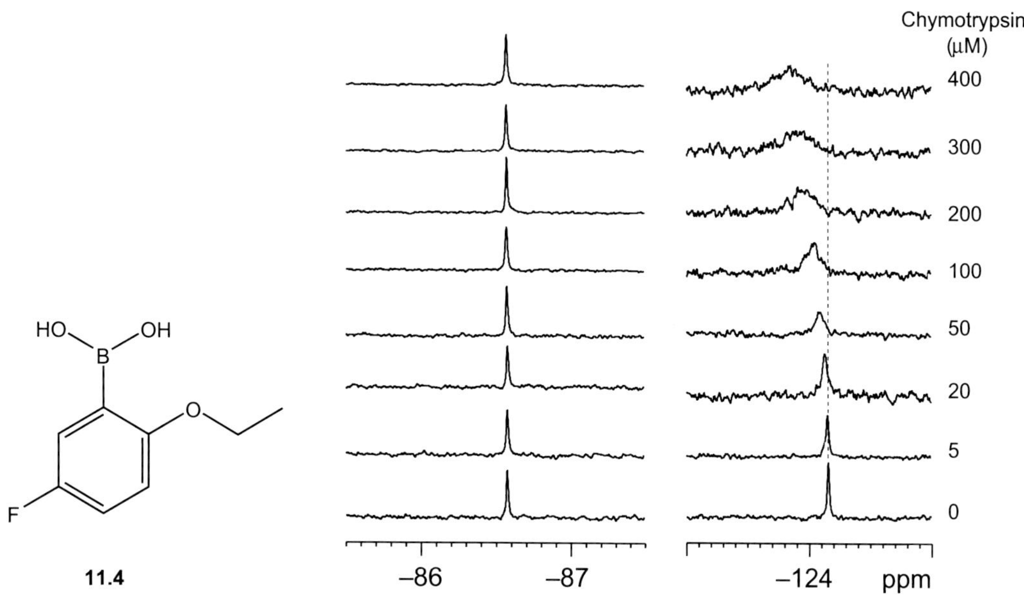

FIGURE 11.9 **^{19}F NMR monitoring the binding of the fluorinated boronic acid 11.4 (200 μM) to α-chymotrypsin as a function of increasing protein concentration.** Changes in chemical shift and significant resonance broadening are both apparent for the ligand (~–124 ppm) whereas the non-binding internal control 1,1,1-trifluoroacetone (–86.6 ppm) remains invariant to protein concentration.

The wide chemical shift range of ^{19}F also makes it more sensitive to binding events than the proton in direct observe experiments [12]. This advantage arises, first, because the fluorine nucleus is likely to exhibit significant chemical shift differences between the free- and the receptor-bound ligand states meaning the exchange broadening described above (the R_2^{ex} term of Eq. 11.8) is likely to be greater than that of protons. In other words, the exchange is more likely to be intermediate on the ^{19}F shift timescale, yielding exchange-broadened resonances whereas this may be fast exchange on the proton shift timescale due to the more similar ^1H shifts between the free and bound states. Second, the large chemical shift anisotropy (CSA) associated with ^{19}F means its relaxation in the receptor-bound state will be enhanced, further increasing line broadening and hence sensitivity to binding. As CSA scales with the square of the applied magnetic field (B_0), the use of higher fields will further enhance this contribution. Finally, fluorine chemical shift changes for exchange-averaged ligands in the presence of the protein receptor may be significant, making this parameter also a useful reporter on binding events. These effects are clearly manifested in the binding of the fluorinated boronic acid **11.4** to α-chymotrypsin, in which boron forms a reversible covalent adduct with the active site serine residue of the protease (Fig. 11.9).

For the reasons given above, it is not surprising that ^{19}F-based experiments have become popular for ligand screening in pharmaceutical lead discovery [14]. The use of ^{19}F-optimised cryogenic probes further increases the attractiveness of fluorine detection as a rapid method for the screening of individual compounds or of ligand mixtures. Experiments may either employ direct observation of the fluorinated ligand or of a selected fluorinated reporter ligand of appropriate affinity in the so-called 'fluorine chemical shift anisotropy and exchange for screening' (FAXS) competition approach (Section 11.7.3), in either case optionally employing CPMG relaxation editing of the spectra.

11.3.3 Paramagnetic Relaxation Enhancement

A further route to enhancing the relaxation rates of ligand nuclei in the bound state, and thus increasing sensitivity toward binding, is through the use of paramagnetic spin labels attached to the protein [15]. This additional sensitivity can reduce the protein concentration required for screening by at least an order of magnitude relative to conventional screening conditions, since lower populations of bound ligands may now be detected. The unpaired electron of the spin label acts as a powerful dipolar relaxation source, adding a further term R_2^{para} to the nuclear spin relaxation rate expressed in Eq. 11.8, leading to:

$$R_2^{av} = F_F^L R_2^F + F_B^L R_2^B + R_2^{ex} + F_B^L R_2^{para} \tag{11.11}$$

This last term is observed as *paramagnetic relaxation enhancement* (PRE), the magnitude of which is dependent upon the inverse sixth power of the distance between the electron and nuclear spin (r^{-6}), the square of the magnetogyric ratio of the spins, and on the overall correlation time of the vector connecting the electron and nuclear spin. Since the magnetogyric ratio of the electron is 658 times that of the proton, relaxation rates for protons in the vicinity of the paramagnetic centre are therefore greatly enhanced. This effect can operate over long distances and remain significant out to ~15 Å, meaning ligands need only bind in the approximate vicinity of the label to experience some degree of PRE. This in turn provides some flexibility as to where a spin label may be placed on the protein to influence the relaxation of a bound ligand nucleus.

Spin-labelling schemes typically employ either organic radicals, for example those derived from the nitroxide radical TEMPO (2,2,6,6-tetramethyl-1-piperidine-*N*-oxyl) [16], or paramagnetic metal ions [17], which may be native to a metalloprotein of interest or may be substituted in place of a diamagnetic native metal ion. Attachment of organic radicals may be achieved through chemical ligation methods which mostly involve addition to reactive amino acid residues such as lysine, cysteine, tyrosine or histidine, this approach being termed *SLAPSTIC* (spin labels attached to protein sidechains as a tool to identify interacting compounds) [16]. This modification procedure will likely lead to non-specific labelling of the protein and may require further analyses, typically by liquid chromatography–mass spectrometry (LC-MS) combined with protein digestion, if the labelled sites are to be identified. It is also necessary to ensure labelling does not adversely affect the structure or function of the protein receptor so that the spin-labelled species provides a representative analogue of the unmodified protein for screening. In the case of metalloproteins, the presence of a native paramagnetic metal ion provides an alternative and potentially less disruptive route to exploit PRE. Otherwise, this may be achieved through suitable substitution of a diamagnetic metal ion with an appropriate paramagnetic surrogate, for example exchanging zinc for cobalt(II). In such cases, metals with long electronic relaxation times are best suited to enhance relaxation, manganese(II) being a good example [18]. In any of these labelling scenarios, it is likely that ligand nuclei may demonstrate differing degrees of attenuation from PRE on binding to the receptor, due to their differing distances from the spin label. This may also provide some insight on ligand orientation relative to the label and may thus be considered beneficial. As described above, the use of relaxation filters will further enhance signal attenuation for bound ligand resonances.

An alternative approach to utilizing PRE is to combine this with the reporter ligand–screening concept described in Section 11.7.2. One interesting approach is to use proton relaxation of *bulk water* to report on binding events. This concept has been used for many years to investigate metal binding to proteins [19] and has more recently served to interrogate ligand binding to paramagnetic metalloproteins [20,21]. The principle being exploited is that for any protein in an aqueous environment, water molecules will invariably exist in the vicinity of the paramagnetic centre (and may even coordinate directly to this) and their proton relaxation rates will be greatly enhanced as a result. Following exchange of these receptor-bound water molecules with the bulk solvent water that surrounds the protein, net enhancement of the bulk water longitudinal relaxation rate is observed relative to that measured in the absence of the protein. Any ligand that binds in the vicinity of the paramagnetic center may act to inhibit access to this by water molecules, leading to net reduction in the bulk water relaxation rate. Hence, the behaviour of the water resonance itself serves as an indirect and generic reporter of ligand binding in this assay and provides the additional benefit that the binding of very high–affinity ligands may also be detected. Water relaxation may be measured through inversion (or saturation) recovery experiments (Section 2.4) and, due to the intense water resonance being observed (the H_2O:D_2O ratio may be varied), experiments may be performed on very small sample volumes, so reducing the total amount of material required. The method also operates at low magnetic fields and would be well suited to the recent generation of low-field (<100 MHz), bench-top NMR instruments.

11.4 SATURATION TRANSFER DIFFERENCE

The saturation transfer difference (STD) method [22] has become a widely employed technique to detect the binding of ligands to receptors and is suitable for studies of single ligands or of ligand mixtures in compound screening assays. The method relies on direct observation of the free ligand in solution and again requires the ligand to be in fast exchange with the receptor molecule. It also demands that the ligand be present in large excess, typically 50- to 100-fold. The use of large ligand excess means rather low protein concentrations may be employed (typically 10–50 μM) and this, together with the relative simplicity of the experimental technique and associated data processing, has led to the popularity of STD as an NMR screening tool.

The method relies on saturation of the resonances of the protein receptor through selective irradiation of a narrow region of the protein ^{1}H spectrum. The saturation dissipates rapidly through the network of protons in the macromolecule via dipolar cross-relaxation (ie interproton NOEs), a process known as *spin diffusion* that occurs rapidly in large molecules. Any ligand molecule that binds to the receptor with a significant residence lifetime will receive saturation of its protons through the same process. This saturation will then be carried with the ligand as it passes back into solution through chemical exchange with the receptor. A suitable analogy would be heat transfer between metal objects in contact. Here the application of direct heat to a metal bowl (the receptor) will result in the whole object becoming hot through conduction

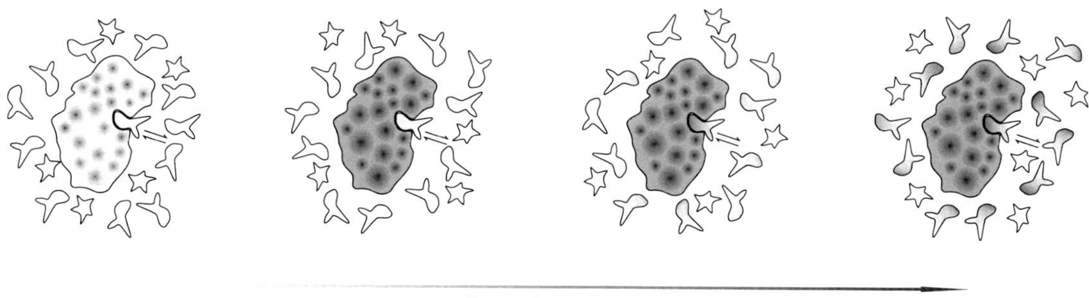

FIGURE 11.10 **Schematic representation of the saturation transfer process between protein receptor and ligand.** Irradiation of receptor protons (grey circles) leads to protein saturation throughout by rapid spin diffusion. Ligand on –off exchange during the ongoing saturation period leads to the progressive saturation of the ligand pool for receptor binders, whereas those that do not bind (stars) receive no saturation.

(spin diffusion), and any metal ball (the binding ligand) that sits in the bowl will likewise become warm and will carry this energy with it as is leaves the bowl; warming of the ball is thus indicative of its extended residence time with the bowl. Ligands that do not associate with the receptor will not experience such saturation transfer (Fig. 11.10) and hence only the resonance intensities of ligand binders will display attenuated intensities in the resulting ^{1}H spectrum. To clearly illustrate these perturbations, a control ^{1}H spectrum is acquired in which saturation is applied *off-resonance* (ie far away from the protein and ligand resonances, but under otherwise identical conditions) and the *on-resonance* protein saturation spectrum subtracted from this. The resulting *saturation transfer difference* spectrum displays resonances of only those ligands that have bound to the saturated receptor and the broad background spectrum of the saturated protein itself, with the resonances of any non-binding molecules nulled (Fig. 11.11).

As illustrated in Fig. 11.10, during the saturation period multiple ligands may enter the receptor binding site and carry saturation into the free ligand state, leading to amplification of the saturation effect, one of the principal advantages of the STD method. Thus, longer saturation times and larger ligand excess will amplify the STD response. The magnitude of this response may be quantified as a fractional intensity change in a manner similar to NOE experiments (Section 9.2) as:

$$\eta_{STD} = \frac{I_{REF} - I_{SAT}}{I_{REF}} = \frac{I_{STD}}{I_{REF}} \tag{11.12}$$

where I_{REF} and I_{SAT} are the peak intensities in the off- and on-resonance saturation experiments, respectively, and I_{STD} (= $I_{REF} - I_{SAT}$) reflects observed intensity in the final STD spectrum.

Fig. 11.12 shows the ^{1}H STD spectrum of a mixture of L-tryptophan and sucrose (**11.2** and **11.3**) with BSA. While the tryptophan binds to the albumin and is thus present in the STD spectrum, sucrose does not bind and its signals are not apparent. Similarly, when screening multiple compound mixtures in the presence of a receptor, only those that bind with appropriate affinity will give rise to an STD response.

The ability to detect an STD response is correlated to the off-rate of the ligand which must be sufficiently high to efficiently carry the saturated ligand back into bulk solution for detection, but not so rapid so as to prevent the transfer of

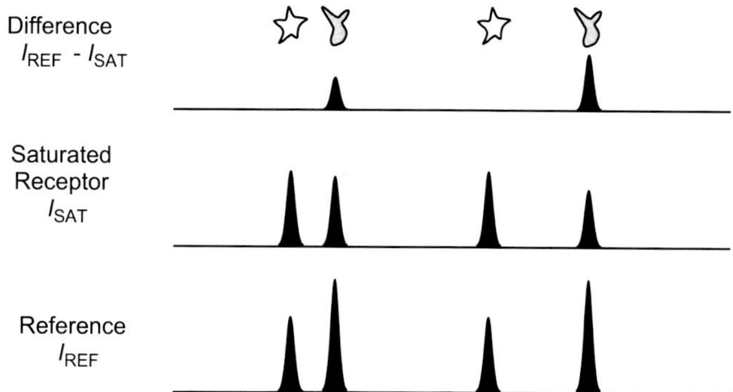

FIGURE 11.11 **Schematic representation of the ^{1}H STD spectrum arising from the processes illustrated in Fig. 11.10.** In the difference spectrum only responses from the ligands that bind to the receptor are observed.

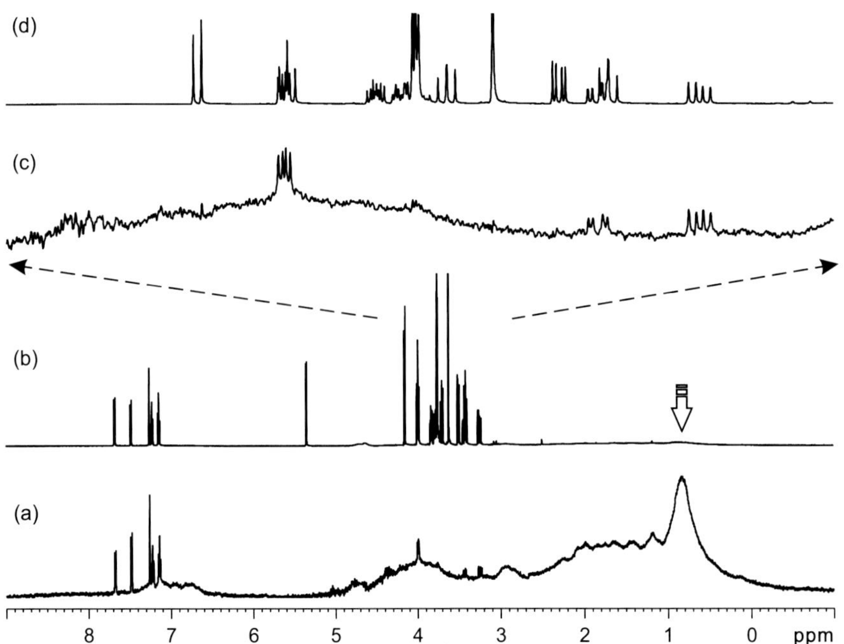

FIGURE 11.12 **The 600 MHz ^{1}H saturation transfer difference spectrum of L-tryptophan (2 mM) and sucrose (2 mM) in the presence of 20 μM BSA (M_r 66.5 kDa) in D_2O.** (a) the STD spectrum (3-s saturation), (b) the reference ^{1}H spectrum, (c and d) expansions showing the aliphatic region, highlighting the cancellation of non-binding sucrose peaks. The arrow indicates the location of protein saturation (0.83 ppm).

saturation from receptor to ligand. This means that very strong or rather weak binders can give rise to vanishingly small responses in STD spectra, and thus the lack of an STD response does not always imply the ligand does not bind. In practice, this means the STD method can be used to detect binding for ligand affinities in the K_D range of high nanomolar to low millimolar, although these limits are also dependent on the ligand on-rate which is rarely known but often assumed to be diffusion-limited. In cases where no response is observed yet the ligand is expected to bind, the use of competition experiments (Section 11.7) may be better suited to identify a high-affinity ligand. It should also be noted that the STD method benefits from the use of larger protein receptors since the longer rotational correlation times (slower tumbling) associated with these enhances the efficiency of spin diffusion and thus the transfer of saturation onto the ligand. Conversely, smaller proteins are likely to give poorer STD responses and the process may be ineffectual below ~15 kDa.

11.4.1 The STD Sequence and Practicalities

The essential requirement for operation of the STD experiment is selective saturation of proton resonances belonging to the protein and complete avoidance of direct saturation of the ligand itself. Such protein saturation could be achieved through application of long, low-power irradiation in a manner similar to the classic steady-state NOE difference experiments. However, to avoid undesirable spillover of the saturating field the standard protocol instead utilizes the repeated application of shaped selective pulses since these should have no significant effect outside of their operating bandwidths. The typical procedure is to use a series of 90 degree Gaussian-shaped pulses each of 50-ms duration (these have a total effective bandwidth of ~100 Hz; see Section 12.4) repeated to give the desired total saturation period, followed by a hard excitation pulse; the sequence is represented in Fig. 11.13. In a standard application the recovery delay (RD) will not be necessary as relaxation of non-irradiated protons will occur during the period in which the protein is being saturated, although this delay is required for the generation of STD buildup profiles, as discussed below.

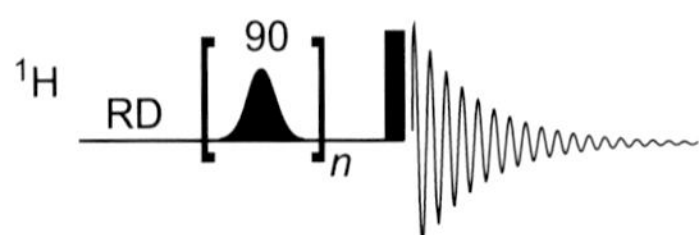

FIGURE 11.13 **The standard STD sequence.** A train of *n* shaped 90 degree pulses are applied to achieve selective protein saturation.

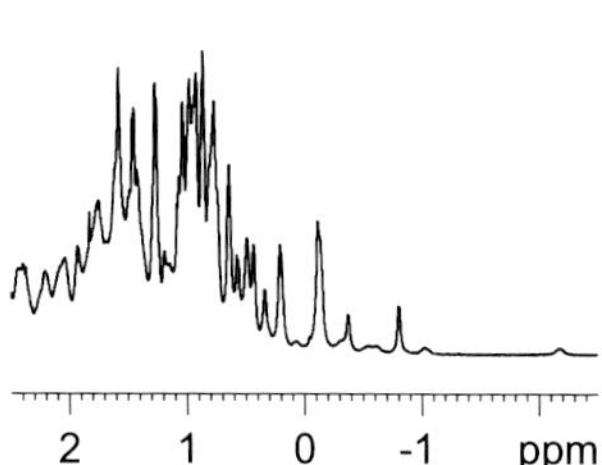

FIGURE 11.14 **A region of the protein spectrum of lysozyme (M_r 14,4 kDa).** This shows the prevalence of shielded resonances around 0 ppm that would be suitable targets for selective protein saturation. This region will likely be even more crowded in larger proteins.

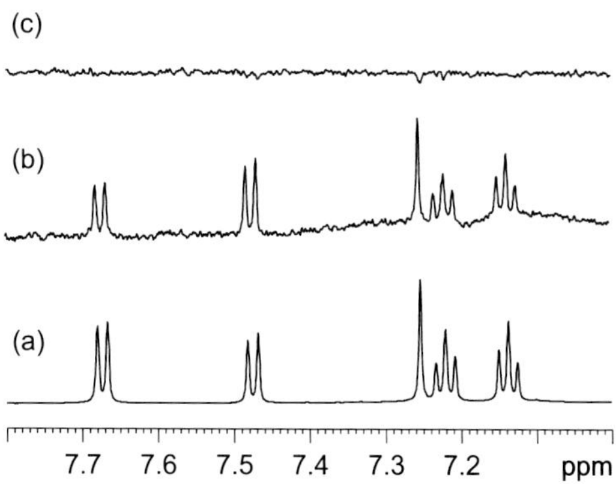

FIGURE 11.15 **STD difference spectra.** These show (a) the reference ^{1}H spectrum, (b) on-resonance vs off- resonance STD spectrum and (c) the off-resonance(1) vs off-resonance(2) blank difference spectrum. Trace (b) shows the usual ligand-binding response while the lack of peaks in (c) demonstrates excellent instrument stability throughout the experiment, leading to complete ligand cancellation in the absence of protein saturation.

The saturating rf field most commonly targets the protein methyl region below 1 ppm so as to avoid ligand resonances, and it is even feasible to irradiate slightly below 0 ppm in many instances since shielded methyl resonances often fall in this region (Fig. 11.14). In cases where methyl saturation is not possible due to the presence of ligand resonances, an alternative approach is to saturate the aromatic region of the protein spectrum. For the off-resonance reference spectrum, the irradiation frequency is set to >30 ppm where no resonances for any species would be anticipated (an exception may be protons close to paramagnetic centres in metalloproteins, where strongly paramagnetically shifted peaks may occur). Comparison of the peak intensities in the difference spectrum ($I_{REF} - I_{SAT}$) with those in the reference spectrum (I_{REF}) directly yield the STD enhancements η_{STD}.

To improve the quality of suppression in the final difference spectrum, modern STD sequences collate data by interleaving acquisitions for the on- and off-resonance experiments so that any long-term perturbations have minimal effect on the quality of the difference spectrum arising from subtraction of the two data sets. A useful test when undertaking STD experiments for the first time on any spectrometer is to collect not one but two off-resonance control spectra with the saturation offset at different frequencies (say, ±30 ppm). When these two control experiments are subtracted the resulting difference spectrum should contain no responses, indicating good instrument and environmental stability over the course of the STD experiment (Fig. 11.15). Any artefact responses that do appear are indicative of instability and these are likely to appear in the actual STD spectrum also. Alternatively, such artefacts may arise from sample degradation over the course of the experiment and the collection of a proton spectrum before and after STD may provide a useful check for this, especially if extended STD experiments are required to detect weak responses.

In the typical STD screening setup a total saturation period of 2–3 s is employed (requiring 40–60 50-ms shaped pulses), representing a balance between STD intensity buildup and total experiment duration, since maximum STD intensities are observed at longer saturation periods. An example of saturation time-dependent STD spectra may be seen in Fig. 11.16 for L-tryptophan binding to BSA with the corresponding STD intensity buildup profiles plotted in Fig. 11.17. Initial protein spin diffusion is a rapid process leading to receptor saturation within hundreds of milliseconds, which in this case is complete after ~3 s, with no further increase in intensity of the protein signals. The ligand STD buildup is a more gradual function of saturation time until a steady-state plateau is reached, at which point the STD enhancement is balanced by the spin relaxation processes of the free ligand. The generation of such profiles can be a useful indicator that observed responses do indeed arise from saturation transfer processes and find use in more quantitative applications of STD, described below. In generating these profiles the total recovery delay (RD + T_{SAT}; Fig. 11.13) is held fixed and only the saturation time T_{SAT} increased so as to ensure all spectra are recorded under conditions of equal spin relaxation between experiments. As discussed further in Section 11.4.2, the final level of STD signal enhancement varies significantly across the protons in a molecule due to a number of factors, including proton proximity to the receptor surface and to proton spin relaxation rates, and their interpretation demands some caution.

It is also the case the STD intensities can be influenced by sample temperature due to the effect this has on the exchange kinetics of the receptor–ligand complex, meaning temperature variation can assist in the search for ligand binders. Thus, if weak or negligible STD responses are observed, repeating the experiment at a different temperature may improve results, although STD temperature dependence for any receptor–ligand combination can be difficult to predict when kinetic

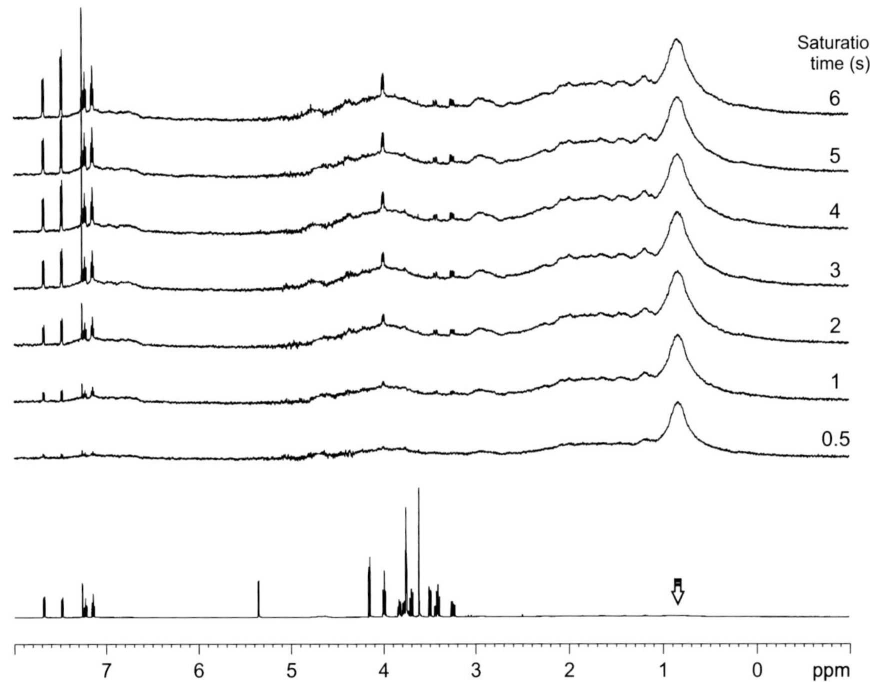

FIGURE 11.16 **Saturation time-dependent STD spectra of the same L-tryptophan/BSA sample used in Fig. 11.12.** The arrow indicates the location of protein saturation.

parameters are unknown, which is invariably the case. Reducing the sample temperature would be expected to reduce the exchange k_{off} rate, so is likely to be beneficial in the search for weaker binders, whereas high-affinity ligands may be better detected at higher temperatures, provided protein stability allows. Lower temperatures will also reduce receptor correlation times and so enhance spin diffusion, potentially enhancing STD intensities; this may prove especially beneficial for smaller protein receptors. Similarly, use of higher magnetic fields also promotes spin diffusion and aids detection sensitivity.

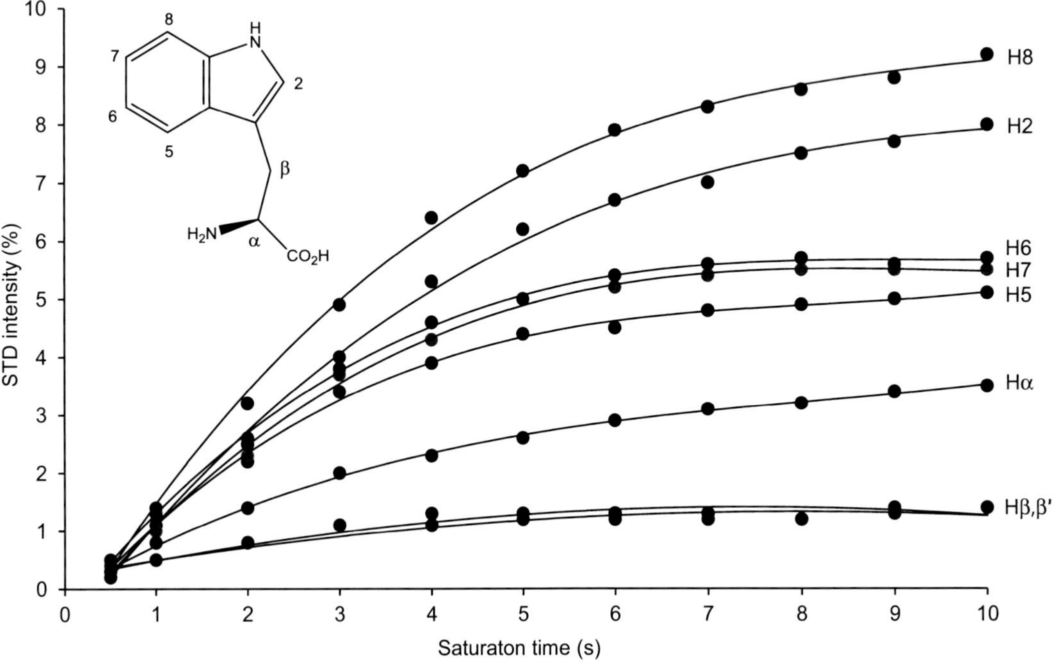

FIGURE 11.17 **Saturation time-dependent STD buildup profiles of L-tryptophan binding to BSA.** These data are derived from the spectra of Fig. 11.16, extended for spectra with up to 10 s saturation times. The η_{STD} values are expressed as percentages.

11.4.1.1 Sample Preparation

Typical solution conditions for STD experiments require protein concentrations in the range of 10–100 μM and ligand excesses of 20–100 corresponding to ligand concentrations in the low-millimolar range. While concentrations chosen for a study may be restricted by practical limitations, such as protein availability or ligand solubility, the considerations described in Section 11.2 should be borne in mind. For example using lower protein concentrations will favour detection of the binding of higher affinity ligands over those of lower affinity when studying ligand mixtures. Likewise, if STD responses are not apparent in an initial screen, using higher protein concentrations may reveal the presence of low-affinity ligands in the sample. It is also advantageous to use *deuterated* aqueous solution for STD studies when possible, rather than protonated water. This reduces relaxation of protein and ligand protons through intermolecular ^{1}H–^{1}H interactions with the water and aids the retention of STD enhancements in the free ligand. Solution buffers are also often required for protein stability and their selection can be important since many commonly used buffers have multiple proton resonances that may overlap with ligand resonances and readily mask information at the concentrations typically employed (tens of millimolar). Proton-free buffers such as phosphate are attractive, otherwise deuterated buffers are commercially available, although can be expensive. The presence of other small molecules in protein preparations is also surprisingly common and often revealed only in NMR studies, suggesting the ^{1}H NMR spectrum of the protein solution in the absence of ligands as a useful reference. A common impurity from protein purification made apparent by NMR (at least in this author's experience) is glycerol, used as a coating for molecular weight cutoff filters used in spin concentrators or as a stabilizer added to proteins during cryogenic storage, the appearance of which surprises many protein chemists. This can mask a substantial window around 3.4–3.8 ppm in the proton spectrum.

11.4.1.2 Extended Sequences

Although, as stated above, it is optimal to collect STD data on samples in deuterated solvent, it is often the case that a significant water resonance remains. This may be because it is impractical to provide protein in deuterated solutions or simply because the materials available are intrinsically wet. In either case, suppression of the protonated water resonance can be desirable and the use of gradient suppression schemes such as WATERGATE or excitation sculpting appended to the STD sequence prove effective. Fig. 11.18a illustrates the addition of the excitation sculpting element in STD; for details on excitation sculpting see Section 12.5.3. It is also sometimes the case that removal of the broad protein background responses seen in STD can be desirable, either to make the ligand responses clearer to observe or to allow their quantification without interference from protein resonance overlap. As described in Section 11.3.1, the suppression of protein resonances can be achieved through the use of a spin lock $R_{1\rho}$ relaxation filter applied for typically tens of milliseconds (30–50 ms being suitable in many cases) at low rf power. This may be included in the STD sequence as illustrated in Fig. 11.18b, which shows this combined with the water suppression element. While the protein responses are significantly attenuated, there will also be some smaller loss of ligand intensity leading to a slight sensitivity penalty when using this element.

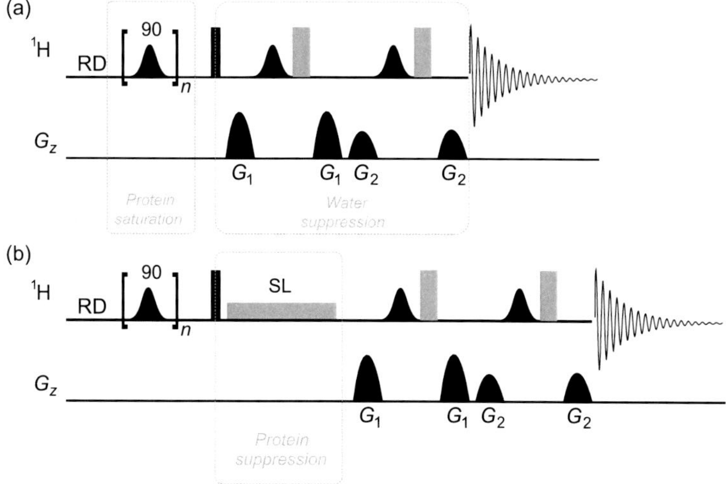

FIGURE 11.18 **Extended STD sequences.** These incorporate (a) water suppression using the excitation sculpting scheme and (b) a low-power spin-lock (SL) protein suppression period (~30–50 ms) to attenuate broad receptor resonances, in addition to the water suppression element.

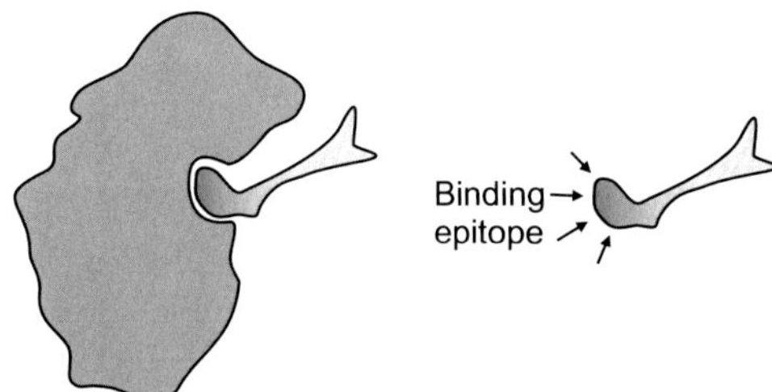

FIGURE 11.19 **The principle of group epitope-mapping through STD measurements.** The parts of the ligand that experience the greatest saturation transfer (darkest grey) are those considered to be making most intimate contact with the receptor (see text).

11.4.2 Epitope Mapping by STD and DIRECTION

Beyond using STD experiments to qualitatively assess whether a ligand binds to a receptor, in favourable cases it is possible to gain further information on the binding event. One such example is identification of the ligand protons most closely associated with the receptor binding site, the so-called *group epitope map* of the ligand (the definition of ligand–receptor interactions more generally is sometimes referred to as pharmacophore mapping). The principle is rather simple to follow: those ligand protons that sit closest to the receptor surface at the binding site will experience a greater degree of saturation transfer from the receptor protons and will therefore display the greatest STD intensities. This notion is represented schematically in Fig. 11.19. Those parts of the ligand in intimate contact with the receptor receive significant direct enhancements (dark grey) while those in parts of the ligand that sit away from the receptor surface show lesser perturbation that arises from spin diffusion through the ligand while bound (light grey). Meaningful epitope maps require the ligand dissociation off-rates k_{off} to be more rapid than the diffusion of magnetization through the ligand in the bound state, and thus can be obtained only for moderate to weak binding ligands. One further important limiting issue is the fact that the relaxation rates for protons of the ligand in the free state can vary significantly, meaning the ability of the ligand protons to retain their saturation when free from the receptor also vary, so contributing to the observed STD intensities [23]. Protons that relax quickly (those with shorter T_1 values) will lose their STD enhancements more readily and so produce weaker net responses. For example methylene CH_2 protons generally experience faster relaxation rates owing to the near geminal neighbour and tend to exhibit weaker STD intensities relative to more slowly relaxing methine CH protons. This influence is clearly apparent in the significantly lower STD intensities observed for the tryptophan Hβ protons in Fig. 11.17. Thus, STD intensities recorded with long saturation times, as typically used in screening experiments, will be weighted by the spin relaxation rates of the free ligand in particular and may not accurately represent its binding epitope. This suggests caution when interpreting STD data used for ligand screening in terms of its possible epitope map.

A more appropriate analysis for epitope mapping is to consider the initial STD buildup rates such that relaxation of the free ligand is insignificant; there is a direct analogy here to using NOE buildup rates to determine internuclear distances within molecules (Section 9.4.2). Under conditions of short saturation times there is also reduced likelihood of spin diffusion through the bound ligand biasing the observed intensities. Unfortunately, these conditions also mean absolute STD intensities are also intrinsically weak, meaning accurate epitope-mapping investigations can demand extended data collection times. Presentation of the epitope map is typically made by classifying the most intense STD response (or more correctly the buildup rate) as 100% and scaling all others relative to this, providing a readily interpreted map (Fig. 11.20).

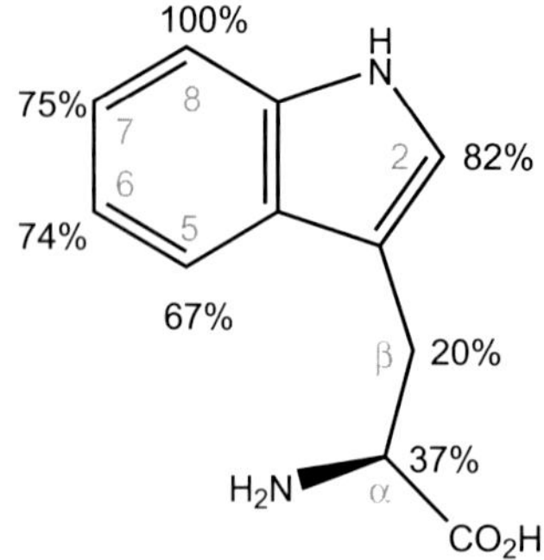

FIGURE 11.20 **Representation of the group epitope map for tryptophan binding to BSA.** Epitope intensities were derived from initial STD buildup rates and normalised to the largest value.

An alternative method held to provide more reliable epitope maps compares the ligand proton longitudinal relaxation rates (as measured with the inversion recovery experiment; Section 2.4.2) determined in the presence of the protein, recorded either without or with continuous irradiation of the protein resonances. In the absence of protein irradiation, the relaxation rates of ligand protons that sit at the surface of the protein will be dictated by intermolecular cross-relaxation with the receptor protons, in addition to intramolecular relaxation caused by dipolar interactions with other ligand protons. However, saturation of all protein resonances, caused by protein irradiation exactly as for the STD experiment, prevents the receptor protons from acting as relaxation sources, meaning any protons that were relaxed by these at the receptor surface will now show a reduced overall relaxation rate. The relative change in ligand relaxation rates may be correlated with their proximity to the receptor surface and therefore yields the ligand epitope map. The method is referred to as DIRECTION (difference of inversion recovery rate with and without target irradiation) [24] and has similar requirements to the STD experiment itself, including fast ligand off-rates, so is also not applicable to the study of high-affinity ligands.

More thorough methods for the quantitative analysis of STD intensities have also been described, including those based on the *complete relaxation and conformational exchange matrix for STD* (CORCEMA-STD) computational algorithm [25]. This enables the prediction of STD intensities from the atomic coordinates of the receptor–ligand complex, typically derived from a crystal structure of the complex or from ligands computationally docked into receptor crystal structures. CORCEMA calculations can therefore be used to help define bound ligand poses from experimental STD data [25–27].

11.4.3 K_D Measurement by STD

A second area in which STD experiments can play a more quantitative role is in the determination of ligand binding affinities through ligand titration experiments. In principle, following the change in STD response as a function of ligand concentration while the receptor concentration is held fixed should allow fitting of the responses as discussed in Section 11.2 and yield experimental K_D values. For this it is useful to define the STD amplification factor [28] STD^{AF}:

$$STD^{AF} = \eta_{STD}\ \varepsilon \tag{11.13}$$

where the measured STD intensities η_{STD} are scaled according to the ligand excess being employed ε (= $[L_T]/[P_T]$) such that the STD^{AF} provides a measure of the effective amount of receptor–ligand complex being generated throughout the STD saturation period. This also allows a comparison to be made of experimental data between samples of differing receptor concentrations. Since it is a measure of the receptor–ligand complex formed that is required for K_D determination, consideration of the variation in STD^{AF} versus ligand concentration should yield the dissociation constant through standard data fitting (Section 11.2 and Eq. 11.9). However, it has been observed that dissociation constants determined from such STD titration analyses are influenced strongly by the experimental parameters employed, including saturation times and receptor concentrations, and tend to overestimate the value of K_D (see below). Thus, while direct titration may indicate an upper limit for K_D, it may not provide a value that could be considered accurate.

There are two principal solutions to this limitation. The simplest and often preferable approach is to use STD measurement in competition assays whereby the STD^{AF} response of a ligand is observed and its variation measured as a second ligand is titrated into the solution. Attenuation of the STD^{AF} response of the observed ligand indicates its direct displacement by the competitor from which the K_D of one ligand may be derived provided that of the other is known; such competition assays are considered further in Section 11.7.

A second solution is appropriate when a suitable binding molecule of known affinity does not exist, and direct determination of K_D is required. To understand the basis of this approach, it is necessary to consider why K_D determination by direct STD titration fails to yield accurate values. The principal problem is that STD responses can be perturbed by rebinding of ligands during the extended saturation time required by the method [29]. Any ligand that has previously bound with a receptor and received some degree of saturation will, following rebinding and subsequent release, carry a reduced level of saturation back into solution relative to a freshly saturated ligand binding for the first time. This tends to attenuate the overall buildup of the observed STD response and leads to biasing of the 'apparent' dissociation constant. STD experiments used purely for the screening of binding tend to use large ligand excess and the likelihood of ligand rebinding is correspondingly low. However, titrations for determining dissociation constants will demand the use of low ligand concentrations at early points, and these conditions tend to favour ligand rebinding and explain why direct titration experiments yield inaccurate K_Ds. Ligand rebinding is also more likely to occur when longer saturation times and high receptor concentrations are employed, both cases leading to higher 'apparent' K_D values being measured; unfortunately, these conditions also correspond to the generation of more intense STD effects favoured experimentally. These perturbing effects are also enhanced for ligands of higher affinity, meaning their measured K_Ds will likely be less accurate.

To overcome these problems it is necessary to record STD intensities under conditions where ligand rebinding is insignificant, a situation that occurs when saturation times tend towards zero and only single binding events influence STD intensity development [29]. Unfortunately, these conditions lead to only weak STD responses that may be difficult to measure accurately, so in practice it is necessary to record the initial buildup rates of STD amplification factors $STD^{AF}{}_0$ (these are analogous to the STD initial buildup rates considered for epitope mapping, as described earlier). The variation in $STD^{AF}{}_0$ as a function of ligand concentration can then yield accurate K_D values. Thus, STD^{AF} values are recorded over a range of saturation times T_{SAT} for a fixed receptor–ligand ratio and initial buildup rates $STD^{AF}{}_0$ determined from a fit of these data points. Such $STD^{AF}{}_0$ values are likewise recorded for a series of ligand concentrations, thus allowing K_D determination through conventional data fitting. Clearly, this is potentially a lengthy process suggesting direct STD titrations may not be the optimum approach to accurate determination of dissociation constants in many cases.

11.5 WATER-LOGSY

The water-LOGSY experiment (*water–ligand observed with gradient spectroscopy*) is in many ways complementary to STD and again relies on the transfer of magnetisation from the receptor to the ligand that is present in excess and undergoing rapid on–off exchange with the macromolecule. In this case, the source of magnetisation is not the direct saturation of the receptor itself, but of the bulk water in solution. This saturation may then be transferred via NOE processes onto ligands that interact with the receptor through a number of magnetisation transfer pathways (Fig. 11.21). These include transfer from water molecules that reside at the receptor surface onto the protein and then via spin diffusion to a binding ligand or through direct transfer onto a ligand that binds in the vicinity of receptor-bound water molecules. Other possible pathways involve chemical exchange of saturated water protons with exchangeable protons on the protein surface, principally OH and NH groups, such that saturation may be then transferred to any ligand at the protein surface, either directly or mediated by spin diffusion through the receptor. In any case, the transfer of saturation onto the ligand occurs while it experiences the slow tumbling associated with the large receptor molecule giving rise to negative NOE effects. These are then carried by the ligand into free solution where they are observed as a reduction in ligand signal intensity which is therefore indicative of its binding to the receptor. As for STD, the ligand off-rate k_{off} is critical to the success of this process and must be rapid relative to ligand proton relaxation rates in the bound state.

An essential requirement for assessing whether binding has occurred is a control water-LOGSY experiment performed in the absence of the receptor to provide a reference against which to judge intensity changes. This is required because the free ligand in solution will itself experience direct magnetisation transfer from the water but this will be observed as positive NOE enhancements since the ligand will experience rapid tumbling in solution. Since this effect also occurs in the experiment with the receptor present, the reference provides a measure of the direct-water-to-free-ligand enhancement. Thus, differences in the experiments recorded in the presence and absence of receptor will reveal evidence for ligand binding. The control experiment also serves to expose ligands that may undergo aggregation under the assay conditions employed since these will then behave as macromolecular species and so give rise to negative enhancements from water

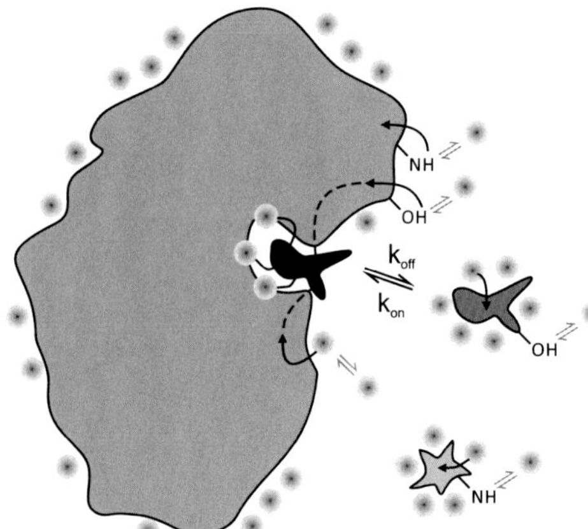

FIGURE 11.21 **Magnetisation transfer processes from water to ligand operating in the water-LOGSY experiment.** Saturated water molecules (grey spheres) may transfer magnetisation to the ligand through a number of pathways including direct NOEs (solid arrows) to the receptor-bound ligand (a negative NOE) and to the free ligand (a positive NOE), spin diffusion through the receptor (dashed arrows) and via direct chemical exchange with labile protons (double arrows). Ligands that do not bind to the receptor experience only direct positive NOEs from the water (pink).

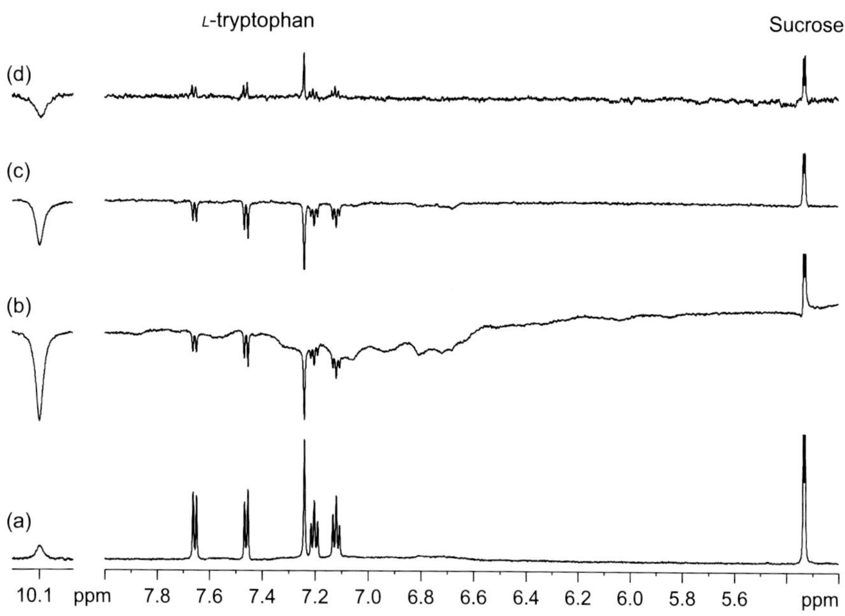

FIGURE 11.22 **The 600 MHz ^{1}H water-LOGSY spectra of L-tryptophan (1.3 mM) and sucrose (2 mM) in the presence of 100 μM BSA (M_r 66.5 kDa) in 90% H_2O: 10% D_2O.** (a) Reference ^{1}H spectrum, (b) water-LOGSY with $\tau_m = 0.8$ s, (c) as (b) with an additional 30 ms spin-lock to attenuate the receptor background and (d) water-LOGSY control in the absence of receptor $\tau_m = 0.8$ s.

saturation. These effects, if observed with the receptor present, could mistakenly be interpreted as evidence for ligand binding, so giving rise to 'false positives' in a ligand screening assay. The common use of small, relatively hydrophobic fragments in compound libraries means evidence for ligand aggregation is quite frequently observed in water-LOGSY controls, testifying to their importance.

Using the BSA–tryptophan–sucrose system once more, it is apparent (Fig. 11.22) that in the receptor-free control the magnetisation transfer from the water to ligands gives rise to weak positive responses arising from direct NOEs between these species. In the presence of the receptor, the response of the binding ligand *L*-tryptophan inverts and matches that seen for the background protein resonances, while those of the non-binding sucrose remain unchanged (Fig. 11.22b). The broad receptor background can often prove problematic when seeking weak ligand responses, but may be effectively removed through the incorporation of a short (~30 ms) spin-lock prior to the water suppression element (Fig. 11.22c), as described below. In the water-LOGSY experiment there will always exist competition between the slowly developing positive enhancements arising from direct water–ligand interactions of the free ligand and opposing effects originating from the rapidly developing ligand–protein complex enhancements, with the net result manifest in the receptor present water-LOGSY spectrum. It is important to realise that ligand binding does not necessarily present such a clear negative response; the response may simply be weaker than in the control spectrum or may even be nulled completely if the competing ligand-bound and ligand-free enhancements happen to cancel. Only by comparison with the receptor-free control run under otherwise identical experiment conditions can changes due to binding be properly assessed. Fig. 11.22 also illustrates the appearance of strong negative peaks arising for exchangeable protons (here the tryptophan indole NH at 10.1 ppm) from direct exchange with water, an 'artefact' commonly observed in water-LOGSY spectra.

The experimental requirements for performing water-LOGSY experiments are in many ways similar to those of STD described above, in that low protein concentrations (10–50 μM) and large ligand excesses are employed. Typically for water-LOGSY, ligand excesses of only 10- to 20-fold are employed (invariably below 50-fold) so as to favour the development of the bound ligand response over that of the free ligand. An essential difference relative to STD is that samples must be prepared in predominantly *protonated* water (typically 90% H_2O: 10% D_2O) since these protons are the source of the magnetisation that is transferred to the ligand. This demands rigorous attenuation of the water resonance so that the substantially weaker ligand responses may be observed, for which gradient suppression methods are mandatory, as described below. Water-LOGSY experiments prove effective in binding studies of low to moderate affinity binders (millimolar or less) and have become popular for the screening of small-fragment libraries. As spin diffusion throughout the receptor is not a requirement for the transfer of saturation in the water-LOGSY mechanism (unlike STD), it also has greater potential for studying ligand binding to proton-sparse macromolecules such as nucleic acids. As for STD, water-LOGSY can be effectively employed in competition assays whereby the change in response of a binder is monitored to reflect the binding of a competitive ligand; see Section 11.7.

11.5.1 The Water-LOGSY Sequence

The primary requirement for the execution of the water-LOGSY experiment is perturbation of the H_2O resonance via either saturation or inversion, again following the requirements of NOE experiments. While early implementations employed direct water saturation by extended irradiation, all current implementations instead employ selective inversion of the water resonance to initiate magnetisation transfer. This is followed by a mixing period in which direct and chemical exchange–mediated transfer processes operate, followed by recording of the ^{1}H spectrum after suppression of the intense water peak. Note that early sequences were developed for the study of protein hydration, so these are also referred to as ePHOGSY (*enhanced protein hydration observed through gradient spectroscopy*).

The key elements of the basic water-LOGSY sequence (Fig. 11.23a) are an initial gradient echo that selects only the H_2O resonance but dephases all other magnetisation, such that the water is inverted by the subsequent 90 degree pulse. The inverted water thus acts as the source for NOE transfer and/or chemical exchange, which occurs during the mixing time τ_m. During this period it is also beneficial to apply a weak continuous gradient pulse (~1% of maximum) throughout so that any recovery of the water signal into the transverse plane does not induce perturbations arising from radiation damping (a deleterious effect associated with very intense resonances) and remains dephased. After the mixing time the NOEs that have developed from the water are 'read' by a 90 degree pulse followed by the final and essential element that is a gradient water suppression scheme such as excitation sculpting. Selection of the desired magnetisation is achieved by stepping the phase of the selective pulse by 90 degrees with inversion of the receiver phase on each step following the four-step EXORCYCLE procedure (Section 8.2.2). The initial water-selective shaped pulse is chosen to have a reasonably narrow inversion profile so that resonances close to the water are not themselves inverted and Gaussian pulses of duration 5–15 ms are appropriate. For the water suppression element, pulses with wider profiles and thus shorter durations are preferred for complete attenuation of the water, as for conventional excitation-sculpting suppression. Due to the multiple transfer pathways used to develop the water-LOGSY signals, the method benefits from the use of rather long mixing times [30] with τ_m values of 0.5–1.5 s being commonly employed.

A more refined version of the water-LOGSY sequence more likely to be employed nowadays is illustrated in Fig. 11.23b. While the key elements described above are retained, this also employs an additional inversion pulse at the midpoint of the mixing time to counter the recovery of ligand magnetisation during the long mixing times employed and so prevent artefacts appearing in the final spectrum; this is exactly equivalent to the use of inversion pulses in the mixing times of 1D NOE spectroscopy (NOESY) experiments to help null signal recovery, as described in Section 9.6.2. A further refinement is the use of a water-selective 90 degree 'flip-back' pulse prior to the water suppression element. This is employed to help return the transverse water signal back to the +z-axis and so avoid the need for lengthy recovery delays between transients that

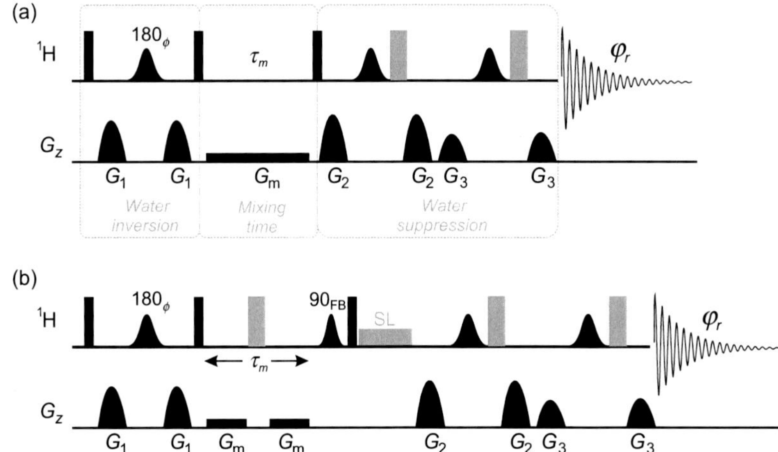

FIGURE 11.23 **Water-LOGSY sequences.** (a) A standard sequence illustrating basic elements of the technique and (b) a more refined sequence yielding improved spectra. In (b) the selective 90 degree flip-back pulse 90_{FB} acts to return the water resonance to the +z-axis and the optional low-power spin lock period SL (greyed) may be employed to attenuate the protein background. The basic phase cycle is to EXORCYCLE the water-selective 180 degree pulse $\phi = x, y, -x, -y$ with corresponding inversion of the receiver phase $\varphi = x, -x, x, -x$.

would otherwise be required for water relaxation. The final addition is an optional spin lock period to attenuate resonances of the macromolecule in a manner similar to that used in STD experiments (Fig. 11.18b) and as exemplified in Fig. 11.22c above; for routine application this may be omitted. In situations where large compound libraries are to be screened, further refinements relating to the recovery of water have been described for the polarisation-optimised water-LOGSY sequence which has greater overall time efficiency [31].

11.5.2 Water-LOGSY Practicalities

Other than the need for protonated aqueous solutions and moderate ligand excess as described above, the requirements for water-LOGSY are not overly stringent and the screening of multiple ligands in solution is possible provided their resonances are sufficiently resolved. The presence of a small molecule in solution other than the ligand(s) of interest can be helpful in the analysis of the spectrum as this should give only a positive 'reference' response indicative of no binding with the receptor. Often solution buffers or even residual dimethylsulfoxide (DMSO) used to prepare ligand stock solutions can serve as such reference compounds, otherwise materials such as trimethylsilyl propionic acid (TSP) can be added for this purpose (although this may itself bind to some protein receptors). It is also the case that these experiments often benefit from being recorded at reduced sample temperatures of 4–10°C where stronger water-LOGSY responses are frequently seen. This can arise due to the slower off-rate of weak binding ligands at lower temperatures, slower exchange of protons with bulk water, slower tumbling of the ligand–receptor complex, or likely from a combination of these effects. It is certainly the case that if initial water-LOGSY experiments fail to yield meaningful responses at ambient temperatures, then repeating these at reduced sample temperatures may greatly improve results. Care must then be taken so as to not to freeze samples in the probe, and probe temperature calibration (Section 3.5.3) is advisable in helping to avoid this since actual sample temperatures can often be lower than those reported by uncorrected spectrometer temperature units.

Beyond these practical considerations, other effects can potentially add complications to the interpretation of water-LOGSY spectra. A common observation is a strong response from the exchangeable protons of the ligand (often acid or amide protons; see Fig. 11.22) arising from direct chemical exchange with the inverted water. These should not be mistaken as being indicative of binding, and indeed should be observed in the receptor-free control also. As indicated above, aggregation of the ligand can also yield responses that can be mistaken as arising from receptor binding and must also be assessed in the receptor-free control. The direct inversion of a ligand resonance that sits under that of the water can also cause complications through the generation of conventional NOEs within the ligand in either the free or bound states, and these should also be considered when reviewing screening results. Direct inversion of the receptor resonances under the water (amino acid Hα protons) appears to play a negligible role in the overall response likely because of the low receptor concentrations used.

Finally we note that the water-LOGSY experiment does not readily provide information on the binding epitope of the ligand due to the multiple and competing pathways that operate to yield the observed response. Furthermore, while in principle ligand titrations may yield ligand dissociation constants the reliability of this approach has not been rigorously demonstrated. Current results suggest estimated K_Ds are strongly dependent on the solution conditions employed for such measurements [32], including receptor concentrations, and it seems likely that this method suffers similar limitations to those described above for STD. As such, water-LOGSY cannot at present be recommended for the *direct* quantitative measurement of accurate binding affinities. However, the use of this method in competitive binding assays employing known reporter ligands is valid and offers a method for K_D quantification (Section 11.7).

11.6 EXCHANGE-TRANSFERRED NUCLEAR OVERHAUSER EFFECTS

Exchange-transferred NOE experiments [33,34] rely upon the generation of intramolecular NOEs between protons within a ligand while it is bound to the macromolecular receptor. These NOEs are subsequently transferred to the free ligand state by chemical exchange, as the name suggests (the term *transferred NOEs* is also widely used). Since the NOEs observed for the free ligand arise while it is in the bound state, they have a negative sign characteristic of the slowly tumbling macromolecule so are opposite in sign and readily distinguished from the positive NOEs observed for small molecules free in solution in the absence of a receptor. The technique again relies on fast exchange between the bound and free ligand states such that the strong, rapidly growing negative NOEs of the bound state dominate any weaker, slowly developing free ligand NOEs. In favourable cases the intramolecular NOEs also provide information on the conformation of the ligand *while bound* (Fig. 11.24). The ability to probe bound-state conformations provides a significant advantage over methods such as STD and water-LOGSY.

Thus, while exchange-transferred NOEs may not be as widely used for primary screening as STD and water-LOGSY, which are generally more sensitive approaches, they have proved more useful in the study of the conformations of ligands while in the receptor-bound state [35–37]. This is especially true for larger ligands with significant conformational flexibility

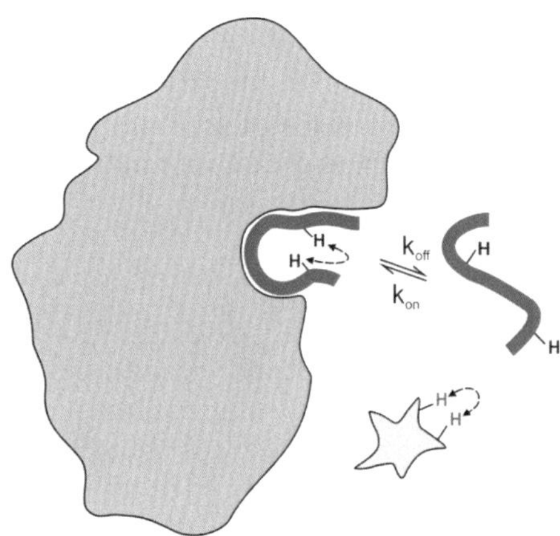

FIGURE 11.24 **Schematic representation of the exchange-transferred NOE process for conformational analysis of a bound ligand.** For binding ligands (dark ribbon) negative NOEs generated in the bound state (between red protons) are observed on the free ligand while for those ligands that do not bind (light star) only positive NOEs for the free molecule are detected (grey protons).

in solution for which substantial conformational changes may occur on binding to the receptor, with peptides and oligosaccharides being the most commonly studied ligands. Such ligands also often have the advantage (in this situation) of giving rise to negligible or weak NOEs in free solution due to their intermediate correlation times (around the NOE 'zero-crossing point'; see Section 9.3) making very apparent any negative NOEs arising from receptor binding. Furthermore, the use of STD for peptide binding can be limited by the absolute need for selective protein saturation while avoiding peptide irradiation, making exchange-transferred NOE studies a useful alternative.

Experimentally, sample requirements are similar to those of STD or water-LOGSY. The ligand must be present in excess (typically 10- to 30-fold) and must have an off-rate k_{off} that is again greater than the ligand proton relaxation rate in the bound state, meaning it is again applicable to mid- to low-affinity binders ($K_D \geq$ μM). It also benefits from larger receptors where stronger negative NOEs would be expected. Most commonly, exchange-transferred NOEs are observed in conventional 2D NOESY experiments, although for screening 1D NOESY can prove equally effective and quicker. When used to study bound ligand conformations, experiments are typically acquired with short mixing times (≤400 ms) to strongly favour faster growing bound-state NOEs, but in ligand screening the requirements are less stringent and longer mixing times may be employed, although may offer little gain. Fig. 11.25 shows the NOESY spectrum of the disaccharide **11.1** introduced earlier, recorded in the presence and absence of WGA. In the absence of the protein only rather weak

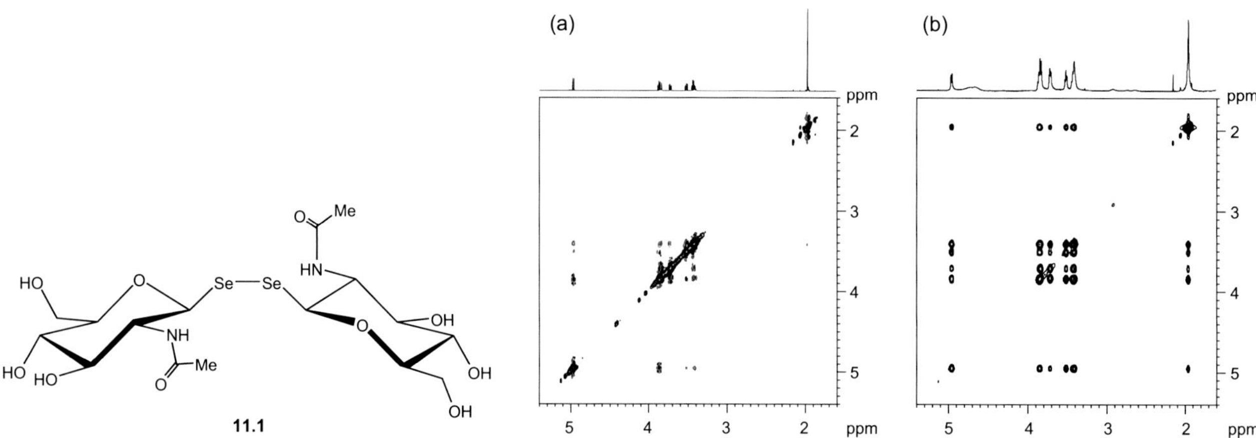

FIGURE 11.25 **Exchange-transferred NOESY experiments for the disaccharide 11.1 (5 mM) at 700 MHz.** (a) in the absence of protein and (b) in the presence of 100 μM of the lectin WGA. A mixing time τ_m of 400 ms was used for both experiments. Red crosspeaks are of opposite sign to the diagonal and indicate positive NOEs whereas black crosspeaks indicate negative NOEs.

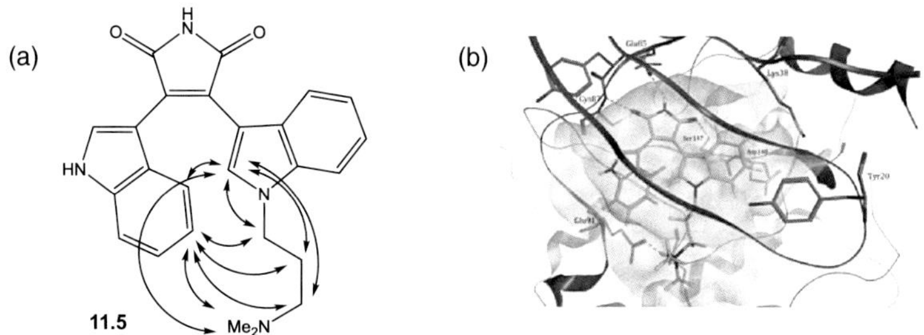

FIGURE 11.26 **Applying exchange transferred NOEs.** (a) NOEs observed for **11.5** only in the presence of the protein kinase Chk 1 and (b) the model for the bound ligand structure derived from transferred NOE data and molecular modeling. *(Source: Reproduced with permission from Ref. [38] copyright Elsevier).*

positive NOEs are observed, unsurprising for a molecule of mass 566 Da in a 700 MHz spectrometer. In clear contrast, strong negative NOEs are apparent in the presence of the receptor and stand testament to ligand binding.

The use of exchange-transferred NOE data in the definition of bound ligand conformations is illustrated in the case of the inhibitor **11.5** (known as GF109203X) binding to the protein kinase checkpoint 1 (Chk 1) [38] (Fig. 11.26). The protein is involved in the mammalian response to DNA damage and presents a therapeutic target for cancer treatment for which selective Chk 1 inhibitors are required. In studies of **11.5**, unique transferred NOEs observed only in the presence of the receptor protein were used to select binding conformations for the bound ligand from 10 structures that had been generated through molecular docking of the ligand into the protein crystal structure. Only one of the potential structures was compatible with NOE data and allowed a structure for the protein–ligand complex to be proposed, which was also compatible with the binding epitope data obtained from STD experiments.

Optimum sample conditions for the observation of transferred NOEs can be difficult to predict, and variation in ligand excess and/or receptor concentration or sample temperature may be required for clear evidence of the bound ligand state being present. The use of 1D NOESY experiments can be employed for sample optimisation as a quicker route to detecting transferred NOEs, but may have more limited use in conformational analysis due to their reduced information content. The quantitative estimation of interproton distances in the bound ligand conformation is possible through the study of time-dependent NOE buildup profiles using the isolated spin pair approximation (Section 9.4). However, caution is required here due to the possible interference of protein-mediated NOE transfer pathways occurring at the receptor surface and from contributions from positive free ligand NOEs that may perturb observed signal intensities. While interference from protein spin diffusion is thought not to be too problematic for proton-rich ligands [37], this may not always be the case and comparison of data from NOESY and ROESY experiments may be required to assess this; see discussions in [39]. The likelihood of such complications may be reduced by ensuring low levels of bound receptor exist, meaning reduced protein concentrations can be advantageous, although longer mixing times are then required to aid sensitivity. The relevant theory for quantitative studies has been extensively reviewed but is beyond further discussion here [33,34].

11.7 COMPETITION LIGAND SCREENING

Although the ligand observe methods described earlier find extensive use in ligand screening, they can suffer from limitations in some circumstances which may be addressed through the use of competition studies. It has been stated already that high-affinity ligands can escape detection due to their low off-rates, leading to false negatives in screening assays. It is also the case that the methods described so far provide no direct evidence of the site at which a ligand binds. Furthermore, complications can arise in the determination of ligand-binding affinities, meaning accurate determination of K_Ds may not be possible for all methods through direct observation. In situations when the ligand of interest has low solubility, methods that employ large ligand excess, including STD and water-LOGSY, may not be applicable and the use of competition experiments where a known binder of higher solubility is observed offers an alternative route to assess their binding.

We shall define two approaches to interrogate the binding of a molecule of interest based on the use of an established ligand whose site of binding and affinity to the target receptor is already known. The first is referred to as *competitive displacement*. In this the NMR-binding response of the novel ligand is perturbed by the presence of the known inhibitor molecule, indicating the ligands compete for the same binding site on the receptor (in the absence of allosteric effects). In the second approach it is the binding response of the known ligand that is monitored and perturbation of this in the presence of the novel ligand is indicative of competitive binding. We shall refer to this approach as *reporter ligand screening*, since the known ligand is used to report on the behaviour of the novel ligand(s) and requires only NMR observation of the known ligand, which can offer a number of advantages that will be outlined later.

In either of these approaches, any of the techniques described above are suitable for monitoring binding and hence the influence of a competing ligand. The presence of this competitive inhibitor ligand (I) will perturb the dynamic equilibrium between the observed ligand (L) and protein receptor (P) leading to a modification of Eq. 11.2 for single-site binding [3]:

$$\mathrm{PI} \underset{k_{\mathrm{off}}^{\mathrm{I}}}{\overset{k_{\mathrm{on}}^{\mathrm{I}}}{\rightleftharpoons}} \mathrm{P} + \mathrm{L} + \mathrm{I} \underset{k_{\mathrm{off}}^{\mathrm{L}}}{\overset{k_{\mathrm{on}}^{\mathrm{L}}}{\rightleftharpoons}} \mathrm{PL} \tag{11.14}$$

This leads to a reduced apparent binding affinity for the ligand L of $K_{\mathrm{D,app}}$:

$$K_{\mathrm{D,\,app}} = \frac{([\mathrm{P}]+[\mathrm{PI}])[\mathrm{L}]}{[\mathrm{PL}]} \tag{11.15}$$

which reflects the reduced availability of the receptor to the ligand L due to its partial occupancy by the competitive inhibitor I. Consequently, the fraction of the protein bound by the ligand in the presence of the inhibitor is given, by analogy with Eq. 11.5, as:

$$F_{\mathrm{B+I}}^{\mathrm{P}} = \frac{[\mathrm{L}]}{[\mathrm{L}] + K_{\mathrm{D,app}}} \tag{11.16}$$

where the apparent K_{D} equates to the free ligand concentration when half of the total protein receptor is occupied by the ligand and will be higher than the true ligand K_{D} because of the need to compete with the inhibitor for receptor sites. As will be described later, knowledge of $K_{\mathrm{D,app}}$ can be used to derive the K_{D} of one ligand when the binding affinity of the other is known, which is especially relevant to reporter ligand assays.

11.7.1 Competitive Displacement

The competitive displacement of a bound ligand by a second (inhibitor) molecule is the classic approach to determining site specificity of binding (Fig. 11.27). Once a ligand of interest has been shown to bind by any of the ligand observe NMR techniques presented above (or indeed by any biophysical method that reports on ligand binding), attenuation of the binding response in the presence of the inhibitor is interpreted as being indicative of the compounds being ligands for the same receptor site. One caveat to this is the possibility that the binding of the inhibitor may occur at a remote site to the ligand but through allosteric influences may lead to its displacement, leading to false interpretation of the observed displacement. The experiment requires that sufficient inhibitor be added to compete with and effectively 'knock off' the ligand from the receptor, and a relatively high-affinity inhibitor is therefore suitable to test site specificity, else a large excess of the inhibitor may be required for effective displacement. Appropriate control experiments will also be required, such as the addition of only the solvent in which the inhibitor was dissolved (most frequently DMSO) to rule out this as being responsible for any observed displacement through, for example, protein disruption.

Fig. 11.28 shows the binding of the glutathione analogue **11.6** to the bacterial potassium efflux protein Kef (which it is known to activate) as detected by CPMG-edited ^{1}H NMR. Titration to give increasing receptor concentrations against a fixed concentration of **11.6** leads to progressive reduction in resonance intensity, here illustrated by the *tert*-butyl group,

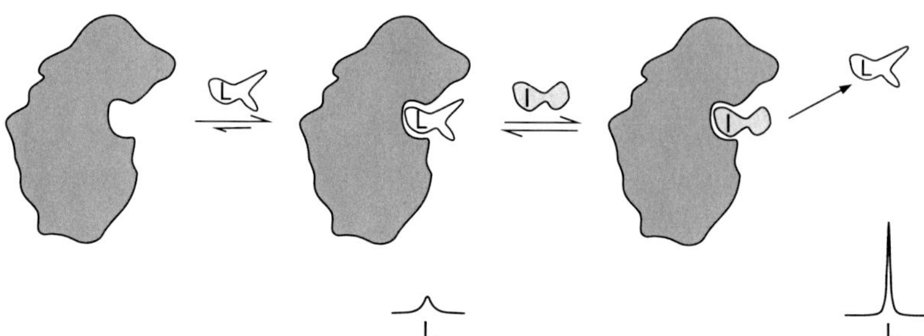

FIGURE 11.27 **Schematic representation of the competition displacement experiment.** The binding response of the ligand L is modulated by the presence of an inhibitor I if it competes for the same receptor binding site. Here the resonance lineshape reflects the environment of the observed ligand, although any suitable NMR response may be monitored.

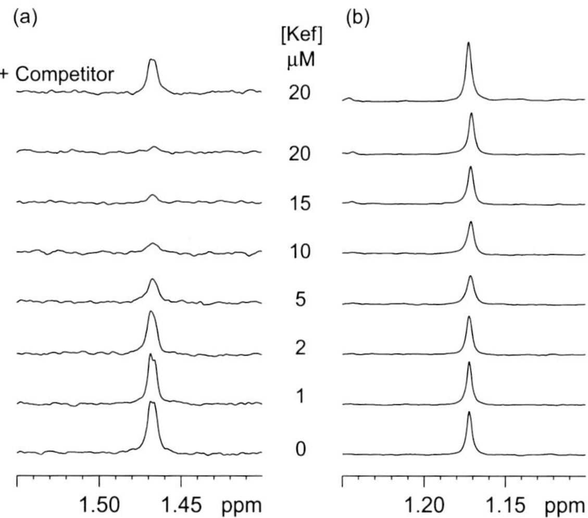

FIGURE 11.28 **Competitive displacement of the glutathione analogue 11.6 by 11.7 as observed by ^{1}H NMR.** (a) Titration of the protein Kef to ligand **11.6** at fixed concentration (10 μM), followed by addition of excess competitor (1 mM) and (b) response of the internal control *tert*-butanol. The *t*-butyl group only is shown, and an 88 ms CPMG filter was used to suppress the protein background for all spectra.

consistent with the ligand being in slow exchange with Kef. Addition of the known binder **11.7** causes a reversal in signal loss, indicating competitive displacement of the initially bound ligand **11.6**. In contrast, the resonance of the internal non-binding control *tert*-butanol shows no change in response throughout the titration.

11.6 **11.7**

Through titration of the inhibitor and monitoring of the progressive loss of binding response of the ligand it becomes possible to determine the K_D of the ligand provided that of the inhibitor is known. From the classic work of Cheng and Prusoff on enzyme kinetics [40], equivalent terms can be derived for competitive ligand binding such that:

$$K_D = \frac{[L]K_I}{K_{D,app} - K_I} \tag{11.17}$$

where K_I is the known binding affinity of the inhibitor. Through titration of the inhibitor against a fixed ligand and protein concentration an apparent binding K_D can be obtained ($K_{D,app}$) and from this the true ligand K_D determined. This assumes that the ligand is present in excess over the receptor such that $[L] \approx [L_T]$ (ie the total ligand concentration approximates the free ligand concentration); it is also then the case that $K_{D,app}$ may be replaced by IC_{50} (the absolute ligand concentration when the protein is 50% inhibited) which is likewise obtained by titration. The assumption of high ligand excess may not always be applicable and if not valid will lead to an overestimate of K_D. The STD experiment benefits from a large ligand excess and is an appropriate technique for this type of analysis for which the STD amplification factor is plotted against the titrated inhibitor concentration to yield the required IC_{50} [28].

11.7.2 Reporter Ligand Screening

Reporter ligand assays are a variation on competitive displacement in which the NMR response of an already known and characterised ligand is observed and its perturbation monitored in the presence of a novel ligand or even a pool of ligands

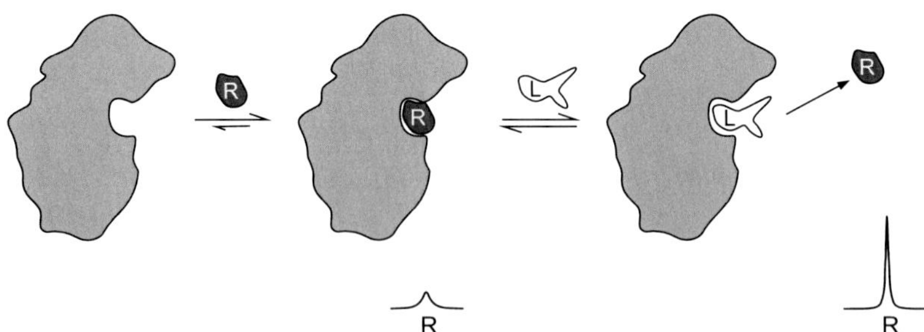

FIGURE 11.29 **Schematic representation of the reporter ligand–screening method.** The NMR binding response of the known reporter molecule R is influenced by the presence of the novel ligand L that competes for the same receptor binding site. Here the resonance lineshape reflects the environment of the observed reporter, although any suitable NMR response may be monitored; the ligand need not be observed as its binding is monitored indirectly through the reporter molecule.

whose binding with the receptor is to be assessed (Fig. 11.29). The ligand that is observed is herein referred to as the *reporter*, but has also been termed the *spy*, *reference*, *probe* or *indicator* molecule. This approach is especially effective for the identification of high-affinity ligands that give no measurable response in ligand observe screening methods since it is only their influence on the observed reporter that is monitored. Likewise, poorly soluble ligands can be studied provided they are able to compete effectively with the reporter at low concentrations, meaning they become amenable to assays requiring high ligand excess such as STD and water-LOGSY. Furthermore, ligands that are not favourable to NMR detection (eg they may be proton sparse) may also be studied indirectly in this manner.

Critical to the success of these assays is the selection of a suitable reporter molecule. It should have a clear NMR signal that can be monitored by the technique of choice and have moderate binding affinity (typically in the range 1–100 μM), so that it may be effectively displaced by competing ligands yet yield a strong binding response in their absence. Since reporters of lower affinity will be displaced by weaker competitors whereas higher affinity reporters will be most sensitive to relatively stronger binders, it may possible to tune the selectivity of screening experiments by appropriate selection of the reporter molecule.

Most of the primary ligand-screening methods have been demonstrated as being applicable to competition reporter assays including direct proton observation and the use of relaxation editing [41–43], STD [44] and water-LOGSY [45]. By monitoring the displacement of the reporter when the novel ligand is titrated into a fixed receptor and reporter concentration, this approach is also applicable to the determination of ligand affinities [46]. As described earlier, this will lead to an apparent dissociation constant for the ligand $K_{D,app}$ which, through knowledge of the dissociation constant of the reporter molecule $K_{D,rep}$, may be translated into the desired ligand K_D according to Eq. 11.7, again provided the reporter molecule is present in large excess [12]. However, when the reporter ligand is *not* in large excess over the receptor, the correct procedure for this translation must account for the true concentrations of all species in the solution, meaning the commonly employed Cheng–Prusoff equation (Eq. 11.17) is no longer applicable. In such cases, more appropriate correlations have been described [47,48], from which the K_D of the titrated ligand may be derived:

$$K_D = \frac{[K_{D,app}]}{\left[\left(\frac{[R_{50}]}{K_{D,rep}}\right) + \left(\frac{[P_0]}{K_{D,rep}}\right) + 1\right]} \tag{11.18}$$

where $[P_0]$ represents the free protein concentration in the presence of the reporter molecule only (ie prior to the ligand being titrated). This may be calculated from the standard solution to the binding equilibrium expressed in Eq. 11.6 since the reporter K_D, protein and ligand concentrations are all known quantities and $[P_0] = [P_T] - [PR]$ (here we use PR to represent the protein–reporter molecule complex in place of PL above). $[R_{50}]$ is the concentration of the *free* reporter when it is 50% displaced which may be calculated as:

$$[R_{50}] = [R_T] - [PR_{50}] = [R_T] - \frac{[P_T] - [P_0]}{2} \tag{11.19}$$

where $[PR_{50}]$ is the bound reporter concentration at 50% displacement, $[R_T]$ and $[P_T]$ are the total reporter and protein concentrations, respectively, and $[P_0]$ is as defined earlier. From these, Eq. 11.18 may be solved explicitly.

This approach is likely to be demanded when using relaxation-edited 1D spectra, for example, since large reporter ligand excesses are not typically employed. As an example, Fig. 11.30 shows the competitive displacement of a reporter

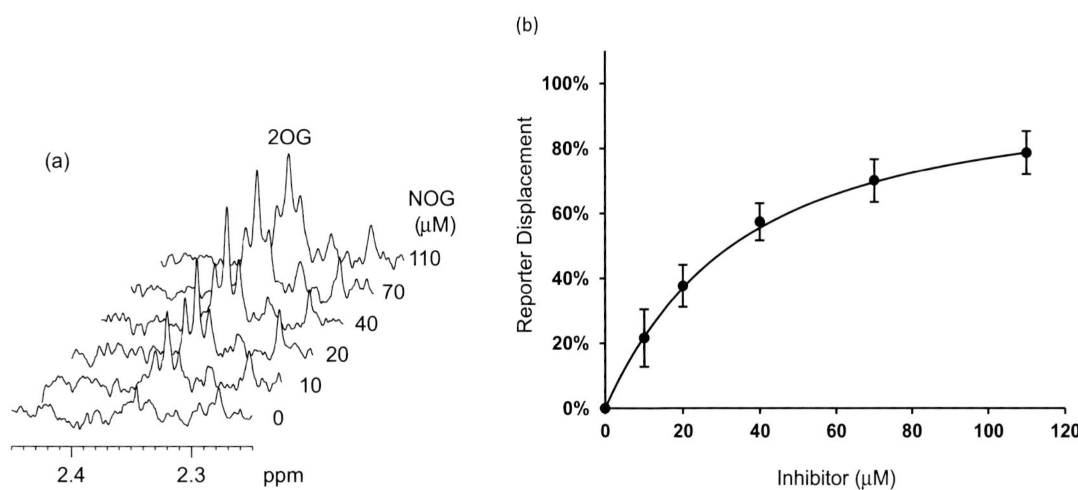

FIGURE 11.30 **Reporter ligand displacement for the measurement of ligand dissociation constants.** (a) Progressive displacement of the observed reporter molecule 2OG **11.8** as a function of the competitive ligand NOG **11.9**, and (b) data fitting yields an *apparent* K_D for the ligand that can be corrected to yield the true value. Data were recorded at 700 MHz for 10 μM 2OG with 10 μM protein using 48 ms CPMG filter for protein background suppression. Samples were prepared in triplicate and error bars reflect standard deviations.

ligand observed by ^{1}H NMR when using CPMG editing to suppress the protein background, together with the associated response curve. The reporter molecule is 2-oxoglutarate (2OG) **11.8**, a cosubstrate in the hydroxylation of proline residues by the Fe(II)-dependent oxygenase prolyl hydroxylase domain 2 (PHD2), and the competitive ligand is the catalytically inactive analogue *N*-oxalylglycine (NOG) **11.9**. For the assay, the enzyme was inactivated by replacement of Fe(II) by Zn(II). Direct displacement of the reporter yields an apparent dissociation constant of 28 μM for NOG, and from the known reporter affinity (0.9 μM) the actual ligand K_D was determined to be 2.6 μM [49].

11.8 **11.9**

One of the potential limitations to the reporter-screening method occurs when the resonances of the reporter are masked by other species in the solution, including the competing ligand(s), such that its displacement cannot be detected. In such cases the use of isotopically labelled reporters has been proposed and demonstrated [50]. Here, ^{13}C labels may be selected through proton-detected isotope-edited techniques such that only these are observed, to the limit of carbon-13 natural abundance, and the unwanted (unlabelled) species are effectively removed from the ^{1}H spectrum. The 1D gradient-selected HSQC is a suitable editing method, equating to the first row of a 2D HSQC where $t_1 = 0$. However, one needs to pay careful attention to the acquisition time if carbon-13 decoupling is employed, as is standard for the 2D experiment. Since 1D experiments tend to be acquired with longer free induction decay (FID) acquisition times, significantly more rf energy will be fed to the probe and sample, potentially leading to undesirable sample heating or, worse still, hardware damage. Thus, either the acquisition times need to remain short or carbon decoupling be removed completely, thus yielding $^1J_{CH}$ doublets in the ^{1}H spectrum. Other nuclei may also be observed, with ^{19}F being most beneficial, as explained in the following section.

11.7.3 ^{19}F *FAXS*

An alternative and now widely used 'labelling' scheme employs reporter molecules containing ^{19}F in the so-called *FAXS* (fluorine chemical shift anisotropy and exchange for screening) experiment [12,51]. Here direct ^{19}F observation is employed and since typically only the fluorinated reporter is observed its behaviour in the presence of a competing ligand or ligand library is readily apparent. Molecules containing CF or CF_3 moieties are most commonly employed, such fragments finding

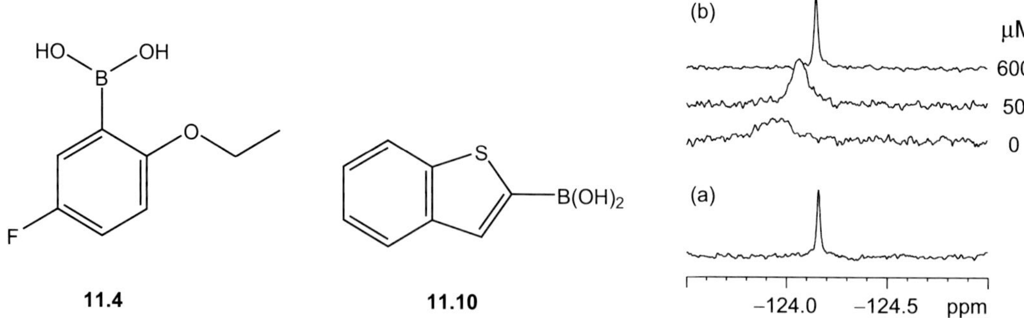

FIGURE 11.31 **^{19}F FAXS reporter screening.** This is demonstrated for the interaction of boronic acid **11.4** with α-chymotrypsin (200 μM). The ^{19}F reporter ligand resonance (200 μM) in (a) the absence and (b) the presence of the protein, and with addition of the competitive inhibitor **11.10** at the concentrations shown.

widespread use in pharmaceuticals and agrochemicals. The FAXS method benefits from the many advantages associated with ^{19}F observation, as outlined in Section 11.3.2 [12], and will often employ relaxation editing of the ^{19}F spectrum with a CPMG filter to enhance the binding response of the reporter. It is also good practice to include a second fluorinated molecule that does not bind to the receptor to act as an internal control reflecting solution conditions, so that, for example, significant viscosity changes associated with ligand additions are made apparent. Since the ligands under test need not themselves contain fluorine atoms, high-throughput screening of fragment libraries is possible under automated conditions [14] providing a robust and readily interpretable analytical method. Fig. 11.31 shows *FAXS* analysis using the fluorinated boronic acid **11.4** as the reporter for ligand binding to α-chymotrypsin. In the presence of the protein, substantial resonance broadening and peak shifting is observed relative to the free ligand. The binding of the competitive, non-fluorinated ligand benzo[*b*]thiophene-2-boronic acid **11.10** is clearly apparent due to its displacement of the reporter and the associated recovery of its response.

11.8 PROTEIN OBSERVE METHODS

While methods for the detection of receptor–ligand interactions have so far included only those employing ligand observation, it is also appropriate to consider approaches where the receptor itself is observed as these are also widely used. In most cases the receptor is a protein and the discussions below will focus on methods for assessing ligand binding with this class of macromolecule but will not address issues of protein resonance assignment or structure elucidation, topics which are well covered in other texts [52–54]. Protein observation offers a number of advantages over ligand observe methods, as described below, and requires largely standard NMR techniques. Ligand binding may induce chemical shift changes of the protein resonances through direct interaction with protein residues at the binding site, or through conformational changes that may influence the environment of residues remote from the binding site (allosteric effects). These experiments may provide direct evidence for the location of ligand binding on the receptor if its resonance assignments are known, information not provided by ligand observe techniques. Changes in chemical shifts of the receptor on ligand binding are known as *chemical shift perturbations*, and the mapping of these is now a well-established method for locating ligand-binding sites [55].

11.8.1 ^{1}H–^{15}N Mapping

The study of chemical shift perturbations using ^{15}N HSQC is the most common protein observe method, its primary limitation being the need for ^{15}N isotopically labelled protein which is usually produced biosynthetically from ^{15}N-enriched bacterial growth media [56]. Assuming uniformly ^{15}N-labelled protein is available, its ^{15}N HSQC spectrum will be expected to yield only a single backbone response from each amino acid residue (with the exception of proline), and thus often provides a well-dispersed 'fingerprint' of the protein structure. Fig. 11.32 shows an example from the N-terminal domain of the eukaryotic elongation release factor 1 protein (eRF1) which has a role in terminating protein translation at the ribosome. The protein consists of 142 amino acids and the well-dispersed crosspeaks are indicative of a well-structured protein fold (Fig. 11.33). In addition to the backbone resonances, peaks from a smaller number of sidechains will be observed, including most commonly the amide groups of asparagine and glutamine and the unsaturated rings of tryptophan or histidine. Ligand binding may induce shift changes to both ^{1}H and ^{15}N resonances as evidenced by a change in position of crosspeaks on addition of the ligand. These

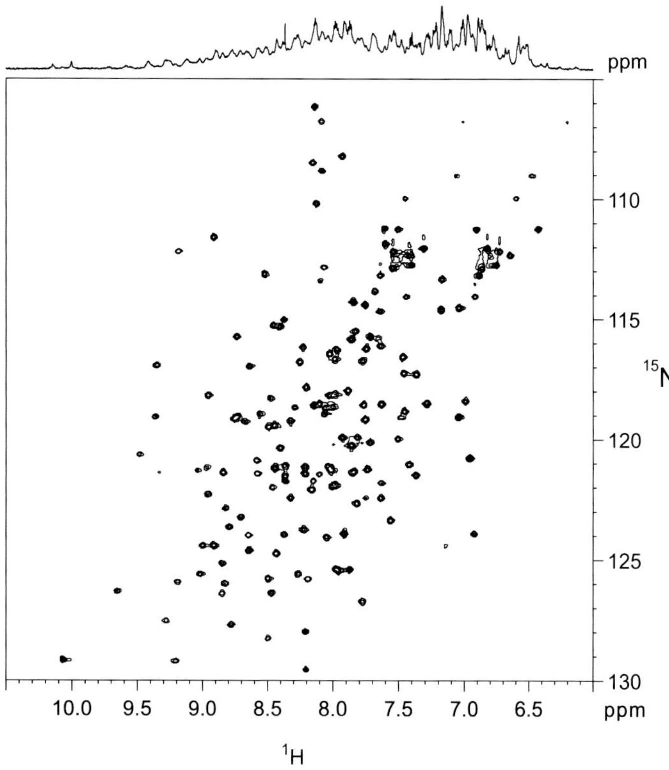

FIGURE 11.32 **The 700 MHz ^{1}H–^{15}N HSQC spectrum of the 15.5-kDa protein eRF1.** The sample was prepared at 300 μM in pH 6.8 Tris buffer, 150 mM NaCl, 0.7 mM 2-mercaptoethanol, 90% H_2O:10% D_2O, 298 K.

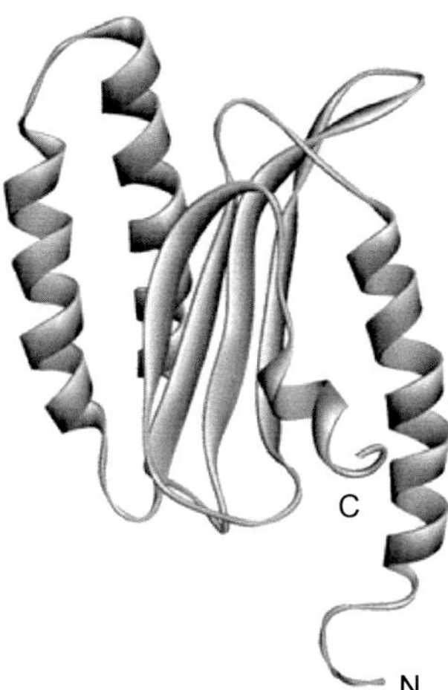

FIGURE 11.33 **Solid ribbon representation of the NMR-derived backbone fold of eRF1 demonstrating the presence of well-defined helical and β-sheet regions, linked by less well-defined flexible loops.** N and C mark the N-terminus and C-terminus of the protein, respectively. *(Source: Reproduced with permission from the BioMagResBank [59] from data reported in reference [60].)*

changes may be observed for strong and weak binders and for both fast- and slow-exchange scenarios, meaning HSQC screening is well suited to the study of high-affinity binders for which ligand observe techniques may give no detectable response. In situations where protein resonance assignments are known, shift perturbation can also be used to localise the ligand-binding site which can be mapped onto the 3D structure of the protein (which may be derived from NMR studies or more commonly will be a crystal structure). The assignment of protein spectra nowadays follows well-established protocols although the techniques employed fall outside the scope of this text and the interested reader is referred elsewhere [52–54,57,58]. Table 11.1 provides a guide as to the likely requirements for the assignment of proteins of differing molecular masses.

TABLE 11.1 Summary of Isotope Labelling Schemes Typically Required for Protein Resonance Assignment Providing an Approximate Guide to the Feasibility of Assigning Proteins of Varying Sizes

Protein Mass Range (kDa)	Labelling Required [56]	Nuclei Assigned	Methods Employed	References
≤8	None	^{1}H	Homonuclear ^{1}H 2D experiments sufficient	
≤15	^{15}N	^{1}H and ^{15}N	^{15}N-edited ^{1}H 3D experiments	[53]
≤25	^{15}N and ^{13}C	^{1}H, ^{15}N and ^{13}C	^{1}H, ^{15}N and ^{13}C–correlated 3D experiments	[53]
≤40	^{15}N, ^{13}C and ^{2}H	^{1}H, ^{15}N and ^{13}C	^{1}H, ^{15}N and ^{13}C–correlated 3D experiments (^{2}H labelling attenuates relaxation losses present in protonated macromolecules)	[61]
≥40	^{13}C-methyl, ^{2}H	Selected ^{1}H and ^{13}C	^{1}H–^{13}C 2D correlations for selective ^{13}C-labelled methyl groups (^{2}H labelling or all other positions attenuate relaxation losses present in protonated macromolecules)	[62]

Even in the absence of any protein assignments, the HSQC experiments can provide a qualitative indication of binding as well as yielding ligand dissociation constants from titration data. The ligand in question is most commonly a small molecule, but could be a metal ion or even a second (unlabeled) protein, thus also providing insights into the nature of protein–protein interactions. The apparent changes observed on ligand binding are dependent on whether ligand exchange is in the fast- or slow-exchange regimes. In situations of a rapidly exchanging ligand, the protein resonance positions will be a population-weighted average of its free and ligand-complexed forms that will move progressively through a titration in which the ligand concentration is increased (Fig. 11.34a). Absolute shift changes of up to 0.5 ppm for ^{1}H and 3 ppm for ^{15}N are most commonly observed, but it is standard practice to report shift variation as a scaled geometric average of the ^{1}H and ^{15}N changes $\Delta\delta H$ and $\Delta\delta N$ according to:

$$\Delta\delta = \sqrt[2]{\tfrac{1}{2}\left[(\Delta\delta \mathrm{H})^2 + (0.14\Delta\delta \mathrm{N})^2\right]} \tag{11.20}$$

so that the total contribution from the two nuclei is comparable. The scaling factor of 0.14 (1/7) for ^{15}N has no theoretical justification but represents a ratio of the total chemical shift ranges occupied by protein amide resonances (~22 ppm for

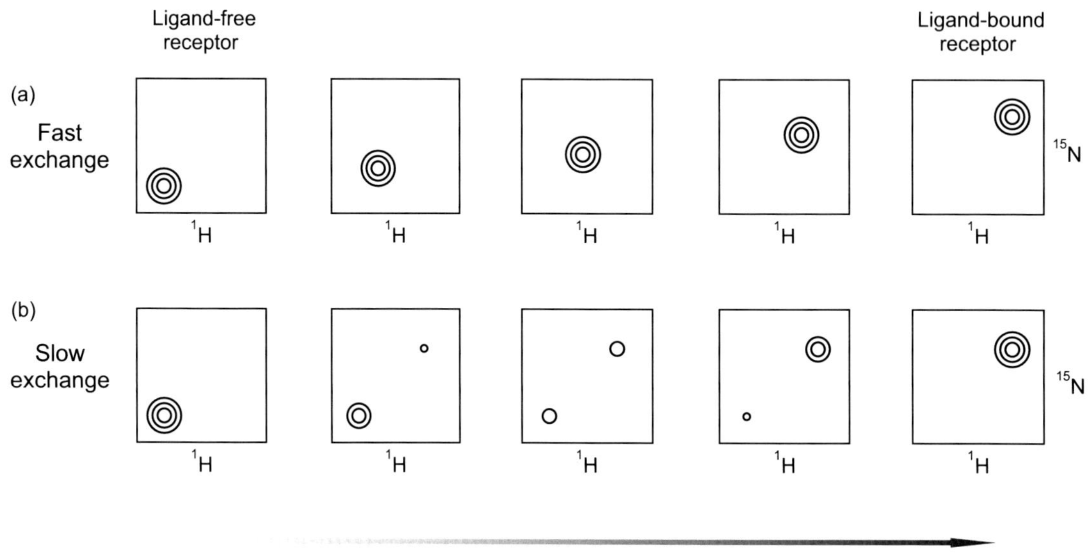

FIGURE 11.34 **Schematic representation of the protein ^{1}H–^{15}N crosspeak behaviour in ligand titration experiments.** In the fast-exchange regime (a) a progressive shift of the averaged peak is observed; when rates tend towards intermediate some broadening of the peak around the titration midpoint may also be seen. In the slow-exchange regime (b) a loss of the free receptor peak is observed with corresponding growth of that from the bound receptor.

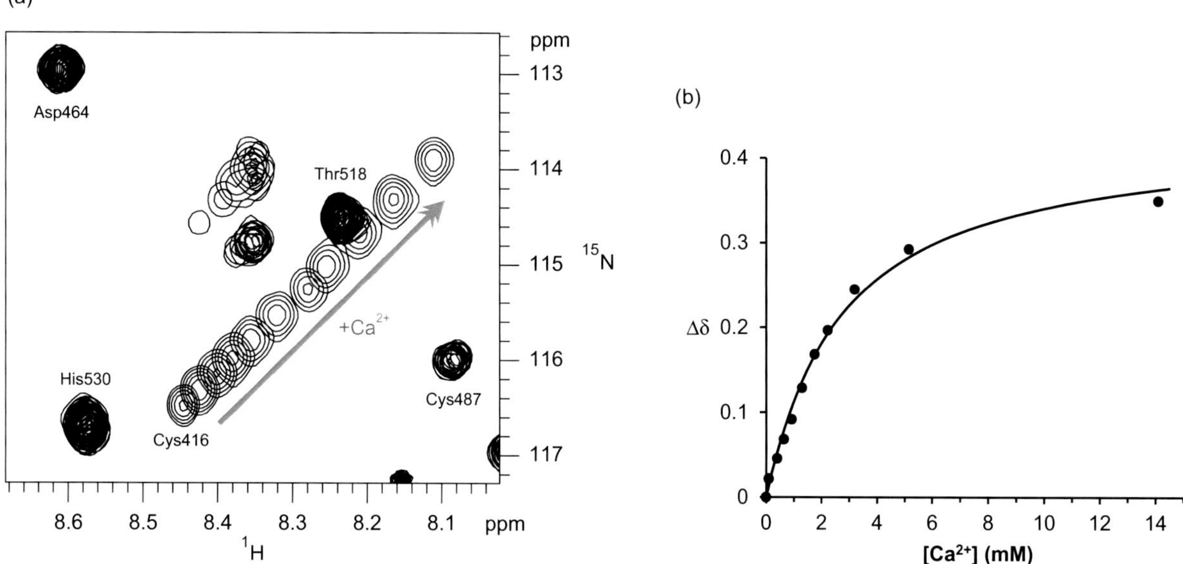

FIGURE 11.35 **HSQC titration data.** (a) The overlay of a selected region of 11 ^{15}N HSQC spectra used to determine the calcium-binding affinity of an EGF protein domain (600 MHz, 300 μM protein). (b) The K_D can be determined by monitoring the averaged chemical shift change Δδ as a function of added calcium chloride (K_D = 2.6 mM). *(Source: Data provided by courtesy of Prof. Christina Redfield, Department of Biochemistry, University of Oxford).*

^{15}N vs. ~3.1 ppm for ^{1}H) [63]. Note that differing (yet largely similar) values for this scaling factor have been employed based on alternative justifications, although 0.14 appears reasonable; in any case consistency in one's methodology is most critical. As for any titration experiment, the influence of other solvent factors must also be considered when interpreting resonance changes. It is common practice to utilise stock ligand solutions in DMSO, for example, and the influence of additions of DMSO alone also need to be assessed and accounted for where appropriate. Binding affinities may be determined for titration data following similar procedures to those described in Section 11.2 using non-linear fitting. In such titrations multiple peaks should ideally be monitored and their shift dependencies fitted to better assess the accuracy of the measured dissociation constant. Fig. 11.35 illustrates the HSQC titration profile for the binding of calcium ions to an epidermal growth factor (EGF)-like domain of a 15-kDa protein. The progressive response of Cys416 on addition of the metal indicates the fast-exchange behavior of the calcium ion, which may be fitted to determine its binding affinity (Fig. 11.35b). Other residues are clearly unaffected by calcium binding and their shifts remain invariant throughout the titration.

In the case of slow on–off ligand exchange, separate resonances will be observed for the free and complexed forms and their intensities will change through a titration, directly reflecting the population of the two states (Fig. 11.34b). In this scenario, K_D affinities may be determined directly from the relative peak intensities provided the protein concentration is known accurately. One complication in the case of slow exchange is the potential difficulty in assigning the new peaks associated with the bound state, since one cannot know *a priori* where these are likely to appear. Complete reassignment of the protein resonances is possible but presents a significant burden, so in most cases the assumption of *minimal shift changes* is made and the closest crosspeak to that newly formed is taken to be the parent. Clearly, such an approach becomes more prone to error in more crowded regions of the spectrum demanding caution, but this does at least place a lower limit on peak shift changes and avoids overestimation of these.

11.8.1.1 Practical Aspects

The nitrogen-15 correlation experiment relies upon the detection of backbone amide protons which in turn requires the use of protonated water as solvent so that these are not lost by exchange with deuterium, as would occur when using D_2O as solvent. In practice, it is thus common to use only 5–10% D_2O to provide the spectrometer lock, with the remainder a suitably buffered H_2O solution. Experimentally, this then demands suppression of the intense solvent resonance to be included in the HSQC sequence. Direct water saturation methods are not appropriate here as these could again lead to amide signal loss through exchange (ie via saturation transfer), and while in principle the use of gradients for coherence selection should attenuate the water resonance, it proves beneficial to add additional steps to minimize the contribution from water. One of the more popular ways to achieve this is to selectively return the water to the transverse plane so that it may be destroyed by an accompanying gradient pulse while all required proton magnetisation is longitudinal; that is along the ±*z*-axis and thus insensitive to the *z*

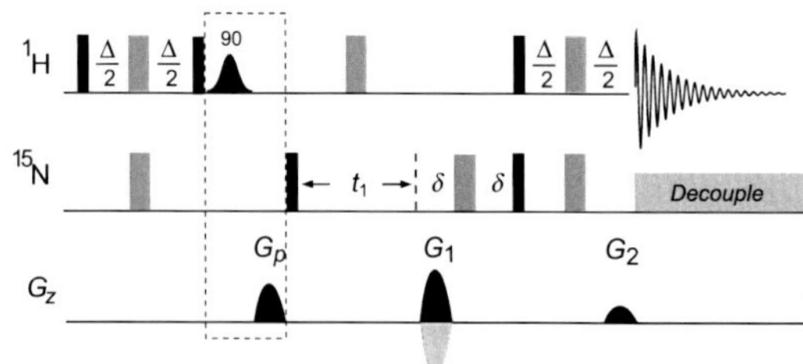

FIGURE 11.36 **^{1}H–^{15}N HSQC sequence (Section 7.3.1) suitable for proteins in ~90% H_2O buffer solutions.** The initial INEPT transfer step incorporates a water-selective 90 degree flip-back pulse to return only the water magnetisation to the transverse plane for destruction by the following purge gradient G_p (boxed). Gradients G_1 and G_2 select for ^{15}N coherences and have the ratio γ_H/γ_N (10:1). The period $\Delta = 1/2{}^1J_{NH}$. The sequence can optionally be extended using the PEP scheme (Section 7.3.1) for enhanced sensitivity.

gradient. A suitable scheme for this is illustrated in Fig. 11.36, which combines a so-called water 'flip-back' pulse [64] as part of the initial INEPT transfer followed by a strong purge gradient G_p to attenuate the water to an acceptable level. The Δ delays are set equal to $1/2{}^1J_{NH}$ and for backbone amide protons $^1J_{NH} = 90$ Hz.

The use of high-field spectrometers with probes optimised for proton detection means protein concentrations of a few hundred micromolar enable the collection of ^{15}N HSQC spectra within about 30 min, and the use of cryoprobes reduces this further, meaning titration by HSQC provides a realistic route to quantify ligand-binding affinities (Fig. 11.35). Similarly, the use of automated probe tuning and the high quality of magnetic field optimisation achievable through gradient shim routines (in this case efficient ^{1}H shim mapping can be used) means that these methods can be run under full automation, further enhancing their utility for ligand screening. Typically, protein and ligand concentrations will be similar or the ligand will be in excess for screening applications. For the determination of accurate binding affinities, it has been demonstrated that for 1:1 complexes an optimal titration will use fixed protein concentrations $[P_T]$ of $\sim 0.5K_D$ and ligand concentrations ranging from 0.5 to 10 times that of the protein [65], so an initial estimate of K_D is desirable, although clearly these optimal conditions may not always be achievable for high-affinity ligands. The term *SAR by NMR* (structure–activity relationships by NMR) has been coined for the use of ^{15}N HSQC to guide high-affinity ligand development [66,67] and is nowadays used more informally with reference to HSQC-based screening in general.

11.8.2 ^{1}H–^{13}C Mapping

Here we consider briefly the use of ^{1}H–^{13}C correlation experiments to detect ligand binding, arguably a more obvious option for those from an organic chemistry background. In principle, this approach is as valid as the use of ^{1}H–^{15}N experiments but in practice tends not to be as useful. Aside from technical issues of producing biosynthetically ^{13}C-labelled proteins and the associated expense, it is the case that ^{1}H–^{13}C 2D spectra tend to be less well resolved than their ^{15}N counterparts and that ^{13}C chemical shift changes on ligand binding tend to be rather smaller than those of ^{15}N. Furthermore, many potentially informative ^{1}H–^{13}C resonances may be hidden under the water resonance, Hα resonances in particular (although D_2O solutions become feasible in this case). Thus, ^{15}N correlations remain by far the favoured protein observe method for protein–ligand binding studies. Nevertheless, ^{1}H–^{13}C correlations of ^{13}C-methyl–labelled proteins have been demonstrated for ligand screening [68] and may be more appropriate than ^{15}N methods for protein targets with masses greater than ~40 kDa. They also find use (when employed with simultaneous protein perdeuteration) in the structural studies of very large proteins for which ^{15}N correlation spectra prove uninformative, often due to excessive crowding [61,62]. Here, selective ^{13}C-methyl–labelling strategies based on biosynthetic incorporation have been successfully employed for sidechains belonging to valine, leucine and isoleucine residues, producing relatively simple spectra where only these methyl group correlations are observed [62,69]. The use of ^{13}C correlation spectra at *natural* carbon abundance for the detection of ligand binding has also been exemplified for the interaction of tryptophan and BSA [70]. Although this requires the use of a cryoprobe at 700 MHz and rather high protein concentrations (~1 mM) it does serve to illustrate the potential for quantifying ligand binding with modern instrumentation without the need for isotopic labelling of the macromolecule.

11.8.3 ^{19}F Mapping

In this final section on methods utilising protein observation we consider a more focused labelling scheme that is gaining in popularity: the use of fluorine labelling. Provided the fluorine label can be located in a suitable position the sensitivity of the

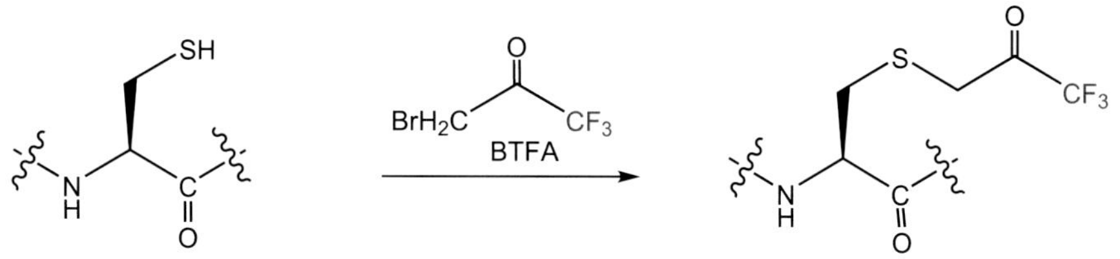

FIGURE 11.37 **Protein cysteine fluorine labelling using BTFA.**

^{19}F nucleus to its surrounding environment make it a useful tool to report on ligand binding. Typically, this requires a crystal structure of the protein to be available so that a suitable site for the label can be identified and an appropriate labelling scheme employed. This may be achieved either by biosynthetic incorporation of a fluorinated amino acid into the protein or by chemical modification of either the native protein or a modified variant generated by site-directed mutagenesis. Biosynthetic incorporation has most often utilised fluorinated phenylalanine, although aliphatic amino acids in which CH_3 has been replaced by CF_3 have also found use [71]. Chemical modification comes in myriad forms most of which require sulfhydryl or amino groups for attachment of the label [72]. Most often this is the CF_3 group as this provides a simple, intense resonance often devoid of significant proton–fluorine couplings. One such example is in the labelling of a cysteine residue with 3-bromo-1,1,1-trifluoroacetone (BTFA; Fig. 11.37).

The modified cysteine residue(s) may be native to the protein (provided it is not involved in a disulphide bond) or may be introduced by mutagenesis, through which it may be placed in the optimum position for reporting on ligand binding. The residue need also be sufficiently exposed for the modifying reagent to gain access. Effective modification is most readily monitored through mass spectrometry and any activity of the modified protein should also be checked to ensure this matches that of the unmodified (wild-type) protein.

The fluorine label offers a number of specific advantages relative to other schemes including simplicity, lack of background resonances, no interference from water, high sensitivity to environmental change and, in favourable situations, site specificity. This approach is also applicable to very large proteins and is well suited to the measurement of ligand-binding affinities by direct titration. The high detection sensitivity of the fluorine nucleus can also be enhanced by the use of cryogenic probes suitable for fluorine observation which are becoming increasingly available. The exquisite sensitivity of the fluorine chemical shift may also be employed to provide an indication of the solvent exposure of the fluorine label [71,73]. A change in the solvent from 100% H_2O to 100% D_2O leads to a deshielding shift of a solvent-exposed ^{19}F resonance of approximately 0.2 ppm, so a measure of this *solvent-induced isotope shift* can report on the label's accessibility to solvent.

Fig. 11.38 shows the binding of the pharmaceutical D-captopril **11.11** to the zinc-dependent New Delhi metallo-β-lactamase (NDM1) labelled with a single CF_3 group [74]. Labelling was achieved by chemical modification by BTFA of a single cysteine following its introduction through site-directed mutagenesis of a methionine, producing the M67C mutant. Titration of the ligand, which coordinates directly through its sulphur to an active site zinc, leads to a progressive change in the ^{19}F chemical shift which may be fitted to yield a binding affinity for the ligand of 6 μM.

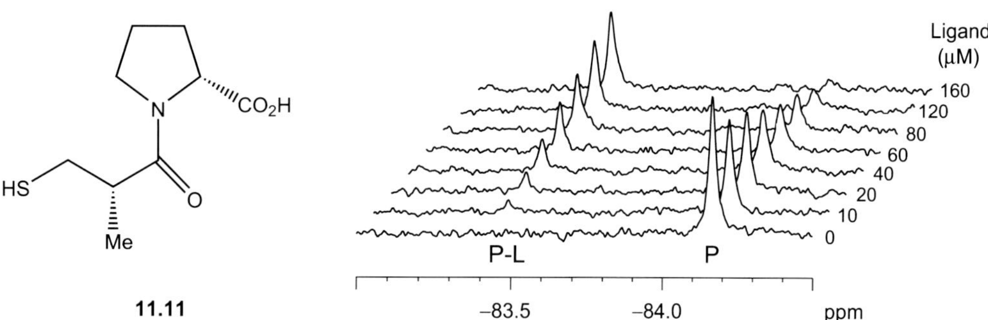

FIGURE 11.38 **Monitoring ligand binding through ^{19}F NMR of a protein site specifically labelled with a CF_3 group.** Resonance P is from NDM1 protein and P–L the bound D-captopril/NDM1 complex.

REFERENCES

[1] Meyer B, Peters T. Angew Chem Int Ed 2003;42:864–90.
[2] Lepre CA, Moore JM, Peng JW. Chem Rev 2004;104:3641–76.
[3] Peng JW, Moore J, Abdul-Manan N. Prog Nucl Magn Reson Spectrosc 2004;44:225–56.
[4] Cala O, Guillière F, Krimm I. Anal Bioanal Chem 2014;406:943–56.
[5] Pellecchia M, Bertini I, Cowburn D, Dalvit C, Giralt E, Jahnke W, James TL, Homans SW, Kessler H, Luchinat C, Meyer B, Oschkinat H, Peng J, Schwalbe H, Siegal G. Nat Rev Drug Discov 2008;7:738–45.
[6] Dalvit C. Drug Discov Today 2009;14:1051–7.
[7] Lucas LH, Larive CL. Concept Magn Reson A 2004;20A:24–41.
[8] Fielding L. Prog Nucl Magn Reson Spectrosc 2007;51:219–42.
[9] Fielding L. Curr Top Med Chem 2003;3:39–53.
[10] Kronis KA, Carver JP. Biochemistry 1982;21:3050–7.
[11] Hajduk PJ, Olejniczak ET, Fesik SW. J Am Chem Soc 1997;119:12257–61.
[12] Dalvit C. Prog Nucl Magn Reson Spectrosc 2007;51:243–71.
[13] Vulpetti A, Hommel U, Landrum G, Lewis R, Dalvit C. J Am Chem Soc 2009;131:12949–59.
[14] Jordan JB, Poppe L, Xia X, Cheng AC, Sun Y, Michelsen K, Eastwood H, Schnier PD, Nixey T, Zhong W. J Med Chem 2011;55:678–87.
[15] Jahnke W. ChemBioChem 2002;3:167–73.
[16] Jahnke W, Rudisser S, Zurini M. J Am Chem Soc 2001;123:3149–50.
[17] Bertini I, Fragai M, Lee Y-M, Luchinat C, Terni B. Angew Chem Int Ed 2004;43:2254–6.
[18] Burton DR, Forsen S, Karlstrom G, Dwek RA. Prog Nucl Magn Reson Spectrosc 1979;13:1–45.
[19] Sherry AD, Birnbaum ER, Darnall DW. Anal Biochem 1973;52:415–20.
[20] Bertini I, Fragai M, Luchinat C, Talluri E. Angew Chem Int Ed 2008;47:4533–7.
[21] Leung IKH, Flashman E, Yeoh KK, Schofield CJ, Claridge TDW. J Med Chem 2010;53:867–75.
[22] Mayer M, Meyer B. Angew Chem Int Ed 1999;38:1784–8.
[23] Yan J, Kline AD, Mo H, Shapiro MJ, Zartler ER. J Magn Reson 2003;163:270–6.
[24] Mizukoshi Y, Abe A, Takizawa T, Hanzawa H, Fukunishi Y, Shimada I, Takahashi H. Angew Chem Int Ed 2012;51:1362–5.
[25] Krishna NR, Jayalakshmi V. Prog Nucl Magn Reson Spectrosc 2006;49:1–25.
[26] Angulo J, Nieto PM. Eur Biophys J 2011;40:1357–69.
[27] Kemper S, Patel MK, Errey JC, Davis BG, Jones JA, Claridge TDW. J Magn Reson 2010;203:1–10.
[28] Mayer M, Meyer B. J Am Chem Soc 2001;123:6108–17.
[29] Angulo J, Enríquez-Navas PM, Nieto PM. Chem Eur J 2010;16:7803–12.
[30] Dalvit C, Fogliatto G, Stewart A, Veronesi M, Stockman B. J Biomol NMR 2001;21:349–59.
[31] Gossert A, Henry C, Blommers MJ, Jahnke W, Fernández C. J Biomol NMR 2009;43:211–7.
[32] Fielding L, Rutherford S, Fletcher D. Magn Reson Chem 2005;43:463–70.
[33] Campbell AP, Sykes BD. Annu Rev Bioph Biom 1993;22:99–122.
[34] Ni F. Prog Nucl Magn Reson Spectrosc 1994;26:517–606.
[35] Albrand JP, Birdsall B, Feeney J, Roberts GCK, Burgen ASV. Int J Biol Macromol 1979;1:37–41.
[36] Gronenborn AM, Clore GM. Biochem Pharmacol 1990;40:115–9.
[37] Post CB. Curr Opin Struct Biol 2003;13:581–8.
[38] Lancelot N, Piotto M, Theret I, Lesur B, Hennig P. J Pharm Biomed Anal 2014;93:125–35.
[39] Gizachew D, Dratz E. Chem Biol Drug Design 2011;78:14–24.
[40] Cheng Y, Prusoff WH. Biochem Pharmacol 1973;22:3099–108.
[41] Jahnke W, Floersheim P, Ostermeier C, Zhang X, Hemmig R, Hurth K, Uzunov DP. Angew Chem Int Ed 2002;41:3420–3.
[42] Siriwardena AH, Tian F, Noble S, Prestegard JH. Angew Chem Int Ed 2002;41:3454–7.
[43] Dalvit C, Flocco M, Knapp S, Mostardini M, Perego R, Stockman BJ, Veronesi M, Varasi M. J Am Chem Soc 2002;124:7702–9.
[44] Wang Y-S, Liu D, Wyss DF. Magn Reson Chem 2004;42:485–9.
[45] Dalvit C, Fasolini M, Flocco M, Knapp S, Pevarello P, Veronesi M. J Med Chem 2002;45:2610–4.
[46] Dalvit C. Concept Magn Reson A 2008;32A:341–72.
[47] Nikolovska-Coleska Z, Wang R, Fang X, Pan H, Tomita Y, Li P, Roller PP, Krajewski K, Saito NG, Stuckey JA, Wang S. Anal Biochem 2004;332: 261–73.
[48] Cer RZ, Mudunuri U, Stephens R, Lebeda FJ. Nucleic Acids Res 2009;37:W441–5.
[49] Leung IKH, Demetriades M, Hardy AP, Lejeune C, Smart TJ, Szöllössi A, Kawamura A, Schofield CJ, Claridge TDW. J Med Chem 2013;56: 547–55.
[50] Swann SL, Song D, Sun C, Hajduk PJ, Petros AM. Med Chem Lett 2010;1:295–9.
[51] Dalvit C, Flocco M, Veronesi M, Stockman BJ. Com Chem High T Scr 2002;5:605–11.
[52] Zerbe O. BioNMR in drug research, 16. Weinheim: Wiley-VCH; 2003.
[53] Cavanagh J, Fairbrother WJ, Palmer AG, Skelton NJ, Rance M. Protein NMR spectroscopy: principles and practice. 2nd ed. San Diego: Academic Press (Elsevier); 2006.

[54] Roberts GCK, Lian L-Y. Protein NMR spectroscopy: practical techniques and applications. Weinheim: Wiley; 2011.
[55] Williamson MP. Prog Nucl Magn Reson Spectrosc 2013;73:1–16.
[56] Ohki S-Y, Kainosho M. Prog Nucl Magn Reson Spectrosc 2008;53:208–26.
[57] Kwan AH, Mobli M, Gooley PR, King GF, Mackay JP. FEBS J 2011;278:687–703.
[58] Bieri M, Kwan AH, Mobli M, King GF, Mackay JP, Gooley PR. FEBS J 2011;278:704–15.
[59] Ulrich EL, Akutsu H, Doreleijers JF, Harano Y, Ioannidis YE, Lin J, Livny M, Mading S, Maziuk D, Miller Z, Nakatani E, Schulte CF, Tolmie DE, Kent Wenger R, Yao H, Markley JL. Nucleic Acids Res 2008;36:D402–8.
[60] Polshakov VI, Eliseev BD, Birdsall B, Frolova LY. Protein Sci 2012;21:896–903.
[61] Kay LE, Gardner KH. Curr Opin Struct Biol 1997;7:722–31.
[62] Frueh DP. Prog Nucl Magn Reson Spectrosc 2014;78:47–75.
[63] Williamson RA, Carr MD, Frenkiel TA, Feeney J, Freedman RB. Biochemistry 1997;36:13882–9.
[64] Grzesiek S, Bax A. J Am Chem Soc 1993;115:12593–4.
[65] Granot J. J Magn Reson 1983;55:216–24.
[66] Shuker SB, Hajduk PJ, Meadows RP, Fesik SW. Science 1996;274:1531–4.
[67] Hajduk PJ, Meadows RP, Fesik SW. Science 1997;278:497–9.
[68] Hajduk PJ, Augeri DJ, Mack J, Mendoza R, Yang J, Betz SF, Fesik SW. J Am Chem Soc 2000;122:7898–904.
[69] Goto NK, Kay LE. Curr Opin Struct Biol 2000;10:585–92.
[70] Quinternet M, Starck J-P, Delsuc M-A, Kieffer B. Chem Eur J 2012;18:3969–74.
[71] Gerig JT. Prog Nucl Magn Reson Spectrosc 1994;26:293–370.
[72] Kitevski-LeBlanc JL, Prosser RS. Prog Nucl Magn Reson Spectrosc 2012;62:1–33.
[73] Hansen PE, Dettman HD, Sykes BD. J Magn Reson 1985;62:487–96.
[74] Rydzik AM, Brem J, van Berkel SS, Pfeffer I, Makena A, Claridge TDW, Schofield CJ. Angew Chem Int Ed 2014;53:3129–33.

Chapter 12

Experimental Methods

Chapter Outline

12.1 Composite Pulses **457**
12.1.1 A Myriad of Pulses 459
12.1.2 Inversion Versus Refocusing 460
12.2 Adiabatic and Broadband Pulses **461**
12.2.1 Common Adiabatic Pulses 462
12.2.2 Broadband Inversion Pulses: BIPs 464
12.3 Broadband Decoupling and Spin Locking **465**
12.3.1 Broadband Adiabatic Decoupling 467
12.3.2 Spin Locking 468
12.4 Selective Excitation and Soft Pulses **468**
12.4.1 Shaped Soft Pulses 469
12.4.2 Excitation Sculpting 473
12.4.3 Chemical Shift Selective Filters 475
12.4.4 DANTE Sequences 477
12.4.5 Practical Considerations 478
12.5 Solvent Suppression **480**
12.5.1 Presaturation 480
12.5.2 Zero Excitation 482
12.5.3 Pulsed Field Gradients 483
12.6 Suppression of Zero-Quantum Coherences **486**
12.6.1 The Variable-Delay Z-Filter 486
12.6.2 Zero-Quantum Dephasing 487
12.7 Heterogeneous Samples and Magic Angle Spinning **489**
12.8 Hyperpolarisation **491**
12.8.1 *Para*-Hydrogen–Induced Polarisation 491
12.8.2 Dynamic Nuclear Polarisation 493
References **496**

This chapter considers a collection of experimental methods that find widespread use in modern high-resolution NMR yet cannot be considered as individual techniques in their own right. Rather they are sequence segments that are used within, or may be added to, the techniques already encountered to enhance their information content or to overcome a range of experimental limitations; they are the components used to construct modern NMR experiments. Depending on context, they may be considered as essential to the correct execution of the desired experiment or viewed as an optional extra to enhance performance. For example the broadband decoupling of protons during carbon acquisition routinely makes use of so-called 'composite pulse decoupling' schemes to achieve efficient removal of all proton couplings, and is nowadays considered essential. Pulsed field gradients (PFGs) could equally well have been included in this chapter since in many cases they serve as an alternative means of signal selection to traditional phase-cycling procedures. The fact that PFGs have been described earlier lays testament to the manner in which these pervade modern NMR sequences and are now considered routine for many experiments. The chapter concludes by considering some experimental methods which largely find use in specific areas of application. These include the use of magic-angle sample spinning for heterogeneous samples, and techniques for boosting sensitivity through hyperpolarisation.

12.1 COMPOSITE PULSES

The plethora of NMR multipulse sequences used throughout modern chemical research all depend critically on nuclear spins experiencing rf pulses of precise flip angles for their successful execution. Careful pulse width calibration is essential in this context yet despite this a number of factors invariably conspire against the experimentalist and produce pulses that deviate from these ideals, so leading to degraded experimental performance. There are two notable contributions to pulse imperfections that can be addressed experimentally, namely rf (B_1) inhomogeneity and off-resonance effects.

Inhomogeneity in the applied rf field means not all nuclei within the sample volume experience the desired pulse flip angle (Fig. 12.1b), notably those at the sample periphery. This is similar in effect to the (localised) poor calibration of pulse widths and references to 'rf (or B_1) inhomogeneity' below could equally read 'pulse width miscalibration'. Modifications that make sequences more tolerant of rf inhomogeneity therefore also provide the experimentalist with some leeway when

High-Resolution NMR Techniques in Organic Chemistry. http://dx.doi.org/10.1016/B978-0-08-099986-9.00012-9

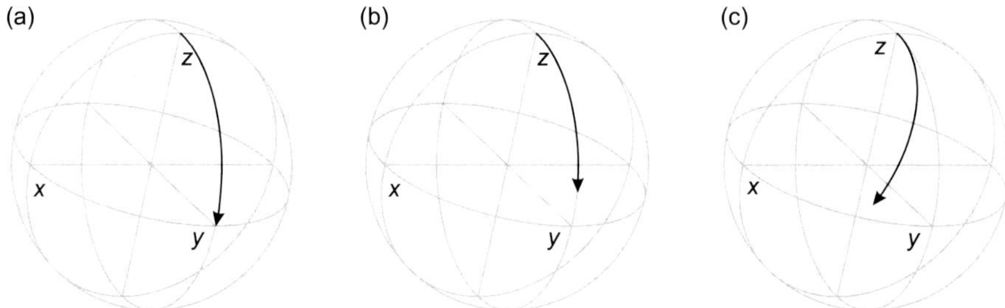

FIGURE 12.1 **Imperfections in pulse excitation.** (a) A perfect $90°_x$ pulse applied at equilibrium, (b) the effect of B_1 inhomogeneity and (c) the effect of off-resonance excitation.

setting up pulse experiments. Off-resonance excitation arises when the transmitter frequency does not exactly match the Larmor frequency of the spin, and has already been introduced in Section 3.2. To recap briefly, off-resonance magnetisation vectors experience an effective rf field in the rotating frame that is tilted out of the x–y plane, rather than the applied B_1 field itself (Fig. 12.2). These vectors are driven about this effective field and thus do not follow the trajectory of the ideal on-resonance case (Fig. 12.1c). The problem stems from the fact that pulse NMR instruments typically use monochromatic rf radiation when polychromatic excitation is required.

Combating the effects of B_1 inhomogeneity lies very much with the design of the rf coil and this is in the hands of probe manufacturers who have made considerable progress. Overcoming off-resonance effects directly requires the use of higher power transmitters that are able to excite over wider bandwidths, but there are two principle problems with this. First is the inability of probe circuitry to sustain such high powers without being damaged and, second, is the insidious effect of rf sample heating which may damage precious samples. Improvements in probe circuitry that minimise this heating have largely been paralleled by increases in static field strengths so despite technical developments the problem remains significant. This has greater relevance at higher field strengths and for those nuclei that exhibit a wide chemical shift range, such as ^{13}C or ^{19}F. Consider data acquisition with a simple 90 degree pulse which has a uniform excitation bandwidth of around $\pm \gamma B_1$ Hz or $2\gamma B_1$ Hz in total, where γB_1 is the rf field strength (and where γB_1 symbolises $\gamma B_1/2\pi$; Section 3.5). Taking a typical value of 10 μs for the 90 degree pulse, this corresponds to a γB_1 of 25 kHz. For ^1H excitation at 400 MHz, the total excited bandwidth is thus 125 ppm which is clearly ample, and for ^{13}C is a satisfactory 500 ppm. In contrast, a 180 degree inversion pulse has a total bandwidth (here taken to mean > 90% effective) of only ca. $0.4\gamma B_1$, which in this example corresponds to 25 ppm for proton, again acceptable, but only 100 ppm for ^{13}C. Carbon nuclei that resonate outside this bandwidth will experience reduced inversion efficiency. Off-resonance errors for a 180 degree inversion pulse are illustrated in Fig. 12.3a in which it is clear that the trajectories of the magnetisation vectors are far from ideal and terminate further from the 'South Pole' as the offset increases, so generating unwanted transverse magnetisation. The combined effects of such pulse imperfections in multi-pulse sequences are to degrade sensitivity and to potentially introduce spurious artefacts, and so it is advantageous to somehow compensate for these.

An elegant approach to compensating deficiencies arising from B_1 inhomogeneity or resonance offset is the use of a *composite pulse* [1]. This is, in fact, a cluster of pulses of varying duration and phase that are used in place of a single pulse and have a net rotation angle equal to this but have greater overall tolerance to these errors. The first composite pulse [2] is the three-pulse cluster $90_x180_y90_x$, equivalent to a $180°_y$ pulse. The self-compensating abilities of this sequence for spin

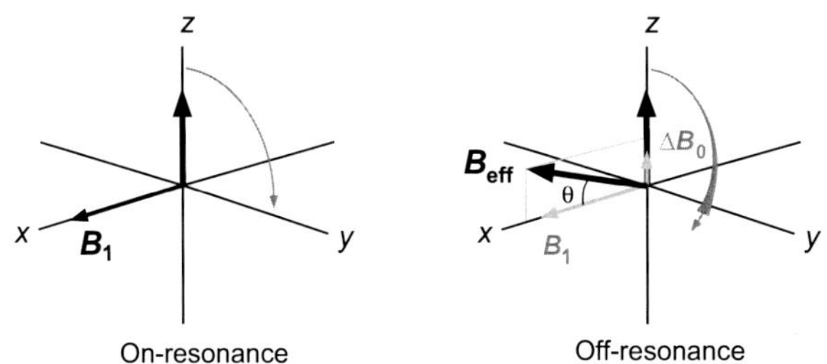

FIGURE 12.2 **Off-resonance excitation.** This causes the bulk vector to rotate about an effective rf field, tipped out of the transverse plane.

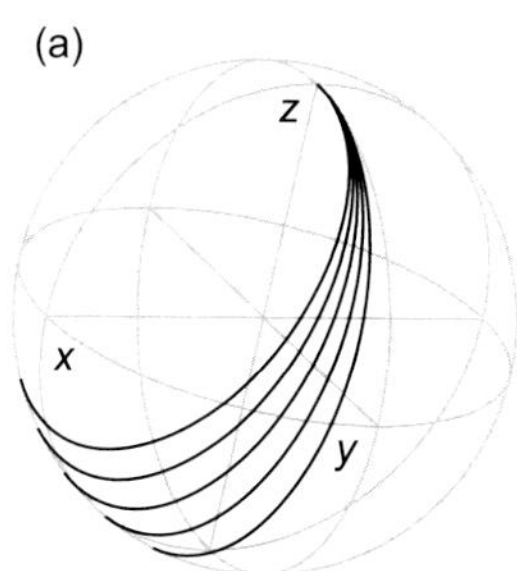

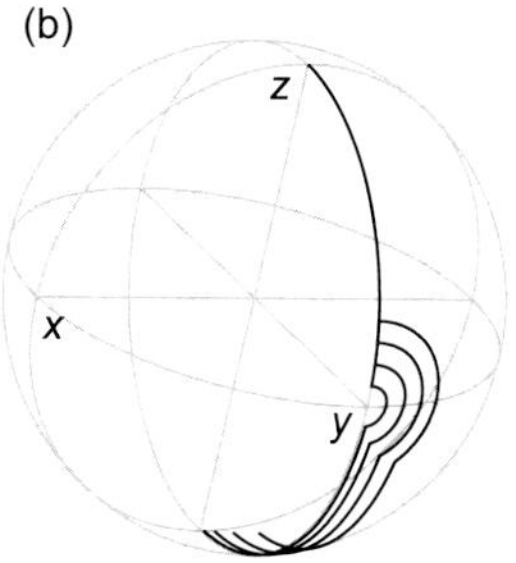

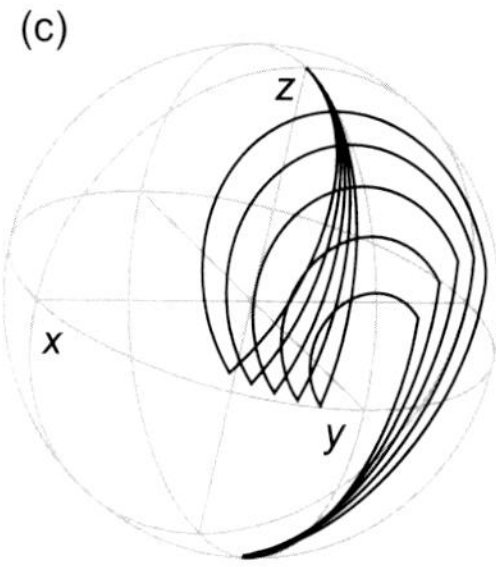

FIGURE 12.3 **Composite pulses.** Simulated effect of (a) resonance offset (of 0.2–0.6γB_1) on the inversion properties of a single 180°$_x$ pulse, and the compensating effect of the $90_x180_y90_x$ composite 180 degree pulse on (b) B_1 inhomogeneity and (c) resonance offset.

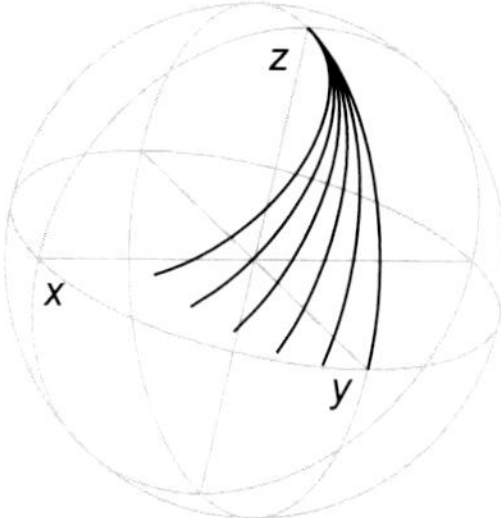

FIGURE 12.4 **The 90 degree pulse.** This is self-compensating with respect to resonance offset in its ability to generate transverse magnetisation, although there is no compensation for phase errors.

FIGURE 12.5 **The 90_x90_y composite 90 degree pulse.** This compensates for B_1 inhomogeneity by placing the vector closer to the transverse plane.

inversion are illustrated by the trajectories of Fig. 12.3. For both B_1 inhomogeneity (Fig. 12.3b) and for resonance offset (Fig. 12.3c) the improved performance is apparent from the clustering of vectors close to the South Pole. The effective bandwidth of the $90_x180_y90_x$ composite pulse is ca. $2\gamma B_1$, five times greater than the simple 180 degree pulse, a considerable improvement from such a simple modification. A single 90 degree excitation pulse, in contrast, is itself effective over ca. $2\gamma B_1$ owing to a degree of 'in-built' self-compensation for off-resonance effects when judged by its ability to generate transverse magnetisation. Here, the increased effective field experienced by off-resonance spins tends to drive the vectors further towards the transverse plane (as described in Section 3.2.1) at the expense of frequency dependent phase errors (Fig. 12.4). For high-power pulses, these errors are approximately a linear function of frequency and are readily removed through phase correction of the spectrum. Compensation for B_1 inhomogeneity may be achieved with the composite 90_x90_y (or better still 90_x110_y [1]) sequence which places the magnetisation vectors closer to the transverse plane than a single pulse (Fig. 12.5). This may prove beneficial in situations where the elimination of z-magnetisation is of utmost importance, such as the nuclear Overhauser effect (NOE) difference experiment (see discussions in Section 9.8; the $270_x360_{-x}90_y$ composite 90 degree pulse has been suggested as a better alternative in this case [3]).

A composite pulse may be included within a pulse sequence directly in place of a single pulse. The *relative* phase relationships of pulses within each cluster *must* be maintained but are otherwise stepped according to the phase cycling associated with the single pulse they have replaced. The selection of a suitable composite pulse is not always a trivial process and they are now increasingly superseded by adiabatic pulses which offer a number of advantages as described in the following section. Composite pulses do however still find routine use within broadband decoupling schemes such as those discussed in Section 12.3 and composite adiabatic pulses follow similar principles of self-compensation.

12.1.1 A Myriad of Pulses

The design of composite pulses has been a major area of research in NMR for many years and a huge variety of sequences have been published and many reviewed [4]. In reality, only a rather small subset of these has found widespread use in routine NMR applications, some of which are summarised in Table 12.1. One general limitation is that the sequences which provide the best compensation tend to be the longest and most complex, often many times longer than the simple pulse they have replaced, and as such may not be readily implemented or are unsuitable for use within a pulse sequence.

Furthermore, most composite pulses are effective against *either* B_1 inhomogeneity *or* offset effects, but not both simultaneously, so some compromise must be made, usually with the emphasis on offset compensation in high-resolution NMR (in other applications, such as *in vivo* spectroscopy, B_1 compensation has far greater importance because of the heterogeneous nature of the samples studied). In some cases, dual compensation has been included [5], although again at the expense of longer sequences (Table 12.1). Numerical optimisation methods have also been applied to the design of more advanced sequences which aim to keep the total duration acceptably small, most of which also make use of small rf phase shifts, that is those other than 90 degrees. A comparison of the performance of some composite inversion pulses is illustrated in Fig. 12.6, and clearly demonstrates the offset compensation they provide relative to a single 180 degree pulse. Nevertheless, their performance is still inferior to that attainable with frequency swept pulses which are now routinely available on modern spectrometers; see Section 12.2.

12.1.2 Inversion Versus Refocusing

One complicating factor associated with the implementation of composite pulses arises from the fact that many composite sequences have been designed with a particular initial magnetisation state in mind and may not perform well, or give the expected result, when the pulses act on other states. The two principle applications are the use of composite 180 degree pulses for *inversion* or for *refocusing*, in which they act on longitudinal and transverse magnetisation respectively. Thus, for example the $90_y180_x90_y$ sequence provides little offset compensation when used as a refocusing pulse (Table 12.1). Although it offers compensation for B_1 inhomogeneity and thus makes the magnitude of the echoes less sensitive to this by returning vectors closer to the x–y plane, it introduces errors in the *phase* of the echoes which may be detrimental to the overall performance of the experiment. The $90_y240_x90_y$ sequence has been proposed as a better refocusing element for offset compensation [9], and more sophisticated sequences have been generated that exhibit low phase-distortion [10,11].

TABLE 12.1 Properties of a Single 180 Degree Pulse and Some Composite 180 Degree Pulses

Composite Pulse	Duration (× 180 degree)	Bandwidth[a] (γB_1)	Properties	References
Inversion				
180_x	1	0.4		
$90_y180_x90_y$	2	2.0		[2]
$90_y240_x90_y$	2.3	1.2	More uniform inversion profile than $90_y180_x90_y$	[1]
$360_x270_{-x}90_y360_{-y}270_y90_x$	8	2.0	Compensated for B_1 inhomogeneity (±25%) and resonance offset	[5]
$38_x111_{-x}159_x250_{-x}$	3.1	2.6		[6]
$151_{247}342_{182}180_{320}342_{182}151_{247}$	6.5	2.0	Requires small rf phase shifts[b]	[7,8]
Refocusing				
180_x	1	0.5		
$90_y180_x90_y$	2	0.6	Introduces phase errors to spin-echoes	
$90_y240_x90_y$	2.3	1.0	Introduces phase errors to spin-echoes	[9]
$360_x270_{-x}90_y360_{-y}270_y90_x$	8	1.0	Compensated for B_1 inhomogeneity (±25%) and resonance offset	[5]
$336_x246_{-x}10_y74_{-y}10_y246_{-x}336_x$	7	1.2	Phase distortionless spin-echoes	[10]
$151_{247}342_{182}180_{320}342_{182}151_{247}$	6.5	2.0	Phase distortionless spin-echoes, requires small rf phase shifts[b]	[7,8]

Only a selection of available pulses is presented with the emphasis on compensation for resonance offset effects and on sequences of short total duration. Individual pulses are presented in the form XX_{yy} where XX represents the nominal pulse flip angle and *yy* its relative phase, either in units of 90° (x, y, $-x$, $-y$) or directly in degrees where appropriate. Sequences are split into those suitable for (A) population inversion (act on M_z) or (B) spin-echo generation (act on M_{xy}).

[a] *Bandwidths are given as fractions of the rf field strength and represent the total region over which the pulses have ca. 90% efficacy.*

[b] *Small rf phase shifts refer to those other than multiples of 90 degree.*

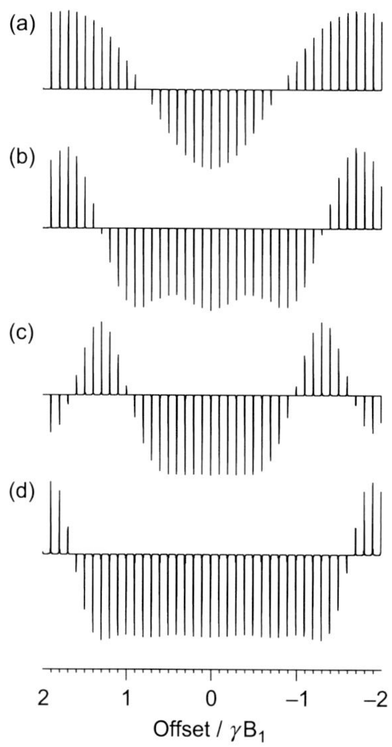

FIGURE 12.6 **Simulated inversion profiles.** These are shown for (a) 180_x, (b) $90_y180_x90_y$, (c) $90_y240_x90_y$ and (d) $38_x111_{-x}159_x250_{-x}$ (see Fig. 12.3 and Table 12.1).

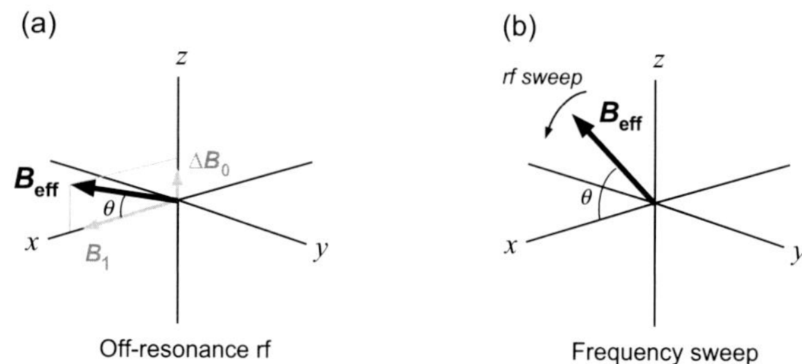

FIGURE 12.7 **Adiabatic inversion pulse.** (a) The effective field B_{eff} resulting from an rf pulse applied off-resonance and (b) the rf frequency sweep during the adiabatic pulse causes the effective field experienced by the spins to trace an arc from the $+z$ to the $-z$-axis, dragging with it the bulk magnetisation vector.

Sequences designed by numerical optimisation methods have been developed without a predefined initial state and are therefore equally effective as inversion or phase-distortionless refocusing elements [7,8].

The moral of all this is that considerable care must be taken when composite pulses are introduced into pulse sequences. It is wise to test the compensated pulse sequences versus the original uncompensated sequences on a known sample to see if improvements are achieved in reality. This offers a check on whether the composite elements have been correctly introduced and on whether the spectrometer is capable of correctly executing the desired sequences, many of which demand accurate control of rf amplitudes and phases for extended periods. The experimental performance of the composite pulse itself is best tested with a simple experiment, such as an inversion sequence ($180_{x,-x}$–90_x–FID$_x$) or spin-echo sequence (90_x–$180_{x,y,-x,-y}$–FID$_{x,-x}$) with a composite 180 degree pulse, and by analysing the results for a single resonance. Offset compensation may be investigated by stepping the transmitter frequency for the 180 degree composite pulse away from the initial on-resonance position (but setting it to be on-resonance for the 90 degree pulse) and B_1 inhomogeneity may be simulated by reducing pulse flip angles.

12.2 ADIABATIC AND BROADBAND PULSES

An alternative and highly efficient approach to spin inversion is offered by the so-called adiabatic pulses [12,13]. Rather than pulses applied at a single frequency, these employ a frequency sweep *during* the pulse that begins far from resonance, for example at a negative offset, passes through the resonance condition and finally terminates far from resonance with, in this case, a positive offset. During this, the effective rf field B_{eff} experienced by the spins (see Section 3.2.1 for discussions of off-resonance effects) begins along the $+z$-axis, traces an arc which passes through the x–y plane at the on-resonance condition and finishes along the $-z$-axis (Fig. 12.7). If the sweep is sufficiently slow, magnetisation vectors initially at equilibrium will continually circle about B_{eff} during the pulse and will be 'dragged along' by the effective field thus also terminating at the South Pole and experiencing the desired inversion. More formally, the sweep must be slow enough to satisfy the adiabatic condition:

$$\left|\frac{d\theta}{dt}\right| << \varpi_{\text{eff}} = \gamma B_{\text{eff}} \tag{12.1}$$

(where θ represents the angle between B_{eff} and the x-axis) but should be fast with respect to spin relaxation (hence it is referred to as *adiabatic fast passage*). So long as the adiabatic condition is met, precise inversion is achieved regardless of the magnitude of the applied rf field and hence these pulses are very tolerant of both B_1 inhomogeneities and power mis-settings. In practice, this simply requires the pulse power to be set at or above the required minimum for the inversion to be effective and that accurate pulse power calibrations are not essential. Furthermore, they are effective over very wide bandwidths and are thus well suited to broadband inversion or refocusing and also have application to ultra-broadband decoupling (Section 12.3.1). The ability to manipulate spins over a very wide frequency window makes these pulses especially desirable for the inversion and refocusing of heteronuclear spins that exhibit large chemical shift ranges on high-field spectrometers where conventional 180 degree pulses have poor offset performance. The application of adiabatic pulses in this context has been mentioned in numerous sections of preceding chapters, notably in the context of manipulating carbon-13 spins.

Adiabatic pulses are described by their frequency sweep and amplitude profile which, when combined with the peak rf amplitude $\omega_{1(max)}$, defines the total power of the pulse. The total frequency range, ΔF, over which the pulse sweeps is commonly many tens of kilohertz and pulse durations, T, are typically of the order of 1 ms, corresponding to frequency sweep rates of 10–100 MHz/s. The degree to which the adiabatic condition is satisfied for the pulse is quantified by the adiabaticity factor Q:

$$Q = \frac{\varpi_{eff}}{|d\theta / dt|} \tag{12.2}$$

this being most critical when the pulse passes through resonance as it is here when ω_{eff} is smallest since there is no contribution to this from $\Delta\omega$ when on-resonance. As the rf field passes through resonance, the expression below can be written for the on-resonance adiabaticity factor Q_0 [14,15]:

$$Q_0 = \frac{\varpi_1^2}{|d\Delta\varpi / dt|} \tag{12.3}$$

In the case of the commonly employed linear (ie constant rate) frequency sweep the term $d\Delta\omega/dt$ may be equated with the sweep rate of the pulse, $\Delta F/T$, and the on-resonance rf field defines the maximum amplitude for the pulse $\omega_{1(max)}$ rad/s, which one can write more usefully in terms of $\gamma B_{1(max)}$ as:

$$\left(\gamma B_{1(max)}\right)^2 = Q_0 \frac{\Delta F}{2\pi T} \quad \text{(Hz)} \tag{12.4}$$

From this expression one calculates the maximum peak amplitude of the adiabatic pulse $\gamma B_{1(max)}$ for a chosen value of Q_0 which must usually be significantly larger than unity and is typically set to 3–5 for most applications. Thus, a 0.5 ms adiabatic pulse sweeping over a 60 kHz window requires a peak amplitude of 9.77 kHz with Q_0 of 5, which corresponds to the power of a 25.6 μs hard 90 degree pulse. Such calculations are performed most conveniently within the tools provided with spectrometer software and their experimental implementation most readily achieved on instruments with linear amplifiers where the necessary rf attenuation may be calculated directly from hard pulse calibrations (see Section 3.5.1).

Eq. 12.4 also demonstrates that the frequency window of an adiabatic pulse scales with the *square* of the applied rf amplitude whereas simple hard pulses exhibit only a linear dependence on this and it is this feature that explains the efficiency of the swept pulses as broadband inversion pulses.

12.2.1 Common Adiabatic Pulses

The development of efficient adiabatic pulses has been an active area of research within the NMR community for some time with the classical hyperbolic secant pulse leading the way. This makes use of modulation of both the rf amplitude (Fig. 12.8a) and frequency but, while providing an excellent inversion profile, has limited application due to its high peak pulse amplitude making it unsuitable for application over wide bandwidths. Nowadays two adiabatic pulses find most widespread use in high-resolution NMR and hence we shall limit discussions to these. The simplest pulse has a rectangular amplitude profile and makes use of a linear frequency sweep and has been termed a CHIRP pulse [16]. The performance of this

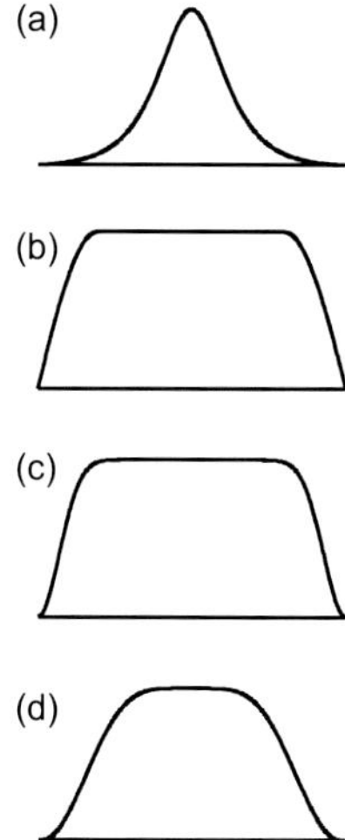

FIGURE 12.8 **Amplitude profiles for adiabatic pulses.** (a) Hyperbolic secant, (b) CHIRP with smoothing of the first and last 20%, (c) WURST with n = 20 (WURST-20) and (d) WURST with n = 5 (WURST-5).

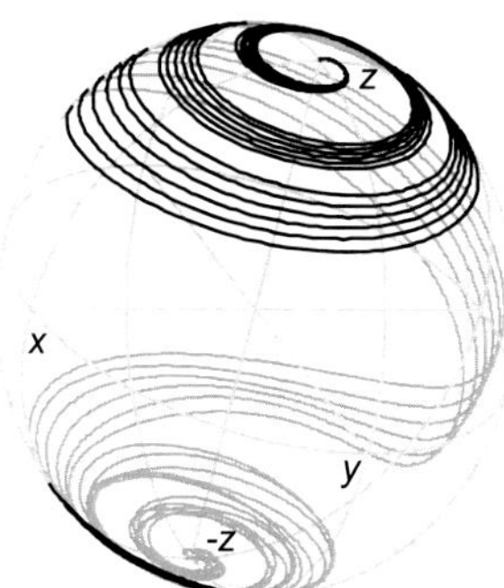

FIGURE 12.9 **Inversion trajectories for a 0.5-ms WURST-20 pulse swept over 60 KHz with $\gamma B_{1(max)}/2\pi$ = 10 kHz (Q = 5.24).** Trajectories are shown for offsets of 0–5 kHz in 1 kHz steps.

at the edges of the frequency sweep can be improved by gently truncating the rectangular amplitude envelope such as with a half-sine profile, so that the rf amplitude begins and ends at zero, leading to the smoothed CHIRP profile of Fig. 12.8b. This smoothing ensures B_{eff} genuinely begins and ends along the ± z-axis at the extremities of the sweep. The second widely used but independently developed pulse has a rather similar amplitude profile to the smoothed CHIRP and again defines a linear frequency sweep having the name WURST (wideband uniform rate and smooth truncation) [14,15,17]. The shape of the amplitude profile is in this case defined by the function:

$$\varpi_1(t) = \varpi_{1(max)}(1-|\sin(\beta t)|^n) \tag{12.5}$$

where $-\pi/2 \leq \beta t \leq +\pi/2$, and defines a sausage-shaped profile suggestive of its name (Fig. 12.8c and d). The exponent n defines the steepness of the truncation with larger values leading to a more rapid power attenuation at the edges; typically $n = 20-40$. The inversion trajectory for a WURST-20 pulse is shown in Fig. 12.9 and shows the swooping paths followed by the magnetisation vectors as they are pulled down toward the South Pole, terminating along the $-z$ axis as required. Initially the vectors circle the North Pole when the rf field is far from resonance, pass through the $x-y$ plane at the on-resonance condition, and are then swept southward as the rf field moves away.

Clearly it can be seen from Fig. 12.8 that the CHIRP and WURST profiles are very similar and not surprisingly their performance characteristics are likewise closely matched. Fig. 12.10 illustrates the inversion profiles for a CHIRP smoothed at 20% and the WURST-20 both simulated for a 60 kHz sweep over 0.5 ms under similar adiabatic conditions (Q = 5.24, $\gamma B_{1(max)}$ = 10 kHz). The total bandwidth over which any inversion occurs, ±25 kHz, is ~80% of the total sweep range although the bandwidth over which the pulses have > 90% inversion efficiency is somewhat less than this at ±17 kHz. Nevertheless, this corresponds to a total 270 ppm window for ^{13}C at 125 MHz (^1H = 500 MHz), clearly ample for complete inversion over the whole chemical shift range one typically encounters. In contrast, a conventional hard 180 degree pulse of 25 μs ($\gamma B_{1(max)}$ = 20 kHz, typical for ^{13}C) exhibits a *total* bandwidth of only 32 kHz and, worse still, displays non-uniform intensity over the whole of this region, despite employing twice the rf amplitude relative to the adiabatic pulses (Fig. 12.10c). This ability to use the adiabatic pulses at low powers yet retain broadband inversion efficiency makes them especially attractive for broadband decoupling; see Section 12.3.1.

The single adiabatic pulses presented so far are suitable for use as inversion pulses (when acting on longitudinal magnetisation), but are less well behaved for refocusing in spin-echoes (when acting on transverse magnetisation) due to the introduction of an offset-dependent phase role across the spectrum. Transverse magnetisation vectors may not experience the expected 180 degree flip but instead effectively undergo rotations in the transverse plane giving rise to the phase errors. One solution to this phase problem is to further elaborate a single pulse to a *composite adiabatic pulse* of sequential sweeps of relative duration 1:2:1 wherein the central pulse flips the effective field B_{eff}, enabling the third pulse to refocus the phase errors created by the first [18]. Thus, the 0.5 ms inversion pulse extends to a 2 ms adiabatic

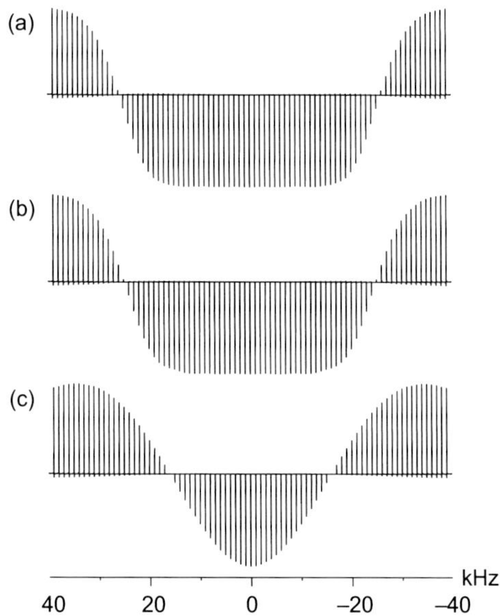

FIGURE 12.10 **Simulated inversion profiles.** These are shown for (a) 20% smoothed CHIRP, (b) WURST-20 and (c) 25 μs hard 180 degree pulse. For both (a) and (b) a 60 kHz sweep was employed over a 0.5 ms pulse with $\gamma B_{1(max)}/2\pi$ = 10 kHz (Q = 5.24). The profiles were generated with a $180^{\circ}_{x,-x}$–$90^{\circ}_{(hard)x}$ sequence over two transients to reveal the inverted M_z component.

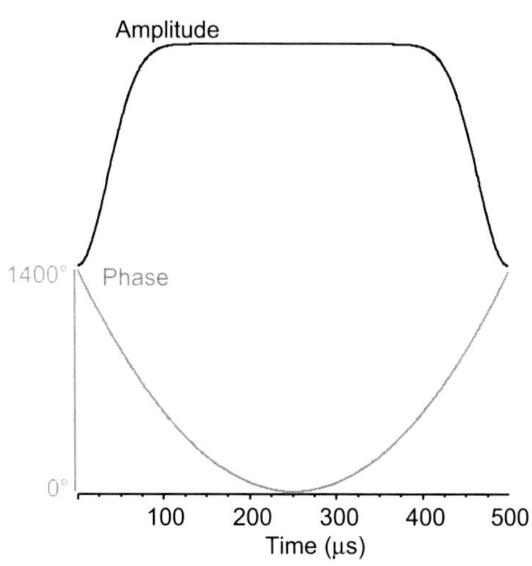

FIGURE 12.11 **The amplitude and phase modulation of a WURST-20 pulse (60 kHz sweep in 500 μs).**

pulse of sequential sweeps of 0.5–1.0–0.5 ms durations, somewhat reminiscent of the 90-180-90 composite hard pulses. Thus, single adiabatic pulses may be used for inversion whereas composite adiabatic pulses should be used for refocusing. The inclusion of adiabatic pulses offers an effective means for extending the operating bandwidth for 180 degree inversion and refocusing pulses for nuclei with wide shift ranges and especially at higher magnetic fields and they now find routine use within 2D experiments. Their inclusion can present additional complications, however, because of the long duration of these elements, in the order of milliseconds. This means that spins with different chemical shift offsets will experience inversion at differing times and this can in turn degrade the refocusing efficiency of these pulses and give rise to phase errors in spectra. These errors may not prove too problematic for many experiments and are unlikely to prevent the use of these pulses, but various approaches have been proposed to compensate for these. For the multiplicity edited HSQC experiment one recommendation is to synchronise the frequency sweep with the magnitude of the $^1J_{CH}$ coupling constant to improve refocusing efficiency on the editing step [19]. The use of matched pairs of adiabatic pulses in HSQC has also been suggested [20].

Finally, we note that in the execution of these swept pulses, spectrometers actually encode the necessary modulation not as a variation of the applied frequency but as a ramp in the *phase* of the applied rf. This is more readily implemented with the very accurate phase control found within NMR instruments. To see why this equality holds one could imagine the behaviour of a rf sine wave of constant frequency passing you by for which the phase of the wave is steadily increased as it does so. The appearance would be to accelerate the passage of the passing wave and thus make it behave as if it was of higher frequency. In effect, the frequency of the rf pulse has been increased by incrementing its phase which then causes excitation at a frequency away from that of the applied transmitter rf; this is how spectrometers generate all shaped pulses at arbitrary offsets from the transmitter frequency, such as those described in Section 12.4. Extending this idea to a pulse whose phase is incremented in a non-linear fashion leads to this having the equivalent effect as one in which the frequency is swept. To exemplify this, the amplitude and phase profile of a WURST-20 pulse swept through 60 kHz in 500 μs is illustrated in Fig. 12.11.

12.2.2 Broadband Inversion Pulses: BIPs

An alternative family of pulses effective over very wide bandwidths are the so-called broadband inversion pulses (BIPs) [21]. These have been designed through numerical optimisation to offer highly effective inversion pulses of higher power

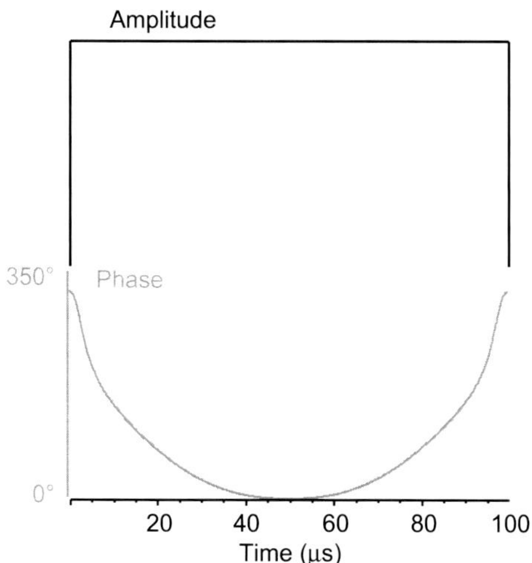

FIGURE 12.12 **The BIP.** The amplitude and phase profiles of the BIP-720-50-20 pulse of 100 μs duration (γB_1 = 20 kHz).

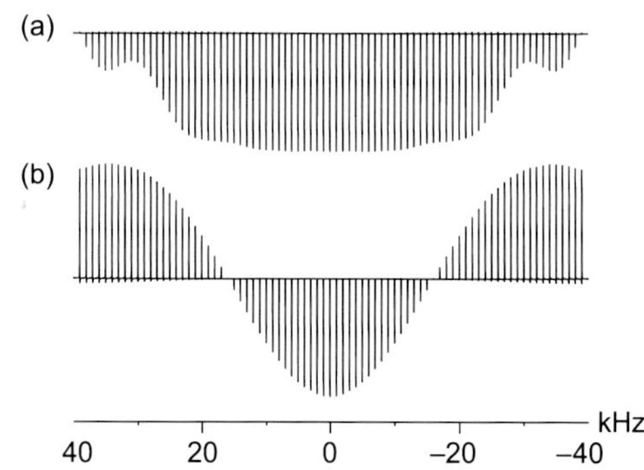

FIGURE 12.13 **The BIP simulated inversion profile.** This compares (a) the 100 μs BIP-720-50-20 and (b) 25 μs hard 180 degree pulse. Both pulses use γB_1= 20 kHz. The profiles were generated with a $180^\circ_{x,-x}$–$90^\circ_{(\text{hard})x}$ sequence over two transients to reveal the inverted M_z component.

and shorter duration than the adiabatic pulses described above and are aimed specifically for use as single pulses within multidimensional sequences rather than for repeated, sequential application as required in broadband decoupling. Surprisingly, it turns out that the resultant pulses do not require shaping of the rf amplitude profile (and in this respect are reminiscent of the early CHIRP pulses) and their properties are dictated by the phase ramping alone. They are defined, and named, according to their individual performance characteristics which encompass:

- Pulse duration: defined as the length (in degrees) of a hard pulse executed at the same rf amplitude.
- Inversion bandwidth: expressed as a percentage of the applied rf field γB_1 over which the pulse has > 95% efficacy.
- Tolerance to B_1 inhomogeneity: expressed as a percentage of the inhomogeneity.

Thus, the BIP-720-50-20 (Fig. 12.12) has an equivalent duration as would a 720 degree hard pulse, has an effective bandwidth of ± 0.5 γB_1 and is tolerant to pulse inhomogeneity, or miscalibration, of 0.2 γB_1. Therefore, applying this with a γB_1 of 20 kHz (which equates to a 25 μs hard 180 degree pulse) defines a 100 μs pulse (720 degree/180 degree × 25 μs) which has >95% inversion efficiency over a total 20 kHz bandwidth. This equates to a 40 ppm or 160 ppm window for ^{1}H and ^{13}C respectively at 500 MHz. The inversion profile of this pulse is shown in Fig. 12.13 and is compared to a 180 degree hard pulse applied at the same rf amplitude. Not only does the BIP easily outperform the hard pulse but its useable bandwidth considerably exceeds the ± 0.5 γB_1 range defined for this pulse when applying the stringent 95% inversion requirement, having good efficacy over approximately twice this range. Still greater bandwidths are available with other members of the BIP family; the 192 μs BIP-1382-250-15 provides complete inversion over a 100 kHz bandwidth.

The BIPs therefore provide for broadband pulses of shorter duration and higher power than the adiabatic pulses and may be better suited when longer pulses can be problematic, as mentioned above. They are suitable as single inversion pulses in multidimensional NMR experiments and can also be used for refocusing of spin-echoes whereby they must be used in pairs to cancel phase errors generated by a single pulse. In a separate development, constant-amplitude, phase-modulated pulses have also been introduced for broadband *excitation*, the so-called BEBOP pulses (broadband excitation by optimised pulses) [22], which have relevance to excitation of, say, ^{13}C at *very* high fields when even conventional 90 degree pulses show inadequate bandwidths.

12.3 BROADBAND DECOUPLING AND SPIN LOCKING

An area in which composite pulses have been applied routinely and continue to be used with great success is broadband heteronuclear decoupling. Their use in this area arises from the realisation that heteronuclear decoupling may be achieved by the continuous application of a train of 180 degree inversion pulses on the decoupled spin. This is illustrated in Fig. 12.14 for a X–H pair in the presence of proton decoupling. Following excitation of X, doublet vectors diverge for a period τ according to J_{XH}. A 180 degree pulse applied to protons at this time inverts the proton α and β states thus reversing the sense of precession of the X-spin vectors which refocus after a further period τ. If this procedure is repeated during

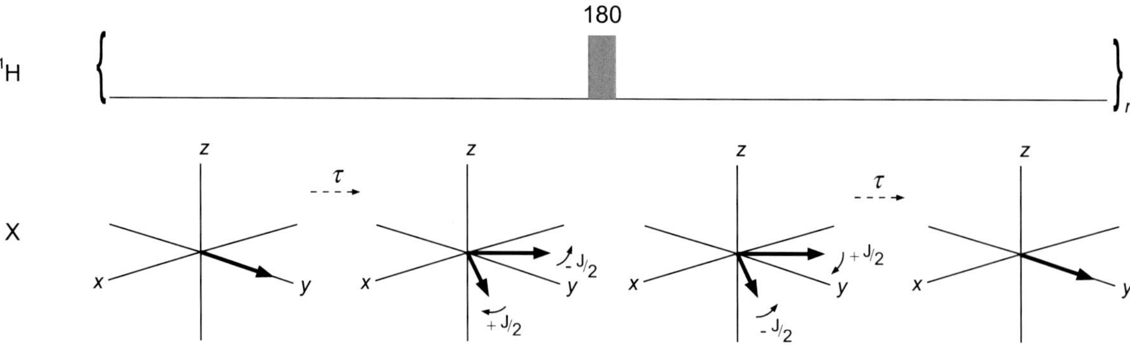

FIGURE 12.14 **Heteronuclear broadband proton decoupling with a sequence of 180 degree proton pulses.**

the observation of X at a fast rate relative to the magnitude of J_{XH}, the action of the coupling is suppressed and hence the spectrum appears decoupled.

From discussions in the previous section it should be apparent that the application of a sequence of simple 180 degree pulses is unlikely to give effective decoupling over a wide bandwidth or in the presence of B_2 inhomogeneity or pulse miscalibration (note the use of the symbol B_2 with reference to a *decoupling* rf field). Further, by reducing the period τ to be infinitely small, the sequence becomes a single, continuous decoupler pulse (so-called *continuous wave* decoupling) and by the same arguments this is also a poor decoupling sequence, particularly off-resonance. One solution to achieving decoupling over wider bandwidths is to employ composite inversion pulses that display superior off-resonance performance, and modern *composite pulse decoupling* (CPD) has evolved from the original $90_x180_y90_x$ and $90_x240_y90_x$ sequences.

Simply inserting these in place of single 180 degree pulses turns out not to be the full answer, however. Defining the composite inversion cluster $90_x180_y90_x$ as an element R, then we can see that the sequence RR (two sequential composite pulses) should return spin vectors back to the $+z$ axis for the rotation sequence to begin again. However, these elements are themselves not perfect (see Fig. 12.3), leaving small errors which will accumulate if the sequence is simply repeated. The trick is to repeat the process but in reverse, with all rf phases inverted so as to counteract these errors, giving rise to the element $\bar{R}$ $(90_{-x}180_{-y}90_{-x})$ and the so-called 'magic-cycle' $RR\bar{R}\bar{R}$. This original decoupling sequence is termed MLEV-4. It was subsequently realised that small residual errors from the magic-cycle could be compensated by further nesting of these elements to produce 'super-cycles', such as $RR\bar{R}\bar{R}\ \bar{R}RR\bar{R}\ \bar{R}\bar{R}RR\ R\bar{R}\bar{R}R$, giving, in this case, MLEV-16. These cycles prove more effective over greater bandwidths without the requirement of excessive rf powers.

Numerous such composite pulse decoupling sequences have been developed along these lines over the years (Table 12.2), the most widely used being WALTZ-16 whose basic inversion element is $90_x180_{-x}270_x$ (or more succinctly $1\bar{2}3$,

TABLE 12.2 Selected Composite-Pulse Sequences for Broadband Decoupling and Spin Locking

Sequence	Bandwidth (γB_2)	Application	References
Decoupling			
Continuous wave	<0.1	Selective decoupling only	
MLEV-16	1.5	^{1}H decoupling	[23]
WALTZ-16	2.0	High-resolution ^{1}H decoupling	[24,25]
DIPSI-2	1.2	Very high-resolution ^{1}H decoupling	[26]
GARP	4.8	X-nucleus decoupling	[27]
Spin locking			
MLEV-17	0.6	TOCSY mixing scheme	[28]
DIPSI-2	1.2	TOCSY mixing scheme	[29]
FLOPSY-8	1.9	TOCSY mixing scheme	[30]

The technicalities of broadband decoupling are extensively discussed in [31].

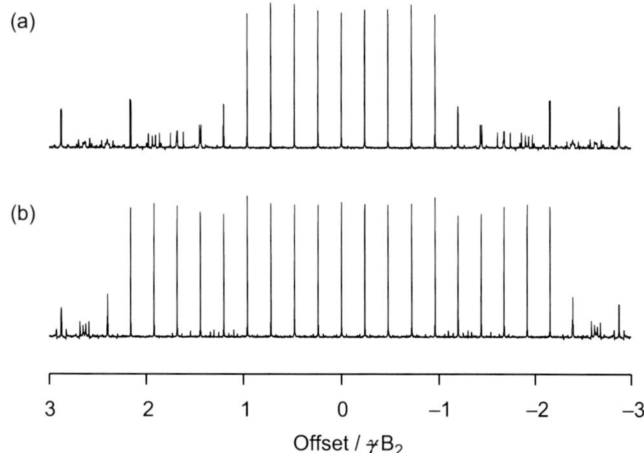

FIGURE 12.15 **Composite pulse decoupling.** Experimental comparison of the ^{1}H decoupling offset profiles of (a) WALTZ-16 and (b) GARP relative to the decoupling rf field strength γB_2. The sample was carbon-13 labelled methanoic acid in D_2O ($^{1}J_{CH}$ = 195 Hz).

hence the name). This provides very effective decoupling over a bandwidth of $2\gamma B_2$ leaving only small residual splittings (< 0.1 Hz) and is thus favoured in high-resolution applications such as broadband proton decoupling of heteronuclear spectra. This bandwidth can, however, be limiting when decoupling nuclei with greater chemical shift dispersion, for example ^{13}C or ^{31}P, as is required in proton-detected heteronuclear shift correlation sequences such as HSQC or HMQC. For such cases, other CPD sequences have been developed through computational optimisation, which have less stringent requirements for residual linewidths (<1.5 Hz). The most popular of these has been GARP (Table 12.2) which is effective over $4.8\gamma B_2$ (Fig. 12.15). With the continued increase in static fields, even greater decoupling bandwidths become necessary and the most effective approach so far appears to be the use of adiabatic decoupling methods.

12.3.1 Broadband Adiabatic Decoupling

As broadband decoupling relies upon the repeated application of 180 degree inversion pulses, it is rational to suppose that greater decoupling bandwidths can be achieved through the use of adiabatic inversion pulses within decoupling schemes. Indeed, such methods have resulted in decoupling schemes that are effective over far greater bandwidths (> 20 γB_2) than those attainable with conventional composite pulse sequences [14,32,33]. As for composite pulse decoupling, these schemes benefit from the use of suitable supercycles, one common scheme employing a five-step 0, 150, 60, 150, 0 degree phase cycle [34] nested with the MLEV-4 cycle described earlier. It is also common practice to employ adiabatic pulses with lower adiabaticity factors Q in order to reduce power requirements and so avoid sample heating; typically a Q of 2–3 is suitable for decoupling applications. The use of such a decoupling scheme is illustrated in Fig. 12.16 for ^{13}C decoupling using a 42 kHz CHIRP sweep of 1.5 ms duration (20% smoothing), with Q reduced to 2.5 and $\gamma B_{2(max)}$ = 3.34 kHz, corresponding to a hard 90 degree pulse of 75 μs. The decoupling is effective over ~30 kHz, (around 70% of the total sweep range) equivalent to a 240 ppm bandwidth for ^{13}C at 500 MHz and ample for most applications. This is contrasted with the decoupling bandwidth afforded by GARP composite pulse decoupling applied at the same rf amplitude, a typical power for heteronuclear decoupling. Clearly the adiabatic decoupling has a useable bandwidth that is in this case, approximately twice that of GARP.

One of the more significant drawbacks to the use of adiabatic decoupling schemes is the appearance of unwelcome cycling sidebands, modulation sidebands that flank the main decoupled resonance and may give rise to interfering peaks in multidimensional spectra [35,36]. These can be minimised by employing a so-called *bi-level decoupling* scheme that starts initially with a brief high-power spin lock applied prior to the adiabatic decoupling itself to momentarily delay the onset of the modulation that gives rise to the sidebands. The duration of the spin lock is varied according to the number of transients collected such that addition of all transients acts to attenuate the sideband responses.

The use of adiabatic schemes has most significance for the decoupling of heteronuclear X-spins at higher fields, although they may still have relevance to decoupling at more modest fields, for example in the decoupling of ^{19}F over a large

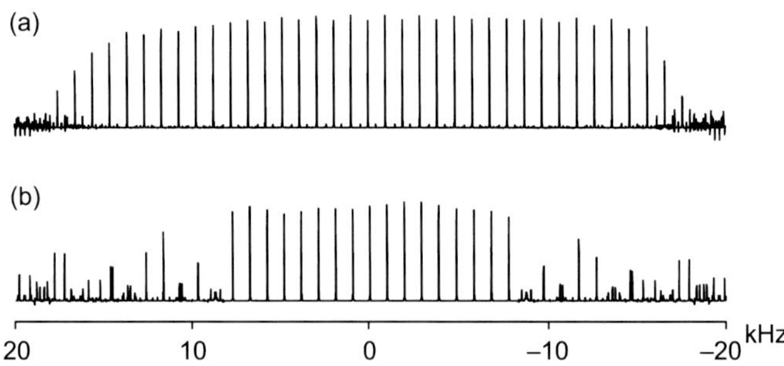

FIGURE 12.16 **Adiabatic decoupling.** Experimental $^{1}H\{^{13}C\}$ broadband decoupling profiles of (a) smoothed CHIRP (20%, 42 kHz, 1.5 ms, $Q = 2.5$) and (b) GARP. In both cases the $\gamma B_{2(max)}$ was 3.34 kHz. The sample was carbon-13 labelled methanoic acid in D_2O ($^{1}J_{CH} = 195$ Hz).

chemical shift range. They have also been suggested for improved ^{1}H decoupling when very accurate quantification of ^{13}C spectra is required [37].

12.3.2 Spin Locking

The ability to 'spin lock' magnetisation along a predefined axis plays an essential role in the TOCSY experiment, and Section 6.2.1 has already introduced the idea that a single continuous pulse or a series of closely spaced 180 degree pulses can, in principle, be used in this context to repeatedly refocus chemical shift evolution while allowing *homonuclear* couplings to evolve. Following from the above discussions, a better approach is clearly to use a series of 180 degree composite pulses which offer compensation for resonance offset and rf inhomogeneity and many of the sequences originally designed as heteronuclear decoupling sequences (which require repeated spin inversions) have since been applied as spin lock sequences (which require repeated spin refocusing). The original and still widely used mixing scheme is based on the MLEV-16 sequence, in which each cycle is followed by a 60 degree pulse to compensate errors that would otherwise accrue during extended mixing, producing the popular MLEV-17 spin lock (Table 12.1). More sophisticated sequences have since been developed that allow for the influence of homonuclear couplings, a factor not considered during the design of the early heteronuclear decoupling sequences. In particular, the DIPSI-2 sequence has better performance than MLEV or WALTZ in these circumstances, and is thus widely used for isotropic mixing in the TOCSY experiment.

12.4 SELECTIVE EXCITATION AND SOFT PULSES

So far discussions in this chapter have dealt with the use of so-called *hard* pulses, that is pulses which (ideally) are equally effective over the whole chemical shift range, or with *adiabatic* pulses that have similar goals. We have also seen composite pulses that help one approach more closely these ideals. In some instances, however, it can be a distinct advantage if only a selected region of the spectrum is influenced by a so-called *soft* pulse, and a number of examples utilising such *selective excitation* have been presented in many of the preceding chapters. Principle applications in a chemical context include:

- Reduced dimensionality of *n*D sequences, for example 1D analogues of 2D experiments.
- Selective removal of unwanted resonances, for example solvent suppression.
- Extraction of specific pieces of information, for example the measurement of specific long-range ^{1}H–^{13}C coupling constants.
- 1D sequences which intrinsically require the selection of a single resonance, for example inversion (saturation) transfer experiments in studies of chemical dynamics.

Replacing hard pulses with their selective counterparts within a multidimensional sequence leads to a spectrum of lower dimensionality having a number of advantages over its fully fledged cousin [38,39]. 1D analogues of more conventional 2D experiments allow greater digital resolution and thus a more detailed insight into fine structures; these equate to high-resolution slices through the related 2D experiment at the shift of the selected spin, for example the 1D TOCSY experiment of Section 6.2.4 or the 1D NOESY of Section 9.6.2. They may also benefit by being quicker to

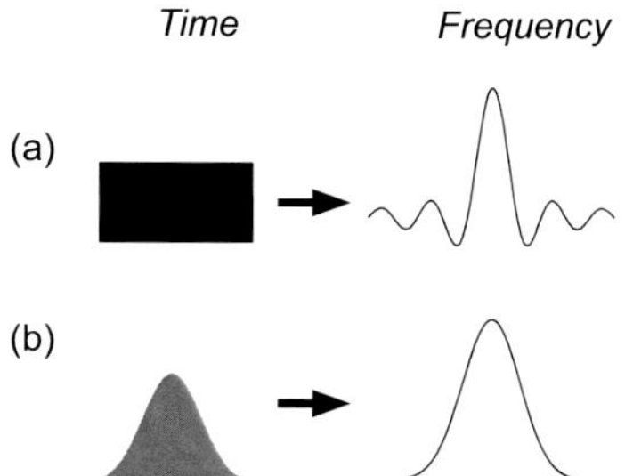

FIGURE 12.17 **Schematic excitation profiles.** These are shown for (a) a low-power rectangular pulse and (b) a smoothly truncated *shaped* pulse.

acquire and analyse. In short, such methods may provide faster and more detailed answers when a specific structural question is being addressed, and discussions on the use of selective pulses in high-resolution NMR have featured in a number of extensive reviews [38,40–42].

The basic approach to producing a *selective* pulse is to reduce the rf power and increase the duration of the pulse. By reducing B_1, the frequency spread over which the pulse is effective likewise diminishes. At the same time, the speed of rotation of vectors being driven by the rf also decreases meaning longer pulses are required to achieve the desired tip angle. The simplest selective pulse is therefore a long, low-power rectangular pulse, by analogy with the usual rectangular hard pulse. Unfortunately, this has a rather undesirable excitation profile in the form of side-lobes that extend far from the principle excitation window (Fig. 12.17a), resulting in rather poor selectivity. The origin of the undesirable side-lobes lies in the sharp edges of the pulse which introduce a sinc-like oscillation to the profile; notice the similarity with the 'sinc wiggles' in spectra caused by the premature truncation of a free induction deday (FID) (Section 3.2.3). These may be suppressed by smoothing the edges of the pulse (Fig. 12.17b), just as apodisation of the truncated FID reduces the wiggles, and such *pulse shaping* has given rise to a multitude of selective pulses with various characteristics that find use in modern NMR. A small selection of these is considered below, which have been chosen for their general applicability and/or otherwise desirable properties for high-resolution spectroscopy.

12.4.1 Shaped Soft Pulses

Associated with the use of selective pulses are a number of experimental factors that have a considerable bearing on the selection (and design) of a soft pulse. In short, the key features are its:

- duration;
- frequency profile; and
- phase behaviour.

Soft pulses are typically 1–100 ms long, three orders of magnitude longer than hard pulses, which may have implications with regard to chemical shift and coupling evolution and to relaxation losses during the pulse. Therefore, it is desirable to have a soft pulse that is as short as possible but still able to deliver the desired selection. This selection is defined by the frequency bandwidth over which the pulse is effective. Ideally, the bandwidth profile should be rectangular, with everything outside the desired window insensitive to the pulse; such a profile is sometimes referred to [43] as the 'top-hat' for obvious reasons. The phase of the excited resonances should also be uniform over the whole excitation window, although in reality this can be difficult to achieve since off-resonance effects are severe with the weak B_1 fields used. Unless corrected, this results in considerable phase distortion away from exact resonance that can be detrimental to multipulse sequences where the precise control of magnetisation is often required. These primary considerations are addressed below for some commonly encountered shaped pulses, while their practical implementation is described in Section 12.4.5.

12.4.1.1 Gaussian Pulses

The original and experimentally simplest shaped pulse has the smooth Gaussian envelope [44] (Fig. 12.18). The (absolute value) frequency excitation profile is also Gaussian, tailing away rapidly with offset and, although not an ideal top-hat profile, is clearly superior to the rectangular pulse (Fig. 12.19). Just outside the principle excitation window, magnetisation vectors also feel the effect of the pulse but tend to be driven back toward the starting position so experience no *net* effect, as can be seen in the trajectories of Fig. 12.20a. This figure also illustrates the problem of phase dispersion across the

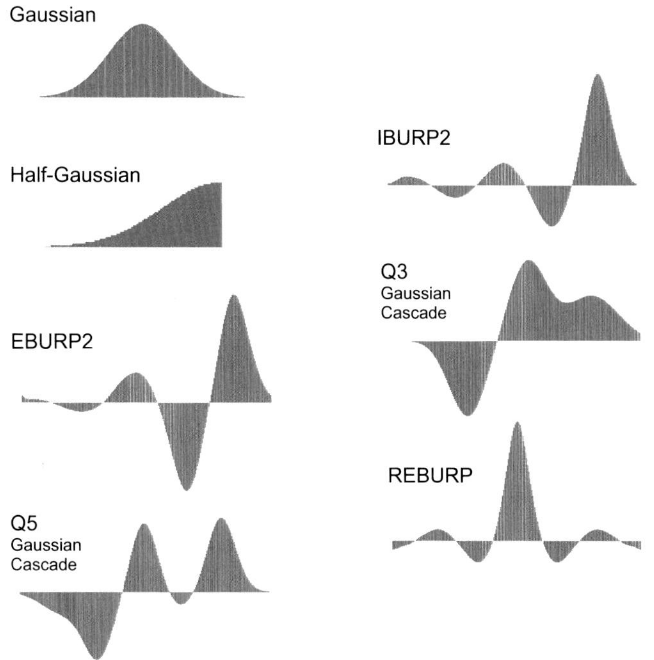

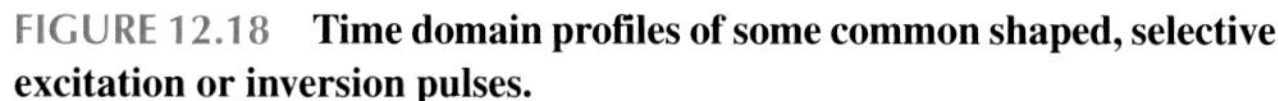

FIGURE 12.18 **Time domain profiles of some common shaped, selective excitation or inversion pulses.**

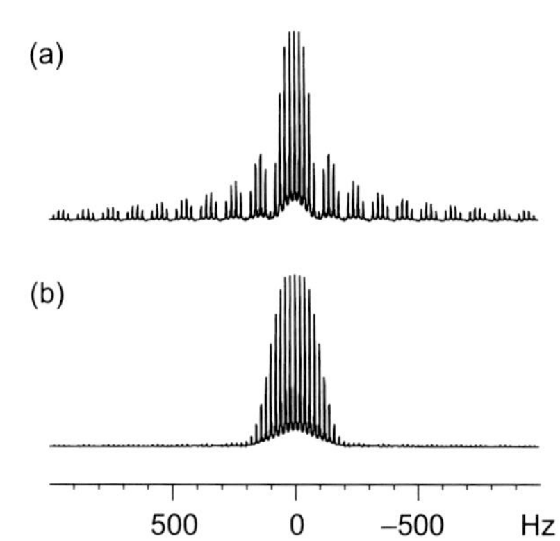

FIGURE 12.19 **Absolute value frequency domain excitation profiles.** These are shown for (a) a rectangular pulse and (b) a Gaussian shaped pulse.

bandwidth for different offsets. For a pulse of phase x both the desired M_y and unwanted M_x components (absorptive and dispersive responses, respectively) are created by the pulse, producing a significant phase gradient which can be corrected in the simplest of experiments, but in general may be problematic. In addition, inverted responses are also generated (those terminating in the $-y$ hemisphere) and thus the profile retains some undesirable oscillatory behaviour (this is hidden in Fig. 12.19 because an absolute-value display was used, but can be seen in Fig. 12.21).

Better alternatives for the excitation of longitudinal magnetisation are the half-Gaussian [45] which, as the name suggests, is simply a Gaussian profile terminated at the midpoint and the 270 degree Gaussian [46]. The half-Gaussian pulse does not produce negative side-lobes because vectors never reach the $-y$ hemisphere, although it still generates a considerable dispersive component, M_x, as is apparent from the trajectories of Fig. 12.20b. The dispersive responses may be removed by a phase-alternated hard 'purge' pulse applied orthogonally to and immediately after the half-Gaussian. This cancels any M_x components, retaining only the desired M_y components and so removing the phase gradient. The improved phase properties of this scheme give rise to the name 'purged half-Gaussian' [47] (Fig. 12.21). The 270 degree Gaussian pulse equates to a net -90 degree rotation and is identical to the 90 degree pulse except for a threefold amplitude difference. The interesting feature of this pulse is that for spins close to resonance it has a self-refocusing effect on both chemical shifts and coupling constants (Fig. 12.20c) and thus has better phase properties than the 90 degree Gaussian pulse (Fig. 12.21). As it does not require purging it is simple to implement and is the excitation pulse of choice when high selectivity is not critical.

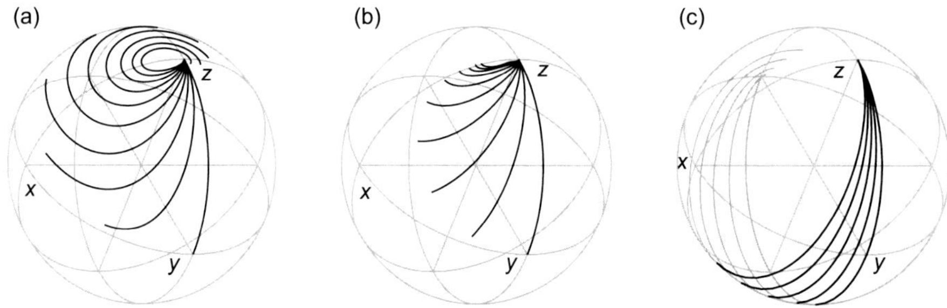

FIGURE 12.20 **Simulated excitation trajectories as a function of resonance offset.** These are shown for (a) the Gaussian, (b) the half-Gaussian and (c) the Gaussian-270.

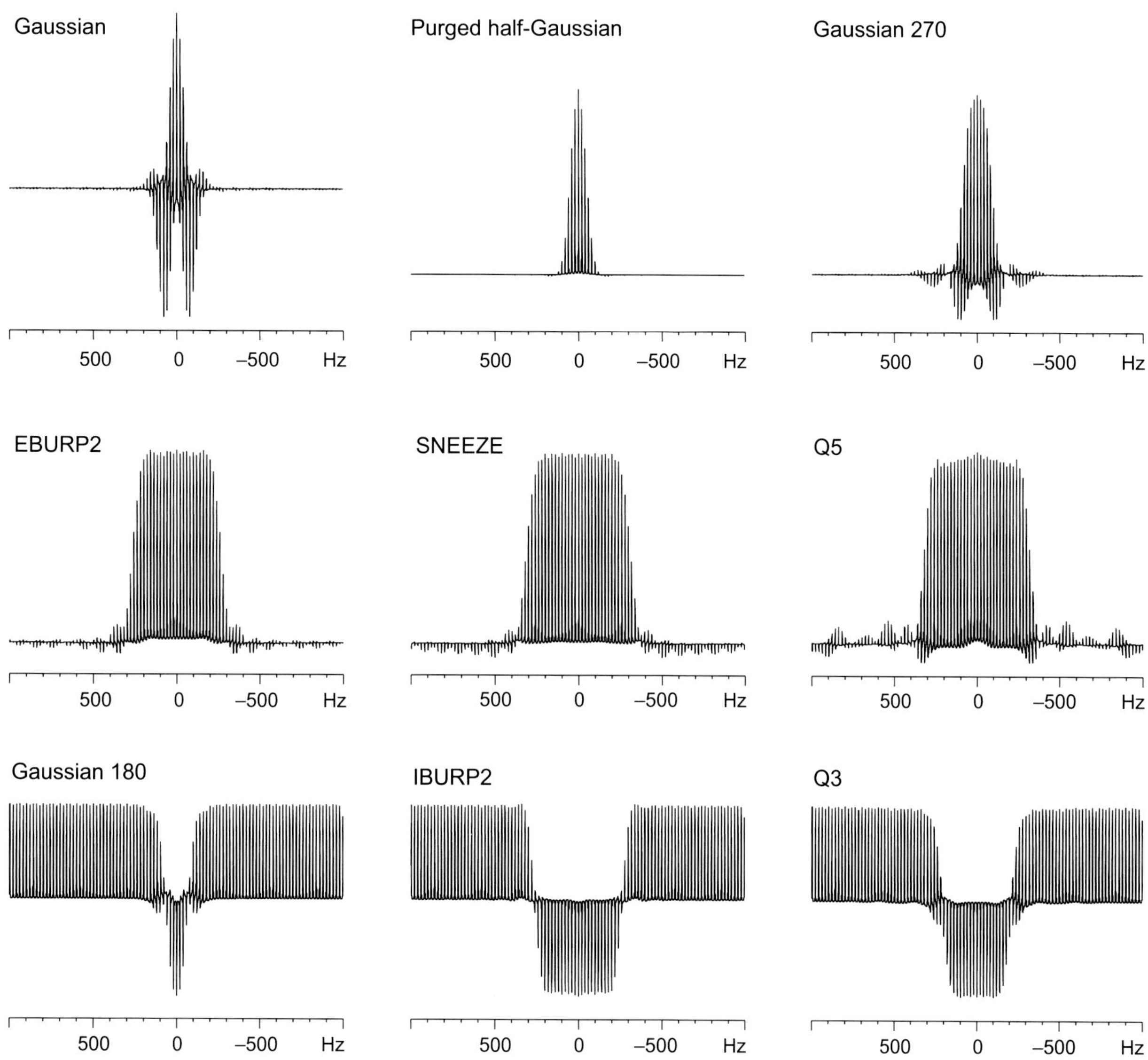

FIGURE 12.21 **Simulated excitation profiles of selected shaped pulses (of 10 ms duration); see Table 12.3.** The inversion profiles (lower trace) were simulated with a 180(soft)–90(hard) sequence.

12.4.1.2 Pure-Phase Pulses

More elaborate pulse shapes have been developed over the years which aim to produce a near top-hat profile yet retain uniform phase for all excited resonances within a predefined frequency window. These operate without the need for purging pulses or further modifications, allowing them to be used directly in place of hard pulses. They are typically generated by computerised procedures which result in more exotic pulse envelopes (Fig. 12.18) that drive magnetisation vectors along rather more tortuous trajectories than the simpler Gaussian-shaped cousins. Trajectories are shown in Fig. 12.22 for two members of the BURP family of pulses (Band-selective, Uniform Response, Pure phase pulses) [48,49] namely the EBURP2 excitation pulse and the IBURP2 inversion pulse (see Section 12.4.5 and Table 12.3). Despite the globetrotting journeys undertaken, trajectories for all offsets within the effective window terminate close to the ideal endpoint with only small intensity or phase errors. The corresponding profiles for these pulses are also shown in Fig. 12.21 and illustrate the uniform, pure-phase behaviour (although experimentally these depend *critically* on accurate pulse calibrations). An experimental demonstration of selective excitation with the EBURP2 pulse is shown in Fig. 12.23 for two different excitation bandwidths. Interestingly, a different design approach has led to a similar family of pulses with largely similar properties, the so-called Gaussian cascades [50,51] (Table 12.3 and Fig. 12.18). As the name implies, these are composed of clusters

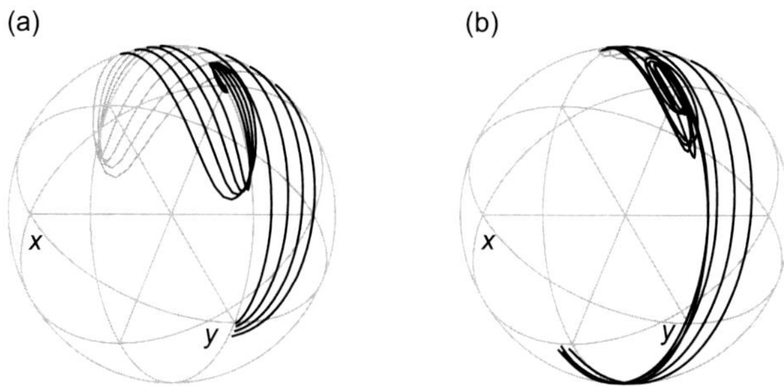

FIGURE 12.22 **BURP trajectories.** These are simulated for nuclear spin vectors as a function of resonance offset during the (a) EBURP-2 90 degree excitation pulse and (b) the IBURP2 180 degree inversion pulse.

TABLE 12.3 Properties of Some Shaped Soft Pulses

Shaped Pulse	Application	Bandwidth Factor	Attenuation Factor (-dB)	References
Rectangular	Universal	1.1	—	—
90 degree pulses				
Gaussian	Universal	2.1	7.7	[44]
Purged half-Gaussian	Excitation only	0.8	7.7	[45,47]
Gaussian-270	Excitation only	1.3	16.7	[46]
EBURP2	Excitation only, Pure phase	4.9	24.3	[49]
SNEEZE[a]	Excitation only, Pure phase	5.8	26.6	[52]
UBURP	Universal, Pure phase	4.7	32.2	[49]
E-SNOB[b]	Excitation only, Pure phase	1.9	7.0	[53]
G4 Gaussian Cascade	Excitation only, Pure phase	7.8	25.4	[50]
Q5 Gaussian Cascade	Universal, Pure phase	6.2	25.3	[51]
180 degree pulses				
Gaussian-180	Universal	0.7	13.7	[44]
IBURP2	Inversion only, Pure phase	4.5	25.9	[49]
REBURP	Universal, Pure phase	4.6	28.0	[49]
I-SNOB2[b]	Inversion only, Pure phase	1.6	13.4	[53]
R-SNOB[b]	Universal, Pure phase	2.3	13.4	[53]
G3 Gaussian Cascade	Inversion only, Pure phase	3.6	23.1	[50]
Q3 Gaussian Cascade	Universal, Pure phase	3.4	22.4	[51]

Universal pulses act equally on any initial magnetisation state whereas excitation and inversion pulses are designed to act on longitudinal magnetisation only. The bandwidth factor is the product of the pulse duration, Δt, and the excitation bandwidth, Δf, which is here defined as the excitation window over which the pulse is at least 70% effective (net pulse amplitude within 3 dB of the maximum; other publications may define this value for higher levels and so quote smaller bandwidth factors). Use this factor to estimate the appropriate pulse duration for the desired bandwidth. The attenuation factor is used for approximate power calibration and represents the amount by which the transmitter output should be increased over that of a soft rectangular pulse of equal duration. The Gaussian-based profiles are truncated at the 1% level.

[a]*The SNEEZE pulse produces more uniform excitation of M_z than EBURP2.*
[b]*The SNOB family are designed to have short pulse durations.*

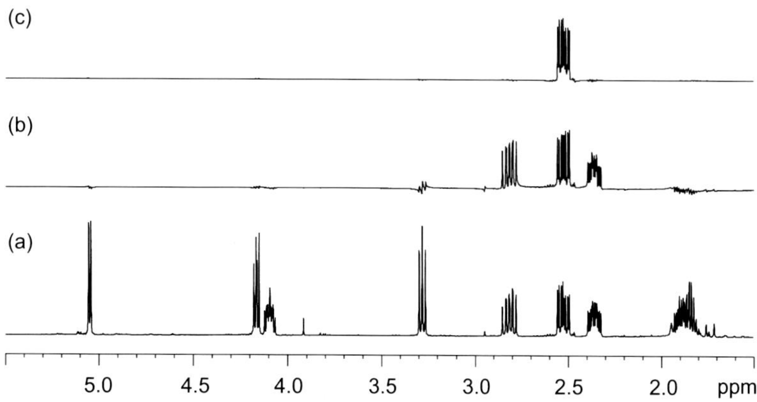

FIGURE 12.23 **Selective excitation with the EBURP2 band-selective pulse.** Spectra were recorded with a pulse duration of (b) 14 ms and (c) 100 ms, shown above the conventional 1D spectrum (a).

of Gaussian pulses whose net effect is closer to the ideal than any single element, reminiscent of the composite hard pulses of Section 12.1.

One feature of these more elaborate envelopes is that they have often been designed with a particular function in mind and may not perform well at anything else. For example the EBURP pulses are 90 degree *excitation pulses* designed to act on longitudinal magnetisation and will not perform as desired on transverse magnetisation. Likewise, selective 180 degree inversion pulses may not work as refocusing elements in the generation of spin-echoes. Those designed to act on *any* initial magnetisation state are referred to as *universal pulses*. More complex profiles also tend to be longer than their simpler counterparts for a given excitation bandwidth and tip angle, meaning relaxation effects may be problematic, particularly for the case of larger molecules or for very long, highly selective pulses. The influence of relaxation on their performance has been addressed [54,55] with more tolerant profiles suggested (SLURPs [56]) and shorter pulses developed with larger molecules in mind (SNOBs [53]).

12.4.1.3 Implementing Shaped Pulses

On modern high-resolution instruments the control of pulse amplitudes 'on-the-fly' is a standard feature. Typically the pulse envelope is defined in a series of discrete steps as a histogram, with each element having a defined amplitude and phase. To provide a close match to the smooth theoretical pulse shape a sufficient number of elements must be defined, typically 1000 or more (but is instrument dependent). Many pulse envelopes come predefined on modern instruments, allowing their direct implementation.

12.4.2 Excitation Sculpting

Modern methods of selective excitation now routinely combine shaped pulses with PFGs to produce experimentally robust excitation sequences with a number of desirable properties [41]. These sequences are based on either single PFG spin-echoes (SPFGSEs) or double PFG spin-echoes (DPFGSEs) (Fig. 12.24), which may be understood with reference to the single echo sequence (Fig. 12.24a). This may be represented G_1-S-G_1 where S represents *any* selective 180 degree pulse (or pulse train) and the bracketing gradients G_1 are identical. For those spins that experience the selective inversion pulse the two gradients act in opposition and thus refocus this selected magnetisation, hence this is known as a gradient-echo. Spins that do not experience this pulse, that is those outside its effective bandwidth, only feel the cumulative effect of both gradients so remain fully dephased in the transverse plane and thus unobservable. This single gradient-echo therefore achieves clean resonance selection according to the profile of pulse S.

The *phase* profile of the selected resonance(s) is also dictated by the phase properties of the selective pulse S, which may not be ideal. Repeating the gradient-echo once again with a different gradient strength G_2 (to avoid accidental refocusing of the previously dephased unwanted magnetisation, Fig. 12.24b) exactly cancels any remaining phase errors and the resulting *pure-phase* excitation profile depends only on the inversion properties of the selective pulse. Experimentally this is an enormous benefit because it makes implementation of the selective sequence straightforward and because the field gradients ensure excellent suppression of unwanted resonances. It is also very much easier to select (and design) a pulse with a desirable 'top-hat' inversion profile when its phase behaviour is of no concern. The

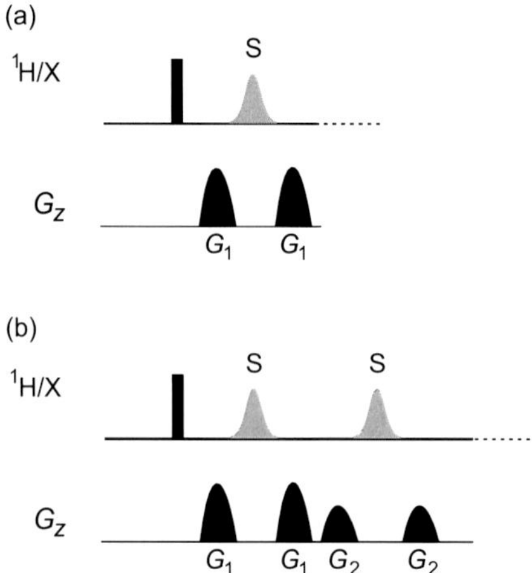

FIGURE 12.24 **Selective excitation PFG sequences.** These are based on (a) a single and (b) a double pulsed field gradient spin-echo. The element S represents any selective 180 degree inversion pulse or pulse cluster.

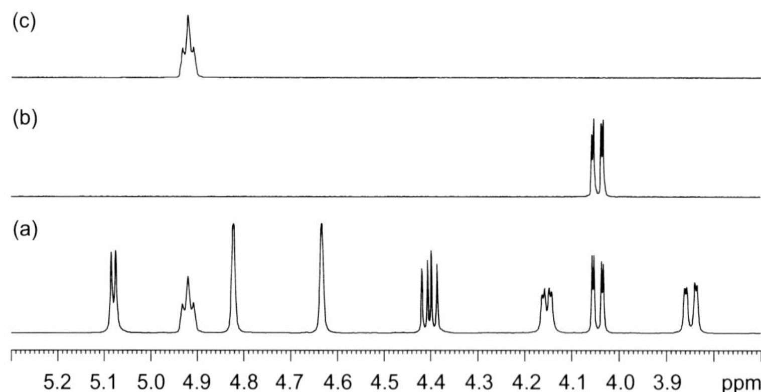

FIGURE 12.25 **Clean selective excitation with the double pulsed field gradient spin-echo sequence.** This used a 40 ms Gaussian 180 degree pulse and gradients of 0.07:0.07:0.03:0.03 T m^{-1}.

resulting excitation profile of the DPFGSE sequence is dictated by the cumulative effect of the repeated inversion pulses, resulting in a 'chipping away' of magnetisation by the series of gradient-echoes, hence the term *excitation sculpting* [57,58]. An example of the clean, pure-phase selective excitation that can be achieved with this sequence is illustrated in Fig. 12.25. This could represent the starting point for a variety of selective 1D experiments, including TOCSY [59,60] and NOESY [58,61]; see for example the 1D gradient TOCSY or NOESY experiments of Sections 6.2.4 and 9.6.2, respectively. The use of a second DPFGSE sequence after one transfer step leads to more elaborate 'doubly selective' experiments also described in Section 9.6.2, while a slight modification produces an extremely effective solvent suppression scheme, described below in Section 12.5. Alternatively, the shorter single-echo excitation may be more appropriate for the study of larger molecules where T_2 relaxation losses can be significant and more generally may be preferable when dealing with broader resonances or fast relaxing spins.

For use in the laboratory, it is convenient to choose a simple, robust inversion pulse as the element S, and the Gaussian pulse is well suited to routine use. Example excitation profiles for this are illustrated in Fig. 12.26 and offer guidance on the selection of pulse duration for a desired excitation window. For proton spectroscopy, a Gaussian pulse of around 40 ms proves suitable for many applications. With increasing ease of implementation of shaped pulses on spectrometers with linearised amplifier outputs, the use of more sophisticated shapes becomes more attractive since their power attenuations may be computed and low-power pulse calibrations are not required. The excitation profile of the Q3 Gaussian cascade is compared with the simpler Gaussian pulse in Fig. 12.26. This demonstrates the more uniform profile one would expect but at the expense of greater pulse duration, as predicted by the corresponding pulse bandwidth factors (Table 12.3).

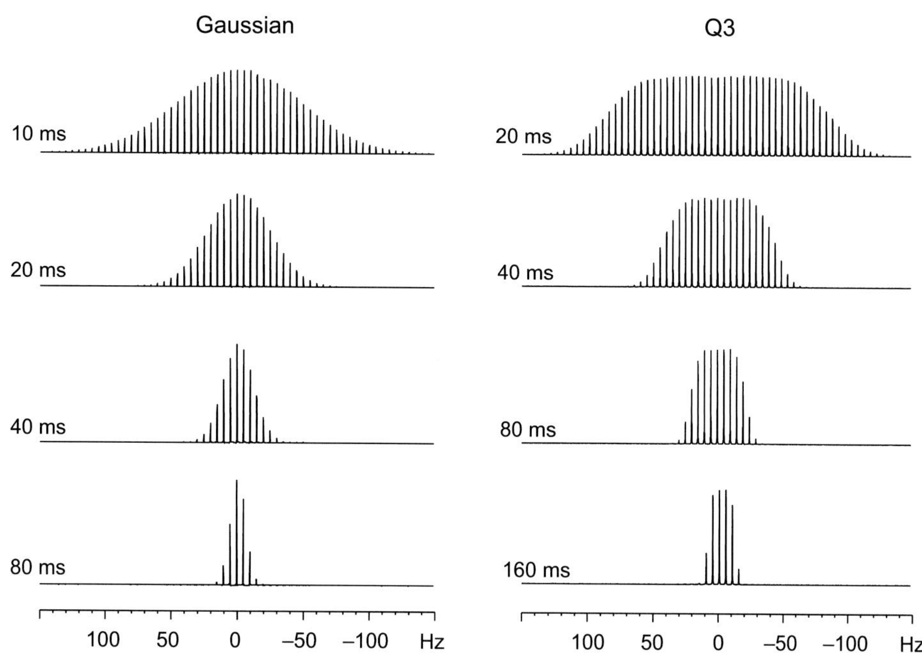

FIGURE 12.26 **Experimental excitation profiles for the double pulsed field gradient spin-echo excitation sequences.** These were recorded with the element S as either a 180 degree Gaussian pulse (truncated at 1%) or a 180 degree Q3 pulse of the durations shown. Line spacings are 5 Hz.

12.4.3 Chemical Shift Selective Filters

Despite the fact that shaped pulses can nowadays produce very clean signal selection, especially when combined with PFGs, they face limitations when a multiplet to be selected overlaps with a neighbouring resonance. This results in the simultaneous selection of multiple peaks and will compromise the experimental results one obtains. This problem is exemplified in Fig. 12.27b which shows excitation sculpting selective excitation of the triplet resonance at 3.415 ppm that is

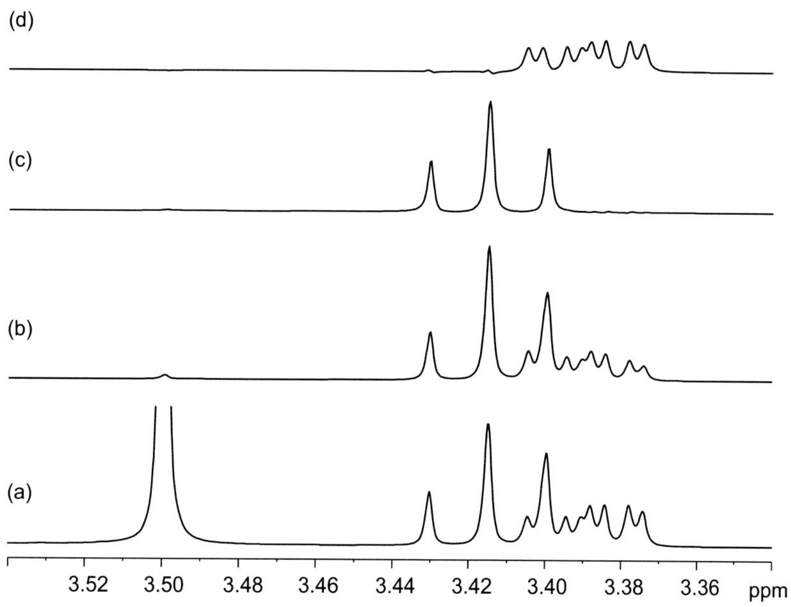

FIGURE 12.27 **The application of chemical shift selective filters.** (a) Partial ^{1}H spectrum, (b) excitation sculpting selection of the triplet at 3.415 ppm using a 40 ms Gaussian pulse, (c) CSSF excitation of the resonance at 3.415 ppm and at (d) 3.390 ppm. The CSSF was optimised for $\Delta\nu = 15.1$ Hz and collected over 8 experiments using a 40-ms Gaussian selective pulse.

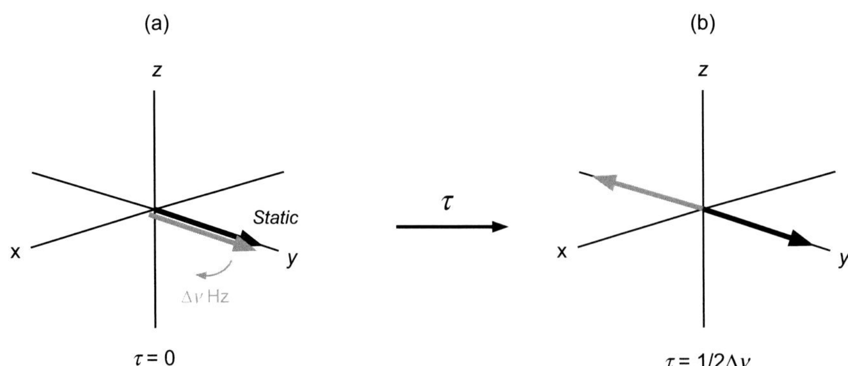

FIGURE 12.28 **The principle of chemical shift selective filters.** The summation of two experiments recorded with (a) $\tau = 0$ and (b) $\tau = 1/2\Delta\nu$, where $\Delta\nu$ is the frequency separation between two spins, will retain the on-resonance spin (black) but cancel the off-resonance spin (grey).

overlapped with another multiplet whose chemical shift differs by only 15 Hz, resulting in the unwanted (partial) excitation of the neighbouring peak. In such instances, it is still possible to select a single resonance, even in the presence of multiplet overlap, through the use of chemical shift selective filters (CSSFs). As the name suggests, these filters will retain the resonance peak at an exactly specified chemical shift but remove those peaks of differing shifts, despite them also falling within the bandwidth of the selective pulse (Fig. 12.27c and d).

Chemical shift selective filtration utilises the differential evolution of chemical shifts that develop during a fixed filter period τ, between the desired spin, which is placed *exactly* on-resonance, and an off-resonance neighbouring spin that is to be suppressed. During the τ delay period, the on-resonance spin will remain static in the rotating frame, while the off-resonance spin will evolve according to its chemical shift offset ($\Delta\nu$ Hz) from the selected resonance (Fig. 12.28). If $\tau = 1/2\Delta\nu$ the off-resonance spin will experience a net rotation of 180 degrees. The co-addition of this data set with one for which $\tau = 0$ (for which no shift evolution occurs for either spin) leads to the constructive summation of the desired on-resonance peak but the destructive cancellation of that which is off-resonance, thus filtering according to chemical shifts.

For this scheme to produce undistorted, in-phase multiplets it is also necessary to suppress the evolution of scalar couplings during the filter period. The CSSFs achieve this by the use of non-selective hard pulses applied subsequent to the selective 180 degree pulse at the centre of a spin-echo, as in the basic CSSF of Fig. 12.29a. Those resonances selected by the shaped pulse are retained in the resulting spectrum, as for excitation sculpting selection, but experience a net 360 degree (0 degree) pulse (the sum of soft and hard 180 degree pulses), resulting in the required chemical shift evolution throughout the filter period for any excited off-resonance spin. Those spins resonating outside the selected bandwidth of the shaped pulse experience only the 180 degree hard pulse, leading to the net refocussing of spin-coupling evolution at the end of the filter delay for all the spin multiplets selected by the shaped pulse. This refocusing of homonuclear couplings shows close similarity to the band-selective pure shift methods of Section 8.5. This J refocussing demands that no spin within the selected region is itself coupled to the on-resonance spin, this requirement being the principle limitation in the application of CSSFs. In the direct acquisition of selective 1D spectra, the sequence also requires a z-filter to produce pure-phase absorptive spectra (the boxed sections in Fig. 12.29), although this would typically be replaced by an appropriate mixing scheme (TOCSY, NOESY or ROESY) when used for selective 1D experiments.

Although in theory the filter should work by summing two experiments recorded with $\tau = 0$ and $1/2\Delta\nu$, in reality this may not produce optimum cancellation of the unwanted resonance due to spin relaxation losses during the $\tau > 0$ filter period. It may also be the case that multiple unwanted off-resonance peaks fall within the bandwidth of the shaped pulse meaning a single optimised τ value cannot be chosen for their suppression. Thus, CSSFs are typically implemented by recording the $\tau = 0$ data and then repeating the experiment N times, progressively incrementing the filter period in smaller steps Δ, and co-adding all data sets. The maximum filter duration $\tau_{max} = N\Delta$ defines the resulting selectivity of the filter and is chosen to be $1/2\Delta\nu$, where $\Delta\nu$ now represents the frequency separation (in Hz) of the *nearest* peak to be suppressed. Typically, Δ = 5–10 ms for ^{1}H spectra and $N \leq 16$. Thus, for a nearest peak offset of 10 Hz, τ_{max} would be 50 ms and one might select Δ = 5 ms and collect 10 filter increments. Since the on-resonance multiplet is co-added from each experiment, the total number of transients required to obtain adequate signal-to-noise may be spread equally over all individual experiments, avoiding the need for lengthy data collection. However, highly selective filters require long total filter periods which may lead to significant signal losses through spin relaxation, and may compromise signal-to-noise relative to the shorter

(a)

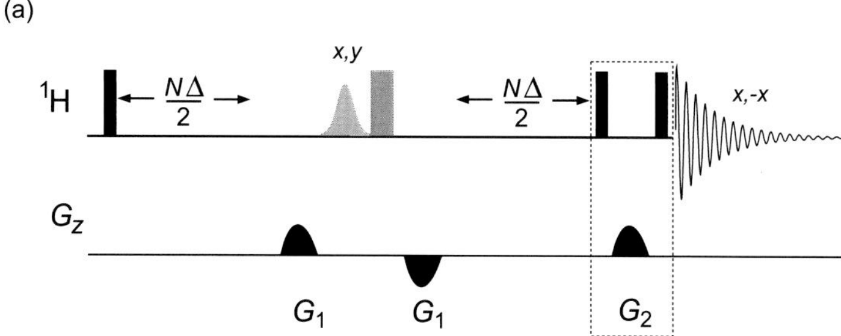

(b)

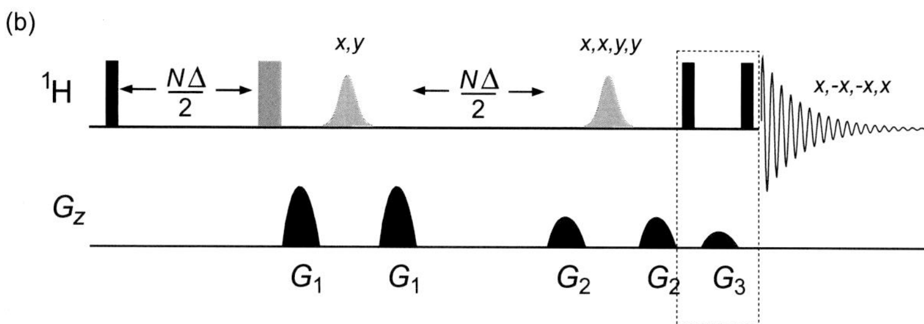

FIGURE 12.29 **CSSF sequences.** (a) The basic single-echo filter sequence and (b) the double-echo filter sequence. The filter delay Δ is held constant and the sequences are repeated with increasing total filter time by incrementing *N* and all resulting FIDs co-added. The boxed z-filter element is required only for demonstrating selective excitation and would typically be replaced with the mixing scheme of choice.

excitation sculpting schemes. The CSSF scheme may also be implemented as the double-gradient spin-echo variant and so benefit from the cleaner phase properties described for excitation sculpting above (Fig. 12.29b).

12.4.4 DANTE Sequences

In the early days of selective excitation, spectrometers were not equipped to generate amplitude modulated rf pulses and the DANTE (delays alternating with nutation for tailored excitation) method was devised, [62] requiring only short, hard pulses. Although largely superseded by the amplitude modulated soft pulses, DANTE may still be the method of choice on older instrumentation.

The basic DANTE sequence is composed of a series of *N* short hard pulses of tip angle α where $\alpha << 90$ degree, interspersed with fixed delays, τ, for free precession:

$$\text{DANTE:} \quad [\alpha - \tau -]_N \quad \text{or more correctly} \quad [\tau/2 - \alpha - \tau/2]_N$$

The total length of the selective pulse is the product $N\tau t_p$ where t_p is the duration of each hard pulse, and the net on-resonance tip angle is the sum of the individual pulses. The effect of such a sequence is illustrated in Fig. 12.30. Following each hard pulse, spins are able to precess during the short delay τ according to their offset from resonance before being pulsed once again. On-resonance trajectories show no precession and are thus driven directly towards the +*y* axis while those close to resonance to follow a zig-zag path. If *N* is sufficiently large (typically ≥ 20) the result closely resembles the smooth path taken under the influence of a soft rectangular pulse of equivalent total duration, and similar selectivity results.

Pulse 'shaping' in the DANTE approach is achieved by keeping the pulse *amplitude* constant but varying the *duration* of each hard pulse throughout the sequence to match the desired envelope. Thus for example a Gaussian envelope may be emulated by varying pulse durations according to a Gaussian profile. Limitations with this approach arise when very small pulse durations are required (<1 μs) since pulse transmitters may then be unable to deliver the necessary precision. In such cases it may be necessary to add fixed attenuation to the transmitter output to allow the use of longer but more accurate hard pulses.

The major difference between soft shaped pulses and DANTE methods is the occurrence of strong sideband excitation windows either side of the principle window with DANTE. These occur at offsets from the transmitter at multiples of the

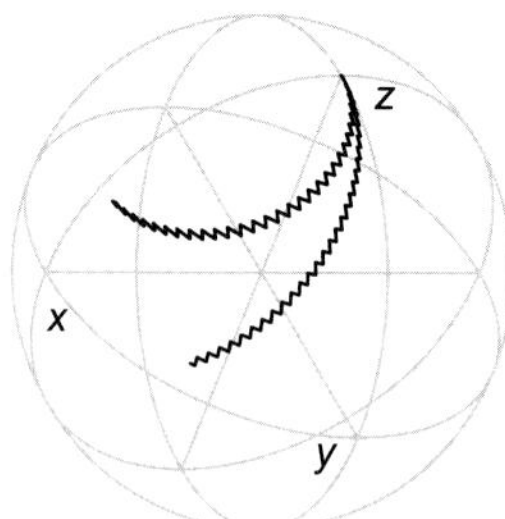

FIGURE 12.30 **Simulated zig-zag excitation trajectories of nuclear spin vectors at different resonance offsets for the DANTE sequence.** Here, 30 pulses of 3 degrees each were separated by delays of 0.33 ms, producing an effective 10-ms selective pulse.

hard-pulse frequency, $1/\tau$. They arise from magnetisation vectors that are far from resonance and which precess full circle during the τ period. Since this behaviour is precisely equivalent to no precession, they are excited as if on-resonance. Further sidebands at $\pm 2/\tau$, $3/\tau$ and so on also occur by virtue of trajectories completing multiple full circles during τ. Such multi-site excitation can at times be desirable [63,64] but if only a single excitation window is required, the hard-pulse repetition frequency must be adjusted by varying τ to ensure the sideband excitations do not coincide with other resonances.

12.4.5 Practical Considerations

This section addresses some of the practical consequences of employing soft pulses which, in general, demand rather more care and attention in their implementation than hard pulses and so typically require operator intervention for their success. The general rule for soft-pulse selection is to choose the simplest pulse that will provide the performance characteristics required. More complex pulse shapes, although providing improved profiles, tend to require the greatest care in calibration and may prove more awkward to use. For example the BURP pulses are extremely sensitive to pulse width miscalibration which leads to significant distortion of otherwise 'uniform' profiles [49]. From a purely practical perspective, the simple Gaussian pulses are most robust and easiest to employ and represent a suitable initial choice for many applications [38] or for novice users. If very high selectivity is required, the pure-phase pulses (BURP or Gaussian cascade families) prove more suitable and the choice of a specific or a universal pulse (Table 12.3) becomes significant, according to the application. The linearised amplifiers found on modern instruments makes implementation of all shaped pulses a more straightforward process since their power settings can be derived by calculation from hard-pulse calibrations and tedious low-power calibrations can be avoided.

The excitation profile of soft pulses is defined by the duration of the pulse, these two factors sharing an inverse proportionality. More precisely, pulse shapes have associated with them a dimensionless *bandwidth factor* which is the product of the pulse duration, Δt, and its effective excitation bandwidth, Δf, for a correctly calibrated pulse. This is fixed for any given pulse envelope, and represents its time efficiency. It is used to estimate the required pulse duration for a desired effective bandwidth; Table 12.3 summarises these factors for some common pulse envelopes. Thus, an excitation bandwidth of 100 Hz requires a 21 ms, 90 degree Gaussian pulse but a 49 ms EBURP2 pulse; clearly the Gaussian pulse is more time efficient but has a poorer excitation profile.

Having determined the necessary pulse duration, the transmitter *power* must be calibrated so that the pulse delivers the appropriate tip angle. As already alluded to, this can be avoided on instruments with linearised amplifier outputs *provided accurate hard-pulse calibrations are known*. The calibration of soft pulses differs from that for hard pulses where one uses a fixed pulse amplitude but varies its duration. For practical convenience, amplitude calibration is usually based on previously recorded calibrations for a soft rectangular pulse (as described later), from which an estimate of the required power change is calculated. Table 12.3 also summarises the necessary changes in transmitter attenuation for various envelopes of equivalent duration, with the more elaborate pulse shapes invariably requiring increased rf peak amplitudes (ie *decreased* attenuation of transmitter output).

As an illustration, suppose one wished to excite a window of 100 Hz with a SNEEZE pulse and had previously determined that a soft 90 degree rectangular pulse of 10 ms required an attenuation of 60 dB. From the bandwidth factor, one can determine the pulse duration must be 58 ms. From the power ratio equations given in Section 3.5.1, one may calculate that a soft *rectangular* pulse of this duration requires 15.3 dB *greater attenuation* than the 10 ms pulse (20 log 5.8), simply because it is longer. Table 12.3 shows that the SNEEZE envelop requires 26.6 dB *less* attenuation than a rectangular pulse of equal duration. The SNEEZE pulse therefore requires 11.3 dB (26.6−15.3) *less attenuation* than that of the reference 10 ms soft rectangular pulse, and the transmitter amplitude setting becomes 48.7 dB. Fine tuning may then be required for

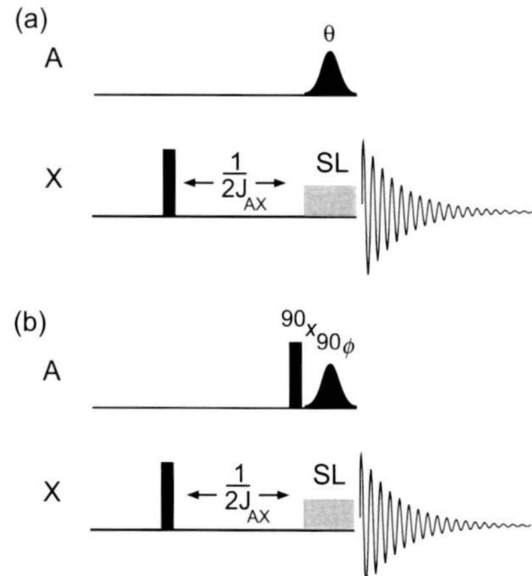

FIGURE 12.31 **Soft-pulse calibration sequences.** These are designed to calibrate (a) amplitude and (b) hard/soft phases differences for selective pulses applied on the indirect (decoupler) channel. SL is a spin lock applied to decouple A and X during the soft pulse.

optimum results without linearised amplifiers, but when these are available the power may be determined directly from the high-power pulse 90 degree calibration. In this example, with a 90 degree pulse of 10 μs at 0 dB attenuation, the calculation becomes 20(log 58000/10) − 26.6 = 48.7 dB. Fortunately, instrument software tools are now provided to perform such calculations, making the implementation of shaped pulses reasonably straightforward.

12.4.5.1 Amplitude Calibration

In the absence of linearised amplifiers, the amplitude calibration of a soft pulse essentially follows the procedures introduced for hard pulses in Section 3.5.1 and the descriptions below assume familiarity with these. For soft pulses on the observe channel, the transmitter frequency should be placed on-resonance for the target spin and the 90 degree or 180 degree condition sought directly by variation of transmitter power, starting with very low values (high attenuations) and progressively increasing (note there is a logarithmic not linear dependence on dB). If calibrations must be performed indirectly, for example on the decoupler channel of older instruments, a slight variation on the method of Fig. 3.62 for a 2-spin AX pair is used (Fig. 12.31a). In this an additional spin-lock pulse is applied to the observe channel during the soft pulse to decouple the AX interaction [65]. This collapse of the AX doublet means the soft pulse can be applied to the centre of the A-spin doublet. Variation of the transmitter power produces results similar to those of Fig. 3.63 when the 90 degree or 180 degree condition is satisfied. With modern instrumentation, however, calibrations determined directly on the observe channel are equally valid if this is later used as the indirect (decoupler) channel in subsequent experiments, so bypassing the need for such indirect calibrations.

12.4.5.2 Phase Calibration

Precise control of the *relative* phase between pulses is crucial to the success of many multi-pulse NMR experiments and some correction to the phase of a soft pulse may be required to maintain these relationships when both hard and soft pulses are to be applied to the same nucleus. When soft pulses are used on the observe channel the phase difference (which may arise because of the potentially different rf paths used for high and low power pulses) may be determined by direct inspection of two separate 1D pulse-acquire spectra recorded with high- and low-power pulses but under otherwise identical conditions. Using only zero-order (frequency independent) phase correction of each spectrum, the difference in the resulting phase constants (soft *minus* hard) represents the phase difference between the high- and low-power rf routes. Adding this as a constant offset to the soft-pulse phase should yield spectra of identical phase to that of the hard-pulse spectrum when processed identically and this correction can be used in all subsequent experiments, *provided the soft pulse power remains unchanged.*

If phase differences must be determined indirectly, the sequence of Fig. 12.31b is suitable in which an additional hard pulse is used on the indirect channel prior to the soft. When two 90 degree pulses are of the same phase their effect is additive and the sequence behaves as if a net 180 degree pulse has been applied to the selected spin, resulting in an inversion of the anti-phase AX doublet. Similarly, two 180 degree pulses of similar phase behave as a net 360 degree pulse and have no effect on the doublet.

12.5 SOLVENT SUPPRESSION

Fortunately the majority of solvents used in solution NMR spectroscopy are readily available in the deuterated form. For proton spectroscopy in particular, this allows the chemist to focus on the solute spectrum, undisturbed by the solvent that is present in vast excess. Unfortunately, most molecules of biochemical or medicinal interest, notably biological macromolecules, must be studied in water and in order to observe all protons of interest within these molecules, including the often vitally important labile protons, protonated water must be employed as the solvent (usually containing 5–10% D_2O to maintain the field-frequency lock). Whereas H_2O is 110 M in protons, solute concentrations are more typically in the millimolar region and the 10^4–10^5 concentration difference imposes severe experimental difficulties which demand the attenuation of the solvent resonance. In addition, liquid chromatography (LC)-NMR may favour the use of protonated solvents for reasons of economy, also making efficient and robust solvent suppression essential. At a more mundane level, when using D_2O or CD_3OD for routine spectroscopy a considerable residual HDO resonance may remain that may limit the useable receiver amplification and appear aesthetically unappealing.

The principle reason to suppress a large solvent resonance in the presence of far smaller solute resonances is to ensure that the dynamic range of the NMR signals lies within the dynamic range of the receiver and analogue-to-digital converter (ADC) (Section 3.2.6). Further concerns include the baseline distortions, t_1-noise in 2D experiments, radiation damping and potential spurious responses that are associated with very intense signals. Radiation damping leads to severe and undesirable broadening of the water resonance which may mask the solute resonances. It arises because the intense NMR signal produced on excitation of the water generates a strong rf current in the detection coil (which ultimately produces the observed resonance). This current in turn generates its own rf field which drives the water magnetisation back towards the equilibrium position; if you like, it acts as a water-selective pulse. The loss of transverse magnetisation is therefore accelerated, producing a reduced apparent T_2 and hence a broadened resonance. A test for radiation damping involves detuning the proton coil and reacquiring the proton spectrum. This degrades the rf coupling so reducing the back electromotive force (emf) and sharpening the water resonance. In general, good lineshape is a prerequisite for successful solvent suppression and while efforts to optimise shimming pay dividends, the intense water singlet of H_2O is ideally suited to gradient shimming (Section 3.4.5). Improved results have also been reported when microscopic air bubbles have been removed from the sample by ultrasonic mixing.

The goal of solvent suppression is therefore to reduce the magnitude of the solvent resonance *before* the NMR signal reaches the receiver. This seemingly simple requirement has generated an enormous research area [66,67], emphasising the fact that this is by no means a trivial exercise. The more widely used approaches can be broadly classified into three areas:

- methods that saturate the water resonance;
- methods that produce zero net excitation of the water resonance; and
- methods that destroy the water resonance with PFGs.

The following sections illustrate examples from these areas that have proved most popular but represent only a small subsection of available methods [68]. All schemes inevitably involve some loss of signal intensity for those resonances close to that of the solvent, and careful adjustment of solution conditions, the simplest being sample temperature, can prove useful in avoiding signal loses by shifting the water resonance relative to solute signals (Fig. 12.32).

12.5.1 Presaturation

The simplest and most robust technique is presaturation of the solvent [69]. This is simple to implement, may be readily added to existing experiments and leaves (non-exchangeable) resonances away from the presaturation frequency unperturbed. It involves the application of continuous, weak rf irradiation at the solvent frequency prior to excitation and acquisition (Fig. 12.33a), rendering the solvent spins saturated and therefore unobservable (Fig. 12.34). Invariably resonances close to the solvent frequency also experience some loss in intensity, with weaker irradiation leading to less spillover but reduced saturation of the solvent. Longer presaturation periods improve the suppression at the expense of extended experiments so a compromise is required and typically 1–3 s are used; trial and error usually represents the best approach to optimisation. Wherever possible the same rf channel should be used for both the presaturation and subsequent proton pulsing, with appropriate transmitter power switching.

A feature of simple presaturation in protonated solutions (particularly with older probes which may also lack appropriate shielding of the coil leads [70]) is a residual 'hump' in the 1D spectrum that originates from peripheral regions of the sample that suffer B_0 and B_1 inhomogeneity. An effective means of suppressing this for 1D acquisitions is the so-called NOESY-presaturation sequence which simply employs a *non-selective* 1D NOESY with short mixing time,

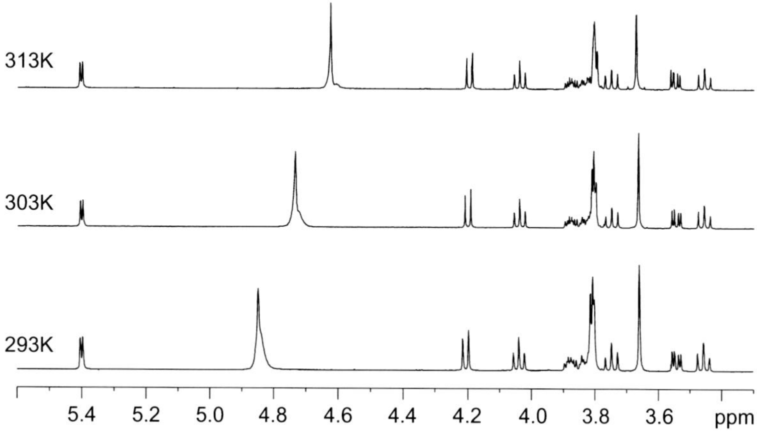

FIGURE 12.32 **The temperature dependence of HDO (here partially suppressed).** The shift corresponds to approx. 5 Hz/K at 500 MHz. Spectra are referenced to internal TSP.

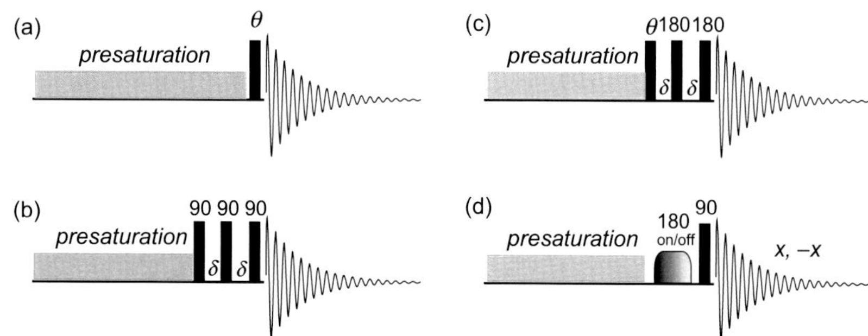

FIGURE 12.33 **Solvent suppression schemes.** These are based on (a) presaturation alone, (b) 1D NOESY, (c) FLIPSY and (d) Pre-SAT180. Sequence (b) makes use of the conventional NOESY phase cycle whereas FLIPSY uses EXORCYCLE on one (or both) of the 180 degree pulses (ie pulse = x, y, $-x$, $-y$; receiver = x, $-x$, x, $-x$). Sequence (d) applies the 180 degree adiabatic sweep on alternate scans combined with receiver inversion; optional opposing purging gradients may be applied either side of the 180 degree pulse.

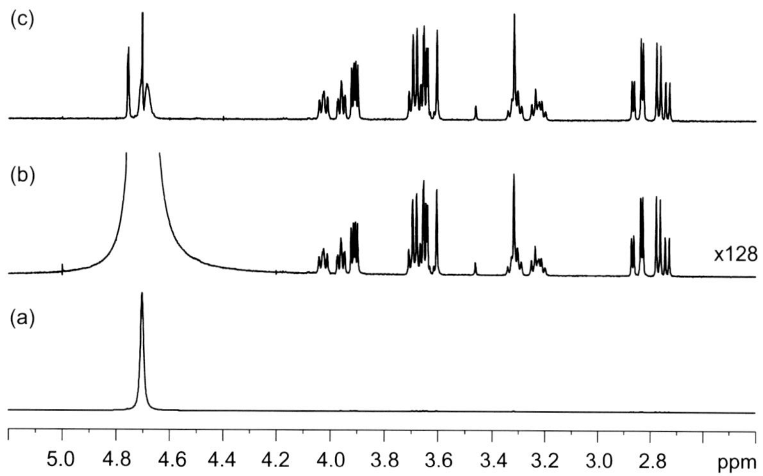

FIGURE 12.34 **Solvent presaturation (2 s).** This reveals a resonance at 4.75 ppm in (c) previously masked by the HDO resonance in (a) and its expansion (b).

or in other words, the first increment of a 2D NOESY experiment with mixing times of around 0–10 ms (Fig. 12.33b). The usual NOESY phase-cycle suppresses the hump, but the sequence is restricted to acquisition with 90 degree pulses which is not optimal for signal averaging or spectrum integration. A variation, termed FLIPSY [71] (flip-angle adjustable 1D NOESY, Fig. 12.33c), allows the use of an arbitrary excitation pulse tip angle θ and is thus better suited to routine use.

A more recent approach [72] to cancelling the solvent signal arising from peripheral regions of the sample that minimises the loss of solute signal intensity employs an alternating (on/off) 180 degree adiabatic sweep after simple presaturation but immediately prior to the observe pulse (Fig. 12.33d). Subtraction of the data sets acquired with and without the adiabatic pulse leads to the co-addition of all signals in the bulk of the sample as these are inverted by the swept pulse, but cancellation of the peripheral solvent signal, which does not sense the inversion. The so-called Pre-SAT180 method may also be incorporated within standard 2D sequences as an alternative to simple presaturation alone.

The principle disadvantage of all presaturation schemes is that they also lead to the suppression of exchangeable protons by the process of saturation transfer. In favourable cases solution conditions may be altered (temperature and pH) to slow the exchange sufficiently to reduce signal attenuation, but this approach has limited applicability. The methods presented below largely avoid such losses.

12.5.2 Zero Excitation

The second general approach strives to produce no net excitation of the solvent resonance. In other words, the sequence ultimately returns the solvent magnetisation to the $+z$ axis while at the same time placing all other magnetisation in the transverse plane prior to acquisition. The following section describes a related approach commonly used in conjunction with PFGs.

12.5.2.1 Jump-Return

To appreciate the principle behind zero excitation, consider the simplest example, the 'jump-return' sequence [73] 90_x–τ–90_{-x} which subsequently spawned many others. The transmitter frequency is placed on the solvent resonance and all magnetisation is tipped into the transverse plane by the first hard 90 degree pulse. During the subsequent delay τ (typically a few hundred μs long) all vectors fan out in the transverse plane according to their offsets, except that of the solvent which, being on-resonance, has zero frequency in the rotating frame. Thus, the solvent resonance is tipped back to the $+z$-axis as the second 90 degree pulse rotates the xy plane into the xz plane. The only remaining transverse magnetisation is the $\pm x$-component of all vectors prior to the second pulse. Hence, this produces a sine-shaped excitation profile (Fig. 12.35a) with maximum amplitude at offsets from the transmitter frequency of $\pm 1/4\tau$ Hz and with resonances either side of the transmitter displaying opposite phase (owing to their $\pm x$ orientations) but without additional phase dispersion. In practice this simple scheme produces a rather narrow null at the transmitter offset and a significant solvent resonance typically remains, so more sophisticated sequences have been investigated.

12.5.2.2 Binomial Sequences

The binomial sequences aim to improve the zero-excitation profile and provide schemes that are less sensitive to spectrometer imperfections. The series may be written $1\text{–}\bar{1}$, $1\text{–}\bar{2}\text{–}1$, $1\text{–}\bar{3}\text{–}3\text{–}\bar{1}$ .. and so on, where the numbers indicate the relative pulse widths, each separated by a delay τ, and the overbar indicates phase inversion of the pulse. For *off-resonance spins* the pulse elements are *additive* at the excitation maximum so for example should one require 90 degree off-resonance excitation, $1\text{–}\bar{1}$ corresponds to the sequence 45_x–τ–45_{-x}. Of this binomial series, it turns out that the $1\text{–}\bar{3}\text{–}3\text{–}\bar{1}$ sequence [74] has good performance and is most tolerant of pulse imperfections by virtue of its symmetry [75]. The trajectory of spins with frequency offset $1/2\tau$ from the transmitter for a net 90 degree pulse ($1 \equiv 11.25$ degree) is shown in Fig. 12.36. During each τ period the spins precess in the rotating frame by half a revolution so the effect of the phase inverted pulses is additive and the magnetisation vector is driven stepwise into the transverse plane. As before, the on-resonance solvent vector shows no precession so is simply tipped back and forth, finally terminating at the North Pole.

While maximum excitation occurs at $\pm$ $1/2\tau$ Hz from the transmitter offset, further nulls occur at offsets of $\pm$ n/τ (n= 1, 2, 3,... corresponding to complete revolutions during each τ) so a judicious choice of τ is required to provide excitation over the desired bandwidth. The excitation profiles of the $1\text{–}\bar{1}$ and $1\text{–}\bar{3}\text{–}3\text{–}\bar{1}$ sequences are shown in Fig. 12.35b and c. Clearly the excitation is non-uniform, so places limits on quantitative measurements, and once again there exists a phase inversion either side of the solvent. Both provide an effective null at the transmitter offset and suppression ratios in excess of 1000-fold can be achieved.

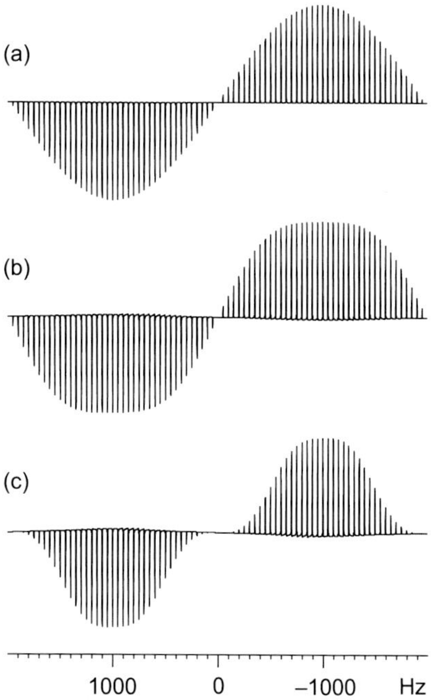

FIGURE 12.35 **Simulated excitation profiles.** These are calculated for (a) the jump-return sequence, (b) the $1\text{-}\bar{1}$ and (c) $1\text{-}\bar{3}\text{-}3\text{-}\bar{1}$ binomial sequences. The jump-return sequence used $\tau = 250$ μs and the binomial sequences used $\tau = 500$ μs.

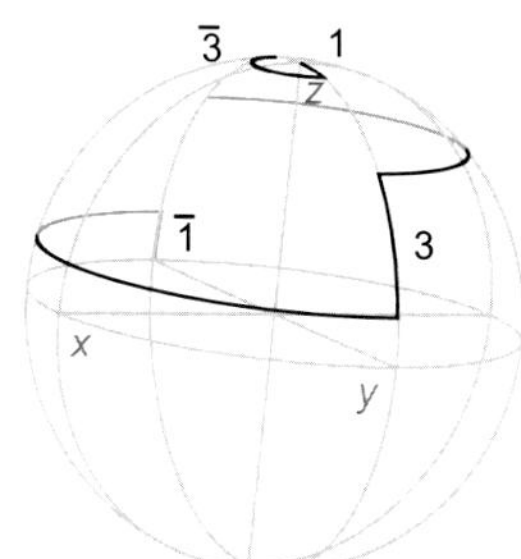

FIGURE 12.36 **The trajectory of a magnetisation vector during the $1\text{-}\bar{3}\text{-}3\text{-}\bar{1}$ binomial sequence.** The trace is shown for a spin with an offset from the transmitter of $+1/2\tau$ Hz, corresponding to the excitation maximum.

When implementing this sequence it may be necessary to attenuate the transmitter output to increase the duration of each pulse so that the shorter elements do not demand very short (<1 μs) pulses (note the similarity with the requirements for the DANTE hard-pulse selective excitation described earlier). The binomial sequences can be adjusted to provide an arbitrary overall tip angle by suitable adjustment of the tip angles for each element. For example *inversion* of all off-resonance signals can be achieved by doubling all elements relative to the net 90 degree condition. Exactly this approach has been exploited in the very effective gradient-echo methods described in the following section.

12.5.3 Pulsed Field Gradients

The most effective approach to date to solvent suppression is the destruction of the net solvent magnetisation by PFGs, so ensuring nothing of this remains observable immediately prior to acquisition. The gradient schemes described here act on transverse magnetisation so are readily appended to existing 1D and multi-dimensional sequences [76].

12.5.3.1 WATERGATE

One popular approach, termed WATERGATE [77,78] (water suppression by gradient-tailored excitation) makes use of a single PFG spin-echo, G_1-S-G_1, (Fig. 12.37a) in which both gradients are applied in the same sense. The element S is chosen to provide zero net rotation of the solvent resonance but to provide a 180 degree *inversion* to all others. This results in the solvent magnetisation experiencing a cumulative dephasing by the two gradients leading to its destruction while all others are refocused in the spin-echo by the second gradient and are therefore retained.

As apparent from the previous section, a binomial sequence has a suitably tailored profile for the element S, and the series $3\alpha\text{-}9\alpha\text{-}19\alpha\text{-}\overline{19\alpha}\text{-}\overline{9\alpha}\text{-}\overline{3\alpha}$ (Fig. 12.38a, with $26\alpha = 180$ degrees and a delay τ between pulses, here termed W3 [79]) has a desirable off-resonance inversion profile for this purpose. The WATERGATE excitation profile for this is shown in Fig. 12.39a. Once again characteristic nulls also occur at offsets of $\pm n/\tau$ Hz, but between these the excitation is quite uniform and does not suffer the phase inversion of the unaccompanied 90 degree binomials. Extended binomial sequences have been shown to provide a narrower notch at the transmitter offset, so reducing the attenuation of solute signals close to

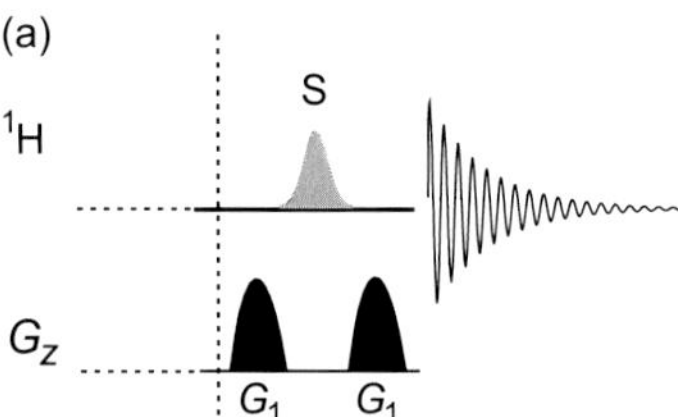

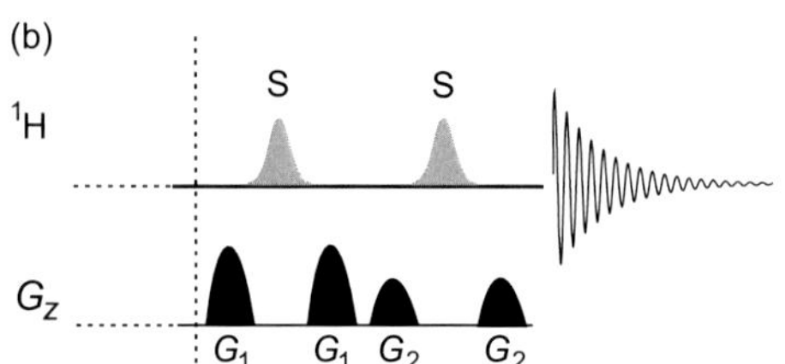

FIGURE 12.37 **Solvent suppression schemes employing pulsed field gradients.** These are based on (a) WATERGATE (single-echo) and (b) excitation sculpting (double-echo) principles. The pulse element S has zero net effect on the solvent resonance but inverts all others.

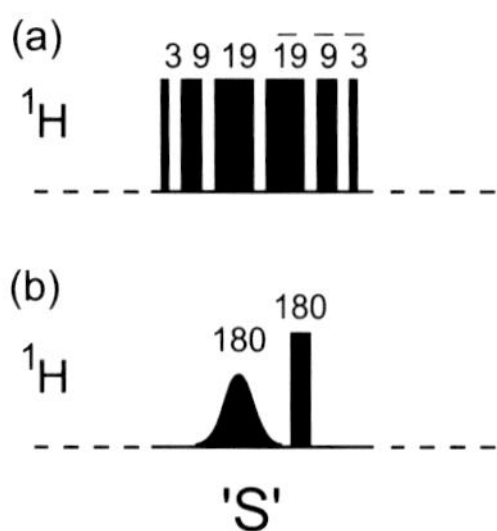

FIGURE 12.38 **Off-resonance inversion elements.** Two possible versions of the element S are shown based on (a) a binomial-type hard-pulse sequence and (b) a combination of soft and hard pulses.

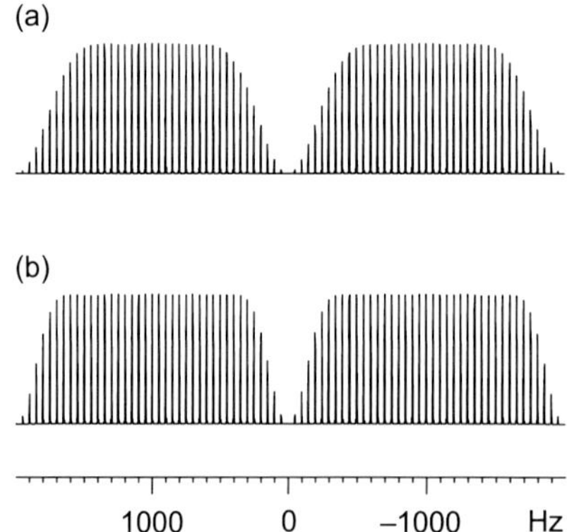

FIGURE 12.39 **Excitation profiles for WATERGATE.** These were simulated using (a) the W3 and (b) W5 binomial-type sequences to provide a null at the solvent shift (τ = 500 μs).

the solvent, and have a wider region of uniform excitation [79]. The improved profile of the so-called W5 sequence is shown in Fig. 12.39b. This sequence lacks the elegant shorthand form of its predecessor and is thus represented 7.8–18.5–37.2–70.0–134.2–$\overline{134.2}$–$\overline{70.0}$–$\overline{37.2}$–$\overline{18.5}$–$\overline{7.8}$, where each number represents the pulse tip angle in degrees.

The element S may also be applied as a combination of hard and soft 180 degree pulses [57,77] (Fig. 12.38b). Here, the soft pulse acts only on the water resonance, causing it to experience a net 360 degree rotation, while all others undergo the desired inversion by virtue of the hard pulse. In this approach, the suppression profile is dictated by the inversion properties of the soft-pulse combination, and off-resonance nulls no longer occur (although correction for any phase difference between the hard and soft pulses may be required). Typically the soft 180 degree pulse has a short duration (~2–8 ms) and a rectangular or, better still, smoothed profile. This method has the notable advantage that the suppression may be applied away from the centre of the spectrum simply by shifting the location of the selective pulse and allows for optimal transmitter positioning for multidimensional experiments.

12.5.3.2 Excitation Sculpting

An improved approach to gradient suppression employs the method of 'excitation sculpting' described in Section 12.4.2 and applies a double PFG spin-echo instead of one [57]. This has the advantage of refocusing the evolution of homonuclear couplings and can produce spectra with improved phase properties and less baseline distortion. The elements S are exactly

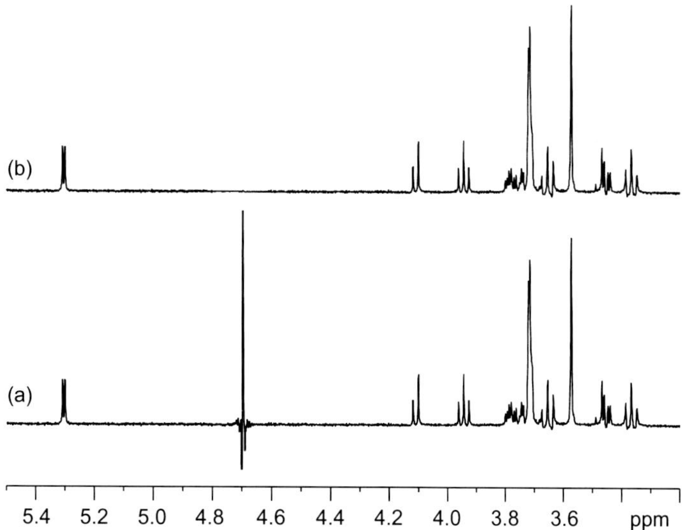

FIGURE 12.40 **Solvent suppression with the excitation sculpting scheme.** This uses the approach of Figure 12.38b with a 8-ms soft pulse (Gaussian) and gradients of 0.1:0.1:0.03:0.03 T m^{-1}. The sample is 2 mM sucrose in 9:1 H_2O:D_2O. In (b) the small residual solvent signal has been completely removed through additional processing of the FID (see text).

as above, except with the double-echo the suppression notch is wider since it is applied twice. The water suppression spectrum of the 2 mM sucrose in 90% H_2O using this approach is shown in Fig. 12.40a, and illustrates the impressive results now routinely available. The residual solvent signal can be further removed (Fig. 12.40b) by subtracting low-frequency components from the FID prior to Fourier transformation [80], which requires that the transmitter frequency be placed on the solvent resonance. Alternatively, the spectrometer software may be able to remove user-defined, higher frequency components of the FID such as those corresponding to the residual solvent resonance, allowing arbitrary positioning of the transmitter. The excitation sculpting sequence can also be readily tailored to achieve multi-site suppression [81,82].

12.5.3.3 Perfect Echo Suppression

One of the features of the gradient suppression methods described above is the potentially wide suppression notch they generate, leading to the loss of resonances close to that of the suppressed water. The use of shaped pulses in the suppression element *S* allows improved selectivity by the use of longer pulses, and helps retain nearby solute resonances. However, scalar coupling evolution during the shaped pulse will give rise to *J*-modulation distortions in the final spectrum that become more severe the more selective the shaped pulse; such perturbations can be seen in the multiplets of Fig. 12.40. To suppress these distortions, it is possible to employ the so-called 'perfect echo' scheme introduced in Section 2.4.4 in the periodic refocusing of J evolution by coherence transfer (PROJECT)/Carr–Purcell–Meiboom–Gill (CPMG) sequence. In this, its role was to suppress J-modulation effects in CPMG spin-echo trains, and it may be used to similar effect with the echoes used for water suppression schemes [83]. For this, it is sufficient to prepend a single-echo to the desired suppression element of Fig. 12.37 and separate them with an orthogonal 90 degree pulse which serves to exchange coherences between spins and so reverse the apparent sense of J modulation (Fig. 12.41). Thus, J evolution occurring in the preparatory echo refocuses during the water suppression echo, so removing J-modulation distortions to multiplet structures. This improvement is clearly apparent in the proton multiplets of Fig. 12.42 generated in spectra recorded with a relatively long water-selective pulse.

Finally, we note the PURGE method (presaturation utilizing relaxation gradients and echoes) [84] offers excellent baseline and phase properties while providing efficient and highly selective suppression. This combines solvent presaturation with gradient-purged echo periods prior to acquisition and has been recommended for when quantitative or statistical analyses of 1D spectra are required. Its implementation within multidimensional experiments is possible but potentially limited by the requirements for phase cycling.

One final point regarding PFGs is that their use in certain experiments for coherence selection also leads to solvent suppression without further modification. For example the DQF-COSY experiment inherently filters out uncoupled spins (ie singlets) and when PFGs are used to provide this filtering the singlet water resonance is also removed at source. In fact, the original publication that stimulated widespread use of field gradients in high-resolution NMR impressively demonstrated such suppression [85]. Similarly, heteronuclear correlation experiments such as HSQC and HMQC intrinsically select

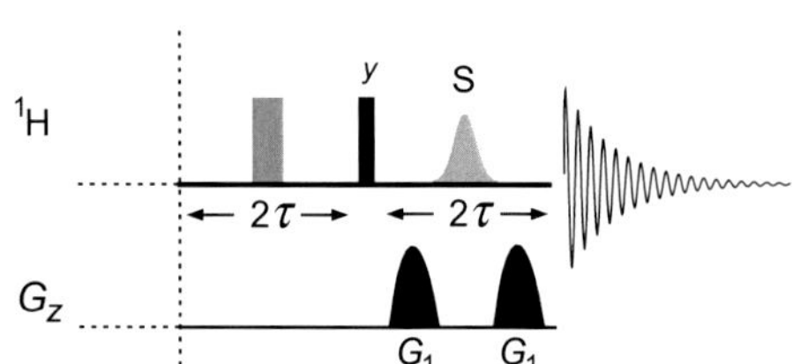

FIGURE 12.41 **Solvent suppression using the perfect echo scheme to remove distortions caused by J modulation.** The suppression element may replace the gradient-echo elements in either of the schemes shown in Figure 12.37.

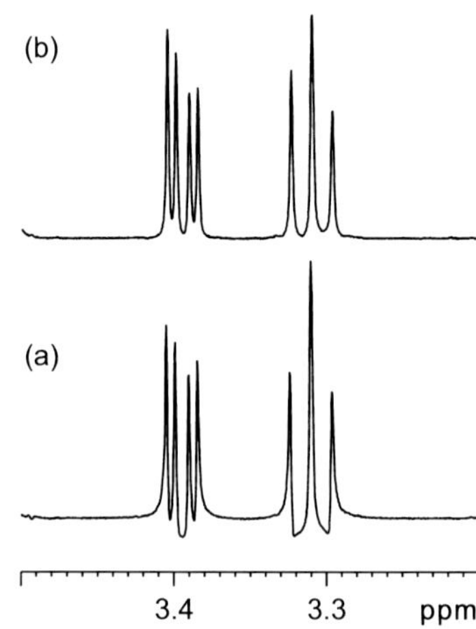

FIGURE 12.42 **Use of perfect echo suppression.** A comparison of multiplet structures seen for (a) the standard excitation sculpting scheme and (b) the perfect echo excitation sculpting variant. A 4-ms rectangular pulse was used to selectively invert water.

protons bound to, say, ^{13}C or ^{15}N and thus also reject water. While these methods on their own are unlikely to prove sufficient for 90% H_2O solutions (in which case the above gradient-echo methods may be appended; see also the flip-back HSQC in Section 11.8.1), they are often sufficient for 'wet' D_2O samples.

12.6 SUPPRESSION OF ZERO-QUANTUM COHERENCES

Zero-quantum coherence (ZQC) was described in Section 5.3 as a form of multiple-quantum coherence that may arise between J-coupled spins and is often generated in multiple-pulse NMR sequences. As with all forms of multiple-quantum coherence it does not give rise to observable signals but can nevertheless be reconverted into observable single-quantum coherence and so give rise to responses in an NMR spectrum that are often considered undesirable artefacts. Commonly employed experiments in which this can be a problem include NOESY and TOCSY in which the zero-quantum contributions give rise to anti-phase dispersive signals that complicate or even mask the appearance of the desired peaks; see for example Fig. 9.38. Unfortunately, since the ZQC has a coherence order $p = 0$, it is insensitive to both phase cycling and PFGs and so cannot be suppressed by either of these methods alone. In this section we review a number of approaches to zero-quantum suppression that can be applied to a wide range of pulse techniques, as has been demonstrated in previous chapters. All methods rely on the fact that ZQC is one manifestation of the general notion of transverse magnetisation and that this can evolve during periods of free precession according to the frequency *differences* of the chemical shifts of the coupled spins A and X ($|\nu_A - \nu_X|$; expressed in Hz). The manipulation of the ZQC evolution provides the key to its suppression.

12.6.1 The Variable-Delay Z-Filter

The idea behind the z-filter is that only desired in-phase magnetisation present during a pulse sequence can pass through the filter and return unaffected while all other components are suppressed. The basic filter is illustrated in Fig. 12.43a and works by temporarily 'storing' the desired magnetisation along the *z*-axis while other components are somehow destroyed (by phase cycling or PFGs) and then returning this to the transverse plane for detection free from undesirable contributions. Considering the case of ZQCs, the interpulse delay τ_{zf} allows any coherence generated to evolve as described earlier and to obtain a phase dictated by the zero-quantum evolution frequency. If one were to repeat the experiment but this time use a different filter delay $\tau_{zf(2)}$ then the extent of evolution and hence the phase acquired will be different from that in the first experiment using delay $\tau_{zf(1)}$. If the filter timings could be set such that the phase difference between the two experiments was exactly 180 degrees, that is when $\tau_{zf(1)} - \tau_{zf(2)} = (2|\nu_A - \nu_X|)^{-1}$, then addition of the two data sets would lead to cancellation of the zero-quantum contributions. In reality, there will exist a range of zero-quantum frequencies evolving, arising from different coupled spin pairs, so it is usually necessary to repeat the experiment many times using systematic variation of the τ_{zf} delays and co-add all the data sets to achieve sufficient artefact suppression [86,87]. The amount of variation required

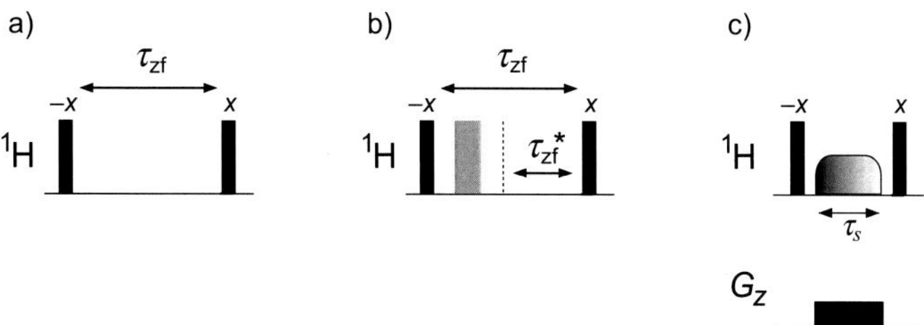

FIGURE 12.43 **Approaches to suppressing zero-quantum coherences.** These employ (a) the basic z-filter, (b) the spin-echo z-filter and (c) the zero-quantum dephasing element employing a rf frequency sweep/gradient pulse combination; see text for details.

clearly depends on the shift differences between coupled spins whereby large differences demand a smaller spread in τ_{zf}. Since higher field magnets also give rise to larger shift separations (expressed in Hz) correspondingly smaller variations are again required. Typical values of τ_{zf} will range from 0–20 ms which should suppress all zero-quantum contributions for J-coupled spins with $\nu_A-\nu_X > 25$ Hz. The need to repeat experiments to achieve suppression can dictate lengthy acquisitions and represents the principle limitation to this approach, leading to the development of more efficient alternatives.

12.6.2 Zero-Quantum Dephasing

The most recent, successful and widely applicable approach to suppressing zero-quantum contributions employs a PFG method for the net dephasing of these coherences *in a single scan* with no requirement for experiment repetition [88]. Since PFGs alone cannot manipulate zero-quantum coherences, the gradients must be applied in a rather ingenious manner to be effective.

To begin we again consider the approach described above for generating differential evolution of ZQCs by variation of the total duration of the *z*-filter delay, τ_{zf}. An alternative approach to this would be to maintain τ_{zf} at a fixed duration and to place within this a 180 degree pulse that moves throughout the delay on each repetition of the experiment [89] (Fig. 12.43b). This pulse introduces a spin-echo to the *z*-filter in which the ZQ evolution is refocused such that the net evolution is restricted only to the residual period τ_{zf}*. Relocation of the 180 degree pulse for each experiment thus leads to differential evolution between them.

Now consider the action of the zero-quantum dephasing sequence of Fig. 12.43c. This employs a frequency swept (adiabatic) 180 degree pulse simultaneously with a field gradient applied along the *z*-axis of the sample. The gradient imposes a frequency spread across the length of the sample tube and so imparts the usual spatial encoding such that responses from spins at the top of the tube would correspond to, say, the high-frequency end and those at the bottom to the lower frequency end. At the same time, the 180 degree adiabatic pulse sweeps through this frequency window (Section 12.2) meaning it inverts spins in different spatial regions of the sample *at different times*. Assuming this sweeps from high to low frequency, this would in the above scenario, correspond to the application of a refocusing pulse initially to the top of the sample tube followed by a steady progression down the tube, finally refocusing at the bottom as the swept pulse ends. In a manner similar to that described above for the scheme of Fig. 12.43b, this results in differential net evolution of the ZQCs but now in *different spatial regions* of the sample (Fig. 12.44). The detected NMR response is clearly the sum of all these regions along the length of the NMR sample leading to a destructive summation for all zero-quantum coherences *on every scan* and hence their removal from the final spectrum.

The degree of attenuation A_{ZQC} depends upon the zero-quantum frequency Ω_{ZQC} (here in rad/s) and the duration of the frequency sweep τ_s according to [88]:

$$A_{ZQC} = \frac{\sin \Omega_{ZQC} \tau_S}{\Omega_{ZQC} \tau_S} \tag{12.6}$$

This approximates to $1/\Omega_{ZQC}\tau_s$ when the sine term oscillations are ignored and allows an estimate of the sweep durations that need to be employed. Thus, in the case of a spectrum recorded at 600 MHz with J-coupled spins separated by 0.5 ppm (ie ZQC frequencies equivalent to 300 Hz or approximately 2000 rad/s) a 50 ms sweep will attenuate ZQC by ~99 % ($A_{ZQC} \sim 0.01$). For spins separated by 1 ppm at this field strength a 25-ms sweep would be sufficient to produce similar attenuation. In general, the larger the frequency separation between the coupled spins, due to greater chemical shift

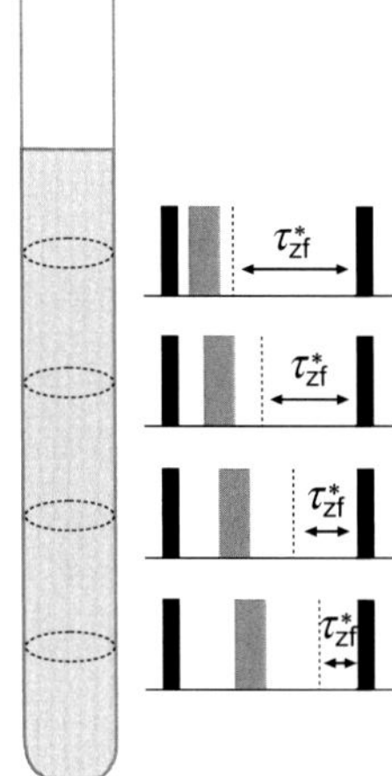

FIGURE 12.44 **Schematic representation of events during the zero-quantum dephasing element.** The effective timing of the 180 degree refocusing pulse and hence the net ZQ evolution time τ_{zf}* becomes dependent upon spatial location along the *z*-axis meaning the zero-quantum coherences evolve to differing extents according to the position of the participating spins in the sample.

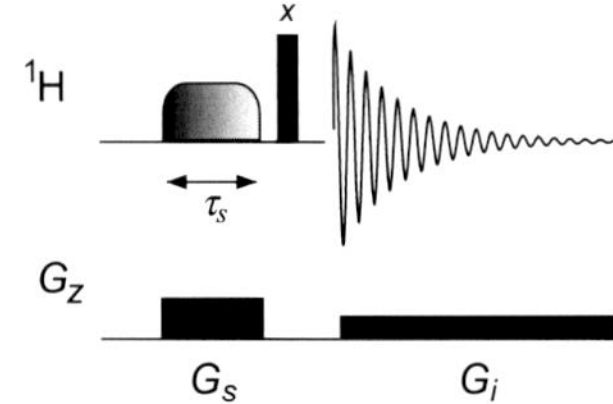

FIGURE 12.45 **The zero-quantum dephasing calibration sequence.**

differences or the use of higher B_0 fields, the shorter the sweep durations that need be employed. Conversely, coupled spins with similar chemical shifts will present the most demanding cases for suppression.

It has also been shown that greater degrees of attenuation can be achieved without the need for very long dephasing elements by cascading two or more discrete elements that employ orthogonal PFG pulses [90] such that the first uses a *z*-gradient, as above, and the second an *x*-gradient, and so on. The resulting attenuation is the product of that given by each element and hence is time efficient but as this demands probes with triple axis (*x*, *y*, *z*) gradient capabilities it has less general applicability. An earlier approach employed the simultaneous application of a field gradient pulse and a short spin lock to dephase ZQC, relying upon the rf (B_1) inhomogeneity of the probe to produce the dephasing [91]. Despite some success, this method is likely to become less useful as probe developments lead to improved B_1 homogeneity.

12.6.2.1 Implementing Zero-Quantum Dephasing

The application of the zero-quantum dephasing element has been illustrated in earlier chapters within sequences such as TOCSY (Section 6.2) and NOESY (Section 9.6) so here we consider only the steps necessary for the calibration of the element itself. For the dephasing to perform correctly one needs to match the frequency profile of the swept pulse with the frequency spread imposed on the spectrum by the simultaneous field gradient, and the calibration sequence of Fig. 12.45 is designed to achieve this.

The process begins with selection of a suitable dephasing duration for the swept pulse τ_s according to Eq. 12.6 above. Since all zero-quantum frequencies may not be known this is likely to provide a guideline only but typical values for the sweep duration would be 20–50 ms. Next the total frequency sweep width for the rf pulse is defined which, as a rule of thumb, may be set to ~10 times the normal spectrum window typically yielding sweep widths of 40–60 kHz. From these figures the radiofrequency field strength for the chosen pulse profile must be determined such that the adiabatic condition is satisfied, a procedure most conveniently performed by direct calculation (see Section 12.2 above on adiabatic pulses). Suitable pulse profiles include WURST and the smoothed CHIRP. Initially, the sequence of Fig. 12.45 is executed with the gradient G_s and the swept-pulse rf field strength both set to zero and the imaging gradient G_i set to a low level such that the resonance lines are broadened slightly (Fig. 12.46b). Applying this gradient during the acquisition time of the FID imposes spatial encoding on the sample effectively providing a 1D image of this along the *z*-axis. Now the sequence is repeated with the adiabatic frequency sweep operational which should lead to complete inversion of the whole spectrum (Fig. 12.46c). The final calibration procedure then involves increasing the filter gradient strength G_s in stages aiming to make this as large as possible while retaining full inversion of all resonances. Should G_s become too high the very edges of each resonance

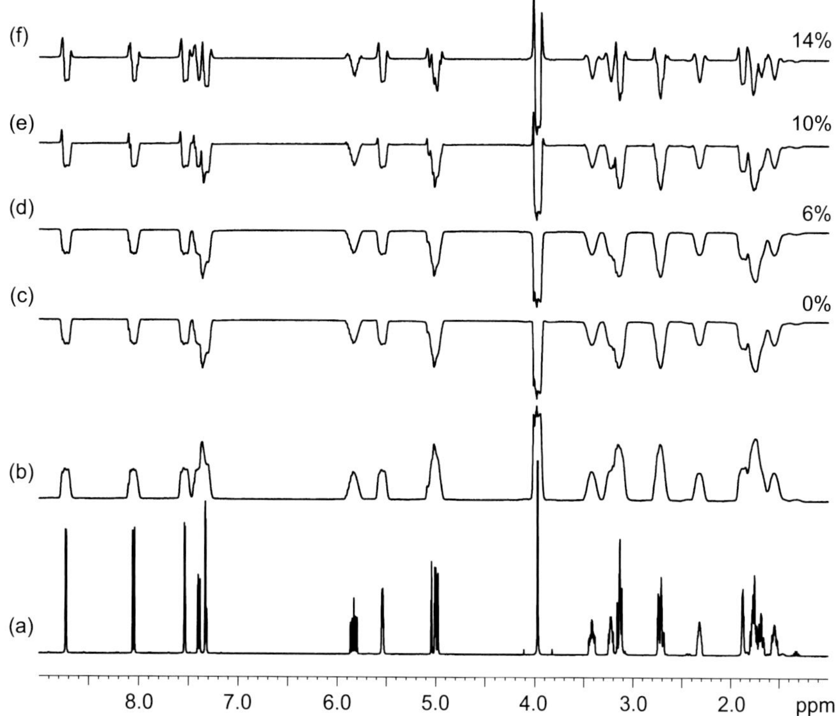

FIGURE 12.46 **Execution of the zero-quantum dephasing calibration routine; see text for discussions.** (a) 1D spectrum, (b) with imaging gradient only applied, (c) with additional application of adiabatic inversion pulse, and then with the filter gradient applied at (d) 6%, (e) 10% and (f) 14% of maximum. A filter gradient a little under 10% represents the optimum setting.

are no longer inverted (Fig. 12.46f); under these conditions the frequency spread imposed by the gradient is now larger than that being covered by the frequency swept pulse and the ends of the sample no longer experience the desired 180 degree inversion.

This calibration process must be performed for all dephasing elements of different durations but the optimum parameters so determined may then be used directly within appropriate sequences containing this element.

12.7 HETEROGENEOUS SAMPLES AND MAGIC ANGLE SPINNING

Most of the methods considered in the preceding chapters have been applied to samples that exist as isotropic solutions, the exception being samples placed deliberately in alignment media. However, there are circumstances when the samples of interest are inherently inhomogeneous and that exist in a semisolid state, for which alternative experimental protocols are required if reasonably high-resolution spectra are to be obtained. For example despite the fact that most synthetic organic chemistry takes place in solution, interest remains in solid-phase organic synthesis protocols for the production of libraries of new molecules [92–94]. Other examples, which lie beyond the scope of this text, include polymers, foodstuffs, whole cells and tissue samples, an area of particular relevance to metabolic profiling [95].

In the area of solid-supported synthesis, a particular challenge is the analysis of the newly synthesised materials for which, ideally, direct structural analysis of the material would be carried out while still attached to the solid support and thus still available for further chemistry. The direct NMR analysis of such materials, even when solvated, is complicated by two principle factors which can severely degrade spectrum resolution:

- restricted motion of the tethered analyte; and
- physical heterogeneity within the sample.

The restricted motion of the tethered analytes relative to that of free molecules in solution may mean that dipolar couplings, D, are not fully averaged to zero, a theme already explored in considering residual dipolar couplings of samples in alignment media (Section 9.12). As described, these have a time-averaged angular dependent:

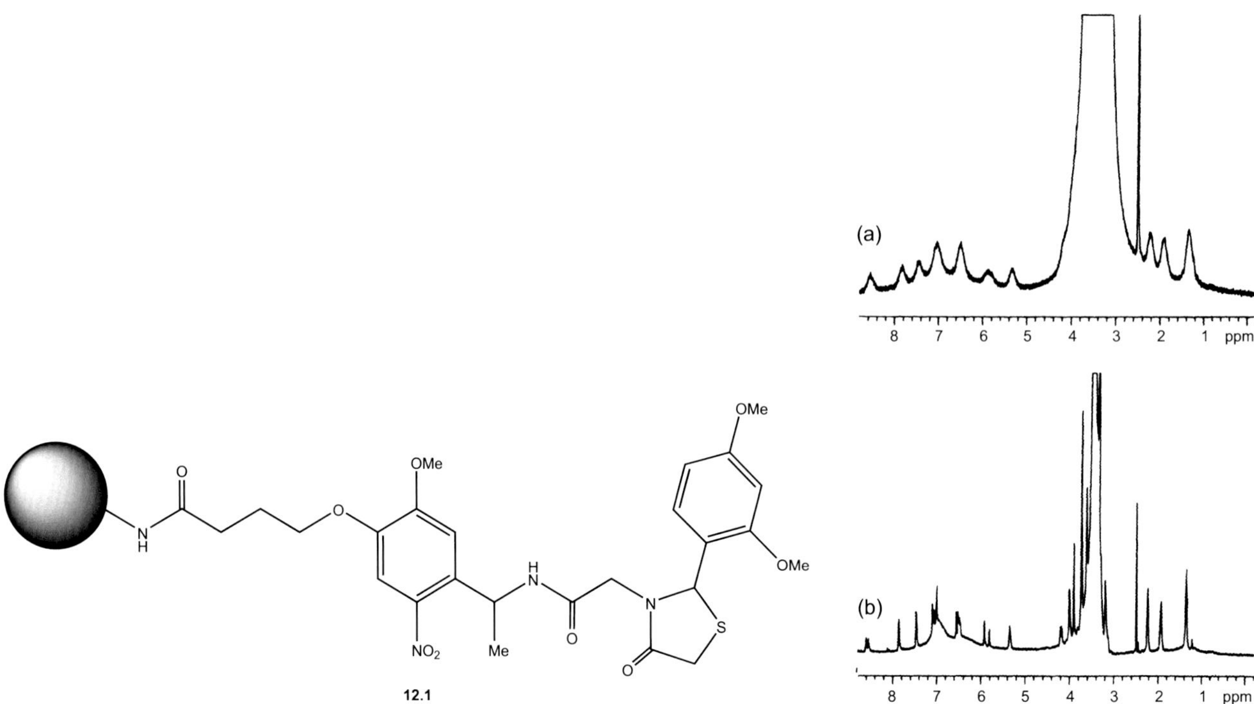

FIGURE 12.47 **Magic angle spinning.** (a) The conventional ^{1}H spectrum of the analyte **12.1** bound to TentaGel resin beads solvated in DMSO and (b) the MAS (2 kHz) spectrum of the same sample. One tenth of the mass of the sample in (a) was used in (b) and both were collected in 16 transients. The large truncated signals arise from the resin itself. *(Source: Reproduced with permission from Ref. [98], copyright American Chemical Society.)*

$$D \propto r^{-3}(3\cos^2\theta - 1) \tag{12.7}$$

where r is the internuclear separation and θ the angle between the static field and the internuclear vector between the dipolar coupled spins. For isotropically tumbling molecules in low-viscosity solutions, the angular term is averaged to zero as the molecule freely rotates, meaning dipolar couplings are not observed under these conditions. When motion is sufficiently restricted, some residual dipolar coupling may be reintroduced, which will contribute to broadening of the NMR resonances. The use of long linker chains allows greater motion of the terminal analyte in solution and can therefore reduce this effect. A potentially greater problem lies in the sample heterogeneity. Solutions containing solid-phase resins, usually in the form of beads, suffer from local field inhomogeneity at the bead–solvent interfaces due to magnetic susceptibility differences and this leads to severe line broadening (Fig. 12.47a). Although some success has been reported with gel-phase NMR of carbon-13 and other nuclei [96,97], these deleterious effects mean high-resolution spectra of solvated resins cannot be obtained with conventional solutions probes, a particular problem for proton spectroscopy.

Magnetic susceptibility has a similar angular dependence to that described above, and line-broadening from both dipolar couplings and susceptibility discontinuities can be eliminated by rapid rotation of the sample about the so-called 'magic angle' $\theta = 54.7$ degrees, such that the above angular term becomes zero and the overall molecular motion emulates that in solution. This *magic angle spinning* (MAS) (Fig. 12.48) has long been used in solid-state NMR spectroscopy to average the effects of chemical shift anisotropy [99] and has now become a standard technique for the direct analysis of resin-bound entities [94,100] and has also found application in the analysis of biomedical tissue extracts [101]. The substantial gain in resolution and sensitivity provided by MAS is illustrated in Fig. 12.47 [98]. The MAS spectrum shows a three-fold

B_0

$\theta = 54.7°$

Rotor

FIGURE 12.48 **MAS involves rapid rotation of a sample inclined at the magic angle of 54.7 degrees to the static field.**

signal-to-noise gain yet uses one-tenth of the sample relative to the conventional solution spectrum, and has considerably greater information content. To achieve such results, samples are typically spun at rates of 2–5 kHz, in purpose-built rotors requiring access to the appropriate high-resolution MAS probehead [102] and spin regulation hardware. The full range of high-resolution 1D and 2D techniques can now be applied to resin-bound samples under MAS conditions for the complete, non-destructive identification of new synthetic products [103–107]. The influence of various solvent and resin combinations has also been investigated and shown to have considerable bearing on the quality of spectra one obtains [108]. Various sequences have also been optimised for the investigation of these heterogeneous samples. The TOCSY experiment benefits from the use of mixing schemes based on adiabatic pulses owing to their greater tolerance of rf inhomogeneity and those based on WURST-2 pulses have been suggested as optimum [109,110]. Variants of the inverse $^{1}H–^{13}C$ correlation experiments HMQC and HMBC have also been presented that avoid the need for gradient pulses since PFGs are not always available on HR-MAS probes [111].

12.8 HYPERPOLARISATION

It is undeniably the case that NMR experiments suffer from poor sensitivity when compared to all other spectrometric methods and this arises primarily from the rather small energy differences, and hence population differences, across the associated spin transitions. These population differences correspond to less than 1 part in 10^4 with conventional NMR spectrometers, or in other words to polarisation levels of $< 0.01\%$. A direct approach to boosting the sensitivity of an NMR experiment is therefore to enhance the initial polarisation levels of the observed nuclei, this generation of non-Boltzmann populations of spin states being referred to as *hyperpolarisation.* This can yield theoretical sensitivity enhancements in excess of 10,000-fold, a massive gain when compared to alternative methods such as cryogenic probes. A number of approaches to generating hyperpolarised samples have been proposed and demonstrated over the years, of which two have most relevance to high-resolution NMR measurements. The first of these, *para*-hydrogen–induced polarisation (PHIP) uses the large population differences associated with the *para*-H_2 molecule and transfers this onto a substrate molecule for observation, typically via a catalytic hydrogenation dependent process. The second approach relies on the transfer of polarisation from unpaired electrons onto nuclear spins through the process known as dynamic nuclear polarisation (DNP).

12.8.1 *Para*-Hydrogen–Induced Polarisation

Hyperpolarised states that are derived from *para*-hydrogen have largely been employed for investigations in which the hydrogen atoms are chemically incorporated into the molecule of interest, and these have provided insights into many aspects of the mechanism and kinetics of inorganic catalytic processes in particular [112–114]. The theory and experimental requirements pertaining to PHIP are now well established and have been reviewed in detail [115]. This methodology was initially demonstrated and presented as *PASADENA* (*para*-hydrogen and synthesis allow dramatically enhanced nuclear alignment [116]) but is now considered to be part of the PHIP phenomenon. At the heart of this process lies the behaviour of the spin states of molecular hydrogen.

Diatomic molecular hydrogen may exist with differing nuclear spin configurations giving rise to four spin isomers. Three of these have symmetric triplet spin states each with total spin angular momentum 1, denoted $\alpha\alpha$, $\beta\beta$ and $\alpha\beta+\beta\alpha$, which correspond to what are known as the *ortho* spin isomers. The fourth has the antisymmetric singlet spin state $\alpha\beta-\beta\alpha$ which has zero total spin angular momentum and is known as the *para* isomer. These spin states (represented symbolically on the left in Fig. 12.49) have their origins in the rotational energy states of the isomers which in turn means *para*-hydrogen is more stable than *ortho*-hydrogen since it has access to lower rotational energy states. Thus, as hydrogen gas is cooled, the *para* state becomes energetically favoured such that the *para*-hydrogen content increases from ~25% at ambient temperature to ~50% at liquid nitrogen temperatures and further to ~99% at 20 K. Since the direct interconversion between the triplet *ortho* and singlet *para* states is forbidden, it requires a suitable catalyst to promote this process, such as paramagnetic materials containing iron. When removed from the catalyst, the *para*-H_2 enriched gas may be stored as a source of enhanced proton spin polarisation for use in subsequent NMR experiments, provided it is maintained in an environment free from paramagnetic sources, including oxygen, that would otherwise allow the undesired equilibration back to thermal populations.

To utilise this entrapped polarisation, it is necessary to react the dihydrogen gas with a substrate molecule so that the two hydrogen nuclei become inequivalent, since it is only then that the hyperpolarised spin states become accessible for detection. This conversion must occur as a spin-correlated process, meaning spin coupling between the nuclei must remain while the hydrogen centres are transferred to the substrate. This results in a product in which the $\alpha\beta$ and $\beta\alpha$ proton spin states are selectively populated, as represented schematically in Fig. 12.49, which compares the conventional Boltzmann populations of a J-coupled spin pair with those resulting from the selective spin state populations derived from

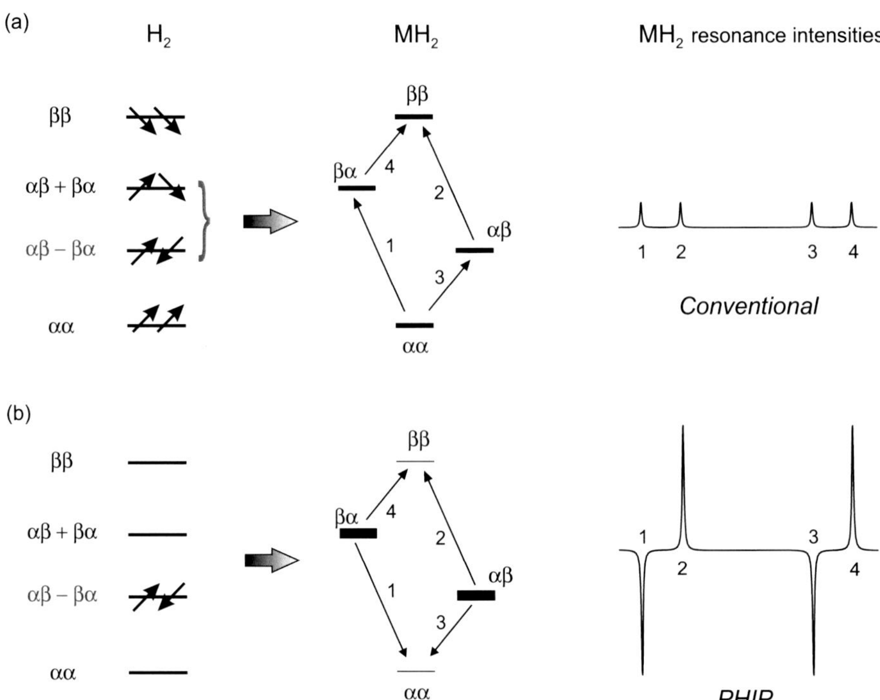

FIGURE 12.49 **Schematic illustration of PHIP.** Energy level population diagrams are shown for (a) standard Boltzmann-derived populations and (b) *para*-hydrogen derived populations. Left: symbolic representations of the spin state isomers present for *ortho*- and *para*-hydrogen gas. Centre: population diagrams that result when molecular hydrogen reacts at the metal centre of the substrate yielding product MH_2, with relative spin state populations represented by line thickness. Right: the resultant proton resonance intensities in the MH_2 product. The αβ+βα and αβ−βα combinations contribute equally to the αβ and βα states of the product.

para-H_2. The exclusive population of only the αβ and βα states corresponds to significantly enhanced population differences and hence greatly increased resonance intensities in NMR experiments detecting these protons, this being the origin of PHIP-based experiments. Although theoretical sensitivity enhancements of the order of 10^4 are predicted, in reality enhancements of 10^2–10^3 are observed. The preferential population of the αβ and βα states also means that the PHIP-derived NMR resonances display a characteristic anti-phase structure (Fig. 12.49). It is also the case that conventional NMR experimental requirements must be altered to be appropriate for PHIP experiments. For example due to the unconventional initial spin populations, the level of excitation caused by a rf pulse of tip angle ϑ has a $\cos\vartheta\sin\vartheta$ dependence, meaning maximum excitation is provided by a 45 degree pulse rather than a 90 degree pulse (which, perhaps unexpectedly, would yield no signal from PHIP). An example of the PHIP spectrum of a hyperpolarised ruthenium complex **12.2** is shown in Fig. 12.50 and displays the expected anti-phase pattern originating from the 5 Hz geminal proton couplings.

Once transferred to the substrate, the H_2 derived protons follow conventional spin relaxation processes through which the enhanced polarisation levels are lost, meaning the amplified sensitivity is short-term and typically sampled only in one transient, in common with dynamic nuclear polarisation (DNP) below. Replenishing the supply of *para*-H_2, such as by over pressurising the NMR tube with the gas or preferably through the use of a continuous capillary flow system in the NMR tube, can allow more substrate to react, so refreshing the hyperpolarised sample. In this manner, PHIP has found greatest use in the study of transition metal hydrides and the investigation of catalytic mechanisms involving hydrogenation and hydroformylation since this enables the detection of low-level and/or transient species forming within catalytic cycles that evade detection by conventional NMR methods. An example of complexes generated by PHIP, for use in such investigations, include those of **12.3**, and many examples may be found in reviews [113–115]. More recent studies have demonstrated the potential to transfer the high levels of polarisation from the initially formed metal hydride substrates to other organic small molecules that can directly but reversibly coordinate with the substrate, an early example being pyridine. The SABRE process (signal amplification by reversible exchange [117]) which operates at relatively low magnetic fields, provides for the exciting possibility of using PHIP for sensitivity enhancement of a wider range of small molecules and nuclei other than protons, so broadening its potential application. This also includes the area of magnetic resonance imaging (MRI), an

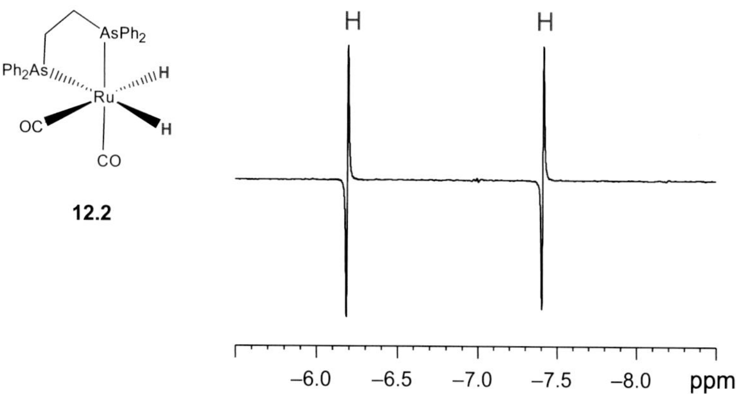

FIGURE 12.50 **The 400 MHz PHIP ^{1}H spectrum of $RuH_2(CO)_2$(bis(diphenylarsino)ethane 12.2.** *(Source: Data courtesy of Prof Simon Duckett, University of York.)*

application also of relevance to DNP. It is noteworthy that hardware for the automated production of *para*-H_2 is now commercially available from at least one NMR instrument vendor, which may lead to the wider adoption of PHIP in appropriate research environments.

Ph_3P CO M Cl PPh_3 — *para*-H_2 → PPh_3 H CO M H Cl Ph_3P

M = Ir, Ru, Rh

12.3

12.8.2 Dynamic Nuclear Polarisation

The method of DNP relies on the transfer of polarisation from unpaired electrons onto nuclear spins in the solid-state, and in this form can directly boost sensitivity in solid-state NMR spectroscopy. However, it has also been further developed to enhance significantly the sensitivity of solution-state NMR, as described briefly here. The technique has been applied to boost the polarisation levels for low-γ heteronuclear spins in particular and shows most promise to date for ^{13}C and ^{15}N observation; the initial report on the technique described a gain in signal-to-noise in excess of 10,000-fold for both these elements when compared to data collected at thermal equilibrium [118]. The hyperpolarisation of a range of nuclei including ^{31}P and ^{29}Si has also been demonstrated.

The method requires a sample to be doped with a suitable radical (organic trityl radicals have proved especially effective) and for this to be dispersed homogeneously within an amorphous frozen matrix to allow intimate contact between the analyte and radical molecules. The electron spins are then polarised by placing the sample in a suitable magnetic field while at low temperature, typically only 1–2 K, and the DNP process initiated by irradiation of the sample with microwaves. The microwave frequency is chosen to match the difference between the electron and nuclear Larmor frequencies and nuclear polarisation levels build over some hours during the irradiation according to the solid-state T_1 relaxation time constants. Carbon-13 polarisation levels in the solid-state can typically reach 20–40%. The mechanisms involved in the polarisation transfer are complex [119] but may be thought of as solid-state Overhauser effects; the original reports of Overhauser were in fact for electron-nuclear transfers. The large solid-state polarisation that develops can be exploited directly in solid-state NMR measurements [119,120] but to be useful for high-resolution NMR these must be transferred into the solution-state. A number of approaches may be envisaged, including melting and dissolution, and it is the second of these that has received greater application to date. In a technically remarkable process [121], the frozen, hyperpolarised pellet is sequentially

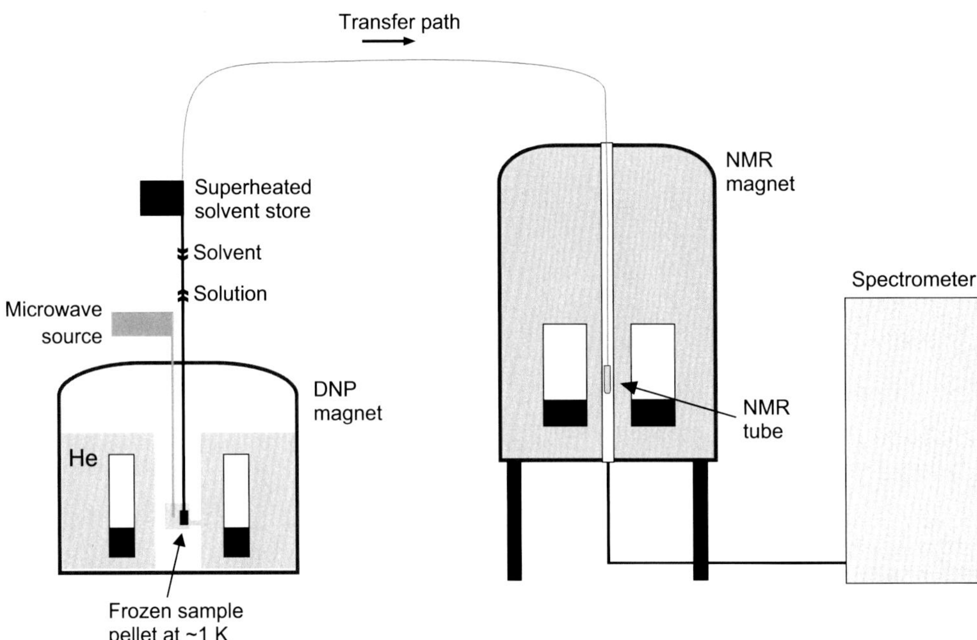

FIGURE 12.51 **A schematic illustration of the DNP-NMR spectrometer assembly.**

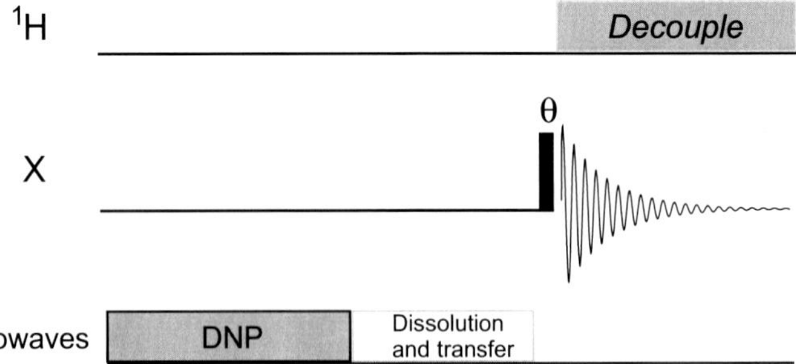

FIGURE 12.52 **A pulse sequence suitable for acquiring 1D heteronuclear spectra from DNP-NMR.**

dissolved in a super-heated solvent, typically methanol, and passed under pressure through a transfer line directly into an NMR tube held within a conventional NMR magnet (Fig. 12.51). This whole process occurs very rapidly and must be fast on the relaxation timescale since polarisation is lost through the usual spin relaxation processes as soon as the sample enters the solution-state. The sample must therefore pass from the solid-state at ~1 K into solution at ~300 K and equilibrate sufficiently for a high-resolution experiment within a matter of seconds! Once in the NMR tube the hyperpolarised sample may be subject to investigation by pulse techniques but again within the constraint that the large polarisation levels will be lost through normal spin-relaxation processes. This means, in essence, that the technique is limited in most part to single-scan acquisitions, typical examples being 1D proton-decoupled heteronuclear spectra. A pulse sequence encapsulating the complete DNP acquisition process is illustrated in Fig. 12.52. While signal averaging is not possible, the high initial polarisation levels means detection sensitivity can be sufficient to observe all resonances in a single-scan even on very small sample quantities. Fig. 12.53 compares the conventional ^{13}C spectrum of 60 μmol ethyl acetoacetate **12.4** collected as 3000 transients over three hours with the single-scan DNP-enhanced spectrum at the same concentration recorded after three hours of polarisation. The labelled resonance at ~39 ppm is missing in trace (a) as the carbon centre becomes deuterated in CD_3OD through *keto-enol* tautomerism whereas this does not occur for the DNP-enhanced sample which used CH_3OH as the dissolution solvent; the minor *enol* form gives rise to the smaller resonances apparent only in the DNP-enhanced spectrum. One notable feature of DNP-enhanced spectra that is apparent in Fig. 12.53 is that, in contrast to conventional heteronuclear spectra, the resonances of non-protonated centres appear with highest intensity, often greater than those of proton-bearing atoms. This is most apparent for the carbonyl resonances above 160 ppm and is a direct consequence of the slower relaxation associated with non-protonated centres (since attached protons act as dominant relaxation sources) meaning these retain more of the initial polarisation during the dissolution and transfer process.

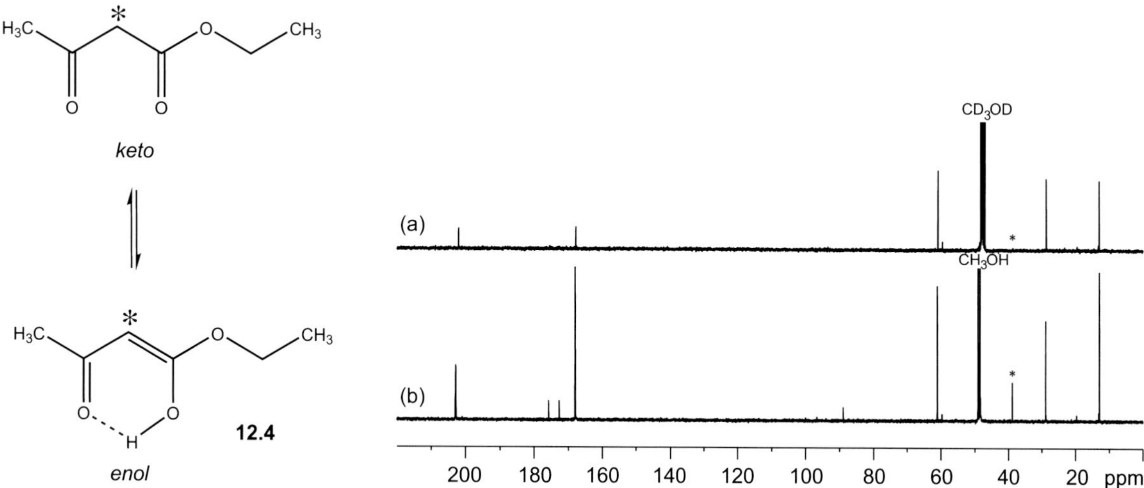

FIGURE 12.53 **DNP enhancement.** A comparison of (a) the conventional 1D ^{13}C spectrum of ethyl acetoacetate **12.4** (60 μmol, 3000 scans, 400 MHz) with (b) the DNP enhanced ^{13}C spectrum (60 μmol, single-scan, 400 MHz). The resonance labelled with an asterix is missing in trace (a) due to deuteration of the centre as a result of *keto-enol* tautomerism; see text. *(Source: Data courtesy of Oxford Instruments Molecular Biotools.)*

The initial motivation for developing the solution-state DNP technology was for *in-vivo* studies employing ^{13}C magnetic resonance imaging [122] and applications of DNP in high-resolution NMR remain limited. The detection of samples of limited availability and perhaps lifetime offer one potential area of application. The simultaneous polarisation of both ^{13}C and ^{15}N has also been demonstrated, with the two nuclei excited and detected sequentially to yield independent DNP enhanced spectra [123]. A more exciting development that shows interesting potential is to marry the DNP method with the single-scan acquisition methods introduced in Section 5.4.5 and so open the way to 2D (or greater) NMR experiments on hyperpolarised samples. This has been demonstrated for single-scan ^{1}H–^{13}C heteronuclear correlation spectrum of 0.47 mM hyperpolarised pyridine [124] (Fig. 12.54). Data were collected with a modified HSQC experiment that exploits the initial ^{13}C polarisation and the total experiment time was a fleeting 110 ms. With current technology one limiting factor in the use

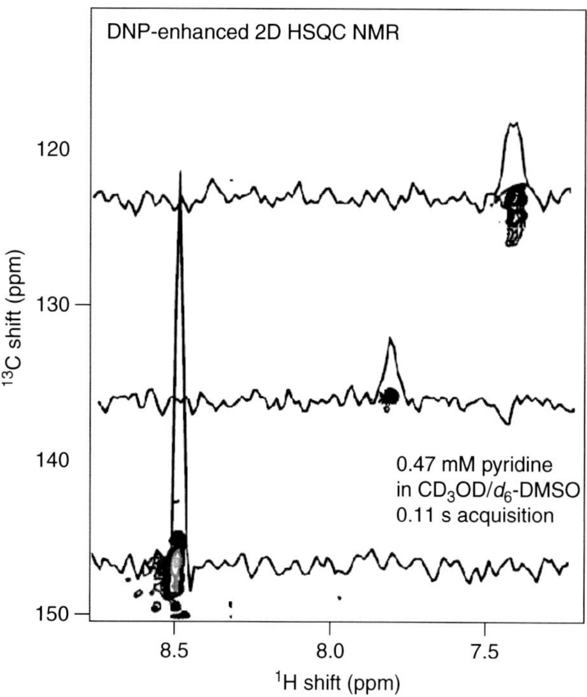

FIGURE 12.54 **The 2D single-scan HSQC spectrum of 0.47 mM ^{13}C-hyperpolarised pyridine.** *(Source: Reproduced with permission from Ref. [124], copyright Nature Publishing Group.)*

of DNP-NMR is the loss of polarisation on transfer from the polariser to the NMR magnet restricting the method to more slowly relaxing samples, usually meaning smaller molecules. This also means signal-to-noise gains of around two orders of magnitude relative to conventional single-scan acquisitions appear to reflect more realistic expectations for small molecules at present. However, recent technological developments may reduce these restrictions [125,126] and extend the scope of DNP-NMR. For example, transfer lines have been placed within "magnetic tunnels" [127] and both DNP and NMR magnets have been combined within a single dewar to reduce the loss of polarization between the DNP and NMR magnets. The use of supercritical fluids as solvents for DNP has also been proposed as a means of enhancing performance [128].

REFERENCES

[1] Freeman R, Kempsell SP, Levitt MH. J Magn Reson 1980;38:453–79.
[2] Levitt MH, Freeman R. J Magn Reson 1979;33:473–6.
[3] Shaka AJ, Bauer C, Freeman R. J Magn Reson 1984;60:479–85.
[4] Levitt MH. Prog Nucl Magn Reson Spectrosc 1986;18:61–122.
[5] Shaka AJ, Freeman R. J Magn Reson 1983;55:487–93.
[6] Keniry M, Sanctuary BC. J Magn Reson 1992;97:382–4.
[7] Bai NS, Ramakrishna M, Ramachandran R. J Magn Reson A 1993;102:235–40.
[8] Ramachandran R. J Magn Reson A 1993;105:328–9.
[9] Levitt MH, Freeman R. J Magn Reson 1981;43:65–80.
[10] Tycko R, Cho HM, Schneider E, Pines A. J Magn Reson 1985;61:90–101.
[11] Derome AE. J Magn Reson 1988;78:113–22.
[12] Garwood M, DelaBarre L. J Magn Reson 2001;153:155–77.
[13] Tannús A, Garwood M. NMR Biomed 1997;10:423–34.
[14] Kupče E, Freeman R. J Magn Reson A 1995;115:273–6.
[15] Kupče E, Freeman R. J Magn Reson A 1996;118:299–303.
[16] Bohlen JM, Bodenhausen G. J Magn Reson A 1993;102:293–301.
[17] Kupče E, Freeman R. J Magn Reson A 1995;117:246–56.
[18] Hwang T-L, van Zijl PCM, Garwood M. J Magn Reson 1997;124:250–4.
[19] Boyer RD, Johnson R, Krishnamurthy K. J Magn Reson 2003;165:253–9.
[20] Kupče E, Freeman R. J Magn Reson 2007;187:258–65.
[21] Smith MA, Hu H, Shaka AJ. J Magn Reson 2001;151:269–83.
[22] Skinner TE, Kobzar K, Luy B, Bendall MR, Bermel W, Khaneja N, Glaser SJ. J Magn Reson 2006;179:241–9.
[23] Levitt MH, Freeman R, Frenkiel T. J Magn Reson 1982;47:328–30.
[24] Shaka AJ, Keeler J, Freeman R. J Magn Reson 1983;53:313–40.
[25] Shaka AJ, Keeler J, Frenkiel T, Freeman R. J Magn Reson 1983;52:335–8.
[26] Shaka AJ, Lee CJ, Pines A. J Magn Reson 1988;77:274–93.
[27] Shaka AJ, Barker PB, Freeman R. J Magn Reson 1985;64:547–52.
[28] Bax A, Davis DG. J Magn Reson 1985;65:355–60.
[29] Rucker SP, Shaka AJ. Mol Phys 1989;68:509–17.
[30] Kadkhodaie M, Rivas O, Tan M, Mohebbi A, Shaka AJ. J Magn Reson 1991;91:437–43.
[31] Shaka AJ, Keeler J. Prog Nucl Magn Reson Spectrosc 1987;19:47–129.
[32] Fu R, Bodenhausen G. J Magn Reson A 1995;117:324–5.
[33] Fu R, Bodenhausen G. J Magn Reson A 1996;119:129–33.
[34] Tycko R, Pines A. Chem Phys Lett 1984;111:462–7.
[35] Kupče E, Freeman R, Wider G, Wuthrich K. J Magn Reson A 1996;122:81–4.
[36] Kupče E, Freeman R, Wider G, Wuthrich K. J Magn Reson A 1996;120:264–8.
[37] Tenailleau E, Akoka S. J Magn Reson 2007;185:50–8.
[38] Kessler H, Mronga S, Gemmecker G. Magn Reson Chem 1991;29:527–57.
[39] Kessler H, Oschkinat H, Griesinger C, Bermel W. J Magn Reson 1986;70:106–33.
[40] Freeman R. Chem Rev 1991;91:1397–412.
[41] Berger S. Prog Nucl Magn Reson Spectrosc 1997;30:137–56.
[42] Freeman R. Prog Nucl Magn Reson Spectrosc 1998;32:59–106.
[43] Geen H, Wimperis S, Freeman R. J Magn Reson 1989;85:620–7.
[44] Bauer C, Freeman R, Frenkiel T, Keeler J, Shaka AJ. J Magn Reson 1984;58:442–57.
[45] Friedrich J, Davies S, Freeman R. J Magn Reson 1987;75:390–5.
[46] Emsley L, Bodenhausen G. J Magn Reson 1989;82:211–21.
[47] Kessler H, Anders U, Gemmecker G, Steuernagel S. J Magn Reson 1989;85:1–14.
[48] Geen H, Freeman R. J Magn Reson 1990;87:415–21.

[49] Geen H, Freeman R. J Magn Reson 1991;93:93–141.
[50] Emsley L, Bodenhausen G. Chem Phys Lett 1990;165:469–76.
[51] Emsley L, Bodenhausen G. J Magn Reson 1992;97:135–48.
[52] Nuzillard J-M, Freeman R. J Magn Reson A 1994;110:252–6.
[53] Kupče E, Boyd J, Campbell ID. J Magn Reson B 1995;106:300–3.
[54] Hajduk PJ, Horita DA, Lerner LE. J Magn Reson A 1993;103:40–52.
[55] Horita DA, Hajduk PJ, Lerner LE. J Magn Reson A 1993;103:53–60.
[56] Nuzillard J-M, Freeman R. J Magn Reson A 1994;107:113–8.
[57] Hwang TL, Shaka AJ. J Magn Reson A 1995;112:275–9.
[58] Stott K, Stonehouse J, Keeler J, Hwang TL, Shaka AJ. J Am Chem Soc 1995;117:4199–200.
[59] Xu GZ, Evans JS. J Magn Reson B 1996;111:183–5.
[60] Kövér KE, Uhrín D, Hruby VJ. J Magn Reson 1998;130:162–8.
[61] Stott K, Keeler J, Van QN, Shaka AJ. J Magn Reson 1997;125:302–24.
[62] Morris GA, Freeman R. J Magn Reson 1978;29:433–62.
[63] Patt SL. J Magn Reson 1992;96:94–102.
[64] Kupče E, Feeman R. J Magn Reson A 1993;105:234–8.
[65] Bernassau J-M, Nuzillard J-M. J Magn Reson B 1994;103:77–81.
[66] Hore PJ. Method Enzymol 1989;176:64–77.
[67] Guéron M, Plateau P, Decorps M. Prog Nucl Magn Reson Spectrosc 1991;23:135–209.
[68] Hull WE. In: Croasmun WR, Carlson RMK, editors. Two-dimensional NMR spectroscopy: applications for chemists and biochemists. New York: VCH; 1994.
[69] Hoult DI. J Magn Reson 1976;21:337–47.
[70] Dykstra RW. J Magn Reson 1987;72:162–7.
[71] Neuhaus D, Ismail IM, Chung C-W. J Magn Reson A 1996;118:256–63.
[72] Mo H, Raftery D. J Magn Reson 2008;190:1–6.
[73] Plateau P, Guéron M. J Am Chem Soc 1982;104:7310–1.
[74] Hore PJ. J Magn Reson 1983;54:539–42.
[75] Hore PJ. J Magn Reson 1983;55:283–300.
[76] Callihan D, West J, Kumar S, Schweitzer BI, Logan TM. J Magn Reson B 1996;112:82–5.
[77] Piotto M, Saudek V, Slenár V. J Biomol NMR 1992;2:661–5.
[78] Sklenár V, Piotto M, Leppik R, Saudek V. J Magn Reson A 1993;102:241–5.
[79] Liu M, Mao X, Ye C, Nicholson JK, Lindon JC. J Magn Reson 1998;132:125–9.
[80] Marion D, Ikura M, Bax A. J Magn Reson 1989;84:425–30.
[81] Parella T, Adell P, Sánchez-Ferrando F, Virgili A. Magn Reson Chem 1998;36:245–9.
[82] Dalvit C, Shapiro G, Böhlen J-M, Parella T. Magn Reson Chem 1999;37:7–14.
[83] Adams RW, Holroyd CM, Aguilar JA, Nilsson M, Morris GA. Chem Commun 2013;49:358–60.
[84] Simpson AJ, Brown SA. J Magn Reson 2005;175:340–6.
[85] Hurd RE. J Magn Reson 1990;87:422–8.
[86] Macura S, Huang Y, Suter D, Ernst RR. J Magn Reson 1981;43:259–81.
[87] Macura S, Wüthrich K, Ernst RR. J Magn Reson 1982;46:269–82.
[88] Thrippleton MJ, Keeler J. Angew Chem Int Ed 2003;42:3938–41.
[89] Rance M, Bodenhausen G, Wagner G, Wuthrich K, Ernst RR. J Magn Reson 1985;62:497–510.
[90] Cano KE, Thrippleton MJ, Keeler J, Shaka AJ. J Magn Reson 2004;167:291–7.
[91] Davis AL, Estcourt G, Keeler J, Laue ED, Titman JJ. J Magn Reson A 1993;105:167–83.
[92] Lowe G. Chem Soc Rev 1995;24:309–17.
[93] Obrecht D, Villalgordo JM. Solid-supported combinatorial and parallel synthesis of small-molecular-weight compound libraries. Oxford: Pergamon; 1998.
[94] Power WP, Webb GA. Ann Rep NMR Spectrosc 2003;51:261–95.
[95] Power WP. Ann Rep NMR Spectrosc 2011;72:111–56.
[96] Giralt E, Rizo J, Pedroso E. Tetrahedron 1984;40:4141–52.
[97] Look GC, Holmes CP, Chinn JP, Gallop MA. J Org Chem 1994;59:7588–90.
[98] Fitch WL, Detre G, Holmes CP, Shoolery JN, Keifer PA. J Org Chem 1994;59:7955–6.
[99] Andrew ER. Prog Nucl Magn Reson Spectrosc 1971;8:1–39.
[100] Shapiro MJ, Gounarides JS. Prog Nucl Magn Reson Spectrosc 1999;35:153–200.
[101] Bollard ME, Murray AJ, Clarke K, Nicholson JK, Griffin JL. FEBS Lett 2003;553:73–8.
[102] Keifer PA, Baltusis L, Rice DM, Tymiak AA, Shoolery JN. J Magn Reson A 1996;119:65–75.
[103] Anderson RC, Jarema MA, Shapiro MJ, Stokes JP, Ziliox M. J Org Chem 1995;60:2650–1.
[104] Anderson RC, Stokes JP, Shapiro MJ. Tetrahedron Lett 1995;36:5311–4.
[105] Jelinek R, Valente AP, Valentine KG, Opella SJ. J Magn Reson 1997;125:185–7.

[106] Dhalluin C, Boutillon C, Tartar A, Lippens G. J Am Chem Soc 1997;119:10494–500.
[107] Rousselot-Pailley P, Maux D, Wieruszeski J-M, Aubagnac J-L, Martinez J, Lippens G. Tetrahedron 2000;56:5163–7.
[108] Keifer PA. J Org Chem 1996;61:1558–9.
[109] Kupče E, Keifer PA, Delepierre M. J Magn Reson 2001;148:115–20.
[110] Kupče E, Hiller W. Magn Reson Chem 2001;39:231–5.
[111] Ramadhar TR, Amador F, Ditty MJT, Power WP. Magn Reson Chem 2008;46:30–5.
[112] Duckett SB, Sleigh CJ. Prog Nucl Magn Reson Spectrosc 1999;34:71–92.
[113] Blazina D, Duckett SB, Dunne JP, Godard C. Dalton Trans 2004;2601–9.
[114] Duckett SB, Wood NJ. Coord Chem Rev 2008;252:2278–91.
[115] Green RA, Adams RW, Duckett SB, Mewis RE, Williamson DC, Green GGR. Prog Nucl Magn Reson Spectrosc 2012;67:1–48.
[116] Bowers CR, Weitekamp DP. J Am Chem Soc 1987;109:5541–2.
[117] Adams RW, Aguilar JA, Atkinson KD, Cowley MJ, Elliott PIP, Duckett SB, Green GGR, Khazal IG, López-Serrano J, Williamson DC. Science 2009;323:1708–11.
[118] Ardenkjaer-Larsen JH, Fridlund B, Gram A, Hansson G, Hansson L, Lerche MH, Servin R, Thaning M, Golman K. Proc Natl Acad Sci USA 2003;100:10158–63.
[119] Wind RA, Duijvestijn MJ, van der Lugt C, Manenschijn A, Vriend J. Prog Nucl Magn Reson Spectrosc 1985;17:33–67.
[120] Bajaj VS, Farrar CT, Hornstein MK, Mastovsky I, Vieregg J, Bryant J, Eléna B, Kreischer KE, Temkin RJ, Griffin RG. J Magn Reson 2003;160: 85–90.
[121] Wolber J, Ellner F, Fridlund B, Gram A, Johannesson H, Hansson G, Hansson LH, Lerche MH, Mansson S, Servin R, Thaning M, Golman K, Ardenkjaer-Larsen JH. Nucl Instrum Meth A 2004;526:173–81.
[122] Golman K, Ardenkjaer-Larsen JH, Petersson JS, Mansson S, Leunbach I. Proc Natl Acad Sci USA 2003;100:10435–9.
[123] Day IJ, Mitchell JC, Snowden MJ, Davis AL. Magn Reson Chem 2007;45:1018–21.
[124] Frydman L, Blazina D. Nat Phys 2007;3:415–9.
[125] Ardenkjaer-Larsen JH. J Magn Reson 2016;264:3–12.
[126] Bornet A, Jannin S. J Magn Reson 2016;264:13–21.
[127] Jähnig F, Kwiatkowski G, Ernst M. J Magn Reson 2016;264:22–9.
[128] van Bentum J, van Meerten B, Sharma M, Kentgens A. J Magn Reson 2016;264:59–67.

Chapter 13

Structure Elucidation and Spectrum Assignment

Chapter Outline

13.1 ^{1}H NMR	**500**
13.2 ^{1}H–^{13}C Edited HSQC	**501**
13.3 ^{1}H–^{1}H COSY and Variants	**503**
13.3.1 Double-Quantum Filtered COSY	505
13.4 ^{1}H–^{1}H TOCSY and Variants	**506**
13.4.1 HSQC-TOCSY	508
13.5 ^{13}C NMR	**508**
13.6 HMBC and Variants	**510**
13.6.1 ^{1}H–^{13}C HMBC	510
13.6.2 ^{31}P and ^{1}H–^{31}P HMBC	512
13.6.3 ^{1}H–^{13}C HMBC Again	513
13.6.4 ^{19}F and ^{19}F–^{13}C HMBC	515
13.7 Nuclear Overhauser Effects	**517**
13.7.1 2D NOESY	517
13.7.2 1D NOESY	521
13.7.3 1D ^{19}F HOESY	522
13.8 Rationalization of ^{1}H–^{1}H Coupling Constants	**523**
13.9 Summary	**525**

In this final chapter we look at the application of the more commonly used techniques, supported by some less commonplace methods, to the structure elucidation and spectrum assignment of a single compound. The goal is to illustrate how these may be combined in a sequential fashion to yield a complete definition of the molecular skeleton and associated stereochemistry. Previous chapters have given many isolated examples of how the techniques being introduced may be employed, but here we aim to bring the most important of these together and provide greater clarity on how these might be used to address structure assignment problems in the laboratory.

The illustrative molecule considered throughout this chapter is the commercially available glucocorticoid dexamethasone-21-phosphate **13.1**, a drug used for its anti-inflammatory and immunosuppressive properties. It was chosen to be representative of the types of compounds produced in synthetic chemistry laboratories or isolated as natural products. It has a reasonably high molecular mass for a 'small' molecule (470 Da) and a proton spectrum that has good resonance dispersion but is sufficiently complex that it cannot be directly interpreted to identify a structure or to yield spectrum assignments without recourse to further 1D and 2D methods. The compound also contains a number of centres requiring stereochemical definition so allows consideration of techniques relevant to this. An additional feature is that it contains NMR active nuclei other than ^{1}H and ^{13}C, allowing the illustration of techniques pertinent to these commonly encountered nuclei. Indeed, ^{1}H, ^{13}C, ^{19}F and ^{31}P may be considered the 'big four' nuclei encountered in the NMR of small organic molecules and are all represented here. The other nucleus of high interest is likely to be ^{15}N which is not found in this compound. However, it is worth noting that most of the methods described for ^{13}C–^{1}H experiments are equally applicable to ^{15}N–^{1}H studies, and may usually be readily adapted for ^{15}N from the ^{13}C-based experimental parameters.

High-Resolution NMR Techniques in Organic Chemistry. http://dx.doi.org/10.1016/B978-0-08-099986-9.00013-0

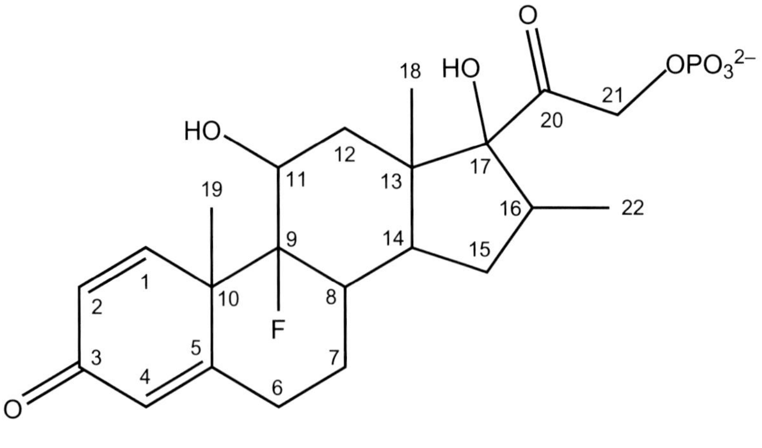

13.1

The aim of the descriptions that follow is to define the structure of **13.1** from the NMR data alone, starting only with the molecular formula ($C_{22}H_{28}FO_8P^{2-}$) that would typically be derived from high-resolution mass spectrometry measurements. The structure is shown here to aid clarity in these discussions, although would not usually be known. Similarly, in these discussions we shall also use the numbering scheme for the final structure, as shown here, as I believe this also provides greater clarity for the reader. The use of the prime notation (eg H^{21} and $H^{21'}$) is used to differentiate diastereotopic methylene protons, with the prime indicating that of lower chemical shift. In reality, with an unknown structure, it would be necessary to define an arbitrary labelling scheme for all resonances, prior to them being correlated, and a systematic approach would be required for this. For example one might label proton resonances a, b, c, etc., from the highest chemical shift, and carbons A, B, C in a similar fashion. Of course, there are multiple approaches to data interpretation and no single correct route, but that described here is illustrative of the approach one might take, and follows a logically consistent process. It is certainly the case that not all of the methods illustrated for **13.1** in what follows would be essential for its structure assignment. Rather, a broader range of methods have been included to illustrate the manner in which these could be applied to resolve chemical structures, and specifically to provide these examples in the context of a single chemical structure. The data presented have all been acquired at 500 MHz (with one exception at 600 MHz noted in the text), a common mid-range field for chemical laboratories.

When applying NMR techniques to solve chemical problems, one may differentiate between structure elucidation and structure verification. Elucidation will likely demand the definition of a structure without prior knowledge of what this might be, so may include the identification of a novel natural product, for example. Verification (or confirmation) may be considered to be the proof of an anticipated structure, so is relevant to the product of a planned chemical synthesis. The first of these is likely to demand the application of a broader range of techniques to lead to an unambiguous structure definition, whereas the second may demand a smaller amount of data be collected to define sufficient consistency with the proposed structure. The process of building a molecular skeleton from NMR correlation data may nowadays be aided by computational algorithms developed for structure elucidation. A number of these are now available from NMR software vendors and may be attractive tools to assist in this process. They serve to generate a range of structures that are consistent with the supplied NMR data, which typically comprise ^1H–^1H correlation data from COSY, and one-bond and long-range ^1H–^{13}C data from HSQC and HMBC experiments as a minimum. Not only do these programs help to manage the significant amounts of correlation data one may be dealing with when building a potential structure, they also help to generate molecular skeletons that might otherwise not have been anticipated by the chemist interpreting the data. The following sections take a stepwise approach to structure elucidation and, as usual, begin with the 1D proton spectrum.

13.1 ^{1}H NMR

The proton spectrum of **13.1** recorded in D_2O is shown in Fig. 13.1 and is referenced to internal trimethylsilyl propionic acid (TSP-d_4) at 0.00 ppm. Resonances are labelled beneath (or beside) according to their final assignments and their associated integral values are shown below the axis. In total, 26 protons are observed, which suggests two are exchangeable and

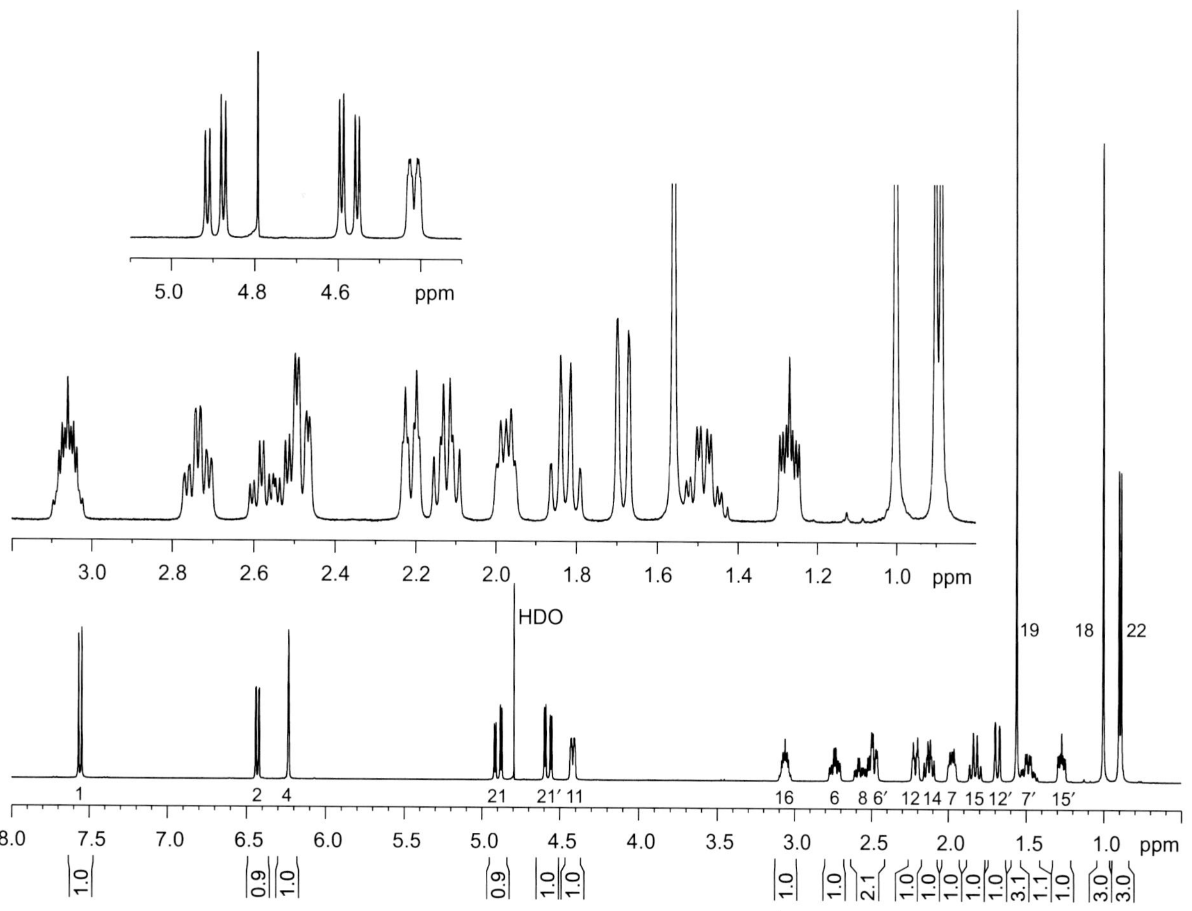

FIGURE 13.1 **The ^{1}H NMR spectrum of 13.1 (500 MHz).** This was recorded with suppression of the HDO resonance using a 1D NOESY presaturation sequence with 2.5-μW irradiation for 2s.

have been lost for deuterium, so are likely OH groups. Three resonances above 6 ppm are suggestive of unsaturation in the structure, and the similar magnitudes of coupling (10.0 Hz) suggest that H^1 and H^2 are adjacent, with the slightly broadened H^4 being more isolated but sharing longer range couplings to other protons. Three resonances appear in the 4–5 ppm region, their shifts suggestive of them likely being adjacent to oxygen atoms and possibly additional features that are deshielding. The roofing of the pair of resonances H^{21} and $H^{21'}$ shows these share a mutual coupling (a rather large 19.0 Hz suggestive of geminal coupling), each with one other coupled spin partner. The remaining 20 protons resonate below 3 ppm and clearly include three methyl resonances Me^{19}, Me^{18} and Me^{22}. Two of these are attached to fully substituted centres and thus appear as singlets, while the doublet structure of Me^{22} demonstrates it is either adjacent to a single proton or shares a heteronuclear coupling. While some progress may be made through direct analysis of the largely resolved multiplet structures of the remaining 11 resonances, a more expedient route is to now consider 2D correlations.

13.2 ^{1}H–^{13}C EDITED HSQC

Consideration of the ^{13}C edited HSQC in Fig. 13.2 provides an immediate overview of the ^{13}C shift dispersion in the molecule (for protonated centres at least) and readily distinguishes CH and CH_2 groups from the signs of crosspeaks. For the three protons H^1, H^2 and H^4, only the carbon resonance of H^1 stands out as being unusual at 156.6 ppm, suggesting the presence of a heteroatom or resonance delocalised deshielding of this centre. The carbon shifts at 68.5 ppm for $H^{21-21'}$ (now clearly a diastereotopic geminal pair as both correlate to the same carbon), and H^{11} at 71.3 ppm are consistent with these being adjacent to oxygen atoms. In the region below 50 ppm (Fig. 13.3), it is now readily possible to identify the presence of four diastereotopic CH_2 groups: $H^{12-12'}$ (δ_C 35.6), $H^{15-15'}$ (δ_C 31.6), $H^{6-6'}$ (δ_C 30.8) and $H^{7-7'}$ (δ_C 27.4), and three CH groups:

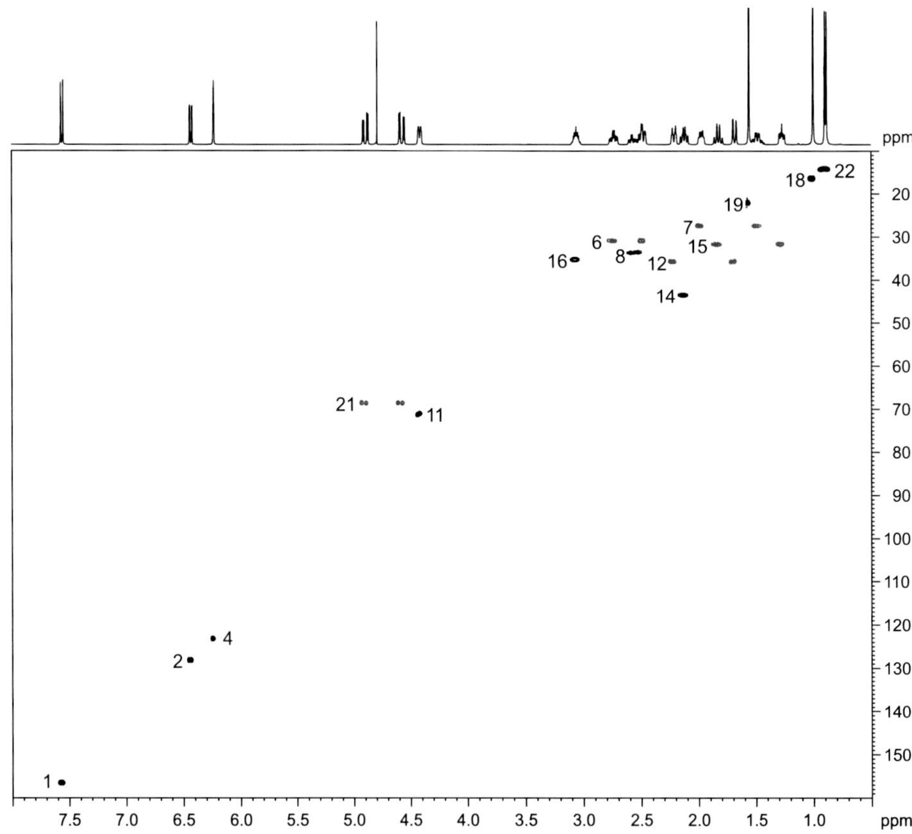

FIGURE 13.2 **The full ^{1}H–^{13}C edited HSQC spectrum of 13.1.**

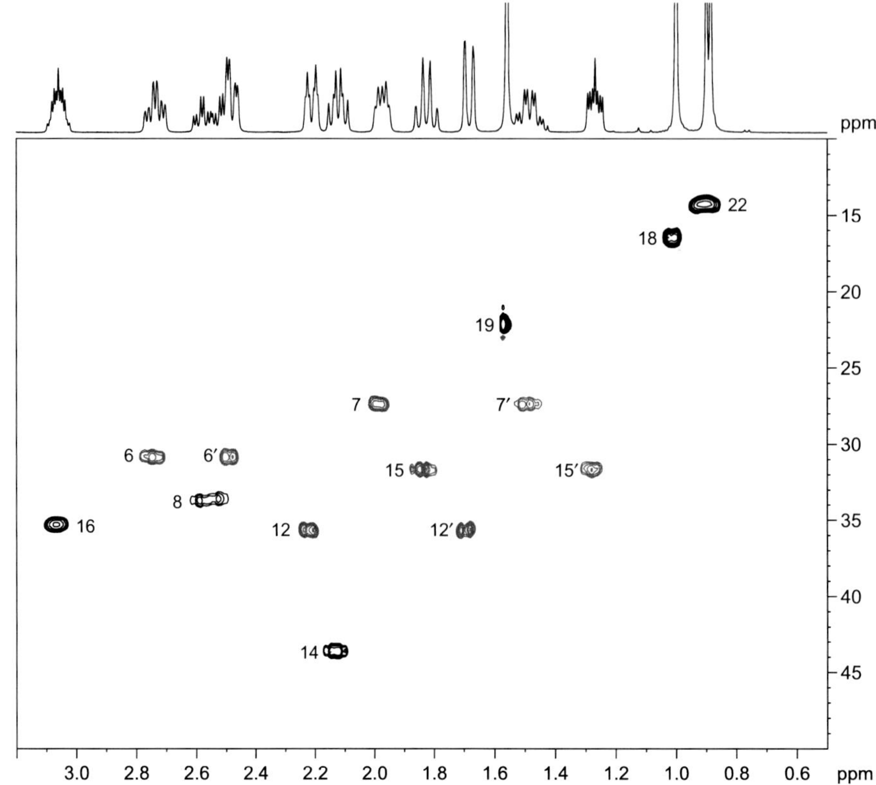

FIGURE 13.3 **An expanded region of the edited HSQC spectrum shown in Fig. 13.2.**

H^{14} (δ_C 43.6), H^{16} (δ_C 35.3) and H^8 (δ_C 33.8), in addition to the three previously identified methyl groups Me^{19} (δ_C 22.2), Me^{18} (δ_C 16.5) and Me^{22} (δ_C 14.3). We also note at this point the rather wide crosspeak observed for H^8, and the titling of the crosspeak structure along f_1, a point we return to later to explain this observation. On a practical point, also notice the slight broadening of the resonances for Me^{18} and Me^{22} in the proton dimension. This arises from a slight breakdown in the efficiency of the carbon-13 decoupling for these resonances that sit at the outer edges of the carbon spectrum and fall just beyond the bandwidth of the ^{13}C GARP decoupling sequence (see Section 12.3) employed during proton detection. Such issues become more significant at higher spectrometer field strengths due to the greater frequency bandwidths encountered.

13.3 ^{1}H–^{1}H COSY AND VARIANTS

Having differentiated the various proton groups in the structure, these may now be correlated through the COSY experiment (Fig. 13.4) as the first stage of piecing together fragments of the molecule's core. Correlations for the unsaturated protons confirm the previous connectivity indicated by coupling constants and suggest a long-range correlation between H^2 and H^4. Combining this with the carbon shift data suggests a possible fragment **13.2**, although the precise nature of the unsaturation remains unresolved at present since an aromatic ring or conjugated alkenes are feasible. Correlations from H^{11} link to a single CH_2 group only ($H^{12-12'}$), and suggest the small fragment **13.3.** Analysis of the region below 3 ppm requires a suitable starting point and in this case it is sensible to start with the methyl group Me^{22} as we now see this must sit at the end of a coupled spin system. Correlations from this are traced in Fig. 13.5 and define the continuous coupled chain of protons CH_3^{22}–CH^{16}–$CH_2^{15-15'}$–CH^{14}–CH^8–$CH_2^{7-7'}$–$CH_2^{6-6'}$ summarized in the fragment **13.4**. A further correlation seen links Me^{18} to H^{12} and this long-range correlation likely arises from a four-bond coupling pathway that is yet to be identified, but suggestive of **13.5**. A final weak coupling correlates H^4 with H^6 (Fig. 13.4) and may arise from a

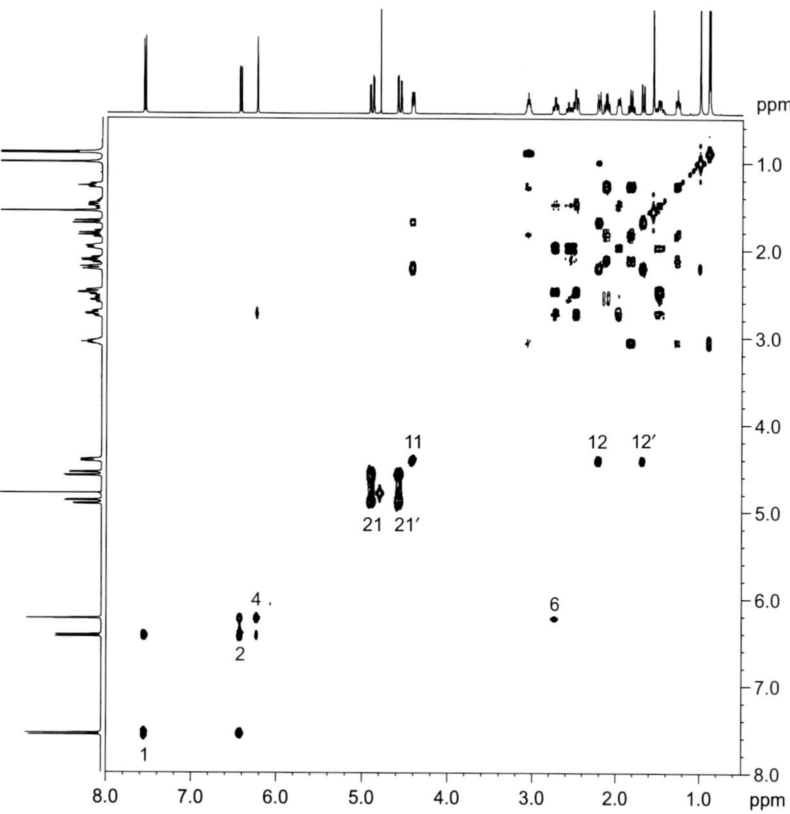

FIGURE 13.4 **The ^{1}H–^{1}H COSY-90 spectrum of 13.1 (magnitude mode display).**

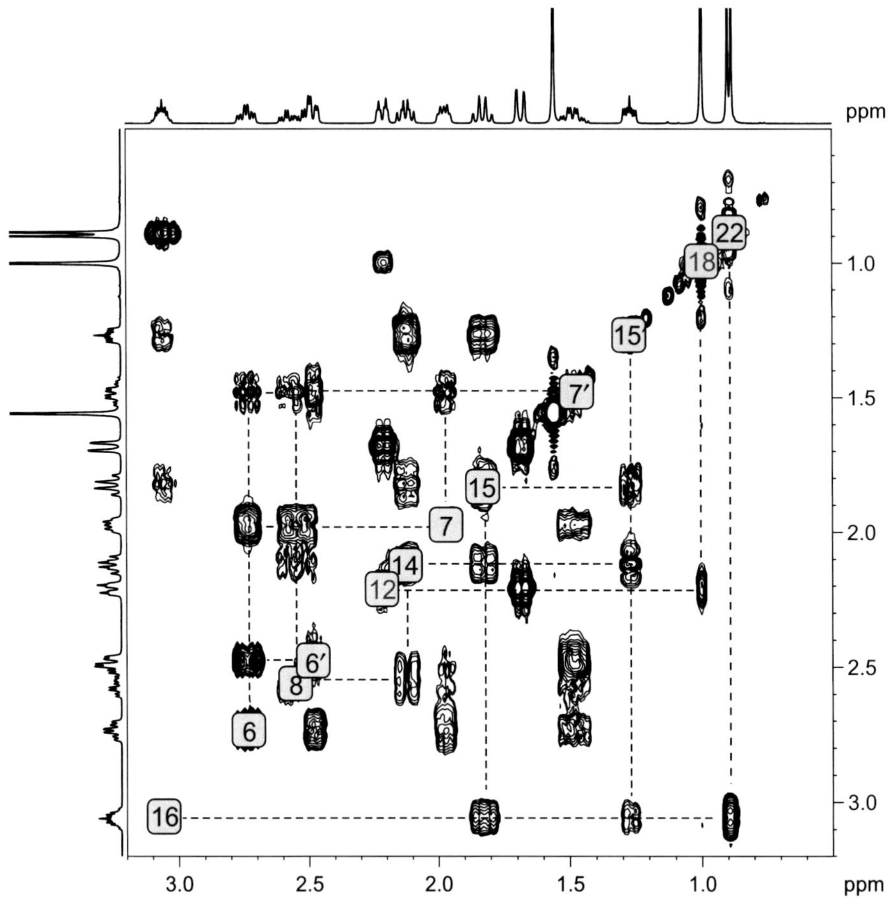

FIGURE 13.5 **An expansion of the COSY spectrum shown in Fig. 13.4.** Correlations connecting the sequence CH_3^{22}–CH^{16}–$CH_2^{15-15'}$–CH^{14}–CH^{8}–$CH_2^{7-7'}$–$CH_2^{6-6'}$ are traced by dashed black lines. The long-range correlation from Me^{18} is traced by the dashed red line.

small vicinal coupling or a longer range interaction between these, enabling a tentative connection to be made, as in **13.6**. We note that protons H^{21} and $H^{21'}$ couple only to each other, despite each appearing as a doublet–doublet, indicating the second coupling in each case is to a heteroatom of high abundance (J_{XH} ~5 Hz each), and both ^{19}F and ^{31}P are potential neighbours at this stage.

13.2 **13.3** **13.4**

13.5 **13.6**

13.3.1 Double-Quantum Filtered COSY

Having now built up a small selection of fragments that account for all (non-exchangeable) protons in the molecule, we take a small digression to consider other proton correlation methods that could be applied at this stage of an investigation. Firstly, we compare the magnitude-mode COSY employed above (Fig. 13.4) with the phase-sensitive double-quantum filtered (DQF)-COSY variant (Fig. 13.6). Unsurprisingly, we see essentially the same correlations in both spectra, yet there are some differences in detail that are worthy of note. Notice first that the DQF-COSY spectrum shows no significant diagonal peaks for the two methyl singlet protons Me^{19} and Me^{18} (or indeed for the singlet water resonance). As explained in Section 6.1.5 this is because these are unable to generate coherence that can pass the double-quantum filter and hence are

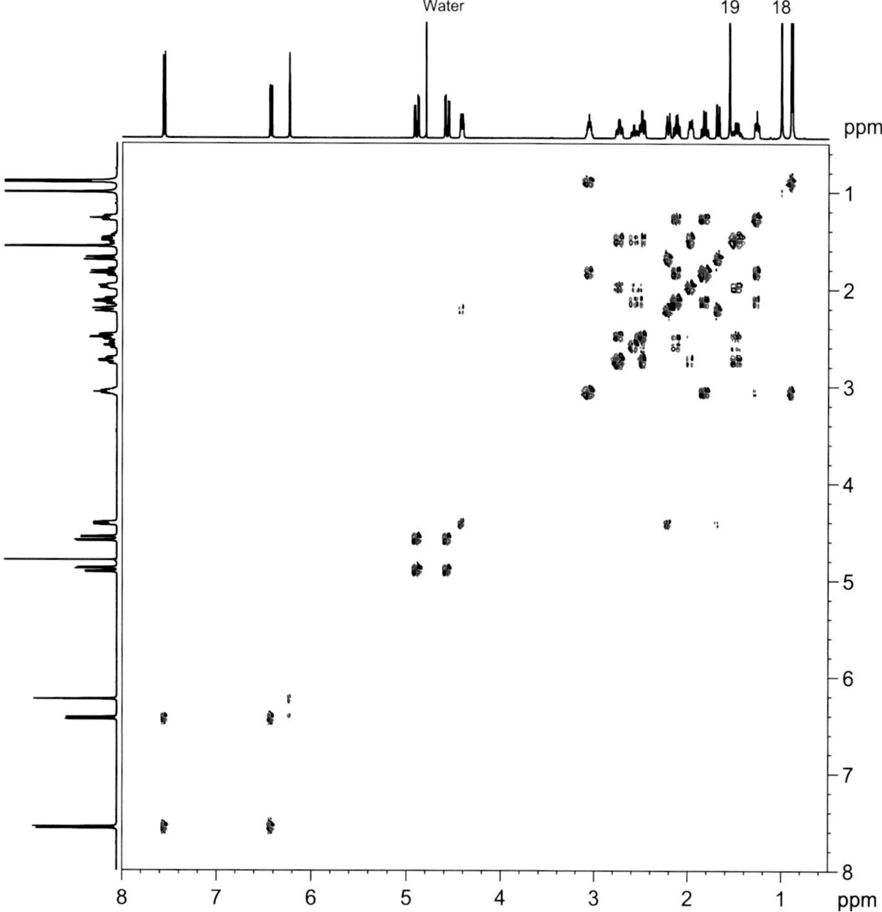

FIGURE 13.6 **The ^1H–^1H DQF-COSY spectrum of 13.1.**

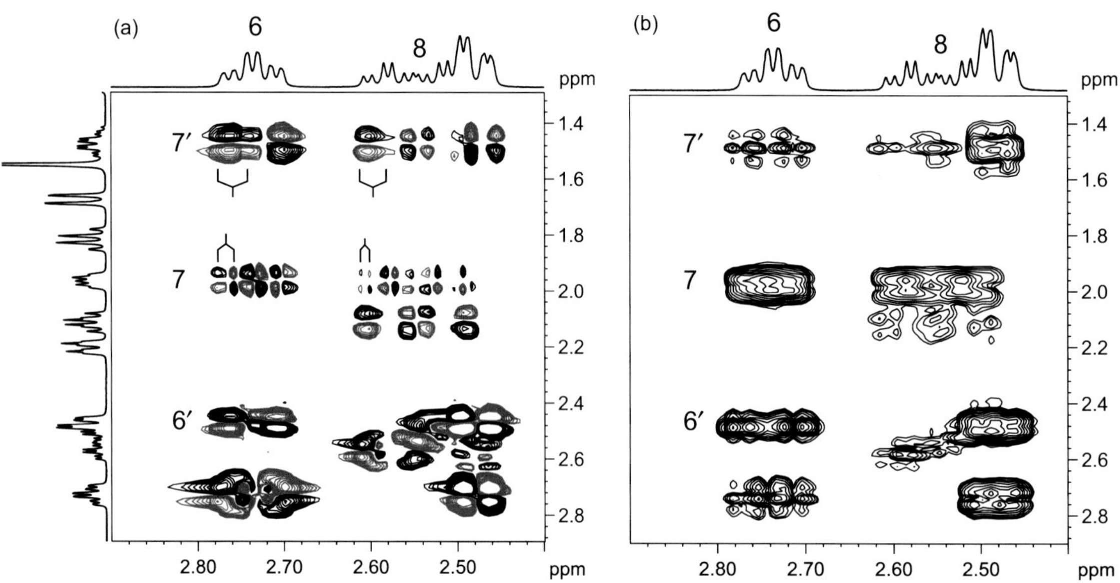

FIGURE 13.7 **An expansion of (a) the DQF-COSY spectrum (Fig. 13.6) and (b) the magnitude-mode COSY spectrum (Fig. 13.4) showing selected correlations from H6 and H8**. Stick figures highlight the anti-phase active couplings for the H7 and H7′ correlations.

removed from the final COSY spectrum. In fact, the very weak diagonal that may be observed for Me^{18} on closer inspection appears because of the small long-range couplings these protons share with H^{12}, leading to an unresolved doublet structure for the methyl resonance. Other differences may be appreciated by consideration of the expansion of Fig. 13.7. Here we see correlations from H^6 to protons 6', 7 and 7' in both spectra. However, in the DQF-COSY the crosspeak structure provides additional information on the magnitudes of the coupling constants giving rise to these peaks. The ^1H multiplet of H^6 is a double triplet with two couplings of 13.5 Hz and one of 5.8 Hz. From the anti-phase crosspeak structures of the DQF-COSY we can see the smallest active coupling is associated with H^7, indicating this proton shares the smaller 5.8 Hz coupling. The larger active coupling for the $H^{7'}$ crosspeak leads to partial signal cancellation at its centre. The crosspeaks of the magnitude COSY spectrum mask this level of detail, leaving this assignment unobtainable. Similar features may be observed for the crosspeaks of H^8 also in Fig. 13.7, for which the smaller active coupling (4.9 Hz) is again associated with the correlation to H^7. The additional detail in DQF-COSY crosspeak structure can also help locate and identify the crosspeak when this suffers overlap with other peaks, although this process does benefit from some experience of interpreting DQF-COSY peak structures. Unravelling overlapping crosspeaks can be yet more challenging in the magnitude-mode COSY spectrum. The level of peak detail afforded by the DQF-COSY can prove helpful in the analysis of more heavily overlapped 2D spectra or in the interpretation of data from more complex unknown structures.

13.4 ^1H–^1H TOCSY AND VARIANTS

With the analysis of more complex data sets in mind, we continue the digression by considering the data provided by the total correlation spectroscopy (TOCSY) spectrum, implemented as both the full 2D experiment and as the 1D analogue. The 2D spectrum is presented in Fig. 13.8 for the region below 5.0 ppm, and the large number of crosspeaks observed demonstrates extensive magnetisation transfer throughout the previously identified fragments. H^{11} clearly shows correlations to its directly coupled neighbours $H^{12-12'}$, but in addition we now see a further relayed correlation to Me^{18} via H^{12} that was not apparent in COSY. Furthermore, fragment **13.4** is now comprehensively mapped from any of the protons in this spin system. This extensive correlation map arises from the long mixing time used to generate these data (100 ms), allowing multi-stage transfer of magnetisation to occur. This process may be further appreciated in the high-resolution 1D TOCSY traces of Fig. 13.9, recorded with progressively greater mixing times of up to 120 ms, using Me^{22} as the target (source) resonance in each case. In the spectrum with the longer mixing periods, all protons within the fragment can be identified and their multiplet structures interrogated with the resolution associated with 1D spectra. This is possible even for resonances that were previously (partially) overlapped with others, as can be seen for the multiplet of H^8 that is partly hidden by that of $H^{6'}$ in the 1D proton spectrum but which is fully exposed in the earlier 1D TOCSY traces (eg 60 ms). The clean multiplet traces are ensured through the use of zero-quantum suppression in the experiment.

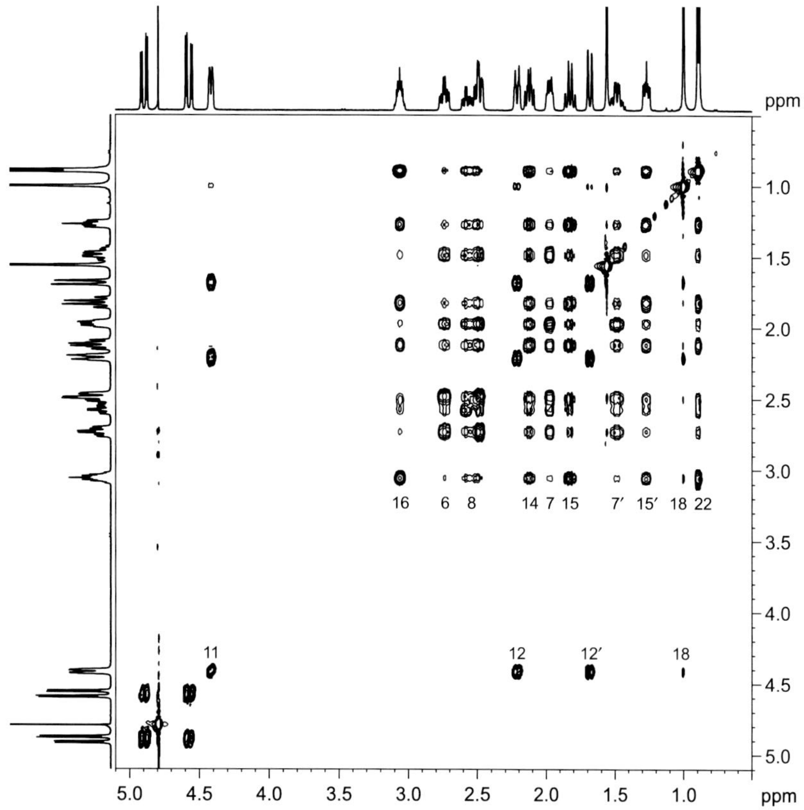

FIGURE 13.8 **A ^{1}H–^{1}H TOCSY spectrum of 13.1 recorded with a mixing time τ_m of 100 ms.** Data were recorded using a DIPSI2 mixing scheme and the sequence incorporated zero-quantum suppression.

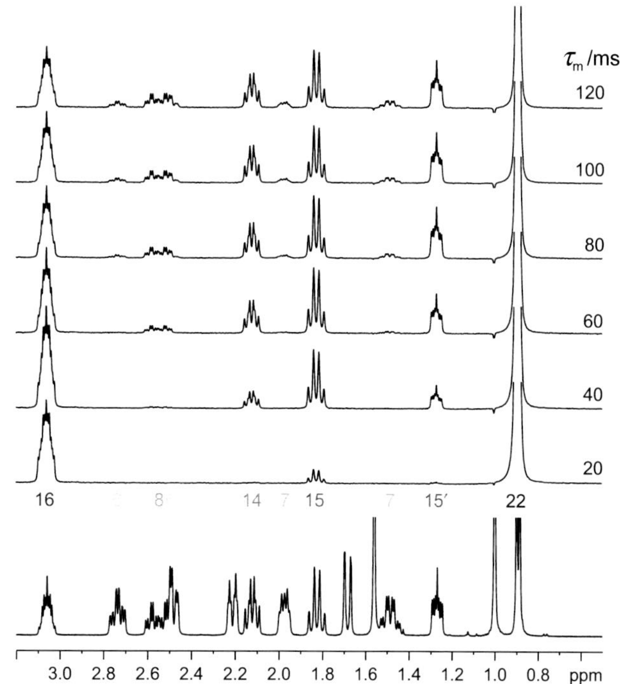

FIGURE 13.9 **A series of 1D TOCSY spectra of 13.1 recorded with increasing mixing times τ_m.** Me^{22} was the target resonance selected by a gradient spin-echo comprising an 80-ms Gaussian inversion pulse. Data were recorded using a DIPSI2 mixing scheme and the sequence incorporated zero-quantum suppression.

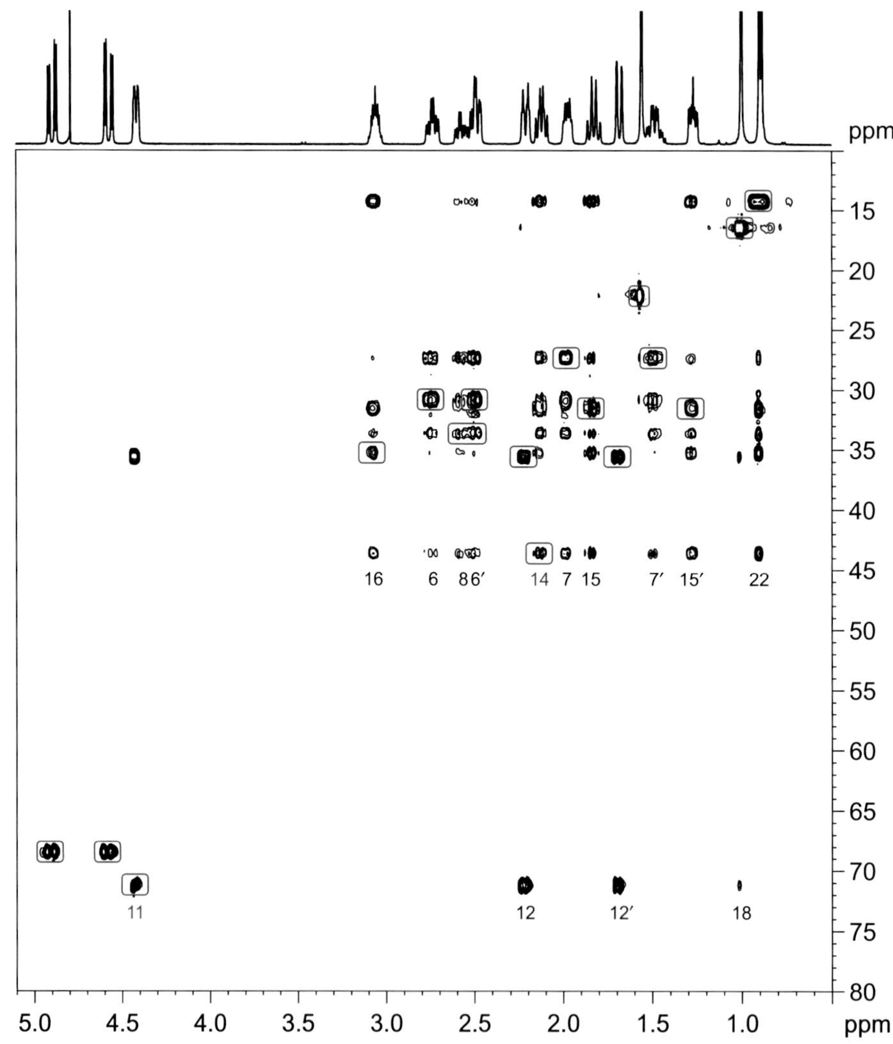

FIGURE 13.10 **A HSQC-TOCSY spectrum of 13.1 recorded with a 100-ms DIPSI2 mixing scheme.** Red boxes highlight the equivalent HSQC direct correlation peaks, with all others arising from magnetisation transfer through the proton coupling networks.

13.4.1 HSQC-TOCSY

An alternative approach to presenting TOCSY correlations is through the hybrid HSQC-TOCSY experiment, in which the proton correlations observed in the conventional TOCSY are dispersed according to chemical shifts of a heteroatom, mostly commonly carbon-13. This additional dispersion may again prove most beneficial in case of severe proton overlap when the homonuclear TOCSY becomes too crowded for unambiguous assignment. However, the penalty is the sensitivity loss associated with selection of the ^{13}C satellites via the HSQC step. To illustrate its appearance relative to the standard TOCSY, the HSQC-TOCSY spectrum recorded with 100 ms mixing time is presented in Fig. 13.10, again for the ^{1}H region below 5 ppm. The correlations seen match those of the TOCSY spectrum, as most easily observed for H^{11}, and are most readily identified from the corresponding HSQC correlation peaks (boxed red in figure). So, for example from the clearly resolved HSQC crosspeak of H^{14} it becomes possible to trace the complete proton network of fragment **13.4** at this carbon chemical shift.

13.5 ^{13}C NMR

We shall now return to the structure elucidation of **13.1,** and to progress further we must identify the remaining elements of the structure and how these are linked to the fragments already identified. Thus far we have assigned 15 protonated carbon centres, yet we know there to be 22 in the structure. Furthermore, there exists another 10 heteroatoms to be accounted for including both NMR active (^{19}F, ^{31}P) and inactive (^{16}O) nuclei. Identification of the non-protonated centres may proceed either by the use of direct 1D carbon observation or through the use of long-range heteronuclear correlation methods,

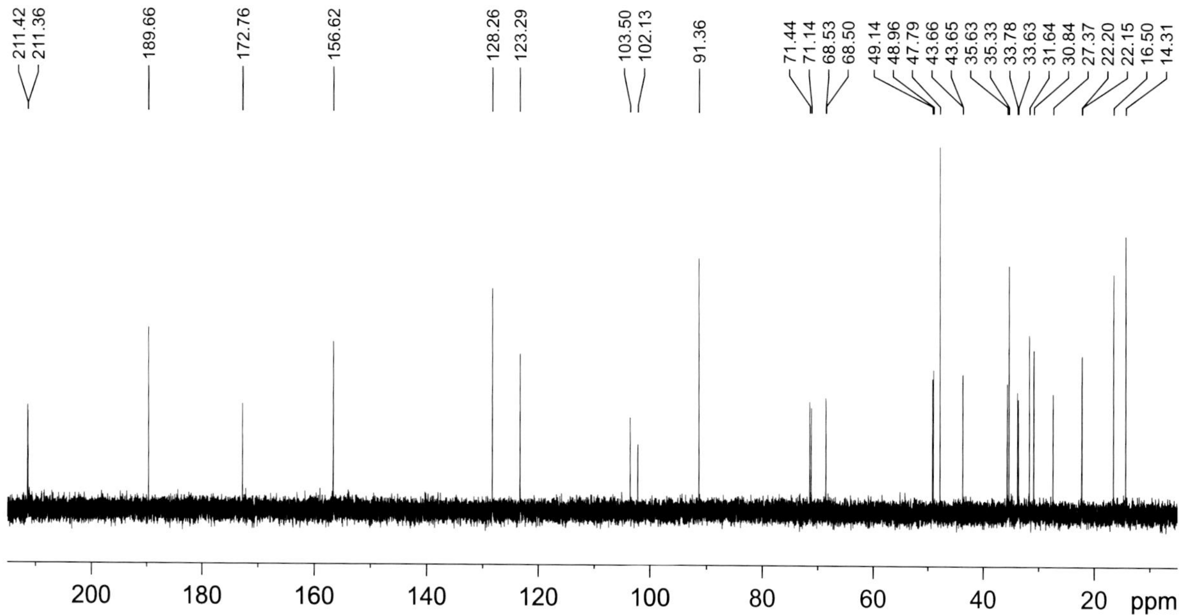

FIGURE 13.11 **The 1D proton decoupled ^{13}C spectrum of 13.1.**

primarily HMBC, or by a combination of these. To date, the presentation of 1D carbon spectra as proof of structure of synthetic materials remains a common requirement for journal publications, and their simple appearance provides a ready fingerprint for an organic structure. In situations where sample quantities are limiting, then direct carbon spectra can prove unobtainable, in which case the necessary assignments may be obtained indirectly from HMBC correlation data. Hence, the HMBC experiment serves a dual role, proving valuable for providing carbon assignments and also for piecing together molecular fragments. In the following discussion, the combined analyses of 1D carbon and 2D HMBC data are considered.

The carbon spectrum of **13.1** is presented in Fig. 13.11, for which a count of resolved peaks identifies 30 in total. There is no solvent resonance and the sample has high purity as evidenced by its ^{1}H spectrum, so we can assume the additional peaks identified above the 22 expected arise from carbon coupling to the other NMR active nuclei in the structure. In the absence of a molecular formula or at least a molecular mass, one may have been drawn to the conclusion that the compound contains 30 carbon centres, so highlighting the care that is needed when correlating the carbon peak count with carbon numbers. Initially therefore we correlate known crosspeaks in the HSQC spectrum with carbon shifts to identify those originating from protonated centres, mindful of the fact that closely neighbouring peaks in the 1D spectrum may originate from one carbon that is split by heteronuclear coupling to ^{19}F and/or ^{31}P.

In the region above 120 ppm, four peaks are observed of which the higher two (211.4 and 211.3 ppm) are indicative of a ketone; either two similar ketone environments or a single ketone with one heteronuclear coupling. The peaks at 189.7 and 172.8 ppm might suggest other carbonyl functionalities in the structure. Unusually, we see three peaks at 103.5, 102.1 and 91.4 ppm, in a region not commonly occupied in ^{13}C spectra, but often associated with the anomeric carbons of carbohydrates. However, in this case we know the centres involved are not protonated. The relatively high chemical shift is likely indicative of the influence of one or more nearby heteroatoms, or alternatively for the peaks above 100 ppm, a highly shielded unsaturated carbon centre. Of the two resonances close to 70 ppm known from the HSQC data, both appear to be split in the carbon spectrum, or happen to have close quaternary neighbours. Earlier analysis indicated protons $H^{21-21'}$ to share coupling with a heteroatom, so coupling to their connected carbon ($\delta_C \sim 68.5$) would also seem likely and would account for the appearance of two peaks ($J_{XC} = 4$ Hz). Remaining resonances occur below 50 ppm, and of these the Me^{19} resonance at 22.1 ppm is observed to carry a heteronuclear coupling of 6 Hz, since it is unlikely that a quaternary resonance would appear this low. Although other resonances in this region must also carry heteronuclear couplings, these are less readily identified, so we now turn our attention to long-range correlation data from HMBC. From this we shall aim to build the remainder of the carbon skeleton of the structure in a stepwise process, and support this with further experiments where required.

For small molecules obtained from a synthetic procedure, the data considered thus far may well be adequate to verify a structure, provided all data were self-consistent and a complete rationalisation of carbon shift data could be made. In this regard, the use of spectrum prediction routines can prove helpful, and a small number of highly developed commercial packages are now available. It should be stressed that predictions of carbon-13 spectra from molecular structures are generally regarded as being more reliable than proton predictions, which should be taken as suggestive of potential assignments at best. At the basic level, proton environments are highly sensitive to solvent effects, stereochemistry and intermolecular interactions making their reliable prediction highly challenging. Carbon chemical shifts are rather less perturbed by such effects and can be predicted more accurately, especially when closely related structures have contributed to data underlying the predictive algorithms. When applying these, it is always sensible to run trial predictions on known compounds to assess their capabilities for the types of structures being studied.

13.6 HMBC AND VARIANTS

The ^{1}H–^{13}C HMBC experiment is undoubtedly the most widely used implementation of HMBC found in the chemical laboratory. However, the experiment may be readily adapted to other combinations of nuclei and some illustrative examples are included in this section. As noted earlier, these can often be derived from a standard ^{13}C parameter set, with appropriate adjustments made for pulse frequencies and calibrations, chemical shift ranges and coupling constants for the nuclei and for appropriate pulsed field gradient amplitude ratios, if used.

13.6.1 ^{1}H–^{13}C HMBC

The HMBC long-range correlation experiment can be expected to yield a wealth of information, with each proton potentially correlating to numerous carbon centres that exist nearby (Fig. 13.12). The analysis of these data require a well-documented and systematic approach, and again commercially software packages that assist with this process are available and may prove worthwhile when working with complex data sets. Whatever record-keeping method is employed, one needs to define a rational start point for the analysis. It is often the case that methyl groups provide a large number of intense resonances in HMBC and since we have two methyl groups already correlated in the above interpretation, these provide a sensible place to begin our analysis.

It is observed in the spectrum of Fig. 13.13 that Me^{22} correlates to three resonances, of which two are to be anticipated (C^{15} and C^{16}) and a further correlation to a quaternary centre that must attach to C^{16} (91.4 ppm, C^{17}). Similarly, Me^{18}

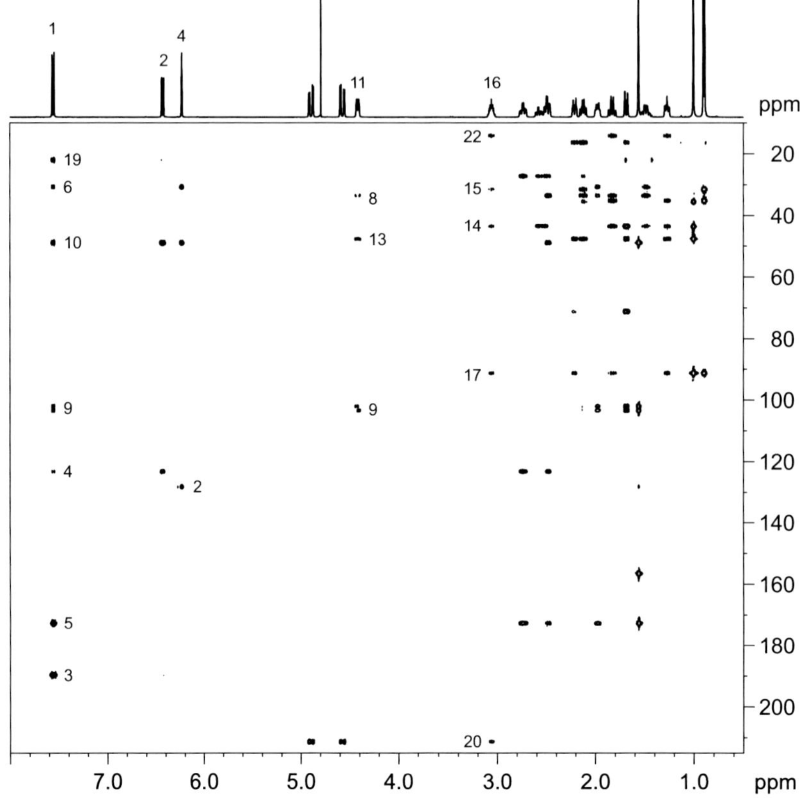

FIGURE 13.12 **The ^{1}H–^{13}C HMBC experiment for 13.1 optimised for J_{CH} long-range couplings of 8 Hz.** The experiment was collected as the phase-sensitive version with three-step low-pass filtration of one-bond correlations, and was processed with magnitude calculation in the proton dimension.

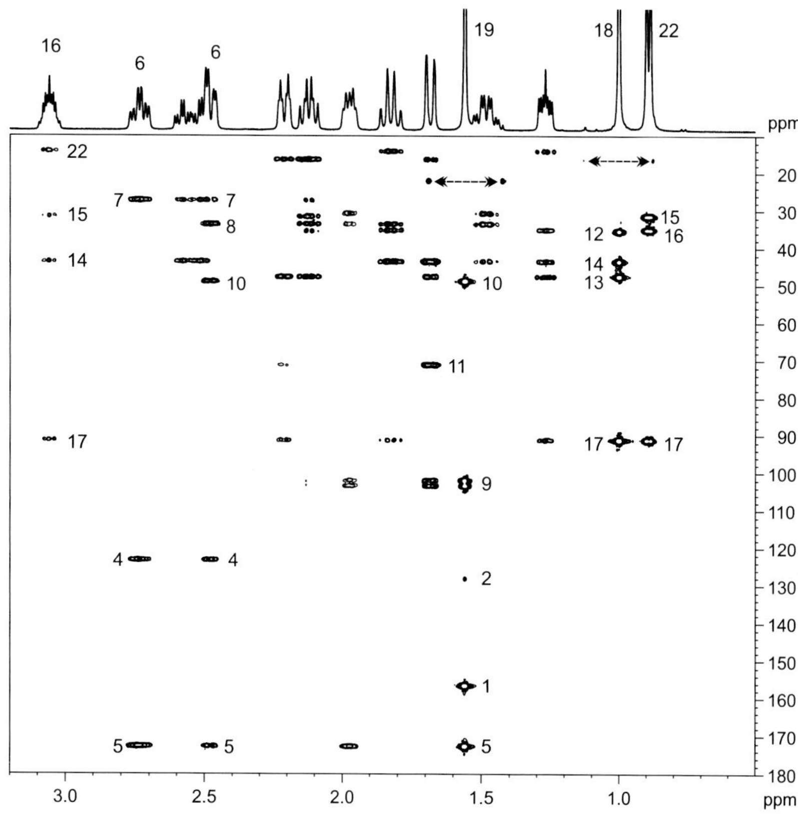

FIGURE 13.13 **Expansion of the HMBC spectrum for the region below 3.2 ppm, with selective labelling of peaks as discussed in the text.** The dashed arrows indicate the appearance of one-bond correlations that have not been completely filtered.

correlates to the previously identified methylene C^{12} and methine C^{14}, and additionally to a new quaternary centre (47.8 ppm) and to the same quaternary of high chemical shift (91.4 ppm) to which Me^{22} was correlated, suggesting these methyl groups exist in neighbouring parts of the structure. However, Me^{18} shows no correlation to C^{16} suggesting it likely does not attach directly to the 91.4 ppm quaternary. Combing these data with the fragments **13.4** and **13.5** identified through proton–proton correlations we may construct the more substantial fragment **13.7**, in which key HMBC derived connections are marked in bold (and X denotes an unidentified atom). To define anything of the structure joined to C^{17} we may consider HMBC correlations from H^{16} (Fig. 13.12). Of the five observed, four are consistent with the ring structure of **13.7** and a fifth links to the highest ^{13}C shift at 211.4 ppm, and thus places a ketone functionality on C^{17}. The only other correlations in HMBC to this ketone arise from protons $H^{21-21'}$ which have already been identified as an isolated pair coupled to another NMR active nucleus, either ^{19}F or ^{31}P. We also note that these protons show a rather weak correlation to C^{17} at 91.4 ppm (not seen in Fig. 13.12), confirming attachment of the side group to the core of the structure.

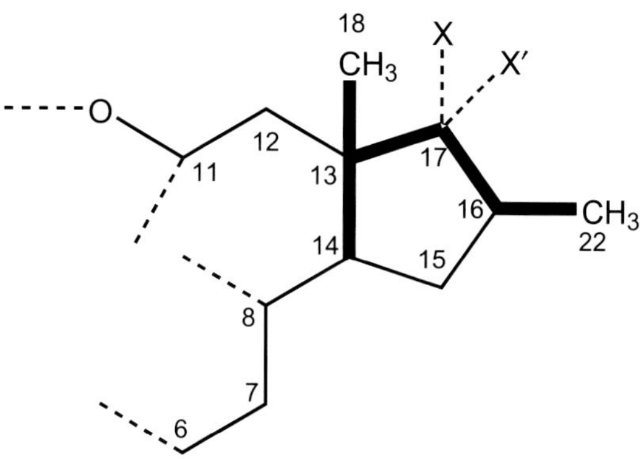

13.7

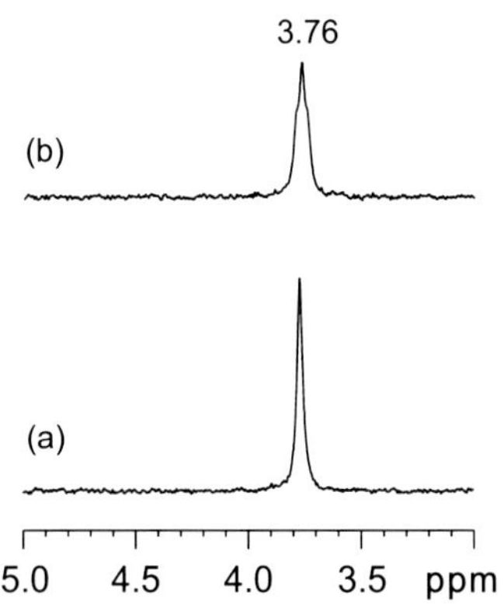

FIGURE 13.14 The ^{31}P NMR spectrum of 13.1 recorded (a) with and (b) without proton decoupling.

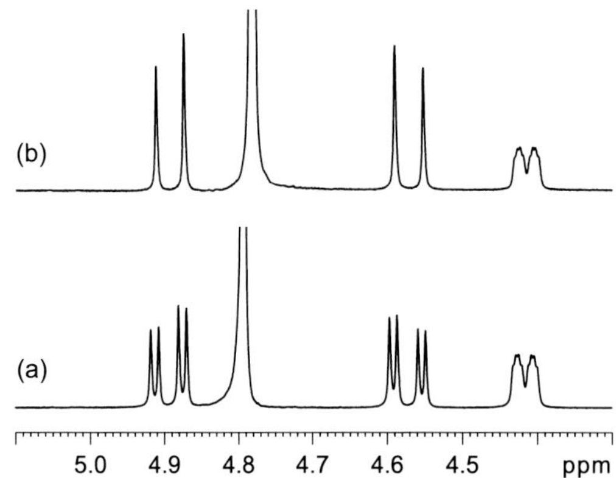

FIGURE 13.15 A region of the ^{1}H NMR spectrum of 13.1. This shows the H^{21}, $H^{21'}$ methylene protons recorded (a) without and (b) with ^{31}P decoupling.

13.6.2 ^{31}P and ^{1}H–^{31}P HMBC

At this stage it is appropriate to consider the influence of the heteroatoms on the NMR spectra. The ^{31}P spectrum of **13.1** contains a single, rather broad resonance at 3.76 ppm (linewidth $\Delta\upsilon_{1/2}$= 6 Hz) which in the absence of proton decoupling appears as a broadened triplet of ~5 Hz, consistent with the ^{1}H coupling structure observed for $H^{21-21'}$ (Fig. 13.14). To prove unambiguously the correlation with ^{31}P, there are a number of possible approaches one might consider. One would be to record the ^{31}P-decoupled ^{1}H spectrum and observe collapse of any multiplet structure associated with the phosphorus coupling (Fig. 13.15). Here the loss of the 5 Hz couplings for the $H^{21-21'}$ resonances prove these to arise from ^{31}P. An alternative would be to record the ^{1}H–^{31}P HMBC spectrum and seek direct evidence of these heteronuclear correlations. This has the potential advantage of very readily revealing other correlations to ^{31}P that arise from smaller, possibly unresolved couplings. Since phosphorus-31 is 100% abundant, its HMBC experiment has very high sensitivity and can typically be acquired very rapidly. The ^{31}P HMBC spectrum of **13.1** (Fig. 13.16) clearly shows only a pair of correlations to the single ^{31}P centre, suggesting this to be remote from all other protons in the molecule. The phosphorus chemical shift is itself suggestive of a phosphate group and the broad resonance is also a common feature of this functional group in aqueous media. Furthermore, the direct attachment of a phosphate group to C^{21} would be consistent with this chemical shift (68.5 ppm). Combining this information with fragment **13.7**, yields **13.8**. This also rationalises the doubling of the C^{20} ketone resonance at 211.4 ppm as arising from $^{3}J_{PC}$ coupling of 7.7 Hz, and of the C^{21} resonance as $^{2}J_{PC}$ of 3.7 Hz, as determined from the ^{13}C spectrum. Finally, the high chemical shift of C^{17} (91.4 ppm) and its singlet structure suggest the attached group is likely to be an oxygen, and most probably a hydroxyl given all HMBC correlations to C^{17} have been assigned, although this remains tentative at this stage. We shall return to the influence of the ^{19}F nucleus shortly.

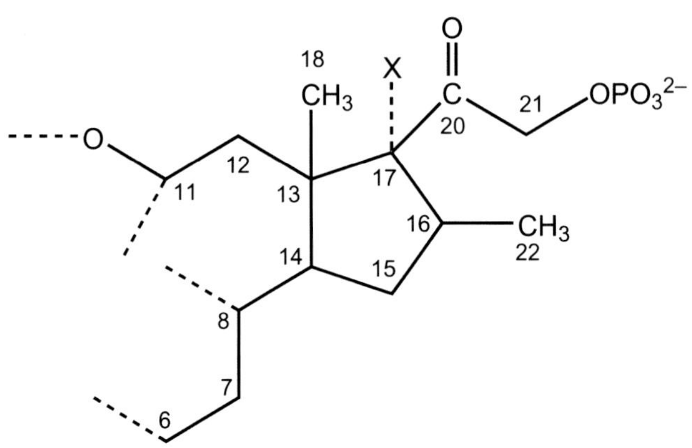

13.8

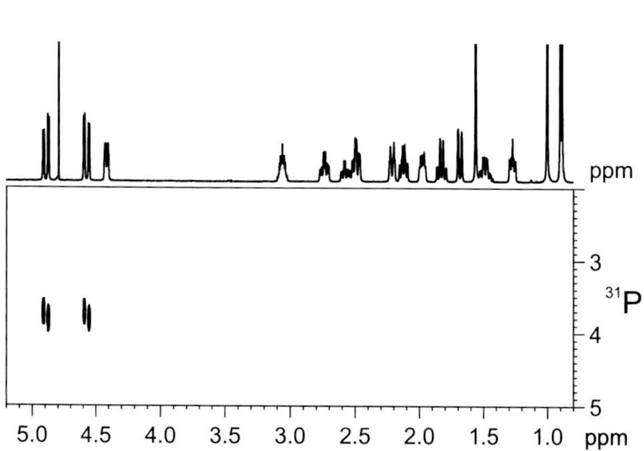

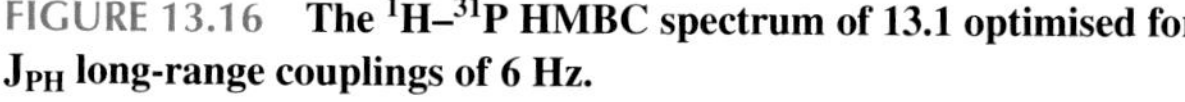

FIGURE 13.16 **The ^{1}H–^{31}P HMBC spectrum of 13.1 optimised for J_{PH} long-range couplings of 6 Hz.**

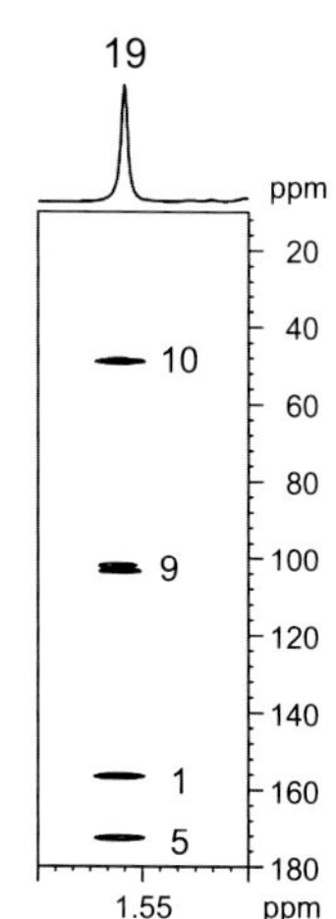

FIGURE 13.17 **An expansion of the ^{1}H–^{13}C HMBC spectrum of 13.1 showing correlations from Me^{19}.**

13.6.3 ^{1}H–^{13}C HMBC Again

Having built substantial parts of the structure, we now return to the remaining methyl group, Me^{19}, and consider its correlations. These link the group to the unsaturated carbon resonance of C^{1} of fragment **13.2** and to a further *sp*^{2} centre C^{5}. Further correlations are seen to 49.0 ppm and to two neighbouring peaks at 103.5 and 102.1 ppm (Fig. 13.17). The appearance of five correlations suggests either that one or more correlations are observed across more than three bonds, or that at least some of these are in fact linking to a carbon split by coupling to another NMR active nucleus so that two peaks are observed in the carbon dimension for a single correlation. The only possible heteronuclear correlation might be for the peaks at 103.5 and 102.1 ppm (in which case J_{XC} = 172 Hz; we shall eventually show these to belong to C^{9}, split by ^{19}F coupling) since all others are too widely separated. We may tentatively propose fragment **13.9** as being consistent with these data.

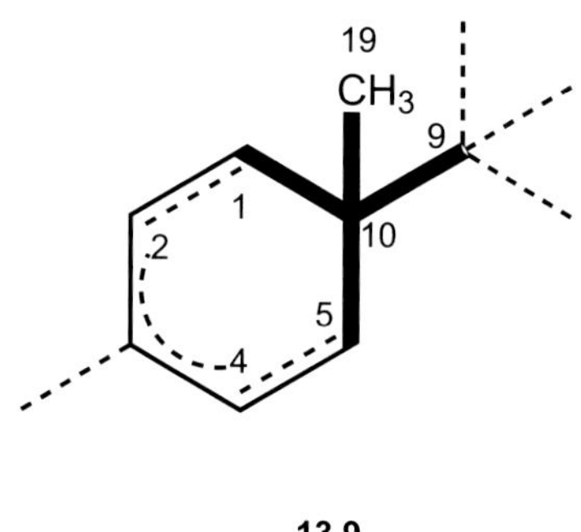

13.9

With a route into the unsaturated core of the structure, we shall next focus on correlations from these protons. H^{1} is observed to yield numerous correlations (Fig. 13.12), including to the carbon of Me^{19}, placing this on the adjacent quaternary centre C^{10} (49.0 ppm) which is observed to be two closely spaced lines in the 1D carbon spectrum (J_{XC} = 22 Hz). H^{1} has many correlations in common with Me^{19}, including to the split peaks at ~103 ppm. Significantly, a correlation is also seen to 189.7 ppm suggesting this to be a carbonyl group and likely a ketone having a rather low shift due to α,β unsaturation (C^{3}). Neither protons H^{2} nor H^{4} correlate to C^{3} over two bonds. Similarly H^{1} shows only a rather weak correlation to C^{2}. Such observations are common for unsaturated systems and weak two-bond correlations are often seen for aromatic rings due to the small magnitude of the associated $^2J_{CH}$ coupling constants. In contrast, correlations from H^1 to C^3 and C^5 are some of the most intense in the HMBC spectrum owing to the *trans* coupling geometry giving rise to large $^3J_{CH}$ values that match well to the optimised 8 Hz coupling. In accordance with this, clear correlations are observed from H^2 and H^4 to 49.0 ppm (C^{10}), leading to the definition of fragment **13.10** for the unsaturated ring structure.

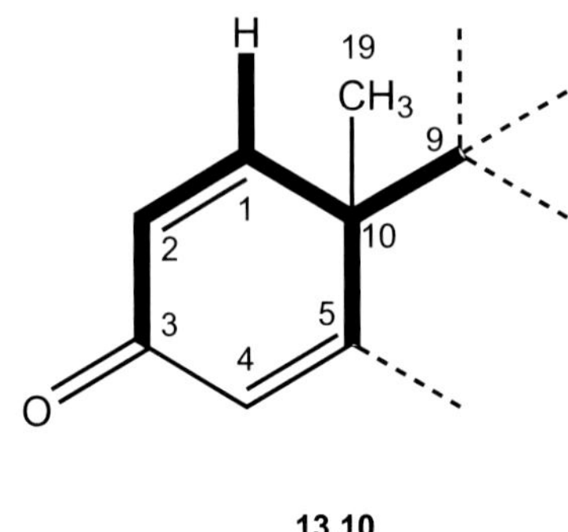

13.10

The remaining unassigned correlations from the alkene protons link both H^1 and H^4 to the previously identified aliphatic C^6. These are important observations since they allow us to connect two otherwise discretely defined fragments of the structure. Since it is known that C^5 is a fully substituted sp^2 centre, a connection to C^6 at this junction would make logical sense. Note that this then defines the H^1–C^6 correlation as arising from a four-bond C–H coupling pathway. This is consistent with the planar 'zig-zag' geometry between these nuclei that would enhance the associated coupling constant.

Combining this HMBC information, allows fragments **13.7** and **13.10** to be connected to yield **13.11**. This connection is further supported through correlations from both H^6 and $H^{6'}$ to C^4 and C^5, and additionally from $H^{6'}$ to C^{10}. With the bulk of the structure now defined, we turn our attention to the remaining core and additional correlations to the peak centred at ~103 ppm. Significant new correlations to this arise from H^7, H^{11} and $H^{12'}$, and taken with the previous correlations to this centre, allows the final ring of the core to be defined, as in **13.12**. Additional correlations from H^{11} to C^9 and C^8 further confirm the ring closure at C^9. From the molecular composition, it is reasonable to assign the missing functionality X on this carbon as being fluorine (see later). Finally we return to the tentative hydroxyl substitution at C^{17} and confirm this as being the correct assignment as all other features have been accounted for. A summary of all HMBC correlations for **13.1** may be found in Table 13.1.

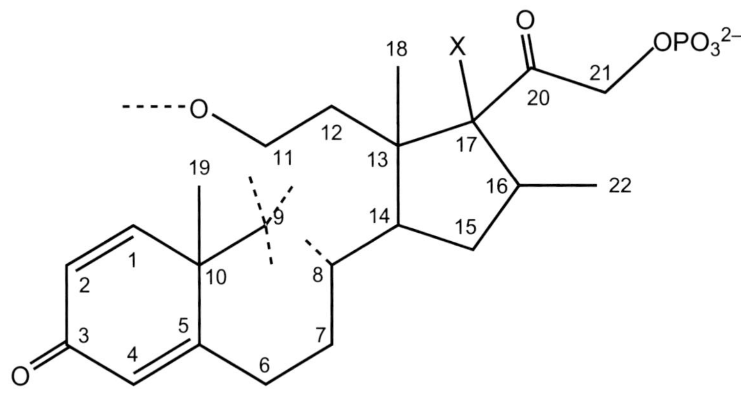

13.11

TABLE 13.1 Correlations Observed in the ^1H–^{13}C HMBC Experiment of 13.1 Optimised for J_{CH} = 8 Hz

Proton	Carbon Correlations
1	3, 4, 5, 6, 9, 10, 19
2	4, 10
4	2, 6, 10
6	4, 5, 7
6′	4, 5, 7, 8, 10
7	5, 6, 8, 9
7′	6, 8, 14
8	7, 14
11	8, 19, 13
12	11, 13, 17, 18
12′	9, 11, 13, 14, 18
14	7, 13, 15, 16, 18
15	8, 14, 16, 17, 22
15′	13, 14, 16, 17, 22
16	14, 15, 17, 20, 22
18	12, 13, 14, 17
19	1, 2, 5, 9, 10
21	20
21′	20
22	15, 16, 17

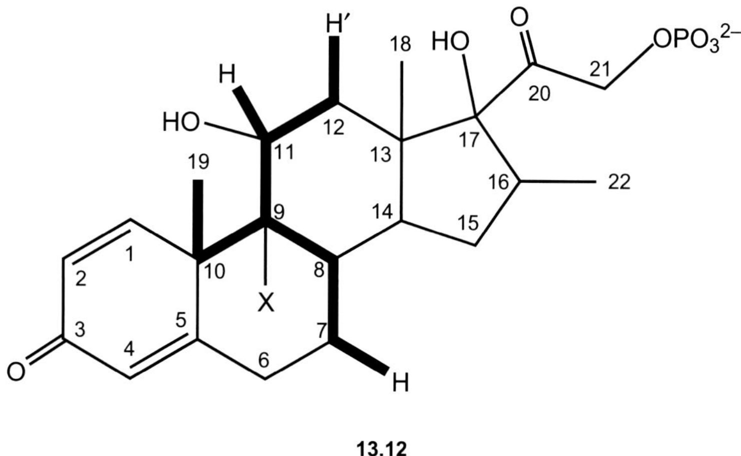

13.12

13.6.4 ^{19}F and ^{19}F–^{13}C HMBC

The ^{19}F spectrum shows a single resonance (-164.5 ppm) which in the presence of proton coupling appears as a double doublet with J_{FH} = 32 and 10 Hz (Fig. 13.18), further broadened likely by other smaller unresolved coupling. Evidence for the location of the fluorine may again been derived from the use of heteronuclear ^{19}F decoupling of the proton spectrum

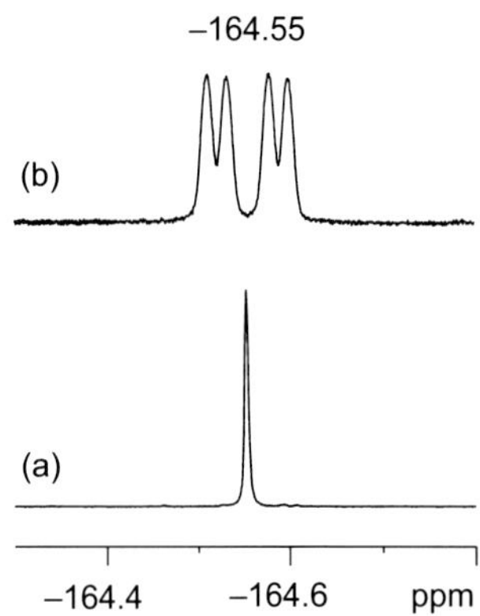

FIGURE 13.18 **The ^{19}F NMR spectrum of 13.1 recorded (a) with and (b) without proton decoupling.** The resolved couplings measure 10 and 32 Hz.

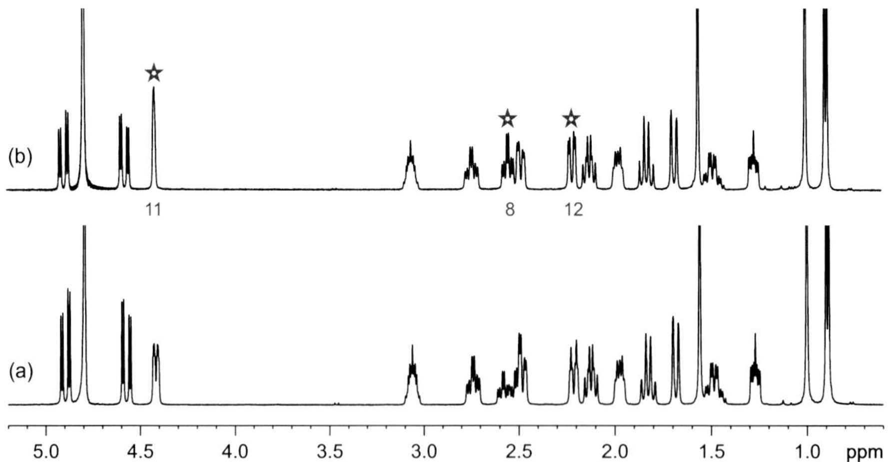

FIGURE 13.19 **A region of the ^{1}H NMR spectrum of 13.1 recorded (a) without and (b) with ^{19}F decoupling (stars highlight regions of multiplet change).** The heteronuclear couplings are observed to be 32 Hz (H^8), 10 Hz (H^{11}) and 3 Hz (H^{12}).

(Fig. 13.19). Comparison of the ^{1}H spectrum with and without this decoupling shows collapse of the multiplet structure for protons H^{11} (10 Hz), H^8 (32 Hz) and H^{12} (3 Hz), identifying couplings across as many as 4-bonds, a common feature for F–H coupling (these couplings were also revealed in the pure shift ^{1}H spectrum of this compound: see Fig. 8.17 in Section 8.5). Of these, the loss of the large coupling to H^8 is very apparent since the multiplet structure then no longer overlaps with the neighbouring $H^{6'}$. The fact that this coupling is in excess of 30 Hz is especially noteworthy since this large value indicates the fluorine and proton share an *anti* orientation with a dihedral angle close to 180 degrees and so provides our first definition of stereochemistry in the structure. With definition of the location of the fluorine centre, it is now possible to rationalise the additional peaks observed in the carbon spectrum, and to assign these as being due to coupling to this heteroatom. In all, six fluorine–carbon couplings can be identified: C^9 ($^{1}J_{FC}$ 172.4 Hz), C^8 ($^{2}J_{FC}$ 19.2 Hz), C^{10} ($^{2}J_{FC}$ 22.4 Hz), C^{11} ($^{2}J_{FC}$ 38.3 Hz), C^{14} ($^{3}J_{FC}$ 1.6 Hz) and C^{19} ($^{3}J_{FC}$ 6.0 Hz). These doublets, in addition to the two identified as arising from phosphorus coupling, account for the eight additional carbon peaks observed beyond those expected from the molecular composition, as noted earlier. The presence of the large fluorine couplings to ^{1}H and ^{13}C also explains the tilted structure observed for the H^8 HSQC crosspeak noted in Section 13.2 (Fig. 13.3) since these heteronuclear couplings are retained in both dimensions for the $^{3}J_{FH}$ and $^{2}J_{FC}$ interactions.

The ^{19}F coupling partners may also be identified more directly through the ^{19}F–^{13}C HMBC experiment. This requires that this combination of nuclei can be tuned simultaneously on a suitable probe, which is not always possible depending on available probe configurations (the spectrum below was recorded using a broadband probe for which the proton channel could alternatively be tuned to fluorine, and employed a dual H/F preamplifier). The spectrum of Fig. 13.20 shows all

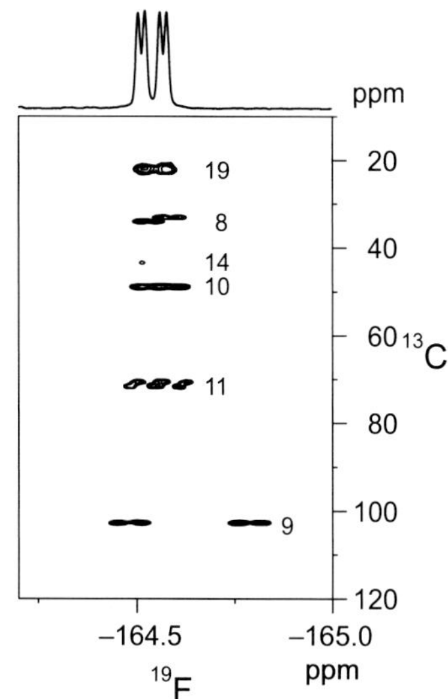

FIGURE 13.20 **The ^{19}F–^{13}C HMBC spectrum of 13.1 optimised for J_{FC} long-range couplings of 8 Hz (recorded at 600 MHz).**

expected correlations consistent with the coupling partners defined above, although the 1.6 Hz coupling to C^{14} yields a peak of very weak intensity. Note also that the one-bond coupling to C^9 is also apparent in this spectrum since no low-pass filtration was employed; the observed offset from the parent fluorine resonance of this correlation is due to the large one-bond $^{12}C/^{13}C$ isotope shift (0.096 ppm).

13.7 NUCLEAR OVERHAUSER EFFECTS

Nuclear Overhauser experiments play a key role in stereochemical definition and nowadays most commonly record the appearance of transient nuclear Overhauser effects (NOEs). These may be detected in both the 2D and 1D version of the NOE spectroscopy (NOESY) experiment, or as the alternative rotating frame NOE spectroscopy (ROESY) experiment when this is more appropriate. Here we illustrate both the 2D and 1D NOESY and demonstrate that equivalent information can be derived from either approach. Also shown is an example of a heteronuclear NOE experiment (HOESY) that compliments the homonuclear version.

13.7.1 2D NOESY

Having defined the gross structure of **13.1** (as summarized in **13.12**), we now consider the stereochemical definition within the structure. While the proton homonuclear vicinal coupling constants may provide insights through consideration of the Karplus relationship between dihedral angle and coupling magnitude, NOE experiments will provide a more direct interrogation of the spatial relationships between protons in the molecule. In a structure of moderate complexity, establishing a complete map of these is most efficiently achieved through consideration of the 2D NOESY spectrum (Fig. 13.21). The observed NOEs are summarized in the grid format of Fig. 13.22; we again begin our analysis with the correlations from the methyl groups.

In Fig. 13.23 methyl-18 is seen to correlate strongly with H^8, $H^{12'}$, H^{15} and H^{16} suggesting these sit on the same face of the structure, which shall be presented as the upper face. Further, weaker NOEs are observed to Me^{19} and to both protons $H^{21–21'}$, which we return to shortly (Fig. 13.21). The lack of a significant correlation to H^{14} suggests this exists on the opposite face to Me^{18}. H^{14} itself correlates strongly with H^{12} and $H^{15'}$, presenting a consistent picture of these occupying the lower side, as in **13.13**. Protons H^{12} and $H^{12'}$ both give an NOE to H^{11}, with that from H^{12} moderately more intense. This, coupled with the lack of NOEs from Me^{18} or from H^8, places H^{11} as equatorial on the ring, sitting approximately midway between H^{12} and $H^{12'}$, and therefore defines the C^{11} hydroxyl group as being axial **13.14**. Continuing the assignment from

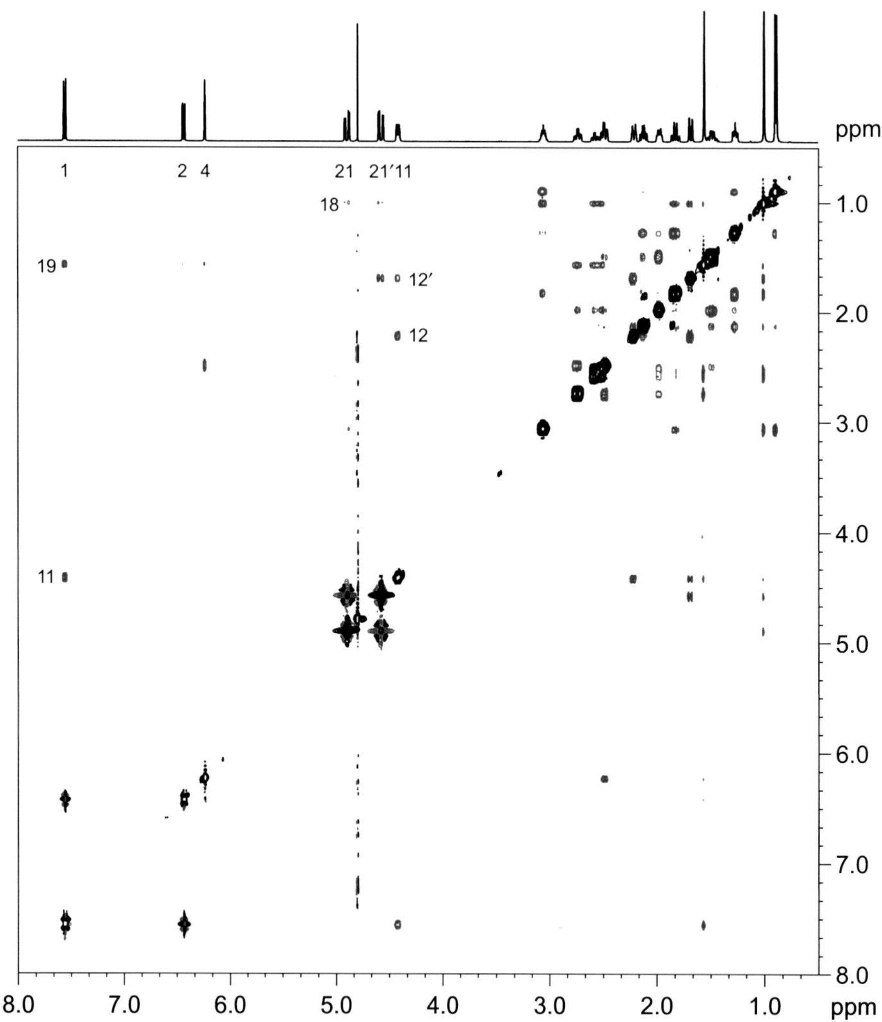

FIGURE 13.21 **The 2D NOESY spectrum of 13.1 recorded with a 500 ms mixing time (with selective labelling of peaks; see text).**

H^8, we see this correlates to only H^7 on the neighbouring carbon and across to Me^{19} suggesting this methyl occupies an axial orientation similar to Me^{18}. Me^{19} show strong NOEs to H^1 and H^6, with additional weaker effects observed to H^2, H^4 and H^{11} (unseen at this threshold). These patterns suggest both $H^{7'}$ and $H^{6'}$ sit on the lower face of the structure, with NOEs from $H^{7'}$ to $H^{6'}$ and H^{14} confirming this geometry, as in **13.15**.

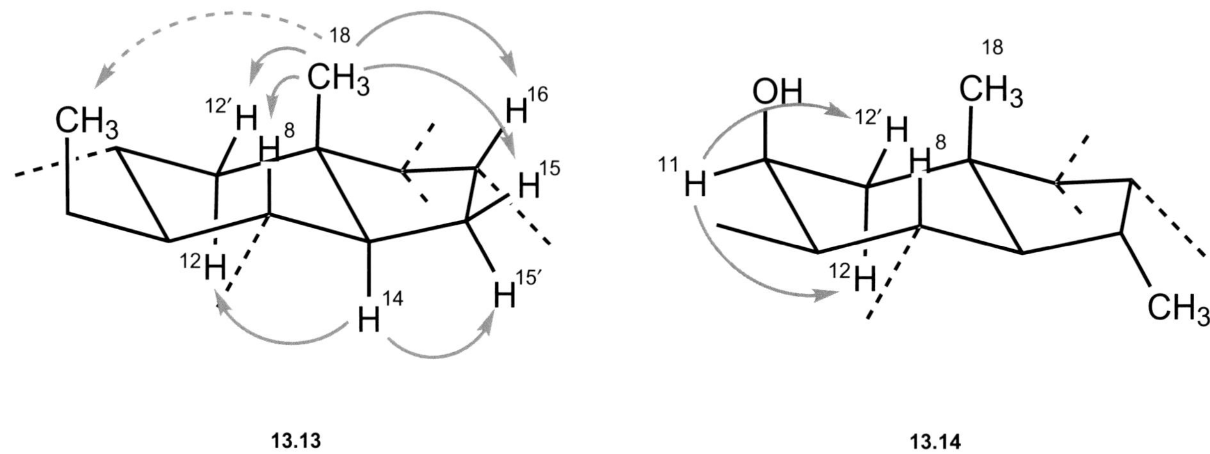

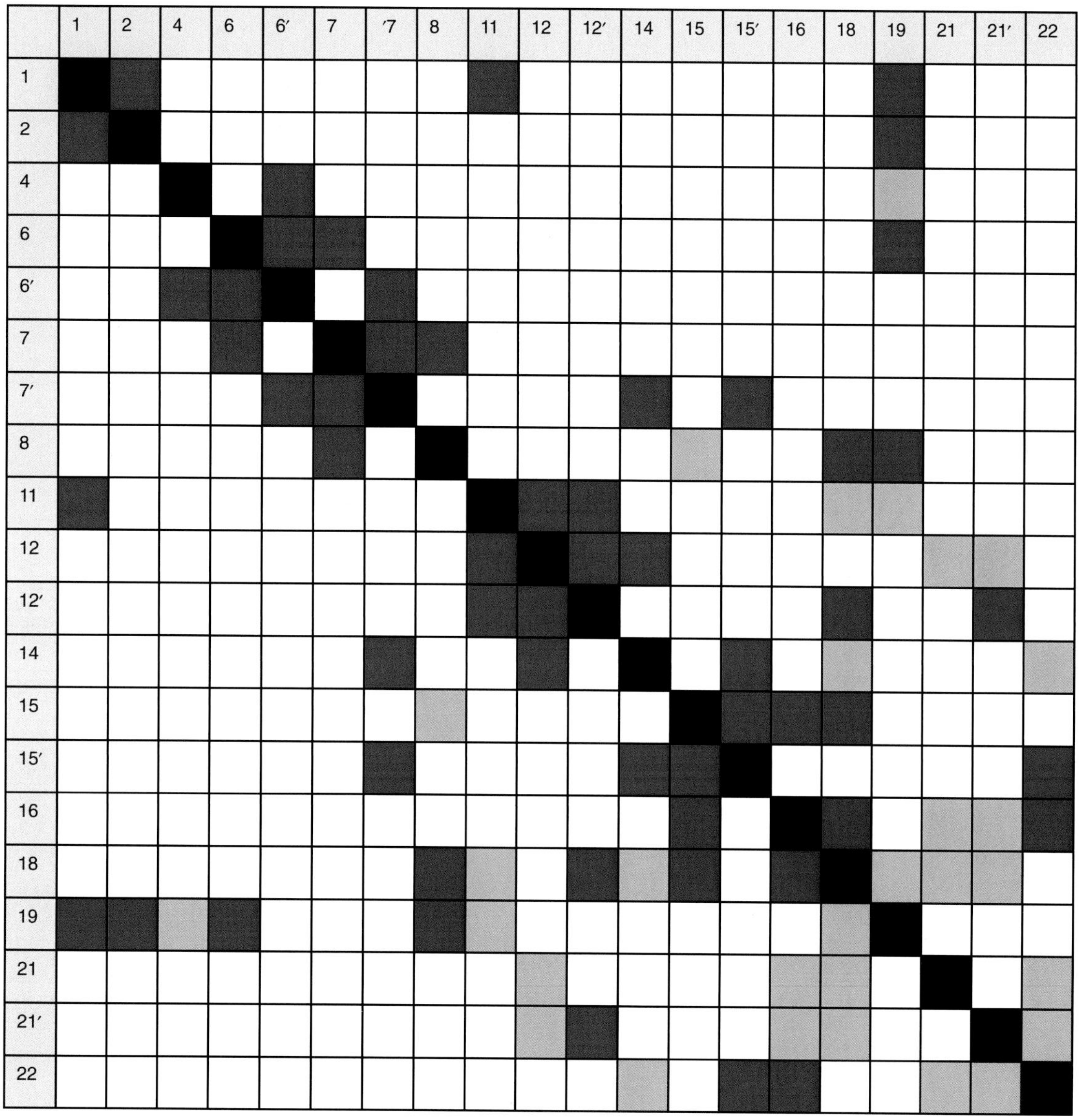

1 2 4 6 6′ 7 ′7 8 11 12 12′ 14 15 15′ 16 18 19 21 21′ 22
1
2
4
6
6′
7
7′
8
11
12
12′
14
15
15′
16
18
19
21
21′
22

FIGURE 13.22 A grid presentation summarising the NOEs observed for 13.1. NOEs of significantly weaker intensity are signified by lighter pink shading.

19
H_3C
H^8
^1H
H^6
H^7
H^{14}
F
6′ H
$H^{7'}$

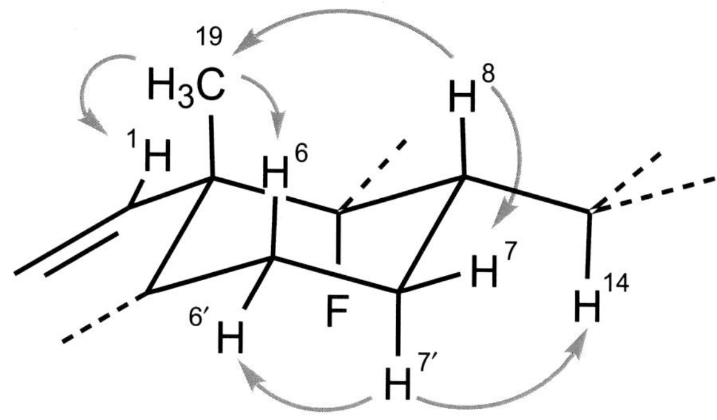

13.15

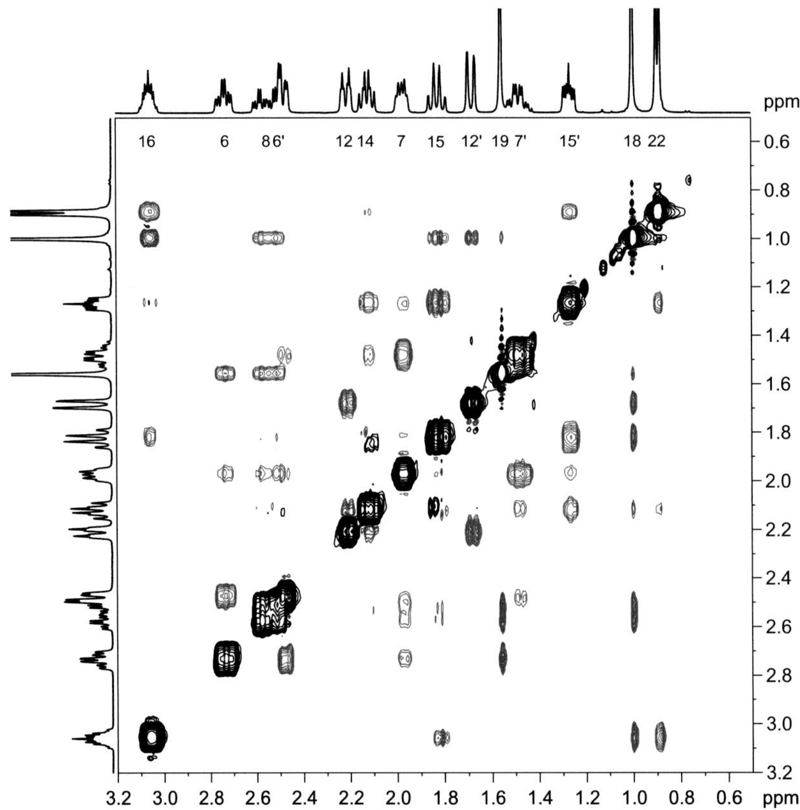

FIGURE 13.23 **The expanded region of the 2D NOESY spectrum of 13.1 below 3.2 ppm.**

Finally, having defined the geometry of the ring system, we return to determine the stereochemistry of the two substituents at the C^{17} centre. We have already noted weak NOEs from Me^{18} equally to H^{21} and $H^{21'}$, and see similarly weak NOEs to both from H^{16} (also unseen at this threshold). However, the most intense NOE for the C^{21} methylene protons occurs between $H^{21'}$ and $H^{12'}$. These data indicate the phosphorylated ketone substituent sits on the top side of the molecule but has considerable conformational flexibility, with weaker NOEs also observed from the C^{21} methylene protons to Me^{22} and H^{12}. Thus, the complete relative stereochemical assignment for the structure is summarised in **13.16**, where the prime notation for diastereotopic pairs again indicates the proton of lower chemical shift, and is represented as a 3D model in Fig. 13.24.

13.16

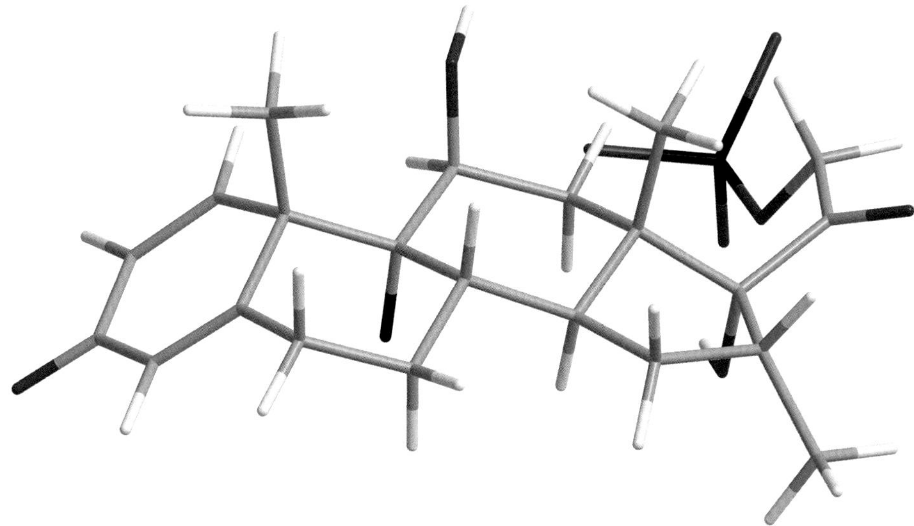

FIGURE 13.24 **A 3D energy minimised representation of 13.1 displaying stereochemistry derived from the NOE analysis.** The fluorine and phosphorus nuclei are shown in black.

13.7.2 1D NOESY

It is worth noting that for this example most of the stereochemical definitions for **13.1** could also have been defined from 1D NOESY experiments specifically by targeting the informative Me resonances. By way of illustration, the relevant 1D NOESY spectra are shown in Fig. 13.25 with all enhancements labelled and highlighted in **13.17**. Particularly for smaller molecules, a few judiciously chosen 1D NOESY experiments may be all that are required to answer a stereochemical problem and these can provide a more rapid solution than a full 2D experiment.

13.17

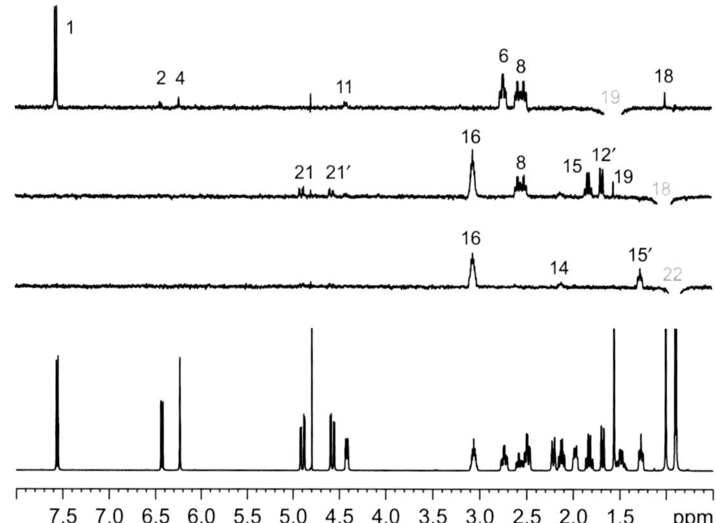

FIGURE 13.25 **1D NOESY spectra of 13.1 recorded with the methyl groups as targets (600 ms mixing time).** Selective inversion employed an 80-ms Gaussian inversion pulse in a single pulsed field gradient spin echo.

13.7.3 1D ^{19}F HOESY

Finally we consider the stereochemical assignment of the fluorine substituent. This was inferred above from the patterns of ^{1}H–^{1}H NOES that were observed, but additional evidence could be derived from the more direct observation of NOEs from the fluorine itself. The ^{1}H observed ^{19}F HOESY spectrum of **13.1** is provided in Fig. 13.26 and clearly maps NOEs to neighbouring protons on the lower face of the structure **13.18**. The lack of an NOE to the adjacent H^8 is consistent with the *anti* geometry predicted from the large $^3J_{FH}$ coupling constant noted above.

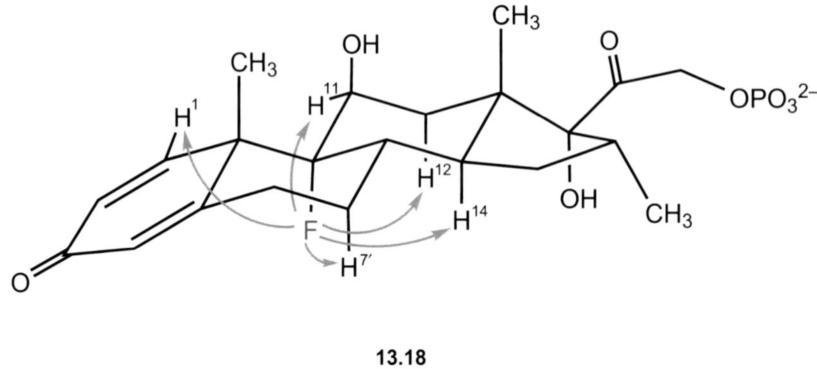

These HOESY experiments can provide useful supporting and complimentary information in the stereochemical assignment of fluorinated molecules. They do require probes capable of simultaneous tuning to ^{1}H and ^{19}F, which is not the default configuration for all types of broadband probes. However, where available, these ^{19}F experiments have a sensitivity

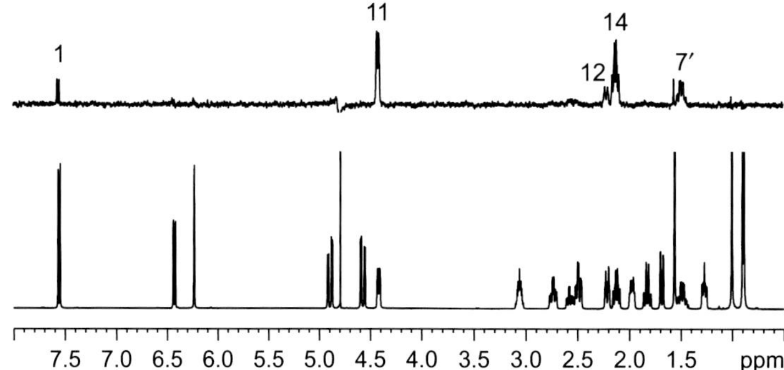

FIGURE 13.26 **1D ^{19}F–^{1}H HOESY spectrum of 13.1 recorded with proton detection (500 ms mixing time).**

similar to that of ^1H–^1H NOEs owing to the similar properties of these nuclei, and can prove to be equally valuable in stereochemical assignment.

13.8 RATIONALIZATION OF ^1H–^1H COUPLING CONSTANTS

Although the structure elucidation of **13.1** has employed proton–proton couplings in many of the correlation experiments presented above, we have so far largely ignored the magnitudes of these coupling constants. Therefore, having defined a structure, it is appropriate to finish by rationalising these values to ensure consistency with this structure, as reflected in **13.16** and Fig. 13.24. From the 500 MHz proton spectrum (Fig. 13.1) it is possible to determine the coupling constants by direct inspection since the multiplets are largely resolved, and these values are summarized in Fig. 13.27 in a manner that allows a more ready visualisation of the data than a conventional listed (journal) format. All geminal couplings (which will appear immediately adjacent to the 'diagonal') fall within the 12–14 Hz range, quite typical for saturated systems, with the exception of the rather high 19 Hz between the more heavily functionalised $H^{21-21'}$ geminal pair. Magnitudes of the ^1H

	1	2	4	6	6′	7	7′	8	11	12	12′	14	15	15′	16	18	19	21	21′	22
1		10.0																		
2	10.0		1.5																	
4		1.5		1.5																
6			1.5		13.5	6.0	13.0													
6′				13.5			4.5													
7				6.0			13.0	5.0												
7′				13.0	4.5	13.0		12.0												
8						5.0	12.0					12.0								
11										3.5	2.0									
12									3.5		14.0									
12′									2.0	14.0										
14								12.0					12.0	8.0						
15												12.0		12.0	12.0					
15′												8.0	12.0		4.0					
16													12.0	4.0						7.0
18																				
19																				
21																			19.0	
21′																		19.0		
22															7.0					

FIGURE 13.27 **Summary of ^1H–^1H couplings constants for 13.1 (values listed to 0.5 Hz).**

vicinal couplings can be correlated with the proton–proton dihedral angles observed within the core of the structure and are consistent with the proposed stereochemistry. It is seen that H^{11} shares rather small couplings of 4.0 and 2.0 Hz with H^{12} and $H^{12'}$ respectively, matching its equatorial orientation. In contrast, H^8 displays large 12 Hz couplings to both H^{14} and $H^{7'}$, indicative of the *anti* relationship for these axial protons; the larger 32 Hz coupling to the *anti* fluorine on C^9 has already been highlighted in previous discussions and may be contrasted with the 10 Hz $^3J_{FH}$ coupling of the equatorial H^{11}. The *gauche* relationship of H^8 and H^7 correlates with a smaller 5 Hz coupling. The diaxial orientations for $H^{7'}$ and H^6 match their large 13 Hz *anti* coupling, with the smaller 6 Hz coupling between H^7 and $H^{6'}$ also consistent with their *gauche* dispositions. H^{14} shares a further large 12 Hz coupling with H^{15} again suggestive of a largely *anti* relationship, with the 8 Hz coupling to $H^{15'}$ correlating with a rather small dihedral angle (~40 degrees in the energy minimized structure). The coupling between H^{15}

TABLE 13.2 Summary of Assignments for 13.1

	Chemical Shifts (ppm)				Heteronuclear Couplings (Hz)				Homonuclear ^1H Couplings (Hz)	
Position	δ_C	δ_H	δ_F	δ_P	$^nJ_{PC}$	$^nJ_{FC}$	$^nJ_{PH}$	$^nJ_{FH}$	H	H′
1	156.62	7.557							10.0	
2	128.26	6.429							10.0, 1.5	
3	189.66	—							—	
4	123.29	6.232							1.5, 1.5	
5	172.76	—							—	
6	30.84	2.736, 2.548							13.5, 13.0, 6.0, 1.5	13.5, 4.5
7	27.37	1.973, 1.559							13.0, 6.0, 5.0	13.0, 13.0, 12.0, 4.5
8	33.78	2.548				19.2		31.8	12.0, 12.0, 5.0	
9	102.80	—	−164.55			172.4			—	
10	49.05	—				22.4		10.1	—	
11	71.29	4.416				38.3			3.5, 2.0	
12	35.63	2.211, 1.683						3.0 (to H12)	14.0, 3.5	14.0, 2.0
13	47.79	—							—	
14	43.65	2.122				1.6			12.0, 12.0, 8.0	
15	31.64	1.825, 1.269							12.0, 12.0, 12.0	12.0, 8.0, 4.0
16	35.33	3.059							12.0, 7.0, 4.0	
17	91.36	—							—	
18	16.50	1.000							—	
19	22.17	1.559				6.0			—	
20	211.39	—			7.6				—	
21	68.52	4.894, 4.572			3.7		5.6, 5.0		19.0	19.0
22	14.31	0.893							7.0	
23				3.76						

Data are referenced to internal TSP (^1H and ^{13}C), external $CFCl_3$ (^{19}F), and external 85% H_3PO_4 (^{31}P) at 0.00 ppm.

and H^{16} is large (12 Hz) which in this case originates from the *syn* relationship and a dihedral angle close to zero degrees. The far smaller $H^{15'}$–H^{16} 4 Hz coupling in turn corresponds to a dihedral angle of ~115 degrees in the modelled structure. Couplings within the unsaturated ring are largely unremarkable, although some longer range four-bond couplings are resolved between H^4–H^6 (1.5 Hz) and H^4–H^2 (1.5 Hz). Finally, we note that the long-range coupling between Me^{18} and H^{12} evidenced in the COSY spectrum of Fig. 13.5 is unresolved, leading only to a slight broadening of the Me^{18} resonance when compared to that of Me^{19}.

13.9 SUMMARY

The previous sections in this chapter have presented a typical protocol for spectrum assignment and structure elucidation. The assignments for **13.1** are summarized in Table 13.2 and the structure as **13.19**. It is certainly the case that not all the experiments presented in this chapter were necessary to define this structure, but rather were included to illustrate how these might be employed in a structure elucidation protocol, and to serve as a reminder that a diverse range of methods now exist to address complex problems. For routine structure assignment the core techniques that one would typically employ are 1D ^{1}H, ^{1}H–^{13}C edited-HSQC, ^{1}H–^{1}H COSY, ^{1}H–^{13}C HMBC and (optionally) 1D ^{13}C, with (1D or 2D) NOESY employed to address issues of stereochemistry. The methods illustrated in this chapter may (in the most part) be readily implemented on a standard two-channel NMR spectrometer equipped with a multinuclear broadband probe. The 500 MHz data presented here used an instrument that was able to acquire fluorine on its broadband channel (as opposed to having this dual tunable on the proton channel), and was thus capable of all ^{1}H–X combinations, where X = ^{13}C, ^{19}F and ^{31}P. All such techniques could, in principle, be obtained under complete automation when automatic tuning of the probe is available, a common configuration on modern probeheads, making all these data available in a time efficient manner.

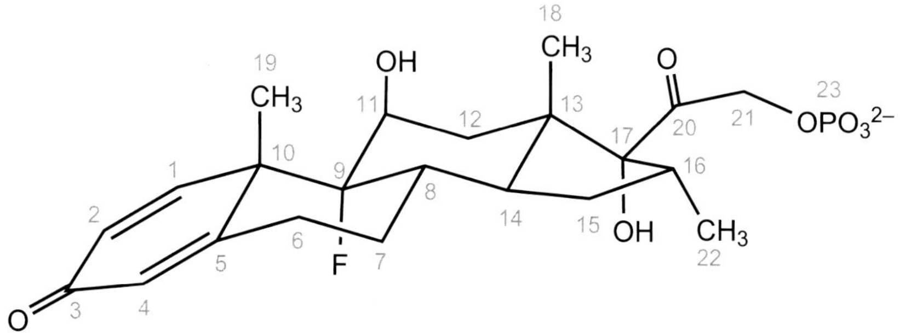

13.19

The need to reach beyond the primary techniques when addressing problems of spectrum assignment and structure verification or elucidation will be dictated by the nature of the problem(s) to be resolved, what information is required and what experiments are practically feasible given the sample amounts and the type of the instrumentation available. Coupled with this is the requirement to have adequate awareness of what is possible with the extensive range of modern NMR techniques that now exist, and an appreciation of the capabilities afforded by modern NMR instrumentation. Hopefully this book goes some way to raising this awareness and provides the necessary theoretical and practical guidance for these NMR techniques to be employed.

Appendix

GLOSSARY OF ACRONYMS

Acronym/term	Translation	Section
ACCORDION	A method of increasing one time period in concert with another	7.4.3
ACOUSTIC	Alternate compound 180s used to suppress transients in the coil (for quadrupolar nuclei)	4.5
ADEQUATE	Adequate sensitivity double-quantum spectroscopy	6.4
APT	Attached proton test (spectrum editing)	4.3.2
BASH	Band-selective homodecoupling	8.5.4
BEBOP	Broadband excitation by optimized pulses	12.2.2
BIP	Broadband inversion pulse	12.2.2
BIRD	Bilinear rotation decoupling	7.3.3
BPP	Bipolar pair: gradient pulse pair of opposite sign	10.2.3
BURP	Band-selective, uniform-response, pure-phase (selective) pulse	12.4.1
CALIS	Calibration for indirect spins	3.5.1
CAMELSPIN	Cross-relaxation appropriate for minimolecules emulated by locked spins (aka ROESY)	9.7
CHIRP	Frequency swept pulse (adiabatic pulse).	12.2.1
CLIP	Clean in-phase (crosspeak structure)	7.45
COLOC	Correlation through long-range coupling	7.5.2
CORCEMA	Complete relaxation and conformational exchange matrix	11.4.2
COSY	Correlation spectroscopy	6.1
CPD	Composite pulse decoupling	12.3
CPMG	Carr-Purcell-Meiboom-Gill (T_2-dependent spin-echo sequence)	2.4.4
CSA	Chemical shift anisotropy	2.5.3
CSSF	Chemical shift selective filter	12.4.3
CYCLOPS	Cyclically-ordered phase-sequence (for suppressing quadrature artefacts)	3.2.5
DANTE	Delays alternating with nutation for tailored (selective) excitation	12.4.4
DEPT	Distortionless enhancement by polarisation transfer (spectrum editing)	4.4.3
DEPTQ	Distortionless enhancement by polarisation transfer retaining quaternary centres	4.4.4
DIPSI	Decoupling in the presence of scalar interactions (isotropic mixing)	12.3.2
DIRECTION	Difference of inversion recovery rate with and without target irradiation	11.4.2
DNMR	Dynamic NMR	2.6
DNP	Dynamic nuclear polarisation	12.8.2
DOSY	Diffusion-ordered spectroscopy	10.2.5
DPFGSE	Double pulsed field gradient spin-echo (selective excitation)	12.4.2
DQF-COSY	Double-quantum filtered correlation spectroscopy	6.1.5
DSTE	Double stimulated-echo	10.3.1
ePHOGSY	Enhanced protein hydration observed through gradient spectroscopy	11.5.1

High-Resolution NMR Techniques in Organic Chemistry. http://dx.doi.org/10.1016/B978-0-08-099986-9.00014-2

Acronym/term	Translation	Section
ERETIC	Electronic reference to access *in-vivo* concentrations	4.1.3
EXORCYCLE	Phase cycle to suppress "ghost" and "phantom" artefacts in spin-echo sequences	8.2.2
EXSY	Exchange spectroscopy	9.11
FAXS	Fluorine chemical shift anisotropy and exchange for screening	11.7.3
FID	Free induction decay	2.2.2
FLIPSY	Flip-angle adjustable one-dimensional NOESY (solvent suppression)	12.5.1
FLOPSY	Flip-flop mixing sequence (for total correlation spectroscopy)	12.3.2
GARP	Globally-optimised, alternating phase, rectangular pulses (for broadband decoupling)	12.3
HEHAHA	Heteronuclear Hartmann-Hahn spectroscopy	6.2
HETCOR	Heteronuclear correlation	7.5.1
H2BC	Heteronuclear two-bond correlation	7.4.4
HMBC	Heteronuclear multiple-bond correlation	7.4
HMQC	Heteronuclear multiple-quantum correlation	7.3.3
HOBS	Homonuclear band-selective decoupling	8.5.4
HOESY	Heteronuclear Overhauser effect spectroscopy	9.9
HOHAHA	Homonuclear Hartmann-Hahn spectroscopy (aka TOCSY)	6.2
HSQC	Heteronuclear single-quantum correlation	7.3.1
HSQMBC	Heteronuclear single-quantum multiple-bond correlation	7.4.5
IMPEACH-MBC	Improved performance accordion heteronuclear multiple-bond correlation	7.4.3
INADEQUATE	Incredible natural-abundance double-quantum transfer experiment	6.3
INEPT	Insensitive nuclei enhanced by polarisation transfer	4.4.2
J-MOD	J-modulated spin-echo	4.3.1
LED	Longitudinal eddy-current delay	10.2.3
LP	Linear prediction	3.2.3
LR-HSQMBC	Long-range heteronuclear single-quantum multiple-bond correlation	7.4.6
MAS	Magic angle spinning	12.7
MLEV	Broadband decoupling cycle (from M. Levitt)	12.3
MQF	Multiple-quantum filter	6.1.5
NOE	Nuclear Overhauser effect	9.2
NOESY	Nuclear Overhauser effect spectroscopy	9.6
NUS	Non-uniform sampling	5.2.4
PANACEA	Parallel acquisition NMR, an all-in-one combination of experimental applications	7.7
PANSY	Parallel acquisition NMR spectroscopy	7.7
PEP	Preservation of equivalent pathways	7.3.1
PFG	Pulsed field gradient	5.4
PIP	Pure in-phase (crosspeak structure)	7.4.5
PHIP	*Para*-hydrogen induced polarisation	12.8.1
PRE	Paramagnetic relaxation enhancement	11.3.3
PROJECT	Periodic refocussing of J evolution by coherence transfer	2.4.4
PSYCHE	Pure shift yielded by CHIRP excitation	8.5.4
PUFSY	Parallel ultra-fast spectroscopy	7.7
PULCON	Pulse-length based concentration determination	4.1.4
PURGE	Presaturation utilizing relaxation gradients and echoes (solvent suppression)	12.5.3
QNMR	Quantitative NMR	4.1.2

Acronym/term	Translation	Section
RDC	Residual dipolar coupling	9.12
RIDE	Ring-down delay sequence (for quadrupolar nuclei)	4.5
ROESY	Rotating-frame Overhauser effect spectroscopy	9.7
SEFT	Spin-echo Fourier transform (spectrum editing)	4.3.1
SLAPSTIC	Spin labels attached to protein sidechains as a tool to identify interacting compounds	11.3.3
SPFGSE	Single pulsed field gradient spin-echo (selective excitation)	12.4.2
SPT	Selective population transfer	4.4.1
STD	Saturation transfer difference	11.4
STE	Stimulated-echo	10.2.2
STEP	Selective TOCSY edited preparation	9.6.2
TOCSY	Total correlation spectroscopy	6.2
TPPI	Time-proportional phase incrementation	5.2.1
Tr-ROESY	Transverse rotating-frame Overhauser effect spectroscopy	9.7
WALTZ	Wideband, alternating-phase, low-power technique for zero-residual splitting (broadband decoupling sequence)	12.3
WATERGATE	Water suppression through gradient tailored excitation	12.5.3
WaterLOGSY	Water-ligand (binding) observed through gradient spectroscopy	11.5
WATR	Water attenuation by transverse relaxation	2.4.4
WURST	Wideband, uniform rate and smooth truncation (adiabatic pulse).	12.2.1

Subject Index

A

Absolute configurations, 376
Absolute-value
  display, 2D, 389
Absorption mode, 88, 212, 226
ACCORD-HMBC, 271
ACCORDION optimisation
  in ADEQUATE, 239
  in HMBC, 270
N-Acetyl chitooligosaccharides
  $(GlcNAc)_{1-6}$, 413
Acetylisomontanolide
  ^{13}C INADEQUATE, 232
ACOUSTIC, 167
Acoustic ringing, 167
Acquisition times, 69–71, 183–185
Acronyms
  glossary of, 8
Active coupling, 211
Active spin resonances, 311
Active volume, 98, 99, 107
ADC. *See* Analogue-to-digital
  converter (ADC)
ADEQUATE, 9, 203, 235
Adiabatic, decoupling, 147
Adiabaticity factor, 462
Adiabatic pulses, 461–463
  amplitude profiles for, 463
Aggregation, 411
Aliasing
  2D NMR, 83
  folded signals, 178
  wrapped signals, 76
Alignment media, for RDCs, 372
Alkynes, unusual behaviour
  DEPT, 164
  HMQC/HSQC, 248
Allylic coupling
  in COSY, 219
Ammonium chloride, 30
Amplitude modulation, 173, 296, 300
Analogue filters, 68, 69
Analogue-to-digital converter (ADC), 63, 480
  Overload, 81
  resolution, 80
Andrographolide
  DQF-COSY, 217
  NOEs, 333
  NOESY, 339
Angular momentum, 11
Angular velocity, 13
Anisotropy
  in chemical bonds, 34
Anti-phase
  splittings, in COSY, 215
  vectors, 285
Apodisation, 72
APT. *See* Attached proton test
Artefacts
  axial peaks, 178
  clipped FID, 81
  ghosts, 300
  phantoms, 300
  quad images, 77
  SPT, 363
  subtraction, 144
  t_1-noise, 251
  truncated FID, 72
ASTM
  test sample, 128
Asymmetrical double STE (aDSTE)
  sequence, 392
Attached proton test (APT), 152
Audio frequencies, 67
Axial peaks, 181–182

B

^{11}B
  coupling, 38
  2D EXSY, 369
  linewidths, 36
Backbone amide protons, nitrogen-15
  correlation, 451
Bacterial potassium efflux protein Kef, 444
Bandpass filter, 68
Band selective homonuclear decoupling
  (BASH), 309
Band-selective method, 309
Bandwidth
  decoupling, 146, 465–468
  excitation, 311, 458
  inversion, 311, 460
Bar magnet analogy, 33
BASH. *See* Band selective homonuclear
  decoupling (BASH)
BEBOP pulses, 465
B_{eff}, effective rf field, 65, 458, 461
Benzene
  test sample, 128
Benzo[*b*]thiophene-2-boronic acid
  ^{19}F FAXS reporter screening, 448
Bi-level decoupling, 467
Bilinear rotation decoupling (BIRD), 311
  HETCOR, 311
  HMBC, 257
  HMQC, 257
  J-resolved spectroscopy, 417
  refocussing element, 312
Bimolecular rate constant, 423
Binomial sequence, 483
Bipolar pulse pair (BPP), 385
BIPs. *See* Broadband inversion pulses
BIRD. *See* Bilinear rotation decoupling
Bloch, Felix, 1
Bloch-Siegert shift, 144
Bloch vector model, 13
^{11}B NMR spectrum, of
  dansylamidophenylboronic acid, 37
Boiling points
  deuterated solvents, 89
Boltzmann constant, 13, 53, 382
Boltzmann distribution, 13, 317
Boronic acids, 36, 37, 448
  fluorinated, 429
Bovine serum albumin (BSA), 427, 432,
  434, 439
BPP. *See* Bipolar pulse pair
BPP-STE experiment, 385
B_2, rf magnetic field, decoupler
  calibrating, 120
B_1, rf magnetic field, transmitter, 14–17
  calibrating, 115
Broadband heteronuclear decoupling,
  465–468
Broadband homonuclear decoupling,
  306–313
Broadband inversion pulses (BIPs), 464
Bromopyrrole alkaloids, 412
3-Bromo-1, 1, 1-trifluoroacetone
  (BTFA), 452
BSA–tryptophan–sucrose system, 427, 432,
  434, 439
B_0, static magnetic field
  inhomogeneity, 107
  optimising, 107–112
Bulk magnetisation vector, 13
  off-resonance excitation, 64–66, 458
  rotation of, 65, 458
BURP pulses, 471–473, 478
Butanol
  ^{13}C INADEQUATE, 232
n-Butyllithium, 412

C

^{13}C
satellites in ^{13}C spectra, 230
satellites in ^{1}H spectra, 248
satellites, selecting, 230, 248–256
Calibrations
decoupler pulses, 118
pulsed field gradients, 122
pulse widths, 113–122
radiofrequency fields, 120–122
shaped pulses, 479
temperature, 124
Calibration standards, 137, 138
CALIS
pulse calibration, 119
CAMELSPIN. *See* ROESY
Carbopeptoids
COSY, 205
HMBC, 269
HSQC, ^{15}N, 250
TOCSY, 354
Tr-ROESY, 354
Carousel analogy
rotating frame, 15
Carr–Purcell–Meiboom–Gill (CPMG), 27–30
PROJECT variant, 29
sequence, 28, 485
spin lock filters, 426
Cartesian coordinates, 13
CAT (computer of average transients), 1
CF_3 group, 428
^{19}F NMR of protein site for monitoring ligand binding, 453
Chemical exchange, 38–58
2D EXSY, 368–371
saturation transfer, 328
solvent attenuation, 32
Chemical shift
anisotropy, 34, 429
refocusing, 18
selective filtration, 476
timescales, 42
visualising in vector model, 17
Chemical shift anisotropy (CSA), 34, 429
Chemical shift selective filter (CSSF), 347–349, 475–476
Cheng–Prusoff equation, 446
CHIRP, 468
Chloroform
lineshape test sample, 130
Chromium(III) acetylacetonate, 134
^{13}C-Hyperpolarised pyridine
2D single-scan HSQC spectrum, 495
CIFIT analysis, 56
Cleaning, NMR tubes, 96
Clipping, FID, 80
^{13}C-Methyl–labelled proteins
^{1}H-^{13}C correlations of, 452
^{13}C-Methyl–labelling strategies, 452
Coherence, 188–191
multiple-quantum coherence in an AX system, 190
net transverse magnetisation, 189
transfer, 188
coherence level diagram, 190
pathways for COSY, 191
Coherence transfer
in COSY, 188
echoes, 2D, 179
in TOCSY, 221
Coherence transfer pathways, 171, 190
in COSY, 190
in DQF-COSY, 212
in HMQC, 257
in HSQC, 247
in INADEQUATE, 231
in NOESY, 341
COLOC, 285
Competitive displacement, 443
Complete relaxation and conformational exchange matrix for STD (CORCEMA-STD), 437
Composite adiabatic pulse, 463
Composite pulse decoupling (CPD), 465–468
Composite pulses, 66, 457–461
broadband decoupling, 121, 465
and spin locking, 466
180 degree pulse, 460
inversion *vs.* refocusing, 460
properties of, selected, 460
Conformational averaging, of NOE, 336
Constant-time experiments
in COLOC, 285
H2BC, 274
HMBC, 268, 273
Continuous wave decoupling, 466
Contour plot, 176
Convection, in diffusion NMR, 390–397
compensating, 395
dealing with, 392
diagnosing, 390
Hadley, 392
Rayleigh-Benard, 390
Cooley-Tukey algorithm
fast Fourier transform, 20
CORCEMA, 437
Correlating dilute spins via INADEQUATE, 230–235
implementing INADEQUATE, 233
one-dimensional INADEQUATE, 233, 234
two-dimensional INADEQUATE, 230–233
Correlating dilute spins via ADEQUATE, 235–239
enhancements to ADEQUATE, 237–239
1, 1-ADEQUATE sequence, 238
Correlation spectroscopy (COSY), 203, 220, 500
COSY-β, 211
double-quantum filtered COSY (DQF-COSY), 212
DQF sequence, 212–213
higher order multiple-quantum filters, 218
interpreting multiplet structure, 214–216
measuring coupling constants, 216–218
DQF-COSY, 212
interpreting, 204
crosspeaks, 204, 216
diagonal peaks, 204
long-range COSY, 219
magnitude-mode COSY, 210
-β variant, 210
peak fine structure, 207
^{31}P–^{31}P COSY, 208
relayed-COSY, 220
signal phases in phase-sensitive, 209
Correlation time, τ_c, 32, 320
COSY. *See* Correlation spectroscopy
COSY-IDOSY sequence, 416
Coupling constants
in cyclohexanes, 217
from DQF-COSY, 214
CPD. *See* Composite pulse decoupling
CPMG. *See* Carr-Purcell-Meiboom-Gill
CPMG filter, 30–32, 427
^{1}H signal attenuation, 428
CPMG-filtered spectra, 30–32, 427–428
Crosspeak
disappearance, in COSY, 220
tilting, in COSY, 211
Cross-polarisation, 220
Cross-relaxation, 317–322, 337
rates, 337, 355
scalar, of the first kind, 342
Cryo-probes
helium cooled, 102
nitrogen cooled, 103
temperature calibration for, 127
CSA. *See* Chemical shift anisotropy
CSSF. *See* Chemical shift selective filter
Cyclodextrins
diffusion measurements, 406
NOESY, 358
ROESY, 358
Cyclohexanes
couplings, from COSY, 217
CYCLOPS
in 2D NMR, 79

D

3-Dansylamidophenylboronic acid, 36
DANTE (Delays alternating with nutation for tailored excitation), 477–479
Data acquisition, 67, 138
carbon decoupling during, 258
fast, 199
phase-sensitive, 184
and processing, 64
real-time pure shift, 309
relative S/N, 136
times, 100, 184, 436
influence of, 70
two-dimensional, 183
Data processing, 139, 153
two-dimensional, 186
dB scale, 114
Decimation, 83
Deconvolution, of lineshapes, 139
Decoupler
calibrating, 118, 144

gating, 144
leakage, 144
Decoupling, 145–148, 465–468
adiabatic, 146, 467
bandwidths, 114, 466
bi-level, 467
broadband, 143, 465
composite pulse, 146, 465
Degassing
for NOE, 325
techniques for, 95
Delayed-COSY. *See* Long-range COSY
DEPT. *See* Distorsionless enhancement by polarisation transfer
DEPTQ, 163, 165
Detection
2D NMR, 83
single channel, 73
two channel, quadrature, 73
Deuterated solvents
changing, 435
properties, 90
Dexamethasone phosphate
assignment of, 499–525
pure shift spectrum of, 306
Diagonal peak, 176, 207
N, N-Dialkyl *p*-nitrosoaniline systems, 48
Diaminoalkanes
diffusion coefficients, 407
Diastereoisomer, 376
Diastereotopic protons
^1H-^{13}C correlations, 245
Difference spectroscopy
NOE, 346
Diffusion coefficients, 381, 384, 388, 393
$CDCl_3$, 391, 405
for free diaminoalkanes, 407
for ionic complexes, 408
for tetrabutylammonium tetrahydroborate, 408
Diffusion decay profiles, 396, 398–401, 403
Diffusion NMR spectroscopy, 381–419
applications of, 403–414
aggregation, 411
host–guest complexes, 405
hydrogen bonding, 405
ion pairing, 408
macromolecular characterisation, 413
mixture separation, 412
signal suppression, 403
supramolecular assemblies, 409
calibrating gradient amplitudes, 397
data analysis
pseudo-2D presentation, 389
regression fitting, 388
hybrid diffusion sequences, 414–417
hydrodynamic radii/molecular weights, 401
molecular size, 382, 401
optimising diffusion parameters, 397–401
practical aspects, 390
convection
dealing with, 390
diagnosing, 390
convection-compensating sequences, 395
sample spinning, 393
solvent viscosity, 393
temperature gradients/decoupling, 393
tube diameters, 393
self-diffusion, measuring, 382
BPP-LED sequence, 386
BPP-STE sequence, 385
enhancements to stimulated-echo, 385
'one-shot' sequence, 387
pulsed field gradient spin-echo, 383
pulsed field gradient stimulated-echo, 384
Diffusion ordered spectroscopy (DOSY), 306, 381, 389, 404. *See also* Diffusion NMR spectroscopy
heteronuclear methods, 415
Diffusion periods, 393
Digital filters, 67, 76, 178
Digital quadrature detection, 74
Digital resolution (DR), 69, 70
Digital signal processor, 67
Digitisation, 80–83
2D NMR, 82, 183
noise, 82
Digitiser. *See* Analogue-to-digital converter
Dilute spins
correlating, 230
N, N-Dimethyl acetamide, 57
Dimethylsulfoxide (DMSO), 50, 393, 441
1, 4-Dinitrobenzene, 137
Dioxane
shift reference, 91
test sample, 128
Diphenyldiazetidinone, 47
Dipolar couplings, 172, 315, 319–322, 489, 490
Dipolar interactions, 33, 324, 325, 437
coupling, 319
residual coupling, 371–377
Dipole-dipole relaxation, 7, 33–34, 324
DIPSI-2, 226, 466
DIRECTION, 437
Distorsionless enhancement by polarisation transfer (DEPT), 162, 171
DEPT-135 *vs.* DEPTQ, 166
editing with, 162
errors in DEPT editing, 164
limitations, 166
optimising sensitivity, 165
sequence, 162
signs of multiplet resonances in, 163
DNMR. *See* Dynamic effects in NMR
2D NMR techniques, 171–188
practical aspects of, 177
absolute value presentations, 179–180
aliasing in two dimensions, 178
phase-sensitive presentations, 177–178
quadrature detection, 177
two-dimensional lineshapes, 177
DNP. *See* Dynamic nuclear polarisation
D_2O exchange, 91
DOSY. *See* Diffusion ordered spectroscopy
DOSY toolbox software, 389
Double PFG spin-echo (DPFGSE), 473
experimental excitation profiles, 475
selective excitation sequences, 474
Double-quantum filter
1D, 214
2D, 212–214
INADEQUATE, 230–235
Double-quantum filtered (DQF)-COSY, 212–218, 505
crosspeak structure, 214–216, 505
'Double-resonance' experiment, 143
Double STE (DSTE) diffusion sequence, 395
Doubly selective
1D TOCSY NOESY, 348
DPFGSE. *See* Double PFG spin-echo
DQD. *See* Digital quadrature detection
DQF. *See* Double-quantum filter
DQF-COSY. *See* Double-quantum filtered COSY
Drying, samples, 94
DSP. *See* Digital signal processor
Dummy scans, 130
Duty cycle, 117
Dwell time, 67
Dynamic effects, in NMR, 38–58
influence of dynamic exchange, 39
intermolecular exchange, 49–50
J coupling, 49–50
practicalities regarding NMR timescales, 50–52
scalar coupling to exchanging sites, 48
two-site exchange
equal populations, 40–43
between scalar coupled nuclei, 47–48
unequal populations, 43–45
Dynamic exchange regimes, 43
Dynamic nuclear polarisation (DNP), 491, 492
Dynamic range, 80

E

EBURP2. *See* BURP pulses
Echo-antiecho method, 250
Eddy currents, 387
Editing
background suppression, 447
diffusion filter, 404
1D carbon spectra, 153–167
1D proton spectra, 256
heteronuclear 2D correlations, 243
by molecular size, 30
multiplicity, 161–165
T_2 filter, 31
Electric field gradients, 36
Electromagnetic spectrum, 13
Electromotive force (emf), 480
Enantiomeric excess
improving accuracy, 71
in liquid crystals, 372
Enthalpy, 53
Entropy, 53, 54
ePHOGSY. *See* Water-LOGSY
Epitope mapping, 421
Equilibrium
establishing, 22
Equilibrium NOE. *See* Steady-state NOE
ERETIC. *See* Quantitative NMR
Ernst angle, 134, 166
for optimum sensitivity, 135

Ethanediol
temperature calibration, 126
Ethanol, first high-resolution ^{1}H spectrum, 2
Ethyl acetate, 167
Ethyl acetoacetate
^{13}C DNP spectrum of, 495
Ethylbenzene
test sample, 129
Ethylene glycol. *See* Ethanediol
Eukaryotic elongation release factor 1 protein (eRF1), 449
Evolution time, 2D NMR, 172–176
Exchange decoupling, 50
Exchange/dynamic broadening of resonances, 42
Exchange/dynamic narrowing of resonances, 42
Exchange spectroscopy (EXSY), 58, 316, 368–371
Exchange-transferred NOE experiments, 441–443
Excitation
bandwidth, 65
sculpting, 346, 435, 473–475
trajectories, 66
Excitation sculpting
solvent suppression, 440
Excitation trajectories, 65–66, 458–459
binomial pulses, 484
composite pulses, 468
DANTE, 483
imperfect pulses, 252
shaped pulses, 475
WURST, 464
EXORCYCLE, 234, 300–301, 440
Experimental methods, 457
adiabatic pulses, 461–462
amplitude calibration, 479
broadband adiabatic decoupling, 467
broadband decoupling, 465–466
broadband inversion pulses (BIPs), 464–465
broadband pulses, 461–462
chemical shift selective filters, 475–476
composite pulses, 457
DANTE, 477
90 degree pulse, 459
180 degree pulse, 460
excitation sculpting, 473–474
heterogeneous samples, 489
hyperpolarisation, 491
dynamic nuclear polarisation, 493–495
para-hydrogen–induced polarisation, 491–492
inversion *vs.* refocusing, 460
magic angle spinning, 489
off-resonance excitation, 458
phase calibration, 479
pulsed field gradients, 483
excitation sculpting, 484
perfect echo suppression, 485
WATERGATE, 483
pulse excitation, imperfections in, 458
selective excitation, 468
simulated effect of, 459
soft pulses, 468
shaped soft pulses, 469
Gaussian pulses, 469
implementing shaped pulses, 473
pure-phase pulses, 471
solvent suppression, 480
1D NOESY presat, 501
excitation sculpting, 473–474
FLIPSY, 481
presaturation, 480
WATERGATE, 483–484
spin locking, 465–466, 468
zero excitation, 482
binomial sequences, 482
jump-return, 482
zero-quantum coherences, suppression of, 486
variable-delay z-filter, 486
zero-quantum dephasing, 487
Experiment selection, guidelines for, 7
Exponential
decay, 110
multiplication, 84
recovery, 23
EXSY. *See* Exchange spectroscopy (EXSY)
Extreme narrowing limit, 32
Eyring relationship, 53

F

^{19}F
coupling to ^{11}B, 38
FAXS, 429
ligand screening with, 429
quantitative, 52
Fast data acquisition, 199. *See also* Gradient-selected spectroscopy
FAXS. *See* Fluorine chemical shift anisotropy and exchange for screening
^{19}F–^{13}C HMBC spectrum, 517
^{19}F coupling partners, 516
^{57}Fe
^{31}P HMQC, 286
^{19}F HOESY spectrum, 522
FID. *See* Free induction decay
Field-gradient pulses. *See* Pulsed field gradients
Filling factor, 96, 102
Filtering
1D proton spectra, 256
low-pass, HMBC, 267
noise, 68
sample solutions, 93
FLIPSY, 481
FLOCK, 285
Floor vibrations, 63
FLOPSY-8, 466
Flow cells, 63
Fluorinated amino acid
biosynthetic incorporation of, 452
Fluorinated boronic acid, 429
Fluorinated compounds, 428
Fluorine–carbon couplings, 515
Fluorine chemical shift anisotropy and exchange for screening (FAXS)
competition approach, 429
Folding. *See* Aliasing
Fourier transformation, 1, 67, 71, 484
complex, 177
in 2D NMR, 20
introduction, to NMR, 140
mathematical expression, 20
real, 22
Free induction decay (FID), 7, 17, 469
processing a truncated, 72
Freeze-pump-thaw method, 367
Frequency domain, 20
Frequency labelling, 176
Frequency sweep
adiabatic, 417
^{19}F shift timescale, 429
^{19}F spectrum
of fluorinated boronic acid binding to α-chymotrypsin, 429
of sodium borofluoride in D_2O, 38

G

γ. *See* Magnetogyric ratio
Gadolinium triethylene-tetraamine-hexaacetate, 234
GARP composite pulse decoupling, 467
Gated decoupling
inverse, 117
in J-resolved spectroscopy, 301
power, 146
Gauche dispositions, 523
Gaussian cascades, 471
Gaussian shaped pulse, 432, 470
Gel-phase NMR, 489
Geminal couplings
in COSY, 204
Ghost
artefact, 300
Gradient-accelerated spectroscopy. *See* Gradient-selected spectroscopy
Gradient coils, 96
Gradient echo
gradient calibrations, 404
selective excitation, 269
shimming, 111
signal selection, 122
Gradient image profiles, 123
Gradient-selected spectroscopy, 191–198
phase-sensitive experiments
Echo–Antiecho selection, 195–196
practical implementation of pulsed field gradients, 198
pulsed field gradients in high-resolution NMR, 196
advantages of, 197
defocusing and refocusing with, 192
limitations of, 197
selective refocusing with, 193–195
signal selection with, 192
Gradient shimming, 111
Gramicidin-S
NOESY, 351
TOCSY, 223
Gyromagnetic ratio. *See* Magnetogyric ratio

H

^{2}H
coupling, 37
lock, 198
Hadley convection, 390
Hard pulses, 468
Hartmann–Hahn match, 223
H2BC, 274
^{1}H–^{13}C correlation experiments, 243, 452
^{1}H–^{13}C–^{15}N correlations in biological macromolecules, 244
HDO
resonance, 480
solvent presaturation, 481
temperature dependence of, 481
Heating, rf induced, 146
HEHAHA. *See* Heteronuclear Hartmann-Hahn
Heisenberg Uncertainty Principle
pulse excitation, 64
relaxation, 22
Helium, 3
HETCOR, 283
Heterogeneous samples
Heteronuclear correlation, 63, 243–291, 501–503, 510–517
Heteronuclear decoupling, 145–148, 465–468
^{1}H{X} decoupling, 147
selective proton decoupling, 149
X{^{1}H} decoupling, 145
Heteronuclear DOSY methods, 415
Heteronuclear Hartmann–Hahn (HEHAHA) spectroscopy, 223
Heteronuclear J-resolved spectroscopy, 295
practical considerations, 300
experimental setup, 301
proton–carbon coupling constants measurement, 298
Heteronuclear multiple-bond correlation (HMBC), 261–282, 490, 500
application, 264–266
long-range proton–carbon coupling constants, 265
practical set-up, 266
extensions and variants, 266
ACCORDION optimisation, 270
broadband $^nJ_{XH}$ detection, 270–274
band-selective, 269
constant time, 268
H2BC, 274
differentiating $^2J_{CH}$ and $^3J_{CH}$ HMBC correlations, 274–275
low-pass J filtration, 267
long-range HSQMBC, 281
interrogating proton-sparse molecules, 281–282
measuring long-range $^nJ_{XH}$ coupling constants, 275
using HMBC, 276–277
using HSQMBC, 278–281
CLIP-selHSQMBC, 278
PIP-HSQMBC, 279
phase-sensitive, 263
sequence, 263
suppressing parent resonances, 263
Heteronuclear multiple quantum coherence, 9, 163, 466
Heteronuclear multiple-quantum correlation (HMQC), 257–261
BIRD-HMQC, 259
suppressing parent resonances without gradient pulses, 259–261
influence of homonuclear proton couplings, 259
comparison of experimental crosspeaks, 259
sequence, 257–258
for X-Y correlations, 286–290
Heteronuclear Overhauser effect spectroscopy (HOESY), 364
applications, 366
experiments, 522
one-dimensional, 365
two-dimensional, 364
Heteronuclear single-bond correlations, 246–261
Heteronuclear single-quantum coherence, 9, 299, 374, 447, 463, 500
Heteronuclear single-quantum correlation (HSQC), 246–257, 501–503
experiment and associated coherence transfer pathway, 247
interference from parent resonances, 248–251
practical set-up, 252–253
sensitivity improvement, 251
preservation of equivalent pathways (PEP), 251–252
sequence, 247–248
Heteronuclear single-quantum multiple-bond correlation (HSQMBC)
CLIP variant, 278
long-range optimised, 279
for measuring $^nJ_{CH}$, 278
PIP variant, 279
Heteronuclear spectra, 134
Heteronuclear X-detected correlations, 282–285
multiple-bond correlations and small couplings, 285
COLOC, 285
single-bond heteronuclear correlations, 283–284
homonuclear decoupling in f_1, 284
Heteronuclear X–Y correlations, 286–290
direct X–Y correlations, 286–287
^{19}F–^{13}C HMBC spectrum, 286
indirect ^{1}H-detected X–Y correlations, 288–290
triple-resonance sequence, 289
High-performance liquid chromatography (HPLC), 63
High-resolution NMR
development of, 1–3
HMBC. *See* Heteronuclear multiple-bond correlation
para-H_2 molecule, 491
HMQC. *See* Heteronuclear multiple-quantum correlation
^{1}H-^{15}N HSQC sequence for proteins, 452
HOBS. *See* Homodecoupled band selective decoupling
H_2O buffer solutions, 452
HOESY. *See* Heteronuclear Overhauser effect spectroscopy
HOHAHA. *See* Total correlation spectroscopy
Homodecoupled band selective decoupling (HOBS), 309
Homonuclear correlation, 203–239, 339–359, 368–371, 503–508
Homonuclear decoupling, 143–145
Bloch–Siegert shifts, 144
broadband, 143
experimental implementation, 144
J-resolved, 301–304
pure shift, 306–313
Homonuclear Hartmann–Hahn (HOHAHA) spectroscopy, 220. *See also* Total correlation spectroscopy
Homonuclear J-resolved spectroscopy, 301
applications, 303
EXORCYCLE, 301–306
phase modulation, 301
projections, 302
symmetrisation, 302
tilting, 302
Homospoil pulse, 196
Host–guest complexes, 406
^{1}H–^{77}Se dipole–dipole mechanism, 34
HSQC. *See* Heteronuclear single-quantum correlation
HSQC-IDOSY sequence, 418
HSQC-TOCSY experiment, 244, 508
HSQMBC. *See* Heteronuclear single-quantum multiple-bond correlation
Hybrid diffusion sequences, 414–417
Hybrid experiments, 243
Hybrid HSQC experiments, 253–257
2D multiplicity editing, 253
editing and filtering 1D proton spectra, 256
utilising X-spin shift dispersion, 253–255
Hydrodynamic radius, 382
Hydrogen bonding, 405
para-Hydrogen–induced polarisation (PHIP), 491–493
Hyperbolic secant, pulse, 462
Hyperpolarisation, 491–496
Hyperpolarised pyridine, 495
Hyperpolarised ruthenium complex
PHIP spectrum of, 491
Hyperpolarised states, 491

I

IBURP2. *See* BURP pulses
IDOSY, 415
Imaginary spectrum, 88
Imaging
gradient shimming, 111
IMPEACH-MBC, 273
INADEQUATE. *See* Incredible natural abundance double-quantum transfer experiment

Incredible natural abundance double-quantum transfer experiment (INADEQUATE)
2D, 230–235
1D, measuring J_{CC}, 230
proton detected, 235–239
INEPT. *See* Insensitive nuclei enhanced by polarisation transfer
$INEPT^{+}$, 158
INEPT-INADEQUATE, 235
Inhomogeneity
B_0 field, 107
B_1 field, 354
Insensitive nuclei enhanced by polarisation transfer (INEPT)
DOSY, 414
editing with, 161
in ^{1}H-X correlations, 247
in ^{1}H-X-Y correlations, 288
multiplet intensities in, 158
optimising sensitivity, 165
refocused, 157
signal averaging, 171
INSIPID, 235
Instrumental artefacts, 182
f_2 quadrature artefacts, 182
t_1 -noise, 182
Instrumentation, 7, 61–64
Integration, 137
Interferogram, 73
Intramolecular dynamic exchange equilibria, 38–58
Inverse shift correlations, 244
Inverse spectroscopy, 129
Inversion recovery
in BIRD-HMQC, 259–261
for measuring T_1, 24–26
Ionic complexes
diffusion coefficients, 408
Ionic liquids, 372
Ion pairing, 408
Isochromats, 27
Isotropic mixing
schemes, 226
in TOCSY, 226. *See also* Spin-lock
IUPAC recommendations, 11, 12, 206

J

$^{1}J_{CC}$
measuring, 275
typical values, 235
$^{n}J_{CC}$
typical values, 235
$^{1}J_{CH}$
typical values, 151, 247
$^{n}J_{CH}$
measuring, 275–281, 298–300
typical values, 265
J-Coupled spins, 47, 176, 343, 487
undergoing interconversion, 48
J Coupling, 9, 18, 40, 49–50, 204, 235, 332, 417
J-IDOSY sequence, 417
J-modulated spin-echo, 149–152
2D analogue, J-resolved, 296
editing in HSQC, 253
J-modulation effects, 29, 485
J-resolved spectroscopy, 295–299
heteronuclear, 295
homonuclear, 301
indirect homonuclear, 304
'Jump-return' sequence, 482

K

Karplus relationship, 517
Keto-enol tautomerism, 493
Ketone functionality, 510

L

Laboratory frame of reference, 14
Larmor
frequency, 13, 14, 66
precession, 13
LC-NMR. *See* Liquid-chromatography and NMR
LED. *See* Longitudinal eddy-current delay
Leu-enkephalin
1D DQF, 213
DQF-COSY, 216
Ligand
binding, 448
affinities, 443
detection, by relaxation editing, 427
characteristic NMR parameters, 422
dissociation constants
reporter ligand displacement for, 447
observe methods, 421–422
screening, 428
titration, 425
data fitting, 426
Linear prediction, 73
Line-broadening functions, 83–86
Lineshape
absorption mode, 178
analysis, for exchange rates, 53
chemical exchange, influence on, 38
2D, 178
deconvolution, 112
defects, 93
dispersion mode, 106
Gaussian, 86
Lorentzian, 84
tests, 126
Lineshape analysis
practical considerations in, 53–55
and thermodynamic parameters, 53
Linewidth
half-height, 26
Liquid chromatography, 480
mass spectrometry (LC-MS), 430
and NMR (LC-NMR), 97
Liquid cryogens, 61
Liquid crystals, 371
Lock
optimising, 96
parameters, 106
system, 112
Longitudinal eddy-current delay, 386
Longitudinal spin relaxation, 22–26, 56
Long-range couplings. *See also* $^{n}J_{CH}$
in COLOC, 285
in COSY, 204
^{1}H-^{13}C, 265
in HMBC, 264–265
measuring $^{n}J_{CH}$, 275–281, 298–300
Long-range heteronuclear single-quantum multiple-bond correlation (LR-HSQMBC), 281–282
Lorentz–Gauss transformation, 187
Low-pass J-filter, in HMBC, 267
Low-power rectangular pulse, 469
LP. *See* Linear prediction
LR-HSQMBC. *See* Long-range heteronuclear single-quantum multiple-bond correlation
L-tryptophan, 427, 432, 434, 436
STD spectrum, 432, 434
water-LOGSY spectrum, 439

M

Macromolecular characterisation, 413
Macromolecule-ligand binding, 422–424
Magic angle spinning (MAS), 490
Magic-cycle, decoupling, 466
Magnetic field, 3
homogeneity, 107, 127
optimisation, 107–112
Magnetic flux density, 12
Magnetic moment, 11
molecular, 35
nuclear, 319
Magnetic resonance imaging, 171
Magnetic susceptibility, 490
matched NMR tubes, 94
probe coils, 107
sample preparation, 435
Magnetisation transfer techniques, 56. *See also* Exchange spectroscopy
under slow-exchange conditions, 55–58
for measuring rate constants, 56
practicalities, 58
Magnetogyric ratio, 12
Magnets, superconducting, 61
Magnitude calculation, 180
Manganese(II) chloride, 33
Marine alkaloids, 350
MAS. *See* Magic angle spinning
Matching
probehead, 103
Melting points
deuterated solvents, 90
Menthol
INEPT-INADEQUATE, 236
J-resolved, 297
Methanol
temperature calibration, 126
α-Methoxy-α-trifluoromethylphenylacetic acid (MTPA), 70
Microprobes, 100
Mixing
2D NMR, 172–176
Mixing time, τ_m
in NOESY, 337
in ROESY, 349
in TOCSY, 225

MLEV-16, 226, 468
MLEV-17, 226, 468
Molecular motion, 22, 32, 319, 321, 322
Molecular sieves
drying samples, 90
Monochromatic radiation
pulsed excitation with, 64
Monopolar gradient pulse, 386
Mosher's acid, 70
MQF. *See* Multiple-quantum filter
Multiple-quantum coherence, 188–191
DEPT, 162
HMQC, 257
simplified picture, 190
Multiple-quantum filter, 218
Multiple receivers, 291
Multiplets
visualising in vector model, 19, 171
Multiplicities from
DEPT, 166
HSQC, 253
INEPT, 161
J-resolved spectroscopy, 297

N

^{14}N
coupling, 38
linewidths, 36
^{15}N
HSQC, 250, 448, 452
INEPT, 160
Natural abundance
of quadrupolar nuclei, selected, 36
of spin-1/2 nuclei, selected, 12
Net magnetisation, 13
New Delhi metallo-β-lactamase (NDM1), 453
NMR spectrometer, schematic, 62
NOE. *See* Nuclear Overhauser effect
NOE difference, 359–363
heteronuclear, 364
optimising, 361
quantifying enhancements, 363
SPT artefacts, 363
subtraction artefacts, 361
NOE spectroscopy (NOESY), 339–353. *See also* Nuclear Overhauser effect
applications, 349
1D sequence, 346
chemical shift selective filters, 347
gradient, 346
interpretation, 349
solvent presaturation sequence, 130
2D sequence, 339
chemical exchange crosspeaks, 342
distance measurement, 345
mixing time, 344
zero-quantum interference, 342
presaturation sequence, 480
NOESY. *See* NOE spectroscopy
Noise
t_1-, 182
filtering, 69
quantisation-, 66
Nomenclature, pulse sequence, 5
Non-uniform sampling, 185
N-type
echoes, 179
pathway, 226
Nuclear Overhauser effect (NOE), 34, 317–330. *See also* NOE Spectroscopy
Nuclear spin, 11
Numerical optimisation methods, 459
NUS. *See* Non-uniform sampling
Nutation angle, 15, 116
Nyquist condition, 67, 68
2D NMR, 82
oversampling, 82

O

^{17}O
backward linear prediction, 168
RIDE, 167
Off-resonance
effects, 64, 457, 469
excitation, 65
One-dimensional (1D) spectra, 133–167, 227, 233, 306, 346, 355, 360, 365, 384, 432, 439
'One-shot' diffusion sequence, 388
On–off exchange equilibria, 422–424
On-resonance
adiabaticity factor, 462
excitation, 65, 458
Oversampling
decimation, 82–83
Overview
of modern NMR methods, 7–10
of NMR spectrometer, 61–64
N-Oxalylglycine (NOG), 446

P

^{31}P
^{13}C HMQC, 287
decoupling, 147
^{57}Fe HMQC, 288
^{31}P COSY, 207
Palladium phosphine complexes
diffusion, 306
HETCOR, 304
^{1}H{^{1}H} decoupled, 305
^{1}H{^{31}P} decoupled, 305
NOESY, 306
ROESY, 353
PANACEA, 291
PANSY. *See* Parallel acquisition NMR
Para-hydrogen induced polarisation (PHIP), 491–493
Parallel acquisition NMR, with multiple receivers, 291–292
PANACEA, 291
PANSY sequence, 291
Parallel evolution periods, 199
Parallel ultra-fast spectroscopy (PUFSY), 291
Paramagnetic relaxation
agents for, 30
enhancement (PRE), 429
mechanism, 33
NOE, quenching, 367
removing O_2, 33
Parent resonance, 233, 259–261, 305
PASADENA (*para*-hydrogen and synthesis allow dramatically enhanced nuclear alignment), 491
Pascal triangle, for coupling, 215, 216
Passive coupling, 211, 215
PBLG, liquid crystal, 372
PEP. *See* Preservation of equivalent pathways
β-Peptides
ROESY, 356
'Perfect ' echoes, 29, 279, 485, 486
Periodic refocusing of J evolution by coherence transfer (PROJECT), 28, 29, 485
PFGs. *See* Pulsed field gradients
Phantom
artefact, 300
in gradient calibrations, 397
Phase
coherence, 16
correction, 1D, 88
correction, 2D, 188
real and imaginary data, 226
Phase cycling, 15–17, 78–80
axial peak suppression, 184
CYCLOPS, 80, 233
double-difference, 77
EXORCYCLE, 234, 300
multiple-quantum filtration, 233
satellite selection, 249
Phase errors
aliased signals, 67
correction, 88, 188
zero and first order, 88
Phase modulation
in J-resolved spectroscopy, 417
Phase-sensitive
detector, 75
display, 2D, 179
gradient spectroscopy, 290
Phase-twist lineshape
in COSY, 180
in J-resolved spectroscopy, 302
Phenol, ^{1}H DOSY spectra of, 405
PHIP. *See Para*-hydrogen induced polarisation
α-Pinene
-^{13}C spectra, 148
proton T_1s, 25
Planck's constant, 13
PMMA. *See* Polymethyl methacrylate
Polarisation transfer, 154–155
in COSY, 207
in DEPT, 162
in HETCOR, 283
in INEPT, 156
versus NOE, sensitivity, 297
Polymethyl methacrylate (PMMA) gel, 375
Population differences, 13, 25, 153, 154, 188, 317, 318, 335, 491
equalising, 188
inverting, 154
Population inversion, 155, 329

Pre-acquisition delay
phase errors, 88
quadrupolar nuclei, 167
Precession
gyroscope, 11
Larmor, 11
Preparation period
2D NMR, 199
Presaturation, solvent, 480–482
NOESY presat, 480
PRESAT-180, 481
PURGE, 485
Presaturation utilizing relaxation gradients and echoes (PURGE), 485
Preservation of equivalent pathways, 236
Probeheads, 97–103
acoustic ringing, 167
actively shielded, 125
cryogenic, 101
flow, 96
magic angle spinning, 490
micro, 100
Q-modulation, 368
tuning and matching, 103
Processing parameters
1D NMR, 64
2D NMR, 186
PROJECT. *See* Periodic refocussing of J evolution by coherence transfer
Prolyl hydroxylase domain 2 (PHD2), 446
Propagation, of magnetisation
with spin-lock, 223
Protein backbone amides
^{15}N labelling of, 421
Protein digestion, 430
Protein ^{1}H–^{15}N crosspeak behaviour
in ligand titration experiments, 450
Protein–ligand screening, 421–453
binding equilibria, 422–424
characteristic NMR parameters, 422
competition ligand screening, 443–447
competitive displacement, 444–445
^{19}F FAXS, 447
reporter ligand screening, 445–447
exchange-transferred NOEs, 441–443
ligand titration, 425
data fitting, 426
protein observe methods, 448–453
^{19}F mapping, 452–453
^{1}H–^{13}C mapping, 452
^{1}H–^{15}N mapping, 448–452
resonance lineshapes/relaxation editing, 424–430
^{19}F NMR, 428–429
^{1}H relaxation-edited NMR, 426–428
paramagnetic relaxation enhancement, 429–430
saturation transfer difference (STD), 430–438
epitope mapping, 436–437
K_D values, 437–438
sequence and practicalities, 432–435
water-LOGSY, 438–441
practicalities, 441
sequence, 440
Protein-mediated NOE transfer pathways, 443
Protein resonances, 426
assignment, isotope labelling schemes, 450
shift changes of, 426
Protein saturation spectrum, 430
Protocol, for structure confirmation, 10
Proton–carbon coupling constants, 298
$^{1}J_{CH}$ values, 151, 247
$^{n}J_{CH}$ values, 265
measuring, 275–281, 298–300
Proton coupling constant timescale, 50
Proton exchange process, in coupled CH–OH pair, 50
Proton-free buffers, 435
Proton–proton couplings
in dexamethasone phosphate, 523
removing, 307
Pseudo-diagonal, 231
PSYCHE. *See* Pure shift yielded by CHIRP excitation
^{195}Pt
field dependence, satellites, 323
^{1}H-^{13}C-^{195}Pt correlation, 289
P-type
anti-echoes, 387
pathway, 250
PUFSY. *See* Parallel ultra-fast spectroscopy
PULCON. *See* Quantitative NMR
Pulsed field gradients (PFGs), 191–198
calibrating, 123
coherence selection, 197
implementing, 122
introduction of, to NMR, 123
shimming with, 127
solvent suppression, 483
symbols used for, 6
Pulse excitation, 64–66, 457–459
bandwidth, 65
off-resonance effects, 64–65, 458
tip or flip angle, 16
vector model, 18
Pulse imperfections
B_1 inhomogeneity, 458
compensating for, 363, 457–461
excitation trajectories, 66
in HSQC, 299
in spin-echoes, 27–30
off-resonance effects, 300
Pulses
adiabatic, 462
broadband, 464
gradient, 191
hard, rf, 64
shaped/selective, rf, 468–473
soft, rf, 64, 468
symbols used for, 6
Pulse sequence elements, 6
Pulse sequence nomenclature, 5–6
Pulse width
calibration, 113
definition, 64
miscalibration, 300
Purcell, Edward, 1
Pure-phase excitation profile, 473
Pure-phase pulse, shaped, 471
Pure shift broadband-decoupling, 306–313
pseudo-2D, 307
real-time, 309
refocussing elements, 309
band selective, 309
BIRD, 311
PSYCHE, 312
Zangger–Sterk, 310
refocussing pulse, 307
Pure shift spectra, 306
Pure shift yielded by CHIRP excitation (PSYCHE), 312
PURGE (presaturation utilizing relaxation gradients and echoes), 485

Q

Q3, shaped pulse, 470
Q5, shaped pulse, 470
Q-modulation, 368
QNMR. *See* Quantitative NMR
Quad-images
compensating for, CYCLOPS, 79, 182
Quadrature detection
aliasing, 77
in 2D NMR, 177
echo-antiecho method, 2D, 178
images, 77
simultaneous and sequential sampling, 74
States method, 2D, 177
States-TPPI method, 2D, 178
TPPI method, 2D, 178
Quadrupolar nuclei, 167–168
coupling with spin-½ nuclei, 37
Quadrupolar relaxation, 35–37
Quantitative NMR, 137–143
calibration samples for, 137
deconvolution of lineshapes, 139
with ERETIC, 137
^{19}F, 147
with PULCON, 137
Quinine
diffusion coefficients, 394
diffusion decay profiles, 396

R

Radiation damping, 480
Radiofrequency (RF), 300
amplitude profile, 462
pulses, 6, 7
Rapid on–off exchange, 422
Rate constant determination
through lineshape analysis, 54
through magnetisation exchange, 56, 368
Rayleigh–Bénard convection, 392
Rayleigh number, 392
RDCs. *See* Residual dipolar couplings
Real spectrum, 88
REBURP. *See* BURP pulses
Receiver gain, 82
Receiver imbalance
compensating for, 79
Receptor-bound water molecules, 430
Receptor–ligand interactions, 421–453
Recovery delay, 134. *See also* Relaxation delay

Reference compounds, 91
Reference deconvolution, 112
Reference distances, NOE, 345
Reference frequency, 67
Referencing spectra, 91–92
Refocusing
  chemical shifts and couplings, 162
  with gradient pulses, 192
Relative sensitivity
  of quadrupolar nuclei, 36
  of spin-½ nuclei, 12
Relaxation, 22–38
  free induction decay, 17
  longitudinal, 22
  mechanisms, 31–38
  spin-lattice, 23
  spin-spin, 27
  time constants, 23, 26
  transverse, 26
Relaxation delay, 134
Relayed- COSY, 220
Relaying of information
  in HSQC-TOCSY, 243
  in TOCSY, 220
Repetition rate
  Ernst angle, 134–135
  optimum, in 1D NMR, 134
  optimum, in 2D NMR, 185
  optimum, with 90° pulse, 136
Reporter ligand-screening, 445–447
Residual dipolar couplings (RDCs), 316, 371–377
  applying, 375
  measuring, 372
    alignment media, 372
Resolution
  tests, 128
Resolution enhancement, 86–88
  linear prediction, 266
Resonance offset
  90 degree pulse, 16, 459
  simulated effect, 459
RF. *See* Radiofrequency
RIDE
  quadrupolar nuclei, 167
Robotic sample changer, 63
ROE. *See* Rotating frame NOE
ROESY. *See* Rotating frame NOE spectroscopy
Rotating frame, 14, 337
Rotating frame NOE, 337–339
  appearance in TOCSY, 338
  applications, examples, 337
  definition, 337
  distance dependence, 325
  experimental aspects, 338
  false, 354
  kinetics, 342
  mixing time, 338
  spin-lock, 338
Rotating frame NOE spectroscopy (ROESY), 353, 359
  1D, 355
  2D, 353
  spin-locks, 355
  transverse, Tr-, 355
Rotaxanes
  NOESY, 351
$RuH_2(CO)_2$ bis(diphenylarsino)ethane
  PHIP ^1H spectrum of, 493

S

SABRE process (signal amplification by reversible exchange), 492
Saccharides
  1D TOCSY-NOESY, 347
  HSQC, edited, 253
  HSQC-TOCSY, 255
  J-resolved, 347
  $^nJ_{CH}$ measurement in, 299
Sample preparation, 89
  filtering and degassing, 94
  reference compounds, 91
  selecting solvent, 89–91
  tubes and sample volumes, 92–94
Sample spinning
  in diffusion NMR, 393
  2D NMR, 393
  shimming, 61
  sidebands, 109
Sample temperature, 124–126
  calibration, 124
Sampling, data
  filtering noise, 68
  Nyquist condition, 67
  sequential, 74
  simultaneous, 75
Satellites, 35, 246
Saturation, 134
  transfer, 56, 328
Saturation transfer difference (STD), 430–438
  amplification factors, 438
  group epitope-mapping, 436
  K_D measurement, 437
  L-tryptophan, 432
    BSA sample, 434, 436
  receptor–ligand complex, 437
  STD sequence, 432
Saturation transfer process
  schematic representation of, 431
Scalar couplings
  $^1J_{CC}$ values, 235
  $^nJ_{CC}$ values, 235
  $^1J_{CH}$ values, 152, 247
  $^nJ_{CH}$ values, 265
Scalar relaxation
  of the 1st kind, in NOESY, 342
Sealing tubes, 95
^{77}Se
  CSA relaxation, 34
  longitudinal relaxation times, 34
SEFT. *See* Spin-echo Fourier transform
Selective detection, 83
Selective excitation, 468–479
  CSSF, 475
  DANTE, 483
  excitation sculpting, 473
  shaped pulses, 469
Selective population inversion, 155
Selective population transfer, 154–155
  artefacts, 362
Selective pulses. *See* Shaped pulses
Selective TOCSY edited preparation (STEP), 347
Self-diffusion coefficient, 381
SELTRIP-gs, 290
Sensitivity, 133, 244
  enhancement, 153
  heteronuclear correlations, 244
  optimising, in 1D NMR, 133
  oversampling, 83
  probe, 97
  tests, 128
Sequential assignment, peptides, 351
Shaped pulse, 469–473
  amplitude calibration of, 479
  bandwidth factor, 472
  calibrating, 479
  calibration sequences, 479
  DANTE, 477
  Gaussian, 469
  properties of, selected, 472
  pure phase, 471
Shift correlation
  chemical exchange, 368–371
  dipolar couplings (NOE), 339
  heteronuclear, 243
  homonuclear, 203
Shim coils, 61
Shimming, 107–112
  defects, 110
  on the FID, 110
  gradient, 111, 480
^{29}Si
  HMBC, 266
Signal averaging, 80, 134
  NOE difference, 361
Signal detection, 66
  digitisation process, 66
Signal-to-noise ratio
  comparing probes, 99
  measuring, tests, 97
  in signal averaging, 80
Single-channel detection, 74
  TPPI, 178
Single PFG spin echo (SPFGSE), 473–474
Single-pulse experiment, 133–136
  essential elements, 134
  optimising sensitivity, 133
  optimum pulse repetition time, 136
Single-scan 2D NMR, 199
SLAPSTIC (spin labels attached to protein sidechains as a tool to identify interacting compounds), 430
SLURP, shaped pulse, 473
SNEEZE, shaped pulse, 472, 478
Sodium dodecyl sulphate (SDS), 412
Soft pulse. *See* Shaped pulse
Solid-state NMR spectroscopy, 493
Solvent-induced isotope shift, 453
Solvents
  melting and boiling points, 90
  viscosity, 394

Solvent suppression, 480–486
 binomial sequences, 482
 jump-return, 482
 presaturation, 480
 pulsed field gradients, 483
 test sample, sucrose, 439
 via chemical exchange, 368
 via diffusion, 403
Spatially selective excitation, 199
Spectral density, 32
Spectral width, 83
Spectrometer calibrations, 113–126
 heteronuclear decoupling field strength, 121
 homonuclear decoupling field strength, 120
 pulsed field gradients, 122
  gradient recovery times, 124
  gradient strengths, 123–124
 radiofrequency pulses, 113
  indirect pulses, 118
   on high-abundance nuclides, 118
   on low-abundance nuclides, 119–120
  observe pulses
   high sensitivity, 115
   low sensitivity, 117
   single-scan nutation spectroscopy, 116–117
  Rf field strengths, 114
Spectrometer performance tests, 126–130
 lineshape, 127
 resolution, 127
 sensitivity, 128–129
 solvent presaturation, 130
 standard sensitivity test samples of some common nuclei, 129
Spectrometer preparation, 95–113
 field frequency lock, 105
  acquiring data unlocked, 106
  lock system, 105
  optimising the lock, 106
 optimising field homogeneity, 107
  common lineshape defects, 109
  gradient shimming, 111–112
  shimming, 107–109
  shimming using the FID or spectrum, 110
  shim system, 107
 probe, 95, 103
  flow, 96
  tuning and matching, 104
 probe design and sensitivity, 97
  cryogenic probes, 101–103
  detection sensitivity, 97–99
  mass and concentration sensitivity, 99
  micro-flow probes, 101
  microprobes, 100
Spectrum assignment, 499–523
Spectrum deconvolution, 140
Spectrum editing, 153
 with spin-echoes, 148
  attached proton test (APT), 152
  J-modulated spin-echo, 149–152
SPFGSE. *See* Single PFG spin echo
Spin coupling
 visualising in vector model, 18
Spin diffusion, 326, 328, 430, 436
Spin-echoes, 18–20
 diffusion NMR, 417
 experimental observation, 29
 heteronuclear, 19
 homonuclear, 19
 J-modulated, 149
 in J-resolved spectroscopy, 295
 in spin-locks, 221
 T_2 measurement, 29
Spin-echo Fourier transform (SEFT), 149
Spin-lattice relaxation. *See* Longitudinal spin relaxation
Spin-lock, 466
 in ROESY, 354
 in TOCSY, 468
Spinning sidebands, 109
Spin quantum number, I, 11, 35
Spin relaxation, 22–38
 longitudinal relaxation, 22–23
 measuring T_2 with a spin-echo sequence, 27
  experimental observation of spin echoes, 29
  operation of the CPMG sequence, 30
  T_2 filter, 31
  T_2 spectrum editing, 30
 measuring T_1 with the inversion recovery sequence, 24
  inversion recovery process, 24
  inversion recovery sequence, 24
  quick T_l estimation, 25
 transverse relaxation, 26–27
Spin rotation relaxation, 35
Spin-spin coupling
 vector model, 18, 176
Spin-spin relaxation. *See* Transverse spin relaxation
Spontaneous emission, 23
SPT. *See* Selective population transfer
Stacked plot, 180
States method
 axial peaks, 178
 2D quad detection, 178
States-TPPI method
 axial peaks, 178
 2D quad detection, 181
STD. *See* Saturation transfer difference
STE. *See* Stimulated echo
Steady-state
 2D NMR, 315
 magnetisation, 133
Steady-state NOE, 322. *See also* Nuclear Overhauser effect
 applications, examples, 317
 distance dependence, 319
 indirect, 3-spin, 326
 measuring, 327
 multispin system, 325
 origin, 317
 spin-diffusion, 322
 summary of, key points, 329
 versus transient NOE, 328
 two-spin system, 317
Stejskal–Tanner equation, 397, 399
STEP. *See* Selective TOCSY edited preparation
STEP-DOSY sequence, 416
Steroids
 example structure elucidation, 352
 NOESY, 352
 ROESY, 353
Stimulated echo (STE), 384–388
 diffusion NMR, 385
 double, 392
 double, asymmetrical, 392
Stimulated emission, 23
Stokes–Einstein equation, 382
Stretched induced alignment in gels (SAG), 373
Strong coupling
 in J-resolved spectroscopy, 304
 spin-locks, TOCSY, 466
Structure elucidation, 499–525
 ^{13}C NMR, 508
 DQF-COSY, 505
 ^{19}F–^{13}C HMBC, 515, 517
 ^{1}H–^{13}C edited HSQC, 501
 ^{1}H–^{13}C HMBC, 513
 ^{1}H–^{1}H COSY, 503
 ^{1}H–^{1}H TOCSY, 506
 ^{1}H NMR, 500
 ^{1}H–^{31}P HMBC, 512
 hybrid HSQC-TOCSY experiment, 508
 proton decoupling
  ^{19}F NMR spectrum, 516
  ^{1}H NMR spectrum, 512, 516
  ^{31}P NMR spectrum, 512
Sucrose
 test sample, 130, 427
Sudlow site II on serum albumin, 426
Superconducting magnets. *See* Magnets
Supercritical fluids
 as NMR solvents, 36
Super-cycle, decoupling, 466
Supramolecular assemblies, 409
Symmetrisation
 in COSY, 183
 in J-resolved spectroscopy, 302
Symmetry
 in complexes, 183
 quadrupolar lineshapes, 187
Synthetic organic materials
 protocol for routine structure confirmation of, 10

T

Tables of $^{1}J_{CC}$, 235
Tables of $^{1}J_{CH}$, 247
Tables of $^{n}J_{CC}$, 235
Tables of $^{n}J_{CH}$, 265
t_1, 2D NMR, 173
 noise, 182
t_2, 2D NMR, 173
Temperature calibration, 124
TentaGel resin, 490
Terakis(trimethylsilyl)silane (TMSS), 402
Terpene, andrographolide
 bicyclic, 163
 carbon and DEPT spectra, 166
 decoupling schemes, 146
 2D NOESY spectrum, 340
 proton spectrum, 50

Tests, spectrometer, 126–130
Tetrabutylammonium tetrahydroborate
diffusion coefficients, 408
Tetramethylsilane (TMS)
diffusion reference, 381
shift reference, 91
Thermal expansivity, 392
Three-dimensional NMR
DOSY, 415
^{1}H-^{13}C-^{15}N correlation of protein, 288
schematic sequence, 67, 175
Three-spin effects, NOE, 327
Tilting
within COSY crosspeaks, 215
of J-resolved spectra, 301
Time domain
extension, linear prediction, 73
extension, zero-filling, 71
Time domain profiles
common shaped, selective excitation pulses, 470
Time proportional phase incrementation (TPPI)
axial peaks, 341
2D quad detection, 340
τ_m. *See* Mixing time
TMS. *See* Tetramethylsilane
τ_{null}
inversion recovery, 25
TOCSY. *See* Total Correlation Spectroscopy
TOE. *See* Truncated driven NOE
Top-hat profile, 469
Total Correlation Spectroscopy (TOCSY), 220–230
applications, 229
breakthrough in ROESY, 338
1D, selective, 227
1D, TOCSY-NOESY, 227
gradient-selected, 226
HSQC-TOCSY, 243
implementing, 225
mixing schemes for, 225
spin-locks, 221
ZQ suppression, 229
TPPI. *See* Time proportional phase incrementation
Transferred NOEs, for ligand binding, 441
Transient NOE, 335–337
applications, examples, 342
initial rate approximation, 335
internuclear separations, 335
interpreting, comments, 349
mixing time, 349
versus steady-state NOE, 359
Transition probabilities, 318
Transitions
directly connected, 211
remotely connected, 211
Transmitter
attenuation, 115
radiofrequency, 63
Transverse magnetisation
loss of, 26
observable, 26
Transverse spin relaxation, 26–31. *See also* Spin relaxation
T_1, relaxation time constant
definition, 23
dependence on tumbling rates, 32
measuring, 24
T_2*, relaxation time constant
definition, 26
T_2, relaxation time constant, 26–31
definition, 26
measuring, 26
3, 4, 5-Trichloropyridine, 137
1, 3, 5-Trimethoxybenzene, 137
Trimethylsilyl propionic acid (TSP), 137, 500
Trim pulses, 226
Triple-quantum filter, 218
Triple-resonance experiment, 143, 286–290
Tris(trimethylsilyl)methane, 53
Trityl radicals, for DNP, 153
Tr-ROESY. *See* Rotating frame NOE spectroscopy
Truncated driven NOE, 335
Truncation artefacts
sinc wiggles, 72
suppressing, 183
TSP. *See* Trimethylsilyl propionic acid
Tubes
cleaning, 96
micro, 103
sapphire, 393
sealing, 95
susceptibility matched, 94
Tumbling rates
correlation time, 320
in solution, 32
Tuning
probehead, 104
Two-dimensional data acquisition, 183–185
non-uniform sampling, 185–186
Two-dimensional data processing, 186–187
linear prediction, 71
phase correction, 188
presentation, 188
window functions, 186
zero-filling, 71
Two-dimensional (2D) experiments, 172–188
generating second dimension, 172–176
practical aspects, 177–188

U

Ultrafast 2D NMR, 199
Ultrasonic mixing, 480
Universal pulse, shaped, 310

V

van der Waals volumes, 409
Vector model, of NMR, 14–20
chemical shifts and couplings, 17–18
pulses, 15–17
phase coherence, 16
rotating frame of reference, 14–15
spin-echoes, 18–20
chemical shift evolution, 19
scalar coupling evolution, 20
Verification of proposed structures, 9
Vibration damping system, 63
Vicinal couplings
in COSY, 204
Viscosities, of protonated solvents, 394
Viscosity–density ratio, 392
Volume integrals, 2D, 345
Volumes
NMR samples, 93

W

WALTZ-16, 129, 226, 393, 466
Water
suppression, 480–486
temperature dependence, 40
Water attenuation by transverse relaxation (WATR), 30
WATERGATE. *See* Water suppression by gradient tailored excitation
Water-ligand observed with gradient spectroscopy (Water-LOGSY), 438–443
magnetisation transfer, 438
practicalities, 441
sequences, 440
Water-LOGSY. *See* Water-ligand observed with gradient spectroscopy
Water suppression by gradient tailored excitation (WATERGATE), 483
WATR. *See* Water attenuation by transverse relaxation
Wheat germ agglutinin (WGA), 424
Window functions, 83–88
2D NMR, 187
WURST pulse, 462, 464
inversion trajectories, 463

X

XCORFE, 285
X-nucleus, definition, 6
X-Y heteronuclear correlations, 286–290
^{1}H detected, 288

Z

Zangger–Sterk, 310
refocussing element, 310
shaped pulse gradient combination, 310
Zero-crossing, in NOE, 337
Zero-excitation profile, 482
Zero-filling, 71, 139, 187, 252, 301
Zero-quantum coherence (ZQC), 188, 193, 228, 312
suppression, 486
in NOESY, 342
Zero-quantum dephasing, 487
calibration routine, 488
in NOESY, 342
in TOCSY, 229, 230
in TOCSY-NOESY, 488
Zero-quantum frequency, 343, 487
Z-filtration
TOCSY, 226